Digital Design
Principles and Practices

DIGITAL DESIGN
Principles and Practices

Third Edition

John F. Wakerly
Cisco Systems, Inc.
Stanford University

Prentice Hall, Upper Saddle River, New Jersey 07458

Library of Congress Cataloging-in-Publication Data

Wakerly, John F.
    Digital design: principles and practices, third edition.
/John F. Wakerly.
    p. cm.
  Includes bibliographical references and index.
  ISBN 0-13-769191-2
  1. Digital integrated circuits--Design and construction.
I. Title.
TK7874.65.W34 2000
621.39'5--dc21   99-36681
                                                  CIP

Publisher: TOM ROBBINS
Editor-in-chief: MARCIA HORTON
Production editor: IRWIN ZUCKER
Executive managing editor: VINCE O'BRIEN
Assistant managing editor: EILEEN CLARK
Manufacturing buyer: PAT BROWN
Copy editor: ROBERT LENTZ

Director of production and manufacturing: DAVID W. RICCARDI
Art director: ANN FRANCE
Associate creative director: AMY ROSEN
Cover art: ROBERT MCFADDEN
Cover designer: JOHN CHRISTIANA
Interior designer: DONNA WICKES
Editorial assistant: DAN DEPASQUALE

The author and publisher of this book have used their best efforts in preparing this book. These efforts include the development, research, and testing of the theories and programs to determine their effectiveness. The author and publisher make no warranty of any kind, expressed or implied, with regard to these programs or the documentation contained in this book. The author and publisher shall not be liable in any event for incidental or consequential damages in connection with, or arising out of, the furnishing, performance, or use of these programs.

We wish to thank Aldec, Inc. and Synopsys® for allowing the use of their software in the Xilinx Student Edition.

Trademark Information: Verilog is a trademark of Cadence Design Systems, Inc. Silos III is a trademark of Simucad Inc. Synopsys, and Foundation Express are trademarks of Synopsys, Inc. Xilinx is a trademark of Xilinx Corp. Aldec is a trademark of Aldec.

Printed in the United States of America

10 9 8 7 6 5 4 3 2

ISBN 0-13-769191-2

Prentice-Hall International (UK) Limited, London
Prentice-Hall of Australia Pty. Limited, Sydney
Prentice-Hall Canada Inc., Toronto
Prentice-Hall Hispanoamericana, S.A., Mexico
Prentice-Hall of India Private Limited, New Delhi
Prentice-Hall of Japan, Inc., Tokyo
Pearson Education Asia Pte. Ltd., Singapore
Editora Prentice-Hall do Brasil, Ltda., Rio de Janeiro

To my teachers

CONTENTS

FOREWORD

Moore's Law, which observes that semiconductor technology advances exponentially, has been valid for over three decades. Experts predict it will continue to hold for at least one more. When integrated circuits were introduced, logic packages had a dozen or so transistors. Today, with exponential increases in circuit density, microprocessor chips have passed the 10-million-transistor mark. In less than another decade they will reach 100 million transistors per chip.

To keep up with Moore's Law, design techniques have changed drastically. Hand-crafted logic circuits were once the norm. Now designers generate circuits from high-level descriptions. Connections on printed-circuit boards have been absorbed within chips. With programmable logic, on-chip logic functions and connections can be updated within the user environment.

How does education keep up with Moore's Law? What can we do to enable students to practice their skills today and adapt them tomorrow to new generations of devices? This is the challenge John Wakerly faced when he began this work.

His approach is multifaceted. It is grounded in basic principles of digital design that do not change with technology, such as combinational logic, sequential logic, and state machines. Wakerly weds these principles with tools and practical techniques that teach how to design for today's technology. These include how to use the ABEL and VHDL design languages, how to structure designs with large building blocks, and how to implement designs with programmable logic devices. These techniques are essential for successful design.

The most difficult goal is to help the student adapt to the inevitable changes to come. Wakerly does this by revealing what is happening underneath the logic. For example, he gives transistor models of gates and uses them to expose issues related to timing and noise. Gates may become faster and denser and may use different control voltages, but how to assure correct and reliable operation will be a continuing concern. We learn the characteristics, constraints, and failure modes and how to design to them. We learn through examples of alternate designs how to judge design quality and evaluate tradeoffs. As new technology emerges, we will be able to design to it.

Wakerly enhances the approach with presentation skills that are rare in college texts. The reader will quickly appreciate the effective graphics, entertaining writing style, and instructive exercises.

Moore's Law condemns textbooks in this field to short lives. Nevertheless, Wakerly's text is a classic.

Harold S. Stone
Princeton, New Jersey

PREFACE

This book is for everyone who wants to design and build real digital circuits. It is based on the idea that, in order to do this, you have to grasp the fundamentals, but at the same time you need to understand how things work in the real world. Hence, the "principles and practices" theme.

The material in this book is appropriate for introductory courses on digital logic design in electrical or computer engineering or computer science curricula. Computer science students who are unfamiliar with basic electronics concepts or who just aren't interested in the electrical behavior of digital devices may wish to skip Chapter 3; the rest of the book is written to be independent of this material as much as possible. On the other hand, *anyone* with a basic electronics background who wants to get up to speed on digital electronics can do so by reading Chapter 3. In addition, students with *no* electronics background can get the basics by reading Bruce M. Fleischer's "Electrical Circuits Review," a freely reproducible 20-page electronics tutorial available on this book's Web site, www.ddpp.com.

introductory courses

electronics concepts

Although this book's level is introductory, it contains much more material than can be taught in a typical introductory course. Once I started writing, I found that I had many important things to say that wouldn't fit into a one-quarter course at Stanford or a 400-page book. Therefore, I have followed my usual practice of including *everything* that I think is at least moderately important, and leaving it up to the instructor or reader to decide what is most important in a particular environment. To help these decisions along, though, I've marked the headings of *optional sections* with an asterisk. In general, these sections can be skipped without any loss of continuity in the non-optional sections that follow.

optional sections

Undoubtedly, some people will use this book in advanced courses and in laboratory courses. Advanced students will want to skip the basics and get right into the fun stuff. Once you know the basics, the most important and fun stuff in

advanced courses
laboratory courses
fun stuff

this book is in the sections on hardware description languages ABEL and VHDL, where you'll discover that your programming courses actually helped prepare you to design hardware.

working digital designers

Another use of this book is as a self-study reference for a working digital designer, who may be either of two kinds:

Novice   If you're just getting started as a working digital designer, and you took a very "theoretical" logic design course in school, you should concentrate on Chapters 3, 5, 6, and 8–11 to get prepared for the real world.

Old pro   If you're experienced, you may not need all of the "practices" material in this book, but the principles in Chapters 2, 4, and 7 can help you organize your thinking, and the discussions there of what's important and what's not might relieve the guilt you feel for not having used a Karnaugh map in 10 years. The examples in Chapters 6, 8, and 9 should give you additional insights into and appreciation for a variety of design methods. Finally, the ABEL and VHDL language descriptions and examples sprinkled throughout Chapters 4 through 9 may serve as your first organized introduction to HDL-based design.

marginal notes marginal pun

All readers should make good use of the comprehensive index and of the *marginal notes* throughout the text that call attention to definitions and important topics. Maybe the highlighted topics in *this* section were more marginal than important, but I just wanted to show off my text formatting system.

Chapter Descriptions

What follows is a list of short descriptions of this book's eleven chapters. This may remind you of the section in software guides, "For People Who Hate Reading Manuals." If you read this list, then maybe you don't have to read the rest of the book.

- Chapter 1 gives a few basic definitions and lays down the ground rules for what we think is and is not important in this book.

- Chapter 2 is an introduction to binary number systems and codes. Readers who are already familiar with binary number systems from a software course should still read Sections 2.10–2.13 to get an idea of how binary codes are used by hardware. Advanced students can get a nice introduction to error-detecting codes by reading Sections 2.14 and 2.15. The material in Section 2.16.1 should be read by everyone; it is used in some design examples in Chapter 8.

- Chapter 3 describes "everything you ever wanted to know about" digital circuit operation, placing primary emphasis on the external electrical characteristics of logic devices. The starting point is a basic electronics background including voltage, current, and Ohm's law; readers unfamiliar with these concepts may wish to consult the "Electrical Circuits Review" mentioned earlier. This chapter may be omitted by readers who aren't interested in how to make real circuits work, or who have the luxury of having someone else to do the dirty work.

- Chapter 4 teaches combinational logic design principles, including switching algebra and combinational-circuit analysis, synthesis, and minimization. Introductions to ABEL and VHDL appear at the end of this chapter.

- Chapter 5 begins with a discussion of digital-system documentation standards, probably the most important practice for aspiring designers to start practicing. Next, this chapter introduces programmable logic devices (PLDs), focusing on their capability to realize combinational logic functions. The rest of the chapter describes commonly used combinational logic functions and applications. For each function, it describes standard MSI building blocks, ABEL programs for PLD realizations, and VHDL models.

- Chapter 6 is a collection of larger combinational-circuit design examples. For each example, it shows how the design can be carried out with MSI building blocks (if appropriate), ABEL and PLDs, or VHDL that can be targeted to a CPLD or FPGA.

- Chapter 7 teaches sequential logic design principles, starting with latches and flip-flops. The main emphasis in this chapter is on the analysis and design of clocked synchronous state machines. However, for the brave and daring, the chapter includes an introduction to fundamental-mode circuits and the analysis and design of feedback sequential circuits. The chapter ends with sections on ABEL and VHDL features that support sequential-circuit design.

- Chapter 8 is all about the practical design of sequential circuits. Like Chapter 5 before it, this chapter focuses on commonly used functions and gives examples using MSI building blocks, ABEL and PLDs, and VHDL. Sections 8.8 and 8.9 discuss the inevitable impediments to the ideal of fully synchronous design and address the important problem of how to live synchronously in an asynchronous world.

- Chapter 9 is a collection of state-machine and larger sequential-circuit design examples. Each example is carried out both using ABEL for a PLD and using VHDL that can be targeted to a CPLD or FPGA.

- Chapter 10 is an introduction to memory devices, CPLDs, and FPGAs. Memory coverage includes read-only memory and static and dynamic read-write memories from the points of view of both internal circuitry and functional behavior. The last two sections introduce CPLD and FPGA architecture.

- Chapter 11 discusses several miscellaneous real-world topics that are of interest to digital designers. When I started writing what I thought would be a 300-page book, I included this chapter in the outline to pad out the "core" material to a more impressive length. Well, the book is obviously long enough without it, but this material is useful just the same.

Most of the chapters contain references, drill problems, and exercises. Drill problems are typically short-answer or turn-the-crank questions that can be answered directly based on the text material, while exercises may require a little more thinking. The drill problems in Chapter 3 are particularly extensive and are designed to allow non-EE types to ease into this material.

Xilinx Foundation Tools

Xilinx, Inc. (San Jose, CA 95124) has kindly allowed us to package their Foundation Express digital-design tools on two CD-ROMs at the end of this book (in most domestic and international printings). These tools are quite comprehensive, including an ABEL compiler, VHDL and Verilog language processors, a schematic drawing package, and a simulator. Much of the software in the package is based on the popular Active-CAD™ and Active-HDL™ tools from Aldec, Inc. The package also includes FPGA Express™ software from Synopsys, which allows ABEL, VHDL, and Verilog designs to be targeted to CPLDs and FPGAs; popular Xilinx parts are supported by the included version.

The Foundation tools were very useful to me as an author. Using them, I was able to write and test all of the example programs in the text. I trust that the tools will be even more useful to the students who use the text. They will allow you to write and test your own hardware designs and download them into Xilinx CPLDs and FPGAs in a lab environment. There are no artificial design-size limits, as long as your design fits into a single device. An excellent lab-oriented manual for getting started with Xilinx devices and tools is David Van den Bout's *The Practical Xilinx Designer: Version 1.5* (Prentice Hall, 1999).

Even if you're not ready to do your own original designs, you can use the Foundation tools to try out and modify any of the examples in the text, since the source code for all of them is available at the book's Web site, discussed next.

WWW.DDPP.COM

Support materials for this book are available at the book's own dedicated Web site, www.ddpp.com. A key resource for students is the set of source listings for all of the example C, ABEL, and VHDL programs in the book. Also available are ZIP'ed Foundation project directories, including not only ABEL and VHDL source files but also schematics that are set up to use and simulate some of the example designs.

During the preparation of this edition, I was surprised and delighted to see how much digital design reference material is available on the Web, especially from device manufacturers. The DDPP Web site contains a "living" references section with links to many useful sites that you can use as a jumping-off point for your own independent study.

A couple of appendices from previous editions are available on the Web site—"Electrical Circuits Review" by Bruce M. Fleischer and "IEEE Standard Symbols." Students taking lab courses may also appreciate the four pages of handy IC pinout guides, which appeared on the inside-cover pages of previous editions.

One thing that students may or may not like is a collection of new exercises that I expect to build up as I continue to teach digital design at Stanford and as I receive contributions from others.

For Instructors

The DDPP Web site has additional materials for instructors only. This part of the site is password protected; if you plan to use it, please allow up to a week to obtain a login name and password via the procedure published there.

The instructors' area contains files with all of the figures and tables in the book. You can use these files to make presentation slides directly, or you can insert selected materials into your own customized presentations.

The site also contains answers to selected exercises—more than half of the exercises in the book, equivalent to over 200 printed pages. There are also several sample exams and solutions.

Another important resource for instructors is Xilinx' University Program (www.xilinx.com/programs/univ). The site offers a variety of product materials, course materials, and discounts on chips and boards that you can use in digital-design lab courses.

How This Book Was Prepared

The text of the third edition of this book was converted from the original second-edition TEX version into Adobe FrameMaker®. Figures from previous editions were converted from Cricket Draw into Adobe Illustrator® EPS files.

All of the writing, editing, drawing, and circuit designing was done on a PC running Windows 95 or 98 with 384 Mbytes of memory, which, regrettably, would still crash if too many programs or files were open at once. The good news is that this edition's use of standard programs and tools has allowed me to provide readers and instructors with a large collection of useful materials on the book's Web site, as described earlier.

Errors

Warning: This book may contain errors. The author assumes no liability for any damage—incidental, brain, or otherwise—caused by errors.

There, that should make the lawyers happy. Now, to make *you* happy, let me assure you that a great deal of care has gone into the preparation of this manuscript to make it as error free as possible. I am anxious to learn of the remaining errors so that they may be fixed in future printings, editions, and spin-offs. Therefore I will pay $5 to the first finder of each undiscovered error, be it technical, typographical, or otherwise. Please email your comments by using the link on the Web site, or by writing to me at `john@wakerly.com`.

Any reader can obtain an up-to-date list of discovered errors using the link at the book's Web site. It will be a very short file transfer, I hope.

Acknowledgements

Many people helped make this book possible. Most of them helped with the first and second editions and are acknowledged there. Preparation of the third edition has been a lonelier task, but it was made easier by my colleagues Mario Mazzola and Prem Jain at Cisco Systems. They and the company made it possible for me to cut back my commitment at Cisco to less than half time for the eight months that it took to prepare this edition.

For the ideas on the "principles" side of this book, I still owe great thanks to my teacher, research advisor, and friend Ed McCluskey. On the "practices" side, my personal "Digital Designers Hall of Fame" includes (in chronological order): Ed Davidson, Jim McClure, Courtenay Heater, Sam Wood, Curt Widdoes, Prem Jain, Ted Tracy, Dave Raaum, Akhil Duggal, Des Young, and Tom Edsall.

The seed that got me started writing this book and many others was planted in the early 70s by Harold Stone at Stanford. He put me to work reviewing and indexing his books, and his computer organization books inspired me to write my first software book. Now, I'd like to offer Harold my belated thanks for getting me started, and I give him special thanks for helping to pad out this edition by another two pages!

In the summer of 1997, during the early stages of this edition's planning, friend and colleague Jean-Pierre Steger took a sabbatical from the Burgdorf School of Engineering near Bern, Switzerland to help me get jump-started with VHDL, the Xilinx Foundation tools, and other topics. A number of other people contributed review comments or other materials to this edition, including John Birkner, Rebecca Farley, Don Gaubatz, John Gill, Linley Gwennap, Jesse Jenkins, and Jeff Purnell.

Xilinx, Inc. naturally deserves credit for providing the Foundation tools that are an important adjunct to this edition. On the people side, their original

University Program leader Jason Feinsmith was very helpful, and their recently appointed leader Patrick Kane has supported our efforts enthusiastically.

Since the second edition was published, I have received many helpful comments from readers. In addition to suggesting or otherwise motivating many improvements, readers have spotted dozens of typographical and technical errors whose fixes are incorporated in this third edition.

My sponsoring editor at Prentice Hall, Tom Robbins, deserves great thanks for his patience. He's the second editor to have been lured to Prentice Hall in part by the (falsely attractive) prospect of working on a project with me, only to find out upon his arrival that the project was very late. In Tom's case, though, I had known him since the early 80s when he first tried to sign me up for a project with another publisher, and we'd been trying to find a way to work together ever since. We're finally there, starting the third decade of our friendship. Tom contributed more than patience—among other things, you have him to thank for getting you the free Xilinx software at the back of this book (most printings).

Production editor Irwin Zucker also deserves credit for providing a very smooth interface with the production side of the house and for putting in long hours to help me during the final "crunch" stage of the project. If not for him, I would not be able to leave on a long-planned three-week family vacation later this morning. (I'm told that if I hadn't finished in time, our 90-pound dog would have gone to Europe in my seat instead!) Robert Lentz also did a great job as a copy editor which I could live with :-) .

Special thanks go to artist Robert McFadden, whose trippy cover painting is hanging in my home along with several other of his far-out works. His painting, which I commissioned and he completed well over a year ago, provided me with the motivation to actually get the *inside* of the book done.

It seems like some disaster always strikes just as I am completing one of these projects. For the first edition, it was the World-Series earthquake of 1989. For the second edition, it was surgery four days before the book's completion for a ruptured and very yucky-looking appendix. This time, I seemed to have dodged the bullet so far. Actually, I won't be completely done until I've finished the index, which I'll be preparing on my laptop as we ride the rails through Europe over the next few weeks. Let's hope I don't forget it on some train!

As always, I must thank my wife Kate for putting up with the late hours, frustration, crabbiness, preoccupation, and phone calls from weird people that occur when I'm engaged in a writing project like this. We hope you enjoy starting this book as much as we enjoy finishing it!

John F. Wakerly
Mountain View, California

Digital Design
Principles and Practices

Hi, I'm John . . . .

Introduction

Welcome to the world of digital design. Perhaps you're a computer science student who knows all about computer software and programming, but you're still trying to figure out how all that fancy hardware could possibly work. Or perhaps you're an electrical engineering student who already knows something about analog electronics and circuit design, but you wouldn't know a bit if it bit you. No matter. Starting from a fairly basic level, this book will show you how to design digital circuits and subsystems.

We'll give you the basic principles that you need to figure things out, and we'll give you lots of examples. Along with principles, we'll try to convey the flavor of real-world digital design by discussing current, practical considerations whenever possible. And I, the author, will often refer to myself as "we" in the hope that you'll be drawn in and feel that we're walking through the learning process together.

1.1 About Digital Design

Some people call it "logic design." That's OK, but ultimately the goal of design is to build systems. To that end, we'll cover a whole lot more in this text than just logic equations and theorems.

This book claims to be about principles and practices. Most of the principles that we present will continue to be important years from now; some may be applied in ways that have not even been discovered yet. As for practices, they may be a little different from what's presented here by the time you start working in the field, and they will certainly continue to change throughout your career. So you should treat the "practices" material in this book as a way to reinforce principles, and as a way to learn design methods by example.

One of the book's goals is to present enough about basic principles for you to know what's happening when you use software tools to turn the crank for you. The same basic principles can help you get to the root of problems when the tools happen to get in your way.

Listed in the box on this page are several key points that you should learn through your studies with this text. Most of these items probably make no sense to you right now, but you should come back and review them later.

Digital design is engineering, and engineering means "problem solving." My experience is that only 5%–10% of digital design is "the fun stuff"—the creative part of design, the flash of insight, the invention of a new approach. Much of the rest is just "turning the crank." To be sure, turning the crank is much easier now than it was 20 or even 10 years ago, but you still can't spend 100% or even 50% of your time on the fun stuff.

IMPORTANT THEMES IN DIGITAL DESIGN

- Good tools do not guarantee good design, but they help a lot by taking the pain out of doing things right.
- Digital circuits have analog characteristics.
- Know when to worry and when not to worry about the analog aspects of digital design.
- Always document your designs to make them understandable by yourself and others.
- Associate active levels with signal names and practice bubble-to-bubble logic design.
- Understand and use standard functional building blocks.
- Design for minimum cost at the system level, including your own engineering effort as part of the cost.
- State-machine design is like programming; approach it that way.
- Use programmable logic to simplify designs, reduce cost, and accommodate last-minute modifications.
- Avoid asynchronous design. Practice synchronous design until a better methodology comes along.
- Pinpoint the unavoidable asynchronous interfaces between different subsystems and the outside world, and provide reliable synchronizers.
- Catching a glitch in time saves nine.

Besides the fun stuff and turning the crank, there are many other areas in which a successful digital designer must be competent, including the following:

- *Debugging.* It's next to impossible to be a good designer without being a good troubleshooter. Successful debugging takes planning, a systematic approach, patience, and logic: if you can't discover where a problem *is*, find out where it *is not*!

- *Business requirements and practices.* A digital designer's work is affected by a lot of non-engineering factors, including documentation standards, component availability, feature definitions, target specifications, task scheduling, office politics, and going to lunch with vendors.

- *Risk-taking.* When you begin a design project you must carefully balance risks against potential rewards and consequences, in areas ranging from new-component selection (will it be available when I'm ready to build the first prototype?) to schedule commitments (will I still have a job if I'm late?).

- *Communication.* Eventually, you'll hand off your successful designs to other engineers, other departments, and customers. Without good communication skills, you'll never complete this step successfully. Keep in mind that communication includes not just transmitting but also receiving; learn to be a good listener!

In the rest of this chapter, and throughout the text, I'll continue to state some opinions about what's important and what is not. I think I'm entitled to do so as a moderately successful practitioner of digital design. Of course, you are always welcome to share your own opinions and experience (send email to john@wakerly.com).

1.2 Analog versus Digital

Analog devices and systems process time-varying signals that can take on any value across a continuous range of voltage, current, or other metric. So do *digital* circuits and systems; the difference is that we can pretend that they don't! A digital signal is modeled as taking on, at any time, only one of two discrete values, which we call *0* and *1* (or LOW and HIGH, FALSE and TRUE, negated and asserted, Sam and Fred, or whatever).

analog
digital

0
1

Digital computers have been around since the 1940s and have been in widespread commercial use since the 1960s. Yet only in the past 10 to 20 years has the "digital revolution" spread to many other aspects of life. Examples of once-analog systems that have now "gone digital" include the following:

- *Still pictures.* The majority of cameras still use silver-halide film to record images. However, the increasing density of digital memory chips has allowed the development of digital cameras which record a picture as a

640×480 or larger array of pixels, where each pixel stores the intensities of its red, green, and blue color components as 8 bits each. This large amount of data, over seven million bits in this example, may be processed and compressed in a format called JPEG down to as little as 5% of the original storage size, depending on the amount of picture detail. So, digital cameras rely on both digital storage and digital processing.

- *Video recordings.* A digital versatile disc (DVD) stores video in a highly compressed digital format called MPEG-2. This standard encodes a small fraction of the individual video frames in a compressed format similar to JPEG, and encodes each other frame as the difference between it and the previous one. The capacity of a single-layer, single-sided DVD is about 35 billion bits, sufficient for about 2 hours of high-quality video, and a two-layer, double-sided disc has four times that capacity.

- *Audio recordings.* Once made exclusively by impressing analog waveforms onto vinyl or magnetic tape, audio recordings now commonly use digital compact discs (CDs). A CD stores music as a sequence of 16-bit numbers corresponding to samples of the original analog waveform, one sample per stereo channel every 22.7 microseconds. A full-length CD recording (73 minutes) contains over six billion bits of information.

- *Automobile carburetors.* Once controlled strictly by mechanical linkages (including clever "analog" mechanical devices that sensed temperature, pressure, etc.), automobile engines are now controlled by embedded microprocessors. Various electronic and electromechanical sensors convert engine conditions into numbers that the microprocessor can examine to determine how to control the flow of fuel and oxygen to the engine. The microprocessor's output is a time-varying sequence of numbers that operate electromechanical actuators which, in turn, control the engine.

- *The telephone system.* It started out a hundred years ago with analog microphones and receivers connected to the ends of a pair of copper wires (or was it string?). Even today, most homes still use analog telephones, which transmit analog signals to the phone company's central office (CO). However, in the majority of COs, these analog signals are converted into a digital format before they are routed to their destinations, be they in the same CO or across the world. For many years the private branch exchanges (PBXs) used by businesses have carried the digital format all the way to the desktop. Now many businesses, COs, and traditional telephony service providers are converting to integrated systems that combine digital voice with data traffic over a single IP (Internet Protocol) network.

- *Traffic lights.* Stop lights used to be controlled by electromechanical timers that would give the green light to each direction for a predetermined amount of time. Later, relays were used in controllers that could activate the lights according to the pattern of traffic detected by sensors embedded in the pavement. Today's controllers use microprocessors and can control the lights in ways that maximize vehicle throughput or, in some California cities, frustrate drivers in all kinds of creative ways.

- *Movie effects.* Special effects used to be created exclusively with miniature clay models, stop action, trick photography, and numerous overlays of film on a frame-by-frame basis. Today, spaceships, bugs, other-worldly scenes, and even babies from hell (in Pixar's animated feature *Tin Toy*) are synthesized entirely using digital computers. Might the stunt man or woman someday no longer be needed, either?

The electronics revolution has been going on for quite some time now, and the "solid-state" revolution began with analog devices and applications like transistors and transistor radios. So why has there now been a *digital* revolution? There are in fact many reasons to favor digital circuits over analog ones:

- *Reproducibility of results.* Given the same set of inputs (in both value and time sequence), a properly designed digital circuit always produces exactly the same results. The outputs of an analog circuit vary with temperature, power-supply voltage, component aging, and other factors.

- *Ease of design.* Digital design, often called "logic design," is logical. No special math skills are needed, and the behavior of small logic circuits can be visualized mentally without any special insights about the operation of capacitors, transistors, or other devices that require calculus to model.

- *Flexibility and functionality.* Once a problem has been reduced to digital form, it can be solved using a set of logical steps in space and time. For example, you can design a digital circuit that scrambles your recorded voice so that it is absolutely indecipherable by anyone who does not have your "key" (password), but it can be heard virtually undistorted by anyone who does. Try doing that with an analog circuit.

- *Programmability.* You're probably already quite familiar with digital computers and the ease with which you can design, write, and debug programs for them. Well, guess what? Much of digital design is carried out today by writing programs, too, in *hardware description languages (HDLs)*. These languages allow both structure and function of a digital circuit to be specified or *modeled*. Besides a compiler, a typical HDL also comes with simulation and synthesis programs. These software tools are used to test the hardware model's behavior before any real hardware is built, and then to synthesize the model into a circuit in a particular component technology.

hardware description language (HDL)

hardware model

- *Speed.* Today's digital devices are very fast. Individual transistors in the fastest integrated circuits can switch in less than 10 picoseconds, and a complete, complex device built from these transistors can examine its

SHORT TIMES A *microsecond (μs)* is 10^{-6} second. A *nanosecond (ns)* is just 10^{-9} second, and a *picosecond (ps)* is 10^{-12} second. In a vacuum, light travels about a foot in a nanosecond, and an inch in 85 picoseconds. With individual transistors in the fastest integrated circuits now switching in less than 10 picoseconds, the speed-of-light delay between these transistors across a half-inch-square silicon chip has become a limiting factor in circuit design.

inputs and produce an output in less than 2 nanoseconds. This means that such a device can produce 500 million or more results per second.

- *Economy.* Digital circuits can provide a lot of functionality in a small space. Circuits that are used repetitively can be "integrated" into a single "chip" and mass-produced at very low cost, making possible throw-away items like calculators, digital watches, and singing birthday cards. (You may ask, "Is this such a good thing?" Never mind!)

- *Steadily advancing technology.* When you design a digital system, you almost always know that there will be a faster, cheaper, or otherwise better technology for it in a few years. Clever designers can accommodate these expected advances during the initial design of a system, to forestall system obsolescence and to add value for customers. For example, desktop computers often have "expansion sockets" to accommodate faster processors or larger memories than are available at the time of the computer's introduction.

So, that's enough of a sales pitch on digital design. The rest of this chapter will give you a bit more technical background to prepare you for the rest of the book.

1.3 Digital Devices

gate

The most basic digital devices are called *gates*, and no, they were not named after the founder of a large software company. Gates originally got their name from their function of allowing or retarding ("gating") the flow of digital information. In general, a gate has one or more inputs and produces an output that is a function of the current input value(s). While the inputs and outputs may be analog conditions such as voltage, current, even hydraulic pressure, they are modeled as taking on just two discrete values, 0 and 1.

AND gate

Figure 1-1 shows symbols for the three most important kinds of gates. A 2-input *AND gate*, shown in (a), produces a 1 output if both of its inputs are 1; otherwise it produces a 0 output. The figure shows the same gate four times, with the four possible combinations of inputs that may be applied to it and the result-

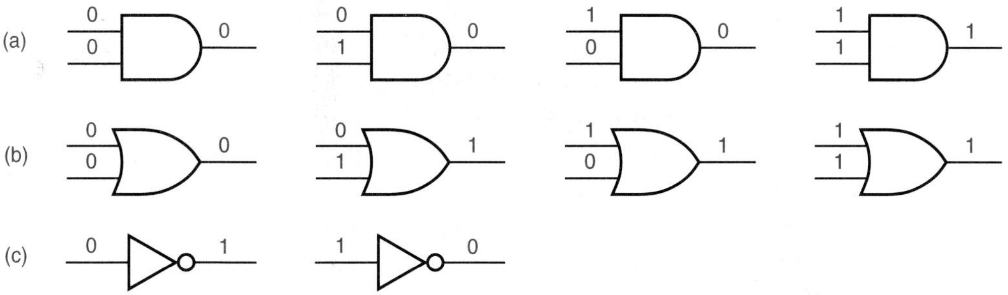

Figure 1-1 Digital devices: (a) AND gate; (b) OR gate; (c) NOT gate or inverter.

ing outputs. A gate is called a *combinational circuit* because its output depends only on the current input combination.

combinational circuit

A 2-input *OR gate*, shown in (b), produces a 1 output if one or both of its inputs are 1; it produces a 0 output only if both inputs are 0. Once again, there are four possible input combinations, resulting in the outputs shown in the figure.

OR gate

A *NOT gate*, more commonly called an *inverter*, produces an output value that is the opposite of the input value, as shown in (c).

NOT gate
inverter

We called these three gates the most important for good reason. Any digital function can be realized using just these three kinds of gates. In Chapter 3 we'll show how gates are realized using transistor circuits. You should know, however, that gates have been built or proposed using other technologies, such as relays, vacuum tubes, hydraulics, and molecular structures.

A *flip-flop* is a device that stores either a 0 or 1. The *state* of a flip-flop is the value that it currently stores. The stored value can be changed only at certain times determined by a "clock" input, and the new value may further depend on the flip-flop's current state and its "control" inputs. A flip-flop can be built from a collection of gates hooked up in a clever way, as we'll show in Section 7.2.

flip-flop
state

A digital circuit that contains flip-flops is called a *sequential circuit*, because its output at any time depends not only on its current input but also on the past sequence of inputs that have been applied to it. In other words, a sequential circuit has *memory* of past events.

sequential circuit

memory

1.4 Electronic Aspects of Digital Design

Digital circuits are not exactly a binary version of alphabet soup—with all due respect to Figure 1-1, they don't have little 0s and 1s floating around in them. As we'll see in Chapter 3, digital circuits deal with analog voltages and currents and are built with analog components. The "digital abstraction" allows analog behavior to be ignored in most cases, so circuits can be modeled as if they really did process 0s and 1s.

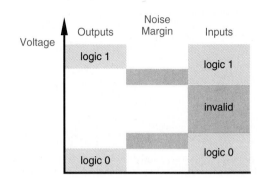

Figure 1-2
Logic values and
noise margins.

One important aspect of the digital abstraction is to associate a *range* of analog values with each logic value (0 or 1). As shown in Figure 1-2, a typical gate is not guaranteed to have a precise voltage level for a logic 0 output. Rather, it may produce a voltage somewhere in a range that is a *subset* of the range guaranteed to be recognized as a 0 by other gate inputs. The difference between the range boundaries is called *noise margin*—in a real circuit, a gate's output can be corrupted by this much noise and still be correctly interpreted at the inputs of other gates.

noise margin

Behavior for logic 1 outputs is similar. Note in the figure that there is an "invalid" region between the input ranges for logic 0 and logic 1. Although any given digital device operating at a particular voltage and temperature will have a fairly well defined boundary (or threshold) between the two ranges, different devices may have different boundaries. Still, all properly operating devices have their boundary *somewhere* in the "invalid" range. Therefore, any signal that is within the defined ranges for 0 and 1 will be interpreted identically by different devices. This characteristic is essential for reproducibility of results.

It is the job of an *electronic* circuit designer to ensure that logic gates produce and recognize logic signals that are within the appropriate ranges. This is an analog circuit-design problem; we touch upon some of its aspects in Chapter 3. It is not possible to design a circuit that has the desired behavior under every possible condition of power-supply voltage, temperature, loading, and other factors. Instead, the electronic circuit designer or device manufacturer provides *specifications* that define the conditions under which correct behavior is guaranteed.

specifications

As a *digital* designer, then, you need not delve into the detailed analog behavior of a digital device to ensure its correct operation. Rather, you need only examine enough about the device's operating environment to determine that it is operating within its published specifications. Granted, some analog knowledge is needed to perform this examination, but not nearly what you'd need to design a digital device starting from scratch. In Chapter 3 we'll give you just what you need.

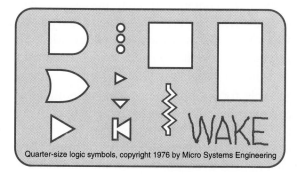

Figure 1-3
A logic-design
template.

1.5 Software Aspects of Digital Design

Digital design need not involve any software tools. For example, Figure 1-3 shows the primary tool of the "old school" of digital design—a plastic template for drawing logic symbols in schematic diagrams by hand (the designer's name was engraved into the plastic with a soldering iron).

Today, however, software tools are an essential part of digital design. Indeed, the availability and practicality of hardware description languages (HDLs) and accompanying circuit simulation and synthesis tools have changed the entire landscape of digital design over the past several years. We'll make extensive use of HDLs throughout this book.

In *computer-aided design (CAD)* various software tools improve the designer's productivity and help to improve the correctness and quality of designs. In a competitive world, the use of software tools is mandatory to obtain high-quality results on aggressive schedules. Important examples of software tools for digital design are listed below:

computer-aided design (CAD)

- *Schematic entry.* This is the digital designer's equivalent of a word processor. It allows schematic diagrams to be drawn "on-line," instead of with paper and pencil. The more advanced schematic-entry programs also check for common, easy-to-spot errors, such as shorted outputs, signals that don't go anywhere, and so on. Such programs are discussed in greater detail in Section 11.1.2.

- *HDLs.* Hardware description languages, originally developed for circuit modeling, are now being used more and more for hardware *design*. They can be used to design anything from individual function modules to large, multichip digital systems. We'll introduce two HDLs, ABEL and VHDL, at the end of Chapter 4, and we'll provide examples in both languages in the chapters that follow.

- *HDL compilers, simulators, and synthesis tools.* A typical HDL software package contains several components. In a typical environment, the designer writes a text-based "program," and the HDL compiler analyzes the program for syntax errors. If it compiles correctly, the designer has the option of handing it over to a synthesis tool that creates a corresponding circuit design targeted to a particular hardware technology. Most often, before synthesis the designer will use the compiler's results as input to a "simulator" to verify the behavior of the design.

- *Simulators.* The design cycle for a customized, single-chip digital integrated circuit is long and expensive. Once the first chip is built, it's very difficult, often impossible, to debug it by probing internal connections (they are really tiny), or to change the gates and interconnections. Usually, changes must be made in the original design database and a new chip must be manufactured to incorporate the required changes. Since this process can take months to complete, chip designers are highly motivated to "get it right" (or almost right) on the first try. Simulators help designers predict the electrical and functional behavior of a chip without actually building it, allowing most if not all bugs to be found before the chip is fabricated.

- Simulators are also used in the design of "programmable logic devices," introduced later, and in the overall design of systems that incorporate many individual components. They are somewhat less critical in this case because it's easier for the designer to make changes in components and interconnections on a printed-circuit board. However, even a little bit of simulation can save time by catching simple but stupid mistakes.

- *Test benches.* Digital designers have learned how to formalize circuit simulation and testing into software environments called "test benches." The idea is to build a set of programs around a design to automatically exercise its functions and check both its functional and its timing behavior. This is especially useful when small design changes are made—the test bench can be run to ensure that bug fixes or "improvements" in one area do not break something else. Test-bench programs may be written in the same HDL as the design itself, in C or C++, or in combination of languages including scripting languages like PERL.

- *Timing analyzers and verifiers.* The time dimension is very important in digital design. All digital circuits take time to produce a new output value in response to an input change, and much of a designer's effort is spent ensuring that such output changes occur quickly enough (or, in some cases, not too quickly). Specialized programs can automate the tedious task of drawing timing diagrams and specifying and verifying the timing relationships between different signals in a complex system.

- *Word processors.* Let's not forget the lowly text editor and word processor. These tools are obviously useful for creating the source code for HDL-based designs, but they have an important use in every design—to create documentation!

PROGRAMMABLE LOGIC DEVICES VERSUS SIMULATION

Later in this book you'll learn how programmable logic devices (PLDs) and field-programmable gate arrays (FPGAs) allow you to design a circuit or subsystem by writing a sort of program. PLDs and FPGAs are now available with up to millions of gates, and the capabilities of these technologies are ever increasing. If a PLD- or FPGA-based design doesn't work the first time, you can often fix it by changing the program and physically reprogramming the device, without changing any components or interconnections at the system level. The ease of prototyping and modifying PLD- and FPGA-based systems can eliminate the need for simulation in board-level design; simulation is required only for chip-level designs.

The most widely held view in industry trends says that as chip technology advances, more and more design will be done at the chip level, rather than the board level. Therefore, the ability to perform complete and accurate simulation will become increasingly important to the typical digital designer.

However, another view is possible. If we extrapolate trends in PLD and FPGA capabilities, in the next decade we will witness the emergence of devices that include not only gates and flip-flops as building blocks, but also higher-level functions such as processors, memories, and input/output controllers. At this point, most digital designers will use complex on-chip components and interconnections whose basic functions have already been tested by the device manufacturer.

In this future view, it is still possible to misapply high-level programmable functions, but it is also possible to fix mistakes simply by changing a program; detailed simulation of a design before simply "trying it out" could be a waste of time. Another, compatible view is that the PLD or FPGA is merely a full-speed simulator for the program, and this full-speed simulator is what gets shipped in the product!

Does this extreme view have any validity? To guess the answer, ask yourself the following question. How many software programmers do you know who debug a new program by "simulating" its operation rather than just trying it out?

In any case, modern digital systems are much too complex for a designer to have any chance of testing every possible input condition, with or without simulation. As in software, correct operation of digital systems is best accomplished through practices that ensure that the systems are "correct by design." It is a goal of this text to encourage such practices.

In addition to using the tools above, designers may sometimes write specialized programs in high-level languages like C or C++, or scripts in languages like PERL, to solve particular design problems. For example, Section 10.1.6 gives a couple of examples of C programs that generate the "truth tables" for complex combinational logic functions.

Although CAD tools are important, they don't make or break a digital designer. To take an analogy from another field, you couldn't consider yourself to be a great writer just because you're a fast typist or very handy with a word processor. During your study of digital design, be sure to learn and use all the tools that are available to you, such as schematic-entry programs, simulators, and HDL compilers. But remember that learning to use tools is no guarantee that you'll be able to produce good results. Please pay attention to what you're producing with them!

1.6 Integrated Circuits

integrated circuit (IC)

A collection of one or more gates fabricated on a single silicon chip is called an *integrated circuit (IC)*. Large ICs with tens of millions of transistors may be half an inch or more on a side, while small ICs may be less than one-tenth of an inch on a side.

wafer

Regardless of its size, an IC is initially part of a much larger, circular *wafer*, up to ten inches in diameter, containing dozens to hundreds of replicas of the same IC. All of the IC chips on the wafer are fabricated at the same time, like pizzas that are eventually sold by the slice, except in this case, each piece (IC chip) is called a *die*. After the wafer is fabricated, the dice are tested in place on the wafer and defective ones are marked. Then the wafer is sliced up to produce the individual dice, and the marked ones are discarded. (Compare with the pizza-maker who sells all the pieces, even the ones without enough pepperoni!) Each unmarked die is mounted in a package, its pads are connected to the package pins, and the packaged IC is subjected to a final test and is shipped to a customer.

die

Some people use the term "IC" to refer to a silicon die. Some use "chip" to refer to the same thing. Still others use "IC" or "chip" to refer to the combination of a silicon die and its package. Digital designers tend to use the two terms interchangeably, and they really don't care what they're talking about. They don't require a precise definition, since they're only looking at the functional and electrical behavior of these things. In the balance of this text, we'll use the term *IC* to refer to a packaged die.

IC

A DICEY DECISION

A reader of the second edition wrote to me to collect a $5 reward for pointing out my "glaring" misuse of "dice" as the plural of "die." According to the dictionary, she said, the plural form of "die" is "dice" *only* when describing those little cubes with dots on each side; otherwise it's "dies," and she produced the references to prove it.

Being stubborn, I asked my friends at the *Microprocessor Report* about this issue. According to the editor,

> There is, indeed, much dispute over this term. We actually stopped using the term "dice" in *Microprocessor Report* more than four years ago. I actually prefer the plural "die," … but perhaps it is best to avoid using the plural whenever possible.

So there you have it, even the experts don't agree with the dictionary! Rather than cop out, I rolled the dice and decided to use "dice" anyway.

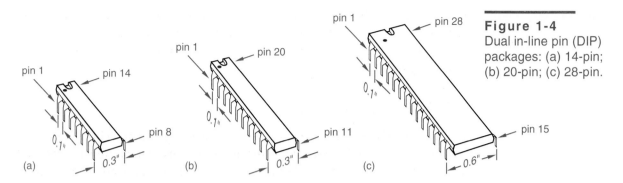

Figure 1-4
Dual in-line pin (DIP)
packages: (a) 14-pin;
(b) 20-pin; (c) 28-pin.

In the early days of integrated circuits, ICs were classified by size—small, medium, or large—according to how many gates they contained. The simplest type of commercially available ICs are still called *small-scale integration (SSI)* and contain the equivalent of 1 to 20 gates. SSI ICs typically contain a handful of gates or flip-flops, the basic building blocks of digital design.

    The SSI ICs that you're likely to encounter in an educational lab come in a 14-pin *dual in-line-pin (DIP)* package. As shown in Figure 1-4(a), the spacing between pins in a column is 0.1 inch and the spacing between columns is 0.3 inch. Larger DIP packages accommodate functions with more pins, as shown in (b) and (c). A *pin diagram* shows the assignment of device signals to package pins, or *pinout*. Figure 1-5 shows the pin diagrams for a few common SSI ICs. Such diagrams are used only for mechanical reference, when a designer needs to determine the pin numbers for a particular IC. In the schematic diagram for a

small-scale integration (SSI)

dual in-line-pin (DIP) package

pin diagram
pinout

Figure 1-5 Pin diagrams for a few 7400-series SSI ICs.

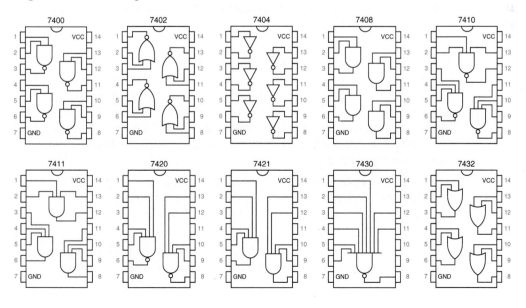

digital circuit, pin diagrams are not used. Instead, the various gates are grouped functionally, as we'll show in Section 5.1.

Although SSI ICs are still sometimes used as "glue" to tie together larger-scale elements in complex systems, they have been largely supplanted by programmable logic devices (PLDs), which we'll study in Sections 5.3 and 8.3.

medium-scale integration (MSI)

The next larger commercially available ICs are called *medium-scale integration (MSI)* and contain the equivalent of about 20 to 200 gates. An MSI IC typically contains a functional building block, such as a decoder, register, or counter. In Chapters 5 and 8 we'll place a strong emphasis on these building blocks. Even though the use of discrete MSI ICs is declining, the equivalent building blocks are used extensively in the design of larger ICs.

large-scale integration (LSI)

Large-scale integration (LSI) ICs are bigger still, containing the equivalent of 200 to 200,000 gates or more. LSI parts include small memories, microprocessors, programmable logic devices, and customized devices.

The dividing line between LSI and *very large-scale integration (VLSI)* is fuzzy and tends to be stated in terms of transistor count rather than gate count. Any IC with over 1,000,000 transistors is definitely VLSI, and that includes most microprocessors and memories nowadays, as well as larger programmable logic devices and customized devices. In 1999, VLSI ICs with as many as 50 million transistors were being designed.

very large-scale integration (VLSI)

1.7 Programmable Logic Devices

There are a wide variety of ICs that can have their logic function "programmed" into them after they are manufactured. Most of these devices use technology that also allows the function to be *re*programmed, which means that if you find a bug in your design, you may be able to fix it without physically replacing or rewiring the device. In this book, we'll frequently refer to the design opportunities and methods for such devices.

Historically, *programmable logic arrays (PLAs)* were the first programmable logic devices. PLAs contained a two-level structure of AND and OR gates with user-programmable connections. Using this structure, a designer could accommodate any logic function up to a certain level of complexity using the well-known theory of logic synthesis and minimization that we'll present in Chapter 4.

programmable logic array (PLA)

PLA structure was enhanced and PLA costs were reduced with the introduction of *programmable array logic (PAL) devices*. Today, such devices are generically called programmable logic devices (PLDs) and are the "MSI" of the programmable logic industry. We'll have a lot to say about PLD architecture and technology in Sections 5.3 and 8.3.

programmable array logic (PAL) device
programmable logic device (PLD)

The ever-increasing capacity of integrated circuits created an opportunity for IC manufacturers to design larger PLDs for larger digital-design applications. However, for technical reasons that we'll discuss in Section 10.5, the basic two-level AND-OR structure of PLDs could not be scaled to larger sizes. Instead, IC manufacturers devised *complex PLD (CPLD)* architectures to achieve the required scale. A typical CPLD is merely a collection of multiple PLDs and an interconnection structure, all on the same chip. In addition to the individual PLDs, the on-chip interconnection structure is also programmable, providing a rich variety of design possibilities. CPLDs can be scaled to larger sizes by increasing the number of individual PLDs and the richness of the interconnection structure on the CPLD chip.

complex PLD (CPLD)

At about the same time that CPLDs were being invented, other IC manufacturers took a different approach to scaling the size of programmable logic chips. Compared to a CPLD, a field-programmable gate array (FPGA) contains a much larger number of smaller individual logic blocks and provides a large, distributed interconnection structure that dominates the entire chip. Figure 1-6 illustrates the difference between the two chip-design approaches.

field-programmable gate array (FPGA)

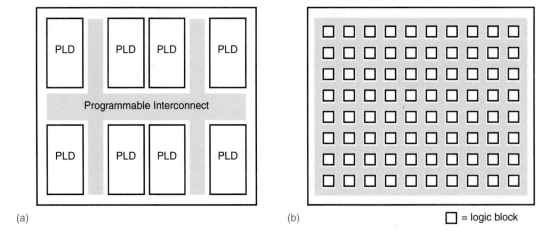

Figure 1-6 Large programmable-logic-device scaling approaches: (a) CPLD; (b) FPGA.

Proponents of one approach or the other used to get into "religious" arguments over which way was better, but the largest manufacturer of large programmable logic devices, Xilinx, Inc., acknowledges that there is a place for both approaches and manufactures both types of devices. What's more important than chip architecture is that both approaches support a style of design in which products can be moved from design concept to prototype and production in a very short time.

Also important in achieving short "time-to-market" for all kinds of PLD-based products is the use of HDLs in their design. Languages like ABEL and VHDL, and their accompanying software tools, allow a design to be compiled, synthesized, and downloaded into a PLD, CPLD, or FPGA literally in minutes. The power of highly structured, hierarchical languages like VHDL is especially important in helping designers utilize the hundreds of thousands or millions of gates that are provided in the largest CPLDs and FPGAs.

1.8 Application-Specific ICs

Perhaps the most interesting developments in IC technology for the average digital designer are not the ever-increasing chip sizes, but the ever-increasing opportunities to "design your own chip." Chips designed for a particular, limited product or application are called *semicustom ICs* or *application-specific ICs (ASICs)*. ASICs generally reduce the total component and manufacturing cost of a product by reducing chip count, physical size, and power consumption, and they often provide higher performance.

The *nonrecurring engineering (NRE) cost* for designing an ASIC can exceed the cost of a discrete design by $5,000 to $250,000 or more. NRE charges are paid to the IC manufacturer and others who are responsible for designing the

semicustom IC
application-specific IC (ASIC)

nonrecurring engineering (NRE) cost

internal structure of the chip, creating tooling such as the metal masks for manufacturing the chips, developing tests for the manufactured chips, and actually making the first few sample chips.

The NRE cost for a typical, medium-complexity ASIC with about 100,000 gates is $30,000–$50,000. An ASIC design normally makes sense only when the NRE cost can be offset by the per-unit savings over the expected sales volume of the product.

The NRE cost to design a *custom LSI* chip—a chip whose functions, internal architecture, and detailed transistor-level design is tailored for a specific customer—is very high, $250,000 or more. Thus, full custom LSI design is done only for chips that have general commercial application or that will enjoy very high sales volume in a specific application (e.g., a digital watch chip, a network interface, or a bus-interface circuit for a PC).

custom LSI

To reduce NRE charges, IC manufacturers have developed libraries of *standard cells* including commonly used MSI functions such as decoders, registers, and counters and commonly used LSI functions such as memories, programmable logic arrays, and microprocessors. In a *standard-cell design*, the logic designer interconnects functions in much the same way as in a multichip MSI/LSI design. Custom cells are created (at added cost, of course) only if absolutely necessary. All of the cells are then laid out on the chip, optimizing the layout to reduce propagation delays and minimize the size of the chip. Minimizing the chip size reduces the per-unit cost of the chip, since it increases the number of chips that can be fabricated on a single wafer. The NRE cost for a standard-cell design is typically on the order of $150,000.

standard cells

standard-cell design

Well, $150,000 is still a lot of money for most folks, so IC manufacturers have gone one step further to bring ASIC design capability to the masses. A *gate array* is an IC whose internal structure is an array of gates whose interconnections are initially unspecified. The logic designer specifies the gate types and interconnections. Even though the chip design is ultimately specified at this very low level, the designer typically works with "macrocells," the same high-level functions used in multichip MSI/LSI and standard-cell designs; software expands the high-level design into a low-level one.

gate array

The main difference between standard-cell and gate-array design is that the macrocells and the chip layout of a gate array are not as highly optimized as those in a standard-cell design, so the chip may be 25% or more larger and therefore may cost more. Also, there is no opportunity to create custom cells in the gate-array approach. On the other hand, a gate-array design can be finished faster and at lower NRE cost, ranging from about $5000 (what you're told initially) to $75,000 (what you find you've spent when you're all done).

The basic digital design methods that you'll study throughout this book apply very well to the functional design of ASICs. However, there are additional opportunities, constraints, and steps in ASIC design, which usually depend on the particular ASIC vendor and design environment.

1.9 Printed-Circuit Boards

printed-circuit board (PCB)

printed-wiring board (PWB)

PCB traces

mil

fine-line

An IC is normally mounted on a *printed-circuit board (PCB)* [or *printed-wiring board (PWB)*] that connects it to other ICs in a system. The multilayer PCBs used in typical digital systems have copper wiring etched on multiple, thin layers of fiberglass that are laminated into a single board about 1/16 inch thick.

Individual wire connections, or *PCB traces*, are usually quite narrow, 10 to 25 mils in typical PCBs. (A *mil* is one-thousandth of an inch.) In *fine-line* PCB technology, the traces are extremely narrow, as little as 4 mils wide with 4-mil spacing between adjacent traces. Thus, up to 125 connections may be routed in a one-inch-wide band on a single layer of the PCB. If higher connection density is needed, then more layers are used.

surface-mount technology (SMT)

Most of the components in modern PCBs use *surface-mount technology (SMT)*. Instead of having the long pins of DIP packages that poke through the board and are soldered to the underside, the leads of SMT IC packages are bent to make flat contact with the top surface of the PCB. Before such components are mounted on the PCB, a special "solder paste" is applied to contact pads on the PCB using a stencil whose hole pattern matches the contact pads to be soldered. Then the SMT components are placed (by hand or by machine) on the pads, where they are held in place by the solder paste (or in some cases, by glue). Finally, the entire assembly is passed through an oven to melt the solder paste, which then solidifies when cooled.

Surface-mount component technology, coupled with fine-line PCB technology, allows extremely dense packing of integrated circuits and other components on a PCB. This dense packing does more than save space. For very high-speed circuits, dense packing goes a long way toward minimizing adverse analog phenomena, including transmission-line effects and speed-of-light limitations.

multichip module (MCM)

To satisfy the most stringent requirements for speed and density, *multichip modules (MCMs)* have been developed. In this technology, IC dice are not mounted in individual plastic or ceramic packages. Instead, the IC dice for a high-speed subsystem (say, a processor and its cache memory) are bonded directly to a substrate that contains the required interconnections on multiple layers. The MCM is hermetically sealed and has its own external pins for power, ground, and just those signals that are required by the system that contains it.

1.10 Digital-Design Levels

Digital design can be carried out at several different levels of representation and abstraction. Although you may learn and practice design at a particular level, from time to time you'll need to go up or down a level or two to get the job done. Also, the industry itself and most designers have been steadily moving to higher levels of abstraction as circuit density and functionality have increased.

The lowest level of digital design is device physics and IC manufacturing processes. This is the level that is primarily responsible for the breathtaking advances in IC speed and density that have occurred over the past decades. The effects of these advances are summarized in *Moore's Law*, first stated by Intel founder Gordon Moore in 1965: that the number of transistors per square inch in an IC doubles every year. In recent years, the rate of advance has slowed down to doubling about every 18 months, but it is important to note that with each doubling of density has also come a doubling of speed.

Moore's Law

This book does not reach down to the level of device physics and IC processes, but you need to recognize the importance of that level. Being aware of likely technology advances and other changes is important in system and product planning. For example, decreases in chip geometries have recently forced a move to lower logic-power-supply voltages, causing major changes in the way designers plan and specify modular systems and upgrades.

In this book, we jump into digital design at the transistor level and go all the way up to the level of logic design using HDLs. We stop short of the next level, which includes computer design and overall system design. The "center" of our discussion is at the level of functional building blocks.

To get a preview of the levels of design that we'll cover, consider a simple design example. Suppose you are to build a "multiplexer" with two data input bits, A and B, a control input bit S, and an output bit Z. Depending on the value of S, 0 or 1, the circuit is to transfer the value of either A or B to the output Z. This idea is illustrated in the "switch model" of Figure 1-7. Let us consider the design of this function at several different levels.

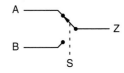

Figure 1-7
Switch model for multiplexer function.

Although logic design is usually carried out at higher level, for some functions it is advantageous to optimize them by designing at the transistor level. The multiplexer is such a function. Figure 1-8 shows how the multiplexer can be designed in "CMOS" technology using specialized transistor circuit structures

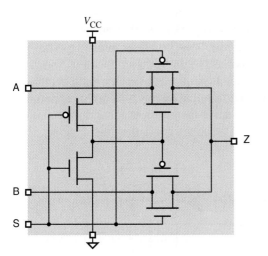

Figure 1-8
Multiplexer design using CMOS transmission gates.

Table 1-1
Truth table for the
multiplexer function.

S	A	B	Z
0	0	0	0
0	0	1	0
0	1	0	1
0	1	1	1
1	0	0	0
1	0	1	1
1	1	0	0
1	1	1	1

called "transmission gates," discussed in Section 3.7.1. Using this approach, the multiplexer can be built with just six transistors. Any of the other approaches that we describe requires at least 14 transistors.

In the traditional study of logic design, we would use a "truth table" to describe the multiplexer's logic function. A truth table lists all possible combinations of input values and the corresponding output values for the function. Since the multiplexer has three inputs, it has 2^3 or 8 possible input combinations, as shown in the truth table in Table 1-1.

Once we have a truth table, traditional logic design methods, described in Section 4.3, use Boolean algebra and well-understood minimization algorithms to derive an "optimal" two-level **AND-OR** equation from the truth table. For the multiplexer truth table, we would derive the following equation:

$$Z = S' \cdot A + S \cdot B$$

This equation is read "Z equals not S and A or S and B." Going one step further, we can convert the equation into a corresponding set of logic gates that perform the specified logic function, as shown in Figure 1-9. This circuit requires 14 transistors if we use standard CMOS technology for the four gates shown.

A multiplexer is a very commonly used function, and most digital logic technologies provide predefined multiplexer building blocks. For example, the 74x157 is an MSI chip that performs multiplexing on two 4-bit inputs simultaneously. Figure 1-10 is a logic diagram that shows how we can hook up just one bit of this 4-bit building block to solve the problem at hand. The numbers in color are pin numbers of a 16-pin DIP package containing the device.

Figure 1-9
Gate-level logic diagram
for multiplexer function.

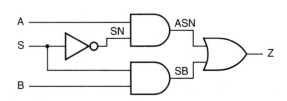

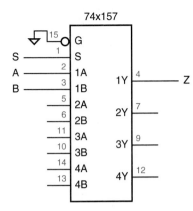

74x157

Figure 1-10
Logic diagram for a
multiplexer using an
MSI building block.

We can also realize the multiplexer function as part of a programmable logic device. Languages like ABEL allow us to specify outputs using Boolean equations similar to the one on the previous page, but it's usually more convenient to use "higher-level" language elements. For example, Table 1-2 is an ABEL program for the multiplexer function. The first three lines define the name of the program module and specify the type of PLD in which the function will be realized. The next two lines specify the device pin numbers for inputs and output. The "WHEN" statement specifies the actual logic function in a way that's very easy to understand, even though we haven't covered ABEL yet.

An even higher-level language, VHDL, can be used to specify the multiplexer function in a way that is very flexible and hierarchical. Table 1-3 is an example VHDL program for the multiplexer. The first two lines specify a standard library and set of definitions to use in the design. The next four lines specify only the inputs and outputs of the function, purposely hiding any details about the way the function is realized internally. The "architecture" section of the program specifies the function's behavior. VHDL syntax takes a little getting used to, but the single "when" statement says basically the same thing that the ABEL version did. A VHDL "synthesis tool" can start with this behavioral

Table 1-2
ABEL program for
the multiplexer.

```
module chap1mux
title 'Two-input multiplexer example'
CHAP1MUX device 'P16V8'

A, B, S        pin 1, 2, 3;
Z              pin 13 istype 'com';

equations

WHEN S == 0 THEN Z = A;  ELSE Z = B;

end chap1mux
```

Table 1-3
VHDL program for
the multiplexer.

```
library IEEE;
use IEEE.std_logic_1164.all;

entity Vchap1mux is
    port ( A, B, S: in  STD_LOGIC;
            Z:         out STD_LOGIC );
end Vchap1mux;

architecture Vchap1mux_arch of Vchap1mux is
begin
  Z <= A when S = '0' else B;
end Vchap1mux_arch;
```

description and produce a circuit that has this behavior in a specified target digital-logic technology.

By explicitly enforcing a separation of input/output definitions ("entity") and internal realization ("architecture"), VHDL makes it easy for designers to define alternate realizations of functions without having to make changes elsewhere in the design hierarchy. For example, a designer could specify an alternate, structural architecture for the multiplexer as shown in Table 1-4. This architecture is basically a text equivalent of the logic diagram in Figure 1-9.

Going one step further, VHDL is powerful enough that we could actually define operations that model functional behavior at the transistor level (though we won't explore such capabilities in this book). Thus, we could come full circle by writing a VHDL program that specifies a transistor-level realization of the multiplexer equivalent to Figure 1-8.

Table 1-4
"Structural" VHDL
program for the
multiplexer.

```
architecture Vchap1mux_gate_arch of Vchap1mux is
signal SN, ASN, SB: STD_LOGIC;
begin
  U1: INV (S, SN);
  U2: AND2 (A, SN, ASN);
  U3: AND2 (S, B, SB);
  U4: OR2 (ASN, SB, Z);
end Vchap1mux_gate_arch;
```

1.11 The Name of the Game

board-level design

Given the functional and performance requirements for a digital system, the name of the game in practical digital design is to minimize cost. For *board-level designs*—systems that are packaged on a single PCB—this usually means minimizing the number of IC packages. If too many ICs are required, they won't all fit on the PCB. "Well, just use a bigger PCB," you say. Unfortunately, PCB sizes are usually constrained by factors such as preexisting standards (e.g., add-in boards for PCs), packaging constraints (e.g., it has to fit in a toaster), or edicts from above (e.g., in order to get the project approved three months ago, you foolishly told your manager that it would all fit on a 3 × 5 inch PCB, and now you've got to deliver!). In each of these cases, the cost of using a larger PCB or multiple PCBs may be unacceptable.

Minimizing the number of ICs is usually the rule even though individual IC costs vary. For example, a typical SSI or MSI IC may cost 25 cents, while a small PLD may cost a dollar. It may be possible to perform a particular function with three SSI and MSI ICs (75 cents) or one PLD (a dollar). In most situations, the more expensive PLD solution is used, not because the designer owns stock in the IC company, but because the PLD solution uses less PCB area and is also a lot easier to change if it's not right the first time.

In *ASIC design*, the name of the game is a little different, but the importance of structured, functional design techniques is the same. Although it's easy to burn hours and weeks creating custom macrocells and minimizing the total gate count of an ASIC, only rarely is this advisable. The per-unit cost reduction achieved by having a 10% smaller chip is negligible except in high-volume applications. In applications with low to medium volume (the majority), two other factors are more important: design time and NRE cost.

ASIC design

A shorter design time allows a product to reach the market sooner, increasing revenues over the lifetime of the product. A lower NRE cost also flows right to the "bottom line" and in small companies may be the only way the project can be completed before the company runs out of money (believe me, I've been there!). If the product is successful, it's always possible and profitable to "tweak" the design later to reduce per-unit costs. The need to minimize design time and NRE cost argues in favor of a structured, as opposed to highly optimized, approach to ASIC design, using standard building blocks provided in the ASIC manufacturer's library.

The considerations in PLD, CPLD, and FPGA design are a combination of the above. The choice of a particular PLD technology and device size is usually made fairly early in the design cycle. Later, as long as the design "fits" in the selected device, there's no point in trying to optimize gate count or board area—the device has already been committed. However, if new functions or bug fixes push the design beyond the capacity of the selected device, that's when you must work very hard to modify the design to make it fit.

1.12 Going Forward

This concludes the introductory chapter. As you continue reading this book, keep in mind two things. First, the ultimate goal of digital design is to build systems that solve problems for people. While this book will give you the basic tools for design, it's still your job to keep "the big picture" in the back of your mind. Second, cost is an important factor in every design decision; and you must consider not only the cost of digital components, but also the cost of the design activity itself.

Finally, as you get deeper into the text, if you encounter something that you think you've seen before but don't remember where, please consult the index. I've tried to make it as helpful and complete as possible.

Drill Problems

1.1 Suggest some better-looking chapter-opening artwork to put on page 1 of the next edition of this book.

1.2 Give three different definitions for the word "bit" as used in this chapter.

1.3 Define the following acronyms: ASIC, CAD, CD, CO, CPLD, DIP, DVD, FPGA, HDL, IC, IP, LSI, MCM, MSI, NRE, OK, PBX, PCB, PLD, PWB, SMT, SSI, VHDL, VLSI.

1.4 Research the definitions of the following acronyms: ABEL, CMOS, JPEG, MPEG, OK, PERL, VHDL. (Is OK really an acronym?)

1.5 Excluding the topics in Section 1.2, list three once-analog systems that have "gone digital" since you were born.

1.6 Draw a digital circuit consisting of a 2-input AND gate and three inverters, where an inverter is connected to each of the AND gate's inputs and its output. For each of the four possible combinations of inputs applied to the two primary inputs of this circuit, determine the value produced at the primary output. Is there a simpler circuit that gives the same input/output behavior?

1.7 When should you use the pin diagrams of Figure 1-5 in the schematic diagram of a circuit?

1.8 What is the relationship between "die" and "dice"?

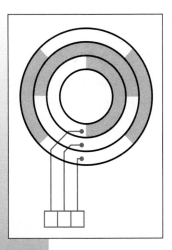

Number Systems and Codes

igital systems are built from circuits that process binary digits—
0s and 1s—yet very few real-life problems are based on binary
numbers or any numbers at all. As a result, a digital system
designer must establish some correspondence between the
binary digits processed by digital circuits and real-life numbers,
events, and conditions. The purpose of this chapter is to show you how
familiar numeric quantities can be represented and manipulated in a digital
system, and how nonnumeric data, events, and conditions also can be
represented.

The first nine sections describe binary number systems and show how
addition, subtraction, multiplication, and division are performed in these
systems. Sections 2.10–2.13 show how other things, such as decimal
numbers, text characters, mechanical positions, and arbitrary conditions,
can be encoded using strings of binary digits.

Section 2.14 introduces "n-cubes," which provide a way to visualize
the relationship between different bit strings. The n-cubes are especially
useful in the study of error-detecting codes in Section 2.15. We conclude the
chapter with an introduction to codes for transmitting and storing data one
bit at a time.

2.1 Positional Number Systems

positional number system

weight

The traditional number system that we learned in school and use every day in business is called a *positional number system*. In such a system, a number is represented by a string of digits, where each digit position has an associated *weight*. The value of a number is a weighted sum of the digits, for example:

$$1734 = 1 \cdot 1000 + 7 \cdot 100 + 3 \cdot 10 + 4 \cdot 1$$

Each weight is a power of 10 corresponding to the digit's position. A decimal point allows negative as well as positive powers of 10 to be used:

$$5185.68 = 5 \cdot 1000 + 1 \cdot 100 + 8 \cdot 10 + 5 \cdot 1 + 6 \cdot 0.1 + 8 \cdot 0.01$$

In general, a number D of the form $d_1 d_0 . d_{-1} d_{-2}$ has the value

$$D = d_1 \cdot 10^1 + d_0 \cdot 10^0 + d_{-1} \cdot 10^{-1} + d_{-2} \cdot 10^{-2}$$

base

radix

Here, 10 is called the *base* or *radix* of the number system. In a general positional number system, the radix may be any integer $r \geq 2$, and a digit in position i has weight r^i. The general form of a number in such a system is

$$d_{p-1} d_{p-2} \cdots d_1 d_0 . d_{-1} d_{-2} \cdots d_{-n}$$

radix point

where there are p digits to the left of the point and n digits to the right of the point, called the *radix point*. If the radix point is missing, it is assumed to be to the right of the rightmost digit. The value of the number is the sum of each digit multiplied by the corresponding power of the radix:

$$D = \sum_{i=-n}^{p-1} d_i \cdot r^i$$

high-order digit

most significant digit

low-order digit

least significant digit

Except for possible leading and trailing zeroes, the representation of a number in a positional number system is unique. (Obviously, 0185.6300 equals 185.63, and so on.) The leftmost digit in such a number is called the *high-order* or *most significant digit*; the rightmost is the *low-order* or *least significant digit*.

binary digit

bit

binary radix

As we'll learn in Chapter 3, digital circuits have signals that are normally in one of only two conditions—low or high, charged or discharged, off or on. The signals in these circuits are interpreted to represent *binary digits* (or *bits*) that have one of two values, 0 and 1. Thus, the *binary radix* is normally used to represent numbers in a digital system. The general form of a binary number is

$$b_{p-1} b_{p-2} \cdots b_1 b_0 . b_{-1} b_{-2} \cdots b_{-n}$$

and its value is

$$B = \sum_{i=-n}^{p-1} b_i \cdot 2^i$$

In a binary number, the radix point is called the *binary point*. When dealing with *binary point* binary and other nondecimal numbers, we use a subscript to indicate the radix of each number, unless the radix is clear from the context. Examples of binary numbers and their decimal equivalents are given below.

$$10011_2 \ = \ 1 \cdot 16 + 0 \cdot 8 + 0 \cdot 4 + 1 \cdot 2 + 1 \cdot 1 \ = \ 19_{10}$$

$$100010_2 \ = \ 1 \cdot 32 + 0 \cdot 16 + 0 \cdot 8 + 0 \cdot 4 + 1 \cdot 2 + 0 \cdot 1 \ = \ 34_{10}$$

$$101.001_2 \ = \ 1 \cdot 4 + 0 \cdot 2 + 1 \cdot 1 + 0 \cdot 0.5 + 0 \cdot 0.25 + 1 \cdot 0.125 \ = \ 5.125_{10}$$

The leftmost bit of a binary number is called the *high-order* or *most significant* *MSB* *bit (MSB)*; the rightmost is the *low-order* or *least significant bit (LSB)*. *LSB*

2.2 Octal and Hexadecimal Numbers

Radix 10 is important because we use it in everyday business, and radix 2 is important because binary numbers can be processed directly by digital circuits. Numbers in other radices are not often processed directly but may be important for documentation or other purposes. In particular, the radices 8 and 16 provide convenient shorthand representations for multibit numbers in a digital system.

The *octal number system* uses radix 8, while the *hexadecimal number* *octal number system* *system* uses radix 16. Table 2-1 shows the binary integers from 0 to 1111 and *hexadecimal number* their octal, decimal, and hexadecimal equivalents. The octal system needs 8 *system* digits, so it uses digits 0–7 of the decimal system. The hexadecimal system needs 16 digits, so it supplements decimal digits 0–9 with the letters *A–F*. *hexadecimal digits*
A–F

The octal and hexadecimal number systems are useful for representing multibit numbers because their radices are powers of 2. Since a string of three bits can take on eight different combinations, it follows that each 3-bit string can be uniquely represented by one octal digit, according to the third and fourth columns of Table 2-1. Likewise, a 4-bit string can be represented by one hexadecimal digit according to the fifth and sixth columns of the table.

Thus, it is very easy to convert a binary number to octal. Starting at the *binary-to-octal* binary point and working left, we simply separate the bits into groups of three *conversion* and replace each group with the corresponding octal digit:

$$100011001110_2 \ = \ 100\ 011\ 001\ 110_2 \ = \ 4316_8$$

$$1110110111010101001_2 \ = \ 011\ 101\ 101\ 110\ 101\ 001_2 \ = \ 355651_8$$

The procedure for binary-to-hexadecimal conversion is similar, except we use *binary-to-hexadecimal* groups of four bits: *conversion*

$$100011001110_2 \ = \ 1000\ 1100\ 1110_2 \ = \ 8CE_{16}$$

$$1110110111010101001_2 \ = \ 0001\ 1101\ 1011\ 1010\ 1001_2 \ = \ 1DBA9_{16}$$

In these examples we have freely added zeroes on the left to make the total number of bits a multiple of 3 or 4 as required.

Table 2-1
Binary, decimal, octal, and hexadecimal numbers.

Binary	Decimal	Octal	3-Bit String	Hexadecimal	4-Bit String
0	0	0	000	0	0000
1	1	1	001	1	0001
10	2	2	010	2	0010
11	3	3	011	3	0011
100	4	4	100	4	0100
101	5	5	101	5	0101
110	6	6	110	6	0110
111	7	7	111	7	0111
1000	8	10	—	8	1000
1001	9	11	—	9	1001
1010	10	12	—	A	1010
1011	11	13	—	B	1011
1100	12	14	—	C	1100
1101	13	15	—	D	1101
1110	14	16	—	E	1110
1111	15	17	—	F	1111

If a binary number contains digits to the right of the binary point, we can convert them to octal or hexadecimal by starting at the binary point and working right. Both the lefthand and righthand sides can be padded with zeroes to get multiples of three or four bits, as shown in the example below:

$$10.1011001011_2 = 010 \, . \, 101 \ 100 \ 101 \ 100_2 = 2.5454_8$$
$$= 0010 \, . \, 1011 \ 0010 \ 1100_2 = 2.B2C_{16}$$

octal- or hexadecimal-to-binary conversion

Converting in the reverse direction, from octal or hexadecimal to binary, is very easy. We simply replace each octal or hexadecimal digit with the corresponding 3- or 4-bit string, as shown below:

$$1357_8 = 001 \ 011 \ 101 \ 111_2$$
$$2046.17_8 = 010 \ 000 \ 100 \ 110 \, . \, 001 \ 111_2$$
$$BEAD_{16} = 1011 \ 1110 \ 1010 \ 1101_2$$
$$9F.46C_{16} = 1001 \ 1111 \, . \, 0100 \ 0110 \ 1100_2$$

The octal number system was quite popular 25 years ago because of certain minicomputers that had their front-panel lights and switches arranged in groups of three. However, the octal number system is not used much today, because of the preponderance of machines that process 8-bit *bytes*. It is difficult to extract individual byte values in multibyte quantities in the octal representation; for

byte

WHEN I'M 64   As you grow older, you'll find that the hexadecimal number system is useful for more than just computers. When I turned 40, I told friends that I had just turned 28_{16}. The "$_{16}$" was whispered under my breath, of course. At age 50, I'll be only 32_{16}.

People get all excited about decennial birthdays like 20, 30, 40, 50, ..., but you should be able to convince your friends that the decimal system is of no fundamental significance. More significant life changes occur around birthdays 2, 4, 8, 16, 32, and 64, when you add a most significant bit to your age. Why do you think the Beatles sang "When I'm sixty-*four*"?

example, what are the octal values of the four 8-bit bytes in the 32-bit number with octal representation 12345670123_8?

In the hexadecimal system, two digits represent an 8-bit byte, and $2n$ digits represent an n-byte word; each pair of digits constitutes exactly one byte. For example, the 32-bit hexadecimal number $5678ABCD_{16}$ consists of four bytes with values 56_{16}, 78_{16}, AB_{16}, and CD_{16}. In this context, a 4-bit hexadecimal digit is sometimes called a *nibble*; a 32-bit (4-byte) number has eight nibbles. Hexadecimal numbers are often used to describe a computer's memory address space. For example, a computer with 16-bit addresses might be described as having read/write memory installed at addresses $0–EFFF_{16}$, and read-only memory at addresses $F000–FFFF_{16}$. Many computer programming languages use the prefix "0x" to denote a hexadecimal number, for example, `0xBFC0000`.

nibble

0x prefix

2.3 General Positional-Number-System Conversions

In general, conversion between two radices cannot be done by simple substitutions; arithmetic operations are required. In this section, we show how to convert a number in any radix to radix 10 and vice versa, using radix-10 arithmetic.

In Section 2.1, we indicated that the value of a number in any radix is given by the formula

radix-r-to-decimal conversion

$$D = \sum_{i=-n}^{p-1} d_i \cdot r^i$$

where r is the radix of the number and there are p digits to the left of the radix point and n to the right. Thus, the value of the number can be found by converting each digit of the number to its radix-10 equivalent and expanding the formula using radix-10 arithmetic. Some examples are given below:

$$1CE8_{16} = 1\cdot16^3 + 12\cdot16^2 + 14\cdot16^1 + 8\cdot16^0 = 7400_{10}$$
$$F1A3_{16} = 15\cdot16^3 + 1\cdot16^2 + 10\cdot16^1 + 3\cdot16^0 = 61859_{10}$$
$$436.5_8 = 4\cdot8^2 + 3\cdot8^1 + 6\cdot8^0 + 5\cdot8^{-1} = 286.625_{10}$$
$$132.3_4 = 1\cdot4^2 + 3\cdot4^1 + 2\cdot4^0 + 3\cdot4^{-1} = 30.75_{10}$$

nested expansion formula

A shortcut for converting whole numbers to radix 10 can be obtained by rewriting the expansion formula in a nested fashion:

$$D = ((\cdots((d_{p-1}) \cdot r + d_{p-2}) \cdot r + \cdots) \cdot r + d_1) \cdot r + d_0$$

That is, we start with a sum of 0; beginning with the leftmost digit, we multiply the sum by r and add the next digit to the sum, repeating until all digits have been processed. For example, we can write

$$F1AC_{16} = (((15) \cdot 16 + 1) \cdot 16 + 10) \cdot 16 + 12$$

decimal-to-radix-r conversion

This formula is used in iterative, programmed conversion algorithms (such as Table 4-38 on page 279). It is also the basis of a very convenient method of converting a decimal number D to a radix r. Consider what happens if we divide the formula by r. Since the parenthesized part of the formula is evenly divisible by r, the quotient will be

$$Q = (\cdots((d_{p-1}) \cdot r + d_{p-2}) \cdot r + \cdots) \cdot r + d_1$$

and the remainder will be d_0. Thus, d_0 can be computed as the remainder of the long division of D by r. Furthermore, the quotient Q has the same form as the original formula. Therefore, successive divisions by r yield successive digits of D from right to left, until all the digits of D have been derived. Examples are given below:

$$179 \div 2 = 89 \text{ remainder } 1 \quad \text{(LSB)}$$
$$\div 2 = 44 \text{ remainder } 1$$
$$\div 2 = 22 \text{ remainder } 0$$
$$\div 2 = 11 \text{ remainder } 0$$
$$\div 2 = 5 \text{ remainder } 1$$
$$\div 2 = 2 \text{ remainder } 1$$
$$\div 2 = 1 \text{ remainder } 0$$
$$\div 2 = 0 \text{ remainder } 1 \quad \text{(MSB)}$$

$$179_{10} = 10110011_2$$

$$467 \div 8 = 58 \text{ remainder } 3 \quad \text{(least significant digit)}$$
$$\div 8 = 7 \text{ remainder } 2$$
$$\div 8 = 0 \text{ remainder } 7 \quad \text{(most significant digit)}$$

$$467_{10} = 723_8$$

$$3417 \div 16 = 213 \text{ remainder } 9 \quad \text{(least significant digit)}$$
$$\div 16 = 13 \text{ remainder } 5$$
$$\div 16 = 0 \text{ remainder } 13 \quad \text{(most significant digit)}$$

$$3417_{10} = D59_{16}$$

Table 2-2 summarizes methods for converting among the most common radices.

■ **Table 2-2** Conversion methods for common radices.

Conversion	Method	Example
Binary to		
Octal	Substitution	$10111011001_2 = 10\ 111\ 011\ 001_2 = 2731_8$
Hexadecimal	Substitution	$10111011001_2 = 101\ 1101\ 1001_2 = 5D9_{16}$
Decimal	Summation	$10111011001_2 = 1 \cdot 1024 + 0 \cdot 512 + 1 \cdot 256 + 1 \cdot 128 + 1 \cdot 64$ $+ 0 \cdot 32 + 1 \cdot 16 + 1 \cdot 8 + 0 \cdot 4 + 0 \cdot 2 + 1 \cdot 1 = 1497_{10}$
Octal to		
Binary	Substitution	$1234_8 = 001\ 010\ 011\ 100_2$
Hexadecimal	Substitution	$1234_8 = 001\ 010\ 011\ 100_2 = 0010\ 1001\ 1100_2 = 29C_{16}$
Decimal	Summation	$1234_8 = 1 \cdot 512 + 2 \cdot 64 + 3 \cdot 8 + 4 \cdot 1 = 668_{10}$
Hexadecimal to		
Binary	Substitution	$C0DE_{16} = 1100\ 0000\ 1101\ 1110_2$
Octal	Substitution	$C0DE_{16} = 1100\ 0000\ 1101\ 1110_2 = 1\ 100\ 000\ 011\ 011\ 110_2 = 140336_8$
Decimal	Summation	$C0DE_{16} = 12 \cdot 4096 + 0 \cdot 256 + 13 \cdot 16 + 14 \cdot 1 = 49374_{10}$
Decimal to		
Binary	Division	$108_{10} \div 2 = 54$ remainder 0 (LSB) $\div 2 = 27$ remainder 0 $\div 2 = 13$ remainder 1 $\div 2 = 6$ remainder 1 $\div 2 = 3$ remainder 0 $\div 2 = 1$ remainder 1 $\div 2 = 0$ remainder 1 (MSB) $108_{10} = 1101100_2$
Octal	Division	$108_{10} \div 8 = 13$ remainder 4 (least significant digit) $\div 8 = 1$ remainder 5 $\div 8 = 0$ remainder 1 (most significant digit) $108_{10} = 154_8$
Hexadecimal	Division	$108_{10} \div 16 = 6$ remainder 12 (least significant digit) $\div 16 = 0$ remainder 6 (most significant digit) $108_{10} = 6C_{16}$

Table 2-3
Binary addition and
subtraction table.

c_{in} **or** b_{in}	x	y	c_{out}	s	b_{out}	d
0	0	0	0	0	0	0
0	0	1	0	1	1	1
0	1	0	0	1	0	1
0	1	1	1	0	0	0
1	0	0	0	1	1	1
1	0	1	1	0	1	0
1	1	0	1	0	0	0
1	1	1	1	1	1	1

2.4 Addition and Subtraction of Nondecimal Numbers

Addition and subtraction of nondecimal numbers by hand uses the same technique that we learned in grammar school for decimal numbers; the only catch is that the addition and subtraction tables are different.

binary addition   Table 2-3 is the addition and subtraction table for binary digits. To add two binary numbers X and Y, we add together the least significant bits with an initial carry (c_{in}) of 0, producing carry (c_{out}) and sum (s) bits according to the table. We continue processing bits from right to left, adding the carry out of each column into the next column's sum.

Two examples of decimal additions and the corresponding binary additions are shown in Figure 2-1, using a colored arrow to indicate a carry of 1. The same examples are repeated below along with two more, with the carries shown as a bit string C:

C		101111000		C		001011000
X	190	10111110		X	173	10101101
Y	+141	+ 10001101		Y	+ 44	+ 00101100
$X+Y$	331	101001011		$X+Y$	217	11011001

C		011111110		C		000000000
X	127	01111111		X	170	10101010
Y	+ 63	+ 00111111		Y	+ 85	+ 01010101
$X+Y$	190	10111110		$X+Y$	255	11111111

binary subtraction   Binary subtraction is performed similarly, using borrows (b_{in} and b_{out}) instead of carries between steps, and producing a difference bit d. Two examples of decimal subtractions and the corresponding binary subtractions are shown in *minuend* Figure 2-2. As in decimal subtraction, the binary minuend values in the columns *subtrahend* are modified when borrows occur, as shown by the colored arrows and bits. The

			1	1	1	1								1		1	1		
X	190	1 0	1	1	1	1 1 0				X	173	1 0	1 0	1	1 0 1

$$
\begin{array}{lr@{\quad}l}
 & & \;1\;1\;1\;1 \\
X & 190 & 1\;0\;1\;1\;1\;1\;1\;0 \\
Y & +141 & +\,1\;0\;0\;0\;1\;1\;0\;1 \\
\hline
X+Y & 331 & 1\;0\;1\;0\;0\;1\;0\;1\;1
\end{array}
\qquad
\begin{array}{lr@{\quad}l}
 & & \;1\;\;\;\;1\;1 \\
X & 173 & 1\;0\;1\;0\;1\;1\;0\;1 \\
Y & +44 & +\,0\;0\;1\;0\;1\;1\;0\;0 \\
\hline
X+Y & 217 & 1\;1\;0\;1\;1\;0\;0\;1
\end{array}
$$

Figure 2-1   Examples of decimal and corresponding binary additions.

examples from the figure are repeated below along with two more, this time showing the borrows as a bit string B:

$$
\begin{array}{lrl}
B & & 001111100 \\
X & 229 & 11100101 \\
Y & -46 & -\,00101110 \\
\hline
X-Y & 183 & 10110111
\end{array}
\qquad
\begin{array}{lrl}
B & & 011011010 \\
X & 210 & 11010010 \\
Y & -109 & -\,01101101 \\
\hline
X-Y & 101 & 01100101
\end{array}
$$

$$
\begin{array}{lrl}
B & & 010101010 \\
X & 170 & 10101010 \\
Y & -85 & -\,01010101 \\
\hline
X-Y & 85 & 01010101
\end{array}
\qquad
\begin{array}{lrl}
B & & 000000000 \\
X & 221 & 11011101 \\
Y & -76 & -\,01001100 \\
\hline
X-Y & 145 & 10010001
\end{array}
$$

A very common use of subtraction in computers is to compare two numbers. For example, if the operation $X - Y$ produces a borrow out of the most significant bit position, then X is less than Y; otherwise, X is greater than or equal to Y. The relationship between carries and borrows in adders and subtractors will be explored in Section 5.10.

comparing numbers

Addition and subtraction tables can be developed for octal and hexadecimal digits, or any other desired radix. However, few computer engineers bother to memorize these tables. If you rarely need to manipulate nondecimal numbers,

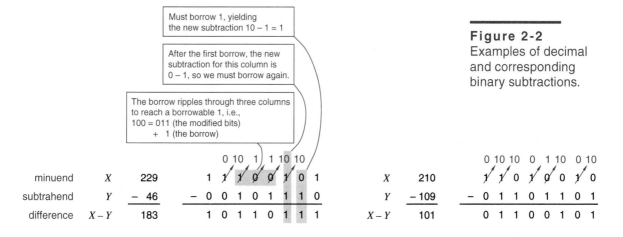

Must borrow 1, yielding the new subtraction 10 − 1 = 1

After the first borrow, the new subtraction for this column is 0 − 1, so we must borrow again.

The borrow ripples through three columns to reach a borrowable 1, i.e., 100 = 011 (the modified bits) + 1 (the borrow)

Figure 2-2
Examples of decimal and corresponding binary subtractions.

then it's easy enough on those occasions to convert them to decimal, calculate results, and convert back. On the other hand, if you must perform calculations in binary, octal, or hexadecimal frequently, then you should ask Santa for a programmer's "hex calculator" from Texas Instruments or Casio.

If the calculator's battery wears out, some mental shortcuts can be used to facilitate nondecimal arithmetic. In general, each column addition (or subtraction) can be done by converting the column digits to decimal, adding in decimal, and converting the result to corresponding sum and carry digits in the non-decimal radix. (A carry is produced whenever the column sum equals or exceeds the radix.) Since the addition is done in decimal, we rely on our knowledge of the decimal addition table; the only new thing that we need to learn is the conversion from decimal to nondecimal digits and vice versa. The sequence of steps for

hexadecimal addition   mentally adding two hexadecimal numbers is shown below:

C	1 1 0 0		1	1	0	0
X	1 9 B 9 $_{16}$		1	9	11	9
Y	+ C 7 E 6 $_{16}$		+12	7	14	6
$X+Y$	E 1 9 F $_{16}$		14	17	25	15
			14	16+1	16+9	15
			E	1	9	F

2.5 Representation of Negative Numbers

So far, we have dealt only with positive numbers, but there are many ways to represent negative numbers. In everyday business we use the signed-magnitude system, discussed next. However, most computers use one of the complement number systems that we introduce later.

2.5.1 Signed-Magnitude Representation

signed-magnitude   In the *signed-magnitude system*, a number consists of a magnitude and a symbol
system   indicating whether the magnitude is positive or negative. Thus, we interpret decimal numbers +98, −57, +123.5, and −13 in the usual way, and we also assume that the sign is "+" if no sign symbol is written. There are two possible representations of zero, "+0" and "−0", but both have the same value.

sign bit   The signed-magnitude system is applied to binary numbers by using an extra bit position to represent the sign (the *sign bit*). Traditionally, the most significant bit (MSB) of a bit string is used as the sign bit (0 = plus, 1 = minus), and the lower-order bits contain the magnitude. Thus, we can write several 8-bit signed-magnitude integers and their decimal equivalents:

$$01010101_2 = +85_{10} \qquad 11010101_2 = -85_{10}$$
$$01111111_2 = +127_{10} \qquad 11111111_2 = -127_{10}$$
$$00000000_2 = +0_{10} \qquad 10000000_2 = -0_{10}$$

The signed-magnitude system has an equal number of positive and negative integers. An n-bit signed-magnitude integer lies within the range $-(2^{n-1} - 1)$ through $+(2^{n-1} - 1)$, and there are two possible representations of zero.

Now suppose that we wanted to build a digital logic circuit that adds signed-magnitude numbers. The circuit must examine the signs of the addends to determine what to do with the magnitudes. If the signs are the same, it must add the magnitudes and give the result the same sign. If the signs are different, it must compare the magnitudes, subtract the smaller from the larger, and give the result the sign of the larger. All of these "ifs," "adds," "subtracts," and "compares" translate into a lot of logic-circuit complexity. Adders for complement number systems are much simpler, as we'll show next. Perhaps the one redeeming feature of a signed-magnitude system is that, once we know how to build a signed-magnitude adder, a signed-magnitude subtractor is almost trivial to build—it need only change the sign of the subtrahend and pass it along with the minuend to an adder.

signed-magnitude adder

signed-magnitude subtractor

2.5.2 Complement Number Systems

While the signed-magnitude system negates a number by changing its sign, a *complement number system* negates a number by taking its complement as defined by the system. Taking the complement is more difficult than changing the sign, but two numbers in a complement number system can be added or subtracted directly without the sign and magnitude checks required by the signed-magnitude system. We shall describe two complement number systems, called the "radix complement" and the "diminished radix-complement."

complement number system

In any complement number system, we normally deal with a fixed number of digits, say n. (However, we can increase the number of digits by "sign extension" as shown in Exercise 2.23, and decrease the number by truncating high-order digits as shown in Exercise 2.24.) We further assume that the radix is r, and that numbers have the form

$$D = d_{n-1}d_{n-2}\cdots d_1 d_0.$$

The radix point is on the right and so the number is an integer. If an operation produces a result that requires more than n digits, we throw away the extra high-order digit(s). If a number D is complemented twice, the result is D.

2.5.3 Radix-Complement Representation

In a *radix-complement system*, the complement of an n-digit number is obtained by subtracting it from r^n. In the decimal number system, the radix complement is called the *10's complement*. Some examples using 4-digit decimal numbers (and subtraction from 10,000) are shown in Table 2-4.

radix-complement system

10's complement

By definition, the radix complement of an n-digit number D is obtained by subtracting it from r^n. If D is between 1 and $r^n - 1$, this subtraction produces

Table 2-4
Examples of 10's and
9s' complements.

Number	10's Complement	9s' Complement
1849	8151	8150
2067	7933	7932
100	9900	9899
7	9993	9992
8151	1849	1848
0	10000 (= 0)	9999

another number between 1 and $r^n - 1$. If D is 0, the result of the subtraction is r^n, which has the form $100 \cdots 00$, where there are a total of $n + 1$ digits. We throw away the extra high-order digit and get the result 0. Thus, there is only one representation of zero in a radix-complement system.

computing the radix complement

It seems from the definition that a subtraction operation is needed to compute the radix complement of D. However, this subtraction can be avoided by rewriting r^n as $(r^n - 1) + 1$ and $r^n - D$ as $((r^n - 1) - D) + 1$. The number $r^n - 1$ has the form $mm \cdots mm$, where $m = r - 1$ and there are n m's. For example, 10,000 equals 9,999 + 1. If we define the complement of a digit d to be $r - 1 - d$, then $(r^n - 1) - D$ is obtained by complementing the digits of D. Therefore, the radix complement of a number D is obtained by complementing the individual

Table 2-5
Digit complements.

Digit	Complement Binary	Octal	Decimal	Hexadecimal
0	1	7	9	F
1	0	6	8	E
2	–	5	7	D
3	–	4	6	C
4	–	3	5	B
5	–	2	4	A
6	–	1	3	9
7	–	0	2	8
8	–	–	1	7
9	–	–	0	6
A	–	–	–	5
B	–	–	–	4
C	–	–	–	3
D	–	–	–	2
E	–	–	–	1
F	–	–	–	0

digits of D and adding 1. For example, the 10's complement of 1849 is 8150 + 1, or 8151. You should confirm that this trick also works for the other 10's-complement examples above. Table 2-5 lists the digit complements for binary, octal, decimal, and hexadecimal numbers.

2.5.4 Two's-Complement Representation

For binary numbers, the radix complement is called the *two's complement*. The MSB of a number in this system serves as the sign bit; a number is negative if and only if its MSB is 1. The decimal equivalent for a two's-complement binary number is computed the same way as for an unsigned number, except that the weight of the MSB is -2^{n-1} instead of $+2^{n-1}$. The range of representable numbers is $-(2^{n-1})$ through $+(2^{n-1} - 1)$. Some 8-bit examples are shown below:

two's complement

weight of MSB

$17_{10} =$ 00010001_2 $-99_{10} =$ 10011101_2

                   $\Downarrow$ complement bits $\Downarrow$ complement bits

              11101110 01100010

                 $+1$ $+1$

             $11101111_2 = -17_{10}$ $01100011_2 = 99_{10}$

$119_{10} =$ 01110111_2 $-127_{10} =$ 10000001_2

                   $\Downarrow$ complement bits $\Downarrow$ complement bits

              10001000_2 01111110_2

                 $+1$ $+1$

             $10001001_2 = -119_{10}$ $01111111_2 = 127_{10}$

$0_{10} =$ 00000000_2 $-128_{10} =$ 10000000_2

                   $\Downarrow$ complement bits $\Downarrow$ complement bits

              11111111 01111111

                 $+1$ $+1$

       $1\ 00000000_2 = 0_{10}$ $10000000_2 = -128_{10}$

A carry out of the MSB position occurs in one case, as shown in color above. As in all two's-complement operations, this bit is ignored and only the low-order n bits of the result are used.

In the two's-complement number system, zero is considered positive because its sign bit is 0. Since two's complement has only one representation of zero, we end up with one extra negative number, -2^{n-1}, that doesn't have a positive counterpart.

extra negative number

We can convert an n-bit two's-complement number X into an m-bit one, but some care is needed. If $m > n$, we must append $m - n$ copies of X's sign bit to the left of X (see Exercise 2.23). That is, we pad a positive number with 0s and a negative one with 1s; this is called *sign extension*. If $m < n$, we discard X's $n - m$

sign extension

leftmost bits; however, the result is valid only if all of the discarded bits are the same as the sign bit of the result (see Exercise 2.24).

Most computers and other digital systems use the two's-complement system to represent negative numbers. However, for completeness, we'll also describe the diminished radix-complement and ones'-complement systems.

*2.5.5 Diminished Radix-Complement Representation

diminished radix-complement system

In a *diminished radix-complement system*, the complement of an n-digit number D is obtained by subtracting it from $r^n - 1$. This can be accomplished by complementing the individual digits of D, *without* adding 1 as in the radix-complement system. In decimal, this is called the *9s' complement*; some examples are given in the last column of Table 2-4 on page 36.

9s' complement

*2.5.6 Ones'-Complement Representation

ones' complement

The diminished radix-complement system for binary numbers is called the *ones' complement*. As in two's complement, the most significant bit is the sign, 0 if positive and 1 if negative. Thus there are two representations of zero, positive zero $(00 \cdots 00)$ and negative zero $(11 \cdots 11)$. Positive-number representations are the same for both ones' and two's complements. However, negative-number representations differ by 1. A weight of $-(2^{n-1} - 1)$, rather than -2^{n-1}, is given to the most significant bit when computing the decimal equivalent of a ones'-complement number. The range of representable numbers is $-(2^{n-1} - 1)$ through $+(2^{n-1} - 1)$. Some 8-bit numbers and their ones' complements are shown below:

$$17_{10} = 00010001_2 \qquad\qquad -99_{10} = 10011100_2$$
$$\Downarrow \qquad\qquad\qquad\qquad \Downarrow$$
$$11101110_2 = -17_{10} \qquad\qquad 01100011_2 = 99_{10}$$

$$119_{10} = 01110111_2 \qquad\qquad -127_{10} = 10000000_2$$
$$\Downarrow \qquad\qquad\qquad\qquad \Downarrow$$
$$10001000_2 = -119_{10} \qquad\qquad 01111111_2 = 127_{10}$$

$$0_{10} = 00000000_2 \text{ (positive zero)}$$
$$\Downarrow$$
$$11111111_2 = 0_{10} \text{ (negative zero)}$$

The main advantages of the ones'-complement system are its symmetry and the ease of complementation. However, the adder design for ones'-complement numbers is somewhat trickier than a two's-complement adder (see Exercise 7.72). Also, zero-detecting circuits in a ones'-complement system

---

* Throughout this book, *optional sections* are marked with an asterisk.

either must check for both representations of zero, or must always convert $11\cdots11$ to $00\cdots00$.

*2.5.7 Excess Representations

Yes, the number of different systems for representing negative numbers *is* excessive, but there's just one more for us to cover. In *excess-B representation*, an *m*-bit string whose unsigned integer value is M ($0 \leq M < 2^m$) represents the signed integer $M - B$, where B is called the *bias* of the number system.

excess-B representation

bias

For example, an *excess-2^{m-1} system* represents any number X in the range -2^{m-1} through $+2^{m-1} - 1$ by the *m*-bit binary representation of $X + 2^{m-1}$ (which is always nonnegative and less than 2^m). The range of this representation is exactly the same as that of *m*-bit two's-complement numbers. In fact, the representations of any number in the two systems are identical except for the sign bits, which are always opposite. (Note that this is true only when the bias is 2^{m-1}.)

excess-2^{m-1} system

The most common use of excess representations is in floating-point number systems (see References).

2.6 Two's-Complement Addition and Subtraction

2.6.1 Addition Rules

A table of decimal numbers and their equivalents in different number systems, Table 2-6, reveals why the two's complement is preferred for arithmetic operations. If we start with 1000_2 (-8_{10}) and count up, we see that each successive two's-complement number all the way to 0111_2 ($+7_{10}$) can be obtained by adding 1 to the previous one, ignoring any carries beyond the fourth bit position. The same cannot be said of signed-magnitude and ones'-complement numbers. Because ordinary addition is just an extension of counting, two's-complement numbers can thus be added by ordinary binary addition, ignoring any carries beyond the MSB. The result will always be the correct sum as long as the range of the number system is not exceeded. Some examples of decimal addition and the corresponding 4-bit two's-complement additions confirm this:

two's-complement addition

$$
\begin{array}{rl}
+3 & \quad 0011 \\
+\ +4 & \quad +\ 0100 \\
\hline
+7 & \quad 0111
\end{array}
\qquad
\begin{array}{rl}
-2 & \quad 1110 \\
+\ -6 & \quad +\ 1010 \\
\hline
-8 & \quad 1\,1000
\end{array}
$$

$$
\begin{array}{rl}
+6 & \quad 0110 \\
+\ -3 & \quad +\ 1101 \\
\hline
+3 & \quad 1\,0011
\end{array}
\qquad
\begin{array}{rl}
+4 & \quad 0100 \\
+\ -7 & \quad +\ 1001 \\
\hline
-3 & \quad 1101
\end{array}
$$

Table 2-6 Decimal and 4-bit numbers.

Decimal	Two's Complement	Ones' Complement	Signed Magnitude	Excess 2^{m-1}
−8	1000	—	—	0000
−7	1001	1000	1111	0001
−6	1010	1001	1110	0010
−5	1011	1010	1101	0011
−4	1100	1011	1100	0100
−3	1101	1100	1011	0101
−2	1110	1101	1010	0110
−1	1111	1110	1001	0111
0	0000	1111 or 0000	1000 or 0000	1000
1	0001	0001	0001	1001
2	0010	0010	0010	1010
3	0011	0011	0011	1011
4	0100	0100	0100	1100
5	0101	0101	0101	1101
6	0110	0110	0110	1110
7	0111	0111	0111	1111

2.6.2 A Graphical View

Another way to view the two's-complement system uses the 4-bit "counter" shown in Figure 2-3. Here we have shown the numbers in a circular or "modular" representation. The operation of this counter very closely mimics that of a real up/down counter circuit, which we'll study in Section 8.4. Starting

Figure 2-3
A modular counting representation of 4-bit two's-complement numbers.

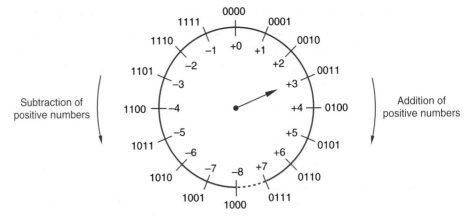

Subtraction of positive numbers

Addition of positive numbers

with the arrow pointing to any number, we can add $+n$ to that number by counting up n times, that is, by moving the arrow n positions clockwise. It is also evident that we can subtract n from a number by counting down n times, that is, by moving the arrow n positions counterclockwise. Of course, these operations give correct results only if n is small enough that we don't cross the discontinuity between -8 and $+7$.

What is most interesting is that we can also subtract n (or add $-n$) by moving the arrow $16 - n$ positions clockwise. Notice that the quantity $16 - n$ is what we defined to be the 4-bit two's complement of n, that is, the two's-complement representation of $-n$. This graphically supports our earlier claim that a negative number in two's-complement representation may be added to another number simply by adding the 4-bit representations using ordinary binary addition. Adding a number in Figure 2-3 is equivalent to moving the arrow a corresponding number of positions clockwise.

2.6.3 Overflow

If an addition operation produces a result that exceeds the range of the number system, *overflow* is said to occur. In the modular counting representation of Figure 2-3, overflow occurs during addition of positive numbers when we count past $+7$. Addition of two numbers with different signs can never produce overflow, but addition of two numbers of like sign can, as shown by the following examples:

overflow

$$
\begin{array}{rl}
-3 & 1101 \\
+ \ -6 & +\ 1010 \\
\hline
-9 & 10111 = +7
\end{array}
\qquad
\begin{array}{rl}
+5 & 0101 \\
+ \ +6 & +\ 0110 \\
\hline
+11 & 1011 = -5
\end{array}
$$

$$
\begin{array}{rl}
-8 & 1000 \\
+ \ -8 & +\ 1000 \\
\hline
-16 & 10000 = +0
\end{array}
\qquad
\begin{array}{rl}
+7 & 0111 \\
+ \ +7 & +\ 0111 \\
\hline
+14 & 1110 = -2
\end{array}
$$

Fortunately, there is a simple rule for detecting overflow in addition: An addition overflows if the signs of the addends are the same and the sign of the sum is different from the addends' sign. The overflow rule is sometimes stated in terms of carries generated during the addition operation: An addition overflows if the carry bits c_{in} into and c_{out} out of the sign position are different. Close examination of Table 2-3 on page 32 shows that the two rules are equivalent—there are only two cases where $c_{in} \neq c_{out}$, and these are the only two cases where $x = y$ and the sum bit is different.

overflow rules

2.6.4 Subtraction Rules

Two's-complement numbers may be subtracted as if they were ordinary unsigned binary numbers, and appropriate rules for detecting overflow may be formulated. However, most subtraction circuits for two's-complement numbers

two's-complement subtraction

do not perform subtraction directly. Rather, they negate the subtrahend by taking its two's complement, and then add it to the minuend using the normal rules for addition.

Negating the subtrahend and adding the minuend can be accomplished with only one addition operation as follows: Perform a bit-by-bit complement of the subtrahend and add the complemented subtrahend to the minuend with an initial carry (c_{in}) of 1 instead of 0. Examples are given below:

$$
\begin{array}{rll}
 & & 1 \longleftarrow c_{in} \\
+4 & 0100 & 0100 \\
-\ +3 & -\ 0011 & +\ 1100 \\
\hline
+1 & & 1\,0001
\end{array}
\qquad
\begin{array}{rll}
 & & 1 \longleftarrow c_{in} \\
+3 & 0011 & 0011 \\
-\ +4 & -\ 0100 & +\ 1011 \\
\hline
-1 & & 1111
\end{array}
$$

$$
\begin{array}{rll}
 & & 1 \longleftarrow c_{in} \\
+3 & 0011 & 0011 \\
-\ -4 & -\ 1100 & +\ 0011 \\
\hline
+7 & & 0111
\end{array}
\qquad
\begin{array}{rll}
 & & 1 \longleftarrow c_{in} \\
-3 & 1101 & 1101 \\
-\ -4 & -\ 1100 & +\ 0011 \\
\hline
+1 & & 1\,0001
\end{array}
$$

Overflow in subtraction can be detected by examining the signs of the minuend and the *complemented* subtrahend, using the same rule as in addition. Or, using the technique in the preceding examples, the carries into and out of the sign position can be observed and overflow detected irrespective of the signs of inputs and output, again using the same rule as in addition.

An attempt to negate the "extra" negative number results in overflow according to the rules above, when we add 1 in the complementation process:

$$
\begin{array}{rl}
-(-8) = -1000 = & 0111 \\
 & +\ 0001 \\
\hline
 & 1000 \ = \ -8
\end{array}
$$

However, this number can still be used in additions and subtractions as long as the final result does not exceed the number range:

$$
\begin{array}{rll}
+4 & 0100 \\
+\ -8 & +\ 1000 \\
\hline
-4 & 1100
\end{array}
\qquad
\begin{array}{rll}
 & & 1 \longleftarrow c_{in} \\
-3 & 1101 & 1101 \\
-\ -8 & -\ 1000 & +\ 0111 \\
\hline
+5 & & 1\,0101
\end{array}
$$

2.6.5 Two's-Complement and Unsigned Binary Numbers

Since two's-complement numbers are added and subtracted by the same basic binary addition and subtraction algorithms as unsigned numbers of the same length, a computer or other digital system can use the same adder circuit to handle numbers of both types. However, the results must be interpreted differently, depending on whether the system is dealing with signed numbers (e.g., −8 through +7) or unsigned numbers (e.g., 0 through 15).

signed vs. unsigned numbers

We introduced a graphical representation of the 4-bit two's-complement system in Figure 2-3. We can relabel this figure as shown in Figure 2-4 to obtain a representation of the 4-bit unsigned numbers. The binary combinations occupy the same positions on the wheel, and a number is still added by moving the arrow a corresponding number of positions clockwise, and subtracted by moving the arrow counterclockwise.

An addition operation can be seen to exceed the range of the 4-bit unsigned-number system in Figure 2-4 if the arrow moves clockwise through the discontinuity between 0 and 15. In this case a *carry* out of the most significant bit position is said to occur.

carry

Likewise a subtraction operation exceeds the range of the number system if the arrow moves counterclockwise through the discontinuity. In this case a *borrow* out of the most significant bit position is said to occur.

borrow

From Figure 2-4 it is also evident that we may subtract an unsigned number *n* by counting *clockwise* $16 - n$ positions. This is equivalent to *adding* the 4-bit two's-complement of *n*. The subtraction produces a borrow if the corresponding addition of the two's complement *does not* produce a carry.

In summary, in unsigned addition the carry or borrow in the most significant bit position indicates an out-of-range result. In signed, two's-complement addition the overflow condition defined earlier indicates an out-of-range result. The carry from the most significant bit position is irrelevant in signed addition in the sense that overflow may or may not occur independently of whether or not a carry occurs.

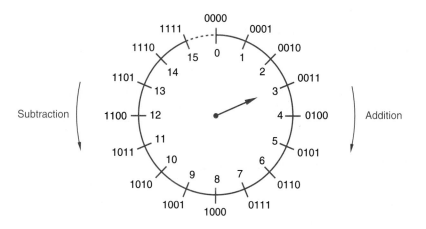

Figure 2-4
A modular counting representation of 4-bit unsigned numbers.

*2.7 Ones'-Complement Addition and Subtraction

Another look at Table 2-6 helps to explain the rule for adding ones'-complement numbers. If we start at 1000_2 (-7_{10}) and count up, we obtain each successive ones'-complement number by adding 1 to the previous one, *except* at the transition from 1111_2 (negative 0) to 0001_2 $(+1_{10})$. To maintain the proper count, we must add 2 instead of 1 whenever we count past 1111_2. This suggests a technique for adding ones'-complement numbers: Perform a standard binary addition, but add an extra 1 whenever we count past 1111_2.

Counting past 1111_2 during an addition can be detected by observing the carry out of the sign position. Thus, the rule for adding ones'-complement numbers can be stated quite simply:

ones'-complement addition

- Perform a standard binary addition; if there is a carry out of the sign position, add 1 to the result.

end-around carry

This rule is often called *end-around carry*. Examples of ones'-complement addition are given below; the last three include an end-around carry:

+3	0011		+4	0100		+5	0101
+ +4	+ 0100		+ −7	+ 1000		+ −5	+ 1010
+7	0111		−3	1100		−0	1111

−2	1101		+6	0110		−0	1111
+ −5	+ 1010		+ −3	+ 1100		+ −0	+ 1111
−7	10111		+3	10010		−0	11110
	+ 1			+ 1			+ 1
	1000			0011			1111

Following the two-step addition rule above, the addition of a number and its ones' complement produces negative 0. In fact, an addition operation using this rule can never produce positive 0 unless both addends are positive 0.

ones'-complement subtraction

As with two's complement, the easiest way to do ones'-complement subtraction is to complement the subtrahend and add. Overflow rules for ones'-complement addition and subtraction are the same as for two's complement.

Table 2-7 summarizes the rules that we presented in this and previous sections for negation, addition, and subtraction in binary number systems.

Table 2-7 Summary of addition and subtraction rules for binary numbers.

Number System	Addition Rules	Negation Rules	Subtraction Rules
Unsigned	Add the numbers. Result is out of range if a carry out of the MSB occurs.	Not applicable	Subtract the subtrahend from the minuend. Result is out of range if a borrow out of the MSB occurs.
Signed magnitude	(same sign) Add the magnitudes; overflow occurs if a carry out of MSB occurs; result has the same sign. (opposite sign) Subtract the smaller magnitude from the larger; overflow is impossible; result has the sign of the larger.	Change the number's sign bit.	Change the sign bit of the subtrahend and proceed as in addition.
Two's complement	Add, ignoring any carry out of the MSB. Overflow occurs if the carries into and out of MSB are different.	Complement all bits of the number; add 1 to the result.	Complement all bits of the subtrahend and add to the minuend with an initial carry of 1.
Ones' complement	Add; if there is a carry out of the MSB, add 1 to the result. Overflow occurs if carries into and out of MSB are different.	Complement all bits of the number.	Complement all bits of the subtrahend and proceed as in addition.

*2.8 Binary Multiplication

In grammar school we learned to multiply by adding a list of shifted multiplicands computed according to the digits of the multiplier. The same method can be used to obtain the product of two unsigned binary numbers. Forming the shifted multiplicands is trivial in binary multiplication, since the only possible values of the multiplier digits are 0 and 1. An example is shown below:

shift-and-add multiplication

unsigned binary multiplication

$$
\begin{array}{rr}
 & 11 \\
\times & 13 \\
\hline
 & 33 \\
 & 11 \\
\hline
 & 143
\end{array}
\qquad
\begin{array}{rr}
 & 1011 \\
\times & 1101 \\
\hline
 & 1011 \\
 & 0000 \\
 & 1011 \\
 & 1011 \\
\hline
 & 10001111
\end{array}
$$

multiplicand
multiplier

shifted multiplicands

product

partial product

Instead of listing all the shifted multiplicands and then adding, in a digital system it is more convenient to add each shifted multiplicand as it is created to a *partial product*. Applying this technique to the previous example, four additions and partial products are used to multiply 4-bit numbers:

$$
\begin{array}{rl}
11 & \\
\times\ 13 & \\
\end{array}
\qquad
\begin{array}{rl}
1011 & \text{multiplicand} \\
\times\quad 1101 & \text{multiplier} \\
\hline
0000 & \text{partial product} \\
1011 & \text{shifted multiplicand} \\
\hline
01011 & \text{partial product} \\
0000\!\downarrow & \text{shifted multiplicand} \\
\hline
001011 & \text{partial product} \\
1011\!\downarrow\!\downarrow & \text{shifted multiplicand} \\
\hline
0110111 & \text{partial product} \\
1011\!\downarrow\!\downarrow\!\downarrow & \text{shifted multiplicand} \\
\hline
10001111 & \text{product}
\end{array}
$$

In general, when we multiply an n-bit number by an m-bit number, the resulting product requires at most $n + m$ bits to express. The shift-and-add algorithm requires m partial products and additions to obtain the result, but the first addition is trivial, since the first partial product is zero. Although the first partial product has only n significant bits, after each addition step the partial product gains one more significant bit, since each addition may produce a carry. At the same time, each step yields one more partial product bit, starting with the rightmost and working toward the left, that does not change. The shift-and-add algorithm can be performed by a digital circuit that includes a shift register, an adder, and control logic, as shown in Section 8.7.2.

signed multiplication

Multiplication of signed numbers can be accomplished using unsigned multiplication and the usual grammar-school rules: Perform an unsigned multiplication of the magnitudes and make the product positive if the operands had the same sign, negative if they had different signs. This is very convenient in signed-magnitude systems, since the sign and magnitude are separate.

In the two's-complement system, obtaining the magnitude of a negative number and negating the unsigned product are nontrivial operations. This leads us to seek a more efficient way of performing two's-complement multiplication, described next.

two's-complement multiplication

Conceptually, unsigned multiplication is accomplished by a sequence of unsigned additions of the shifted multiplicands; at each step, the shift of the multiplicand corresponds to the weight of the multiplier bit. The bits in a two's-complement number have the same weights as in an unsigned number, except for the MSB, which has a negative weight (see Section 2.5.4). Thus, we can perform two's-complement multiplication by a sequence of two's-complement additions of shifted multiplicands, except for the last step, in which the shifted multiplicand corresponding to the MSB of the multiplier must be negated before it is added to the partial product. Our previous example is repeated below, this time interpreting the multiplier and multiplicand as two's-complement numbers:

−5	1011	multiplicand
× −3	× 1101	multiplier
	00000	partial product
	11011	shifted multiplicand
	111011	partial product
	00000↓	shifted multiplicand
	1111011	partial product
	11011↓↓	shifted multiplicand
	11100111	partial product
	00101↓↓↓	shifted and negated multiplicand
	00001111	product

Handling the MSBs is a little tricky because we gain one significant bit at each step and we are working with signed numbers. Therefore, before adding each shifted multiplicand and k-bit partial product, we change them to $k + 1$ significant bits by sign extension, as shown in color above. Each resulting sum has $k + 1$ bits; any carry out of the MSB of the $k + 1$-bit sum is ignored.

*2.9 Binary Division

The simplest binary division algorithm is based on the shift-and-subtract method that we learned in grammar school. Table 2-8 gives examples of this method for unsigned decimal and binary numbers. In both cases, we mentally compare the reduced dividend with multiples of the divisor to determine which multiple of

shift-and-subtract division

unsigned division

19	10011	quotient
11)217	1011)11011001	dividend
11	1011	shifted divisor
107	0101	reduced dividend
99	0000	shifted divisor
8	1010	reduced dividend
	0000	shifted divisor
	10100	reduced dividend
	1011	shifted divisor
	10011	reduced dividend
	1011	shifted divisor
	1000	remainder

Table 2-8
Example of long division.

the shifted divisor to subtract. In the decimal case, we first pick 11 as the greatest multiple of 11 less than 21, and then pick 99 as the greatest multiple less than 107. In the binary case, the choice is somewhat simpler, since the only two choices are zero and the divisor itself.

Division methods for binary numbers are somewhat complementary to binary multiplication methods. A typical division algorithm takes an $(n + m)$-bit dividend and an n-bit divisor, and produces an m-bit quotient and an n-bit remainder. A division *overflows* if the divisor is zero or the quotient would take more than m bits to express. In most computer division circuits, $n = m$.

division overflow

signed division

Division of signed numbers can be accomplished using unsigned division and the usual grammar school rules: Perform an unsigned division of the magnitudes and make the quotient positive if the operands had the same sign, negative if they had different signs. The remainder should be given the same sign as the dividend. As in multiplication, there are special techniques for performing division directly on two's-complement numbers; these techniques are often implemented in computer division circuits (see References).

2.10 Binary Codes for Decimal Numbers

Even though binary numbers are the most appropriate for the internal computations of a digital system, most people still prefer to deal with decimal numbers. As a result, the external interfaces of a digital system may read or display decimal numbers, and some digital devices actually process decimal numbers directly.

The human need to represent decimal numbers doesn't change the basic nature of digital electronic circuits—they still process signals that take on one of only two states that we call 0 and 1. Therefore, a decimal number is represented in a digital system by a string of bits, where different combinations of bit values in the string represent different decimal numbers. For example, if we use a 4-bit string to represent a decimal number, we might assign bit combination 0000 to decimal digit 0, 0001 to 1, 0010 to 2, and so on.

code
code word

A set of n-bit strings in which different bit strings represent different numbers or other things is called a *code*. A particular combination of n bit-values is called a *code word*. As we'll see in the examples of decimal codes in this section, there may or may not be an arithmetic relationship between the bit values in a code word and the thing that it represents. Furthermore, a code that uses n-bit strings need not contain 2^n valid code words.

At least four bits are needed to represent the ten decimal digits. There are billions and billions of different ways to choose ten 4-bit code words, but some of the more common decimal codes are listed in Table 2-9.

binary-coded decimal (BCD)

Perhaps the most "natural" decimal code is *binary-coded decimal (BCD)*, which encodes the digits 0 through 9 by their 4-bit unsigned binary representations, 0000 through 1001. The code words 1010 through 1111 are not used.

Table 2-9 Decimal codes.

Decimal digit	BCD (8421)	2421	Excess-3	Biquinary	1-out-of-10
0	0000	0000	0011	0100001	1000000000
1	0001	0001	0100	0100010	0100000000
2	0010	0010	0101	0100100	0010000000
3	0011	0011	0110	0101000	0001000000
4	0100	0100	0111	0110000	0000100000
5	0101	1011	1000	1000001	0000010000
6	0110	1100	1001	1000010	0000001000
7	0111	1101	1010	1000100	0000000100
8	1000	1110	1011	1001000	0000000010
9	1001	1111	1100	1010000	0000000001
Unused code words					
	1010	0101	0000	0000000	0000000000
	1011	0110	0001	0000001	0000000011
	1100	0111	0010	0000010	0000000101
	1101	1000	1101	0000011	0000000110
	1110	1001	1110	0000101	0000000111
	1111	1010	1111	· · ·	· · ·

Conversions between BCD and decimal representations are trivial, a direct substitution of four bits for each decimal digit. Some computer programs place two BCD digits in one 8-bit byte in *packed-BCD representation*; thus, one byte may represent the values from 0 to 99 as opposed to 0 to 255 for a normal unsigned 8-bit binary number. BCD numbers with any desired number of digits may be obtained by using one byte for each two digits.

packed-BCD representation

As with binary numbers, there are many possible representations of negative BCD numbers. Signed BCD numbers have one extra digit position for

---

BINOMIAL COEFFICIENTS

The number of different ways to choose m items from a set of n items is given by a *binomial coefficient*, denoted $\binom{n}{m}$, whose value is $\dfrac{n!}{m! \cdot (n-m)!}$. For a 4-bit decimal code, there are $\binom{16}{10}$ different ways to choose 10 out of 16 4-bit code words, and 10! ways to assign each different choice to the 10 digits. So there are $\dfrac{16!}{10! \cdot 6!} \cdot 10!$ or 29,059,430,400 different 4-bit decimal codes.

the sign. Both the signed-magnitude and 10's-complement representations are popular. In signed-magnitude BCD, the encoding of the sign bit string is arbitrary; in 10's-complement, 0000 indicates plus and 1001 indicates minus.

BCD addition

Addition of BCD digits is similar to adding 4-bit unsigned binary numbers, except that a correction must be made if a result exceeds 1001. The result is corrected by adding 6; examples are shown below:

$$
\begin{array}{rl}
5 & 0101 \\
+\ 9 & +\ 1001 \\
\hline
14 & 1110 \\
 & +\ 0110 \quad\text{— correction} \\
\hline
10+4 & 1\,0100
\end{array}
\qquad
\begin{array}{rl}
4 & 0100 \\
+\ 5 & +\ 0101 \\
\hline
9 & 1001
\end{array}
$$

$$
\begin{array}{rl}
8 & 1000 \\
+\ 8 & +\ 1000 \\
\hline
16 & 1\,0000 \\
 & +\ 0110 \quad\text{— correction} \\
\hline
10+6 & 1\,0110
\end{array}
\qquad
\begin{array}{rl}
9 & 1001 \\
+\ 9 & +\ 1001 \\
\hline
18 & 1\,0010 \\
 & +\ 0110 \quad\text{— correction} \\
\hline
10+8 & 1\,1000
\end{array}
$$

Notice that the addition of two BCD digits produces a carry into the next digit position if either the initial binary addition or the correction-factor addition produces a carry. Many computers perform packed-BCD arithmetic using special instructions that handle the carry correction automatically.

weighted code

Binary-coded decimal is a *weighted code* because each decimal digit can be obtained from its code word by assigning a fixed weight to each code-word bit. The weights for the BCD bits are 8, 4, 2, and 1, and for this reason the code

8421 code
2421 code
self-complementing code

is sometimes called the *8421 code*. Another set of weights results in the *2421 code* shown in Table 2-9. This code has the advantage that it is *self-complementing*, that is, the code word for the 9s' complement of any digit may be obtained by complementing the individual bits of the digit's code word.

excess-3 code

Another self-complementing code shown in Table 2-9 is the *excess-3 code*. Although this code is not weighted, it has an arithmetic relationship with the BCD code—the code word for each decimal digit is the corresponding BCD code word plus 0011_2. Because the code words follow a standard binary counting sequence, standard binary counters can easily be made to count in excess-3 code, as we'll show in Figure 8-37 on page 700.

biquinary code

Decimal codes can have more than four bits; for example, the *biquinary code* in Table 2-9 uses seven. The first two bits in a code word indicate whether the number is in the range 0–4 or 5–9, and the last five bits indicate which of the five numbers in the selected range is represented.

One potential advantage of using more than the minimum number of bits in a code is an error-detecting property. In the biquinary code, if any one bit in a code word is accidentally changed to the opposite value, the resulting code word does not represent a decimal digit and can therefore be flagged as an error. Out of 128 possible 7-bit code words, only 10 are valid and recognized as decimal digits; the rest can be flagged as errors if they appear.

A *1-out-of-10 code*, such as the one shown in the last column of Table 2-9, is the sparsest encoding for decimal digits, using 10 out of 1024 possible 10-bit code words.

1-out-of-10 code

2.11 Gray Code

In electromechanical applications of digital systems—such as machine tools, automotive braking systems, and copiers—it is sometimes necessary for an input sensor to produce a digital value that indicates a mechanical position. For example, Figure 2-5 is a conceptual sketch of an encoding disk and a set of contacts that produce one of eight 3-bit binary-coded values depending on the rotational position of the disk. The dark areas of the disk are connected to a signal source corresponding to logic 1, and the light areas are unconnected, which the contacts interpret as logic 0.

The encoder in Figure 2-5 has a problem when the disk is positioned at certain boundaries between the regions. For example, consider the boundary between the 001 and 010 regions of the disk; two of the encoded bits change here. What value will the encoder produce if the disk is positioned right on the theoretical boundary? Since we're on the border, both 001 and 010 are acceptable. However, because the mechanical assembly is not perfect, the two righthand contacts may both touch a "1" region, giving an incorrect reading of 011. Likewise, a reading of 000 is possible. In general, this sort of problem can occur at any boundary where more than one bit changes. The worst problems occur when all three bits are changing, as at the 000–111 and 011–100 boundaries.

The encoding-disk problem can be solved by devising a digital code in which only one bit changes between each pair of successive code words. Such a code is called a *Gray code*; a 3-bit Gray code is listed in Table 2-10. We've

Gray code

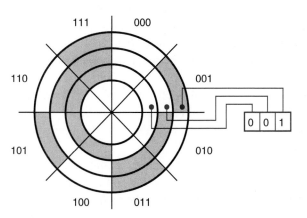

Figure 2-5
A mechanical encoding disk using a 3-bit binary code.

Table 2-10
A comparison of 3-bit binary code and Gray code.

Decimal Number	Binary Code	Gray Code
0	000	000
1	001	001
2	010	011
3	011	010
4	100	110
5	101	111
6	110	101
7	111	100

redesigned the encoding disk using this code as shown in Figure 2-6. Only one bit of the new disk changes at each border, so borderline readings give us a value on one side or the other of the border.

There are two convenient ways to construct a Gray code with any desired number of bits. The first method is based on the fact that Gray code is a *reflected code*; it can be defined (and constructed) recursively using the following rules:

reflected code

1. A 1-bit Gray code has two code words, 0 and 1.

2. The first 2^n code words of an $(n+1)$-bit Gray code equal the code words of an n-bit Gray code, written in order with a leading 0 appended.

3. The last 2^n code words of an $(n+1)$-bit Gray code equal the code words of an n-bit Gray code, but written in reverse order with a leading 1 appended.

If we draw a line between rows 3 and 4 of Table 2-10, we can see that rules 2 and 3 are true for the 3-bit Gray code. Of course, to construct an n-bit Gray code for an arbitrary value of n with this method, we must also construct a Gray code of each length smaller than n.

Figure 2-6
A mechanical encoding disk using a 3-bit Gray code.

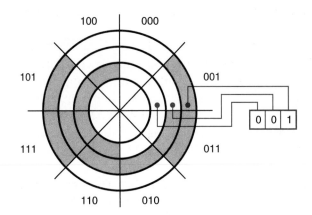

The second method allows us to derive an n-bit Gray-code code word directly from the corresponding n-bit binary code word:

1. The bits of an n-bit binary or Gray-code code word are numbered from right to left, from 0 to $n - 1$.
2. Bit i of a Gray-code code word is 0 if bits i and $i + 1$ of the corresponding binary code word are the same, else bit i is 1. (When $i + 1 = n$, bit n of the binary code word is considered to be 0.)

Again, inspection of Table 2-10 shows that this is true for the 3-bit Gray code.

*2.12 Character Codes

As we showed in the preceding section, a string of bits need not represent a number, and in fact most of the information processed by computers is nonnumeric. The most common type of nonnumeric data is *text*, strings of characters from some character set. Each character is represented in the computer by a bit string according to an established convention.

text

The most commonly used character code is *ASCII* (pronounced *ASS key*), the American Standard Code for Information Interchange. ASCII represents each character with a 7-bit string, yielding a total of 128 different characters shown in Table 2-11. The code contains the uppercase and lowercase alphabet, numerals, punctuation, and various nonprinting control characters. Thus, the text string "Yeccch!" is represented by a rather innocuous-looking list of seven 7-bit numbers:

ASCII

1011001   1100101   1100011   1100011   1100011   1101000   0100001

2.13 Codes for Actions, Conditions, and States

The codes that we've described so far are generally used to represent things that we would probably consider to be "data"—things like numbers, positions, and characters. Programmers know that dozens of different data types can be used in a single computer program.

In digital system design, we often encounter nondata applications where a string of bits must be used to control an action, to flag a condition, or to represent the current state of the hardware. Probably the most commonly used type of code for such an application is a simple binary code. If there are n different actions, conditions, or states, we can represent them with a b-bit binary code with $b = \lceil \log_2 n \rceil$ bits. (The brackets $\lceil \ \rceil$ denote the *ceiling function*—the smallest integer greater than or equal to the bracketed quantity. Thus, b is the smallest integer such that $2^b \geq n$.)

$\lceil \ \rceil$

ceiling function

For example, consider a simple traffic-light controller. The signals at the intersection of a north-south (N-S) and an east-west (E-W) street might be in any

■ **Table 2-11** American Standard Code for Information Interchange (ASCII),
Standard No. X3.4-1968 of the American National Standards Institute.

$b_3b_2b_1b_0$	Row (hex)	000 0	001 1	010 2	011 3	100 4	101 5	110 6	111 7
0000	0	NUL	DLE	SP	0	@	P	`	p
0001	1	SOH	DC1	!	1	A	Q	a	q
0010	2	STX	DC2	"	2	B	R	b	r
0011	3	ETX	DC3	#	3	C	S	c	s
0100	4	EOT	DC4	$	4	D	T	d	t
0101	5	ENQ	NAK	%	5	E	U	e	u
0110	6	ACK	SYN	&	6	F	V	f	v
0111	7	BEL	ETB	'	7	G	W	g	w
1000	8	BS	CAN	(	8	H	X	h	x
1001	9	HT	EM	)	9	I	Y	i	y
1010	A	LF	SUB	*	:	J	Z	j	z
1011	B	VT	ESC	+	;	K	[	k	{
1100	C	FF	FS	,	<	L	\	l	|
1101	D	CR	GS	–	=	M	]	m	}
1110	E	SO	RS	.	>	N	^	n	~
1111	F	SI	US	/	?	O	_	o	DEL

The top header row reads: $b_6b_5b_4$ (column)

Control codes

NUL	Null	DLE	Data link escape
SOH	Start of heading	DC1	Device control 1
STX	Start of text	DC2	Device control 2
ETX	End of text	DC3	Device control 3
EOT	End of transmission	DC4	Device control 4
ENQ	Enquiry	NAK	Negative acknowledge
ACK	Acknowledge	SYN	Synchronize
BEL	Bell	ETB	End transmitted block
BS	Backspace	CAN	Cancel
HT	Horizontal tab	EM	End of medium
LF	Line feed	SUB	Substitute
VT	Vertical tab	ESC	Escape
FF	Form feed	FS	File separator
CR	Carriage return	GS	Group separator
SO	Shift out	RS	Record separator
SI	Shift in	US	Unit separator
SP	Space	DEL	Delete or rubout

Table 2-12 States in a traffic-light controller.

	Lights						
State	**N-S Green**	**N-S Yellow**	**N-S Red**	**E-W Green**	**E-W Yellow**	**E-W Red**	**Code Word**
N-S go	ON	off	off	off	off	ON	000
N-S wait	off	ON	off	off	off	ON	001
N-S delay	off	off	ON	off	off	ON	010
E-W go	off	off	ON	ON	off	off	100
E-W wait	off	off	ON	off	ON	off	101
E-W delay	off	off	ON	off	off	ON	110

of the six states listed in Table 2-12. These states can be encoded in three bits, as shown in the last column of the table. Only six of the eight possible 3-bit code words are used, and the assignment of the six chosen code words to states is arbitrary, so many other encodings are possible. An experienced digital designer chooses a particular encoding to minimize circuit cost or to optimize some other parameter (like design time—there's no need to try billions and billions of possible encodings).

Another application of a binary code is illustrated in Figure 2-7(a). Here, we have a system with n devices, each of which can perform a certain action. The characteristics of the devices are such that they may be enabled to operate only one at a time. The control unit produces a binary-coded "device-select" word with $\lceil \log_2 n \rceil$ bits to indicate which device is enabled at any time. The "device-select code word is applied to each device, which compares it with its own "device ID" to determine whether it is enabled. Although its code words have the minimum number of bits, a binary code isn't always the best choice for encoding actions, conditions, or states. Figure 2-7(b) shows how to control n devices with a *1-out-of-n code*, an n-bit code in which valid code words have one bit equal to 1 and the rest of the bits equal to 0. Each bit of the 1-out-of-n code word is connected directly to the enable input of a corresponding device. This simplifies the design of the devices, since they no longer have device IDs; they need only a single "enable" input bit.

1-out-of-n code

The code words of a 1-out-of-10 code were listed in Table 2-9. Sometimes an all-0s word may also be included in a 1-out-of-n code, to indicate that no device is selected. Another common code is an *inverted 1-out-of-n code*, in which valid code words have one 0 bit and the rest of the bits equal to 1.

inverted 1-out-of-n code

In complex systems, a combination of coding techniques may be used. For example, consider a system similar to Figure 2-7(b), in which each of the n devices contains up to s subdevices. The control unit could produce a device-

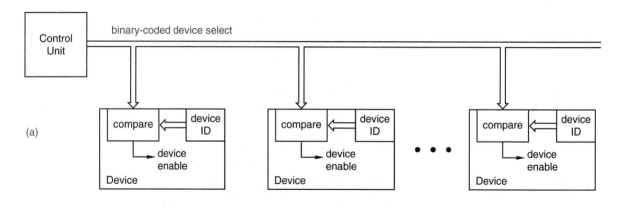

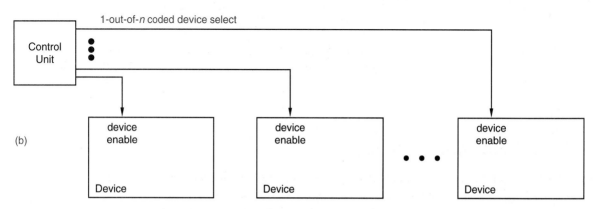

Figure 2-7 Control structure for a digital system with *n* devices:
(a) using a binary code; (b) using a 1-out-of-*n* code.

select code word with a 1-out-of-*n* coded field to select a device, and a $\lceil \log_2 s \rceil$-bit binary-coded field to select one of the *s* subdevices of the selected device.

m-out-of-n code

An *m-out-of-n code* is a generalization of the 1-out-of-*n* code in which valid code words have *m* bits equal to 1 and the rest of the bits equal to 0. An *m*-out-of-*n* code word can be detected with an *m*-input **AND** gate, which produces a 1 output if all of its inputs are 1. This is fairly simple and inexpensive to do, yet for most values of *m*, an *m*-out-of-*n* code typically has far more valid code words than a 1-out-of-*n* code. The total number of code words is given by the binomial coefficient $\binom{n}{m}$, which has the value $\frac{n!}{m! \cdot (n-m)!}$. Thus, a 2-out-of-4 code has 6 valid code words, and a 3-out-of-10 code has 120.

8B10B code

An important variation of an *m*-out-of-*n* code is the *8B10B code* used in the 802.3z Gigabit Ethernet standard. This code uses 10 bits to represent 256 valid code words, or 8 bits worth of data. Most code words use a 5-out-of-10 coding. However, since $\binom{5}{10}$ is only 252, some 4- and 6-out-of-10 words are also used to complete the code in a very interesting way; more on this in Section 2.16.2.

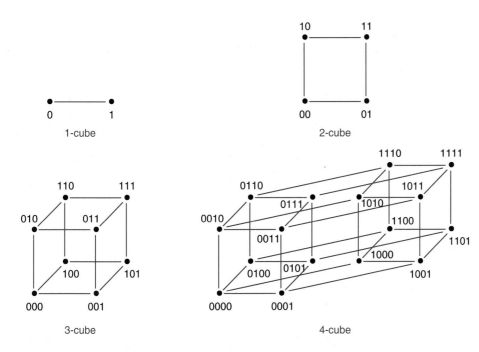

Figure 2-8
n-cubes for $n = 1, 2,$
3, and 4.

*2.14 *n*-Cubes and Distance

An *n*-bit string can be visualized geometrically, as a vertex of an object called an   *n-cube*
n-cube. Figure 2-8 shows *n*-cubes for $n = 1, 2, 3, 4$. An *n*-cube has 2^n vertices,
each of which is labeled with an *n*-bit string. Edges are drawn so that each vertex
is adjacent to *n* other vertices whose labels differ from the given vertex in only
one bit. Beyond $n = 4$, *n*-cubes are really tough to draw.

For reasonable values of *n*, *n*-cubes make it easy to visualize certain coding
and logic-minimization problems. For example, the problem of designing an
n-bit Gray code is equivalent to finding a path along the edges of an *n*-cube, a
path that visits each vertex exactly once. The paths for 3- and 4-bit Gray codes
are shown in Figure 2-9.

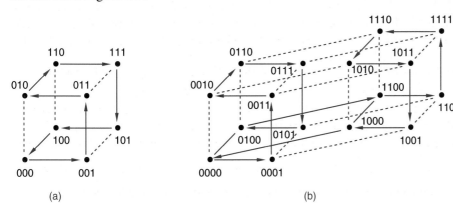

Figure 2-9
Traversing *n*-cubes
in Gray-code order:
(a) 3-cube;
(b) 4-cube.

distance
Hamming distance

Cubes also provide a geometrical interpretation for the concept of *distance*, also called *Hamming distance*. The distance between two *n*-bit strings is the number of bit positions in which they differ. In terms of an *n*-cube, the distance is the minimum length of a path between the two corresponding vertices. Two adjacent vertices have distance 1; vertices 001 and 100 in the 3-cube have distance 2. The concept of distance is crucial in the design and understanding of error-detecting codes, discussed in the next section.

m-subcube

An *m-subcube* of an *n*-cube is a set of 2^m vertices in which $n - m$ of the bits have the same value at each vertex, and the remaining *m* bits take on all 2^m combinations. For example, the vertices (000, 010, 100, 110) form a 2-subcube of the 3-cube. This subcube can also be denoted by a single string, xx0, where "x" denotes that a particular bit is a *don't-care*; any vertex whose bits match in the non-x positions belongs to this subcube. The concept of subcubes is particularly useful in visualizing algorithms that minimize the cost of combinational logic functions, as we'll show in Section 4.4.

don't-care

*2.15 Codes for Detecting and Correcting Errors

error
failure
temporary failure
permanent failure

An *error* in a digital system is the corruption of data from its correct value to some other value. An error is caused by a physical *failure*. Failures can be either temporary or permanent. For example, a cosmic ray or alpha particle can cause a temporary failure of a memory circuit, changing the value of a bit stored in it. Letting a circuit get too hot or zapping it with static electricity can cause a permanent failure, so that it never works correctly again.

error model
independent error model
single error
multiple error

The effects of failures on data are predicted by *error models*. The simplest error model, which we consider here, is called the *independent error model*. In this model, a single physical failure is assumed to affect only a single bit of data; the corrupted data is said to contain a *single error*. Multiple failures may cause *multiple errors*—two or more bits in error—but multiple errors are normally assumed to be less likely than single errors.

2.15.1 Error-Detecting Codes

Recall from our definitions in Section 2.10 that a code that uses *n*-bit strings need not contain 2^n valid code words; this is certainly the case for the codes that we now consider. An *error-detecting code* has the property that corrupting or garbling a code word will likely produce a bit string that is not a code word (a *noncode word*).

error-detecting code

noncode word

A system that uses an error-detecting code generates, transmits, and stores only code words. Thus, errors in a bit string can be detected by a simple rule—if the bit string is a code word, it is assumed to be correct; if it is a noncode word, it contains an error.

An *n*-bit code and its error-detecting properties under the independent error model are easily explained in terms of an *n*-cube. A code is simply a subset

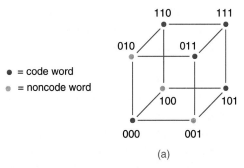

= code word
= noncode word

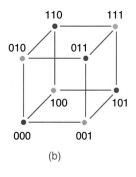

Figure 2-10
Code words in two
different 3-bit codes:
(a) minimum distance
= 1, does not detect
all single errors;
(b) minimum distance
= 2, detects all single
errors.

of the vertices of the *n*-cube. In order for the code to detect all single errors, no code-word vertex can be immediately adjacent to another code-word vertex.

For example, Figure 2-10(a) shows a 3-bit code with five code words. Code word 111 is immediately adjacent to code words 110, 011 and 101. Since a single failure could change 111 to 110, 011 or 101 this code does not detect all single errors. If we make 111 a noncode word, we obtain a code that does have the single-error-detecting property, as shown in (b). No single error can change one code word into another.

The ability of a code to detect single errors can be stated in terms of the concept of distance introduced in the preceding section:

- A code detects all single errors if the *minimum distance* between all possible pairs of code words is 2.

minimum distance

In general, we need $n + 1$ bits to construct a single-error-detecting code with 2^n code words. The first *n* bits of a code word, called *information bits*, may be any of the 2^n *n*-bit strings. To obtain a minimum-distance-2 code, we add one more bit, called a *parity bit*, that is set to 0 if there are an even number of 1s among the information bits, and to 1 otherwise. This is illustrated in the first two columns of Table 2-13 for a code with three information bits. A valid $(n+1)$-bit code word has an even number of 1s, and this code is called an *even-parity code*.

information bit

parity bit

even-parity code

Information Bits	Even-parity Code	Odd-parity Code
000	000 0	000 1
001	001 1	001 0
010	010 1	010 0
011	011 0	011 1
100	100 1	100 0
101	101 0	101 1
110	110 0	110 1
111	111 1	111 0

Table 2-13
Distance-2 codes with
three information bits.

We can also construct a code in which the total number of 1s in a valid $(n+1)$-bit code word is odd; this is called an *odd-parity code* and is shown in the third column of the table. These codes are also sometimes called *1-bit parity codes*, since they each use a single parity bit.

The 1-bit parity codes do not detect 2-bit errors, since changing two bits does not affect the parity. However, the codes can detect errors in any *odd* number of bits. For example, if three bits in a code word are changed, then the resulting word has the wrong parity and is a noncode word. This doesn't help us much, though. Under the independent error model, 3-bit errors are much less likely than 2-bit errors, which are not detectable. Thus, practically speaking, the 1-bit parity codes' error-detection capability stops after 1-bit errors. Other codes, with minimum distance greater than 2, can be used to detect multiple errors.

2.15.2 Error-Correcting and Multiple-Error-Detecting Codes

check bits

By using more than one parity bit, or *check bits*, according to some well-chosen rules, we can create a code whose minimum distance is greater than 2. Before showing how this can be done, let's look at how such a code can be used to correct single errors or detect multiple errors.

Suppose that a code has a minimum distance of 3. Figure 2-11 shows a fragment of the *n*-cube for such a code. As shown, there are at least two noncode words between each pair of code words. Now suppose we transmit code words

Figure 2-11
Some code words and noncode words in a 7-bit, distance-3 code.

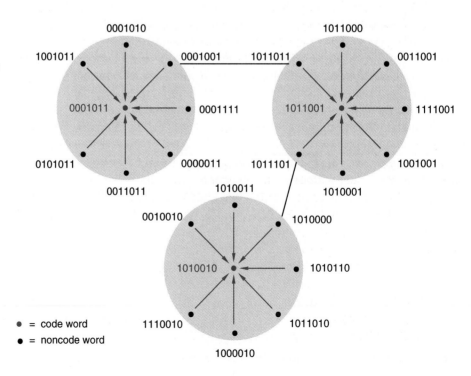

and assume that failures affect at most one bit of each received code word. Then a received noncode word with a 1-bit error will be closer to the originally transmitted code word than to any other code word. Therefore, when we receive a noncode word, we can *correct* the error by changing the received noncode word to the nearest code word, as indicated by the arrows in the figure. Deciding which code word was originally transmitted to produce a received word is called *decoding*, and the hardware that does this is an error-correcting *decoder*.

error correction

decoding
decoder

A code that is used to correct errors is called an *error-correcting code*. In general, if a code has minimum distance $2c + 1$, it can be used to correct errors that affect up to c bits ($c = 1$ in the preceding example). If a code's minimum distance is $2c + d + 1$, it can be used to correct errors in up to c bits and to detect errors in up to d additional bits.

error-correcting code

For example, Figure 2-12(a) shows a fragment of the n-cube for a code with minimum distance 4 ($c = 1$, $d = 1$). Single-bit errors that produce noncode words 00101010 and 11010011 can be corrected. However, an error that produces 10100011 cannot be corrected, because no single-bit error can produce this noncode word, and either of two 2-bit errors could have produced it. So the code can detect a 2-bit error, but it cannot correct it.

When a noncode word is received, we don't know which code word was originally transmitted; we only know which code word is closest to what we've received. Thus, as shown in Figure 2-12(b), a 3-bit error may be "corrected" to the wrong value. The possibility of making this kind of mistake may be acceptable if 3-bit errors are very unlikely to occur. On the other hand, if we are concerned about 3-bit errors, we can change the decoding policy for the code. Instead of trying to correct errors, we just flag all noncode words as uncorrectable errors. Thus, as shown in (c), we can use the same distance-4 code to detect up to 3-bit errors but correct no errors ($c = 0$, $d = 3$).

2.15.3 Hamming Codes

In 1950, R. W. Hamming described a general method for constructing codes with a minimum distance of 3, now called *Hamming codes*. For any value of i, his method yields a $(2^i - 1)$-bit code with i check bits and $2^i - 1 - i$ information bits. Distance-3 codes with a smaller number of information bits are obtained by deleting information bits from a Hamming code with a larger number of bits.

Hamming code

The bit positions in a Hamming code word can be numbered from 1 through $2^i - 1$. In this case, any position whose number is a power of 2 is a check bit, and the remaining positions are information bits. Each check bit is grouped with a subset of the information bits, as specified by a *parity-check matrix*. As

parity-check matrix

DECISIONS, DECISIONS   The names *decoding* and *decoder* make sense, since they are just distance-1 perturbations of *deciding* and *decider*.

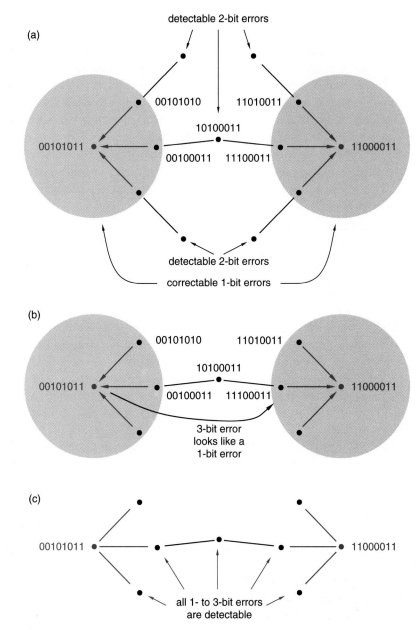

Figure 2-12
Some code words and
noncode words in an 8-bit,
distance-4 code:
(a) correcting 1-bit and
detecting 2-bit errors;
(b) incorrectly "correcting"
a 3-bit error;
(c) correcting no errors but
detecting up to 3-bit errors.

shown in Figure 2-13(a), each check bit is grouped with the information positions whose numbers have a 1 in the same bit when expressed in binary. For
example, check bit 2 (010) is grouped with information bits 3 (011), 6 (110), and
7 (111). For a given combination of information-bit values, each check bit is
chosen to produce even parity, that is, so the total number of 1s in its group is
even.

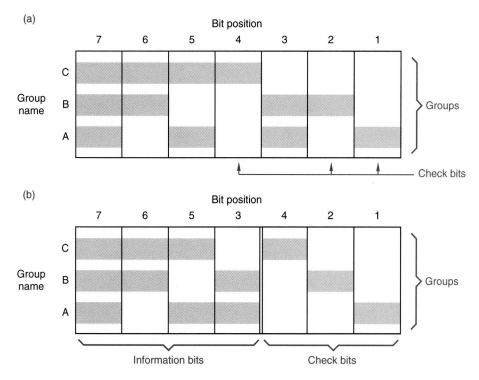

Figure 2-13
Parity-check matrices for 7-bit Hamming codes: (a) with bit positions in numerical order; (b) with check bits and information bits separated.

Traditionally, the bit positions of a parity-check matrix and the resulting code words are rearranged so that all of the check bits are on the right, as in Figure 2-13(b). The first two columns of Table 2-14 list the resulting code words.

We can prove that the minimum distance of a Hamming code is 3 by proving that at least a 3-bit change must be made to a code word to obtain another code word. That is, we'll prove that a 1-bit or 2-bit change in a code word yields a noncode word.

If we change one bit of a code word, in position j, then we change the parity of every group that contains position j. Since every information bit is contained in at least one group, at least one group has incorrect parity, and the result is a noncode word.

What happens if we change two bits, in positions j and k? Parity groups that contain both positions j and k will still have correct parity, since parity is not affected when an even number of bits are changed. However, since j and k are different, their binary representations differ in at least one bit, corresponding to one of the parity groups. This group has only one bit changed, resulting in incorrect parity and a noncode word.

If you understand this proof, you should also understand how the position-numbering rules for constructing a Hamming code are a simple consequence of the proof. For the first part of the proof (1-bit errors), we required that the position numbers be nonzero. And for the second part (2-bit errors), we required

■ **Table 2-14** Code words in distance-3 and distance-4 Hamming codes with four information bits.

Minimum-Distance-3 Code		Minimum-Distance-4 Code	
Information Bits	Parity Bits	Information Bits	Parity Bits
0000	000	0000	0000
0001	011	0001	0111
0010	101	0010	1011
0011	110	0011	1100
0100	110	0100	1101
0101	101	0101	1010
0110	011	0110	0110
0111	000	0111	0001
1000	111	1000	1110
1001	100	1001	1001
1010	010	1010	0101
1011	001	1011	0010
1100	001	1100	0011
1101	010	1101	0100
1110	100	1110	1000
1111	111	1111	1111

*error-correcting
decoder*

syndrome

that no two positions have the same number. Thus, with an i-bit position number, you can construct a Hamming code with up to $2^i - 1$ bit positions.

The proof also suggests how we can design an *error-correcting decoder* for a received Hamming code word. First, we check all of the parity groups; if all have even parity, then the received word is assumed to be correct. If one or more groups have odd parity, then a single error is assumed to have occurred. The pattern of groups that have odd parity (called the *syndrome*) must match one of the columns in the parity-check matrix; the corresponding bit position is assumed to contain the wrong value and is complemented. For example, using the code defined by Figure 2-13(b), suppose we receive the word 0101011. Groups B and C have odd parity, corresponding to position 6 of the parity-check matrix (the syndrome is 110, or 6). By complementing the bit in position 6 of the received word, we determine that the correct word is 0001011.

A distance-3 Hamming code can easily be modified to increase its minimum distance to 4. We simply add one more check bit, chosen so that the parity of all the bits, including the new one, is even. As in the 1-bit even-parity code, this bit ensures that all errors affecting an odd number of bits are detectable. In particular, any 3-bit error is detectable. We already showed that 1- and 2-bit errors are detected by the other parity bits, so the minimum distance of the modified code must be 4.

Distance-3 and distance-4 Hamming codes are commonly used to detect and correct errors in computer memory systems, especially in large mainframe computers where memory circuits account for the bulk of the system's failures. These codes are especially attractive for very wide memory words, since the required number of parity bits grows slowly with the width of the memory word, as shown in Table 2-15.

Table 2-15 Word sizes of distance-3 and distance-4 Hamming codes.

	Minimum-Distance-3 Codes		Minimum-Distance-4 Codes	
Information Bits	Parity Bits	Total Bits	Parity Bits	Total Bits
1	2	3	3	4
≤ 4	3	≤ 7	4	≤ 8
≤ 11	4	≤ 15	5	≤ 16
≤ 26	5	≤ 31	6	≤ 32
≤ 57	6	≤ 63	7	≤ 64
≤ 120	7	≤ 127	8	≤ 128

2.15.4 CRC Codes

Beyond Hamming codes, many other error-detecting and -correcting codes have been developed. The most important codes, which happen to include Hamming codes, are the *cyclic-redundancy-check (CRC) codes*. A rich theory has been developed for these codes, focused both on their error-detecting and correcting properties and on the design of inexpensive encoders and decoders for them (see References).

cyclic-redundancy-check (CRC) code

Two important applications of CRC codes are in disk drives and in data networks. In a disk drive, each block of data (typically 512 bytes) is protected by a CRC code, so that errors within a block can be detected and, in some drives, corrected. In a data network, each packet of data ends with check bits in a CRC code. The CRC codes for both applications were selected because of their burst-error detecting properties. In addition to single-bit errors, they can detect multibit errors that are clustered together within the disk block or packet. Such errors are more likely than errors of randomly distributed bits, because of the likely physical causes of errors in the two applications—surface defects in disk drives and noise bursts in communication links.

2.15.5 Two-Dimensional Codes

two-dimensional code

Another way to obtain a code with large minimum distance is to construct a *two-dimensional code*, as illustrated in Figure 2-14(a). The information bits are conceptually arranged in a two-dimensional array, and parity bits are provided to check both the rows and the columns. A code C_{row} with minimum distance d_{row} is used for the rows, and a possibly different code C_{col} with minimum distance d_{col} is used for the columns. That is, the row-parity bits are selected so that each row is a code word in C_{row} and the column-parity bits are selected so that each column is a code word in C_{col}. (The "corner" parity bits can be chosen according to either code.) The minimum distance of the two-dimensional code is the product of d_{row}

product code

and d_{col}; in fact, two-dimensional codes are sometimes called *product codes*.

As shown in Figure 2-14(b), the simplest two-dimensional code uses 1-bit even-parity codes for the rows and columns and has a minimum distance of $2 \cdot 2$,

(a)

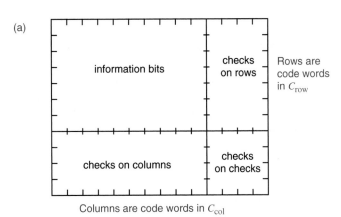

Columns are code words in C_{col}

Figure 2-14
Two-dimensional codes:
(a) general structure;
(b) using even parity for both the row and column codes to obtain minimum distance 4;
(c) typical pattern of an undetectable error.

(b)

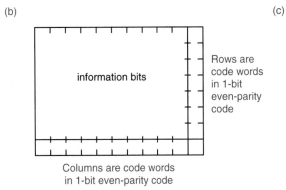

Columns are code words
in 1-bit even-parity code

(c)

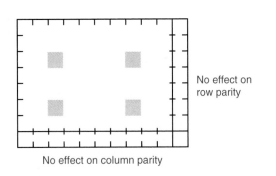

No effect on column parity

or 4. You can easily prove that the minimum distance is 4 by convincing yourself that any pattern of one, two, or three bits in error causes incorrect parity of a row or a column or both. In order to obtain an undetectable error, at least four bits must be changed in a rectangular pattern as in (c).

The error-detecting and correcting procedures for this code are straightforward. Assume we are reading information one row at a time. As we read each row, we check its row code. If an error is detected in a row, we can't tell which bit is wrong from the row check alone. However, assuming only one row is bad, we can reconstruct it by forming the bit-by-bit Exclusive OR of the columns, omitting the bad row, but including the column-check row.

To obtain an even larger minimum distance, a distance-3 or -4 Hamming code can be used for the row or column code or both. It is also possible to construct a code in three or more dimensions, with minimum distance equal to the product of the minimum distances in each dimension.

An important application of two-dimensional codes is in RAID storage systems. *RAID* stands for "redundant array of inexpensive disks." In this scheme, $n + 1$ identical disk drives are used to store n disks worth of data. For example, eight 8-gigabyte drives could be used to store 64 gigabytes of nonredundant data, and a ninth 8-gigabyte drive would be used to store checking information.

RAID

Figure 2-15 shows the general scheme of a two-dimensional code for a RAID system; each disk drive is considered to be a row in the code. Each drive stores m blocks of data, where a block typically contains 512 bytes. For example, an 8-gigabyte drive would store about 16 million blocks. As shown in the figure, each block includes its own check bits in a CRC code, to detect errors within that block. The first n drives store the nonredundant data. Each block in drive $n + 1$ stores parity bits for the corresponding blocks in the first n drives. That is, each

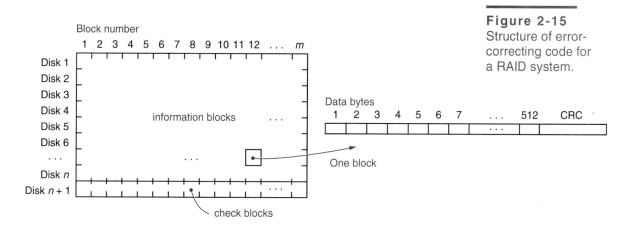

Figure 2-15
Structure of error-correcting code for a RAID system.

bit i in drive $n+1$, block b, is chosen so that there are an even number of 1s in block b, bit position i, across all the drives.

In operation, errors in the information blocks are detected by the CRC code. Whenever an error is detected in a block on one of the drives, the correct contents of that block can be constructed simply by computing the parity of the corresponding blocks in all the other drives, including drive $n+1$. Although this requires n extra disk-read operations, it's better than losing your data! Write operations require extra disk accesses as well, to update the corresponding check block when an information block is written (see Exercise 2.46). Since disk writes are much less frequent than reads in typical applications, this overhead usually is not a problem.

2.15.6 Checksum Codes

The parity-checking operation that we've used in the previous subsections is essentially modulo-2 addition of bits—the sum modulo 2 of a group of bits is 0 if the number of 1s in the group is even, and 1 if it is odd. The technique of modular addition can be extended to other bases besides 2 to form check digits.

For example, a computer stores information as a set of 8-bit bytes. Each byte may be considered to have a decimal value from 0 to 255. Therefore, we can use modulo-256 addition to check the bytes. We form a single check byte, called

checksum
checksum code

a *checksum*, that is the sum modulo 256 of all the information bytes. The resulting *checksum code* can detect any single *byte* error, since such an error will cause a recomputed sum of bytes to disagree with the checksum.

Checksum codes can also use a different modulus of addition. In particular,

ones'-complement
checksum code

checksum codes using modulo-255, ones'-complement addition are important because of their special computational and error-detecting properties, and because they are used to check packet headers in the ubiquitous Internet Protocol (IP) (see References).

2.15.7 *m*-out-of-*n* Codes

The 1-out-of-n and m-out-of-n codes that we introduced in Section 2.13 have a minimum distance of 2, since changing only one bit changes the total number of 1s in a code word and therefore produces a noncode word.

These codes have another useful error-detecting property—they detect

unidirectional error

unidirectional multiple errors. In a *unidirectional error*, all of the erroneous bits change in the same direction (0s change to 1s, or vice versa). This property is very useful in systems where the predominant error mechanism tends to change all bits in the same direction.

2.16 Codes for Serial Data Transmission and Storage

2.16.1 Parallel and Serial Data

Most computers and other digital systems transmit and store data in a *parallel* format. In parallel data transmission, a separate signal line is provided for each bit of a data word. In parallel data storage, all of the bits of a data word can be written or read simultaneously.

parallel data

    Parallel formats are not cost effective for some applications. For example, parallel transmission of data bytes over the telephone network would require eight phone lines, and parallel storage of data bytes on a magnetic disk would require a disk drive with eight separate read/write heads. *Serial* formats allow data to be transmitted or stored one bit at a time, reducing system cost in many applications.

serial data

    Figure 2-16 illustrates some of the basic ideas in serial data transmission. A repetitive clock signal, named CLOCK in the figure, defines the rate at which bits are transmitted, one bit per clock cycle. Thus, the *bit rate* in bits per second (bps) numerically equals the clock frequency in cycles per second (hertz, or Hz).

bit rate, bps

    The reciprocal of the bit rate is called the *bit time* and numerically equals the clock period in seconds (s). This amount of time is reserved on the serial data line (named SERDATA in the figure) for each bit that is transmitted. The time occupied by each bit is sometimes called a *bit cell*. The format of the actual signal that appears on the line during each bit cell depends on the *line code*. In the simplest line code, called *Non-Return-to-Zero (NRZ)*, a 1 is transmitted by placing a 1 on the line for the entire bit cell, and a 0 is transmitted as a 0. More complex line codes have other rules, as discussed in the next subsection.

bit time

bit cell
line code
Non-Return-to-Zero (NRZ)

Figure 2-16 Basic concepts for serial data transmission.

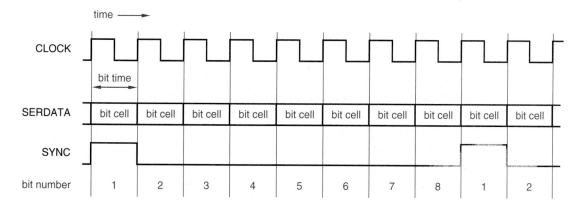

Regardless of the line code, a serial data-transmission or storage system needs some way of identifying the significance of each bit in the serial stream. For example, suppose that 8-bit bytes are transmitted serially. How can we tell which *synchronization signal* is the first bit of each byte? A *synchronization signal*, named SYNC in Figure 2-16, provides the necessary information; it is 1 for the first bit of each byte.

Evidently, we need a minimum of three signals to recover a serial data stream: a clock to define the bit cells, a synchronization signal to define the word boundaries, and the serial data itself. In some applications, like the interconnection of modules in a computer or telecommunications system, a separate wire is used for each of these signals, since reducing the number of wires per connection from *n* to three is savings enough. We'll give an example of a 3-wire serial data system in Section 8.5.4.

In many applications, the cost of having three separate signals is still too high (e.g., three phone lines, three read/write heads). Such systems typically combine all three signals into a single serial data stream and use sophisticated analog and digital circuits to recover the clock and synchronization information from the data stream.

*2.16.2 Serial Line Codes

The most commonly used line codes for serial data are illustrated in Figure 2-17. In the NRZ code, each bit value is sent on the line for the entire bit cell. This is the simplest and most reliable coding scheme for short-distance transmission. However, it generally requires a clock signal to be sent along with the data to define the bit cells. Otherwise, it is not possible for the receiver to determine how many 0s or 1s are represented by a continuous 0 or 1 level. For example, without a clock to define the bit cells, the NRZ waveform in Figure 2-17 might be erroneously interpreted as 01010.

Figure 2-17
Commonly used line codes for serial data.

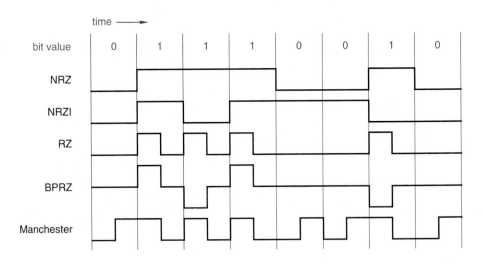

A *digital phase-locked loop (DPLL)* is an analog/digital circuit that can be used to recover a clock signal from a serial data stream. The DPLL works only if the serial data stream contains enough 0-to-1 and 1-to-0 transitions to give the DPLL "hints" about when the original clock transitions took place. With NRZ-coded data, the DPLL works only if the data does not contain any long, continuous streams of 1s or 0s.

digital phase-locked loop (DPLL)

Some serial transmission and storage media are *transition sensitive*; they cannot transmit or store absolute 0 or 1 levels, only transitions between two discrete levels. For example, a magnetic disk or tape stores information by changing the polarity of the medium's magnetization in regions corresponding to the stored bits. When the information is recovered, it is not feasible to determine the absolute magnetization polarity of a region, only that the polarity changes between one region and the next.

transition-sensitive media

Data stored in NRZ format on transition-sensitive media cannot be recovered unambiguously; the data in Figure 2-17 might be interpreted as 01110010 or 10001101. The *Non-Return-to-Zero Invert-on-1s (NRZI)* code overcomes this limitation by sending a 1 as the opposite of the level that was sent during the previous bit cell, and a 0 as the same level. A DPLL can recover the clock from NRZI-coded data as long as the data does not contain any long, continuous streams of 0s.

Non-Return-to-Zero Invert-on-1s (NRZI)

The Return-to-Zero (RZ) code is similar to NRZ except that, for a 1 bit, the 1 level is transmitted only for a fraction of the bit time, usually 1/2. With this code, data patterns that contain a lot of 1s create lots of transitions for a DPLL to use to recover the clock. However, as in the other line codes, a string of 0s has no transitions, and a long string of 0s makes clock recovery impossible.

Return-to-Zero (RZ)

Another requirement of some transmission media, such as high-speed fiber-optic links, is that the serial data stream be *DC balanced*. That is, it must have an equal number of 1s and 0s; any long-term DC component in the stream (created by having a lot more 1s than 0s or vice versa) creates a bias at the receiver that reduces its ability to distinguish reliably between 1s and 0s.

DC balance

Ordinarily, NRZ, NRZI or RZ data has no guarantee of DC balance; there's nothing to prevent a user data stream from having a long string of words with more than 1s than 0s or vice versa. However, DC balance can still be achieved by using a few extra bits to code the user data in a *balanced code*, in which each code word has an equal number of 1s and 0s, and then sending these code words in NRZ format.

balanced code

For example, in Section 2.13 we introduced the 8B10B code, which codes 8 bits of user data into 10 bits in a mostly 5-out-of-10 code. Recall that there are only 252 5-out-of-10 code words, but there are another $\binom{4}{10} = 210$ 4-out-of-10 code words and an equal number of 6-out-of-10 code words. Of course, these code words aren't quite DC balanced. The 8B10B code solves this problem by associating with each 8-bit value to be encoded a *pair* of unbalanced code words,

---

KILO-, MEGA-,
GIGA-, TERA-

The prefixes K (kilo-), M (mega-), G (giga-), and T (tera-) mean 10^3, 10^6, 10^9, and 10^{12}, respectively, when referring to bps, hertz, ohms, watts, and most other engineering quantities. However, when referring to memory sizes, the prefixes mean 2^{10}, 2^{20}, 2^{30}, and 2^{40}. Historically, the prefixes were co-opted for this purpose because memory sizes are normally powers of 2, and 2^{10} (1024) is very close to 1000,

Now, when somebody offers you 50 kilobucks a year for your first engineering job, it's up to you to negotiate what the prefix means!

---

running disparity

one 4-out-of-10 ("light") and the other 6-out-of-10 ("heavy"). The coder also keeps track of the *running disparity*, a single bit of information indicating whether the last unbalanced code word that it transmitted was heavy or light. When it comes time to transmit another unbalanced code word, the coder selects the one of the pair with the opposite weight. This simple trick makes available $252 + 210 = 462$ code words for the 8B10B to encode 8 bits of user data. Some of the "extra" code words are used to conveniently encode nondata conditions on the serial line, such as IDLE, SYNC, and ERROR. Not all the unbalanced code words are used. Also, some of the balanced code words, such as 0000011111, are not used either, in favor of unbalanced pairs that contain more transitions.

Bipolar Return-to-Zero (BPRZ)

Alternate Mark Inversion (AMI)

All of the preceding codes transmit or store only two signal levels. The *Bipolar Return-to-Zero (BPRZ)* code transmits three signal levels: $+1$, 0, and -1. The code is like RZ except that 1s are alternately transmitted as $+1$ and -1; for this reason, the code is also known as *Alternate Mark Inversion (AMI)*.

The big advantage of BPRZ over RZ is that it's DC balanced. This makes it possible to send BPRZ streams over transmission media that cannot tolerate a DC component, such as transformer-coupled phone lines. In fact, the BPRZ code has been used in T1 digital telephone links for decades, where analog speech signals are carried as streams of 8000 8-bit digital samples per second that are transmitted in BPRZ format on 64-Kbps serial channels.

As with RZ, it is possible to recover a clock signal from a BPRZ stream as long as there aren't too many 0s in a row. Although TPC (The Phone Company) has no control over what you say (at least, not yet), they still have a simple way of limiting runs of 0s. If one of the 8-bit bytes that results from sampling your analog speech pattern is all 0s, they simply change second-least significant bit to 1! This is called *zero-code suppression* and I'll bet you never noticed it. And this is also why, in many data applications of T1 links, you get only 56 Kbps of usable data per 64-Kbps channel; the LSB of each byte is always set to 1 to prevent zero-code suppression from changing the other bits.

zero-code suppression

Manchester

diphase

The last code in Figure 2-17 is called *Manchester* or *diphase* code. The major strength of this code is that, regardless of the transmitted data pattern, it provides at least one transition per bit cell, making it very easy to recover the clock. As shown in the figure, a 0 is encoded as a 0-to-1 transition in the middle

ABOUT TPC   Watch the 1967 James Coburn movie, *The President's Analyst,* for an amusing view of TPC. With the growing pervasiveness of digital technology and cheap wireless communications, the concept of universal, *personal* connectivity to the phone network presented in the movie's conclusion has become much less far-fetched.

of the bit cell, and a 1 is encoded as a 1-to-0 transition. The Manchester code's major strength is also its major weakness. Since it has more transitions per bit cell than other codes, it also requires more media bandwidth to transmit a given bit rate. Bandwidth is not a problem in coaxial cable, however, which was used in the original Ethernet local area networks to carry Manchester-coded serial data at the rate of 10 Mbps (megabits per second).

References

The presentation in the first nine sections of this chapter is based on Chapter 4 of *Microcomputer Architecture and Programming*, by John F. Wakerly (Wiley, 1981). Precise, thorough, and entertaining discussions of these topics can also be found in Donald E. Knuth's *Seminumerical Algorithms*, third edition (Addison-Wesley, 1997). Mathematically inclined readers will find Knuth's analysis of the properties of number systems and arithmetic to be excellent, and all readers should enjoy the insights and history sprinkled throughout the text.

Descriptions of digital logic circuits for arithmetic operations, as well as an introduction to properties of various number systems, appear in *Computer Arithmetic* by Kai Hwang (Wiley, 1979). *Decimal Computation* by Hermann Schmid (Wiley, 1974) contains a thorough description of techniques for BCD arithmetic.

An introduction to algorithms for binary multiplication and division and to floating-point arithmetic appears in *Microcomputer Architecture and Programming: The 68000 Family* by John F. Wakerly (Wiley, 1989). A more thorough discussion of arithmetic techniques and floating-point number systems can be found in *Introduction to Arithmetic for Digital Systems Designers* by Shlomo Waser and Michael J. Flynn (Holt, Rinehart and Winston, 1982).

CRC codes are based on the theory of *finite fields,* which was developed by   *finite fields* French mathematician Évariste Galois (1811–1832) shortly before he was killed in a duel with a political opponent. The classic book on error-detecting and error-correcting codes is *Error-Correcting Codes* by W. W. Peterson and E. J. Weldon, Jr. (MIT Press, 1972, second edition); however, this book is recommended only for mathematically sophisticated readers. A more accessible introduction to coding can be found in *Error Control Coding: Fundamentals and Applications* by S. Lin and D. J. Costello, Jr. (Prentice Hall, 1983). Another communication-oriented introduction to coding theory can be found in *Error-*

Control Techniques for Digital Communication by A. M. Michelson and A. H. Levesque (Wiley-Interscience, 1985). Hardware applications of codes in computer systems are discussed in *Error-Detecting Codes, Self-Checking Circuits, and Applications* by John F. Wakerly (Elsevier/North-Holland, 1978).

As shown in the above reference by Wakerly, ones'-complement checksum codes have the ability to detect long bursts of unidirectional errors; this is useful in communication channels where errors all tend to be in the same direction. The special computational properties of these codes also make them quite amenable to efficient checksum calculation by software programs, important for their use in the Internet Protocol; see RFC-1071 and RFC-1141. RFCs (Requests for Comments) are archived in many places on the Web; just search for "RFC".

An introduction to coding techniques for serial data transmission, including mathematical analysis of the performance and bandwidth requirements of several codes, appears in *Introduction to Communications Engineering* by R. M. Gagliardi (Wiley-Interscience, 1988, second edition). A nice introduction to the serial codes used in magnetic disks and tapes is given in *Computer Storage Systems and Technology* by Richard Matick (Wiley-Interscience, 1977).

The structure of the 8B10B code and the rationale behind it is explained nicely in the original IBM patent by Peter Franaszek and Albert Widmer, U.S. patent number 4,486,739 (1984). This and almost all U.S. patents issued after 1971 can be found on the Web at `www.patents.ibm.com`.

Drill Problems

2.1   Perform the following number system conversions:

    (a) $1101011_2 = ?_{16}$ (b) $174003_8 = ?_2$

    (c) $10110111_2 = ?_{16}$ (d) $67.24_8 = ?_2$

    (e) $10100.1101_2 = ?_{16}$ (f) $F3A5_{16} = ?_2$

    (g) $11011001_2 = ?_8$ (h) $AB3D_{16} = ?_2$

    (i) $101111.0111_2 = ?_8$ (j) $15C.38_{16} = ?_2$

2.2   Convert the following octal numbers into binary and hexadecimal:

    (a) $1023_8 = ?_2 = ?_{16}$ (b) $761302_8 = ?_2 = ?_{16}$

    (c) $163417_8 = ?_2 = ?_{16}$ (d) $552273_8 = ?_2 = ?_{16}$

    (e) $5436.15_8 = ?_2 = ?_{16}$ (f) $13705.207_8 = ?_2 = ?_{16}$

2.3   Convert the following hexadecimal numbers into binary and octal:

    (a) $1023_{16} = ?_2 = ?_8$ (b) $7E6A_{16} = ?_2 = ?_8$

    (c) $ABCD_{16} = ?_2 = ?_8$ (d) $C350_{16} = ?_2 = ?_8$

    (e) $9E36.7A_{16} = ?_2 = ?_8$ (f) $DEAD.BEEF_{16} = ?_2 = ?_8$

2.4 What are the octal values of the four 8-bit bytes in the 32-bit number with octal representation 123456701238_8?

2.5 Convert the following numbers into decimal:

(a) $1101011_2 = ?_{10}$ (b) $174003_8 = ?_{10}$

(c) $10110111_2 = ?_{10}$ (d) $67.24_8 = ?_{10}$

(e) $10100.1101_2 = ?_{10}$ (f) $F3A5_{16} = ?_{10}$

(g) $12010_3 = ?_{10}$ (h) $AB3D_{16} = ?_{10}$

(i) $7156_8 = ?_{10}$ (j) $15C.38_{16} = ?_{10}$

2.6 Perform the following number-system conversions:

(a) $125_{10} = ?_2$ (b) $3489_{10} = ?_8$

(c) $209_{10} = ?_2$ (d) $9714_{10} = ?_8$

(e) $132_{10} = ?_2$ (f) $23851_{10} = ?_{16}$

(g) $727_{10} = ?_5$ (h) $57190_{10} = ?_{16}$

(i) $1435_{10} = ?_8$ (j) $65113_{10} = ?_{16}$

2.7 Add the following pairs of binary numbers, showing all carries:

(a)   110101   (b)   101110   (c)   11011101   (d)   1110010

    +   11001   + 100101   +   1100011   + 1101101

2.8 Repeat Drill 2.7 using subtraction instead of addition, and showing borrows instead of carries.

2.9 Add the following pairs of octal numbers:

(a)   1372   (b)   47135   (c)   175214   (d)   110321

    +   4631   +   5125   + 152405   + 56573

2.10 Add the following pairs of hexadecimal numbers:

(a)   1372   (b)   4F1A5   (c)   F35B   (d)   1B90F

    +   4631   +   B8D5   + 27E6   +   C44E

2.11 Write the 8-bit signed-magnitude, two's-complement, and ones'-complement representations for each of these decimal numbers: +18, +115, +79, −49, −3, −100.

2.12 Indicate whether or not overflow occurs when adding the following 8-bit two's-complement numbers:

(a)   11010100   (b)   10111001   (c)   01011101   (d)   00100110

    + 10101011   + 11010110   + 00100001   + 01011010

2.13 How many errors can be detected by a code with minimum distance d?

2.14 What is the minimum number of parity bits required to obtain a distance-4, two-dimensional code with n information bits?

Exercises

2.15   Here's a problem to whet your appetite. What is the hexadecimal equivalent of 61453_{10}?

2.16   Each of the following arithmetic operations is correct in at least one number system. Determine possible radices of the numbers in each operation.

    (a) $1234 + 5432 = 6666$ (b) $41 / 3 = 13$

    (c) $33/3 = 11$ (d) $23+44+14+32 = 223$

    (e) $302/20 = 12.1$ (f) $\sqrt{41} = 5$

2.17   The first expedition to Mars found only the ruins of a civilization. From the artifacts and pictures, the explorers deduced that the creatures who produced this civilization were four-legged beings with a tentacle that branched out at the end with a number of grasping "fingers." After much study, the explorers were able to translate Martian mathematics. They found the following equation:

$$5x^2 - 50x + 125 = 0$$

with the indicated solutions $x = 5$ and $x = 8$. The value $x = 5$ seemed legitimate enough, but $x = 8$ required some explanation. Then the explorers reflected on the way in which Earth's number system developed, and found evidence that the Martian system had a similar history. How many fingers would you say the Martians had? (From *The Bent of Tau Beta Pi*, February 1956.)

2.18   Suppose a $4n$-bit number B is represented by an n-digit hexadecimal number H. Prove that the two's complement of B is represented by the 16's complement of H. Make and prove true a similar statement for octal representation.

2.19   Repeat Exercise 2.18 using the ones' complement of B and the 15s' complement of H.

2.20   Given an integer x in the range $-2n^{-1} \le x \le 2n^{-1} - 1$, we define $[x]$ to be the two's-complement representation of x, expressed as a positive number: $[x] = x$ if $x \ge 0$ and $[x] = 2n - |x|$ if $x < 0$, where $|x|$ is the absolute value of x. Let y be another integer in the same range as x. Prove that the two's-complement addition rules given in Section 2.6 are correct by proving that the following equation is always true:

$$[x + y] = [x] + [y] \text{ modulo } 2^n$$

(*Hints:* Consider four cases based on the signs of x and y. Without loss of generality, you may assume that $|x| \ge |y|$.)

2.21   Repeat Exercise 2.20 using appropriate expressions and rules for ones'-complement addition.

2.22   State an overflow rule for addition of two's-complement numbers in terms of counting operations in the modular representation of Figure 2-3.

2.23   Show that a two's-complement number can be converted to a representation with more bits by *sign extension*. That is, given an n-bit two's-complement number X, show that the m-bit two's-complement representation of X, where $m > n$, can be obtained by appending $m - n$ copies of X's sign bit to the left of the n-bit representation of X.

2.24 Show that a two's-complement number can be converted to a representation with fewer bits by removing higher-order bits. That is, given an n-bit two's-complement number X, show that the m-bit two's-complement number Y obtained by discarding the d leftmost bits of X represents the same number as X if and only if the discarded bits all equal the sign bit of Y.

2.25 Why is the punctuation of "two's complement" and "ones' complement" inconsistent? (See the first two citations in the References.)

2.26 A n-bit binary adder can be used to perform an n-bit unsigned subtraction operation $X - Y$, by performing the operation $X + \overline{Y} + 1$, where X and Y are n-bit unsigned numbers and $\overline{Y}$ represents the bit-by-bit complement of Y. Demonstrate this fact as follows. First, prove that $(X - Y) = (X + \overline{Y} + 1) - 2^n$. Second, prove that the carry out of the n-bit adder is the opposite of the borrow from the n-bit subtraction. That is, show that the operation $X - Y$ produces a borrow out of the MSB position if and only if the operation $X + \overline{Y} + 1$ *does not* produce a carry out of the MSB position.

2.27 In most cases, the product of two n-bit two's-complement numbers requires fewer than $2n$ bits to represent it. In fact, there is only one case in which $2n$ bits are needed—find it.

2.28 Prove that a two's-complement number can be multiplied by 2 by shifting it one bit position to the left, with a carry of 0 into the least significant bit position and disregarding any carry out of the most significant bit position, assuming no overflow. State the rule for detecting overflow.

2.29 State and prove correct a technique similar to the one described in Exercise 2.28, for multiplying a ones'-complement number by 2.

2.30 Show how to subtract BCD numbers, by stating the rules for generating borrows and applying a correction factor. Show how your rules apply to each of the following subtractions: $9 - 3, 5 - 7, 4 - 9, 1 - 8$.

2.31 How many different 3-bit binary state encodings are possible for the traffic-light controller of Table 2-12?

2.32 List all of the "bad" boundaries in the mechanical encoding disk of Figure 2-5, where an incorrect position may be sensed.

2.33 As a function of n, how many "bad" boundaries are there in a mechanical encoding disk that uses an n-bit binary code?

2.34 On-board altitude transponders on commercial and private aircraft use Gray code to encode the altitude readings that are transmitted to air traffic controllers. Why?

2.35 An incandescent light bulb is stressed every time it is turned on, so in some applications the lifetime of the bulb is limited by the number of on/off cycles rather than the total time it is illuminated. Use your knowledge of codes to suggest a way to double the lifetime of 3-way bulbs in such applications.

2.36 As a function of n, how many different distinct subcubes of an n-cube are there?

2.37 Find a way to draw a 3-cube on a sheet of paper (or other two-dimensional object) so that none of the lines cross, or prove that it's impossible.

2.38 Repeat Exercise 2.37 for a 4-cube.

2.39 Write a formula that gives the number of m-subcubes of an n-cube for a specific value of m. (Your answer should be a function of n and m.)

2.40 Define parity groups for a distance-3 Hamming code with 11 information bits.

2.41 Write the code words of a Hamming code with one information bit.

2.42 Exhibit the pattern for a 3-bit error that is not detected if the "corner" parity bits are not included in the two-dimensional codes of Figure 2-14.

2.43 The *rate of a code* is the ratio of the number of information bits to the total number of bits in a code word. High rates, approaching 1, are desirable for efficient transmission of information. Construct a graph comparing the rates of distance-2 parity codes and distance-3 and -4 Hamming codes for up to 100 information bits.

2.44 Which type of distance-4 code has a higher rate—a two-dimensional code or a Hamming code? Support your answer with a table in the style of Table 2-15, including the rate as well as the number of parity and information bits of each code for up to 100 information bits.

2.45 Show how to construct a distance-6 code with four information bits. Write a list of its code words.

2.46 Describe the operations that must be performed in a RAID system to write new data into information block b in drive d, so the data can be recovered in the event of an error in block b in any drive. Minimize the number of disk accesses required.

2.47 In the style of Figure 2-17, draw the waveforms for the bit pattern 10101110 when sent serially using the NRZ, NRZI, RZ, BPRZ, and Manchester codes, assuming that the bits are transmitted in order from left to right.

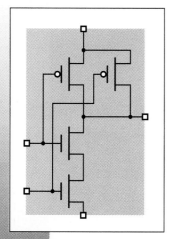

Digital Circuits

Marketing hype notwithstanding, we live in an analog world, not a digital one. Voltages, currents, and other physical quantities in real circuits take on values that are infinitely variable, depending on properties of the real devices that comprise the circuits. Because real values are continuously variable, we could use a physical quantity such as a signal voltage in a circuit to represent a real number (e.g., 3.14159265358979 volts represents the mathematical constant *pi* to 14 decimal digits of precision).

Unfortunately, stability and accuracy in physical quantities are difficult to obtain in real circuits. They are affected by manufacturing tolerances, temperature, power-supply voltage, cosmic rays, and noise created by other circuits, among other things. If we used an analog voltage to represent *pi*, we might find that instead of being an absolute mathematical constant, *pi* varied over a range of 10% or more.

Also, many mathematical and logical operations can be difficult or impossible to perform with analog quantities. While it is possible with some cleverness to build an analog circuit whose output voltage is the square root of its input voltage, no one has ever built a 100-input, 100-output analog circuit whose outputs are a set of voltages identical to the set of input voltages, but sorted arithmetically.

The purpose of this chapter is to give you a solid working knowledge of the electrical aspects of digital circuits, enough for you to understand and build real circuits and systems. We'll see in later chapters that with modern software tools, it's possible to "build" circuits in the abstract, using hardware design languages to specify their design and simulators to test their operation. Still, to build real, production-quality circuits, either at the board level or the chip level, you need to understand most of the material in this chapter. However, if you're anxious to start designing and simulating abstract circuits, you can just read the first section of this chapter and come back to the rest of it later.

3.1 Logic Signals and Gates

digital logic

logic values

Digital logic hides the pitfalls of the analog world by mapping the infinite set of real values for a physical quantity into two subsets corresponding to just two possible numbers or *logic values*—0 and 1. As a result, digital logic circuits can be analyzed and designed functionally, using switching algebra, tables, and other abstract means to describe the operation of well-behaved 0s and 1s in a circuit.

binary digit
bit

A logic value, 0 or 1, is often called a *binary digit*, or *bit*. If an application requires more than two discrete values, additional bits may be used, with a set of n bits representing 2^n different values.

Examples of the physical phenomena used to represent bits in some modern (and not-so-modern) digital technologies are given in Table 3-1. With most phenomena, there is an undefined region between the 0 and 1 states (e.g., voltage = 1.8 V, dim light, capacitor slightly charged, etc.). This undefined region is needed so that the 0 and 1 states can be unambiguously defined and reliably detected. Noise can more easily corrupt results if the boundaries separating the 0 and 1 states are too close.

LOW
HIGH

When discussing electronic logic circuits such as CMOS and TTL, digital designers often use the words "LOW" and "HIGH" in place of "0" and "1" to remind them that they are dealing with real circuits, not abstract quantities:

LOW   A signal in the range of algebraically lower voltages, which is interpreted as a logic 0.

HIGH   A signal in the range of algebraically higher voltages, which is interpreted as a logic 1.

Note that the assignments of 0 and 1 to LOW and HIGH are somewhat arbitrary. Assigning 0 to LOW and 1 to HIGH seems most natural and is called *positive logic*. The opposite assignment, 1 to LOW and 0 to HIGH, is not often used and is called *negative logic*.

positive logic
negative logic

buffer amplifier

Because a wide range of physical values represent the same binary value, digital logic is highly immune to component and power-supply variations and noise. Furthermore, *buffer-amplifier* circuits can be used to regenerate "weak" values into "strong" ones, so that digital signals can be transmitted over arbitrary distances without loss of information. For example, a buffer amplifier for CMOS

Table 3-1 Physical states representing bits in different computer logic and memory technologies.

Technology	State Representing Bit	
	0	1
Pneumatic logic	Fluid at low pressure	Fluid at high pressure
Relay logic	Circuit open	Circuit closed
Complementary metal-oxide semiconductor (CMOS) logic	0–1.5 V	3.5–5.0 V
Transistor-transistor logic (TTL)	0–0.8 V	2.0–5.0 V
Fiber optics	Light off	Light on
Dynamic memory	Capacitor discharged	Capacitor charged
Nonvolatile, erasable memory	Electrons trapped	Electrons released
Bipolar read-only memory	Fuse blown	Fuse intact
Bubble memory	No magnetic bubble	Bubble present
Magnetic tape or disk	Flux direction "north"	Flux direction "south"
Polymer memory	Molecule in state A	Molecule in state B
Read-only compact disc	No pit	Pit
Rewriteable compact disc	Dye in crystalline state	Dye in noncrystalline state

logic converts any HIGH input voltage into an output very close to 5.0 V, and any LOW input voltage into an output very close to 0.0 V.

A logic circuit can be represented with a minimum amount of detail simply as a "black box" with a certain number of inputs and outputs. For example, Figure 3-1 shows a logic circuit with three inputs and one output. However, this representation does not describe how the circuit responds to input signals.

From the point of view of electronic circuit design, it takes a lot of information to describe the precise electrical behavior of a circuit. However, since the inputs of a digital logic circuit can be viewed as taking on only discrete 0 and 1 values, the circuit's "logical" operation can be described with a table that ignores electrical behavior and lists only discrete 0 and 1 values.

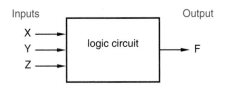

Figure 3-1
"Black-box" representation of a three-input, one-output logic circuit.

combinational circuit

truth table

A logic circuit whose outputs depend only on its current inputs is called a *combinational circuit*. Its operation is fully described by a *truth table* that lists all combinations of input values and the output value(s) produced by each one. Table 3-2 is the truth table for a logic circuit with three inputs X, Y, and Z and a single output F.

Table 3-2
Truth table for a combinational logic circuit.

X	Y	Z	F
0	0	0	0
0	0	1	1
0	1	0	0
0	1	1	0
1	0	0	0
1	0	1	0
1	1	0	1
1	1	1	1

sequential circuit

state table

A circuit with memory, whose outputs depend on the current input *and* the sequence of past inputs, is called a *sequential circuit*. The behavior of such a circuit may be described by a *state table* that specifies its output and next state as functions of its current state and input. Sequential circuits will be introduced in Chapter 7.

As we'll show in Section 4.1, just three basic logic functions, AND, OR, and NOT, can be used to build any combinational digital logic circuit. Figure 3-2 shows the truth tables and symbols for logic "gates" that perform these functions. The symbols and truth tables for AND and OR may be extended to gates with any number of inputs. The gates' functions are easily defined in words:

AND gate
- An *AND gate* produces a 1 output if and only if all of its inputs are 1.

OR gate
- An *OR gate* produces a 1 if and only if one or more of its inputs are 1.

NOT gate

inverter
- A *NOT gate,* usually called an *inverter,* produces an output value that is the opposite of its input value.

Figure 3-2
Basic logic elements:
(a) AND; (b) OR;
(c) NOT (inverter).

(a)

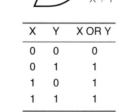

X	Y	X AND Y
0	0	0
0	1	0
1	0	0
1	1	1

(b)

X	Y	X OR Y
0	0	0
0	1	1
1	0	1
1	1	1

(c)

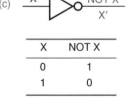

X	NOT X
0	1
1	0

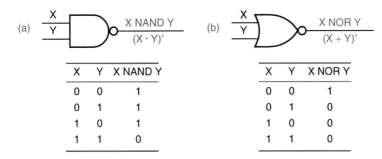

X	Y	X NAND Y
0	0	1
0	1	1
1	0	1
1	1	0

X	Y	X NOR Y
0	0	1
0	1	0
1	0	0
1	1	0

Figure 3-3
Inverting gates:
(a) NAND; (b) NOR.

The circle on the inverter symbol's output is called an *inversion bubble* and is used in this and other gate symbols to denote "inverting" behavior.   *inversion bubble*

Notice that in the definitions of AND and OR functions, we only had to state the input conditions for which the output is 1, because there is only one possibility when the output is not 1—it must be 0.

Two more logic functions are obtained by combining NOT with an AND or OR function in a single gate. Figure 3-3 shows the truth tables and symbols for these gates; Their functions are also easily described in words:

- A *NAND gate* produces the opposite of an AND gate's output, a 0 if and   *NAND gate*
  only if all of its inputs are 1.

- A *NOR gate* produces the opposite of an OR gate's output, a 0 if and only   *NOR gate*
  if one or more of its inputs are 1.

As with AND and OR gates, the symbols and truth tables for NAND and NOR may be extended to gates with any number of inputs.

Figure 3-4 is a logic circuit using AND, OR, and NOT gates that functions according to the truth table of Table 3-2. In Chapter 4 you'll learn how to go from a truth table to a logic circuit, and vice versa, and you'll also learn about the switching-algebra notation used in Figures 3-2 through 3-4.

Real logic circuits also function in another analog dimension—time. For example, Figure 3-5 is a *timing diagram* that shows how the circuit of Figure 3-4   *timing diagram* might respond to a time-varying pattern of input signals. The timing diagram shows that the logic signals do not change between 0 and 1 instantaneously, and also that there is a lag between an input change and the corresponding output

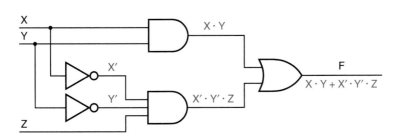

Figure 3-4
Logic circuit with the truth table of Table 3-2.

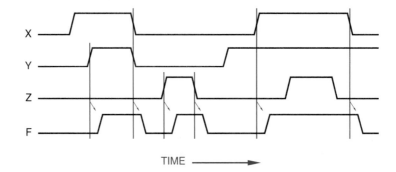

Figure 3-5
Timing diagram for a
logic circuit.

change. Later in this chapter you'll learn some of the reasons for this timing behavior, and how it is specified and handled in real circuits. And once again, you'll learn in a later chapter how this analog timing behavior can be generally ignored in most sequential circuits, and instead the circuit can be viewed as moving between discrete states at precise intervals defined by a clock signal.

Thus, even if you know nothing about analog electronics, you should be able to understand the logical behavior of digital circuits. However, there comes a time in design and debugging when every digital logic designer must throw out "the digital abstraction" temporarily and consider the analog phenomena that limit or disrupt digital performance. The rest of this chapter prepares you for that day by discussing the electrical characteristics of digital logic circuits.

THERE'S HOPE FOR NON-EE'S	If all of this electrical "stuff" bothers you, don't worry, at least for now. The rest of this book is written to be as independent of this stuff as possible. But you'll need it later, if you ever have to design and build digital systems in the real world.

3.2 Logic Families

There are many, many ways to design an electronic logic circuit. The first electrically controlled logic circuits, developed at Bell Laboratories in 1930s, were based on relays. In the mid-1940s, the first electronic digital computer, the Eniac, used logic circuits based on vacuum tubes. The Eniac had about 18,000 tubes and a similar number of logic gates, not a lot by today's standards of microprocessor chips with tens of millions of transistors. However, the Eniac could hurt you a lot more than a chip could if it fell on you—it was 100 feet long, 10 feet high, 3 feet deep, and consumed 140,000 watts of power!

semiconductor diode
bipolar junction
transistor

The inventions of the *semiconductor diode* and the *bipolar junction transistor* allowed the development of smaller, faster, and more capable computers in the late 1950s. In the 1960s, the invention of the *integrated circuit (IC)* allowed multiple diodes, transistors, and other components to be fabricated on a single chip, and computers got still better.

integrated circuit (IC)

The 1960s also saw the introduction of the first integrated-circuit logic families. A *logic family* is a collection of different integrated-circuit chips that have similar input, output, and internal circuit characteristics, but that perform different logic functions. Chips from the same family can be interconnected to perform any desired logic function. On the other hand, chips from differing families may not be compatible; they may use different power-supply voltages or may use different input and output conditions to represent logic values.

logic family

The most successful *bipolar logic family* (one based on bipolar junction transistors) is *transistor-transistor logic (TTL)*. First introduced in the 1960s, TTL now is actually a family of logic families that are compatible with each other but differ in speed, power consumption, and cost. Digital systems can mix components from several different TTL families, according to design goals and constraints in different parts of the system. Although TTL was largely replaced by CMOS in the 1990s, you're still likely to encounter TTL components in academic labs; therefore, we introduce TTL families in Section 3.10.

bipolar logic family
transistor-transistor logic (TTL)

Ten years *before* the bipolar junction transistor was invented, the principles of operation were patented for another type of transistor, called the *metal-oxide semiconductor field-effect transistor (MOSFET)*, or simply *MOS transistor*. However, MOS transistors were difficult to fabricate in the early days, and it wasn't until the 1960s that a wave of developments made MOS-based logic and memory circuits practical. Even then, MOS circuits lagged bipolar circuits considerably in speed and were attractive only in selected applications because of their lower power consumption and higher levels of integration.

metal-oxide semiconductor field-effect transistor (MOSFET)

MOS transistor

Beginning in the mid-1980s, advances in the design of MOS circuits, in particular *complementary MOS (CMOS)* circuits, vastly increased their performance and popularity. By far the majority of new large-scale integrated circuits, such as microprocessors and memories, use CMOS. Likewise, small- to medium-scale applications, for which TTL was once the logic family of choice, are now likely to use CMOS devices with equivalent functionality but higher speed and lower power consumption. CMOS circuits now account for the vast majority of the worldwide IC market.

complementary MOS (CMOS)

CMOS logic is both the most capable and the easiest to understand commercial digital logic technology. Beginning in the next section, we describe the basic structure of CMOS logic circuits and introduce the most commonly used commercial CMOS logic families.

GREEN STUFF   Nowadays, the acronym "MOS" is usually spoken as "moss," rather than spelled out. And "CMOS" has always been spoken as "sea moss."

As a consequence of the industry's transition from TTL to CMOS over a long period of time, many CMOS families were designed to be somewhat compatible with TTL. In Section 3.12 we show how TTL and CMOS families can be mixed within a single system.

3.3 CMOS Logic

The functional behavior of a CMOS logic circuit is fairly easy to understand, even if your knowledge of analog electronics is not particularly deep. The basic (and typically only) building blocks in CMOS logic circuits are MOS transistors, described shortly. Before introducing MOS transistors and CMOS logic circuits, we must talk about logic levels.

3.3.1 CMOS Logic Levels

Abstract logic elements process binary digits, 0 and 1. However, real logic circuits process electrical signals such as voltage levels. In any logic circuit, there is a range of voltages (or other circuit conditions) that is interpreted as a logic 0, and another, nonoverlapping range that is interpreted as a logic 1.

A typical CMOS logic circuit operates from a 5-volt power supply. Such a circuit may interpret any voltage in the range 0–1.5 V as a logic 0, and in the range 3.5–5.0 V as a logic 1. Thus, the definitions of LOW and HIGH for 5-volt CMOS logic are as shown in Figure 3-6. Voltages in the intermediate range (1.5–3.5 V) are not expected to occur except during signal transitions, and yield undefined logic values (i.e., a circuit may interpret them as either 0 or 1). CMOS circuits using other power-supply voltages, such as 3.3 or 2.7 volts, partition the voltage range similarly.

3.3.2 MOS Transistors

A MOS transistor can be modeled as a 3-terminal device that acts like a voltage-controlled resistance. As suggested by Figure 3-7, an input voltage applied to one terminal controls the resistance between the remaining two terminals. In digital logic applications, a MOS transistor is operated so its resistance is always either very high (and the transistor is "off") or very low (and the transistor is "on").

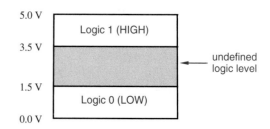

Figure 3-6
Logic levels for typical
CMOS logic circuits.

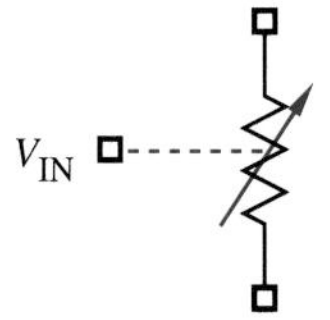

Figure 3-7
The MOS transistor as
a voltage-controlled
resistance.

There are two types of MOS transistors, *n*-channel and *p*-channel; the names refer to the type of semiconductor material used for the resistance-controlled terminals. The circuit symbol for an *n-channel MOS (NMOS) transistor* is shown in Figure 3-8. The terminals are called *gate, source,* and *drain.* (Note that the "gate" of a MOS transistor has nothing to do with a "logic gate.") As you might guess from the orientation of the circuit symbol, the drain is normally at a higher voltage than the source.

n-channel MOS
 (NMOS) transistor
gate
source
drain

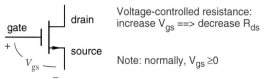

Figure 3-8
Circuit symbol for an
n-channel MOS (NMOS)
transistor.

The voltage from gate to source (V_{gs}) in an NMOS transistor is normally zero or positive. If $V_{gs} = 0$, then the resistance from drain to source (R_{ds}) is very high, at least a megohm (10^6 ohms) or more. As we increase V_{gs} (i.e., increase the voltage on the gate), R_{ds} decreases to a very low value, 10 ohms or less in some devices.

The circuit symbol for a *p-channel MOS (PMOS) transistor* is shown in Figure 3-9. Operation is analogous to that of an NMOS transistor, except that the source is normally at a higher voltage than the drain, and V_{gs} is normally zero or negative. If V_{gs} is zero, then the resistance from source to drain (R_{ds}) is very high. As we algebraically decrease V_{gs} (i.e., *decrease* the voltage on the gate), R_{ds} decreases to a very low value.

p-channel MOS
 (PMOS) transistor

The gate of a MOS transistor has a very high impedance. That is, the gate is separated from the source and the drain by an insulating material with a very high resistance. However, the gate voltage creates an electric field that enhances or retards the flow of current between source and drain. This is the "field effect" in the "MOSFET" name.

Regardless of gate voltage, almost no current flows from the gate to source, or from the gate to drain for that matter. The resistance between the gate and the

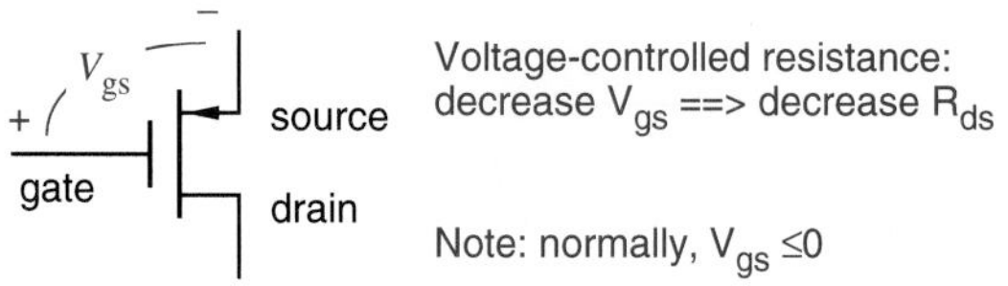

Figure 3-9
Circuit symbol for a
p-channel MOS (PMOS)
transistor.

| IMPEDANCE VS. RESISTANCE | Technically, there's a difference between the words "impedance" and "resistance," but electrical engineers often use the terms interchangeably. So do we in this text. |

leakage current

other terminals of the device is extremely high, well over a megohm. The small amount of current that flows across this resistance is very small, typically less than one microampere (μA, 10^{-6} A), and is called a *leakage current.*

The MOS transistor symbol itself reminds us that there is no connection between the gate and the other two terminals of the device. However, the gate of a MOS transistor is capacitively coupled to the source and drain, as the symbol might suggest. In high-speed circuits, the power needed to charge and discharge this capacitance on each input-signal transition accounts for a nontrivial portion of a circuit's power consumption.

3.3.3 Basic CMOS Inverter Circuit

CMOS logic

NMOS and PMOS transistors are used together in a complementary way to form *CMOS logic.* The simplest CMOS circuit, a logic inverter, requires only one of each type of transistor, connected as shown in Figure 3-10(a). The power-supply voltage, V_{DD}, typically may be in the range 2–6 V and is most often set at 5.0 V for compatibility with TTL circuits.

Ideally, the functional behavior of the CMOS inverter circuit can be characterized by just two cases tabulated in Figure 3-10(b):

1. V_{IN} is 0.0 V. In this case, the bottom, *n*-channel transistor $Q1$ is off, since its V_{gs} is 0, but the top, *p*-channel transistor $Q2$ is on, since its V_{gs} is a large negative value (−5.0 V). Therefore, $Q2$ presents only a small resistance between the power-supply terminal (V_{DD}, +5.0 V) and the output terminal (V_{OUT}), and the output voltage is 5.0 V.

Figure 3-10
CMOS inverter:
(a) circuit diagram;
(b) functional behavior;
(c) logic symbol.

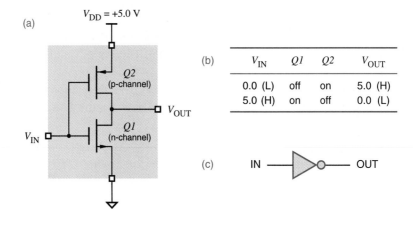

(b)

V_{IN}	$Q1$	$Q2$	V_{OUT}
0.0 (L)	off	on	5.0 (H)
5.0 (H)	on	off	0.0 (L)

2. V_{IN} is 5.0 V. Here, $Q1$ is on, since its V_{gs} is a large positive value (+5.0 V), but $Q2$ is off, since its V_{gs} is 0. Thus, $Q1$ presents a small resistance between the output terminal and ground, and the output voltage is 0 V.

With the foregoing functional behavior, the circuit clearly behaves as a logical inverter, since a 0-volt input produces a 5-volt output, and vice versa.

Another way to visualize CMOS operation uses switches. As shown in Figure 3-11(a), the *n*-channel (bottom) transistor is modeled by a normally-open switch, and the *p*-channel (top) transistor by a normally-closed switch. Applying a HIGH voltage changes each switch to the opposite of its normal state, as shown in (b).

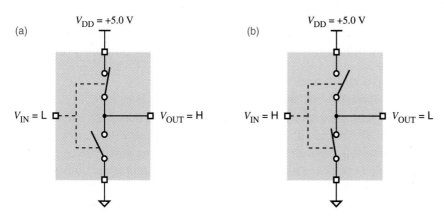

Figure 3-11
Switch model for CMOS inverter: (a) LOW input; (b) HIGH input.

The switch model gives rise to a way of drawing CMOS circuits that makes their logical behavior more readily apparent. As shown in Figure 3-12, different symbols are used for the *p*- and *n*-channel transistors to reflect their logical behavior. The *n*-channel transistor ($Q1$) is switched "on," and current flows between source and drain, when a HIGH voltage is applied to its gate; this seems natural enough. The *p*-channel transistor ($Q2$) has the opposite behavior. It is

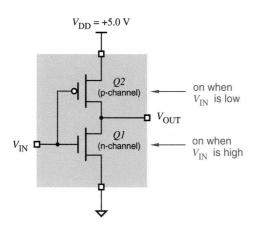

Figure 3-12
CMOS inverter logical operation.

"on" when a LOW voltage is applied; the inversion bubble on its gate indicates this inverting behavior.

3.3.4 CMOS NAND and NOR Gates

Both NAND and NOR gates can be constructed using CMOS. A k-input gate uses k p-channel and k n-channel transistors. Figure 3-13 shows a 2-input CMOS NAND gate. If either input is LOW, the output Z has a low-impedance connection to V_{DD} through the corresponding "on" p-channel transistor, and the path to ground is blocked by the corresponding "off" n-channel transistor. If both inputs are HIGH, the path to V_{DD} is blocked, and Z has a low-impedance connection to ground. Figure 3-14 shows the switch model for the NAND gate's operation.

Figure 3-15 shows a CMOS NOR gate. If both inputs are LOW, then the output Z has a low-impedance connection to V_{DD} through the "on" p-channel transistors, and the path to ground is blocked by the "off" n-channel transistors. If either input is HIGH, the path to V_{DD} is blocked, and Z has a low-impedance connection to ground.

Figure 3-13
CMOS 2-input
NAND gate:
(a) circuit diagram;
(b) function table;
(c) logic symbol.

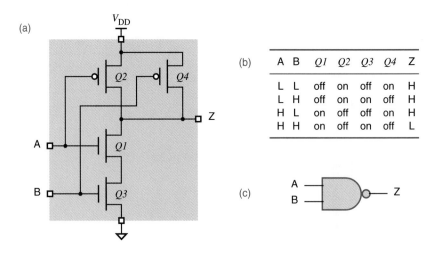

(b)

A	B	Q1	Q2	Q3	Q4	Z
L	L	off	on	off	on	H
L	H	off	on	on	off	H
H	L	on	off	off	on	H
H	H	on	off	on	off	L

NAND VS. NOR   CMOS NAND and NOR gates do not have identical performance. For a given silicon area, an *n*-channel transistor has lower "on" resistance than a *p*-channel transistor. Therefore, when transistors are put in series, *k* *n*-channel transistors have lower "on" resistance than do *k* *p*-channel ones. As a result, a *k*-input NAND gate is generally faster than and preferred over a *k*-input NOR gate.

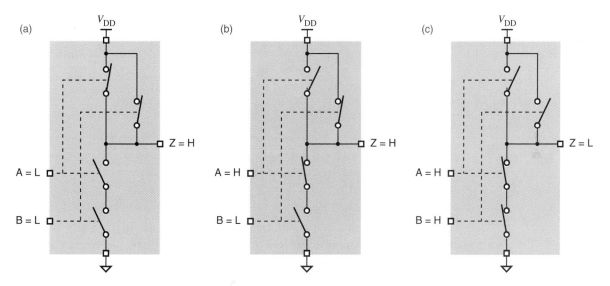

Figure 3-14 Switch model for CMOS 2-input NAND gate: (a) both inputs LOW; (b) one input HIGH; (c) both inputs HIGH.

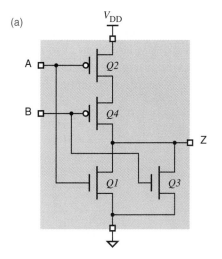

A	B	*Q1*	*Q2*	*Q3*	*Q4*	Z
L	L	off	on	off	on	H
L	H	off	on	on	off	L
H	L	on	off	off	on	L
H	H	on	off	on	off	L

Figure 3-15
CMOS 2-input
NOR gate:
(a) circuit diagram;
(b) function table;
(c) logic symbol.

(a)

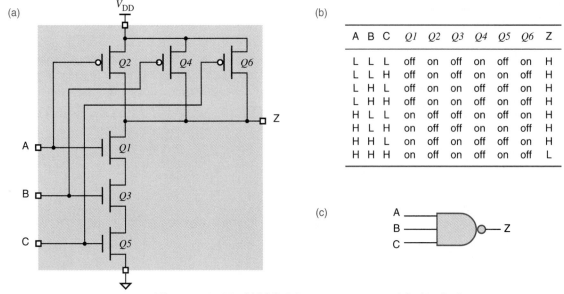

(b)

A	B	C	$Q1$	$Q2$	$Q3$	$Q4$	$Q5$	$Q6$	Z
L	L	L	off	on	off	on	off	on	H
L	L	H	off	on	off	on	on	off	H
L	H	L	off	on	on	off	off	on	H
L	H	H	off	on	on	off	on	off	H
H	L	L	on	off	off	on	off	on	H
H	L	H	on	off	off	on	on	off	H
H	H	L	on	off	on	off	off	on	H
H	H	H	on	off	on	off	on	off	L

(c)

Figure 3-16 CMOS 3-input NAND gate: (a) circuit diagram; (b) function table; (c) logic symbol.

3.3.5 Fan-In

fan-in

The number of inputs that a gate can have in a particular logic family is called the logic family's *fan-in*. CMOS gates with more than two inputs can be obtained by extending series-parallel designs on Figures 3-13 and 3-15 in the obvious manner. For example, Figure 3-16 shows a 3-input CMOS NAND gate.

In principle, you could design a CMOS NAND or NOR gate with a very large number of inputs. In practice, however, the additive "on" resistance of series transistors limits the fan-in of CMOS gates, typically to 4 for NOR gates and 6 for NAND gates.

As the number of inputs is increased, designers of CMOS gate circuits may compensate by increasing the size of the series transistors to reduce their resistance and the corresponding switching delay. However, at some point this becomes inefficient or impractical. Gates with a large number of inputs can be made faster and smaller by cascading gates with fewer inputs. For example,

Figure 3-17
Logic diagram equivalent to the internal structure of an 8-input CMOS NAND gate.

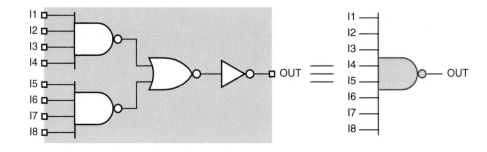

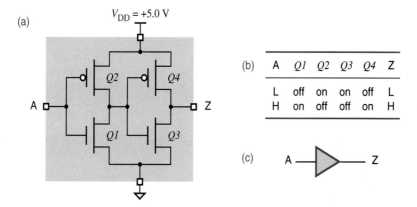

Figure 3-18
CMOS noninverting
buffer:
(a) circuit diagram;
(b) function table;
(c) logic symbol.

Figure 3-17 shows the logical structure of an 8-input CMOS NAND gate. The total delay through a 4-input NAND, a 2-input NOR, and an inverter is typically less than the delay of a one-level 8-input NAND circuit.

3.3.6 Noninverting Gates

In CMOS, and in most other logic families, the simplest gates are inverters, and the next simplest are NAND gates and NOR gates. A logical inversion comes "for free," and it typically is not possible to design a noninverting gate with a smaller number of transistors than an inverting one.

CMOS noninverting buffers and AND and OR gates are obtained by connecting an inverter to the output of the corresponding inverting gate. Thus, Figure 3-18 shows a noninverting buffer and Figure 3-19 shows an AND gate. Combining Figure 3-15(a) with an inverter yields an OR gate.

Figure 3-19 CMOS 2-input AND gate: (a) circuit diagram; (b) function table; (c) logic symbol.

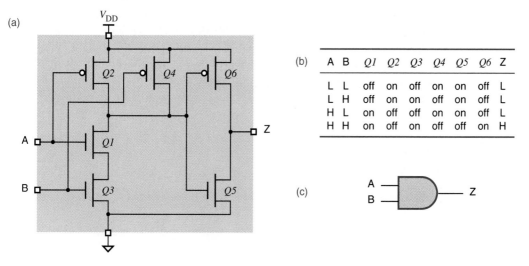

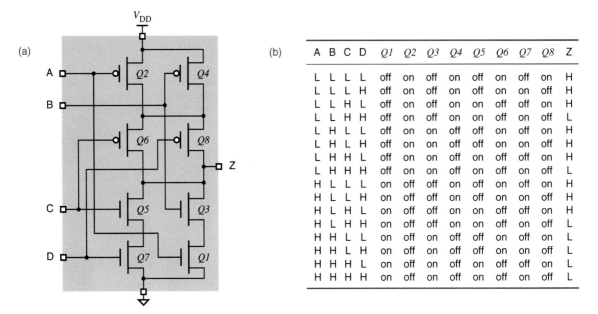

A	B	C	D	Q1	Q2	Q3	Q4	Q5	Q6	Q7	Q8	Z
L	L	L	L	off	on	off	on	off	on	off	on	H
L	L	L	H	off	on	off	on	off	on	on	off	H
L	L	H	L	off	on	off	on	on	off	off	on	H
L	L	H	H	off	on	off	on	on	off	on	off	L
L	H	L	L	off	on	on	off	off	on	off	on	H
L	H	L	H	off	on	on	off	off	on	on	off	H
L	H	H	L	off	on	on	off	on	off	off	on	H
L	H	H	H	off	on	on	off	on	off	on	off	L
H	L	L	L	on	off	off	on	off	on	off	on	H
H	L	L	H	on	off	off	on	off	on	on	off	H
H	L	H	L	on	off	off	on	on	off	off	on	H
H	L	H	H	on	off	off	on	on	off	on	off	L
H	H	L	L	on	off	on	off	off	on	off	on	L
H	H	L	H	on	off	on	off	off	on	on	off	L
H	H	H	L	on	off	on	off	on	off	off	on	L
H	H	H	H	on	off	on	off	on	off	on	off	L

Figure 3-20 CMOS AND-OR-INVERT gate: (a) circuit diagram; (b) function table.

3.3.7 CMOS AND-OR-INVERT and OR-AND-INVERT Gates

AND-OR-INVERT (AOI) gate

CMOS circuits can perform two levels of logic with just a single "level" of transistors. For example, the circuit in Figure 3-20(a) is a two-wide, two-input CMOS *AND-OR-INVERT (AOI) gate*. The function table for this circuit is shown in (b) and a logic diagram for this function using AND and NOR gates is shown in Figure 3-21. Transistors can be added to or removed from this circuit to obtain an AOI function with a different number of ANDs or a different number of inputs per AND.

The contents of each of the $Q1$–$Q8$ columns in Figure 3-20(b) depends only on the input signal connected to the corresponding transistor's gate. The last column is constructed by examining each input combination and determining whether Z is connected to V_{DD} or ground by "on" transistors for that input combination. Note that Z is never connected to *both* V_{DD} and ground for any input combination; in such a case the output would be a nonlogic value

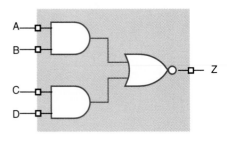

Figure 3-21
Logic diagram for CMOS
AND-OR-INVERT gate.

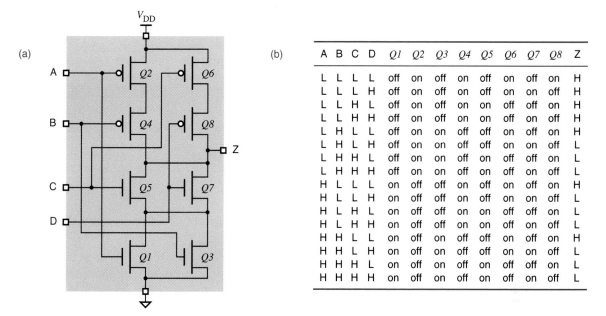

(a)

(b)

A	B	C	D	Q1	Q2	Q3	Q4	Q5	Q6	Q7	Q8	Z
L	L	L	L	off	on	off	on	off	on	off	on	H
L	L	L	H	off	on	off	on	off	on	on	off	H
L	L	H	L	off	on	off	on	on	off	off	on	H
L	L	H	H	off	on	off	on	on	off	on	off	H
L	H	L	L	off	on	on	off	off	on	off	on	H
L	H	L	H	off	on	on	off	off	on	on	off	L
L	H	H	L	off	on	on	off	on	off	off	on	L
L	H	H	H	off	on	on	off	on	off	on	off	L
H	L	L	L	on	off	off	on	off	on	off	on	H
H	L	L	H	on	off	off	on	off	on	on	off	L
H	L	H	L	on	off	off	on	on	off	off	on	L
H	L	H	H	on	off	off	on	on	off	on	off	L
H	H	L	L	on	off	on	off	off	on	off	on	H
H	H	L	H	on	off	on	off	off	on	on	off	L
H	H	H	L	on	off	on	off	on	off	off	on	L
H	H	H	H	on	off	on	off	on	off	on	off	L

Figure 3-22 CMOS OR-AND-INVERT gate: (a) circuit diagram; (b) function table.

somewhere between LOW and HIGH, and the output structure would consume excessive power due to the low-impedance connection between V_{DD} and ground.

A circuit can also be designed to perform an OR-AND-INVERT function. For example, Figure 3-22(a) is a two-wide, two-input CMOS *OR-AND-INVERT (OAI) gate*. The function table for this circuit is shown in (b); the values in each column are determined just as we did for the CMOS AOI gate. A logic diagram for the OAI function using OR and NAND gates is shown in Figure 3-23.

OR-AND-INVERT (AOI) gate

The speed and other electrical characteristics of a CMOS AOI or OAI gate are quite comparable to those of a single CMOS NAND or NOR gate. As a result, these gates are very appealing because they can perform two levels of logic (AND-OR or OR-AND) with just one level of delay. Most digital designers don't bother to use AOI gates in their discrete designs. However, CMOS VLSI devices often use these gates internally, since many HDL synthesis tools can automatically convert AND/OR logic into AOI gates when appropriate.

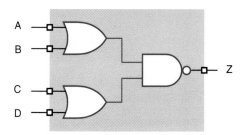

Figure 3-23
Logic diagram for CMOS OR-AND-INVERT gate.

3.4 Electrical Behavior of CMOS Circuits

The next three sections discuss electrical, not logical, aspects of CMOS circuit operation. It's important to understand this material when you design real circuits using CMOS or other logic families. Most of this material is aimed at providing a framework for ensuring that the "digital abstraction" is really valid for a given circuit. In particular, a circuit or system designer must provide in a number of areas adequate engineering design margins—insurance that the circuit will work properly even under the worst of conditions.

engineering design margins

3.4.1 Overview

The topics that we discuss in Sections 3.5–3.7 include the following:

- *Logic voltage levels.* CMOS devices operating under normal conditions are guaranteed to produce output voltage levels within well-defined LOW and HIGH ranges. And they recognize LOW and HIGH input voltage levels over somewhat wider ranges. CMOS manufacturers specify these ranges and operating conditions very carefully to ensure compatibility among different devices in the same family and to provide a degree of interoperability (if you're careful) among devices in different families.

- *DC noise margins.* Nonnegative DC noise margins ensure that the highest LOW voltage produced by an output is always lower than the highest voltage that an input can reliably interpret as LOW, and that the lowest HIGH voltage produced by an output is always higher than the lowest voltage that an input can reliably interpret as HIGH. A good understanding of noise margins is especially important in circuits that use devices from a number of different families.

- *Fanout.* This refers to the number and type of inputs that are connected to a given output. If too many inputs are connected to an output, the DC noise margins of the circuit may be inadequate. Fanout may also affect the speed at which the output changes from one state to another.

- *Speed.* The time that it takes a CMOS output to change from the LOW state to the HIGH state, or vice versa, depends on both the internal structure of the device and the characteristics of the other devices that it drives, even to the extent of being affected by the wire or printed-circuit-board traces connected to the output. We'll look at two separate components of "speed"—transition time and propagation delay.

- *Power consumption.* The power consumed by a CMOS device depends on a number of factors, including not only its internal structure, but also the input signals that it receives, the other devices that it drives, and how often its output changes between LOW and HIGH.

- *Noise.* The main reason for providing engineering design margins is to ensure proper circuit operation in the presence of noise. Noise can be generated by a number of sources; several of them are listed below, from the least likely to the (perhaps surprisingly) most likely:
  - Cosmic rays.
  - Magnetic fields from nearby machinery.
  - Power-supply disturbances.
  - The switching action of the logic circuits themselves.

- *Electrostatic discharge.* Would you believe that you can destroy a CMOS device just by touching it?

- *Open-drain outputs.* Some CMOS outputs omit the usual *p*-channel pull-up transistors. In the HIGH state, such an output behaves essentially like a "no-connection," which is useful in some applications.

- *Three-state outputs.* Some CMOS devices have an extra "output enable" control input that can be used to disable both the *p*-channel pull-up transistors and the *n*-channel pull-down transistors. Many such device outputs can be tied together to create a multisource bus, as long as the control logic is arranged so that at most one output is enabled at a time.

3.4.2 Data Sheets and Specifications

The manufacturers of real-world devices provide *data sheets* that specify the   *data sheet*
devices' logical and electrical characteristics. The electrical specifications portion of a minimal data sheet for a simple CMOS device, the 54/74HC00 quadruple NAND gate, is shown in Table 3-3. Different manufacturers typically specify additional parameters, and they may vary in how they specify even the "standard" parameters shown in the table. Thus, they usually also show the test circuits and waveforms that they use to define various parameters, for example as shown in Figure 3-24. Note that this figure contains information for some parameters in addition to those used with the 54/74HC00.

Most of the terms in the data sheet and the waveforms in the figure are probably meaningless to you at this point. However, after reading the next three sections you should know enough about the electrical characteristics of CMOS circuits that you'll be able to understand the salient points of this or any other data sheet. As a logic designer, you'll need this knowledge to create reliable and robust real-world circuits and systems.

DON'T BE AFRAID   Computer science students and other non-EE readers should not have undue fear of the material in the next three sections. Only a basic understanding of electronics, at about the level of Ohm's law, is required.

Table 3-3 Manufacturer's data sheet for a typical CMOS device, a 54/74HC00 quad NAND gate.

DC ELECTRICAL CHARACTERISTICS OVER OPERATING RANGE

The following conditions apply unless otherwise specified:

Commercial: $T_A = -40°C$ to $+85°C$, $V_{CC} = 5.0V \pm 5\%$;   Military: $T_A = -55°C$ to $+125°C$, $V_{CC} = 5.0$ V $\pm 10\%$

Sym.	Parameter	Test Conditions[1]		Min.	Typ.[2]	Max.	Unit
V_{IH}	Input HIGH level	Guaranteed logic HIGH level		3.15	—	—	V
V_{IL}	Input LOW level	Guaranteed logic LOW level		—	—	1.35	V
I_{IH}	Input HIGH current	$V_{CC} = $ Max., $V_I = V_{CC}$		—	—	1	μA
I_{IL}	Input LOW current	$V_{CC} = $ Max., $V_I = 0$ V		—	—	−1	μA
V_{IK}	Clamp diode voltage	$V_{CC} = $ Min., $I_N = -18$ mA		—	−0.7	−1.2	V
I_{IOS}	Short-circuit current	$V_{CC} = $ Max.,[3] $V_O = $ GND		—	—	−35	mA
V_{OH}	Output HIGH voltage	$V_{CC} = $ Min., $V_{IN} = V_{IL}$	$I_{OH} = -20\ \mu$A	4.4	4.499	—	V
			$I_{OH} = -4$ mA	3.84	4.3	—	V
V_{OL}	Output LOW voltage	$V_{CC} = $ Min. $V_{IN} = V_{IH}$	$I_{OL} = 20\ \mu$A	—	.001	0.1	V
			$I_{OL} = 4$ mA		0.17	0.33	
I_{CC}	Quiescent power supply current	$V_{CC} = $ Max. $V_{IN} = $ GND or V_{CC}, $I_O = 0$		—	2	10	μA

SWITCHING CHARACTERISTICS OVER OPERATING RANGE, $C_L = 50$ pF

Sym.	Parameter[4]	Test Conditions	Min.	Typ.	Max.	Unit
t_{PD}	Propagation delay	A or B to Y	—	9	19	ns
C_I	Input capacitance	$V_{IN} = 0$ V	—	3	10	pF
C_{pd}	Power dissipation capacitance per gate	No load	—	22	—	pF

NOTES:

1. For conditions shown as Max. or Min., use appropriate value specified under Electrical Characteristics.

2. Typical values are at $V_{CC} = 5.0$ V, $+25°C$ ambient.

3. Not more than one output should be shorted at a time. Duration of short-circuit test should not exceed one second.

4. This parameter is guaranteed but not tested.

WHAT'S IN A NUMBER?   Two different prefixes, "74" and "54," are used in the part numbers of CMOS and TTL devices. These prefixes simply distinguish between commercial and military versions. A 74HC00 is the commercial part and the 54HC00 is the military version.

TEST CIRCUIT FOR ALL OUTPUTS

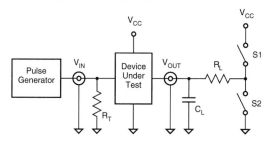

LOADING

Parameter		R_L	C_L	S1	S2
t_{en}	t_{pZH}	1 KΩ	50 pF or 150 pF	Open	Closed
	t_{pZL}			Closed	Open
t_{dis}	t_{pHZ}	1 KΩ		Open	Closed
	t_{pLZ}			Closed	Open
t_{pd}		—	50 pF or 150 pF	Open	Open

DEFINITIONS:
C_L = Load capacitance, includes jig and probe capacitance.
R_T = Termination resistance, should equal Z_{OUT} of the Pulse Generator.

SETUP, HOLD, AND RELEASE TIMES

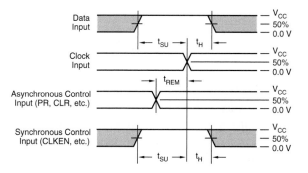

PULSE WIDTH

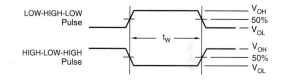

PROPAGATION DELAY

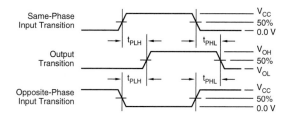

THREE-STATE ENABLE AND DISABLE TIMES

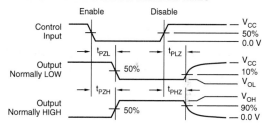

Figure 3-24 Test circuits and waveforms for HC-series logic.

3.5 CMOS Steady-State Electrical Behavior

This section discusses the steady-state behavior of CMOS circuits, that is, the circuits' behavior when inputs and outputs are not changing. The next section discusses dynamic behavior, including speed and power dissipation.

3.5.1 Logic Levels and Noise Margins

The table in Figure 3-10(b) on page 88 defined the CMOS inverter's behavior only at two discrete input voltages; other input voltages may yield different output voltages. The complete input-output transfer characteristic can be

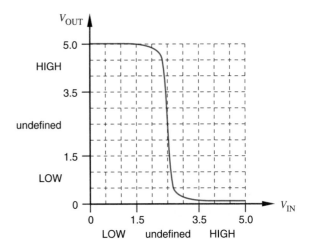

Figure 3-25
Typical input-output
transfer characteristic
of a CMOS inverter.

described by a graph such as Figure 3-25. In this graph, the input voltage is varied from 0 to 5 V, as shown on the X axis; the Y axis plots the output voltage.

If we believed the curve in Figure 3-25, we could define a CMOS LOW input level as any voltage under 2.4 V, and a HIGH input level as anything over 2.6 V. Only when the input is between 2.4 and 2.6 V does the inverter produce a nonlogic output voltage under this definition.

Unfortunately, the typical transfer characteristic shown in Figure 3-25 is just that—typical, but not guaranteed. It varies greatly under different conditions of power-supply voltage, temperature, and output loading. The transfer characteristic may even vary depending on when the device was fabricated. For example, after months of trying to figure out why gates made on some days were good and on other days were bad, legend has it that one manufacturer discovered that the bad gates were victims of airborne contamination by a particularly noxious perfume worn by one of its production-line workers!

Sound engineering practice dictates that we use more conservative specifications for LOW and HIGH. The conservative specs for a typical CMOS logic family (HC-series) are depicted in Figure 3-26. These parameters are specified by CMOS device manufacturers in data sheets like Table 3-3, and are defined as follows:

V_{OHmin} The minimum output voltage in the HIGH state.

V_{IHmin} The minimum input voltage guaranteed to be recognized as a HIGH.

V_{ILmax} The maximum input voltage guaranteed to be recognized as a LOW.

V_{OLmax} The maximum output voltage in the LOW state.

The input voltages are determined mainly by switching thresholds of the two transistors, while the output voltages are determined mainly by the "on" resistance of the transistors.

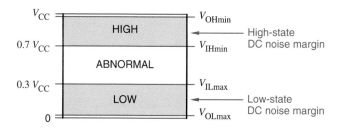

Figure 3-26
Logic levels and
noise margins
for the HC-series
CMOS logic family.

All of the parameters in Figure 3-26 are guaranteed by CMOS manu-facturers over a range of temperature and output loading. Parameters are also guaranteed over a range of power-supply voltage V_{CC}, typically 5.0 V±10%.

The data sheet in Table 3-3 on page 98 specifies values for each of these parameters for HC-series CMOS. Notice that there are two values specified for V_{OHmin} and V_{OLmax}, depending on whether the output current (I_{OH} or I_{OL}) is large or small. When the device outputs are connected only to other CMOS inputs, the output current is low (e.g., $I_{OL} \leq 20\,\mu A$), so there's very little voltage drop across the output transistors. In the next few subsections we'll focus on these "pure" CMOS applications.

The power-supply voltage V_{CC} and ground are often called the *power-* *supply rails*. CMOS levels are typically a function of the power-supply rails:

power-supply rails

V_{OHmin}	$V_{CC} - 0.1$ V
V_{IHmin}	70% of V_{CC}
V_{ILmax}	30% of V_{CC}
V_{OLmax}	ground + 0.1 V

Notice in Table 3-3 that V_{OHmin} is specified as 4.4 V. This is only a 0.1-V drop from V_{CC}, since the worst-case number is specified with V_{CC} at its minimum value of 5.0 − 10% = 4.5 V.

DC noise margin is a measure of how much noise it takes to corrupt a worst-case output voltage into a value that may not be recognized properly by an input. For HC-series CMOS in the LOW state, V_{ILmax} (1.35 V) exceeds V_{OLmax} (0.1 V) by 1.25 V so the LOW-state DC noise margin is 1.25 V. Likewise, there is DC noise margin of 1.25 V in the HIGH state. In general, CMOS outputs have excellent DC noise margins when driving other CMOS inputs.

DC noise margin

Regardless of the voltage applied to the input of a CMOS inverter, the input consumes very little current, only the leakage current of the two transistors' gates. The maximum amount of current that can flow is also specified by the device manufacturer:

I_{IH} The maximum current that flows into the input in the HIGH state.

I_{IL} The maximum current that flows into the input in the LOW state.

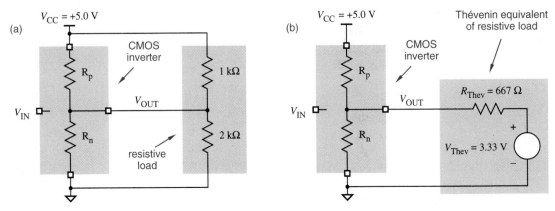

Figure 3-27 Resistive model of a CMOS inverter with a resistive load: (a) showing actual load circuit; (b) using Thévenin equivalent of load.

The input current shown in Table 3-3 for the 'HC00 is only $\pm 1\ \mu A$. Thus, it takes very little power to maintain a CMOS input in one state or the other. This is in sharp contrast to bipolar logic circuits like TTL and ECL, whose inputs may consume significant current (and power) in one or both states.

3.5.2 Circuit Behavior with Resistive Loads

resistive load
DC load

As mentioned previously, CMOS gate inputs have very high impedance and consume very little current from the circuits that drive them. There are other devices, however, which require nontrivial amounts of current to operate. When such a device is connected to a CMOS output, we call it a *resistive load* or a *DC load*. Here are some examples of resistive loads:

- Discrete resistors may be included to provide transmission-line termination, discussed in Section 11.4.
- Discrete resistors may not really be present in the circuit, but the load presented by one or more TTL or other non-CMOS inputs may be modeled by a simple resistor network.
- The resistors may be part of or may model a current-consuming device such as a light-emitting diode (LED) or a relay coil.

When the output of a CMOS circuit is connected to a resistive load, the output behavior is not nearly as ideal as we described previously. In either logic state, the CMOS output transistor that is "on" has a nonzero resistance, and a load connected to the output terminal will cause a voltage drop across this resistance. Thus, in the LOW state, the output voltage may be somewhat higher than 0.1 V, and in the HIGH state it may be lower than 4.4 V. The easiest way to see how this happens is look at a resistive model of the CMOS circuit and load.

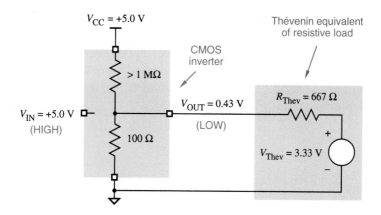

Figure 3-28
Resistive model for
CMOS LOW output
with resistive load.

Figure 3-27(a) shows the resistive model. The *p*-channel and *n*-channel transistors have resistances R_p and R_n, respectively. In normal operation, one resistance is high (> 1 MΩ) and the other is low (perhaps 100 Ω), depending on whether the input voltage is HIGH or LOW. The load in this circuit consists of two resistors attached to the supply rails; a real circuit may have any resistor values, or an even more complex resistive network. In any case, a resistive load, consisting only of resistors and voltage sources, can always be modeled by a Thévenin equivalent network, such as the one shown in Figure 3-27(b).

When the CMOS inverter has a HIGH input, the output should be LOW; the actual output voltage can be predicted using the resistive model shown in Figure 3-28. The *p*-channel transistor is "off" and has a very high resistance, high enough to be negligible in the calculations that follow. The *n*-channel

REMEMBERING THÉVENIN   Any two-terminal circuit consisting of only voltage sources and resistors can be modeled by a *Thévenin equivalent* consisting of a single voltage source in series with a single resistor. The *Thévenin voltage* is the open-circuit voltage of the original circuit, and the *Thévenin resistance* is the Thévenin voltage divided by the short-circuit current of the original circuit.

In the example of Figure 3-27, the Thévenin voltage of the resistive load, including its connection to V_{CC}, is established by the 1-kΩ and 2-kΩ resistors, which form a voltage divider:

$$V_{Thev} = \frac{2\,k\Omega}{2\,k\Omega + 1\,k\Omega} \cdot 5.0\,V = 3.33\,V$$

The short-circuit current is (5.0 V)/(1 kΩ) = 5 mA, so the Thévenin resistance is (3.33 V)/(5 mA) = 667 Ω. Experienced readers may recognize this as the parallel resistance of the 1-kΩ and 2-kΩ resistors.

transistor is "on" and has a low resistance, which we assume to be 100 Ω. (The actual "on" resistance depends on the CMOS family and other characteristics such as operating temperature and whether or not the device was manufactured on a good day.) The "on" transistor and the Thévenin-equivalent resistor R_{Thev} in Figure 3-28 form a simple voltage divider. The resulting output voltage can be calculated as follows:

$$V_{OUT} = 3.33 \text{ V} \cdot [100/(100 + 667)]$$

$$= 0.43 \text{ V}$$

Similarly, when the inverter has a LOW input, the output should be HIGH, and the actual output voltage can be predicted with the model in Figure 3-29. We'll assume that the *p*-channel transistor's "on" resistance is 200 Ω. Once again, the "on" transistor and the Thévenin-equivalent resistor R_{Thev} in the figure form a simple voltage divider, and the resulting output voltage can be calculated as follows:

$$V_{OUT} = 3.33 \text{ V} + (5 \text{ V} - 3.33 \text{ V}) \cdot [667/(200 + 667)]$$

$$= 4.61 \text{ V}$$

In practice, it's seldom necessary to calculate output voltages as in the preceding examples. In fact, IC manufacturers usually don't specify the equivalent resistances of the "on" transistors, so you wouldn't have the necessary information to make the calculation anyway. Instead, IC manufacturers specify a maximum load for the output in each state (HIGH or LOW), and guarantee a worst-case output voltage for that load. The load is specified in terms of current:

I_{OLmax} The maximum current that the output can sink in the LOW state while still maintaining an output voltage no greater than V_{OLmax}.

I_{OHmax} The maximum current that the output can source in the HIGH state while still maintaining an output voltage no less than V_{OHmin}.

Figure 3-29
Resistive model for
CMOS HIGH output
with resistive load.

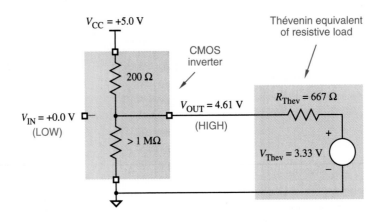

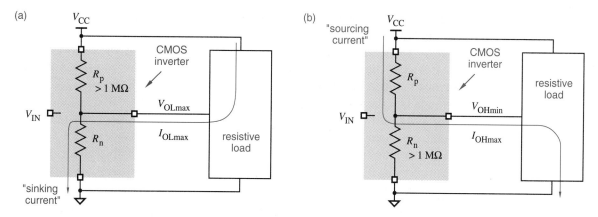

Figure 3-30 Circuit definitions of (a) I_{OLmax}; (b) I_{OHmax}.

These definitions are illustrated in Figure 3-30. A device output is said to *sink current* when current flows from the power supply, through the load, and through the device output to ground as in (a). The output is said to *source current* when current flows from the power supply, out of the device output, and through the load to ground as in (b).

sinking current

sourcing current

Most CMOS devices have two sets of loading specifications. One set is for "CMOS loads," where the device output is connected to other CMOS inputs, which consume very little current. The other set is for "TTL loads," where the output is connected to resistive loads such as TTL inputs or other devices that consume significant current. For example, the specifications for HC-series CMOS outputs were shown in Table 3-3 and are repeated in Table 3-4.

Notice in the table that the output current in the HIGH state is shown as a negative number. By convention, the *current flow* measured at a device terminal is positive if positive current flows *into* the device; in the HIGH state, current flows *out* of the output terminal.

current flow

Table 3-4 Output loading specifications for HC-series CMOS with a 5-volt supply.

Parameter	CMOS Load		TTL Load	
	Name	Value	Name	Value
Maximum LOW-state output current (mA)	I_{OLmaxC}	0.02	I_{OLmaxT}	4.0
Maximum LOW-state output voltage (V)	V_{OLmaxC}	0.1	V_{OLmaxT}	0.33
Maximum HIGH-state output current (mA)	I_{OHmaxC}	−0.02	I_{OHmaxT}	−4.0
Minimum HIGH-state output voltage (V)	V_{OHminC}	4.4	V_{OHminT}	3.84

As the table shows, with CMOS loads, the CMOS gate's output voltage is maintained within 0.1 V of the power-supply rail. With TTL loads, the output voltage may degrade quite a bit. Also notice that for the same output current (± 4 mA) the maximum voltage drop with respect to the power-supply rail is twice as much in the HIGH state (0.66 V) as in the LOW state (0.33 V). This suggests that the p-channel transistors in HC-series CMOS have a higher "on" resistance than the n-channel transistors do. This is natural, since in any CMOS circuit, a p-channel transistor has over twice the "on" resistance of an n-channel transistor with the same area. Equal voltage drops in both states could be obtained by making the p-channel transistors much larger than the n-channel transistors, but for various reasons this was not done.

Ohm's law can be used to determine how much current an output sources or sinks in a given situation. In Figure 3-28 on page 103, the "on" n-channel transistor modeled by a 100-Ω resistor has a 0.43-V drop across it; therefore it sinks (0.43 V)/(100 Ω) = 4.3 mA of current. Similarly, the "on" p-channel transistor in Figure 3-29 sources (0.39 V)/(200 Ω) = 1.95 mA.

The actual "on" resistances of CMOS output transistors usually aren't published, so it's not always possible to use the exact models of the previous paragraphs. However, you can estimate "on" resistances using the following equations, which rely on specifications that are always published:

$$R_{\text{p(on)}} = \frac{V_{\text{DD}} - V_{\text{OHminT}}}{|I_{\text{OHmaxT}}|}$$

$$R_{\text{n(on)}} = \frac{V_{\text{OLmaxT}}}{I_{\text{OLmaxT}}}$$

These equations use Ohm's law to compute the "on" resistance as the voltage drop across the "on" transistor divided by the current through it with a worst-case resistive load. Using the numbers given for HC-series CMOS in Table 3-4, we can calculate $R_{\text{p(on)}}$ = 175 Ω and $R_{\text{n(on)}}$ = 82.5 Ω.

Very good *worst-case* estimates of output current can be made by assuming that there is *no* voltage drop across the "on" transistor. This assumption simplifies the analysis, and yields a conservative result that is almost always good enough for practical purposes. For example, Figure 3-31 shows a CMOS inverter driving the same Thévenin-equivalent load that we've used in previous examples. The resistive model of the output structure is not shown, because it is no longer needed; we assume that there is no voltage drop across the "on" CMOS transistor. In (a), with the output LOW, the entire 3.33-V Thévenin-equivalent voltage source appears across R_{Thev}, and the estimated sink current is (3.33 V)/(667 Ω) = 5.0 mA. In (b), with the output HIGH and assuming a 5.0-V supply, the voltage drop across R_{Thev} is 1.67 V, and the estimated source current is (1.67 V)/(667 Ω) = 2.5 mA.

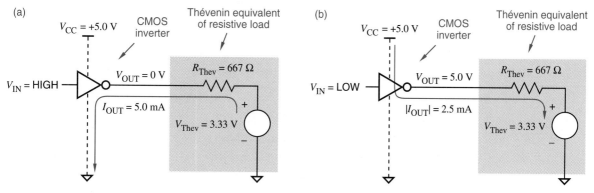

Figure 3-31 Estimating sink and source current: (a) output LOW; (b) output HIGH.

An important feature of the CMOS inverter (or any CMOS circuit) is that the output structure by itself consumes very little current in either state, HIGH or LOW. In either state, one of the transistors is in the high-impedance "off" state. All of the current flow that we've been talking about occurs when a resistive load is connected to the CMOS output. If there's no load, then there's no current flow, and the power consumption is zero. With a load, however, current flows through both the load and the "on" transistor, and power is consumed in both.

THE TRUTH ABOUT POWER CONSUMPTION	As we've stated elsewhere, an "off" transistor's resistance is over one megohm, but it's not infinite. Therefore, a very tiny leakage current actually does flow in "off" transistors, and the CMOS output structure does have a correspondingly tiny but nonzero power consumption. In most applications, this power consumption is tiny enough to ignore. It is usually significant only in "standby mode" in battery-powered devices, such as the laptop computer on which this chapter was first prepared.

3.5.3 Circuit Behavior with Nonideal Inputs

So far, we have assumed that the HIGH and LOW inputs to a CMOS circuit are ideal voltages, very close to the power-supply rails. However, the behavior of a real CMOS inverter circuit depends on the input voltage as well as on the characteristics of the load. If the input voltage is not close to the power-supply rail, then the "on" transistor may not be fully "on" and its resistance may increase. Likewise, the "off" transistor may not be fully "off" and its resistance may be quite a bit less than one megohm. These two effects combine to move the output voltage away from the power-supply rail.

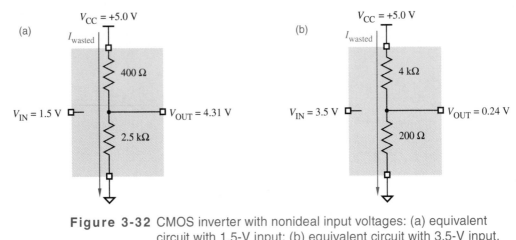

Figure 3-32 CMOS inverter with nonideal input voltages: (a) equivalent circuit with 1.5-V input; (b) equivalent circuit with 3.5-V input.

For example, Figure 3-32(a) shows a CMOS inverter's possible behavior with a 1.5-V input. The *p*-channel transistor's resistance has doubled at this point, and that the *n*-channel transistor is beginning to turn on. (These values are simply assumed for the purposes of illustration; the actual values depend on the detailed characteristics of the transistors.)

In the figure, the output at 4.31 V is still well within the valid range for a HIGH signal, but not quite the ideal of 5.0 V. Similarly, with a 3.5-V input in (b), the LOW output is 0.24 V, not 0 V. The slight degradation of output voltage is generally tolerable; what's worse is that the output structure is now consuming a nontrivial amount of power. The current flow with the 1.5-V input is

$$I_{wasted} = 5.0 \text{ V}/(400 \text{ } \Omega + 2.5 \text{ k}\Omega) = 1.72 \text{ mA}$$

and the power consumption is

$$P_{wasted} = 5.0 \text{ V} \cdot I_{wasted} = 8.62 \text{ mW}$$

The output voltage of a CMOS inverter deteriorates further with a resistive load. Such a load may exist for any of a variety of reasons discussed previously. Figure 3-33 shows a CMOS inverter's possible behavior with a resistive load. With a 1.5-V input, the output at 3.98 V is still within the valid range for a HIGH signal, but it is far from the ideal of 5.0 V. Similarly, with a 3.5-V input as shown in Figure 3-34, the LOW output is 0.93 V, not 0 V.

In "pure" CMOS systems, all of the logic devices in a circuit are CMOS. Since CMOS inputs have a very high impedance, they present very little resistive load to the CMOS outputs that drive them. Therefore, the CMOS output levels all remain very close to the power-supply rails (0 V and 5 V), and none of the devices waste power in their output structures. On the other hand, if TTL outputs or other nonideal logic signals are connected to CMOS inputs, then the CMOS

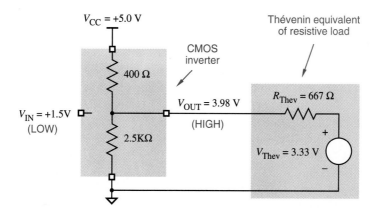

Figure 3-33
CMOS inverter with load and nonideal 1.5-V input.

outputs use power in the way depicted in this subsection; this is formalized in the box at the top of page 139. In addition, if TTL inputs or other resistive loads are connected to CMOS outputs, then the CMOS outputs use power in the way depicted in the preceding subsection.

3.5.4 Fanout

The *fanout* of a logic gate is the number of inputs that the gate can drive without exceeding its worst-case loading specifications. The fanout depends not only on the characteristics of the output, but also on the inputs that it is driving. Fanout must be examined for both possible output states, HIGH and LOW.

fanout

For example, we showed in Table 3-4 on page 105 that the maximum LOW-state output current I_{OLmaxC} for an HC-series CMOS gate driving CMOS inputs is 0.02 mA (20 μA). We also stated previously that the maximum input current I_{Imax} for an HC-series CMOS input in any state is ± 1 μA. Therefore, the *LOW*-state fanout for an HC-series output driving HC-series inputs is 20. Table 3-4 also showed that the maximum HIGH-state output current I_{OHmaxC} is

LOW-state fanout

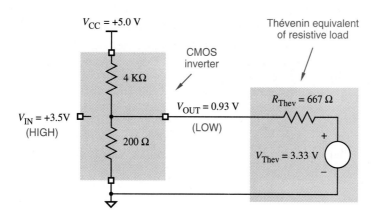

Figure 3-34
CMOS inverter with load and nonideal 3.5-V input.

HIGH-state fanout

−0.02 mA (−20 μA) Therefore, the *HIGH-state fanout* for an HC-series output driving HC-series inputs is also 20.

Note that the HIGH-state and LOW-state fanouts of a gate aren't necessarily equal. In general, the *overall fanout* of a gate is the minimum of its HIGH-state and LOW-state fanouts, 20 in the foregoing example.

overall fanout

In the fanout example that we just completed, we assumed that we needed to maintain the gate's output at CMOS levels, that is, within 0.1 V of the power-supply rails. If we were willing to live with somewhat degraded, TTL output levels, then we could use I_{OLmaxT} and I_{OHmaxT} in the fanout calculation. Table 3-4 shows that these specifications are 4.0 mA and −4.0 mA, respectively. Therefore, the fanout of an HC-series output driving HC-series inputs at TTL levels is 4000—virtually unlimited, apparently.

DC fanout

Well, not quite. The calculations that we've just carried out give the *DC fanout,* defined as the number of inputs that an output can drive *with the output in a constant state* (HIGH or LOW). Even if the DC fanout specification is met, a CMOS output driving a large number of inputs may not behave satisfactorily on transitions, LOW-to-HIGH or vice versa.

During transitions, the CMOS output must charge or discharge the stray capacitance associated with the inputs that it drives. If this capacitance is too large, the transition from LOW to HIGH (or vice versa) may be too slow, causing improper system operation.

AC fanout

The ability of an output to charge and discharge stray capacitance is sometimes called *AC fanout*, though it is seldom calculated as precisely as DC fanout. As you'll see in Section 3.6.1, it's more a matter of deciding how much speed degradation you're willing to live with.

3.5.5 Effects of Loading

Loading an output beyond its rated fanout has several effects:

- In the LOW state, the output voltage (V_{OL}) may increase beyond V_{OLmax}.

- In the HIGH state, the output voltage (V_{OH}) may fall below V_{OHmin}.

- Propagation delay to the output may increase beyond specifications.

- Output rise and fall times may increase beyond their specifications.

- The operating temperature of the device may increase, thereby reducing the reliability of the device and eventually causing device failure.

The first four effects reduce the DC noise margins and timing margins of the circuit. Thus, a slightly overloaded circuit may work properly in ideal conditions, but experience says that it will fail once it's out of the friendly environment of the engineering lab.

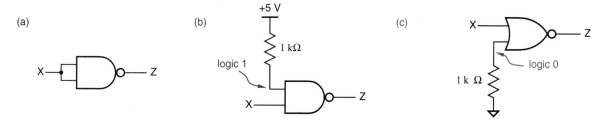

Figure 3-35 Unused inputs: (a) tied to another input; (b) NAND pulled up; (c) NOR pulled down.

3.5.6 Unused Inputs

Sometimes not all of the inputs of a logic gate are used. In a real design problem, you may need an n-input gate but have only an $(n+1)$-input gate available. Tying together two inputs of the $(n+1)$-input gate gives it the functionality of an n-input gate. You can convince yourself of this fact intuitively now, or use switching algebra to prove it after you've studied Section 4.1. Figure 3-35(a) shows a NAND gate with its inputs tied together.

You can also tie unused inputs to a constant logic value. An unused AND or NAND input should be tied to logic 1, as in (b), and an unused OR or NOR input should be tied to logic 0, as in (c). In high-speed circuit design, it's usually better to use method (b) or (c) rather than (a), which increases the capacitive load on the driving signal and may slow things down. In (b) and (c), a resistor value in the range 1–10 kΩ is typically used, and a single pull-up or pull-down resistor can serve multiple unused inputs. It is also possible to tie unused inputs directly to the appropriate power-supply rail.

Unused CMOS inputs should never be left unconnected (or *floating*). On *floating input* one hand, such an input will behave as if it had a LOW signal applied to it and will normally show a value of 0 V when probed with an oscilloscope or volt-meter. So you might think that an unused OR or NOR input can be left floating, because it will act as if a logic 0 is applied and not affect the gate's output.

SUBTLE BUGS   Floating CMOS inputs are often the cause of mysterious circuit behavior, as an unused input erratically changes its effective state based on noise and conditions elsewhere in the circuit. When you're trying to debug such a problem, the extra capacitance of an oscilloscope probe touched to the floating input is often enough to damp out the noise and make the problem go away. This can be especially baffling if you don't realize that the input is floating!

However, since CMOS inputs have such high impedance, it takes only a small amount of circuit noise to temporarily make a floating input look HIGH, creating some very nasty intermittent circuit failures.

3.5.7 Current Spikes and Decoupling Capacitors

When a CMOS output switches between LOW and HIGH, current flows from V_{CC} to ground through the partially-on p- and n-channel transistors. These *current spikes* currents, often called *current spikes* because of their brief duration, may show up as noise on the power-supply and ground connections in a CMOS circuit, especially when multiple outputs are switched simultaneously.

decoupling capacitors   For this reason, systems that use CMOS circuits require *decoupling capacitors* between V_{CC} and ground. These capacitors must be distributed throughout the circuit, at least one within an inch or so of each chip, to supply *filtering capacitors* current during transitions. The large *filtering capacitors* typically found in the power supply itself don't satisfy this requirement, because stray wiring inductance prevents them from supplying the current fast enough, hence the need for a *physically distributed* system of decoupling capacitors.

3.5.8 How to Destroy a CMOS Device

Hit it with a sledge hammer. Or simply walk across a carpet and then touch an input pin with your finger. Because CMOS device inputs have such high imped-
electrostatic discharge   ance, they are subject to damage from *electrostatic discharge (ESD)*.
(ESD)   ESD occurs when a buildup of charge on one surface arcs through a dielectric to another surface with the opposite charge. In the case of a CMOS input, the dielectric is the insulation between an input transistor's gate and its source and drain. ESD may damage this insulation, causing a short-circuit between the device's input and output.

The input structures of modern CMOS devices use various measures to reduce their susceptibility to ESD damage, but no device is completely immune. Therefore, to protect CMOS devices from ESD damage during shipment and handling, manufacturers normally package their devices in conductive bags, tubes, or foam. To prevent ESD damage when handling loose CMOS devices, circuit assemblers and technicians usually wear conductive wrist straps that are connected by a coil cord to earth ground; this prevents a static charge from building up on their bodies as they move around the factory or lab.

latch-up   Once a CMOS device is installed in a system, another possible source of damage is *latch-up*. The physical input structure of just about any CMOS device contains parasitic bipolar transistors between V_{CC} and ground configured as a "silicon-controlled rectifier (SCR)." In normal operation, this "parasitic SCR" has no effect on device operation. However, an input voltage that is less than ground or more than V_{CC} can "trigger" the SCR, creating a virtual short-circuit between V_{CC} and ground. Once the SCR is triggered, the only way to turn it off

---

ELIMINATE RUDE, SHOCKING BEHAVIOR!

Some design engineers consider themselves above such inconveniences, but to be safe you should follow several ESD precautions in the lab:

- Before handling a CMOS device, touch the grounded metal case of a plugged-in instrument or another source of earth ground.

- Before transporting a CMOS device, insert it in conductive foam.

- When carrying a circuit board containing CMOS devices, handle the board by the edges, and touch a ground terminal on the board to earth ground before poking around with it.

- When handing over a CMOS device to a partner, especially on a dry winter day, touch the partner first. He or she will thank you for it.

---

is to turn off the power supply. Before you have a chance to do this, enough power may be dissipated to destroy the device (i.e., you may see smoke).

One possible trigger for latch-up is "undershoot" on high-speed HIGH-to-LOW signal transitions, discussed in Section 11.4. In this situation, the input signal may go several volts below ground for several nanoseconds before settling into the normal LOW range. However, modern CMOS logic circuits are fabricated with special structures that prevent latch-up in this transient case.

Latch-up can also occur when CMOS inputs are driven by the outputs of another system or subsystem with a separate power supply. If a HIGH input is applied to a CMOS gate before power is present, the gate may come up in the "latched-up" state when power is applied. Again, modern CMOS logic circuits are fabricated with special structures that prevent this in most cases. However, if the driving output is capable of sourcing lots of current (e.g., tens of mA), latch-up is still possible. One solution to this problem is to apply power before hooking up input cables.

3.6 CMOS Dynamic Electrical Behavior

Both the speed and the power consumption of a CMOS device depend to a large extent on AC or dynamic characteristics of the device and its load, that is, what happens when the output changes between states. As part of the internal design of CMOS ASICs, logic designers must carefully examine the effects of output loading and redesign where the load is too high. Even in board-level design, the effects of loading must be considered for clocks, buses, and other signals that have high fanout or long interconnections.

Speed depends on two characteristics, transition time and propagation delay, discussed in the next two subsections. Power dissipation is discussed in the third subsection.

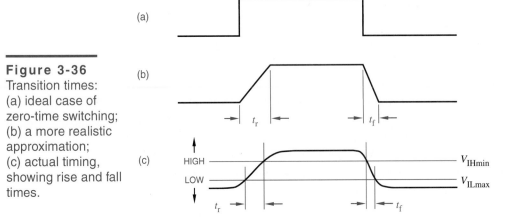

Figure 3-36
Transition times:
(a) ideal case of
zero-time switching;
(b) a more realistic
approximation;
(c) actual timing,
showing rise and fall
times.

3.6.1 Transition Time

transition time

The amount of time that the output of a logic circuit takes to change from one state to another is called the *transition time*. Figure 3-36(a) shows how we might like outputs to change state—in zero time. However, real outputs cannot change instantaneously, because they need time to charge the stray capacitance of the wires and other components that they drive. A more realistic view of a circuit's output is shown in (b). An output takes a certain time, called the *rise time* (t_r), to change from LOW to HIGH, and a possibly different time, called the *fall time* (t_f), to change from HIGH to LOW.

rise time (t_r)
fall time (t_f)

Even Figure 3-36(b) is not quite accurate, because the rate of change of the output voltage does not change instantaneously, either. Instead, the beginning and the end of a transition are smooth, as shown in (c). To avoid difficulties in defining the endpoints, rise and fall times are normally measured at the boundaries of the valid logic levels as indicated in the figure.

With the convention in (c), the rise and fall times indicate how long an output voltage takes to pass through the "undefined" region between LOW and HIGH. The initial part of a transition is not included in the rise- or fall-time number. Instead, the initial part of a transition contributes to the "propagation delay" number discussed in the next subsection.

The rise and fall times of a CMOS output depend mainly on two factors, the "on" transistor resistance and the load capacitance. A large capacitance increases transition times; since this is undesirable, it is very rare for a logic designer to purposely connect a capacitor to a logic circuit's output. However, *stray capacitance* is present in every circuit; it comes from at least three sources:

stray capacitance

1. Output circuits, including a gate's output transistors, internal wiring, and packaging, have some capacitance associated with them, in the range of 2–10 picofarads (pF) in typical logic families, including CMOS.

2. The wiring that connects an output to other inputs has capacitance, about 1 pF per inch or more, depending on the wiring technology.

3. Input circuits, including transistors, internal wiring, and packaging, have capacitance, from 2 to 15 pF per input in typical logic families.

Stray capacitance is sometimes called a *capacitive load* or an *AC load*.   *capacitive load*
   A CMOS output's rise and fall times can be analyzed using the equivalent   *AC load*
circuit shown in Figure 3-37. As in the preceding section, the *p*-channel and
n-channel transistors are modeled by resistances R_p and R_n, respectively. In
normal operation, one resistance is high and the other is low, depending on the
output's state. The output's load is modeled by an *equivalent load circuit* with   *equivalent load circuit*
three components:

R_L, V_L These two components represent the DC load. They determine the
voltages and currents that are present when the output has settled into a
stable **HIGH** or **LOW** state. The DC load doesn't have too much effect on
transition times when the output changes states.

C_L This capacitance represents the AC load. It determines the voltages and
currents that are present while the output is changing, and how long it
takes to change from one state to the other

When a CMOS output drives only CMOS inputs, the DC load is negligible. To
simplify matters, we'll analyze only this case, with $R_L = \infty$ and $V_L = 0$, in the
remainder of this subsection. The presence of a nonnegligible DC load would
affect the results, but not dramatically (see Exercise 3.68).
   We can now analyze the transition times of a CMOS output. For the
purpose of this analysis, we'll assume $C_L = 100$ pF, a moderate capacitive load.
Also, we'll assume that the "on" resistances of the *p*-channel and *n*-channel
transistors are 200 Ω and 100 Ω, respectively, as in the preceding subsection.
The rise and fall times depend on how long it takes to charge or discharge the
capacitive load C_L.

Figure 3-37
Equivalent circuit for
analyzing transition
times of a CMOS output.

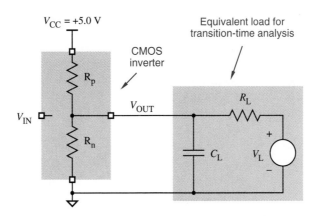

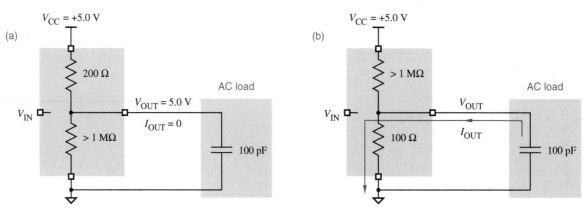

Figure 3-38 Model of a CMOS HIGH-to-LOW transition: (a) in the HIGH state; (b) after *p*-channel transistor turns off and *n*-channel transistor turns on.

First, we'll look at fall time. Figure 3-38(a) shows the electrical conditions in the circuit when the output is in a steady HIGH state. (R_L and V_L are not drawn; they have no effect, since we assume $R_L = \infty$.) For the purposes of our analysis, we'll assume that when CMOS transistors change between "on" and "off," they do so instantaneously. We'll assume that at time $t = 0$ the CMOS output changes to the LOW state, resulting in the situation depicted in (b).

At time $t = 0$, V_{OUT} is still 5.0 V. (A useful electrical engineering maxim is that the voltage across a capacitor cannot change instantaneously.) At time $t = \infty$, the capacitor must be fully discharged and V_{OUT} will be 0 V. In between, the value of V_{OUT} is governed by an exponential law:

$$V_{OUT} = V_{DD} \cdot e^{-t/R_n C_L}$$

$$= 5.0 \cdot e^{-t(100 \cdot 100 \cdot 10^{-12})} \text{ V}$$

$$= 5.0 \cdot e^{-t/(10 \cdot 10^{-9})} \text{ V}$$

RC time constant

The factor $R_n C_L$ has units of seconds and is called an *RC time constant*. The preceding calculation shows that the *RC* time constant for HIGH-to-LOW transitions is 10 nanoseconds (ns).

Figure 3-39 plots V_{OUT} as a function of time. To calculate fall time, recall that 1.5 V and 3.5 V are the defined boundaries for LOW and HIGH levels for CMOS inputs being driven by the CMOS output. To obtain the fall time, we must solve the preceding equation for $V_{OUT} = 3.5$ and $V_{OUT} = 1.5$, yielding:

$$t = -R_n C_L \cdot \ln \frac{V_{OUT}}{V_{DD}} = -10 \cdot 10^{-9} \cdot \ln \frac{V_{OUT}}{5.0}$$

$$t_{3.5} = 3.57 \text{ ns}$$

$$t_{1.5} = 12.04 \text{ ns}$$

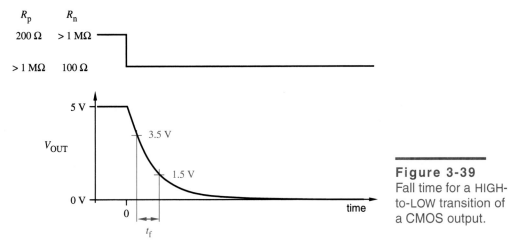

Figure 3-39
Fall time for a HIGH-
to-LOW transition of
a CMOS output.

The fall time t_f is the difference between these two numbers, or about 8.5 ns.

Rise time can be calculated in a similar manner. Figure 3-40(a) shows the conditions in the circuit when the output is in a steady **LOW** state. If at time $t = 0$ the CMOS output changes to the **HIGH** state, the situation depicted in (b) results. Once again, V_{OUT} cannot change instantly, but at time $t = \infty$, the capacitor will be fully charged and V_{OUT} will be 5.0 V. Once again, the value of V_{OUT} in between is governed by an exponential law:

$$
\begin{aligned}
V_{OUT} &= V_{DD} \cdot (1 - e^{-t/R_p C_L}) \\
&= 5.0 \cdot (1 - e^{-t(200 \cdot 100 \cdot 10^{-12})}) \text{ V} \\
&= 5.0 \cdot (1 - e^{-t/(20 \cdot 10^{-9})}) \text{ V}
\end{aligned}
$$

Figure 3-40 Model of a CMOS LOW-to-HIGH transition: (a) in the LOW state; (b) after *n*-channel transistor turns off and *p*-channel transistor turns on.

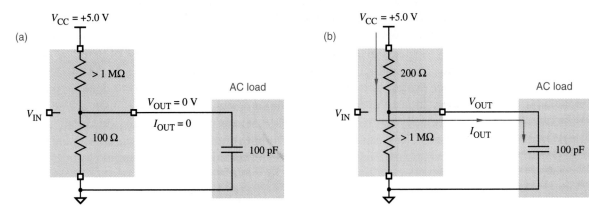

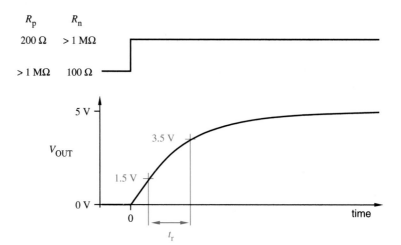

R_p	R_n
200 Ω	> 1 MΩ
> 1 MΩ	100 Ω

Figure 3-41
Rise time for a LOW-
to-HIGH transition of
a CMOS output.

The *RC* time constant in this case is 20 ns. Figure 3-41 plots V_OUT as a function of time. To obtain the rise time, we must solve the preceding equation for $V_\mathrm{OUT} = 1.5$ and $V_\mathrm{OUT} = 3.5$, yielding

$$t = -RC \cdot \ln \frac{V_\mathrm{DD} - V_\mathrm{OUT}}{V_\mathrm{DD}}$$

$$= -20 \cdot 10^{-9} \cdot \ln \frac{5.0 - V_\mathrm{OUT}}{5.0}$$

$$t_{1.5} = 7.13 \text{ ns}$$

$$t_{3.5} = 24.08 \text{ ns}$$

The rise time t_r is the difference between these two numbers, or about 17 ns.

The foregoing example assumes that the *p*-channel transistor has twice the resistance of the *n*-channel one, and as a result the rise time is twice as long as the fall time. It takes longer for the "weak" *p*-channel transistor to pull the output up than it does for the "strong" *n*-channel transistor to pull it down; the output's drive capability is "asymmetric." High-speed CMOS devices are sometimes fabricated with larger *p*-channel transistors to make the transition times more nearly equal and output drive more symmetric.

Regardless of the transistors' characteristics, an increase in load capacitance causes an increase in the *RC* time constant and a corresponding increase in the transition times of the output. Thus, it is a goal of high-speed circuit designers to minimize load capacitance, especially on the most timing-critical signals. This can be done by minimizing the number of inputs driven by the signal, by creating multiple copies of the signal, and by careful physical layout of the circuit.

When working with real digital circuits, it's often useful to estimate transition times, without going through a detailed analysis. A useful rule of thumb is that the transition time approximately equals the *RC* time constant of the charging or discharging circuit. For example, estimates of 10 and 20 ns for fall and rise time in the preceding example would have been pretty much on target, especially considering that most assumptions about load capacitance and transistor "on" resistances are approximate to begin with.

Manufacturers of commercial CMOS circuits typically do not specify transistor "on" resistances on their data sheets. If you search carefully, you might find this information published in the manufacturers' application notes. In any case, you can estimate an "on" resistance as the voltage drop across the "on" transistor divided by the current through it with a worst-case resistive load, as we showed in Section 3.5.2:

$$R_{p(on)} = \frac{V_{DD} - V_{OHminT}}{|I_{OHmaxT}|}$$

$$R_{n(on)} = \frac{V_{OLmaxT}}{|I_{OLmaxT}|}$$

THERE'S A CATCH!   Calculated transition times are actually quite sensitive to the choice of logic levels. In the examples in this subsection, if we used 2.0 V and 3.0 V instead of 1.5 V and 3.5 V as the thresholds for LOW and HIGH, we would calculate shorter transition times. On the other hand, if we used 0.0 and 5.0 V, the calculated transition times would be infinity! You should also be aware that in some logic families (most notably TTL), the thresholds are not symmetric around the voltage midpoint. Still, it is the author's experience that the "time-constant-equals-transition-time" rule of thumb usually works for practical circuits.

3.6.2 Propagation Delay

Rise and fall times only partially describe the dynamic behavior of a logic element; we need additional parameters to relate output timing to input timing. A *signal path* is the electrical path from a particular input signal to a particular output signal of a logic element. The *propagation delay* t_p of a signal path is the amount of time that it takes for a change in the input signal to produce a change in the output signal.

signal path

propagation delay t_p

A complex logic element with multiple inputs and outputs may specify a different value of t_p for each different signal path. Also, different values may be specified for a particular signal path, depending on the direction of the output change. Ignoring rise and fall times, Figure 3-42(a) shows two different

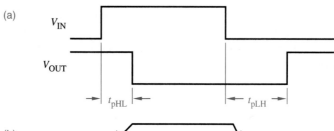

(a)

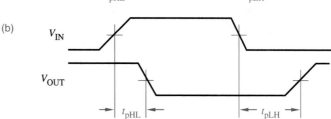

(b)

Figure 3-42
Propagation delays
for a CMOS inverter:
(a) ignoring rise and
fall times; (b) measured at
midpoints of transitions.

propagation delays for the input-to-output signal path of a CMOS inverter, depending on the direction of the output change:

t_{pHL}

t_{pHL} The time between an input change and the corresponding output change when the output is changing from HIGH to LOW.

t_{pLH}

t_{pLH} The time between an input change and the corresponding output change when the output is changing from LOW to HIGH.

Several factors lead to nonzero propagation delays. In a CMOS device, the rate at which transistors change state is influenced both by the semiconductor physics of the device and by the circuit environment, including input-signal transition rate, input capacitance, and output loading. Multistage devices such as noninverting gates or more complex logic functions may require several internal transistors to change state before the output can change state. And even when the output begins to change state, with nonzero rise and fall times it takes quite some time to cross the region between states, as we showed in the preceding subsection. All of these factors are included in propagation delay.

To factor out the effect of rise and fall times, manufacturers usually specify propagation delays at the midpoints of input and output transitions, as shown in Figure 3-42(b). However, sometimes the delays are specified at the logic-level boundary points, especially if the device's operation may be adversely affected by slow rise and fall times. For example, Figure 3-43 shows how the minimum input pulse width for an SR latch (discussed in Section 7.2.1) might be specified.

Figure 3-43
Worst-case timing
specified using logic-
level boundary points.

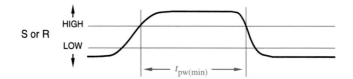

In addition, a manufacturer may specify absolute maximum input rise and fall times that must be satisfied to guarantee proper operation. High-speed CMOS circuits may consume excessive current or oscillate if their input transitions are too slow.

3.6.3 Power Consumption

The power consumption of a CMOS circuit whose output is not changing is called *static power dissipation* or *quiescent power dissipation.* (The words *consumption* and *dissipation* are used pretty much interchangeably when discussing how much power a device uses.) Most CMOS circuits have very low static power dissipation. This is what makes them so attractive for laptop computers and other low-power applications—when computation pauses, very little power is consumed. A CMOS circuit consumes significant power only during transitions; this is called *dynamic power dissipation.*

static power dissipation

quiescent power dissipation

dynamic power dissipation

One source of dynamic power dissipation is the partial short-circuiting of the CMOS output structure. When the input voltage is not close to one of the power supply rails (0 V or V_{CC}), both the *p*-channel and *n*-channel output transistors may be partially "on," creating a series resistance of 600 Ω or less. In this case, current flows through the transistors from V_{CC} to ground. The amount of power consumed in this way depends on both the value of V_{CC} and the rate at which output transitions occur, according to the formula

$$P_T = C_{PD} \cdot V_{CC}^2 \cdot f$$

The following variables are used in the formula:

P_T The circuit's internal power dissipation due to output transitions.

V_{CC} The power-supply voltage. As all electrical engineers know, power dissipation across a resistive load (the partially-on transistors) is proportional to the *square* of the voltage.

f The *transition frequency* of the output signal. This specifies the number of power-consuming output transitions per second. (But note that frequency is defined as the number of transitions divided by 2.)

transition frequency

C_{PD} The *power-dissipation capacitance.* This constant, normally specified by the device manufacturer, completes the formula. C_{PD} turns out to have units of capacitance, but does not represent an actual output capacitance. Rather, it embodies the dynamics of current flow through the changing output-transistor resistances during a single pair of output transitions, HIGH-to-LOW and LOW-to-HIGH. For example, C_{PD} for HC-series CMOS gates is typically 20–24 pF, even though the actual output capacitance is much less.

power-dissipation capacitance

The P_T formula is valid only if input transitions are fast enough, leading to fast output transitions. If the input transitions are too slow, then the output transistors stay partially on for a longer time, and power consumption increases. Device manufacturers usually recommend a maximum input rise and fall time, below which the value specified for C_{PD} is valid.

C_L

A second, and often more significant, source of CMOS power consumption is the capacitive load (C_L) on the output. During a LOW-to-HIGH transition, current flows through a p-channel transistor to charge C_L. Likewise, during a HIGH-to-LOW transition, current flows through an n-channel transistor to discharge C_L. In each case, power is dissipated in the "on" resistance of the

P_L

transistor. We'll use P_L to denote the total amount of power dissipated by charging and discharging C_L.

The units of P_L are power, or energy usage per unit time. The energy for one transition could be determined by calculating the current through the charging transistor as a function of time (using the RC time constant as in Section 3.6.1), squaring this function, multiplying by the "on" resistance of the charging transistor, and integrating over time. An easier way is described below.

During a transition, the voltage across the load capacitance C_L changes by $\pm V_{CC}$. According to the definition of capacitance, the total amount of charge that must flow to make a voltage change of V_{CC} across C_L is $C_L \cdot V_{CC}$. The total amount of energy used in one transition is charge times the average voltage change. The first little bit of charge makes a voltage change of V_{CC}, while the last bit of charge makes a vanishingly small voltage change, hence the average change is $V_{CC}/2$. The total energy per transition is therefore $C_L \cdot V_{CC}^2/2$. If there are $2f$ transitions per second, the total power dissipated due to the capacitive load is

$$P_L = C_L \cdot (V_{CC}^2/2) \cdot 2f$$
$$= C_L \cdot V_{CC}^2 \cdot f$$

The total dynamic power dissipation of a CMOS circuit is the sum of P_T and P_L:

$$P_D = P_T + P_L$$
$$= C_{PD} \cdot V_{CC}^2 \cdot f + C_L \cdot V_{CC}^2 \cdot f$$
$$= (C_{PD} + C_L) \cdot V_{CC}^2 \cdot f$$

CV²f power

Based on this formula, dynamic power dissipation is often called *CV²f power*. In most applications of CMOS circuits, *CV²f* power is by far the major contributor to total power dissipation. Note that *CV²f* power is also consumed by bipolar logic circuits like TTL and ECL, but at low to moderate frequencies it is insignificant compared to the static (DC or quiescent) power dissipation of bipolar circuits.

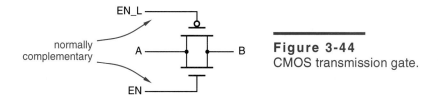

Figure 3-44
CMOS transmission gate.

3.7 Other CMOS Input and Output Structures

Circuit designers have modified the basic CMOS circuit in many ways to produce gates that are tailored for specific applications. This section describes some of the more common variations in CMOS input and output structures.

3.7.1 Transmission Gates

A *p*-channel and *n*-channel transistor pair can be connected together to form a logic-controlled switch. Shown in Figure 3-44(a), this circuit is called a CMOS *transmission gate*.

transmission gate

    A transmission gate is operated so that its input signals EN and EN_L are always at opposite levels. When EN is HIGH and EN_L is LOW, there is a low-impedance connection (as low as 2–5 Ω) between points A and B. When EN is LOW and EN_L is HIGH, points A and B are disconnected.

    Once a transmission gate is enabled, the propagation delay from A to B (or vice versa) is very short. Because of their short delays and conceptual simplicity, transmission gates are often used internally in larger-scale CMOS devices such as multiplexers and flip-flops. For example, Figure 3-45 shows how transmission gates can be used to create a "2-input multiplexer." When S is LOW, the X "input" is connected to the Z "output"; when S is HIGH, Y is connected to Z.

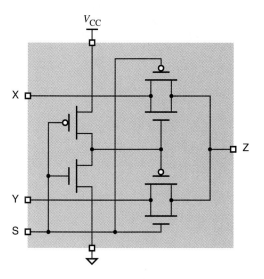

Figure 3-45
Two-input multiplexer using CMOS transmission gates.

At least one commercial manufacturer (Integrated Device Technology) makes a variety of logic functions based on transmission gates. In their multiplexer devices, it takes several nanoseconds for a change in the "select" inputs (such as in Figure 3-45) to affect the input-output path (X or Y to Z). Once a path is set up, however, the propagation delay from input to output is specified to be at most 0.25 ns; this is the fastest discrete CMOS multiplexer you can buy.

3.7.2 Schmitt-Trigger Inputs

Schmitt-trigger input

The input-output transfer characteristic for a typical CMOS gate was shown in Figure 3-25 on page 100. The corresponding transfer characteristic for a gate with *Schmitt-trigger inputs* is shown in Figure 3-46(a). A Schmitt trigger is a special circuit that uses feedback internally to shift the switching threshold depending on whether the input is changing from LOW to HIGH or from HIGH to LOW.

For example, suppose the input of a Schmitt-trigger inverter is initially at 0 V, a solid LOW. Then the output is HIGH, close to 5.0 V. If the input voltage is increased, the output will not go LOW until the input voltage reaches about 2.9 V. However, once the output is LOW, it will not go HIGH again until the input is decreased to about 2.1 V. Thus, the switching threshold for positive-going input changes, denoted V_{T+}, is about 2.9 V, and for negative-going input changes, denoted V_{T-}, is about 2.1 V. The difference between the two thresholds is called *hysteresis*. The Schmitt-trigger inverter provides about 0.8 V of hysteresis.

hysteresis

To demonstrate the usefulness of hysteresis, Figure 3-47(a) shows an input signal with long rise and fall times and about 0.5 V of noise on it. An ordinary inverter, without hysteresis, has the same switching threshold for both positive-going and negative-going transitions, $V_T \approx 2.5$ V. Thus, the ordinary inverter responds to the noise as shown in (b), producing multiple output changes each time the noisy input voltage crosses the switching threshold. However, as shown in (c), a Schmitt-trigger inverter does not respond to the noise, because its hysteresis is greater than the noise amplitude.

Figure 3-46
A Schmitt-trigger inverter: (a) input-output transfer characteristic; (b) logic symbol.

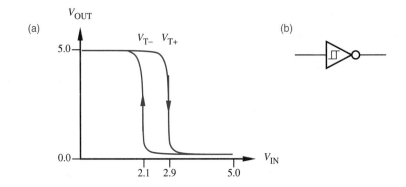

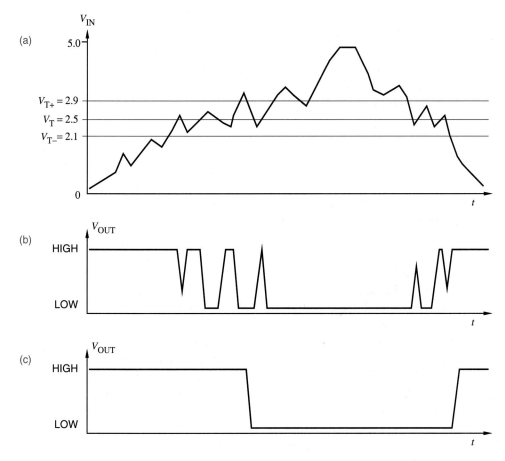

Figure 3-47 Device operation with slowly changing inputs: (a) a noisy, slowly changing input; (b) output produced by an ordinary inverter; (c) output produced by an inverter with 0.8 V of hysteresis.

FIXING YOUR TRANSMISSION   Schmitt-trigger inputs have better noise immunity than ordinary gate inputs for signals that contain transmission-line reflections, discussed in Section 11.4, or that have long rise and fall times. Such signals typically occur in physically long connections, such as input-output buses and computer interface cables. Noise immunity is important in these applications, since long signal lines are more likely to have reflections or to pick up noise from adjacent signal lines, circuits, and appliances.

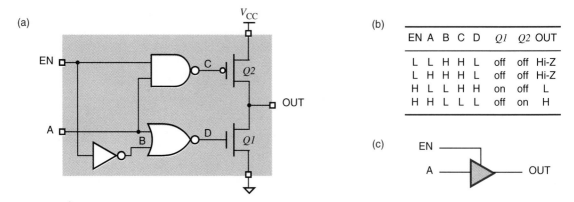

Figure 3-48 CMOS three-state buffer: (a) circuit diagram; (b) function table; (c) logic symbol.

3.7.3 Three-State Outputs

Logic outputs have two normal states, LOW and HIGH, corresponding to logic values 0 and 1. However, some outputs have a third electrical state that is not a logic state at all, called the *high-impedance, Hi-Z,* or *floating state.* In this state, the output behaves as if it isn't even connected to the circuit, except for a small leakage current that may flow into or out of the output pin. Thus, an output can have one of three states—logic 0, logic 1, and Hi-Z.

high-impedance state
Hi-Z state
floating state

An output with three possible states is called (surprise!) a *three-state output* or, sometimes, a *tri-state output.* Three-state devices have an extra input, usually called "output enable" or "output disable," for placing the device's output(s) in the high-impedance state.

three-state output
tri-state output

A *three-state bus* is created by wiring several three-state outputs together. Control circuitry for the "output enables" must ensure that at most one output is enabled (not in its Hi-Z state) at any time. The single enabled device can transmit logic levels (HIGH and LOW) on the bus. Examples of three-state bus design are given in Section 5.6.

three-state bus

A circuit diagram for a CMOS *three-state buffer* is shown in Figure 3-48(a). To simplify the diagram, the internal NAND, NOR, and inverter functions are shown in functional rather than transistor form; they actually use a total of 10 transistors (see Exercise 3.80). As shown in the function table (b), when the enable (EN) input is LOW, both output transistors are off, and the output is in the Hi-Z state. Otherwise, the output is HIGH or LOW as controlled by

three-state buffer

the "data" input A. Logic symbols for three-state buffers and gates are normally drawn with the enable input coming into the top, as shown in (c).

In practice, the three-state control circuit may be different from what we have shown, in order to provide proper dynamic behavior of the output transistors during transitions to and from the Hi-Z state. In particular, devices with three-state outputs are normally designed so that the output-enable delay (Hi-Z to LOW or HIGH) is somewhat longer than the output-disable delay (LOW or HIGH to Hi-Z). Thus, if a control circuit activates one device's output-enable input at the same time that it deactivates a second's, the second device is guaranteed to enter the Hi-Z state before the first places a HIGH or LOW level on the bus.

If two three-state outputs on the same bus are enabled at the same time and try to maintain opposite states, the situation is similar to tying standard active-pull-up outputs together as in Figure 3-56 on page 133—a nonlogic voltage is produced on the bus. If fighting is only momentary, the devices probably will not be damaged, but the large current drain through the tied outputs can produce noise pulses that affect circuit behavior elsewhere in the system.

There is a leakage current of up to 10 μA associated with a CMOS three-state output in its Hi-Z state. This current, as well as the input currents of receiving gates, must be taken into account when calculating the maximum number of devices that can be placed on a three-state bus. That is, in the LOW or HIGH state, an enabled three-state output must be capable of sinking or sourcing up to 10 μA of leakage current for every other three-state output on the bus, as well as handling the current required by every input on the bus. As with standard CMOS logic, separate LOW-state and HIGH-state calculations must be made to ensure that the fanout requirements of a particular circuit configuration are met.

*3.7.4 Open-Drain Outputs

The *p*-channel transistors in CMOS output structures are said to provide *active pull-up*, since they actively pull up the output voltage on a LOW-to-HIGH transition. These transistors are omitted in gates with *open-drain outputs*, such as the NAND gate in Figure 3-49(a). The drain of the topmost *n*-channel transistor is left unconnected internally, so if the output is not LOW it is "open," as indicated in (b). The underscored diamond in the symbol in (c) is sometimes used to indicate an open-drain output. A similar structure, called an "open-collector output," is provided in TTL logic families as described in Section 3.10.5.

An open-drain output requires an external *pull-up resistor* to provide *passive pull-up* to the HIGH level. For example, Figure 3-50 shows an open-drain CMOS NAND gate, with its pull-up resistor, driving a load.

For the highest possible speed, an open-drain output's pull-up resistor should be as small as possible; this minimizes the *RC* time constant for LOW-to-

active pull-up
open-drain output

pull-up resistor
passive pull-up

---

*Throughout this book, optional sections are marked with an asterisk.

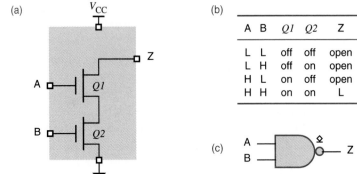

Figure 3-49
Open-drain CMOS NAND gate: (a) circuit diagram; (b) function table; (c) logic symbol.

(b)

A	B	Q1	Q2	Z
L	L	off	off	open
L	H	off	on	open
H	L	on	off	open
H	H	on	on	L

HIGH transitions (rise time). However, the pull-up resistance cannot be arbitrarily small; the minimum resistance is determined by the open-drain output's maximum sink current, I_{OLmax}. For example, in HC- and HCT-series CMOS, I_{OLmax} is 4 mA, and the pull-up resistor can be no less than 5.0 V/4 mA, or 1.25 kΩ. Since this is an order of magnitude greater than the "on" resistance of the p-channel transistors in a standard CMOS gate, the LOW-to-HIGH output transitions are much slower for an open-drain gate than for standard gate with active pull-up.

As an example, let us assume that the open-drain gate in Figure 3-50 is HC-series CMOS, the pull-up resistance is 1.5 kΩ, and the load capacitance is 100 pF. We showed in Section 3.5.2 that the "on" resistance of an HC-series CMOS output in the LOW state is about 80 Ω. Thus, the RC time constant for a HIGH-to-LOW transition is about 80 Ω · 100 pF = 8 ns, and the output's fall time is about 8 ns. However, the RC time constant for a LOW-to-HIGH transition is about 1.5 kΩ · 100 pF = 150 ns, and the rise time is about 150 ns. This relatively slow rise time is contrasted with the much faster fall time in Figure 3-51. A friend of the author calls such slow rising transitions *ooze*.

ooze

Figure 3-50
Open-drain CMOS NAND gate driving a load.

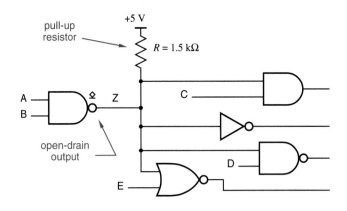

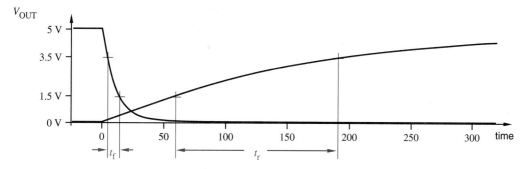

Figure 3-51 Rising and falling transitions of an open-drain CMOS output.

So why use open-drain outputs? Despite slow rise times, they can be useful in at least three applications: driving light-emitting diodes (LEDs) and other devices; performing wired logic; and driving multisource buses.

*3.7.5 Driving LEDs

An open-drain output can drive an LED as shown in Figure 3-52. If either input A or B is LOW, the corresponding n-channel transistor is off and the LED is off. When A and B are both HIGH, both transistors are on, the output Z is LOW, and the LED is on. The value of the pull-up resistor R is chosen so that the proper amount of current flows through the LED in the "on" state.

Typical LEDs require 10 mA for normal brightness. HC- and HCT-series CMOS outputs are only specified to sink or source 4 mA and are not normally used to drive LEDs. However, the outputs in advanced CMOS families such as 74AC and 74ACT can sink 24 mA or more and can be used quite effectively to drive LEDs.

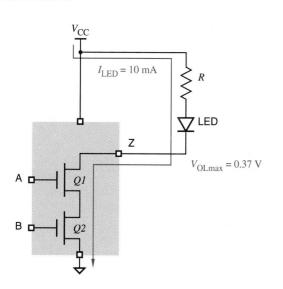

Figure 3-52
Driving an LED with an open-drain output.

Three pieces of information are needed to calculate the proper value of the pull-up resistor R:

1. The LED current I_{LED} needed for the desired brightness, 10 mA for typical LEDs.

2. The voltage drop V_{LED} across the LED in the "on" condition, about 1.6 V for typical LEDs.

3. The output voltage V_{OL} of the open-drain output that sinks the LED current. In the 74AC and 74ACT CMOS families, V_{OLmax} is 0.37 V. If an output can sink I_{LED} and maintain a lower voltage, say 0.2 V, then the calculation below yields a resistor value that is a little too low, but normally with no harm done. A little more current than I_{LED} will flow and the LED will be just a little brighter than expected.

Using the above information, we can write the following equation:

$$V_{OL} + V_{LED} + (I_{LED} \cdot R) = V_{CC}$$

Assuming $V_{CC} = 5.0$ V and the other typical values above, we can solve for the required value of R:

$$R = \frac{V_{CC} - V_{OL} - V_{LED}}{I_{LED}}$$
$$= (5.0 - 0.37 - 1.6) \text{ V}/10 \text{ mA} = 303 \ \Omega$$

Note that you don't have to use an open-drain output to drive an LED. Figure 3-53(a) shows an LED driven by an ordinary CMOS NAND-gate output with active pull-up. If both inputs are HIGH, the bottom (*n*-channel) transistors pull the output LOW as in the open-drain version. If either input is LOW, the output is HIGH; although one or both of the top (*p*-channel) transistors is on, no current flows through the LED.

With some CMOS families, you can turn an LED "on" when the output is in the HIGH state, as shown in Figure 3-53(b). This is possible if the output can source enough current to satisfy the LED's requirements. However, method (b) isn't used as often as method (a), because most CMOS and TTL outputs cannot source as much current in the HIGH state as they can sink in the LOW state.

RESISTOR VALUES   In most applications, the precise value of LED series resistors is unimportant, as long as groups of nearby LEDs have similar drivers and resistors to give equal apparent brightness. In the example in this subsection, one might use an off-the-shelf resistor value of 270, 300, or 330 ohms, whatever is readily available.

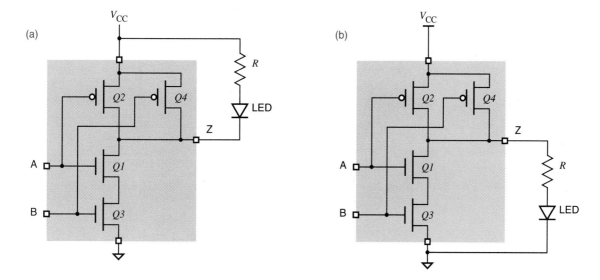

Figure 3-53 Driving an LED with an ordinary CMOS output: (a) sinking current, "on" in the LOW state; (b) sourcing current, "on" in the HIGH state.

*3.7.6 Multisource Buses

Open-drain outputs can be tied together to allow several devices, one at a time,   *open-drain bus*
to put information on a common bus. At any time all but one of the outputs on
the bus are in their HIGH (open) state. The remaining output either stays in the
HIGH state or pulls the bus LOW, depending on whether it wants to transmit a
logical 1 or a logical 0 on the bus. Control circuitry selects the particular device
that is allowed to drive the bus at any time.

For example, in Figure 3-54, eight 2-input open-drain NAND-gate outputs
drive a common bus. The top input of each NAND gate is a data bit, and the

Figure 3-54 Eight open-drain outputs driving a bus.

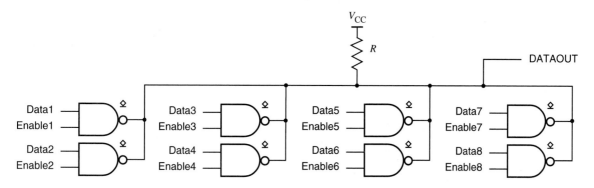

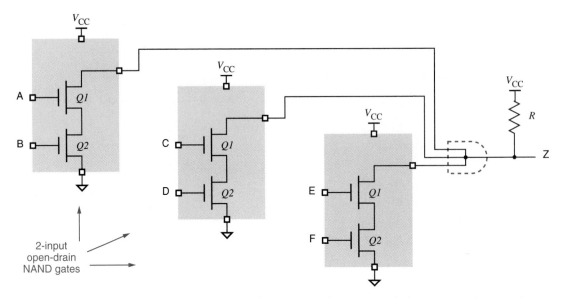

Figure 3-55 Wired-AND function on three open-drain NAND-gate outputs.

bottom input of each is a control bit. At most one control bit is HIGH at any time, enabling the corresponding data bit to be passed through to the bus. (Actually, the complement of the data bit is placed on the bus.) The other gate outputs are HIGH, that is, "open," so the data input of the enabled gate determines the value on the bus.

*3.7.7 Wired Logic

wired logic

wired AND

fighting

If the outputs of several open-drain gates are tied together with a single pull-up resistor, then *wired logic* is performed. (That's *wired*, not *weird*!) An AND function is obtained, since the wired output is HIGH if and only if all of the individual gate outputs are HIGH (actually, open); any output going LOW is sufficient to pull the wired output LOW. For example, a three-input *wired AND* function is shown in Figure 3-55. If any of the individual 2-input NAND gates has both inputs HIGH, it pulls the wired output LOW; otherwise, the pull-up resistor R pulls the wired output HIGH.

Note that wired logic cannot be performed using gates with active pull-up. Two such outputs wired together and trying to maintain opposite logic values result in a very high current flow and an abnormal output voltage. Figure 3-56 shows this situation, which is sometimes called *fighting*. The exact output voltage depends on the relative "strengths" of the fighting transistors, but with 5-V CMOS devices it is typically about 1–2 V, almost always a nonlogic voltage. Worse, if outputs fight continuously for more than a few seconds, the chips can get hot enough to sustain internal damage *and* to burn your fingers!

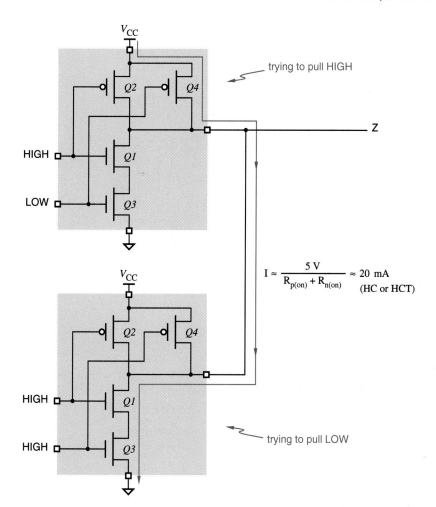

$$I \approx \frac{5\ V}{R_{p(on)} + R_{n(on)}} \approx 20\ mA$$
(HC or HCT)

Figure 3-56
Two CMOS outputs
trying to maintain
opposite logic values
on the same line.

*3.7.8 Pull-Up Resistors

A proper choice of value for the pull-up resistor R must be made in open-drain *pull-up- resistor* applications. Two calculations are made to bracket the allowable values of R: *calculation*

Minimum The sum of the current through R in the LOW state and the LOW-state input currents of the gates driven by the wired outputs must not exceed the LOW-state driving capability of the active output, 4 mA for HC and HCT, 24 mA for AC and ACT.

Maximum The voltage drop across R in the HIGH state must not reduce the output voltage below 2.4 V, which is V_{IHmin} for typical driven gates plus a 400-mV noise margin. This drop is produced by the HIGH-state output leakage current of the wired outputs and the HIGH-state input currents of the driven gates.

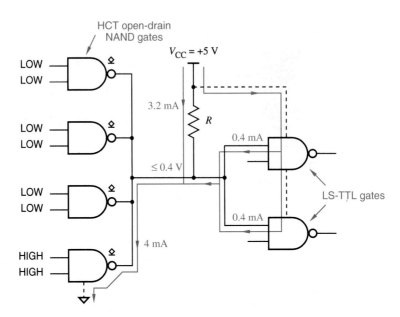

Figure 3-57
Four open-drain
outputs driving two
inputs in the LOW
state.

For example, suppose that four HCT open-drain outputs are wired together
and drive two LS-TTL inputs (Section 3.11) as shown in Figure 3-57. A LOW
output must sink 0.4 mA from each LS-TTL input as well as sink the current
through the pull-up resistor R. For the total current to stay within the HCT I_{OLmax}
spec of 4 mA, the current through R may be no more than

$$I_{R(max)} = 4 - (2 \cdot 0.4) = 3.2 \text{ mA}$$

Figure 3-58
Four open-drain
outputs driving two
inputs in the HIGH
state.

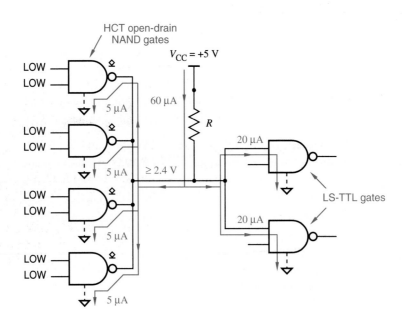

Assuming that V_{OL} of the open-drain output is 0.0 V, the minimum value of R is

$$R_{min} = (5.0 - 0.0)/I_{R(max)} = 1562.5 \ \Omega$$

In the HIGH state, typical open-drain outputs have a maximum leakage current of 5 μA, and typical LS-TTL inputs require 20 μA of source current. Hence, the HIGH-state current requirement as shown in Figure 3-58 is

$$I_{R(leak)} = (4 \cdot 5) + (2 \cdot 20) = 60 \ \mu A$$

This current produces a voltage drop across R, and must not lower the output voltage below $V_{OHmin} = 2.4$ V; thus the maximum value of R is

$$R_{max} = (5.0 - 2.4)/I_{R(leak)} = 43.3 \ k\Omega$$

Hence, any value of R between 1562.5 Ω and 43.3 $k\Omega$ may be used. Higher values reduce power consumption and improve the LOW-state noise margin, while lower values increase power consumption but improve both the HIGH-state noise margin and the speed of LOW-to-HIGH output transitions.

OPEN-DRAIN ASSUMPTION   In our open-drain resistor calculations, we assume that the output voltage can be as low as 0.0 V rather than 0.4 V (V_{OLmax}) in order to obtain a worst-case result. That is, even if the open-drain output is so strong that it can pull the output voltage all the way down to 0.0 V (it's only required to pull down to 0.4 V), we'll never allow it to sink more than 4 mA, so it doesn't get overstressed. Some designers prefer to use 0.4 V in this calculation, figuring that if the output is so good that it can pull lower than 0.4 V, a little bit of excess sink current beyond 4 mA won't hurt it.

3.8 CMOS Logic Families

The first commercially successful CMOS family was *4000-series CMOS*. Although 4000-series circuits offered the benefit of low power dissipation, they were fairly slow and were not easy to interface with the most popular logic family of the time, bipolar TTL. Thus, the 4000 series was supplanted in most applications by the more capable CMOS families discussed in this section.

4000-series CMOS

All of the CMOS devices that we discuss have part numbers of the form "74FAM*nn*," where "FAM" is an alphabetic family mnemonic and *nn* is a numeric function designator. Devices in different families with the same value of *nn* perform the same function. For example, the 74HC30, 74HCT30, 74AC30, 74ACT30, and 74AHC30 are all 8-input NAND gates.

The prefix "74" is simply a number that was used by an early, popular supplier of TTL devices, Texas Instruments. The prefix "54" is used for identical parts that are specified for operation over a wider range of temperature and power-supply voltage, for use in military applications. Such parts are usually

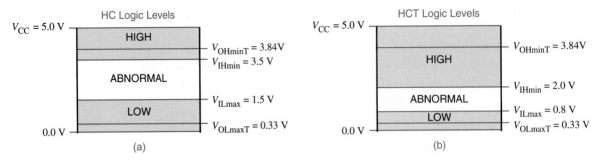

Figure 3-59 Input and output levels for CMOS devices using a 5-V supply: (a) HC; (b) HCT.

fabricated in the same way as their 74-series counterparts, except that they are tested, screened, and marked differently, a lot of extra paperwork is generated, and a higher price is charged, of course.

3.8.1 HC and HCT

HC (High-speed CMOS)

HCT (High-speed CMOS, TTL compatible)

The first two 74-series CMOS families are *HC (High-speed CMOS)* and *HCT (High-speed CMOS, TTL compatible)*. Compared with the original 4000 family, HC and HCT both have higher speed and better current sinking and sourcing capability. The HCT family uses a power-supply voltage V_{CC} of 5 V and can be intermixed with TTL devices, which also use a 5-V supply.

The HC family is optimized for use in systems that use CMOS logic exclusively, and can use any power-supply voltage between 2 and 6 V. A higher voltage is used for higher speed, and a lower voltage for lower power dissipation. Lowering the supply voltage is especially effective, since most CMOS power dissipation is proportional to the square of the voltage (CV^2f power).

Even when used with a 5-V supply, HC devices are not quite compatible with TTL. In particular, HC circuits are designed to recognize CMOS input levels. Assuming a supply voltage of 5.0 V, Figure 3-59(a) shows the input and output levels of HC devices. The output levels produced by TTL devices do not quite match this range, so HCT devices use the different input levels shown in (b). These levels are established in the fabrication process by making transistors with different switching thresholds, producing the different transfer characteristics shown in Figure 3-60.

Figure 3-60
Transfer characteristics of HC and HCT circuits under typical conditions.

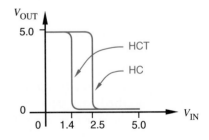

We'll have more to say about CMOS/TTL interfacing in Section 3.12. For now, it is useful simply to note that HC and HCT are essentially identical in their output specifications; only their input levels differ.

3.8.2 VHC and VHCT

Several new CMOS families were introduced in the 1980s and the 1990s. Two of the most recent and probably the most versatile are *VHC (Very High-Speed CMOS)* and *VHCT (Very High-Speed CMOS, TTL compatible)*. These families are about twice as fast as HC/HCT while maintaining backward compatibility with their predecessors. Like HC and HCT, the VHC and VHCT families differ from each other only in the input levels that they recognize; their output characteristics are the same.

VHC (Very High-speed CMOS)

VHCT (Very High-speed CMOS, TTL compatible)

Also like HC/HCT, VHC/VHCT outputs have *symmetric output drive*. That is, an output can sink or source equal amounts of current; the output is just as "strong" in both states. Other logic families, including the FCT and TTL families introduced later, have *asymmetric output drive;* they can sink much more current in the LOW state than they can source in the HIGH state.

symmetric output drive

asymmetric output drive

3.8.3 HC, HCT, VHC, and VHCT Electrical Characteristics

Electrical characteristics of the HC, HCT, VHC, and VHCT families are summarized in this subsection. The specifications assume that the devices are used with a nominal 5-V power supply, although (derated) operation is possible with any supply voltage in the range 2–5.5 V (up to 6 V for HC/HCT). We'll take a closer look at low-voltage and mixed-voltage operation in Section 3.13.

Commercial (74-series) parts are intended to be operated at temperatures between 0°C and 70°C, while military (54-series) parts are characterized for operation between −55°C and 125°C. The specs in Table 3-5 assume an operating temperature of 25°C. A full manufacturer's data sheet provides additional specifications for device operation over the entire temperature range.

Most devices within a given logic family have the same electrical specifications for inputs and outputs, typically differing only in power consumption and propagation delay. Table 3-5 includes specifications for a 74x00 two-input NAND gate and a 74x138 3-to-8 decoder in the HC, HCT, VHC, and VHCT families. The '00 NAND gate is included as the smallest logic-design building block in each family, while the '138 is a "medium-scale" part containing the equivalent of about 15 NAND gates. (The '138 spec is included to allow

---

VERY = ADVANCED, SORT OF	The VHC and VHCT logic families are manufactured by several companies, including Motorola, Fairchild, and Toshiba. Compatible families with similar but not identical specifications are manufactured by Texas Instruments and Philips; they are called AHC and AHCT, where the "A" stands for "Advanced."

■ **Table 3-5** Speed and power characteristics of CMOS families operating at 5 V.

Description	Part	Symbol	Condition	Family HC	HCT	VHC	VHCT
Typical propagation delay (ns)	'00	t_{PD}		9	10	5.2	5.5
	'138			18	20	7.2	8.1
Quiescent power-supply current (μA)	'00	I_{CC}	$V_{in} = 0$ or V_{CC}	2.5	2.5	5.0	5.0
	'138		$V_{in} = 0$ or V_{CC}	40	40	40	40
Quiescent power dissipation (mW)	'00		$V_{in} = 0$ or V_{CC}	0.0125	0.0125	0.025	0.025
	'138		$V_{in} = 0$ or V_{CC}	0.2	0.2	0.2	0.2
Power-dissipation capacitance (pF)	'00	C_{PD}		22	15	19	17
	'138	C_{PD}		55	51	34	49
Dynamic power dissipation (mW/MHz)	'00			0.55	0.38	0.48	0.43
	'138			1.38	1.28	0.85	1.23
Total power dissipation (mW)	'00		$f = 100$ kHz	0.068	0.050	0.073	0.068
	'00		$f = 1$ MHz	0.56	0.39	0.50	0.45
	'00		$f = 10$ MHz	5.5	3.8	4.8	4.3
	'138		$f = 100$ kHz	0.338	0.328	0.285	0.323
	'138		$f = 1$ MHz	1.58	1.48	1.05	1.43
	'138		$f = 10$ MHz	14.0	13.0	8.7	12.5
Speed-power product (pJ)	'00		$f = 100$ kHz	0.61	0.50	0.38	0.37
	'00		$f = 1$ MHz	5.1	3.9	2.6	2.5
	'00		$f = 10$ MHz	50	38	25	24
	'138		$f = 100$ kHz	6.08	6.55	2.05	2.61
	'138		$f = 1$ MHz	28.4	29.5	7.56	11.5
	'138		$f = 10$ MHz	251	259	63	101

comparison with the faster FCT family in Section 3.8.4; '00 gates are not manufactured in the FCT family.)

The first row of Table 3-5 specifies propagation delay. As discussed in Section 3.6.2, two numbers, t_{pHL} and t_{pLH}, may be used to specify delay; the number in the table is the worst case of the two. Skipping ahead to Table 3-11 on page 167, you can see that HC and HCT are about the same speed as LS TTL, and that VHC and VHCT are almost as fast as ALS TTL. The propagation delay

NOTE ON NOTATION   The "x" in the notation "74x00" takes the place of a family designator such as HC, HCT, VHC, VHCT, FCT, LS, ALS, AS, or F. We may also refer to such a generic part simply as a " '00" and leave off the "74x."

QUIETLY GETTING
MORE DISS'ED

HCT and VHCT circuits can also be driven by TTL devices, which may produce HIGH output levels as low as 2.4 V. As we explained in Section 3.5.3, a CMOS output may draw additional current from the power supply if any of the inputs are nonideal. In the case of an HCT or VHCT inverter with a HIGH input of 2.4 V, the bottom, n-channel output transistor is fully "on." However, the top, p-channel transistor is also partially "on." This allows the additional quiescent current flow, specified as ΔI_{CC} or I_{CCT} in the data sheet, which can be as much as 2–3 mA per nonideal input in HCT and VHCT devices.

---

for the '138 is somewhat longer than for the '00, since signals must travel through three or four levels of gates internally.

The second and third rows of the table show that the quiescent power dissipation of these CMOS devices is practically nil, well under a milliwatt (mW) if the inputs have CMOS levels—0 V for LOW and V_{CC} for HIGH. (Note that in the table, the quiescent power dissipation numbers given for the '00 are per gate, while for the '138 they apply to the entire MSI device.)

As we discussed in Section 3.6.3, the dynamic power dissipation of a CMOS gate depends on the voltage swing of the output (usually V_{CC}), the output transition frequency (f), and the capacitance that is being charged and discharged on transitions, according to the formula

$$P_D = (C_L + C_{PD}) \cdot V_{DD}^2 \cdot f$$

Here, C_{PD} is the power-dissipation capacitance of the device and C_L is the capacitance of the load attached to the CMOS output in a given application. The table lists both C_{PD} and an equivalent dynamic power-dissipation factor in units of milliwatts per megahertz, assuming that $C_L = 0$. Using this factor, the total power dissipation is computed at various frequencies as the sum of the dynamic power dissipation at that frequency and the quiescent power dissipation.

Shown next in the table, the *speed-power product* is simply the product of *speed-power product* the propagation delay and power consumption of a typical gate; the result is measured in picojoules (pJ). Recall from physics that the joule is a unit of energy, so the speed-power product measures a sort of efficiency—how much energy a logic gate uses to switch its output. In this day and age, it's obvious that the lower the energy usage, the better.

SAVING ENERGY

There are practical as well as geopolitical reasons for saving energy in digital systems. Lower energy consumption means lower cost of power supplies and cooling systems. Also, a digital system's reliability is improved more by running it cooler than by any other single reliability improvement strategy.

Table 3-6 Input specifications for CMOS families with V_{CC} between 4.5 and 5.5 V.

Description	Symbol	Condition	Family HC	HCT	VHC	VHCT
Input leakage current (μA)	I_{Imax}	V_{in} = any	±1	±1	±1	±1
Maximum input capacitance (pF)	C_{INmax}		10	10	10	10
LOW-level input voltage (V)	V_{ILmax}		1.35	0.8	1.35	0.8
HIGH-level input voltage (V)	V_{IHmin}		3.85	2.0	3.85	2.0

Table 3-6 gives the input specs of typical CMOS devices in each of the families. Some of the specs assume that the 5-V supply has a ±10% margin; that is, V_{CC} can be anywhere between 4.5 and 5.5 V. These parameters were discussed in previous sections, but for reference purposes their meanings are summarized here:

I_{Imax} The maximum input current for any value of input voltage. This spec states that the current flowing into or out of a CMOS input is 1 μA or less for any value of input voltage. In other words, CMOS inputs create almost no DC load on the circuits that drive them.

C_{INmax} The maximum capacitance of an input. This number can be used when figuring the AC load on an output that drives this and other inputs. Most manufacturers also specify a lower, typical input capacitance of about 5 pF, which gives a good estimate of AC load if you're not unlucky.

V_{ILmax} The maximum voltage that an input is guaranteed to recognize as LOW. Note that the values are different for HC/VHC versus HCT/VHCT. The "CMOS" value, 1.35 V, is 30% of the minimum power-supply voltage, while the "TTL" value is 0.8 V for compatibility with TTL families.

CMOS VS. TTL POWER DISSIPATION

At high transition frequencies (f), CMOS families actually use more power than TTL. For example, compare HCT CMOS in Table 3-5 at f = 10 MHz with LS TTL in Table 3-11; a CMOS gate uses three times as much power as a TTL gate at this frequency. Both HCT and LS may be used in systems with maximum "clock" frequencies of up to about 20 MHz, so you might think that CMOS is not so good for high-speed systems. However, the transition frequencies of most outputs in typical systems are much less than the maximum frequency present in the system (e.g., see Exercise 3.77). Thus, typical CMOS systems have a lower total power dissipation than they would have if they were built with TTL.

Table 3-7 Output specifications for CMOS families operating with V_{CC} between 4.5 and 5.5 V.

Description	Symbol	Condition	Family HC	HCT	VHC	VHCT
LOW-level output current (mA)	I_{OLmaxC}	CMOS load	0.02	0.02	0.05	0.05
	I_{OLmaxT}	TTL load	4.0	4.0	8.0	8.0
LOW-level output voltage (V)	V_{OLmaxC}	$I_{out} \leq I_{OLmaxC}$	0.1	0.1	0.1	0.1
	V_{OLmaxT}	$I_{out} \leq I_{OLmaxT}$	0.33	0.33	0.44	0.44
HIGH-level output current (mA)	I_{OHmaxC}	CMOS load	−0.02	−0.02	−0.05	−0.05
	I_{OHmaxT}	TTL load	−4.0	−4.0	−8.0	−8.0
HIGH-level output voltage (V)	V_{OHminC}	$|I_{out}| \leq |I_{OHmaxC}|$	4.4	4.4	4.4	4.4
	V_{OHminT}	$|I_{out}| \leq |I_{OHmaxT}|$	3.84	3.84	3.80	3.80

V_{IHmin} The minimum voltage that an input is guaranteed to recognize as HIGH. The "CMOS" value, 3.85 V, is 70% of the maximum power-supply voltage, while the "TTL" value is 2.0 V for compatibility with TTL families. (Unlike CMOS levels, TTL input levels are not symmetric with respect to the power-supply rails.)

The specifications for TTL-compatible CMOS outputs usually have two sets of output parameters; one set or the other is used depending on how an output is loaded. A *CMOS load* is one that requires the output to sink and source very little DC current, 20 μA for HC/HCT and 50 μA for VHC/VHCT. This is, of course, the case when the CMOS outputs drive only CMOS inputs. With CMOS loads, CMOS outputs maintain an output voltage within 0.1 V of the supply rails, 0 and V_{CC}. (A worst-case $V_{CC} = 4.5$ V is used for the table entries; hence, $V_{OHminC} = 4.4$ V.)

CMOS load

A *TTL load* can consume much more sink and source current, up to 4 mA from an HC/HCT output and 8 mA from a VHC/VHCT output. In this case, a higher voltage drop occurs across the "on" transistors in the output circuit, but the output voltage is still guaranteed to be within the normal range of TTL output levels.

TTL load

Table 3-7 lists CMOS output specifications for both CMOS and TTL loads. These parameters have the following meanings:

I_{OLmaxC} The maximum current that an output can supply in the LOW state while driving a CMOS load. Since this is a positive value, current flows *into* the output pin.

I_{OLmaxT} The maximum current that an output can supply in the LOW state while driving a TTL load.

V_{OLmaxC} The maximum voltage that a LOW output is guaranteed to produce while driving a CMOS load, that is, as long as I_{OLmaxC} is not exceeded.

V_{OLmaxT} The maximum voltage that a LOW output is guaranteed to produce while driving a TTL load, that is, as long as I_{OLmaxT} is not exceeded.

I_{OHmaxC} The maximum current that an output can supply in the HIGH state while driving a CMOS load. Since this is a negative value, positive current flows out of the output pin.

I_{OHmaxT} The maximum current that an output can supply in the HIGH state while driving a TTL load.

V_{OHminC} The minimum voltage that a HIGH output is guaranteed to produce while driving a CMOS load, that is, as long as I_{OHmaxC} is not exceeded.

V_{OHminT} The minimum voltage that a HIGH output is guaranteed to produce while driving a TTL load, that is, as long as I_{OHmaxT} is not exceeded.

The voltage parameters above determine DC noise margins. The LOW-state DC noise margin is the difference between V_{OLmax} and V_{ILmax}. This depends on the characteristics of both the driving output and the driven inputs. For example, the LOW-state DC noise margin of HCT driving a few HCT inputs (a CMOS load) is $0.8 - 0.1 = 0.7$ V. With a TTL load, the noise margin for the HCT inputs drops to $0.8 - 0.33 = 0.47$ V. Similarly, the HIGH-state DC noise margin is the difference between V_{OHmin} and V_{IHmin}. In general, when different families are interconnected, you have to compare the appropriate V_{OLmax} and V_{OHmin} of the driving gate with V_{ILmax} and V_{IHmin} of all the driven gates to determine the worst-case noise margins.

The I_{OLmax} and I_{OHmax} parameters in the table determine fanout capability and are especially important when an output drives inputs in one or more different families. Two calculations must be performed to determine whether an output is operating within its rated fanout capability:

HIGH-state fanout The I_{IHmax} values for all of the driven inputs are added. The sum must be less than I_{OHmax} of the driving output.

LOW-state fanout The I_{ILmax} values for all of the driven inputs are added. The sum must be less than I_{OLmax} of the driving output

Note that the input and output characteristics of specific components may vary from the representative values given in Table 3-7, so you must always consult the manufacturers' data sheets when analyzing a real design.

*3.8.4 FCT and FCT-T

FCT (Fast CMOS, TTL compatible)

In the early 1990s, yet another CMOS family was launched. The key benefit of the *FCT (Fast CMOS, TTL compatible)* family was its ability to meet or exceed the speed and the output drive capability of the best TTL families while reducing power consumption and maintaining full compatibility with TTL.

The original FCT family had the drawback of producing a full 5-V CMOS V_{OH}, creating enormous CV^2f power dissipation and circuit noise as its outputs swung from 0 V to almost 5 V in high-speed (25 MHz+) applications. A variation of the family, *FCT-T (Fast CMOS, TTL compatible with TTL V_{OH})*, was quickly introduced with circuit innovations to reduce the HIGH-level output voltage, thereby reducing both power consumption and switching noise while maintaining the same high operating speed as the original FCT. A suffix of "T" is used on part numbers to denote the FCT-T output structure, for example, 74FCT138T versus 74FCT138.

FCT-T (Fast CMOS, TTL compatible with TTL V_{OH}

The FCT-T family remains very popular today. A key application of FCT-T is driving buses and other heavy loads. Compared with other CMOS families, it can source or sink gobs of current, up to 64 mA in the LOW state.

*3.8.5 FCT-T Electrical Characteristics

Electrical characteristics of the 5-V FCT-T family are summarized in Table 3-8. The family is specifically designed to be intermixed with TTL devices, so its operation is only specified with a nominal 5-V supply and TTL logic levels. Some manufacturers are beginning to sell parts with similar capabilities using a 3.3-V supply, and using the FCT designation. However, they are different devices with different part numbers.

Individual logic gates are not manufactured in the FCT family. Perhaps the simplest FCT logic element is a 74FCT138T decoder, which has six inputs, eight outputs, and contains the equivalent of about a dozen 4-input gates internally. (This function is described later, in Section 5.4.4.) Comparing its propagation delay and power consumption in Table 3-8 with the corresponding HCT and VHCT numbers in Table 3-5 on page 138, you can see that the FCT-T family is superior in both speed and power dissipation. When comparing, note that FCT-T manufacturers specify only maximum, not typical propagation delays.

Unlike other CMOS families, FCT-T does not have a C_{PD} specification. Instead, it has an I_{CCD} specification:

I_{CCD} Dynamic power-supply current, in units of mA/MHz. This is the amount of additional power-supply current that flows when one input is changing at the rate of 1 MHz.

EXTREME SWITCHING   Device outputs in the FCT and FCT-T families have very low impedance and as a consequence extremely fast rise and fall times. In fact, they are so fast that they are often a major source of "analog" problems, including switching noise and "ground bounce," so extra care must be taken in the analog and physical design of printed-circuit boards using these and other extremely high-speed parts. To reduce the effects of transmission-line reflections (Section 11.4), another high-speed design worry, some FCT-T outputs have built-in 25-Ω series resistors.

Table 3-8 Specifications for a 74FCT138T decoder in the FCT-T logic family.

Description	Symbol	Condition	Value
Maximum propagation delay (ns)	t_{PD}		5.8
Quiescent power-supply current (µA)	I_{CC}	$V_{in} = 0$ or V_{CC}	200
Quiescent power dissipation (mW)		$V_{in} = 0$ or V_{CC}	1.0
Dynamic power-supply current (mA/MHz)	I_{CCD}	Outputs open, one input changing	0.12
Quiescent power-supply current per TTL input (mA)	ΔI_{CC}	$V_{in} = 3.4$ V	2.0
Total power dissipation (mW)		$f = 100$ kHz $f = 1$ MHz $f = 10$ MHz	0.60 1.06 1.6
Speed-power product (pJ)		$f = 100$ kHz $f = 1$ MHz $f = 10$ MHz	6.15 9.3 41
Input leakage current (µA)	I_{Imax}	$V_{in} =$ any	±5
Typical input capacitance (pF)	C_{INtyp}		5
LOW-level input voltage (V)	V_{ILmax}		0.8
HIGH-level input voltage (V)	V_{IHmin}		2.0
LOW-level output current (mA)	I_{OLmax}		64
LOW-level output voltage (V)	V_{OLmax}	$I_{out} \leq I_{OLmax}$	0.55
HIGH-level output current (mA)	I_{OHmax}		−15
HIGH-level output voltage (V)	V_{OHmin} V_{OHtyp}	$|I_{out}| \leq |I_{OHmax}|$ $|I_{out}| \leq |I_{OHmax}|$	2.4 3.3

The I_{CCD} specification gives the same information as C_{PD}, but in a different way. The circuit's internal power dissipation due to transitions at a given frequency f can be calculated by the formula

$$P_T = V_{CC} \cdot I_{CCD} \cdot f$$

Thus, I_{CCD}/V_{CC} is algebraically equivalent to the C_{PD} specification of other CMOS families (see Exercise 3.83). FCT-T also has a ΔI_{CC} specification for the extra quiescent current that is consumed with nonideal HIGH inputs (see box at the top of page 139).

3.9 Bipolar Logic

Bipolar logic families use semiconductor diodes and bipolar junction transistors as the basic building blocks of logic circuits. The simplest bipolar logic elements use diodes and resistors to perform logic operations; this is called diode logic. *diode logic* Most TTL logic gates use diode logic internally and boost their output drive capability using transistor circuits. Some TTL gates use parallel configurations of transistors to perform logic functions. ECL gates, described in Section 3.14, use transistors as current switches to achieve very high speed.

This section covers the basic operation of bipolar logic circuits made from diodes and transistors, and the next section covers TTL circuits in detail. Although TTL is the most commonly used bipolar logic family, it has been largely supplanted by the CMOS families that we studied in previous sections.

Still, it is useful to study basic TTL operation for the occasional application that requires TTL/CMOS interfacing, discussed in Section 3.12. Also, an understanding of TTL may give you insight into the fortuitous similarity of logic levels that allowed the industry to migrate smoothly from TTL to 5-V CMOS logic, and now to lower-voltage, higher-performance 3.3-V CMOS logic, as described in Section 3.13. If you're not interested in all the gory details of TTL, you can skip to Section 3.11 for an overview of TTL families.

3.9.1 Diodes

A *semiconductor diode* is fabricated from two types of semiconductor material, *semiconductor diode* called *p*-type and *n*-type, that are brought into contact with each other as shown *p-type material* in Figure 3-61(a). This is basically the same material that is used in *p*-channel *n-type material* and *n*-channel MOS transistors. The point of contact between the *p* and *n* materials is called a *pn junction*. (Actually, a diode is normally fabricated from a *pn junction* single monolithic crystal of semiconductor material in which the two halves are "doped" with different impurities to give them *p*-type and *n*-type properties.)

The physical properties of a *pn* junction are such that positive current can easily flow from the *p*-type material to the *n*-type. Thus, if we build the circuit shown in Figure 3-61(b), the *pn* junction acts almost like a short circuit. However, the physical properties also make it very difficult for positive current to

Figure 3-61 Semiconductor diodes: (a) the *pn* junction; (b) forward-biased junction allowing current flow; (c) reverse-biased junction blocking current flow.

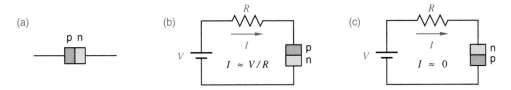

Figure 3-62 Diodes: (a) symbol; (b) transfer characteristic of an ideal diode; (c) transfer characteristic of a real diode.

diode action

flow in the opposite direction, from n to p. Thus, in the circuit of Figure 3-61(c), the pn junction behaves almost like an open circuit. This is called *diode action*.

diode

anode

cathode

Although it's possible to build vacuum tubes and other devices that exhibit diode action, modern systems use pn junctions—semiconductor diodes—which we'll henceforth call simply *diodes*. Figure 3-62(a) shows the schematic symbol for a diode. As we've shown, in normal operation significant amounts of current can flow only in the direction indicated by the two arrows, from *anode* to *cathode*. In effect, the diode acts like a short circuit as long as the voltage across the anode-to-cathode junction is nonnegative. If the anode-to-cathode voltage is negative, the diode acts like an open circuit and no current flows.

reverse-biased diode

forward-biased diode

The transfer characteristic of an ideal diode shown in Figure 3-62(b) further illustrates this principle. If the anode-to-cathode voltage, V, is negative, the diode is said to be *reverse biased* and the current I through the diode is zero. If V is nonnegative, the diode is said to be *forward biased* and I can be an arbitrarily large positive value. In fact, V can never get larger than zero, because an ideal diode acts like a zero-resistance short circuit when forward biased.

A nonideal, real diode has a resistance that is less than infinity when reverse biased, and greater than zero when forward biased, so the transfer characteristic looks like Figure 3-62(c). When forward biased, the diode acts

YES, THERE ARE TWO ARROWS   . . . in Figure 3-62(a). The second arrow is built into the diode symbol to help you remember the direction of current flow. Once you know this, there are many ways to remember which end is called the anode and which is the cathode. Aficionados of vacuum-tube hi-fi amplifiers may remember that electrons travel from the hot cathode to the anode, and therefore positive current flow is from anode to cathode. Those of us who were still watching "Sesame Street" when most vacuum tubes went out of style might like to think in terms of the alphabet—current flows alphabetically from A to C. (Thanks to Big-Bird-fan Rick Casey for suggesting this memory crutch.)

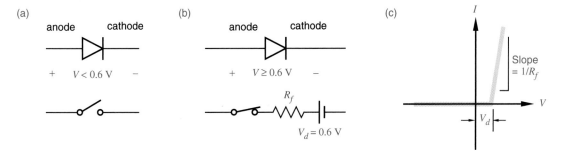

Figure 3-63 Model of a real diode: (a) reverse biased; (b) forward biased; (c) transfer characteristic of forward-biased diode.

like a small nonlinear resistance; its voltage drop increases as current increases, but not strictly proportionally. When the diode is reverse biased, a small amount of negative *leakage current* flows. If the voltage is made too negative, the diode *breaks down,* and large amounts of negative current can flow; in most applications, this type of operation is avoided.

leakage current
diode breakdown

    A real diode can be modeled very simply as shown in Figure 3-63(a) and (b). When the diode is reverse biased, it acts like an open circuit; we ignore leakage current. When the diode is forward biased, it acts like a small resistance, R_f, in series with V_d, a small voltage source. R_f is called the *forward resistance* of the diode, and V_d is called a *diode-drop.*

forward resistance
diode-drop

    Careful choice of values for R_f and V_d yields a reasonable piecewise-linear approximation to the real diode transfer characteristic, as in Figure 3-63(c). In a typical small-signal diode such as a 1N914, the forward resistance R_f is about 25 Ω and the diode-drop V_d is about 0.6 V.

    In order to get a feel for diodes, you should remember that a real diode does not actually contain the 0.6-V source that appears in the model. It's just that, due to the nonlinearity of the real diode's transfer characteristic, significant amounts of current do not begin to flow until the diode's forward voltage V has reached about 0.6 V. Also note that in typical applications, the 25-Ω forward resistance of the diode is small compared to other resistances in the circuit, so that very little additional voltage drop occurs across the forward-biased diode once V has reached 0.6 V. Thus, for practical purposes, a forward-biased diode may be considered to have a fixed drop of 0.6 V or so.

ZENER DIODES   *Zener diodes* take advantage of diode breakdown, in particular the steepness of the V–I slope in the breakdown region. A Zener diode can function as a voltage regulator when used with a resistor to limit the breakdown current. A wide variety of Zeners with different breakdown voltages are produced for voltage-regulator applications.

Table 3-9
Logic levels in a simple diode logic system.

Signal Level	Designation	Binary Logic Value
0–2 volts	LOW	0
2–3 volts	noise margin	undefined
3–5 volts	HIGH	1

3.9.2 Diode Logic

LOW

HIGH

diode AND gate

Diode action can be exploited to perform logical operations. Consider a logic system with a 5-V power supply and the characteristics shown in Table 3-9. Within the 5-volt range, signal voltages are partitioned into two ranges, LOW and HIGH, with a 1-volt noise margin between. A voltage in the LOW range is considered to be a logic 0, and a voltage in the HIGH range is a logic 1.

With these definitions, a *diode AND gate* can be constructed as shown in Figure 3-64(a). In this circuit, suppose that both inputs X and Y are connected to HIGH voltage sources, say 4 V, so that V_X and V_Y are both 4 V as in (b). Then both diodes are forward biased, and the output voltage V_Z is one diode-drop above 4 V, or about 4.6 V. A small amount of current, determined by the value of R, flows from the 5-V supply through the two diodes and into the 4-V sources. The colored arrows in the figure show the path of this current flow.

Figure 3-64
Diode AND gate:
(a) electrical circuit;
(b) both inputs HIGH;
(c) one input HIGH, one LOW; (d) function table; (e) truth table.

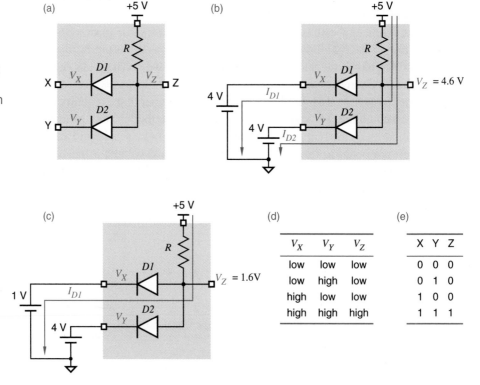

V_X	V_Y	V_Z
low	low	low
low	high	low
high	low	low
high	high	high

X	Y	Z
0	0	0
0	1	0
1	0	0
1	1	1

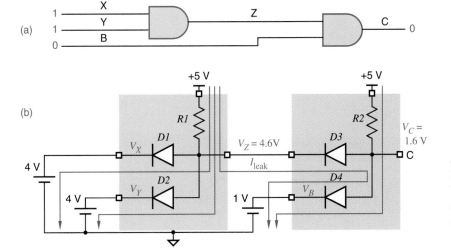

Figure 3-65
Two AND gates:
(a) logic diagram;
(b) electrical circuit.

Now suppose that V_X drops to 1 V as in Figure 3-64(c). In the diode AND gate, the output voltage equals the lower of the two input voltages plus a diode-drop. Thus, V_Z drops to 1.6 V, and diode $D2$ is reverse biased (the anode is at 1.6 V and the cathode is still at 4 V). The single LOW input "pulls down" the output of the diode AND gate to a LOW value. Obviously, two LOW inputs create a LOW output as well. This functional operation is summarized in (d) and is repeated in terms of binary logic values in (e); clearly, this is an AND gate.

Figure 3-65(a) shows a logic circuit with two AND gates connected together; Figure 3-65(b) shows the equivalent electrical circuit with a particular set of input values. This example shows the necessity of diodes in the AND circuit: $D3$ allows the output Z of the first AND gate to remain HIGH while the output C of the second AND gate is being pulled LOW by input B through $D4$.

When diode logic gates are cascaded as in Figure 3-65, the voltage levels of the logic signals move away from the power-supply rails and toward the undefined region. Thus, in practice, a diode AND gate normally must be followed by a transistor amplifier to restore the logic levels; this is the scheme used in TTL NAND gates, described in Section 3.10.1. However, logic designers are occasionally tempted to use discrete diodes to perform logic under special circumstances; for example, see Exercise 3.95.

3.9.3 Bipolar Junction Transistors

A *bipolar junction transistor* is a three-terminal device that, in most logic circuits, acts like a current-controlled switch. If we put a small current into one of the terminals, called the *base*, then the switch is "on"—current may flow between the other two terminals, called the *emitter* and the *collector*. If no current is put into the base, then the switch is "off"—no current flows between the emitter and the collector.

bipolar junction transistor

base

emitter

collector

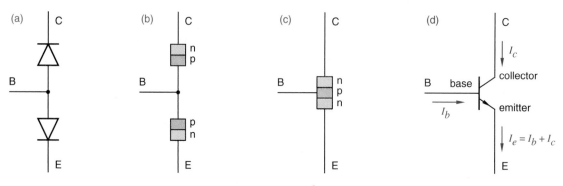

Figure 3-66 Development of an *npn* transistor: (a) back-to-back diodes;
(b) equivalent *pn* junctions; (c) structure of an *npn* transistor;
(d) *npn* transistor symbol.

To study the operation of a transistor, we first consider the operation of a pair of diodes connected as shown in Figure 3-66(a). In this circuit, current can flow from node B to node C or node E, when the appropriate diode is forward biased. However, no current can flow from C to E, or vice versa, since for any choice of voltages on nodes B, C, and E, one or both diodes will be reverse biased. The *pn* junctions of the two diodes in this circuit are shown in (b).

Now suppose that we fabricate the back-to-back diodes so that they share a common *p*-type region, as shown in Figure 3-66(c). The resulting structure is called an *npn transistor* and has an amazing property. (At least, the physicists working on transistors back in the 1950s thought it was amazing!) If we put current across the base-to-emitter *pn* junction, then current is also enabled to flow across the collector-to-base *np* junction (which is normally impossible) and from there to the emitter.

npn transistor

The circuit symbol for the *npn* transistor is shown in Figure 3-66(d). Notice that the symbol contains a subtle arrow in the direction of positive current flow. This also reminds us that the base-to-emitter junction is a *pn* junction, the same as a diode whose symbol has an arrow pointing in the same direction.

pnp transistor

It is also possible to fabricate a *pnp transistor,* as shown in Figure 3-67. However, *pnp* transistors are seldom used in digital circuits, so we won't discuss them any further.

The current I_e flowing out of the emitter of an *npn* transistor is the sum of the currents I_b and I_c flowing into the base and the collector. A transistor is often used as a signal *amplifier*, because over a certain operating range (the *active region*) the collector current is equal to a fixed constant times the base current ($I_c = \beta \cdot I_b$). However, in digital circuits, we normally use a transistor as a simple switch that's always fully "on" or fully "off," as explained next.

amplifier
active region

Figure 3-68 shows the *common-emitter configuration* of an *npn* transistor, which is most often used in digital switching applications. This configuration

common-emitter
configuration

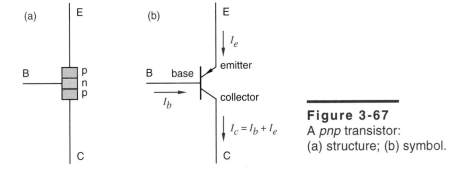

Figure 3-67
A *pnp* transistor:
(a) structure; (b) symbol.

uses two discrete resistors, *R1* and *R2*, in addition to a single *npn* transistor. In this circuit, if V_{IN} is 0 or negative, then the base-to-emitter diode junction is reverse biased, and no base current (I_b) can flow. If no base current flows, then no collector current (I_c) can flow, and the transistor is said to be *cut off (OFF)*.

cut off (OFF)

Since the base-to-emitter junction is a *real* diode, as opposed to an ideal one, V_{IN} must reach at least +0.6 V (one diode-drop) before any base current can flow. Once this happens, Ohm's law tells us that

$$I_b = (V_{IN} - 0.6) / R1$$

(We ignore the forward resistance R_f of the forward-biased base-to-emitter junction, which is usually small compared to the base resistor *R1*.) When base current flows, then collector current can flow in an amount proportional to I_b, that is,

$$I_c = \beta \cdot I_b$$

The constant of proportionality, β, is called the *gain* of the transistor, and is in the range of 10 to 100 for typical transistors.

β

gain

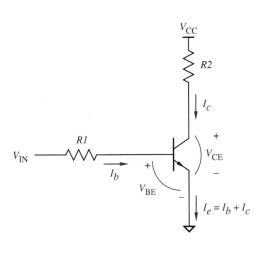

Figure 3-68
Common-emitter
configuration of an
npn transistor.

Although the base current I_b controls the collector current flow I_c, it also indirectly controls the voltage V_{CE} across the collector-to-emitter junction, since V_{CE} is just the supply voltage V_{CC} minus the voltage drop across resistor $R2$:

$$V_{CE} = V_{CC} - I_c \cdot R2$$
$$= V_{CC} - \beta \cdot I_b \cdot R2$$
$$= V_{CC} - \beta \cdot (V_{IN} - 0.6) \cdot R2 / R1$$

However, in an ideal transistor V_{CE} can never be less than zero (the transistor cannot just create a negative potential), and in a real transistor V_{CE} can never be less than $V_{CE(sat)}$, a transistor parameter that is typically about 0.2 V.

If the values of V_{IN}, β, $R1$, and $R2$ are such that the above equation predicts a value of V_{CE} that is less than $V_{CE(sat)}$, then the transistor cannot be operating in the active region and the equation does not apply. Instead, the transistor is *saturation region* operating in the *saturation region,* and is said to be *saturated (ON).* No matter *saturated (ON)* how much current I_b we put into the base, V_{CE} cannot drop below $V_{CE(sat)}$, and the collector current I_c is determined mainly by the load resistor $R2$:

$$I_c = (V_{CC} - V_{CE(sat)}) / (R2 + R_{CE(sat)})$$

saturation resistance   Here, $R_{CE(sat)}$ is the *saturation resistance* of the transistor. Typically, $R_{CE(sat)}$ is 50 Ω or less and is insignificant compared with $R2$.

transistor simulation   Computer scientists might like to imagine an *npn* transistor as a device that continuously looks at its environment and executes the program in Table 3-10.

3.9.4 Transistor Logic Inverter

Figure 3-69 shows that we can make a logic inverter from an *npn* transistor in the common-emitter configuration. When the input voltage is LOW, the output voltage is HIGH, and vice versa.

In digital switching applications, bipolar transistors are often operated so they are always either cut off or saturated. That is, digital circuits such as the

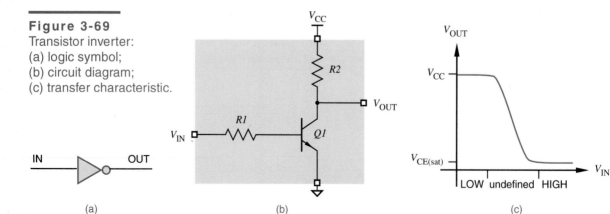

Figure 3-69
Transistor inverter:
(a) logic symbol;
(b) circuit diagram;
(c) transfer characteristic.

(a)   (b)   (c)

Table 3-10 A C program that simulates the function of an *npn* transistor in the common-emitter configuration.

```c
/* Transistor parameters */
#define DIODEDROP 0.6 /* volts */
#define BETA 10;
#define VCE_SAT 0.2    /* volts */
#define RCE_SAT 50     /* ohms  */

main()
{
    float Vcc, Vin, R1, R2;   /* circuit parameters */
    float Ib, Ic, Vce;        /* circuit conditions */

    if (Vin < DIODEDROP) {    /* cut off */
        Ib  = 0.0;
        Ic  = 0.0;
        Vce = Vcc;
    }
    else {                        /* active or saturated */
        Ib = (Vin - DIODEDROP) / R1;
        if ((Vcc - ((BETA * Ib) * R2)) >= VCE_SAT) {    /* active */
            Ic  = BETA * Ib;
            Vce = Vcc - (Ic * R2);
        }
        else {                    /* saturated */
            Vce = VCE_SAT;
            Ic  = (Vcc - Vce) / (R2 + RCE_SAT);
        }
    }
}
```

Figure 3-70 Normal states of an *npn* transistor in a digital switching circuit: (a) transistor symbol and currents; (b) equivalent circuit for a cut-off (OFF) transistor; (c) equivalent circuit for a saturated (ON) transistor.

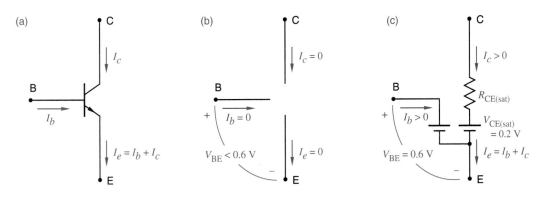

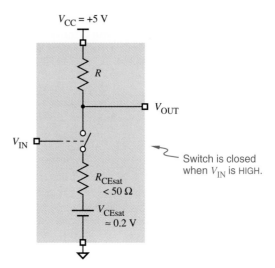

Figure 3-71
Switch model for a
transistor inverter.

inverter in Figure 3-69 are designed so that their transistors are always (well, almost always) in one of the states depicted in Figure 3-70. When the input voltage V_{IN} is LOW, it is low enough that I_b is zero and the transistor is cut off; the collector-emitter junction looks like an open circuit. When V_{IN} is HIGH, it is high enough (and $R1$ is low enough and β is high enough) that the transistor will be saturated for any reasonable value of $R2$; the collector-emitter junction looks almost like a short circuit. Input voltages in the undefined region between LOW and HIGH are not allowed, except during transitions. This undefined region corresponds to the noise margin that we discussed in conjunction with Table 3-1.

Another way to visualize the operation of a transistor inverter is shown in Figure 3-71. When V_{IN} is HIGH, the transistor switch is closed, and the output terminal is connected to ground, definitely a LOW voltage. When V_{IN} is LOW, the transistor switch is open and the output terminal is pulled to +5 V through a resistor; the output voltage is HIGH unless the output terminal is too heavily loaded (i.e., improperly connected through a low impedance to ground).

3.9.5 Schottky Transistors

When the input of a saturated transistor is changed, the output does not change immediately; it takes extra time, called *storage time*, to come out of saturation. In fact, storage time accounts for a significant portion of the propagation delay in the original TTL logic family.

storage time

Storage time can be eliminated and propagation delay can be reduced by ensuring that transistors do not saturate in normal operation. Contemporary TTL logic families do this by placing a *Schottky diode* between the base and collector of each transistor that might saturate, as shown in Figure 3-72. The resulting transistors, which do not saturate, are called *Schottky-clamped transistors* or *Schottky transistors* for short.

Schottky diode
Schottky-clamped transistor
Schottky transistor

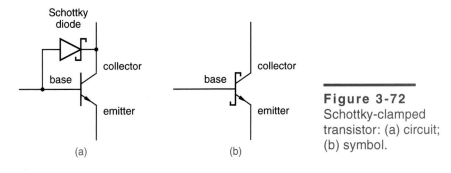

Figure 3-72
Schottky-clamped transistor: (a) circuit; (b) symbol.

When forward biased, a Schottky diode's voltage drop is much less than a standard diode's, 0.25 V vs. 0.6 V. In a standard saturated transistor, the base-to-collector voltage is 0.4 V, as shown in Figure 3-73(a). In a Schottky transistor, the Schottky diode shunts current from the base into the collector before the transistor goes into saturation, as shown in (b). Figure 3-74 is the circuit diagram of a simple inverter using a Schottky transistor.

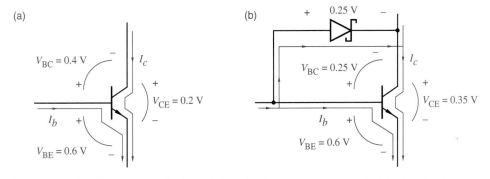

Figure 3-73 Operation of a transistor with large base current: (a) standard saturated transistor; (b) transistor with Schottky diode to prevent saturation.

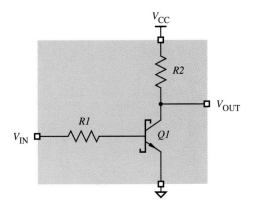

Figure 3-74
Inverter using Schottky transistor.

3.10 Transistor-Transistor Logic

The most commonly used bipolar logic family is transistor-transistor logic. Actually, there are many different TTL families, with a range of speed, power consumption, and other characteristics. The circuit examples in this section are based on a representative TTL family, Low-power Schottky (LS or LS-TTL).

TTL families use basically the same logic levels as the TTL-compatible CMOS families in previous sections. We'll use the following definitions of LOW and HIGH in our discussions of TTL circuit behavior:

LOW   0–0.8 volts.

HIGH   2.0–5.0 volts.

3.10.1 Basic TTL NAND Gate

The circuit diagram for a two-input LS-TTL NAND gate, part number 74LS00, is shown in Figure 3-75. The NAND function is obtained by combining a diode AND gate with an inverting buffer amplifier. The circuit's operation is best understood by dividing it into the three parts that are shown in the figure and discussed in the next three paragraphs:

- Diode AND gate and input protection.
- Phase splitter.
- Output stage.

diode AND gate

clamp diode

Diodes *D1X* and *D1Y* and resistor *R1* in Figure 3-75 form a *diode AND gate,* as in Section 3.9.2. *Clamp diodes D2X* and *D2Y* do nothing in normal operation, but limit undesirable negative excursions on the inputs to a single diode-drop. Such negative excursions may occur on HIGH-to-LOW input transitions as a result of transmission-line effects, discussed in Section 11.4.

phase splitter

Transistor *Q2* and the surrounding resistors form a *phase splitter* that controls the output stage. Depending on whether the diode AND gate produces a "low" or a "high" voltage at V_A, *Q2* is either cut off or turned on.

WHERE IN THE WORLD IS *Q1*?

Notice that there is no transistor *Q1* in Figure 3-75, but the other transistors are named in a way that's traditional; some TTL devices do in fact have a transistor named *Q1*. Instead of diodes like *D1X* and *D1Y*, these devices use a multiple-emitter transistor *Q1* to perform logic. This transistor has one emitter per logic input, as shown in the figure to the right. Pulling any one of the emitters LOW is sufficient to turn the transistor ON and thus pull V_A LOW.

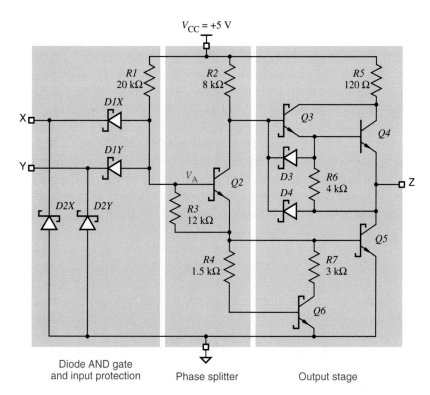

$V_{CC} = +5$ V

Diode AND gate
and input protection

Phase splitter

Output stage

Figure 3-75
Circuit diagram of
two-input LS-TTL
NAND gate.

The *output stage* has two transistors, *Q4* and *Q5*, only one of which is on at
any time. The TTL output stage is sometimes called a *totem-pole* or *push-pull*
output. Similar to the *p*-channel and *n*-channel transistors in CMOS, *Q4* and *Q5*
provide active pull-up and pull-down to the HIGH and LOW states, respectively.

The functional operation of the TTL NAND gate is summarized in
Figure 3-76(a). The gate does indeed perform the NAND function, with the truth
table and logic symbol shown in (b) and (c). TTL NAND gates can be designed
with any desired number of inputs simply by changing the number of diodes in

output stage
totem-pole output
push-pull output

(a)

X	Y	V_A	Q2	Q3	Q4	Q5	Q6	V_Z	Z
L	L	≤1.05	off	on	on	off	off	2.7	H
L	H	≤1.05	off	on	on	off	off	2.7	H
H	L	≤1.05	off	on	on	off	off	2.7	H
H	H	1.2	on	off	off	on	on	≤0.35	L

Figure 3-76
Functional operation
of a TTL two-input
NAND gate:
(a) function table;
(b) truth table;
(c) logic symbol.

(b)

X	Y	Z
0	0	1
0	1	1
1	0	1
1	1	0

(c)

X ———
Y ——— Z

the diode AND gate in the figure. Commercially available TTL NAND gates have as many as 13 inputs. A TTL inverter is designed as a 1-input NAND gate, omitting diodes *D1Y* and *D2Y* in Figure 3-75.

Since the output transistors *Q4* and *Q5* are normally complementary—one ON and the other OFF—you might question the purpose of the 120-Ω resistor *R5* in the output stage. A value of 0 Ω would give even better driving capability in the HIGH state. This is certainly true from a DC point of view. However, when the TTL output is changing from HIGH to LOW or vice versa, there is a short time when both transistors may be on. The purpose of *R5* is to limit the amount of current that flows from VCC to ground during this time. Even with a 120 Ω resistor in the TTL output stage, higher-than-normal currents called current spikes flow when TTL outputs are switched. These are similar to the current spikes that occur when high-speed CMOS outputs switch.

So far we have shown the input signals to a TTL gate as ideal voltage sources. Figure 3-77 shows the situation when a TTL input is driven LOW by the output of another TTL gate. Transistor *Q5A* in the driving gate is ON, and thereby provides a path to ground for the current flowing out of the diode *D1XB* in the driven gate. When current flows *into* a TTL output in the LOW state, as in *sinking current* this case, the output is said to be *sinking current*.

Figure 3-78 shows the same circuit with a HIGH output. In this case, *Q4A* in the driving gate is turned on enough to supply the small amount of leakage current flowing through reverse-biased diodes *D1XB* and *D2XB* in the driven gate. When current flows out of a TTL output in the HIGH state, the output is *sourcing current* said to be *sourcing current*.

3.10.2 Logic Levels and Noise Margins

At the beginning of this section, we indicated that we would consider TTL signals between 0 and 0.8 V to be LOW, and signals between 2.0 and 5.0 V to be HIGH. Actually, we can be more precise by defining TTL input and output levels in the same way as we did for CMOS:

V_{OHmin} The minimum output voltage in the HIGH state, 2.7 V for most TTL families.

V_{IHmin} The minimum input voltage guaranteed to be recognized as a HIGH, 2.0 V for all TTL families.

CURRENT SPIKES AGAIN   Current spikes can show up as noise on the power-supply and ground connections in TTL and CMOS circuits, especially when multiple outputs are switched simultaneously. For this reason, reliable circuits require decoupling capacitors between V_{CC} and ground, distributed throughout the circuit so that there is a capacitor within an inch or so of each chip. Decoupling capacitors supply the instantaneous current needed during transitions.

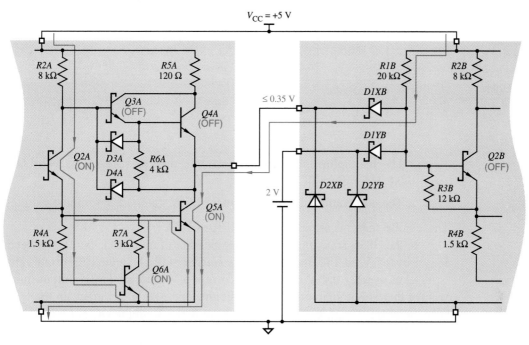

Figure 3-77 A TTL output driving a TTL input LOW.

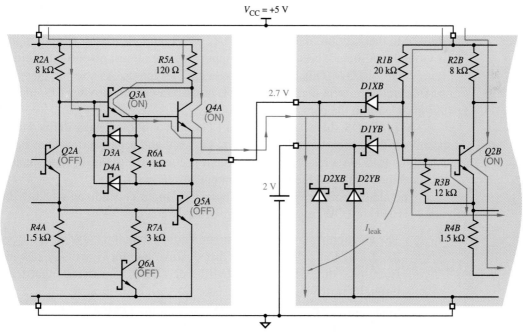

Figure 3-78 A TTL output driving a TTL input HIGH.

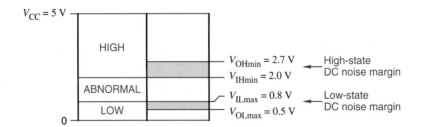

Figure 3-79
Noise margins for
popular TTL logic
families (74LS, 74S,
74ALS, 74AS, 74F).

V_{ILmax} The maximum input voltage guaranteed to be recognized as a **LOW**,
0.8 V for most TTL families.

V_{OLmax} The maximum output voltage in the **LOW** state, 0.5 V for most families.

These noise margins are illustrated in Figure 3-79.

DC noise margin

In the **HIGH** state, the V_{OHmin} specification of most TTL families exceeds V_{IHmin} by 0.7 V, so TTL has a *DC noise margin* of 0.7 V in the **HIGH** state. That is, it takes at least 0.7 V of noise to corrupt a worst-case **HIGH** output into a voltage that is not guaranteed to be recognizable as a **HIGH** input. In the **LOW** state, however, V_{ILmax} exceeds V_{OLmax} by only 0.3 V, so the DC noise margin in the **LOW** state is only 0.3 V. In general, TTL and TTL-compatible circuits tend to be more sensitive to noise in the **LOW** state than in the **HIGH** state.

3.10.3 Fanout

fanout

As we defined it previously in Section 3.5.4, *fanout* is a measure of the number of gate inputs that are connected to (and driven by) a single gate output. As we showed in that section, the DC fanout of CMOS outputs driving CMOS inputs is virtually unlimited, because CMOS inputs require almost no current in either state, **HIGH** or **LOW**. This is not the case with TTL inputs. As a result, there are very definite limits on the fanout of TTL or CMOS outputs driving TTL inputs, as you'll learn in the paragraphs that follow.

current flow

As in CMOS, the *current flow* in a TTL input or output lead is defined to be positive if the current actually flows *into* the lead, and negative if current flows *out* of the lead. As a result, when an output is connected to one or more inputs, the algebraic sum of all the input and output currents is 0.

The amount of current required by a TTL input depends on whether the input is **HIGH** or **LOW**, and is specified by two parameters:

I_{ILmax} The maximum current that an input requires to pull it **LOW**. Recall from the discussion of Figure 3-77 that positive current is actually flowing from V_{CC}, through *R1B*, through diode *D1XB*, out of the input lead, through the driving output transistor *Q5A*, and into ground.

Since current flows out of a TTL input in the **LOW** state, I_{ILmax} has a negative value. Most LS-TTL inputs have $I_{\text{ILmax}} = -0.4$ mA, which is sometimes called a *LOW-state unit load* for LS-TTL.

LOW-state unit load

I_{IHmax} The maximum current that an input requires to pull it HIGH. As shown in Figure 3-78 on page 159, positive current flows from V_{CC}, through *R5A* and *Q4A* of the driving gate, and *into* the driven input, where it leaks to ground through reverse-biased diodes *D1XB* and *D2XB*.

Since current flows *into* a TTL input in the HIGH state, I_{IHmax} has a positive value. Most LS-TTL inputs have $I_{\text{IHmax}} = 20\ \mu\text{A}$, which is sometimes called a *HIGH-state unit load* for LS-TTL.

HIGH-state unit load

Like CMOS outputs, TTL outputs can source or sink a certain amount of current depending on the state, HIGH or LOW:

I_{OLmax} The maximum current an output can sink in the LOW state while maintaining an output voltage no more than V_{OLmax}. Since current flows into the output, I_{OLmax} has a positive value, 8 mA for most LS-TTL outputs.

I_{OHmax} The maximum current an output can source in the HIGH state while maintaining an output voltage no less than V_{OHmin}. Since current flows out of the output, I_{OHmax} has a negative value, $-400\ \mu\text{A}$ for most LS-TTL outputs.

Notice that the value of I_{OLmax} for typical LS-TTL outputs is exactly 20 times the absolute value of I_{ILmax}. As a result, LS-TTL is said to have a *LOW-state fanout* of 20, because an output can drive up to 20 inputs in the LOW state. Similarly, the absolute value of I_{OHmax} is exactly 20 times I_{IHmax}, so LS-TTL is said to have a *HIGH-state fanout* of 20 also. The *overall fanout* is the lesser of the LOW- and HIGH-state fanouts.

LOW-state fanout

HIGH-state fanout
overall fanout

Loading a TTL output with more than its rated fanout has the same deleterious effects that were described for CMOS devices in Section 3.5.5 on page 110. That is, DC noise margins may be reduced or eliminated, transition times and delays may increase, and the device may overheat.

TTL OUTPUT ASYMMETRY   Although LS-TTL's numerical fanouts for HIGH and LOW states are equal, LS-TTL and other TTL families have a definite asymmetry in current driving capability—an LS-TTL output can sink 8 mA in the LOW state, but can source only 400 μA in the HIGH state.

This asymmetry is no problem when TTL outputs drive other TTL inputs, because it is matched by a corresponding asymmetry in TTL input current requirements (I_{ILmax} is large, while I_{IHmax} is small). However, it is a limitation when TTL is used to drive LEDs, relays, solenoids, or other devices requiring large amounts of current, often tens of milliamperes. Circuits using these devices must be designed so that current flows (and the driven device is "on") when the TTL output is in the LOW state, and so little or no current flows in the HIGH state. Special TTL buffer/driver gates are made that can sink up to 60 mA in the LOW state, but that still have a rather puny current-sourcing capability in the HIGH state (2.4 mA).

BURNED FINGERS If a TTL or CMOS output is forced to sink a lot more than I_{OLmax}, the device may be damaged, especially if high current is allowed to flow for more than a second or so. For example, suppose that a TTL output in the **LOW** state is short-circuited directly to the 5 V supply. The **ON** resistance, $R_{CE(sat)}$, of the saturated *Q5* transistor in a typical TTL output stage is less than 10 Ω. Thus, *Q5* must dissipate about $5^2/10$ or 2.5 watts. Don't try this yourself unless you're prepared to deal with the consequences! That's enough heat to destroy the device (and burn your finger) in a very short time.

In general, two calculations must be carried out to confirm that an output is not being overloaded:

HIGH state The I_{IHmax} values for all of the driven inputs are added. This sum must be less than or equal to the absolute value of I_{OHmax} for the driving output.

LOW state The I_{ILmax} values for all of the driven inputs are added. The absolute value of this sum must be less than or equal to I_{OLmax} for the driving output.

For example, suppose you designed a system in which a certain LS-TTL output drives ten LS-TTL and three S-TTL gate inputs. In the HIGH state, a total of $10 \cdot 20 + 3 \cdot 50\ \mu A = 350\ \mu A$ is required. This is within an LS-TTL output's HIGH-state current-sourcing capability of 400 μA. But in the LOW state, a total of $10 \cdot 0.4 + 3 \cdot 2.0\ \text{mA} = 10.0\ \text{mA}$ is required. This is more than an LS-TTL output's LOW-state current-sinking capability of 8 mA, so the output is overloaded.

3.10.4 Unused Inputs

Unused inputs of TTL gates can be handled in the same way as we described for CMOS gates in Section 3.5.6 on page 111. That is, unused inputs may be tied to used ones, or unused inputs may be pulled HIGH or LOW as is appropriate for the logic function.

The resistance value of a pull-up or pull-down resistor is more critical with TTL gates than CMOS gates, because TTL inputs draw significantly more current, especially in the LOW state. If the resistance is too large, the voltage drop across the resistor may result in a gate input voltage beyond the normal LOW or HIGH range.

For example, consider the pull-down resistor shown in Figure 3-80. The pull-down resistor must sink 0.4 mA of current from each of the unused LS-TTL inputs that it drives. Yet the voltage drop across the resistor must be no more than 0.5 V in order to have a LOW input voltage no worse than that produced by a normal gate output. If the resistor drives *n* LS-TTL inputs, then we must have

$$n \cdot 0.4\ \text{mA} \cdot R_{pd} < 0.5\ \text{V}$$

FLOATING TTL INPUTS	Analysis of the TTL input structure shows that unused inputs left unconnected (or *floating*) behave as if they have a HIGH voltage applied—they are pulled HIGH by base resistor *R1* in Figure 3-75 on page 157. However, *R1*'s pull-up is much weaker than that of a TTL output driving the input. As a result, a small amount of circuit noise, such as that produced by other gates when they switch, can make a floating input spuriously behave like it's LOW. Therefore, for the sake of reliability, unused TTL inputs should be tied to a stable HIGH or LOW voltage source.

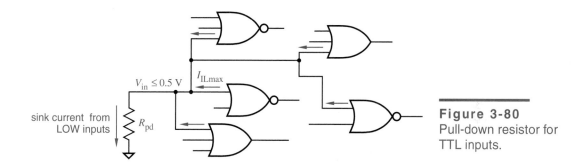

Figure 3-80
Pull-down resistor for TTL inputs.

Thus, if the resistor must pull 10 LS-TTL inputs LOW, then we must have $R_{pd} < 0.5 / (10 \cdot 4 \cdot 10^{-3})$, or $R_{pd} < 125\ \Omega$.

Similarly, consider the pull-up resistor shown in Figure 3-81. It must source 20 μA of current to each unused input while producing a HIGH voltage no worse than that produced by a normal gate output, 2.7 V. Therefore, the voltage drop across the resistor must be no more than 2.3 V; if *n* LS-TTL input are driven, we must have

$$n \cdot 20\ \mu A \cdot R_{pu} < 2.3\ V$$

Thus, if 10 LS-TTL inputs are pulled up, then $R_{pu} < 2.3 / (10 \cdot 20 \cdot 10^{-6})$, or $R_{pu} < 11.5\ K\Omega$.

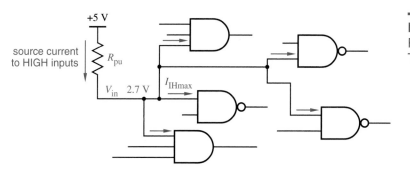

Figure 3-81
Pull-up resistor for TTL inputs.

**WHY USE A
RESISTOR?**

You might be asking yourself, "Why use a pull-up or pull-down resistor, when a direct connection to ground or the 5-V power supply should be a perfectly good source of LOW or HIGH?"

Well, for a HIGH source, a direct connection to the 5-V power supply is not recommended, since an input transient of over 5.5 V can damage some TTL devices, ones that use a multiple-emitter transistor in the input stage. The pull-up resistor limits current and prevents damage in this case.

For a LOW source, a direct connection to ground without the pull-down resistor is actually OK in most cases. You'll see many examples of this sort of connection throughout this book. However, as explained in Section 11.2.2 on page 904, the pull-down resistor is still desirable in some cases, so that the "constant" LOW signal it produces can be overridden and driven HIGH for system-testing purposes.

3.10.5 Additional TTL Gate Types

Although the NAND gate is the "workhorse" of the TTL family, other types of gates can be built with the same general circuit structure.

The circuit diagram for an LS-TTL NOR gate is shown in Figure 3-82. If either input X or Y is HIGH, the corresponding phase-splitter transistor $Q2X$ or $Q2Y$ is turned on, which turns off $Q3$ and $Q4$ while turning on $Q5$ and $Q6$, and the output is LOW. If both inputs are LOW, then both phase-splitter transistors are off, and the output is forced HIGH. This functional operation is summarized in Figure 3-83.

The LS-TTL NOR gate's input circuits, phase splitter, and output stage are almost identical to those of an LS-TTL NAND gate. The difference is that an LS-TTL NAND gate uses diodes to perform the AND function, while an LS-TTL NOR gate uses parallel transistors in the phase splitter to perform the OR function.

The speed, input, and output characteristics of a TTL NOR gate are comparable to those of a TTL NAND. However, an n-input NOR gate uses more transistors and resistors and is thus more expensive in silicon area than an n-input NAND. Also, internal leakage current limits the number of $Q2$ transistors that can be placed in parallel, so NOR gates have poor fan-in. (The largest discrete TTL NOR gate has only 5 inputs, compared with a 13-input NAND.) As a result, NOR gates are less commonly used than NAND gates in TTL designs.

The most "natural" TTL gates are inverting gates like NAND and NOR. Noninverting TTL gates include an extra inverting stage, typically between the input stage and the phase splitter. As a result, noninverting TTL gates are typically larger and slower than the inverting gates on which they are based.

Like CMOS, TTL gates can be designed with three-state outputs. Such gates have an "output-enable" or "output-disable" input that allows the output to be placed in a high-impedance state where neither output transistor is turned on.

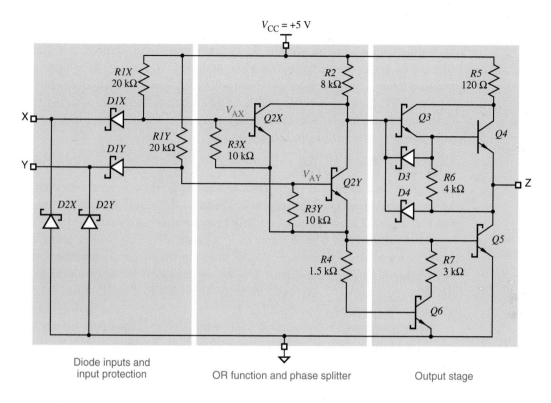

Figure 3-82 Circuit diagram of a two-input LS-TTL NOR gate.

(a)

X	Y	V_{AX}	Q2X	V_{AY}	Q2Y	Q3	Q4	Q5	Q6	V_Z	Z
L	L	≤ 1.05	off	≤ 1.05	off	on	on	off	off	≥ 2.7	H
L	H	≤ 1.05	off	1.2	on	off	off	on	on	≤ 0.35	L
H	L	1.2	on	≤ 1.05	off	off	off	on	on	≤ 0.35	L
H	H	1.2	on	1.2	on	off	off	on	on	≤ 0.35	L

Figure 3-83
Two-input LS-TTL
NOR gate:
(a) function table;
(b) truth table;
(c) logic symbol.

(b)

X	Y	Z
0	0	1
0	1	0
1	0	0
1	1	0

(c)

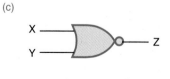

Some TTL gates are also available with *open-collector outputs*. Such gates omit the entire upper half of the output stage in Figure 3-75, so that only passive pull-up to the HIGH state is provided by an external resistor. The applications and required calculations for TTL open-collector gates are similar to those for CMOS gates with open-drain outputs.

open-collector output

3.11 TTL Families

TTL families have evolved over the years in response to the demands of digital designers for better performance. As a result, three TTL families have come and gone, and today's designers have five surviving families from which to choose. All of the TTL families are compatible in that they use the same power-supply voltage and logic levels, but each family has its own advantages in terms of speed, power consumption, and cost.

3.11.1 Early TTL Families

The original TTL family of logic gates was introduced by Sylvania in 1963. It was popularized by Texas Instruments, whose "7400-series" part numbers for gates and other TTL components quickly became an industry standard.

As in 7400-series CMOS, devices in a given TTL family have part numbers of the form 74FAMnn, where "FAM" is an alphabetic family mnemonic and nn is a numeric function designator. Devices in different families with the same value of nn perform the same function. In the original TTL family, "FAM" is null and the family is called *74-series TTL*.

74-series TTL

Resistor values in the original TTL circuit were changed to obtain two more TTL families with different performance characteristics. The *74H (High-speed TTL)* family used lower resistor values to reduce propagation delay at the expense of increased power consumption. The *74L (Low-power TTL)* family used higher resistor values to reduce power consumption at the expense of propagation delay.

74H (High-speed TTL)

74L (Low-power TTL)

The availability of three TTL families allowed digital designers in the 1970s to make a choice between high speed and low power consumption for their circuits. However, like many people in the 1970s, they wanted to "have it all, now." The development of Schottky transistors provided this opportunity and made 74, 74H, and 74L TTL obsolete. The characteristics of better-performing, contemporary TTL families are discussed in the rest of this section.

3.11.2 Schottky TTL Families

74S (Schottky TTL)

Historically, the first family to make use of Schottky transistors was *74S (Schottky TTL)*. With Schottky transistors and low resistor values, this family has much higher speed, but higher power consumption, than the original 74-series TTL.

74LS (Low-power Schottky TTL)

Perhaps the most widely used and certainly the least expensive TTL family is *74LS (Low-power Schottky TTL)*, introduced shortly after 74S. By combining Schottky transistors with higher resistor values, 74LS TTL matches the speed of 74-series TTL but has about one-fifth of its power consumption. Thus, 74LS is a preferred logic family for new TTL designs.

74AS (Advanced Schottky TTL)

Subsequent IC processing and circuit innovations gave rise to two more Schottky logic families. The *74AS (Advanced Schottky TTL)* family offers speeds approximately twice as fast as 74S with approximately the same power

Table 3-11 Characteristics of gates in TTL families.

Description	Symbol	Family 74S	74LS	74AS	74ALS	74F
Maximum propagation delay (ns)		3	9	1.7	4	3
Power consumption per gate (mW)		19	2	8	1.2	4
Speed-power product (pJ)		57	18	13.6	4.8	12
LOW-level input voltage (V)	V_{ILmax}	0.8	0.8	0.8	0.8	0.8
LOW-level output voltage (V)	V_{OLmax}	0.5	0.5	0.5	0.5	0.5
HIGH-level input voltage (V)	V_{IHmin}	2.0	2.0	2.0	2.0	2.0
HIGH-level output voltage (V)	V_{OHmin}	2.7	2.7	2.7	2.7	2.7
LOW-level input current (mA)	I_{ILmax}	−2.0	−0.4	−0.5	−0.2	−0.6
LOW-level output current (mA)	I_{OLmax}	20	8	20	8	20
HIGH-level input current (μA)	I_{IHmax}	50	20	20	20	20
HIGH-level output current (μA)	I_{OHmax}	−1000	−400	−2000	−400	−1000

consumption. The *74ALS (Advanced Low-power Schottky TTL)* family offers both lower power and higher speeds than 74LS, and rivals 74LS in popularity for general-purpose requirements in new TTL designs. The *74F (Fast TTL)* family is positioned between 74AS and 74ALS in the speed/power tradeoff, and is probably the most popular choice for high-speed requirements in new TTL designs.

74ALS (Advanced Low-power Schottky TTL)

74F (Fast TTL)

3.11.3 Characteristics of TTL Families

The important characteristics of contemporary TTL families are summarized in Table 3-11. The first two rows of the table list the propagation delay (in nanoseconds) and the power consumption (in milliwatts) of a typical 2-input NAND gate in each family.

One figure of merit of a logic family is its *speed-power product* listed in the third row of the table. As discussed previously, this is simply the product of the propagation delay and power consumption of a typical gate. The speed-power product measures a sort of efficiency—how much energy a logic gate uses to switch its output.

The remaining rows in Table 3-11 describe the input and output parameters of typical TTL gates in each of the families. Using this information, you can analyze the external behavior of TTL gates without knowing the details of the internal TTL circuit design. These parameters were defined and discussed in Sections 3.10.2 and 3.10.3. As always, the input and output characteristics of specific components may vary from the representative values given in Table 3-11, so you must always consult the manufacturer's data book when analyzing a real design.

3.11.4 A TTL Data Sheet

Table 3-12 shows the part of a typical manufacturer's data sheet for the 74LS00. The 54LS00 listed in the data sheet is identical to the 74LS00, except that it is specified to operate over the full "military" temperature and voltage range, and it costs more. Most TTL parts have corresponding 54-series (military) versions. Three sections of the data sheet are shown in the table:

recommended operating conditions

- *Recommended operating conditions* specify power-supply voltage, input-voltage ranges, DC output loading, and temperature values under which the device is normally operated.

electrical characteristics

- *Electrical characteristics* specify additional DC voltages and currents that are observed at the device inputs and output when it is operated under the recommended conditions:

  I_I Maximum input current for a very high HIGH input voltage.

  I_{OS} Output current with HIGH output shorted to ground.

  I_{CCH} Power-supply current when all outputs (on four NAND gates) are HIGH. (The number given is for the entire package, which contains four NAND gates, so the current per gate is one-fourth of the specified amount.)

  I_{CCL} Power-supply current when all outputs (on four NAND gates) are LOW.

switching characteristics

- *Switching characteristics* give maximum and typical propagation delays under "typical" operating conditions of $V_{CC} = 5$ V and $T_A = 25°C$. A conservative designer must increase these delays by 5%–10% to account for different power-supply voltages and temperatures, and even more under heavy loading conditions.

A fourth section is also included in the manufacturer's data book:

absolute maximum ratings

- *Absolute maximum ratings* indicate the worst-case conditions for operating or storing the device without damage.

A complete data book also shows test circuits that are used to measure the parameters when the device is manufactured, and graphs that show how the typical parameters vary with operating conditions such as power-supply voltage (V_{CC}), ambient temperature (T_A), and load (R_L, C_L).

Table 3-12 Typical manufacturer's data sheet for the 74LS00.

RECOMMENDED OPERATING CONDITIONS

Parameter	Description	SN54LS00			SN74LS00			Unit
		Min.	Nom.	Max.	Min.	Nom.	Max.	
V_{CC}	Supply voltage	4.5	5.0	5.5	4.75	5.0	5.25	V
V_{IH}	High-level input voltage	2.0			2.0			V
V_{IL}	Low-level input voltage			0.7			0.8	V
I_{OH}	High-level output current			−0.4			−0.4	mA
I_{OL}	Low-level output current			4			8	mA
T_A	Operating free-air temperature	−55		125	0		70	°C

ELECTRICAL CHARACTERISTICS OVER RECOMMENDED FREE-AIR TEMPERATURE RANGE

Parameter	Test Conditions[1]	SN54LS00			SN74LS00			Unit
		Min.	Typ.[2]	Max.	Min.	Typ.[2]	Max.	
V_{IK}	$V_{CC} = $ Min., $I_N = −18$ mA			−1.5			−1.5	V
V_{OH}	$V_{CC} = $ Min., $V_{IL} = $ Max., $I_{OH} = −0.4$ mA	2.5	3.4		2.7	3.4		V
V_{OL}	$V_{CC} = $ Min., $V_{IH} = 2.0$ V, $I_{OL} = 4$ mA		0.25	0.4		0.25	0.4	V
	$V_{CC} = $ Min., $V_{IH} = 2.0$ V, $I_{OL} = 8$ mA					0.35	0.5	V
I_I	$V_{CC} = $ Max., $V_I = 7.0$ V			0.1			0.1	mA
I_{IH}	$V_{CC} = $ Max. $V_I = 2.7$ V			20			20	μA
I_{IL}	$V_{CC} = $ Max. $V_I = 0.4$ V			−0.4			−0.4	mA
I_{IOS} [3]	$V_{CC} = $ Max.	−20		−100	−20		−100	mA
I_{CCH}	$V_{CC} = $ Max., $V_I = 0$ V		0.8	1.6		0.8	1.6	mA
I_{CCL}	$V_{CC} = $ Max., $V_I = 4.5$ V		2.4	4.4		2.4	4.4	mA

SWITCHING CHARACTERISTICS, $V_{CC} = 5.0$ V, $T_A = 25$°C

Parameter	From (Input)	To (Output)	Test Conditions	Min.	Typ.	Max.	Unit
t_{PLH}	A or B	Y	$R_L = 2$ kΩ, $C_L = 15$ pF		9	15	ns
t_{PHL}					10	15	ns

NOTES:
1. For conditions shown as Max. or Min., use appropriate value specified under Recommended Operating Conditions.
2. All typical values are at $V_{CC} = 5.0$ V, $T_A = 25$°C.
3. Not more than one output should be shorted at a time; duration of short-circuit should not exceed one second.

*3.12 CMOS/TTL Interfacing

A digital designer selects a "default" logic family to use in a system, based on general requirements of speed, power, cost, and so on. However, the designer may select devices from other families in some cases because of availability or other special requirements. (For example, not all 74LS part numbers are available in 74HCT, and vice versa.) Thus, it's important for a designer to understand the implications of connecting TTL outputs to CMOS inputs, and vice versa.

There are several factors to consider in TTL/CMOS interfacing, and the first is noise margin. The LOW-state DC noise margin depends on V_{OLmax} of the driving output and V_{ILmax} of the driven input, and equals $V_{ILmax} - V_{OLmax}$. Similarly, the HIGH-state DC noise margin equals $V_{OHmin} - V_{IHmin}$. Figure 3-84 shows the relevant numbers for TTL and CMOS families.

For example, the LOW-state DC noise margin of HC or HCT driving TTL is $0.8 - 0.33 = 0.47$ V, and the HIGH-state is $3.84 - 2.0 = 1.84$ V. On the other hand, the HIGH-state margin of TTL driving HC or VHC is $2.7 - 3.85 = -1.15$ V. In other words, TTL driving HC or AC doesn't work, unless the TTL HIGH output happens to be higher and the CMOS HIGH input threshold happens to be lower by a total of 1.15 V compared to their worst-case specs. To drive CMOS inputs properly from TTL outputs, the CMOS devices should be HCT, VHCT. or FCT rather than HC or VHC.

The next factor to consider is fanout. As with pure TTL (Section 3.10.3), a designer must sum the input current requirements of devices driven by an output and compare with the output's capabilities in both states. Fanout is not a problem when TTL drives CMOS, since CMOS inputs require almost no current in either state. On the other hand, TTL inputs, especially in the LOW state, require

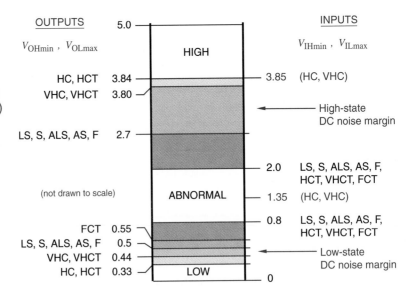

Figure 3-84
Output and input levels for interfacing TTL and CMOS families. (Note that HC and VHC inputs are not TTL compatible.)

substantial current, especially compared to HC and HCT output capabilities. For example, an HC or HCT output can drive 10 LS or only two S-TTL inputs.

The last factor is capacitive loading. We've seen that load capacitance increases both the delay and the power dissipation of logic circuits. Increases in delay are especially noticeable with HC and HCT outputs, whose transition times increase about 1 ns for each 5 pF of load capacitance. The transistors in FCT outputs have very low "on" resistances, so their transition times increase only about 0.1 ns for each 5 pF of load capacitance.

For a given load capacitance, power-supply voltage, and application, all of the CMOS families have similar dynamic power dissipation, since each variable in the CV^2f equation is the same. On the other hand, TTL outputs have somewhat lower dynamic power dissipation, since the voltage swing between TTL HIGH and LOW levels is smaller.

*3.13 Low-Voltage CMOS Logic and Interfacing

Two important factors have led the IC industry to move toward lower power-supply voltages in CMOS devices:

- In most applications, CMOS output voltages swing from rail to rail, so the V in the CV^2f equation is the power-supply voltage. Cutting power-supply voltage reduces dynamic power dissipation more than proportionally.

- As the industry moves toward ever-smaller transistor geometries, the oxide insulation between a CMOS transistor's gate and its source and drain is getting ever thinner, and thus incapable of insulating voltage potentials as "high" as 5 V.

As a result, JEDEC, an IC industry standards group, selected $3.3\,V \pm 0.3V$, $2.5\,V \pm 0.2V$, and $1.8\,V \pm 0.15V$ as the next "standard" logic power-supply voltages. JEDEC standards specify the input and output logic voltage levels for devices operating with these power-supply voltages.

The migration to lower voltages has occurred in stages and will continue to do so. For discrete logic families, the trend has been to produce parts that operate and produce outputs at the lower voltage but can also tolerate inputs at the higher voltage. This approach has allowed 3.3-V CMOS families to operate with 5-V CMOS and TTL families, as we'll see in the next section.

Many ASICs and microprocessors have followed a similar approach, but another approach is often used as well. These devices are large enough that it can make sense to provide them with two power-supply voltages. A low voltage, such as 2.5 V, is supplied to operate the chip's internal gates, or *core logic*. A *core logic* higher voltage, such as 3.3 V, is supplied to operate the external input and output circuits, or *pad ring*, for compatibility with older-generation devices in the *pad ring* system. Special buffer circuits are used internally to translate safely and quickly between the core-logic and the pad-ring logic voltages.

*3.13.1 3.3-V LVTTL and LVCMOS Logic

The relationships among signal levels for standard TTL and low-voltage CMOS devices operating at their nominal power-supply voltages are illustrated nicely in Figure 3-85, adapted from a Texas Instruments application note. The original, symmetric signal levels for pure 5-V CMOS families such as HC and VHC are shown in (a). TTL-compatible CMOS families such as HCT, VHCT, and FCT shift the voltage levels downward for compatibility with TTL as shown in (b).

The first step in the progression of lower CMOS power-supply voltages was 3.3 V. The JEDEC standard for 3.3-V logic actually defines two sets of levels. *LVCMOS (low-voltage CMOS)* levels are used in pure CMOS applications where outputs have light DC loads (less than 100 µA), so V_{OL} and V_{OH} are maintained within 0.2 V of the power-supply rails. *LVTTL (low-voltage TTL)* levels, shown in (c), are used in applications where outputs have significant DC loads, so V_{OL} can be as high as 0.4 V and V_{OH} can be as low as 2.4 V.

LVCMOS (low-voltage CMOS)

LVTTL (low-voltage TTL)

The positioning of TTL's logic levels at the low end of the 5-V range was really quite fortuitous. As shown in Figure 3-85(b) and (c), it was possible to define the LVTTL levels to match up with TTL levels exactly. Thus, an LVTTL output can drive a TTL input with no problem, as long as its output current specifications (I_{OLmax}, I_{OHmax}) are respected. Similarly, a TTL output can drive an LVTTL input, except for the problem of driving it beyond LVTTL's 3.3-V V_{CC}, as discussed next.

Figure 3-85 Comparison of logic levels: (a) 5-V CMOS; (b) 5-V TTL, including 5-V TTL-compatible CMOS; (c) 3.3-V LVTTL; (d) 2.5-V CMOS; (e) 1.8-V CMOS.

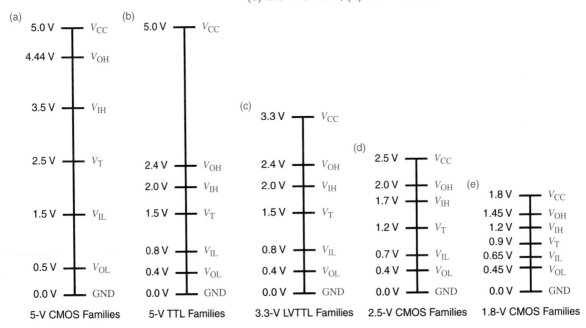

*3.13.2 5-V Tolerant Inputs

The inputs of a gate won't necessarily tolerate voltages greater than V_{CC}. This is a problem when both 5-V and 3.3-V logic families are used in a system. For example, 5-V CMOS devices easily produce 4.9-V outputs when lightly loaded, and both CMOS and TTL devices routinely produce 4.0-V outputs even when moderately loaded. The inputs of 3.3-V devices may not like these high voltages.

The maximum voltage V_{Imax} that can be tolerated by an input is listed in the "absolute maximum ratings" section of the manufacturer's data sheet. For HC devices, V_{Imax} equals V_{CC}. Thus, if an HC device is powered by a 3.3-V supply, its inputs cannot be driven by any 5-V CMOS or TTL outputs. For VHC devices, on the other hand, V_{Imax} is 7 V; thus, VHC devices with a 3.3-V power supply may be used to convert 5-V outputs to 3.3-V levels for use with 3.3-V microprocessors, memories, and other devices in a pure 3.3-V subsystem.

Figure 3-86 explains why some inputs are 5-V tolerant and others are not. As shown in (a), the HC and HCT input structure actually contains two reverse-biased *clamp diodes*, which we haven't shown before, between each input signal and V_{CC} and ground. The purpose of these diodes is specifically to shunt any transient input signal value less than 0 through *D1* or greater than V_{CC} through *D2* to the corresponding power-supply rail. Such transients can occur as a result of transmission-line reflections, as described in Section 11.4. Shunting the so-called "undershoot" or "overshoot" to ground or V_{CC} reduces the magnitude and duration of reflections.

clamp diode

Of course, diode *D2* can't distinguish between transient overshoot and a persistent input voltage greater than V_{CC}. Hence, if a 5-V output is connected to one of these inputs, it will not see the very high impedance normally associated with a CMOS input. Instead, it will see a relatively low impedance path to V_{CC} through the now forward-biased diode *D2*, and excessive current will flow.

Figure 3-86(b) shows a 5-V tolerant CMOS input. This input structure simply omits *D2*; diode *D1* is still provided to clamp undershoot. The VHC and AHC families use this input structure.

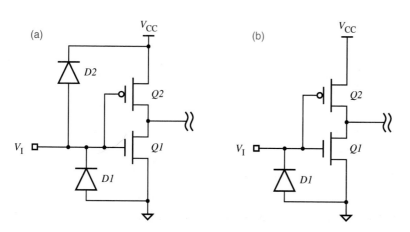

(a)

(b)

Figure 3-86
CMOS input structures:
(a) non-5-V tolerant HC;
(b) 5-V tolerant VHC.

The kind of input structure shown in Figure 3-86(b) is necessary but not sufficient to create 5-V tolerant inputs. The transistors in a device's particular fabrication process must also be able to withstand voltage potentials higher than V_{CC}. On this basis, V_{Imax} in the VHC family is limited to 7.0 V. In many 3.3-V ASIC processes, it's not possible to get 5-V tolerant inputs, even if you're willing to give up the transmission-line benefits of diode $D2$.

*3.13.3 5-V Tolerant Outputs

Five-volt tolerance must also be considered for outputs, in particular, when both 3.3-V and 5-V three-state outputs are connected to a bus. When the 3.3-V output is in the disabled, Hi-Z state, a 5-V device may be driving the bus, and a 5-V signal may appear on the 3.3-V device's *output*.

In this situation, Figure 3-87 explains why some outputs are 5-V tolerant and others are not. As shown in (a), the standard CMOS three-state output has an *n*-channel transistor $Q1$ to ground and a *p*-channel transistor $Q2$ to V_{CC}. When the output is disabled, circuitry (not shown) holds the gate of $Q1$ near 0 V, and the gate of $Q2$ near V_{CC}, so both transistors are off and Y is Hi-Z.

Now consider what happens if V_{CC} is 3.3 V and a different device applies a 5-V signal to the output pin Y in (a). Then the drain of $Q2$ (Y) is at 5 V while the gate (V_2) is still at only 3.3 V. With the gate at a lower potential than the drain, $Q2$ will begin to conduct and provide a relatively low-impedance path from Y to V_{CC}, and excessive current will flow. Both HC and VHC three-state outputs have this structure and therefore are not 5-V tolerant.

Figure 3-87(b) shows a 5-V tolerant output structure. An extra *p*-channel transistor $Q3$ is used to prevent $Q2$ from turning on when it shouldn't. When V_{OUT} is greater than V_{CC}, $Q3$ turns on. This forms a relatively low impedance path from Y to the gate of $Q2$, which now stays off because its gate voltage V_2 can no longer be below the drain voltage. This output structure is used in Texas Instruments' LVC (Low-Voltage CMOS) family.

Figure 3-87
CMOS three-state output structures:
(a) non-5-V tolerant HC and VHC;
(b) 5-V tolerant LVC.

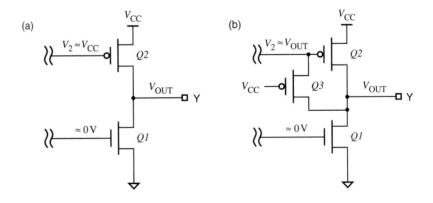

*3.13.4 TTL/LVTTL Interfacing Summary

Based on the information in the preceding subsections, TTL (5-V) and LVTTL (3.3-V) devices can be mixed in the same system subject to just three rules:

1. LVTTL outputs can drive TTL inputs directly, subject to the usual constraints on output current (I_{OLmax}, I_{OHmax}) of the driving devices.

2. TTL outputs can drive LVTTL inputs if the inputs are 5-V tolerant.

3. TTL and LVTTL three-state outputs can drive the same bus if the LVTTL outputs are 5-V tolerant.

*3.13.5 2.5-V and 1.8-V Logic

The transition from 3.3-V to 2.5-V logic will not be so easy. It is true that 3.3-V outputs can drive 2.5-V inputs as long as the inputs are 3.3-V tolerant. However, a quick look at Figure 3-85(c) and (d) on page 172 shows that V_{OH} of a 2.5-V output equals V_{IH} of a 3.3-V input. In other words, there is zero HIGH-state DC noise margin when a 2.5-V output drives a 3.3-V input—not a good situation.

The solution to this problem is to use a *level translator* or *level shifter*, a device which is powered by both supply voltages and which internally boosts the lower logic levels (2.5 V) to the higher ones (3.3 V). Many of today's ASICs and microprocessors contain level translators internally, allowing them to operate with a 2.5-V or 2.7-V core and a 3.3-V pad ring, as we discussed at the beginning of this section. If and when 2.5-V discrete devices become popular, we can expect the major semiconductor vendors to produce level translators as stand-alone components as well.

level translator
level shifter

The next step will be a transition from 2.5-V to 1.8-V logic. Referring to Figure 3-85(d) and (e), you can see that the HIGH-state DC noise margin is actually negative when a 1.8-V output drives a 2.5-V input, so level translators will be needed in this case also.

*3.14 Emitter-Coupled Logic

The key to reducing propagation delay in a bipolar logic family is to prevent a gate's transistors from saturating. In Section 3.9.5, we learned how Schottky diodes prevent saturation in TTL gates. However, it is also possible to prevent saturation by using a radically different circuit structure, called *current-mode logic (CML)* or *emitter-coupled logic (ECL)*.

current-mode logic (CML)

emitter-coupled logic (ECL)

Unlike the other logic families in this chapter, CML does not produce a large voltage swing between the LOW and HIGH levels. Instead, it has a small voltage swing, less than a volt, and it internally switches current between two possible paths, depending on the output state.

The first CML logic family was introduced by General Electric in 1961. The concept was soon refined by Motorola and others to produce the still popular 10K and 100K *emitter-coupled logic (ECL)* families. These families are

emitter-coupled logic (ECL)

extremely fast, offering propagation delays as short as 1 ns. The newest ECL family, ECLinPS (literally, ECL in picoseconds), offers maximum delays under 0.5 ns (500 ps), including the signal delay getting on and off of the IC package. Throughout the evolution of digital circuit technology, some type of ECL has always been the fastest technology for discrete, packaged logic components.

Still, commercial ECL families aren't nearly as popular as CMOS and TTL, mainly because they consume much more power. In fact, high power consumption made the design of ECL supercomputers, such as the Cray-1 and Cray-2, as much of a challenge in cooling technology as in digital design. Also, ECL has a poor speed-power product, does not provide a high level of integration, has fast edge rates requiring design for transmission-line effects in most applications, and is not directly compatible with TTL and CMOS. Nevertheless, ECL still finds its place as a logic and interface technology in very high-speed communications gear, including fiber-optic transceiver interfaces for gigabit Ethernet and Asynchronous Transfer Mode (ATM) networks.

*3.14.1 Basic CML Circuit

differential amplifier

The basic idea of current-mode logic is illustrated by the inverter/buffer circuit in Figure 3-88. This circuit has both an inverting output (OUT1) and a noninverting output (OUT2). Two transistors are connected as a *differential amplifier* with a common emitter resistor. The supply voltages for this example are $V_{CC} = 5.0$, $V_{BB} = 4.0$, and $V_{EE} = 0$ V, and the input LOW and HIGH levels are defined to be 3.6 and 4.4 V. This circuit actually produces output LOW and HIGH levels that are 0.6 V higher (4.2 and 5.0 V), but this is corrected in real ECL circuits.

Figure 3-88
Basic CML inverter/buffer circuit with input HIGH.

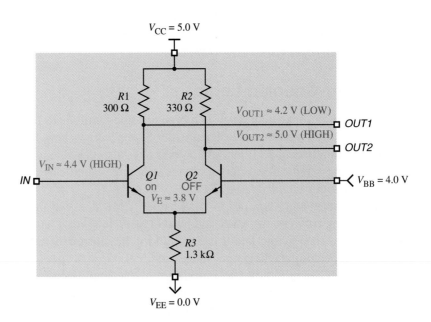

When V_{IN} is HIGH, as shown in the figure, transistor $Q1$ is on, but not saturated, and transistor $Q2$ is OFF. This is true because of a careful choice of resistor values and voltage levels. Thus, V_{OUT2} is pulled to 5.0 V (HIGH) through $R2$, and it can be shown that the voltage drop across $R1$ is about 0.8 V, so that V_{OUT1} is about 4.2 V (LOW).

When V_{IN} is LOW, as shown in Figure 3-89, transistor $Q2$ is on, but not saturated, and transistor $Q1$ is OFF. Thus, V_{OUT1} is pulled to 5.0 V through $R1$, and it can be shown that V_{OUT2} is about 4.2 V.

The outputs of this inverter are called *differential outputs* because they are always complementary, and it is possible to determine the output state by looking at the difference between the output voltages ($V_{OUT1} - V_{OUT2}$) rather than their absolute values. That is, the output is 1 if ($V_{OUT1} - V_{OUT2}$) > 0, and it is 0 if ($V_{OUT1} - V_{OUT2}$) > 0. It is possible to build input circuits with two wires per logical input that define the logical signal value in this way; these are called *differential inputs*.

differential outputs

differential inputs

Differential signals are used in most ECL "interfacing" and "clock distribution" applications because of their low skew and high noise immunity. They are "low skew" because the timing of a 0-to-1 or 1-to-0 transition does not depend critically on voltage thresholds, which may change with temperature or between devices. Instead, the timing depends only on when the voltages cross over relative to each other. Similarly, the "relative" definition of 0 and 1 provides outstanding noise immunity, since noise created by variations in the power supply or coupled from external sources tends to be a *common-mode signal* that affect both differential signals similarly, leaving the difference value unchanged.

common-mode signal

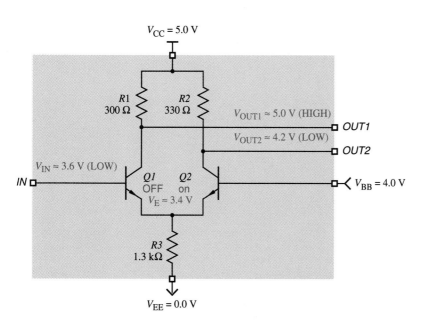

Figure 3-89
Basic CML inverter/buffer circuit with input LOW.

It is also possible, of course, to determine the logic value by sensing the absolute voltage level of one input signal, called a *single-ended input*. Single-ended signals are used in most ECL "logic" applications to avoid the obvious expense of doubling the number of signal lines. The basic CML inverter in Figure 3-89 has a single-ended input. It always has both "outputs" available internally; the circuit is actually either an inverter or a noninverting buffer, depending on whether we use OUT1 or OUT2.

To perform logic with the basic circuit of Figure 3-89, we simply place additional transistors in parallel with $Q1$, similar to the approach in a TTL NOR gate. For example, Figure 3-90 shows a 2-input CML OR/NOR gate. If any input is HIGH, the corresponding input transistor is active, and V_{OUT1} is LOW (NOR output). At the same time, $Q3$ is OFF, and V_{OUT2} is HIGH (OR output).

Recall that the input levels for the inverter/buffer are defined to be 3.6 and 4.4 V, while the output levels that it produces are 4.2 and 5.0 V. This is obviously

Figure 3-90 CML 2-input OR/NOR gate: (a) circuit diagram; (b) function table; (c) logic symbol; (d) truth table.

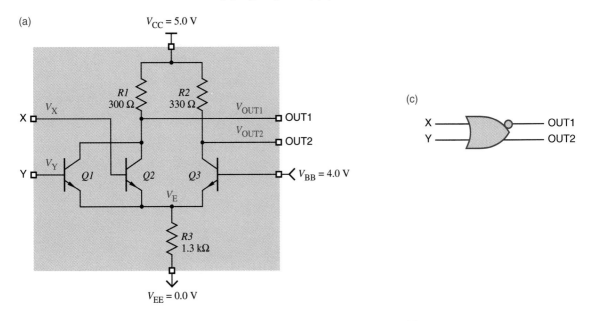

X	Y	V_X	V_Y	$Q1$	$Q2$	$Q3$	V_E	V_{OUT1}	V_{OUT2}	OUT1	OUT2
L	L	3.6	3.6	OFF	OFF	on	3.4	5.0	4.2	H	L
L	H	3.6	4.4	OFF	on	OFF	3.8	4.2	5.0	L	H
H	L	4.4	3.6	on	OFF	OFF	3.8	4.2	5.0	L	H
H	H	4.4	4.4	on	on	OFF	3.8	4.2	5.0	L	H

X	Y	OUT1	OUT2
0	0	1	0
0	1	0	1
1	0	0	1
1	1	0	1

a problem. We could put a diode in series with each output to lower it by 0.6 V to match the input levels, but that still leaves another problem—the outputs have poor fanout. A HIGH output must supply base current to the inputs that it drives, and this current creates an additional voltage drop across $R1$ or $R2$, reducing the output voltage (and we don't have much margin to work with). These problems are solved in commercial ECL families, such as the 10K family described next.

*3.14.2 ECL 10K/10H Families

The packaged components in today's most popular ECL family have 5-digit part numbers of the form "10xxx" (e.g., 10102, 10181, 10209), so the family is generically called *ECL 10K*. This family has several improvements over the basic CML circuit described previously:

ECL 10K family

- An emitter-follower output stage shifts the output levels to match the input levels and provides very high current-driving capability, up to 50 mA per output. It is also responsible for the family's name, "emitter-coupled" logic.

- An internal bias network provides V_{BB} without the need for a separate, external power supply.

- The family is designed to operate with $V_{CC} = 0$ (ground) and $V_{EE} = -5.2$ V. In most applications, ground signals are more noise-free than the power-supply signals. In ECL, the logic signals are referenced to the algebraically higher power-supply voltage rail, so the family's designers decided to make that 0 V (the "clean" ground) and use a negative voltage for V_{EE}. The power-supply noise that does appear on V_{EE} is a "common-mode" signal that is rejected by the input structure's differential amplifier.

- Parts with a 10H prefix (the *ECL 10H family*) are fully voltage compensated, so they will work properly with power-supply voltages other than $V_{EE} = -5.2$ V, as we'll discuss in Section 3.14.4.

ECL 10H family

Logic LOW and HIGH levels are defined in the ECL 10K family as shown in Figure 3-91. Note that even though the power supply is negative, ECL assigns the names LOW and HIGH to the *algebraically* lower and higher voltages, respectively.

DC noise margins in ECL 10K are much less than in CMOS and TTL, only 0.155 V in the LOW state and 0.125 V in the HIGH state. However, ECL gates do not need as much noise margin as these families. Unlike CMOS and TTL, an ECL gate generates very little power-supply and ground noise when it changes state; its current requirement remains constant as it merely steers current from one path to another. Also, ECL's emitter-follower outputs have very low impedance in either state, and it is difficult to couple noise from an external source into a signal line driven by such a low-impedance output.

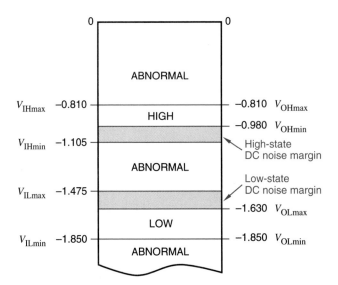

Figure 3-91
ECL 10K logic levels.

Figure 3-92(a) is the circuit for an ECL OR/NOR gate, one section of a quad OR/NOR gate with part number 10102. A pull-down resistor on each input ensures that if the input is left unconnected, it is treated as LOW. The bias network has component values selected to generate $V_{BB} = -1.29$ V for proper operation of the differential amplifier. Each output transistor, using the emitter-follower configuration, maintains its emitter voltage at one diode-drop below its base voltage, thereby achieving the required output-level shift. Figure 3-92(b) summarizes the electrical operation of the gate.

The emitter-follower outputs used in ECL 10K require external pull-down resistors, as shown in the figure. The 10K family is designed to use external rather than internal pull-down resistors for good reason. The rise and fall times of ECL output transitions are so fast (typically 2 ns) that any connection longer than a few inches must be treated as a transmission line and must be terminated as discussed in Section 11.4. Rather than waste power with an internal pull-down resistor, ECL 10K allows the designer to select an external resistor that satisfies both pull-down and transmission-line termination requirements. The simplest termination, sufficient for short connections, is to connect a resistor in the range of 270 Ω to 2 kΩ from each output to V_{EE}.

A typical ECL 10K gate has a propagation delay of 2 ns, comparable to 74AS TTL. With its outputs left unconnected, a 10K gate consumes about 26 mW of power, also comparable to a 74AS TTL gate, which consumes about 20 mW. However, the termination required by ECL 10K also consumes power, from 10 to 150 mW per output depending on the type of termination circuit. A 74AS TTL output may or may not require a power-consuming termination circuit, depending on the physical characteristics of the application.

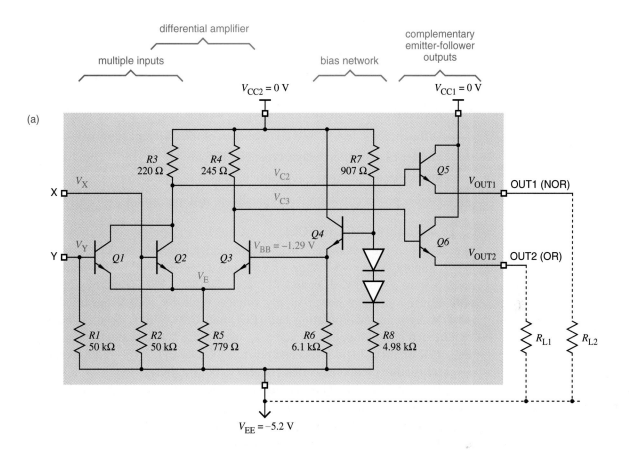

(a)

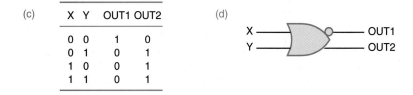

(b)

X Y	V_X	V_Y	Q1	Q2	Q3	V_E	V_{C2}	V_{C3}	V_{OUT1}	V_{OUT2}	OUT1	OUT2
L L	−1.8	−1.8	OFF	OFF	on	−1.9	−0.2	−1.2	−0.9	−1.8	H	L
L H	−1.8	−0.9	OFF	on	OFF	−1.5	−1.2	−0.2	−1.8	−0.9	L	H
H L	−0.9	−1.8	on	OFF	OFF	−1.5	−1.2	−0.2	−1.8	−0.9	L	H
H H	−0.9	−0.9	on	on	OFF	−1.5	−1.2	−0.2	−1.8	−0.9	L	H

(c)

X Y	OUT1	OUT2
0 0	1	0
0 1	0	1
1 0	0	1
1 1	0	1

(d)

Figure 3-92 Two-input 10K ECL OR/NOR gate: (a) circuit diagram;
(b) function table; (c) truth table; (d) logic symbol.

*3.14.3 ECL 100K Family

ECL 100K family

Members of the *ECL 100K family* have 6-digit part numbers of the form "100xxx" (e.g., 100101, 100117, 100170), but in general their functions are different from those of 10K parts with similar numbers. The 100K family has the following major differences from the 10K family:

- Reduced power-supply voltage, $V_{EE} = -4.5$ V.
- Different logic levels, as a consequence of the different supply voltage.
- Shorter propagation delays, typically 0.75 ns.
- Shorter transition times, typically 0.70 ns.
- Higher power consumption, typically 40 mW per gate.

*3.14.4 Positive ECL (PECL)

We described the advantage of noise immunity provided by ECL's negative power supply ($V_{EE} = -5.2$ V or -4.5 V), but there's also a big disadvantage—today's most popular CMOS and TTL logic families, ASICs, and microprocessors all use a positive power-supply voltage, typically +5.0 V but trending to +3.3 V. Systems incorporating both ECL and CMOS/TTL devices therefore require two power supplies. In addition, interfacing between standard, negative ECL 10K or 100K logic levels and positive CMOS/TTL levels requires special level-translation components that connect to both supplies.

positive ECL (PECL)

Positive ECL (*PECL*, pronounced "peckle") uses a standard +5.0-V power supply. Note that there's nothing in the ECL 10K circuit design of Figure 3-92 that requires V_{CC} to be grounded and V_{EE} to be connected to a -5.2-V supply. The circuit will function exactly the same with V_{EE} connected to ground, and V_{CC} to a +5.2-V supply.

Thus, PECL components are nothing more than standard ECL components with V_{EE} connected to ground and V_{CC} to a +5.0-V supply. The voltage between V_{EE} and V_{CC} is a little less than with standard 10K ECL and more than with standard 100K ECL, but the 10H-series and 100K parts are voltage compensated, designed to still work well with the supply voltage being a little high or low.

Like ECL logic levels, PECL levels are referenced to V_{CC}, so the PECL HIGH level is about $V_{CC} - 0.9$ V, and LOW is about $V_{CC} - 1.7$ V, or about 4.1 V and 3.3 V with a nominal 5-V V_{CC}. Since these levels are referenced to V_{CC}, they move up and down with any variations in V_{CC}. Thus, PECL designs require particularly close attention to power-distribution issues, to prevent noise on V_{CC} from corrupting the logic levels transmitted and received by PECL devices.

Recall that CML/ECL devices produce differential outputs and can have differential inputs. A differential input is relatively insensitive to the absolute voltage levels of an input-signal pair, and sensitive only to their difference. Therefore, differential signals can be used quite effectively in PECL applications to ease the noise concerns raised in the preceding paragraph.

It is also quite common to provide differential PECL-compatible inputs and outputs on CMOS devices, allowing a direct interface between the CMOS device and a device such as a fiber-optic transceiver that expects ECL or PECL levels. In fact, as CMOS circuits have migrated to 3.3-V power supplies, it has even been possible to build PECL-like differential inputs and outputs that are simply referenced to the 3.3-V supply instead of a 5-V supply.

References

Students who need to study the basics may wish to consult "Electrical Circuits Review" by Bruce M. Fleischer. This 20-page tutorial covers all of the basic circuit concepts that are used in this chapter. It appears both as an appendix in this book's first edition and as a .pdf file on this book's Web page, www.ddpp.com.

After seeing the results of last few decades' amazing pace of development in digital electronics, it's easy to forget that logic circuits had an important place in technologies that came before the transistor. In Chapter 5 of *Introduction to the Methodology of Switching Circuits* (Van Nostrand, 1972), George J. Klir shows how logic can be (and has been) performed by a variety of physical devices, including relays, vacuum tubes, and pneumatic systems.

If you're interested in more history, a nice introduction to all of the early bipolar logic families can be found in *Logic Design with Integrated Circuits* by William E. Wickes (Wiley-Interscience, 1968). The classic introduction to TTL electrical characteristics appeared in *The TTL Applications Handbook,* edited by Peter Alfke and Ib Larsen (Fairchild Semiconductor, 1973). Early logic designers also enjoyed *The TTL Cookbook* by Don Lancaster.

For another perspective on the electronics material in this chapter, you can consult almost any modern electronics text. Many contain a much more analytical discussion of digital circuit operation; for example, see *Microelectronics* by J. Millman and A. Grabel (McGraw-Hill, 1987, second edition). Another good introduction to ICs and important logic families can be found in *VLSI System Design* by Saburo Muroga (Wiley, 1982). For NMOS and CMOS circuits in particular, two good books are *Introduction to VLSI Systems* by Carver Mead and Lynn Conway (Addison-Wesley, 1980) and *Principles of CMOS VLSI Design* by Neil H. E. Weste and Kamran Eshraghian (Addison-Wesley, 1993).

A light-hearted and very readable introduction to digital circuits can be found in Clive Maxfield's *Bebop to the Boolean Boogie* (LLH Technology Publishing, www.llh-publishing.com). Some people think that the seafood gumbo recipe in Appendix H is alone worth the price! Even without the recipe, the book is a well-illustrated classic that guides you through the basics of digital electronics fundamentals, components, and processes.

A sound understanding of the electrical aspects of digital circuit operation is mandatory for successful high-speed circuit design. Unquestionably the best book on this subject is *High-Speed Digital Design: A Handbook of Black Magic,*

Table 3-13 Manufacturers' logic data books.

Manufacturer	Order Number	Topics	Title	Year
Texas Instruments	SDLD001	74, 74S, 74LS TTL	*TTL Logic Data Book*	1988
Texas Instruments	SDAD001C	74AS, 74ALS TTL	*ALS/AS Logic Data Book*	1995
Texas Instruments	SDFD001B	74F TTL	*F Logic Data Book*	1994
Texas Instruments	SCLD001D	74HC, 74HCT CMOS	*HC/HCT Logic Data Book*	1997
Texas Instruments	SCAD001D	74AC, 74ACT CMOS	*AC/ACT Logic Data Book*	1997
Texas Instruments	SCLD003A	74AHC, 74AHCT CMOS	*AHC/AHCT Logic Data Book*	1997
Motorola	DL121/D	74F, 74LSTTL	*Fast and LSTTL Data*	1989
Motorola	DL129/D	74HC, 74HCT	*High-Speed CMOS Data*	1988
Motorola	DL138/D	74AC, 74ACT	*FACT Data*	1988
Motorola	DL122/D	10K ECL	*MECL Device Data*	1989

by Howard Johnson and Martin Graham (Prentice Hall, 1993). It combines solid electronics principles with tremendous insight and experience in the design of practical digital systems.

Characteristics of today's logic families can be found in the data books published by the device manufacturers. Both Texas Instruments and Motorola publish comprehensive data books for TTL and CMOS devices, as listed in Table 3-13. Both manufacturers keep up-to-date versions of their data books on the Web, at www.ti.com and www.mot.com. Motorola also provides a nice introduction to ECL design in the *MECL System Design Handbook* (publ. HB205, rev. 1, 1988).

The JEDEC (Joint Electron Device Engineering Council) standards for digital logic levels can be found on JEDEC's Web site, www.jedec.org. The JEDEC standards for 3.3-V, 2.5-V, and 1.8-V logic were published in 1994, 1995, and 1997, respectively.

Drill Problems

3.1   A particular logic family defines a LOW signal to be in the range 0.0–0.8 V and a HIGH signal to be in the range 2.0–3.3 V. Under a positive-logic convention, indicate the logic value associated with each of the following signal levels:

   (a)   0.0 V   (b)   3.0 V   (c)   0.8 V   (d)   1.9 V

   (e)   2.0 V   (f)   5.0 V   (g)   −0.7 V   (h)   −3.0 V

3.2   Repeat Drill 3.1 using a negative-logic convention.

3.3   Discuss how a logic buffer amplifier is different from an audio amplifier.

3.4   Is a buffer amplifier equivalent to a 1-input AND gate or a 1-input OR gate?

3.5   True or false: For a given set of input values, a NAND gate produces the opposite output as a NOR gate.

3.6   Write two completely different definitions of "gate" used in this chapter.

3.7   What kind of transistors are used in CMOS gates?

3.8   (Hobbyists only.) Draw an equivalent circuit for a CMOS inverter using a single-pole, double-throw relay.

3.9   For a given silicon area, which is likely to be faster, a CMOS NAND gate or a CMOS NOR?

3.10   Define "fan-in" and "fanout." Which one are you likely to have to calculate?

3.11   Draw the circuit diagram, function table, and logic symbol for a 3-input CMOS NOR gate in the style of Figure 3-16.

3.12   Draw switch models in the style of Figure 3-14 for a 2-input CMOS NOR gate for all four input combinations.

3.13   Draw a circuit diagram, function table, and logic symbol for a CMOS OR gate in the style of Figure 3-19.

3.14   Which has fewer transistors, a CMOS inverting gate or a noninverting gate?

3.15   Name and draw the logic symbols of four different 4-input CMOS gates that each use 8 transistors.

3.16   The circuit in Figure X3.16(a) is a type of CMOS AND-OR-INVERT gate. Write a function table for this circuit in the style of Figure 3-15(b), and a corresponding logic diagram using AND and OR gates and inverters.

3.17   The circuit in Figure X3.16(b) is a type of CMOS OR-AND-INVERT gate. Write function table for this circuit in the style of Figure 3-15(b), and a corresponding logic diagram using AND and OR gates and inverters.

3.18   How is it that perfume can be bad for digital designers?

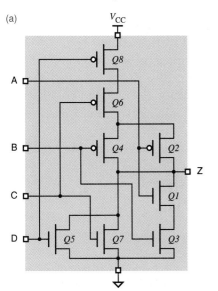 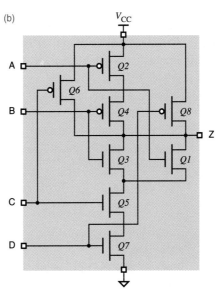

Figure X3.16

3.19   How much high-state DC noise margin is available in a CMOS inverter whose transfer characteristic under worst-case conditions looks like Figure 3-25? How much low-state DC noise margin is available? (Assume standard 1.5-V and 3.5-V thresholds for LOW and HIGH.)

3.20   Using the data sheet in Table 3-3, determine the worst-case LOW-state and HIGH-state DC noise margins of the 74HC00. State any assumptions required by your answer.

3.21   Section 3.5 defines seven different electrical parameters for CMOS circuits. Using the data sheet in Table 3-3, determine the worst-case value of each of these for the 74HC00. State any assumptions required by your answer.

3.22   Based on the conventions and definitions in Section 3.4, if the current at a device output is specified as a negative number, is the output sourcing current or sinking current?

3.23   For each of the following resistive loads, determine whether the output drive specifications of the 74HC00 over the commercial operating range are exceeded. (Refer to Table 3-3, and use $V_{OHmin} = 3.84$ V and $V_{CC} = 5.0$ V.)

(a)   120 Ω to V_{CC} (b)   270 Ω to V_{CC} and 330 Ω to GND

(c)   1 KΩ to GND   (d)   150 Ω to V_{CC} and 150 Ω to GND

(e)   100 Ω to V_{CC} (f)   75 Ω to V_{CC} and 150 Ω to GND

(g)   75 Ω to V_{CC} (h)   270 Ω to V_{CC} and 150 Ω to GND

3.24   Across the range of valid HIGH input levels, 2.0–5.0 V, at what input level would you expect the 74HC00 (Table 3-3) to consume the most power?

3.25   Determine the LOW-state and HIGH-state DC fanout of the 74HC00 when it drives 74LS00-like inputs. (Refer to Tables 3-3 and 3-12.)

3.26   Estimate the "on" resistances of the p-channel and n-channel output transistors of the 74HC00 using information in Table 3-3.

3.27   Under what circumstances is it safe to allow an unused CMOS input to float?

3.28   Explain "latch up" and the circumstances under which it occurs.

3.29   Explain why putting all the decoupling capacitors in one corner of a printed-circuit board is not a good idea.

3.30   When is it important to hold hands with a friend?

3.31   Name the two components of CMOS logic gate's delay. Which one is most affected by load capacitance?

3.32   Determine the RC time constant for each of the following resistor-capacitor combinations:

(a) $R = 100$ Ω, $C = 50$ pF   (b) $R = 330$ Ω, $C = 150$ pF

(c) $R = 1$ KΩ, $C = 30$ pF   (d) $R = 4.7$ KΩ, $C = 100$ pF

3.33   Besides delay, what other characteristic(s) of a CMOS circuit are affected by load capacitance?

3.34   Explain why the number of CMOS inputs connected to the output of a CMOS gate generally is not limited by DC fanout considerations.

3.35   It is possible to operate 74VHC CMOS devices with a 3.3-volt power supply. How much power does this typically save, compared to 5-volt operation?

3.36   A particular Schmitt-trigger inverter has $V_{ILmax} = 0.8$ V, $V_{IHmin} = 2.0$ V, $V_{T+} = 1.6$ V, and $V_{T-} = 1.3$ V. How much hysteresis does it have?

3.37   Why are three-state outputs usually designed to turn off faster than they turn on?

3.38   Discuss the pros and cons of larger vs. smaller pull-up resistors for open-drain CMOS outputs or open-collector TTL outputs.

3.39   A particular LED has a voltage drop of about 2.0 V in the "on" state and requires about 5 mA of current for normal brightness. Determine an appropriate value for the pull-up resistor when the LED is connected as shown in Figure 3-52.

3.40   How does the answer for Drill 3.39 change if the LED is connected as shown in Figure 3-53(a)?

3.41   A wired-AND function is obtained simply by tying two open-drain or open-collector outputs together, without going through another level of transistor circuitry. How is it, then, that a wired-AND function can actually be slower than a discrete AND gate? (*Hint:* Recall the title of a Teenage Mutant Ninja Turtles movie.)

3.42   Which CMOS or TTL logic family in this chapter has the strongest output driving capability?

3.43   Concisely summarize the difference between HC and HCT logic families. The same concise statement should apply to AC versus ACT.

3.44   Why don't the specifications for FCT devices include parameters like V_{OLmaxC} that apply to CMOS loads, as HCT and ACT specifications do?

3.45   How do FCT-T devices reduce power consumption compared to FCT devices?

3.46   How many diodes are required for an n-input diode AND gate?

3.47   True or false: A TTL NOR gate uses diode logic.

3.48   Are TTL outputs more capable of sinking current or sourcing current?

3.49   Compute the maximum fanout for each of the following cases of a TTL output driving multiple TTL inputs. Also indicate how much "excess" driving capability is available in the LOW or HIGH state for each case.

   (a)   74LS driving 74LS   (b)   74LS driving 74S

   (c)   74S driving 74AS   (d)   74F driving 74S

   (e)   74AS driving 74AS   (f)   74AS driving 74F

   (g)   74ALS driving 74F   (h)   74AS driving 74ALS

3.50   Which resistor dissipates more power, the pull-down for an unused LS-TTL NOR-gate input, or the pull-up for an unused LS-TTL NAND-gate input? Use the maximum allowable resistor value in each case.

3.51   Which would you expect to be faster, a TTL AND gate or a TTL AND-OR-INVERT gate? Why?

3.52   Describe the key benefit and the key drawback of Schottky transistors in TTL.

3.53   Using the data sheet in Table 3-12, determine the worst-case LOW-state and HIGH-state DC noise margins of the 74LS00.

3.54 Sections 3.10.2 and 3.10.3 define eight different electrical parameters for TTL circuits. Using the data sheet in Table 3-12, determine the worst-case value of each of these for the 74LS00.

3.55 For each of the following resistive loads, determine whether the output drive specifications of the 74LS00 over the commercial operating range are exceeded. (Refer to Table 3-12, and use $V_{OLmax} = 0.5$ V and $V_{CC} = 5.0$ V.)

(a)   470 Ω to V_{CC} (b)   330 Ω to V_{CC} and 470 Ω to GND

(c)   10 KΩ to GND   (d)   390 Ω to V_{CC} and 390 Ω to GND

(e)   600 Ω to V_{CC} (f)   510 Ω to V_{CC} and 510 Ω to GND

(g)   4.7 KΩ to GND   (h)   220 Ω to V_{CC} and 330 Ω to GND

3.56 Compute the LOW-state and HIGH-state DC noise margins for each of the following cases of a TTL output driving a TTL-compatible CMOS input, or vice versa.

(a)   74HCT driving 74LS   (b)   74VHCT driving 74AS

(c)   74LS driving 74HCT   (d)   74S driving 74VHCT

3.57 Compute the maximum fanout for each of the following cases of a TTL-compatible CMOS output driving multiple inputs in a TTL logic family. Also indicate how much "excess" driving capability is available in the LOW or HIGH state for each case.

(a)   74HCT driving 74LS   (b)   74HCT driving 74S

(c)   74VHCT driving 74AS   (d)   74VHCT driving 74LS

3.58 For a given load capacitance and transition rate, which logic family in this chapter has the lowest dynamic power dissipation?

Exercises

3.59 Design a CMOS circuit that has the functional behavior shown in Figure X3.59. (*Hint:* Only six transistors are required.)

3.60 Design a CMOS circuit that has the functional behavior shown in Figure X3.60. (*Hint:* Only six transistors are required.)

3.61 Draw a circuit diagram, function table, and logic symbol in the style of Figure 3-19 for a CMOS gate with two inputs A and B and an output Z, where Z=1 if A=0 and B=1, and Z=0 otherwise. (*Hint:* Only six transistors are needed.)

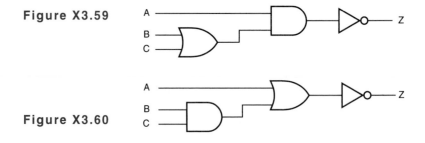

Figure X3.59

Figure X3.60

3.62   Draw a circuit diagram, function table, and logic symbol in the style of Figure 3-19 for a CMOS gate with two inputs A and B and an output Z, where Z=0 if A=1 and B=0, and Z=1 otherwise. (*Hint:* Only six transistors are needed.)

3.63   Draw a figure showing the logical structure of an 8-input CMOS NOR gate, assuming that at most 4-input gate circuits are practical. Using your general knowledge of CMOS characteristics, select a circuit structure that minimizes the NOR gate's propagation delay for a given silicon area, and explain why this is so.

3.64   The circuit designers of TTL-compatible CMOS families presumably could have made the voltage drop across the "on" transistor under load in the HIGH state as little as it is in the LOW state, simply by making the *p*-channel transistors bigger. Why do you suppose they didn't bother to do this?

3.65   How much current and power are "wasted" in Figure 3-32(b)?

3.66   Perform a detailed calculation of V_{OUT} in Figures 3-34 and 3-33. (*Hint:* Create a Thévenin equivalent for the CMOS inverter in each figure.)

3.67   Consider the dynamic behavior of a CMOS output driving a given capacitive load. If the resistance of the charging path is double the resistance of the discharging path, is the rise time exactly twice the fall time? If not, what other factors affect the transition times?

3.68   Analyze the fall time of the CMOS inverter output of Figure 3-37, with $R_L=1$ kΩ and $V_L=2.5$ V. Compare your result with the result in Section 3.6.1 and explain.

3.69   Repeat Exercise 3.68 for rise time.

3.70   Assuming that the transistors in an FCT CMOS three-state buffer are perfect zero-delay on-off devices that switch at an input threshold of 1.5 V, determine the value of t_{PLZ} for the test circuit and waveforms in Figure 3-24. (*Hint:* You have to determine the time using an RC time constant.) Explain the difference between your result and the specifications in Table 3-3.

3.71   Repeat Exercise 3.70 for t_{PHZ}.

3.72   Using the specifications in Table 3-7, estimate the "on" resistances of the *p*-channel and *n*-channel transistors in 74VHC-series CMOS logic.

3.73   Create a $4 \times 4 \times 2 \times 2$ matrix of worst-case DC noise margins for the following CMOS interfacing situations: an (HC, HCT, VHC, or VHCT) output driving an (HC, HCT, VHC, or VHCT) input with a (CMOS, TTL) load in the (LOW, HIGH) state; Figure X3.73 illustrates. (*Hints:* There are 64 different combinations, but many give identical results. Some combinations yield negative margins.)

Figure X3.73

		HC		HCT		VHC		VHCT	
Output									
	HC	CL	TL	CL	TL	CL	TL	CL	TL
		CH	TH	CH	TH	CH	TH	CH	TH
	HCT	CL	TL	CL	TL	CL	TL	CL	TL
		CH	TH	CH	TH	CH	TH	CH	TH
	VHC	CL	TL	CL	TL	CL	TL	CL	TL
		CH	TH	CH	TH	CH	TH	CH	TH
	VHCT	CL	TL	CL	TL	CL	TL	CL	TL
		CH	TH	CH	TH	CH	TH	CH	TH

Key:
CL = CMOS load, LOW
CH = CMOS load, HIGH
TL = TTL load, LOW
TH = TTL load, HIGH

3.74   Using Figure 3-85, determine the DC noise margins for 5-V-tolerant, 3.3-V CMOS driving 5-V CMOS logic with TTL input levels, and vice versa.

3.75   Using Figure 3-85, determine the DC noise margins for 3.3-V-tolerant, 2.5-V CMOS driving 3.3-V CMOS, and vice versa.

3.76   In the LED example in Section 3.7.5, a designer chose a resistor value of 300 Ω and found that the open-drain gate was able to maintain its output at 0.1 V while driving the LED. How much current flows through the LED, and how much power is dissipated by the pull-up resistor in this case?

3.77   Consider a CMOS 8-bit binary counter (Section 8.4) clocked at 16 MHz. For the purpose of computing the counter's dynamic power dissipation, what is the transition frequency of the least significant bit? Of the most significant bit? For the purpose of determining the dynamic power dissipation of the eight output bits, what frequency should be used?

3.78   Using only AND and NOR gates, draw a logic diagram for the logic function performed by the circuit in Figure 3-55.

3.79   Calculate the approximate output voltage at Z in Figure 3-56, assuming that the gates are HCT-series CMOS.

3.80   Redraw the circuit diagram of a CMOS 3-state buffer in Figure 3-48 using actual transistors instead of NAND, NOR, and inverter symbols. Can you find a circuit for the same function that requires a smaller total number of transistors? If so, draw it.

3.81   Modify the CMOS 3-state buffer circuit in Figure 3-48 so that the output is in the High-Z state when the enable input is HIGH. The modified circuit should require no more transistors than the original.

3.82   Using information in Table 3-3, estimate how much current can flow through each output pin if the outputs of two different 74HC00s are fighting.

3.83   Show that at a given power-supply voltage, an FCT-type I_{CCD} specification can be derived from an HCT/ACT-type C_{PD} specification, and vice versa.

3.84   If both V_Z and V_B in Figure 3-65(b) are 4.6 V, can we get $V_C = 5.2$ V? Explain.

3.85   Modify the program in Table 3-10 to account for leakage current in the OFF state.

3.86   Assuming "ideal" conditions, what is the minimum voltage that will be recognized as a HIGH in the TTL NAND gate in Figure 3-75 with one input LOW and the other HIGH?

3.87   Assuming "ideal" conditions, what is the maximum voltage that will be recognized as a LOW in the TTL NAND gate in Figure 3-75 with both inputs HIGH?

3.88   Find a commercial TTL part that can source 40 mA in the HIGH state. What is its application?

3.89   What happens if you try to drive an LED with its cathode grounded and its anode connected to a TTL totem-pole output, analogous to Figure 3-53(b) for CMOS?

3.90   What happens if you try to drive a 12-volt relay with a TTL totem-pole output?

3.91   Suppose that a single pull-up resistor to +5 V is used to provide a constant-1 logic source to 15 different 74LS00 inputs. What is the maximum value of this resistor? How much HIGH-state DC noise margin are you providing in this case?

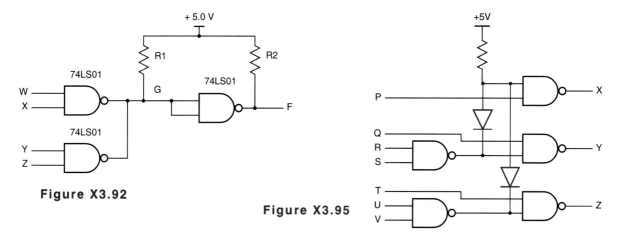

Figure X3.92

Figure X3.95

3.92   The circuit in Figure X3.92 uses open-collector NAND gates to perform "wired logic." Write a truth table for output signal F and, if you've read Section 4.2, a logic expression for F as a function of the circuit inputs.

3.93   What is the maximum allowable value for *R1* in Figure X3.92? Assume that a 0.7-V HIGH-state noise margin is required. The 74LS01 has the specs shown in the 74LS column of Table 3-11, except that I_{OHmax} is 100 μA, a leakage current that flows *into* the output in the HIGH state.

3.94   Suppose that the output signal F in Figure X3.92 drives the inputs of two 74S04 inverters. Compute the minimum and maximum allowable values of *R2*, assuming that a 0.7 V HIGH-state noise margin is required.

3.95   A logic designer found a problem in a certain circuit's function after the circuit had been released to production and 1000 copies of it built. A portion of the circuit is shown in Figure X3.95 in black; all of the gates are 74LS00 NAND gates. The logic designer fixed the problem by adding the two diodes shown in color. What do the diodes do? Describe both the logical effects of this change on the circuit's function and the electrical effects on the circuit's noise margins.

3.96   A 74LS125 is a buffer with a three-state output. When enabled, the output can sink 24 mA in the LOW state and source 2.6 mA in the HIGH state. When disabled, the output has a leakage current of ±20 μA (the sign depends on the output voltage—plus if the output is pulled HIGH by other devices, minus if it's LOW). Suppose a system is designed with multiple modules connected to a bus, where each module has a single 74LS125 to drive the bus and one 74LS04 to receive information on the bus. What is the maximum number of modules that can be connected to the bus without exceeding the 74LS125's specs?

3.97   Repeat Exercise 3.97, this time assuming that a single pull-up resistor is connected from the bus to +5 V to guarantee that the bus is HIGH when no device is driving it. Calculate the maximum possible value of the pull-up resistor, and the number of modules that can be connected to the bus.

3.98   Find the circuit design in a TTL data book for an actual three-state gate, and explain how it works.

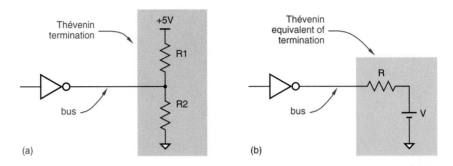

Figure X3.99   (a)   (b)

3.99   A *Thévenin termination* for an open-collector or three-state bus has the structure shown in Figure X3.99(a). The idea is that, by selecting appropriate values of $R1$ and $R2$, a designer can obtain a circuit equivalent to the termination in (b) for any desired values of V and R. The value of V determines the voltage on the bus when no device is driving it, and the value of R is selected to match the characteristic impedance of the bus for transmission-line purposes (Section 11.4). For each of the following pairs of V and R, determine the required values of $R1$ and $R2$.

(a) $V = 2.75, R = 148.5$ (b) $V = 2.7, R = 180$

(c) $V = 3.0, R = 130$ (d) $V = 2.5, R = 75$

3.100   For each of the $R1$ and $R2$ pairs in Exercise 3.99, determine whether the termination can be properly driven by a three-state output in each of the following logic families: 74LS, 74S, 74FCT-T. For proper operation, the family's I_{OL} and I_{OH} specs must not be exceeded when $V_{OL} = V_{OLmax}$ and $V_{OH} = V_{OHmin}$, respectively.

3.101   Using the graphs in a TTL data book, develop some rules of thumb for derating the maximum-propagation-delay specification of LS-TTL under nonoptimal conditions of power-supply voltage, temperature, and loading.

3.102   Determine the total power dissipation of the circuit in Figure X3.102 as function of transition frequency f for two realizations: (a) using 74LS gates; (b) using 74HC gates. Assume that input capacitance is 3 pF for a TTL gate and 7 pF for a CMOS gate, that a 74LS gate has an internal power-dissipation capacitance of 20 pF, and that there is an additional 20 pF of stray wiring capacitance in the circuit. Also assume that the X, Y, and Z inputs are always HIGH, and that input C is driven with a CMOS-level square wave with frequency f. Other information that you need for this problem can be found in Tables 3-5 and 3-11. State any other assumptions that you make. At what frequency does the TTL circuit dissipate less power than the CMOS circuit?

Figure X3.102

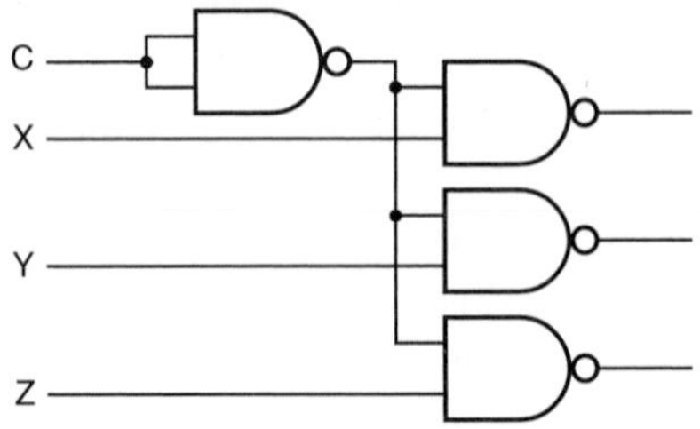

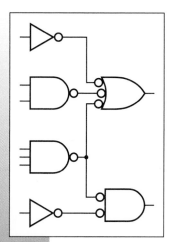

Combinational Logic Design Principles

L ogic circuits are classified into two types, "combinational" and "sequential." A *combinational logic circuit* is one whose outputs depend only on its current inputs. The rotary channel selector knob on an old-fashioned television is like a combinational circuit—its "output" selects a channel based only on the current position of the knob ("input").

The outputs of a *sequential logic circuit* depend not only on the current inputs but also on the past sequence of inputs, possibly arbitrarily far back in time. The channel selector controlled by the up and down pushbuttons on a TV or VCR is a sequential circuit—the channel selection depends on the past sequence of up/down pushes, at least since when you started viewing 10 hours before, and perhaps as far back as when you first powered up the device. Sequential circuits are discussed in Chapters 7 and 8.

A combinational circuit may contain an arbitrary number of logic gates and inverters but no feedback loops. A *feedback loop* is a signal path of a circuit that allows the output of a gate to propagate back to the input of that same gate; such a loop generally creates sequential circuit behavior.

In combinational circuit *analysis* we start with a logic diagram and proceed to a formal description of the function performed by that circuit, such as a truth table or a logic expression. In *synthesis* we do the reverse, starting with a formal description and proceeding to a logic diagram.

193

SYNTHESIS VS.
DESIGN

SYNTHESIS VS. DESIGN   Logic circuit design is a superset of synthesis, since in a real design problem we usually start out with an informal (word or thought) description of the circuit. Often the most challenging and creative part of design is to formalize the circuit description, defining the circuit's input and output signals and specifying its functional behavior by means of truth tables and equations. Once we've created the formal circuit description, we can usually follow a "turn-the-crank" synthesis procedure to obtain a logic diagram for a circuit with the required functional behavior. The material in the first four sections of this chapter is the basis for "turn-the-crank" procedures, whether the crank is turned by hand or by a computer. The last two sections describe actual design languages, ABEL and VHDL. When we create a design using one of these languages, a computer program can perform the synthesis steps for us. In later chapters we'll encounter many examples of the real design process.

Combinational circuits may have one or more outputs. Most analysis and synthesis techniques can be extended in an obvious way from single-output to multiple-output circuits (e.g., "Repeat these steps for each output"). We'll also point out how some techniques can be extended in a not-so-obvious way for improved effectiveness in the multiple-output case.

The purpose of this chapter is to give you a solid theoretical foundation for the analysis and synthesis of combinational logic circuits, a foundation that will be doubly important later when we move on to sequential circuits. Although most of the analysis and synthesis procedures in this chapter are automated nowadays by computer-aided design tools, you need a basic understanding of the fundamentals to use the tools and to figure out what's wrong when you get unexpected or undesirable results.

With the fundamentals well in hand, it is appropriate next to understand how combinational functions can be expressed and analyzed using hardware description languages (HDLs). So, the last two sections of this chapter introduce basic features of ABEL and VHDL, which we'll use to design all kinds of logic circuits throughout the balance of the text.

Before launching into a discussion of combinational logic circuits, we must introduce switching algebra, the fundamental mathematical tool for analyzing and synthesizing logic circuits of all types.

4.1 Switching Algebra

Boolean algebra

Formal analysis techniques for digital circuits have their roots in the work of an English mathematician, George Boole. In 1854, he invented a two-valued algebraic system, now called *Boolean algebra*, to "give expression … to the fundamental laws of reasoning in the symbolic language of a Calculus." Using this system, a philosopher, logician, or inhabitant of the planet Vulcan can formulate propositions that are true or false, combine them to make new propositions, and determine the truth or falsehood of the new propositions. For example, if we agree that "People who haven't studied this material are either failures or not nerds," and "No computer designer is a failure," then we can answer questions like "If you're a nerdy computer designer, then have you already studied this?"

Long after Boole, in 1938, Bell Labs researcher Claude E. Shannon showed how to adapt Boolean algebra to analyze and describe the behavior of circuits built from relays, the most commonly used digital logic elements of that time. In Shannon's *switching algebra*, the condition of a relay contact, open or closed, is represented by a variable X that can have one of two possible values, 0 or 1. In today's logic technologies, these values correspond to a wide variety of physical conditions—voltage HIGH or LOW, light off or on, capacitor discharged or charged, fuse blown or intact, and so on—as we detailed in Table 3-1 on page 81.

switching algebra

In the remainder of this section we develop the switching algebra directly, using "first principles" and what we already know about the behavior of logic elements (gates and inverters). For more historical and/or mathematical treatments of this material, consult the References.

4.1.1 Axioms

In switching algebra we use a symbolic variable, such as X, to represent the condition of a logic signal. A logic signal is in one of two possible conditions— low or high, off or on, and so on, depending on the technology. We say that X has the value "0" for one of these conditions and "1" for the other.

For example, with the CMOS and TTL logic circuits in Chapter 3, the *positive-logic convention* dictates that we associate the value "0" with a LOW voltage and "1" with a HIGH voltage. The *negative-logic convention* makes the opposite association: 0 = HIGH and 1 = LOW. However, the choice of positive or negative logic has no effect on our ability to develop a consistent algebraic description of circuit behavior; it only affects details of the physical-to-algebraic abstraction, as we'll explain later in our discussion of "duality." For the moment, we may ignore the physical realities of logic circuits and pretend that they operate directly on the logic symbols 0 and 1.

positive-logic convention

negative-logic convention

The *axioms* (or *postulates*) of a mathematical system are a minimal set of basic definitions that we assume to be true, from which all other information about the system can be derived. The first two axioms of switching algebra embody the "digital abstraction" by formally stating that a variable X can take on only one of two values:

axiom

postulate

$$(\text{A1}) \quad X = 0 \quad \text{if } X \neq 1 \qquad (\text{A1}') \quad X = 1 \quad \text{if } X \neq 0$$

Notice that we stated these axioms as a pair, the only difference between A1 and A1′ being the interchange of the symbols 0 and 1. This is a characteristic of all the axioms of switching algebra and is the basis of the "duality" principle that we'll study later.

complement
prime (′)

In Section 3.3.3 we showed the design of an inverter, a logic circuit whose output signal level is the opposite (or *complement*) of its input signal level. We use a *prime (′)* to denote an inverter's function. That is, if a variable X denotes an inverter's input signal, then X′ denotes the value of a signal on the inverter's output. This notation is formally specified in the second pair of axioms:

(A2)   If X = 0, then X′ = 1   (A2′)   If X = 1, then X′ = 0

algebraic operator
expression
NOT operation

As shown in Figure 4-1, the output of an inverter with input signal X may have an arbitrary signal name, say Y. Algebraically, however, we write Y = X′ to say "signal Y always has the opposite value as signal X." The prime (′) is an *algebraic operator*, and X′ is an *expression*, which you can read as "X prime" or "NOT X." This usage is analogous to what you've learned in programming languages, where if J is an integer variable, then −J is an expression whose value is 0 − J. Although this may seem like a small point, you'll learn that the distinction between signal names (X, Y), expressions (X′), and equations (Y = X′) is very important when we study documentation standards and software tools for logic design. In the logic diagrams in this book we maintain this distinction by writing signal names in black and expressions in color.

Figure 4-1
Signal naming and algebraic notation for an inverter.

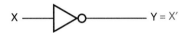

In Section 3.3.6 we showed how to build a 2-input CMOS AND gate, a circuit whose output is 1 if both of its inputs are 1. The function of a 2-input AND gate is sometimes called *logical multiplication* and is symbolized algebraically by a *multiplication dot (·)*. That is, an AND gate with inputs X and Y has an output signal whose value is X · Y, as shown in Figure 4-2(a). Some authors, especially mathematicians and logicians, denote logical multiplication with a wedge (X ∧ Y). We follow standard engineering practice by using the dot (X ·Y). When

logical multiplication
multiplication dot (·)

NOTE ON NOTATION   The notations X, ~X, and ¬X are also used by some authors to denote the complement of X. The overbar notation ($\overline{X}$) is probably the most widely used and the best looking typographically. However, we use the prime notation to get you used to writing logic expressions on a single text line without the more graphical overbar, and to force you to parenthesize complex complemented subexpressions—because this is what you'll have to do when you use HDLs and other tools.

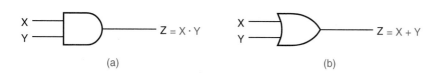

Figure 4-2
Signal naming and
algebraic notation:
(a) AND gate;
(b) OR gate.

we study hardware description languages (HDLs), we'll encounter several other symbols that are used to denote the same thing.

We also described in Section 3.3.6 how to build a 2-input CMOS OR gate, a circuit whose output is 1 if either of its inputs is 1. The function of a 2-input OR gate is sometimes called *logical addition* and is symbolized algebraically by a plus sign (+). An OR gate with inputs X and Y has an output signal whose value is X + Y, as shown in Figure 4-2(b). Some authors denote logical addition with a vee (X ∨ Y), but we follow the standard engineering practice of using the plus sign (X + Y). Once again, other symbols may be used in HDLs. By convention, in a logic expression involving both multiplication and addition, multiplication has *precedence*, just as in integer expressions in conventional programming languages. That is, the expression W · X + Y · Z is equivalent to (W · X) + (Y · Z).

The last three pairs of axioms state the formal definitions of the AND and OR operations by listing the output produced by each gate for each possible input combination:

logical addition

precedence

AND operation
OR operation

(A3)	$0 \cdot 0 = 0$	(A3′)	$1 + 1 = 1$
(A4)	$1 \cdot 1 = 1$	(A4′)	$0 + 0 = 0$
(A5)	$0 \cdot 1 = 1 \cdot 0 = 0$	(A5′)	$1 + 0 = 0 + 1 = 1$

The five pairs of axioms, A1–A5 and A1′–A5′, completely define switching algebra. All other facts about the system can be proved using these axioms as a starting point.

JUXT A MINUTE...   Older texts use simple *juxtaposition* (XY) to denote logical multiplication, but we don't. In general, juxtaposition is a clear notation only when signal names are limited to a single character. Otherwise, is XY a logical product or a two-character signal name? One-character variable names are common in algebra, but in real digital design problems we prefer to use multicharacter signal names that mean something. Thus, we need a separator between names, and the separator might just as well be a multiplication dot rather than a space. The HDL equivalent of the multiplication dot (often * or &) is absolutely required when logic formulas are written in a hardware-description language.

Table 4-1
Switching-algebra
theorems with one
variable.

(T1)	$X + 0 = X$	(T1′) $X \cdot 1 = X$	(Identities)
(T2)	$X + 1 = 1$	(T2′) $X \cdot 0 = 0$	(Null elements)
(T3)	$X + X = X$	(T3′) $X \cdot X = X$	(Idempotency)
(T4)	$(X')' = X$		(Involution)
(T5)	$X + X' = 1$	(T5′) $X \cdot X' = 0$	(Complements)

4.1.2 Single-Variable Theorems

theorem

During the analysis or synthesis of logic circuits, we often write algebraic expressions that characterize a circuit's actual or desired behavior. Switching-algebra *theorems* are statements, known to be always true, that allow us to manipulate algebraic expressions to allow simpler analysis or more efficient synthesis of the corresponding circuits. For example, the theorem $X + 0 = X$ allows us to substitute every occurrence of $X + 0$ in an expression with X.

Table 4-1 lists switching-algebra theorems involving a single variable X. How do we know that these theorems are true? We can either prove them ourselves or take the word of someone who has. OK, we're in college now, let's learn how to prove them.

perfect induction

Most theorems in switching algebra are exceedingly simple to prove using a technique called *perfect induction*. Axiom A1 is the key to this technique— since a switching variable can take on only two different values, 0 and 1, we can prove a theorem involving a single variable X by proving that it is true for both $X = 0$ and $X = 1$. For example, to prove theorem T1, we make two substitutions:

$$[X = 0] \quad 0 + 0 = 0 \quad \text{true, according to axiom A4}'$$
$$[X = 1] \quad 1 + 0 = 1 \quad \text{true, according to axiom A5}'$$

All of the theorems in Table 4-1 can be proved using perfect induction, as you're asked to do in the Drills 4.2 and 4.3.

4.1.3 Two- and Three-Variable Theorems

Switching-algebra theorems with two or three variables are listed in Table 4-2. Each of these theorems is easily proved by perfect induction, by evaluating the theorem statement for the four possible combinations of X and Y, or the eight possible combinations of X, Y, and Z.

The first two theorem pairs concern commutativity and associativity of logical addition and multiplication and are identical to the commutative and associative laws for addition and multiplication of integers and reals. Taken together, they indicate that the parenthesization or order of terms in a logical sum or logical product is irrelevant. For example, from a strictly algebraic point of view, an expression such as $W \cdot X \cdot Y \cdot Z$ is ambiguous; it should be written as $(W \cdot (X \cdot (Y \cdot Z)))$ or $(((W \cdot X) \cdot Y) \cdot Z)$ or $(W \cdot X) \cdot (Y \cdot Z)$ (see Exercise 4.34).

Table 4-2 Switching-algebra theorems with two or three variables.

(T6)	$X + Y = Y + X$		(T6')	$X \cdot Y = Y \cdot X$	(Commutativity)
(T7)	$(X + Y) + Z = X + (Y + Z)$		(T7')	$(X \cdot Y) \cdot Z = X \cdot (Y \cdot Z)$	(Associativity)
(T8)	$X \cdot Y + X \cdot Z = X \cdot (Y + Z)$		(T8')	$(X + Y) \cdot (X + Z) = X + Y \cdot Z$	(Distributivity)
(T9)	$X + X \cdot Y = X$		(T9')	$X \cdot (X + Y) = X$	(Covering)
(T10)	$X \cdot Y + X \cdot Y' = X$		(T10')	$(X + Y) \cdot (X + Y') = X$	(Combining)
(T11)	$X \cdot Y + X' \cdot Z + Y \cdot Z = X \cdot Y + X' \cdot Z$				(Consensus)
(T11')	$(X + Y) \cdot (X' + Z) \cdot (Y + Z) = (X + Y) \cdot (X' + Z)$				

But the theorems tell us that the ambiguous form of the expression is OK because we get the same results in any case. We even could have rearranged the order of the variables (e.g., $X \cdot Z \cdot Y \cdot W$) and gotten the same results.

As trivial as this discussion may seem, it is very important, because it forms the theoretical basis for using logic gates with more than two inputs. We defined $\cdot$ and $+$ as *binary operators*—operators that combine *two* variables. Yet we use 3-input, 4-input, and larger AND and OR gates in practice. The theorems tell us we can connect gate inputs in any order; in fact, many printed-circuit-board and ASIC layout programs take advantage of this. We can use either one n-input gate or $(n-1)$ 2-input gates interchangeably, though propagation delay and cost are likely to be higher with multiple 2-input gates.

binary operator

Theorem T8 is identical to the distributive law for integers and reals—that is, logical multiplication distributes over logical addition. Hence, we can "multiply out" an expression to obtain a sum-of-products form, as in the example below:

$$V \cdot (W + X) \cdot (Y + Z) = V \cdot W \cdot Y + V \cdot W \cdot Z + V \cdot X \cdot Y + V \cdot X \cdot Z$$

However, switching algebra also has the unfamiliar property that the reverse is true—logical addition distributes over logical multiplication—as demonstrated by theorem T8'. Thus, we can also "add out" an expression to obtain a product-of-sums form:

$$(V \cdot W \cdot X) + (Y \cdot Z) = (V + Y) \cdot (V + Z) \cdot (W + Y) \cdot (W + Z) \cdot (X + Y) \cdot (X + Z)$$

Theorems T9 and T10 are used extensively in the minimization of logic functions. For example, if the subexpression $X + X \cdot Y$ appears in a logic expression, the *covering theorem* T9 says that we need only include X in the expression; X is said to *cover* $X \cdot Y$. The *combining theorem* T10 says that if the subexpression $X \cdot Y + X \cdot Y'$ appears in an expression, we can replace it with X. Since Y must be 0 or 1, either way the original subexpression is 1 if and only if X is 1.

covering theorem
cover
combining theorem

Although we could easily prove T9 by perfect induction, the truth of T9 is more obvious if we prove it using the other theorems that we've proved so far:

$$
\begin{aligned}
X + X \cdot Y &= X \cdot 1 + X \cdot Y && \text{(according to T1')} \\
&= X \cdot (1 + Y) && \text{(according to T8)} \\
&= X \cdot 1 && \text{(according to T2)} \\
&= X && \text{(according to T1')}
\end{aligned}
$$

Likewise, the other theorems can be used to prove T10, where the key step is to use T8 to rewrite the lefthand side as $X \cdot (Y + Y')$.

consensus theorem

consensus

Theorem T11 is known as the *consensus theorem*. The $Y \cdot Z$ term is called the *consensus* of $X \cdot Y$ and $X' \cdot Z$. The idea is that if $Y \cdot Z$ is 1, then either $X \cdot Y$ or $X' \cdot Z$ must also be 1, since Y and Z are both 1 and either X or X' must be 1. Thus, the $Y \cdot Z$ term is redundant and may be dropped from the righthand side of T11. The consensus theorem has two important applications. It can be used to eliminate certain timing hazards in combinational logic circuits, as we'll see in Section 4.5. And it also forms the basis of the iterative-consensus method of finding prime implicants (see References).

In all of the theorems, it is possible to replace each variable with an arbitrary logic expression. A simple replacement is to complement one or more variables:

$$(X + Y') + Z' = X + (Y' + Z') \quad \text{(based on T7)}$$

But more complex expressions may be substituted as well:

$$(V' + X) \cdot (W \cdot (Y' + Z)) + (V' + X) \cdot (W \cdot (Y' + Z))' = V' + X \quad \text{(based on T10)}$$

4.1.4 *n*-Variable Theorems

Several important theorems, listed in Table 4-3, are true for an arbitrary number of variables, *n*. Most of these theorems can be proved using a two-step method called *finite induction*—first proving that the theorem is true for $n = 2$ (the *basis step*) and then proving that if the theorem is true for $n = i$, then it is also true for $n = i + 1$ (the *induction step*). For example, consider the generalized idempotency theorem T12. For $n = 2$, T12 is equivalent to T3 and is therefore true. If it is true for a logical sum of *i* X's, then it is also true for a sum of $i + 1$ X's, according to the following reasoning:

finite induction

basis step

induction step

$$
\begin{aligned}
X + X + X + \cdots + X &= X + (X + X + \cdots + X) && (i + 1 \text{ X's on either side}) \\
&= X + (X) && (\text{if T12 is true for } n = i) \\
&= X && (\text{according to T3})
\end{aligned}
$$

Thus, the theorem is true for all finite values of *n*.

DeMorgan's theorems

DeMorgan's theorems (T13 and T13') are probably the most commonly used of all the theorems of switching algebra. Theorem T13 says that an *n*-input

Table 4-3 Switching-algebra theorems with n variables.

(T12)	$X + X + \cdots + X = X$	(Generalized idempotency)
(T12′)	$X \cdot X \cdot \ \cdots\ \cdot X = X$	
(T13)	$(X_1 \cdot X_2 \cdot\ \cdots\ \cdot X_n)' = X_1' + X_2' + \cdots + X_n'$	(DeMorgan's theorems)
(T13′)	$(X_1 + X_2 + \cdots + X_n)' = X_1' \cdot X_2' \cdot\ \cdots\ \cdot X_n'$	
(T14)	$[F(X_1, X_2, ..., X_n, +, \cdot)]' = F(X_1', X_2', ..., X_n', \cdot, +)$	(Generalized DeMorgan's theorem)
(T15)	$F(X_1, X_2, ..., X_n) = X_1 \cdot F(1, X_2, ..., X_n) + X_1' \cdot F(0, X_2, ..., X_n)$	(Shannon's expansion theorems)
(T15′)	$F(X_1, X_2, ..., X_n) = [X_1 + F(0, X_2, ..., X_n)] \cdot [X_1' + F(1, X_2, ..., X_n)]$	

AND gate whose output is complemented is equivalent to an n-input OR gate whose inputs are complemented. That is, the circuits of Figure 4-3(a) and (b) are equivalent.

In Section 3.3.4 we showed how to build a CMOS NAND gate. The output of a NAND gate for any set of inputs is the complement of an AND gate's output for the same inputs, so a NAND gate can have the logic symbol in Figure 4-3(c). However, the CMOS NAND circuit is not designed as an AND gate followed by a transistor inverter (NOT gate); it's just a collection of transistors that happens to perform the AND-NOT function. In fact, theorem T13 tells us that the logic symbol in (d) denotes the same logic function (bubbles on the OR-gate inputs indicate logical inversion). That is, a NAND gate may be viewed as performing a NOT-OR function.

By observing the inputs and output of a NAND gate, it is impossible to determine whether it has been built internally as an AND gate followed by an inverter, as inverters followed by an OR gate, or as a direct CMOS realization, *because all NAND circuits perform precisely the same logic function.* Although the choice of symbol has no bearing on the functionality of a circuit, we'll show in Section 5.1 that the proper choice can make the circuit's function much easier to understand.

Figure 4-3 Equivalent circuits according to DeMorgan's theorem T13: (a) AND-NOT; (b) NOT-OR; (c) logic symbol for a NAND gate; (d) equivalent symbol for a NAND gate.

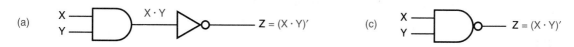

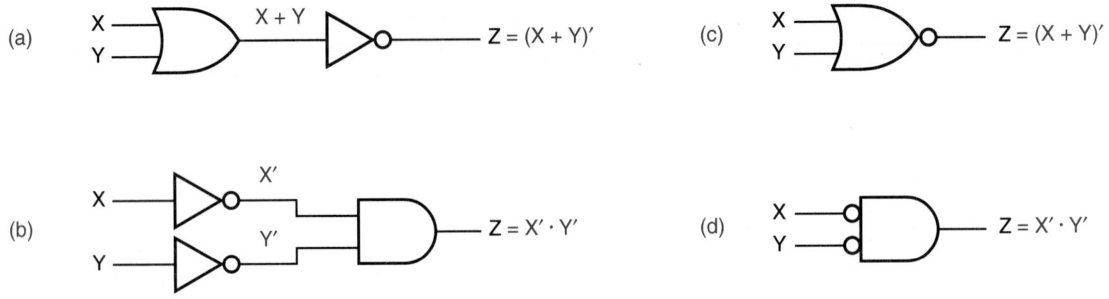

(a) X, Y → OR gate → X + Y → inverter → Z = (X + Y)′

(b) X → inverter → X′, Y → inverter → Y′ → AND gate → Z = X′ · Y′

(c) X, Y → NOR gate → Z = (X + Y)′

(d) X, Y → gate with inverted inputs → Z = X′ · Y′

Figure 4-4 Equivalent circuits according to DeMorgan's theorem T13′: (a) OR-NOT; (b) NOT-AND; (c) logic symbol for a NOR gate; (d) equivalent symbol for a NOR gate.

generalized DeMorgan's theorem

complement of a logic expression

A similar symbolic equivalence can be inferred from theorem T13′. As shown in Figure 4-4, a NOR gate may be realized as an OR gate followed by an inverter, or as inverters followed by an AND gate.

Theorems T13 and T13′ are just special cases of a *generalized DeMorgan's theorem*, T14, that applies to an arbitrary logic expression F. By definition, the *complement of a logic expression*, denoted (F)′, is an expression whose value is the opposite of F's for every possible input combination. Theorem T14 is very important because it gives us a way to manipulate and simplify the complement of an expression.

Theorem T14 states that, given any n-variable logic expression, its complement can be obtained by swapping + and · and complementing all variables. For example, suppose that we have

$$F(W, X, Y, Z) = (W' \cdot X) + (X \cdot Y) + (W \cdot (X' + Z'))$$
$$= ((W)' \cdot X) + (X \cdot Y) + (W \cdot ((X)' + (Z)'))$$

In the second line we have enclosed complemented variables in parentheses to remind you that the ′ is an operator, not part of the variable name. By applying theorem T14, we obtain

$$[F(W, X, Y, Z)]' = ((W')' + X') \cdot (X' + Y') \cdot (W' + ((X')' \cdot (Z')'))$$

Using theorem T4, this can be simplified to

$$[F(W, X, Y, Z)]' = (W + X') \cdot (X' + Y') \cdot (W' + (X \cdot Z))$$

In general, we can use theorem T14 to complement a parenthesized expression by swapping + and ·, complementing all uncomplemented variables, and uncomplementing all complemented ones.

The generalized DeMorgan's theorem T14 can be proved by showing that all logic functions can be written as either a sum or a product of subfunctions,

and then applying T13 and T13′ recursively. However, a much more enlighten-
ing and satisfying proof can be based on the principle of duality, explained next.

4.1.5 Duality

We stated all of the axioms of switching algebra in pairs. The primed version of
each axiom (e.g., A5′) is obtained from the unprimed version (e.g., A5) by
simply swapping 0 and 1 and, if present, · and +. As a result, we can state the
following *metatheorem*, a theorem about theorems: *metatheorem*

Principle of Duality Any theorem or identity in switching algebra remains true
                        if 0 and 1 are swapped and · and + are swapped throughout.

The metatheorem is true because the duals of all the axioms are true, so duals of
all switching-algebra theorems can be proved using duals of the axioms.

After all, what's in a name, or in a symbol for that matter? If the software
that was used to typeset this book had a bug, one that swapped $0 \leftrightarrow 1$ and $\cdot \leftrightarrow +$
throughout this chapter, you still would have learned exactly the same switching
algebra; only the nomenclature would have been a little weird, using words like
"product" to describe an operation that uses the symbol "+".

Duality is important because it doubles the usefulness of everything that
you learn about switching algebra and manipulation of switching functions.
Stated more practically, from a student's point of view, it halves the amount that
you have to learn! For example, once you learn how to synthesize two-stage
AND-OR logic circuits from sum-of-products expressions, you automatically
know a dual technique to synthesize OR-AND circuits from product-of-sums
expressions.

There is just one convention in switching algebra where we did not treat ·
and + identically, so duality does not necessarily hold true—can you figure out
what it is before reading the answer below? Consider the following statement of
theorem T9 and its clearly absurd "dual":

$$X + X \cdot Y = X \qquad \text{(theorem T9)}$$
$$X \cdot X + Y = X \qquad \text{(after applying the principle of duality)}$$
$$X + Y = X \qquad \text{(after applying theorem T3′)}$$

Obviously the last line above is false—where did we go wrong? The problem is
in operator precedence. We were able to write the lefthand side of the first line
without parentheses because of our convention that · has precedence. However,
once we applied the principle of duality, we should have given precedence to +
instead, or written the second line as $X \cdot (X + Y) = X$. The best way to avoid
problems like this is to parenthesize an expression fully before taking its dual.

Let us formally define the *dual of a logic expression*. If   *dual of a logic*
$F(X_1, X_2, ..., X_n, +, \cdot, ')$ is a fully parenthesized logic expression involving the   *expression*

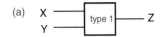

(a) X, Y — type 1 — Z

X	Y	Z
LOW	LOW	LOW
LOW	HIGH	LOW
HIGH	LOW	LOW
HIGH	HIGH	HIGH

(b) X, Y — type 1 — Z $= X \cdot Y$

X	Y	Z
0	0	0
0	1	0
1	0	0
1	1	1

(c) X, Y — type 1 — Z $= X + Y$

X	Y	Z
1	1	1
1	0	1
0	1	1
0	0	0

Figure 4-5 A "type-1"logic gate: (a) electrical function table; (b) logic function table and symbol with positive logic; (c) logic function table and symbol with negative logic.

variables $X_1, X_2, ..., X_n$ and the operators $+$, $\cdot$, and $'$, then the dual of F, written F^D, is the same expression with $+$ and $\cdot$ swapped:

$$F^D(X_1, X_2, ..., X_n, +, \cdot, ') = F(X_1, X_2, ..., X_n, \cdot, +, ')$$

You already knew this, of course, but we wrote the definition in this way just to highlight the similarity between duality and the generalized DeMorgan's theorem T14, which may now be restated as follows:

$$[F(X_1, X_2, ..., X_n)]' = F^D(X_1', X_2', ..., X_n')$$

Let's examine this statement in terms of a physical network.

Figure 4-5(a) shows the electrical function table for a logic element that we'll simply call a "type-1" gate. Under the positive-logic convention (LOW = 0 and HIGH = 1), this is an AND gate, but under the negative-logic convention (LOW = 1 and HIGH = 0), it is an OR gate, as shown in (b) and (c). We can also imagine a "type-2" gate, shown in Figure 4-6, that is a positive-logic OR or a negative-logic AND. Similar tables can be developed for gates with more than two inputs.

Figure 4-6 A "type-2" logic gate: (a) electrical function table; (b) logic function table and symbol with positive logic; (c) logic function table and symbol with negative logic.

(a) X, Y — type 2 — Z

X	Y	Z
LOW	LOW	LOW
LOW	HIGH	HIGH
HIGH	LOW	HIGH
HIGH	HIGH	HIGH

(b) X, Y — type 2 — Z $= X + Y$

X	Y	Z
0	0	0
0	1	1
1	0	1
1	1	1

(c) X, Y — type 2 — Z $= X \cdot Y$

X	Y	Z
1	1	1
1	0	0
0	1	0
0	0	0

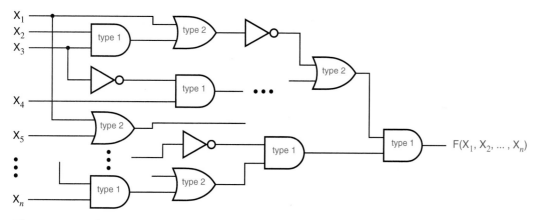

Figure 4-7 Circuit for a logic function using inverters and type-1 and type-2 gates under a positive-logic convention.

Suppose that we are given an arbitrary logic expression, $F(X_1, X_2, ..., X_n)$. Following the positive-logic convention, we can build a circuit corresponding to this expression using inverters for **NOT** operations, type-1 gates for **AND**, and type-2 gates for **OR**, as shown in Figure 4-7. Now suppose that, without changing this circuit, we simply change the logic convention from positive to negative. Then we should redraw the circuit as shown in Figure 4-8. Clearly, for every possible combination of input voltages (**HIGH** and **LOW**), the circuit still produces the same output voltage. However, from the point of view of switching algebra, the output value—0 or 1—is the opposite of what it was under the positive-logic convention. Likewise, each input value is the opposite of what it was. Therefore, for each possible input combination to the circuit in Figure 4-7, the output is the opposite of that produced by the opposite input combination applied to the circuit in Figure 4-8:

$$F(X_1, X_2, ..., X_n) = [F^D(X_1', X_2', ..., X_n')]'$$

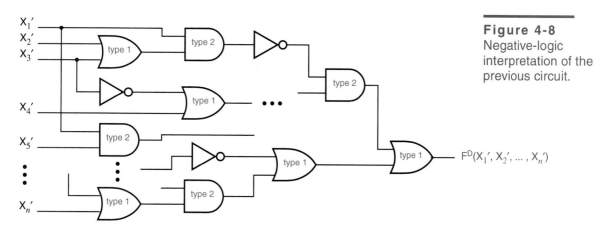

Figure 4-8
Negative-logic interpretation of the previous circuit.

By complementing both sides, we get the generalized DeMorgan's theorem:

$$[F(X_1, X_2, ..., X_n)]' \ = \ F^D(X_1', X_2', ..., X_n')$$

Amafing!

DUALITY IS GOOD FOR STUDENTS AND AUTHORS	You've seen that duality is the basis for the generalized DeMorgan's theorem. Going forward, it will halve the number of methods you must learn to manipulate and simplify logic functions. It also halved the amount of original material I had to write in these sections!

4.1.6 Standard Representations of Logic Functions

Before moving on to analysis and synthesis of combinational logic functions we'll introduce some necessary nomenclature and notation.

truth table

The most basic representation of a logic function is the *truth table*. Similar in philosophy to the perfect-induction proof method, this brute-force representation simply lists the output of the circuit for every possible input combination. Traditionally, the input combinations are arranged in rows in ascending binary counting order, and the corresponding output values are written in a column next to the rows. The general structure of a 3-variable truth table is shown below in Table 4-4.

The rows are numbered 0–7, corresponding to the binary input combinations, but this numbering is not an essential part of the truth table. The truth table for a particular 3-variable logic function is shown in Table 4-5. Each distinct pattern of 0s and 1s in the output column yields a different logic function; there are 2^8 such patterns. Thus, the logic function in Table 4-5 is one of 2^8 different logic functions of three variables.

The truth table for an *n*-variable logic function has 2^n rows. Obviously, truth tables are practical to write only for logic functions with a small number of variables, say, 10 for students and about 4–5 for everyone else.

Table 4-4 General truth table structure for a 3-variable logic function, F(X, Y, Z).	Row	X	Y	Z	F
	0	0	0	0	F(0,0,0)
	1	0	0	1	F(0,0,1)
	2	0	1	0	F(0,1,0)
	3	0	1	1	F(0,1,1)
	4	1	0	0	F(1,0,0)
	5	1	0	1	F(1,0,1)
	6	1	1	0	F(1,1,0)
	7	1	1	1	F(1,1,1)

Row	X	Y	Z	F
0	0	0	0	1
1	0	0	1	0
2	0	1	0	0
3	0	1	1	1
4	1	0	0	1
5	1	0	1	0
6	1	1	0	1
7	1	1	1	1

Table 4-5
Truth table for a particular 3-variable logic function, $F(X, Y, Z)$.

The information contained in a truth table can also be conveyed algebraically. To do so, we first need some definitions:

- A *literal* is a variable or the complement of a variable. Examples: X, Y, X', Y'. *literal*

- A *product term* is a single literal or a logical product of two or more literals. Examples: Z', W · X · Y, X · Y' · Z, W' · Y' · Z. *product term*

- A *sum-of-products expression* is a logical sum of product terms. Example: Z' + W · X · Y + X · Y' · Z + W' · Y' · Z. *sum-of-products expression*

- A *sum term* is a single literal or a logical sum of two or more literals. Examples: Z', W + X + Y, X + Y' + Z, W' + Y' + Z. *sum term*

- A *product-of-sums expression* is a logical product of sum terms. Example: Z' · (W + X + Y) · (X + Y' + Z) · (W' + Y' + Z). *product-of-sums expression*

- A *normal term* is a product or sum term in which no variable appears more than once. A nonnormal term can always be simplified to a constant or a normal term using one of theorems T3, T3', T5, or T5'. Examples of nonnormal terms: W · X · X · Y', W + W + X' + Y, X · X' · Y. Examples of normal terms: W · X · Y', W + X' + Y. *normal term*

- An *n*-variable *minterm* is a normal product term with *n* literals. There are 2^n such product terms. Examples of 4-variable minterms: W' · X' · Y' · Z', W · X · Y' · Z, W' · X' · Y · Z'. *minterm*

- An *n*-variable *maxterm* is a normal sum term with *n* literals. There are 2^n such sum terms. Examples of 4-variable maxterms: W' + X' + Y' + Z', W + X' + Y' + Z, W' + X' + Y + Z'. *maxterm*

There is a close correspondence between the truth table and minterms and maxterms. A minterm can be defined as a product term that is 1 in exactly one row of the truth table. Similarly, a maxterm can be defined as a sum term that is 0 in exactly one row of the truth table. Table 4-6 shows this correspondence for a 3-variable truth table.

Table 4-6
Minterms and maxterms for a 3-variable logic function, $F(X, Y, Z)$.

Row	X	Y	Z	F	*Minterm*	*Maxterm*
0	0	0	0	$F(0,0,0)$	$X' \cdot Y' \cdot Z'$	$X + Y + Z$
1	0	0	1	$F(0,0,1)$	$X' \cdot Y' \cdot Z$	$X + Y + Z'$
2	0	1	0	$F(0,1,0)$	$X' \cdot Y \cdot Z'$	$X + Y' + Z$
3	0	1	1	$F(0,1,1)$	$X' \cdot Y \cdot Z$	$X + Y' + Z'$
4	1	0	0	$F(1,0,0)$	$X \cdot Y' \cdot Z'$	$X' + Y + Z$
5	1	0	1	$F(1,0,1)$	$X \cdot Y' \cdot Z$	$X' + Y + Z'$
6	1	1	0	$F(1,1,0)$	$X \cdot Y \cdot Z'$	$X' + Y' + Z$
7	1	1	1	$F(1,1,1)$	$X \cdot Y \cdot Z$	$X' + Y' + Z'$

minterm number
minterm i

An n-variable minterm can be represented by an n-bit integer, the *minterm number*. We'll use the name *minterm i* to denote the minterm corresponding to row i of the truth table. In minterm i, a particular variable appears complemented if the corresponding bit in the binary representation of i is 0; otherwise, it is uncomplemented. For example, row 5 has binary representation 101 and the corresponding minterm is $X \cdot Y' \cdot Z$. As you might expect, the correspondence

maxterm i

for maxterms is just the opposite: in *maxterm i*, a variable is complemented if the corresponding bit in the binary representation of i is 1. Thus, maxterm 5 (101) is $X' + Y + Z'$. Note that all of this makes sense only if we know the number of variables in the truth table, three in the examples.

Based on the correspondence between the truth table and minterms, we can easily create an algebraic representation of a logic function from its truth table.

canonical sum

The *canonical sum* of a logic function is a sum of the minterms corresponding to truth-table rows (input combinations) for which the function produces a 1 output. For example, the canonical sum for the logic function in Table 4-5 on page 207 is

$$F = \Sigma_{X,Y,Z}(0, 3, 4, 6, 7)$$
$$= X' \cdot Y' \cdot Z' + X' \cdot Y \cdot Z + X \cdot Y' \cdot Z' + X \cdot Y \cdot Z' + X \cdot Y \cdot Z$$

minterm list

Here, the notation $\Sigma_{X,Y,Z}(0, 3, 4, 6, 7)$ is a *minterm list* and means "the sum of minterms 0, 3, 4, 6, and 7 with variables X, Y, and Z." The minterm list is also

on-set

known as the *on-set* for the logic function. You can visualize that each minterm "turns on" the output for exactly one input combination. Any logic function can be written as a canonical sum.

canonical product

The *canonical product* of a logic function is a product of the maxterms corresponding to input combinations for which the function produces a 0 output. For example, the canonical product for the logic function in Table 4-5 is

$$F = \Pi_{X,Y,Z}(1, 2, 5)$$
$$= (X + Y + Z') \cdot (X + Y' + Z) \cdot (X' + Y + Z')$$

Here, the notation $\prod_{X,Y,Z}(1,2,5)$ is a *maxterm list* and means "the product of maxterms 1, 2, and 5 with variables X, Y, and Z." The maxterm list is also known as the *off-set* for the logic function. You can visualize that each maxterm "turns off" the output for exactly one input combination. Any logic function can be written as a canonical product.

maxterm list

off-set

It's easy to convert between a minterm list and a maxterm list. For a function of n variables, the possible minterm and maxterm numbers are in the set $\{0, 1, \ldots, 2^n - 1\}$; a minterm or maxterm list contains a subset of these numbers. To switch between list types, take the set complement, for example,

$$\Sigma_{A,B,C}(0,1,2,3) = \prod_{A,B,C}(4,5,6,7)$$
$$\Sigma_{X,Y}(1) = \prod_{X,Y}(0,2,3)$$
$$\Sigma_{W,X,Y,Z}(0,1,2,3,5,7,11,13) = \prod_{W,X,Y,Z}(4,6,8,9,10,12,14,15)$$

We have now learned five possible representations for a combinational logic function:

1. A truth table.
2. An algebraic sum of minterms, the canonical sum.
3. A minterm list using the Σ notation.
4. An algebraic product of maxterms, the canonical product.
5. A maxterm list using the $\prod$ notation.

Each one of these representations specifies exactly the same information; given any one of them, we can derive the other four using a simple mechanical process.

4.2 Combinational-Circuit Analysis

We analyze a combinational logic circuit by obtaining a formal description of its logic function. Once we have a description of the logic function, a number of other operations are possible:

- We can determine the behavior of the circuit for various input combinations.

- We can manipulate an algebraic description to suggest different circuit structures for the logic function.

- We can transform an algebraic description into a standard form corresponding to an available circuit structure. For example, a sum-of-products expression corresponds directly to the circuit structure used in PLDs (programmable logic devices).

- We can use an algebraic description of the circuit's functional behavior in the analysis of a larger system that includes the circuit.

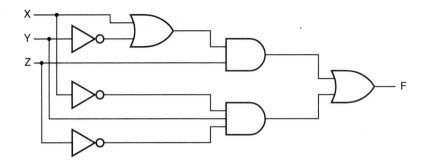

Figure 4-9
A three-input, one-output logic circuit.

Given a logic diagram for a combinational circuit, such as Figure 4-9, there are a number of ways to obtain a formal description of the circuit's function. The most primitive functional description is the truth table.

Using only the basic axioms of switching algebra, we can obtain the truth table of an n-input circuit by working our way through all 2^n input combinations. For each input combination, we determine all of the gate outputs produced by that input, propagating information from the circuit inputs to the circuit outputs. Figure 4-10 applies this "exhaustive" technique to our example circuit. Written on each signal line in the circuit is a sequence of eight logic values, the values present on that line when the circuit inputs XYZ are 000, 001, ..., 111. The truth table can be written by transcribing the output sequence of the final OR gate, as

Figure 4-10 Gate outputs created by all input combinations.

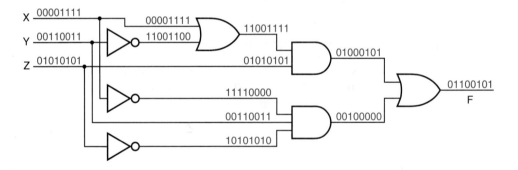

A LESS EXHAUSTING WAY TO GO	You can easily obtain the results in Figure 4-10 with typical logic design tools that include a logic simulator. First, you draw the schematic. Then, you apply the outputs of a 3-bit binary counter to the X, Y, and Z inputs. (Most simulators have such counter outputs built in for just this sort of exercise.) The counter repeatedly cycles through the eight possible input combinations, in the same order that we've shown in the figure. The simulator allows you to graph the resulting signal values at any point in the schematic, including the intermediate points as well as the output.

Row	X	Y	Z	F
0	0	0	0	0
1	0	0	1	1
2	0	1	0	1
3	0	1	1	0
4	1	0	0	0
5	1	0	1	1
6	1	1	0	0
7	1	1	1	1

Table 4-7
Truth table for the logic circuit of Figure 4-9.

shown in Table 4-7. Once we have the truth table for the circuit, we can also directly write a logic expression—the canonical sum or product—if we wish.

The number of input combinations of a logic circuit grows exponentially with the number of inputs, so the exhaustive approach can quickly become exhausting. Instead, we normally use an algebraic approach whose complexity is more linearly proportional to the size of the circuit. The method is simple—we build up a parenthesized logic expression corresponding to the logic operators and structure of the circuit. We start at the circuit inputs and propagate expressions through gates toward the output. Using the theorems of switching algebra, we may simplify the expressions as we go, or we may defer all algebraic manipulations until an output expression is obtained.

Figure 4-11 applies the algebraic technique to our example circuit. The output function is given on the output of the final OR gate:

$$F = ((X+Y') \cdot Z) + (X' \cdot Y \cdot Z')$$

No switching-algebra theorems were used to obtain this expression. However, we can use theorems to transform this expression into another form. For example, a sum of products can be obtained by "multiplying out":

$$F = X \cdot Z + Y' \cdot Z + X' \cdot Y \cdot Z'$$

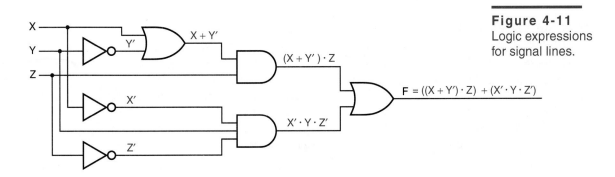

Figure 4-11
Logic expressions for signal lines.

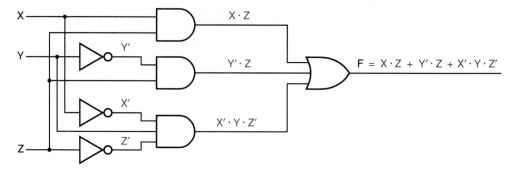

Figure 4-12 Two-level AND-OR circuit.

The new expression corresponds to a different circuit for the same logic function, as shown in Figure 4-12.

Similarly, we can "add out" the original expression to obtain a product of sums:

$$
\begin{aligned}
F &= ((X + Y') \cdot Z) + (X' \cdot Y \cdot Z') \\
&= (X + Y' + X') \cdot (X + Y' + Y) \cdot (X + Y' + Z') \cdot (Z + X') \cdot (Z + Y) \cdot (Z + Z') \\
&= 1 \cdot 1 \cdot (X + Y' + Z') \cdot (X' + Z) \cdot (Y + Z) \cdot 1 \\
&= (X + Y' + Z') \cdot (X' + Z) \cdot (Y + Z)
\end{aligned}
$$

The corresponding logic circuit is shown in Figure 4-13.

Our next example of algebraic analysis uses a circuit with NAND and NOR gates, shown in Figure 4-14. This analysis is a little messier than the previous example, because each gate produces a complemented subexpression, not just a simple sum or product. However, the output expression can be simplified by repeated application of the generalized DeMorgan's theorem:

Figure 4-13 Two-level OR-AND circuit.

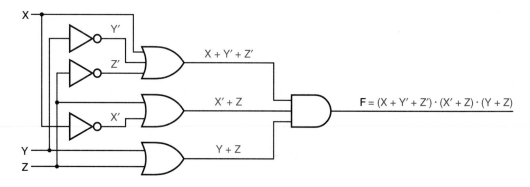

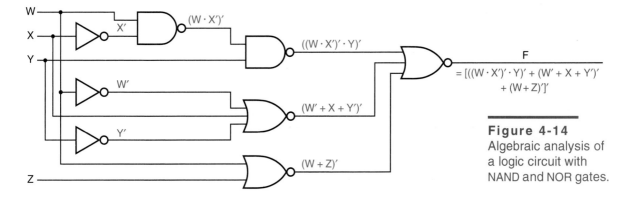

Figure 4-14
Algebraic analysis of
a logic circuit with
NAND and NOR gates.

$$
\begin{aligned}
F &= [((W \cdot X')' \cdot Y)' + (W' + X + Y')' + (W + Z)']' \\
  &= ((W' + X)' + Y')' \cdot (W \cdot X' \cdot Y)' \cdot (W' \cdot Z')' \\
  &= ((W \cdot X')' \cdot Y) \cdot (W' + X + Y') \cdot (W + Z) \\
  &= ((W' + X) \cdot Y) \cdot (W' + X + Y') \cdot (W + Z)
\end{aligned}
$$

Quite often, DeMorgan's theorem can be applied *graphically* to simplify algebraic analysis. Recall from Figures 4-3 and 4-4 that NAND and NOR gates each have two equivalent symbols. By judiciously redrawing Figure 4-14, we make it possible to cancel out some of the inversions during the analysis by using theorem T4 $[(X')' = X]$, as shown in Figure 4-15. This manipulation leads us to a simplified output expression directly:

$$F = ((W' + X) \cdot Y) \cdot (W' + X + Y') \cdot (W + Z)$$

Figures 4-14 and 4-15 were just two different ways of drawing the same physical logic circuit. However, when we simplify a logic expression using the theorems of switching algebra, we get an expression corresponding to a different physical circuit. For example, the simplified expression above corresponds to the circuit of Figure 4-16, which is physically different from the one in the previous two figures. Furthermore, we could multiply out and add out the

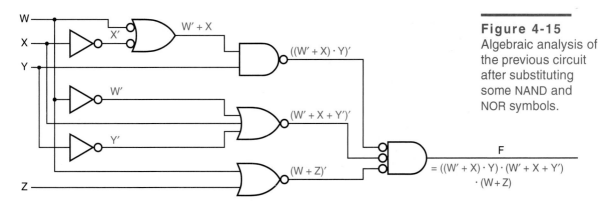

Figure 4-15
Algebraic analysis of
the previous circuit
after substituting
some NAND and
NOR symbols.

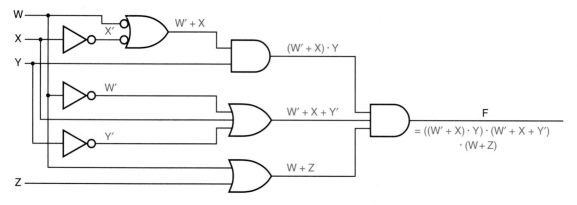

Figure 4-16 A different circuit for same logic function.

expression to obtain sum-of-products and product-of-sums expressions corresponding to two more physically different circuits for the same logic function.

Although we used logic expressions above to convey information about the physical structure of a circuit, we don't always do this. For example, we might use the expression $G(W, X, Y, Z) = W \cdot X \cdot Y + Y \cdot Z$ to describe any one of the circuits in Figure 4-17. Normally, the only sure way to determine a circuit's structure is to look at its schematic drawing. However, for certain restricted classes of circuits, structural information can be inferred from logic expressions. For example, the circuit in (a) could be described without reference to the drawing as "a two-level AND-OR circuit for $W \cdot X \cdot Y + Y \cdot Z$," while the circuit in (b) could be described as "a two-level NAND-NAND circuit for $W \cdot X \cdot Y + Y \cdot Z$."

Figure 4-17 Three circuits for $G(W, X, Y, Z) = W \cdot X \cdot Y + Y \cdot Z$: (a) two-level AND-OR; (b) two-level NAND-NAND; (c) ad hoc.

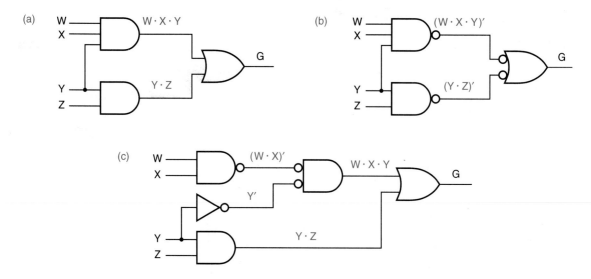

4.3 Combinational-Circuit Synthesis

4.3.1 Circuit Descriptions and Designs

What is the starting point for designing combinational logic circuits? Usually, we are given a word description of a problem or we develop one ourselves. Occasionally, the description is a list of input combinations for which a signal should be on or off, the verbal equivalent of a truth table or the Σ or Π notation introduced previously. For example, the description of a 4-bit prime-number detector might be, "Given a 4-bit input combination $N = N_3N_2N_1N_0$, this function produces a 1 output for $N = 1, 2, 3, 5, 7, 11, 13$, and 0 otherwise." A logic function described in this way can be designed directly from the canonical sum or product expression. For the prime-number detector, we have

$$\begin{aligned}
F &= \Sigma_{N_3,N_2,N_1,N_0}(1, 2, 3, 5, 7, 11, 13) \\
&= N_3' \cdot N_2' \cdot N_1' \cdot N_0 + N_3' \cdot N_2' \cdot N_1 \cdot N_0' + N_3' \cdot N_2' \cdot N_1 \cdot N_0 + N_3' \cdot N_2 \cdot N_1' \cdot N_0 \\
&\quad + N_3' \cdot N_2 \cdot N_1 \cdot N_0 + N_3 \cdot N_2' \cdot N_1 \cdot N_0 + N_3 \cdot N_2 \cdot N_1' \cdot N_0
\end{aligned}$$

The corresponding circuit is shown in Figure 4-18.

More often, we describe a logic function using the English-language connectives "and," "or," and "not." For example, we might describe an alarm circuit by saying, "The ALARM output is 1 if the PANIC input is 1, or if the ENABLE input is 1, the EXITING input is 0, and the house is not secure; the

Figure 4-18 Canonical-sum design for 4-bit prime-number detector.

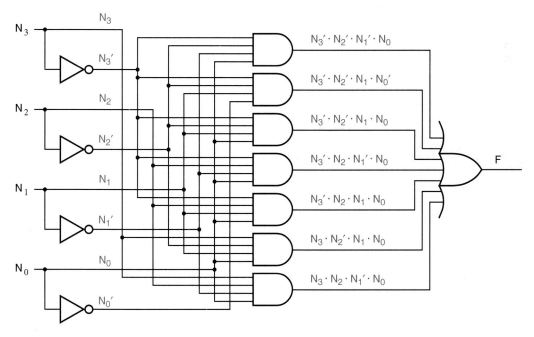

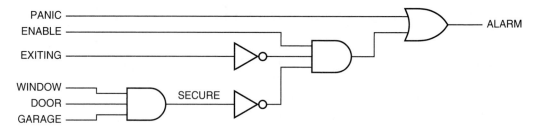

Figure 4-19 Alarm circuit derived from logic expression.

house is secure if the WINDOW, DOOR, and GARAGE inputs are all 1." Such a description can be translated directly into algebraic expressions:

ALARM = PANIC + ENABLE · EXITING' · SECURE'

SECURE = WINDOW · DOOR · GARAGE

ALARM = PANIC + ENABLE · EXITING' · (WINDOW · DOOR · GARAGE)'

Notice that we used the same method in switching algebra as in ordinary algebra to formulate a complicated expression—we defined an auxiliary variable SECURE to simplify the first equation, developed an expression for SECURE, and used substitution to get the final expression. We can easily draw a circuit using AND, OR, and NOT gates that realizes the final expression, as shown in Figure 4-19. A circuit *realizes* ["makes real"] an expression if its output function equals that expression, and the circuit is called a *realization* of the function.

realize
realization

Once we have an expression, any expression, for a logic function, we can do other things besides building a circuit directly from the expression. We can manipulate the expression to get different circuits. For example, the ALARM expression above can be multiplied out to get the sum-of-products circuit in Figure 4-20. Or, if the number of variables is not too large, we can construct the truth table for the expression and use any of the synthesis methods that apply to truth tables, including the canonical sum or product method described earlier and the minimization methods described later.

Figure 4-20 Sum-of-products version of alarm circuit.

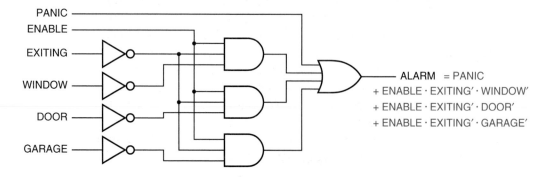

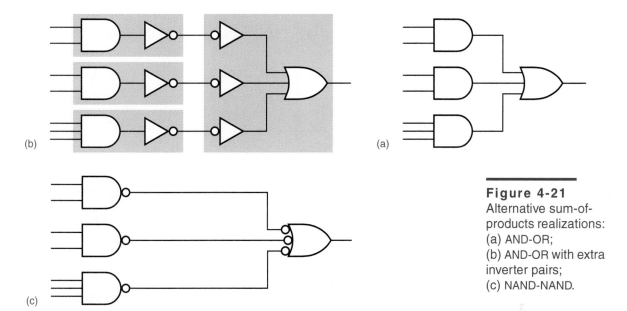

(b)

(a)

(c)

Figure 4-21
Alternative sum-of-products realizations:
(a) AND-OR;
(b) AND-OR with extra inverter pairs;
(c) NAND-NAND.

In general, it's easier to describe a circuit in words using logical connectives and to write the corresponding logic expressions than it is to write a complete truth table, especially if the number of variables is large. However, sometimes we have to work with imprecise word descriptions of logic functions, for example, "The ERROR output should be 1 if the GEARUP, GEARDOWN, and GEARCHECK inputs are inconsistent." In this situation, the truth-table approach is best because it allows us to determine the output required for every input combination, based on our knowledge and understanding of the problem environment (e.g., the brakes cannot be applied unless the gear is down).

4.3.2 Circuit Manipulations

The design methods that we've described so far use AND, OR, and NOT gates. We might like to use NAND and NOR gates, too—they're faster than ANDs and ORs in most technologies. However, most people don't develop logical propositions in terms of NAND and NOR connectives. That is, you probably wouldn't say, "I won't date you if you're not clean or not wealthy and also you're not smart or not friendly." It would be more natural for you to say, "I'll date you if you're clean and wealthy, or if you're smart and friendly." So, given a "natural" logic expression, we need ways to translate it into other forms.

We can translate any logic expression into an equivalent sum-of-products expression, simply by multiplying it out. As shown in Figure 4-21(a), such an expression may be realized directly with AND and OR gates. The inverters required for complemented inputs are not shown.

As shown in Figure 4-21(b), we may insert a pair of inverters between each AND-gate output and the corresponding OR-gate input in a two-level AND-OR

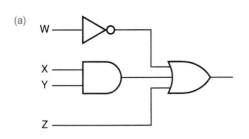

(a)

(b)

Figure 4-22
Another two-level
sum-of-products
circuit: (a) AND-OR;
(b) AND-OR with extra
inverter pairs;
(c) NAND-NAND.

(c)

circuit. According to theorem T4, these inverters have no effect on the output function of the circuit. In fact, we've drawn the second inverter of each pair with its inversion bubble on its input to provide a graphical reminder that the inverters cancel. However, if these inverters are absorbed into the AND and OR gates, we wind up with AND-NOT gates at the first level and a NOT-OR gate at the second level. These are just two different symbols for the same type of gate—a NAND gate. Thus, a two-level AND-OR *circuit* may be converted to a two-level NAND-NAND *circuit* simply by substituting gates.

AND-OR circuit

NAND-NAND circuit

Figure 4-23
Realizations of a
product-of-sums
expression:
(a) OR-AND;
(b) OR-AND with extra
inverter pairs;
(c) NOR-NOR.

(c)

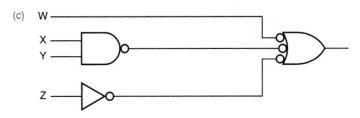

(a)

(b)

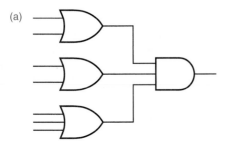

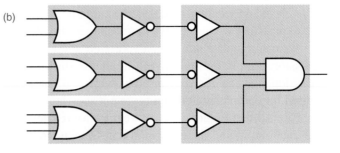

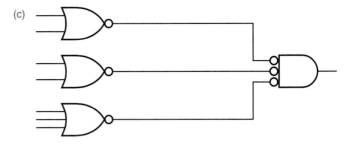

If any product terms in the sum-of-products expression contain just a single literal, then we may gain or lose inverters in the transformation from AND-OR to NAND-NAND. For example, Figure 4-22 is an example where an inverter on the W input is no longer needed, but an inverter must be added to the Z input.

We have shown that any sum-of-products expression can be realized in either of two ways—as an AND-OR circuit or as a NAND-NAND circuit. The dual of this statement is also true: any product-of-sums expression can be realized as an OR-AND *circuit* or as a NOR-NOR *circuit*. Figure 4-23 shows an example. Any logic expression can be translated into an equivalent product-of-sums expression by adding it out, and hence has both OR-AND and NOR-NOR circuit realizations.

OR-AND circuit
NOR-NOR circuit

The same kind of manipulations can be applied to arbitrary logic circuits. For example, Figure 4-24(a) shows a circuit built from AND and OR gates. After adding pairs of inverters, we obtain the circuit in (b). However, one of the gates, a 2-input AND gate with a single inverted input, is not a standard type. We can use a discrete inverter as shown in (c) to obtain a circuit that uses only standard gate types—NAND, AND, and inverters. Actually, a better way to use the inverter is shown in (d); one level of gate delay is eliminated, and the bottom gate becomes a NOR instead of AND. In most logic technologies, inverting gates like NAND and NOR are faster than noninverting gates like AND and OR.

Figure 4-24 Logic-symbol manipulations: (a) original circuit; (b) transformation with a nonstandard gate; (c) inverter used to eliminate nonstandard gate; (d) preferred inverter placement.

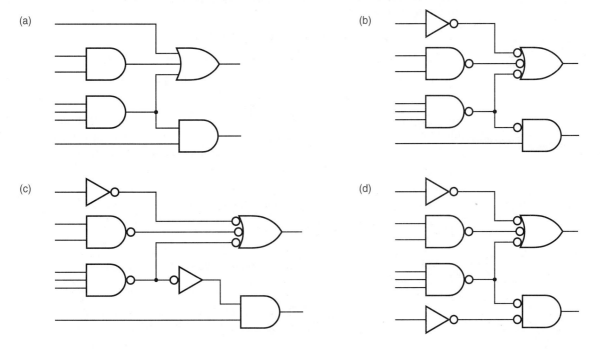

WHY MINIMIZE?   Minimization is an important step in both ASIC design and in PLD-based design. Extra gates and gate inputs require more area in an ASIC chip and thereby increase cost. The number of gates in a PLD is fixed, so you might think that extra gates are free—and they are, until you run out of them and have to upgrade to a bigger, slower, more expensive PLD. Fortunately, most software tools for both ASIC and PLD design have a minimization program built in. The purpose of Sections 4.3.3 through 4.3.8 is to give you a feel for how minimization works.

4.3.3 Combinational-Circuit Minimization

It's often uneconomical to realize a logic circuit directly from the first logic expression that pops into your head. Canonical sum and product expressions are especially expensive because the number of possible minterms or maxterms (and hence gates) grows exponentially with the number of variables. We *minimize* a combinational circuit by reducing the number and size of gates that are needed to build it.

minimize

The traditional combinational-circuit-minimization methods that we'll study have as their starting point a truth table or, equivalently, a minterm list or maxterm list. If we are given a logic function that is not expressed in this form, then we must convert it to an appropriate form before using these methods. For example, if we are given an arbitrary logic expression, then we can evaluate it for every input combination to construct the truth table. The minimization methods reduce the cost of a two-level AND-OR, OR-AND, NAND-NAND, or NOR-NOR circuit in three ways:

1. By minimizing the number of first-level gates.
2. By minimizing the number of inputs on each first-level gate.
3. By minimizing the number of inputs on the second-level gate. This is actually a side effect of the first reduction.

However, the minimization methods do not consider the cost of input inverters; they assume that both true and complemented versions of all input variables are available. While this is not always the case in gate-level or ASIC design, it's very appropriate for PLD-based design; PLDs have both true and complemented versions of all input variables available "for free."

Most minimization methods are based on a generalization of the combining theorems, T10 and T10':

$$\text{given product term} \cdot \text{Y} + \text{given product term} \cdot \text{Y}' = \text{given product term}$$

$$(\text{given sum term} + \text{Y}) \cdot (\text{given sum term} + \text{Y}') = \text{given sum term}$$

That is, if two product or sum terms differ only in the complementing or not of one variable, we can combine them into a single term with one less variable. So we save one gate and the remaining gate has one fewer input.

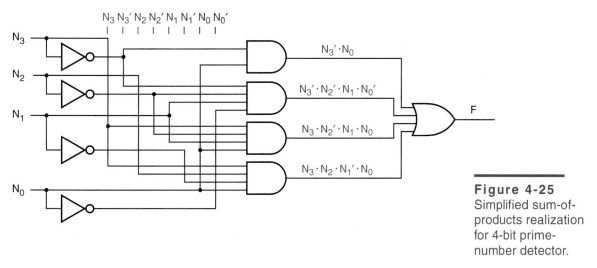

Figure 4-25
Simplified sum-of-products realization for 4-bit prime-number detector.

We can apply this algebraic method repeatedly to combine minterms 1, 3, 5, and 7 of the prime-number detector shown in Figure 4-18 on page 215:

$$
\begin{aligned}
F &= \Sigma_{N_3,N_2,N_1,N_0}(1, 3, 5, 7, 2, 11, 13) \\
&= N_3' \cdot N_2' N_1' N_0 + N_3' \cdot N_2' \cdot N_1 \cdot N_0 + N_3' \cdot N_2 \cdot N_1' \cdot N_0 + N_3' \cdot N_2 \cdot N_1 \cdot N_0 + \ldots \\
&= (N_3' \cdot N_2' \cdot N_1' \cdot N_0 + N_3' \cdot N_2' \cdot N_1 \cdot N_0) + (\cdot N_3' \cdot N_2 \cdot N_1' \cdot N_0 + N_3' \cdot N_2 \cdot N_1 \cdot N_0) + \ldots \\
&= N_3' N_2' \cdot N_0 + N_3' \cdot N_2 \cdot N_0 + \ldots \\
&= N_3' \cdot N_0 + \ldots
\end{aligned}
$$

The resulting circuit is shown in Figure 4-25; it has three fewer gates, and one of the remaining gates has two fewer inputs.

If we had worked a little harder on the preceding expression, we could have saved a couple more first-level gate inputs, though not any gates. It's difficult to find terms that can be combined in a jumble of algebraic symbols. In the next subsection we'll begin to explore a minimization method that is more fit for human consumption. Our starting point will be the graphical equivalent of a truth table.

4.3.4 Karnaugh Maps

A *Karnaugh map* is a graphical representation of a logic function's truth table.   *Karnaugh map*
Figure 4-26 shows Karnaugh maps for logic functions of 2, 3, and 4 variables. The map for an *n*-input logic function is an array with 2^n cells, one for each possible input combination or minterm.

The rows and columns of a Karnaugh map are labeled so that the input combination for any cell is easily determined from the row and column headings for that cell. The small number inside each cell is the corresponding minterm number in the truth table, assuming that the truth-table inputs are labeled alphabetically from left to right (e.g., X, Y, Z) and the rows are numbered in binary

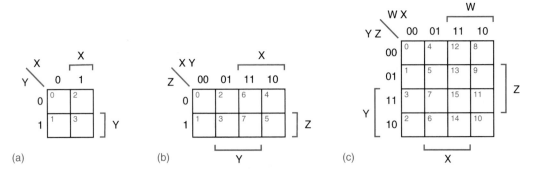

Figure 4-26 Karnaugh maps: (a) 2-variable; (b) 3-variable; (c) 4-variable.

counting order, like all the examples in this text. For example, cell 13 in the 4-variable map corresponds to the truth-table row in which $W \, X \, Y \, Z = 1101$.

When we draw the Karnaugh map for a given function, each cell of the map contains the information from the like-numbered row of the function's truth table—a 0 if the function is 0 for that input combination, a 1 otherwise.

In this text we use two redundant labelings for map rows and columns. For example, consider the 4-variable map in Figure 4-26(c). The columns are labeled with the four possible combinations of W and X, $W \, X = 00$, 01, 11, and 10. Similarly, the rows are labeled with the $Y \, Z$ combinations. These labels give us all the information we need. However, we also use brackets to associate four regions of the map with the four variables. Each bracketed region is the part of the map in which the indicated variable is 1. Obviously, the brackets convey the same information that is given by the row and column labels.

When we draw a map by hand, it is much easier to draw the brackets than to write out all of the labels. However, we retain the labels in the text's Karnaugh maps as an additional aid to understanding. In any case, you must be sure to label the rows and columns in the proper order to preserve the correspondence between map cells and truth-table row numbers shown in Figure 4-26.

To represent a logic function on a Karnaugh map, we simply copy 1s and 0s from the truth table or equivalent to the corresponding cells of the map. Figures 4-27(a) and (b) show the truth table and Karnaugh map for a logic function that we analyzed (beat to death?) in Section 4.2. From now on, we'll reduce the clutter in maps by copying only the 1s or the 0s, not both.

4.3.5 Minimizing Sums of Products

By now you must be wondering about the "strange" ordering of the row and column numbers in a Karnaugh map. There is a very important reason for this ordering—each cell corresponds to an input combination that differs from each of its immediately adjacent neighbors in only one variable. For example, cells 5 and 13 in the 4-variable map differ only in the value of W. In the 3- and

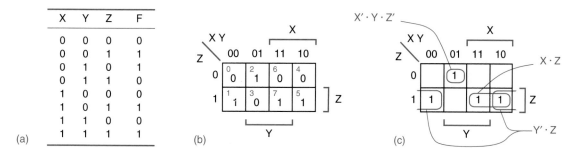

Figure 4-27 $F = \Sigma_{X,Y,Z}(1,2,5,7)$: (a) truth table; (b) Karnaugh map; (c) combining adjacent 1-cells.

4-variable maps, corresponding cells on the left/right or top/bottom borders are less obvious neighbors; for example, cells 12 and 14 in the 4-variable map are adjacent because they differ only in the value of Y.

Each input combination with a "1" in the truth table corresponds to a minterm in the logic function's canonical sum. Since pairs of adjacent "1" cells in the Karnaugh map have minterms that differ in only one variable, the minterm pairs can be combined into a single product term using the generalization of theorem T10, term $\cdot$ Y + term $\cdot$ Y' = term. Thus, we can use a Karnaugh map to simplify the canonical sum of a logic function.

For example, consider cells 5 and 7 in Figure 4-27(b) and their contribution to the canonical sum for this function:

$$
\begin{aligned}
F &= \cdots + X \cdot Y' \cdot Z + X \cdot Y \cdot Z \\
&= \cdots + (X \cdot Z) \cdot Y' + (X \cdot Z) \cdot Y \\
&= \cdots + X \cdot Z
\end{aligned}
$$

Remembering wraparound, we see that cells 1 and 5 in Figure 4-27(b) are also adjacent and can be combined:

$$
\begin{aligned}
F &= X' \cdot Y' \cdot Z + X \cdot Y' \cdot Z + \cdots \\
&= X' \cdot (Y' \cdot Z) + X \cdot (Y' \cdot Z) + \cdots \\
&= Y' \cdot Z + \ldots
\end{aligned}
$$

In general, we can simplify a logic function by combining pairs of adjacent 1-cells (minterms) whenever possible and writing a sum of product terms that covers all of the 1-cells. Figure 4-27(c) shows the result for our example logic function. We circle a pair of 1s to indicate that the corresponding minterms are combined into a single product term. The corresponding **AND-OR** circuit is shown in Figure 4-28.

In many logic functions the cell-combining procedure can be extended to combine more than two 1-cells into a single product term. For example, consider

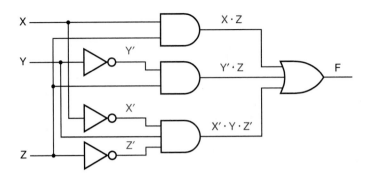

Figure 4-28
Minimized AND-OR circuit.

the canonical sum for the logic function $F = \Sigma_{X,Y,Z}(0, 1, 4, 5, 6)$. We can use the algebraic manipulations of the previous examples iteratively to combine four of the five minterms:

$$
\begin{aligned}
F &= X' \cdot Y' \cdot Z' + X' \cdot Y' \cdot Z + X \cdot Y' \cdot Z' + X \cdot Y' \cdot Z + X \cdot Y \cdot Z' \\
&= [(X' \cdot Y') \cdot Z' + (X' \cdot Y') \cdot Z] + [(X \cdot Y') \cdot Z' + (X \cdot Y') \cdot Z] + X \cdot Y \cdot Z' \\
&= X' \cdot Y' + X \cdot Y' + X \cdot Y \cdot Z' \\
&= [X' \cdot (Y') + X \cdot (Y')] + X \cdot Y \cdot Z' \\
&= Y' + X \cdot Y \cdot Z'
\end{aligned}
$$

In general, 2^i 1-cells may be combined to form a product term containing $n - i$ literals, where n is the number of variables in the function.

A precise mathematical rule determines how 1-cells may be combined and the form of the corresponding product term:

- A set of 2^i 1-cells may be combined if there are i variables of the logic function that take on all 2^i possible combinations within that set, while the remaining $n - i$ variables have the same value throughout that set. The corresponding product term has $n - i$ literals, where a variable is complemented if it appears as 0 in all of the 1-cells, and uncomplemented if it appears as 1.

rectangular sets of 1s   Graphically, this rule means that we can circle *rectangular* sets of 2^i 1s, literally as well as figuratively stretching the definition of rectangular to account for wraparound at the edges of the map. We can determine the literals of the corresponding product terms directly from the map; for each variable we make the following determination:

- If a circle covers only areas of the map where the variable is 0, then the variable is complemented in the product term.

- If a circle covers only areas of the map where the variable is 1, then the variable is uncomplemented in the product term.

- If a circle covers both areas of the map where the variable is 0 and areas where it is 1, then the variable does not appear in the product term.

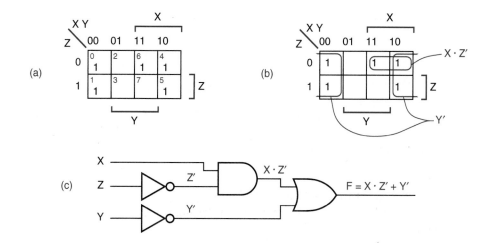

Figure 4-29
$F = \Sigma_{X,Y,Z}(0,1,4,5,6)$:
(a) initial Karnaugh map; (b) Karnaugh map with circled product terms;
(c) AND/OR circuit.

A sum-of-products expression for a function must contain product terms (circled sets of 1-cells) that cover all of the 1s and none of the 0s on the map.

The Karnaugh map for our most recent example, $F = \Sigma_{X,Y,Z}(0, 1, 4, 5, 6)$, is shown in Figure 4-29(a) and (b). We have circled one set of four 1s, corresponding to the product term Y', and a set of two 1s corresponding to the product term $X \cdot Z'$. Notice that the second product term has one less literal than the corresponding product term in our algebraic solution $(X \cdot Y \cdot Z')$. By circling the largest possible set of 1s containing cell 6, we have found a less expensive realization of the logic function, since a 2-input AND gate should cost less than a 3-input one. The fact that two different product terms now cover the same 1-cell (4) does not affect the logic function, since for logical addition $1 + 1 = 1$, not 2! The corresponding two-level AND/OR circuit is shown in (c).

As another example, the prime-number detector circuit that we introduced in Figure 4-18 on page 215 can be minimized as shown in Figure 4-30.

At this point, we need some more definitions to clarify what we're doing:

- A *minimal sum* of a logic function $F(X_1,...,X_n)$ is a sum-of-products   *minimal sum*
  expression for F such that no sum-of-products expression for F has fewer product terms, and any sum-of-products expression with the same number of product terms has at least as many literals.

That is, the minimal sum has the fewest possible product terms (first-level gates and second-level gate inputs) and, within that constraint, the fewest possible literals (first-level gate inputs). Thus, among our three prime-number detector circuits, only the one in Figure 4-30 on the next page realizes a minimal sum.

The next definition says precisely what the word "imply" means when we talk about logic functions:

- A logic function $P(X_1,...,X_n)$ *implies* a logic function $F(X_1,...,X_n)$ if for every   *imply*
  input combination such that $P = 1$, then $F = 1$ also.

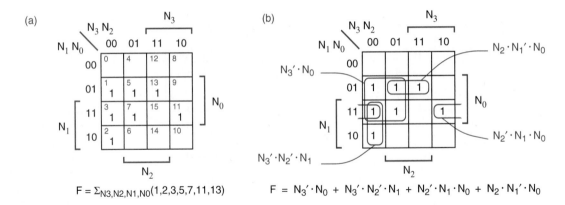

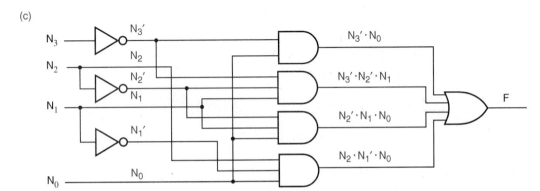

Figure 4-30 Prime-number detector: (a) initial Karnaugh map; (b) circled product terms; (c) minimized circuit.

That is, if P implies F, then F is 1 for every input combination that P is 1, and maybe some more. We may write the shorthand $P \Rightarrow F$. We may also say that "F *includes* P," or that "F *covers* P."

includes

covers

prime implicant

- A *prime implicant* of a logic function $F(X_1,...,X_n)$ is a normal product term $P(X_1,...,X_n)$ that implies F, such that if any variable is removed from P, then the resulting product term does not imply F.

In terms of a Karnaugh map, a prime implicant of F is a circled set of 1-cells satisfying our combining rule, such that if we try to make it larger (covering twice as many cells), it covers one or more 0s.

Now comes the most important part, a theorem that limits how much work we must do to find a minimal sum for a logic function:

Prime-Implicant Theorem   A minimal sum is a sum of prime implicants.

That is, to find a minimal sum, we need not consider any product terms that are not prime implicants. This theorem is easily proved by contradiction. Suppose

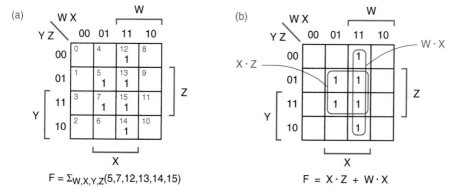

F = X·Z + W·X

Figure 4-31 $F = \Sigma_{W,X,Y,Z}(5,7,12,13,14,15)$: (a) Karnaugh map; (b) prime implicants.

that a product term P in a "minimal" sum is *not* a prime implicant. Then according to the definition of prime implicant, if P is not one, it is possible to remove some literal from P to obtain a new product term P* that still implies F. If we replace P with P* in the presumed "minimal" sum, the resulting sum still equals F but has one fewer literal. Therefore, the presumed "minimal" sum was not minimal after all.

Another minimization example, this time a 4-variable function, is shown in Figure 4-31. There are just two prime implicants, and it's quite obvious that both of them must be included in the minimal sum in order to cover all of the 1-cells on the map. We didn't draw the logic diagram for this example because you should know how to do that yourself by now.

The sum of all the prime implicants of a logic function is called the *complete sum*. Although the complete sum is always a legitimate way to realize a logic function, it's not always minimal. For example, consider the logic function shown in Figure 4-32. It has five prime implicants, but the minimal sum includes

complete sum

Figure 4-32 $F = \Sigma_{W,X,Y,Z}(1,3,4,5,9,11,12,13,14,15)$: (a) Karnaugh map; (b) prime implicants and distinguished 1-cells.

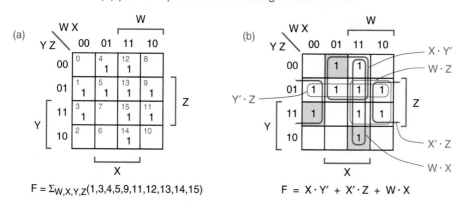

F = $\Sigma_{W,X,Y,Z}(1,3,4,5,9,11,12,13,14,15)$ F = X·Y' + X'·Z + W·X

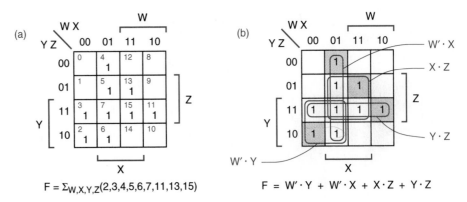

Figure 4-33 F $=\Sigma_{W,X,Y,Z}(2,3,4,5,6,7,11,13,15)$: (a) Karnaugh map; (b) prime implicants and distinguished 1-cells.

only three of them. So, how can we systematically determine which prime implicants to include and which to leave out? Two more definitions are needed:

distinguished 1-cell

- A *distinguished 1-cell* of a logic function is an input combination that is covered by only one prime implicant.

essential prime implicant

- An *essential prime implicant* of a logic function is a prime implicant that covers one or more distinguished 1-cells.

Since an essential prime implicant is the *only* prime implicant that covers some 1-cell, it *must* be included in every minimal sum for the logic function. So, the first step in the prime-implicant selection process is simple—we identify distinguished 1-cells and the corresponding prime implicants and include the essential prime implicants in the minimal sum. Then we need only determine how to cover the 1-cells, if any, that are not covered by the essential prime implicants. In the example of Figure 4-32, the three distinguished 1-cells are shaded, and the corresponding essential prime implicants are circled with heavier lines. All of the 1-cells in this example are covered by essential prime implicants, so we need go no further. Likewise, Figure 4-33 shows an example where all of the prime implicants are essential, and so all are included in the minimal sum.

A logic function in which not all the 1-cells are covered by essential prime implicants is shown in Figure 4-34. By removing the essential prime implicants and the 1-cells they cover, we obtain a reduced map with only a single 1-cell and two prime implicants that cover it. The choice in this case is simple—we use the $W' \cdot Z$ product term because it has fewer inputs and therefore lower cost.

For more complex cases, we need yet another definition:

eclipse

- Given two prime implicants P and Q in a reduced map, P is said to eclipse Q (written P ... Q) if P covers at least all the 1-cells covered by Q.

If P costs no more than Q and eclipses Q, then removing Q from consideration cannot prevent us from finding a minimal sum; that is, P is at least as good as Q.

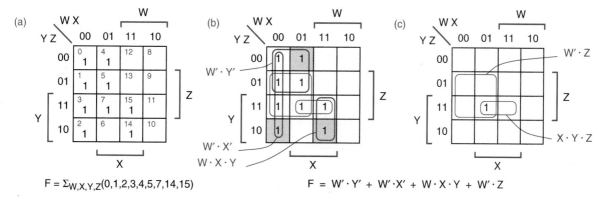

F = Σ_W,X,Y,Z(0,1,2,3,4,5,7,14,15)

F = W′·Y′ + W′·X′ + W·X·Y + W′·Z

Figure 4-34 $F = \Sigma_{W,X,Y,Z}(0,1,2,3,4,5,7,14,15)$: (a) Karnaugh map; (b) prime implicants and distinguished 1-cells; (c) reduced map after removal of essential prime implicants and covered 1-cells.

An example of eclipsing is shown in Figure 4-35. After removing essential prime implicants, we are left with two 1-cells, each of which is covered by two prime implicants. However, $X \cdot Y \cdot Z$ eclipses the other two prime implicants, which therefore may be removed from consideration. The two 1-cells are then covered only by $X \cdot Y \cdot Z$, which is a *secondary essential prime implicant* that must be included in the minimal sum.

secondary essential prime implicant

Figure 4-36 shows a more difficult case—a logic function with no essential prime implicants. By trial and error we can find two different minimal sums for this function.

We can also approach the problem systematically using the *branching method*. Starting with any 1-cell, we arbitrarily select one of the prime implicants that covers it, and we include it as if it were essential. This simplifies the

branching method

Figure 4-35 $F = \Sigma_{W,X,Y,Z}(2,6,7,9,13,15)$: (a) Karnaugh map; (b) prime implicants and distinguished 1-cells; (c) reduced map after removal of essential prime implicants and covered 1-cells.

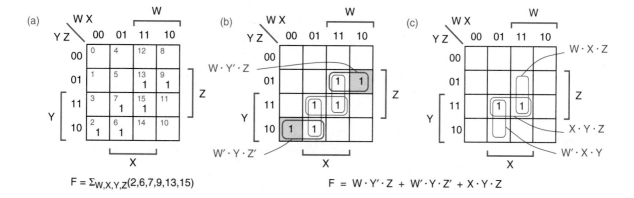

F = Σ_W,X,Y,Z(2,6,7,9,13,15)

F = W·Y′·Z + W′·Y·Z′ + X·Y·Z

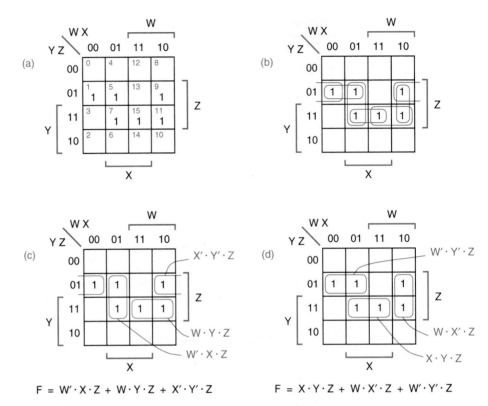

Figure 4-36 $F = \Sigma_{W,X,Y,Z}(1,5,7,9,11,15)$: (a) Karnaugh map; (b) prime implicants; (c) a minimal sum; (d) another minimal sum.

remaining problem, which we can complete in the usual way to find a tentative minimal sum. We repeat this process starting with all other prime implicants that cover the starting 1-cell, generating a different tentative minimal sum from each starting point. We may get stuck along the way and have to apply the branching method recursively. Finally, we examine all of the tentative minimal sums that we generated in this way and select one that is truly minimal.

4.3.6 Simplifying Products of Sums

Using the principle of duality, we can minimize product-of-sums expressions by looking at the 0s on a Karnaugh map. Each 0 on the map corresponds to a maxterm in the canonical product of the logic function. The entire process in the preceding subsection can be reformulated in a dual way, including the rules for writing sum terms corresponding to circled sets of 0s, in order to find a *minimal product*.

minimal product

Fortunately, once we know how to find minimal sums, there's an easier way to find the minimal product for a given logic function F. The first step is to complement F to obtain F′. Assuming that F is expressed as a minterm list or a truth table, complementing is very easy; the 1s of F' are just the 0s of F. Next, we find a minimal sum for F' using the method of the preceding subsection. Finally, we complement the result using the generalized DeMorgan's theorem, which yields a minimal product for $(F')' = F$. (Note that if you simply "add out" the minimal-sum expression for the original function, the resulting product-of-sums expression is not necessarily minimal; for example, see Exercise 4.61.)

In general, to find the lowest-cost two-level realization of a logic function, we have to find both a minimal sum and a minimal product and compare them. If a minimal sum for a logic function has many terms, then a minimal product for the same function may have few terms. As a trivial example of this tradeoff, consider a 4-input OR function:

$$F = (W) + (X) + (Y) + (Z) \text{ (a sum of four trivial product terms)}$$
$$= (W + X + Y + Z) \text{ (a product of one sum term)}$$

For a nontrivial example, you're invited to find the minimal product for the function that we minimized in Figure 4-33 on page 228; it has just two sum terms.

The opposite situation is also sometimes true, as trivially illustrated by a 4-input AND:

$$F = (W) \cdot (X) \cdot (Y) \cdot (Z) \text{ (a product of four trivial sum terms)}$$
$$= (W \cdot X \cdot Y \cdot Z) \text{ (a sum of one product term)}$$

A nontrivial example with a higher-cost product-of-sums is the function in Figure 4-29 on page 225.

For some logic functions, both minimal forms are equally costly. For example, consider a 3-input "Exclusive OR" function; both minimal expressions have four terms, and each term has three literals:

$$F = \Sigma_{X,Y,Z}(1,2,4,7)$$
$$= (X' \cdot Y' \cdot Z) + (X' \cdot Y \cdot Z') + (X \cdot Y' \cdot Z') + (X \cdot Y \cdot Z)$$
$$= (X + Y + Z) \cdot (X + Y' + Z') \cdot (X' + Y + Z') \cdot (X' + Y' + Z)$$

Still, in most cases, one form or the other will give better results. Looking at both forms is especially useful in PLD-based designs.

PLD MINIMIZATION   Typical PLDs have an AND-OR array corresponding to a sum-of-products form, so you might think that only the minimal sum-of-products is relevant to a PLD-based design. However, most PLDs also have a programmable inverter/buffer at the output of the AND-OR array, which can either invert or not. Thus, the PLD can utilize the equivalent of the minimal sum by using the AND-OR array to realize the complement of the desired function and then programming the inverter/buffer to invert. Most logic-minimization programs for PLDs automatically find both the minimal sum and the minimal product and select the one that requires fewer terms.

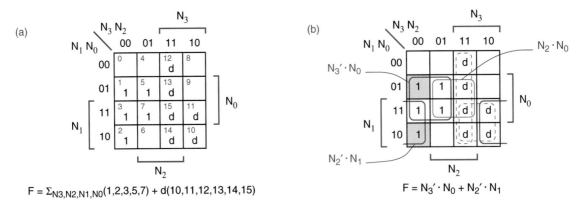

$$F = \Sigma_{N3,N2,N1,N0}(1,2,3,5,7) + d(10,11,12,13,14,15)$$

$$F = N_3{}' \cdot N_0 + N_2{}' \cdot N_1$$

Figure 4-37 Prime BCD-digit detector: (a) initial Karnaugh map;
(b) Karnaugh map with prime implicants and distinguished 1-cells.

*4.3.7 "Don't-Care" Input Combinations

don't-care

Sometimes the specification of a combinational circuit is such that its output doesn't matter for certain input combinations, called *don't-cares*. This may be true because the outputs really don't matter when these input combinations occur, or because these input combinations never occur in normal operation. For example, suppose we wanted to build a prime-number detector whose 4-bit input $N = N_3N_2N_1N_0$ is always a BCD digit; then minterms 10–15 should never occur. A prime BCD-digit detector function may therefore be written as follows:

$$F \;=\; \Sigma_{N_3,N_2,N_1,N_0}(1,2,3,5,7) + d(10,11,12,13,14,15)$$

d-set

The d(...) list specifies the don't-care input combinations for the function, also known as the *d-set*. Here F must be 1 for input combinations in the on-set $(1,2,3,5,7)$, F can have any values for inputs in the d-set $(10,11,12,13,14,15)$, and F must be 0 for all other input combinations (in the 0-set).

Figure 4-37 shows how to find a minimal sum-of-products realization for the prime BCD-digit detector, including don't-cares. The d's in the map denote the don't-care input combinations. We modify the procedure for circling sets of 1s (prime implicants) as follows:

- Allow d's to be included when circling sets of 1s, to make the sets as large as possible. This reduces the number of variables in the corresponding prime implicants. Two such prime implicants ($N_2 \cdot N_0$ and $N_2{}' \cdot N_1$) appear in the example.

- Do not circle any sets that contain only d's. Including the corresponding product term in the function would unnecessarily increase its cost. Two such product terms ($N_3 \cdot N_2$ and $N_3 \cdot N_1$) are circled in the example.

- Just a reminder: As usual, do not circle any 0s.

*Throughout this book, optional sections are marked with an asterisk.

The remainder of the procedure is the same. In particular, we look for distinguished 1-cells and *not* distinguished d-cells, and we include only the corresponding essential prime implicants and any others that are needed to cover all the 1s on the map. In Figure 4-37, the two essential prime implicants are sufficient to cover all of the 1s on the map. Two of the d's also happen to be covered, so F will be 1 for don't-care input combinations 10 and 11, and 0 for the other don't-cares.

Some HDLs, including ABEL, provide a means for the designer to specify don't-care inputs, and the logic-minimization program takes these into account when computing a minimal sum.

*4.3.8 Multiple-Output Minimization

Most practical combinational logic circuits require more than one output. We can always handle a circuit with n outputs as n independent single-output design problems. However, in doing so, we may miss some opportunities for optimization. For example, consider the following two logic functions:

$$F = \Sigma_{X,Y,Z}(3,6,7)$$
$$G = \Sigma_{X,Y,Z}(0,1,3)$$

Figure 4-38 shows the design of F and G as two independent single-output functions. However, as shown in Figure 4-39, we can also find a pair of sum-of-products expressions that share a product term, such that the resulting circuit has one fewer gate than our original design.

Figure 4-38 Treating a 2-output design as two independent single-output designs: (a) Karnaugh maps; (b) "minimal" circuit.

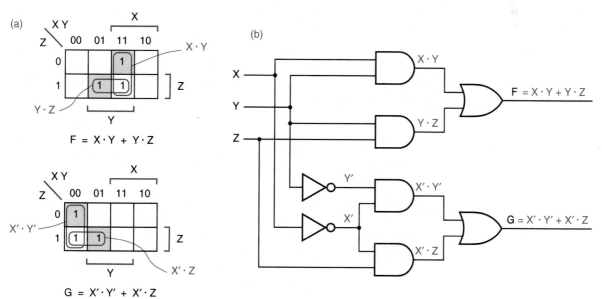

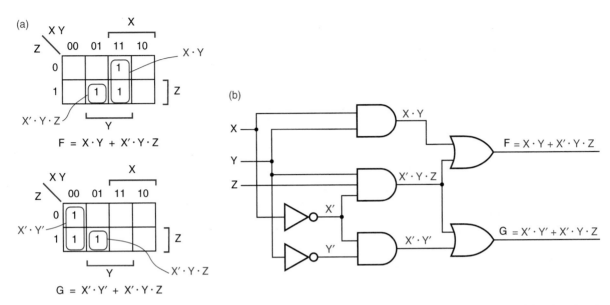

Figure 4-39 Multiple-output minimization for a 2-output circuit: (a) minimized maps including a shared term; (b) minimal multiple-output circuit

When we design multiple-output combinational circuits using discrete gates, as in an ASIC, product-term sharing obviously reduces circuit size and cost. In addition, PLDs contain multiple copies of the sum-of-products structure that we've been learning how to minimize, one per output, and some PLDs allow product terms to be shared among multiple outputs. Thus, the ideas introduced in this subsection are used in many logic-minimization programs.

You probably could have "eyeballed" the Karnaugh maps for F and G in Figure 4-39 and discovered the minimal solution. However, larger circuits can be minimized only with a formal multiple-output minimization algorithm. We'll outline the ideas in such an algorithm here; details can be found in the References.

m-product function

The key to successful multiple-output minimization of a set of n functions is to consider not only the n original single-output functions, but also "product functions." An *m-product function* of a set of n functions is the product of m of the functions, where $2 \leq m \leq n$. There are $2^n - n - 1$ such functions. Fortunately, $n = 2$ in our example and there is only one product function, $F \cdot G$, to consider. The Karnaugh maps for F, G, and F · G are shown in Figure 4-40; in general, the map for an *m*-product function is obtained by ANDing the maps of its *m* components.

multiple-output prime implicant

A *multiple-output prime implicant* of a set of n functions is a prime implicant of one of the n functions or of one of the product functions. The

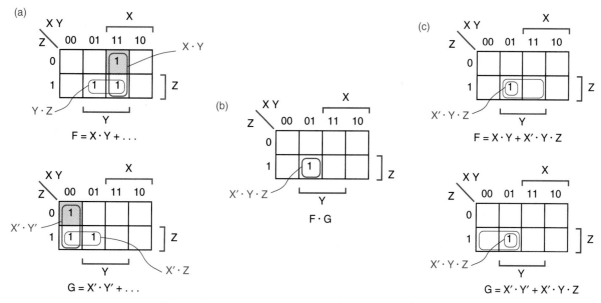

Figure 4-40 Karnaugh maps for a set of two functions: (a) maps for F and G;
(b) 2-product map for F · G; (c) reduced maps for F and G after
removal of essential prime implicants and covered 1-cells.

first step in multiple-output minimization is to find all of the multiple-output
prime implicants. Each prime implicant of an m-product function is a possible
term to include in the corresponding m outputs of the circuit. If we were trying to
minimize a set of 8 functions, we would have to find the prime implicants for
$2^8 - 8 - 1 = 247$ product functions as well as for the 8 given functions. Obviously,
multiple-output minimization is not for the faint-hearted!

Once we have found the multiple-output prime implicants, we try to
simplify the problem by identifying the essential ones. A *distinguished 1-cell* of
a particular single-output function F is a 1-cell that is covered by exactly one
prime implicant of F or of the product functions involving F. The distinguished
1-cells in Figure 4-40 are shaded. An *essential prime implicant* of a particular
single-output function is one that contains a distinguished 1-cell. As in single-
output minimization, the essential prime implicants must be included in a
minimum-cost solution. Only the 1-cells that are not covered by essential prime
implicants are considered in the remainder of the algorithm.

The final step is to select a minimal set of prime implicants to cover the
remaining 1-cells. In this step we must consider all n functions simultaneously,
including the possibility of sharing; details of this procedure are discussed in the
References. In the example of Figure 4-40(c), we see that there exists a single,
shared product term that covers the remaining 1-cell in both F and G.

distinguished 1-cell
*essential prime
  implicant*

*4.4 Programmed Minimization Methods

Obviously, logic minimization can be a very involved process. In real logic-design applications you are likely to encounter only two kinds of minimization problems: functions of a few variables that you can "eyeball" using the methods of the previous section, and more complex, multiple-output functions that are hopeless without the use of a minimization program.

Quine-McCluskey
algorithm

We know that minimization can be performed visually for functions of a few variables using the Karnaugh-map method. We'll show in this section that the same operations can be performed for functions of an arbitrarily large number of variables (at least in principle) using a tabular method called the *Quine-McCluskey algorithm*. Like all algorithms, the Quine-McCluskey algorithm can be translated into a computer program. And like the map method, the algorithm has two steps: (a) finding all prime implicants of the function, and (b) selecting a minimal set of prime implicants that covers the function.

The Quine-McCluskey algorithm is often described in terms of handwritten tables and manual check-off procedures. However, since no one ever uses these procedures manually, it's more appropriate for us to discuss the algorithm in terms of data structures and functions in a high-level programming language. The goal of this section is to give you an appreciation for computational complexity involved in a large minimization problem. We consider only fully specified, single-output functions; don't-cares and multiple-output functions can be handled by fairly straightforward modifications to the single-output algorithms, as discussed in the References.

*4.4.1 Representation of Product Terms

The starting point for the Quine-McCluskey minimization algorithm is the truth table or, equivalently, the minterm list of a function. If the function is specified differently, it must first be converted into this form. For example, an arbitrary n-variable logic expression can be multiplied out (perhaps using DeMorgan's theorem along the way) to obtain a sum-of-products expression. Once we have a sum-of-products expression, each p-variable product term produces 2^{n-p} minterms in the minterm list.

We showed in Section 4.1.6 that a minterm of an n-variable logic function can be represented by an n-bit integer (the minterm number), where each bit indicates whether the corresponding variable is complemented or uncomplemented. However, a minimization algorithm must also deal with product terms that are not minterms, where some variables do not appear at all. Thus, we must represent three possibilities for each variable in a general product term:

1   Uncomplemented.

0   Complemented.

x   Doesn't appear.

These possibilities are represented by a string of *n* of the above digits in the *cube* *representation* of a product term. For example, if we are working with product terms of up to eight variables, X7, X6, ..., X1, X0, we can write the following product terms and their cube representations:

cube representation

$$X7' \cdot X6 \cdot X5 \cdot X4' \cdot X3 \cdot X2 \cdot X1 \cdot X0' \equiv 01101110$$

$$X3 \cdot X2 \cdot X1 \cdot X0' \equiv \text{xxxx}1110$$

$$X7 \cdot X5' \cdot X4 \cdot X3 \cdot X2' \cdot X1 \equiv 1\text{x}01101\text{x}$$

$$X6 \equiv \text{x}1\text{xxxxxx}$$

Notice that for convenience, we named the variables just like the bit positions in *n*-bit binary integers.

In terms of the *n*-cube and *m*-subcube nomenclature of Section 2.14, the string 1x01101x represents a 2-subcube of an 8-cube, and the string 01101110 represents a 0-subcube of an 8-cube. However, in the minimization literature, the maximum dimension *n* of a cube or subcube is usually implicit, and an *m*-subcube is simply called an "*m*-cube" or a "cube" for short; we'll follow this practice in this section.

To represent a product term in a computer program, we can use a data structure with *n* elements, each of which has three possible values. In C, we might make the following declarations:

```
typedef enum {complemented, uncomplemented, doesntappear} TRIT;
typedef TRIT[16] CUBE;  /* Represents a single product
                           term with up to 16 variables */
```

However, these declarations do not lead to a particularly efficient internal representation of cubes. As we'll see, cubes are easier to manipulate using conventional computer instructions if an *n*-variable product term is represented by two *n*-bit computer words, as suggested by the following declarations:

```
#define MAX_VARS 16     /* Max # of variables in a product term */
typedef unsigned short WORD;   /* Use 16-bit words */
struct cube {
  WORD t; /* Bits 1 for uncomplemented variables. */
  WORD f; /* Bits 1 for complemented variables.   */
};
typedef struct cube CUBE;
CUBE P1, P2, P3;    /* Allocate three cubes for use by program. */
```

Here, a WORD is a 16-bit integer, and a 16-variable product term is represented by a record with two WORDs, as shown in Figure 4-41(a). The first word in a CUBE has a 1 for each variable in the product term that appears uncomplemented (or "true," t), and the second has a 1 for each variable that appears complemented (or "false," f). If a particular bit position has 0s in both WORDs, then the corresponding variable does not appear, while the case of a particular bit

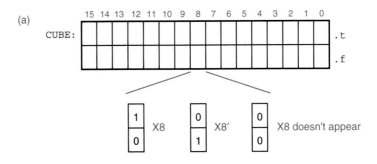

(a)

Figure 4-41
Internal representation
of 16-variable product terms
in a Pascal program:
(a) general format; (b) P1 =
X15·X12′·X10′·X9·X4′·X1·X0

(b)

position having 1s in both WORDs is not used. Thus, the program variable P1 in (b) represents the product term P1 = X15 · X12′ · X10′ · X9 · X4′ · X1 · X0. If we wished to represent a logic function F of up to 16 variables, containing up to 100 product terms, we could declare an array of 100 CUBEs:

```
CUBE F[100];      /* Storage for a logic function
                     with up to 100 product terms. */
```

Using the foregoing cube representation, it is possible to write short, efficient C functions that manipulate product terms in useful ways. Table 4-8 shows several such functions. Corresponding to two of the functions, Figure 4-42 depicts how two cubes can be compared and combined if possible

Figure 4-42 Cube manipulations: (a) determining whether two cubes are combinable using theorem T10, term · X + term · X′ = term; (b) combining cubes using theorem T10.

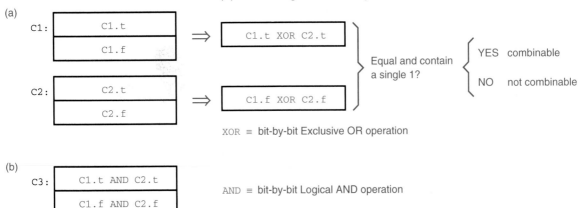

Table 4-8 Cube comparing and combining functions used in minimization program.

```
int EqualCubes(CUBE C1, CUBE C2)         /* Returns true if C1 and C2 are identical.  */
{
  return ( (C1.t == C2.t) && (C1.f == C2.f) );
}


int Oneone(WORD w)            /* Returns true if w has exactly one 1 bit.           */
{                             /* Optimizing the speed of this routine is critical   */
  int ones, b;                /*    and is left as an exercise for the hacker.      */
  ones = 0;
  for (b=0; b<MAX_VARS; b++) {
    if (w & 1) ones++;
    w = w>>1;
  }
  return((ones==1));
}


int Combinable(CUBE C1, CUBE C2)
{                             /* Returns true if C1 and C2 differ in only one variable, */
  WORD twordt, twordf;        /* which appears true in one and false in the other.      */

  twordt = C1.t ^ C2.t;
  twordf = C1.f ^ C2.f;
  return( (twordt==twordf) && Oneone(twordt) );
}


void Combine(CUBE C1, CUBE C2, CUBE *C3)
                             /* Combines C1 and C2 using theorem T10, and stores the   */
{                            /*    result in C3.  Assumes Combinable(C1,C2) is true.    */
  C3->t = C1.t & C2.t;
  C3->f = C1.f & C2.f;
}
```

using theorem T10, $term \cdot X + term \cdot X' = term$. This theorem says that two product terms can be combined if they differ in only one variable that appears complemented in one term and uncomplemented in the other. Combining two m-cubes yields an $(m + 1)$-cube. Using cube representation, we can apply the combining theorem to a few examples:

$$010 + 000 = 0x0$$
$$00111001 + 00111000 = 0011100x$$
$$101xx0x0 + 101xx1x0 = 101xxxx0$$
$$x111xx00110x000x + x111xx00010x000x = x111xx00x10x000x$$

*4.4.2 Finding Prime Implicants by Combining Product Terms

The first step in the Quine-McCluskey algorithm is to determine all of the prime implicants of the logic function. With a Karnaugh map, we do this visually by identifying "largest possible rectangular sets of 1s." In the algorithm, this is done by systematic, repeated application of theorem T10 to combine minterms, then 1-cubes, 2-cubes, and so on, creating the largest possible cubes (smallest possible product terms) that cover only 1s of the function.

The C program in Table 4-9 applies the algorithm to functions with up to 16 variables. It uses 2-dimensional arrays, cubes[m][j] and covered[m][j], to keep track of MAX_VARS m-cubes. The 0-cubes (minterms) are supplied by the user. Starting with the 0-cubes, the program examines every pair of cubes at each level and combines them when possible into cubes at the next level. Cubes that are combined into a next-level cube are marked as "covered"; cubes that are not covered are prime implicants.

Even though the program in Table 4-9 is short, an experienced programmer could become very pessimistic just looking at its structure. The inner for loop is nested four levels deep, and the number of times it might be executed is of the order of MAX_VARS · MAX_CUBES^3. That's right, that's an exponent, not a footnote! We picked the value maxCubes = 1000 somewhat arbitrarily (in fact, too optimistically for many functions), but if you believe this number, then the inner loop can be executed *billions and billions* of times.

The maximum number of minterms of an n-variable function is 2^n, of course, and so by all rights the program in Table 4-9 should declare maxCubes to be 2^{16}, at least to handle the maximum possible number of 0-cubes. Such a declaration would not be overly pessimistic. If an n-variable function has a product term equal to a single variable, then 2^{n-1} minterms are in fact needed to cover that product term.

For larger cubes, the situation is actually worse. The number of possible m-subcubes of an n-cube is $\binom{n}{m} \times 2^{n-m}$, where the binomial coefficient $\binom{n}{m}$ is the number of ways to choose m variables to be x's, and 2^{n-m} is the number of ways to assign 0s and 1s to the remaining variables. For 16-variable functions, the worst case occurs with $m = 5$; there are 8,945,664 possible 5-subcubes of a 16-cube. The total number of distinct m-subcubes of an n-cube, over all values of m, is 3^n. So a general minimization program might require a *lot* more memory than we've allocated in Table 4-9.

There are a few things that we can do to optimize the storage space and execution time required in Table 4-9 (see Exercises 4.77–4.80), but they are piddling compared to the overwhelming combinatorial complexity of the problem. Thus, even with today's fast computers and huge memories, direct application of the Quine-McCluskey algorithm for generating prime implicants is generally limited to functions with only a few variables (fewer than 15–20).

Table 4-9 A C program that finds prime implicants using the Quine-McCluskey algorithm.

```
#define TRUE    1
#define FALSE   0
#define MAX_CUBES 50

void main()
{
  CUBE cubes[MAX_VARS+1][MAX_CUBES];
  int covered[MAX_VARS+1][MAX_CUBES];
  int numCubes[MAX_VARS+1];
  int m;           /* Value of m in an m-cube, i.e., ''level m.'' */
  int j, k, p;     /* Indices into the cubes or covered array.    */
  CUBE tempCube;
  int found;

  /* Initialize number of m-cubes at each level m. */
  for (m=0; m<MAX_VARS+1; m++) numCubes[m] = 0;

  /* Read a list of minterms (0-cubes) supplied by the user, storing them   */
  /* in the cubes[0,j] subarray, setting covered[0,j] to false for each     */
  /* minterm, and setting numCubes[0] to the total number of minterms read. */
  ReadMinterms;

  for (m=0; m<MAX_VARS; m++)           /* Do for all levels except the last */
    for (j=0; j<numCubes[m]; j++)        /* Do for all cubes at this level   */
      for (k=j+1; k<numCubes[m]; k++)    /* Do for other cubes at this level */
        if (Combinable(cubes[m][j], cubes[m][k])) {
          /* Mark the cubes as covered. */
          covered[m][j] = TRUE;   covered[m][k] = TRUE;
          /* Combine into an (m+1)-cube, store in tempCube. */
          Combine(cubes[m][j], cubes[m][k], &tempCube);
          found = FALSE;   /* See if we've generated this one before. */
          for (p=0; p<numCubes[m+1]; p++)
            if (EqualCubes(cubes[m+1][p],tempCube)) found = TRUE;
          if (!found) {  /* Add the new cube to the next level. */
            numCubes[m+1] = numCubes[m+1] + 1;
            cubes[m+1][numCubes[m+1]-1] = tempCube;
            covered[m+1][numCubes[m+1]-1] = FALSE;
          }
        }
  for (m=0; m<MAX_VARS; m++)         /* Do for all levels                */
    for (j=0; j<numCubes[m]; j++)  /* Do for all cubes at this level */
      /* Print uncovered cubes -- these are the prime implicants. */
      if (!covered[m][j]) PrintCube(cubes[m][j]);
}
```

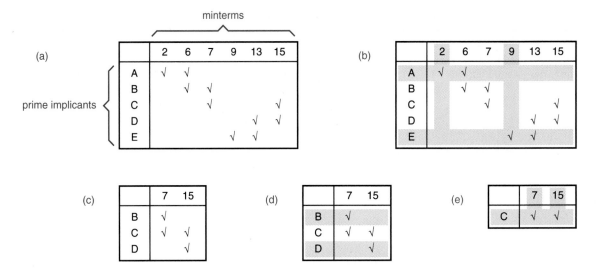

Figure 4-43 Prime-implicant tables: (a) original table; (b) showing distinguished 1-cells and essential prime implicants; (c) after removal of essential prime implicants; (d) showing eclipsed rows; (e) after removal of eclipsed rows, showing secondary essential prime implicant.

*4.4.3 Finding a Minimal Cover Using a Prime-Implicant Table

prime-implicant table

The second step in minimizing a combinational logic function, once we have a list of all its prime implicants, is to select a minimal subset of them to cover all the 1s of the function. The Quine-McCluskey algorithm uses a two-dimensional array called a *prime-implicant table* to do this. Figure 4-43(a) shows a small but representative prime-implicant table, corresponding to the Karnaugh-map minimization problem of Figure 4-35. There is one column for each minterm of the function, and one row for each prime implicant. Each entry is a bit that is 1 if and only if the prime implicant for that row covers the minterm for that column (shown in the figure as a check).

The steps for selecting prime implicants with the table are analogous to the steps that we used in Section 4.3.5 with Karnaugh maps:

1. Identify distinguished 1-cells. These are easily identified in the table as columns with a single 1, as shown in Figure 4-43(b).

2. Include all essential prime implicants in the minimal sum. A row that contains a check in one or more distinguished-1-cell columns corresponds to an essential prime implicant.

3. Remove from consideration the essential prime implicants and the 1-cells (minterms) that they cover. In the table, this is done by deleting the corresponding rows and columns, marked in color in Figure 4-43(b). If any

rows have no checks remaining, they are also deleted; the corresponding prime implicants are *redundant*, that is, completely covered by essential prime implicants. This step leaves the reduced table shown in (c).

redundant prime implicant

4. Remove from consideration any prime implicants that are "eclipsed" by others with equal or lesser cost. In the table, this is done by deleting any rows whose checked columns are a proper subset of another row's, and deleting all but one of a set of rows with identical checked columns. This is shown in color in (d) and leads to the further reduced table in (e).

   When a function is realized in a PLD, all of its prime implicants may be considered to have equal cost, because all of the AND gates in a PLD have all of the inputs available. Otherwise, the prime implicants must be sorted and selected according to the number of AND-gate inputs.

5. Identify distinguished 1-cells and include all secondary essential prime implicants in the minimal sum. As before, any row that contains a check in one or more distinguished-1-cell columns corresponds to a secondary essential prime implicant.

6. If all remaining columns are covered by the secondary essential prime implicants, as in (e), we're done. Otherwise, if any secondary essential prime implicants were found in the previous step, we go back to step 3 and iterate. Otherwise, the branching method must be used, as described in Section 4.3.5. This involves picking rows one at a time, treating them as if they were essential, and recursing (and cursing) on steps 3–6.

Although a prime-implicant table allows a fairly straightforward prime-implicant selection algorithm, the data structure required in a corresponding computer program is huge, since it requires on the order of $p \cdot 2^n$ bits, where p is the number of prime implicants and n is the number of input bits (assuming that the given function produces a 1 output for most input combinations). Worse, executing the steps that we so blithely described in a few sentences above requires a huge amount of computation.

*4.4.4 Other Minimization Methods

Although the previous subsections form an introduction to logic-minimization algorithms, the methods they describe are by no means the latest and greatest. Spurred on by the ever-increasing density of VLSI chips, many researchers have discovered more effective ways to minimize combinational logic functions. Their results fall roughly into three categories:

1. *Computational improvements.* Improved algorithms typically use clever data structures or rearrange the order of the steps to reduce the memory requirements and execution time of the classical algorithms.

2. *Heuristic methods.* Some minimization problems are just too big to be solved using an "exact" algorithm. These problems can be attacked using

shortcuts and well-educated guesses to reduce memory size and execution time to a fraction of what an "exact" algorithm would require. However, rather than finding a provably minimal expression for a logic function, heuristic methods attempt to find an "almost minimal" one.

Even for problems that can be solved by an "exact" method, a heuristic method typically finds a good solution ten times faster. The most successful heuristic program, Espresso-II, does in fact produce minimal or near-minimal results for the majority of problems (within one or two product terms), including problems with dozens of inputs and hundreds of product terms.

3. *Looking at things differently.* As we mentioned earlier, multiple-output minimization can be handled by straightforward, fairly mechanical modifications to single-output minimization methods. However, by looking at multiple-output minimization as a problem in multivalued (nonbinary) logic, the designers of the Espresso-MV algorithm were able to make substantial performance improvements over Espresso-II.

More information on these methods can be found in the References.

*4.5 Timing Hazards

steady-state behavior

The analysis methods that we developed in Section 4.2 ignore circuit delay and predict only the *steady-state behavior* of combinational logic circuits. That is, they predict a circuit's output as a function of its inputs under the assumption that the inputs have been stable for a long time, relative to the delays in the circuit's electronics. However, we showed in Section 3.6 that the actual delay from an input change to the corresponding output change in a real logic circuit is nonzero and depends on many factors.

transient behavior

glitch

hazard

Because of circuit delays, the *transient behavior* of a logic circuit may differ from what is predicted by a steady-state analysis. In particular, a circuit's output may produce a short pulse, often called a *glitch*, at a time when steady-state analysis predicts that the output should not change. A *hazard* is said to exist when a circuit has the possibility of producing such a glitch. Whether or not the glitch actually occurs depends on the exact delays and other electrical characteristics of the circuit. Since such parameters are difficult to control in production circuits, a logic designer must be prepared to eliminate hazards (the *possibility* of a glitch) even though a glitch may occur only under a worst-case combination of logical and electrical conditions.

*4.5.1 Static Hazards

static-1 hazard

A *static-1 hazard* is the possibility of a circuit's output producing a 0 glitch when we would expect the output to remain at a nice steady 1 based on a static analysis of the circuit function. A formal definition is given as follows.

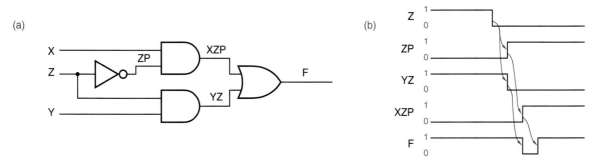

Figure 4-44 Circuit with a static-1 hazard: (a) logic diagram; (b) timing diagram.

Definition: A static-1 hazard is a pair of input combinations that: (a) differ in only one input variable and (b) both give a 1 output; such that it is possible for a momentary 0 output to occur during a transition in the differing input variable.

For example, consider the logic circuit in Figure 4-44(a). Suppose that X and Y are both 1 and that Z is changing from 1 to 0. Then (b) shows the timing diagram, assuming that the propagation delay through each gate or inverter is one unit time. Even though "static" analysis predicts that the output is 1 for both input combinations $X,Y,Z = 111$ and $X,Y,Z = 110$, the timing diagram shows that F goes to 0 for one unit time during a 1-0 transition on Z, because of the delay in the inverter that generates Z'.

A *static-0 hazard* is the possibility of a 1 glitch when we expect the circuit   *static-0 hazard*
to have a steady 0 output:

Definition: A static-0 hazard is a pair of input combinations that: (a) differ in only one input variable and (b) both give a 0 output; such that it is possible for a momentary 1 output to occur during a transition in the differing input variable.

Since a static-0 hazard is just the dual of a static-1 hazard, an OR-AND circuit that is the dual of Figure 4-44(a) would have a static-0 hazard.

An OR-AND circuit with four static-0 hazards is shown in Figure 4-45(a). One of the hazards occurs when $W,X,Y = 000$ and Z is changed, as shown in (b). You should be able to find the other three hazards and eliminate all of them after studying the next subsection.

*4.5.2 Finding Static Hazards Using Maps

A Karnaugh map can be used to detect static hazards in a two-level sum-of-products or product-of-sums circuit. The existence or nonexistence of static hazards depends on the circuit design for a logic function.

A properly designed two-level sum-of-products (AND-OR) circuit has no static-0 hazards. A static-0 hazard would exist in such a circuit only if both a

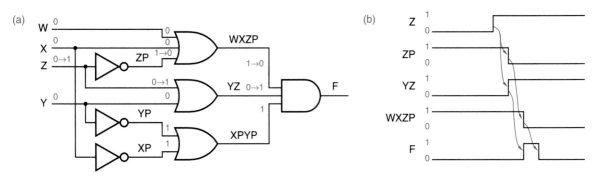

Figure 4-45 Circuit with static-0 hazards: (a) logic diagram; (b) timing diagram.

variable and its complement were connected to the same AND gate, which would be silly. However, the circuit *may* have static-1 hazards. Their existence can be predicted from a Karnaugh map where the product terms corresponding to the AND gates in the circuit are circled.

Figure 4-46(a) shows the Karnaugh map for the circuit of Figure 4-44. It is clear from the map that there is no single product term that covers both input combinations X,Y,Z = 111 and X,Y,Z = 110. Thus, intuitively, it is possible for the output to "glitch" momentarily to 0 if the AND gate output that covers one of the combinations goes to 0 before the AND gate output covering the other input combination goes to 1. The way to eliminate the hazard is also quite apparent: Simply include an extra product term (AND gate) to cover the hazardous input pair, as shown in Figure 4-46(b). The extra product term, it turns out, is the *consensus* of the two original terms; in general, we must add consensus terms to eliminate hazards. The corresponding hazard-free circuit is shown in Figure 4-47.

Another example is shown in Figure 4-48. In this example, three product terms must be added to eliminate the static-1 hazards.

A properly designed two-level product-of-sums (OR-AND) circuit has no static-1 hazards. It *may* have static-0 hazards, however. These hazards can be detected and eliminated by studying the adjacent 0s in the Karnaugh map, in a manner dual to the foregoing.

consensus

Figure 4-46
Karnaugh map for the circuit of Figure 4-44:
(a) as originally designed;
(b) with static-1 hazard eliminated.

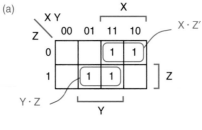

(a)

$$F = X \cdot Z' + Y \cdot Z$$

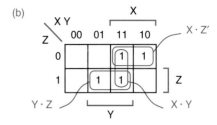

(b)

$$F = X \cdot Z' + Y \cdot Z + X \cdot Y$$

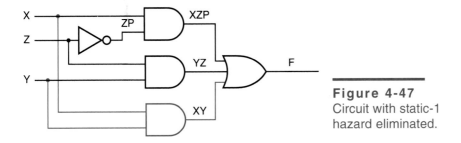

Figure 4-47
Circuit with static-1
hazard eliminated.

*4.5.3 Dynamic Hazards

A *dynamic hazard* is the possibility of an output changing more than once as the   *dynamic hazard*
result of a single input transition. Multiple output transitions can occur if there
are multiple paths with different delays from the changing input to the changing
output.

For example, consider the circuit in Figure 4-49; it has three different paths
from input X to the output F. One of the paths goes through a slow OR gate, and
another goes through an OR gate that is even slower. If the input to the circuit is
W,X,Y,Z = 0,0,0,1, then the output will be 1, as shown. Now suppose we change
the X input to 1. Assuming that all of the gates except the two marked "slow" and
"slower" are very fast, the transitions shown in black occur next, and the output
goes to 0. Eventually, the output of the "slow" OR gate changes, creating the
transitions shown in nonitalic color, and the output goes to 1. Finally, the output
of the "slower" OR gate changes, creating the transitions shown in italic color,
and the output goes to its final state of 0.

Dynamic hazards do not occur in a properly designed two-level AND-OR
or OR-AND circuit, that is, one in which no variable and its complement are con-

Figure 4-48 Karnaugh map for another sum-of-products circuit: (a) as originally
designed; (b) with extra product terms to cover static-1 hazards.

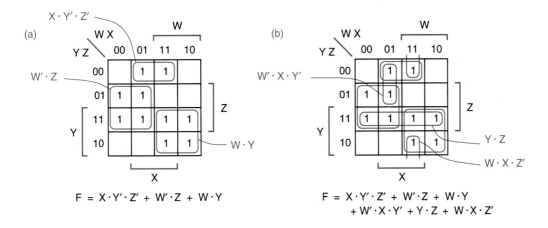

$$F = X \cdot Y' \cdot Z' + W' \cdot Z + W \cdot Y$$

$$F = X \cdot Y' \cdot Z' + W' \cdot Z + W \cdot Y$$
$$+ W' \cdot X \cdot Y' + Y \cdot Z + W \cdot X \cdot Z'$$

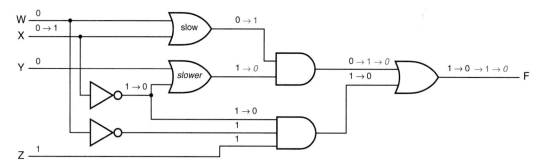

Figure 4-49 Circuit with a dynamic hazard.

nected to the same first-level gate. In multilevel circuits, dynamic hazards can be discovered using a method described in the References.

*4.5.4 Designing Hazard-Free Circuits

Only a few situations, such as the design of feedback sequential circuits, require hazard-free combinational circuits. Techniques for finding hazards in arbitrary circuits, described in the References, are rather difficult to use. So, when you require a hazard-free design, it's best to use a circuit structure that is easy to analyze.

In particular, we have indicated that a properly designed two-level AND-OR circuit has no static-0 or dynamic hazards. Static-1 hazards may exist in such a circuit, but they can be found and eliminated using the map method described earlier. If cost is not a problem, then a brute-force method of obtaining a hazard-free realization is to use the complete sum—the sum of all of the prime implicants of the logic function (see Exercise 4.84). In a dual manner, a hazard-free two-level OR-AND circuit can be designed for any logic function. Finally, note that everything we've said about AND-OR circuits naturally applies to the corresponding NAND-NAND designs, and about OR-AND applies to NOR-NOR.

MOST HAZARDS ARE NOT HAZARDOUS!	Any combinational circuit can be analyzed for the presence of hazards. However, a well-designed, *synchronous* digital system is structured so that hazard analysis is not needed for most of its circuits. In a synchronous system, all of the inputs to a combinational circuit are changed at a particular time, and the outputs are not "looked at" until they have had time to settle to a steady-state value. Hazard analysis and elimination are typically needed only in the design of asynchronous sequential circuits, such as the feedback sequential circuits discussed in Section 7.9. You'll rarely have reason to design such a circuit, but if you do, an understanding of hazards will be absolutely essential for a reliable result.

4.6 The ABEL Hardware Description Language

ABEL is a hardware description language (HDL) that was invented to allow designers to specify logic functions for realization in PLDs. An ABEL program is a text file containing several elements:

- Documentation, including program name and comments.

- Declarations that identify the inputs and outputs of the logic functions to be performed.

- Statements that specify the logic functions to be performed.

- Usually, a declaration of the type of PLD or other targeted device in which the specified logic functions are to be performed.

- Usually, "test vectors" that specify the logic functions' expected outputs for certain inputs.

ABEL is supported by an *ABEL language processor*, which we'll simply call an *ABEL compiler*. The compiler's job is to translate the ABEL text file into a "fuse pattern" that can be downloaded into a physical PLD. Even though most PLDs can be physically programmed only with patterns corresponding to sum-of-products expressions, ABEL allows PLD functions to be expressed with truth tables or nested "IF" statements as well as by any algebraic expression format. The compiler manipulates these formats and minimizes the resulting equations to fit, if possible, into the available PLD structure.

ABEL language processor
ABEL compiler

   We'll talk about PLD structures, fuse patterns, and related topics later, in Section 5.3, and we'll show how to target ABEL programs to specific PLDs. In the meantime, we'll show how ABEL can be used to specify combinational logic functions without necessarily having to declare the targeted device type. Later, in Section 7.11, we'll do the same for sequential logic functions.

4.6.1 ABEL Program Structure

Table 4-10 shows the typical structure of an ABEL program, and Table 4-11 shows an actual program exhibiting the following language features:

- *Identifiers* must begin with a letter or underscore, may contain up to 31 letters, digits, and underscores, and are case sensitive.

identifier

- A program file begins with a `module` statement, which associates an identifier (`Alarm_Circuit`) with the program module. Large programs can have multiple modules, each with its own local title, declarations, and equations. Note that keywords such as "`module`" are not case sensitive.

module

LEGAL NOTICE   ABEL (Advanced Boolean Equation Language) is a trademark of Data I/O Corporation (Redmond, WA 98073).

Table 4-10
Typical structure of an
ABEL program.

```
module module name
title string
deviceID device deviceType;
pin declarations
other declarations
equations
equations
test_vectors
test vectors
end module name
```

title
- The *title* statement specifies a title string that will be inserted into the documentation files that are created by the compiler.

string
- A *string* is a series of characters enclosed by single quotes.

device
- The optional *device* declaration includes a device identifier (ALARMCKT) and a string that denotes the device type ('P16V8C' for a GAL16V8). The compiler uses the device identifier in the names of documentation files that it generates, and it uses the device type to determine whether the device can really perform the logic functions specified in the program.

comment
- *Comments* begin with a double quote and end with another double quote or the end of the line, whichever comes first.

pin declarations
- *Pin declarations* tell the compiler about symbolic names associated with the device's external pins. If the signal name is preceded with the **NOT** prefix (!), then the complement of the named signal will appear on the pin. Pin declarations may or may not include pin numbers; if none are given, the compiler assigns them based on the capabilities of the targeted device.

istype
com
- The *istype* keyword precedes a list of one or more properties, separated by commas. This tells the compiler the type of output signal. The "com" keyword indicates a combinational output. If no istype keyword is given, the compiler generally assumes that the signal is an input unless it appears on the lefthand side of an equation, in which case it tries to figure out the output's properties from the context. For your own protection, it's best just to use the istype keyword for all outputs!

other declarations
- *Other declarations* allow the designer to define constants and expressions to improve program readability and to simplify logic design.

equations
- The *equations* statement indicates that logic equations defining output signals as functions of input signals will follow.

equations
- *Equations* are written like assignment statements in a conventional programming language. Each equation is terminated by a semicolon. ABEL uses the following symbols for logical operations:

Table 4-11 An ABEL program for the alarm circuit of Figure 4-11.

```
module Alarm_Circuit
title 'Alarm Circuit Example
J. Wakerly, Micro Systems Engineering'
ALARMCKT device 'P16V8C';

" Input pins
PANIC, ENABLEA, EXITING      pin 1, 2, 3;
WINDOW, DOOR, GARAGE         pin 4, 5, 6;
" Output pins
ALARM                        pin 11 istype 'com';

" Constant definition
X = .X.;

" Intermediate equation
SECURE = WINDOW & DOOR & GARAGE;

equations
ALARM = PANIC # ENABLEA & !EXITING & !SECURE;

test_vectors
([PANIC,ENABLEA,EXITING,WINDOW,DOOR,GARAGE] -> [ALARM])
[    1,    .X.,    .X.,    .X., .X.,    .X.] -> [    1];
[    0,     0,     .X.,    .X., .X.,    .X.] -> [    0];
[    0,     1,      1,     .X., .X.,    .X.] -> [    0];
[    0,     1,      0,      0, .X.,    .X.] -> [    1];
[    0,     1,      0,     .X.,  0,    .X.] -> [    1];
[    0,     1,      0,     .X., .X.,     0] -> [    1];
[    0,     1,      0,      1,   1,     1] -> [    0];

end Alarm_Circuit
```

&	AND.		*& (AND)*
#	OR.		*# (OR)*
!	NOT (used as a prefix).		*! (NOT)*
$	XOR.		*$ (XOR)*
!$	XNOR.		*!$ (XNOR)*

As in conventional programming languages, AND (&) has precedence over OR (#) in expressions. The *@ALTERNATE* directive makes the compiler recognize an alternate set of symbols for these operations: *, +, /, :+:, and :*:, respectively. This book uses the default symbols throughout.   *@ALTERNATE*

- The optional *test_vectors* statement indicates that test vectors follow.   *test_vectors*

- *Test vectors* associate input combinations with expected output values; they are used for simulation and testing as explained in Section 4.6.7.   *test vectors*

.X.

• The compiler recognizes several special constants, including .X., a single bit whose value is "don't-care."

end

• The *end* statement marks the end of the module.

unclocked assignment operator, =

Equations for combinational outputs use the *unclocked assignment operator, =*. The lefthand side of an equation normally contains a signal name. The righthand side is a logic expression, not necessarily in sum-of-products form. The signal name on the lefthand side of an equation may be optionally preceded by the NOT operator !; this is equivalent to complementing the right-hand side. The compiler's job is to generate a fuse pattern such that the signal named on the lefthand side realizes the logic expression on the righthand side.

4.6.2 ABEL Compiler Operation

The program in Table 4-11 realizes the alarm function that we described on page 215. The signal named ENABLE has been coded as ENABLEA because ENABLE is a reserved word in ABEL.

intermediate equation

Notice that not all of the equations appear under the equations statement. An intermediate equation for an identifier SECURE appears earlier. This equation is merely a definition that associates an expression with the identifier SECURE. The ABEL compiler substitutes this expression for the identifier SECURE in every place that SECURE appears after its definition.

In Figure 4-19 on page 216 we realized the alarm circuit directly from the SECURE and ALARM expressions, using multiple levels of logic. The ABEL compiler doesn't use expressions to interconnect gates in this way. Rather, it "crunches" the expressions to obtain a minimal two-level sum-of-products result appropriate for realization in a PLD. Thus, when compiled, Table 4-11 should yield a result equivalent to the AND-OR circuit that we showed in Figure 4-20 on page 216, which happens to be minimal.

In fact, it does. Table 4-12 shows the synthesized equations file created by the ABEL compiler. Notice that the compiler creates equations only for the ALARM signal, the only output. The SECURE signal does not appear anywhere.

The compiler finds a minimal sum-of-products expression for both ALARM and its complement, !ALARM. As mentioned previously, many PLDs have the ability selectively to invert or not to invert their AND-OR output. The "reverse-polarity equation" in Table 4-12 is a sum-of-products realization of !ALARM and would be used if output inversion were selected.

In this example, the reverse-polarity equation has one less product term than the normal-polarity equation for ALARM, so the compiler will select this equation if the targeted device has selectable output inversion. A user can also force the compiler to use either normal or reverse polarity for a signal by including the keyword "buffer" or "invert," respectively, in the signal's istype property list. (With some ABEL compilers, keywords "pos" and "neg" can be used for this purpose, but see Section 4.6.6.)

Table 4-12 Synthesized equations file produced by ABEL for program in Table 4-11.

```
ABEL 6.30

Design alarmckt created Tue Nov 24 1998

Title: Alarm Circuit Example
Title: J. Wakerly, Micro Systems Engineering

P-Terms  Fan-in  Fan-out  Type  Name (attributes)
---------  ------  -------  ----  -----------------
   4/3       6        1     Pin   ALARM
=========
   4/3            Best P-Term Total: 3
                        Total Pins: 7
                       Total Nodes: 0
              Average P-Term/Output: 3


Equations:

ALARM = (ENABLEA & !EXITING & !DOOR
      # ENABLEA & !EXITING & !WINDOW
      # ENABLEA & !EXITING & !GARAGE
      # PANIC);


Reverse-Polarity Equations:

!ALARM = (!PANIC & WINDOW & DOOR & GARAGE
      # !PANIC & EXITING
      # !PANIC & !ENABLEA);
```

4.6.3 WHEN Statements and Equation Blocks

In addition to equations, ABEL provides the *WHEN statement* as another means   *WHEN statement*
to specify combinational logic functions in the *equations* section of an ABEL
program. Table 4-13 shows the general structure of a WHEN statement, similar to
an IF statement in a conventional programming language. The ELSE clause is
optional. Here *LogicExpression* is an expression which results in a value of true
(1) or false (0). Either *TrueEquation* or *FalseEquation* is "executed" depending

WHEN *LogicExpression* THEN
TrueEquation;
ELSE
FalseEquation;

Table 4-13
Structure of an ABEL
WHEN statement.

on the value of *LogicExpression*. But we need to be a little more precise about what we mean by "executed," as discussed below.

In the simplest case, *TrueEquation* and the optional *FalseEquation* are assignment statements, as in the first two WHEN statements in Table 4-14 (for X1 and X2). In this case, *LogicExpression* is logically ANDed with the righthand side of *TrueEquation*, and the complement of *LogicExpression* is ANDed with the righthand side of *FalseEquation*. Thus, the equations for X1A and X2A produce the same results as the corresponding WHEN statements but do not use WHEN.

Notice in the first example that X1 appears in the *TrueEquation*, but there is no *FalseEquation*. So, what happens to X1 when *LogicExpression* (!A#B) is false? You might think that X1's value should be don't-care for these input combinations, but it's not, as explained below.

Formally, the unclocked assignment operator, =, specifies input combinations that should be added to the on-set for the output signal appearing on the lefthand side of the equation. An output's on-set starts out empty and is augmented each time that the output appears on the lefthand side of an equation. That is, the righthand sides of all equations for the same (uncomplemented) output are ORed together. (If the output appears complemented on the lefthand side, the righthand side is complemented before being ORed.) Thus, the value of X1 is 1 only for the input combinations for which *LogicExpression* (!A#B) is true and the righthand side of *TrueEquation* (C&!D) is also true.

In the second example, X2 appears on the lefthand side of two equations, so the equivalent equation shown for X2A is obtained by ORing two righthand sides after ANDing each with the appropriate condition.

The *TrueEquation* and the optional *FalseEquation* in a WHEN statement can be any equation. In addition, WHEN statements can be "nested" by using another WHEN statement as the *FalseEquation*. When statements are nested, all of the conditions leading to an "executed" statement are ANDed. The equation for X3 and its WHEN-less counterpart for X3A in Table 4-14 illustrate the concept.

The *TrueEquation* can be another WHEN statement if it's enclosed in braces, as shown in the X4 example in the table. This is just one instance of the general use of braces described shortly.

Although all of our WHEN examples have assigned values to the same output within each part of a given WHEN statement, this does not have to be the case. The second-to-last WHEN statement in Table 4-14 is such an example.

It's often useful to make more than one assignment in *TrueEquation* or *FalseEquation* or both. For this purpose, ABEL supports equation blocks anywhere that it supports a single equation. An *equation block* is just a sequence of statements enclosed in braces, as shown in the last WHEN statement in the table. The individual statements in the sequence may be simple assignment statements, or they may be WHEN statements or nested equation blocks. A semicolon is not used after a block's closing brace. Just for fun, Table 4-15 shows the equations that the ABEL compiler produces for the entire example program.

equation block

Table 4-14 Examples of WHEN statements.

```
module WhenEx
title 'WHEN Statement Examples'

" Input pins
A, B, C, D, E, F                pin;

" Output pins
X1, X1A, X2, X2A, X3, X3A, X4    pin istype 'com';
X5, X6, X7, X8, X9, X10          pin istype 'com';

equations

WHEN (!A # B) THEN X1 = C & !D;

X1A = (!A # B) & (C & !D);

WHEN (A & B) THEN X2 = C # D;
ELSE X2 = E # F;

X2A = (A & B) & (C # D)
    # !(A & B) & (E # F);

WHEN (A) THEN X3 = D;
ELSE WHEN (B) THEN X3 = E;
ELSE WHEN (C) THEN X3 = F;

X3A = (A) & (D)
    # !(A) & (B) & (E)
    # !(A) & !(B) & (C) & (F);

WHEN (A) THEN
  {WHEN (B) THEN X4 = D;}
ELSE X4 = E;

WHEN (A & B) THEN X5 = D;
ELSE WHEN (A # !C) THEN X6 = E;
ELSE WHEN (B # C) THEN X7 = F;

WHEN (A) THEN {
    X8 = D & E & F;
    WHEN (B) THEN X8 = 1; ELSE {X9 = D; X10 = E;}
} ELSE {
    X8 = !D # !E;
    WHEN (D) THEN X9 = 1;
    {X10 = C & D;}
}

end WhenEx
```

■ **Table 4-15** Synthesized equations file produced by ABEL for program in Table 4-14.

ABEL 6.30

Design whenex created Wed Dec 2 1998

Title: WHEN Statement Examples

P-Terms	Fan-in	Fan-out	Type	Name
2/3	4	1	Pin	X1
2/3	4	1	Pin	X1A
6/3	6	1	Pin	X2
6/3	6	1	Pin	X2A
3/4	6	1	Pin	X3
3/4	6	1	Pin	X3A
2/3	4	1	Pin	X4
1/3	3	1	Pin	X5
2/3	4	1	Pin	X6
1/3	3	1	Pin	X7
4/4	5	1	Pin	X8
2/2	3	1	Pin	X9
2/4	5	1	Pin	X10

=========

36/42   Best P-Term Total: 30
              Total Pins: 19
              Total Nodes: 0
    Average P-Term/Output: 2

Equations:

```
X1 = (C & !D & !A
     # C & !D & B);

X1A = (C & !D & !A
      # C & !D & B);

X2 = (D & A & B
     # C & A & B
     # !B & E
     # !A & E
     # !B & F
     # !A & F);

X2A = (D & A & B
      # C & A & B
      # !B & E
      # !A & E
      # !B & F
      # !A & F);

X3 = (C & !A & !B & F
     # !A & B & E
     # D & A);

X3A = (C & !A & !B & F
      # !A & B & E
      # D & A);

X4 = (D & A & B
     # !A & E);

X5 = (D & A & B);

X6 = (A & !B & E
     # !C & !A & E);

X7 = (C & !A & F);

X8 = (D & A & E & F
     # A & B
     # !A & !E
     # !D & !A);

X9 = (D & !A
     # D & !B);

X10 = (C & D & !A
      # A & !B & E);
```

Reverse-Polarity Eqns:

```
!X1 = (A & !B
      # D
      # !C);

!X1A = (A & !B
       # D
       # !C);

!X2 = (!C & !D & A & B
      # !B & !E & !F
      # !A & !E & !F);

!X2A = (!C & !D & A & B
       # !B & !E & !F
       # !A & !E & !F);

!X3 = (!C & !A & !B
      # !A & B & !E
      # !D & A
      # !A & !B & !F);

!X3A = (!C & !A & !B
       # !A & B & !E
       # !D & A
       # !A & !B & !F);

!X4 = (A & !B
      # !D & A
      # !A & !E);

!X5 = (!A
      # !D
      # !B);

!X6 = (A & B
      # C & !A
      # !E);

!X7 = (A
      # !C
      # !F);

!X8 = (A & !B & !F
      # D & !A & E
      # A & !B & !E
      # !D & A & !B);

!X9 = (!D
      # A & B);

!X10 = (A & B
       # !D & !A
       # !C & !A
       # A & !E);
```

truth_table (*input-list -> output-list*)
input-value -> output-value ;
. . .
input-value -> output-value ;

Table 4-16
Structure of an ABEL
truth table.

4.6.4 Truth Tables

ABEL provides one more way to specify combinational logic functions—
the *truth table*, with the general format shown in Table 4-16. The keyword
`truth_table` introduces a truth table. The *input-list* and *output-list* give the
names of the input signals and the outputs that they affect. Each of these lists is
either a single signal name or a *set*; sets are described fully in Section 4.6.5.
Following the truth-table introduction are a series of statements, each of which
specifies an input value and a required output value using the "->" operator. For
example, the truth table for an inverter is shown below:

truth table
truth_table
input-list
output-list

unclocked truth-table
  operator, ->

```
truth_table (X -> NOTX)
          0 -> 1;
          1 -> 0;
```

The list of input values does not need to be complete; only the on-set of the
function needs to be specified unless don't-care processing is enabled (see
Section 4.6.6). Table 4-17 shows how the prime-number-detector function
described on page 215 can be specified using an ABEL program. For convenience, the identifier NUM is defined as a synonym for the set of four input bits
[N3,N2,N1,N0], allowing a 4-bit input value to be written as a decimal integer.

Table 4-17 An ABEL program for the prime-number detector.

```
module PrimeDet
title '4-Bit Prime Number Detector'

" Input and output pins
N0, N1, N2, N3                   pin;
F                                pin istype 'com';

" Definition
NUM = [N3,N2,N1,N0];

truth_table (NUM -> F)
             1 -> 1;
             2 -> 1;
             3 -> 1;
             5 -> 1;
             7 -> 1;
            11 -> 1;
            13 -> 1;
end PrimeDet
```

Both truth tables and equations can be used within the same ABEL program. The equations keyword introduces a sequence of equations, while the truth_table keyword introduces a single truth table.

4.6.5 Ranges, Sets, and Relations

Most digital systems include buses, registers, and other circuits that handle a group of two or more signals in an identical fashion. ABEL provides several shortcuts for conveniently defining and using such signals.

range

The first shortcut is for naming similar, numbered signals. As shown in the pin definitions in Table 4-18, a *range* of signal names can be defined by stating the first and last names in the range, separated by "..". For example, writing "N3..N0" is the same as writing "N3,N2,N1,N0." Notice in the table that the range can be ascending or descending.

set

Next, we need a facility for writing equations more compactly when a group of signals are all handled identically, in order to reduce the chance of errors and inconsistencies. An ABEL *set* is simply a defined collection of signals that is handled as a unit. When a logical operation such as AND, OR, or assignment is applied to a set, it is applied to each element of the set.

Each set is defined at the beginning of the program by associating a set name with a bracketed list of the set elements (e.g., N=[N3,N2,N1,N0] in Table 4-18). The set element list may use shortcut notation (YOUT=[Y1..Y4]), but the element names need not be similar or have any correspondence with the set name (COMP=[EQ,GE]). Set elements can also be constants (GT=[0,1]). In any case, the number and order of elements in a set are significant, as we'll see.

Most of ABEL's operators, can be applied to sets. When an operation is applied to two or more sets, all of the sets must have the same number of elements, and the operation is applied individually to set elements in like positions, regardless of their names or numbers. Thus, the equation "YOUT = N & M" is equivalent to four equations:

```
Y1 = N3 & M3;
Y2 = N2 & M2;
Y3 = N1 & M1;
Y4 = N0 & M0;
```

When an operation includes both set and nonset variables, the nonset variables are combined individually with set elements in each position. Thus, the equation "ZOUT = (SEL & N) # (!SEL & M)" is equivalent to four equations of the form "Z_i = (SEL & N_i) # (!SEL & M_i)" for i equal 0 to 3.

relation
relational operator

Another important feature is ABEL's ability to convert "relations" into logic expressions. A *relation* is a pair of operands combined with one of the *relational operators* listed in Table 4-19. The compiler converts a relation into a logic expression that is 1 if and only if the relation is true.

The operands in a relation are treated as unsigned integers, and either operand may be an integer or a set. If the operand is a set, it is treated as an unsigned

Table 4-18 Examples of ABEL ranges, sets, and relations.

```
module SetOps
title 'Set Operation Examples'

" Input and output pins
N3..N0, M3..M0, SEL                          pin;
Y1..Y4, Z0..Z3, EQ, GE, GTR, LTH, UNLUCKY    pin istype 'com';

" Definitions
N    = [N3,N2,N1,N0];
M    = [M3,M2,M1,M0];
YOUT = [Y1..Y4];
ZOUT = [Z3..Z0];

COMP = [EQ,GE];
GT   = [ 0, 1];
LT   = [ 0, 0];

equations

YOUT = N & M;
ZOUT = (SEL & N) # (!SEL & M);
EQ = (N == M);
GE = (N >= M);
GTR = (COMP == GT);
LTH = (COMP == LT);
UNLUCKY = (N == 13) # (M == ^hD) # ((N + M) == ^b1101);

end SetOps
```

binary integer with the leftmost variable representing the most significant bit. By default, numbers in ABEL programs are assumed to be base-10. Hexadecimal and binary numbers are denoted by a prefix of "$^\wedge$h" or "$^\wedge$b," respectively, as shown in the last equation in Table 4-18.

$^\wedge$h hexadecimal prefix
$^\wedge$b binary prefix

ABEL sets and relations allow a lot of functionality to be expressed in very few lines of code. For example, the equations in Table 4-18 generate minimized equations with 69 product terms, as shown in the summary in Table 4-20.

Symbol	Relation
==	equal
!=	not equal
<	less than
<=	less than or equal
>	greater than
>=	greater than or equal

Table 4-19
Relational operators in ABEL.

Table 4-20 Synthesized equations summary produced by ABEL for program in Table 4-18.

P-Terms	Fan-in	Fan-out	Type	Name (attributes)
1/2	2	1	Pin	Y1
1/2	2	1	Pin	Y2
1/2	2	1	Pin	Y3
1/2	2	1	Pin	Y4
2/2	3	1	Pin	Z0
2/2	3	1	Pin	Z1
2/2	3	1	Pin	Z2
2/2	3	1	Pin	Z3
16/8	8	1	Pin	EQ
23/15	8	1	Pin	GE
1/2	2	1	Pin	GTR
1/2	2	1	Pin	LTH
16/19	8	1	Pin	UNLUCKY

```
=========
  69/62              Best P-Term Total: 53
                        Total Pins: 22
                        Total Nodes: 0
             Average P-Term/Output: 4
```

*4.6.6 Don't-Care Inputs

Some versions of the ABEL compiler have a limited ability to handle don't-care inputs. As mentioned previously, ABEL equations specify input combinations that belong to the on-set of a logic function; the remaining combinations are assumed to belong to the off-set. If some input combinations can instead be assigned to the d-set, then the program may be able to use these don't-care inputs to do a better job of minimization.

@DCSET

dc

?= don't-care unclocked assignment operator

The ABEL language defines two mechanisms for assigning input combinations to the d-set. In order to use either mechanism, you must include the compiler directive @DCSET in your program or include "dc" in the istype property list of the outputs for which you want don't-cares to be considered.

The first mechanism is the *don't-care unclocked assignment operator*, ?=. This operator is used instead of = in equations to indicate that input combinations matching the righthand side should be put into the d-set instead of the on-set. Although this operator is documented in the ABEL compiler that I use, unfortunately it seems to be broken, so I'm not going to talk about it any more.

The second mechanism is the truth table. When don't-care processing is enabled, any input combinations that are not explicitly listed in the truth table are put into the d-set. Thus, the prime BCD-digit detector described on page 232 can be specified in ABEL as shown in Table 4-21. A don't-care value is implied for input combinations 10–15 because these combinations do not appear in the truth table and the @DCSET directive is in effect.

```
module DontCare
title 'Dont Care Examples'
@DCSET

" Input and output pins
N3..N0, A, B                    pin;
F, Y                           pin istype 'com';

NUM = [N3..N0];
X = .X.;

truth_table (NUM->F)
              0->0;
              1->1;
              2->1;
              3->1;
              4->0;
              5->1;
              6->0;
              7->1;
              8->0;
              9->0;

truth_table ([A,B]->Y)
              [0,0]->0;
              [0,1]->X;
              [1,0]->X;
              [1,1]->1;


end DontCare
```

Table 4-21
ABEL program using
don't-cares.

It's also possible to specify don't-care combinations explicitly, as shown in the second truth table. As introduced at the very beginning of this section, ABEL recognizes .X. as a special one-bit constant whose value is "don't-care." In Table 4-21, the identifier "X" has been equated to this constant just to make it easier to type don't-cares in the truth table. The minimized equations resulting from Table 4-21 are shown in Table 4-22. Notice that the two equations for F are not equal; the compiler has selected different values for the don't-cares.

```
Equations:
F = (!N2 & N1
     # !N3 & N0);
Y = (B);

Reverse-Polarity Equations:
!F = (N2 & !N0
      # N3
      # !N1 & !N0);
!Y = (!B);
```

Table 4-22
Minimized equations
derived from
Table 4-21.

Table 4-23
Structure of ABEL
test vectors.

```
test_vectors  (input-list -> output-list)
              input-value -> output-value;
              . . .
              input-value -> output-value;
```

4.6.7 Test Vectors

test_vectors

input-list
output-list

ABEL programs may contain optional test vectors, as we showed in Table 4-11 on page 251. The general format of test vectors is very similar to a truth table and is shown in Table 4-23. The keyword *test_vectors* introduces a truth table. The *input-list* and *output-list* give the names of the input signals and the outputs that they affect. Each of these lists is either a single signal name or a set. Following the test-vector introduction are a series of statements, each of which specifies an input value and an expected output value using the "->" operator.

ABEL test vectors have two main uses and purposes:

1. After the ABEL compiler translates the program into "fuse pattern" for a particular device, it simulates the operation of the final programmed device by applying the test-vector inputs to a software model of the device and comparing its outputs with the corresponding test-vector outputs. The designer may specify a series of test vectors in order to double-check that device will behave as expected for some or all input combinations.

2. After a PLD is physically programmed, the programming unit applies the test-vector inputs to the physical device and compares the device outputs with the corresponding test-vector outputs. This is done to check for correct device programming and operation.

Unfortunately, ABEL test vectors seldom do a very good job at either one of these tasks, as we'll explain.

The test vectors from Table 4-11 are repeated in Table 4-24, except that for readability we've assumed that the identifier X has been equated to the don't-care constant .X., and we've added comments to number the test vectors.

Table 4-24 actually appears to be a pretty good set of test vectors. From the designer's point of view, these vectors fully cover the expected operation of the alarm circuit, as itemized vector-by-vector below:

1. If PANIC is 1, then the alarm output (F) should be on regardless of the other input values. All of the remaining vectors cover cases where PANIC is 0.

2. If the alarm is not enabled, then the output should be off.

3. If the alarm is enabled but we're exiting, then the output should be off.

4. If the alarm is enabled and we're not exiting, then the output should be on if any of the sensor signals WINDOW, DOOR, or GARAGE is 0.

5. If the alarm is enabled, we're not exiting, and all of the sensor signals are 1, then the output should be off.

```
test_vectors
([PANIC,ENABLEA,EXITING,WINDOW,DOOR,GARAGE] -> [ALARM])
[     1,       X,       X,       X,   X,       X] -> [    1];   "1
[     0,       0,       X,       X,   X,       X] -> [    0];   "2
[     0,       1,       1,       X,   X,       X] -> [    0];   "3
[     0,       1,       0,       0,   X,       X] -> [    1];   "4
[     0,       1,       0,       X,   0,       X] -> [    1];   "5
[     0,       1,       0,       X,   X,       0] -> [    1];   "6
[     0,       1,       0,       1,   1,       1] -> [    0];   "7
```

Table 4-24
Test vectors for the
alarm circuit program
in Table 4-11.

The problem is that ABEL doesn't handle don't-cares in test-vector inputs the way that it should. For example, by all rights, test vector 1 should test 32 distinct input combinations corresponding to all 32 possible combinations of don't-care inputs ENABLEA, EXITING, WINDOW, DOOR, and GARAGE. But it doesn't. In this situation, the ABEL compiler interprets "don't care" as "the user doesn't care what input value I use," and it just assigns 0 to all don't-care inputs in a test vector. In this example, you could have incorrectly written the output equation as "F = PANIC & !ENABLEA # ENABLEA & ..."; the test vectors would still pass, even though the panic button would work only when the system is disabled.

The second use of test vectors is in physical device testing. Most physical defects in logic devices can be detected using the *single stuck-at fault model*, which assumes that any physical defect is equivalent to having a single gate input or output stuck at a logic 0 or 1 value. Just putting together a set of test vectors that seems to exercise a circuit's functional specifications, as we did in Table 4-24, doesn't guarantee that all single stuck-at faults can be detected. The test vectors have to be chosen so that every possible stuck-at fault causes an incorrect value at the circuit output for some test-vector input combination.

single stuck-at fault model

Table 4-25 shows a complete set of test vectors for the alarm circuit when it is realized as a two-level sum-of-products circuit. The first four vectors check for stuck-at-1 faults on the OR gate, and the last three check for stuck-at-0 faults on the AND gates; it turns out that this is sufficient to detect all single stuck-at faults. If you know something about fault testing you can generate test vectors for small circuits by hand (as I did in this example), but most designers use automated third-party tools to create high-quality test vectors for their PLD designs.

```
test_vectors
([PANIC,ENABLEA,EXITING,WINDOW,DOOR,GARAGE] -> [ALARM])
[     1,       0,       1,       1,   1,       1] -> [    1];   "1
[     0,       1,       0,       0,   1,       1] -> [    1];   "2
[     0,       1,       0,       1,   0,       1] -> [    1];   "3
[     0,       1,       0,       1,   1,       0] -> [    1];   "4
[     0,       0,       0,       0,   0,       0] -> [    0];   "5
[     0,       1,       1,       0,   0,       0] -> [    0];   "6
[     0,       1,       0,       1,   1,       1] -> [    0];   "7
```

Table 4-25
Single stuck-at fault
test vectors for the
minimal sum-of-
products realization
of the alarm circuit.

4.7 The VHDL Hardware Description Language

VHDL

In the mid-1980s, the U.S. Department of Defense (DoD) and the IEEE sponsored the development of a highly capable hardware-description language called *VHDL*. The language started out with and still has the following features:

- Designs may be decomposed hierarchically.

- Each design element has both a well-defined interface (for connecting it to other elements) and a precise behavioral specification (for simulating it).

- Behavioral specifications can use either an algorithm or an actual hardware structure to define an element's operation. For example, an element can be defined initially by an algorithm, to allow design verification of higher-level elements that use it; later, the algorithmic definition can be replaced by a hardware structure.

- Concurrency, timing, and clocking can all be modeled. VHDL handles asynchronous as well as synchronous sequential-circuit structures.

- The logical operation and timing behavior of a design can be simulated.

Thus, VHDL started out as a documentation and modeling language, allowing the behavior of digital-system designs to be precisely specified and simulated.

VHDL synthesis tools

While the VHDL language and simulation environment were important innovations by themselves, VHDL's utility and popularity took a quantum leap with the commercial development of *VHDL synthesis tools*. These programs can create logic-circuit structures directly from VHDL behavioral descriptions. Using VHDL, you can design, simulate, and synthesize anything from a simple combinational circuit to a complete microprocessor system on a chip.

VHDL-87
VHDL-93

VHDL was standardized by the IEEE in 1987 (*VHDL-87*) and extended in 1993 (*VHDL-93*). In this section we'll describe a subset of language features that are legal under either standard. We'll describe additional features for sequential logic design in Section 7.12.

4.7.1 Design Flow

design flow

It's useful to understand the overall VHDL design environment before jumping into the language itself. There are several steps in a VHDL-based design process, often called the *design flow*. These steps are applicable to any HDL-based design process and are outlined in Figure 4-50 on page 266.

THE MEANING OF VHDL   "VHDL" stands for "VHSIC Hardware Description Language." VHSIC, in turn, stands for "Very High Speed Integrated Circuit," which was a U.S. Department of Defense program to encourage research on high-performance IC technology (using Very Healthy Sums of Instant Cash!).

VERILOG AND VHDL

At about the same time that VHDL was developing, a different hardware design language appeared on the scene. *Verilog HDL*, or simply *Verilog*, was introduced by Gateway Design Automation in 1984 as a proprietary hardware description and simulation language. The subsequent introduction of Verilog-based synthesis tools in 1988 by then-fledgling Synopsys and the 1989 acquisition of Gateway by Cadence Design Systems was a winning combination that led to widespread use of the language.

Today, VHDL and Verilog both enjoy widespread use and share the logic synthesis market roughly 50/50. Verilog has its syntactic roots in C and is in some respects an easier language to learn and use, while VHDL is more like Ada (a DoD-sponsored software programming language) and has more features that support large project development.

Comparing the pros and cons of starting out with one language versus the other, David Pellerin and Douglas Taylor probably put it best in their book, *VHDL Made Easy!* (Prentice Hall, 1997):

> Both languages are easy to learn and hard to master. And once you have learned one of these languages, you will have no trouble transitioning to the other.

The so-called "front end" begins with figuring out the basic approach and building blocks at the block-diagram level. Large logic designs, like software programs, are usually hierarchical, and VHDL gives you a good framework for defining modules and their interfaces and filling in the details later.

The next step is the actual writing of VHDL code for modules, their interfaces, and their internal details. Since VHDL is a text-based language, in principle you can use any text editor for this part of the job. However, most design environments include a specialized *VHDL text editor* that makes the job a little easier. Such editors include features like automatic highlighting of VHDL keywords, automatic indenting, built-in templates for frequently used program structures, and built-in syntax checking and one-click access to the compiler.

VHDL text editor

Once you've written some code, you will want to compile it, of course. A *VHDL compiler* analyzes your code for syntax errors and also checks it for compatibility with other modules on which it relies. It also creates the internal information that is needed for a simulator to process your design later. As in other programming endeavors, you probably shouldn't wait until the very end of coding to compile all of your code. Doing a piece at a time can prevent you from

THE MEANING OF VERILOG

"Verilog" isn't an acronym, but it has some interesting anagrams, including "I, Glover" (Danny?), "G.I. lover," "Go, liver!" and "I grovel." Oh, I suppose it could also be a contraction of "VERIfy LOGic."

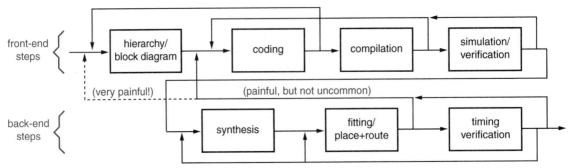

Figure 4-50 Steps in a VHDL or other HDL-based design flow.

proliferating syntax errors, inconsistent names, and so on, and can certainly give you a much-needed sense of progress when the project end is far from sight!

VHDL simulator

Perhaps the most satisfying step comes next—simulation. A *VHDL simulator* allows you to define and apply inputs to your design, and to observe its outputs, without ever having to build the physical circuit. In small projects, the kind you might do as homework in a digital-design class, you would probably generate inputs and observe outputs manually. But for larger projects, VHDL gives you the ability to create "test benches" that automatically apply inputs and compare them with expected outputs.

verification

Actually, simulation is just one piece of a larger step called *verification*. Sure, it is satisfying to watch your simulated circuit produce simulated outputs, but the purpose of simulation is larger—it is to *verify* that the circuit works as desired. In a typical large project, a substantial amount of effort is expended both during and after the coding stage to define test cases that exercise the circuit over a wide range of logical operating conditions. Finding design bugs at this stage has a high value; if bugs are found later, all of the so-called "back-end" steps must typically be repeated.

functional verification

Note that there are at least two dimensions to verification. In *functional verification*, we study the circuit's logical operation independent of timing considerations; gate delays and other timing parameters are considered to be zero. In *timing verification*, we study the circuit's operation including estimated delays, and we verify that the setup, hold, and other timing requirements for sequential devices like flip-flops are met. It is customary to perform thorough *functional* verification before starting the back-end steps. However, our ability to do *timing* verification at this stage is often limited, since timing may be heavily dependent on the results of synthesis and fitting. We may do preliminary timing verification to gain some comfort with the overall design approach, but detailed timing verification must wait until the end.

timing verification

After verification, we are ready to move into the "back-end" stage. The nature of and tools for this stage vary somewhat, depending on the target technology for the design, but there are three basic steps. The first is *synthesis*, converting the VHDL description into a set of primitives or components that can

synthesis

be assembled in the target technology. For example, with PLDs or CPLDs, the synthesis tool may generate two-level sum-of-products equations. With ASICs, it may generate a list of gates and a *netlist* that specifies how they should be interconnected. The designer may "help" the synthesis tool by specifying certain technology-specific *constraints*, such as the maximum number of logic levels or the strength of logic buffers to use.

netlist

constraints

In the *fitting* step, a fitting tool or *fitter* maps the synthesized primitives or components onto available device resources. For a PLD or CPLD, this may mean assigning equations to available AND-OR elements. For an ASIC, it may mean laying down individual gates in a pattern and finding ways to connect them within the physical constraints of the ASIC die; this is called the *place-and-route* process. The designer can usually specify additional constraints at this stage, such as the placement of modules with a chip or the pin assignments of external input and output pins.

fitting
fitter

place and route

The "final" step is timing verification of the fitted circuit. It is only at this stage that the actual circuit delays due to wire lengths, electrical loading, and other factors can be calculated with reasonable precision. It is usual during this step to apply the same test cases that were used in functional verification, but in this step they are run against the circuit as it will actually be built.

As in any other creative process, you may occasionally take two steps forward and one step back (or worse!). As suggested in the figure, during coding you may encounter problems that force you to go back and rethink your hierarchy, and you will almost certainly have compilation and simulation errors that force you to rewrite parts of the code.

The most painful problems are the ones that you encounter in the back end of the design flow. For example, if the synthesized design doesn't fit into an available FPGA or doesn't meet timing requirements, you may have to go back as far as rethinking your whole design approach. That's worth remembering— excellent tools are still no substitute for careful thought at the outset of a design.

IT WORKS!?   As a long-time logic designer and system builder, I always thought I knew what it means when someone says about their circuit, "It works!" It means you can go into the lab, power-up a prototype without seeing smoke, and push a reset button and use an oscilloscope or logic analyzer to watch the prototype go through its paces.

But over the years, the meaning of "It works" has changed, at least for some people. When I took a new job a few years ago, I was very pleased to hear that several key ASICs for an important new product were all "working." But later (just a short time later) I figured out that the ASICs were working only in simulation, and that the design team still had to do several arduous months of synthesis, fitting, timing verification, and repeating, before they could order any prototypes. "It works!"—sure. Just like my kids' homework—"It's done!"

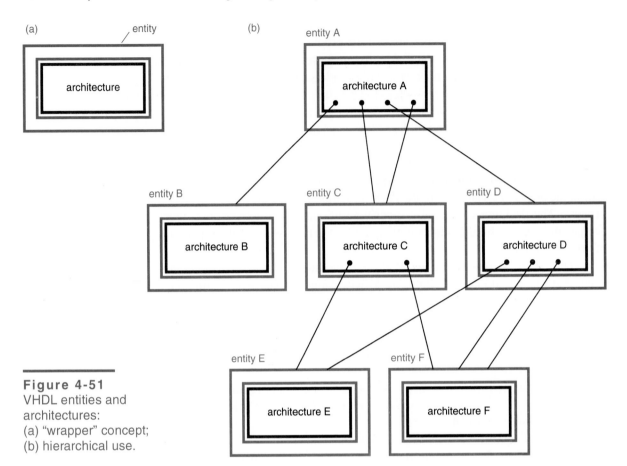

Figure 4-51
VHDL entities and
architectures:
(a) "wrapper" concept;
(b) hierarchical use.

4.7.2 Program Structure

VHDL was designed with principles of structured programming in mind, borrowing ideas from the Pascal and Ada software programming languages. A key idea is to define the interface of a hardware module while hiding its internal details. Thus, a VHDL *entity* is simply a declaration of a module's inputs and outputs, while a VHDL *architecture* is a detailed description of the module's internal structure or behavior.

entity
architecture

Figure 4-51(a) illustrates the concept. Many designers like to think of a VHDL entity declaration as a "wrapper" for the architecture, hiding the details of what's inside while providing the "hooks" for other modules to use it. This forms the basis for hierarchical system design—the architecture of a top-level entity may use (or "instantiate") other entities, while hiding the architectural details of lower-level entities from the higher-level ones. As shown in (b), a higher-level architecture may use a lower-level entity multiple times, and multiple top-level architectures may use the same lower-level one. In the figure, architectures B, E and F stand alone; they do not use any other entities.

GO CONFIGURE! VHDL actually allows you to define multiple architectures for a single entity, and it provides a configuration management facility that allows you to specify which one to use during a particular compilation or synthesis run. This lets you try out a different architectural approach without throwing away or hiding your other efforts. However, we won't use or further discuss this facility in this text.

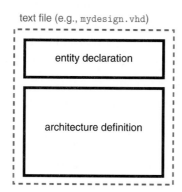

Figure 4-52
VHDL program file structure.

In the text file of a VHDL program, the *entity declaration* and *architecture definition* are separated, as shown in Figure 4-52. For example, Table 4-26 is a very simple VHDL program for a 2-input "inhibit" gate. In large projects, entities and architectures are sometimes defined in separate files, which the compiler matches up according to their declared names.

entity declaration
architecture definition

Like other high-level programming languages, VHDL generally ignores spaces and line breaks, and these may be provided as desired for readability. *Comments* begin with two hyphens (--) and end at the end of a line.

comments

VHDL defines many special character strings, called *reserved words* or *keywords*. Our example includes several—entity, port, is, in, out, end, architecture, begin, when, else, and not. User-defined *identifiers* begin with a letter and contain letters, digits, and underscores. (An underscore may not follow another underscore or be the last character in an identifier.) Identifiers in the example are Inhibit, X, Y, BIT, Z, and Inhibit_arch. "BIT" is a built-in identifier for a predefined type; it's not considered a reserved word because it can be redefined. Reserved words and identifiers are not case sensitive.

reserved words
keywords
identifiers

```
entity Inhibit is      -- also known as 'BUT-NOT'
  port (X,Y: in BIT;   --  as in 'X but not Y'
        Z:  out BIT);  --  (see [Klir, 1972])
end Inhibit;

architecture Inhibit_arch of Inhibit is
begin
  Z <= '1' when X='1' and Y='0' else '0';
end Inhibit_arch;
```

Table 4-26
VHDL program for an "inhibit" gate.

Table 4-27
Syntax of a VHDL
entity declaration.

```
entity entity-name is
   port (signal-names : mode signal-type;
         signal-names : mode signal-type;
         . . .
         signal-names : mode signal-type);
end entity-name;
```

A basic entity declaration has the syntax shown in Table 4-27. Besides naming the entity, the purpose of the entity declaration is to define its external interface signals or *ports* in its *port declaration* part. In addition to the keywords *entity*, *is*, *port*, and *end*, an entity declaration has the following elements:

port
port declaration

entity-name   A user-selected identifier to name the entity.

signal-names   A comma-separated list of one or more user-selected identifiers to name external-interface signals.

mode   One of four reserved words, specifying the signal direction:

in   The signal is an input to the entity.

out   The signal is an output of the entity. Note that the value of such a signal cannot be "read" inside the entity's architecture, only by other entities that use it.

buffer   The signal is an output of the entity, and its value can also be read inside the entity's architecture.

inout   The signal can be used as an input or an output of the entity. This mode is typically used for three-state input/output pins on PLDs.

signal-type   A built-in or user-defined signal type. We'll have a lot to say about types in the next subsection.

Note that there is no semicolon after the final *signal-type*; swapping the closing parenthesis with the semicolon after it is a common syntax error for beginning VHDL programmers.

An entity's ports and their modes and types are all that is seen by other modules that use it. The entity's internal operation is specified in its *architecture definition*, whose general syntax is shown in Table 4-28. The *entity-name* in this definition must be the same as the one given previously in the entity declaration. The *architecture-name* is a user-selected identifier, usually related to the entity name; it can be the same as the entity name if desired.

architecture definition

An architecture's external interface signals (ports) are inherited from the port-declaration part of its corresponding entity declaration. An architecture may also include signals and other declarations that are local to that architecture, similar to other high-level languages. Declarations common to multiple entities can be made in a separate "package" used by all entities, as discussed later.

```
architecture architecture-name of entity-name is
    type declarations
    signal declarations
    constant declarations
    function definitions
    procedure definitions
    component declarations
begin
    concurrent-statement
    ...
    concurrent-statement
end architecture-name;
```

Table 4-28
Syntax of a VHDL
architecture definition.

The declarations in Table 4-28 can appear in any order. In due course we'll discuss many different kinds of declarations and statements that can appear in the architecture definition, but the easiest to start with is the *signal declaration*. It gives the same information about a signal as in a port declaration, except that no mode is specified:

signal declaration

```
signal signal-names : signal-type;
```

Zero or more signals can be defined within an architecture, and they roughly correspond to named wires in a logic diagram. They can be read or written within the architecture definition and, like other local objects, can be referenced only within the encompassing architecture definition.

VHDL *variables* are similar to signals, except that they usually don't have physical significance in a circuit. In fact, notice that Table 4-28 has no provision for "variable declarations" in an architecture definition. Rather, variables are used in VHDL functions, procedures, and processes, each of which we'll discuss later. Within these program elements, the syntax of a *variable declaration* is just like that of a signal declaration, except that the `variable` keyword is used:

variable

variable declaration

```
variable variable-names : variable-type;
```

4.7.3 Types and Constants

All signals, variables, and constants in a VHDL program must have an associated "type." The *type* specifies the set or range of values that the object can take on, and there is also typically a set of operators (such as add, **AND**, and so on) associated with a given type.

type

VHDL has just a few *predefined types*, listed in Table 4-29. In the rest of this book, the only predefined types that we'll use are `integer`, `character`, and `boolean`. You would think that types with names "bit" and "bit_vector"

predefined types

bit	character	severity_level	
bit_vector	integer	string	
boolean	real	time	

Table 4-29
VHDL predefined types.

Table 4-30
Predefined operators
for VHDL's `integer`
and `boolean` types.

integer **Operators**		boolean **Operators**	
+	addition	and	AND
–	subtraction	or	OR
*	multiplication	nand	NAND
/	division	nor	NOR
mod	modulo division	xor	Exclusive OR
rem	modulo remainder	xnor	Exclusive NOR
abs	absolute value	not	complementation
**	exponentiation		

would be essential in digital design, but it turns out that user-defined versions of these types are more useful, as discussed shortly.

integer

Type *integer* is defined as the range of integers including at least the range $-2{,}147{,}483{,}647$ through $+2{,}147{,}483{,}647$ ($-2^{31}+1$ through $+2^{31}-1$); VHDL implementations may extend this range. Type *boolean* has two values, *true* and *false*. The *character* type contains all of the characters in the ISO 8-bit character set; the first 128 are the ASCII characters. Built-in operators for the `integer` and `boolean` types are listed in Table 4-30.

boolean
true, false
character

user-defined type
enumerated type

The most commonly used types in typical VHDL programs are *user-defined types*, and the most common of these are *enumerated types*, which are defined by listing their values. Predefined types `boolean` and `character` are enumerated types. A type declaration for an enumerated type has the format shown in the first line of Table 4-31. Here, *value-list* is a comma-separated list (enumeration) of all possible values of the type. The values may be user-defined identifiers or characters (where a "character" is an ISO character enclosed in

Table 4-31
Syntax of VHDL
type and constant
declarations.

```
type type-name is (value-list);

subtype subtype-name is type-name start to end;
subtype subtype-name is type-name start downto end;

constant constant-name: type-name := value;
```

**STRONG
TYPING**

Unlike C, VHDL is a strongly typed language. This means that the compiler does not allow you to assign a value to a signal or variable unless the type of the value precisely matches the declared type of the signal or variable.

Strong typing is both a blessing and a curse. It makes your programs more reliable and easier to debug, because it makes it difficult for you to make "dumb errors" where you assign a value of the wrong type or size. On the other hand, it can be exasperating at times. Even simple operations, such as reinterpreting a 2-bit signal as an integer (for example, to select one of four outcomes in a "case" statement), may require you to call a type-conversion function explicitly.

```
type STD_ULOGIC is ( 'U',   -- Uninitialized
                     'X',   -- Forcing   Unknown
                     '0',   -- Forcing   0
                     '1',   -- Forcing   1
                     'Z',   -- High Impedance
                     'W',   -- Weak      Unknown
                     'L',   -- Weak      0
                     'H',   -- Weak      1
                     '-'    -- Don't care
                   );
subtype STD_LOGIC is resolved STD_ULOGIC;
```

Table 4-32
Definition of VHDL
`std_logic` type
(see Section 5.6.4
for discussion of
"`resolved`").

single quotes). The first style is used most often to define cases or states for a state machine, for example,

```
type traffic_light_state is (reset, stop, wait, go);
```

The second style is used in the very important case of a standard user-defined logic type *std_logic*, shown in Table 4-32 and part of the IEEE 1164 standard package, discussed in Section 4.7.5. This type includes not only '0' and '1', but seven other values that have been found useful in simulating a logic signal (bit) in a real logic circuit, as explained in more detail in Section 5.6.4.

  VHDL also allows users to create *subtypes* of a type, using the syntax shown in Table 4-31. The values in the subtype must be a contiguous range of values of the base type, from *start* to *end*. For an enumerated type, "contiguous" refers to positions in the original, defining *value-list*. Some examples of subtype definitions are shown below:

std_logic

subtypes

```
subtype twoval_logic is std_logic range '0' to '1';
subtype fourval_logic is std_logic range 'X' to 'Z';
subtype negint is integer range -2147483647 to -1;
subtype bitnum is integer range 31 downto 0;
```

Notice that the order of a range may be specified in ascending or descending order, depending on whether *to* or *downto* is used. There are certain attributes of subtypes for which this distinction is significant, but we don't use them in this book and we won't discuss this further.

to

downto

WHAT A CHARACTER! You may be wondering why the values in the `std_logic` type are defined as characters rather than one-letter identifiers. Certainly "U", "X", and so on would be easier to type than "'U'", "'X'", and so on. Well, that would require a one-letter identifier other than "–" to be used for don't-care, but that's no big deal. The main reason for using characters is that "0" and "1" could not be used, because they're already recognized as integer constants. This goes back to VHDL's strong typing; it was not deemed advisable to let the compiler perform an automatic type conversion depending on the context.

Table 4-33
Syntax of VHDL
array declarations.

```
type type-name is array (start to end) of element-type;

type type-name is array (start downto end) of element-type;

type type-name is array (range-type) of element-type;

type type-name is array (range-type range start to end) of element-type;

type type-name is array (range-type range start downto end) of element-type;
```

VHDL has two predefined `integer` subtypes, defined below:

```
subtype natural is integer range 0 to highest-integer;
subtype positive is integer range 1 to highest-integer;
```

constants
constant declaration

Constants contribute to the readability, maintainability, and portability of programs in any language. The syntax of a *constant declaration* in VHDL is shown in the last line of Table 4-31; examples are shown below:

```
constant BUS_SIZE: integer := 32;     -- width of component
constant MSB: integer := BUS_SIZE-1;  -- bit number of MSB
constant Z: character := 'Z';         -- synonym for Hi-Z value
```

Notice that the value of a constant can be a simple expression. Constants can be used anywhere the corresponding value can be used, and they can be put to especially good use in type definitions, as we'll soon show.

array types
array
array index

Another very important category of user-defined types are *array types*. Like other languages, VHDL defines an *array* as an ordered set of elements of the same type, where each element is selected by an *array index*. Table 4-33 shows several versions of the syntax for declaring an array in VHDL. In the first two versions, *start* and *end* are integers that define the possible range of the array index and hence the total number of array elements. In the last three versions, all or a subset of the values of an existing type (*range-type*) are the range of the array index.

Examples of array declarations are given in Table 4-34. The first pair of examples are very ordinary and show both ascending and descending ranges. The next example shows how a constant, `WORD_LEN`, can be used with an array declaration, showing also that a range value can be a simple expression. The

UNNATURAL ACTS

Although VHDL defines the subtype "natural" as being nonnegative integers starting with 0, most mathematicians consider and define the natural numbers to begin with 1. After all, in early history people began counting with 1, and the concept of "0" arrived much later. Still, there is some discussion and perhaps controversy on the subject, especially as the computer age has led more of us to think with 0 as a starting number. For the latest thinking, search the Web for "natural numbers."

```
type monthly_count is array (1 to 12) of integer;
type byte is array (7 downto 0) of STD_LOGIC;

constant WORD_LEN: integer := 32;
type word is array (WORD_LEN-1 downto 0) of STD_LOGIC;

constant NUM_REGS: integer := 8;
type reg_file is array (1 to NUM_REGS) of word;

type statecount is array (traffic_light_state) of integer;
```

Table 4-34
Examples of VHDL
array declarations.

third example shows that an array element may itself be an array, thus creating a two-dimensional array. The last example shows that an enumerated type (or a subtype) may be specified as the array element range; the array in this example has four elements, based on our previous definition of `traffic_light_state`.

Array elements are considered to be ordered from left to right, in the same direction as index range. Thus, the leftmost elements of arrays of types `monthly_count`, `byte`, `word`, `reg_file`, and `statecount` have indices 1, 7, 31, 1, and `reset`, respectively.

Within VHDL program statements, individual array elements are accessed using the array name and the element's index in parentheses. For example, if `M`, `B`, `W`, `R`, and `S` are signals or variables of the five array types defined in Table 4-34, then `M(11)`, `B(5)`, `W(WORD_LEN-5)`, `R(0,0)`, `R(0)`, and `S(reset)` are all valid elements.

Array literals can be specified by listing the element values in parentheses. For example, the `byte` variable B could be set to all ones by the statement

array literal

```
B := ('1','1','1','1','1','1','1','1');
```

VHDL also has a shorthand notation that allows you to specify values by index. For example, to set `word` variable W to all ones except for zeroes in the LSB of each byte, you can write

```
W := (0=>'0',8=>'0',16=>'0',24=>'0',others=>'1');
```

others

The methods just described work for arrays with any *element-type*, but the easiest way write a literal of type STD_LOGIC is to use a "string." A VHDL *string* is a sequence of ISO characters enclosed in double quotes, such as `"Hi there"`. A string is just an array of characters; as a result, a STD_LOGIC array of a given length can be assigned the value of a string of the same length, as long as the characters in the string are taken from the set of nine characters defined as the possible values of the STD_LOGIC elements—`'0'`, `'1'`, `'U'`, and so on. Thus, the previous two examples can be rewritten as follows:

string

```
B := "11111111";
W := "11111110111111101111111011111110";
```

array slice

It is also possible to refer to a contiguous subset or *slice* of an array by specifying the starting and ending indices of the subset, for example, M(6 to 9), B(3 downto 0), W(15 downto 8), R(0,7 downto 0), R(1 to 2), S(stop to go). Notice that the slice's direction must be the same as the original array's.

&, concatenation operator

Finally, you can combine arrays or array elements using the *concatenation operator &*, which joins arrays and elements in the order written, from left to right. For example, '0'&'1'&"1Z" is equivalent to "011Z", and the expression B(6 downto 0) & B(7) yields a 1-bit left circular shift of the 8-bit array B.

std_logic_vector

The most important array type in typical VHDL programs is the IEEE 1164 standard user-defined logic type std_logic_vector, which defines an ordered set of std_logic bits. The definition of this type is:

```
type STD_LOGIC_VECTOR is array ( natural range <> ) of STD_LOGIC;
```

unconstrained array type

This is an example of an *unconstrained array type*—the range of the array is unspecified, except that it must be a subrange of a defined type, in this case, natural. This VHDL feature allows us to develop architectures, functions, and other program elements in a more general way, somewhat independent of the array size or its range of index values. An actual range is specified when a signal or variable is assigned this type. We'll see examples in the next subsection.

4.7.4 Functions and Procedures

function
arguments
result
function definition
formal parameters
actual parameters

Like a function in a high-level programming language, a VHDL *function* accepts a number of *arguments* and returns a *result*. Each of the arguments and the result in a VHDL function definition or function call have a predetermined type.

The syntax of a *function definition* is shown in Table 4-35. After giving the name of the function, it lists zero or more *formal parameters* which are used within the function body. When the function is called, the *actual parameters* in the function call are substituted for the formal parameters. Following VHDL's

Table 4-35
Syntax of a VHDL function definition.

```
function function-name (
      signal-names : signal-type;
      signal-names : signal-type;
      . . .
      signal-names : signal-type
) return return-type is
      type declarations
      constant declarations
      variable declarations
      function definitions
      procedure definitions
begin
      sequential-statement
      . . .
      sequential-statement
end function-name;
```

strong-typing policy, the actual parameters must be the same type or a subtype of the formal parameters. When the function is called from within an architecture, a value of the type *return-type* is returned in place of the function call.

As shown in the table, a function may define its own local types, constants, variables, and nested functions and procedures. The keywords begin and end enclose a series of "sequential statements" that are executed when the function is called. We'll take a closer look at different kinds of sequential statements and their syntax later, but you should be able to understand the examples here based on your previous programming experience.

The VHDL "inhibit-gate" architecture of Table 4-26 on page 269 is modified in Table 4-36 to use a function. Within the function definition, the keyword return indicates when the function should return to the caller, and it is followed *return* by an expression with the value to be returned to the caller. The type resulting from this expression must match the *return-type* in the function declaration.

The IEEE 1164 standard logic package defines many functions that operate on the standard types std_logic and std_logic_vector. Besides specifying a number of user-defined types, the package also defines the basic logic operations on these types such as and and or. This takes advantage of VHDL's ability to *overload* operators. This facility allows the user to specify a *operator overloading* function that is invoked whenever a built-in operator symbol (and, or, +, etc.) is used with a matching set of operand types. There may be several definitions for a given operator symbol; the compiler automatically picks the definition that matches the operand types in each use of the operator.

For example, Table 4-37 contains code, taken from the IEEE package, that shows how the "and" operation is defined for std_logic operands. This code may look complicated, but we've already introduced all of the basic language elements that it uses (except for "resolved", which we describe in connection with three-state logic in Section 5.6.4).

The inputs to the function may be of type std_ulogic or its subtype std_logic. Another subtype UX01 is defined to be used as the function's return type; even if one of the "and" inputs is a nonlogic value ('Z', 'W', etc.), the

Table 4-36
VHDL program for an "inhibit" function.

```
architecture Inhibit_archf of Inhibit is

function ButNot (A, B: bit) return bit is
begin
  if B = '0' then return A;
  else return '0';
  end if;
end ButNot;

begin
  Z <= ButNot(X,Y);
end Inhibit_archf;
```

Table 4-37
Definitions relating to the "and" operation on STD_LOGIC values in IEEE 1164.

```
SUBTYPE UX01 IS resolved std_ulogic RANGE 'U' TO '1';
                                 -- ('U','X','0','1')
TYPE stdlogic_table IS ARRAY(std_ulogic, std_ulogic) OF std_ulogic;

-- truth table for "and" function
CONSTANT and_table : stdlogic_table := (
--      -------------------------------------------------------
--      | U     X     0     1     Z     W     L     H     -    |    |
--      -------------------------------------------------------
       ( 'U',  'U',  '0',  'U',  'U',  'U',  '0',  'U',  'U' ),  -- | U |
       ( 'U',  'X',  '0',  'X',  'X',  'X',  '0',  'X',  'X' ),  -- | X |
       ( '0',  '0',  '0',  '0',  '0',  '0',  '0',  '0',  '0' ),  -- | 0 |
       ( 'U',  'X',  '0',  '1',  'X',  'X',  '0',  '1',  'X' ),  -- | 1 |
       ( 'U',  'X',  '0',  'X',  'X',  'X',  '0',  'X',  'X' ),  -- | Z |
       ( 'U',  'X',  '0',  'X',  'X',  'X',  '0',  'X',  'X' ),  -- | W |
       ( '0',  '0',  '0',  '0',  '0',  '0',  '0',  '0',  '0' ),  -- | L |
       ( 'U',  'X',  '0',  '1',  'X',  'X',  '0',  '1',  'X' ),  -- | H |
       ( 'U',  'X',  '0',  'X',  'X',  'X',  '0',  'X',  'X' )   -- | - |
);

FUNCTION "and" ( L : std_ulogic; R : std_ulogic ) RETURN UX01 IS
BEGIN
     RETURN (and_table(L, R));
END "and";
```

function will return one of only four possible values. Type stdlogic_table defines a two-dimensional, 9×9 array indexed by a pair of std_ulogic values. For the and_table, the table entries are arranged so that if either index is '0' or 'L' (a weak '0'), the entry is '0'. A '1' entry is found only if both inputs are '1' or 'H' (a weak '1'). Otherwise, a 'U' or 'X' entry appears.

In the function definition itself, double quotes around the function name indicate operator overloading. The "executable" part of the function is just a single statement that returns the table element indexed by the two inputs, L and R, of the "and" function.

Because of VHDL's strong typing requirements, it's often necessary to convert a signal from one type to another, and the IEEE 1164 package contains several conversion functions—for example, from BIT to STD_LOGIC or vice versa. A commonly needed conversion is from STD_LOGIC_VECTOR into a corresponding integer value. IEEE 1164 does not include such a conversion function, because different designs may need to use different number interpretations—for example, signed versus unsigned. However, we can write our own conversion function as shown in Table 4-38.

The CONV_INTEGER function uses a simple iterative algorithm equivalent to the nested expansion formula on page 30. We won't be describing the FOR, CASE, and WHEN statements that it uses until Section 4.7.8, but you should get the idea. The *null statement* is easy—it means "do nothing." The range of the FOR loop is specified by "X'range", where the single quote after a signal name

null statement

```
function CONV_INTEGER (X: STD_LOGIC_VECTOR) return INTEGER is
  variable RESULT: INTEGER;
begin
  RESULT := 0;
  for i in X'range loop
    RESULT := RESULT * 2;
    case X(i) is
      when '0' | 'L'  => null;
      when '1' | 'H'  => RESULT := RESULT + 1;
      when others     => null;
    end case;
  end loop;
  return RESULT;
end CONV_INTEGER;
```

Table 4-38
VHDL function
for converting
STD_LOGIC_VECTOR
to INTEGER.

means "attribute," and *range* is a built-in attribute identifier that applies only to arrays and means "range of this array's index, from left to right."

range attribute

In the other direction, we can convert an integer to a STD_LOGIC_VECTOR as shown in Table 4-39. Here we must specify not only the integer value to be converted (ARG), but also the number of bits in the desired result (SIZE). Notice that the function declares a local variable "result", a STD_LOGIC_VECTOR whose index range is dependent on SIZE. For this reason, SIZE must be a constant or other value that is known when CONV_STD_LOGIC_VECTOR is compiled. To perform the conversion, the function uses the successive-division algorithm that was described on page 30.

A VHDL *procedure* is similar to a function, except it does not return a result. Whereas a function call can be used in the place of an expression, a procedure call can be used in the place of a statement. VHDL procedures allow their arguments to be specified with type out or inout, so it is actually possible for a procedure to "return" a result. However, we don't use VHDL procedures in the rest of this book, so we won't discuss them further.

procedure

```
function CONV_STD_LOGIC_VECTOR(ARG: INTEGER; SIZE: INTEGER)
    return STD_LOGIC_VECTOR is
  variable result: STD_LOGIC_VECTOR (SIZE-1 downto 0);
  variable temp: integer;
begin
  temp := ARG;
  for i in 0 to SIZE-1 loop
    if (temp mod 2) = 1 then result(i) := '1';
    else result(i) := '0';
    end if;
    temp := temp / 2;
  end loop;
  return result;
end;
```

Table 4-39
VHDL function
for converting
INTEGER to
STD_LOGIC_VECTOR.

4.7.5 Libraries and Packages

library

A VHDL *library* is a place where the VHDL compiler stores information about a particular design project, including intermediate files that are used in the analysis, simulation, and synthesis of the design. The location of the library within a host computer's file system is implementation dependent. For a given VHDL design, the compiler automatically creates and uses a library named "`work`".

A complete VHDL design usually has multiple files, each containing different design units including entities and architectures. When the VHDL compiler analyzes each file in the design, it places the results in the "`work`" library, and it also searches this library for needed definitions, such as other entities. Because of this feature, a large design can be broken up into multiple files, yet the compiler will find external references as needed.

Not all of the information needed in a design may be in the "`work`" library. For example, a designer may rely on common definitions or functional modules across a family of different projects. Each project has its own "`work`" library (typically a subdirectory within that project's overall directory), but it must also refer to a common library containing the shared definitions. Even small projects may use a standard library such as the one containing IEEE standard definitions.

library clause

The designer can specify the name of such a library using a `library` *clause* at the beginning of the design file. For example, we can specify the IEEE library:

```
library ieee;
```

The clause "`library work;`" is included implicitly at the beginning of every VHDL design file.

Specifying a library name in a design gives it access to any previously analyzed entities and architectures stored in the library, but it does not give access to type definitions and the like. This is the function of "packages" and "use clauses," described next.

package

A VHDL *package* is a file containing definitions of objects that can be used in other programs. The kind of objects that can be put into a package include signal, type, constant, function, procedure, and component declarations.

Signals that are defined in a package are "global" signals, available to any VHDL entity that uses the package. Types and constants defined in a package are known in any file that uses the package. Likewise, functions and procedures defined in a package can be called in files that use the package, and components (described in the next subsection) can be "instantiated" in architectures that use the package.

use clause

A design can "use" a package by including a *use clause* at the beginning of the design file. For example, to use all of the definitions in the IEEE standard 1164 package, we would write

```
use ieee.std_logic_1164.all;
```

Here, "ieee" is the name of a library which has been previously given in a library clause. Within this library, the file named "std_logic_1164" contains the desired definitions. The suffix "all" tells the compiler to use all of the definitions in this file. Instead of "all", you can write the name of a particular object to use just its definition, for example,

```
use ieee.std_logic_1164.std_ulogic
```

This clause would make available just the definition of the std_ulogic type in Table 4-32 on page 273, without all of the related types and functions. However, multiple "use" clauses can be written to use additional definitions.

Defining packages is not limited to standards bodies. Anyone can write a package, using the syntax shown in Table 4-40. All of the objects declared between "package" and the first "end" statement are visible in any design file that uses the package; objects following the "package body" keyword are local. In particular, notice that the first part includes "function declarations," not definitions. A *function declaration* lists only the function name, arguments, and type, up to but not including the "is" keyword in Table 4-35 on page 276. The complete function definition is given in the package body and is not visible to function users.

function declaration

IEEE VHDL STANDARDS

VHDL has excellent capabilities for extending its data types and functions. This is important, because the language's built-in BIT and BIT_VECTOR actually are quite inadequate for modeling real circuits that also handle three-state, unknown, don't-care, and varying-strength signals.

As a result, soon after the language was formalized as IEEE standard 1076, commercial vendors began to introduce their own built-in data types to deal with logic values other than 0 and 1. Of course, each vendor had different definitions for these extended types, creating a potential "Tower of Babel."

To avoid this situation, the IEEE developed the 1164 standard logic package (std_logic_1164) with a nine-valued logic system that satisfies most designers' needs. This was later followed by standard 1076-3, discussed in Section 5.9.6, which includes several packages with standard types and operations for vectors of STD_LOGIC components that are interpreted as signed or unsigned integers. The packages include std_logic_arith, std_logic_signed, and std_logic_unsigned.

By using IEEE standards, designers can ensure a high degree of portability and interoperability among their designs. This is increasingly important, as the deployment of very large ASICs necessitates cooperation not only among multiple designers but also among multiple vendors who may each contribute different pieces of a "system-on-a-chip" design.

Table 4-40
Syntax of a VHDL
package definition.

```
package package-name is
    type declarations
    signal declarations
    constant declarations
    component declarations
    function declarations
    procedure declarations
end package-name;
package body package-name is
    type declarations
    constant declarations
    function definitions
    procedure definitions
end package-name;
```

4.7.6 Structural Design Elements

We're finally ready to look at the guts of a VHDL design, the "executable" portion of an architecture. Recall from Table 4-28 on page 271 that the body of an architecture is a series of concurrent statements. In VHDL, each *concurrent statement* executes simultaneously with the other concurrent statements in the same architecture body.

concurrent statement

This behavior is markedly different from that of statements in conventional software programming languages, where statements execute sequentially. Concurrent statements are necessary to simulate the behavior of hardware, where connected elements affect each other continuously, not just at particular, ordered time steps. Thus, in a VHDL architecture body, if the last statement updates a signal that is used by the first statement, then the simulator will go back to that first statement and update its results according to the signal that just changed. In fact, the simulator will keep propagating changes and updating results until the simulated circuit stabilizes; we'll discuss this in more detail in Section 4.7.9.

VHDL has several different concurrent statements, as well as a mechanism for bundling a set of sequential statements to operate as a single concurrent statement. Used in different ways, these statements give rise to three somewhat distinct styles of circuit design and description, which we cover in this and the next two subsections.

The most basic of VHDL's concurrent statements is the `component` *statement*, whose basic syntax is shown in Table 4-41. Here, *component-name* is the name of a previously defined entity that is to be used, or *instantiated*, within the current architecture body. One instance of the named entity is created for

instantiate

Table 4-41 Syntax of a VHDL component statement.

label: *component-name* port map(*signal1*, *signal2*, ..., *signaln*);

label: *component-name* port map(*port1*=>*signal1*, *port2*=>*signal2*, ..., *portn*=>*signaln*);

```
component component-name
   port (signal-names : mode signal-type;
         signal-names : mode signal-type;
         ...
         signal-names : mode signal-type);
end component;
```

Table 4-42
Syntax of a
VHDL component
declaration.

each component statement that invokes its name, and each instance must be named by a unique *label*.

The `port map` keywords introduce a list that associates ports of the named entity with signals in the current architecture. The list may be written in either of two different styles. The first is a positional style; as in conventional programming languages, the signals in the list are associated with the entity's ports in the same order in which they appear in the entity's definition. The second is an explicit style; each of the entity's ports is connected to a signal using the "=>" operator, and these associations may be listed in any order.

port map

Before being instantiated in an architecture, a component must be declared in a *component declaration* in the architecture's definition (see Table 4-28 on page 271). As shown in Table 4-42, a component declaration is essentially the same as the port-declaration part of the corresponding entity declaration—it lists the name, mode, and type of each of its ports.

component declaration

The components used in an architecture may be ones that were previously defined as part of a design, or they may be part of a library. Table 4-43 is an example of a VHDL entity and its architecture that uses components, a "prime-number detector" that is structurally identical to the gate-level circuit in Figure 4-30(c) on page 226. The entity declaration names the inputs and the output of the circuit. The declarations section of the architecture defines all of the signal names and the components that are used internally. The components, INV, AND2, AND3, and OR4, are predefined in the design environment in which this example was created and compiled (Xilinx Foundation 1.5, see References).

Note that component statements in Table 4-43 execute *concurrently*. Even if the statements were listed in a different order, the same circuit would be synthesized, and the simulated circuit operation would be the same.

A VHDL architecture that uses components is often called a *structural description* or *structural design*, because it defines the precise interconnection structure of signals and entities that realize the entity. In this regard, a pure structural description is equivalent to a schematic or a net list for the circuit.

structural description
structural design

In some applications it is necessary to create multiple copies of a particular structure within an architecture. For example, we'll see in Section 5.10.2 that an *n*-bit "ripple adder" can be created by cascading *n* "full adders." VHDL includes a *generate statement* that allows you to create such repetitive structures using a kind of "for loop," without having to write out all of the component instantiations individually.

Table 4-43 Structural VHDL program for a prime-number detector.

```
library IEEE;
use IEEE.std_logic_1164.all;

entity prime is
    port ( N: in STD_LOGIC_VECTOR (3 downto 0);
            F: out STD_LOGIC );
end prime;

architecture prime1_arch of prime is
signal N3_L, N2_L, N1_L: STD_LOGIC;
signal N3L_N0, N3L_N2L_N1, N2L_N1_N0, N2_N1L_N0: STD_LOGIC;
component INV port (I: in STD_LOGIC; O: out STD_LOGIC); end component;
component AND2 port (I0,I1: in STD_LOGIC; O: out STD_LOGIC); end component;
component AND3 port (I0,I1,I2: in STD_LOGIC; O: out STD_LOGIC); end component;
component OR4 port (I0,I1,I2,I3: in STD_LOGIC; O: out STD_LOGIC); end component;
begin
  U1: INV port map (N(3), N3_L);
  U2: INV port map (N(2), N2_L);
  U3: INV port map (N(1), N1_L);
  U4: AND2 port map (N3_L, N(0), N3L_N0);
  U5: AND3 port map (N3_L, N2_L, N(1), N3L_N2L_N1);
  U6: AND3 port map (N2_L, N(1), N(0), N2L_N1_N0);
  U7: AND3 port map (N(2), N1_L, N(0), N2_N1L_N0);
  U8: OR4 port map (N3L_N0, N3L_N2L_N1, N2L_N1_N0, N2_N1L_N0, F);
end prime1_arch;
```

The syntax of a simple iterative `generate` loop is shown in Table 4-44. The *identifier* is implicitly declared as a variable with type compatible with the *range*. The *concurrent statement* is executed once for each possible value of the *identifier* within the range, and *identifier* may be used within the concurrent statement. For example, Table 4-45 shows how an 8-bit inverter can be created.

The value of a constant must be known at the time that a VHDL program is compiled. In many applications it is useful to design and compile an entity and its architecture while leaving some of its parameters, such as bus width, unspecified. VHDL's "generic" facility lets you do this.

generic constant
generic declaration

One or more *generic constants* can be defined in an entity declaration with a *generic declaration* before the port declaration, using the syntax shown in Table 4-46. Each of the named constants can be used within the architecture definition for the entity, and the value of the constant is deferred until the entity is instantiated using a component statement within another architecture. Within that component statement, values are assigned to the generic constants using a

generic map

`generic map` clause in the same style as the port map clause. Table 4-47 is an example that combines generic and generate statements to define a "bus inverter" with a user-specifiable width. Multiple copies of this inverter, each with a different width, are instantiated in the program in Table 4-48.

```
label: for identifier in range generate
            concurrent-statement
        end generate;
```

Table 4-44
Syntax of a VHDL
for-generate loop.

```
library IEEE;
use IEEE.std_logic_1164.all;

entity inv8 is
    port ( X: in STD_LOGIC_VECTOR (1 to 8);
           Y: out STD_LOGIC_VECTOR (1 to 8) );
end inv8;

architecture inv8_arch of inv8 is
component INV port (I: in STD_LOGIC; O: out STD_LOGIC); end component;
begin
  g1: for b in 1 to 8 generate
        U1: INV port map (X(b), Y(b));
      end generate;
end inv8_arch;
```

Table 4-45
VHDL entity and
architecture for an
8-bit inverter.

```
entity entity-name is
    generic (constant-names : constant-type;
             constant-names : constant-type;
             . . .
             constant-names : constant-type);
    port (signal-names : mode signal-type;
          signal-names : mode signal-type;
          . . .
          signal-names : mode signal-type);
end entity-name;
```

Table 4-46
Syntax of a VHDL ge-
neric declaration with-
in an entity
declaration.

```
library IEEE;
use IEEE.std_logic_1164.all;

entity businv is
    generic (WIDTH: positive);
    port ( X: in STD_LOGIC_VECTOR (WIDTH-1 downto 0);
           Y: out STD_LOGIC_VECTOR (WIDTH-1 downto 0) );
end businv;

architecture businv_arch of businv is
component INV port (I: in STD_LOGIC; O: out STD_LOGIC); end component;
begin
  g1: for b in WID-1 downto 0 generate
        U1: INV port map (X(b), Y(b));
      end generate;
end businv_arch;
```

Table 4-47
VHDL entity and
architecture for an
arbitrary-width bus
inverter.

Table 4-48
VHDL entity and
architecture that use
the arbitrary-width
bus inverter.

```
library IEEE;
use IEEE.std_logic_1164.all;

entity businv_example is
    port ( IN8: in STD_LOGIC_VECTOR (7 downto 0);
           OUT8: out STD_LOGIC_VECTOR (7 downto 0);
           IN16: in STD_LOGIC_VECTOR (15 downto 0);
           OUT16: out STD_LOGIC_VECTOR (15 downto 0);
           IN32: in STD_LOGIC_VECTOR (31 downto 0);
           OUT32: out STD_LOGIC_VECTOR (31 downto 0) );
end businv_example;

architecture businv_ex_arch of businv_example is
component businv
    generic (WIDTH: positive);
    port ( X: in STD_LOGIC_VECTOR (WIDTH-1 downto 0);
           Y: out STD_LOGIC_VECTOR (WIDTH-1 downto 0) );
end component;
begin
U1: businv generic map (WIDTH=>8) port map (IN8, OUT8);
U2: businv generic map (WIDTH=>16) port map (IN16, OUT16);
U3: businv generic map (WIDTH=>32) port map (IN32, OUT32);
end businv_ex_arch;
```

4.7.7 Dataflow Design Elements

If component statements were its only concurrent statements, then VHDL would be little more than a strongly typed, hierarchical net-list description language. Several additional concurrent statements allow VHDL to describe a circuit in terms of the flow of data and operations on it within the circuit. This style is called a *dataflow description* or *dataflow design*.

dataflow description
dataflow design
*concurrent signal-
  assignment statement*

Two additional concurrent statements used in dataflow designs are shown in Table 4-49. The first of these is the most often used and is called a *concurrent signal-assignment statement.* You can read this as "*signal-name* gets *expression.*" Because of VHDL's strong typing, the type of *expression* must be compatible with that of *signal-name*. In general, this means that either the types must be identical or *expression*'s type is a subtype of *signal-name*'s. In the case of arrays, both the element type and the length must match; however, the index range and direction need not match.

Table 4-49
Syntax of VHDL
concurrent signal-
assignment statements.

signal-name <= *expression*;

signal-name <= *expression* when *boolean-expression* else
                  expression when *boolean-expression* else
                    . . .
                  expression when *boolean-expression* else
                  expression;

```
architecture prime2_arch of prime is
signal N3L_N0, N3L_N2L_N1, N2L_N1_N0, N2_N1L_N0: STD_LOGIC;
begin
  N3L_N0     <= not N(3)                            and N(0);
  N3L_N2L_N1 <= not N(3) and not N(2) and     N(1)         ;
  N2L_N1_N0  <=              not N(2) and     N(1) and N(0);
  N2_N1L_N0  <=                        N(2) and not N(1) and N(0);
  F <= N3L_N0 or N3L_N2L_N1 or N2L_N1_N0 or N2_N1L_N0;
end prime2_arch;
```

Table 4-50
Dataflow VHDL architecture for the prime-number detector.

Table 4-50 shows an architecture for the prime-number detector entity (Table 4-43 on page 284) written in dataflow style. In this style we don't show the explicit gates and their connections; rather, we use VHDL's built-in and, or, and not operators. (Actually, these operators are not built in for signals of type STD_LOGIC, but they are defined and overloaded by the IEEE 1164 package.) Note that the not operator has the highest precedence, so no parentheses are required around subexpressions like "not N(3)" to get the intended result.

We can also use the second, *conditional* form of the concurrent signal-assignment statement, using the keywords when and else as shown in Table 4-49. Here, a *boolean-expression* combines individual boolean terms using VHDL's built-in boolean operators such as and, or, and not. Boolean terms are typically boolean variables or results of comparisons using *relational operators* =, /= (not equal), >, >=, <, and <=.

conditional signal-assignment statement
when
else
relational operators
=, /=, >, >=, <, <=

Table 4-51 is an example using conditional concurrent assignment statements. Each of the comparisons of a individual STD_LOGIC bit such as N(3) is made against a character literal '1' or '0' and returns a value of type boolean. These comparison results are combined in the boolean expression between the when and else keywords in each statement. The else clauses are generally required; the combined set of conditions in a single statement should cover all possible input combinations.

Another kind of concurrent assignment statement is the *selected signal assignment*, whose syntax is shown in Table 4-52. This statement evaluates the given *expression*, and when the value matches one of the *choices*, it assigns the corresponding *signal-value* to *signal-name*. The *choices* in each when clause

selected signal-assignment statement

```
architecture prime3_arch of prime is
signal N3L_N0, N3L_N2L_N1, N2L_N1_N0, N2_N1L_N0: STD_LOGIC;
begin
  N3L_N0     <= '1' when N(3)='0' and N(0)='1' else '0';
  N3L_N2L_N1 <= '1' when N(3)='0' and N(2)='0' and N(1)='1' else '0';
  N2L_N1_N0  <= '1' when N(2)='0' and N(1)='1' and N(0)='1' else '0';
  N2_N1L_N0  <= '1' when N(2)='1' and N(1)='0' and N(0)='1' else '0';
  F <= N3L_N0 or N3L_N2L_N1 or N2L_N1_N0 or N2_N1L_N0;
end prime3_arch;
```

Table 4-51
Prime-number-detector architecture using conditional assignments.

Table 4-52
Syntax of VHDL
selected signal-
assignment statement.

```
with expression select
  signal-name <= signal-value when choices,
                 signal-value when choices,
                 . . .
                 signal-value when choices;
```

may be a single value of *expression* or a list of values separated by vertical bars
(|). The *choices* for the entire statement must be mutually exclusive and all
inclusive. The keyword *others* can be used in the last when clause to denote all
values of *expression* that have not yet been covered.

others

Table 4-53 is an architecture for the prime-number detector that uses a
selected signal-assignment statement. All of the *choices* for which F is '1' could
have been written in a single when clause, but multiple clauses are shown just for
instructional purposes. In this example, the selected signal-assignment statement reads somewhat like a listing of the on-set of the function F.

We can modify the previous architecture slightly to take advantage of the
numeric interpretation of N in the function definition. Using the CONV_INTEGER
function that we defined previously, Table 4-54 writes the *choices* in terms of
integers, which we can readily see are prime as required. We can think of this
version of the architecture as a "behavioral" description, because it describes the
desired function in such a way that its behavior is quite evident.

Table 4-53
Prime-number
detector architecture
using selected signal
assignment.

```
architecture prime4_arch of prime is
begin
  with N select
    F <= '1' when "0001",
         '1' when "0010",
         '1' when "0011" | "0101" | "0111",
         '1' when "1011" | "1101",
         '0' when others;
end prime4_arch;
```

**COVERING
ALL THE
CASES**

Conditional and selected signal assignments require all possible conditions to be
covered. In a conditional signal assignment, the final "else *expression*" covers
missing conditions. In a selected signal assignment, "others" can be used in the
final when clause to pick up the remaining conditions.

In Table 4-53, you might think that instead of writing "others" in the
final when clause, we could have written the nine remaining 4-bit combinations,
"0000", "0100", and so on. But that's not true! Remember that STD_LOGIC is
a nine-valued system, so a 4-bit STD_LOGIC_VECTOR actually has 9^4 possible
values. So "others" in this example is really covering 6,554 cases!

```
architecture prime5_arch of prime is
begin
  with CONV_INTEGER(N) select
    F <= '1' when 1 | 2 | 3 | 5 | 7 | 11 | 13,
         '0' when others;
end prime5_arch;
```

Table 4-54
A more behavioral description of the prime-number detector.

4.7.8 Behavioral Design Elements

As we saw in the last example, it is sometimes possible to directly describe a desired logic-circuit behavior using a concurrent statement. This is a good thing, as the ability to create a *behavioral design* or *behavioral description* is one of the key benefits of hardware-description languages in general and VHDL in particular. However, for most behavioral descriptions, we need to employ some additional language elements described in this subsection.

behavioral design
behavioral description

VHDL's key behavioral element is the "process." A *process* is a collection of "sequential" statements (described shortly) that executes in parallel with other concurrent statements and other processes. Using a process, you can specify a complex interaction of signals and events in a way that executes in essentially zero simulated time during simulation and that gives rise to a synthesized combinational or sequential circuit that performs the modeled operation directly.

process

A VHDL *process statement* can be used anywhere that a concurrent statement can be used. A process statement is introduced by the keyword *process* and has the syntax shown in Table 4-55. Since a process statement is written within the scope of an enclosing architecture, it has visibility of the types, signals, constants, functions, and procedures that are declared or are otherwise visible in the enclosing architecture. However, you can also define types, signals, constants, functions, and procedures that are local to the process.

process statement
process

Note that a process may not declare signals, only "variables." A VHDL *variable* keeps track of the state within a process and is not visible outside of the process. Depending on its use, it may or may not give rise to a corresponding signal in a physical realization of the modeled circuit. The syntax for defining a

variable

```
process (signal-name, signal-name, ..., signal-name)
    type declarations
    variable declarations
    constant declarations
    function definitions
    procedure definitions
begin
    sequential-statement
    ...
    sequential-statement
end process;
```

Table 4-55
Syntax of a VHDL process statement.

variable

variable within a process is similar to the syntax for a signal declaration within an architecture, except that the keyword *variable* is used:

variable *variable-names* : *variable-type*;

running process
suspended process
sensitivity list

A VHDL process is always either *running* or *suspended*. The list of signals in the process definition, called the *sensitivity list*, determines when the process runs. A process initially is suspended; when any signal in its sensitivity list changes value, the process resumes execution, starting with its first sequential statement and continuing until the end. If any signal in the sensitivity list change value as a result of running the process, it runs again. This continues until the process runs without any of these signals changing value. In simulation, all of this happens in zero simulated time.

Upon resumption, a properly written process will suspend after one or a few runs. However, it is possible to write an incorrect process that never suspends. For example, consider a process with just one sequential statement, "X <= not X" and a sensitivity list of "(X)". Since X changes on every pass, the process will run forever in zero simulated time—not very useful! In practice, simulators have safeguards that normally can detect such unwanted behavior, terminating the misbehaving process after a thousand or so passes.

The sensitivity list is optional; a process without a sensitivity list starts running at time zero in simulation. One application of such a process is to generate input waveforms in a test bench, as in Table 4-65 on page 296.

*sequential signal-
   assignment statement*

VHDL has several kinds of sequential statements. The first is a *sequential signal-assignment statement*; this has the same syntax as the concurrent version (*signal-name <= expression;*), but it occurs within the body of a process rather than an architecture. An analogous statement for variables is the *variable-assignment statement*, which has the syntax "*variable-name := expression;*". Notice that a different assignment operator, :=, is used for variables.

*variable-assignment
   statement*

:=

For instruction purposes, the dataflow architecture of the prime-number detector in Table 4-50 is rewritten as a process in Table 4-56. Notice that we're still working off the same original entity declaration of prime that appeared in

Table 4-56
Process-based
dataflow VHDL
architecture for the
prime-number
detector.

```
architecture prime6_arch of prime6 is
begin
  process(N)
    variable N3L_N0, N3L_N2L_N1, N2L_N1_N0, N2_N1L_N0: STD_LOGIC;
  begin
    N3L_N0      := not N(3)                            and N(0);
    N3L_N2L_N1 := not N(3) and not N(2) and     N(1)          ;
    N2L_N1_N0  :=               not N(2) and     N(1) and N(0);
    N2_N1L_N0  :=                   N(2) and not N(1) and N(0);
    F <= N3L_N0 or N3L_N2L_N1 or N2L_N1_N0 or N2_N1L_N0;
  end process;
end prime6_arch;
```

WEIRD BEHAVIOR	Remember that the statements within a process are executed *sequentially*. Suppose that for some reason we wrote the last statement in Table 4-56 (the signal assignment to F) as the first. Then we would see rather weird behavior from this process.

The first time the process was run, the simulator would complain that the values of the variables were being read before any value was assigned to them. On subsequent resumptions, a value would be assigned to F based on the *previous* values of the variables, which are remembered while the process is suspended. New values would then be assigned to the variables and remembered until the next resumption. So the circuit's output value would always be one input-change behind.

Table 4-43. Within the new architecture (prime6_arch), we have just one concurrent statement, which is a process. The process sensitivity list contains just N, the primary inputs of the desired combinational logic function. The AND-gate outputs must be defined as variables rather than signals, since signal definitions are not allowed within a process. Otherwise, the body of the process is very similar to that of the original architecture. In fact, a typical synthesis tool would probably create the same circuit from either description.

Other sequential statements, beyond simple assignment, can give us more creative control in expressing circuit behavior. The *if statement*, with the syntax shown in Table 4-57, is probably the most familiar of these. In the first and simplest form of the statement, a *boolean-expression* is tested, and a *sequential-statement* is executed if the expression's value is true. In the second form,

if statement

```
if boolean-expression then sequential-statement
end if;

if boolean-expression then sequential-statement
else sequential-statement
end if;

if boolean-expression then sequential-statement
elsif boolean-expression then sequential-statement
 . . .
elsif boolean-expression then sequential-statement
end if;

if boolean-expression then sequential-statement
elsif boolean-expression then sequential-statement
 . . .
elsif boolean-expression then sequential-statement
else sequential-statement
end if;
```

Table 4-57
Syntax of a VHDL
if statement.

Table 4-58
Prime-number-detector architecture using an `if` statement.

```
architecture prime7_arch of prime is
begin
  process(N)
    variable NI: INTEGER;
  begin
    NI := CONV_INTEGER(N);
    if NI=1 or NI=2 then F <= '1';
    elsif NI=3 or NI=5 or NI=7 or NI=11 or NI=13 then F <= '1';
    else F <= '0';
    end if;
  end process;
end prime7_arch;
```

else

we've added an "*else*" clause with another *sequential-statement* that's executed if the expression's value is `false`.

elsif

To create nested `if-then-else` statements, VHDL uses a special keyword *elsif*, which introduces the "middle" clauses. An `elsif` clause's *sequential-statement* is executed if its *boolean-expression* is `true` and all of the preceding *boolean-expressions* were `false`. The optional final `else`-clause's *sequential-statement* is executed if all of the preceding *boolean-expressions* were `false`.

Table 4-58 is a version of the prime-number-detector architecture that uses an `if` statement. A local variable `NI` is used to hold a converted, integer version of the input `N`, so that the comparisons in the `if` statement can be written using integer values.

The boolean expressions in Table 4-58 are nonoverlapping; that is, only one of them is `true` at a time. For this application we really didn't need the full power of nested `if` statements. In fact, a synthesis engine might create a circuit that evaluates the boolean expressions in series, with slower operation than might otherwise be possible. When we need to select among multiple alterna-

case statement

tives based on the value of just one signal or expression, a *case statement* is usually more readable and may yield a better synthesized circuit.

Table 4-59 shows the syntax of a `case` statement. This statement evaluates the given *expression*, finds a matching value in one of the *choices*, and executes the corresponding *sequential-statements*. Note that one or more sequential statements can be written for each set of *choices*. The *choices* may take the form of a single value or of multiple values separated by vertical bars (|). The *choices* must be mutually exclusive and include all possible values of *expression*'s type; the keyword `others` can be used as the last *choices* to denote all values that have not yet been covered.

Table 4-59
Syntax of a VHDL `case` statement.

```
case expression is
  when choices => sequential-statements
  . . .
  when choices => sequential-statements
end case;
```

```
architecture prime8_arch of prime is
begin
  process(N)
  begin
    case CONV_INTEGER(N) is
      when 1 => F <= '1';
      when 2 => F <= '1';
      when 3 | 5 | 7 | 11 | 13 => F <= '1';
      when others => F <= '0';
    end case;
  end process;
end prime8_arch;
```

Table 4-60
Prime-number-
detector architecture
using a case statement.

Table 4-60 is yet another architecture for the prime-number detector, this time coded with a case statement. Like the concurrent version, the select statement in Table 4-54 on page 289, the case statement makes it very easy to see the desired functional behavior.

Another important class of sequential statements are the *loop statements*, The simplest of these has the syntax shown in Table 4-61 and creates an infinite loop. Although infinite loops are undesirable in conventional software programming languages, we'll show in Section 7.12 how such a loop can be very useful in hardware modeling.

loop statement

```
loop
   sequential-statement
   . . .
   sequential-statement
end loop;
```

Table 4-61
Syntax of a basic
VHDL loop
statement.

A more familiar loop, one that we've seen before, is the *for loop*, with the syntax shown in Table 4-62. Note that the loop variable, *identifier*, is declared implicitly by its appearance in the for loop and has the same type as *range*. This variable may be used within the loop's sequential statements, and it steps through all of the values in *range*, from left to right, one per iteration.

for loop

Two more useful sequential statements that can be executed within a loop are "*exit*" and "*next*". When executed, *exit* transfers control to the statement immediately following the loop end. On the other hand, *next* causes any remaining statements in the loop to be bypassed and begins the next iteration of the loop.

exit statement
next statement

```
for identifier in range loop
   sequential-statement
   . . .
   sequential-statement
end loop;
```

Table 4-62
Syntax of a VHDL
for loop.

Table 4-63
Prime-number-
detector architecture
using a for statement.

```
library IEEE;
use IEEE.std_logic_1164.all;

entity prime9 is
    port ( N: in STD_LOGIC_VECTOR (15 downto 0);
           F: out STD_LOGIC );
end prime9;

architecture prime9_arch of prime9 is
begin
  process(N)
  variable NI: INTEGER;
  variable prime: boolean;
  begin
    NI := CONV_INTEGER(N);
    prime := true;
    if NI=1 or NI=2 then null; -- take care of boundary cases
    else for i in 2 to 253 loop
          if NI mod i = 0 then
             prime := false; exit;
          end if;
        end loop;
    end if;
    if prime then F <= '1'; else F <= '0'; end if;
  end process;
end prime9_arch;
```

Our good old prime-number detector is coded one more time in Table 4-63, this time using a for loop. The striking thing about this example is that it is truly a behavioral description—we have actually used VHDL to compute whether the input N is a prime number. We've also increased the size of N to 16 bits, just to emphasize the fact that we were able to create a compact model for the circuit without having to explicitly list hundreds of primes.

BAD DESIGN

Table 4-63 has a good example of a for loop, but is a bad example of how to design a circuit. Although VHDL is a powerful programming language, design descriptions that use its full power may be inefficient or unsynthesizable.

The culprit in Table 4-63 is the mod operator. This operation requires an integer division, and most VHDL tools are unable to synthesize division circuits except for special cases, such as division by a power of two (realized as a shift).

Even if the tools could synthesize a divider, we wouldn't want to specify a prime number detector in this way. The description in Table 4-63 implies a combinational circuit, and the tools would have to create 252 combinational dividers, one for each value of i, to "unroll" the for loop and realize the circuit!

Table 4-65
Using the VHDL `wait` statement to generate input waveforms in a test-bench program.

```
entity InhibitTestBench is
end InhibitTestBench;

architecture InhibitTB_arch of InhibitTestBench is
component Inhibit port (X,Y: in BIT; Z: out BIT); end component;
signal XT, YT, ZT: BIT;
begin
  U1: Inhibit port map (XT, YT, ZT);
  process
  begin
    XT <= '0'; YT <= '0';
    wait for 10 ns;
    XT <= '0'; YT <= '1';
    wait for 10 ns;
    XT <= '1'; YT <= '0';
    wait for 10 ns;
    XT <= '1'; YT <= '1';
    wait; -- this suspends the process indefinitely
  end process;
end InhibitTB_arch;
```

simulation time

Simulator operation begins at *simulation time* of zero. At this time, the simulator initializes all signals to a default value (which you shouldn't depend on!). It also initializes any signals or variables for which initial values have been declared explicitly (we haven't shown you how to do this). Next, the simulator begins the execution of all the processes and concurrent statements in the design.

Of course, the simulator can't really simulate all of the processes and concurrent statements simultaneously, but it can pretend that it does, using a time-based "event list" and a "signal-sensitivity matrix." Note that each concurrent statement is equivalent to one process.

At simulation time zero, all of the processes are scheduled for execution, and one of these is selected. All of its sequential statements are executed, including any looping behavior that is specified. When the execution of this process is completed, another one is selected, and so on, until all of the processes have been

simulation cycle

executed. This completes one *simulation cycle*.

event list

During its execution, a process may assign new values to signals. The new values are not assigned immediately; rather, they are placed on the *event list* and scheduled to become effective at a certain time. If the assignment has an explicit simulation time associated with it (for example, after a delay specified by an `after` clause), then it is scheduled on the event list to occur at that time. Otherwise, it is supposed to occur "immediately"; however, it is actually scheduled to

delta delay

occur at the current simulation time plus one "delta delay." The *delta delay* is an infinitesimally short time, such that the current simulation time plus any number of delta delays still equals the current simulation time. This concept allows processes to execute multiple times if necessary in zero simulated time.

``
while *boolean-expression* loop
    sequential-statement
    . . .
    sequential-statement
end loop;
``
	Table 4-64
	Syntax of a VHDL
	`while` loop.

The last kind of `loop` statement is the `while *loop*`, with the syntax shown in Table 4-64. In this form, *boolean-expression* is tested before each iteration of the loop, and the loop is executed only if the value of the expression is `true`.

while loop

We can use processes to write behavioral descriptions of both combinational and sequential circuits. Many more examples of combinational-circuit descriptions appear in the VHDL subsections of Chapter 5. A few additional language features are needed to describe sequential circuits; these are described in Section 7.12, and sequential examples appear in the VHDL subsections of Chapter 8.

4.7.9 The Time Dimension and Simulation

None of the examples that we've dealt with so far models the time dimension of circuit operation—everything happens in zero simulated time. However, VHDL has excellent facilities for modeling the time, and it is indeed another significant dimension of the language. In this book we won't go into detail on this subject, but we'll introduce just a few ideas here.

VHDL allows you to specify a time delay using the keyword `after` in any signal-assignment statement, including sequential, concurrent, conditional, and selected assignments. For example, in the inhibit-gate architecture of Table 4-26 on page 269 you could write

after keyword

```
Z <= '1' after 4 ns when X='1' and Y='0' else '0' after 3 ns;
```

This allows you to model an inhibit gate that has 4 ns of delay on a 0-to-1 output transition and only 3 ns on a 1-to-0 transition. In typical ASIC design environments, the VHDL models for all of the low-level components in the component library include such delay parameters. Using these estimates, a VHDL simulator can predict the approximate timing behavior of a larger circuit that uses these components.

Another way to invoke the time dimension is with `wait`, a sequential statement. This statement can be used to suspend a process for a specified time period. Table 4-65 is an example program that uses `wait` to create simulated input waveforms to test the operation of the inhibit gate for four different input combinations at 10-ns time steps.

wait statement

Once you have a VHDL program whose syntax and semantics are correct, you can use a VHDL simulator to observe its operation. Although we won't go into great detail, it's useful to have a basic understanding of how such a simulator works.

After a simulation cycle completes, the event list is scanned for the signal or signals that change at the next earliest time on the list. This may be as little as one delta delay later, or it may be a real circuit delay later, in which case the simulation time is advanced to this time. In any case, the scheduled signal changes are made. Some processes may be sensitive to the changing signals, as indicated by the signal-sensitivity matrix. It indicates, for each signal, which processes have that signal in their sensitivity list. (The process equivalent of a concurrent statement has *all* of its control and data signals in its sensitivity list.) All of the processes that are sensitive to a signal that just changed are scheduled for execution in the next simulation cycle, which now begins.

The simulator's two-phase operation of a simulation cycle followed by scanning the event list and making the next scheduled signal assignments goes on indefinitely, until the list is empty. At this point the simulation is complete.

The event-list mechanism makes it possible to simulate the operation of concurrent processes even though the simulator runs on a single computer with a single thread of execution. And the delta-delay mechanism ensures correct operation even though a process or set of processes may require multiple executions, spanning several delta delays, before changing signals settle down to a stable value. This mechanism is also used to detect runaway processes (such as "X <= not X"); if a thousand simulation cycles occur over a thousand delta delays without advancing simulation time by any "real" amount, it's most likely that something's amiss.

4.7.10 Synthesis

As we mentioned at the beginning of this section, VHDL was originally invented as a logic circuit description and simulation language and was later adapted to synthesis. The language has many features and constructs that cannot be synthesized. However, the subset of the language and the style of programs that we've presented in this section are generally synthesizable by most tools.

Still, the code that you write can have a big effect on the quality of the synthesized circuits that you get. A few examples are listed below:

- "Serial" control structures like if-elsif-elsif-else can result in a corresponding serial chain of logic gates to test conditions. It's better to use a case or select statement if the conditions are mutually exclusive.

- Loops in processes are generally "unwound" to create multiple copies of combinational logic to execute the statements in the loop. If you want to use just one copy of the combinational logic in a sequence of steps, then you have to design a sequential circuit, as discussed in later chapters.

- When using conditional statements in a process, failing to state an outcome for some input combination will cause the compiler to create a "latch" to hold the old value of a signal that might otherwise change. Such latches are generally not intended.

In addition, some language features and constructs may just be unsynthesizable, depending on the tool. Naturally, you have to consult the documentation to find out what's disallowed, allowed, and recommended for a particular tool.

For the foreseeable future, digital designers who use synthesis tools will need to pay reasonably close attention to their coding style in order to obtain good results. And for the moment, the definition of "good coding style" depends somewhat on both the synthesis tool and the target technology. The examples in the rest of this book, while syntactically and semantically correct, hardly scratch the surface of coding methods for large HDL-based designs. The art and practice of large HDL-based hardware design is still very much evolving.

References

A historical description of Boole's development of "the science of Logic" appears in *The Computer from Pascal to von Neumann* by Herman H. Goldstine (Princeton University Press, 1972). Claude E. Shannon showed how Boole's work could be applied to logic circuits in "A Symbolic Analysis of Relay and Switching Circuits" (*Trans. AIEE,* Vol. 57, 1938, pp. 713–723).

Although the two-valued Boolean algebra is the basis for switching algebra, a Boolean algebra need not have only two values. Boolean algebras with 2^n values exist for every integer n; for example, see *Discrete Mathematical Structures and Their Applications* by Harold S. Stone (SRA, 1973). Such

Huntington postulates

algebras may be formally defined using the so-called *Huntington postulates* devised by E. V. Huntington in 1907; for example, see *Digital Design* by M. Morris Mano (Prentice Hall, 1984). A mathematician's development of Boolean algebra based on a more modern set of postulates appears in *Modern Applied Algebra* by G. Birkhoff and T. C. Bartee (McGraw-Hill, 1970). Our engineering-style, "direct" development of switching algebra follows that of Edward J. McCluskey in his *Introduction to the Theory of Switching Circuits* (McGraw-Hill, 1965) and *Logic Design Principles* (Prentice Hall, 1986).

The prime-implicant theorem was proved by W. V. Quine in "The Problem of Simplifying Truth Functions" (*Am. Math. Monthly,* Vol. 59, No. 8, 1952, pp. 521–531). In fact it is possible to prove a more general prime-implicant theorem showing that there exists at least one minimal sum that is a sum of prime implicants even if we remove the constraint on the number of literals in the definition of "minimal."

A graphical method for simplifying Boolean functions was proposed by E. W. Veitch in "A Chart Method for Simplifying Boolean Functions" (*Proc. ACM,* May 1952, pp. 127–133). His *Veitch diagram,* shown in Figure 4-53, actually reinvented a chart proposed by an English archaeologist, A. Marquand ("On Logical Diagrams for *n* Terms," *Philosophical Magazine* XII, 1881, pp. 266–270). The Veitch diagram or Marquand chart uses "natural" binary counting order for its rows and columns, with the result that some adjacent rows and columns differ in more than one value, and product terms do not always cover adjacent cells. M. Karnaugh showed how to fix the problem in "A Map Method for Synthesis of Combinational Logic Circuits" (*Trans. AIEE, Comm. and Electron.*, Vol. 72, Part I, November 1953, pp. 593–599). On the other hand, George J. Klir, in his book *Introduction to the Methodology of Switching Circuits* (Van Nostrand, 1972), claims that binary counting order is just as good as, perhaps better than, Karnaugh-map order for minimizing logic functions.

At this point, the Karnaugh vs. Veitch argument is of course irrelevant, because no one draws charts any more to minimize logic circuits. Instead, we use computer programs running logic-minimization algorithms. The first of such algorithms was described by Quine in "A Way to Simplify Truth Functions" (*Am. Math. Monthly*, Vol. 62, No. 9, 1955, pp. 627–631) and modified by E. J. McCluskey in "Minimization of Boolean Functions" (*Bell Sys. Tech. J.*, Vol. 35, No. 5, November 1956, pp. 1417–1444). The Quine-McCluskey algorithm is fully described in McCluskey's books cited earlier.

McCluskey's 1965 book also covers the iterative consensus algorithm for finding prime implicants and proves that it works. The starting point for this algorithm is a sum-of-products expression, or equivalently, a list of cubes. The product terms *need not* be minterms or prime implicants, but *may* be either or anything in between. In other words, the cubes in the list may have any and all dimensions, from 0 to *n* in an *n*-variable function. Starting with the list of cubes, the algorithm generates a list of all the prime-implicant cubes of the function, without ever having to generate a full minterm list.

The iterative consensus algorithm was first published by T. H. Mott, Jr., in "Determination of the Irredundant Normal Forms of a Truth Function by Iterated Consensus of the Prime Implicants" (*IRE Trans. Electron. Computers*, Vol. EC-9, No. 2, 1960, pp. 245–252). A generalized consensus algorithm was published by Pierre Tison in "Generalization of Consensus Theory and Application to the Minimization of Boolean Functions" (*IEEE Trans. Electron. Computers*, Vol. EC-16, No. 4, 1967, pp. 446–456). All of these algorithms are described by Thomas Downs in *Logic Design with Pascal* (Van Nostrand Reinhold, 1988).

As we explained in Section 4.4.4, the huge number of prime implicants in some logic functions makes it impractical or impossible deterministically to find them all or select a minimal cover. However, efficient heuristic methods can find

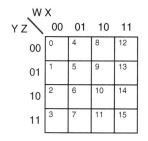

Figure 4-53
A 4-variable Veitch diagram or Marquand chart.

solutions that are close to minimal. The Espresso-II method is described in *Logic Minimization Algorithms for VLSI Synthesis* by R. K. Brayton, C. McMullen, G. D. Hachtel, and A. Sangiovanni-Vincentelli (Kluwer Academic Publishers, 1984). The more recent Espresso-MV and Espresso-EXACT algorithms are described in "Multiple-Valued Minimization for PLA Optimization" by R. L. Rudell and A. Sangiovanni-Vincentelli (*IEEE Trans. CAD*, Vol. CAD-6, No. 5, 1987, pp. 727–750).

In this chapter we described a map method for finding static hazards in two-level AND-OR and OR-AND circuits, but any combinational circuit can be analyzed for hazards. In both his 1965 and 1986 books, McCluskey defines the *0-set* and *1-sets* of a circuit and shows how they can be used to find static hazards. He also defines *P-sets* and *S-sets* and shows how they can be used to find dynamic hazards.

0-set
1-set
P-set
S-set

Many deeper and varied aspects of switching theory have been omitted from this book but have been beaten to death in other books and literature. A good starting point for an academic study of classical switching theory is Zvi Kohavi's book, *Switching and Finite Automata Theory*, second edition (McGraw-Hill, 1978), which includes material on set theory, symmetric networks, functional decomposition, threshold logic, fault detection, and path sensitization. Another area of great academic interest (but little commercial activity) is nonbinary *multiple-valued logic*, in which each signal line can take on more than two values. In his 1986 book, McCluskey gives a good introduction to multiple-valued logic, explaining its pros and cons and why it has seen little commercial development.

multiple-valued logic

Over the years, I've struggled to find a readily accessible and definitive reference on the ABEL language, and I finally found it—Appendix A of *Digital Design Using ABEL*, by David Pellerin and Michael Holley (Prentice Hall, 1994). It makes sense that this would be the definitive work—Pellerin and Holley invented the language and wrote the original compiler code! Pellerin also wrote, with co-author Douglas Taylor, a very nice stand-alone introduction to VHDL, titled *VHDL Made Easy!* (Prentice Hall, 1997).

VHDL is a "Very Huge Design Language," and we cover only a subset of its features and capabilities in this book. There are many language options (such as statement labels) that we don't use, but these are covered in dedicated VHDL tutorial and reference books. One of the best is *The Designer's Guide to VHDL* (Morgan Kaufmann, 1996). Though it's big (700 pages), it covers the whole language quite systematically and includes a very helpful and concise appendix on the complete language syntax of both VHDL-87 and VHDL-93. In this text we use a subset of the language that's compatible with both standards.

A somewhat shorter VHDL introduction is *VHDL for Designers* by Stefan Sjoholm and Lennart Lindh (Prentice Hall, 1997). This book is recommended for its nice emphasis on practical design issues including synthesis and test benches. Another practical book is *VHDL for Programmable Logic*, by Kevin Skahill of Cypress Semiconductor (Addison-Wesley, 1996).

All of the ABEL and VHDL examples in this chapter and throughout the text were compiled and in most cases simulated using Foundation 1.5 Student Edition software from Xilinx, Inc. (San Jose, CA 95124, www.xilinx.com). The Foundation software integrates a schematic editor, HDL editor, compilers for ABEL, VHDL and Verilog, and a simulator from Aldec, Inc. (Henderson, NV 89014, www.aldec.com) along with Xilinx' own specialized tools for CPLD and FPGA design and programming. It also contains an excellent on-line help system, including reference manuals for both ABEL and VHDL. The Foundation software is packaged on CD-ROM with some editions of this book.

The IEEE standard VHDL packages are important elements of any VHDL design environment. A listing of their type and function definitions is included as an appendix in some VHDL texts, but if you're really curious about them, you can find the complete source files in the library subsystem of any VHDL design system, including Foundation software.

We briefly discussed device testing in the context of ABEL test vectors. There is a large, well-established body of literature on digital device testing, and a good starting point for study is McCluskey's 1986 book. Generating a set of test vectors that completely tests a large circuit such as a PLD is a task best left to a program. At least one company's entire business is focused on programs that automatically create test vectors for PLD testing (ACUGEN Software, Inc., Nashua, NH 03063, www.acugen.com).

Drill Problems

4.1   Using variables NERD, DESIGNER, FAILURE, and STUDIED, write a boolean expression that is 1 for successful designers who never studied and for nerds who studied all the time.

4.2   Prove theorems T2–T5 using perfect induction.

4.3   Prove theorems T1′–T3′ and T5′ using perfect induction.

4.4   Prove theorems T6–T9 using perfect induction.

4.5   According to DeMorgan's theorem, the complement of $X + Y \cdot Z$ is $X' \cdot Y' + Z'$. Yet both functions are 1 for $XYZ = 110$. How can both a function and its complement be 1 for the same input combination? What's wrong here?

4.6   Use the theorems of switching algebra to simplify each of the following logic functions:

(a) $F = W \cdot X \cdot Y \cdot Z \cdot (W \cdot X \cdot Y \cdot Z' + W \cdot X' \cdot Y \cdot Z + W' \cdot X \cdot Y \cdot Z + W \cdot X \cdot Y' \cdot Z)$

(b) $F = A \cdot B + A \cdot B \cdot C' \cdot D + A \cdot B \cdot D \cdot E' + A \cdot B \cdot C' \cdot E + C' \cdot D \cdot E$

(c) $F = M \cdot N \cdot O + Q' \cdot P' \cdot N' + P \cdot R \cdot M + Q' \cdot O \cdot M \cdot P' + M \cdot R$

4.7   Write the truth table for each of the following logic functions:

(a) $F = X' \cdot Y + X' \cdot Y' \cdot Z$ (b) $F = W' \cdot X + Y' \cdot Z' + X' \cdot Z$

(c) $F = W + X' \cdot (Y' + Z)$ (d) $F = A \cdot B + B' \cdot C + C' \cdot D + D' \cdot A$

(e) $F = V \cdot W + X' \cdot Y' \cdot Z$ (f) $F = (A' + B' + C \cdot D) \cdot (B + C' + D' \cdot E')$

(g) $F = (W \cdot X)' \cdot (Y' + Z')'$ (h) $F = (((A + B)' + C')' + D)'$

(i) $F = (A' + B + C) \cdot (A + B' + D') \cdot (B + C' + D') \cdot (A + B + C + D)$

4.8   Write the truth table for each of the following logic functions:

(a) $F = X' \cdot Y' \cdot Z' + X \cdot Y \cdot Z + X \cdot Y' \cdot Z$

(b) $F = M' \cdot N' + M \cdot P + N' \cdot P$

(c) $F = A \cdot B + A \cdot B' \cdot C' + A' \cdot B \cdot C$

(d) $F = A' \cdot B \cdot (C \cdot B \cdot A' + B \cdot C')$

(e) $F = X \cdot Y \cdot (X' \cdot Y \cdot Z + X \cdot Y' \cdot Z + X \cdot Y \cdot Z' + X' \cdot Y' \cdot Z)$ (f) $F = M \cdot N + M' \cdot N' \cdot P'$

(g) $F = (A + A') \cdot B + B \cdot A \cdot C' + C \cdot (A + B') \cdot (A' + B)$ (h) $F = X \cdot Y' + Y \cdot Z + Z' \cdot X$

4.9   Write the canonical sum and product for each of the following logic functions:

(a) $F = \Sigma_{X,Y}(1,2)$ (b) $F = \Pi_{A,B}(0,1,2)$

(c) $F = \Sigma_{A,B,C}(2,4,6,7)$ (d) $F = \Pi_{W,X,Y}(0,1,3,4,5)$

(e) $F = X + Y' \cdot Z'$ (f) $F = V' + (W' \cdot X)'$

4.10   Write the canonical sum and product for each of the following logic functions:

(a) $F = \Sigma_{X,Y,Z}(0,3)$ (b) $F = \Pi_{A,B,C}(1,2,4)$

(c) $F = \Sigma_{A,B,C,D}(1,2,5,6)$ (d) $F = \Pi_{M,N,P}(0,1,3,6,7)$

(e) $F = X' + Y \cdot Z' + Y \cdot Z'$ (f) $F = A'B + B'C + A$

4.11   If the canonical sum for an n-input logic function is also a minimal sum, how many literals are in each product term of the sum? Might there be any other minimal sums in this case?

4.12   Give two reasons why the cost of inverters is not included in the definition of "minimal" for logic minimization.

4.13   Using Karnaugh maps, find a minimal sum-of-products expression for each of the following logic functions. Indicate the distinguished 1-cells in each map.

(a) $F = \Sigma_{X,Y,Z}(1,3,5,6,7)$ (b) $F = \Sigma_{W,X,Y,Z}(1,4,5,6,7,9,14,15)$

(c) $F = \Pi_{W,X,Y}(0,1,3,4,5)$ (d) $F = \Sigma_{W,X,Y,Z}(0,2,5,7,8,10,13,15)$

(e) $F = \Pi_{A,B,C,D}(1,7,9,13,15)$ (f) $F = \Sigma_{A,B,C,D}(1,4,5,7,12,14,15)$

4.14   Find a minimal product-of-sums expression for each function in Drill 4.13 using the method of Section 4.3.6.

4.15   Find a minimal product-of-sums expression for the function in each of the following figures and compare its cost with the previously found minimal sum-of-products expression: (a) Figure 4-27; (b) Figure 4-29; (c) Figure 4-33.

4.16   Using Karnaugh maps, find a minimal sum-of-products expression for each of the following logic functions. Indicate the distinguished 1-cells in each map.

(a) $F = \Sigma_{A,B,C}(0,1,2,4)$ 　　　　　　　(b) $F = \Sigma_{W,X,Y,Z}(1,4,5,6,11,12,13,14)$

(c) $F = \Pi_{A,B,C}(1,2,6,7)$ 　　　　　　　　(d) $F = \Sigma_{W,X,Y,Z}(0,1,2,3,7,8,10,11,15)$

(e) $F = \Sigma_{W,X,Y,Z}(1,2,4,7,8,11,13,14)$ 　(f) $F = \Pi_{A,B,C,D}(1,3,4,5,6,7,9,12,13,14)$

4.17   Find a minimal product-of-sums expression for each function in Drill 4.16 using the method of Section 4.3.6.

4.18   Find the complete sum for the logic functions in Drill 4.16(d) and (e).

4.19   Using Karnaugh maps, find a minimal sum-of-products expression for each of the following logic functions. Indicate the distinguished 1-cells in each map.

(a) $F = \Sigma_{W,X,Y,Z}(0,1,3,5,14) + d(8,15)$ 　(b) $F = \Sigma_{W,X,Y,Z}(0,1,2,8,11) + d(3,9,15)$

(c) $F = \Sigma_{A,B,C,D}(1,5,9,14,15) + d(11)$ 　(d) $F = \Sigma_{A,B,C,D}(1,5,6,7,9,13) + d(4,15)$

(e) $F = \Sigma_{W,X,Y,Z}(3,5,6,7,13) + d(1,2,4,12,15)$

4.20   Repeat Drill 4.19, finding a minimal product-of-sums expression for each logic function.

4.21   For each logic function in the two preceding exercises, determine whether the minimal sum-of-products expression equals the minimal product-of-sums expression. Also compare the circuit cost for realizing each of the two expressions.

4.22   For each of the following logic expressions, find all of the static hazards in the corresponding two-level AND-OR or OR-AND circuit, and design a hazard-free circuit that realizes the same logic function.

(a) $F = W \cdot X + W'Y'$ 　　　　　　　　　(b) $F = W \cdot X' \cdot Y' + X \cdot Y' \cdot Z + X \cdot Y$

(c) $F = W' \cdot Y + X' \cdot Y' + W \cdot X \cdot Z$ 　　(d) $F = W' \cdot X + Y' \cdot Z + W \cdot X \cdot Y \cdot Z + W \cdot X' \cdot Y \cdot Z'$

(e) $F = (W + X + Y) \cdot (X' + Z')$ 　　　　　(f) $F = (W + Y' + Z') \cdot (W' + X' + Z') \cdot (X' + Y + Z)$

(g) $F = (W + Y + Z') \cdot (W + X' + Y + Z) \cdot (X' + Y') \cdot (X + Z)$

4.23   Write a set of ABEL test vectors for prime-number detector in Table 4-17.

4.24   Write a structural VHDL program (entity and architecture) for the alarm circuit in Figure 4-19.

4.25   Repeat Drill 4.24, using the dataflow style of description.

4.26   Repeat Drill 4.24, using a process and a behavioral description style.

4.27   Write a structural VHDL program corresponding to the NAND-gate based logic circuit in Figure 5-17.

Exercises

4.28   Design a non-trivial-looking logic circuit that contains a feedback loop but has an output that depends only on its current input.

4.29   Prove the combining theorem T10 without using perfect induction, but assuming that theorems T1–T9 and T1′–T9′ are true.

4.30   Show that the combining theorem T10 is just a special case of consensus (T11) used with covering (T9).

4.31   Prove that $(X + Y') \cdot Y = X \cdot Y$ *without* using perfect induction. You may assume that theorems T1–T11 and T1′–T11′ are true.

4.32   Prove that $(X + Y) \cdot (X' + Z) = X \cdot Z + X' \cdot Y$ *without* using perfect induction. You may assume that theorems T1–T11 and T1′–T11′ are true.

4.33   Show that an n-input AND gate can be replaced by $(n-1)$ 2-input AND gates. Can the same statement be made for NAND gates? Justify your answer.

4.34   How many physically different ways are there to realize $V \cdot W \cdot X \cdot Y \cdot Z$ using four 2-input AND gates (4/4 of a 74x08)? Justify your answer.

4.35   Use switching algebra to prove that tying together two inputs of an $(n+1)$-input AND or OR gate gives it the functionality of an n-input gate.

4.36   Prove DeMorgan's theorems (T13 and T13′) using finite induction.

4.37   Which logic symbol more closely approximates the internal realization of a TTL NOR gate, Figure 4-4(c) or (d)? Why?

4.38   Use the theorems of switching algebra to rewrite the following expression using as few inversions as possible (complemented parentheses are allowed):

$$B' \cdot C + A \cdot C \cdot D' + A' \cdot C + E \cdot B' + E \cdot (A + C) \cdot (A' + D')$$

4.39   Prove or disprove the following propositions:

(a) Let A and B be switching-algebra *variables*. Then $A \cdot B = 0$ and $A + B = 1$ implies that $A = B'$.

(b) Let X and Y be switching-algebra *expressions*. Then $X \cdot Y = 0$ and $X + Y = 1$ implies that $X = Y'$.

4.40   Prove Shannon's expansion theorems. (*Hint:* Don't get carried away; it's easy.)

generalized Shannon-expansion theorems

4.41   Shannon's expansion theorems can be generalized to "pull out" not just one but i variables so that a logic function can be expressed as a sum or product of 2^i terms. State the generalized Shannon expansion theorems.

4.42   Show how the generalized Shannon expansion theorems lead to the canonical sum and canonical product representations of logic functions.

Exclusive OR (XOR) gate

4.43   An *Exclusive OR (XOR) gate* is a 2-input gate whose output is 1 if and only if exactly one of its inputs is 1. Write a truth table, sum-of-products expression, and corresponding AND-OR circuit for the Exclusive OR function.

4.44   From the point of view of switching algebra, what is the function of a 2-input XOR gate whose inputs are tied together? How might the output behavior of a real XOR gate differ?

4.45   After completing the design and fabrication of an SSI-based digital system, a designer finds that one more inverter is required. However, the only spare gates in the system are a 3-input OR, a 2-input AND, and a 2-input XOR. How should the designer realize the inverter function without adding another IC?

4.46   Any set of logic-gate types that can realize any logic function is called a *complete set* of logic gates. For example, 2-input AND gates, 2-input OR gates, and inverters are a complete set, because any logic function can be expressed as a sum of products of variables and their complements, and AND and OR gates with any number of inputs can be made from 2-input gates. Do 2-input NAND gates form a complete set of logic gates? Prove your answer.

complete set

4.47   Do 2-input NOR gates form a complete set of logic gates? Prove your answer.

4.48   Do 2-input XOR gates form a complete set of logic gates? Prove your answer.

4.49   Define a two-input gate, other than NAND, NOR, or XOR, that forms a complete set of logic gates if the constant inputs 0 and 1 are allowed. Prove your answer.

4.50   Some people think that there are *four* basic logic functions, AND, OR, NOT, and *BUT*. Figure X4.50 is a possible symbol for a 4-input, 2-output *BUT gate*. Invent a useful, nontrivial function for the BUT gate to perform. The function should have something to do with the name (BUT). Keep in mind that, due to the symmetry of the symbol, the function should be symmetric with respect to the A and B inputs of each section and with respect to sections 1 and 2. Describe your BUT's function and write its truth table.

BUT
BUT gate

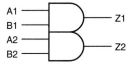

Figure X4.50

4.51   Write logic expressions for the Z1 and Z2 outputs of the BUT gate you designed in the preceding exercise, and draw a corresponding logic diagram using AND gates, OR gates, and inverters.

4.52   Most students have no problem using theorem T8 to "multiply out" logic expressions, but many develop a mental block if they try to use theorem T8′ to "add out" a logic expression. How can duality be used to overcome this problem?

4.53   How many different logic functions are there of n variables?

4.54   How many different 2-variable logic functions $F(X,Y)$ are there? Write a simplified algebraic expression for each of them.

4.55   A *self-dual logic function* is a function F such that $F = F^D$. Which of the following functions are self-dual? (The symbol $\oplus$ denotes the Exclusive OR (XOR) operation.)

self-dual logic function
$\oplus$

(a) $F = X$

(b) $F = \Sigma_{X,Y,Z}(0,3,5,6)$

(c) $F = X \cdot Y' + X' \cdot Y$

(d) $F = W \cdot (X \oplus Y \oplus Z) + W' \cdot (X \oplus Y \oplus Z)'$

(e) A function F of 7 variables such that $F = 1$ if and only if 4 or more of the variables are 1

(f) A function F of 10 variables such that $F = 1$ if and only if 5 or more of the variables are 1

4.56   How many self-dual logic functions of n input variables are there? (*Hint:* Consider the structure of the truth table of a self-dual function.)

irredundant sum

4.57 Prove that any n-input logic function $F(X_1, ..., X_n)$ that can be written in the form $F = X_1 \cdot G(X_2, ..., X_n) + X_1' \cdot G^D(X_2, ..., X_n)$ is self-dual.

4.58 Assuming that an inverting gate has a propagation delay of 5 ns, and that a non-inverting gate has a propagation delay of 8 ns, compare the speeds of the circuits in Figure 4-24(a), (c), and (d).

4.59 Find the minimal product-of-sums expressions for the logic functions in Figures 4-27 and 4-29.

4.60 Use switching algebra to show that the logic functions obtained in Exercise 4.59 equal the AND-OR functions obtained in Figures 4-27 and 4-29.

4.61 Determine whether the product-of-sums expressions obtained by "adding out" the minimal sums in Figure 4-27 and 4-29 are minimal.

4.62 Prove that the rule for combining 2^i 1-cells in a Karnaugh map is true, using the axioms and theorems of switching algebra.

4.63 An *irredundant sum* for a logic function F is a sum of prime implicants for F such that if any prime implicant is deleted, the sum no longer equals F. This sounds a lot like a minimal sum, but an irredundant sum is not necessarily minimal. For example, the minimal sum of the function in Figure 4-35 has only three product terms, but there is an irredundant sum with four product terms. Find the irredundant sum and draw a map of the function, circling only the prime implicants in the irredundant sum.

4.64 Find another logic function in Section 4.3 that has one or more nonminimal irredundant sums, and draw its map, circling only the prime implicants in the irredundant sum.

4.65 Draw a Karnaugh map and assign variables to the inputs of the AND-XOR circuit in Figure X4.65 so that its output is $F = \Sigma_{W,X,Y,Z}(6, 7, 12, 13)$. Note that the output gate is a 2-input XOR rather than an OR.

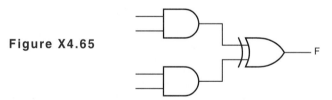

Figure X4.65

4.66 A 3-bit "comparator" circuit receives two 3-bit numbers, $P = P_2 P_1 P_0$ and $Q = Q_2 Q_1 Q_0$. Design a minimal sum-of-products circuit that produces a 1 output if and only if $P > Q$.

4.67 Find minimal multiple-output sum-of-products expressions for $F = \Sigma_{X,Y,Z}(0,1,2)$, $G = \Sigma_{X,Y,Z}(1,4,6)$, and $H = \Sigma_{X,Y,Z}(0,1,2,4,6)$.

4.68 Prove whether or not the following expression is a minimal sum. Do it the easiest way possible (algebraically, not using maps).

$$F = T' \cdot U \cdot V \cdot W \cdot X + T' \cdot U \cdot V' \cdot X \cdot Z + T' \cdot U \cdot W \cdot X \cdot Y' \cdot Z$$

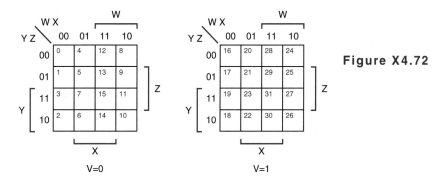

Figure X4.72

4.69 The text states that a truth table or equivalent is the starting point for traditional combinational minimization methods. A Karnaugh map itself contains the same information as a truth table. Given a sum-of-products expression, it is possible to write the 1s corresponding to each product term directly on the map without developing an explicit truth table or minterm list, and then proceed with the map-minimization procedure. In this way, find a minimal sum-of-products expression for each of the following logic functions:

(a) $F = X' \cdot Z + X \cdot Y + X \cdot Y' \cdot Z$

(b) $F = A' \cdot C' \cdot D + B' \cdot C \cdot D + A \cdot C' \cdot D + B \cdot C \cdot D$

(c) $F = W \cdot X \cdot Z' + W \cdot X' \cdot Y \cdot Z + X \cdot Z$

(d) $F = (X' + Y') \cdot (W' + X' + Y) \cdot (W' + X + Z)$

(e) $F = A \cdot B \cdot C' \cdot D' + A' \cdot B \cdot C' + A \cdot B \cdot D + A' \cdot C \cdot D + B \cdot C \cdot D'$

4.70 Repeat Exercise 4.69, finding a minimal product-of-sums expression for each logic function.

4.71 Derive the minimal product-of-sums expression for the prime BCD-digit-detector function of Figure 4-37. Determine whether or not the expression algebraically equals the minimal sum-of-products expression and explain your result.

4.72 A Karnaugh map for a 5-variable function can be drawn as shown in Figure X4.72. In such a map, cells that occupy the same relative position in the $V = 0$ and $V = 1$ submaps are considered to be adjacent. (Many worked examples of 5-variable Karnaugh maps appear in Sections 7.4.4 and 7.4.5.) Find a minimal sum-of-products expression for each of the following functions using a 5-variable map:

5-variable Karnaugh map

(a) $F = \Sigma_{V,W,X,Y,Z}(5,7,13,15,16,20,25,27,29,31)$

(b) $F = \Sigma_{V,W,X,Y,Z}(0,7,8,9,12,13,15,16,22,23,30,31)$

(c) $F = \Sigma_{V,W,X,Y,Z}(0,1,2,3,4,5,10,11,14,20,21,24,25,26,27,28,29,30)$

(d) $F = \Sigma_{V,W,X,Y,Z}(0,2,4,6,7,8,10,11,12,13,14,16,18,19,29,30)$

(e) $F = \Pi_{V,W,X,Y,Z}(4,5,10,12,13,16,17,21,25,26,27,29)$

(f) $F = \Sigma_{V,W,X,Y,Z}(4,6,7,9,11,12,13,14,15,20,22,25,27,28,30) + d(1,5,29,31)$

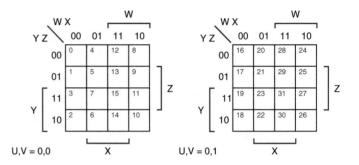

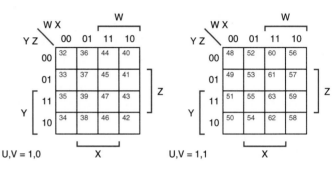

Figure X4.74

4.73 Repeat Exercise 4.72, finding a minimal product-of-sums expression for each logic function.

6-variable Karnaugh map

4.74 A Karnaugh map for a 6-variable function can be drawn as shown in Figure X4.74. In such a map, cells that occupy the same relative position in adjacent submaps are considered to be adjacent. Minimize the following functions using 6-variable maps:

(a) $F = \Sigma_{U,V,W,X,Y,Z}(1,5,9,13,21,23,29,31,37,45,53,61)$

(b) $F = \Sigma_{U,V,W,X,Y,Z}(0,4,8,16,24,32,34,36,37,39,40,48,50,56)$

(c) $F = \Sigma_{U,V,W,X,Y,Z}(2,4,5,6,12-21,28-31,34,38,50,51,60-63)$

4.75 There are $2n$ m-subcubes of an n-cube for the value $m = n - 1$. Show their text representations and the corresponding product terms. (You may use ellipses as required, e.g., 1, 2, ..., n.)

4.76 There is just one m-subcube of an n-cube for the value $m = n$; its text representation is xx...xx. Write the product term corresponding to this cube.

4.77 The C program in Table 4-9 uses memory inefficiently because it allocates memory for a maximum number of cubes at each level, even if this maximum is never used. Redesign the program so that the cubes and used arrays are one-dimensional arrays, and each level uses only as many array entries as needed. (*Hint:* You can still allocate cubes sequentially, but keep track of the starting point in the array for each level.)

4.78   As a function of m, how many times is each distinct m-cube rediscovered in Table 4-9, only to be found in the inner loop and thrown away? Suggest some ways to eliminate this inefficiency.

4.79   The third for-loop in Table 4-9 tries to combine all m-cubes at a given level with all other m-cubes at that level. In fact, only m-cubes with x's in the same positions can be combined, so it is possible to reduce the number of loop iterations by using a more sophisticated data structure. Design a data structure that segregates the cubes at a given level according to the position of their x's, and determine the maximum size required for various elements of the data structure. Rewrite Table 4-9 accordingly.

4.80   Estimate whether the savings in inner-loop iterations achieved in Exercise 4.80 outweighs the overhead of maintaining a more complex data structure. Try to make reasonable assumptions about how cubes are distributed at each level, and indicate how your results are affected by these assumptions.

4.81   Optimize the Oneone function in Table 4-8. An obvious optimization is to drop out of the loop early, but other optimizations exist that eliminate the for loop entirely. One is based on table look-up and another uses a tricky computation involving complementing, Exclusive ORing, and addition.

4.82   Extend the C program in Table 4-9 to handle don't-care conditions. Provide another data structure, dc[MAX_VARS+1][MAX_CUBES], that indicates whether a given cube contains only don't-cares, and update it as cubes are read and generated.

4.83   (*Hamlet circuit.*) Complete the timing diagram and explain the function of the circuit in Figure X4.83. Where does the circuit get its name?

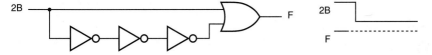

Figure X4.83

4.84   Prove that a two-level AND-OR circuit corresponding to the complete sum of a logic function is always hazard free.

4.85   Find a four-variable logic function whose minimal sum-of-products realization is not hazard free, but for which there exists a hazard-free sum-of-products realization with fewer product terms than the complete sum.

4.86   Starting with the WHEN statements in the ABEL program in Table 4-14, work out the logic equations for variables X4 through X10 in the program. Explain any discrepancies between your results and the equations in Table 4-15.

4.87   Draw a circuit diagram corresponding to the minimal two-level sum-of-products equations for the alarm circuit, as given in Table 4-12. On each inverter, AND gate, and OR gate input and output, write a pair of numbers ($t0, t1$), where $t0$ is the test number from Table 4-25 that detects a stuck-at-0 fault on that line, and $t1$ is the test number that detects a stuck-at-1 fault.

4.88   Write a dataflow-style VHDL program (entity and architecture) corresponding to the full-adder circuit in Figure 5-86.

4.89   Using the entity that you designed in Exercise 4.88, write a structural VHDL program for a 4-bit ripple adder using the structure of Figure 5-87.

4.90   Using the entity that you defined in Exercise 4.88, write a structural VHDL program for a 16-bit ripple adder along the lines of Figure 5-87. Use a generate statement to create the 16 full adders and their signal connection.

4.91   Rewrite the prime-number-detector architecture of Table 4-63 using a while statement.

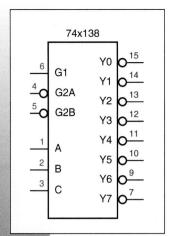

74x138

chapter 5

Combinational Logic Design Practices

The preceding chapter described the theoretical principles used in combinational logic design. In this chapter we'll build on that foundation and describe many of the devices, structures, and methods used by engineers to solve practical digital design problems.

A practical combinational circuit may have dozens of inputs and outputs and could require hundreds, thousands, even millions of terms to describe as a sum of products, and *billions and billions* of rows to describe in a truth table. Thus, most real combinational logic design problems are too large to solve by "brute-force" application of theoretical techniques.

But wait, you say, how could any human being conceive of such a complex logic circuit in the first place? The key is structured thinking. A complex circuit or system is conceived as a collection of smaller subsystems, each of which has a much simpler description.

In combinational logic design, there are several straightforward structures—decoders, multiplexers, comparators, and the like—that turn up quite regularly as building blocks in larger systems. The most important of these structures are described in this chapter. We describe each structure generally and then we give examples and applications using 74-series components, ABEL, and VHDL.

Before launching into these combinational building blocks, we need to discuss several important topics. The first topic is documentation standards

THE IMPORTANCE OF 74-SERIES LOGIC	Later in this chapter we'll look at commonly used 74-series ICs that perform well-structured logic functions. These parts are important building blocks in a digital designer's toolbox, because their level of functionality often matches a designer's level of thinking when partitioning a large problem into smaller chunks.
	Even when you design for PLDs, FPGAs, or ASICs, understanding 74-series MSI functions is important. In PLD-based design, standard MSI functions can be used as a starting point for developing logic equations for more specialized functions. And in FPGA and ASIC design, the basic building blocks (or "standard cells" or "macros") provided by the FPGA or ASIC manufacturer may actually be defined as 74-series MSI functions, even to the extent of having similar descriptive numbers.

that are used by digital designers to ensure that their designs are correct, manufacturable, and maintainable. Next we discuss circuit timing, a crucial element for successful digital design. Third, we describe the internal structure of combinational PLDs, which we use later as "universal" building blocks.

5.1 Documentation Standards

Good documentation is essential for correct design and efficient maintenance of digital systems. In addition to being accurate and complete, documentation must be somewhat instructive, so that a test engineer, maintenance technician, or even the original design engineer (six months after designing the circuit) can figure out how the system works just by reading the documentation.

Although the type of documentation depends on system complexity and the engineering and manufacturing environments, a documentation package should generally contain at least the following six items:

circuit specification

1. A *specification* describes exactly what the circuit or system is supposed to do, including a description of all inputs and outputs ("interfaces") and the functions that are to be performed. Note that the "spec" doesn't have to specify *how* the system achieves its results, just *what* the results are supposed to be. However, in many companies it is common practice also to incorporate one or more of the documents below into the spec to describe how the system works at the same time.

block diagram

2. A *block diagram* is an informal pictorial description of the system's major functional modules and their basic interconnections.

schematic diagram

3. A *schematic diagram* is a formal specification of the electrical components of the system, their interconnections, and all of the details needed to construct the system, including IC types, reference designators, and pin numbers. We've been using the term *logic diagram* for an informal drawing that does not have quite this level of detail. Most schematic

logic diagram

DOCUMENTS ON-LINE Professional engineering documentation nowadays is carefully maintained on corporate intranets, so it's very useful to include URLs in circuit specifications and descriptions so that references can be easily located. On-line documentation is so important and authoritative in one company that the footer on every page of every specification contains the warning that "A printed version of this document is an uncontrolled copy." That is, a printed copy could very well be obsolete.

drawing programs have the ability to generate a *bill of materials (BOM)* from the schematic; this tells the purchasing department what electrical components they have to order to build the system.

bill of materials (BOM)

4. A *timing diagram* shows the values of various logic signals as a function of time, including the cause-and-effect delays between critical signals.

timing diagram

5. A *structured logic device description* describes the internal function of a programmable logic device (PLD), field-programmable gate array (FPGA), or application-specific integrated circuit (ASIC). It is normally written in a hardware description language (HDL) such as ABEL or VHDL, but it may be in the form of logic equations, state tables, or state diagrams. In some cases, a conventional programming language such as C may be used to model the operation of a circuit or to specify its behavior.

structured logic device description

6. A *circuit description* is a narrative text document that, in conjunction with the other documentation, explains how the circuit works internally. The circuit description should list any assumptions and potential pitfalls in the circuit's design and operation, and point out the use of any nonobvious design "tricks." A good circuit description also contains definitions of acronyms and other specialized terms and has references to related documents.

circuit description

You've probably already seen block diagrams in many contexts. We present a few rules for drawing them in the next subsection, and then in the rest of this section we concentrate on schematics for combinational logic circuits. Section 5.2.1 introduces timing diagrams. Structured logic descriptions in the form of ABEL and VHDL programs were covered in Sections 4.6 and 4.7. In Section 10.1.6 we'll show how a C program can be used to generate the contents of a read-only memory.

The last area of documentation, the circuit description, is very important in practice. Just as an experienced programmer creates a program design document before beginning to write code, an experienced logic designer starts writing the circuit description before drawing the schematic. Unfortunately, the circuit description is sometimes the last document to be created, and sometimes it's never written at all. A circuit without a description is difficult to debug, manufacture, test, maintain, modify, and enhance.

> **DON'T FORGET TO WRITE!** In order to create great products, logic designers must develop their language and writing skills, especially in the area of *logical* outlining and organization. The most successful logic designers (and later, project leaders, system architects, and entrepreneurs) are the ones who communicate their ideas, proposals, and decisions effectively to others. Even though it's a lot of fun to tinker in the digital design lab, don't use that as an excuse to shortchange your writing and communications courses and projects!

5.1.1 Block Diagrams

block diagram

A *block diagram* shows the inputs, outputs, functional modules, internal data paths, and important control signals of a system. In general, it should not be so detailed that it occupies more than one page, yet it must not be too vague. A small block diagram may have three to six blocks, while a large one may have 10 to 15, depending on system complexity. In any case, the block diagram must

Figure 5-1
Block diagram for a digital design project.

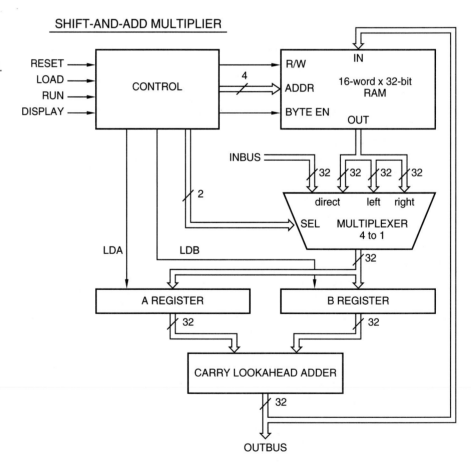

SHIFT-AND-ADD MULTIPLIER

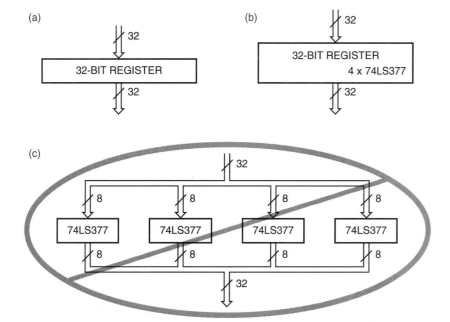

Figure 5-2
A 32-bit register block: (a) realization unspecified; (b) chips specified; (c) too much detail.

show the most important system elements and how they work together. Large systems may require additional block diagrams of individual subsystems, but there should always be a "top-level" diagram showing the entire system.

Figure 5-1 shows a sample block diagram. Each block is labeled with the function of the block, not the individual chips that comprise it. As another example, Figure 5-2(a) shows the block-diagram symbol for a 32-bit register. If the register is to be built using four 74LS377 8-bit registers, and this information is important to someone reading the diagram (e.g., for cost reasons), then it can be conveyed as shown in (b). However, splitting the block to show individual chips as in (c) is incorrect.

A *bus* is a collection of two or more related signal lines. In a block diagram,   *bus*
buses are drawn with a double or heavy line. A slash and a number may indicate how many individual signal lines are contained in a bus. Alternatively, size may be denoted in the bus name (e.g., INBUS[31..0] or INBUS[31:0]). Active levels (defined later) and inversion bubbles may or may not appear in block diagrams; in most cases, they are unimportant at this level of detail. However, important control signals and buses should have names, usually the same names that appear in the more detailed schematic.

The flow of control and data in a block diagram should be clearly indicated. Logic diagrams are generally drawn with signals flowing from left to right, but in block diagrams this ideal is more difficult to achieve. Inputs and outputs may be on any side of a block, and the direction of signal flow may be arbitrary. Arrowheads are used on buses and ordinary signal lines to eliminate ambiguity.

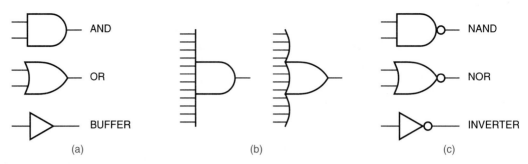

Figure 5-3 Shapes for basic logic gates: (a) AND, OR, and buffers;
(b) expansion of inputs; (c) inversion bubbles.

5.1.2 Gate Symbols

The symbol shapes for AND and OR gates and buffers are shown in Figure 5-3(a). (Recall from Chapter 3 that a buffer is a circuit that simply converts "weak" logic signals into "strong" ones.) To draw logic gates with more than a few inputs, we expand the AND and OR symbols as shown in (b). A small circle, called an *inversion bubble*, denotes logical inversion or complementing and is used in the symbols for NAND and NOR gates and inverters in (c).

inversion bubble

Using the generalized DeMorgan's theorem, we can manipulate the logic expressions for gates with complemented outputs. For example, if X and Y are the inputs of a NAND gate with output Z, then we can write

$$Z = (X \cdot Y)'$$
$$= X' + Y'$$

This gives rise to two different but equally correct symbols for a NAND gate, as we demonstrated in Figure 4-3 on page 201. In fact, this sort of manipulation may be applied to gates with uncomplemented inputs as well. For example, consider the following equations for an AND gate:

$$Z = X \cdot Y$$
$$= ((X \cdot Y)')'$$
$$= (X' + Y')'$$

IEEE STANDARD LOGIC SYMBOLS

Together with the American National Standards Institute (ANSI), the Institute of Electrical and Electronic Engineers (IEEE) has developed a standard set of logic symbols. The most recent revision of the standard is ANSI/IEEE Std 91-1984, *IEEE Standard Graphic Symbols for Logic Functions*, and it allows both rectangular- and distinctive-shape symbols for logic gates. We have been using and will continue to use the distinctive-shape symbols in this book, but the rectangular-shape symbols are described in a guide to IEEE symbols on the web at www.ddpp.com.

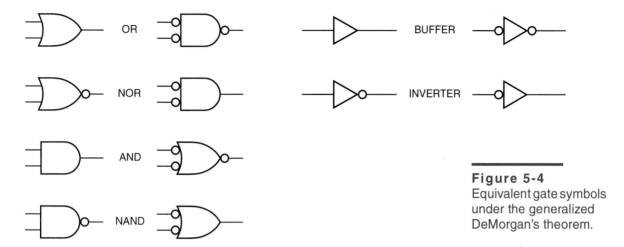

Figure 5-4
Equivalent gate symbols under the generalized DeMorgan's theorem.

Thus, an AND gate may be symbolized as an OR gate with inversion bubbles on its inputs and output.

Equivalent symbols for standard gates that can be obtained by these manipulations are summarized in Figure 5-4. Even though both symbols in a pair represent the same logic function, the choice of one symbol or the other in a logic diagram is not arbitrary, at least not if we are adhering to good documentation standards. As we'll show in the next three subsections, proper choices of signal names and gate symbols can make logic diagrams much easier to use and understand.

5.1.3 Signal Names and Active Levels

Each input and output signal in a logic circuit should have a descriptive alphanumeric label, the signal's name. Most computer-aided design systems for drawing logic circuits also allow certain special characters, such as *, _, and !, to be included in signal names. In the analysis and synthesis examples in Chapter 4, we used mostly single-character signal names (X, Y, etc.) because the circuits didn't do much. However, in a real system, well-chosen signal names convey information to someone reading the logic diagram the same way that variable names in a software program do. A signal's name indicates an action that is controlled (GO, PAUSE), a condition that it detects (READY, ERROR), or data that it carries (INBUS[31:0]).

Each signal name should have an *active level* associated with it. A signal is *active high* if it performs the named action or denotes the named condition when it is HIGH or 1. (Under the positive-logic convention, which we use throughout this book, "HIGH" and "1" are equivalent.) A signal is *active low* if it performs the named action or denotes the named condition when it is LOW or 0. A signal is said to be *asserted* when it is at its active level. A signal is said to be *negated* (or, sometimes, *deasserted*) when it is not at its active level.

active level
active high

active low
assert
negate
deassert

Table 5-1
Each line shows a different naming convention for active levels.

Active Low	Active High
READY–	READY+
ERROR.L	ERROR.H
ADDR15(L)	ADDR15(H)
RESET*	RESET
ENABLE~	ENABLE
~GO	GO
/RECEIVE	RECEIVE
TRANSMIT_L	TRANSMIT

active-level naming convention

The active level of each signal in a circuit is normally specified as part of its name, according to some convention. Examples of several different *active-level naming conventions* are shown in Table 5-1. The choice of one of these or other signal-naming conventions is sometimes just a matter of personal preference, but more often it is constrained by the engineering environment. Since the active-level designation is part of the signal name, the naming convention must be compatible with the input requirements of any computer-aided design tools that will process the signal names, such as schematic editors, HDL compilers, and simulators. In this text, we'll use the last convention in the table: An active-low signal name has a suffix of _L, and an active-high signal has no suffix. The _L suffix may be read as if it were a prefix "not."

_L suffix

signal name
logic expression

logic equation

It's extremely important for you to understand the difference between signal names, expressions, and equations. A *signal name* is just a name—an alphanumeric label. A *logic expression* combines signal names using the operators of switching algebra—AND, OR, and NOT—as we explained and used throughout Chapter 4. A *logic equation* is an assignment of a logic expression to a signal name—it describes one signal's function in terms of other signals.

The distinction between signal names and logic expressions can be related to a concept used in computer programming languages: The lefthand side of an assignment statement contains a variable *name*, and the righthand side contains an *expression* whose value will be given to the named variable (e.g., $Z = -(X+Y)$ in C). In a programming language, you can't put an expression on the lefthand side of an assignment statement. In logic design, you can't use a logic expression as a signal name.

Logic signals may have names like X, READY, and GO_L. The "_L" in GO_L is just part of the signal's name, like an underscore in a variable name in a C program. There is *no* signal whose name is READY′—this is an expression, since ′ is an operator. However, there may be two signals named READY and READY_L such that READY_L = READY′ during normal operation of the circuit. We are very careful in this book to distinguish between signal names, which are always printed in black, and logic expressions, which are always printed in color when they are written near the corresponding signal lines.

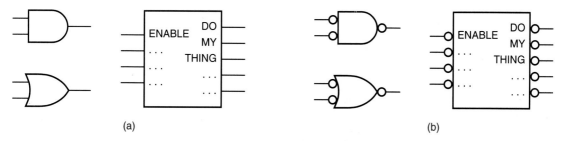

Figure 5-5 Logic symbols: (a) AND, OR, and a larger-scale logic element; (b) the same elements with active-low inputs and outputs.

5.1.4 Active Levels for Pins

When we draw the outline of an AND or OR symbol, or a rectangle representing a larger-scale logic element, we think of the given logic function as occurring *inside* that symbolic outline. In Figure 5-5(a), we show the logic symbols for an AND and OR gate and for a larger-scale element with an ENABLE input. The AND and OR gates have active-high inputs—they require 1s on the input to assert their outputs. Likewise, the larger-scale element has an active-high ENABLE input, which must be 1 to enable the element to do its thing. In (b), we show the same logic elements with active-low input and output pins. Exactly the same logic functions are performed *inside* the symbolic outlines, but the inversion bubbles indicate that 0s must now be applied to the input pins to activate the logic functions, and that the outputs are 0 when they are "doing their thing."

Thus, active levels may be associated with the input and output pins of gates and larger-scale logic elements. We use an inversion bubble to indicate an active-low pin and the absence of a bubble to indicate an active-high pin. For example, the AND gate in Figure 5-6(a) performs the logical AND of two active-high inputs and produces an active-high output: if both inputs are asserted (1), the output is asserted (1). The NAND gate in (b) also performs the AND function, but it produces an active-low output. Even a NOR or OR gate can be construed to perform the AND function using active-low inputs and outputs, as shown in (c) and (d). All four gates in the figure can be said to perform the same function: the output of each gate is asserted if both of its inputs are asserted. Figure 5-7 shows the same idea for the OR function: The output of each gate is asserted if either of its inputs is asserted.

Figure 5-6 Four ways of obtaining an AND function: (a) AND gate (74x08); (b) NAND gate (74x00); (c) NOR gate (74x02); (d) OR gate (74x32).

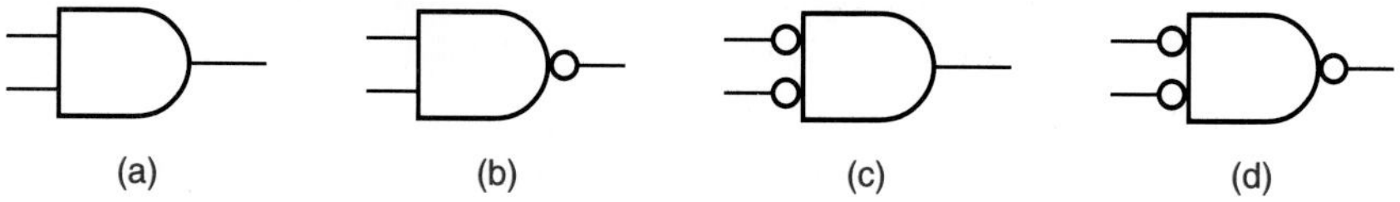

(a)   (b)   (c)   (d)

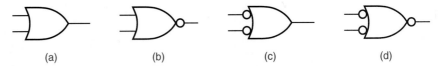

Figure 5-7 Four ways of obtaining an OR function: (a) OR gate (74x32); (b) NOR gate (74x02); (c) NAND gate (74x00); (d) AND gate

Figure 5-8 Alternate logic symbols: (a, b) inverters; (c, d) noninverting buffers.

Sometimes a noninverting buffer is used simply to boost the fanout of a logic signal without changing its function. Figure 5-8 shows the possible logic symbols for both inverters and noninverting buffers. In terms of active levels, all of the symbols perform exactly the same function: Each asserts its output signal if and only if its input is asserted.

5.1.5 Bubble-to-Bubble Logic Design

Experienced logic circuit designers formulate their circuits in terms of the logic functions performed *inside* the symbolic outlines. Whether you're designing with discrete gates or in an HDL like ABEL or VHDL, it's easiest to think of logic signals and their interactions using active-high names. However, once you're ready to realize your circuit, you may have to deal with active-low signals due to the requirements of the environment.

When you design with discrete gates, either at board or ASIC level, a key requirement is often speed. As we showed in Section 3.3.6, inverting gates are typically faster than noninverting ones, so there's often a significant performance payoff in carrying some signals in active-low form.

When you design with larger-scale elements, many of them may be off-the-shelf chips or other existing components that already have some inputs and outputs fixed in active-low form. The reasons that they use active-low signals may range from performance improvement to years of tradition, but in any case, you still have to deal with it.

NAME THAT SIGNAL! Although it is absolutely necessary to name only a circuit's main inputs and outputs, most logic designers find it useful to name internal signals as well. During circuit debugging, it's nice to have a name to use when pointing to an internal signal that's behaving strangely. Most computer-aided design systems automatically generate labels for unnamed signals, but a user-chosen name is preferable to a computer-generated one like XSIG1057.

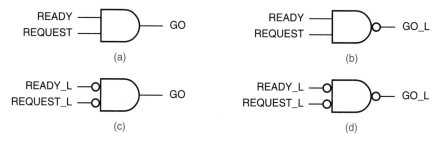

Figure 5-9 Many ways to GO: (a) active-high inputs and output;
(b) active-high inputs, active-low output; (c) active-low
inputs, active-high output; (d) active-low inputs and outputs.

Bubble-to-bubble logic design is the practice of choosing logic symbols
and signal names, including active-level designators, that make the function of a
logic circuit easier to understand. Usually, this means choosing signal names and
gate types and symbols so that most of the inversion bubbles "cancel out" and
the logic diagram can be analyzed as if all of the signals were active high.

bubble-to-bubble logic design

For example, suppose we need to produce a signal that tells a device to
"GO" when we are "READY" and we get a "REQUEST." Clearly from the problem statement, an AND function is required; in switching algebra, we would
write GO = READY · REQUEST. However, we can use different gates to perform
the AND function, depending on the active level required for the GO signal and
the active levels of the available input signals.

Figure 5-9(a) shows the simplest case, where GO must be active-high and
the available input signals are also active-high; we use an AND gate. If, on the
other hand, the device that we're controlling requires an active-low GO_L signal,
we can use a NAND gate as shown in (b). If the available input signals are active-
low, we can use a NOR or OR gate as shown in (c) and (d).

The active levels of available signals don't always match the active levels
of available gates. For example, suppose we are given input signals READY_L
(active-low) and REQUEST (active-high). Figure 5-10 shows two different ways
to generate GO using an inverter to generate the active level needed for the AND
function. The second way is generally preferred, since inverting gates like NOR
are generally faster than noninverting ones like AND. We drew the inverter
differently in each case to make the output's active level match its signal name.

Figure 5-10 Two more ways to GO, with mixed input levels: (a) with an AND
gate; (b) with a NOR gate.

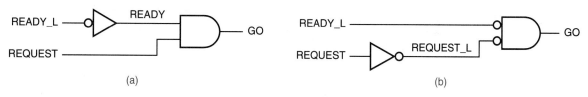

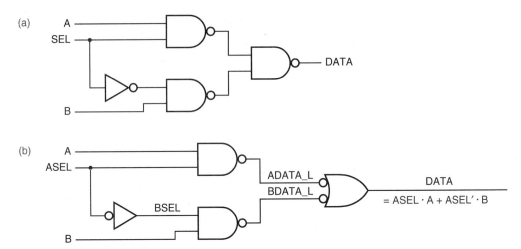

Figure 5-11 A 2-input multiplexer (you're not supposed to know what that is yet): (a) cryptic logic diagram; (b) proper logic diagram using active-level designators and alternate logic symbols.

To understand the benefits of bubble-to-bubble logic design, consider the circuit in Figure 5-11(a). What does it do? In Section 4.2 we showed several ways to analyze such a circuit, and we could certainly obtain a logic expression for the DATA output using these techniques. However, when the circuit is redrawn in Figure 5-11(b), the output function can be read directly from the logic diagram, as follows. The DATA output is asserted when either ADATA_L or BDATA_L is asserted. If ASEL is asserted, then ADATA_L is asserted if and only if A is asserted; that is, ADATA_L is a copy of A. If ASEL is negated, BSEL is asserted and BDATA_L is a copy of B. In other words, DATA is a copy of A if ASEL is asserted, and DATA is a copy of B if ASEL is negated. Even though there are five inversion bubbles in the logic diagram, we mentally had to perform only one negation to understand the circuit—that BSEL is asserted if ASEL is not asserted.

If we wish, we can write an algebraic expression for the DATA output. We use the technique of Section 4.2, simply propagating expressions through gates toward the output. In doing so, we can ignore pairs of inversion bubbles that cancel, and directly write the expression shown in color in the figure.

Figure 5-12 Another properly drawn logic diagram.

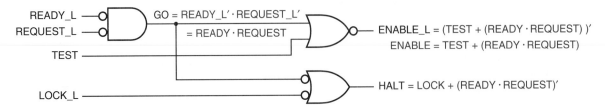

BUBBLE-TO-BUBBLE LOGIC DESIGN RULES

The following rules are useful for performing bubble-to-bubble logic design:

- The signal name on a device's output should have the same active level as the device's output pin—that is, active-low if the device symbol has an inversion bubble on the output pin, active-high if not.

- If the active level of an input signal is the same as that of the input pin to which it is connected, then the logic function inside the symbolic outline is activated when the signal is asserted. This is the most common case in a logic diagram.

- If the active level of an input signal is the opposite of that of the input pin to which it is connected, then the logic function inside the symbolic outline is activated when the signal is negated. This case should be avoided whenever possible because it forces us to keep track mentally of a logical negation to understand the circuit.

Another example is shown in Figure 5-12. Reading directly from the logic diagram, we see that ENABLE_L is asserted if READY_L and REQUEST_L are asserted or if TEST is asserted. The HALT output is asserted if READY_L and REQUEST_L are not both asserted or if LOCK_L is asserted. Once again, this example has only one place where a gate input's active level does not match the input signal level, and this is reflected in the verbal description of the circuit.

We can, if we wish, write algebraic equations for the ENABLE_L and HALT outputs. As we propagate expressions through gates toward the output, we obtain expressions like READY_L' · REQUEST'. However, we can use our active-level naming convention to simplify terms like READY_L'. The circuit contains no signal with the name READY; but if it did, it would satisfy the relationship READY = READY_L' according to the naming convention. This allows us to write the ENABLE_L and HALT equations as shown. Complementing both sides of the ENABLE_L equation, we obtain an equation that describes a hypothetical active-high ENABLE output in terms of hypothetical active-high inputs.

We'll see more examples of bubble-to-bubble logic design in this and later chapters, especially as we begin to use larger-scale logic elements.

5.1.6 Drawing Layout

Logic diagrams and schematics should be drawn with gates in their "normal" orientation with inputs on the left and outputs on the right. The logic symbols for larger-scale logic elements are also normally drawn with inputs on the left and outputs on the right.

A complete schematic page should be drawn with system inputs on the left and outputs on the right, and the general flow of signals should be from left to right. If an input or output appears in the middle of a page, it should be extended to the left or right edge, respectively. In this way, a reader can find all inputs and outputs by looking at the edges of the page only. All signal paths on the page

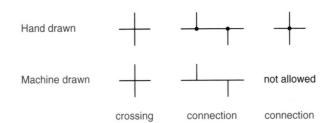

Figure 5-13
Line crossings and
connections.

should be connected when possible; paths may be broken if the drawing gets crowded, but breaks should be flagged in both directions, as described later.

Sometimes block diagrams are drawn without crossing lines for a neater appearance, but this is never done in logic diagrams. Instead, lines are allowed to cross and connections are indicated clearly with a dot. Still, some computer-aided design systems (and some designers) can't draw legible connection dots. To distinguish between crossing lines and connected lines, they adopt the convention that only "T"-type connections are allowed, as shown in Figure 5-13. This is a good convention to follow in any case.

Schematics that fit on a single page are the easiest to work with. The largest practical paper size for a schematic might be E-size (34"×44"). Although its drawing capacity is great, such a large paper size is unwieldy to work with. The best compromise of drawing capacity and practicality is B-size (11"×17"). It can be easily folded for storage and quick reference in standard 3-ring notebooks, and it can be copied on most office copiers. Regardless of paper size, schematics

Figure 5-14 Flat schematic structure.

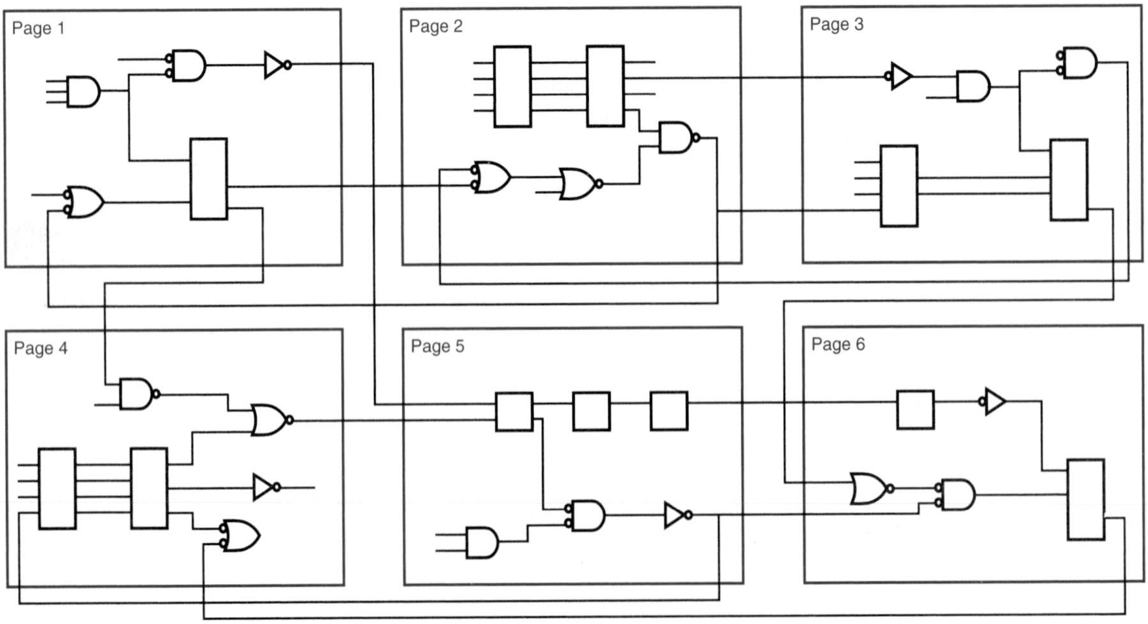

come out best when the page is used in landscape format, that is, with its long dimension oriented from left to right, the direction of most signal flow.

Schematics that don't fit on a single page should be broken up into individual pages in a way that minimizes the connections (and confusion) between pages. They may also use a coordinate system, like that of a road map, to flag the *signal flags* sources and destinations of signals that travel from one page to another. An outgoing signal should have flags referring to all of the destinations of that signal, while an incoming signal should have a flag referring to the source only. That is, an incoming signal should be flagged to the place where it is generated, not to a place somewhere in the middle of a chain of destinations that use the signal.

A multiple-page schematic usually has a "flat" structure. As shown in *flat schematic structure* Figure 5-14, each page is carved out from the complete schematic and can connect to any other page as if all the pages were on one large sheet. However, much like programs, schematics can also be constructed hierarchically, as *hierarchical schematic* illustrated in Figure 5-15. In this approach, the "top-level" schematic is just a *structure* single page that may take the place of a block diagram. Typically, the top-level

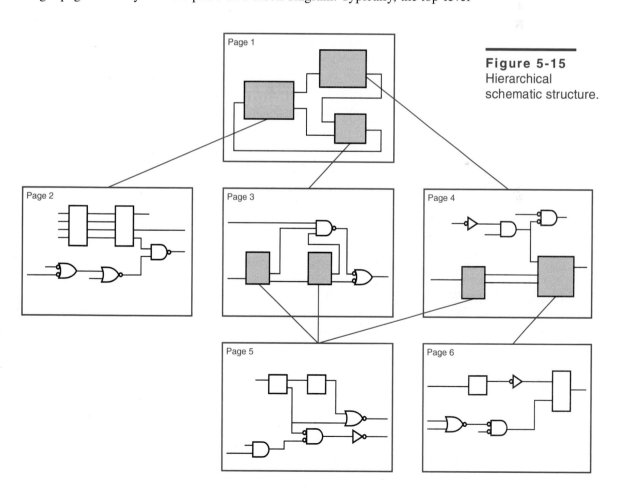

Figure 5-15
Hierarchical
schematic structure.

schematic contains no gates or other logic elements; it only shows blocks corresponding to the major subsystems, and their interconnections. The blocks or subsystems are in turn defined on lower-level pages, which may contain ordinary gate-level descriptions, or which may themselves use blocks defined in lower-level hierarchies. If a particular lower-level hierarchy needs to be used more than once, it may be reused (or "called," in the programming sense) multiple times by the higher-level pages.

Most computer-aided logic design systems support both flat and hierarchical schematics. Proper signal naming is very important in both styles, since there are a number of common errors that can occur:

- Like any other program, a schematic-entry program does what you say, not what you mean. If you use slightly different names for what you intend to be the same signal on different pages, they won't be connected.

- Conversely, if you inadvertently use the same name for different signals on different pages of a flat schematic, many programs will dutifully connect them together, even if you haven't connected them with an off-page flag. (In a hierarchical schematic, reusing a name at different places in the hierarchy is generally OK, because the program qualifies each name with its position in the hierarchy.)

- In a hierarchical schematic, you have to be careful in naming the external interface signals on pages in the lower levels of the hierarchy. These are the names that will appear inside the blocks corresponding to these pages when they are used at higher levels of the hierarchy. It's very easy to transpose signal names or use a name with the wrong active level, yielding incorrect results when the block is used.

- This is not usually a naming problem, but all schematic programs seem to have quirks in which signals that appear to be connected are not. Using the "T" convention in Figure 5-13 can help minimize this problem.

Fortunately, most schematic programs have error-checking facilities that can catch many of these errors, for example, by searching for signal names that have no inputs, no outputs, or multiple outputs associated with them. But most logic designers learn the importance of careful, manual schematic double-checking only through the bitter experience of building a printed-circuit board or an ASIC based on a schematic containing some dumb error.

5.1.7 Buses

As defined previously, a bus is a collection of two or more related signal lines. For example, a microprocessor system might have an address bus with 16 lines, ADDR0–ADDR15, and a data bus with 8 lines, DATA0–DATA7. The signal names in a bus are not necessarily related or ordered as in these first examples. For example, a microprocessor system might have a control bus containing five signals, ALE, MIO, RD_L, WR_L, and RDY.

Logic diagrams use special notation for buses in order to reduce the amount of drawing and to improve readability. As shown in Figure 5-16, a bus has its own descriptive name, such as ADDR[15:0], DATA[7:0], or CONTROL. A bus name may use brackets and a colon to denote a range. Buses are drawn with thicker lines than ordinary signals. Individual signals are put into or pulled out of the bus by connecting an ordinary signal line to the bus and writing the signal name. Often a special connection dot is also used, as in the example.

A computer-aided design system keeps track of the individual signals in a bus. When it actually comes time to build a circuit from the schematic, signal lines in a bus are treated just as though they had all been drawn individually.

The symbols at the righthand edge of Figure 5-16 are interpage signal flags. They indicate that LA goes out to page 2, DB is bidirectional and connects to page 2, and CONTROL is bidirectional and connects to pages 2 and 3.

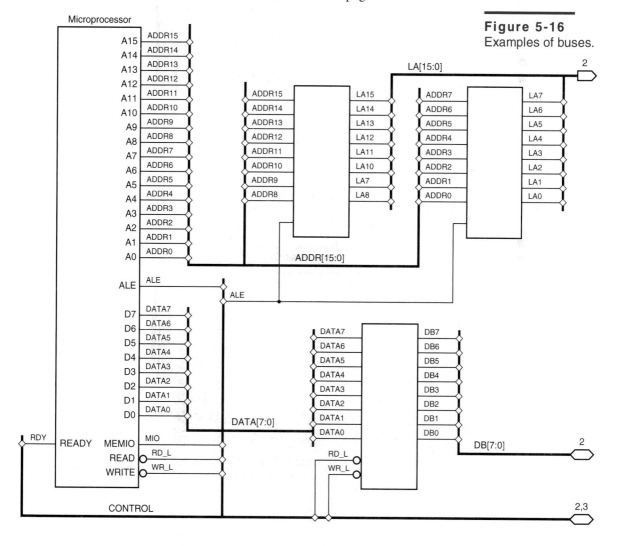

Figure 5-16
Examples of buses.

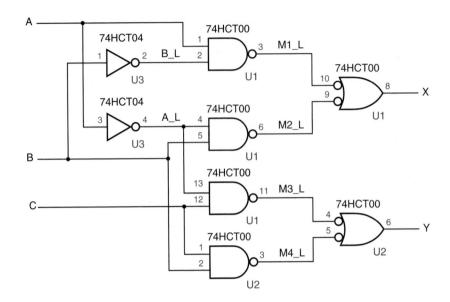

Figure 5-17
Schematic diagram
for a circuit using a
74HCT00.

5.1.8 Additional Schematic Information

Complete schematic diagrams indicate IC types, reference designators, and pin
numbers, as in Figure 5-17. The *IC type* is a part number identifying the integrated circuit that performs a given logic function. For example, a 2-input NAND
gate might be identified as a 74HCT00 or a 74LS00. In addition to the logic
function, the IC type identifies the device's logic family and speed.

IC type

reference designator

The *reference designator* for an IC identifies a particular instance of that IC
type installed in the system. In conjunction with the system's mechanical
documentation, the reference designator allows a particular IC to be located
during assembly, test, and maintenance of the system. Traditionally, reference
designators for ICs begin with the letter U (for "unit").

pin number

Once a particular IC is located, *pin numbers* are used to locate individual
logic signals on its pins. The pin numbers are written near the corresponding
inputs and outputs of the standard logic symbol, as shown in Figure 5-17.

In the rest of this book, just to make you comfortable with properly drawn
schematics, we'll include reference designators and pin numbers for all of the
logic-circuit examples that use SSI and MSI parts.

Figure 5-18 shows the pinouts of many different SSI ICs that are used in
examples throughout this book. Some special graphic elements appear in a few
of the symbols:

- Symbols for the 74x14 Schmitt-trigger inverter have a special element
  inside the symbol to indicate hysteresis.

- Symbols for the 74x03 quad NAND and the 74x266 quad Exclusive NOR
  have a special element to indicate an open-drain or open-collector output.

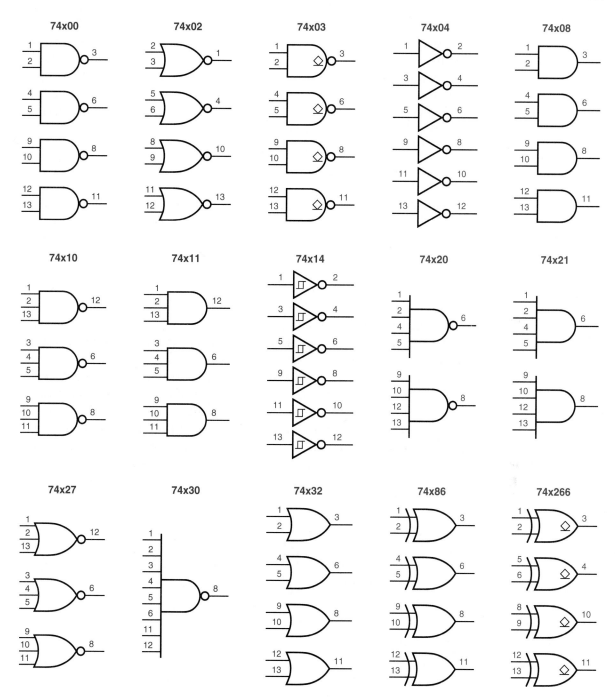

Figure 5-18 Pinouts for SSI ICs in standard dual-inline packages.

When you prepare a schematic diagram for a board-level design using a schematic drawing program, the program automatically provides the pin numbers for the devices that you select from its component library. Note that an IC's pin numbers may differ depending on package type, so you have to be careful to select the right version of the component from the library. Figure 5-18 shows the pin numbers that are used in a dual-inline-pin (DIP) package, the type of package that you would use in a digital design laboratory course or in a low-density, "thru-hole" commercial printed-circuit board.

5.2 Circuit Timing

"Timing is everything"—in investing, in comedy, and yes, in digital design. As we studied in Section 3.6, the outputs of real circuits take time to react to their inputs, and many of today's circuits and systems are so fast that even the speed-of-light delay in propagating an output signal to an input on the other side of a board or chip is significant.

Most digital systems are sequential circuits that operate step-by-step under the control of a periodic clock signal, and the speed of the clock is limited by the worst-case time that it takes for the operations in one step to complete. Thus, digital designers need to be keenly aware of timing behavior in order to build fast circuits that operate correctly under all conditions.

The last several years have seen great advances in the number and quality of CAD tools for analyzing circuit timing. Still, quite often the greatest challenge in completing a board-level or especially an ASIC design is achieving the required timing performance. In this section, we start with the basics, so you can understand what the tools are doing when you use them, and so you can figure out how to fix your circuits when their timing isn't quite making it.

5.2.1 Timing Diagrams

timing diagram

A *timing diagram* illustrates the logical behavior of signals in a digital circuit as a function of time. Timing diagrams are an important part of the documentation of any digital system. They can be used both to explain the timing relationships among signals within a system and to define the timing requirements of external signals that are applied to the system.

Figure 5-19(a) is the block diagram of a simple combinational circuit with two inputs and two outputs. Assuming that the ENB input is held at a constant value, (b) shows the delay of the two outputs with respect to the GO input. In each waveform, the upper line represents a logic 1 and the lower line a logic 0. Signal transitions are drawn as slanted lines to remind us that they do not occur in zero time in real circuits. (Also, slanted lines look nicer than vertical ones.)

causality

Arrows are sometimes drawn, especially in complex timing diagrams, to show *causality*—which input transitions cause which output transitions. In any case, the most important information provided by a timing diagram is a specification of the *delay* between transitions.

delay

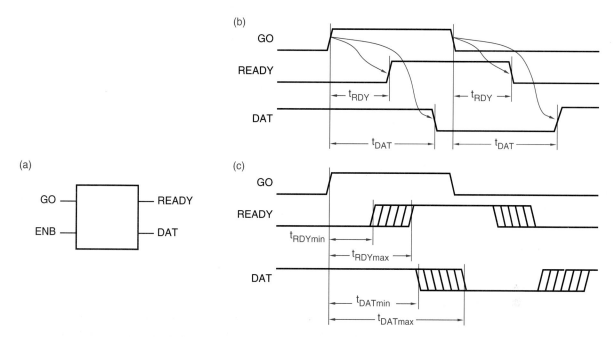

Figure 5-19 Timing diagrams for a combinational circuit: (a) block diagram of circuit; (b) causality and propagation delay; (c) minimum and maximum delays.

Different paths through a circuit may have different delays. For example, Figure 5-19(b) shows that the delay from GO to READY is shorter than the delay from GO to DAT. Similarly, the delays from the ENB input to the outputs may vary, and could be shown in another timing diagram. And, as we'll discuss later, the delay through any given path may vary depending on whether the output is changing from LOW to HIGH or from HIGH to LOW (this phenomenon is not shown in the figure).

Delay in a real circuit is normally measured between the centerpoints of transitions, so the delays in a timing diagram are marked this way. A single timing diagram may contain many different delay specifications. Each different delay is marked with a different identifier, such as t_{RDY} and t_{DAT} in the figure. In large timing diagrams, the delay identifiers are usually numbered for easier reference (e.g., t_1, t_2, …, t_{42}). In either case, the timing diagram is normally accompanied by a *timing table* that specifies each delay amount and the conditions under which it applies.

timing table

Since the delays of real digital components can vary depending on voltage, temperature, and manufacturing parameters, delay is seldom specified as a single number. Instead, a timing table may specify a range of values by giving *minimum*, *typical*, and *maximum* values for each delay. The idea of a range of delays is sometimes carried over into the timing diagram itself by showing the transitions to occur at uncertain times, as in Figure 5-19(c).

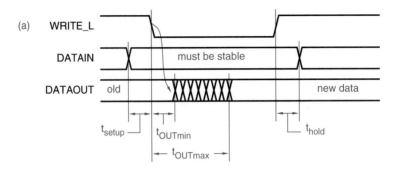

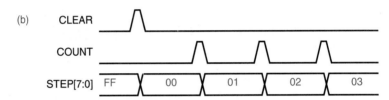

Figure 5-20
Timing diagrams for
"data" signals:
(a) certain and
uncertain transitions;
(b) sequence of values
on an 8-bit bus.

For some signals, the timing diagram needn't show whether the signal changes from 1 to 0 or from 0 to 1 at a particular time, only that a transition occurs then. Any signal that carries a bit of "data" has this characteristic—the actual value of the data bit varies according to circumstances but, regardless of value, the bit is transferred, stored, or processed at a particular time relative to "control" signals in the system. Figure 5-20(a) is a timing diagram that illustrates this concept. The "data" signal is normally at a steady 0 or 1 value, and transitions occur only at the times indicated. The idea of an uncertain delay time can also be used with "data" signals, as shown for the DATAOUT signal.

Quite often in digital systems, a group of data signals in a bus is processed by identical circuits. In this case, all signals in the bus have the same timing, and they can be represented by a single line in the timing diagram and corresponding specifications in the timing table. If the bus bits are known to take on a particular combination at a particular time, this is sometimes shown in the timing diagram using binary, octal, or hexadecimal numbers, as in Figure 5-20(b).

5.2.2 Propagation Delay

propagation delay

In Section 3.6.2 we formally defined the *propagation delay* of a signal path as the time that it takes for a change at the input of the path to produce a change at the output of the path. A combinational circuit with many inputs and outputs has many different paths, and each one may have a different propagation delay. Also, the propagation delay when the output changes from LOW to HIGH (t_{pLH}) may be different from the delay when it changes from HIGH to LOW (t_{pHL}).

The manufacturer of a combinational-logic IC normally specifies all of these different propagation delays, or at least the delays that would be of interest in typical applications. A logic designer who combines ICs in a larger circuit uses the individual device specifications to analyze the overall circuit timing. The delay of a path through the overall circuit is the sum of the delays through subpaths in the individual devices.

5.2.3 Timing Specifications

The timing specification for a device may give minimum, typical, and maximum values for each propagation-delay path and transition direction:

- *Maximum.* This specification is the one that is most often used by experienced designers, since a path "never" has a propagation delay longer than the maximum. However, the definition of "never" varies among logic families and manufacturers. For example, "maximum" propagation delays of 74LS and 74S TTL devices are specified with $V_{CC} = 5$ V, $T_A = 25°C$, and almost no capacitive load. If the voltage or temperature is different, or if the capacitive load is more than 15 pF, the delay may be longer. On the other hand, a "maximum" propagation delay is specified for 74AC and 74ACT devices over the full operating voltage and temperature range and with a heavier capacitive load of 50 pF.

  maximum delay

- *Typical.* This specification is the one most often used by designers who don't expect to be around when their product leaves the friendly environment of the engineering lab and is shipped to customers. The "typical" delay is what you see from a device that was manufactured on a good day and is operating under near-ideal conditions.

  typical delay

- *Minimum.* This is the smallest propagation delay that a path will ever exhibit. Most well-designed circuits don't depend on this number; that is, they will work properly even if the delay is zero. That's good because manufacturers don't specify minimum delay in most moderate-speed logic families, including 74LS and 74S TTL. However, in high-speed families, including ECL and 74AC and 74ACT CMOS, a nonzero minimum delay is specified to help the designer ensure that timing requirements of latches and flip-flops, discussed in Section 7.2, are met.

  minimum delay

Table 5-2 lists the typical and maximum delays of several 74-series CMOS and TTL gates. Table 5-3 does the same thing for most of the CMOS and TTL MSI parts that are introduced later in this chapter.

HOW TYPICAL IS TYPICAL?   Most ICs, perhaps 99%, really are manufactured on "good" days and exhibit delays near the "typical" specifications. However, if you design a system that works only if all of its 100 ICs meet the "typical" timing specs, probability theory suggests that 63% $(1 - .99^{100})$ of the systems won't work. But see the next box ....

◼ **Table 5-2** Propagation delay in nanoseconds of selected 5-V CMOS and TTL SSI parts.

Part number	74HCT Typical t_{pLH}, t_{pHL}	74HCT Maximum t_{pLH}, t_{pHL}	74AHCT Typical t_{pLH}	74AHCT Typical t_{pHL}	74AHCT Maximum t_{pLH}	74AHCT Maximum t_{pHL}	74LS Typical t_{pLH}	74LS Typical t_{pHL}	74LS Maximum t_{pLH}	74LS Maximum t_{pHL}
'00, '10	11	35	5.5	5.5	9.0	9.0	9	10	15	15
'02	9	29	3.4	4.5	8.5	8.5	10	10	15	15
'04	11	35	5.5	5.5	8.5	8.5	9	10	15	15
'08, '11	11	35	5.5	5.5	9.0	9.0	8	10	15	20
'14	16	48	5.5	5.5	9.0	9.0	15	15	22	22
'20	11	35					9	10	15	15
'21	11	35					8	10	15	20
'27	9	29	5.6	5.6	9.0	9.0	10	10	15	15
'30	11	35					8	13	15	20
'32	9	30	5.3	5.3	8.5	8.5	14	14	22	22
'86 (2 levels)	13	40	5.5	5.5	10	10	12	10	23	17
'86 (3 levels)	13	40	5.5	5.5	10	10	20	13	30	22

A COROLLARY OF MURPHY'S LAW

Murphy's law states, "If something can go wrong, it will." A corollary to this is, "If you want something to go wrong, it won't."

In the boxed example on the previous page, you might think that you have a 63% chance of detecting the potential timing problems in the engineering lab. The problems aren't spread out evenly, though, since all ICs from a given batch tend to behave about the same. Murphy's corollary says that *all* of the engineering prototypes will be built with ICs from the same, "good" batches. Therefore, everything works fine for a while, just long enough for the system to get into volume production and for everyone to become complacent and self-congratulatory.

Then, unbeknownst to the production department, a "slow" batch of some IC type arrives from a supplier and gets used in every system that is built, so that *nothing* works. The production engineers scurry around trying to analyze the problem (not easy, because the designer is long gone and didn't bother to write a circuit description), and in the meantime the company loses big bucks because it is unable to ship its product.

Table 5-3 Propagation delay in nanoseconds of selected CMOS and TTL MSI parts.

Part	From	To	74HCT Typical t_{pLH}, t_{pHL}	74HCT Maximum t_{pLH}, t_{pHL}	74AHCT / FCT Typical t_{pLH}, t_{pHL}	74AHCT / FCT Maximum t_{pLH}, t_{pHL}	74LS Typical t_{pLH}	74LS Typical t_{pHL}	74LS Maximum t_{pLH}	74LS Maximum t_{pHL}
'138	any select	output (2)	23	45	8.1 / 5	13 / 9	11	18	20	41
	any select	output (3)	23	45	8.1 / 5	13 / 9	21	20	27	39
	$\overline{G2A}$, $\overline{G2B}$	output	22	42	7.5 / 4	12 / 8	12	20	18	32
	G1	output	22	42	7.1 / 4	11.5 / 8	14	13	26	38
'139	any select	output (2)	14	43	6.5 / 5	10.5 / 9	13	22	20	33
	any select	output (3)	14	43	6.5 / 5	10.5 / 9	18	25	29	38
	enable	output	11	43	5.9 / 5	9.5 / 9	16	21	24	32
'151	any select	Y	17	51	- / 5	- / 9	27	18	43	30
	any select	$\overline{Y}$	18	54	- / 5	- / 9	14	20	23	32
	any data	Y	16	48	- / 4	- / 7	20	16	32	26
	any data	$\overline{Y}$	15	45	- / 4	- / 7	13	12	21	20
	enable	Y	12	36	- / 4	- / 7	26	20	42	32
	enable	$\overline{Y}$	15	45	- / 4	- / 7	15	18	24	30
'153	any select	output	14	43	- / 5	- / 9	19	25	29	38
	any data	output	12	43	- / 4	- / 7	10	17	15	26
	enable	output	11	34	- / 4	- / 7	16	21	24	32
'157	select	output	15	46	6.8 / 7	11.5 / 10.5	15	18	23	27
	any data	output	12	38	5.6 / 4	9.5 / 6	9	9	14	14
	enable	output	12	38	7.1 / 7	12.0 / 10.5	13	14	21	23
'182	any $\overline{Gi}$, $\overline{Pi}$	C1-3	13	41			4.5	4.5	7	7
	any $\overline{Gi}$, $\overline{Pi}$	$\overline{G}$	13	41			5	7	7.5	10.5
	any $\overline{Pi}$	$\overline{P}$	11	35			4.5	6.5	6.5	10
	C0	C1-3	17	50			6.5	7	10	10.5
'280	any input	EVEN	18	53	- / 6	- / 10	33	29	50	45
	any input	ODD	19	56	- / 6	- / 10	23	31	35	50
'283	C0	any Si	22	66			16	15	24	24
	any Ai, Bi	any Si	21	61			15	15	24	24
	C0	C4	19	58			11	11	17	22
	any Ai, Bi	C4	20	60			11	12	17	17
'381	CIN	any Fi					18	14	27	21
	any Ai, Bi	$\overline{G}$					20	21	30	33
	any Ai, Bi	$\overline{P}$					21	33	23	33
	any Ai, Bi	any Fi					20	15	30	23
	any select	any Fi					35	34	53	51
	any select	$\overline{G}$, $\overline{P}$					31	32	47	48
'682	any Pi	$\overline{PEQQ}$	26	69	- / 7	- / 11	13	15	25	25
	any Qi	$\overline{PEQQ}$	26	69	- / 7	- / 11	14	15	25	25
	any Pi	$\overline{PGTQ}$	26	69	- / 9	- / 14	20	15	30	30
	any Qi	$\overline{PGTQ}$	26	69	- / 9	- / 14	21	19	30	30

All inputs of an SSI gate have the same propagation delay to the output. Note that TTL gates usually have different delays for LOW-to-HIGH and HIGH-to-LOW transitions (t_{pLH} and t_{pHL}), but CMOS gates usually do not. CMOS gates have a more symmetrical output driving capability, so any difference between the two cases is usually not worth noting.

The delay from an input transition to the corresponding output transition depends on the internal path taken by the changing signal, and in larger circuits the path may be different for different input combinations. For example, the 74LS86 2-input XOR gate is constructed from four NAND gates as shown in Figure 5-71 on page 412, and has two different-length paths from either input to the output. If one input is LOW, and the other is changed, the change propagates through two NAND gates, and we observe the first set of delays shown in Table 5-2. If one input is HIGH, and the other is changed, the change propagates through *three* NAND gates internally, and we observe the second set of delays. Similar behavior is exhibited by the 74LS138 and 74LS139 in Table 5-3. However, the corresponding CMOS parts do not show these differences; they are small enough to be ignored.

5.2.4 Timing Analysis

To accurately analyze the timing of a circuit containing multiple SSI and MSI devices, a designer may have to study its logical behavior in excruciating detail. For example, when TTL inverting gates (NAND, NOR, etc.) are placed in series, a LOW-to-HIGH change at one gate's output causes a HIGH-to-LOW change at the next one's, and so the differences between t_{pLH} and t_{pHL} tend to average out. On the other hand, when noninverting gates (AND, OR, etc.) are placed in series, a transition causes all outputs to change in the same direction, and so the gap between t_{pLH} and t_{pHL} tends to widen. As a reader, you'll have the privilege of carrying out this sort of analysis in Drills 5.8–5.13.

The analysis gets more complicated if there are MSI devices in the delay path, or if there are multiple paths from a given input signal to a given output signal. Thus, in large circuits, analysis of all of the different delay paths and transition directions can be very complex.

To permit a simplified "worst-case" analysis, designers often use a single *worst-case delay* specification that is the maximum of t_{pLH} and t_{pHL} specifications. The worst-case delay through a circuit is then computed as the sum of the worst-case delays through the individual components, independent of the transition direction and other circuit conditions. This may give an overly pessimistic view of the overall circuit delay, but it saves design time and it's guaranteed to work.

worst-case delay

5.2.5 Timing Analysis Tools

Sophisticated CAD tools for logic design make timing analysis even easier. Their component libraries typically contain not only the logic symbols and functional models for various logic elements, but also their timing models. A simulator allows you to apply input sequences and observe how and when outputs are produced in response. You typically can control whether minimum, typical, maximum, or some combination of delay values are used.

Even with a simulator, you're not completely off the hook. It's usually up the designer to supply the input sequences for which the simulator should produce outputs. Thus, you'll need to have a good feel for what to look for and how to stimulate your circuit to produce and observe the worst-case delays.

Some timing analysis programs can automatically find all possible delay paths in a circuit and print out a sorted list of them, starting with the slowest. These results may be overly pessimistic, however, as some paths may actually not be used in normal operations of the circuit; the designer must still use some intelligence to interpret the results properly.

5.3 Combinational PLDs

5.3.1 Programmable Logic Arrays

Historically, the first PLDs were *programmable logic arrays (PLAs)*. A PLA is a combinational, two-level **AND-OR** device that can be programmed to realize any sum-of-products logic expression, subject to the size limitations of the device. Limitations are

programmable logic array (PLA)

- the number of inputs (n),
- the number of outputs (m), and
- the number of product terms (p).

inputs

outputs

product terms

We might describe such a device as "an $n \times m$ PLA with p product terms." In general, p is far less than the number of n-variable minterms (2^n). Thus, a PLA cannot perform arbitrary n-input, m-output logic functions; its usefulness is limited to functions that can be expressed in sum-of-products form using p or fewer product terms.

An $n \times m$ PLA with p product terms contains p $2n$-input **AND** gates and m p-input **OR** gates. Figure 5-21 shows a small PLA with four inputs, six **AND**

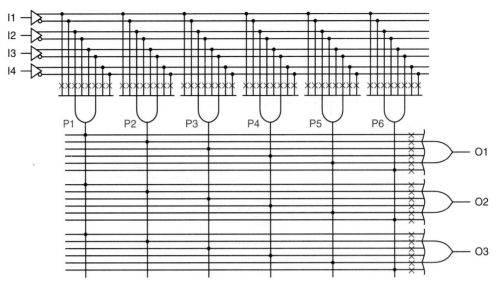

Figure 5-21 A 4×3 PLA with six product terms.

gates, and three OR gates and outputs. Each input is connected to a buffer that produces both a true and a complemented version of the signal for use within the array. Potential connections in the array are indicated by X's; the device is programmed by keeping only the connections that are actually needed. The

PLD fuses

selected connections are made by *fuses*, which are actual fusible links or non-volatile memory cells, depending on technology as we explain in Sections 5.3.4 and 5.3.5. Thus, each AND gate's inputs can be any subset of the primary input signals and their complements. Similarly, each OR gate's inputs can be any subset of the AND-gate outputs.

As shown in Figure 5-22, a more compact diagram can be used to represent

PLA diagram

a PLA. Moreover, the layout of this diagram more closely resembles the actual internal layout of a PLA chip (e.g., Figure 5-28 on page 346).

Figure 5-22
Compact representation of a 4×3 PLA with six product terms.

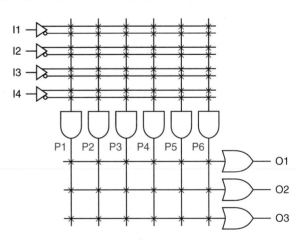

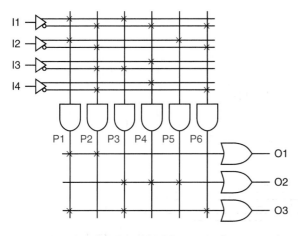

Figure 5-23
A 4 × 3 PLA programmed
with a set of three logic
equations.

The PLA in Figure 5-22 can perform any three 4-input combinational logic functions that can be written as sums of products using a total of six or fewer distinct product terms, for example:

$$O1 = I1 \cdot I2 + I1' \cdot I2' \cdot I3' \cdot I4'$$
$$O2 = I1 \cdot I3' + I1' \cdot I3 \cdot I4 + I2$$
$$O3 = I1 \cdot I2 + I1 \cdot I3' + I1' \cdot I2' \cdot I4'$$

These equations have a total of eight product terms, but the first two terms in the O3 equation are the same as the first terms in the O1 and O2 equations. The programmed connection pattern in Figure 5-23 matches these logic equations.

Sometimes a PLA output must be programmed to be a constant 1 or a constant 0. That's no problem, as shown in Figure 5-24. Product term P1 is always 1 because its product line is connected to no inputs and is therefore always pulled HIGH; this constant-1 term drives the O1 output. No product term drives the O2 output, which is therefore always 0. Another method of obtaining a constant-0 output is shown for O3. Product term P2 is connected to each input variable and its complement; therefore, it's always 0 ($X \cdot X' = 0$).

PLA constant outputs

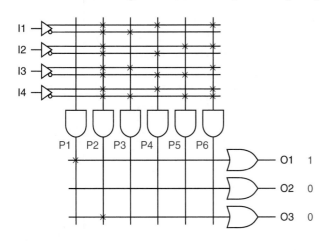

Figure 5-24
A 4 × 3 PLA
programmed to
produce constant
0 and 1 outputs.

Our example PLA has too few inputs, outputs, and AND gates (product terms) to be very useful. An n-input PLA could conceivably use as many as 2^n product terms, to realize all possible n-variable minterms. The actual number of product terms in typical commercial PLAs is far fewer, on the order of 4 to 16 per output, regardless of the value of n.

The Signetics 82S100 was a typical example of the PLAs that were introduced in the mid-1970s. It had 16 inputs, 48 AND gates, and 8 outputs. Thus, it had $2 \times 16 \times 48 = 1536$ fuses in the AND array and $8 \times 48 = 384$ in the OR array. Off-the-shelf PLAs like the 82S100 have since been supplanted by PALs, CPLDs, and FPGAs, but custom PLAs are often synthesized to perform complex combinational logic within a larger ASIC.

5.3.2 Programmable Array Logic Devices

programmable array logic (PAL) device

A special case of a PLA, and today's most commonly used type of PLD, is the *programmable array logic (PAL) device*. Unlike a PLA, in which both the AND and OR arrays are programmable, a PAL device has a *fixed* OR array.

The first PAL devices used TTL-compatible bipolar technology and were introduced in the late 1970s. Key innovations in the first PAL devices, besides the introduction of a catchy acronym, were the use of a fixed OR array and bidirectional input/output pins.

PAL16L8

These ideas are well illustrated by the *PAL16L8*, shown in Figures 5-25 and 5-26 and one of today's most commonly used combinational PLD structures. Its programmable AND array has 64 rows and 32 columns, identified for programming purposes by the small numbers in the figure, and $64 \times 32 = 2048$ fuses. Each of the 64 AND gates in the array has 32 inputs, accommodating 16 variables and their complements; hence, the "16" in "PAL16L8".

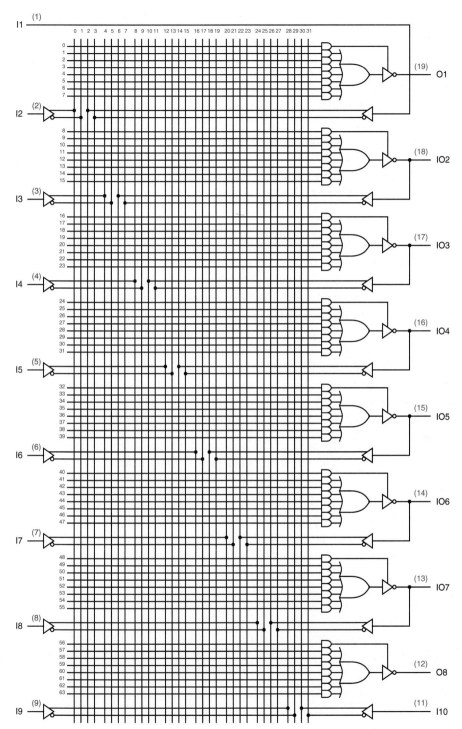

Figure 5-25 Logic diagram of the PAL16L8.

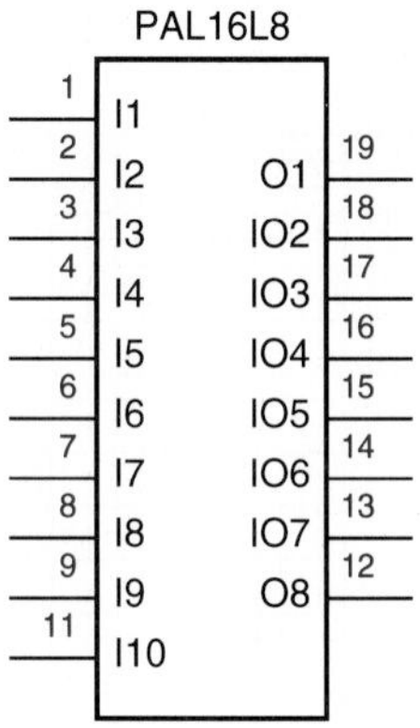

PAL16L8

```
         1  ┌─────────┐
        ────┤ I1      │
         2  │         │        19
        ────┤ I2   O1 ├────
         3  │         │        18
        ────┤ I3  IO2 ├────
         4  │         │        17
        ────┤ I4  IO3 ├────
         5  │         │        16
        ────┤ I5  IO4 ├────
         6  │         │        15
        ────┤ I6  IO5 ├────
         7  │         │        14
        ────┤ I7  IO6 ├────
         8  │         │        13
        ────┤ I8  IO7 ├────
         9  │         │        12
        ────┤ I9   O8 ├────
        11  │         │
        ────┤ I10     │
            └─────────┘
```

Figure 5-26
Traditional logic symbol for
the PAL16L8.

output-enable gate

Eight AND gates are associated with each output pin of the PAL16L8. Seven of them provide inputs to a fixed 7-input OR gate. The eighth, which we call the *output-enable gate*, is connected to the three-state enable input of the output buffer; the buffer is enabled only when the output-enable gate has a 1 output. Thus, an output of the PAL16L8 can perform only logic functions that can be written as sums of seven or fewer product terms. Each product term can be a function of any or all 16 inputs, but only seven such product terms are available.

Although the PAL16L8 has up to 16 inputs and up to 8 outputs, it is housed in a dual in-line package with only 20 pins, including two for power and ground (the corner pins, 10 and 20). This magic is the result of six bidirectional pins (13–18) that may be used as inputs or outputs or both. This and other differences between the PAL16L8 and a PLA structure are summarized below:

- The PAL16L8 has a fixed OR array, with seven AND gates permanently connected to each OR gate. AND-gate outputs cannot be shared; if a product term is needed by two OR gates, it must be generated twice.

- Each output of the PAL16L8 has an individual three-state output enable signal, controlled by a dedicated AND gate (the output-enable gate). Thus, outputs may be programmed as always enabled, always disabled, or enabled by a product term involving the device inputs.

HOW USEFUL ARE SEVEN PRODUCT TERMS?

The worst-case logic function for two-level AND-OR design is an n-input XOR (parity) function, which requires 2^{n-1} product terms. However, less perverse functions with more than seven product terms of a PAL16L8 can often be built by decomposing them into a 4-level structure (AND-OR-AND-OR) that can be realized with two passes through the AND-OR array. Unfortunately, besides using up PLD outputs for the first-pass terms, this doubles the delay, since a first-pass input must pass through the PLD twice to propagate to the output.

COMBINATIONAL, NOT COMBINATORIAL!	A step *backward* in MMI's introduction of PAL devices was their popularization of the word "combinatorial" to describe combinational circuits. *Combinational* circuits have no memory—their output at any time depends on the current input *combination*. For well-rounded computer engineers, the word "combinatorial" should conjure up vivid images of binomial coefficients, problem-solving complexity, and computer-science-great Donald Knuth.

- There is an inverter between the output of each **OR** gate and the external pin of the device.

- Six of the output pins, called *I/O pins*, may also be used as inputs. This provides many possibilities for using each I/O pin, depending on how the device is programmed:

  I/O pin

  - If an I/O pin's output-control gate produces a constant 0, then the output is always disabled and the pin is used strictly as an input.

  - If the input signal on an I/O pin is not used by any gates in the **AND** array, then the pin may be used strictly as an output. Depending on the programming of the output-enable gate, the output may always be enabled, or it may be enabled only for certain input conditions.

  - If an I/O pin's output-control gate produces a constant 1, then the output is always enabled, but the pin may still be used as an input too. In this way, outputs can be used to generate first-pass "helper terms" for logic functions that cannot be performed in a single pass with the limited number of **AND** terms available for a single output. We'll show an example of this case on page 365.

  - In another case with an I/O pin always output enabled, the output may be used as an input to **AND** gates that affect the very same output. That is, we can embed a feedback sequential circuit in a PAL16L8. We'll discuss this case in Section 8.2.6.

The *PAL20L8* is another combinational PLD similar to the PAL16L8, except that its package has four more input-only pins and each of its **AND** gates has eight more inputs to accommodate them. Its output structure is the same as the PAL16L8's.

PAL20L8

5.3.3 Generic Array Logic Devices

In Section 8.3 we'll introduce sequential PLDs, programmable logic devices that provide flip-flops at some or all **OR**-gate outputs. These devices can be programmed to perform a variety of useful sequential-circuit functions.

generic array logic
GAL device
GAL16V8

One type of sequential PLD, first introduced by Lattice Semiconductor, is called *generic array logic* or a *GAL device*, and is particularly popular. A single GAL device type, the *GAL16V8*, can be configured (via programming and a corresponding fuse pattern) to emulate the AND-OR, flip-flop, and output structure of any of a variety of combinational and sequential PAL devices, including the PAL16L8 introduced previously. What's more, the GAL device can be erased electrically and reprogrammed.

Figure 5-27 shows the logic diagram for a GAL16V8 when it has been configured as a strictly combinational device similar to the PAL16L8. This configuration is achieved by programming two "architecture-control" fuses, not shown. In this configuration, the device is called a *GAL16V8C*.

GAL16V8C

The most important thing to note about the GAL16V8C logic diagram, compared to that of a PAL16L8 on page 341, is that an XOR gate has been inserted between each OR output and the three-state output driver. One input of the XOR gate is "pulled up" to a logic 1 value but connected to ground (0) via a fuse. If this fuse is intact, the XOR gate simply passes the OR-gate's output unchanged, but if the fuse is blown, the XOR gate inverts the OR-gate's output. This fuse is said to control the *output polarity* of the corresponding output pin.

output polarity

Output-polarity control is a very important feature of modern PLDs, including the GAL16V8. As we discussed in Section 4.6.2, given a logic function to minimize, an ABEL compiler finds minimal sum-of-products expressions for both the function and its complement. If the complement yields fewer product terms, it can be used if the GAL16V8's output polarity fuse is set to invert. Unless overridden, the compiler automatically makes the best selection and sets up the fuse patterns appropriately.

PALCE16V8
GAL20V8
PALCE20V8

Several companies make a part that is equivalent to the GAL16V8, called the *PALCE16V8*. There is also a 24-pin GAL device, the *GAL20V8* or *PALCE20V8*, that can be configured to emulate the structure of the PAL20L8 or any of a variety of sequential PLDs, as described in Section 8.3.2.

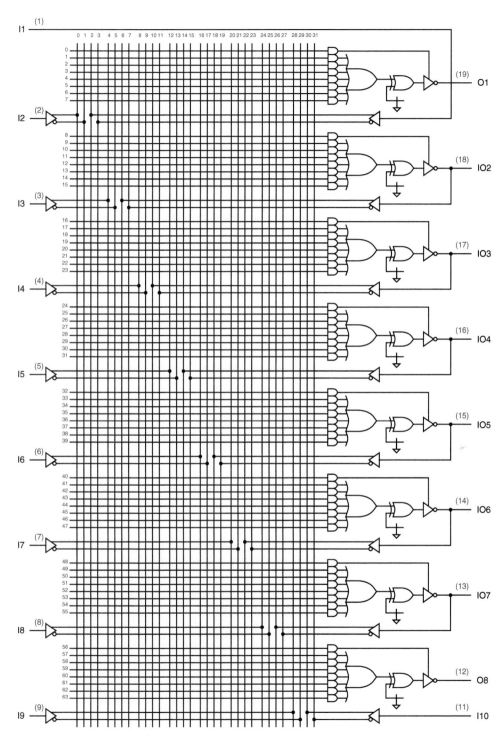

Figure 5-27 Logic diagram of the GAL16V8C.

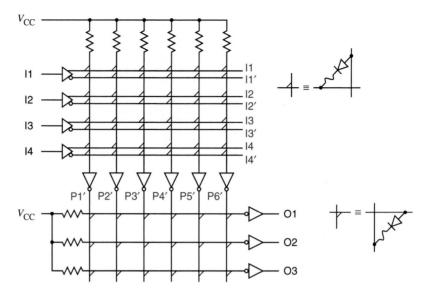

Figure 5-28
A 4 × 3 PLA built using
TTL-like open-collector
gates and diode logic.

*5.3.4 Bipolar PLD Circuits

There are several different circuit technologies for building and physically programming a PLD. Early commercial PLAs and PAL devices used bipolar circuits. For example, Figure 5-28 shows how the example 4 × 3 PLA circuit of the preceding section might be built in a bipolar, TTL-like technology. Each potential connection is made by a diode in series with a metal link that may be present or absent. If the link is present, then the diode connects its input into a diode-AND function. If the link is missing, then the corresponding input has no effect on that AND function.

A diode-AND function is performed because each and every horizontal "input" line that is connected via a diode to a particular vertical "AND" line must be HIGH in order for that AND line to be HIGH. If an input line is LOW, it pulls LOW all of the AND lines to which it is connected. This first matrix of circuit elements that perform the AND function is called the *AND plane.*

AND plane

Each AND line is followed by an inverting buffer, so overall a NAND function is obtained. The outputs of the first-level NAND functions are combined by another set of programmable diode AND functions, once again followed by inverters. The result is a two-level NAND-NAND structure that is functionally equivalent to the AND-OR PLA structure described in the preceding section. The matrix of circuit elements that perform the OR function (or the second NAND function, depending on how you look at it) is called the *OR plane.*

OR plane

A bipolar PLD chip is manufactured with all of its diodes present, but with a tiny *fusible link* in series with each one (the little squiggles in Figure 5-28). By

fusible link

---

* Throughout this book, optional sections are marked with an asterisk.

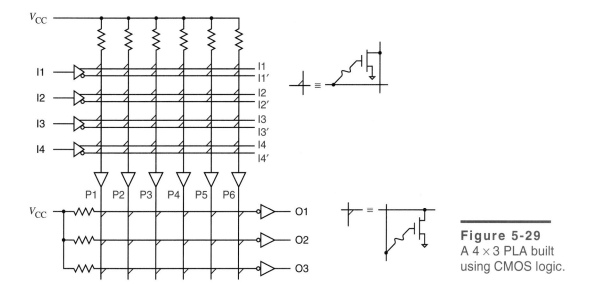

Figure 5-29
A 4 × 3 PLA built
using CMOS logic.

applying special input patterns to the device, it is possible to select individual links, apply a high voltage (10–30 V), and thereby vaporize selected links.

Early bipolar PLDs had reliability problems. Sometimes the stored patterns changed because of incompletely vaporized links that would "grow back," and sometimes intermittent failures occurred because of floating shrapnel inside the IC package. However, these problems have been largely worked out, and reliable fusible-link technology is used in today's bipolar PLDs.

*5.3.5 CMOS PLD Circuits

Although they're still available, bipolar PLDs have been largely supplanted by CMOS PLDs with a number of advantages, including reduced power consumption and reprogrammability. Figure 5-29 shows a CMOS design for the 4 × 3 PLA circuit of Section 5.3.1.

Instead of a diode, an *n*-channel transistor with a programmable connection is placed at each intersection between an input line and a word line. If the input is LOW, then the transistor is "off," but if the input is HIGH, then the transistor is "on," which pulls the AND line LOW. Overall, an inverted-input AND (i.e., NOR) function is obtained. This is similar in structure and function to a normal CMOS *k*-input NOR gate, except that the usual series connection of *k* *p*-channel pull-up transistors has been replaced with a passive pull-up resistor (in practice, the pull-up is a single *p*-channel transistor with a constant bias).

As shown in color on Figure 5-29, the effects of using an *inverted*-input AND gate are canceled by using the opposite (complemented) input lines for each input, compared with Figure 5-28. Also notice that the connection between

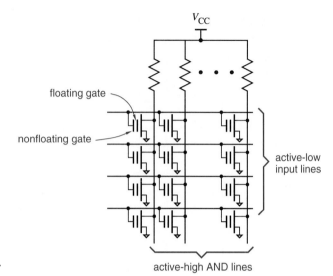

Figure 5-30
AND plane of an
EPLD using floating-
gate MOS transistors.

the AND plane and the OR plane is noninverting, so the AND plane performs a true AND function.

The outputs of the first-level AND functions are combined in the OR plane by another set of NOR functions with programmable connections. The output of each NOR function is followed by an inverter, so a true OR function is realized, and overall the PLA performs an AND-OR function as desired.

In CMOS PLD technologies, the programmable links shown in Figure 5-29 are not normally fuses. In non-field-programmable devices, such as custom VLSI chips, the presence or absence of each link is simply established as part of the metal mask pattern for the manufacture of the device. By far the most common programming technology, however, is used in CMOS EPLDs, as discussed next.

erasable programmable logic device (EPLD)

An *erasable programmable logic device (EPLD)* can be programmed with any desired link configuration, as well as "erased" to its original state, either electronically or by exposing it to ultraviolet light. No, erasing does not cause links to suddenly appear or disappear! Rather, EPLDs use a different technology, called "floating-gate MOS."

floating-gate MOS transistor

As shown in Figure 5-30, an EPLD uses *floating-gate MOS transistors*. Such a transistor has two gates. The "floating" gate is unconnected and is surrounded by extremely high-impedance insulating material. In the original, manufactured state, the floating gate has no charge on it and has no effect on circuit operation. In this state, all transistors are effectively "connected"; that is, there is a logical link present at every crosspoint in the AND and OR planes.

To program an EPLD, the programmer applies a high voltage to the non-floating gate at each location where a logical link is not wanted. This causes a temporary breakdown in the insulating material and allows a negative charge to accumulate on the floating gate. When the high voltage is removed, the negative charge remains on the floating gate. During subsequent operations, the negative charge prevents the transistor from turning "on" when a HIGH signal is applied to the nonfloating gate; the transistor is effectively disconnected from the circuit.

EPLD manufacturers claim that a properly programmed bit will retain 70% of its charge for at least 10 years, even if the part is stored at 125°C, so for most applications the programming can be considered to be permanent. However, EPLDs can also be erased.

Although some early EPLDs were packaged with a transparent lid and used light for erasing, today's most popular devices are *electrically erasable PLDs*. The floating gates in an electrically erasable PLD are surrounded by an extremely thin insulating layer and can be erased by applying a voltage of the opposite polarity as the charging voltage to the nonfloating gate. Thus, the same piece of equipment that is normally used to program a PLD can also be used to erase an EPLD before programming it.

electrically erasable PLD

Larger-scale, "complex" PLDs (CPLDs), also use floating-gate programming technology. Even larger devices, often called *field-programmable gate arrays (FPGAs)*, use read/write memory cells to control the state of each connection. The read/write memory cells are volatile—they do not retain their state when power is removed. Therefore, when power is first applied to the FPGA, all of its read/write memory must be initialized to a state specified by a separate, external nonvolatile memory. This memory typically is either a programmable read-only memory (PROM) chip attached directly to the FPGA or is part of a microprocessor subsystem that initializes the FPGA as part of overall system initialization.

field-programmable gate array (FPGA)

*5.3.6 Device Programming and Testing

A special piece of equipment is used to vaporize fuses, charge up floating-gate transistors, or do whatever else is required to program a PLD. This piece of equipment, found nowadays in almost all digital design labs and production facilities, is called a *PLD programmer* or a *PROM programmer*. (It can be used

PLD programmer
PROM programmer

CHANGING HARDWARE ON THE FLY	PROMs are normally used to supply the connection pattern for a read/write-memory-based FPGA, but there are also applications where the pattern is actually read from a floppy disk. You just received a floppy with a new software version? Guess what, you just got a new hardware version too!
	This concept leads us to the intriguing idea, already being applied in some applications, of "reconfigurable hardware," where a hardware subsystem is redefined, on the fly, to optimize its performance for the particular task at hand.

with programmable read-only memories, "PROMs," as well as for PLDs.) A typical PLD programmer includes a socket or sockets that physically accept the devices to be programmed, and a way to "download" desired programming patterns into the programmer, typically by connecting the programmer to a PC.

PLD programmers typically place a PLD into a special mode of operation in order to program it. For example, a PLD programmer typically programs the PLDs described in this chapter eight fuses at a time as follows:

1. Raise a certain pin to a predetermined, high voltage (such as 14 V) to put the device into programming mode.

2. Select a group of eight fuses by applying a binary "address" to certain inputs of the device. (For example, the 82S100 has 1920 fuses, and it would therefore require 8 inputs to select one of 240 groups of 8 fuses.)

3. Apply an 8-bit value to the *outputs* of the device to specify the desired programming for each fuse (the outputs are used as inputs in programming mode).

4. Raise a second predetermined pin to the high voltage for a predetermined length of time (such as 100 microseconds) to program the eight fuses.

5. Lower the second predetermined pin to a low voltage (such as 0 V) to read out and verify the programming of the eight fuses.

6. Repeat steps 1–5 for each group of eight fuses.

in-system programmability

JTAG port

Many PLDs, especially larger CPLDs, feature *in-system programmability*. This means that the device can be programmed after it is already soldered into the system. In this case, the fuse patterns are applied to the device serially using four extra signals and pins, called the *JTAG port*, defined by IEEE standard 1149.1. These signals are defined so that multiple devices on the same printed-circuit board can be "daisy chained" and selected and programmed during the board manufacturing process using just one JTAG port on a special connector. No special high-voltage power supply is needed; each device uses a charge-pump circuit internally to generate the high voltage needed for programming.

As noted in step 5 above, fuse patterns are verified as they are programmed into a device. If a fuse fails to program properly the first time, the operation can be retried; if it fails to program properly after a few tries, the device is discarded (often with great prejudice and malice aforethought).

While verifying the fuse pattern of a programmed device proves that its fuses are programmed properly, it does not prove that the device will perform the logic function specified by those fuses. This is true because the device may have unrelated internal defects such as missing connections between the fuses and elements of the AND-OR array.

The only way to test for all defects is to put the device into its normal operational mode, apply a set of normal logic inputs, and observe the outputs. The required input and output patterns, called test vectors, can be specified by the designer as we showed in Section 4.6.7, or can be generated automatically by a special test-vector-generation program. Regardless of how the test vectors are generated, most PLD programmers have the ability to apply test-vector inputs to a PLD and to check its outputs against the expected results.

Most PLDs have a *security fuse* which, when programmed, disables the ability to read fuse patterns from the device. Manufacturers can program this fuse to prevent others from reading out the PLD fuse patterns in order to copy the product design. Even if the security fuse is programmed, test vectors still work, so the PLD can still be checked.

security fuse

5.4 Decoders

A *decoder* is a multiple-input, multiple-output logic circuit that converts coded inputs into coded outputs, where the input and output codes are different. The input code generally has fewer bits than the output code, and there is a one-to-one mapping from input code words into output code words. In a *one-to-one mapping*, each input code word produces a different output code word.

decoder

one-to-one mapping

The general structure of a decoder circuit is shown in Figure 5-31. The enable inputs, if present, must be asserted for the decoder to perform its normal mapping function. Otherwise, the decoder maps all input code words into a single, "disabled," output code word.

The most commonly used input code is an n-bit binary code, where an n-bit word represents one of 2^n different coded values, normally the integers from 0 through $2^n - 1$. Sometimes an n-bit binary code is truncated to represent fewer than 2^n values. For example, in the BCD code, the 4-bit combinations 0000 through 1001 represent the decimal digits 0–9, and combinations 1010 through 1111 are not used.

The most commonly used output code is a 1-out-of-m code, which contains m bits, where one bit is asserted at any time. Thus, in a 1-out-of-4 code with active-high outputs, the code words are 0001, 0010, 0100, and 1000. With active-low outputs, the code words are 1110, 1101, 1011, and 0111.

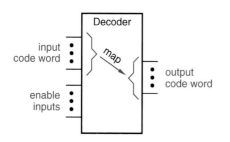

Figure 5-31
Decoder circuit structure.

Table 5-4
Truth table for a 2-to-4
binary decoder.

Inputs			Outputs			
EN	I1	I0	Y3	Y2	Y1	Y0
0	x	x	0	0	0	0
1	0	0	0	0	0	1
1	0	1	0	0	1	0
1	1	0	0	1	0	0
1	1	1	1	0	0	0

5.4.1 Binary Decoders

binary decoder

The most common decoder circuit is an n-to-2^n decoder or *binary decoder*. Such a decoder has an n-bit binary input code and a 1-out-of-2^n output code. A binary decoder is used when you need to activate exactly one of 2^n outputs based on an n-bit input value.

For example, Figure 5-32(a) shows the inputs and outputs and Table 5-4 is the truth table of a 2-to-4 decoder. The input code word 1,I0 represents an integer in the range 0–3. The output code word Y3,Y2,Y1,Y0 has Yi equal to 1 if and only

enable input

if the input code word is the binary representation of i and the *enable input* EN is 1. If EN is 0, then all of the outputs are 0. A gate-level circuit for the 2-to-4

decode

decoder is shown in Figure 5-32(b). Each AND gate *decodes* one combination of the input code word I1,I0.

The binary decoder's truth table introduces a "don't-care" notation for input combinations. If one or more input values do not affect the output values for some combination of the remaining inputs, they are marked with an "x" for that input combination. This convention can greatly reduce the number of rows in the truth table, as well as make the functions of the inputs more clear.

The input code of an n-bit binary decoder need not represent the integers from 0 through 2^n-1. For example, Table 5-5 shows the 3-bit Gray-code output

Figure 5-32
A 2-to-4 decoder:
(a) inputs and outputs;
(b) logic diagram.

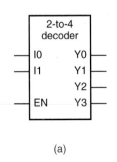

(a)

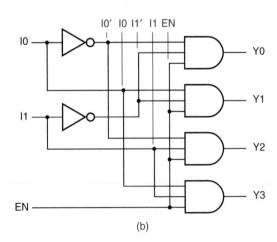

(b)

Disk Position	I2	I1	I0	Binary Decoder Output
0°	0	0	0	Y0
45°	0	0	1	Y1
90°	0	1	1	Y3
135°	0	1	0	Y2
180°	1	1	0	Y6
225°	1	1	1	Y7
270°	1	0	1	Y5
315°	1	0	0	Y4

Table 5-5
Position encoding for a 3-bit mechanical encoding disk.

of a mechanical encoding disk with eight positions. The eight disk positions can be decoded with a 3-bit binary decoder with the appropriate assignment of signals to the decoder outputs, as shown in Figure 5-33.

Also, it is not necessary to use all of the outputs of a decoder, or even to decode all possible input combinations. For example, a *decimal* or *BCD decoder* decodes only the first ten binary input combinations 0000–1001 to produce outputs Y0–Y9.

decimal decoder
BCD decoder

5.4.2 Logic Symbols for Larger-Scale Elements

Before describing some commercially available 74-series MSI decoders, we need to discuss general guidelines for drawing logic symbols for larger-scale logic elements.

The most basic rule is that logic symbols are drawn with inputs on the left and outputs on the right. The top and bottom edges of a logic symbol are not normally used for signal connections. However, explicit power and ground connections are sometimes shown at the top and bottom, especially if these connections are made on "nonstandard" pins. (Most MSI parts have power and ground connected to the corner pins, e.g., pins 8 and 16 of a 16-pin DIP package.)

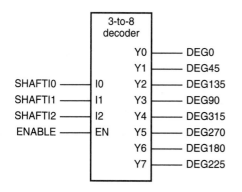

Figure 5-33
Using a 3-to-8 binary decoder to decode a Gray code.

LOGIC FAMILIES   Most logic gates and larger-scale elements are available in a variety of CMOS and families, many of which we described in Sections 3.8 and 3.11. For example, the 74LS139, 74S139, 74ALS139, 74AS139, 74F139, 74HC139, 74HCT139, 74ACT139, 74AC139, 74FCT139 74AHC139, 74AHCT139, 74LC139, 74LVC139, and 74VHC139 are all dual 2-to-4 decoders with the same logic function, but in electrically different TTL and CMOS families and sometimes in different packages. In addition, "macro" logic elements with the same pin names and functions as the '139 and other popular 74-series devices are available as building blocks in most FPGA and ASIC design environments.

Throughout this text, we use "74x" as a generic prefix. And we'll sometimes omit the prefix and write, for example, '139. In a real schematic diagram for a circuit that you are going to build or simulate, you should include the full part number, since timing, loading, and packaging characteristics depend on the family.

Like gate symbols, the logic symbols for larger-scale elements associate an active level with each pin. With respect to active levels, it's important to use a consistent convention to naming the internal signals and external pins.

Larger-scale elements almost always have their signal names defined in terms of the functions performed *inside* their symbolic outline, as explained in Section 5.1.4. For example, Figure 5-34(a) shows the logic symbol for one section of a 74x139 dual 2-to-4 decoder, an MSI part that we'll fully describe in the next subsection. When the G input is asserted, one of the outputs Y0–Y3 is asserted, as selected by a 2-bit code applied to the A and B inputs. It is apparent from the symbol that the G input pin and all of the output pins are active low.

When the 74x139 symbol appears in the logic diagram for a real application, its inputs and outputs have signals connected to other devices, and each such signal has a name that indicates its function in the application. However, when we describe the 74x139 in isolation, we might still like to have a name for the signal on each external pin. Figure 5-34(b) shows our naming convention in this case. Active-high pins are given the same name as the internal signal, while active-low pins have the internal signal name followed by the suffix "_L".

Figure 5-34
Logic symbol for one-half of a 74x139 dual 2-to-4 decoder: (a) conventional symbol; (b) default signal names associated with external pins.

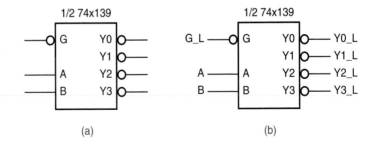

5.4.3 The 74x139 Dual 2-to-4 Decoder

Two independent and identical 2-to-4 decoders are contained in a single MSI part, the *74x139*. The gate-level circuit diagram for this IC is shown in Figure 5-35(a). Notice that the outputs and the enable input of the '139 are active-low. Most MSI decoders were originally designed with active-low outputs, since TTL inverting gates are generally faster than noninverting ones. Also notice that the '139 has extra inverters on its select inputs. Without these inverters, each select input would present three AC or DC loads instead of one, consuming much more of the fanout budget of the device that drives it.

74x139

Figure 5-35 The 74x139 dual 2-to-4 decoder: (a) logic diagram, including pin numbers for a standard 16-pin dual in-line package; (b) traditional logic symbol; (c) logic symbol for one decoder.

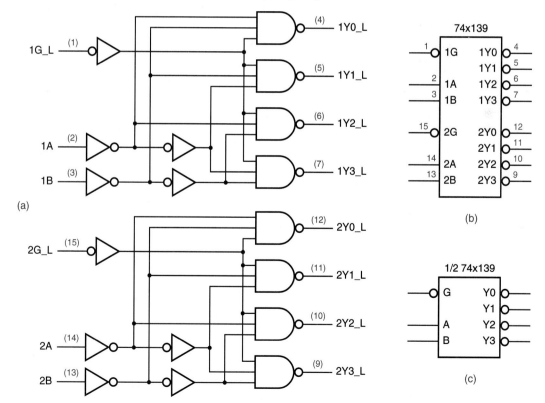

Table 5-6
Truth table for one-half of a 74x139 dual 2-to-4 decoder.

Inputs			Outputs			
G_L	B	A	Y3_L	Y2_L	Y1_L	Y0_L
1	x	x	1	1	1	1
0	0	0	1	1	1	0
0	0	1	1	1	0	1
0	1	0	1	0	1	1
0	1	1	0	1	1	1

A logic symbol for the 74x139 is shown in Figure 5-35(b). Notice that all of the signal names inside the symbol outline are active-high (no "_L"), and that inversion bubbles indicate active-low inputs and outputs. Often a schematic may use a generic symbol for just one decoder, one-half of a '139, as shown in (c). In this case, the assignment of the generic function to one half or the other of a particular '139 package can be deferred until the schematic is completed.

Table 5-6 is the truth table for a 74x139-type decoder. The truth tables in some manufacturers' data books use L and H to denote the input and output signal voltage levels, so there can be no ambiguity about the electrical function of the device; a truth table written this way is sometimes called a *function table*. However, since we use positive logic throughout this book, we can use 0 and 1 without ambiguity. In any case, the truth table gives the logic function in terms of the *external pins* of the device. A truth table for the function performed *inside* the symbol outline would look just like Table 5-4, except that the input signal names would be G, B, A.

Some logic designers draw the symbol for 74x139s and other logic functions without inversion bubbles. Instead, they use an overbar on signal names inside the symbol outline to indicate negation, as shown in Figure 5-36(a). This notation is self-consistent, but it is inconsistent with our drawing standards for bubble-to-bubble logic design. The symbol shown in (b) is absolutely *incorrect*: according to this symbol, a logic 1, not 0, must be applied to the enable pin to enable the decoder.

function table

Figure 5-36
More ways to symbolize a 74x139: (a) correct but to be avoided; (b) incorrect because of double negations.

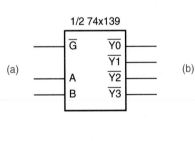

(a)

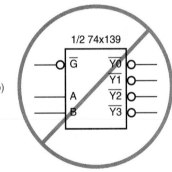

(b)

BAD NAMES   Some manufacturers' data sheets have inconsistencies similar to Figure 5-36(b). For example, Texas Instruments' data sheet for the 74AHC139 uses active-low names like $\overline{1G}$ for the enable inputs, with the overbar indicating an active-low pin, but active-high names like 1Y0 for all the active-low output pins. On the other hand, Motorola's data sheet for the 74VHC139 correctly uses overbars on the names for both the enable inputs and the outputs, but the overbars are barely visible in the device's function table due to a typographical problem.

I've also had the personal experience of building a printed-circuit board with many copies of a new device from a vendor whose documentation clearly indicated that a particular input was active low, only to find out upon the first power-on that the input was active high.

The moral of the story is that you have to study the description of each device to know what's really going on. And if it's a brand-new device, whether from a commercial vendor or your own company's ASIC group, you should double-check all of the signal polarities and pin assignments before committing to a PCB. Rest assured, however, that the signal names in this text *are* consistent and correct.

5.4.4 The 74x138 3-to-8 Decoder

The *74x138* is a commercially available MSI 3-to-8 decoder whose gate-level *74x138* circuit diagram and symbol are shown in Figure 5-37; its truth table is given in Table 5-7. Like the 74x139, the 74x138 has active-low outputs, and it has three enable inputs (G1, /G2A, /G2B), all of which must be asserted for the selected output to be asserted.

The logic function of the '138 is straightforward—an output is asserted if and only if the decoder is enabled and the output is selected. Thus, we can easily write logic equations for an internal output signal such as Y5 in terms of the internal input signals:

$$Y5 = \underbrace{G1 \cdot G2A \cdot G2B}_{\text{enable}} \cdot \underbrace{C \cdot B' \cdot A}_{\text{select}}$$

However, because of the inversion bubbles, we have the following relations between internal and external signals:

$$G2A = G2A\_L'$$
$$G2B = G2B\_L'$$
$$Y5 = Y5\_L'$$

Therefore, if we're interested, we can write the following equation for the external output signal Y5_L in terms of external input signals:

$$Y5\_L = Y5' = (G1 \cdot G2A\_L' \cdot G2B\_L' \cdot C \cdot B' \cdot A)'$$
$$= G1' + G2A\_L + G2B\_L + C' + B + A'$$

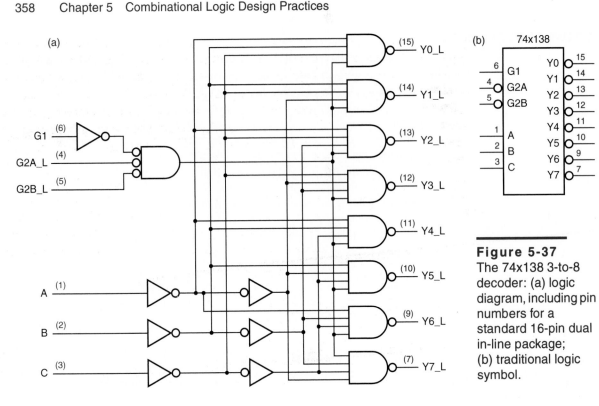

Figure 5-37
The 74x138 3-to-8 decoder: (a) logic diagram, including pin numbers for a standard 16-pin dual in-line package; (b) traditional logic symbol.

Table 5-7 Truth table for a 74x138 3-to-8 decoder.

Inputs						Outputs							
G1	G2A_L	G2B_L	C	B	A	Y7_L	Y6_L	Y5_L	Y4_L	Y3_L	Y2_L	Y1_L	Y0_L
0	x	x	x	x	x	1	1	1	1	1	1	1	1
x	1	x	x	x	x	1	1	1	1	1	1	1	1
x	x	1	x	x	x	1	1	1	1	1	1	1	1
1	0	0	0	0	0	1	1	1	1	1	1	1	0
1	0	0	0	0	1	1	1	1	1	1	1	0	1
1	0	0	0	1	0	1	1	1	1	1	0	1	1
1	0	0	0	1	1	1	1	1	1	0	1	1	1
1	0	0	1	0	0	1	1	1	0	1	1	1	1
1	0	0	1	0	1	1	1	0	1	1	1	1	1
1	0	0	1	1	0	1	0	1	1	1	1	1	1
1	0	0	1	1	1	0	1	1	1	1	1	1	1

On the surface, this equation doesn't resemble what you might expect for a decoder, since it is a logical sum rather than a product. However, if you practice bubble-to-bubble logic design, you don't have to worry about this; you just give the output signal an active-low name and remember that it's active low when you connect it to other inputs.

5.4.5 Cascading Binary Decoders

Multiple binary decoders can be used to decode larger code words. Figure 5-38 shows how two 3-to-8 decoders can be combined to make a 4-to-16 decoder. The availability of both active-high and active-low enable inputs on the 74x138 makes it possible to enable one or the other directly based on the state of the most significant input bit. The top decoder (U1) is enabled when N3 is 0, and the bottom one (U2) is enabled when N3 is 1.

To handle even larger code words, binary decoders can be cascaded hierarchically. Figure 5-39 shows how to use half of a 74x139 to decode the two high-order bits of a 5-bit code word, thereby enabling one of four 74x138s that decode the three low-order bits.

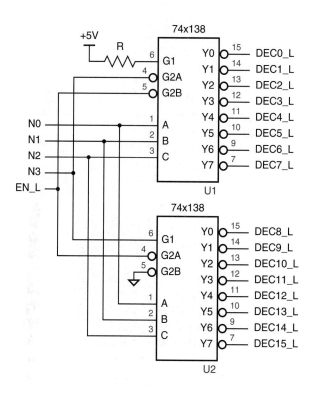

Figure 5-38
Design of a 4-to-16 decoder using 74x138s.

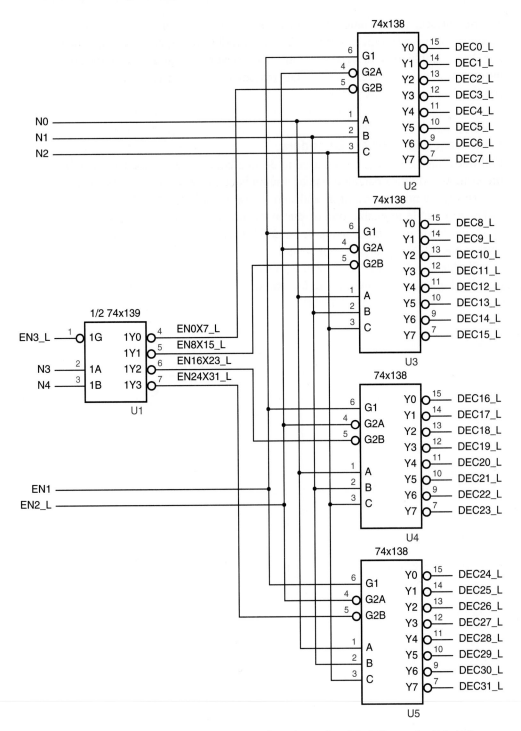

Figure 5-39 Design of a 5-to-32 decoder using 74x138s and a 74x139.

5.4.6 Decoders in ABEL and PLDs

Nothing in logic design is much easier than writing the PLD equations for a decoder. Since the logic expression for each output is typically just a single product term, decoders are very easily targeted to PLDs and use few product-term resources.

For example, Table 5-8 is an ABEL program for a 74x138-like 3-to-8 binary decoder as realized in a PAL16L8. Note that some of the input pins and all of the output pins have active-low names ("_L" suffix) in the pin declarations, corresponding to the logic diagram in Figure 5-37 on page 358. However, the program also defines a corresponding active-high name for each signal so that the equations can all be written "naturally," in terms of active-high signals. An alternate way to achieve the same effect is described in the box on page 363.

Table 5-8 An ABEL program for a 74x138-like 3-to-8 binary decoder.

```
module Z74X138
title '74x138 Decoder PLD
J. Wakerly, Stanford University'
Z74X138 device 'P16L8';

" Input and output pins
A, B, C, G2A_L, G2B_L, G1                        pin 1, 2, 3, 4, 5, 6;
Y0_L, Y1_L, Y2_L, Y3_L, Y4_L, Y5_L, Y6_L, Y7_L   pin 19..12 istype 'com';

" Active-high signal names for readability
G2A = !G2A_L;
G2B = !G2B_L;
Y0 = !Y0_L;
Y1 = !Y1_L;
Y2 = !Y2_L;
Y3 = !Y3_L;
Y4 = !Y4_L;
Y5 = !Y5_L;
Y6 = !Y6_L;
Y7 = !Y7_L;

" Constant expression
ENB = G1 & G2A & G2B;

equations
Y0 = ENB & !C & !B & !A;
Y1 = ENB & !C & !B &  A;
Y2 = ENB & !C &  B & !A;
Y3 = ENB & !C &  B &  A;
Y4 = ENB &  C & !B & !A;
Y5 = ENB &  C & !B &  A;
Y6 = ENB &  C &  B & !A;
Y7 = ENB &  C &  B &  A;

end Z74X138
```

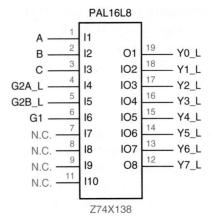

Figure 5-40
Logic diagram for
the PAL16L8 used as
a 74x138 decoder.

Also note that the ABEL program defines a constant expression for ENB. Here, ENB is not an input or output signal, but merely a user-defined name. In the equations section, the compiler substitutes the expression (G1 & G2A & G2B) everywhere that "ENB" appears. Assigning the constant expression to a user-defined name improves this program's readability and maintainability.

If all you needed was a '138, you'd be better off using a real '138 than a more expensive PLD. However, if you need nonstandard functionality, then the PLD can usually achieve it much more cheaply and easily than an MSI/SSI-based solution. For example, if you need the functionality of a '138 but with active-high outputs, you need only to change one line in the pin declarations of Table 5-8:

```
Y0, Y1, Y2, Y3, Y4, Y5, Y6, Y7        pin 19..12 istype 'com';
```

(Also, the original definitions of Y0–Y7 in Table 5-8 must be deleted.) Since each of the equations required a single product of six variables (including the three in the ENB expression), each complemented equation requires a *sum* of six product terms, less than the seven available in a PAL16L8. If you use a PAL16V8 or other device with output polarity selection, then the compiler can select non-inverted output polarity to use only one product term per output.

Table 5-9   Alternate declarations for a 74x138-like 3-to-8 binary decoder.

```
" Input and output pins
A, B, C, !G2A, !G2B, G1                pin 1, 2, 3, 4, 5, 6;
!Y0, !Y1, !Y2, !Y3, !Y4, !Y5, !Y6, !Y7    pin 19..12 istype 'com';

" Constant expression
ENB = G1 & G2A & G2B;
```

ACTIVE-LOW PIN DEFINITIONS

ABEL allows you to use an inversion prefix (!) on signal names in the pin definitions of a program. When a pin name is defined with the inversion prefix, the compiler automatically prepends an inversion prefix to the signal name anywhere it appears elsewhere in the program. If it's already inverted, this results in a double inversion.

This feature can be used to define a different but consistent convention for defining active-low inputs and outputs—give each active-low signal an active-*high* name, but precede it with the inversion prefix in its pin definition. For the 3-to-8 decoder in Table 5-8, we replace the first part of the program with the code shown in Table 5-9; the equations section of the program stays exactly the same.

Which convention to use may be a matter of personal taste, but it can also depend on the capabilities of the CAD tools that you use to draw schematics. Many tools allow you to automatically create a schematic symbol from a logic block that is defined by an ABEL program. If the tool allows you to place inversion bubbles on selected inputs and outputs of the symbol, then the convention in Table 5-9 yields a symbol with active-high signal names inside the function outline. You can then specify external bubbles on the active-low signals to obtain a symbol that matches the conventions described in Section 5.4.2 and shown in Figure 5-41(a).

On the other hand, you may not be able or want to provide inversion bubbles on CAD-created symbols. In that case, you should use the convention in Table 5-8; this yields a CAD-created symbol in which the active level is indicated by the signal name inside the function outline; no external inversion bubbles are needed. This is shown in Figure 5-41(b). Note that unlike Figure 5-36(a) on page 356, we use a text-based convention (_L) rather than an overbar on the signal name to indicate active level. A properly chosen text-based convention provides portability among different CAD tools.

We'll somewhat arbitrarily select one convention or the other in each of the ABEL examples in the rest of this book, just to help you get comfortable with both approaches.

(a)

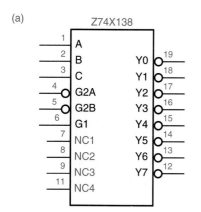

(b)

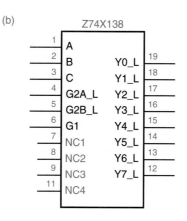

Figure 5-41
Possible CAD-created symbols for the PLD-based, 74x138-like decoder: (a) based on Table 5-9, after manual insertion of inversion bubbles; (b) based on Table 5-8.

Another easy change is to provide alternate enable inputs that are ORed with the main enable inputs. To do this, you need only define additional pins and modify the definition of ENB:

```
EN1, EN2_L                              pin 7, 8;
...
EN2 = !EN2_L;
...
ENB = G1 & G2A & G2B # EN1 # EN2;
```

This change expands the number of product terms per output to three, each having a form similar to

```
Y0 = G1 & G2A & G2B & !C & !B &!A
   # EN1 & !C & !B & !A
   # EN2 & !C & !B & !A;
```

(Remember that the PAL16L8 has a fixed inverter and the PAL16V8 has a selectable inverter between the AND-OR array and the output of the PLD, so the actual output is active low as desired.)

If you add the extra enables to the version of the program with active-high outputs, then the PLD must realize the complement of the sum-of-products expression above. It's not immediately obvious how many product terms this expression will have, and whether it will fit in a PAL16L8, but we can use the ABEL compiler to get the answer for us:

```
!Y0 = C # B # A # !G2B & !EN1 & !EN2
              # !G2A & !EN1 & !EN2
              # !G1  & !EN1 & !EN2;
```

The expression has a total of six product terms, so it fits in a PAL16L8.

As a final tweak, we can add an input to dynamically control whether the output is active high or active low, and modify all of the equations as follows:

```
POL                 pin 9;
...
Y0 = POL $ (ENB & !C & !B & !A);
Y1 = POL $ (ENB & !C & !B &  A);
...
Y7 = POL $ (ENB &  C &  B &  A);
```

As a result of the XOR operation, the number of product terms needed per output increases to 9, in either output-pin polarity. Thus, even a PAL16V8 cannot implement the function as written.

helper output

The function can still be realized if we create a *helper output* to reduce the product-term explosion. As shown in Table 5-10, we allocate an output pin for

Table 5-10 ABEL program fragment showing two-pass logic.

```
...
" Output pins
Y0_L, Y1_L, Y2_L, Y3_L          pin 19, 18, 17, 16 istype 'com';
Y4_L, Y5_L, Y6_L, ENB           pin 15, 14, 13, 12 istype 'com';

equations
ENB = G1 & G2A & G2B # EN1 # EN2;
Y0 = POL $ (ENB & !C & !B & !A);
...
```

the ENB expression (losing the Y7_L output) , and move the ENB equation into the equations section of the program. This reduces the product-term requirement to five in either polarity.

helper output

Besides sacrificing a pin for the helper output, this realization has the disadvantage of being slower. Any changes in the inputs to the helper expression must propagate through the PLD twice before reaching the final output. This is called *two-pass logic*. Many PLD and FPGA synthesis tools can automatically generate logic with two or more passes if a required expression cannot be realized in just one pass through the logic array.

two-pass logic

Decoders can be customized in other ways. A common customization is for a single output to decode more than one input combination. For example, suppose you needed to generate a set of enable signals according to Table 5-11. A 74x138 MSI decoder can be augmented as shown in Figure 5-42 to perform the required function. This approach, while potentially less expensive than a PLD, has the disadvantages that it requires extra components and delay to create the required outputs, and it is not easily modified.

CS_L	RD_L	A2	A1	A0	**Output(s) to Assert**
1	x	x	x	x	none
x	1	x	x	x	none
0	0	0	0	0	BILL_L, MARY_L
0	0	0	0	1	MARY_L, KATE_L
0	0	0	1	0	JOAN_L
0	0	0	1	1	PAUL_L
0	0	1	0	0	ANNA_L
0	0	1	0	1	FRED_L
0	0	1	1	0	DAVE_L
0	0	1	1	1	KATE_L

Table 5-11
Truth table for a customized decoder function.

Figure 5-42
Customized
decoder circuit.

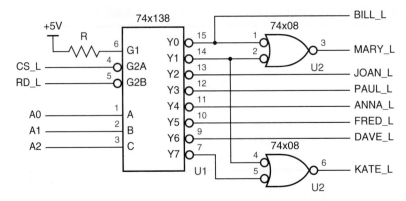

A PLD solution to the same problem is shown in Table 5-12. Note that this program uses the pin active-low pin-naming convention described in the box on page 363 (you should be comfortable with either convention). Each of the last six equations uses a single AND gate in the PLD. The ABEL compiler will also minimize the MARY equation to use just one AND gate. Active-high output signals could be obtained just by changing two lines in the declaration section:

```
BILL, MARY, JOAN, PAUL      pin 19, 18, 17, 16 istype 'com';
ANNA, FRED, DAVE, KATE      pin 15, 14, 13, 12 istype 'com';
```

Table 5-12 ABEL equations for a customized decoder.

```
module CUSTMDEC
title 'Customized Decoder PLD
J. Wakerly, Stanford University'
CUSTMDEC device 'P16L8';

" Input pins
!CS, !RD, A0, A1, A2        pin 1, 2, 3, 4, 5;
" Output pins
!BILL, !MARY, !JOAN, !PAUL  pin 19, 18, 17, 16 istype 'com';
!ANNA, !FRED, !DAVE, !KATE  pin 15, 14, 13, 12 istype 'com';

equations
BILL = CS & RD & (!A2 & !A1 & !A0);
MARY = CS & RD & (!A2 & !A1 & !A0 # !A2 & !A1 &  A0);
KATE = CS & RD & (!A2 & !A1 &  A0 #  A2 &  A1 &  A0);
JOAN = CS & RD & (!A2 &  A1 & !A0);
PAUL = CS & RD & (!A2 &  A1 &  A0);
ANNA = CS & RD & ( A2 & !A1 & !A0);
FRED = CS & RD & ( A2 & !A1 &  A0);
DAVE = CS & RD & ( A2 &  A1 & !A0);

end CUSTMDEC
```

Another way of writing the equations is shown in Table 5-13. In most applications, this style is more clear, especially if the select inputs have numeric significance.

Table 5-13 Equivalent ABEL equations for a customized decoder.

```
ADDR = [A2,A1,A0];

equations
BILL = CS & RD & (ADDR == 0);
MARY = CS & RD & (ADDR == 0) # (ADDR == 1);
KATE = CS & RD & (ADDR == 1) # (ADDR == 7);
JOAN = CS & RD & (ADDR == 2);
PAUL = CS & RD & (ADDR == 3);
ANNA = CS & RD & (ADDR == 4);
FRED = CS & RD & (ADDR == 5);
DAVE = CS & RD & (ADDR == 6);
```

5.4.7 Decoders in VHDL

There are several ways to approach the design of decoders in VHDL. The most primitive approach would be to write a structural equivalent of a decoder logic circuit, as Table 5-14 does for the 2-to-4 binary decoder of Figure 5-32 on page 352. The components and3 and inv are assumed to already exist in the target technology. Of course, this mechanical conversion of an existing design into the equivalent of a netlist defeats the purpose of using VHDL in the first place.

Instead, we would like to write a program that uses VHDL to make our decoder design more understandable and maintainable. Table 5-15 shows one approach to writing code for a 3-to-8 binary decoder equivalent to the 74x138, using the dataflow style of VHDL. The address inputs A(2 downto 0) and the active-low decoded outputs Y_L(0 to 7) are declared using vectors to improve readability. A select statement enumerates the eight decoding cases and assigns the appropriate active-low output pattern to an 8-bit internal signal Y_L_i. This value is assigned to the actual circuit output Y_L only if all of the enable inputs are asserted.

This design is a good start, and it works, but it does have a potential pitfall. The adjustments that handle the fact that two inputs and all the outputs are active-low happen to be buried in the final assignment statement. While it's true that most VHDL programs are written almost entirely with active-high signals, if we're defining a device with active-low external pins, we really should handle them in a more systematic and easily maintainable way.

■ Table 5-14 VHDL structural program for the decoder in Figure 5-32.

```
library IEEE;
use IEEE.std_logic_1164.all;

entity V2to4dec is
  port (I0, I1, EN: in STD_LOGIC;
        Y0, Y1, Y2, Y3: out STD_LOGIC );
end V2to4dec;

architecture V2to4dec_s of V2to4dec is
  signal NOTI0, NOTI1: STD_LOGIC;
  component inv port (I: in STD_LOGIC; O: out STD_LOGIC ); end component;
  component and3 port (I0, I1, I2: in STD_LOGIC; O: out STD_LOGIC ); end component;
begin
  U1: inv port map (I0,NOTI0);
  U2: inv port map (I1,NOTI1);
  U3: and3 port map (NOTI0,NOTI1,EN,Y0);
  U4: and3 port map (   I0,NOTI1,EN,Y1);
  U5: and3 port map (NOTI0,   I1,EN,Y2);
  U6: and3 port map (   I0,   I1,EN,Y3);
end V2to4dec_s;
```

■ Table 5-15 Dataflow-style VHDL program for a 74x138-like 3-to-8 binary decoder.

```
library IEEE;
use IEEE.std_logic_1164.all;

entity V74x138 is
    port (G1, G2A_L, G2B_L: in STD_LOGIC;        -- enable inputs
        A: in STD_LOGIC_VECTOR (2 downto 0);     -- select inputs
        Y_L: out STD_LOGIC_VECTOR (0 to 7) );    -- decoded outputs
  end V74x138;

architecture V74x138_a of V74x138 is
  signal Y_L_i: STD_LOGIC_VECTOR (0 to 7);
begin
  with A select Y_L_i <=
    "01111111" when "000",
    "10111111" when "001",
    "11011111" when "010",
    "11101111" when "011",
    "11110111" when "100",
    "11111011" when "101",
    "11111101" when "110",
    "11111110" when "111",
    "11111111" when others;
  Y_L <= Y_L_i when (G1 and not G2A_L and not G2B_L)='1' else "11111111";
end V74x138_a;
```

Table 5-16 VHDL architecture with a maintainable approach to active-level handling.

```
architecture V74x138_b of V74x138 is
  signal G2A, G2B: STD_LOGIC;              -- active-high version of inputs
  signal Y: STD_LOGIC_VECTOR (0 to 7);    -- active-high version of outputs
  signal Y_s: STD_LOGIC_VECTOR (0 to 7);  -- internal signal
begin
  G2A <= not G2A_L; -- convert inputs
  G2B <= not G2B_L; -- convert inputs
  Y_L <= Y;         -- convert outputs
  with A select Y_s <=
    "10000000" when "000",
    "01000000" when "001",
    "00100000" when "010",
    "00010000" when "011",
    "00001000" when "100",
    "00000100" when "101",
    "00000010" when "110",
    "00000001" when "111",
    "00000000" when others;
  Y <= not Y_s when (G1 and G2A and G2B)='1' else "00000000";
end V74x138_b;
```

Table 5-16 shows such an approach. No changes are made to the `entity` declarations. However, active-high versions of the active-low external pins are defined within the V74x138_a architecture, and explicit assignment statements are used to convert between the active-high and active-low signals. The decoder function itself is defined in terms of only the active-high signals, probably the biggest advantage of this approach. Another advantage is that the design can be easily modified in just a few well-defined places if changes are required in the external active levels.

OUT-OF-ORDER EXECUTION   In Table 5-16, we've grouped all three of the active-level conversion statements together at the beginning of program, even a value isn't assigned to Y_L until *after* a value is assigned to Y, later in the program. Remember that this is OK because the assignment statements in the architecture body are executed concurrently. That is, an assignment to any signal causes all the other statements that use that signal to be reevaluated, regardless of their position in the architecture body.

You could put the "Y_L <= Y" statement at the end of the body if its current position bothers you, but the program is a bit more maintainable in its present form, with all the active-level conversions together.

Table 5-17 Hierarchical definition of 74x138-like decoder with active-level handling.

```
architecture V74x138_c of V74x138 is
  signal G2A, G2B: STD_LOGIC;             -- active-high version of inputs
  signal Y: STD_LOGIC_VECTOR (0 to 7);   -- active-high version of outputs
  component V3to8dec port (G1, G2, G3: in STD_LOGIC;
                           A: in STD_LOGIC_VECTOR (2 downto 0);
                           Y: out STD_LOGIC_VECTOR (0 to 7) ); end component;
begin
  G2A <= not G2A_L;   -- convert inputs
  G2B <= not G2B_L;   -- convert inputs
  Y_L <= not Y;       -- convert outputs
  U1: V3to8dec port map (G1, G2A, G2B, A, Y);
end V74x138_c;
```

Active levels can be handled in an even more structured way. As shown in Table 5-17, the V74x138 architecture can be defined hierarchically, using a fully active-high V3to8dec component that has its own dataflow-style definition in Table 5-18. Once again, no changes are required in the top-level definition of the V74x138 entity. Figure 5-43 shows the relationship between the entities.

Still another approach to decoder design is shown in Table 5-19, which can replace the V3to8dec_a architecture of Table 5-18. Instead of concurrent state-

Table 5-18
Dataflow definition of an active-high 3-to-8 decoder.

```
library IEEE;
use IEEE.std_logic_1164.all;

entity V3to8dec is
    port (G1, G2, G3: in STD_LOGIC;
            A: in STD_LOGIC_VECTOR (2 downto 0);
            Y: out STD_LOGIC_VECTOR (0 to 7) );
end V3to8dec;

architecture V3to8dec_a of V3to8dec is
  signal Y_s: STD_LOGIC_VECTOR (0 to 7);
begin
    with A select Y_s <=
      "10000000" when "000",
      "01000000" when "001",
      "00100000" when "010",
      "00010000" when "011",
      "00001000" when "100",
      "00000100" when "101",
      "00000010" when "110",
      "00000001" when "111",
      "00000000" when others;
    Y <= Y_s when (G1 and G2 and G3)='1'
                else "00000000";
end V3to8dec_a;
```

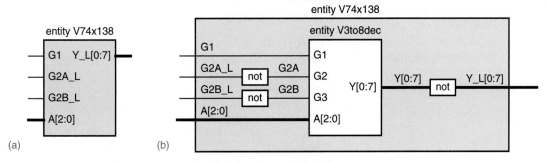

Figure 5-43 VHDL entity V74x138: (a) top level; (b) internal structure using architecture V74x138_c.

NAME MATCHING   In Figure 5-43, the port names of an entity are drawn inside the corresponding box. The names of the signals that are connected to the ports when the entity is used are drawn on the signal lines. Notice that the signal names may match, but they don't have to. The VHDL compiler keeps everything straight, associating a scope with each name. The situation is completely analogous to the way variable and parameter names are handled in structured, procedural programming languages like C.

ments, this architecture uses a process and sequential statements to define the decoder's operation in a behavioral style. However, a close comparison of the two architectures shows that they're really not that different except for syntax.

```
architecture V3to8dec_b of V3to8dec is
  signal Y_s: STD_LOGIC_VECTOR (0 to 7);
begin
process(A, G1, G2, G3, Y_s)
  begin
    case A is
      when "000" => Y_s <= "10000000";
      when "001" => Y_s <= "01000000";
      when "010" => Y_s <= "00100000";
      when "011" => Y_s <= "00010000";
      when "100" => Y_s <= "00001000";
      when "101" => Y_s <= "00000100";
      when "110" => Y_s <= "00000010";
      when "111" => Y_s <= "00000001";
      when others => Y_s <= "00000000";
    end case;
    if (G1 and G2 and G3)='1' then Y <= Y_s;
    else Y <= "00000000";
    end if;
  end process;
end V3to8dec_b;
```

Table 5-19
Behavioral-style architecture definition for a 3-to-8 decoder.

Table 5-20
Truly behavioral architecture definition for a 3-to-8 decoder.

```
architecture V3to8dec_c of V3to8dec is
begin
process (G1, G2, G3, A)
   variable i: INTEGER range 0 to 7;
   begin
     Y <= "00000000";
     if (G1 and G2 and G3) = '1' then
       for i in 0 to 7 loop
         if i=CONV_INTEGER(A) then Y(i) <= '1'; end if;
       end loop;
     end if;
   end process;
end V3to8dec_c;
```

As a final example, a more truly behavioral, process-based architecture for the 3-to-8 decoder is shown in Table 5-20. (Recall that the CONV_INTEGER function was defined in Section 4.7.4.) Of the examples we've given, this is the only one that describes the decoder function without essentially embedding a truth table in the VHDL program. In that respect, it is more flexible because it can be easily adapted to make a binary decoder of any size. In another respect, it is less flexible in that it does not have a truth table that can be easily modified to make custom decoders like the one we specified in Table 5-11 on page 365.

*5.4.8 Seven-Segment Decoders

seven-segment display

Look at your wrist and you'll probably see a *seven-segment display*. This type of display, which normally uses light-emitting diodes (LEDs) or liquid-crystal display (LCD) elements, is used in watches, calculators, and instruments to display decimal data. A digit is displayed by illuminating a subset of the seven line segments shown in Figure 5-44(a).

seven-segment decoder

A *seven-segment decoder* has 4-bit BCD as its input code and the "seven-segment code," which is graphically depicted in Figure 5-44(b), as its output code. Figure 5-45 and Table 5-21 are the logic diagram and truth table for a

74x49

74x49 seven-segment decoder. Except for the strange (clever?) connection of the "blanking input" BI_L, each output of the 74x49 is a minimal product-of-sums

Figure 5-44 Seven-segment display: (a) segment identification; (b) decimal digits.

(a)   (b)

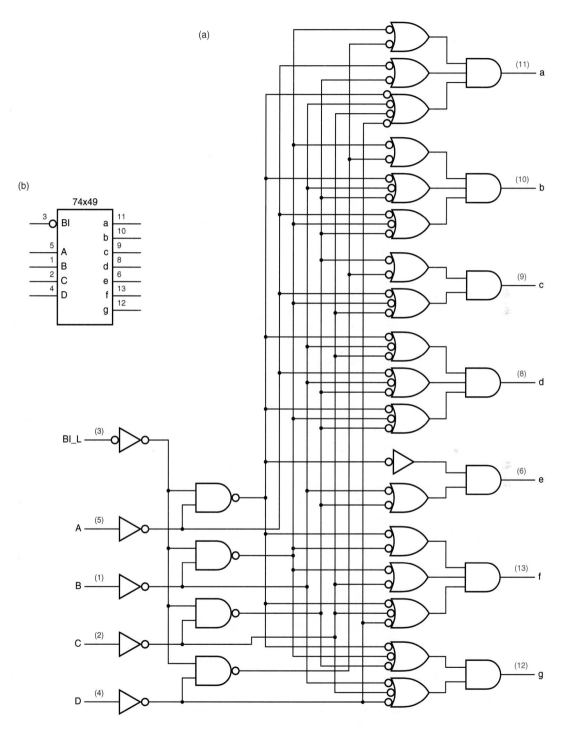

Figure 5-45 The 74x49 seven-segment decoder: (a) logic diagram, including pin numbers; (b) traditional logic symbol.

Table 5-21 Truth table for a 74x49 seven-segment decoder.

Inputs					Outputs						
BI_L	D	C	B	A	a	b	c	d	e	f	g
0	x	x	x	x	0	0	0	0	0	0	0
1	0	0	0	0	1	1	1	1	1	1	0
1	0	0	0	1	0	1	1	0	0	0	0
1	0	0	1	0	1	1	0	1	1	0	1
1	0	0	1	1	1	1	1	1	0	0	1
1	0	1	0	0	0	1	1	0	0	1	1
1	0	1	0	1	1	0	1	1	0	1	1
1	0	1	1	0	0	0	1	1	1	1	1
1	0	1	1	1	1	1	1	0	0	0	0
1	1	0	0	0	1	1	1	1	1	1	1
1	1	0	0	1	1	1	1	0	0	1	1
1	1	0	1	0	0	0	0	1	1	0	1
1	1	0	1	1	0	0	1	1	0	0	1
1	1	1	0	0	0	1	0	0	0	1	1
1	1	1	0	1	1	0	0	1	0	1	1
1	1	1	1	0	0	0	0	1	1	1	1
1	1	1	1	1	0	0	0	0	0	0	0

realization for the corresponding segment, assuming "don't-cares" for the non-decimal input combinations. The INVERT-OR-AND structure used for each output may seem a little strange, but it is equivalent under the generalized DeMorgan's theorem to an AND-OR-INVERT gate, which is a fairly fast and compact structure to build in CMOS or TTL.

Most modern seven-segment display elements have decoders built into them, so that a 4-bit BCD word can be applied directly to the device. Many of the older, discrete seven-segment decoders have special high-voltage or high-current outputs that are well suited for driving large, high-powered display elements.

Table 5-22 is an ABEL program for a seven-segment decoder. Sets are used to define the digit patterns to make the program more readable.

Table 5-22 ABEL program for a 74x49-like seven-segment decoder.

```
module Z74X49H
title 'Seven-Segment_Decoder
J. Wakerly, Micro Design Resources, Inc.'
Z74X49H device 'P16L8';

" Input pins
A, B, C, D                  pin 1, 2, 3, 4;
BI_L                        pin 5;
" Output pins
SEGA, SEGB, SEGC, SEGD      pin 19, 18, 17, 16 istype 'com';
SEGE, SEGF, SEGG            pin 15, 14, 13    istype 'com';

" Definitions
BI = !BI_L
DIGITIN = [D,C,B,A];
SEGOUT = [SEGA,SEGB,SEGC,SEGD,SEGE,SEGF,SEGG];

" Segment encodings for digits
DIG0 = [1,1,1,1,1,1,0];   " 0
DIG1 = [0,1,1,0,0,0,0];   " 1
DIG2 = [1,1,0,1,1,0,1];   " 2
DIG3 = [1,1,1,1,0,0,1];   " 3
DIG4 = [0,1,1,0,0,1,1];   " 4
DIG5 = [1,0,1,1,0,1,1];   " 5
DIG6 = [1,0,1,1,1,1,1];   " 6  'tail' included
DIG7 = [1,1,1,0,0,0,0];   " 7
DIG8 = [1,1,1,1,1,1,1];   " 8
DIG9 = [1,1,1,1,0,1,1];   " 9  'tail' included
DIGA = [1,1,1,0,1,1,1];   " A
DIGB = [0,0,1,1,1,1,1];   " b
DIGC = [1,0,0,1,1,1,0];   " C
DIGD = [0,1,1,1,1,0,1];   " d
DIGE = [1,0,0,1,1,1,1];   " E
DIGF = [1,0,0,0,1,1,1];   " F

equations

SEGOUT = !BI & ( (DIGITIN ==  0) & DIG0 # (DIGITIN ==  1) & DIG1
              # (DIGITIN ==  2) & DIG2 # (DIGITIN ==  3) & DIG3
              # (DIGITIN ==  4) & DIG4 # (DIGITIN ==  5) & DIG5
              # (DIGITIN ==  6) & DIG6 # (DIGITIN ==  7) & DIG7
              # (DIGITIN ==  8) & DIG8 # (DIGITIN ==  9) & DIG9
              # (DIGITIN == 10) & DIGA # (DIGITIN == 11) & DIGB
              # (DIGITIN == 12) & DIGC # (DIGITIN == 13) & DIGD
              # (DIGITIN == 14) & DIGE # (DIGITIN == 15) & DIGF );

end Z74X49H
```

5.5 Encoders

encoder

A decoder's output code normally has more bits than its input code. If the device's output code has *fewer* bits than the input code, the device is usually called an *encoder*. For example, consider a device with eight input bits representing an unsigned binary number, and two output bits indicating whether the number is prime or divisible by 7. We might call such a device a lucky/prime encoder.

2^n-to-n encoder
binary encoder

Probably the simplest encoder to build is a 2^n-to-n or *binary encoder*. As shown in Figure 5-46(a), it has just the opposite function as a binary *decoder*— its input code is the 1-out-of-2^n code and its output code is n-bit binary. The equations for an 8-to-3 encoder with inputs I0–I7 and outputs Y0–Y2 are given below:

$$Y0 = I1 + I3 + I5 + I7$$
$$Y1 = I2 + I3 + I6 + I7$$
$$Y2 = I4 + I5 + I6 + I7$$

The corresponding logic circuit is shown in (b). In general, a 2^n-to-n encoder can be built from n 2^{n-1}-input OR gates. Bit i of the input code is connected to OR gate j if bit j in the binary representation of i is 1.

Figure 5-46
Binary encoder:
(a) general structure;
(b) 8-to-3 encoder.

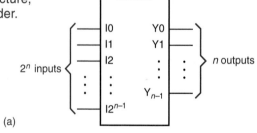

(a)

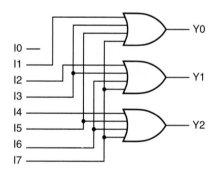

(b)

5.5.1 Priority Encoders

The 1-out-of-2^n coded outputs of an n-bit binary decoder are generally used to control a set of 2^n devices, where at most one device is supposed to be active at any time. Conversely, consider a system with 2^n *inputs*, each of which indicates a request for service, as in Figure 5-47. This structure is often found in microprocessor input/output subsystems, where the inputs might be interrupt requests.

In this situation, it may seem natural to use a binary encoder of the type shown in Figure 5-46 to observe the inputs and indicate which one is requesting service at any time. However, this encoder works properly only if the inputs are guaranteed to be asserted at most one at a time. If multiple requests can be made

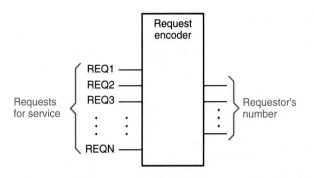

Figure 5-47
A system with 2^n requestors, and a "request encoder" that indicates which request signal is asserted at any time.

simultaneously, the encoder gives undesirable results. For example, suppose that inputs I2 and I4 of the 8-to-3 encoder are both 1; then the output is 110, the binary encoding of 6.

Either 2 or 4, not 6, would be a useful output in the preceding example, but how can the encoding device decide which? The solution is to assign *priority* to the input lines, so that when multiple requests are asserted, the encoding device produces the number of the highest-priority requestor. Such a device is called a *priority encoder*.

priority

priority encoder

The logic symbol for an 8-input priority encoder is shown in Figure 5-48. Input I7 has the highest priority. Outputs A2–A0 contain the number of the highest-priority asserted input, if any. The IDLE output is asserted if no inputs are asserted.

In order to write logic equations for the priority encoder's outputs, we first define eight intermediate variables H0–H7, such that Hi is 1 if and only if Ii is the highest priority 1 input:

$$H7 = I7$$
$$H6 = I6 \cdot I7'$$
$$H5 = I5 \cdot I6' \cdot I7'$$
$$\cdots$$
$$H0 = I0 \cdot I1' \cdot I2' \cdot I3' \cdot I4' \cdot I5' \cdot I6' \cdot I7'$$

Using these signals, the equations for the A2–A0 outputs are similar to the ones for a simple binary encoder:

$$A2 = H4 + H5 + H6 + H7$$
$$A1 = H2 + H3 + H6 + H7$$
$$A0 = H1 + H3 + H5 + H7$$

The IDLE output is 1 if no inputs are 1:

$$IDLE = (I0 + I1 + I2 + I3 + I4 + I5 + I6 + I7)'$$
$$= I0' \cdot I1' \cdot I2' \cdot I3' \cdot I4' \cdot I5' \cdot I6' \cdot I7'$$

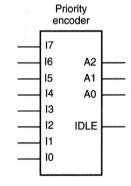

Figure 5-48
Logic symbol for a generic 8-input priority encoder.

5.5.2 The 74x148 Priority Encoder

The *74x148* is a commercially available, MSI 8-input priority encoder. Its logic symbol is shown in Figure 5-49 and its schematic is shown in Figure 5-50. The main difference between this IC and the "generic" priority encoder of Figure 5-48 is that its inputs and outputs are active low. Also, it has an enable input, EI_L, that must be asserted for any of its outputs to be asserted. The complete truth table is given in Table 5-23.

Instead of an IDLE output, the '148 has a GS_L output that is asserted when the device is enabled and one or more of the request inputs are asserted. The manufacturer calls this "Group Select," but it's easier to remember as "Got Something." The EO_L signal is an enable *output* designed to be connected to the EI_L input of another '148 that handles lower-priority requests. EO_L is asserted if EI_L is asserted but no request input is asserted; thus, a lower-priority '148 may be enabled.

Figure 5-51 shows how four 74x148s can be connected in this way to accept 32 request inputs and produce a 5-bit output, RA4–RA0, indicating the highest-priority requestor. Since the A2–A0 outputs of at most one '148 will be enabled at any time, the outputs of the individual '148s can be ORed to produce RA2–RA0. Likewise, the individual GS_L outputs can be combined in a 4-to-2 encoder to produce RA4 and RA3. The RGS output is asserted if any GS output is asserted.

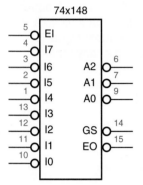

Figure 5-49
Logic symbol for the 74x148 8-input priority encoder.

Table 5-23 Truth table for a 74x148 8-input priority encoder.

	Inputs									Outputs				
EI_L	I0_L	I1_L	I2_L	I3_L	I4_L	I5_L	I6_L	I7_L		A2_L	A1_L	A0_L	GS_L	EO_L
1	x	x	x	x	x	x	x	x		1	1	1	1	1
0	x	x	x	x	x	x	x	0		0	0	0	0	1
0	x	x	x	x	x	x	0	1		0	0	1	0	1
0	x	x	x	x	x	0	1	1		0	1	0	0	1
0	x	x	x	x	0	1	1	1		0	1	1	0	1
0	x	x	x	0	1	1	1	1		1	0	0	0	1
0	x	x	0	1	1	1	1	1		1	0	1	0	1
0	x	0	1	1	1	1	1	1		1	1	0	0	1
0	0	1	1	1	1	1	1	1		1	1	1	0	1
0	1	1	1	1	1	1	1	1		1	1	1	1	0

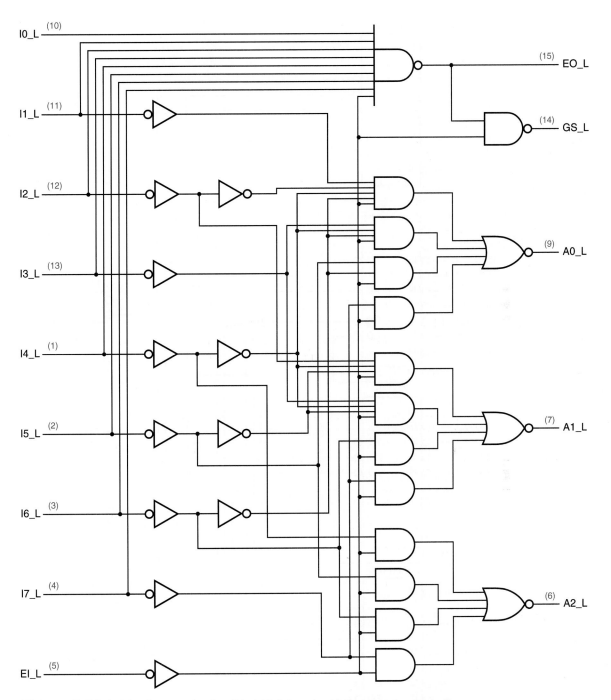

Figure 5-50 Logic diagram for the 74x148 8-input priority encoder, including
pin numbers for a standard 16-pin dual in-line package.

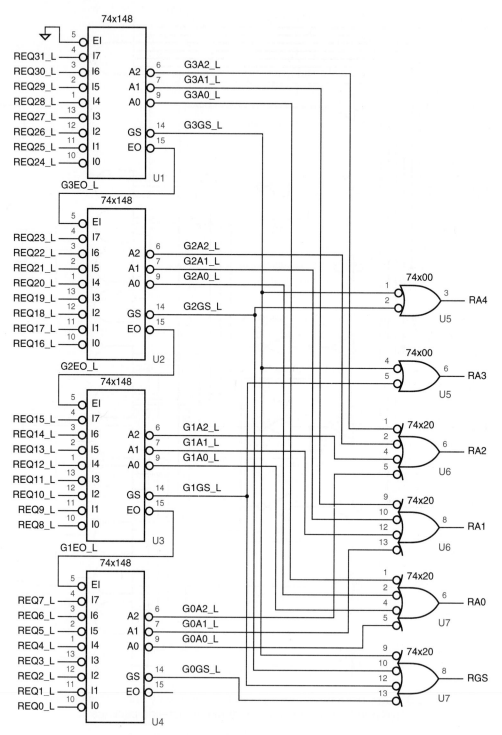

Figure 5-51 Four 74x148s cascaded to handle 32 requests.

5.5.3 Encoders in ABEL and PLDs

Encoders can be designed in ABEL using an explicit equation for each input combination, as in Table 5-8 on page 361, or using truth tables. However, since the number of inputs is usually large, the number of input combinations is very large, and this method often is not practical.

For example, how would we specify a 15-input priority encoder for inputs P0–P14? We obviously don't want to deal with all 2^{15} possible input combinations! One way to do it is to decompose the priority function into two parts. First, we write equations for 15 variables Hi ($0 \leq i \leq 14$) such that Hi is 1 if Pi is the highest-priority asserted input. Since by definition at most one Hi variable is 1 at any time, we can combine the Hi's in a binary encoder to obtain a 4-bit number identifying the highest-priority asserted input.

An ABEL program using this approach is shown in Table 5-24, and a logic diagram for the encoder using a single PAL20L8 or GAL20V8 is given in Figure 5-52. Inputs P0–P14 are asserted to indicate requests, with P14 having the highest priority. If EN_L (Enable) is asserted, then the Y3_L–Y0_L outputs give the number (active low) of the highest-priority request, and GS is asserted if any request is present. If EN_L is negated, then the Y3_L–Y0_L outputs are negated and GS is negated. ENOUT_L is asserted if EN_L is asserted and no request is present.

Notice that in the ABEL program, the equations for the Hi variables are written as "constant expressions," before the equations declaration. Thus, these signals will not be generated explicitly. Rather, they will be incorporated in the subsequent equations for Y0–Y3, which the compiler cranks on to obtain

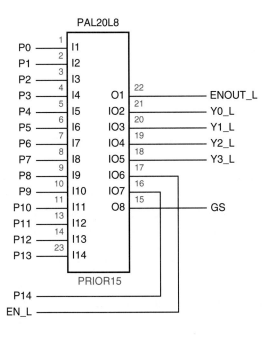

Figure 5-52
Logic diagram for a PLD-based 15-input priority encoder

Table 5-24 An ABEL program for a 15-input priority encoder.

```
module PRIOR15
title '15-Input Priority Encoder '
PRIOR15 device 'P20L8';

" Input and output pins
P0..P14, EN_L                          pin 1..11, 13, 14, 23, 16, 17;
Y3_L, Y2_L, Y1_L, Y0_L, GS, ENOUT_L    pin 18..21, 15, 22 istype 'com';

" Active-level translation
EN = !EN_L; ENOUT = !ENOUT_L;
Y3 = !Y3_L; Y2 = !Y2_L; Y1 = !Y1_L; Y0 = !Y0_L;

" Constant expressions
H14 = EN&P14;
H13 = EN&!P14&P13;
H12 = EN&!P14&!P13&P12;
H11 = EN&!P14&!P13&!P12&P11;
H10 = EN&!P14&!P13&!P12&!P11&P10;
H9  = EN&!P14&!P13&!P12&!P11&!P10&P9;
H8  = EN&!P14&!P13&!P12&!P11&!P10&!P9&P8;
H7  = EN&!P14&!P13&!P12&!P11&!P10&!P9&!P8&P7;
H6  = EN&!P14&!P13&!P12&!P11&!P10&!P9&!P8&!P7&P6;
H5  = EN&!P14&!P13&!P12&!P11&!P10&!P9&!P8&!P7&!P6&P5;
H4  = EN&!P14&!P13&!P12&!P11&!P10&!P9&!P8&!P7&!P6&!P5&P4;
H3  = EN&!P14&!P13&!P12&!P11&!P10&!P9&!P8&!P7&!P6&!P5&!P4&P3;
H2  = EN&!P14&!P13&!P12&!P11&!P10&!P9&!P8&!P7&!P6&!P5&!P4&!P3&P2;
H1  = EN&!P14&!P13&!P12&!P11&!P10&!P9&!P8&!P7&!P6&!P5&!P4&!P3&!P2&P1;
H0  = EN&!P14&!P13&!P12&!P11&!P10&!P9&!P8&!P7&!P6&!P5&!P4&!P3&!P2&!P1&P0;

equations

Y3 = H8 # H9 # H10 # H11 # H12 # H13 # H14;
Y2 = H4 # H5 # H6 # H7 # H12 # H13 # H14;
Y1 = H2 # H3 # H6 # H7 # H10 # H11 # H14;
Y0 = H1 # H3 # H5 # H7 # H9 # H11 # H13;

GS    = EN&(P14#P13#P12#P11#P10#P9#P8#P7#P6#P5#P4#P3#P2#P1#P0);
ENOUT = EN&!P14&!P13&!P12&!P11&!P10&!P9&!P8&!P7&!P6&!P5&!P4&!P3&!P2&!P1&!P0;

end PRIOR15
```

minimal sum-of-products expressions. As it turns out, each Yi output has only seven product terms, as you can see from the structure of the equations.

The priority encoder can be designed even more intuitively using ABEL's WHEN statement. As shown in Table 5-25, a deeply nested series of WHEN statements expresses precisely the logical function of the priority encoder. This program yields exactly the same set of output equations as the previous program.

Table 5-25 Alternate ABEL program for the same 15-input priority encoder.

```
module PRIOR15W
title '15-Input Priority Encoder'
PRIOR15W device 'P20L8';

" Input and output pins
P0..P14, EN_L                          pin 1..11, 13, 14, 23, 16, 17;
Y3_L, Y2_L, Y1_L, Y0_L, GS, ENOUT_L    pin 18..21, 15, 22 istype 'com';

" Active-level translation
EN = !EN_L; ENOUT = !ENOUT_L;

" Set definition
Y_L = [Y3_L, Y2_L, Y1_L, Y0_L];

equations

WHEN !EN THEN !Y_L = 0;              " Note: !Y_L === active-high Y
ELSE WHEN P14 THEN !Y_L = 14;       " Can't define active-high Y set
  ELSE WHEN P13 THEN !Y_L = 13;     "      due to ABEL set quirks
    ELSE WHEN P12 THEN !Y_L = 12;
      ELSE WHEN P11 THEN !Y_L = 11;
        ELSE WHEN P10 THEN !Y_L = 10;
          ELSE WHEN P9 THEN !Y_L = 9;
            ELSE WHEN P8 THEN !Y_L = 8;
              ELSE WHEN P7 THEN !Y_L = 7;
                ELSE WHEN P6 THEN !Y_L = 6;
                  ELSE WHEN P5 THEN !Y_L = 5;
                    ELSE WHEN P4 THEN !Y_L = 4;
                      ELSE WHEN P3 THEN !Y_L = 3;
                        ELSE WHEN P2 THEN !Y_L = 2;
                          ELSE WHEN P1 THEN !Y_L = 1;
                            ELSE WHEN P0 THEN !Y_L = 0;
                              ELSE {!Y_L = 0; ENOUT = 1;};

GS = EN&(P14#P13#P12#P11#P10#P9#P8#P7#P6#P5#P4#P3#P2#P1#P0);

end PRIOR15W
```

5.5.4 Encoders in VHDL

The approach to specifying encoders in VHDL is similar to the ABEL approach. We could embed the equivalent of a truth table or explicit equations into the VHDL program, but a behavioral description is far more intuitive. Since VHDL's IF-THEN-ELSE construct best describes prioritization and is available only within a process, we use the process-based behavioral approach.

Table 5-26 is a behavioral VHDL program for a priority encoder whose function is equivalent to the 74x148. It uses a FOR loop to look for an asserted

Table 5-26 Behavioral VHDL program for a 74x148-like 8-input priority encoder.

```
library IEEE;
use IEEE.std_logic_1164.all;

entity V74x148 is
    port (
        EI_L: in STD_LOGIC;
        I_L: in STD_LOGIC_VECTOR (7 downto 0);
        A_L: out STD_LOGIC_VECTOR (2 downto 0);
        EO_L, GS_L: out STD_LOGIC
    );
end V74x148;

architecture V74x148p of V74x148 is
  signal EI: STD_LOGIC;                        -- active-high version of input
  signal I: STD_LOGIC_VECTOR (7 downto 0); -- active-high version of inputs
  signal EO, GS: STD_LOGIC;                    -- active-high version of outputs
  signal A: STD_LOGIC_VECTOR (2 downto 0); -- active-high version of outputs
begin
  process (EI_L, I_L, EI, EO, GS, I, A)
  variable j: INTEGER range 7 downto 0;
  begin
    EI <= not EI_L; -- convert input
    I <= not I_L;   -- convert inputs
    EO <= '1'; GS <= '0'; A <= "000";
    if (EI)='0' then EO <= '0';
    else for j in 7 downto 0 loop
        if I(j)='1' then
          GS <= '1'; EO <= '0'; A <= CONV_STD_LOGIC_VECTOR(j,3);
          exit;
        end if;
      end loop;
    end if;
    EO_L <= not EO; -- convert output
    GS_L <= not GS; -- convert output
    A_L <= not A;   -- convert outputs
  end process;
end V74x148p;
```

input, starting with the highest-priority input. Like some of our previous programs, it performs explicit active-level conversion at the beginning and end. Also recall that the CONV_STD_LOGIC_VECTOR(j,n) function was defined in Section 4.7.4 to convert from an integer j to a STD_LOGIC_VECTOR of a specified length n. This program is easily modified to use a different priority order or a different number of inputs, or to add more functionality such as finding a second-highest-priority input, as explored in Section 6.2.3.

5.6 Three-State Devices

In Sections 3.7.3 and 3.10.5 we described the electrical design of CMOS and TTL devices whose outputs may be in one of three states—0, 1, or Hi-Z. In this section we'll show how to use them.

5.6.1 Three-State Buffers

The most basic three-state device is a *three-state buffer*, often called a *three-state driver*. The logic symbols for four physically different three-state buffers are shown in Figure 5-53. The basic symbol is that of a noninverting buffer [(a), (b)] or an inverter [(c), (d)]. The extra signal at the top of the symbol is a *three-state enable* input, which may be active high [(a), (c)] or active low [(b), (d)]. When the enable input is asserted, the device behaves like an ordinary buffer or inverter. When the enable input is negated, the device output "floats"; that is, it goes to a high-impedance (Hi-Z), disconnected state and functionally behaves as if it weren't even there.

three-state buffer
three-state driver

three-state enable

    Three-state devices allow multiple sources to share a single "party line," as long as only one device "talks" on the line at a time. Figure 5-54 gives an example of how this can be done. Three input bits, SSRC2–SSRC0, select one of eight sources of data that may drive a single line, SDATA. A 3-to-8 decoder, the 74x138, ensures that only one of the eight SEL lines is asserted at a time, enabling only one three-state buffer to drive SDATA. However, if not all of the EN lines are asserted, then none of the three-state buffers is enabled. The logic value on SDATA is undefined in this case.

Figure 5-53 Various three-state buffers: (a) noninverting, active-high enable; (b) noninverting, active-low enable; (c) inverting, active-high enable; (d) inverting, active-low enable.

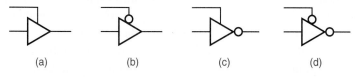

    (a)   (b)   (c)   (d)

DEFINING "UNDEFINED"	The actual voltage level of a floating signal depends on circuit details, such as resistive and capacitive load, and may vary over time. Also, the interpretation of this level by other circuits depends on the input characteristics of those circuits, so it's best not to count on a floating signal as being anything other than "undefined." Sometimes a pull-up resistor is used on three-state party lines to ensure that a floating value is pulled to a HIGH voltage and interpreted as logic 1. This is especially important on party lines that drive CMOS devices, which may consume excessive current when their input voltage is halfway between logic 0 and 1.

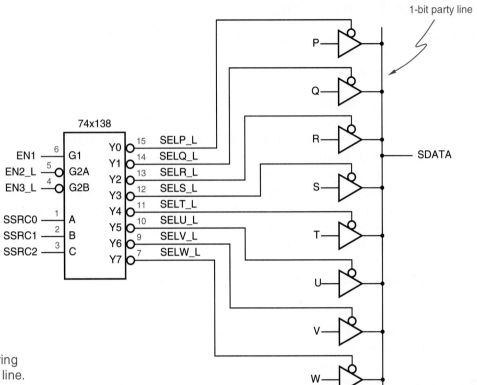

Figure 5-54
Eight sources sharing
a three-state party line.

Typical three-state devices are designed so that they go into the Hi-Z state faster than they come out of the Hi-Z state. (In terms of the specifications in a data book, t_{pLZ} and t_{pHZ} are both less than t_{pZL} and t_{pZH}; also see Section 3.7.3.) This means that if the outputs of two three-state devices are connected to the same party line, and we simultaneously disable one and enable the other, the first device will get off the party line before the second one gets on. This is important because, if both devices were to drive the party line at the same time, and if both were trying to maintain opposite output values (0 and 1), then excessive current would flow and create noise in the system, as discussed in Section 3.7.7. This is often called *fighting*.

fighting

Unfortunately, delays and timing skews in control circuits make it difficult to ensure that the enable inputs of different three-state devices change "simultaneously." Even when this is possible, a problem arises if three-state devices from different-speed logic families (or even different ICs manufactured on different days) are connected to the same party line. The turn-on time (t_{pZL} or t_{pZH}) of a "fast" device may be shorter than the turn-off time (t_{pLZ} or t_{pHZ}) of a "slow" one, and the outputs may still fight.

The only really safe way to use three-state devices is to design control logic that guarantees a *dead time* on the party line during which no one is driving it.

dead time

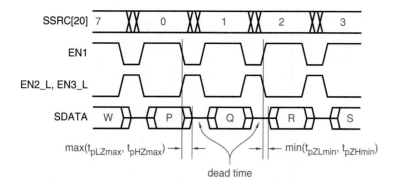

Figure 5-55
Timing diagram for the three-state party line.

The dead time must be long enough to account for the worst-case differences between turn-off and turn-on times of the devices and for skews in the three-state control signals. A timing diagram that illustrates this sort of operation for the party line of Figure 5-54 is shown in Figure 5-55. This timing diagram also illustrates a drawing convention for three-state signals—when in the Hi-Z state, they are shown at an "undefined" level halfway between 0 and 1.

5.6.2 Standard SSI and MSI Three-State Buffers

Like logic gates, several independent three-state buffers may be packaged in a single SSI IC. For example, Figure 5-56 shows the pinouts of *74x125* and *74x126*, each of which contains four independent noninverting three-state buffers in a 14-pin package. The three-state enable inputs in the '125 are active low, and in the '126 they are active high. *74x125* *74x126*

Most party-line applications use a bus with more than one bit of data. For example, in an 8-bit microprocessor system, the data bus is eight bits wide, and peripheral devices normally place data on the bus eight bits at a time. Thus, a peripheral device enables eight three-state drivers to drive the bus, all at the same time. Independent enable inputs, as in the '125 and '126, are not necessary.

Thus, to reduce the package size in wide-bus applications, most commonly used MSI parts contain multiple three-state buffers with common enable inputs. For example, Figure 5-57 shows the logic diagram and symbol for a *74x541*   *74x541* octal noninverting three-state buffer. *Octal* means that the part contains eight

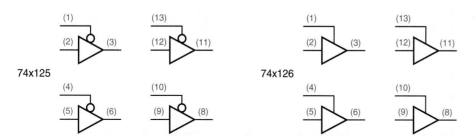

Figure 5-56
Pinouts of the 74x125 and 74x126 three-state buffers.

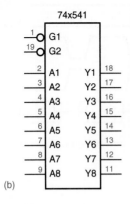

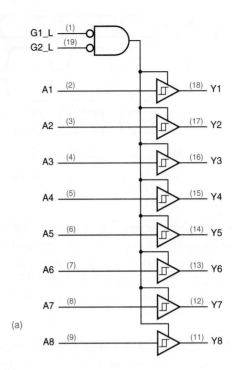

Figure 5-57
The 74x541 octal three-state buffer: (a) logic diagram, including pin numbers for a standard 20-pin dual in-line package; (b) traditional logic symbol.

Figure 5-58
Using a 74x541 as a microprocessor input port.

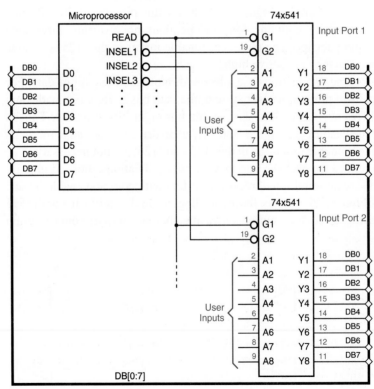

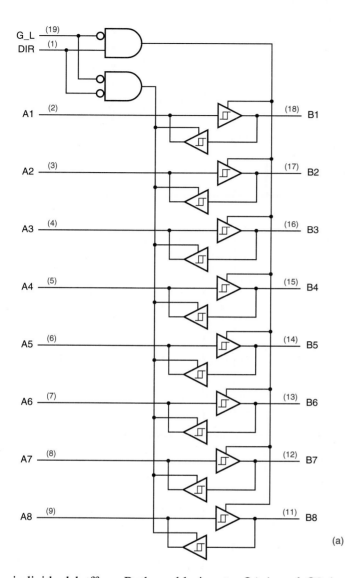

(a)

(b)

Figure 5-59
The 74x245 octal
three-state transceiver:
(a) logic diagram;
(b) traditional logic
symbol.

individual buffers. Both enable inputs, G1_L and G2_L, must be asserted to *octal* enable the device's three-state outputs. The little rectangular symbols inside the buffer symbols indicate *hysteresis*, an electrical characteristic of the inputs that *hysteresis* improves noise immunity, as we explained in Section 3.7.2. The 74x541 inputs typically have 0.4 volts of hysteresis.

Figure 5-58 shows part of a microprocessor system with an 8-bit data bus, DB[0–7], and a 74x541 used as an input port. The microprocessor selects Input Port 1 by asserting INSEL1 and requests a read operation by asserting READ. The selected 74x541 responds by driving the microprocessor data bus with user-supplied input data. Other input ports may be selected when a different INSEL line is asserted along with READ.

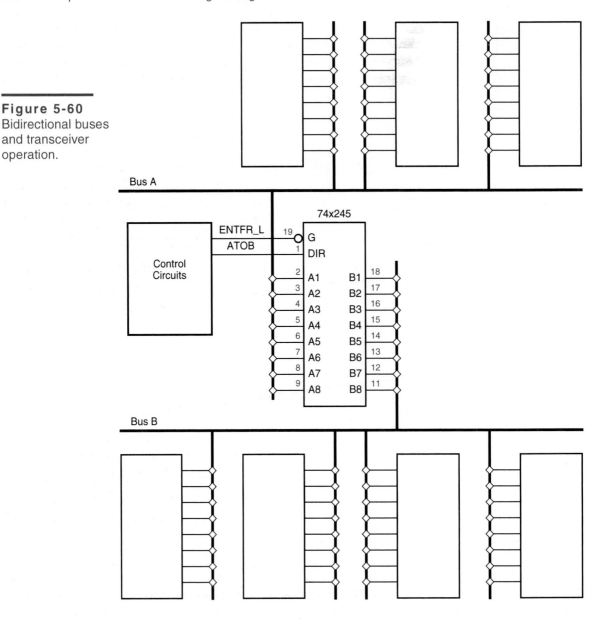

Figure 5-60
Bidirectional buses
and transceiver
operation.

Many other varieties of octal three-state buffers are commercially available. For example, the *74x540* is identical to the 74x541 except that it contains inverting buffers. The *74x240* and *74x241* are similar to the '540 and '541, except that they are split into two 4-bit sections, each with a single enable line.

A *bus transceiver* contains pairs of three-state buffers connected in opposite directions between each pair of pins, so that data can be transferred in either direction. For example, Figure 5-59 on the preceding page shows the logic diagram and symbol for a *74x245* octal three-state transceiver. The DIR input

74x540

74x240

74x241

bus transceiver

74x245

Table 5-27 Modes of operation for a pair of bidirectional buses.

ENTFR_L	ATOB	*Operation*
0	0	Transfer data from a source on bus B to a destination on bus A.
0	1	Transfer data from a source on bus A to a destination on bus B.
1	x	Transfer data on buses A and B independently.

determines the direction of transfer, from A to B (DIR = 1) or from B to A (DIR = 0). The three-state buffer for the selected direction is enabled only if G_L is asserted.

A bus transceiver is typically used between two *bidirectional buses*, as *bidirectional bus* shown in Figure 5-60. Three different modes of operation are possible, depending on the state of G_L and DIR, as shown in Table 5-27. As usual, it is the designer's responsibility to ensure that neither bus is ever driven simultaneously by two devices. However, independent transfers where both buses are driven at the same time may occur when the transceiver is disabled, as indicated in the last row of the table.

5.6.3 Three-State Outputs in ABEL and PLDs

The combinational-PLD applications in previous sections have used the bidirectional I/O pins (IO2–IO7 on a PAL16L8 or GAL16V8) statically, that is, always output-enabled or always output-disabled. In such applications, the compiler can take care of programming the output-enable gates appropriately—all fuses blown, or all fuses intact. By default in ABEL, a three-state output pin is programmed to be always enabled if its signal name appears on the lefthand side of an equation, and always disabled otherwise.

Three-state output pins can also be controlled dynamically, by a single input, by a product term, or, using two-pass logic, by a more complex logic expression. In ABEL, an *attribute suffix* .OE is attached to a signal name on the *.OE attribute suffix* lefthand side of an equation to indicate that the equation applies to the output-enable for the signal. In a PAL16L8 or GAL16V8, the output enable is controlled by a single AND gate, so the righthand side of the enable equation must reduce to a single product term.

Table 5-28 shows a simple PLD program fragment with three-state control. Adapted from the program for a 74x138-like decoder on Table 5-8, this program includes a three-state output control OE for all eight decoder outputs. Notice that a set Y is defined to allow all eight output enables to be specified in a single equation; the .OE suffix is applied to each member of the set.

In the preceding example, the output pins Y0–Y7 are always either enabled or floating, and are used strictly as "output pins." I/O pins (IO2–IO7 in a 16L8 or 16V8) can be used as "bidirectional pins"; that is, they can be used dynamically

■ **Table 5-28** ABEL program for a 74x138-like 3-to-8 binary decoder with three-state output control.

```
module Z74X138T
title '74x138 Decoder with Three-State Output Enable'
Z74X138T device 'P16L8';

" Input pins
A, B, C, !G2A, !G2B, G1, !OE  pin 1, 2, 3, 4, 5, 6, 7;
" Output pins
!Y0, !Y1, !Y2, !Y3            pin 19, 18, 17, 16 istype 'com';
!Y4, !Y5, !Y6, !Y7            pin 15, 14, 13, 12 istype 'com';

" Constant expression
ENB = G1 & G2A & G2B;
Y = [Y0..Y7];

equations
Y.OE = OE;
Y0 = ENB & !C & !B & !A;
...
Y7 = ENB &  C &  B &  A;

end Z74X138T
```

as inputs or outputs depending on whether the output-enable gate is producing a 0 or a 1. An example application of I/O pins is a four-way, 2-bit bus transceiver with the following specifications:

- The transceiver handles four 2-bit bidirectional buses, A[1:2], B[1:2], C[1:2], and D[1:2].

- The source of data to drive the buses is selected by three select inputs, S[2:0], according to Table 5-29. If S2 is 0, the buses are driven with a constant value, otherwise they are driven with one of the other buses. However, when the selected source is a bus, the source bus is driven with 00.

Table 5-29
Bus-selection codes for a four-way bus transceiver.

S2	S1	S0	Source selected
0	0	0	00
0	0	1	01
0	1	0	10
0	1	1	11
1	0	0	A bus
1	0	1	B bus
1	1	0	C bus
1	1	1	D bus

Table 5-30 An ABEL program for four-way, 2-bit bus transceiver.

```
module XCVR4X2
title 'Four-way 2-bit Bus Transceiver'
XCVR4X2 device 'P16L8';

" Input pins
A1I, A2I                        pin 1, 11;
!AOE, !BOE, !COE, !DOE, !MOE    pin 2, 3, 4, 5, 6;
S0, S1, S2                      pin 7, 8, 9;
" Output and bidirectional pins
A1O, A2O                        pin 19, 12                     istype 'com';
B1, B2, C1, C2, D1, D2          pin 18, 17, 16, 15, 14, 13 istype 'com';

" Set definitions
ABUSO = [A1O,A2O];
ABUSI = [A1I,A2I];
BBUS  = [B1,B2];
CBUS  = [C1,C2];
DBUS  = [D1,D2];
SEL   = [S2,S1,S0];
CONST = [S1,S0];
" Constants
SELA = [1,0,0];
SELB = [1,0,1];
SELC = [1,1,0];
SELD = [1,1,1];

equations
ABUSO.OE = AOE & MOE;
BBUS.OE = BOE & MOE;
CBUS.OE = COE & MOE;
DBUS.OE = DOE & MOE;
ABUSO = !S2&CONST # (SEL==SELB)&BBUS # (SEL==SELC)&CBUS # (SEL==SELD)&DBUS;
BBUS = !S2&CONST # (SEL==SELA)&ABUSI # (SEL==SELC)&CBUS # (SEL==SELD)&DBUS;
CBUS = !S2&CONST # (SEL==SELA)&ABUSI # (SEL==SELB)&BBUS # (SEL==SELD)&DBUS;
DBUS = !S2&CONST # (SEL==SELA)&ABUSI # (SEL==SELB)&BBUS # (SEL==SELC)&CBUS;

end XCVR4X2
```

- Each bus has its own output-enable signal, AOE_L, BOE_L, COE_L, or DOE_L. There is also a "master" output-enable signal, MOE_L. The transceiver drives a particular bus if and only if MOE_L and the output-enable signal for that bus are both asserted.

Table 5-30 is an ABEL program that performs the transceiver function. According to the enable (.OE) equations, each bus is output enabled if MOE and its own OE are asserted. Each bus is driven with S1 and S0 if S2 is 0, and with

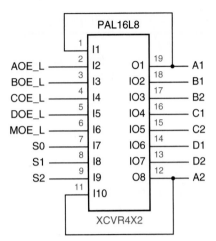

Figure 5-61
PLD inputs and
outputs for a four-way,
2-bit bus transceiver.

the selected bus if a different bus is selected. If the bus itself is selected, the output equation evaluates to 0, and the bus is driven with 00 as required.

Figure 5-61 is a logic diagram for a PAL16L8 (or GAL16V8) with the required inputs and outputs. Since the device has only six bidirectional pins and the specification requires eight, the A bus uses one pair of pins for input and another for output. This is reflected in the program by the use of separate signals and sets for the A-bus input and output.

Table 5-31 IEEE 1164 package declarations for STD_ULOGIC and STD_LOGIC.

```
PACKAGE std_logic_1164 IS
-- logic state system (unresolved)
    TYPE std_ulogic IS ( 'U',  -- Uninitialized
                         'X',  -- Forcing  Unknown
                         '0',  -- Forcing  0
                         '1',  -- Forcing  1
                         'Z',  -- High Impedance
                         'W',  -- Weak     Unknown
                         'L',  -- Weak     0
                         'H',  -- Weak     1
                         '-'   -- Don't care
                       );

-- unconstrained array of std_ulogic
    TYPE std_ulogic_vector IS ARRAY ( NATURAL RANGE <> ) OF std_ulogic;

-- resolution function
    FUNCTION resolved ( s : std_ulogic_vector ) RETURN std_ulogic;

-- *** industry standard logic type ***
    SUBTYPE std_logic IS resolved std_ulogic;
...
```

Table 5-32 IEEE 1164 package body for STD_ULOGIC and STD_LOGIC.

```
PACKAGE BODY std_logic_1164 IS
-- local type
    TYPE stdlogic_table IS ARRAY(std_ulogic, std_ulogic) OF std_ulogic;

-- resolution function
    CONSTANT resolution_table : stdlogic_table := (
--      -------------------------------------------------------------
--      |  U    X    0    1    Z    W    L    H    -        |  |
--      -------------------------------------------------------------
        ( 'U', 'U', 'U', 'U', 'U', 'U', 'U', 'U', 'U' ), -- | U |
        ( 'U', 'X', 'X', 'X', 'X', 'X', 'X', 'X', 'X' ), -- | X |
        ( 'U', 'X', '0', 'X', '0', '0', '0', '0', 'X' ), -- | 0 |
        ( 'U', 'X', 'X', '1', '1', '1', '1', '1', 'X' ), -- | 1 |
        ( 'U', 'X', '0', '1', 'Z', 'W', 'L', 'H', 'X' ), -- | Z |
        ( 'U', 'X', '0', '1', 'W', 'W', 'W', 'W', 'X' ), -- | W |
        ( 'U', 'X', '0', '1', 'L', 'W', 'L', 'W', 'X' ), -- | L |
        ( 'U', 'X', '0', '1', 'H', 'W', 'W', 'H', 'X' ), -- | H |
        ( 'U', 'X', 'X', 'X', 'X', 'X', 'X', 'X', 'X' )  -- | - |
    );

    FUNCTION resolved ( s : std_ulogic_vector ) RETURN std_ulogic IS
        VARIABLE result : std_ulogic := 'Z';  -- weakest state default
    BEGIN
        -- the test for a single driver is essential otherwise the
        -- loop would return 'X' for a single driver of '-' and that
        -- would conflict with the value of a single driver unresolved
        -- signal.
        IF    (s'LENGTH = 1) THEN    RETURN s(s'LOW);
        ELSE
            FOR i IN s'RANGE LOOP
                result := resolution_table(result, s(i));
            END LOOP;
        END IF;
        RETURN result;
    END resolved;
...
```

*5.6.4 Three-State Outputs in VHDL

VHDL itself does not have built-in types and operators for three-state outputs. However, it does have primitives which can be used to create signals and systems with three-state behavior; the IEEE 1164 package uses these primitives. As a start, the IEEE 1164 STD_LOGIC type defines 'Z' as one of its nine possible signal values; this value is used for the high-impedance state. You can assign this value to any STD_LOGIC signal, and the definitions of the standard logic functions account for the possibility of 'Z' inputs (generally a 'Z' input will cause a 'U' output).

Given the availability of three-state signals, how do we create three-state buses in VHDL? A three-state bus generally has two or more drivers, although the mechanisms we discuss work fine with just one driver. In VHDL, there is no explicit language construct for joining three-state outputs into a bus. Instead, the compiler automatically joins together signals that are driven in two or more different processes, that is, signals that appear on the lefthand side of a signal assignment statement in two or more processes. However, the signals must have the appropriate type, as explained below.

subtype
STD_ULOGIC
unresolved type
resolution function

The IEEE 1164 STD_LOGIC type is actually defined as a *subtype* of an unresolved type, *STD_ULOGIC*. In VHDL, an *unresolved type* is used for any signal that may be driven in two or more processes. The definition of an unresolved type includes a *resolution function* that is called every time an assignment is made to a signal having that type. As the name implies, this function resolves the value of the signal when it has multiple drivers.

Tables 5-31 and 5-32 on the preceding pages show the IEEE 1164 definitions of STD_ULOGIC, STD_LOGIC and the resolution function "resolved". This code uses a two-dimensional array resolution_table to determine the final STD_LOGIC value produced by n processes that drive a signal to n possibly different values passed in the input vector s. If, for example, a signal has four drivers, the VHDL compiler automatically constructs a 4-element vector containing the four driven values, and passes this vector to resolved every time that any one of those values changes. The result is passed back to the simulation.

Notice that the order in which the driven signal values appear in s does not affect the result produced by resolved, due to the strong ordering of "strengths" in the resolution_table: 'U'>'X'>'0','1'>'W'>'L','H'>'-'. That is, once a signal is partially resolved to a particular value, it never further resolves to a "weaker" value; and 0/1 and L/H conflicts always resolve to a stronger undefined value ('X' or 'W').

So, do you need to know all of this in order to use three-state outputs in VHDL? Well, usually not, but it can help if your simulations don't match up with reality. All that's normally required to use three-state outputs within VHDL is to declare the corresponding signals as type STD_ULOGIC.

For example, Table 5-33 describes a system that uses four 8-bit three-state drivers (in four processes) to select one of four 8-bit buses, A, B, C, and D, to drive onto a result bus X. Within each process, the IEEE 1164 standard function To_StdULogicVector is used to convert the input type of STD_LOGIC_VECTOR to STD_ULOGIC_VECTOR as required to make a legal assignment to result bus X. Recall that the construct "(others => 'Z')" denotes a vector of all Z's, with whatever length is needed to match the lefthand side of the assignment.

VHDL is flexible enough that you can use it to define other types of bus operation. For example, you could define a subtype and resolution function for open-drain outputs such that a wired-AND function is obtained. However, you'll

Table 5-33 VHDL program with four 8-bit three-state drivers.

```vhdl
library IEEE;
use IEEE.std_logic_1164.all;

entity V3statex is
    port (
        G_L: in STD_LOGIC;                          -- Global output enable
        SEL: in STD_LOGIC_VECTOR (1 downto 0);      -- Input select 0,1,2,3 ==> A,B,C,D
        A, B, C, D: in STD_LOGIC_VECTOR (1 to 8);   -- Input buses
        X: out STD_ULOGIC_VECTOR (1 to 8)           -- Output bus (three-state)
    );
end V3statex;

architecture V3states of V3statex is
begin
  process (G_L, SEL, A)
  begin
    if G_L='0' and SEL = "00" then X <= To_StdULogicVector(A);
    else X <= (others => 'Z');
    end if;
  end process;

  process (G_L, SEL, B)
  begin
    if G_L='0' and SEL = "01" then X <= To_StdULogicVector(B);
    else X <= (others => 'Z');
    end if;
  end process;

  process (G_L, SEL, C)
  begin
    if G_L='0' and SEL = "10" then X <= To_StdULogicVector(C);
    else X <= (others => 'Z');
    end if;
  end process;

  process (G_L, SEL, D)
  begin
    if G_L='0' and SEL = "11" then X <= To_StdULogicVector(D);
    else X <= (others => 'Z');
    end if;
  end process;

end V3states;
```

seldom have to use this feature; the definitions for specific output types in PLDs, FPGAs, and ASICs are usually already done for you in libraries provided by the component vendors.

5.7 Multiplexers

multiplexer

A *multiplexer* is a digital switch—it connects data from one of n sources to its output. Figure 5-62(a) shows the inputs and outputs of an n-input, b-bit multiplexer. There are n sources of data, each of which is b bits wide, and there are b output bits. In typical commercially available multiplexers, $n = 1, 2, 4, 8$, or 16, and $b = 1, 2$, or 4. There are s inputs that select among the n sources, so $s = \lceil \log_2 n \rceil$. An enable input EN allows the multiplexer to "do its thing"; when EN = 0, all of the outputs are 0. A multiplexer is often called a *mux* for short.

mux

Figure 5-62(b) shows a switch circuit that is roughly equivalent to the multiplexer. However, unlike a mechanical switch, a multiplexer is a unidirectional device: information flows only from inputs (on the left) to outputs (on the right).

We can write a general logic equation for a multiplexer output:

$$iY = \sum_{j=0}^{n-1} EN \cdot M_j \cdot iDj$$

Here, the summation symbol represents a logical sum of product terms. Variable iY is a particular output bit ($1 \leq i \leq b$), and variable iDj is input bit i of source j ($0 \leq j \leq n-1$). M_j represents minterm j of the s select inputs. Thus, when the multiplexer is enabled and the value on the select inputs is j, each output iY equals the corresponding bit of the selected input, iDj.

Multiplexers are obviously useful devices in any application in which data must be switched from multiple sources to a destination. A common application in computers is the multiplexer between the processor's registers and its arithmetic logic unit (ALU). For example, consider a 16-bit processor in which

Figure 5-62
Multiplexer structure:
(a) inputs and outputs;
(b) functional equivalent.

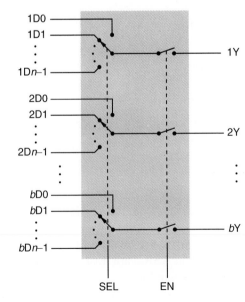

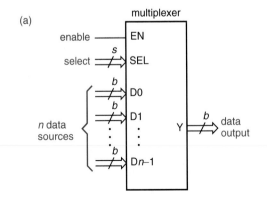

each instruction has a 3-bit field that specifies one of eight registers to use. This 3-bit field is connected to the select inputs of an 8-input, 16-bit multiplexer. The multiplexer's data inputs are connected to the eight registers, and its data outputs are connected to the ALU to execute the instruction using the selected register.

5.7.1 Standard MSI Multiplexers

The sizes of commercially available MSI multiplexers are limited by the number of pins available in an inexpensive IC package. Commonly used muxes come in 16-pin packages. At one extreme is the *74x151*, shown in Figure 5-63, which selects among eight 1-bit inputs. The select inputs are named C, B, and A, where C is most significant numerically. The enable input EN_L is active low; both active-high (Y) and active-low (Y_L) versions of the output are provided.

74x151

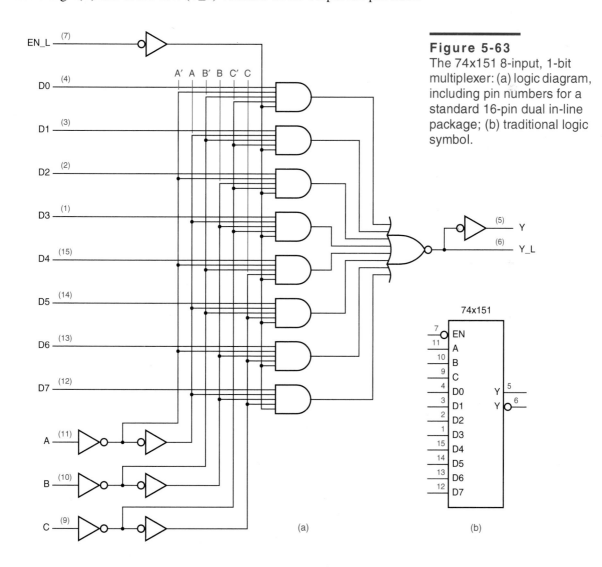

Figure 5-63
The 74x151 8-input, 1-bit multiplexer: (a) logic diagram, including pin numbers for a standard 16-pin dual in-line package; (b) traditional logic symbol.

Table 5-34
Truth table for a
74x151 8-input,
1-bit multiplexer.

Inputs				Outputs	
EN_L	C	B	A	Y	Y_L
1	x	x	x	0	1
0	0	0	0	D0	D0′
0	0	0	1	D1	D1′
0	0	1	0	D2	D2′
0	0	1	1	D3	D3′
0	1	0	0	D4	D4′
0	1	0	1	D5	D5′
0	1	1	0	D6	D6′
0	1	1	1	D7	D7′

Figure 5-64 The 74x157 2-input, 4-bit multiplexer: (a) logic diagram, including pin numbers for a standard 16-pin dual in-line package; (b) traditional logic symbol.

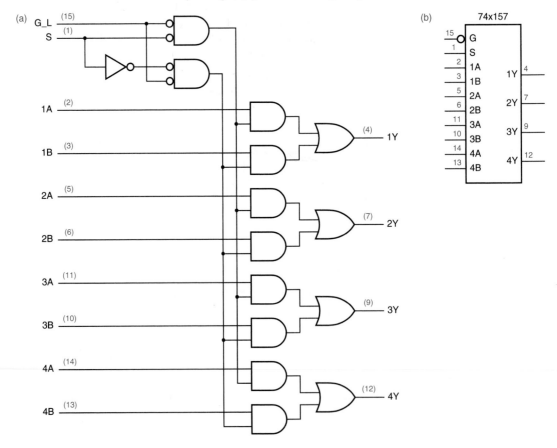

Inputs		Outputs			
G_L	S	1Y	2Y	3Y	4Y
1	x	0	0	0	0
0	0	1A	2A	3A	4A
0	1	1B	2B	3B	4B

Table 5-35
Truth table for a
74x157 2-input,
4-bit multiplexer.

The 74x151's truth table is shown in Table 5-34. Here we have once again extended our notation for truth tables. Up until now, our truth tables have specified an output of 0 or 1 for each input combination. In the 74x151's table, only a few of the inputs are listed under the "Inputs" heading. Each output is specified as 0, 1, or a simple logic function of the remaining inputs (e.g., D0 or D0′). This notation saves eight columns and eight rows in the table and presents the logic function more clearly than a larger table would.

At the other extreme of muxes in 16-pin packages, we have the *74x157*, *74x157* shown in Figure 5-64, which selects between two 4-bit inputs. Just to confuse things, the manufacturer has named the select input S and the active-low enable input G_L. Also note that the data sources are named A and B instead of D0 and D1 as in our generic example. Our extended truth-table notation makes the 74x157's description very compact, as shown in Table 5-35.

Intermediate between the 74x151 and 74x157 is the *74x153*, a 4-input, *74x153* 2-bit multiplexer. This device, whose logic symbol is shown in Figure 5-65, has separate enable inputs (1G, 2G) for each bit. As shown in Table 5-36, its function is very straightforward.

Inputs				Outputs	
1G_L	2G_L	B	A	1Y	2Y
0	0	0	0	1C0	2C0
0	0	0	1	1C1	2C1
0	0	1	0	1C2	2C2
0	0	1	1	1C3	2C3
0	1	0	0	1C0	0
0	1	0	1	1C1	0
0	1	1	0	1C2	0
0	1	1	1	1C3	0
1	0	0	0	0	2C0
1	0	0	1	0	2C1
1	0	1	0	0	2C2
1	0	1	1	0	2C3
1	1	x	x	0	0

Table 5-36
Truth table for a
74x153 4-input, 2-bit
multiplexer.

Figure 5-65
Traditional logic
symbol for the
74x153.

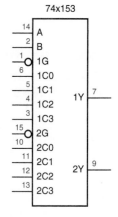

74x251

74x253

74x257

Some multiplexers have three-state outputs. The enable input of such a multiplexer, instead of forcing the outputs to zero, forces them to the Hi-Z state. For example, the *74x251* is identical to the '151 in its pinout and its internal logic design, except that Y and Y_L are three-state outputs. When the EN_L input is negated, instead of forcing the outputs to be negated, it forces the outputs into the Hi-Z state. Similarly, the *74x253* and *74x257* are three-state versions of the '153 and '157. The three-state outputs are especially useful when *n*-input muxes are combined to form larger muxes, as suggested in the next subsection.

5.7.2 Expanding Multiplexers

Seldom does the size of an MSI multiplexer match the characteristics of the problem at hand. For example, we suggested earlier that an 8-input, 16-bit multiplexer might be used in the design of a computer processor. This function could be performed by 16 74x151 8-input, 1-bit multiplexers or equivalent ASIC cells, each handling one bit of all the inputs and the output. The processor's 3-bit register-select field would be connected to the A, B, and C inputs of all 16 muxes, so they would all select the same register source at any given time.

The device that produces the 3-bit register-select field in this example must have enough fanout to drive 16 loads. With 74LS-series ICs this is possible because typical devices have a fanout of 20 LS-TTL loads.

Still, it is fortunate that the '151 was designed so that each of the A, B, and C inputs presents only one LS-TTL load to the circuit driving it. Theoretically, the '151 could have been designed without the first rank of three inverters shown on the select inputs in Figure 5-63, but then each select input would have presented five LS-TTL loads, and the drivers in the register-select application would need a fanout of 80.

Another dimension in which multiplexers can be expanded is the number of data sources. For example, suppose we needed a 32-input, 1-bit multiplexer. Figure 5-66 shows one way to build it. Five select bits are required. A 2-to-4 decoder (one-half of a 74x139) decodes the two high-order select bits to enable one of four 74x151 8-input multiplexers. Since only one '151 is enabled at a time, the '151 outputs can simply be ORed to obtain the final output.

CONTROL-SIGNAL FANOUT IN ASICS   Just the sort of fanout consideration that we described above occurs quite frequently in ASIC design. When a set of control signals, such as the register-select field in the example, controls a large number of bits, the required fanout can be enormous. In CMOS chips, the consideration is not DC loading but capacitive load, which slows down performance. In such an application, the designer must carefully partition the load and select points at which to buffer the control signals to reduce fanout. While inserting the extra buffers, the designer must be careful not increase the chip area significantly or to put so many buffers in series that their delay is unacceptable.

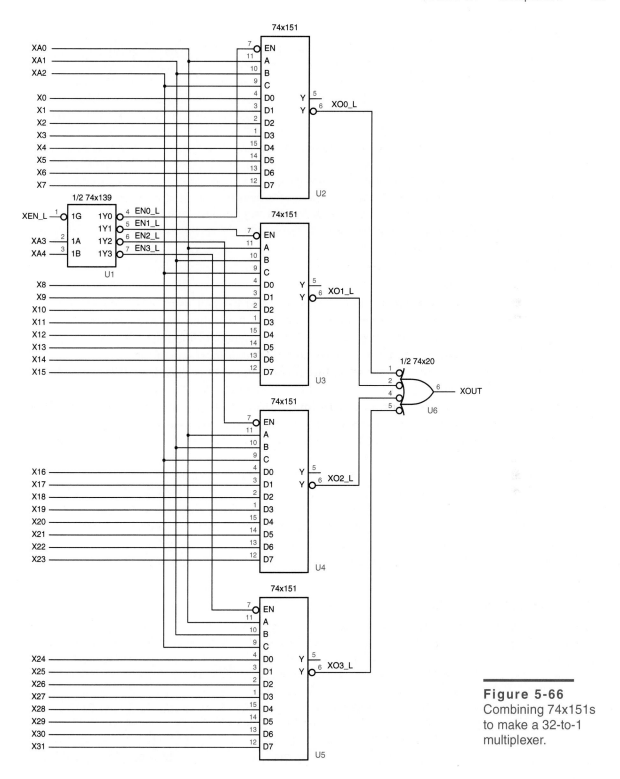

Figure 5-66
Combining 74x151s
to make a 32-to-1
multiplexer.

TURN ON THE BUBBLE MACHINE

The use of bubble-to-bubble logic design should help your understanding of these multiplexer design examples. Since the decoder outputs and the multiplexer enable inputs are all active low, they can be hooked up directly. You can ignore the inversion bubbles when thinking about the logic function that is performed—you just say that when a particular decoder output is asserted, the corresponding multiplexer is enabled.

Bubble-to-bubble design also provides two options for the final OR function in Figure 5-66. The most obvious design would have used a 4-input OR gate connected to the Y outputs. However, for faster operation, we used an inverting gate, a 4-input NAND connected to the /Y outputs. This eliminated the delay of two inverters—the one used inside the '151 to generate Y from /Y, and the extra inverter circuit that is used to obtain an OR function from a basic NOR circuit in a CMOS or TTL OR gate.

The 32-to-1 multiplexer can also be built using 74x251s. The circuit is identical to Figure 5-66, except that the output NAND gate is eliminated. Instead, the Y (and, if desired, Y_L) outputs of the four '251s are simply tied together. The '139 decoder ensures that at most one of the '251s has its three-state outputs enabled at any time. If the '139 is disabled (XEN_L is negated), then all of the '251s are disabled, and the XOUT and XOUT_L outputs are undefined. However, if desired, resistors may be connected from each of these signals to +5 volts to pull the output HIGH in this case.

Figure 5-67
A multiplexer driving a bus and a demultiplexer receiving the bus:
(a) switch equivalent;
(b) block-diagram symbols.

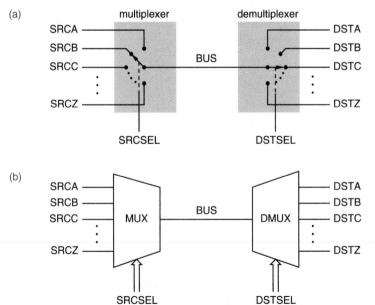

5.7.3 Multiplexers, Demultiplexers, and Buses

A multiplexer can be used to select one of *n* sources of data to transmit on a bus. At the far end of the bus, a *demultiplexer* can be used to route the bus data to one of *m* destinations. Such an application, using a 1-bit bus, is depicted in terms of our switch analogy in Figure 5-67(a). In fact, block diagrams for logic circuits often depict multiplexers and demultiplexers using the wedge-shaped symbols in (b), to suggest visually how a selected one of multiple data sources gets directed onto a bus and routed to a selected one of multiple destinations.

demultiplexer

The function of a demultiplexer is just the inverse of a multiplexer's. For example, a 1-bit, *n*-output demultiplexer has one data input and *s* inputs to select one of $n = 2^s$ data outputs. In normal operation, all outputs except the selected one are 0; the selected output equals the data input. This definition may be generalized for a *b*-bit, *n*-output demultiplexer; such a device has *b* data inputs, and its *s* select inputs choose one of $n = 2^s$ sets of *b* data outputs.

A binary decoder with an enable input can be used as a demultiplexer, as shown in Figure 5-68. The decoder's enable input is connected to the data line, and its select inputs determine which of its output lines is driven with the data bit. The remaining output lines are negated. Thus, the 74x139 can be used as a 2-bit, 4-output demultiplexer with active-low data inputs and outputs, and the 74x138 can be used as a 1-bit, 8-output demultiplexer. In fact, the manu-facturer's catalog typically lists these ICs as "decoders/demultiplexers."

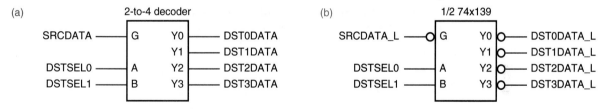

Figure 5-68 Using a 2-to-4 binary decoder as a 1-bit, 4-output demultiplexer: (a) generic decoder; (b) 74x139.

5.7.4 Multiplexers in ABEL and PLDs

Multiplexers are very easy to design using ABEL and combinational PLDs. For example, the function of a 74x153 4-input, 2-bit multiplexer can be duplicated in a PAL16L8 as shown in Figure 5-69 and Table 5-37. Several characteristics of the PLD-based design and program are worth noting:

- Signal names in the ABEL program are changed slightly from the signal names shown for a 74x153 in Figure 5-65 on page 401, since ABEL does not allow a number to be used as the first character of a signal name.

- A 74x153 has twelve inputs, while a PAL16L8 has only ten inputs. There-fore, two of the '153 inputs are assigned to 16L8 I/O pins, which are no longer usable as outputs.

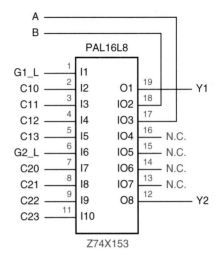

Figure 5-69
Logic diagram for the
PAL16L8 used as a
74x153-like multiplexer.

- The '153 outputs (**1Y** and **2Y**) are assigned to pins 19 and 12 on the 16L8, which are usable *only* as outputs. This is preferred over assigning them to I/O pins; given a choice, it's better to leave I/O pins than output-only pins as spares.

- Although the multiplexer equations in the table are written quite naturally in sum-of-products form, they don't map directly onto the 16L8's structure

Table 5-37
ABEL program for a
74x153-like 4-input,
2-bit multiplexer.

```
module Z74X153
title '74x153-like multiplexer PLD '
Z74X153 device 'P16L8';

" Input and output pins
A, B, G1_L, G2_L          pin 17, 18, 1, 6;
C10, C11, C12, C13        pin 2, 3, 4, 5;
C20, C21, C22, C23        pin 7, 8, 9, 11;
Y1, Y2                    pin 19, 12 istype 'com';

" Active-level conversion
G1 = !G1_L;   G2 = !G2_L;

equations
Y1 = G1 & ( !B & !A & C10
        # !B &  A & C11
        #  B & !A & C12
        #  B &  A & C13);

Y2 = G2 & ( !B & !A & C20
        # !B &  A & C21
        #  B & !A & C22
        #  B &  A & C23);
end Z74X153
```

```
!Y1 = (!B  &  !A  &  !C10
   #  !B  &   A  &  !C11
   #   B  &  !A  &  !C12
   #   B  &   A  &  !C13
   #   G1);
!Y2 = (!B  &  !A  &  !C20
   #  !B  &   A  &  !C21
   #   B  &  !A  &  !C22
   #   B  &   A  &  !C23
   #   G2);
```

Table 5-38
Inverted, reduced equations for 74x153-like 4-input, 2-bit multiplexer.

because of the inverter between the **AND-OR** array and the actual output pins. Therefore, the ABEL compiler must complement the equations in the table and then reduce the result to sum-of-products form. With a GAL16V8, either version of the equations could be used.

Multiplexer functions are even easier to expression using ABEL's sets and relations. For example, Table 5-39 shows the ABEL program for a 4-input, 8-bit multiplexer. No device statement is included, because this function has too many inputs and outputs to fit in any of the PLDs we've described so far. However, it's quite obvious that a multiplexer of any size can be specified in just a few lines of code in this way.

```
module mux4in8b
title '4-input, 8-bit wide multiplexer PLD'

" Input and output pins
!G                 pin;          " Output enable for Y bus
S1..S0             pin;          " Select inputs, 0-3 ==> A-D
A1..A8, B1..B8, C1..C8, D1..D8 pin; " 8-bit input buses A, B, C, D
Y1..Y8             pin istype 'com'; " 8-bit three-state output bus

" Sets
SEL = [S1..S0];
A = [A1..A8];
B = [B1..B8];
C = [C1..C8];
D = [D1..D8];
Y = [Y1..Y8];

equations
Y.OE = G;
WHEN (SEL == 0) THEN Y = A;
ELSE WHEN (SEL == 1) THEN Y = B;
ELSE WHEN (SEL == 2) THEN Y = C;
ELSE WHEN (SEL == 3) THEN Y = D;
end mux4in8b
```

Table 5-39
ABEL program for a 4-input, 8-bit multiplexer.

Table 5-40
Function table for
a specialized 4-input,
18-bit multiplexer.

S2	S1	S0	Input to Select
0	0	0	A
0	0	1	B
0	1	0	A
0	1	1	C
1	0	0	A
1	0	1	D
1	1	0	A
1	1	1	B

Likewise, it is easy to customize multiplexer functions using ABEL. For example, suppose that you needed a circuit that selects one of four 18-bit input buses, A, B, C, or D, to drive an 18-bit output bus F, as specified in Table 5-40 by three control bits. There are more control-bit combinations than multiplexer inputs, so a standard 4-input multiplexer doesn't quite fit the bill (but see Exercise 5.61). A 4-input, 3-bit multiplexer with the required behavior can be designed to fit into a single PAL16L8 or GAL16V8 as shown in Figure 5-70 and Table 5-41, and six copies of this device can be used to make the 18-bit mux. Alternatively, a single, larger PLD could be used. In any case, the ABEL program is very easily modified for different selection criteria.

Since this function uses all of the available pins on the PAL16L8, we had to make the pin assignment in Figure 5-70 carefully. In particular, we had to assign two output signals to the two output-only pins (O1 and O8), in order to maximize the number of input pins available.

Figure 5-70
Logic diagram for the
PAL16L8 used as a
specialized 4-input,
3-bit multiplexer.

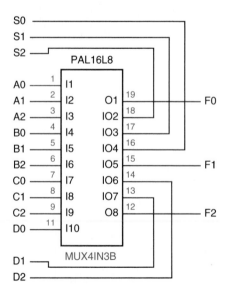

Table 5-41 ABEL program for a specialized 4-input, 3-bit multiplexer.

```
module mux4in3b
title 'Specialized 4-input, 3-bit Multiplexer'
mux4in3b device 'P16L8';

" Input and output pins
S2..S0                       pin 16..18;               " Select inputs
A0..A2, B0..B2, C0..C2, D0..D2  pin 1..9, 11, 13, 14;     " Bus inputs
F0..F2                       pin 19, 15, 12 istype 'com'; " Bus outputs

" Sets
SEL = [S2..S0];
A = [A0..A2];
B = [B0..B2];
C = [C0..C2];
D = [D0..D2];
F = [F0..F2];

equations
WHEN (SEL== 0) # (SEL== 2) # (SEL== 4) # (SEL== 6) THEN F = A;
ELSE WHEN (SEL== 1) # (SEL== 7) THEN F = B;
ELSE WHEN (SEL== 3) THEN F = C;
ELSE WHEN (SEL== 5) THEN F = D;

end mux4in3b
```

EASIEST, BUT NOT CHEAPEST As you've seen, it's very easy to program a PLD to perform decoder and multiplexer functions. Still, if you need the logic function of a standard decoder or multiplexer, it's usually less costly to use a standard MSI chip than it is to use a PLD. The PLD-based approach is best if the multiplexer has some nonstandard functional requirements, or if you think you may have to change its function as a result of debugging.

5.7.5 Multiplexers in VHDL

Multiplexers are very easy to describe in VHDL. In the dataflow style of architecture, the SELECT statement provides the required functionality, as shown in Table 5-42, the VHDL description of 4-input, 8-bit multiplexer.

In a behavioral architecture, a CASE statement is used. For example, Table 5-43 shows a process-based architecture for the same mux4in8b entity.

As in ABEL, it is very easy to customize the selection criteria in a VHDL multiplexer program. For example, Table 5-44 is a behavioral-style program for a specialized 4-input, 18-bit multiplexer with the selection criteria of Table 5-40.

In each example, if the select inputs are not valid (e.g., contain U's or X's), the output bus is set to "unknown" to help catch errors during simulation.

Table 5-42 Dataflow VHDL program for a 4-input, 8-bit multiplexer.

```
library IEEE;
use IEEE.std_logic_1164.all;

entity mux4in8b is
    port (
        S: in STD_LOGIC_VECTOR (1 downto 0);        -- Select inputs, 0-3 ==> A-D
        A, B, C, D: in STD_LOGIC_VECTOR (1 to 8); -- Data bus input
        Y: out STD_LOGIC_VECTOR (1 to 8)          -- Data bus output
    );
end mux4in8b;

architecture mux4in8b of mux4in8b is
begin
    with S select Y <=
      A when "00",
      B when "01",
      C when "10",
      D when "11",
      (others => 'U') when others; -- this creates an 8-bit vector of 'U'
end mux4in8b;
```

Table 5-43 Behavioral architecture for a 4-input, 8-bit multiplexer.

```
architecture mux4in8p of mux4in8b is
begin
process(S, A, B, C, D)
  begin
    case S is
      when "00" => Y <= A;
      when "01" => Y <= B;
      when "10" => Y <= C;
      when "11" => Y <= D;
      when others => Y <= (others => 'U');  -- 8-bit vector of 'U'
    end case;
  end process;
end mux4in8p;
```

5.8 Exclusive-OR Gates and Parity Circuits

5.8.1 Exclusive-OR and Exclusive-NOR Gates

Exclusive OR (XOR)
Exclusive NOR (XNOR)
Equivalence

An *Exclusive-OR (XOR)* gate is a 2-input gate whose output is 1 if exactly one of its inputs is 1. Stated another way, an XOR gate produces a 1 output if its inputs are different. An *Exclusive NOR (XNOR)* or *Equivalence* gate is just the opposite—it produces a 1 output if its inputs are the same. A truth table for these

Table 5-44 Behavioral VHDL program for a specialized 4-input, 3-bit multiplexer.

```
library IEEE;
use IEEE.std_logic_1164.all;

entity mux4in3b is
    port (
        S: in STD_LOGIC_VECTOR (2 downto 0);        -- Select inputs, 0-7 ==> ABACADAB
        A, B, C, D: in STD_LOGIC_VECTOR (1 to 18);  -- Data bus inputs
        Y: out STD_LOGIC_VECTOR (1 to 18)           -- Data bus output
    );
end mux4in3b;

architecture mux4in3p of mux4in3b is
begin
process(S, A, B, C, D)
variable i: INTEGER;
  begin
    case S is
      when "000" | "010" | "100" | "110" => Y <= A;
      when "001" | "111" => Y <= B;
      when "011" => Y <= C;
      when "101" => Y <= D;
      when others => Y <= (others => 'U'); -- 18-bit vector of 'U'
    end case;
  end process;
end mux4in3p;
```

functions is shown in Table 5-45. The XOR operation is sometimes denoted by the symbol "⊕", that is,

$$X \oplus Y = X' \cdot Y + X \cdot Y'$$

Although EXCLUSIVE OR is not one of the basic functions of switching algebra, discrete XOR gates are fairly commonly used in practice. Most switching technologies cannot perform the XOR function directly; instead, they use multigate designs like the ones shown in Figure 5-71.

X	Y	$X \oplus Y$ (XOR)	$(X \oplus Y)'$ (XNOR)
0	0	0	1
0	1	1	0
1	0	1	0
1	1	0	1

Table 5-45
Truth table for XOR and XNOR functions.

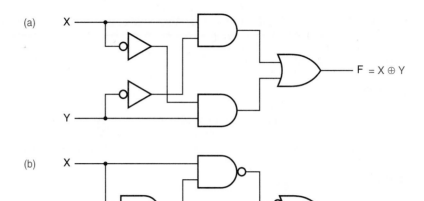

Figure 5-71
Multigate designs for
the 2-input XOR
function: (a) AND-OR;
(b) three-level NAND.

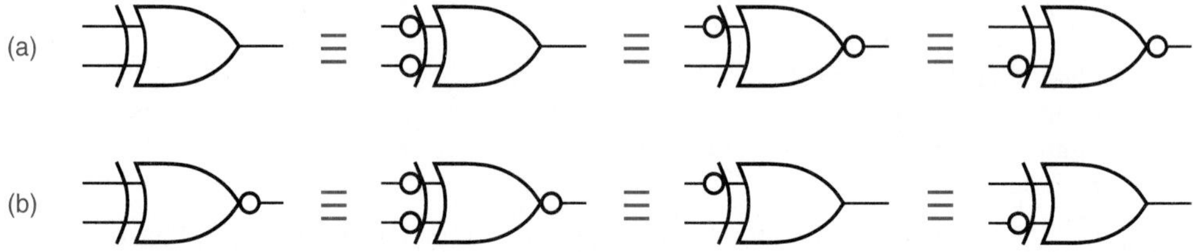

Figure 5-72 Equivalent symbols for (a) XOR gates; (b) XNOR gates.

The logic symbols for **XOR** and **XNOR** functions are shown in Figure 5-72. There are four equivalent symbols for each function. All of these alternatives are a consequence of a simple rule:

- Any two signals (inputs or output) of an **XOR** or **XNOR** gate may be complemented without changing the resulting logic function.

In bubble-to-bubble logic design, we choose the symbol that is most expressive of the logic function being performed.

74x86
Four **XOR** gates are provided in a single 14-pin SSI IC, the *74x86* shown in Figure 5-73. New SSI logic families do not offer **XNOR** gates, although they are readily available in FPGA and ASIC libraries and as primitives in HDLs.

Figure 5-73
Pinouts of the 74x86
quadruple 2-input
XOR gate.

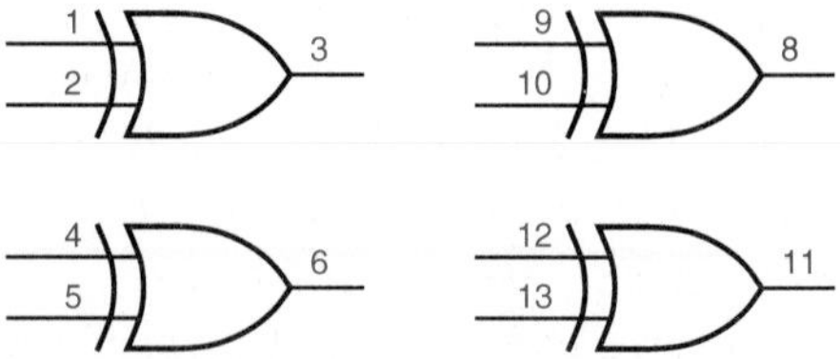

(a)

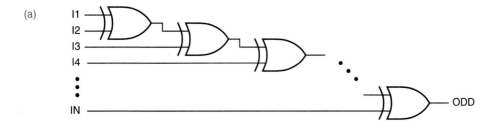

(b)

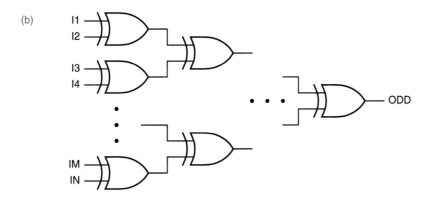

Figure 5-74
Cascading XOR
gates: (a) daisy-chain
connection; (b) tree
structure.

5.8.2 Parity Circuits

As shown in Figure 5-74(a), n XOR gates may be cascaded to form a circuit with $n + 1$ inputs and a single output. This is called an *odd-parity circuit*, because its output is 1 if an odd number of its inputs are 1. The circuit in (b) is also an odd-parity circuit, but it's faster because its gates are arranged in a treelike structure. If the output of either circuit is inverted, we get an *even-parity circuit*, whose output is 1 if an even number of its inputs are 1.

odd-parity circuit

even-parity circuit

5.8.3 The 74x280 9-Bit Parity Generator

Rather than build a multibit parity circuit with discrete XOR gates, it is more economical to put all of the XORs in a single MSI package with just the primary inputs and outputs available at the external pins. The *74x280* 9-bit parity generator, shown in Figure 5-75, is such a device. It has nine inputs and two outputs that indicate whether an even or odd number of inputs are 1.

74x280

5.8.4 Parity-Checking Applications

In Section 2.15 we described error-detecting codes that use an extra bit, called a parity bit, to detect errors in the transmission and storage of data. In an even-parity code, the parity bit is chosen so that the total number of 1 bits in a code word is even. Parity circuits like the 74x280 are used both to generate the correct value of the parity bit when a code word is stored or transmitted and to check the parity bit when a code word is retrieved or received.

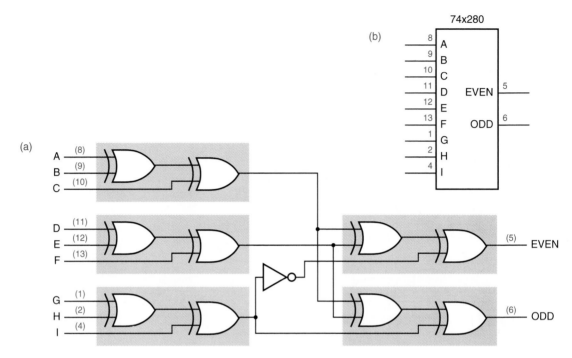

Figure 5-75 The 74x280 9-bit odd/even parity generator: (a) logic diagram, including pin numbers for a standard 16-pin dual in-line package; (b) traditional logic symbol.

Figure 5-76 shows how a parity circuit might be used to detect errors in the memory of a microprocessor system. The memory stores 8-bit bytes, plus a parity bit for each byte. The microprocessor uses a bidirectional bus D[0:7] to transfer data to and from the memory. Two control lines, RD and WR, are used to indicate whether a read or write operation is desired, and an ERROR signal is asserted to indicate parity errors during read operations. Complete details of the memory chips, such as addressing inputs, are not shown; memory chips are described in detail in Chapter 10. For parity checking, we are concerned only with the data connections to the memory.

SPEEDING UP THE XOR TREE

If each XOR gate in Figure 5-75 were built using discrete NAND gates as in Figure 5-71(b), the 74x280 would be pretty slow, having a propagation delay equivalent to $4 \cdot 3 + 1$, or 13, NAND gates. Instead, a typical implementation of the 74x280 uses a 4-wide AND-OR-INVERT gate to perform the function of each shaded pair of XOR gates in the figure with about the same delay as a single NAND gate. The A–I inputs are buffered through two levels of inverters, so that each input presents just one unit load to the circuit driving it. Thus, the total propagation delay through this implementation of the 74x280 is about the same as five inverting gates, not 13.

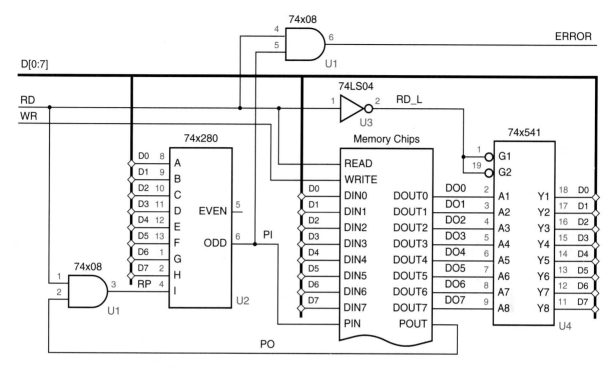

Figure 5-76 Parity generation and checking for an 8-bit-wide memory system.

To store a byte into the memory chips, we specify an address (not shown), place the byte on D[0–7], generate its parity bit on PIN, and assert WR. The AND gate on the I input of the 74x280 ensures that I is 0 except during read operations, so that during writes the '280's output depends only on the parity of the D-bus data. The '280's ODD output is connected to PIN, so that the total number of 1s stored is even.

To retrieve a byte, we specify an address (not shown) and assert RD; the byte value appears on DOUT[0–7] and its parity appears on POUT. A 74x541 drives the byte onto the D bus, and the '280 checks its parity. If the parity of the 9-bit word DOUT[0–7],POUT is odd during a read, the ERROR signal is asserted.

Parity circuits are also used with error-correcting codes such as the Hamming codes described in Section 2.15.3. We showed the parity-check matrix for a 7-bit Hamming code in Figure 2-13 on page 63. We can correct errors in this code as shown in Figure 5-77. A 7-bit word, possibly containing an error, is presented on DU[1–7]. Three 74x280s compute the parity of the three bit-groups defined by the parity-check matrix. The outputs of the '280s form the syndrome, which is the number of the erroneous input bit, if any. A 74x138 is used to decode the syndrome. If the syndrome is zero, the NOERROR_L signal is asserted (this signal also could be named ERROR). Otherwise, the erroneous

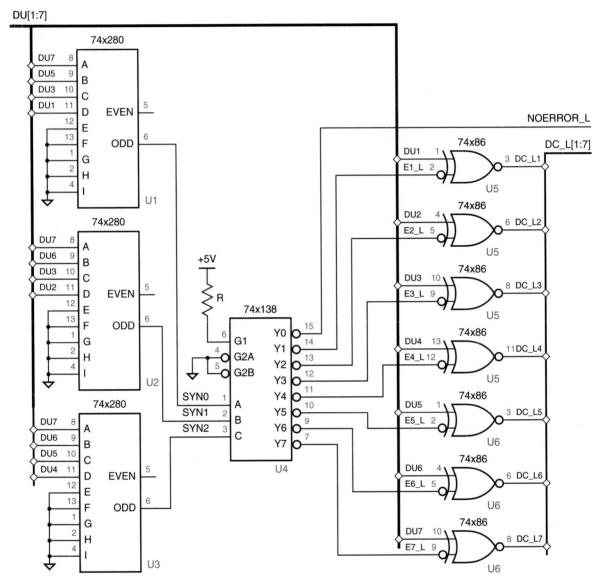

Figure 5-77 Error-correcting circuit for a 7-bit Hamming code.

bit is corrected by complementing it. The corrected code word appears on the DC_L bus.

Note that the active-low outputs of the '138 led us to use an active-low DC_L bus. If we required an active-high DC bus, we could have put a discrete inverter on each XOR input or output, or used a decoder with active-high outputs, or used XNOR gates.

5.8.5 Exclusive-OR Gates and Parity Circuits in ABEL and PLDs

The Exclusive-OR function is denoted in ABEL by the $ operator, and its complement, the Exclusive-NOR function, is denoted by !$. In principle, these operators may be used freely in ABEL expressions. For example, you could specify a PLD output equivalent to the 74x280's EVEN output using the following ABEL equation:

```
EVEN = !(A $ B $ C $ D $ E $ F $ G $ H $ I);
```

However, most PLDs realize expressions using two-level AND-OR logic and have little if any capability of realizing XOR functions directly. Unfortunately, the Karnaugh map of an n-input XOR function is a checkerboard with 2^{n-1} prime implicants. Thus, the sum-of-products realization of the simple equation above requires 256 product terms, well beyond the capability of any PLD.

So-called X-series PLDs can realize a two-input XOR function directly in a three-level structure that uses a single XOR gate ro combine two independent sums of products. This structure turns out to be useful in the design of counters. To create larger XOR functions, however, a board-level designer must normally use a specialized parity generator/checker component like the 74x280, and an ASIC designer must combine individual XOR gates in a multilevel parity tree similar to Figure 5-74(b) on page 413.

5.8.6 Exclusive-OR Gates and Parity Circuits in VHDL

Like ABEL, VHDL provides primitive operators, xor and xnor, for specifying XOR and XNOR functions (xnor is available only in VHDL-93). For example, Table 5-46 is a dataflow-style program for a 3-input XOR device that uses the xor primitive. It's also possible to specify XOR or parity functions behaviorally, as Table 5-47 does for a 9-input parity function similar to the 74x280.

Table 5-46 Dataflow-style VHDL program for a 3-input XOR device.

```
library IEEE;
use IEEE.std_logic_1164.all;

entity vxor3 is
    port (
        A, B, C: in STD_LOGIC;
        Y: out STD_LOGIC
    );
end vxor3;

architecture vxor3 of vxor3 is
begin
  Y <= A xor B xor C;
end vxor3;
```

▌ **Table 5-47** Behavioral VHDL program for a 9-input parity checker.

```vhdl
library IEEE;
use IEEE.std_logic_1164.all;

entity parity9 is
    port (
        I: in STD_LOGIC_VECTOR (1 to 9);
        EVEN, ODD: out STD_LOGIC
    );
end parity9;

architecture parity9p of parity9 is
begin
process (I)
  variable p : STD_LOGIC;
  begin
    p := I(1);
    for j in 2 to 9 loop
      if I(j) = '1' then p := not p; end if;
    end loop;
    ODD <= p;
    EVEN <= not p;
  end process;
end parity9p;
```

When a VHDL program containing large XOR functions is synthesized, the synthesis tool will do the best it can to realize the function in the targeted device technology. There's no magic—if we try to target the VHDL program in Table 5-47 to a 16V8 PLD, it still won't fit!

Typical ASIC and FPGA libraries contain two- and three-input XOR and XNOR functions as primitives. In CMOS ASICs, these primitives are usually realized very efficiently at the transistor level using transmission gates, as shown in Exercises 5.56 and 5.73. Fast and compact XOR trees can be built using these primitives. However, typical VHDL synthesis tools are not smart enough to create an efficient tree structure from a behavioral program like Table 5-47. Instead, we can use a structural program to get exactly what we want.

For example, Table 5-48 is a structural VHDL program for a 9-input XOR function that is equivalent to the 74x280 of Figure 5-75(a) in structure as well as function. In this example, we've used the previously defined vxor3 component as the basic building block of the XOR tree. In an ASIC, we would replace the vxor3 with a 3-input XOR primitive from the ASIC library. Also, if a 3-input XNOR were available, we could eliminate the explicit inversion for Y3N and instead use the XNOR for U5, using the noninverted Y3 signal as its last input.

Our final example is a VHDL version of the Hamming decoder circuit of Figure 5-77 and is shown in Table 5-49. A function syndrome(DU) is defined to

Table 5-48 Structural VHDL program for a 74x280-like parity checker.

```
library IEEE;
use IEEE.std_logic_1164.all;

entity V74x280 is
    port (
        I: in STD_LOGIC_VECTOR (1 to 9);
        EVEN, ODD: out STD_LOGIC
    );
end V74x280;

architecture V74x280s of V74x280 is
component vxor3
 port (A, B, C: in STD_LOGIC; Y: out STD_LOGIC);
end component;
signal Y1, Y2, Y3, Y3N: STD_LOGIC;
begin
  U1: vxor3 port map (I(1), I(2), I(3), Y1);
  U2: vxor3 port map (I(4), I(5), I(6), Y2);
  U3: vxor3 port map (I(7), I(8), I(9), Y3);
  Y3N <= not Y3;
  U4: vxor3 port map (Y1, Y2, Y3, ODD);
  U5: vxor3 port map (Y1, Y2, Y3N, EVEN);
end V74x280s;
```

return the 3-bit syndrome of a 7-bit uncorrected data input vector DU. In the "main" process, the corrected data output vector DC is initially set equal to DU. The CONV_INTEGER function, which we defined in Section 4.7.4, is used to convert the 3-bit syndrome to an integer. If the syndrome is nonzero, the corresponding bit of DC is complemented to correct the assumed 1-bit error. If the syndrome is zero, either no error or an undetectable error has occurred; the output NOERROR is set accordingly.

5.9 Comparators

Comparing two binary words for equality is a commonly used operation in computer systems and device interfaces. For example, in Figure 2-7(a) on page 56, we showed a system structure in which devices are enabled by comparing a "device select" word with a predetermined "device ID." A circuit that compares two binary words and indicates whether they are equal is called a *comparator*.   *comparator*
Some comparators interpret their input words as signed or unsigned numbers and also indicate an arithmetic relationship (greater or less than) between the words. These devices are often called *magnitude comparators*.   *magnitude comparator*

■ Table 5-49 Behavioral VHDL program for Hamming error correction.

```
library IEEE;
use IEEE.std_logic_1164.all;
use IEEE.std_logic_unsigned.all;

entity hamcorr is
  port (
      DU: IN STD_LOGIC_VECTOR (1 to 7);
      DC: OUT STD_LOGIC_VECTOR (1 to 7);
      NOERROR: OUT STD_LOGIC
);
end hamcorr;

architecture hamcorr of hamcorr is

function syndrome (D: STD_LOGIC_VECTOR)
    return STD_LOGIC_VECTOR is
variable SYN: STD_LOGIC_VECTOR (2 downto 0);
begin
  SYN(0) := D(1) xor D(3) xor D(5) xor D(7);
  SYN(1) := D(2) xor D(3) xor D(6) xor D(7);
  SYN(2) := D(4) xor D(5) xor D(6) xor D(7);
  return(SYN);
end syndrome;

begin
process (DU)
variable i: INTEGER;
  begin
    DC <= DU;
    i := CONV_INTEGER(syndrome(DU));
    if i = 0 then NOERROR <= '1';
    else NOERROR <= '0'; DC(i) <= not DU(i); end if;
  end process;
end hamcorr;
```

5.9.1 Comparator Structure

Exclusive-OR and Exclusive-NOR gates may be viewed as 1-bit comparators. Figure 5-78(a) shows an interpretation of the 74x86 XOR gate as a 1-bit comparator. The active-high output, DIFF, is asserted if the inputs are different. The outputs of four XOR gates are ORed to create a 4-bit comparator in (b). The DIFF output is asserted if any of the input-bit pairs are different. Given enough XOR gates and wide enough OR gates, comparators with any number of input bits can be built.

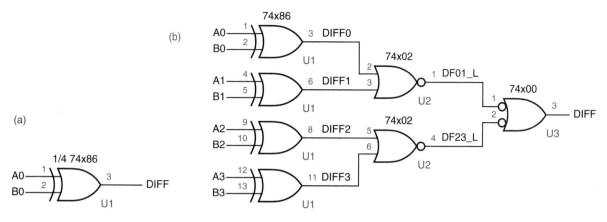

Figure 5-78 Comparators using the 74x86: (a) 1-bit comparator; (b) 4-bit comparator.

AN ITERATIVE
COMPARATOR

The *n*-bit comparators in the preceding subsection might be called *parallel compar-ators* because they look at each pair of input bits simultaneously and deliver the 1-bit comparison results in parallel to an *n*-input OR or AND function. It is also possible to design an "iterative comparator" that looks at its bits one at a time using a small, fixed amount of logic per bit. Before looking at the iterative comparator design, you should understand the general class of "iterative circuits" described in the next subsection.

5.9.2 Iterative Circuits

An *iterative circuit* is a special type of combinational circuit, with the structure shown in Figure 5-79. The circuit contains *n* identical modules, each of which has both *primary inputs and outputs* and *cascading inputs and outputs*. The left-most cascading inputs are called *boundary inputs* and are connected to fixed logic values in most iterative circuits. The rightmost cascading outputs are called *boundary outputs* and usually provide important information.

iterative circuit
primary inputs and outputs
cascading inputs and outputs
boundary outputs

   Iterative circuits are well suited to problems that can be solved by a simple iterative algorithm:

1. Set C_0 to its initial value and set i to 0.
2. Use C_i and PI_i to determine the values of PO_i and C_{i+1}.
3. Increment i.
4. If $i < n$, go to step 2.

In an iterative circuit, the loop of steps 2–4 is "unwound" by providing a separate combinational circuit that performs step 2 for each value of i.

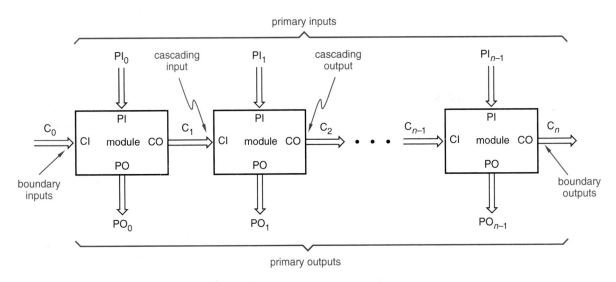

Figure 5-79 General structure of an iterative combinational circuit.

Examples of iterative circuits are the comparator circuit in the next subsection and the ripple adder in Section 5.10.2. The 74x85 4-bit comparator and the 74x283 4-bit adder are examples of MSI circuits that can be used as the individual modules in a larger iterative circuit. In Section 8.6 we'll explore the relationship between iterative circuits and corresponding sequential circuits that execute the 4-step algorithm above in discrete time steps.

5.9.3 An Iterative Comparator Circuit

Two n-bit values X and Y can be compared one bit at a time using a single bit EQ_i at each step to keep track of whether all of the bit-pairs have been equal so far:

1. Set EQ_0 to 1 and set i to 0.
2. If EQ_i is 1 and X_i and Y_i are equal, set EQ_{i+1} to 1. Else set EQ_{i+1} to 0.
3. Increment i.
4. If $i < n$, go to step 2.

Figure 5-80 shows a corresponding iterative circuit. Note that this circuit has no primary outputs; the boundary output is all that interests us. Other iterative circuits, such as the ripple adder of Section 5.10.2, have primary outputs of interest.

Given a choice between the iterative comparator circuit in this subsection and one of the parallel comparators shown previously, you would probably prefer the parallel comparator. The iterative comparator saves little if any cost, and it's very slow because the cascading signals need time to "ripple" from the leftmost to the rightmost module. Iterative circuits that process more than one bit

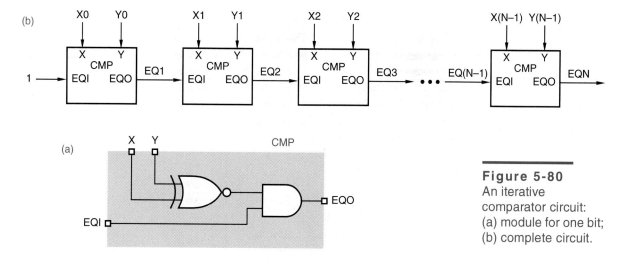

Figure 5-80
An iterative
comparator circuit:
(a) module for one bit;
(b) complete circuit.

at a time, using modules like the 74x85 4-bit comparator and 74x283 4-bit adder, are much more likely to be used in practical designs.

5.9.4 Standard MSI Comparators

Comparator applications are common enough that several MSI comparators have been developed commercially. The *74x85* is a 4-bit comparator with the logic symbol shown in Figure 5-81. It provides a greater-than output (AGTBOUT) and a less-than output (ALTBOUT) as well as an equal output (AEQBOUT). The '85 also has *cascading inputs* (AGTBIN, ALTBIN, AEQBIN) for combining multiple '85s to create comparators for more than four bits. Both the cascading inputs and the outputs are arranged in a 1-out-of-3 code, since in normal operation exactly one input and one output should be asserted.

The cascading inputs are defined so the outputs of an '85 that compares less-significant bits are connected to the inputs of an '85 that compares more-

74x85

cascading inputs

Figure 5-81
Traditional logic symbol for
the 74x85 4-bit comparator.

74x85

2 ALTBIN	ALTBOUT 7	
3 AEQBIN	AEQBOUT 6	
4 AGTBIN	AGTBOUT 5	
10 A0		
9 B0		
12 A1		
11 B1		
13 A2		
14 B2		
15 A3		
1 B3		

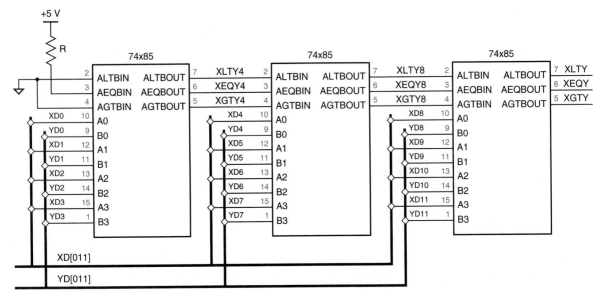

Figure 5-82 A 12-bit comparator using 74x85s.

significant bits, as shown in Figure 5-82 for a 12-bit comparator. This is an iterative circuit according to the definition in Section 5.9.2. Each '85 develops its cascading outputs roughly according to the following pseudo-logic equations:

$$AGTBOUT = (A > B) + (A = B) \cdot AGTBIN$$

$$AEQBOUT = (A = B) \cdot AEQBIN$$

$$ALTBOUT = (A < B) + (A = B) \cdot ALTBIN$$

The parenthesized subexpressions above are not normal logic expressions; rather, they indicate an arithmetic comparison that occurs between the A3–A0 and B3–B0 inputs. In other words, AGTBOUT is asserted if $A > B$ or if $A = B$ and AGTBIN is asserted (if the higher-order bits are equal, we have to look at the lower-order bits for the answer). We'll see this kind of expression again when we look at ABEL comparator design in Section 5.9.5. The arithmetic comparisons can be expressed using normal logic expressions, for example,

$$(A > B) = A3 \cdot B3' +$$
$$(A3 \oplus B3)' \cdot A2 \cdot B2' +$$
$$(A3 \oplus B3)' \cdot (A2 \oplus B2)' \cdot A1 \cdot B1' +$$
$$(A3 \oplus B3)' \cdot (A2 \oplus B2)' \cdot (A1 \oplus B1)' \cdot A0 \cdot B0'$$

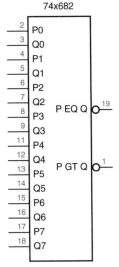

Figure 5-83
Traditional logic symbol for the 74x682 8-bit comparator.

Such expressions must be substituted into the pseudo-logic equations above to obtain genuine logic equations for the comparator outputs.

Several 8-bit MSI comparators are also available. The simplest of these is the *74x682*, whose logic symbol is shown in Figure 5-83 and whose internal

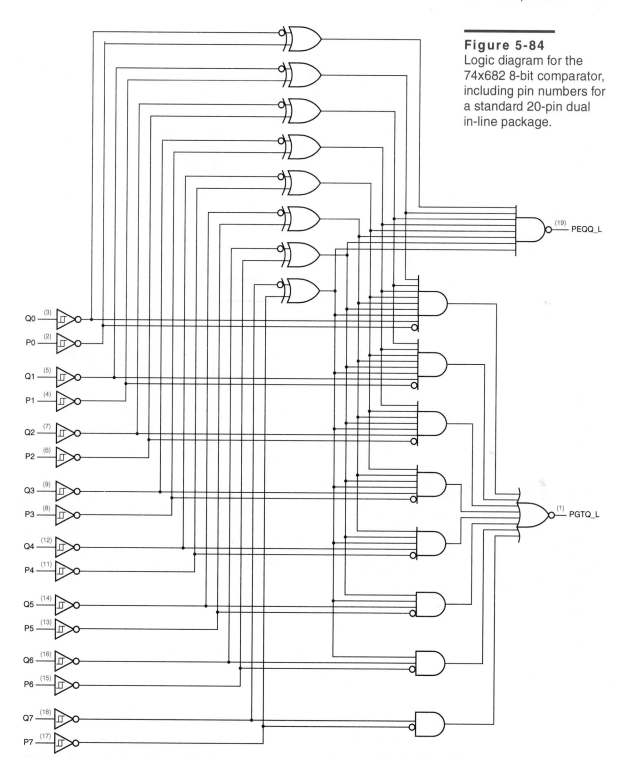

Figure 5-84
Logic diagram for the
74x682 8-bit comparator,
including pin numbers for
a standard 20-pin dual
in-line package.

logic diagram is shown in Figure 5-84. The top half of the circuit checks the two 8-bit input words for equality. Each XNOR-gate output is asserted if its inputs are equal, and the PEQQ_L output is asserted if all eight input-bit pairs are equal. The bottom half of the circuit compares the input words arithmetically and asserts PGTQ_L if P[7–0] > Q[7–0].

Unlike the 74x85, the 74x682 does not have cascading inputs. Also unlike the '85, the '682 does not provide a "less than" output. However, any desired condition, including ≤ and ≥, can be formulated as a function of the PEQQ_L and PGTQ_L outputs, as shown in Figure 5-85.

Figure 5-85
Arithmetic conditions derived from 74x682 outputs.

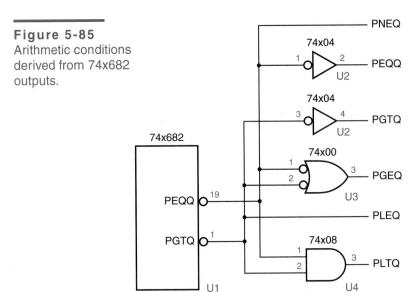

COMPARING COMPARATORS   The individual 1-bit comparators (XNOR gates) in the '682 are drawn in the opposite sense as the examples of the preceding subsection—outputs are asserted for *equal* inputs and then ANDed, rather than asserted for *different* inputs and then ORed. We can look at a comparator's function either way, as long as we're consistent.

5.9.5 Comparators in ABEL and PLDs

Comparing two sets for equality or inequality is very easy to do in ABEL using the "==" or "!=" operator in a relational expression. The only restriction is that the two sets must have an equal number of elements. Thus, given the relational expression "A!=B", where A and B are sets each with n elements, the compiler generates the logic expression

```
(A1 $ B1)  # (A2 $ B2)  # ... # (An $ Bn)
```

The logic expression for "A==B" is just the complement of the one above.

In the preceding logic expression, it takes one 2-input XOR function to compare each bit. Since a 2-input XOR function can be realized as a sum of two product terms, the complete expression can be realized in a PLD as a relatively modest sum of $2n$ product terms:

```
(A1&!B1 # !A1&B1) # (A2&!B2 # !A2&&B2) # ... # (An&!Bn # !An&&Bn)
```

Although ABEL has relational operators for less-than and greater-than comparisons, the resulting logic expressions are not so small or easy to derive. For example, consider the relational expression "A<B", where [An..A1] and [Bn..B1] are sets with n elements. To construct the corresponding logic expression, ABEL first constructs n equations of the form

```
Li = (!Ai & (Bi # Li-1) # (Ai & Bi & Li-1)
```

for $i = 1$ to n and L0 = 0 by definition. This is, in effect, an iterative definition of the less-than function, starting with the least-significant bit. Each Li equation says that, as of bit i, A is less than B if Ai is 0 and Bi is 1 or A was less than B as of the previous bit, or if Ai and Bi are both 1 and A was less than B as of the previous bit.

The logic expression for "A<B" is simply the equation for Ln. So, after creating the n equations above, ABEL collapses them into a single equation for Ln involving only elements of A and B. It does this by substituting the Ln-1 equation into the righthand side of the Ln equation, then substituting the Ln-2 equation into this result, and so on, until substituting 0 for L0. Finally, it derives a minimal sum-of-products expression from the result.

Collapsing an iterative circuit into a two-level sum-of-products realization usually creates an exponential expansion of product terms. The "<" comparison function follows this pattern, requiring $2^n - 1$ product terms for an n-bit comparator. Thus, comparators larger than a few bits cannot be realized practically in one pass through a PLD.

The results for ">" comparators are identical, of course, and logic expressions for ">=" and "<=" are at least as bad, being the complements of the expressions for "<" and ">". If we use a PLD with output polarity control, the inversion is free and the number of product terms is the same; otherwise, the minimal number of product terms after inverting is $2^n + 2^{n-1} - 1$.

5.9.6 Comparators in VHDL

VHDL has comparison operators for all of its built-in types. *Equality (=)* and *inequality (/=)* operators apply to all types; for array and record types, the operands must have equal size and structure, and the operands are compared component by component. We have used the equality operator to compare a signal or signal vector with a constant value in many examples in this chapter. If we compare two signals or variables, the synthesis engine generates equations similar to ABEL's in the preceding subsection.

equality, =
inequality, /=

VHDL's other comparison operators, >, <, >=, and <=, apply only to integer types, enumerated types (such as STD_LOGIC), and one-dimensional arrays of enumeration or integer types. Integer order from smallest to largest is the natural ordering, from minus infinity to plus infinity, and enumerated types use the ordering in which the elements of the type were defined, from first to last (unless you explicitly change the enumeration encoding using a command specific to the synthesis engine, in which case the ordering is that of your encoding).

The ordering for array types is defined iteratively, starting with the *leftmost* element in each array. Arrays are always compared from left to right, regardless of the order of their index range ("to" or "downto"). The order of the leftmost pair of unequal elements is the order of the array. If the arrays have unequal lengths and all the elements of the shorter array match the corresponding elements of the longer one, then the shorter array is considered to be the smaller.

The result of all this is that the built-in comparison operators compare equal-length arrays of type BIT_VECTOR or STD_LOGIC_VECTOR as if they represented unsigned integers. If the arrays have different lengths, then the operators do *not* yield a valid arithmetic comparison, what you'd get by extending the shorter array with zeroes on the left; more on this in a moment.

Table 5-50 is a VHDL program that produces all of the comparison outputs for comparing two 8-bit unsigned integers. Since the two input vectors A and B have equal lengths, the program produces the desired results.

Table 5-50
Behavioral VHDL program for comparing 8-bit unsigned integers.

```
library IEEE;
use IEEE.std_logic_1164.all;

entity vcompare is
    port (
        A, B: in STD_LOGIC_VECTOR (7 downto 0);
        EQ, NE, GT, GE, LT, LE: out STD_LOGIC
    );
end vcompare;

architecture vcompare_arch of vcompare is
begin
process (A, B)
  begin
    EQ <= '0'; NE <= '0'; GT <= '0'; GE <= '0'; LT <= '0'; LE <= '0';
    if A = B then EQ <= '1'; end if;
    if A /= B then NE <= '1'; end if;
    if A > B then GT <= '1'; end if;
    if A >= B then GE <= '1'; end if;
    if A < B then LT <= '1'; end if;
    if A <= B then LE <= '1'; end if;
  end process;
end vcompare_arch
```

To allow more flexible comparisons and arithmetic operations, IEEE standard 1076-3 created a standard package, `std_logic_arith`, which defines two important new types and a host of comparison and arithmetic functions that operate on them. The two new types are `SIGNED` and `UNSIGNED`:

```
type SIGNED is array (NATURAL range <>) of STD_LOGIC;
type UNSIGNED is array (NATURAL range <>) of STD_LOGIC;
```

As you can see, both types are defined just as indeterminate-length arrays of `STD_LOGIC`, no different from `STD_LOGIC_VECTOR`. The important thing is that the package also defines new comparison functions that are invoked when either or both comparison operands have one of the new types. For example, it defines eight new "less-than" functions with the following combinations of parameters:

```
function "<" (L: UNSIGNED; R: UNSIGNED) return BOOLEAN;
function "<" (L: SIGNED; R: SIGNED) return BOOLEAN;
function "<" (L: UNSIGNED; R: SIGNED) return BOOLEAN;
function "<" (L: SIGNED; R: UNSIGNED) return BOOLEAN;
function "<" (L: UNSIGNED; R: INTEGER) return BOOLEAN;
function "<" (L: INTEGER; R: UNSIGNED) return BOOLEAN;
function "<" (L: SIGNED; R: INTEGER) return BOOLEAN;
function "<" (L: INTEGER; R: SIGNED) return BOOLEAN;
```

Thus, the "<" operator can be used with any combination of `SIGNED`, `UNSIGNED`, and `INTEGER` operands; the compiler selects the function whose parameter types match the actual operands. Each of the functions is defined in the package to do the "right thing," including making the appropriate extensions and conversions when operands of different sizes or types are used. Similar functions are provided for the other five relational operators, =, /=, <=, >, and >=.

Using the `IEEE_std_logic_arith` package, you can write programs like the one in Table 5-51. Its 8-bit input vectors, A, B, C, and D, have three different types. In the comparisons involving A, B, and C, the compiler automatically selects the correct version of the comparison function; for example, for "A<B" it selects the first "<" function above, because both operands have type `UNSIGNED`.

In the comparisons involving D, explicit type conversions are used. The assumption is that the designer wants this particular `STD_LOGIC_VECTOR` to be interpreted as `UNSIGNED` in one case and `SIGNED` in another. The important thing to understand here is that the `std_logic_arith` package does not make any assumptions about how `STD_LOGIC_VECTOR`s are to be interpreted; the user must specify the conversion.

Two other packages, `std_logic_signed` and `std_logic_unsigned`, do make assumptions and are useful if all `STD_LOGIC_VECTOR`s are to be interpreted the same way. Each package contains three versions of each comparison function so that `STD_LOGIC_VECTOR`s are interpreted as `SIGNED` or `UNSIGNED`, respectively, when compared with each other or with integers.

Table 5-51 Behavioral VHDL program for comparing 8-bit integers of various types.

```
library IEEE;
use IEEE.std_logic_1164.all;
use IEEE.std_logic_arith.all;

entity vcompa is
    port (
        A, B: in UNSIGNED (7 downto 0);
        C: in SIGNED (7 downto 0);
        D: in STD_LOGIC_VECTOR (7 downto 0);
        A_LT_B, B_GE_C, A_EQ_C, C_NEG, D_BIG, D_NEG: out STD_LOGIC
    );
end vcompa;

architecture vcompa_arch of vcompa is
begin
process (A, B, C, D)
  begin
    A_LT_B <= '0'; B_GE_C <= '0'; A_EQ_C <= '0'; C_NEG <= '0'; D_BIG <= '0'; D_NEG <= '0';
    if A < B then A_LT_B <= '1'; end if;
    if B >= C then B_GE_C <= '1'; end if;
    if A = C then A_EQ_C <= '1'; end if;
    if C < 0 then C_NEG <= '1'; end if;
    if UNSIGNED(D) > 200 then D_BIG <= '1'; end if;
    if SIGNED(D) < 0 then D_NEG <= '1'; end if;
  end process;
end vcompa_arch;
```

When a comparison function is specified in VHDL, realizing the function as a two-level sum of products takes just as many product terms as in ABEL. However, most VHDL synthesis engines will realize the comparator as an iterative circuit with far fewer gates, albeit more levels of logic. Also, better synthesis engines can detect opportunities to eliminate entire comparator circuits. For example, in the program of Table 5-50 on page 428, the NE, GE, and LE outputs could be realized with one inverter each, as the complements of the EQ, LT, and GT outputs, respectively.

*5.10 Adders, Subtractors, and ALUs

adder

subtractor

Addition is the most commonly performed arithmetic operation in digital systems. An *adder* combines two arithmetic operands using the addition rules described in Chapter 2. As we showed in Section 2.6, the same addition rules and therefore the same adders are used for both unsigned and two's-complement numbers. An adder can perform subtraction as the addition of the minuend and the complemented (negated) subtrahend, but you can also build *subtractor* circuits that perform subtraction directly. MSI devices called ALUs, described in Section 5.10.6, perform addition, subtraction, or any of several other operations according to an operation code supplied to the device.

*5.10.1 Half Adders and Full Adders

The simplest adder, called a *half adder*, adds two 1-bit operands X and Y, producing a 2-bit sum. The sum can range from 0 to 2, which requires two bits to express. The low-order bit of the sum may be named HS (half sum), and the high-order bit may be named CO (carry out). We can write the following equations for HS and CO:

half adder

$$HS = X \oplus Y$$
$$= X \cdot Y' + X' \cdot Y$$
$$CO = X \cdot Y$$

To add operands with more than one bit, we must provide for carries between bit positions. The building block for this operation is called a *full adder*. Besides the addend-bit inputs X and Y, a full adder has a carry-bit input, CIN. The sum of the three inputs can range from 0 to 3, which can still be expressed with just two output bits, S and COUT, having the following equations:

full adder

$$S = X \oplus Y \oplus CIN$$
$$= X \cdot Y' \cdot CIN' + X' \cdot Y \cdot CIN' + X' \cdot Y' \cdot CIN + X \cdot Y \cdot CIN$$
$$COUT = X \cdot Y + X \cdot CIN + Y \cdot CIN$$

Here, S is 1 if an odd number of the inputs are 1, and COUT is 1 if two or more of the inputs are 1. These equations represent the same operation that was specified by the binary addition table in Table 2-3 on page 32.

One possible circuit that performs the full-adder equations is shown in Figure 5-86(a). The corresponding logic symbol is shown in (b). Sometimes the symbol is drawn as shown in (c), so that cascaded full adders can be drawn more neatly, as in the next subsection.

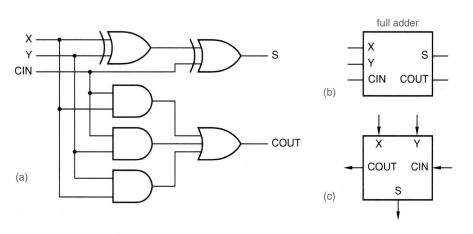

Figure 5-86
Full adder: (a) gate-level circuit diagram; (b) logic symbol; (c) alternate logic symbol suitable for cascading.

*5.10.2 Ripple Adders

Two binary words, each with n bits, can be added using a *ripple adder*—a cascade of n full-adder stages, each of which handles one bit. Figure 5-87 shows the circuit for a 4-bit ripple adder. The carry input to the least significant bit (c_0) is normally set to 0, and the carry output of each full adder is connected to the carry input of the next most significant full adder. The ripple adder is a classic example of an iterative circuit as defined in Section 5.9.2.

A ripple adder is slow, since in the worst case a carry must propagate from the least significant full adder to the most significant one. This occurs if, for example, one addend is 11 … 11 and the other is 00 … 01. Assuming that all of the addend bits are presented simultaneously, the total worst-case delay is

$$t_{ADD} = t_{XYCout} + (n - 2) \cdot t_{CinCout} + t_{CinS}$$

where t_{XYCout} is the delay from X or Y to COUT in the least significant stage, $t_{CinCout}$ is the delay from CIN to COUT in the middle stages, and t_{CinS} is the delay from CIN to S in the most significant stage.

A faster adder can be built by obtaining each sum output s_i with just two levels of logic. This can be accomplished by writing an equation for s_i in terms of x_0–x_i, y_0–y_i, and c_0, "multiplying out" or "adding out" to obtain a sum-of-products or product-of-sums expression, and building the corresponding AND-OR or OR-AND circuit. Unfortunately, beyond s_2 the resulting expressions have too many terms, requiring too many first-level gates and more inputs than typically possible on the second-level gate. For example, even assuming that $c_0 = 0$, a two-level AND-OR circuit for s_2 requires fourteen 4-input ANDs, four 5-input ANDs, and an 18-input OR gate; higher-order sum bits are even worse. Nevertheless, it is possible to build adders with just a few levels of delay using a more reasonable number of gates, as we'll see in Section 5.10.4.

*5.10.3 Subtractors

A binary subtraction operation analogous to binary addition was also specified in Table 2-3 on page 32. A *full subtractor* handles one bit of the binary subtraction algorithm, having input bits X (minuend), Y (subtrahend), and BIN (borrow

Figure 5-87
A 4-bit ripple adder.

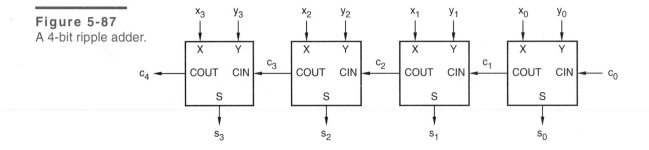

in), and output bits D (difference) and BOUT (borrow out). We can write logic equations corresponding to the binary subtraction table as follows:

$$D = X \oplus Y \oplus BIN$$
$$BOUT = X' \cdot Y + X' \cdot BIN + Y \cdot BIN$$

These equations are very similar to equations for a full adder, which should not be surprising. We showed in Section 2.6 that a two's-complement subtraction operation, $X - Y$, can be performed by an addition operation, namely, by adding the two's complement of Y to X. The two's complement of Y is $\overline{Y} + 1$, where $\overline{Y}$ is the bit-by-bit complement of Y. We also showed in Exercise 2.26 that a binary adder can be used to perform an unsigned subtraction operation $X - Y$ by performing the operation $X + \overline{Y} + 1$. We can now confirm that these statements are true by manipulating the logic equations above:

$$BOUT = X' \cdot Y + X' \cdot BIN + Y \cdot BIN$$
$$BOUT' = (X + Y') \cdot (X + BIN') \cdot (Y' + BIN') \quad \text{(generalized DeMorgan's theorem)}$$
$$= X \cdot Y' + X \cdot BIN' + Y' \cdot BIN' \quad \text{(multiply out)}$$
$$D = X \oplus Y \oplus BIN$$
$$= X \oplus Y' \oplus BIN' \quad \text{(complementing XOR inputs)}$$

For the last manipulation, recall that we can complement the two inputs of an XOR gate without changing the function performed.

Comparing with the equations for a full adder, the above equations tell us that we can build a full subtractor from a full adder as shown in Figure 5-88. Just to keep things straight, we've given the full-adder circuit in (a) a fictitious name, the "74x999." As shown in (c), we can interpret the function of this same physical circuit to be a full subtractor by giving it a new symbol with active-low borrow in, borrow out, and subtrahend signals.

Thus, to build a ripple subtractor for two n-bit active-high operands, we can use n 74x999s and inverters, as shown in (d). Note that for the subtraction operation, the borrow input of the least significant bit should be negated (no borrow), which for an active-low input means that the physical pin must be 1 or HIGH. This is just the opposite as in addition, where the same input pin is an active-high carry-in that is 0 or LOW.

By going back to the math in Chapter 2, we can show that this sort of manipulation works for all adder and subtractor circuits, not just ripple adders and subtractors. That is, any n-bit adder circuit can be made to function as a subtractor by complementing the subtrahend and treating the carry-in and carry-out signals as borrows with the opposite active level. The rest of this section discusses addition circuits only, with the understanding that they can easily be made to perform subtraction.

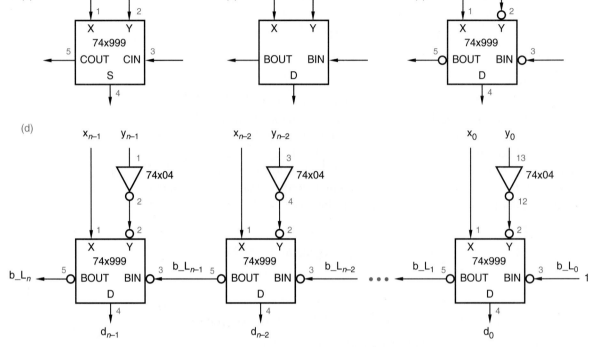

Figure 5-88 Designing subtractors using adders: (a) full adder; (b) full subtractor; (c) interpreting the device in (a) as a full subtractor; (d) ripple subtractor.

*5.10.4 Carry Lookahead Adders

The logic equation for sum bit i of a binary adder can actually be written quite simply:

$$s_i = x_i \oplus y_i \oplus c_i$$

More complexity is introduced when we expand c_i above in terms of $x_0 - x_{i-1}$, $y_0 - y_{i-1}$, and c_0, and we get a real mess expanding the XORs. However, if we're willing to forego the XOR expansion, we can at least streamline the design of c_i logic using ideas of *carry lookahead* discussed in this subsection.

carry lookahead

Figure 5-89 shows the basic idea. The block labeled "Carry Lookahead Logic" calculates c_i in a fixed, small number of logic levels for any reasonable value of i. Two definitions are the key to carry lookahead logic:

carry generate

- For a particular combination of inputs x_i and y_i, adder stage i is said to *generate* a carry if it produces a carry-out of 1 ($c_{i+1} = 1$) independent of the inputs on $x_0 - x_{i-1}$, $y_0 - y_{i-1}$, and c_0.

carry propagate

- For a particular combination of inputs x_i and y_i, adder stage i is said to *propagate* carries if it produces a carry-out of 1 ($c_{i+1} = 1$) in the presence of an input combination of $x_0 - x_{i-1}$, $y_0 - y_{i-1}$, and c_0 that causes a carry-in of 1 ($c_i = 1$).

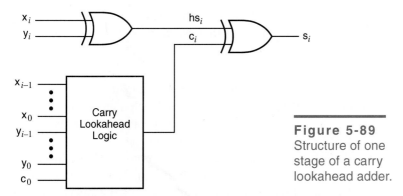

Figure 5-89
Structure of one
stage of a carry
lookahead adder.

Corresponding to these definitions, we can write logic equations for a carry-generate signal, g_i, and a carry-propagate signal, p_i, for each stage of a carry lookahead adder:

$$g_i = x_i \cdot y_i$$
$$p_i = x_i + y_i$$

That is, a stage unconditionally generates a carry if both of its addend bits are 1, and it propagates carries if at least one of its addend bits is 1. The carry output of a stage can now be written in terms of the generate and propagate signals:

$$c_{i+1} = g_i + p_i \cdot c_i$$

To eliminate carry ripple, we recursively expand the c_i term for each stage and multiply out to obtain a 2-level **AND-OR** expression. Using this technique, we can obtain the following carry equations for the first four adder stages:

$$
\begin{aligned}
c_1 &= g_0 + p_0 \cdot c_0 \\
c_2 &= g_1 + p_1 \cdot c_1 \\
&= g_1 + p_1 \cdot (g_0 + p_0 \cdot c_0) \\
&= g_1 + p_1 \cdot g_0 + p_1 \cdot p_0 \cdot c_0 \\
c_3 &= g_2 + p_2 \cdot c_2 \\
&= g_2 + p_2 \cdot (g_1 + p_1 \cdot g_0 + p_1 \cdot p_0 \cdot c_0) \\
&= g_2 + p_2 \cdot g_1 + p_2 \cdot p_1 \cdot g_0 + p_2 \cdot p_1 \cdot p_0 \cdot c_0 \\
c_4 &= g_3 + p_3 \cdot c_3 \\
&= g_3 + p_3 \cdot (g_2 + p_2 \cdot g_1 + p_2 \cdot p_1 \cdot g_0 + p_2 \cdot p_1 \cdot p_0 \cdot c_0) \\
&= g_3 + p_3 \cdot g_2 + p_3 \cdot p_2 \cdot g_1 + p_3 \cdot p_2 \cdot p_1 \cdot g_0 + p_3 \cdot p_2 \cdot p_1 \cdot p_0 \cdot c_0
\end{aligned}
$$

Each equation corresponds to a circuit with just three levels of delay—one for the generate and propagate signals, and two for the sum of products shown. A *carry lookahead adder* uses three-level equations such as these in each adder stage for the block labeled "carry lookahead" in Figure 5-89. The sum output for a stage is produced by combining the carry bit with the two addend bits for the

carry lookahead adder

stage, as we showed in the figure. In the next subsection we'll study some commercial MSI adders and ALUs that use carry lookahead.

*5.10.5 MSI Adders

74x283

74x83

The *74x283* is a 4-bit binary adder that forms its sum and carry outputs with just a few levels of logic, using the carry lookahead technique. Figure 5-90 is a logic symbol for the 74x283. The older *74x83* is identical except for its pinout, which has nonstandard locations for power and ground.

The logic diagram for the '283, shown in Figure 5-91, has just a few differences from the general carry-lookahead design that we described in the preceding subsection. First of all, its addends are named A and B instead of X and Y; no big deal. Second, it produces active-low versions of the carry-generate (g_i') and carry-propagate (p_i') signals, since inverting gates are generally faster than noninverting ones. Third, it takes advantage of the fact that we can algebraically manipulate the half-sum equation as follows:

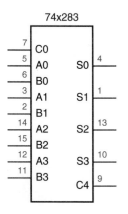

Figure 5-90
Traditional logic symbol for the 74x283 4-bit binary adder.

$$
\begin{aligned}
hs_i &= x_i \oplus y_i \\
&= x_i \cdot y_i' + x_i' \cdot y_i \\
&= x_i \cdot y_i' + x_i \cdot x_i' + x_i' \cdot y_i + y_i \cdot y_i' \\
&= (x_i + y_i) \cdot (x_i' + y_i') \\
&= (x_i + y_i) \cdot (x_i \cdot y_i)' \\
&= p_i \cdot g_i'
\end{aligned}
$$

Thus, an AND gate with an inverted input can be used instead of an XOR gate to create each half-sum bit.

Finally, the '283 creates the carry signals using an INVERT-OR-AND structure (the DeMorgan equivalent of an AND-OR-INVERT), which has about the same delay as a single CMOS or TTL inverting gate. This requires some explaining, since the carry equations that we derived in the preceding subsection are used in a slightly modified form. In particular, the c_{i+1} equation uses the term $p_i \cdot g_i$ instead of g_i. This has no effect on the output, since p_i is always 1 when g_i is 1. However, it allows the equation to be factored as follows:

$$
\begin{aligned}
c_{i+1} &= p_i \cdot g_i + p_i \cdot c_i \\
&= p_i \cdot (g_i + c_i)
\end{aligned}
$$

This leads to the following carry equations, which are used by the circuit :

$$
\begin{aligned}
c_1 &= p_0 \cdot (g_0 + c_0) \\
c_2 &= p_1 \cdot (g_1 + c_1) \\
&= p_1 \cdot (g_1 + p_0 \cdot (g_0 + c_0)) \\
&= p_1 \cdot (g_1 + p_0) \cdot (g_1 + g_0 + c_0)
\end{aligned}
$$

Figure 5-91
Logic diagram for
the 74x283 4-bit
binary adder.

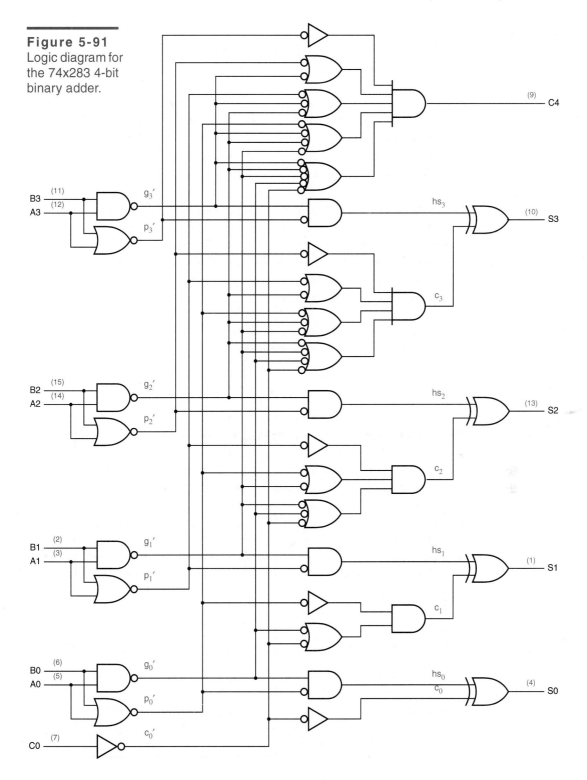

$$c_3 = p_2 \cdot (g_2 + c_2)$$
$$= p_2 \cdot (g_2 + p_1 \cdot (g_1 + p_0) \cdot (g_1 + g_0 + c_0))$$
$$= p_2 \cdot (g_2 + p_1) \cdot (g_2 + g_1 + p_0) \cdot (g_2 + g_1 + g_0 + c_0)$$
$$c_4 = p_3 \cdot (g_3 + c_3)$$
$$= p_3 \cdot (g_3 + p_2 \cdot (g_2 + p_1) \cdot (g_2 + g_1 + p_0) \cdot (g_2 + g_1 + g_0 + c_0))$$
$$= p_3 \cdot (g_3 + p_2) \cdot (g_3 + g_2 + p_1) \cdot (g_3 + g_2 + g_1 + p_0) \cdot (g_3 + g_2 + g_1 + g_0 + c_0)$$

If you've followed the derivation of these equations and can obtain the same ones by reading the '283 logic diagram, then congratulations, you're up to speed on switching algebra! If not, you may want to review Sections 4.1 and 4.2.

group-ripple adder

The propagation delay from the C0 input to the C4 output of the '283 is very short, about the same as two inverting gates. As a result, fairly fast *group-ripple adders* with more than four bits can be made simply by cascading the carry outputs and inputs of '283s, as shown in Figure 5-92 for a 16-bit adder. The total propagation delay from C0 to C16 in this circuit is about the same as that of eight inverting gates.

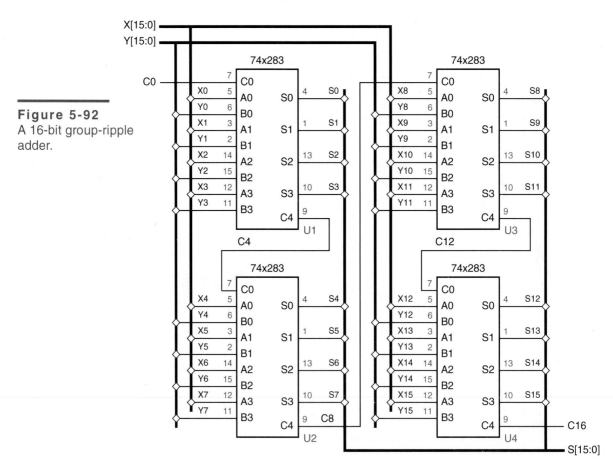

Figure 5-92
A 16-bit group-ripple adder.

*5.10.6 MSI Arithmetic and Logic Units

An *arithmetic and logic unit (ALU)* is a combinational circuit that can perform any of a number of different arithmetic and logical operations on a pair of b-bit operands. The operation to be performed is specified by a set of function-select inputs. Typical MSI ALUs have 4-bit operands and three to five function-select inputs, allowing up to 32 different functions to be performed.

arithmetic and logic unit (ALU)

Figure 5-93 is a logic symbol for the *74x181* 4-bit ALU. The operation performed by the '181 is selected by the M and S3–S0 inputs, as detailed in Table 5-52. Note that the identifiers A, B, and F in the table refer to the 4-bit words A3–A0, B3–B0, and F3–F0; and the symbols · and + refer to logical AND and OR operations.

74x181

The 181's M input selects between arithmetic and logical operations. When M = 1, logical operations are selected, and each output Fi is a function only of the corresponding data inputs, Ai and Bi. No carries propagate between stages, and the CIN input is ignored. The S3–S0 inputs select a particular logical operation; any of the 16 different combinational logic functions on two variables may be selected.

Table 5-52 Functions performed by the 74x181 4-bit ALU.

Inputs				Function	
S3	S2	S1	S0	M = 0 (arithmetic)	M = 1 (logic)
0	0	0	0	F = A minus 1 plus CIN	F = A'
0	0	0	1	F = A · B minus 1 plus CIN	F = A' + B'
0	0	1	0	F = A · B' minus 1 plus CIN	F = A' + B
0	0	1	1	F = 1111 plus CIN	F = 1111
0	1	0	0	F = A plus (A + B') plus CIN	F = A' · B'
0	1	0	1	F = A · B plus (A + B') plus CIN	F = B'
0	1	1	0	F = A minus B minus 1 plus CIN	F = A ⊕ B'
0	1	1	1	F = A + B' plus CIN	F = A + B'
1	0	0	0	F = A plus (A + B) plus CIN	F = A' · B
1	0	0	1	F = A plus B plus CIN	F = A ⊕ B
1	0	1	0	F = A · B' plus (A + B) plus CIN	F = B
1	0	1	1	F = A + B plus CIN	F = A + B
1	1	0	0	F = A plus A plus CIN	F = 0000
1	1	0	1	F = A · B plus A plus CIN	F = A · B'
1	1	1	0	F = A · B' plus A plus CIN	F = A · B
1	1	1	1	F = A plus CIN	F = A

Figure 5-93
Logic symbol for the 74x181 4-bit ALU.

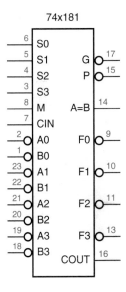

When M = 0, arithmetic operations are selected, carries propagate between the stages, and CIN is used as a carry input to the least significant stage. For operations larger than four bits, multiple '181 ALUs may be cascaded like the group-ripple adder in Figure 5-92, with the carry-out (COUT) of each ALU connected to the carry-in (CIN) of the next most significant stage. The same function-select signals (M, S3–S0) are applied to all the '181s in the cascade.

To perform two's-complement addition, we use S3–S0 to select the operation "A plus B plus CIN." The CIN input of the least-significant ALU is normally set to 0 during addition operations. To perform two's-complement subtraction, we use S3–S0 to select the operation A minus B minus plus CIN. In this case, the CIN input of the least significant ALU is normally set to 1, since CIN acts as the complement of the borrow during subtraction.

The '181 provides other arithmetic operations, such as "A minus 1 plus CIN," that are useful in some applications (e.g., decrement by 1). It also provides a bunch of weird arithmetic operations, such as "A · B′ plus (A + B) plus CIN," that are almost never used in practice, but that "fall out" of the circuit for free.

Notice that the operand inputs A3_L–A0_L and B3_L–B0_L, and the function outputs F3_L–F0_L of the '181 are active low. The '181 can also be used with active-high operand inputs and function outputs. In this case, a different version of the function table must be constructed. When M = 1, logical operations are still performed, but for a given input combination on S3–S0, the function obtained is precisely the dual of the one listed in Table 5-52. When M = 0, arithmetic operations are performed, but the function table is once again different. Refer to a '181 data sheet for more details.

74x381
74x382

Two other MSI ALUs, the *74x381* and *74x382* shown in Figure 5-94, encode their select inputs more compactly, and provide only eight different but useful functions, as detailed in Table 5-53. The only difference between the '381 and '382 is that one provides group-carry lookahead outputs (which we explain next), while the other provides ripple carry and overflow outputs.

Figure 5-94
Logic symbols for 4-bit ALUs: (a) 74x381; (b) 74x382.

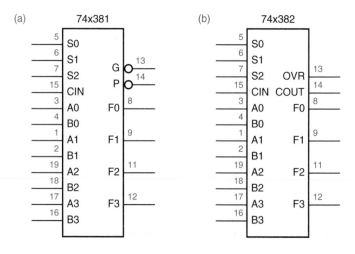

Inputs			
S2	S1	S0	Function
0	0	0	F = 0000
0	0	1	F = B minus A minus 1 plus CIN
0	1	0	F = A minus B minus 1 plus CIN
0	1	1	F = A plus B plus CIN
1	0	0	F = A ⊕ B
1	0	1	F = A + B
1	1	0	F = A · B
1	1	1	F = 1111

Table 5-53
Functions performed by the 74x381 and 74x382 4-bit ALUs.

*5.10.7 Group-Carry Lookahead

The '181 and '381 provide *group-carry lookahead* outputs that allow multiple ALUs to be cascaded without rippling carries between 4-bit groups. Like the 74x283, the ALUs use carry lookahead to produce carries internally. However, they also provide G_L and P_L outputs that are carry lookahead signals for the entire 4-bit group. The G_L output is asserted if the ALU generates a carry—that is, if it will produce a carry-out (COUT = 1) whether or not there is a carry-in (CIN = 1):

group-carry lookahead

$$G\_L = (g_3 + p_3 \cdot g_2 + p_3 \cdot p_2 \cdot g_1 + p_3 \cdot p_2 \cdot p_1 \cdot g_0)'$$

The P_L output is asserted if the ALU propagates a carry—that is, if it will produce a carry-out if there is a carry-in:

$$P\_L = (p_3 \cdot p_2 \cdot p_1 \cdot p_0)'$$

When ALUs are cascaded, the group-carry lookahead outputs may be combined in just two levels of logic to produce the carry input to each ALU. A *lookahead carry circuit*, the *74x182* shown in Figure 5-95, performs this operation. The '182 inputs are C0, the carry input to the least significant ALU ("ALU 0"), and G0–G3 and P0–P3, the generate and propagate outputs of ALUs 0–3. Using these inputs, the '182 produces carry inputs C1–C3 for ALUs 1–3. Figure 5-96 shows the connections for a 16-bit ALU using four '381s and a '182.

lookahead carry circuit
74x182

The 182's carry equations are obtained by "adding out" the basic carry lookahead equation of Section 5.10.4:

$$c_{i+1} = g_i + p_i \cdot c_i$$
$$= (g_i + p_i) \cdot (g_i + c_i)$$

Expanding for the first three values of i, we obtain the following equations:

C1 = (G0+P0) · (G0+C0)
C2 = (G1+P1) · (G1+G0+P0) · (G1+G0+C0)
C3 = (G2+P2) · (G2+G1+P1) · (G2+G1+G0+P0) · (G2+G1+G0+C0)

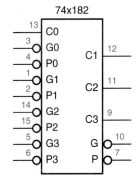

Figure 5-95
Logic symbol for the 74x182 lookahead carry circuit.

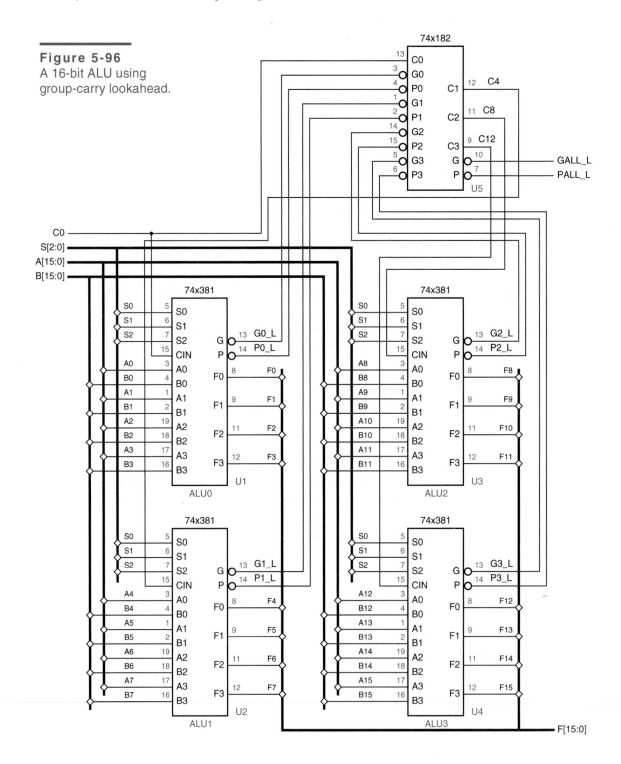

Figure 5-96
A 16-bit ALU using
group-carry lookahead.

The '182 realizes each of these equations with just one level of delay—an INVERT-OR-AND gate.

When more than four ALUs are cascaded, they may be partitioned into "supergroups," each with its own '182. For example, a 64-bit adder would have four supergroups, each containing four ALUs and a '182. The G_L and P_L outputs of each '182 can be combined in a next-level '182, since they indicate whether the supergroup generates or propagates carries:

$$G\_L = ((G3+P3) \cdot (G3+G2+P2) \cdot (G3+G2+G1+P1) \cdot (G3+G2+G1+G0))'$$

$$P\_L = (P0 \cdot P1 \cdot P2 \cdot P3)'$$

*5.10.8 Adders in ABEL and PLDs

ABEL supports addition (+) and subtraction (−) operators which can be applied to sets. Sets are interpreted as unsigned integers; for example, a set with n bits represents an integer in the range of 0 to 2^n-1. Subtraction is performed by negating the subtrahend and adding. Negation is performed in two's complement; that is, the operand is complemented bit-by-bit and then 1 is added.

Table 5-54 shows an example of addition in ABEL. The set definition for SUM was made one bit wider than the addends to accommodate the carry out of the MSB; otherwise this carry would be discarded. The set definitions for the addends were extended on the left with a 0 bit to match the size of SUM.

Even though the adder program is extremely small, it takes a long time to compile and it generates a huge number of terms in the minimal two-level sum of products. While SUM0 has only two product terms, subsequent terms SUMi have $5 \cdot 2^i - 4$ terms, or 636 terms for SUM7! And the carry out (SUM8) has $2^8 - 1 = 255$ product terms. Obviously, adders with more than a few bits cannot be practically realized using two levels of logic.

Table 5-54
ABEL program for an 8-bit adder.

```
module add
title 'Adder Exercise'

" Input and output pins
A7..A0, B7..B0          pin;
SUM8..SUM0              pin istype 'com';

" Set definitions
A = [0, A7..A0];
B = [0, B7..B0];
SUM = [SUM8..SUM0];

equations
SUM = A + B;

end add
```

@CARRY directive

Recognizing that larger adders and comparators are still needed in PLDs from time to time, ABEL provides an *@CARRY directive* which tells the compiler to synthesize a group-ripple adder with *n* bits per group. For example, if the statement "@CARRY 1;" were included in Table 5-54, the compiler would create eight new signals for the carries out of bit positions 0 through 7. The equations for SUM1 through SUM8 would use these internal carries, essentially creating an 8-stage ripple adder with a worst-case delay of eight passes through the PLD.

If the statement "@CARRY 2;" were used, the compiler would compute carries two bits at a time, creating four new signals for carries out of bit positions 1, 3, 5, and 7. In this case, the maximum number of product terms needed for any output is still reasonable, only 7, and the worst-case delay path has just four passes through the PLD. With three bits per group (@CARRY 3;), the maximum number of product terms balloons to 28, which is impractical.

A special case that is often used in ABEL and PLDs is adding or subtracting a constant 1. This operation is used in the definition of counters, where the next state of the counter is just the current state plus 1 for an "up" counter or minus 1 for a "down" counter. The equation for bit *i* of an "up" counter can be stated very simply in words: "Complement bit *i* if counting is enabled and all of the bits lower than *i* are 1." This requires just $i + 2$ product terms for any value of *i*, and can be further reduced to just one product term and an XOR gate in some PLDs and CPLDs.

*5.10.9 Adders in VHDL

Although VHDL has addition (+) and subtraction (−) operators built in, they work only with the integer, real, and physical types. They specifically do *not* work with BIT_VECTOR types or the IEEE standard type STD_LOGIC_VECTOR. Instead, standard packages define these operators.

As we explained in Section 5.9.6, the IEEE_std_logic_arith package defines two new array types, SIGNED and UNSIGNED, and a set of comparison functions for operands of type INTEGER, SIGNED, or UNSIGNED. The package also defines addition and subtraction operations for the same kinds of operands as well as STD_LOGIC and STD_ULOGIC for 1-bit operands.

The large number of overlaid addition and subtraction functions may make it less than obvious what type an addition or subtraction result will have. Normally, if any of the operands is type SIGNED, the result is SIGNED, else the result is UNSIGNED. However, if the result is assigned to a signal or variable of

```
library IEEE;
use IEEE.std_logic_1164.all;
use IEEE.std_logic_arith.all;

entity vadd is
    port (
        A, B: in UNSIGNED (7 downto 0);
        C: in SIGNED (7 downto 0);
        D: in STD_LOGIC_VECTOR (7 downto 0);
        S: out UNSIGNED (8 downto 0);
        T: out SIGNED (8 downto 0);
        U: out SIGNED (7 downto 0);
        V: out STD_LOGIC_VECTOR (8 downto 0)
    );
end vadd;

architecture vadd_arch of vadd is
  begin
    S <= ('0' & A) + ('0' & B);
    T <= A + C;
    U <= C + SIGNED(D);
    V <= C - UNSIGNED(D);
end vadd_arch;
```

Table 5-55
VHDL program for adding and subtracting 8-bit integers of various types.

type STD_LOGIC_VECTOR, then the SIGNED or UNSIGNED result is converted to that type. The length of any result is normally the length of the longest operand. However, when an UNSIGNED operand is combined with a SIGNED or INTEGER operand, its length is increased by 1 to accommodate a sign bit of 0, and then the result's length is determined.

Incorporating these considerations, the VHDL program in Table 5-55 shows 8-bit additions for various operand and result types. The first result, S, is declared to be 9 bits long assuming the designer is interested in the carry from the 8-bit addition of UNSIGNED operands A and B. The concatenation operator & is used to extend A and B so that the addition function will return the carry bit in the MSB of the result.

The next result, T, is also 9 bits long, since the addition function extends the UNSIGNED operand A when combining it with the SIGNED operand C. In the third addition, an 8-bit STD_LOGIC_VECTOR D is type-converted to SIGNED and combined with C to obtain an 8-bit SIGNED result U. In the last statement, D is converted to UNSIGNED, automatically extended by one bit, and subtracted from C to produce a 9-bit result V.

Since addition and subtraction are fairly expensive in terms of the number of gates required, many VHDL synthesis engines will attempt to reuse adder blocks whenever possible. For example, Table 5-56 is a VHDL program that includes two different additions. Rather than building two adders and selecting

Table 5-56
VHDL program that allows adder sharing.

```
library IEEE;
use IEEE.std_logic_1164.all;
use IEEE.std_logic_arith.all;

entity vaddshr is
    port (
        A, B, C, D: in SIGNED (7 downto 0);
        SEL: in STD_LOGIC;
        S: out SIGNED (7 downto 0)
    );
end vaddshr;

architecture vaddshr_arch of vaddshr is
begin
    S <= A + B when SEL = '1' else C + D;
end vaddshr_arch;
```

one's output with a multiplexer, the synthesis engine can build just one adder and select its inputs using multiplexers, potentially creating a smaller overall circuit.

*5.11 Combinational Multipliers

*5.11.1 Combinational Multiplier Structures

In Section 2.8 we outlined an algorithm that uses n shifts and adds to multiply n-bit binary numbers. Although the shift-and-add algorithm emulates the way that we do paper-and-pencil multiplication of decimal numbers, there is nothing inherently "sequential" or "time dependent" about multiplication. That is, given two n-bit input words X and Y, it is possible to write a truth table that expresses the $2n$-bit product $P = X \cdot Y$ as a *combinational* function of X and Y. A *combinational multiplier* is a logic circuit with such a truth table.

combinational multiplier

Most approaches to combinational multiplication are based on the paper-and-pencil shift-and-add algorithm. Figure 5-97 illustrates the basic idea for an 8×8 multiplier for two unsigned integers, multiplicand $X = x_7x_6x_5x_4x_3x_2x_1x_0$ and multiplier $Y = y_7y_6y_5y_4y_3y_2y_1y_0$. We call each row a *product component*, a shifted

product component

Figure 5-97
Partial products in an 8×8 multiplier.

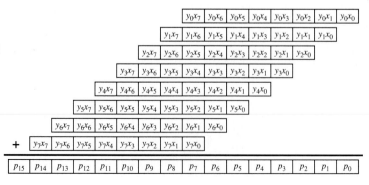

Figure 5-98
Interconnections
for an 8 × 8
combinational
multiplier.

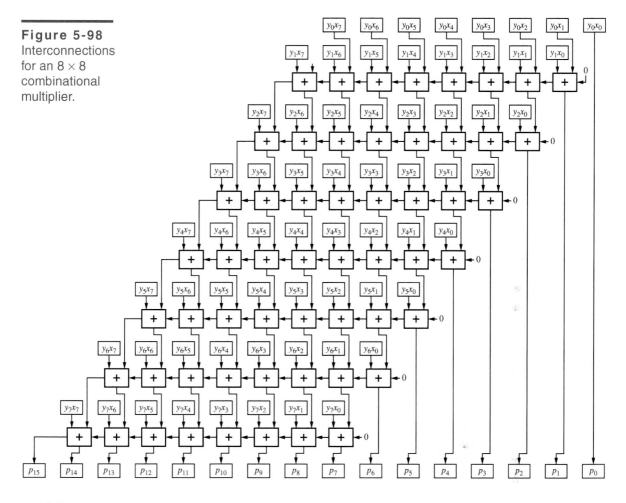

multiplicand that is multiplied by 0 or 1 depending on the corresponding multi-plier bit. Each small box represents one product-component bit $y_i x_j$, the logical **AND** of multiplier bit y_i and multiplicand bit x_j. The product $P = p_{15}p_{14}\ldots p_2 p_1 p_0$ has 16 bits and is obtained by adding together all the product components.

Figure 5-98 shows one way to add up the product components. Here, the product-component bits have been spread out to make space, and each "+" box is a full-adder equivalent to Figure 5-86(c) on page 431. The carries in each row of full adders are connected to make an 8-bit ripple adder. Thus, the first ripple adder combines the first two product components to product the first partial product, as defined in Section 2.8. Subsequent adders combine each partial product with the next product component.

It is interesting to study the propagation delay of the circuit in Figure 5-98. In the worst case, the inputs to the least significant adder ($y_0 x_1$ and $y_1 x_0$) can affect the MSB of the product (p_{15}). If we assume for simplicity that the delays from any input to any output of a full adder are equal, say t_{pd}, then the worst-case

Figure 5-99
Interconnections
for a faster 8×8
combinational
multiplier.

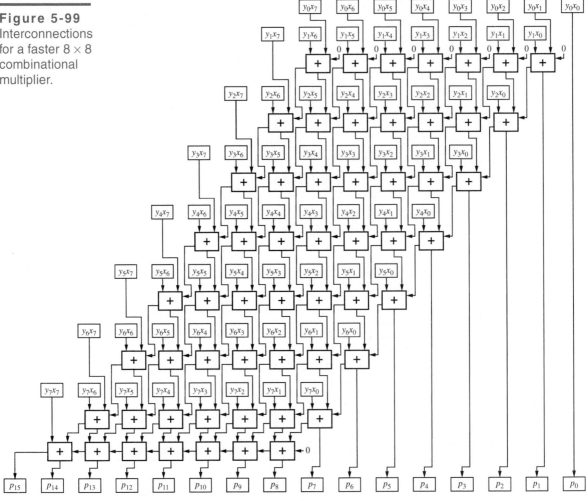

path goes through 20 adders and its delay is $20t_{pd}$. If the delays are different, then the answer depends on the relative delays; see Exercise 5.83.

sequential multiplier

Sequential multipliers use a single adder and a register to accumulate the partial products. The partial-product register is initialized to the first product component, and for an $n \times n$-bit multiplication, $n - 1$ steps are taken and the adder is used $n - 1$ times, once for each of the remaining $n - 1$ product components to be added to the partial-product register.

carry-save addition

Some sequential multipliers use a trick called *carry-save addition* to speed up multiplication. The idea is to break the carry chain of the ripple adder to shorten the delay of each addition. This is done by applying the carry output from bit i during step j to the carry input for bit $i + 1$ during the *next* step, $j + 1$. After the last product component is added, one more step is needed in which the carries are hooked up in the usual way and allowed to ripple from the least to the most significant bit.

The combinational equivalent of an 8×8 multiplier using carry-save addition is shown in Figure 5-99. Notice that the carry out of each full adder in the first seven rows is connected to an input of an adder *below* it. Carries in the eighth row of full adders are connected to create a conventional ripple adder. Although this adder uses exactly the same amount of logic as the previous one (64 2-input AND gates and 56 full adders), its propagation delay is substantially shorter. Its worst-case delay path goes through only 14 full adders. The delay can be further improved by using a carry lookahead adder for the last row.

The regular structure of combinational multipliers makes them ideal for VLSI and ASIC realization. The importance of fast multiplication in microprocessors, digital video, and many other applications has led to much study and development of even better structures and circuits for combinational multipliers; see the References.

*5.11.2 Multiplication in ABEL and PLDs

ABEL provides a multiplication operator $*$, but it can be used only with individual signals, numbers, or special constants, not with sets. Thus, ABEL cannot synthesize a multiplier circuit from a single equation like "$P = X*Y$."

Still, you can use ABEL to specify a combinational multiplier if you break it down into smaller pieces. For example, Table 5-57 shows the design of a 4×4 unsigned multiplier following the same general structure as Figure 5-97 on page page 446. Expressions are used to define the four product components, PC1, PC2, PC3, and PC4, which are then added in the equations section of the program. This does not generate an array of full adders as in Figure 5-98 or 5-99. Rather, the ABEL compiler will dutifully crunch the addition equation to

Table 5-57
ABEL program for a 4×4 combinational multiplier.

```
module mul4x4
title '4x4 Combinational Multiplier'

X3..X0, Y3..Y0 pin;  " multiplicand, multiplier
P7..P0          pin istype 'com';   " product

P = [P7..P0];
PC1 = Y0 & [0, 0, 0, 0,X3,X2,X1,X0];
PC2 = Y1 & [0, 0, 0,X3,X2,X1,X0, 0];
PC3 = Y2 & [0, 0,X3,X2,X1,X0, 0, 0];
PC4 = Y3 & [0,X3,X2,X1,X0, 0, 0, 0];

equations
P = PC1 + PC2 + PC3 + PC4;

end mul4x4
```

produce a minimal sum for each of the eight product output bits. Surprisingly, the worst-case output, P4, has only 36 product terms, a little high but certainly realizable in two passes through a PLD.

*5.11.3 Multiplication in VHDL

VHDL is rich enough to express multiplication in a number of different ways; we'll save the best for last.

Table 5-58 is a behavioral VHDL program that mimics the multiplier structure of Figure 5-99. In order to represent the internal signals in the figure, the program defines a new data type, array8x8, which is a two-dimensional array of STD_LOGIC (recall that STD_LOGIC_VECTOR is a one-dimensional array of STD_LOGIC). Variable PC is declared as such an array to hold the product-component bits, and variables PCS and PCC are similar arrays to hold the sum and carry outputs of the main array of full adders. One-dimensional arrays RAS and RAC hold the sum and carry outputs of the ripple adder. Figure 5-100 shows the variable naming and numbering scheme. Integer variables i and j are used as loop indices for rows and columns, respectively.

The program attempts to illustrate the logic gates that would be used in a faithful realization of Figure 5-99, even though a synthesizer could legitimately create quite a different structure from this behavioral program. If you want to control the structure, then you must use structural VHDL, as we'll show later.

In the program, the first, nested for statement performs 64 AND operations to obtain the product-component bits. The next for loop initializes boundary conditions at the top of the multiplier, using the notion of row-0 "virtual" full

Figure 5-100
VHDL variable names for the 8×8 multiplier.

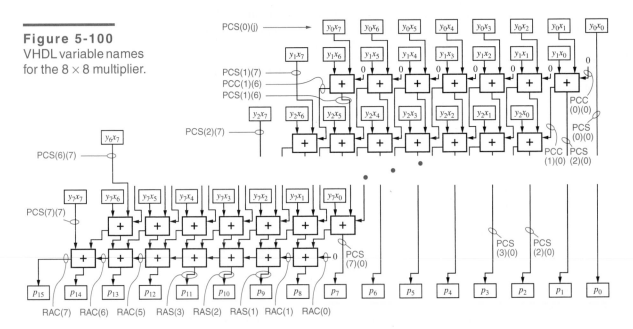

```
library IEEE;
use IEEE.std_logic_1164.all;

entity vmul8x8p is
    port ( X: in STD_LOGIC_VECTOR (7 downto 0);
           Y: in STD_LOGIC_VECTOR (7 downto 0);
           P: out STD_LOGIC_VECTOR (15 downto 0) );
end vmul8x8p;

architecture vmul8x8p_arch of vmul8x8p is
function MAJ (I1, I2, I3: STD_LOGIC) return STD_LOGIC is
  begin
    return ((I1 and I2) or (I1 and I3) or (I2 and I3));
  end MAJ;
begin
process (X, Y)
type array8x8 is array (0 to 7) of STD_LOGIC_VECTOR (7 downto 0);
variable PC: array8x8;        -- product component bits
variable PCS: array8x8;       -- full-adder sum bits
variable PCC: array8x8;       -- full-adder carry output bits
variable RAS, RAC: STD_LOGIC_VECTOR (7 downto 0); -- ripple adder sum
  begin                                            --   and carry bits
    for i in 0 to 7 loop for j in 0 to 7 loop
        PC(i)(j) := Y(i) and X(j); -- compute product component bits
    end loop;  end loop;
    for j in 0 to 7 loop
      PCS(0)(j) := PC(0)(j);  -- initialize first-row "virtual"
      PCC(0)(j) := '0';       --    adders (not shown in figure)
    end loop;
    for i in 1 to 7 loop      -- do all full adders except last row
      for j in 0 to 6 loop
        PCS(i)(j) := PC(i)(j) xor PCS(i-1)(j+1) xor PCC(i-1)(j);
        PCC(i)(j) := MAJ(PC(i)(j), PCS(i-1)(j+1), PCC(i-1)(j));
        PCS(i)(7) := PC(i)(7); -- leftmost "virtual" adder sum output
      end loop;
    end loop;
    RAC(0) := '0';
    for i in 0 to 6 loop  -- final ripple adder
      RAS(i) := PCS(7)(i+1) xor PCC(7)(i) xor RAC(i);
      RAC(i+1) := MAJ(PCS(7)(i+1), PCC(7)(i), RAC(i));
    end loop;
    for i in 0 to 7 loop
      P(i) <= PCS(i)(0);  -- first 8 product bits from full-adder sums
    end loop;
    for i in 8 to 14 loop
      P(i) <= RAS(i-8);   -- next 7 bits from ripple-adder sums
    end loop;
    P(15) <= RAC(7);      -- last bit from ripple-adder carry
  end process;
end vmul8x8p_arch;
```

Table 5-58
Behavioral VHDL
program for an 8×8
combinational
multiplier.

Table 5-59
Structural VHDL
architecture for an
8×8 combinational
multiplier.

```vhdl
architecture vmul8x8s_arch of vmul8x8s is
component AND2
  port( I0, I1: in STD_LOGIC;
        O: out STD_LOGIC );
end component;
component XOR3
  port( I0, I1, I2: in STD_LOGIC;
        O: out STD_LOGIC );
end component;
component MAJ   -- Majority function, O = I0*I1 + I0*I2 + I1*I2
  port( I0, I1, I2: in STD_LOGIC;
        O: out STD_LOGIC );
end component;

type array8x8 is array (0 to 7) of STD_LOGIC_VECTOR (7 downto 0);
signal PC: array8x8;      -- product-component bits
signal PCS: array8x8;     -- full-adder sum bits
signal PCC: array8x8;     -- full-adder carry output bits
signal RAS, RAC: STD_LOGIC_VECTOR (7 downto 0); -- sum, carry
begin
  g1: for i in 0 to 7 generate     -- product-component bits
    g2: for j in 0 to 7 generate
        U1: AND2 port map (Y(i), X(j), PC(i)(j));
    end generate;
  end generate;
  g3: for j in 0 to 7 generate
    PCS(0)(j) <= PC(0)(j);  -- initialize first-row "virtual" adders
    PCC(0)(j) <= '0';
  end generate;
  g4: for i in 1 to 7 generate     -- do full adders except the last row
    g5: for j in 0 to 6 generate
      U2: XOR3 port map (PC(i)(j),PCS(i-1)(j+1),PCC(i-1)(j),PCS(i)(j));
      U3: MAJ  port map (PC(i)(j),PCS(i-1)(j+1),PCC(i-1)(j),PCC(i)(j));
      PCS(i)(7) <= PC(i)(7); -- leftmost "virtual" adder sum output
    end generate;
  end generate;
  RAC(0) <= '0';
  g6: for i in 0 to 6 generate  -- final ripple adder
    U7: XOR3 port map (PCS(7)(i+1), PCC(7)(i), RAC(i), RAS(i));
    U3: MAJ  port map (PCS(7)(i+1), PCC(7)(i), RAC(i), RAC(i+1));
  end generate;
  g7: for i in 0 to 7 generate
    P(i) <= PCS(i)(0); -- get first 8 product bits from full-adder sums
  end generate;
  g8: for i in 8 to 14 generate
    P(i) <= RAS(i-8);  -- get next 7 bits from ripple-adder sums
  end generate;
  P(15) <= RAC(7);     -- get last bit from ripple-adder carry
end vmul8x8s_arch;
```

adders, not shown in the figure, whose sum outputs equal the first row of PC bits and whose carry outputs are 0. The third, nested for loop corresponds to the main array of adders in Figure 5-99, all except the last row, which is handled by the fourth for loop. The last two for loops assign the appropriate adder outputs to the multiplier output signals.

The program in Table 5-58 can be modified to use structural VHDL as shown in Table 5-59. This approach gives the designer complete control over the circuit structure that is synthesized, as might be desired in an ASIC realization. The program assumes that the architectures for AND2, XOR3, and MAJ3 have been defined elsewhere, for example, in an ASIC library.

This program makes good use of the *generate statement* to create the arrays of components used in the multiplier. The generate statement must have a label, and similar to a for-loop statement, it specifies an iteration scheme to control the repetition of the enclosed statements. Within for-generate, the enclosed statements can include any concurrent statements, IF-THEN-ELSE statements, and additional levels of looping constructs. Sometimes generate statements are combined with IF-THEN-ELSE to produce a kind of conditional compilation capability

generate statement

Well, we said we'd save the best for last, and here it is. The IEEE std_logic_arith library that we introduced in Section 5.9.6 defines multiplication functions for SIGNED and UNSIGNED types, and overlays these functions onto the "*" operator. Thus, the program in Table 5-60 can multiply unsigned numbers with a simple one-line assignment statement.

Within the IEEE std_logic_arith library, the multiplication function is defined behaviorally, using the shift-and-add algorithm. We could have showed you this approach at the beginning of this subsection, but then you wouldn't have read the rest of it, would you?

```
library IEEE;
use IEEE.std_logic_1164.all;
use IEEE.std_logic_arith.all;

entity vmul8x8i is
    port (
        X: in UNSIGNED (7 downto 0);
        Y: in UNSIGNED (7 downto 0);
        P: out UNSIGNED (15 downto 0)
    );
end vmul8x8i;

architecture vmul8x8i_arch of vmul8x8i is
begin
  P <= X * Y;
end vmul8x8i_arch;
```

Table 5-60
Truly behavioral VHDL program for an 8×8 combinational multiplier.

SIGNALS VS.
VARIABLES

Variables are used rather than signals in the process in Table 5-58 to make the simulation run faster. Variables are faster because the simulator keeps track of their values only when the process is running. Because variable values are assigned sequentially, the process in Table 5-58 is carefully written to compute values in the proper order. That is, a variable cannot be used until a value has been assigned to it.

Signals, on the other hand, have a value at all times. When a signal value is changed in a process, the simulator schedules a future event in its event list for the value change. If the signal appears on the righthand side of an assignment statement in the process, then the signal must also be included in the process' sensitivity list. If a signal value changes, the process will then execute again, and keep repeating until all of the signals in the sensitivity list are stable.

In Table 5-58, if you wanted to observe internal values or timing during simulation, you could change all the variables (except i and j) to signals and include them in the sensitivity list. To make the program syntactically correct, you would also have to move the type and signal declarations to just after the architecture statement, and change all of the ":=" assignments to "<=".

The simulation will run a lot slower after you make these changes above. As we discussed in connection with Figure 5-99 on page 448, the worst-case signal propagation path through the adder array goes through 14 adders. This is also the number of simulation cycles that must occur for the signals in the changed VHDL program to stabilize, plus one more cycle for the simulator to notice that they've stabilized.

While the choice of signals vs. variables affects the speed of simulation, with most VHDL synthesis engines it does not affect the results of synthesis.

ON THE
THRESHOLD OF
A DREAM

A three-input "majority function," MAJ, is defined at the beginning of Table 5-58 and is subsequently used to compute carry outputs. An *n*-input *majority function* produces a 1 output if the majority of its inputs are 1, two out of three in the case of a 3-input majority function. (If *n* is even, $n/2 + 1$ inputs must be 1.)

Over thirty years ago, there was substantial academic interest in a more general class of *n*-input *threshold functions* which produce a 1 output if *k* or more of their inputs are 1. Besides providing full employment for logic theoreticians, threshold functions could realize many logic functions with a smaller number of elements than could a conventional AND/OR realization. For example, an adder's carry function requires three AND gates and one OR gate, but just one three-input threshold gate.

(Un)fortunately, an economical technology never emerged for threshold gates, and they remain, for now, an academic curiosity.

<table>
<tr><td>SYNTHESIS OF
BEHAVIORAL
DESIGNS</td><td>You've probably heard that compilers for high-level programming languages like C usually generate better code than people do writing in assembly language, even with "hand-tweaking." Most digital designers hope that compilers for behavioral HDLs will also some day produce results superior to a typical hand-tweaked design, be it a schematic or structural VHDL. Better compilers won't put the designers out of work, they will simply allow them to tackle bigger designs.</td></tr>
</table>

We're not quite there yet. However, the more advanced synthesis engines do include some nice optimizations for commonly used behavioral structures. For example, I have to admit that the FPGA synthesis engine that I used to test the VHDL programs in this subsection produced just as fast a multiplier from Table 5-60 as it did from any of the more detailed architectures!

References

Digital designers who want to improve their writing should start by reading the classic *Elements of Style*, third edition, by William Strunk, Jr., and E. B. White (Allyn & Bacon, 1979). Probably the most inexpensive and concise yet very useful guide to technical writing is *The Elements of Technical Writing*, by Gary Blake and Robert W. Bly (Macmillan, 1993). For a more encyclopedic treatment, see *Handbook of Technical Writing*, fifth edition, by Brusaw, Alred, and Oliu (St. Martin's Press, 1997).

The ANSI/IEEE standard for logic symbols is Std 91-1984, *IEEE Standard Graphic Symbols for Logic Functions*. Another standard of interest to logic designers is ANSI/IEEE 991-1986, *Logic Circuit Diagrams*. These two standards and ten others, including standard symbols for 10-inch gongs and maid's-signal plugs, can be found in one handy, five-pound reference, *Electrical and Electronics Graphic and Letter Symbols and Reference Designations Standards Collection*, published by the IEEE in 1996 (www.ieee.org).

Real logic devices are described in data sheets and data books published by the manufacturers. Updated editions of data books used to be published every few years, but recently the trend has been to minimize or eliminate the hardcopy editions and instead to publish up-to-date information on the Web. Two of the largest suppliers with the most comprehensive sites are Texas Instruments (www.ti.com) and Motorola (www.mot.com).

For a given logic family such as 74ALS, all manufacturers list generally equivalent specifications, so you can get by with just one data book per family. Some specifications, especially timing, may vary slightly between manufacturers, so when timing is tight it's best to check a couple of different sources and use the worst case. That's a *lot* easier than convincing your manufacturing department to buy a component only from a single supplier.

The first PAL devices were invented at Monolithic Memories, Inc. (MMI) in 1978 by John Birkner and H. T. Chua. The inventors earned U.S. patent number 4,124,899 for their invention, and MMI rewarded them by buying them a brand new Porsche and Mercedes, respectively! Seeing the value in this technology (PAL devices, not cars), Advanced Micro Devices (AMD) acquired MMI in the early 1980s and became a leading developer and supplier of new PLDs and CPLDs. In 1997, AMD spun off its PLD operations to a subsidiary, Vantis Corporation, which they sold in 1999 to former competitor Lattice Semiconductor.

Some of the best resources for learning about PLD-based design are provided by the PLD manufacturers. Xilinx Corporation, which started in the FPGA business, also makes CPLDs, and publishes a comprehensive *Xilinx Data Book* in paper (San Jose, CA 95124, 1999) and on the Web (`www.xilinx.com`). Similarly, GAL inventor Lattice Semiconductor has a comprehensive *Lattice Data Book* (Hillsboro, OR 97124, 1999) and Web site (`www.latticesemi.com`).

A much more detailed discussion of the internal operation of LSI and VLSI devices, including PLDs, ROMs, and RAMs, can be found in electronics texts such as *Microelectronics*, second edition, by J.Millman and A.Grabel (McGraw-Hill, 1987) and *VLSI Engineering* by Thomas E. Dillinger (Prentice Hall, 1988).

On the technical side of digital design, lots of textbooks cover digital design principles, but only a few cover practical aspects of design. An excellent short book focusing on digital design practices is *The Well-Tempered Digital Design* by Robert B. Seidensticker (Addison-Wesley, 1986). It contains hundreds of readily accessible digital-design "proverbs" in areas ranging from high-level design philosophy to manufacturability.

Drill Problems

5.1 Give three examples of combinational logic circuits that require *billions and billions* of rows to describe in a truth table. For each circuit, describe its inputs and output(s) and indicate exactly how many rows the truth table contains; you need not write out the truth table. (*Hint:* You can find several such circuits right in this chapter.)

5.2 Draw the DeMorgan equivalent symbol for a 74x30 8-input NAND gate.

5.3 Draw the DeMorgan equivalent symbol for a 74x27 3-input NOR gate.

5.4 What's wrong with the signal name "READY′"?

5.5 You may find it annoying to have to keep track of the active levels of all the signals in a logic circuit. Why not use only noninverting gates, so all signals are active high?

5.6 True or false: In bubble-to-bubble logic design, outputs with a bubble can be connected only to inputs with a bubble.

5.7 A digital communication system is being designed with twelve identical network ports. Which type of schematic structure is probably most appropriate for the design?

5.8   Determine the exact maximum propagation delay from IN to OUT of the circuit in Figure X5.8 for both LOW-to-HIGH and HIGH-to-LOW transitions, using the timing information given in Table 5-2. Repeat, using a single worst-case delay number for each gate, and compare and comment on your results.

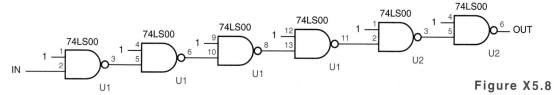

Figure X5.8

5.9   Repeat Drill 5.8, substituting 74HCT00s for the 74LS00s.

5.10   Repeat Drill 5.8, substituting 74LS08s for the 74LS00s.

5.11   Repeat Drill 5.8, substituting 74AHCT02s for the 74LS00s, using constant 0 instead of constant 1 inputs, and using typical rather than maximum timing.

5.12   Estimate the minimum propagation delay from IN to OUT for the circuit shown in Figure X5.12. Justify your answer.

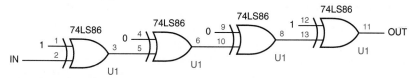

Figure X5.12

5.13   Determine the exact maximum propagation delay from IN to OUT of the circuit in Figure X5.12 for both LOW-to-HIGH and HIGH-to-LOW transitions, using the timing information given in Table 5-2. Repeat, using a single worst-case delay number for each gate, and compare and comment on your results.

5.14   Repeat Drill 5.13, substituting 74HCT86s for the 74LS86s.

5.15   Which would expect to be faster, a decoder with active-high outputs or one with active-low outputs?

5.16   Using the information in Table 5-3 for 74LS components, determine the maximum propagation delay from any input to any output in the 5-to-32 decoder circuit of Figure 5-39. You may use the "worst-case" analysis method.

5.17   Repeat Drill 5.16, performing a detailed analysis for each transition direction, and compare your results.

5.18   Draw the digits created by a 74x49 seven-segment decoder for the nondecimal inputs 1010 through 1111.

5.19   Show how to build each of the following single- or multiple-output logic functions using one or more 74x138 or 74x139 binary decoders and NAND gates. (*Hint:* Each realization should be equivalent to a sum of minterms.)

(a) $F = \Sigma_{X,Y,Z}(2,4,7)$

(b) $F = \Pi_{A,B,C}(3,4,5,6,7)$

(c) $F = \Sigma_{A,B,C,D}(2,4,6,14)$

(d) $F = \Sigma_{W,X,Y,Z}(0,1,2,3,5,7,11,13)$

(e) $F = \Sigma_{W,X,Y}(1,3,5,6)$
     $G = \Sigma_{W,X,Y}(2,3,4,7)$

(f) $F = \Sigma_{A,B,C}(0,4,6)$
     $G = \Sigma_{C,D,E}(1,2)$

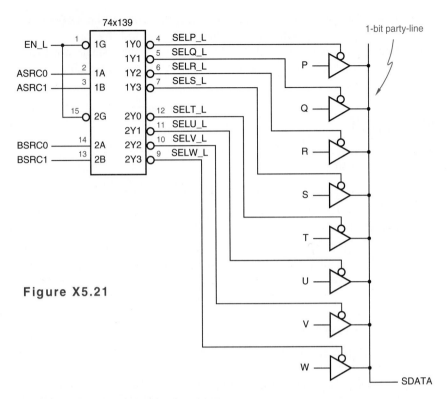

Figure X5.21

5.20 Starting with the logic diagram for the 74x148 priority encoder, write logic equations for its A2_L, A1_L, and A0_L outputs. How do they differ from the "generic" equations given in Section 5.5.1?

5.21 What's terribly wrong with the circuit in Figure X5.21? Suggest a change that eliminates the terrible problem.

5.22 Using the information in Tables 5-2 and 5-3 for 74LS components, determine the maximum propagation delay from any input to any output in the 32-to-1 multiplexer circuit of Figure 5-66. You may use the "worst-case" analysis method.

5.23 Repeat Exercise 5.22 using 74HCT components.

5.24 An n-input parity tree can be built with XOR gates in the style of Figure 5-74(a). Under what circumstances does a similar n-input parity tree built using XNOR gates perform exactly the same function?

5.25 Using the information in Tables 5-2 and 5-3 for 74LS components, determine the maximum propagation delay from the DU bus to the DC bus in the error-correction circuit of Figure 5-77. You may use the "worst-case" analysis method.

5.26 Repeat Exercise 5.25 using 74HCT components.

5.27 Starting with the equations given in Section 5.9.4, write a complete logic expression for the ALTBOUT output of the 74x85.

5.28 Write an algebraic expression for s_2, the third sum bit of a binary adder, as a function of inputs x_0, x_1, x_2, y_0, y_1, and y_2. Assume that $c_0 = 0$, and do not attempt to "multiply out" or minimize the expression.

5.29   Starting with the logic diagram for the 74x682, write a logic expression for the PGTQ_L output in terms of the inputs.

5.30   Using the information in Table 5-3 for 74LS components, determine the maximum propagation delay from any input to any output of the 16-bit group ripple adder of Figure 5-92. You may use the "worst-case" analysis method.

Exercises

5.31   A possible definition of a BUT gate (Exercise 4.50) is "Y1 is 1 if A1 and B1 are 1 *but* either A2 or B2 is 0; Y2 is defined symmetrically." Write the truth table and find minimal sum-of-products expressions for the BUT-gate outputs. Draw the logic diagram for a NAND-NAND circuit for the expressions, assuming that only uncomplemented inputs are available. You may use gates from 74x00, '04, '10, '20, and '30 packages.

5.32   Find a gate-level design for the BUT gate defined in Exercise 5.31 that uses a minimum number of transistors when realized in CMOS. You may use gates from 74x00, '02, '04, '10, '20, and '30 packages. Write the output expressions (which need not be two-level sums of products), and draw the logic diagram.

5.33   For each circuit in the two preceding exercises, compute the worst-case delay from input to output, using the delay numbers for 74HCT components in Table 5-2. Compare the cost (number of transistors), speed, and input loading of the two designs. Which is better?

5.34   Butify the function $F = \Sigma_{W,X,Y,Z}(3,7,11,12,13,14)$. That is, show how to perform F with a single BUT gate as defined in Exercise 5.31 and a single 2-input OR gate.   *butification*

5.35   Suppose that a 74LS138 decoder is connected so that all enable inputs are asserted and C B A = 101. Using the information in Table 5-3 and the '138 internal logic diagram, determine the propagation delay from input to all relevant outputs for each possible single-input change. (*Hint:* There are a total of nine delay numbers, since a change on A, B, or C affects two outputs, and a change on any of the three enable inputs affects one output.)

5.36   Suppose that you are asked to design a new component, a decimal decoder that is optimized for applications in which only decimal input combinations are expected to occur. How can the cost of such a decoder be minimized compared to one that is simply a 4-to-16 decoder with six outputs removed? Write the logic equations for all ten outputs of the minimized decoder, assuming active-high inputs and outputs and no enable inputs.

5.37   How many Karnaugh maps would be required to work Exercise 5.36 using the formal multiple-output minimization procedure described in Section 4.3.8?

5.38   Suppose that a system requires a 5-to-32 binary decoder with a single active-low enable input, similar to Figure 5-39. With the EN1 input pulled HIGH, either the EN2_L or the EN3_L input in the figure could be used as the enable, with the other input grounded. Discuss the pros and cons of using EN2_L versus EN3_L.

5.39   Determine whether the a, b, and c output circuits in the 74x49 seven-segment decoder correspond to minimal product-of-sums expressions for these segments, assuming that the nondecimal input combinations are "don't cares" and BI_L = 1.

Figure X5.40

5.40   Redesign the MSI 74x49 seven-segment decoder so that the digits 6 and 9 have tails as shown in Figure X5.40. Are any of the digit patterns for nondecimal inputs 1010 through 1111 affected by your redesign?

5.41   Starting with the ABEL program in Table 5-22, write a program for a seven-segment decoder with the following enhancements:

- The outputs are all active low.
- Two new inputs, ENHEX and ERRDET, control segment-output decoding.
- If ENHEX = 0, the outputs match the behavior of a 74x49.
- If ENHEX = 1, then the outputs for digits 6 and 9 have tails, and the outputs for digits A–F are controlled by ERRDET.
- If ENHEX = 1 and ERRDET = 0, then the outputs for digits A–F look like the letters A–F, as in the original program.
- If ENHEX = 1 and ERRDET = 1, then digits A–F look like the letter S.

5.42   A famous logic designer decided to quit teaching and make a fortune by licensing the circuit design shown in Figure X5.42.

(a) Label the inputs and outputs of the circuit with appropriate signal names, including active-level indications.

(b) What does the circuit do? Be specific and account for all inputs and outputs.

(c) Draw the logic symbol that would go on the data sheet of this circuit.

(d) Write an ABEL or behavioral VHDL program for the circuit.

(e) With what standard building blocks does the new circuit compete? Do you think it would be successful as an MSI part?

5.43   An FCT three-state buffer drives ten FCT inputs and a 4.7-KΩ pull-up resistor to 5.0 V. When the output changes from LOW to Hi-Z, estimate how long it takes for the FCT inputs to see the output as HIGH. State any assumptions that you make.

5.44   On a three-state bus, ten FCT three-state buffers are driving ten FCT inputs and a 4.7-KΩ pull-up resistor to 5.0 V. Assuming that no other devices are driving the bus, estimate how long the bus signal remains at a valid logic level when an active output enters the Hi-Z state. State any assumptions that you make.

5.45   Design a 10-to-4 encoder with inputs in the 1-out-of-10 code and outputs in BCD.

5.46   Draw the logic diagram for a 16-to-4 encoder using just four 8-input NAND gates. What are the active levels of the inputs and outputs in your design?

5.47   Draw the logic diagram for a circuit that uses the 74x148 to resolve priority among eight active-high inputs, I0–I7, where I7 has the highest priority. The circuit should produce active-high address outputs A2–A0 to indicate the number of the highest-priority asserted input. If no input is asserted, then A2–A0 should be 111 and an IDLE output should be asserted. You may use discrete gates in addition to the '148. Be sure to name all signals with the proper active levels.

5.48   Draw the logic diagram for a circuit that resolves priority among eight active-low inputs, I0_L–I7_L, where I0_L has the highest priority. The circuit should produce active-high address outputs A2–A0 to indicate the number of the highest-priority asserted input. If at least one input is asserted, then an AVALID output should be asserted. Be sure to name all signals with the proper active levels. This circuit can be built with a single 74x148 and no other gates.

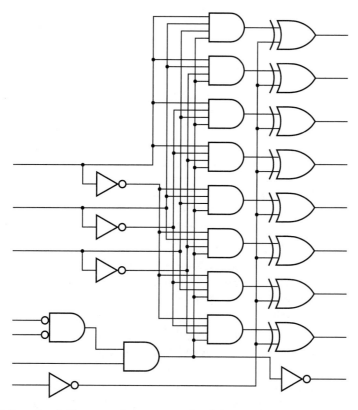

Figure X5.42

5.49 A purpose of Exercise 5.48 was to demonstrate that it is not always possible to maintain consistency in active-level notation unless you are willing to define alternate logic symbols for MSI parts that can be used in different ways. Define an alternate 74x148 symbol that provides this consistency in Exercise 5.48.

5.50 Design a combinational circuit with eight active-low request inputs, R0_L–R7_L, and eight outputs, A2–A0, AVALID, B2–B0, and BVALID. The R0_L–R7_L inputs and A2–A0 and AVALID outputs are defined as in Exercise 5.48. The B2–B0 and BVALID outputs identify the second-highest priority request input that is asserted. You should be able to design this circuit with no more than six SSI and MSI packages, but don't use more than 10 in any case.

5.51 Repeat Exercise 5.50 using ABEL. Does the design fit into a single GAL20V8?

5.52 Repeat Exercise 5.50 using VHDL.

5.53 Create a VHDL type, based on IEEE 1164, that models open-collector outputs, where typing outputs together creates a wired-AND function. You should also model a pull-up resistor element such that if there is no pull-up resistor and no device is driving the bus, then an "unknown" signal is produced. Test your definitions by modeling the circuit in Figure X3.92 for all input combinations, both with and without R1 present.

5.54 Design a 3-input, 5-bit multiplexer that fits in a 24-pin IC package. Write the truth table and draw a logic diagram and logic symbol for your multiplexer.

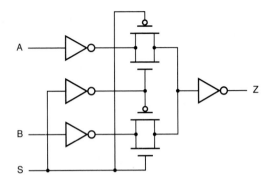

Figure X5.55

5.55 Write the truth table and a logic diagram for the logic function performed by the CMOS circuit in Figure X5.55. (The circuit contains transmission gates, which were introduced in Section 3.7.1.)

5.56 What logic function is performed by the CMOS circuit shown in Figure X5.56?

5.57 A famous logic designer decided to quit teaching and make a fortune by licensing the circuit design shown in Figure X5.57.

(a) Label the inputs and outputs of the circuit with appropriate signal names, including active-level indications.

(b) What does the circuit do? Be specific and account for all inputs and outputs.

(c) Draw the logic symbol that would go on the data sheet of this circuit.

(d) Write an ABEL or behavioral VHDL program for the circuit.

(e) With what standard building blocks does the new circuit compete? Do you think it would be successful as an MSI part?

5.58 Write a VHDL program for 74x157 multiplexer with the function table shown in Table 5-35.

5.59 Write a VHDL program for 74x153 multiplexer with the function table shown in Table 5-36.

5.60 Show how to realize the 4-input, 18-bit multiplexer with the functionality described in Table 5-40 using 18 74x151s.

5.61 Show how to realize the 4-input, 18-bit multiplexer with the functionality of Table 5-40 using 9 74x153s and a "code converter" with inputs S2–S0 and outputs C1,C0 such that [C1,C0] = 00–11 when S2–S0 selects A–D, respectively.

Figure X5.56

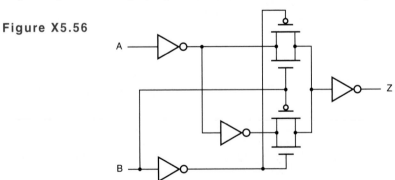

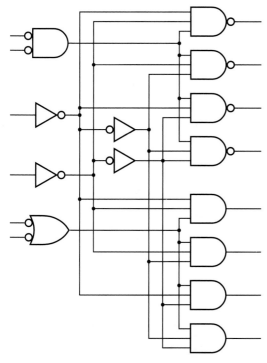

Figure X5.57

5.62   Design a 3-input, 2-output combinational circuit that performs the code conversion specified in the previous exercise, using discrete gates.

5.63   Add a three-state-output control input OE to the VHDL multiplexer program in Table 5-43. Your solution should have only one process.

5.64   A 16-bit *barrel shifter* is a combinational logic circuit with 16 data inputs, 16 data *barrel shifter* outputs, and 4 control inputs. The output word equals the input word, rotated by a number of bit positions specified by the control inputs. For example, if the input word equals ABCDEFGHIJKLMNOP (each letter represents one bit), and the control inputs are 0101 (5), then the output word is FGHIJKLMNOPABCDE. Design a 16-bit barrel shifter using combinational MSI parts discussed in this chapter. Your design should contain 20 or fewer ICs. Do not draw a complete schematic, but sketch and describe your design in general terms and indicate the types and total number of ICs required.

5.65   Write an ABEL program for the barrel shifter in Exercise 5.64.

5.66   Write a VHDL program for the barrel shifter in Exercise 5.64.

5.67   A digital designer who built the circuit in Figure 5-76 accidentally used 74x00s instead of '08s in the circuit and found that the circuit still worked, except for a change in the active level of the ERROR signal. How was this possible?

5.68   An odd-parity circuit with 2^n inputs can be built with $2^n - 1$ XOR gates. Describe two different structures for this circuit, one of which gives a minimum worst-case input to output propagation delay and the other of which gives a maximum. For each structure, state the worst-case number of XOR-gate delays, and describe a situation where that structure might be preferred over the other.

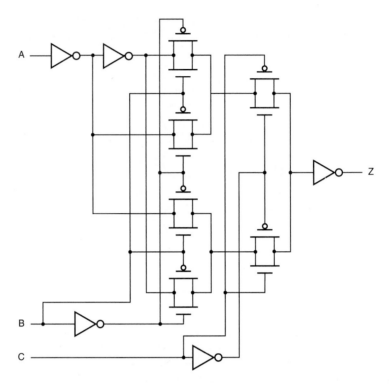

Figure X5.73

5.69 Write a 4-step iterative algorithm corresponding to the iterative comparator circuit of Figure 5-80.

5.70 Write a VHDL program for a 16-bit iterative camparator using the structure of Figure 5-80. Use VHDL's "generate" capability.

5.71 Design a 16-bit comparator using five 74x85s in a treelike structure, such that the maximum delay for a comparison equals twice the delay of one 74x85.

5.72 Write a VHDL program for a device with the functionality of a 74x85.

5.73 Write the truth table and a logic diagram for the logic function performed by the CMOS circuit in Figure X5.73.

5.74 Design a comparator similar to the 74x85 that uses the opposite cascading order. That is, to perform a 12-bit comparison, the cascading outputs of the high-order comparator would drive the cascading inputs of the mid-order comparator, and the mid-order outputs would drive the low-order inputs. You needn't do a complete logic design and schematic; a truth table and an application note showing the interconnection for a 12-bit comparison are sufficient.

5.75 Design a 24-bit comparator using three 74x682s and additional gates as required. Your circuit should compare two 24-bit unsigned numbers P and Q and produce two output bits that indicate whether P = Q or P > Q.

5.76 Using the information in Table 5-3, determine the maximum propagation delay from any A or B bus input to any F bus output of the 16-bit carry lookahead adder of Figure 5-96. You may use the "worst-case" analysis method.

5.77   Starting with the logic diagram for the 74x283 in Figure 5-91, write a logic expression for the S2 output in terms of the inputs, and prove that it does indeed equal the third sum bit in a binary addition as advertised. You may assume that $c_0 = 0$ (i.e., ignore c_0).

5.78   Referring to the data sheet of a 74S182 carry lookahead circuit, determine whether or not its outputs match the equations given in Section 5.10.7.

5.79   Estimate the number of product terms in a minimal sum-of-products expression for the c_{32} output of a 32-bit binary adder. Be more specific than "billions and billions," and justify your answer.

5.80   Draw the logic diagram for a 64-bit ALU using sixteen 74x181s and five 74S182s for full carry lookahead (two levels of '182s). For the '181s, you need show only the CIN inputs and G_L and P_L outputs.

5.81   Write a VHDL model for a 74x181 ALU.

5.82   Show how to build all four of the following functions using one SSI package and one 74x138.

$$F1 = X' \cdot Y' \cdot Z' + X \cdot Y \cdot Z \qquad F2 = X' \cdot Y' \cdot Z + X \cdot Y \cdot Z'$$
$$F3 = X' \cdot Y \cdot Z' + X \cdot Y' \cdot Z \qquad F4 = X \cdot Y' \cdot Z' + X' \cdot Y \cdot Z$$

5.83   Determine the worst-case propagation delay of the multiplier in Figure 5-98, assuming that the propagation delay from any adder input to its sum output is twice as long as the delay to the carry output. Repeat, assuming the opposite relationship. If you were designing the adder cell from scratch, which path would you favor with the shortest delay? Is there an optimal balance?

5.84   Repeat the preceding exercise for the multiplier in Figure 5-99.

5.85   Design a customized decoder with the function table in Table X5.85 using MSI and SSI parts. Minimize the number of IC packages in your design.

Table X5.85

CS_L	A2	A1	A0	*Output to Assert*
1	x	x	x	none
0	0	0	x	BILL_L
0	0	x	0	MARY_L
0	0	1	x	JOAN_L
0	0	x	1	PAUL_L
0	1	0	x	ANNA_L
0	1	x	0	FRED_L
0	1	1	x	DAVE_L
0	1	x	1	KATE_L

5.86   Repeat Exercise 5.85 using ABEL and a single GAL16V8.

5.87   Repeat Exercise 5.85 using VHDL.

5.88   Based on the Hamming code used in the VHDL program in Table 5-77, write a VHDL program for a Hamming encoder entity with 4-bit data inputs and 7-bit encoded data outputs.

5.89 Using ABEL and a single GAL16V8, design a customized multiplexer with four 3-bit input buses P, Q, R, T, and three select inputs S2–S0 that choose one of the buses to drive a 3-bit output bus Y according to Table X5.89.

Table X5.89

S2	S1	S0	*Input to Select*
0	0	0	P
0	0	1	P
0	1	0	P
0	1	1	Q
1	0	0	P
1	0	1	P
1	1	0	R
1	1	1	T

5.90 Design a customized multiplexer with four 4-bit input buses P, Q, R, and T, selecting one of the buses to drive a 4-bit output bus Y according to Table X5.89. Use two 74x153s and a code converter that maps the eight possible values on S2–S0 to four select codes for the 74x153s. Choose a code that minimizes the size and propagation delay of the code converter.

5.91 Design a customized multiplexer with five 4-bit input buses A, B, C, D, and E, selecting one of the buses to drive a 4-bit output bus T according to Table X5.91. You may use no more than three MSI and SSI ICs.

Table X5.91

S2	S1	S0	*Input to Select*
0	0	0	A
0	0	1	B
0	1	0	A
0	1	1	C
1	0	0	A
1	0	1	D
1	1	0	A
1	1	1	E

5.92 Repeat Exercise 5.91 using ABEL and one or more PAL/GAL devices from this chapter. Minimize the number and size of the GAL devices.

5.93 Design a 3-bit equality checker with six inputs, SLOT[2–0] and GRANT[2–0], and one active-low output, MATCH_L. The SLOT inputs are connected to fixed values when the circuit is installed in the system, but the GRANT values are changed on a cycle-by-cycle basis during normal operation of the system. Using only SSI and MSI parts that appear in Tables 5-2 and 5-3, design a comparator with the shortest possible maximum propagation delay from GRANT[2–0] to MATCH_L. (*Note:* The author had to solve this problem "in real life" to shave 2 ns off the critical-path delay in a 25-MHz system design.)

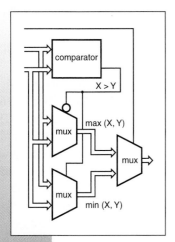

chapter 6

Combinational-Circuit Design Examples

S
o far, we have looked at basic principles in several areas—number systems, digital circuits, and combinational logic—and we have described many of the basic building blocks of combinational design—decoders, multiplexers, and the like. All of that is a needed foundation, but the ultimate goal of studying digital design is eventually to be able to solve real problems by designing digital systems (well, duh...). That usually requires experience beyond what you can get by reading a textbook. We'll try to get you started by presenting a number of larger combinational design examples in this chapter.

The chapter is divided into three sections. The first gives design examples using combinational building blocks. While the examples are written in terms of MSI functions, the same functions are widely used in ASIC and schematic-based FPGA design. The idea of these examples is to show that you can often express a combinational function using a collection of smaller building blocks. This is important for a couple of reasons: a hierarchical approach usually simplifies the overall design task, and the smaller building blocks often have a more efficient, optimized realization in FPGA and ASIC cells than what you'd get if you wrote a larger, monolithic description in an HDL and then just hit the "synthesize" button.

The second section gives design examples using ABEL. These designs are all targeted to small PLDs such as 16V8s and 20V8s. Besides the general use of the ABEL language, some of the examples illustrate the partitioning decisions that a designer must make when an entire circuit does not fit into a single component.

A VHDL-based approach is especially appropriate for larger designs that will be realized in a single CPLD, FPGA, or ASIC, as described in the third section. You may notice that these examples do not target a specific CPLD or FPGA. Indeed, this is one of the benefits of HDL-based design; most or all of the design effort is "portable" and can be targeted to any of a variety of technologies.

The only prerequisites for this chapter are the chapters that precede it. The three sections are written to be pretty much independent of each other, so you don't have to read about ABEL if you're only interested in VHDL, or vice versa. Also, the rest of the book is written so that you can read this chapter now or skip it and come back later.

6.1 Building-Block Design Examples

6.1.1 Barrel Shifter

barrel shifter

A *barrel shifter* is a combinational logic circuit with n data inputs, n data outputs, and a set of control inputs that specify how to shift the data between input and output. A barrel shifter that is part of a microprocessor CPU can typically specify the direction of shift (left or right), the type of shift (circular, arithmetic, or logical), and the amount of shift (typically 0 to $n-1$ bits, but sometimes 1 to n bits).

In this subsection we'll look at the design of a simple 16-bit barrel shifter that does left circular shifts only, using a 4-bit control input S[3:0] to specify the amount of shift. For example, if the input word is ABCDEFGHGIHKLMNOP (where each letter represents one bit), and the control input is 0101 (5), then the output word is FGHGIHKLMNOPABCDE.

From one point of view, this problem is deceptively simple. Each output bit can be obtained from a 16-input multiplexer controlled by the shift-control inputs, where each multiplexer data input is connected to the appropriate data bit. On the other hand, when you look at the details of the design, you'll see that there are tradeoffs in the speed and size of the circuit.

Let us first consider a design that uses off-the-shelf MSI multiplexers. A 16-input, one-bit multiplexer can be built using two 74x151s by applying S3 and its complement to the EN_L inputs and combining the Y_L data outputs with a NAND gate, as we showed in Figure 5-66 for a 32-input multiplexer. The low-order shift-control inputs, S2–S0, connect to the like-named select inputs of the '151s.

We complete the design by replicating this 16-input multiplexer 16 times and hooking up the data inputs appropriately, as shown in Figure 6-1. The top '151 of each pair is enabled by S3_L, and the bottom one by S3; the remaining select bits are connected to all 32 '151s. Data inputs D0–D7 of each '151 are connected to the DIN inputs in the listed order from left to right.

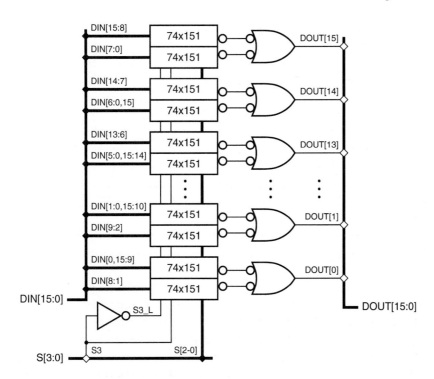

Figure 6-1
One approach to building a 16-bit barrel shifter.

The first row of Table 6-1 shows the characteristics of this first approach. About 36 chips (32 74x151s, 4 74x00s, and 1/6 74x04) are used in the MSI/SSI realization. We can reduce this to 32 chips by replacing the 74x151s with 74x251s and tying their three-state Y outputs together, as tabulated in the second row. Both of these designs have very heavy loading on the control inputs; each of the control bits S[2:0] must be connected to the like-named select input of all 32 multiplexers. The data inputs are also fairly heavily loaded; each data bit must connect to 16 different multiplexer data inputs, corresponding to the 16 possible shift amounts. However, assuming that the heavy control and data loads don't slow things down too much, the 74x251-based approach yields the shortest data delay, with each data bit passing through just one multiplexer.

Alternatively, we could build the barrel shifter using 16 74x157 2-input, 4-bit multiplexers, as tabulated in the last row of the table. We start by using four 74x157s to make a 2-input, 16-bit multiplexer. Then, we can hook up a first set

Multiplexer Component	Data Loading	Data Delay	Control Loading	Total ICs
74x151	16	2	32	36
74x251	16	1	32	32
74x153	4	2	8	16
74x157	2	4	4	16

Table 6-1
Properties of four different barrel-shifter design approaches.

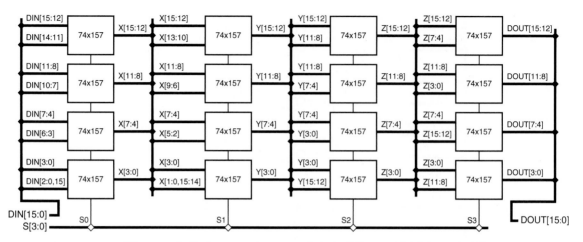

Figure 6-2 A second approach to building a 16-bit barrel shifter.

of four '157s controlled by S0 to shift the input word left by 0 or 1 bit. The data outputs of this set are connected to the inputs of a second set, controlled by S1, which shifts its input word left by 0 or 2 bits. Continuing the cascade, a third and fourth set are controlled by S2 and S3 to shift selectively by 4 and 8 bits, as shown in Figure 6-2. Here, the 1A-4A and 1B-4B inputs and the 1Y-4Y outputs of each '157 are connected to the indicated signals in the listed order from left to right.

The '157-based approach requires only half as many MSI packages and has far less loading on the control and data inputs. On the other hand, it has the longest data-path delay, since each data bit must pass through four 74x157s.

Halfway between the two approaches, we can use eight 74x153 4-input, 2-bit multiplexers to build a 4-input, 16-bit multiplexer. Cascading two sets of these, we can use S[3:2] to shift selectively by 0, 4, 8, or 12 bits, and S[1:0] to shift by 0–3 bits. This approach has the performance characteristics shown in the third row of Table 6-1 and would appear to be the best compromise if you don't need to have the absolutely shortest possible data delay.

The same kind of considerations would apply if you were building the barrel shifter out of ASIC cells instead of MSI parts, except you'd be counting chip area instead of MSI/SSI packages.

Typical ASIC cell libraries have 1-bit-wide multiplexers, usually realized with CMOS transmission gates, with 2 to 8 inputs. To build a larger multiplexer, you have to put together the appropriate combination of smaller cells. Besides the kind of choices we encountered in the MSI example, you have the further complication that CMOS delays are highly dependent on loading. Thus, depending on the approach, you must decide where to add buffers to the control lines, the data lines, or both to minimize loading-related delays. An approach that looks good on paper, before analyzing these delays and adding buffers, may actually turn out to have poorer delay or more chip area than another approach.

6.1.2 Simple Floating-Point Encoder

The previous example used multiple copies of a single building block, a multiplexer, and it was pretty obvious from the problem statement that a multiplexer was the appropriate building block. The next example shows that you sometimes have to look a little harder to see the solution in terms of known building blocks.

Now let's look at a design problem whose MSI solution is not quite so obvious, a "fixed-point to floating-point encoder." An unsigned binary integer B in the range $0 \leq B < 2^{11}$ can be represented by 11 bits in "fixed-point" format, $B = b_{10}b_9...b_1b_0$. We can represent numbers in the same range with less precision using only 7 bits in a floating-point notation, $F = M \cdot 2^E$, where M is a 4-bit mantissa $m_3m_2m_1m_0$ and E is a 3-bit exponent $e_2e_1e_0$. The smallest integer in this format is $0 \cdot 2^0$ and the largest is $(2^4-1) \cdot 2^7$.

Given an 11-bit fixed-point integer B, we can convert it to our 7-bit floating-point notation by "picking off" four high-order bits beginning with the most significant 1, for example,

$$
\begin{aligned}
11010110100 &= 1101 \cdot 2^7 + 0110100 \\
00100101111 &= 1001 \cdot 2^5 + 01111 \\
00000111110 &= 1111 \cdot 2^2 + 10 \\
00000001011 &= 1011 \cdot 2^0 + 0 \\
00000000010 &= 0010 \cdot 2^0 + 0
\end{aligned}
$$

The last term in each equation is a truncation error that results from the loss of precision in the conversion. Corresponding to this conversion operation, we can write the specification for a fixed-point to floating-point encoder circuit:

- A combinational circuit is to convert an 11-bit unsigned binary integer B into a 7-bit floating-point number M,E, where M and E have 4 and 3 bits, respectively. The numbers have the relationship $B = M \cdot 2^E + T$, where T is the truncation error, $0 \leq T < 2^E$.

Starting with a problem statement like the one above, it takes some creativity to come up with an efficient circuit design—the specification gives no clue. However, we can get some ideas by looking at how we converted numbers by hand earlier. We basically scanned each input number from left to right to find the first position containing a 1, stopping at the b_3 position if no 1 was found. We picked off four bits starting at that position to use as the mantissa, and the starting position number determined the exponent. These operations are beginning to sound like MSI building blocks.

"Scanning for the first 1" is what a generic priority encoder does. The output of the priority encoder is a number that tells us the position of the first 1. The position number determines the exponent; first-1 positions of $b_{10} - b_3$ imply exponents of 7–0, and positions of $b_2 - b_0$ or no-1-found imply an exponent of 0. Therefore, we can scan for the first 1 with an 8-input priority encoder with inputs

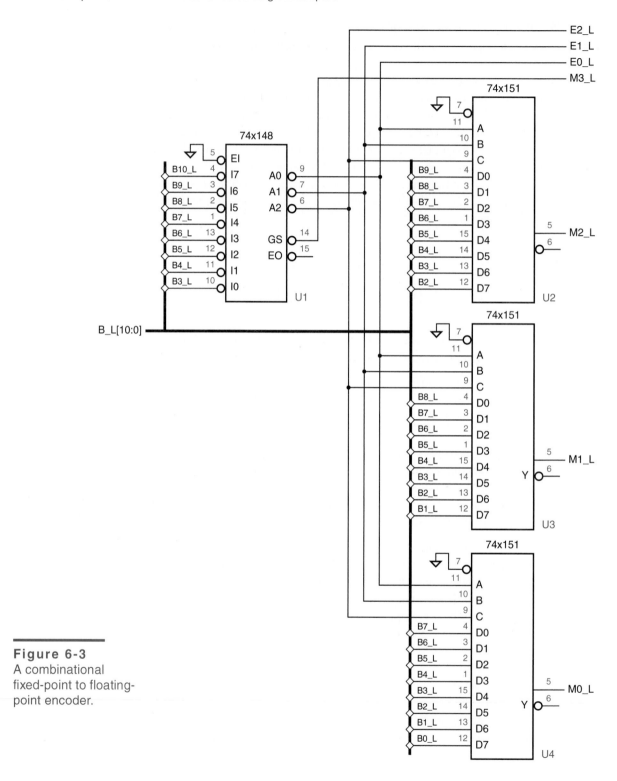

Figure 6-3
A combinational
fixed-point to floating-
point encoder.

I7 (highest priority) through I0 connected to $b_{10} - b_3$. We can use the priority encoder's A2–A0 outputs directly as the exponent, as long as the no-1-found case produces A2–A0 = 000.

"Picking off four bits" sounds like a "selecting" or multiplexing operation. The 3-bit exponent determines which four bits of B we pick off, so we can use the exponent bits to control an 8-input, 4-bit multiplexer that selects the appropriate four bits of B to form M.

An MSI circuit that results from these ideas is shown in Figure 6-3. It contains several optimizations:

- Since the available MSI priority encoder, the 74x148, has active-low inputs, the input number B is assumed to be available on an active-low bus B_L[10:0]. If only an active-high version of B is available, then eight inverters can be used to obtain the active-low version.

- If you think about the conversion operation a while, you'll realize that the most significant bit of the mantissa, m_3, is always 1, except in the no-1-found case. The '148 has a GS_L output that indicates this case, allowing us to eliminate the multiplexer for m_3.

- The '148 has active-low outputs, so the exponent bits (E0_L–E2_L) are produced in active-low form. Naturally, three inverters could be used to produce an active-high version.

- Since everything else is active-low, active-low mantissa bits are used, too. Active-high bits are also readily available on the '148 EO_L and the '151 Y_L outputs.

Strictly speaking, the multiplexers in Figure 6-3 are drawn incorrectly. The 74x151 symbol can be drawn alternatively as shown in Figure 6-4. In words, if the multiplexer's data inputs are active low, then the data outputs have an active level opposite that shown in the original symbol. The "active-low-data" symbol

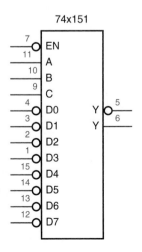

Figure 6-4
Alternate logic symbol
for the 74x151 8-input
multiplexer.

should be preferred in Figure 6-3, since the active levels of the '151 inputs and outputs would then match their signal names. However, in data transfer and storage applications, designers (and the book) don't always "go by the book." It is usually clear from the context that a multiplexer (or a multibit register, in Section 8.2.5) does not alter the active level of its data.

6.1.3 Dual-Priority Encoder

Quite often MSI building blocks need a little help from their friends—ordinary gates—to get the job done. In this example, we'd like to build a priority encoder that identifies not only the highest but also the second-highest-priority asserted signal among a set of eight request inputs.

We'll assume for this example that the request inputs are active low and are named R_L[0:7], where R_L0 has the highest priority. We'll use A[2:0] and AVALID to identify the highest-priority request, where AVALID is asserted only if at least one request input is asserted. We'll use B[2:0] and BVALID to identify the second-highest-priority request, where BVALID is asserted only if at least two request inputs are asserted.

Finding the highest-priority request is easy enough, we can just use a 74x148. To find the second-highest-priority request, we can use another '148, but only if we first "knock out" the highest-priority request before applying the request inputs. This can be done using a decoder to select a signal to knock out, based on A[2:0] and AVALID from the first '148. These ideas are combined in the solution shown in Figure 6-6. A 74x138 decoder asserts at most one of its eight outputs, corresponding to the highest-priority request input. The outputs are fed to a rank of NAND gates to "turn off" the highest-priority request.

A trick is used in this solution is to get active-high outputs from the '148s, as shown in Figure 6-5. We can rename the address outputs A_L[2:0] to be active high if we also change the name of the request input that is associated with each output combination. In particular, we complement the bits of the request number. In the redrawn symbol, request input I0 has the highest priority.

Figure 6-5
Alternate logic symbols for the 74x148 8-input priority encoder.

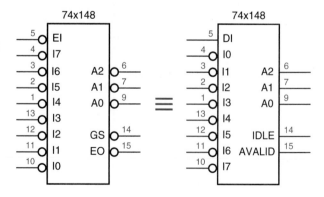

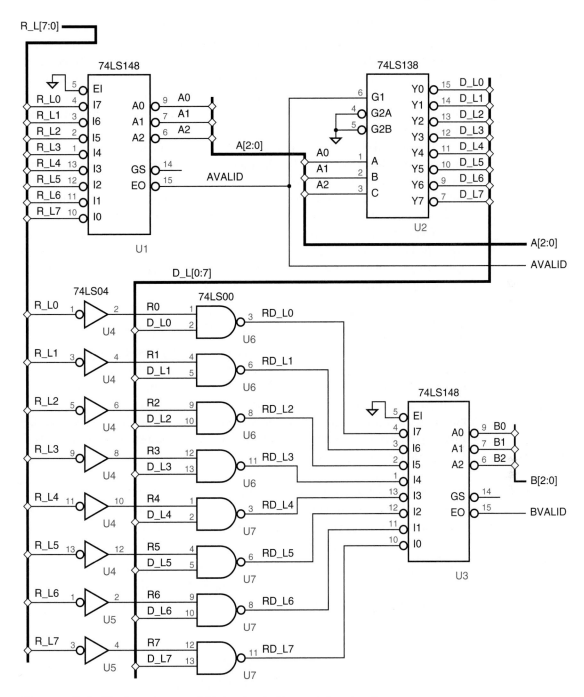

Figure 6-6 First- and second-highest-priority encoder circuit.

6.1.4 Cascading Comparators

In Section 5.9.4, we showed how 74x85 4-bit comparators can be cascaded to create larger comparators. Since the 74x85 uses a serial cascading scheme, it can be used to build arbitrarily large comparators. The 74x682 8-bit comparator, on the other hand, doesn't have any cascading inputs and outputs at all. Thus, at first glance, you might think that it can't be used to build larger comparators. But that's not true.

If you think about the nature of a large comparison, it is clear that two wide inputs, say 32 bits (four bytes) each, are equal only if their corresponding bytes are equal. If we're trying to do a greater-than or less-than comparison, then the corresponding most-significant that are not equal determine the result of the comparison.

Using these ideas, Figure 6-7 uses three 74x682 8-bit comparators to do equality and greater-than comparison on two 24-bit operands. The 24-bit results are derived from the individual 8-bit results using combinational logic for the following equations:

$$PEQQ = EQ2 \cdot EQ1 \cdot EQ0$$
$$PGTQ = GT2 + EQ2 \cdot GT1 + EQ2 \cdot EQ1 \cdot GT0$$

This "parallel" expansion approach is actually faster than the 74x85's serial cascading scheme, because it does not suffer the delay of propagating the cascading signals through a cascade of comparators. The parallel approach can be used to build very wide comparators using two-level AND-OR logic to combine the 8-bit results, limited only by the fan-in constraints of the AND-OR logic. Arbitrary large comparators can be made if you use additional levels of logic to do the combining.

6.1.5 Mode-Dependent Comparator

Quite often, the requirements for a digital-circuit application are specified in a way that makes an MSI or other building-block solution obvious. For example, consider the following problem:

- Design a combinational circuit whose inputs are two 8-bit unsigned binary integers, X and Y, and a control signal MIN/MAX. The output of the circuit is an 8-bit unsigned binary integer Z such that $Z = \min(X,Y)$ if MIN/MAX $= 1$, and $Z = \max(X,Y)$ if MIN/MAX $= 0$.

This circuit is fairly easy to visualize in terms of MSI functions. Clearly, we can use a comparator to determine whether X > Y. We can use the comparator's output to control multiplexers that produce $\min(X,Y)$ and $\max(X,Y)$, and we can use another multiplexer to select one of these results depending on MIN/MAX. Figure 6-8(a) is the block diagram of a circuit corresponding to this approach.

Our first solution approach works, but it's more expensive than it needs to be. Although it has three two-input multiplexers, there are only two input words,

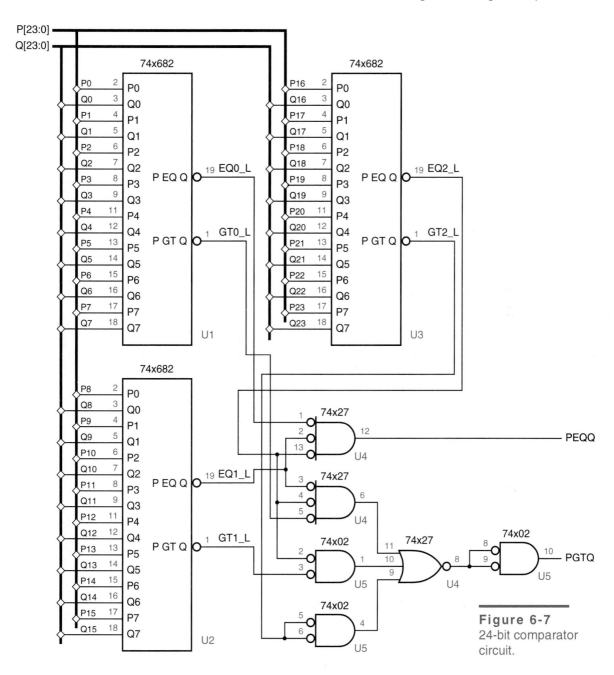

Figure 6-7
24-bit comparator circuit.

X and Y, that may ultimately be selected and produced at the output of the circuit. Therefore, we should be able to use just a single two-input mux, and use some other logic to figure out which input to tell it to select. This approach is shown in Figure 6-8(b) and (c). The "other logic" is very simple indeed, just a single XOR gate.

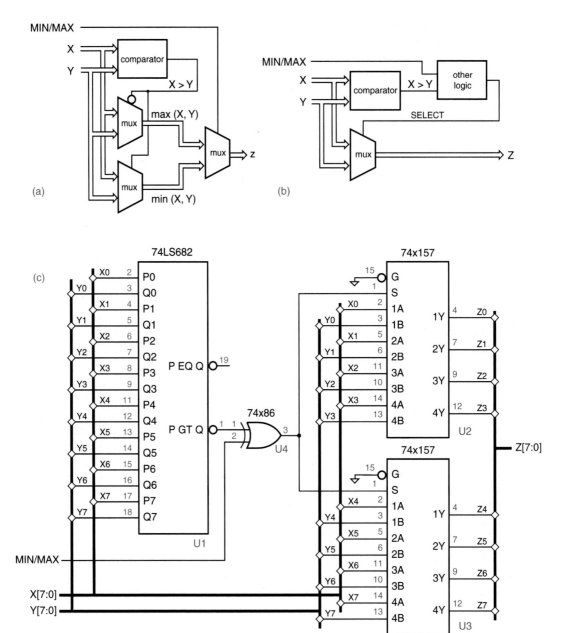

Figure 6-8 Mode-dependent comparator circuit: (a) block diagram of a "first-cut" solution; (b) block diagram of a more cost-effective solution; (c) logic diagram for the second solution.

DON'T BE A
BLOCKHEAD

The wastefulness of our original design approach in Figure 6-8(a) may have been obvious to you from the beginning, but it demonstrates an important approach to designing with building blocks:

- Use standard building blocks to handle data, and look for ways that a single block can perform different functions at different times or in different modes. Design control circuits to select the appropriate functions as needed, to reduce the total parts count of the design.

As Figure 6-8(c) dramatically shows, this approach can save a lot of chips. When designing with IC chips, you should *not* heed the slogan, "Have all you want, we'll make more"!

6.2 Design Examples Using ABEL and PLDs

6.2.1 Barrel Shifter

A barrel shifter, defined on page 468, is good example of something *not* to design using PLDs. However, we can put ABEL to good use to describe a barrel shifter's function, and we'll also see why a typical barrel shifter is not a good fit for a PLD.

Table 6-2 shows the equations for a 16-bit barrel shifter with the same functionality as the example on page 468—it does left circular shifts only, using a 4-bit control input S[3..0] to specify the amount of shift. ABEL makes it easy to specify the functionality of the overall circuit without worrying about how the circuit might be partitioned into multiple chips. Also, ABEL dutifully generates a minimal sum-of-products expression for each output bit. In this case, each output requires 16 product terms.

Partitioning the 16-bit barrel shifter into multiple PLDs is a difficult task in two different ways. First, it should be obvious that the nature of the function is such that every output bit depends on every input bit. A PLD that produces, say, the DOUT0 output must have all 16 DIN inputs and all four S inputs available to it. So, a GAL16V8 definitely cannot be used; it has only 16 inputs.

The GAL20V8 is similar to the GAL16V8, with the addition of four input-only pins. If we use all 20 available inputs, we are left with two output-only pins (corresponding to the top and bottom outputs in Figure 5-27 on page 345). Thus, it seems possible that we could realize the barrel shifter using eight 20V8 chips, producing two output bits per chip.

No, we still have a problem. The second dimension of difficulty in a PLD-based barrel shifter is the number of product terms per output. The barrel shifter requires 16, and the 20V8 provides only 7. We're stuck—any realization of the barrel shifter in 20V8s is going to require multiple-pass logic. At this point, we would be best advised to look at partitioning options along the lines that we did in Section 6.1.1.

Table 6-2

ABEL program for a 16-bit barrel shifter.

```
module barrel16
title '16-bit Barrel Shifter'

" Inputs and Outputs
DIN15..DIN0, S3..S0                      pin;
DOUT15..DOUT0                            pin istype 'com';

S = [S3..S0];

equations

[DOUT15..DOUT0] = (S==0) & [DIN15..DIN0]
               # (S==1) & [DIN14..DIN0,DIN15]
               # (S==2) & [DIN13..DIN0,DIN15..DIN14]
               # (S==3) & [DIN12..DIN0,DIN15..DIN13]
                 ...
               # (S==12) & [DIN3..DIN0,DIN15..DIN4]
               # (S==13) & [DIN2..DIN0,DIN15..DIN3]
               # (S==14) & [DIN1..DIN0,DIN15..DIN2]
               # (S==15) & [DIN0,DIN15..DIN1];

end barrel16
```

The 16-bit barrel shifter can be realized without much difficulty in a larger programmable device, that is, in a CPLD or an FPGA with enough I/O pins. However, imagine that we were trying to design a 32-bit or 64-bit barrel shifter. Clearly, we would need to use a device with even more I/O pins, but that's not all. The number of product terms and the large amount of connectivity (all the inputs connect to all the outputs) would still be challenging.

Indeed, a typical CPLD or FPGA fitter could have difficulty realizing a large barrel shifter with small delay or even at all. There is a critical resource that we took for granted in the partitioned, building-block barrel-shifter designs of Section 6.1.1—wires! An FPGA is somewhat limited in its internal connectivity, and a CPLD is even more so. Thus, even with modern FPGA and CPLD design tools, you may still have to "use your head" to partition the design in a way that helps the tools do their job.

Barrel shifters can be even more complex than what we've shown so far. Just for fun, Table 6-3 shows the design for a barrel shifter that supports six different kinds of shifting. This requires even more product terms, up to 40 per output! Although you'd never build this device in a PLD, CPLD, or small FPGA, the minimized ABEL equations are useful because they can help you understand the effects of some of your design choices. For example, by changing the coding of SLA and SRA to [1,.X.,0] and [1,.X.,1], you can reduce the total number of product terms in the design from 624 to 608. You can save more product terms by changing the coding of the shift amount for some shifts (see Exercise 6.3). The savings from these changes may carry over to other design approaches.

Table 6-3 ABEL program for a multi-mode 16-bit barrel shifter.

```
module barrl16f
Title 'Multi-mode 16-bit Barrel Shifter'

" Inputs and Outputs
DIN15..DIN0, S3..S0, C2..C0          pin;
DOUT15..DOUT0                        pin istype 'com';

S = [S3..S0]; C = [C2..C0]; " Shift amount and mode
L = DIN15; R = DIN0;        " MSB and LSB

ROL = (C == [0,0,0]); " Rotate (circular shift) left
ROR = (C == [0,0,1]); " Rotate (circular shift) right
SLL = (C == [0,1,0]); " Shift logical left (shift in 0s)
SRL = (C == [0,1,1]); " Shift logical right (shift in 0s)
SLA = (C == [1,0,0]); " Shift left arithmetic (replicate LSB)
SRA = (C == [1,0,1]); " Shift right arithmetic (replicate MSB)

equations

[DOUT15..DOUT0] = ROL & (S==0) & [DIN15..DIN0]
               # ROL & (S==1) & [DIN14..DIN0,DIN15]
               # ROL & (S==2) & [DIN13..DIN0,DIN15..DIN14]
               ...
               # ROL & (S==15) & [DIN0,DIN15..DIN1]
               # ROR & (S==0) & [DIN15..DIN0]
               # ROR & (S==1) & [DIN0,DIN15..DIN1]
               ...
               # ROR & (S==14) & [DIN13..DIN0,DIN15..DIN14]
               # ROR & (S==15) & [DIN14..DIN0,DIN15]
               # SLL & (S==0) & [DIN15..DIN0]
               # SLL & (S==1) & [DIN14..DIN0,0]
               ...
               # SLL & (S==14) & [DIN1..DIN0,0,0,0,0,0,0,0,0,0,0,0,0,0,0]
               # SLL & (S==15) & [DIN0,0,0,0,0,0,0,0,0,0,0,0,0,0,0,0]
               # SRL & (S==0) & [DIN15..DIN0]
               # SRL & (S==1) & [0,DIN15..DIN1]
               ...
               # SRL & (S==14) & [0,0,0,0,0,0,0,0,0,0,0,0,0,0,DIN15..DIN14]
               # SRL & (S==15) & [0,0,0,0,0,0,0,0,0,0,0,0,0,0,0,DIN15]
               # SLA & (S==0) & [DIN15..DIN0]
               # SLA & (S==1) & [DIN14..DIN0,R]
               ...
               # SLA & (S==14) & [DIN1..DIN0,R,R,R,R,R,R,R,R,R,R,R,R,R,R]
               # SLA & (S==15) & [DIN0,R,R,R,R,R,R,R,R,R,R,R,R,R,R,R]
               # SRA & (S==0) & [DIN15..DIN0]
               # SRA & (S==1) & [L,DIN15..DIN1]
               ...
               # SRA & (S==14) & [L,L,L,L,L,L,L,L,L,L,L,L,L,L,DIN15..DIN14]
               # SRA & (S==15) & [L,L,L,L,L,L,L,L,L,L,L,L,L,L,L,DIN15];
end barrl16f
```

6.2.2 Simple Floating-Point Encoder

We defined a simple floating-point number format on page 471 and posed the design problem of converting a number from fixed-point to this floating-point format. The I/O-pin requirements of this design are limited—11 inputs and 7 outputs—so we can potentially use a single PLD to replace the four parts that were used in the MSI solution.

An ABEL program for the fixed-to-floating-point converter is given in Table 6-4. The WHEN statement expresses the operation of determining the exponent value E in a very natural way. Then E is used to select the appropriate bits of B to use as the mantissa M.

Despite the deep nesting of the WHEN statement, only four product terms are needed in the minimal sum for each bit of E. The equations for the M bits are not too bad either, requiring only eight product terms each. Unfortunately, the

Table 6-4

An ABEL program for the fixed-point to floating-point PLD.

```
module fpenc
title 'Fixed-point to Floating-point Encoder'
FPENC device 'P20L8';

" Input and output pins
B10..B0                 pin 1..11;
E2..E0, M3..M0          pin 21..15 istype 'com';

" Constant expressions
B = [B10..B0];
E = [E2..E0];
M = [M3..M0];

equations

WHEN B < 16 THEN E = 0;
ELSE WHEN B < 32 THEN E = 1;
ELSE WHEN B < 64 THEN E = 2;
ELSE WHEN B < 128 THEN E = 3;
ELSE WHEN B < 256 THEN E = 4;
ELSE WHEN B < 512 THEN E = 5;
ELSE WHEN B < 1024 THEN E = 6;
ELSE E = 7;

M = (E==0) & [B3..B0]
  # (E==1) & [B4..B1]
  # (E==2) & [B5..B2]
  # (E==3) & [B6..B3]
  # (E==4) & [B7..B4]
  # (E==5) & [B8..B5]
  # (E==6) & [B9..B6]
  # (E==7) & [B10..B7];

end fpenc
```

GAL20V8 has available only seven product terms per output. However, the GAL22V10 (Figure 8-22 on page 688) has more product terms available, so we can use that if we like.

One drawback of the design in Table 6-4 is that the [M3..M0] outputs are slow; since they use [E2..E0], they take two passes through the PLD. A faster approach, if it fits, would be to rewrite the "select" terms (E==0, etc.) as intermediate equations before the equations section, and let ABEL expand the resulting M equations in a single level of logic. Unfortunately, ABEL does not allow WHEN statements outside of the equations section, so we'll have to roll up our sleeves and write our own logic expressions in the intermediate equations.

Table 6-5 shows the modified approach. The expressions for S7–S0 are just mutually-exclusive AND-terms that indicate exponent values of 7–0 depending on the location of the most significant 1 bit in the fixed-point input number. The exponent [E2..E0] is a binary encoding of the select terms, and the mantissa bits [M3..M0] are generated using a select term for each case. It turns out that these M equations still require 8 product terms per output bit, but at least they're a lot faster, since they use just one level of logic.

Table 6-5
Alternative ABEL program for the fixed-point to floating-point PLD.

```
module fpence
title 'Fixed-point to Floating-point Encoder'
FPENCE device 'P20L8';

" Input and output pins
B10..B0                     pin 1..11;
E2..E0, M3..M0              pin 21..15 istype 'com';

" Intermediate equations
S7 = B10;
S6 = !B10 & B9;
S5 = !B10 & !B9 & B8;
S4 = !B10 & !B9 & !B8 & B7;
S3 = !B10 & !B9 & !B8 & !B7 & B6;
S2 = !B10 & !B9 & !B8 & !B7 & !B6 & B5;
S1 = !B10 & !B9 & !B8 & !B7 & !B6 & !B5 & B4;
S0 = !B10 & !B9 & !B8 & !B7 & !B6 & !B5 & !B4;

equations

E2 = S7 # S6 # S5 # S4;
E1 = S7 # S6 # S3 # S2;
E0 = S7 # S5 # S3 # S1;

[M3..M0] = S0 & [B3..B0] # S1 & [B4..B1] # S2 & [B5..B2]
         # S3 & [B6..B3] # S4 & [B7..B4] # S5 & [B8..B5]
         # S6 & [B9..B6] # S7 & [B10..B7];

end fpenc
```

6.2.3 Dual-Priority Encoder

In this example, we'll design a PLD-based priority encoder that identifies both the highest-priority and the second-highest-priority asserted signal among a set of eight active-high request inputs named [R0..R7], where R0 has the highest priority. We'll use [A2..A0] and AVALID to identify the highest-priority request, asserting AVALID only if a highest-priority request is present. Similarly, we'll use [B2:B0] and BVALID to identify the second-highest-priority request.

Table 6-6 shows an ABEL program for the priority encoder. As usual, a nested WHEN statement is perfect for expressing priority behavior. To find the

Table 6-6
ABEL program for a dual-priority encoder.

```
title 'Dual Priority Encoder'
PRIORTWO device 'P16V8';

" Input and output pins
R7..R0                                pin 1..8;
AVALID, A2..A0, BVALID, B2..B0        pin 19..12 istype 'com';

" Set definitions
A = [A2..A0]; B = [B2..B0];

equations

WHEN R0==1 THEN A=0;
ELSE WHEN R1==1 THEN A=1;
ELSE WHEN R2==1 THEN A=2;
ELSE WHEN R3==1 THEN A=3;
ELSE WHEN R4==1 THEN A=4;
ELSE WHEN R5==1 THEN A=5;
ELSE WHEN R6==1 THEN A=6;
ELSE WHEN R7==1 THEN A=7;

AVALID = ([R7..R0] != 0);

WHEN (R0==1) & (A!=0) THEN B=0;
ELSE WHEN (R1==1) & (A!=1) THEN B=1;
ELSE WHEN (R2==1) & (A!=2) THEN B=2;
ELSE WHEN (R3==1) & (A!=3) THEN B=3;
ELSE WHEN (R4==1) & (A!=4) THEN B=4;
ELSE WHEN (R5==1) & (A!=5) THEN B=5;
ELSE WHEN (R6==1) & (A!=6) THEN B=6;
ELSE WHEN (R7==1) & (A!=7) THEN B=7;

BVALID = (R0==1) & (A!=0) # (R1==1) & (A!=1)
       # (R2==1) & (A!=2) # (R3==1) & (A!=3)
       # (R4==1) & (A!=4) # (R5==1) & (A!=5)
       # (R6==1) & (A!=6) # (R7==1) & (A!=7);

end priortwo
```

P-Terms	Fan-in	Fan-out	Type	Name
8/1	8	1	Pin	AVALID
4/5	8	1	Pin	A2
4/5	8	1	Pin	A1
4/5	8	1	Pin	A0
24/8	11	1	Pin	BVALID
24/17	11	1	Pin	B2
20/21	11	1	Pin	B1
18/22	11	1	Pin	B0
=========
106/84				

Best P-Term Total: 76
Total Pins: 16
Average P-Term/Output: 9

Table 6-7
Product-term usage
in the dual priority
encoder PLD.

second-highest priority input, we exclude an input if its input number matches
the highest-priority input number, which is A. Thus, we're using two-pass logic
to compute the B outputs. The equation for AVALID is easy; AVALID is 1 if the
request inputs are not all 0. To compute BVALID, we OR all of the conditions that
set B in the WHEN statement.

Even with two-pass logic, the B outputs use too many product terms to fit in
a 16V8; Table 6-7 shows the product-term usage. The B outputs use too many
terms even for a 22V10, which has 16 terms for two of its output pins and 8–14
for the others. Sometimes you just have to work harder to make things fit!

So, how can we save some product terms? One important thing to notice is
that R0 can never be the second-highest-priority asserted input, and therefore B
can never be valid and 0. Thus, we can eliminate the WHEN clause for the R0==1
case. Making this change reduces the minimum number of terms for B2–B0 to
14, 17, and 15, respectively. We can almost fit the design in a 22V10, if we can
just know one term out of the B1 equation.

Well, let's try something else. The second WHEN clause, for the R0==2 case,
also fails to make use of everything we know. We don't need the full generality

SUMS OF PRODUCTS AND PRODUCTS OF SUMS (SAY THAT 5 TIMES FAST)	You may recall from Section 4.3.6 that the minimal sum-of-products expression for the complement of a function can be manipulated through DeMorgan's theorem to obtain a minimal product-of-sums expression for the original function. You may also recall that the number of product terms in the minimal sum of products may differ from the number of sum terms in the minimal product of sums. The "P-Terms" column in Table 6-7 lists the number of terms in both minimal forms (product/sum terms). If *either* minimal form has less than or equal to the number of product terms available in a 22V10's AND-OR array, then the function can be made to fit.

of A!=0; this case is only important when R0 is 1. So, let us replace the first two lines of the original WHEN statement with

WHEN (R1==1) & (R0==1) THEN B=1;

This subtle change reduces the minimum number of terms for B2–B0 to 12, 16, and 13, respectively. We made it! Can the number of product terms be reduced further, enough to fit into a 16V8 while maintaining the same functionality? It's not likely, but we'll leave that as an exercise (6.4) for the reader!

6.2.4 Cascading Comparators

We showed in Section 5.9.5 that equality comparisons are easy to realize in PLDs but that magnitude comparisons (greater-than or less-than) of more than a few bits are not good candidates for PLD realization, owing to the large number of product terms required. Thus, comparators are best realized using discrete MSI comparator components or as specialized cells within an FPGA or ASIC library. However, PLDs are quite suitable for realizing the combinational logic used in "parallel expansion" schemes that construct wider comparators from smaller ones, as we'll show here.

In Section 5.9.4 we showed how to connect 74x85 4-bit comparators in series to create larger comparators. Although a serial cascading scheme requires no extra logic to build arbitrarily large comparators, it has the major drawback that the delay increases linearly with the length of the cascade.

In Section 6.1.4, on the other hand, we showed how multiple copies of the 74x682 8-bit comparator could be used in parallel along with combinational logic to perform a 24-bit comparison. This scheme can be generalized for comparisons of arbitrary width.

Table 6-8 is an ABEL program that uses a GAL22V10 to perform a 64-bit comparison, using eight 74x682s to combine the equal (EQ) and greater-than (GT) outputs from the individual byte to produce all six possible relations of the two 64-bit input values $(=, \neq, >, \geq, <, \leq)$.

In this program, the PEQQ and PNEQ outputs can be realized with one product term each. The remaining eight outputs use eight product terms each. As we've mentioned previously, the 22V10 provides 8–16 product terms per output, so the design fits.

HAVE IT YOUR WAY Early PLDs such as the PAL16L8s did not have output-polarity control. Designers who used these devices were forced to choose a particular polarity, active high or active low, for some outputs in order to obtain reduced equations that would fit. When a 16V8, 20V8, 22V10, or any of a plethora of modern CPLDs is used, no such restriction exists. If an equation *or its complement* can be reduced to the number of product terms available, then the corresponding output can be made active high or active low by programming the output-polarity fuse appropriately.

```
module compexp
title 'Expansion logic for 64-bit comparator'
COMPEXP device 'P22V10';

" Inputs from the individual comparators, active-low, 7 = MSByte
EQ_L7..EQ_L0, GT_L7..GT_L0              pin 1..11, 13..14, 21..23;

" Comparison outputs
PEQQ, PNEQ, PGTQ, PGEQ, PLTQ, PLEQ        pin 15..20 istype 'com';

" Active-level conversions
EQ7 = !EQ_L7; EQ6 = !EQ_L6; EQ5 = !EQ_L5; EQ4 = !EQ_L4;
EQ3 = !EQ_L3; EQ2 = !EQ_L2; EQ1 = !EQ_L1; EQ0 = !EQ_L0;
GT7 = !GT_L7; GT6 = !GT_L6; GT5 = !GT_L5; GT4 = !GT_L4;
GT3 = !GT_L3; GT2 = !GT_L2; GT1 = !GT_L1; GT0 = !GT_L0;

" Less-than terms
LT7 = !(EQ7 # GT7); LT6 = !(EQ6 # GT6); LT5 = !(EQ5 # GT5);
LT4 = !(EQ4 # GT4); LT3 = !(EQ3 # GT3); LT2 = !(EQ2 # GT2);
LT1 = !(EQ1 # GT1); LT0 = !(EQ0 # GT0);

equations

PEQQ = EQ7 & EQ6 & EQ5 & EQ4 & EQ3 & EQ2 & EQ1 & EQ0;

PNEQ = !(EQ7 & EQ6 & EQ5 & EQ4 & EQ3 & EQ2 & EQ1 & EQ0);

PGTQ = GT7  #  EQ7 & GT6  #  EQ7 & EQ6 & GT5
     # EQ7 & EQ6 & EQ5 & GT4  #  EQ7 & EQ6 & EQ5 & EQ4 & GT3
     # EQ7 & EQ6 & EQ5 & EQ4 & EQ3 & GT2
     # EQ7 & EQ6 & EQ5 & EQ4 & EQ3 & EQ2 & GT1
     # EQ7 & EQ6 & EQ5 & EQ4 & EQ3 & EQ2 & EQ1 & GT0;

PLEQ = !(GT7  #  EQ7 & GT6  #  EQ7 & EQ6 & GT5
     # EQ7 & EQ6 & EQ5 & GT4  #  EQ7 & EQ6 & EQ5 & EQ4 & GT3
     # EQ7 & EQ6 & EQ5 & EQ4 & EQ3 & GT2
     # EQ7 & EQ6 & EQ5 & EQ4 & EQ3 & EQ2 & GT1
     # EQ7 & EQ6 & EQ5 & EQ4 & EQ3 & EQ2 & EQ1 & GT0);

PLTQ = LT7  #  EQ7 & LT6  #  EQ7 & EQ6 & LT5
     # EQ7 & EQ6 & EQ5 & LT4  #  EQ7 & EQ6 & EQ5 & EQ4 & LT3
     # EQ7 & EQ6 & EQ5 & EQ4 & EQ3 & LT2
     # EQ7 & EQ6 & EQ5 & EQ4 & EQ3 & EQ2 & LT1
     # EQ7 & EQ6 & EQ5 & EQ4 & EQ3 & EQ2 & EQ1 & LT0;

PGEQ = !(LT7  #  EQ7 & LT6  #  EQ7 & EQ6 & LT5
     # EQ7 & EQ6 & EQ5 & LT4  #  EQ7 & EQ6 & EQ5 & EQ4 & LT3
     # EQ7 & EQ6 & EQ5 & EQ4 & EQ3 & LT2
     # EQ7 & EQ6 & EQ5 & EQ4 & EQ3 & EQ2 & LT1
     # EQ7 & EQ6 & EQ5 & EQ4 & EQ3 & EQ2 & EQ1 & LT0);

end compexp
```

Table 6-9
Mode-control bits for
the mode-dependent
comparator.

M1	M0	Comparison
0	0	32-bit
0	1	31-bit
1	0	30-bit
1	1	not used

6.2.5 Mode-Dependent Comparator

For the next example, let us suppose we have a system in which we need to compare two 32-bit words under normal circumstances, but where we must sometimes ignore one or two low-order bits of the input words. The operating mode is specified by two mode-control bits, M1 and M0, as shown in Table 6-9.

As we've noted previously, comparing, adding, and other "iterative" operations are usually poor candidates for PLD-based design, because an equivalent two-level sum-of-products expression has far too many product terms. In Section 5.9.5 we calculated how many product terms are needed for an n-bit comparator. Based on these results, we certainly wouldn't be able to build the 32-bit mode-dependent comparator or even an 8-bit slice of it with a PLD; the 74x682 8-bit comparator is just about the most efficient possible single chip we can use to perform an 8-bit comparison. However, a PLD-based design is quite reasonable for handling the mode-control logic and the part of the comparison that is dependent on mode (the two low-order bits).

Figure 6-9 shows a complete circuit design resulting from this idea, and Table 6-11 is the ABEL program for a 16V8 MODECOMP PLD that handles the "random logic." Four '682s are used to compare most of the bits, and the 16V8 combines the '682 outputs and handles the two low-order bits as a function of the mode. Intermediate expressions EQ30 and GT30 are defined to save typing in the equations section of the program.

As shown in Table 6-10, the XEQY and XGTY outputs use 7 and 11 product terms, respectively. Thus, XGTY does not fit into the 7 product terms available on a 16V8 output. However, this is another example where we have some flexibility in our coding choices. By changing the coding of MODE30 to [1,.X.], we can reduce the product-term requirements for XGTY to 7/12 and thereby fit the design into a 16V8.

Table 6-10
Product-term usage
for the MODECOMP
PLD.

P-Terms	Fan-in	Fan-out	Type	Name
7/9	10	1	Pin	XEQY
11/13	14	1	Pin	XGTY
18/22		Best P-Term Total: 18		

Total Pins: 16
Average P-Term/Output: 9

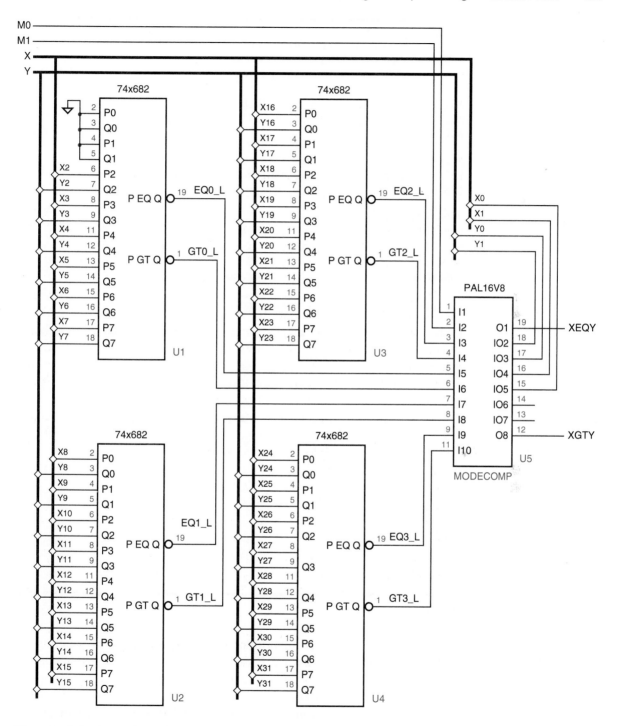

Figure 6-9 A 32-bit mode-dependent comparator.

```
module modecomp
title 'Control PLD for Mode-Dependent Comparator'
MODECOMP device 'P16V8';

" Input and output pins
M0, M1, EQ2_L, GT2_L, EQ0_L, GT0_L              pin 1..6;
EQ1_L, GT1_L, EQ3_L, GT3_L, X0, X1, Y0, Y1     pin 7..9, 10, 15..18;
XEQY, XGTY                                       pin 19, 12 istype 'com';

" Active-level conversions
EQ3 = !EQ3_L; EQ2 = !EQ2_L; EQ1 = !EQ1_L; EQ0 = !EQ0_L;
GT3 = !GT3_L; GT2 = !GT2_L; GT1 = !GT1_L; GT0 = !GT0_L;

" Mode definitions
MODE32 = ([M1,M0] == [0,0]);  " 32-bit comparison
MODE31 = ([M1,M0] == [0,1]);  " 31-bit comparison
MODE30 = ([M1,M0] == [1,0]);  " 30-bit comparison
MODEXX = ([M1,M0] == [1,1]);  " Unused

" Expressions for 30-bit equal and greater-than
EQ30 = EQ3 & EQ2 & EQ1 & EQ0;
GT30 = GT3 # (EQ3 & GT2) # (EQ3 & EQ2 & GT1) # (EQ3 & EQ2 & EQ1 & GT0);

equations

WHEN MODE32 THEN {
   XEQY = EQ30 & (X1==Y1) & (X0==Y0);
   XGTY = GT30 # (EQ30 & (X1>Y1)) # (EQ30 & (X1==Y1) & (X0>Y0));
   }
ELSE WHEN MODE31 THEN {
   XEQY = EQ30 & (X1==Y1);
   XGTY = GT30 # (EQ30 & (X1>Y1));
   }
ELSE WHEN MODE30 THEN {
   XEQY = EQ30;
   XGTY = GT30;
   }

end modecomp
```

6.2.6 Ones Counter

There are several important algorithms that include the step of counting the number of "1" bits in a data word. In fact, some microprocessor instruction sets have been extended recently to include ones counting as a basic instruction.

Counting the ones in a data word can be done easily as an iterative process, where you scan the word from one end to the other and increment a counter each time a "1" is encountered. However, this operation must be done more quickly inside the arithmetic and logic unit of a microprocessor. Ideally, we would like

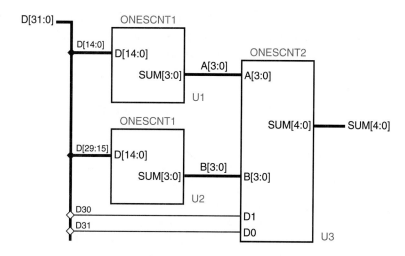

Figure 6-10
Possible partitioning
for the ones-counting
circuit.

ones counting to run as fast as any other arithmetic operation, such as adding two words. Therefore, a combinational circuit is required.

In this example, let us suppose that we have a requirement to build a 32-bit ones counter as part of a larger system. Based on the number of inputs and outputs required, we obviously can't fit the design into a single 22V10-class PLD, but we might be able to partition the design into a reasonably small number of PLDs.

Figure 6-10 shows such a partition. Two copies of a first 22V10, ONESCNT1, are used to count the ones in two 15-bit chunks of the 32-bit input word D[31:0], each producing a 4-bit sum output. A second 22V10, ONESCNT2, is used to add the two 4-bit sums and the last 2 input bits.

The program for ONESCNT1 is deceptively simple, as shown in Table 6-12. The statement "@CARRY 1" is included to limit the carry chain to one stage; as explained in Section 5.10.8, this reduces product-term requirements at the expense of helper outputs and increased delay.

Table 6-12
ABEL program for
counting the 1 bits
in a 15-bit word.

```
module onescnt1
title 'Count the ones in a 15-bit word'
ONESCNT1 device 'P22V10';

" Input and output pins
D14..D0            pin 1..11, 13..15, 23;
SUM3..SUM0         pin 17..20 istype 'com';

equations

@CARRY 1;
[SUM3..SUM0] = D0 + D1 + D2 + D3 + D4 + D5 + D6 + D7
             + D8 + D9 + D10 + D11 + D12 + D13 + D14;

end onescnt1
```

Unfortunately, when I compiled this program, my computer just sat there, CPU-bound, for an hour without producing any results. That gave me time to use my brain, a good exercise for those of us who have become too dependent on CAD tools. I then realized that I could write the logic function for the SUM0 output by hand in just a few seconds,

$$\text{SUM0} = \text{D0} \oplus \text{D1} \oplus \text{D2} \oplus \text{D3} \oplus \text{D4} \oplus \text{D5} \oplus \text{D6} \oplus \text{D7} \oplus \ldots \oplus \text{D13} \oplus \text{D14}$$

The Karnaugh map for this function is a checkerboard, and the minimal sum-of-products expression has 2^{14} product terms. Obviously this is not going to fit in one or a few passes through a 22V10! So, anyway, I killed the ABEL compiler process and rebooted Windows just in case the compiler had gone awry.

Obviously, a partitioning into smaller chunks is required to design the ones-counting circuit. Although we could pursue this further using ABEL and PLDs, it's more interesting to do a structural design using VHDL, as we will in Section 6.3.6. The ABEL and PLD version is left as an exercise (6.6).

6.2.7 Tic-Tac-Toe

In this example, we'll design a combinational circuit that picks a player's next move in the game of Tic-Tac-Toe. The first thing we'll do is decide on a strategy for picking the next move. Let us try to emulate the typical human's strategy by following the decision steps below:

1. Look for a row, column, or diagonal that has two of my marks (X or O, depending on which player I am) and one empty cell. If one exists, place my mark in the empty cell; I win!

2. Else, look for a row, column, or diagonal that has two of my opponent's marks and one empty cell. If one exists, place my mark in the empty cell to block a potential win by my opponent.

3. Else, pick a cell based on experience. For example, if the middle cell is open, it's usually a good bet to take it. Otherwise, the corner cells are good bets. Intelligent players can also notice and block a developing pattern by the opponent or "look ahead" to pick a good move.

TIC-TAC-TOE, IN CASE YOU DIDN'T KNOW   The game of Tic-Tac-Toe is played by two players on a 3×3 grid of cells that are initially empty. One player is "X" and the other is "O". The players alternate in placing their mark in an empty cell; "X" always goes first. The first player to get three of his or her own marks in the same row, column, or diagonal wins. Although the first player to move (X) has a slight advantage, it can be shown that a game between two intelligent players will always end in a draw; neither player will get three in a row before the grid fills up.

row \ column	1	2	3
1	X11,Y11	X13,Y12	X13,Y13
2	X21,Y21	X22,Y22	X23,Y23
3	X21,Y21	X23,Y22	X23,Y23

Figure 6-11
Tic-Tac-Toe grid and ABEL signal names.

Planning ahead, we'll call the second player "Y" to avoid confusion between "O" and "0" in our programs. The next thing to think about is how we might encode the inputs and outputs of the circuit. There are only nine possible moves that a player can make, so the output can be encoded in just four bits. The circuit's input is the current state of the playing grid. There are nine cells, and each cell has one of three possible states (empty, occupied by X, occupied by Y).

There are several choices of how to code the state of one cell. Because the game is symmetric, we'll choose a symmetric encoding that may help us later:

00 Cell is empty.

10 Cell is occupied by X.

01 Cell is occupied by Y.

So, we can encode the 3×3 grid's state in 18 bits. As shown in Figure 6-11, we'll number the grid with row and column numbers and use ABEL signals `Xij` and `Yij` to denote the presence of X or Y in cell `i,j`. We'll look at the output coding later.

With a total of 18 inputs and 4 outputs, the Tic-Tac-Toe circuit could conceivably fit in just one 22V10. However, experience suggests that there's just no way. We're going to have to find a partitioning of the function, and partitioning along the lines of the decision steps on the preceding page seems like a good idea.

In fact, steps 1 and 2 are very similar; they differ only in reversing the roles of the player and the opponent. Here's where our symmetric encoding can pay

COMPACT ENCODING Since each cell in the Tic-Tac-Toe grid can have only three states, not four, the total number of board configurations is 3^9, or 19,683. This is less than 2^{15}, so the board state can be encoded in only 15 bits. However, such an encoding would lead to much larger circuits for picking a move, unless the move-picking circuit was a read-only memory (see Exercise 10.26).

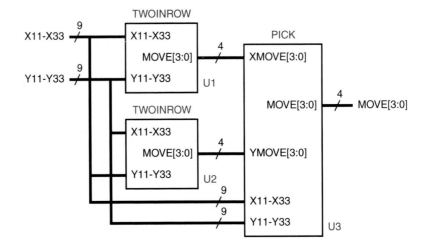

Figure 6-12
Preliminary PLD
partitioning for the
Tic-Tac-Toe game.

off. A PLD that finds me two of my marks in a row along with one empty cell for a winning move (step 1) can find two of my opponent's marks in a row plus an empty cell for a blocking move (step 2). All we have to do is swap the encodings for X and Y. With our selected coding, that doesn't require any logic, just physically swapping the Xij and Yij signals for each cell. With this in mind, we can use two copies of the same PLD, TWOINROW, to perform steps 1 and 2 as shown in Figure 6-12. Notice that the X11–X33 signals are connected to the top inputs of the first TWOINROW PLD but to the bottom inputs of the second.

The moves from the two TWOINROW PLDs can be examined in another PLD, PICK. This device picks a move from the first two PLDs if either found one; else it performs step 3. It looks like PICK has too many inputs and outputs to fit in a 22V10, but we'll come back to that later.

Table 6-13 is a program for the TWOINROW PLD. It looks at the grid's state from the point of view of X; that is, it looks for a move where X can get three in a row. The program makes extensive use of intermediate equations to

Table 6-13
ABEL program to find
two in a row in
Tic-Tac-Toe.

```
module twoinrow
Title 'Find Two Xs and an empty cell in a row, column, or diagonal'
TWOINROW device 'P22V10';

" Inputs and Outputs
X11, X12, X13, X21, X22, X23, X31, X32, X33 pin 1..9;
Y11, Y12, Y13, Y21, Y22, Y23, Y31, Y32, Y33 pin 10,11,13..15,20..23;
MOVE3..MOVE0                                pin 16..19 istype 'com';

" MOVE output encodings
MOVE   = [MOVE3..MOVE0];
MOVE11 = [1,0,0,0]; MOVE12 = [0,1,0,0]; MOVE13 = [0,0,1,0];
MOVE21 = [0,0,0,1]; MOVE22 = [1,1,0,0]; MOVE23 = [0,1,1,1];
MOVE31 = [1,0,1,1]; MOVE32 = [1,1,0,1]; MOVE33 = [1,1,1,0];
NONE   = [0,0,0,0];
```

Table 6-13
(continued)

```
" Find moves in rows.   Rxy ==> a move exists in cell xy
R11 = X12 & X13 & !X11 & !Y11;
R12 = X11 & X13 & !X12 & !Y12;
R13 = X11 & X12 & !X13 & !Y13;
R21 = X22 & X23 & !X21 & !Y21;
R22 = X21 & X23 & !X22 & !Y22;
R23 = X21 & X22 & !X23 & !Y23;
R31 = X32 & X33 & !X31 & !Y31;
R32 = X31 & X33 & !X32 & !Y32;
R33 = X31 & X32 & !X33 & !Y33;

" Find moves in columns.  Cxy ==> a move exists in cell xy
C11 = X21 & X31 & !X11 & !Y11;
C12 = X22 & X32 & !X12 & !Y12;
C13 = X23 & X33 & !X13 & !Y13;
C21 = X11 & X31 & !X21 & !Y21;
C22 = X12 & X32 & !X22 & !Y22;
C23 = X13 & X33 & !X23 & !Y23;
C31 = X11 & X21 & !X31 & !Y31;
C32 = X12 & X22 & !X32 & !Y32;
C33 = X13 & X23 & !X33 & !Y33;

" Find moves in diagonals.  Dxy or Exy ==> a move exists in cell xy
D11 = X22 & X33 & !X11 & !Y11;
D22 = X11 & X33 & !X22 & !Y22;
D33 = X11 & X22 & !X33 & !Y33;
E13 = X22 & X31 & !X13 & !Y13;
E22 = X13 & X31 & !X22 & !Y22;
E31 = X13 & X22 & !X31 & !Y31;

" Combine moves for each cell.  Gxy ==> a move exists in cell xy
G11 = R11 # C11 # D11;
G12 = R12 # C12;
G13 = R13 # C13 # E13;
G21 = R21 # C21;
G22 = R22 # C22 # D22 # E22;
G23 = R23 # C23;
G31 = R31 # C31 # E31;
G32 = R32 # C32;
G33 = R33 # C33 # D33;

equations

WHEN G22 THEN MOVE= MOVE22;
ELSE WHEN G11 THEN MOVE = MOVE11;
ELSE WHEN G13 THEN MOVE = MOVE13;
ELSE WHEN G31 THEN MOVE = MOVE31;
ELSE WHEN G33 THEN MOVE = MOVE33;
ELSE WHEN G12 THEN MOVE = MOVE12;
ELSE WHEN G21 THEN MOVE = MOVE21;
ELSE WHEN G23 THEN MOVE = MOVE23;
ELSE WHEN G32 THEN MOVE = MOVE32;
ELSE MOVE = NONE;

end twoinrow
```

define all possible row, column, and diagonal moves. It combines all of the moves for a cell i,j in an expression for Gij, and finally the equations section uses a WHEN statement to select a move.

Note that a nested WHEN statement must be used rather than nine parallel WHEN statements or assignments, because we can select only one move even if multiple moves are available. Also note that G22, the center cell, is checked first, followed by the corners. This was done hoping that we could minimize the number of terms by putting the most common moves early in the nested WHEN. Alas, the design still requires a ton of product terms, as shown in Table 6-14.

By the way, we still haven't explained why we chose the output coding that we did (as defined by MOVE11, MOVE22, etc. in the program). It's pretty clear that changing the encoding is never going to save us enough product terms to fit the design into a 22V10. But there's still method to this madness, as we'll now show.

Clearly we'll have to split TWOINROW into two or more pieces. As in any design problem, several different strategies are possible. The first strategy I tried was to use two different PLDs, one to find moves in all the rows and one of the diagonals, and the other to work on all the columns and the remaining diagonal. That helped, but not nearly enough to fit each half into a 22V10.

With the second strategy, I tried slicing the problem a different way. The first PLD finds all the moves in cells 11, 12, 13, 21, and 22, and the second PLD finds all the moves in the remaining cells. That worked! The first PLD, named TWOINHAF, is obtained from Table 6-13 simply by commenting out the four lines of the WHEN statement for the moves to cells 23, 31, 32, and 33.

We could obtain the second PLD from TWOINROW in a similar way, but let's wait a minute. In the manufacture of real digital systems, it is always desirable to minimize the number of distinct parts that are used; this saves on inventory costs and complexity. With programmable parts, it is desirable to minimize the number of distinct programs that are used. Even though the physical parts are identical, a different set of test vectors must be devised at some cost for each different program. Also, it's possible that the product will be successful enough for us to save money by converting the PLDs into hard-coded devices, a different one for each program, again encouraging us to minimize programs.

Table 6-14
Product-term usage for the TWOINROW PLD.

P-Terms	Fan-in	Fan-out	Type	Name
61/142	18	1	Pin	MOVE3
107/129	18	1	Pin	MOVE2
77/88	17	1	Pin	MOVE1
133/87	18	1	Pin	MOVE0
=========				
378/446		Best P-Term Total: 332		
		Total Pins: 22		
		Average P-Term/Output: 83		

The Tic-Tac-Toe game is the same game even if we rotate the grid 90° or 180°. Thus, the TWOINHAF PLD can find moves for cells 33, 32, 31, 23, and 22 if we rotate the grid 180°. Because of the way we defined the grid state, with a separate pair of inputs for each cell, we can "rotate the grid" simply by shuffling the input pairs appropriately. That is, we swap 33↔11, 32↔12, 31↔13, and 23↔21.

Of course, once we rearrange inputs, TWOINHAF will still produce output move codes corresponding to cells in the top half of the grid. To keep things straight, we should transform these into codes for the proper cells in the bottom half of the grid. We would like this transformation to take a minimum of logic. This is where our choice of output code comes in. If you look carefully at the MOVE coding defined at the beginning of Table 6-13, you'll see that the code for a given position in the 180°-rotated grid is obtained by complementing and reversing the order of the code bits for the same position in the unrotated grid. In other words, the code transformation can be done with four inverters and a rearrangement of wires. This can be done "for free" in the PLD that looks at the TWOINHAF outputs.

You probably never thought that Tic-Tac-Toe could be so tricky. Well, we're halfway there. Figure 6-13 shows the partitioning of the design as we'll now continue it. Each TWOINROW PLD from our original partition is replaced

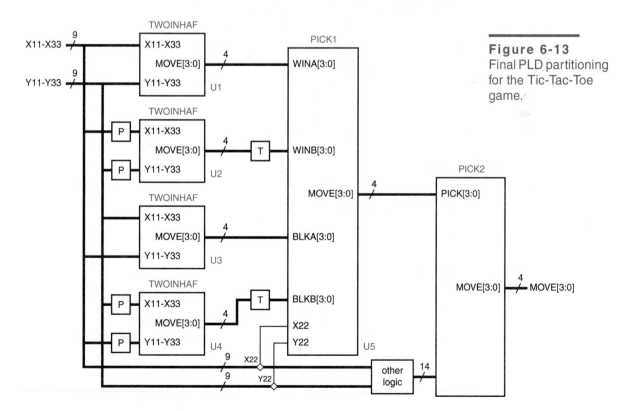

Figure 6-13
Final PLD partitioning for the Tic-Tac-Toe game.

by a pair of TWOINHAF PLDs. The bottom PLD of each pair is preceded by a box labeled "P", which permutes the inputs to rotate the grid 180°, as discussed previously. Likewise, it is followed by a box labeled "T", which compensates for the rotation by transforming the output code; this box will actually be absorbed into the PLD that follows it, PICK1.

The function of PICK1 is pretty straightforward. As shown in Table 6-15, it is simply picks a winning move or a blocking move if one is available. Since there are two extra input pins available on the 22V10, we use them to input the state of the center cell. In this way, we can perform the first part of step 3 of the "human" algorithm on page 492, to pick the center cell if no winning or blocking move is available. The PICK1 PLD uses at most 9 product terms per output.

Table 6-15 ABEL program to pick one move based on four inputs.

```
module pick1
Title 'Pick One Move from Four Possible'
PICK1 device 'P22V10';

" Inputs from TWOINHAF PLDs
WINA3..WINA0        pin 1..4;       "Winning moves in cells 11,12,13,21,22
WINB3..WINB0        pin 5..8;       "Winning moves in cells 11,12,13,21,22 of rotated grid
BLKA3..BLKA0        pin 9..11, 13;  "Blocking moves in cells 11,12,13,21,22
BLKB3..BLKB0        pin 14..16, 21; "Blocking moves in cells 11,12,13,21,22 of rotated grid
" Inputs from grid
X22, Y22            pin 22..23;     "Center cell; pick if no other moves
" Move outputs to PICK2 PLD
MOVE3..MOVE0        pin 17..20 istype 'com';

" Sets
WINA = [WINA3..WINA0];  WINB = [WINB3..WINB0];
BLKA = [BLKA3..BLKA0];  BLKB = [BLKB3..BLKB0];
MOVE = [MOVE3..MOVE0];

" Non-rotated move input and output encoding
MOVE11 = [1,0,0,0]; MOVE12 = [0,1,0,0]; MOVE13 = [0,0,1,0];
MOVE21 = [0,0,0,1]; MOVE22 = [1,1,0,0]; MOVE23 = [0,1,1,1];
MOVE31 = [1,0,1,1]; MOVE32 = [1,1,0,1]; MOVE33 = [1,1,1,0];
NONE   = [0,0,0,0];

equations

WHEN WINA != NONE THEN MOVE = WINA;
ELSE WHEN WINB != NONE THEN MOVE = ![WINB0..WINB3];  " Map rotated coding
ELSE WHEN BLKA != NONE THEN MOVE = BLKA;
ELSE WHEN BLKB != NONE THEN MOVE = ![BLKB0..BLKB3];  " Map rotated coding
ELSE WHEN !X22 & !Y22 THEN MOVE = MOVE22;            " Pick center cell if it's empty
ELSE MOVE = NONE;

end pick1
```

The final part of the design in Figure 6-13 is the PICK2 PLD. This PLD must provide most of the "experience" in step 3 of the human algorithm if PICK1 does not find a move.

We have a little problem with PICK2 in that a 22V10 does not have enough pins to accommodate the 4-bit input from PICK1, its own 4-bit output, and all 18 bits of grid state; it has only 22 I/O pins. Actually, we don't need to connect X22 and Y22, since they were already examined in PICK1, but that still leaves us two pins short. So, the purpose of the "other logic" block in Figure 6-13 is to encode

▌ **Table 6-16** ABEL program to pick one move using "experience."

```
module pick2
Title 'Pick a move using experience'
PICK2 device 'P22V10';

" Inputs from PICK1 PLD
PICK3..PICK0                            pin 1..4;  " Move, if any, from PICK1 PLD
" Inputs from Tic-Tac-Toe grid corners
X11, Y11, X13, Y13, X31, Y31, X33, Y33    pin 5..11, 13;
" Combined inputs from external NOR gates; 1 ==> corresponding cell is empty
E12, E21, E23, E32                      pin 14..15, 22..23;
" Move output
MOVE3..MOVE0                            pin 17..20 istype 'com';

PICK = [PICK3..PICK0];  " Set definition
" Non-rotated move input and output encoding
MOVE = [MOVE3..MOVE0];
MOVE11 = [1,0,0,0]; MOVE12 = [0,1,0,0]; MOVE13 = [0,0,1,0];
MOVE21 = [0,0,0,1]; MOVE22 = [1,1,0,0]; MOVE23 = [0,1,1,1];
MOVE31 = [1,0,1,1]; MOVE32 = [1,1,0,1]; MOVE33 = [1,1,1,0];
NONE   = [0,0,0,0];

" Intermediate equations for empty corner cells
E11 = !X11 & !Y11;  E13 = !X13 & !Y13;  E31 = !X31 & !Y31;  E33 = !X33 & !Y33;

equations

"Simplest approach -- pick corner if available, else side
WHEN PICK != NONE THEN MOVE = PICK;
ELSE WHEN E11 THEN MOVE = MOVE11;
ELSE WHEN E13 THEN MOVE = MOVE13;
ELSE WHEN E31 THEN MOVE = MOVE31;
ELSE WHEN E33 THEN MOVE = MOVE33;
ELSE WHEN E12 THEN MOVE = MOVE12;
ELSE WHEN E21 THEN MOVE = MOVE21;
ELSE WHEN E23 THEN MOVE = MOVE23;
ELSE WHEN E32 THEN MOVE = MOVE32;
ELSE MOVE = NONE;

end pick2
```

some of the information to save two pins. The method that we'll use here is to combine the signals for the middle edge cells 12, 21, 23, and 32 to produce four signals E12, E21, E23, and E32 that are asserted if and only if the corresponding cells are empty. This can be done with four 2-input NOR gates and actually leaves two spare inputs or outputs on the 22V10.

Assuming the four NOR gates as "other logic," Table 6-16 on the preceding page gives a program for the PICK2 PLD. When it must pick a move, this program uses the simplest heuristic possible—it picks a corner cell if one is empty, else it picks a middle edge cell. This program could use some improvement, because it will sometimes lose (see Exercise 6.8). Luckily, the equations resulting from Table 6-16 require only 8 to 10 terms per output, so it's possible to put in more intelligence (see Exercises 6.9 and 6.10).

6.3 Design Examples Using VHDL

6.3.1 Barrel Shifter

On page 468 we defined a barrel shifter as a combinational logic circuit with n data inputs, n data outputs, and a set of control inputs that specify how to shift the data between input and output. We showed in Section 6.1.1 how to build a simple barrel shifter that performs only left circular shifts using MSI building blocks. Later, in Section 6.2.1, we showed how to define a more capable barrel shifter using ABEL, but we also pointed out that PLDs are normally unsuitable for realizing barrel shifters. In this subsection we'll show how VHDL can be used to describe both the behavior and structure of barrel shifters for FPGA or ASIC realization.

Table 6-17 is a behavioral VHDL program for a 16-bit barrel shifter that performs any of six different combinations of shift type and direction. The shift types are circular, logical, and arithmetic, as defined previously in Table 6-3, and the directions are, of course, left and right. As shown in the entity declaration, a 4-bit control input S gives the shift amount, and a 3-bit control input C gives the shift mode (type and direction). We used the IEEE std_logic_arith package and defined the shift amount S to be type UNSIGNED so we could later use the CONV_INTEGER function in that package.

Notice that the entity declaration includes six constant definitions that establish the correspondence between shift modes and the value of C. Although we didn't discuss it in Section 4.7, VHDL allows you to put constant, type, signal, and other declarations within an entity declaration. It makes sense to define such items within the entity declaration only if they must be the same in any architecture. In this case, we are pinning down the shift-mode encodings, so they should go here. Other items should go in the architecture definition.

In the architecture part of the program we define six functions, one for each kind of shift on a 16-bit STD_LOGIC_VECTOR. We defined the subtype DATAWORD to save typing in the function definitions.

Table 6-17 VHDL behavioral description of a 6-function barrel shifter.

```vhdl
library IEEE;
use IEEE.std_logic_1164.all;
use IEEE.std_logic_arith.all;

entity barrel16 is
    port (
        DIN: in STD_LOGIC_VECTOR (15 downto 0);   -- Data inputs
        S: in UNSIGNED (3 downto 0);              -- Shift amount, 0-15
        C: in STD_LOGIC_VECTOR (2 downto 0);      -- Mode control
        DOUT: out STD_LOGIC_VECTOR (15 downto 0)  -- Data bus output
    );
    constant Lrotate:  STD_LOGIC_VECTOR := "000";   -- Define the coding of
    constant Rrotate:  STD_LOGIC_VECTOR := "001";   -- the different shift modes
    constant Llogical: STD_LOGIC_VECTOR := "010";
    constant Rlogical: STD_LOGIC_VECTOR := "011";
    constant Larith:   STD_LOGIC_VECTOR := "100";
    constant Rarith:   STD_LOGIC_VECTOR := "101";
end barrel16;

architecture barrel16_behavioral of barrel16 is
subtype DATAWORD is STD_LOGIC_VECTOR(15 downto 0);

function Vrol (D: DATAWORD; S: UNSIGNED)
      return DATAWORD is
  variable N: INTEGER;
  variable TMPD: DATAWORD;
  begin
    N := CONV_INTEGER(S); TMPD := D;
    for i in 1 to N loop
      TMPD := TMPD(14 downto 0) & TMPD(15);
    end loop;
    return TMPD;
  end Vrol;

...

begin
process(DIN, S, C)
  begin
    case C is
      when Lrotate  => DOUT <= Vrol(DIN,S);
      when Rrotate  => DOUT <= Vror(DIN,S);
      when Llogical => DOUT <= Vsll(DIN,S);
      when Rlogical => DOUT <= Vsrl(DIN,S);
      when Larith   => DOUT <= Vsla(DIN,S);
      when Rarith   => DOUT <= Vsra(DIN,S);
      when others   => null;
    end case;
  end process;
end barrel16_behavioral;
```

ROLLING
YOUR OWN

VHDL-93 actually has built-in array operators, rol, ror, sll, srl, sla, and sra, corresponding to the shift operations that we defined in Table 6-3. Since these operations are not provided in VHDL-87, we've defined our own functions in Table 6-17. Well, actually we've defined only one of them (Vrol); the rest are left as an exercise for the reader (Exercise 6.11).

Table 6-17 shows the details of only the first function (Vrol); the rest are similar with only a one-line change. We define a variable N for converting the shift-amount S into an integer for the for loop. We also assign the input vector D to a local variable TMPD, which is shifted N times in the for loop. In the body of the for loop, a single assignment statement takes a 15-bit slice of the data word [TMPD(14 downto 0)] and uses concatenation [&] to put it back together with the bit that "falls off" the left end [TMPD(15)]. Other shift types can be described with similar operations. Note that the shift functions might not be defined in other, nonbehavioral descriptions of the barrel16 entity, for example in structural architectures.

The "concurrent statements" part of the architecture is a single process that has all of the entity's inputs in its sensitivity list. Within this process, a case statement assigns a result to DOUT by calling the appropriate function based on the value of the mode-control input C.

The process in Table 6-17 is a nice behavioral description of the barrel shifter, but most synthesis tools cannot synthesize a circuit from it. The problem is that most tools require the range of a for loop to be static at the time it is analyzed. The range of the for loop in the Vrol function is dynamic; it depends on the value of input signal S when the circuit is operating.

Well, that's OK; it's hard to predict what kind of circuit the synthesis tool would come up with even if it could handle a dynamic for range. This is an example where as designers we should take a little more control over the circuit structure to obtain a reasonably fast, efficient synthesis result.

In Figure 6-2 on page 470 we showed how to design a 16-bit barrel shifter for left circular shifts using MSI building blocks. We used a cascade of four 16-bit, 2-input multiplexers to shift their inputs by 0 or 1, 2, 4, or 8 positions, depending on the values of S0 through S3, respectively. We can express the same kind of behavior and structure using the VHDL program shown in Table 6-18. Even though the program uses a process and is "behavioral" in style, we can be pretty sure that most synthesis engines will generate a 1-input multiplexer for each "if" statement in the program, thereby creating a cascade similar to Figure 6-2.

Table 6-18 VHDL program for a 16-bit barrel shifter for left circular shifts only.

```
library IEEE;
use IEEE.std_logic_1164.all;

entity rol16 is
    port (
        DIN: in STD_LOGIC_VECTOR(15 downto 0);   -- Data inputs
        S: in STD_LOGIC_VECTOR (3 downto 0);     -- Shift amount, 0-15
        DOUT: out STD_LOGIC_VECTOR(15 downto 0)  -- Data bus output
    );
end rol16;

architecture rol16_arch of rol16 is
begin
process(DIN, S)
  variable X, Y, Z: STD_LOGIC_VECTOR(15 downto 0);
  begin
    if S(0)='1' then X := DIN(14 downto 0) & DIN(15); else X := DIN; end if;
    if S(1)='1' then Y := X(13 downto 0) & X(15 downto 14); else Y := X; end if;
    if S(2)='1' then Z := Y(11 downto 0) & Y(15 downto 12); else Z := Y; end if;
    if S(3)='1' then DOUT <= Z(7 downto 0) & Z(15 downto 8); else DOUT <= Z; end if;
  end process;
end rol16_arch;
```

Of course, our problem statement requires a barrel shifter that can shift both left and right. Table 6-19 revises the previous program to do circular shifts in either direction. An additional input, DIR, specifies the shift direction, 0 for left, 1 for right. Each rank of shifting is specified by a case statement that picks one of four possibilities based on the values of DIR and the bit of S that controls that rank. Notice that we created local 2-bit variables CTRLi to hold the pair of values DIR and S(i); each case statement is controlled by one of these variables. You might like to eliminate these variables and simply control each case statement with a concatenation "DIR & S(i)", but VHDL syntax doesn't allow that because the type of this concatenation would be unknown.

A typical VHDL synthesis tool will generate a 3- or 4-input multiplexer for each of the case statements in Table 6-19. A good synthesis tool will generate only a 2-input multiplexer for the last case statement.

So, now we have a barrel shifter that will do left or right circular shifts, but we're not done yet—we need to take care of the logical and arithmetic shifts in both directions. Figure 6-14 shows our strategy for completing the design. We start out with the ROLR16 component that we just completed, and we use other logic to control the shift direction as a function of C.

Table 6-19 VHDL program for a 16-bit barrel shifter for left and right circular shifts.

```vhdl
library IEEE;
use IEEE.std_logic_1164.all;

entity rolr16 is
    port (
        DIN:  in STD_LOGIC_VECTOR(15 downto 0);   -- Data inputs
        S:    in STD_LOGIC_VECTOR (3 downto 0);   -- Shift amount, 0-15
        DIR:  in STD_LOGIC;                       -- Shift direction, 0=>L, 1=>R
        DOUT: out STD_LOGIC_VECTOR(15 downto 0)   -- Data bus output
    );
end rolr16;

architecture rol16r_arch of rolr16 is
begin
process(DIN, S, DIR)
  variable X, Y, Z: STD_LOGIC_VECTOR(15 downto 0);
  variable CTRL0, CTRL1, CTRL2, CTRL3: STD_LOGIC_VECTOR(1 downto 0);
  begin
    CTRL0 := S(0) & DIR; CTRL1 := S(1) & DIR; CTRL2 := S(2) & DIR; CTRL3 := S(3) & DIR;
    case CTRL0 is
      when "00" | "01" => X := DIN;
      when "10" => X := DIN(14 downto 0) & DIN(15);
      when "11" => X := DIN(0) & DIN(15 downto 1);
      when others => null;  end case;
    case CTRL1 is
      when "00" | "01" => Y := X;
      when "10" => Y := X(13 downto 0) & X(15 downto 14);
      when "11" => Y := X(1 downto 0) & X(15 downto 2);
      when others => null;  end case;
    case CTRL2 is
      when "00" | "01" => Z := Y;
      when "10" => Z := Y(11 downto 0) & Y(15 downto 12);
      when "11" => Z := Y(3 downto 0) & Y(15 downto 4);
      when others => null;  end case;
    case CTRL3 is
      when "00" | "01" => DOUT <= Z;
      when "10" | "11" => DOUT <= Z(7 downto 0) & Z(15 downto 8);
      when others => null;  end case;
  end process;
end rol16r_arch;
```

Next we must "fix up" some of the result bits if we are doing a logical or arithmetic shift. For a left logical or arithmetic *n*-bit shift, we must set the rightmost *n*–1 bits to 0 or the original rightmost bit value, respectively. For a right logical or arithmetic *n*-bit shift, we must set the leftmost *n*–1 bits to 0 or the original leftmost bit value, respectively.

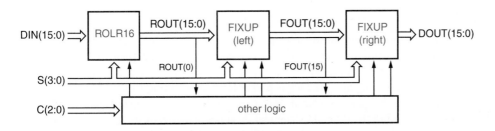

Figure 6-14
Barrel-shifter
components.

As shown in Figure 6-14, our strategy is to follow the circular shifter (ROLR16) with a fix-up circuit (FIXUP) that plugs in appropriate low-order bits for a left logical or arithmetic shift, and follow that with another fix-up circuit that plugs in high-order bits for a right logical or arithmetic shift.

Table 6-20 is a behavioral VHDL program for the left-shift fix-up circuit. The circuit has 16 bits of data input and output, DIN and DOUT. Its control inputs are the shift amount S, an enable input FEN, and the new value FDAT to be plugged into the fixed-up data bits. For each output bit DOUT(i), the circuit puts out the fixed-up bit value if i is less than S and the circuit is enabled; else it puts out the unmodified data input DIN(i).

The for loop in Table 6-20 is readily synthesizable, but you can't really be sure what kind of logic the synthesis tool will generate. In particular, the ">"

Table 6-20 Behavioral VHDL program for left-shift fix-ups.

```
library IEEE;
use IEEE.std_logic_1164.all;
use IEEE.std_logic_arith.all;

entity fixup is
    port (
        DIN:  in STD_LOGIC_VECTOR(15 downto 0); -- Data inputs
        S:    in UNSIGNED(3 downto 0);          -- Shift amount, 0-15
        FEN:  in STD_LOGIC;                     -- Fixup enable
        FDAT: in STD_LOGIC;                     -- Fixup data
        DOUT: out STD_LOGIC_VECTOR(15 downto 0) -- Data bus output
    );
end fixup;

architecture fixup_arch of fixup is
begin
process(DIN, S, FEN, FDAT)
  begin
    for i in 0 to 15 loop
      if (i < CONV_INTEGER(S)) and (FEN = '1') then DOUT(i) <= FDAT;
      else DOUT(i) <= DIN(i);  end if;
    end loop;
  end process;
end fixup_arch;
```

A SERIAL FIX-UP
STRUCTURE

A structural architecture for the fix-up logic is shown in Table 6-21. Here, we have defined what is in effect an iterative circuit to create a 16-bit vector FSEL, where FSEL(i) is 1 if bit i needs fixing up. We start by setting FSEL(15) to 0, since that bit never needs fixing up. Then we note that for the remaining values of i, FSEL(i) should be 1 if S equals i+1 or if FSEL(i+1) is already asserted. Thus, the FSEL assignment within the generate statement creates a serial chain of 2-input OR gates, where one input is asserted if S=i (decoded with a 4-input AND gate), and the other input is connected to the previous OR gate's output. The DOUT(i) assignment statement creates 16 2-input multiplexers that select either DIN(i) or the fix-up data (FDAT), depending on the value of FSEL(i).

Although the serial realization is compact, it is very slow compared to one that realizes each FSEL output as a 2-level sum-of-products circuit. However, the long delay may not matter, because the fix-up circuit appears near the end of the data path. If speed is still a problem, there is a zero-cost trick that cuts the delay in half (see Exercise 6.12).

operation in each step of the loop may cause the synthesis of a general-purpose magnitude comparator, even though one of the operands is a constant and each output could therefore be generated with no more than a handful of gates. (In fact, the logic for "7 < CONV_INTEGER(S)" is just a wire, S(3)!) For a structural version of this function, see the box on this page.

For right shifts, fix-ups start from the opposite end of the data word, so it would appear that we need a second version of the fix-up circuit. However, we can use the original version if we just reverse the order of its input and output bits, as we'll soon see.

Table 6-22 puts together a structural architecture for the complete, 16-bit, 6-function barrel shifter using the design approach of Figure 6-14 on page 505. The entity declaration for barrel16 is unchanged from the original in Table 6-17 on page 501. The architecture declares two components, rolr16 and fixup; these use our previous entity definitions. The statement part of the archi-

Table 6-21 Structural VHDL architecture for left-shift fix-ups.

```
architecture fixup_struc of fixup is
signal FSEL: STD_LOGIC_VECTOR(15 downto 0);        -- Fixup select
begin
  FSEL(15) <= '0'; DOUT(15) <= DIN(15);
  U1: for i in 14 downto 0 generate
    FSEL(i) <= '1' when CONV_INTEGER(S) = i+1 else FSEL(i+1);
    DOUT(i) <= FDAT when (FSEL(i) = '1' and FEN = '1') else DIN(i);
  end generate;
end fixup_struc;
```

tecture instantiates `rolr16` and `fixup` and has several assignment statements
that create needed control signals (the "other logic" in Figure 6-14).

For example, the first assignment statement asserts `DIR_RIGHT` if C specifies one of the right shifts. The enable inputs for the left and right fix-up circuits
are `FIX_LEFT` and `FIX_RIGHT`, asserted for left and right logical and arithmetic
shifts. The fix-up data values are `FIX_LEFT_DAT` and `FIX_RIGHT_DAT`.

While all the statements in the architecture execute concurrently, they are
listed in Table 6-22 in the order of the actual dataflow to improve readability.
First, `rolr16` is instantiated to perform the basic left or right circular shift as
specified. Its outputs are hooked up to the inputs of the first `fixup` component
(U2) to handle fix-ups for left logical and arithmetic shifts. Next comes U3, a
`generate` statement that reverses the order of the data inputs for the next `fixup`

Table 6-22 VHDL structural architecture for the 6-function barrel shifter.

```
architecture barrel16_struc of barrel16 is

component rolr16 port (
        DIN:  in STD_LOGIC_VECTOR(15 downto 0); -- Data inputs
        S:    in UNSIGNED(3 downto 0);          -- Shift amount, 0-15
        DIR:  in STD_LOGIC;                     -- Shift direction, 0=>L, 1=>R
        DOUT: out STD_LOGIC_VECTOR(15 downto 0) -- Data bus output
    ); end component;

component fixup port (
        DIN:  in STD_LOGIC_VECTOR(15 downto 0); -- Data inputs
        S:    in UNSIGNED(3 downto 0);          -- Shift amount, 0-15
        FEN:  in STD_LOGIC;                     -- Fixup enable
        FDAT: in STD_LOGIC;                     -- Fixup data
        DOUT: out STD_LOGIC_VECTOR(15 downto 0) -- Data bus output
    ); end component;

signal DIR_RIGHT, FIX_RIGHT, FIX_RIGHT_DAT, FIX_LEFT, FIX_LEFT_DAT: STD_LOGIC;
signal ROUT, FOUT, RFIXIN, RFIXOUT: STD_LOGIC_VECTOR(15 downto 0);

begin
  DIR_RIGHT <= '1' when C = Rrotate or C = Rlogical or C = Rarith else '0';
  FIX_LEFT <= '1' when DIR_RIGHT='0' and (C = Llogical or C = Larith) else '0';
  FIX_RIGHT <= '1' when DIR_RIGHT='1' and (C = Rlogical or C = Rarith) else '0';
  FIX_LEFT_DAT <= DIN(0) when C = Larith else '0';
  FIX_RIGHT_DAT <= DIN(15) when C = Rarith else '0';
  U1: rolr16 port map (DIN, S, DIR_RIGHT, ROUT);
  U2: fixup port map (ROUT, S, FIX_LEFT, FIX_LEFT_DAT, FOUT);
  U3: for i in 0 to 15 generate RFIXIN(i) <= FOUT(15-i); end generate;
  U4: fixup port map (RFIXIN, S, FIX_RIGHT, FIX_RIGHT_DAT, RFIXOUT);
  U5: for i in 0 to 15 generate DOUT(i) <= RFIXOUT(15-i); end generate;
end barrel16_struc;
```

INFORMATION-HIDING STYLE

Based on the encoding of C, you might like to replace the first assignment statement in Table 6-22 with "DIR_RIGHT <= C(0)", which would be guaranteed to lead to a more efficient realization for that control bit—just a wire! However, this would violate a programming principle of information hiding and lead to possible bugs.

We explicitly wrote the shift encodings using constant definitions in the barrel16 entity declaration. The architecture does not need to be aware of the encoding details. Suppose that we nevertheless made the architecture change suggested above. If somebody else (or we!) came along later and changed the constant definitions in the barrel16 entity to make a different encoding, the architecture would not use the new encodings! Exercise 6.13 asks you to change the definitions so that the cost savings of our suggested change are enabled by the entity definition.

component (U4), which handles fix-ups for right logical and arithmetic shifts. Finally U5, another generate statement, undoes the bit reversing of U3. Note that in synthesis, U3 and U5 are merely permutations of wires.

Many other architectures are possible for the original barrel16 entity. In Exercise 6.14, we suggest an architecture that enables the circular shifting to be done by the rol16 entity, which uses only 2-input multiplexers, rather than the more expensive rolr16.

6.3.2 Simple Floating-Point Encoder

We defined a simple floating-point number format on page 471 and posed the design problem of converting a number from fixed-point to this floating-point format. The problem of determining the exponent of the floating-point number mapped nicely into an MSI priority encoder. In an HDL, the same problem maps into nested "if" statements.

Table 6-23 is a behavioral VHDL program for the floating-point encoder. Within the fpenc_arch architecture, a nested "if" statement checks the range of the input B and sets M and E appropriately. Notice that the program uses the IEEE std_logic_arith package; this is done to get the UNSIGNED type and the comparison operations that go along with it, as we described in Section 5.9.6. Just to save typing, a variable BU is defined to hold the value of B as converted to the UNSIGNED type; alternatively, we could have written "UNSIGNED(B)" in each nested "if" clause.

Although the code in Table 6-23 is fully synthesizable, some synthesis tools may not be smart enough to recognize that the nested comparisons require just one bit to be checked at each level, and might instead generate a full 11-bit comparator at each level. Such logic would be a lot bigger and slower than what otherwise would be possible. If faced with this problem, we can always write the architecture a little differently and more explicitly to help out the tool, as shown in Table 6-24.

```
library IEEE;
use IEEE.std_logic_1164.all;
use IEEE.std_logic_arith.all;

entity fpenc is
  port (
    B: in STD_LOGIC_VECTOR(10 downto 0); -- fixed-point number
    M: out STD_LOGIC_VECTOR(3 downto 0); -- floating-point mantissa
    E: out STD_LOGIC_VECTOR(2 downto 0)  -- floating-point exponent
  );
end fpenc;

architecture fpenc_arch of fpenc is
begin
  process(B)
  variable BU: UNSIGNED(10 downto 0);
  begin
    BU := UNSIGNED(B);
    if    BU < 16   then M <= B( 3 downto 0); E <= "000";
    elsif BU < 32   then M <= B( 4 downto 1); E <= "001";
    elsif BU < 64   then M <= B( 5 downto 2); E <= "010";
    elsif BU < 128  then M <= B( 6 downto 3); E <= "011";
    elsif BU < 256  then M <= B( 7 downto 4); E <= "100";
    elsif BU < 512  then M <= B( 8 downto 5); E <= "101";
    elsif BU < 1024 then M <= B( 9 downto 6); E <= "110";
    else                 M <= B(10 downto 7); E <= "111";
    end if;
  end process;
end fpenc_arch;
```

Table 6-23
Behavioral VHDL program for fixed-point to floating-point conversion.

```
architecture fpence_arch of fpenc is
begin
  process(B)
  begin
    if    B(10) = '1' then M <= B(10 downto 7); E <= "111";
    elsif B(9)  = '1' then M <= B( 9 downto 6); E <= "110";
    elsif B(8)  = '1' then M <= B( 8 downto 5); E <= "101";
    elsif B(7)  = '1' then M <= B( 7 downto 4); E <= "100";
    elsif B(6)  = '1' then M <= B( 6 downto 3); E <= "011";
    elsif B(5)  = '1' then M <= B( 5 downto 2); E <= "010";
    elsif B(4)  = '1' then M <= B( 4 downto 1); E <= "001";
    else                   M <= B( 3 downto 0); E <= "000";
    end if;
  end process;
end fpence_arch;
```

Table 6-24
Alternative VHDL architecture for fixed-point to floating-point conversion.

B'S NOT MY TYPE   In Table 6-23 we used the expression UNSIGNED(B) to convert B, an array of type STD_LOGIC_VECTOR, into an array of type UNSIGNED. This is called an *explicit type conversion*. VHDL lets you convert between related closely related types by writing the desired type followed by the value to be converted in parentheses. Two array types are "closely related" if they have the same element type, the same number of dimensions, and the same index types (typically INTEGER) or ones that can be type converted. The values in the old array are placed in corresponding positions, left to right, in the new array.

On the other hand, we might like to use the real comparators and spend even more gates to improve the functionality of our design. In particular, the present design performs truncation rather than rounding when generating the mantissa bits. A more accurate result is achieved with rounding, but this is a much more complicated design. First, we will need an adder to add 1 to the selected mantissa bits when we round up. However, adding 1 when the mantissa is already 1111 will bump us into the next exponent range, so we need to watch out for this case. Finally, we can never round up if the unrounded mantissa and exponent are 1111 and 111, because there's no higher value in our floating-point representation to round to.

The program in Table 6-25 performs rounding as desired. The function round takes a selected 5-bit slice from the fixed-point number and returns the four high-order bits, adding 1 if the LSB is 1. Thus, if we think of the binary point as being just to the left of the LSB, rounding occurs if the truncated part of the mantissa is 1/2 or more. In each clause in the nested "if" statement in the process, the comparison value is selected so that rounding up will occur only if it does not "overflow," pushing the result into the next exponent range. Otherwise, conversion and rounding occurs in the next clause. In the last clause, we ensure that we do not round up when we're at the end of the floating-point range.

GOBBLE,   The rounding operation does not require a 4-bit adder, only an "incrementer," since
GOBBLE   one of the addends is always 1. Some VHDL tools may synthesize the complete adder, while others may be smart enough to use an incrementer with far fewer gates.

In some cases it may not matter. The most sophisticated tools for FPGA and ASIC design include *gate gobblers*. These programs look for gates with constant inputs and eliminate gates or gate inputs as a result. For example, an AND-gate input with a constant 1 applied to it can be eliminated, and an AND gate with a constant-0 input can be replaced with a constant-0 signal.

A gate-gobbler program propagates the effects of constant inputs as far as possible in a circuit. Thus, it can transform a 4-bit adder with a constant-1 input into a more economical 4-bit incrementer.

```
architecture fpencr_arch of fpenc is
function round (BSLICE: STD_LOGIC_VECTOR(4 downto 0))
    return STD_LOGIC_VECTOR is
  variable BSU: UNSIGNED(3 downto 0);
  begin
    if BSLICE(0) = '0' then return BSLICE(4 downto 1);
    else null;
      BSU := UNSIGNED(BSLICE(4 downto 1)) + 1;
      return STD_LOGIC_VECTOR(BSU);
    end if;
  end;
begin
  process(B)
  variable BU: UNSIGNED(10 downto 0);
  begin
    BU := UNSIGNED(B);
    if    BU < 16       then M <= B( 3 downto 0); E <= "000";
    elsif BU < 32-1     then M <= round(B( 4 downto 0)); E <= "001";
    elsif BU < 64-2     then M <= round(B( 5 downto 1)); E <= "010";
    elsif BU < 128-4    then M <= round(B( 6 downto 2)); E <= "011";
    elsif BU < 256-8    then M <= round(B( 7 downto 3)); E <= "100";
    elsif BU < 512-16   then M <= round(B( 8 downto 4)); E <= "101";
    elsif BU < 1024-32  then M <= round(B( 9 downto 5)); E <= "110";
    elsif BU < 2048-64  then M <= round(B(10 downto 6)); E <= "111";
    else                     M <= "1111";               E <= "111";
    end if;
  end process;
end fpencr_arch;
```

Table 6-25
Behavioral VHDL architecture for fixed-point to floating-point conversion with rounding.

Once again, synthesis results for this behavioral program may or may not be efficient. Besides the multiple comparison statements, we now must worry about the multiple 4-bit adders that might be synthesized as a result of the multiple calls to the round function. Restructuring the architecture so that only a single adder is synthesized is left as an exercise (6.15).

6.3.3 Dual-Priority Encoder

In this example we'll use VHDL to create a behavioral description of a PLD priority encoder that identifies both the highest-priority and the second-highest-priority asserted signal among a set of request inputs R(0 to 7), where R(0) has the highest priority. We'll use A(2 downto 0) and AVALID to identify the highest-priority request, asserting AVALID only if a highest-priority request is present. Similarly, we'll use B(2 downto 0) and BVALID to identify the second-highest-priority request.

Table 6-26
Behavioral VHDL program for a dual-priority encoder.

```
library IEEE;
use IEEE.std_logic_1164.all;
use IEEE.std_logic_arith.all;

entity Vprior2 is
    port (
        R: in STD_LOGIC_VECTOR (0 to 7);
        A, B: out STD_LOGIC_VECTOR (2 downto 0);
        AVALID, BVALID: buffer STD_LOGIC
    );
end Vprior2;

architecture Vprior2_arch of Vprior2 is
begin
  process(R, AVALID, BVALID)
  begin
    AVALID <= '0'; BVALID <= '0'; A <= "000"; B <= "000";
    for i in 0 to 7 loop
      if R(i) = '1' and AVALID = '0' then
        A <= CONV_STD_LOGIC_VECTOR(i,3); AVALID <= '1';
      elsif R(i) = '1' and BVALID = '0' then
        B <= CONV_STD_LOGIC_VECTOR(i,3); BVALID <= '1';
      end if;
    end loop;
  end process;
end Vprior2_arch;
```

Table 6-26 shows a behavioral VHDL program for the priority encoder. Instead of the nested "if" approach of the previous example, we've used a "for" loop. This approach allows us to take care of both the first and the second priorities within the same loop, working our way from highest to lowest priority. Besides std_logic_1164, the program uses the IEEE std_logic_arith package in order to get the CONV_STD_LOGIC_VECTOR function. We also wrote this function explicitly in Table 4-39 on page 279.

Notice in the table that ports AVALID and BVALID are declared as mode buffer, because they are read within the architecture. If you were stuck with an entity definition that declared AVALID and BVALID as mode out, you could still use the same architecture approach, but you would have to declare local variables corresponding to AVALID and BVALID within the process. Notice also that we included AVALID and BVALID in the process sensitivity list. Although this is not strictly necessary, it prevents warnings that the compiler otherwise would give about using the value of a signal that is not on the sensitivity list.

The nested "if" approach can also be used for the dual-priority encoder, but it yields a longer program with more accidents waiting to happen, as shown in Table 6-27. On the other hand, it may yield a better synthesis result; the only

```
architecture Vprior2i_arch of Vprior2 is
begin
  process(R, A, AVALID, BVALID)
  begin
    if    R(0) = '1' then A <= "000"; AVALID <= '1';
    elsif R(1) = '1' then A <= "001"; AVALID <= '1';
    elsif R(2) = '1' then A <= "010"; AVALID <= '1';
    elsif R(3) = '1' then A <= "011"; AVALID <= '1';
    elsif R(4) = '1' then A <= "100"; AVALID <= '1';
    elsif R(5) = '1' then A <= "101"; AVALID <= '1';
    elsif R(6) = '1' then A <= "110"; AVALID <= '1';
    elsif R(7) = '1' then A <= "111"; AVALID <= '1';
    else                  A <= "000"; AVALID <= '0';
    end if;
    if    R(1) = '1' and A /= "001" then B <= "001"; BVALID <= '1';
    elsif R(2) = '1' and A /= "010" then B <= "010"; BVALID <= '1';
    elsif R(3) = '1' and A /= "011" then B <= "011"; BVALID <= '1';
    elsif R(4) = '1' and A /= "100" then B <= "100"; BVALID <= '1';
    elsif R(5) = '1' and A /= "101" then B <= "101"; BVALID <= '1';
    elsif R(6) = '1' and A /= "110" then B <= "110"; BVALID <= '1';
    elsif R(7) = '1' and A /= "111" then B <= "111"; BVALID <= '1';
    else                              B <= "000"; BVALID <= '0';
    end if;
  end process;
end Vprior2i_arch;
```

Table 6-27
Alternative VHDL
architecture for a
dual-priority encoder.

way to know with a particular tool is to synthesize the circuit and analyze the results in terms of delay and cell or gate count.

Both nested "if" statements and "for" statements may lead to long delay chains in synthesis. To guarantee that you get a faster dual-priority encoder, you must follow a structural or semistructural design approach. For example, you can start by writing a dataflow model of a fast 8-input priority encoder using the ideas found in the 74x148 logic diagram (Figure 5-50 on page 379) or in a related ABEL program (Table 5-24 on page 382). Then you can put two of these together in a structure that "knocks out" the highest-priority input in order to find the second, as we showed in Figure 6-6 on page 475.

6.3.4 Cascading Comparators

Cascading comparators is something we typically would not do in a VHDL behavioral model, because the language and the IEEE std_logic_arith package let us define comparators of any desired length directly. However, we may indeed need to write structural or semistructural VHDL programs that hook up smaller comparator components in a specific way to obtain high performance.

Table 6-28
Behavioral VHDL program for a 64-bit comparator.

```
library IEEE;
use IEEE.std_logic_1164.all;
use IEEE.std_logic_unsigned.all;

entity comp64 is
    port ( A, B: in STD_LOGIC_VECTOR (63 downto 0);
           EQ, GT: out STD_LOGIC );
end comp64;

architecture comp64_arch of comp64 is
begin
  EQ <= '1' when A = B else '0';
  GT <= '1' when A > B else '0';
end comp64_arch;
```

Table 6-28 is a simple behavioral model of a 64-bit comparator with equals and greater-than outputs. This program uses the IEEE std_logic_unsigned package, whose built-in comparison functions automatically treat all signals of type STD_LOGIC_VECTOR as unsigned integers. Although the program is fully synthesizable, the speed and size of the result depends on the "intelligence" of the particular tool that is used.

An alternative is to build the comparator by cascading smaller components, such as 8-bit comparators. Table 6-29 is the behavioral model of an 8-bit comparator. A particular tool may or may not synthesize a very fast comparator from this program, but it's sure to be significantly faster than a 64-bit comparator in any case.

Next, we can write a structural program that instantiates eight of these 8-bit comparators and hooks up their individual outputs through additional logic to calculate the overall comparison result. One way to do this is shown Table 6-30. A generate statement creates not only the individual 8-bit comparators but also

Table 6-29
VHDL program for an 8-bit comparator.

```
library IEEE;
use IEEE.std_logic_1164.all;
use IEEE.std_logic_unsigned.all;

entity comp8 is
    port ( A, B: in STD_LOGIC_VECTOR (7 downto 0);
           EQ, GT: out STD_LOGIC );
end comp8;

architecture comp8_arch of comp8 is
begin
  EQ <= '1' when A = B else '0';
  GT <= '1' when A > B else '0';
end comp8_arch;
```

Table 6-30 VHDL structural architecture for a 64-bit comparator.

```
architecture comp64s_arch of comp64 is
component comp8
  port ( A, B: in STD_LOGIC_VECTOR (7 downto 0);
         EQ, GT: out STD_LOGIC);
end component;
signal EQ8, GT8: STD_LOGIC_VECTOR (7 downto 0); -- =, > for 8-bit slice
signal SEQ, SGT: STD_LOGIC_VECTOR (8 downto 0); -- serial chain of slice results
begin
  SEQ(8) <= '1'; SGT(8) <= '0';
  U1: for i in 7 downto 0 generate
      U2: comp8 port map (A(7+i*8 downto i*8), B(7+i*8 downto i*8), EQ8(i), GT8(i));
      SEQ(i) <= SEQ(i+1) and EQ8(i);
      SGT(i) <= SGT(i+1) or (SEQ(i+1) and GT8(i));
  end generate;
  EQ <= SEQ(0); GT <= SGT(0);
end comp64s_arch;
```

cascading logic that serially builds up the overall result from most significant to least significant stage.

An unsophisticated tool could synthesize a slow iterative comparator circuit for our original 64-bit comparator architecture in Table 6-28. In this situation, the architecture in Table 6-30 yields a faster synthesized circuit because it explicitly "pulls" out the cascading information for each 8-bit slice and combines it in a faster combinational circuit (just 8 levels of AND-OR logic, not 64). A more sophisticated tool might flatten the 8-bit comparator into faster, non-iterative structure similar to the 74x682 MSI comparator (Figure 5-84 on page 425), and it might flatten our iterative cascading logic in Table 6-30 into two-level sum-of-products equations similar to the ones in the ABEL solution on page 487.

6.3.5 Mode-Dependent Comparator

For the next example, let us suppose we have a system in which we need to compare two 32-bit words under normal circumstances, but where we must sometimes ignore one or two low-order bits of the input words. The operating mode is specified by two mode-control bits, M1 and M0, as shown in Table 6-9 on page 488.

The desired functionality can be obtained very easily in VHDL using a case statement to select the behavior by mode, as shown in the program in Table 6-31. This is a perfectly good behavioral description that is also fully synthesizable. However, it has one major drawback in synthesis—it will, in all likelihood, cause the creation of three separate equality and magnitude comparators (32-, 31-, and 30-bit), one for each case in the case statement. The

Table 6-31 VHDL behavioral architecture of a 32-bit mode-dependent comparator.

```vhdl
library IEEE;
use IEEE.std_logic_1164.all;
use IEEE.std_logic_unsigned.all;

entity Vmodecmp is
    port ( M: in STD_LOGIC_VECTOR (1 downto 0);      -- mode
           A, B: in STD_LOGIC_VECTOR (31 downto 0); -- unsigned integers
           EQ, GT: out STD_LOGIC );                 -- comparison results
end Vmodecmp;

architecture Vmodecmp_arch of Vmodecmp is
begin
  process (M, A, B)
  begin
    case M is
      when "00" =>
        if A = B then EQ <= '1'; else EQ <= '0'; end if;
        if A > B then GT <= '1'; else GT <= '0'; end if;
      when "01" =>
        if A(31 downto 1) = B(31 downto 1) then EQ <= '1'; else EQ <= '0'; end if;
        if A(31 downto 1) > B(31 downto 1) then GT <= '1'; else GT <= '0'; end if;
      when "10" =>
        if A(31 downto 2) = B(31 downto 2) then EQ <= '1'; else EQ <= '0'; end if;
        if A(31 downto 2) > B(31 downto 2) then GT <= '1'; else GT <= '0'; end if;
      when others => EQ <= '0';  GT <= '0';
    end case;
  end process;
end Vmodecmp_arch;
```

individual comparators may or may not be fast, as discussed in the previous subsection, but we won't worry about speed for this example.

A more efficient alternative is to perform just one comparison for the 30 high-order bits of the inputs, using additional logic that is dependent on mode to give a final result using the low-order bits as necessary. This approach is shown in Table 6-32. Two variables, EQ30 and GT30, are used within the process to hold the results of the comparison of the 30 high-order bits. A case statement similar to the previous architecture's is then used to obtain the final results as a function of the mode. If desired, the speed of the 30-bit comparison can be optimized using the methods discussed in the preceding subsection.

6.3.6 Ones Counter

Several important algorithms include the step of counting the number of "1" bits in a data word. In fact, some microprocessor instruction sets have been extended recently to include ones counting as a basic instruction. In this example, let us suppose that we have a requirement to design a combinational circuit that counts ones in a 32-bit word as part of the arithmetic and logic unit of a microprocessor.

Table 6-32 More efficient architecture for a 32-bit mode-dependent comparator.

```
architecture Vmodecpe_arch of Vmodecmp is
begin
  process (M, A, B)
  variable EQ30, GT30: STD_LOGIC; -- 30-bit comparison results
  begin
    if A(31 downto 2) = B(31 downto 2) then EQ30 := '1'; else EQ30 := '0'; end if;
    if A(31 downto 2) > B(31 downto 2) then GT30 := '1'; else GT30 := '0'; end if;
    case M is
      when "00" =>
        if EQ30='1' and A(1 downto 0) = B(1 downto 0) then
          EQ <= '1'; else EQ <= '0'; end if;
        if GT30='1' or (EQ30='1' and A(1 downto 0) > B(1 downto 0)) then
          GT <= '1'; else GT <= '0'; end if;
      when "01" =>
        if EQ30='1' and A(1) = B(1) then EQ <= '1'; else EQ <= '0'; end if;
        if GT30='1' or (EQ30='1' and A(1) > B(1)) then
          GT <= '1'; else GT <= '0'; end if;
      when "10" =>  EQ <= EQ30;  GT <= GT30;
      when others => EQ <= '0';  GT <= '0';
    end case;
  end process;
end Vmodecpe_arch;
```

Ones counting can be described very easily by a behavioral VHDL program, as shown in Table 6-33. This program is fully synthesizable, but it may generate a very slow, inefficient realization with 32 5-bit adders in series.

Table 6-33
Behavioral VHDL program for a 32-bit ones counter.

```
library IEEE;
use IEEE.std_logic_1164.all;
use IEEE.std_logic_unsigned.all;

entity Vcnt1s is
    port ( D: in STD_LOGIC_VECTOR (31 downto 0);
           SUM: out STD_LOGIC_VECTOR (4 downto 0) );
end Vcnt1s;

architecture Vcnt1s_arch of Vcnt1s is
begin
  process (D)
  variable S: STD_LOGIC_VECTOR(4 downto 0);
  begin
    S := "00000";
    for i in 0 to 31 loop
      if D(i) = '1' then S := S + "00001"; end if;
    end loop;
    SUM <= S;
  end process;
end Vcnt1s_arch;
```

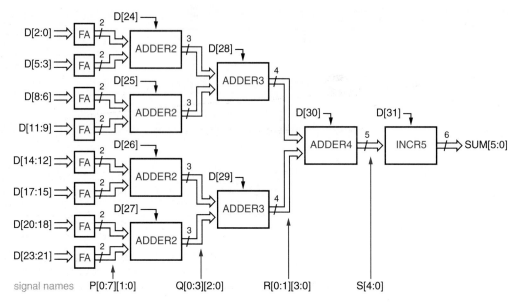

Figure 6-15 Structure of 32-bit ones counter.

To synthesize a more efficient realization of the ones counter, we must come up with an efficient structure and then write an architecture that describes it. Such a structure is the adder tree shown in Figure 6-15. A full adder (FA) adds three input bits to produce a 2-bit sum. Pairs of 2-bit numbers are added by 2-bit adders (ADDER2), each of which also has a carry input that can include add another 1-bit input to its sum. The resulting 3-bit sums are combined by 3-bit adders (ADDER3), and the final pair of 4-bit sums are combined in a 4-bit adder (ADDER4). By making use of the available carry inputs, this tree structure can combine 31 bits. A separate 5-bit incrementer is used at the end to handle the one remaining input bit.

The structure of Figure 6-15 can be created nicely by a structural VHDL architecture, as shown in Table 6-34. The program begins by declaring all of the components that will be used in the design, corresponding to the blocks in the figure.

The letter under each column of signals in Figure 6-15 corresponds to the name used for that signal in the program. Each of signals P, Q, and R is an array with one STD_LOGIC_VECTOR per connection in the corresponding column. The program defines a corresponding type for each of these, followed by the actual signal declaration.

The program in Table 6-34 makes good use of generate statements to create the multiple adder components on the lefthand side of the figure—eight FAs, four ADDER2s, and two ADDER3s. Finally, it instantiates one each of ADDER4 and INCR5.

```
architecture Vcnt1str_arch of Vcnt1str is

component FA port ( A, B, CI: in  STD_LOGIC;
                    S, CO:   out STD_LOGIC );
end component;

component ADDER2 port ( A, B: in  STD_LOGIC_VECTOR(1 downto 0);
                        CI:   in  STD_LOGIC;
                        S:    out STD_LOGIC_VECTOR(2 downto 0) );
end component;

component ADDER3 port ( A, B: in  STD_LOGIC_VECTOR(2 downto 0);
                        CI:   in  STD_LOGIC;
                        S:    out STD_LOGIC_VECTOR(3 downto 0) );
end component;

component ADDER4 port ( A, B: in  STD_LOGIC_VECTOR(3 downto 0);
                        CI:   in  STD_LOGIC;
                        S:    out STD_LOGIC_VECTOR(4 downto 0) );
end component;

component INCR5 port ( A:  in  STD_LOGIC_VECTOR(4 downto 0);
                       CI: in  STD_LOGIC;
                       S:  out STD_LOGIC_VECTOR(5 downto 0) );
end component;

type Ptype is array (0 to 7) of STD_LOGIC_VECTOR(1 downto 0);
type Qtype is array (0 to 3) of STD_LOGIC_VECTOR(2 downto 0);
type Rtype is array (0 to 1) of STD_LOGIC_VECTOR(3 downto 0);
signal P: Ptype;  signal Q: Qtype;  signal R: Rtype;
signal S: STD_LOGIC_VECTOR(4 downto 0);

begin
  U1: for i in 0 to 7 generate
    U1C: FA port map (D(3*i), D(3*i+1), D(3*i+2), P(i)(0), P(i)(1));
  end generate;
  U2: for i in 0 to 3 generate
    U2C: ADDER2 port map (P(2*i), P(2*i+1), D(24+i), Q(i));
  end generate;
  U3: for i in 0 to 1 generate
    U3C: ADDER3 port map (Q(2*i), Q(2*i+1), D(28+i), R(i));
  end generate;
  U4: ADDER4 port map (R(0), R(1), D(30), S);
  U5: INCR5 port map (S, D(31), SUM);
end Vcnt1str_arch;
```

Table 6-34
VHDL structural architecture for a 32-bit ones counter.

The definitions of the ones counter's individual component entities and architectures, from FA to INCR, can be made in separate structural or behavioral programs. For example, Table 6-35 is a structural program for FA. The rest of the components are left as exercises (6.20–6.22).

Table 6-35
Structural VHDL program for a full adder.

```
library IEEE;
use IEEE.std_logic_1164.all;

entity FA is
    port ( A, B, CI: in  STD_LOGIC;
             S, CO:    out STD_LOGIC );
end FA;

architecture FA_arch of FA is
begin
  S <= A xor B xor CI;
  CO <= (A and B) or (A and CI) or (B and CI);
end FA_arch;
```

6.3.7 Tic-Tac-Toe

Our last example is the design of a combinational circuit that picks a player's next move in the game of Tic-Tac-Toe. In case you're not familiar with the game, the rules are explained in the box on page 492. We'll repeat here our strategy for playing and winning the game:

1. Look for a row, column, or diagonal that has two of my marks (X or O, depending on which player I am) and one empty cell. If one exists, place my mark in the empty cell; I win!

2. Else, look for a row, column, or diagonal that has two of my opponent's marks and one empty cell. If one exists, place my mark in the empty cell to block a potential win by my opponent.

3. Else, pick a cell based on experience. For example, if the middle cell is open, it's usually a good bet to take it. Otherwise, the corner cells are good bets. Intelligent players can also notice and block a developing pattern by the opponent or "look ahead" to pick a good move.

To avoid confusion between "O" and "0" in our programs, we'll call the second player "Y". Now we can think about how to encode the inputs and outputs of the circuit. The inputs represent the current state of the playing grid. There are nine cells, and each cell has one of three possible states (empty, occupied by X, occupied by Y). The outputs represent the move to make, assuming that it is X's turn. There are only nine possible moves that a player can make, so the output can be encoded in just four bits.

There are several choices of how to code the state of one cell. Because the game is symmetric, we choose a symmetric encoding that can help us later:

00 Cell is empty.

10 Cell is occupied by X.

01 Cell is occupied by Y.

```
library IEEE;
use IEEE.std_logic_1164.all;

package TTTdefs is

type TTTgrid is array (1 to 3) of STD_LOGIC_VECTOR(1 to 3);
subtype TTTmove is STD_LOGIC_VECTOR (3 downto 0);

constant MOVE11: TTTmove := "1000";
constant MOVE12: TTTmove := "0100";
constant MOVE13: TTTmove := "0010";
constant MOVE21: TTTmove := "0001";
constant MOVE22: TTTmove := "1100";
constant MOVE23: TTTmove := "0111";
constant MOVE31: TTTmove := "1011";
constant MOVE32: TTTmove := "1101";
constant MOVE33: TTTmove := "1110";
constant NONE:   TTTmove := "0000";

end TTTdefs;
```

Table 6-36
VHDL package with definitions for the Tic-Tac-Toe project.

So, we can encode the 3 × 3 grid's state in 18 bits. Since VHDL supports arrays, it is useful to define an array type, `TTTgrid`, that contains elements corresponding to the cells in the grid. Since this type will be used throughout our Tic-Tac-Toe project, it is convenient to put this definition, along with several others that we'll come to, in a VHDL package, as shown in Table 6-36.

It would be natural to define `TTTgrid` as a two-dimensional array of `STD_LOGIC`, but not all VHDL tools support two-dimensional arrays. Instead, we define it as an array of 3-bit `STD_LOGIC_VECTOR`s, which is almost the same thing. To represent the Tic-Tac-Toe grid, we'll use two signals `X` and `Y` of this type, where an element of a variable is 1 if the like-named player has a mark in the corresponding cell. Figure 6-16 shows the correspondence between signal names and cells in the grid.

row \ column	1	2	3
1	X(1)(1) Y(1)(1)	X(1)(2) Y(1)(2)	X(1)(2) Y(1)(2)
2	X(2)(1) Y(2)(1)	X(2)(2) Y(2)(2)	X(2)(2) Y(2)(2)
3	X(3)(1) Y(3)(1)	X(3)(2) Y(3)(2)	X(3)(2) Y(3)(2)

Figure 6-16
Tic-Tac-Toe grid and VHDL signal names.

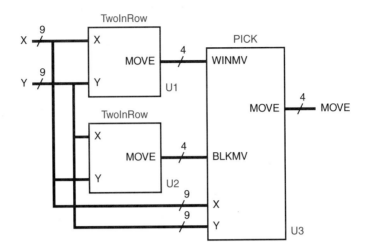

Figure 6-17
Entity partitioning for
the Tic-Tac-Toe game.

The package in Table 6-36 also defines a 4-bit type `TTTmove` for encoded moves. A player has nine possible moves, and one more code is used for the case where no move is possible. This particular coding was chosen and used in the package for no other reason than that it's the same coding that was used in the ABEL version of this example in Section 6.2.7. By defining the coding in the package, we can easily change the definition later without having to change the entities that use it (for example, see Exercise 6.23).

Rather than try to design the Tic-Tac-Toe move-finding circuit as a single monolithic entity, it makes sense for us to try to partition it into smaller pieces. In fact, partitioning it along the lines of the three-step strategy at the beginning of this section seems like a good idea.

We note that steps 1 and 2 of our strategy are very similar; they differ only in reversing the roles of the player and the opponent. An entity that finds a winning move for me can also find a blocking move for my opponent. Looking at this characteristic from another point of view, an entity that finds a winning move for me can find a blocking move for me if the encodings for me and my opponent are swapped. Here's where our symmetric encoding pays off—we can swap players merely by swapping signals X and Y.

With this in mind, we can use two copies of the same entity, TwoInRow, to perform steps 1 and 2 as shown in Figure 6-17. Notice that signal X is connected to the top input of the first TwoInRow entity, but to the bottom input of the second. A third entity, PICK, picks a winning move if one is available from U1, else it picks a blocking move if available from U2, else it uses "experience" (step 3) to pick a move.

Table 6-37 is a structural VHDL program for the top-level entity, GETMOVE. In addition to the IEEE std_logic_1164 package, it uses our TTTdefs package. Notice that the "use" clause for the TTTdefs packages specifies that it is stored in the "work" library, which is automatically created for our project.

```
library IEEE;
use IEEE.std_logic_1164.all;
use work.TTTdefs.all;

entity GETMOVE is
    port ( X, Y: in  TTTgrid;
           MOVE: out TTTmove );
end GETMOVE;

architecture GETMOVE_arch of GETMOVE is

component TwoInRow port ( X, Y: in  TTTgrid;
                          MOVE: out STD_LOGIC_VECTOR(3 downto 0) );
end component;

component PICK port ( X, Y:         in  TTTgrid;
                      WINMV, BLKMV: in  STD_LOGIC_VECTOR(3 downto 0);
                      MOVE:         out STD_LOGIC_VECTOR(3 downto 0) );
end component;

signal WIN, BLK: STD_LOGIC_VECTOR(3 downto 0);

begin
  U1: TwoInRow port map (X, Y, WIN);
  U2: TwoInRow port map (Y, X, BLK);
  U3: PICK port map (X, Y, WIN, BLK, MOVE);
end GETMOVE_arch;
```

Table 6-37
Top-level structural VHDL entity for picking a move in Tic-Tac-Toe.

The architecture in Table 6-37 declares and uses just two components, TwoInRow and PICK, which will be defined shortly. The only internal signals are WIN and BLK, which pass winning and blocking moves from the two instances of TwoInRow to PICK, as in Figure 6-17. The statement part of the architecture has just three statements to instantiate the three blocks in the figure.

Now comes the interesting part, the design of the individual entities in Figure 6-17. We'll start with TwoInRow, since it accounts for two-thirds of the design. Its entity definition is very simple, as shown in Table 6-38. But there's plenty to discuss about its architecture, shown in Table 6-39.

```
library IEEE;
use IEEE.std_logic_1164.all;
use work.TTTdefs.all;

entity TwoInRow is
    port ( X, Y: in  TTTgrid;
           MOVE: out TTTmove );
end TwoInRow;
```

Table 6-38
Declaration of TwoInRow entity.

■ **Table 6-39** Architecture of TwoInRow entity.

```
architecture TwoInRow_arch of TwoInRow is

function R(X, Y: TTTgrid; i, j: INTEGER) return BOOLEAN is
  variable result: BOOLEAN;
  begin                                 -- Find 2-in-row with empty cell i,j
    result := TRUE;
    for jj in 1 to 3 loop
      if jj = j then result := result and X(i)(jj)='0' and Y(i)(jj)='0';
      else result := result and X(i)(jj)='1'; end if;
    end loop;
    return result;
  end R;

function C(X, Y: TTTgrid; i, j: INTEGER) return BOOLEAN is
  variable result: BOOLEAN;
  begin                                 -- Find 2-in-column with empty cell i,j
    result := TRUE;
    for ii in 1 to 3 loop
      if ii = i then result := result and X(ii)(j)='0' and Y(ii)(j)='0';
      else result := result and X(ii)(j)='1'; end if;
    end loop;
    return result;
  end C;

function D(X, Y: TTTgrid; i, j: INTEGER) return BOOLEAN is
  variable result: BOOLEAN;          -- Find 2-in-diagonal with empty cell i,j.
  begin                              -- This is for 11, 22, 33 diagonal.
    result := TRUE;
    for ii in 1 to 3 loop
      if ii = i then result := result and X(ii)(ii)='0' and Y(ii)(ii)='0';
      else result := result and X(ii)(ii)='1'; end if;
    end loop;
    return result;
  end D;

function E(X, Y: TTTgrid; i, j: INTEGER) return BOOLEAN is
  variable result: BOOLEAN;          -- Find 2-in-diagonal with empty cell i,j.
  begin                              -- This is for 13, 22, 31 diagonal.
    result := TRUE;
    for ii in 1 to 3 loop
      if ii = i then result := result and X(ii)(4-ii)='0' and Y(ii)(4-ii)='0';
      else result := result and X(ii)(4-ii)='1'; end if;
    end loop;
    return result;
  end E;
```

Table 6-39
(continued)

```
begin
  process (X, Y)
  variable G11, G12, G13, G21, G22, G23, G31, G32, G33: BOOLEAN;
  begin
    G11 := R(X,Y,1,1) or C(X,Y,1,1) or D(X,Y,1,1);
    G12 := R(X,Y,1,2) or C(X,Y,1,2);
    G13 := R(X,Y,1,3) or C(X,Y,1,3) or E(X,Y,1,3);
    G21 := R(X,Y,2,1) or C(X,Y,2,1);
    G22 := R(X,Y,2,2) or C(X,Y,2,2) or D(X,Y,2,2) or E(X,Y,2,2);
    G23 := R(X,Y,2,3) or C(X,Y,2,3);
    G31 := R(X,Y,3,1) or C(X,Y,3,1) or E(X,Y,3,1);
    G32 := R(X,Y,3,2) or C(X,Y,3,2);
    G33 := R(X,Y,3,3) or C(X,Y,3,3) or D(X,Y,3,3);
    if    G11 then MOVE <= MOVE11;
    elsif G12 then MOVE <= MOVE12;
    elsif G13 then MOVE <= MOVE13;
    elsif G21 then MOVE <= MOVE21;
    elsif G22 then MOVE <= MOVE22;
    elsif G23 then MOVE <= MOVE23;
    elsif G31 then MOVE <= MOVE31;
    elsif G32 then MOVE <= MOVE32;
    elsif G33 then MOVE <= MOVE33;
    else           MOVE <= NONE;
    end if;
  end process;
end TwoInRow_arch;
```

The architecture defines several functions, each of which determines whether there is a winning move (from X's point of view) in a particular cell i,j. A winning move exists if cell i,j is empty and the other two cells in the same row, column, or diagonal contain an X. Functions R and C look for winning moves in cell i,j's row and column, respectively. Functions D and E look in the two diagonals.

Within the architecture's single process, nine BOOLEAN variables G11–G33 are declared to indicate whether each of the cells has a winning move possible. Assignment statements at the beginning of the process set each variable to TRUE if there is such a move, calling and combining all of the appropriate functions for cell i,j.

The rest of the process is a deeply nested "if" statement that looks for a winning move in all possible cells. Although it typically results in slower synthesized logic nested "if" is required rather than some form of "case" statement, because multiple moves may be possible. If no winning move is possible, the value "NONE" is assigned.

**EXPLICIT
IMPURITY**

In addition to a cell index i,j, the functions R, C, D, and E in Table 6-39 are passed the grid state X and Y. This is necessary because VHDL functions are by default *pure*, which means that signals and variables declared in the function's parents are *not* directly visible within the function. However, you can relax this restriction by explicitly declaring a function to be *impure* by placing the keyword impure before the keyword function in its definition.

The PICK entity combines the results of two TwoInRow entities according to the program in Table 6-40. First priority is given to a winning move, followed by a blocking move. Otherwise, function MT is called for each cell, starting with the middle and ending with the side cells, to find an available move. This completes the design of the Tic-Tac-Toe circuit.

Table 6-40
VHDL program to pick a winning or blocking Tic-Tac-Toe move or else use "experience."

```
library IEEE;
use IEEE.std_logic_1164.all;
use work.TTTdefs.all;

entity PICK is
  port ( X, Y:          in  TTTgrid;
         WINMV, BLKMV: in  STD_LOGIC_VECTOR(3 downto 0);
         MOVE:         out STD_LOGIC_VECTOR(3 downto 0) );
end PICK;

architecture PICK_arch of PICK is
function MT(X, Y: TTTgrid; i, j: INTEGER) return BOOLEAN is
  begin                              -- Determine if cell i,j is empty
    return X(i)(j)='0' and Y(i)(j)='0';
  end MT;
begin
  process (X, Y, WINMV, BLKMV)
  begin                                      -- If available, pick:
    if    WINMV /= NONE then MOVE <= WINMV;  -- winning move
    elsif BLKMV /= NONE then MOVE <= BLKMV;  -- else blocking move
    elsif MT(X,Y,2,2)   then MOVE <= MOVE22; -- else center cell
    elsif MT(X,Y,1,1)   then MOVE <= MOVE11; -- else corner cells
    elsif MT(X,Y,1,3)   then MOVE <= MOVE13;
    elsif MT(X,Y,3,1)   then MOVE <= MOVE31;
    elsif MT(X,Y,3,3)   then MOVE <= MOVE33;
    elsif MT(X,Y,1,2)   then MOVE <= MOVE12; -- else side cells
    elsif MT(X,Y,2,1)   then MOVE <= MOVE21;
    elsif MT(X,Y,2,3)   then MOVE <= MOVE23;
    elsif MT(X,Y,3,2)   then MOVE <= MOVE32;
    else                     MOVE <= NONE;   -- else grid is full
    end if;
  end process;
end PICK_arch;
```

Exercises

6.1   Explain how the 16-bit barrel shifter of Section 6.1.1 can be realized with a combination of 74x157s and 74x151s. How does this approach compare with the others in delay and parts count?

6.2   Show how the 16-bit barrel shifter of Section 6.1.1 can be realized in eight identical GAL22V10s.

6.3   Find a coding of the shift amounts (S[3:0]) and modes (C[2:0]) in the barrel shifter of Table 6-3 that further reduces the total number of product terms used by the design.

6.4   Make changes to the dual-priority encoder program of Table 6-6 to further reduce the number of product terms required. State whether your changes increase the delay of the circuit when realized in a GAL22V10. Can you reduce the product terms enough to fit the design into a GAL16V8?

6.5   Here's an exercise where you can use your brain, like the author had to when figuring out the equation for the SUM0 output in Table 6-12. Do each of the SUM1–SUM3 outputs require more terms or fewer terms than SUM0?

6.6   Complete the design of the ABEL and PLD-based ones-counting circuit that was started in Section 6.2.6. Use 22V10 or smaller PLDs and try to minimize the total number of PLDs required. State the total delay of your design in terms of the worst-case number of PLD delays in a signal path from input to output.

6.7   Find another code for the Tic-Tac-Toe moves in Table 6-13 that has the same rotation properties as the original code. That is, it should be possible to compensate for a 180° rotation of the grid using just inverters and wire rearrangement. Determine whether the TWOINHAF equations will still fit in a single 22V10 using the new code.

6.8   Using a simulator, demonstrate a sequence of moves in which the PICK2 PLD in Table 6-16 will lose a Tic-Tac-Toe game, even if X goes first.

6.9   Modify the program in Table 6-16 to give the program a better chance of winning, or at least not losing. Can your new program still lose?

6.10   Modify both the "other logic" in Figure 6-13 and the program in Table 6-16 to give the program a better chance of winning, or at least not losing. Can your new program still lose?

6.11   Write the VHDL functions for Vror, Vsll, Vsrl, Vsla, and Vsra in Table 6-17 using the ror, sll, srl, sla, and sra operations as defined in Table 6-3.

6.12   The iterative-circuit version of fixup in Table 6-20 has a worst-case delay path of 15 OR gates from the first decoded value of i (14) to the FSEL(0) signal. Figure out a trick that cuts this delay path almost in half with no cost (or negative cost) in gates. How can this trick be extended further to save gates or gate inputs?

6.13   Rewrite the barrel16 entity definition in Table 6-17 and the architecture in Table 6-22 so that a single direction-control bit is made explicitly available to the architecture.

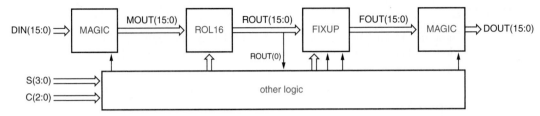

Figure X6.14

6.14   Rewrite the `barrel16` architecture definition in Table 6-22 to use the structure shown in Figure X6.14. Use the existing ROL16 and FIXUP entities; it's up to you to come up with MAGIC and the other logic.

6.15   Write a semibehavioral or structural version of the `fpencr_arch` architecture of Table 6-25 that generates only one adder in synthesis and that does not generate multiple 10-bit comparators for the nested "if" statement.

6.16   Repeat Exercise 6.15, including a structural definition of an efficient rounding circuit that performs the `round` function. Your circuit should require significantly fewer gates than a 4-bit adder.

6.17   Redesign the VHDL dual-priority encoder of Section 6.3.3 to get better, known performance, as suggested in the last paragraph of the section.

6.18   Write a structural VHDL architecture for a 64-bit comparator that is similar to Table 6-30 except that it builds up the comparison result serially from least to most significant stage.

6.19   What significant change occurs in the synthesis of the VHDL program in Table 6-31 if we change the statements in the "when others" case to "null"?

6.20   Write behavioral VHDL programs for the "ADDERx" components used in Table 6-34.

6.21   Write a structural VHDL programs for the "ADDERx" components in Table 6-34. Use a generic definition so that the same entity can be instantiated for ADDER2, ADDER3, and ADDER5, and show what changes must be made in Table 6-34 to do this.

6.22   Write a structural VHDL program for the "INCR5" component in Table 6-34.

6.23   Using an available VHDL synthesis tool, synthesize the Tic-Tac-Toe design of Section 6.3.7, fit it into an available FPGA, and determine how many internal resources it uses. Then try to reduce the resource requirements by specifying a different encoding of the moves in the TTTdefs package.

6.24   The Tic-Tac-Toe program in Section 6.3.7 eventually loses against an intelligent opponent if applied to the grid state shown in Figure X6.24. Use an available VHDL simulator to prove that this is true. Then modify the PICK entity to win in this and similar situations and verify your design using the simulator.

Figure X6.24

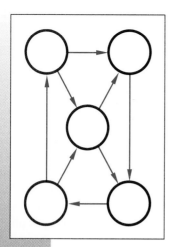

Sequential
Logic Design Principles

L ogic circuits are classified into two types, "combinational" and "sequential." A *combinational* logic circuit is one whose outputs depend only on its current inputs. The rotary channel selector knob on an old-fashioned TV is like a combinational circuit—its "output" selects a channel based only on its current "input"—the position of the knob.

A *sequential* logic circuit is one whose outputs depend not only on its current inputs, but also on the past sequence of inputs, possibly arbitrarily far back in time. The circuit controlled by the channel-up and channel-down pushbuttons on a TV or VCR is a sequential circuit—the channel selection depends on the past sequence of up/down pushes, at least since when you started viewing 10 hours before, and perhaps as far back as when you first plugged the device into the wall.

So it is inconvenient, and often impossible, to describe the behavior of a sequential circuit by means of a table that lists outputs as a function of the input sequence that has been received up until the current time. To know where you're going next, you need to know where you are now. With the TV channel selector, it is impossible to determine what channel is currently selected by looking only at the preceding sequence of presses on the up and down pushbuttons, whether we look at the preceding 10 presses or the preceding 1,000. More information, the current "state" of the channel selector, is needed. Probably the best definition of "state" that I've seen appeared in Herbert Hellerman's book on *Digital Computer System Principles* (McGraw-Hill, 1967):

state
state variable

> The *state* of a sequential circuit is a collection of *state variables* whose values at any one time contain all the information about the past necessary to account for the circuit's future behavior.

In the channel-selector example, the current channel number is the current state. Inside the TV, this state might be stored as seven binary state variables representing a decimal number between 0 and 127. Given the current state (channel number), we can always predict the next state as a function of the inputs (presses of the up/down pushbuttons). In this example, one highly visible output of the sequential circuit is an encoding of the state itself—the channel-number display. Other outputs, internal to the TV, may be combinational functions of the state alone (e.g., VHF/UHF/cable tuner selection) or of both state and input (e.g., turning off the TV if the current state is 0 and the "down" button is pressed).

State variables need not have direct physical significance, and there are often many ways to choose them to describe a particular sequential circuit. For example, in the TV channel selector, the state might be stored as three BCD digits or 12 bits, with many of the bit combinations (4,096 possible) going unused.

In a digital logic circuit, state variables are binary values, corresponding to certain logic signals in the circuit, as we'll see in later sections. A circuit with n binary state variables has 2^n possible states. As large as it might be, 2^n is always

finite-state machine

finite, never infinite, so sequential circuits are sometimes called *finite-state machines*.

clock

The state changes of most sequential circuits occur at times specified by a free-running *clock* signal. Figure 7-1 gives timing diagrams and nomenclature for typical clock signals. By convention, a clock signal is active high if state changes occur at the clock's rising edge or when the clock is HIGH, and active

clock period
clock frequency
clock tick
duty cycle

low in the complementary case. The *clock period* is the time between successive transitions in the same direction, and the *clock frequency* is the reciprocal of the period. The first edge or pulse in a clock period or sometimes the period itself is called a *clock tick*. The *duty cycle* is the percentage of time that the clock signal

---

NON-FINITE-STATE MACHINES   A group of mathematicians recently proposed a non-finite-state machine, but they're still busy listing its states. . . . Sorry, that's just a joke. There *are* mathematical models for infinite-state machines, such as Turing machines. They typically contain a small finite-state-machine control unit, and an infinite amount of auxiliary memory, such as an endless tape.

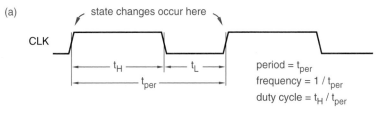

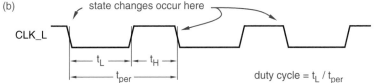

Figure 7-1
Clock signals:
(a) active high;
(b) active low.

is at its asserted level. Typical digital systems, from digital watches to supercomputers, use a quartz-crystal oscillator to generate a free-running clock signal. Clock frequencies might range from 32.768 kHz (for a watch) to 500 MHz (for a CMOS RISC microprocessor with a cycle time of 2 ns); "typical" systems using TTL and CMOS parts have clock frequencies in the 5–150 MHz range.

In this chapter we'll discuss two types of sequential circuits that account for the majority of practical discrete designs. A *feedback sequential circuit* uses ordinary gates and feedback loops to obtain memory in a logic circuit, thereby creating sequential-circuit building blocks such as latches and flip-flops that are used in higher-level designs. A *clocked synchronous state machine* uses these building blocks, in particular edge-triggered D flip-flops, to create circuits whose inputs are examined and whose outputs change in accordance with a controlling clock signal. Other sequential circuit types, such as general fundamental mode, multiple-pulse mode, and multiphase circuits, are sometimes useful in high-performance systems and VLSI and are discussed in advanced texts.

feedback sequential circuit

clocked synchronous state machine

7.1 Bistable Elements

The simplest sequential circuit consists of a pair of inverters forming a feedback loop, as shown in Figure 7-2. It has *no* inputs and two outputs, Q and Q_L.

7.1.1 Digital Analysis

The circuit of Figure 7-2 is often called a *bistable*, since a strictly digital analysis shows that it has two stable states. If Q is HIGH, then the bottom inverter has a HIGH input and a LOW output, which forces the top inverter's output HIGH as we assumed in the first place. But if Q is LOW, then the bottom inverter has a LOW input and a HIGH output, which forces Q LOW, another stable situation. We could use a single state variable, the state of signal Q, to describe the state of the circuit; there are two possible states, Q = 0 and Q = 1.

bistable

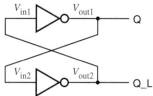

Figure 7-2
A pair of inverters forming
a bistable element.

The bistable element is so simple that it has no inputs and therefore no way of controlling or changing its state. When power is first applied to the circuit, it randomly comes up in one state or the other and stays there forever. Still, it serves our illustrative purposes very well, and we *will* actually show a couple of applications for it in Sections 8.2.3 and 8.2.4.

7.1.2 Analog Analysis

The analysis of the bistable has more to reveal if we consider its operation from an analog point of view. The dark line in Figure 7-3 shows the steady-state (DC) transfer function T for a single inverter; the output voltage is a function of input voltage, $V_{out} = T(V_{in})$. With two inverters connected in a feedback loop as in Figure 7-2, we know that $V_{in1} = V_{out2}$ and $V_{in2} = V_{out1}$; therefore, we can plot the transfer functions for both inverters on the same graph with an appropriate labeling of the axes. Thus, the dark line is the transfer function for the top inverter in Figure 7-2, and the colored line is the transfer function for the bottom one.

Considering only the steady-state behavior of the bistable's feedback loop, and not dynamic effects, the loop is in equilibrium if the input and output voltages of both inverters are constant DC values consistent with the loop connection and the inverters' DC transfer function. That is, we must have

$$
\begin{aligned}
V_{in1} &= V_{out2} \\
&= T(V_{in2}) \\
&= T(V_{out1}) \\
&= T(T(V_{in1}))
\end{aligned}
$$

Figure 7-3
Transfer functions for
inverters in a bistable
feedback loop.

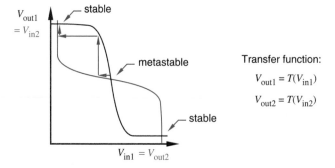

Transfer function:

$$V_{out1} = T(V_{in1})$$

$$V_{out2} = T(V_{in2})$$

Likewise, we must have

$$V_{in2} = T(T(V_{in2}))$$

We can find these equilibrium points graphically from Figure 7-3; they are the points at which the two transfer curves meet. Surprisingly, we find that there are not two but *three* equilibrium points. Two of them, labeled *stable*, correspond to the two states that our "strictly digital" analysis identified earlier, with Q either 0 (LOW) or 1 (HIGH).

stable

The third equilibrium point, labeled *metastable*, occurs with V_{out1} and V_{out2} about halfway between a valid logic 1 voltage and a valid logic 0 voltage; so Q and Q_L are not valid logic signals at this point. Yet the loop equations are satisfied; if we can get the circuit to operate at the metastable point, it could theoretically stay there indefinitely.

metastable

7.1.3 Metastable Behavior

Closer analysis of the situation at the metastable point shows that it is aptly named. It is not truly stable, because random noise will tend to drive a circuit that is operating at the metastable point toward one of the stable operating points, as we'll now demonstrate.

Suppose the bistable is operating precisely at the metastable point in Figure 7-3. Now let us assume that a small amount of circuit noise reduces V_{in1} by a tiny amount. This tiny change causes V_{out1} to *increase* by a small amount. But since V_{out1} produces V_{in2}, we can follow the first horizontal arrow from near the metastable point to the second transfer characteristic, which now demands a lower voltage for V_{out2}, which is V_{in1}. Now we're back where we started, except we have a much larger change in voltage at V_{in1} than the original noise produced, and the operating point is still changing. This "regenerative" process continues until we reach the stable operating point at the upper lefthand corner of Figure 7-3. However, if we perform a "noise" analysis for either of the stable operating points, we find that feedback brings the circuit back toward the stable operating point, rather than away from it.

Metastable behavior of a bistable can be compared to the behavior of a ball dropped onto a hill, as shown in Figure 7-4. If we drop a ball from overhead, it will probably roll down immediately to one side of the hill or the other. But if it lands right at the top, it may precariously sit there for a while before random

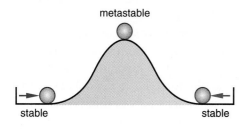

metastable

stable   stable

Figure 7-4
Ball and hill analogy for metastable behavior.

forces (wind, rodents, earthquakes) start it rolling down the hill. Like the ball at the top of the hill, the bistable may stay in the metastable state for an unpredictable length of time before nondeterministically settling into one stable state or the other.

If the *simplest* sequential circuit is susceptible to metastable behavior, you can be sure that *all* sequential circuits are susceptible. And this behavior is not something that only occurs at power-up.

Returning to the ball-and-hill analogy, consider what happens if we try to kick the ball from one side of the hill to the other. Apply a strong force (Arnold Schwarzenegger), and the ball goes right over the top and lands in a stable resting place on the other side. Apply a weak force (Mr. Rogers), and the ball falls back to its original starting place. But apply a wishy-washy force (Charlie Brown), and the ball goes to the top of the hill, teeters, and eventually falls back to one side or the other.

This behavior is completely analogous to what happens to latches and flip-flops under marginal triggering conditions. For example, we'll soon study S-R latches, where a pulse on the S input forces the latch from the 0 state to the 1 state. A minimum pulse width is specified for the S input. Apply a pulse of this width or longer, and the latch immediately goes to the 1 state. Apply a very short pulse, and the latch stays in the 0 state. Apply a pulse just under the minimum width, and the latch may go into the metastable state. Once the latch is in the metastable state, its operation depends on "the shape of its hill." Latches and flip-flops built from high-gain, fast technologies tend to come out of metastability faster than ones built from low-performance technologies.

We'll say more about metastability in the next section in connection with specific latch and flip-flop types, and in Section 8.9 with respect to synchronous design methodology and synchronizer failure.

7.2 Latches and Flip-Flops

Latches and flip-flops are the basic building blocks of most sequential circuits. Typical digital systems use latches and flip-flops that are prepackaged, functionally specified devices in a standard integrated circuit. In ASIC design environments, latches and flip-flops are typically predefined cells specified by the ASIC vendor. However, within a standard IC or an ASIC, each latch or flip-flop cell is typically designed as a feedback sequential circuit using individual logic gates and feedback loops. We'll study these discrete designs for two reasons—to understand the behavior of the prepackaged elements better, and to gain the capability of building a latch or flip-flop "from scratch," as is required occasionally in digital-design practice and often in digital-design exams.

flip-flop   All digital designers use the name *flip-flop* for a sequential device that normally samples its inputs and changes its outputs only at times determined by a

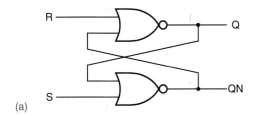

S	R	Q	QN
0	0	last Q	last QN
0	1	0	1
1	0	1	0
1	1	0	0

(a) (b)

Figure 7-5
S-R latch: (a) circuit design using NOR gates; (b) function table.

clocking signal. On the other hand, most digital designers use the name *latch* for a sequential device that watches all of its inputs continuously and changes its outputs at any time, independent of a clocking signal. We follow this standard convention in this text. However, some textbooks and digital designers may (incorrectly) use the name "flip-flop" for a device that we call a "latch."

latch

In any case, because the functional behaviors of latches and flip-flops are quite different, it is important for the logic designer to know which type is being used in a design, either from the device's part number (e.g., 74x374 vs. 74x373) or from other contextual information. We discuss the most commonly used types of latches and flip-flops in the following subsections.

7.2.1 S-R Latch

An *S-R (set-reset) latch* based on NOR gates is shown in Figure 7-5(a). The circuit has two inputs, S and R, and two outputs, labeled Q and QN, where QN is normally the complement of Q. Signal QN is sometimes labeled $\overline{Q}$ or Q_L.

S-R latch

If S and R are both 0, the circuit behaves like the bistable element—we have a feedback loop that retains one of two logic states, Q = 0 or Q = 1. As shown in Figure 7-5(b), either S or R may be asserted to force the feedback loop to a desired state. S *sets* or *presets* the Q output to 1; R *resets* or *clears* the Q output to 0. After the S or R input is negated, the latch remains in the state that it was forced into. Figure 7-6(a) shows the functional behavior of an S-R latch for a typical sequence of inputs. Colored arrows indicate causality, that is, which input transitions cause which output transitions.

set
preset
reset
clear

Figure 7-6 Typical operation of an S-R latch: (a) "normal" inputs; (b) S and R asserted simultaneously.

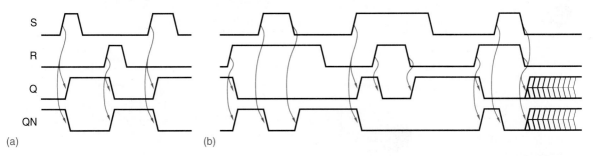

(a) (b)

Q̅ VERSUS QN   In most applications of an S-R latch, the QN (a.k.a. Q̅) output is always the complement of the Q output. However, the Q̅ name is not quite correct, because there is one
case where this output is not the complement of Q. If both S and R are 1, as they are
in several places in Figure 7-6(b), then both outputs are forced to 0. Once we negate
either input, the outputs return to complementary operation as usual. However, if we
negate both inputs simultaneously, the latch goes to an unpredictable next state, and it
may in fact oscillate or enter the metastable state. Metastability may also occur if a 1
pulse that is too short is applied to S or R.

Three different logic symbols for the same S-R latch circuit are shown in
Figure 7-7. The symbols differ in the treatment of the complemented output.
Historically, the first symbol was used, showing the active-low or complemented
signal name inside the function rectangle. However, in bubble-to-bubble logic
design the second form of the symbol is preferred, showing an inversion bubble
outside the function rectangle. The last form of the symbol is obviously wrong.

propagation delay   Figure 7-8 defines timing parameters for an S-R latch. The *propagation
delay* is the time it takes for a transition on an input signal to produce a transition
on an output signal. A given latch or flip-flop may have several different
propagation-delay specifications, one for each pair of input and output signals.
Also, the propagation delay may be different depending on whether the output
makes a LOW-to-HIGH or HIGH-to-LOW transition. With an S-R latch, a
LOW-to-HIGH transition on S can cause a LOW-to-HIGH transition on Q, so a
propagation delay $t_{pLH(SQ)}$ occurs, as shown in transition 1 in the figure.
Similarly, a LOW-to-HIGH transition on R can cause a HIGH-to-LOW transition
on Q, with propagation delay $t_{pHL(RQ)}$ as shown in transition 2. Not shown in the
figure are the corresponding transitions on QN, which would have propagation
delays $t_{pHL(SQN)}$ and $t_{pLH(RQN)}$.

minimum pulse width   Minimum-pulse-width specifications are usually given for the S and R
inputs. As shown in Figure 7-8, the latch may go into the metastable state and
remain there for a random length of time if a pulse shorter than the minimum
width $t_{pw(min)}$ is applied to S or R. The latch can be deterministically brought out
of the metastable state only by applying a pulse to S or R that meets or exceeds
the minimum-pulse-width requirement.

Figure 7-7
Symbols for an S-R latch:
(a) without bubble;
(b) preferred for bubble-
to-bubble design;
(c) incorrect because
of double negation.

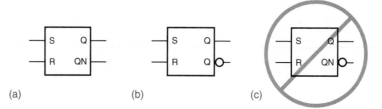

(a)   (b)   (c)

HOW CLOSE IS CLOSE?   As mentioned in the previous note, an S-R latch may go into the metastable state if S and R are negated simultaneously. Often, but not always, a commercial latch's specifications define "simultaneously" (e.g., S and R negated within 20 ns of each other). This parameter is sometimes called the *recovery time*, t_{rec}. It is the minimum delay between negating S and R for them to be considered nonsimultaneous and it is closely related to the minimum-pulse-width specification. Both specifications are measures of how long it takes for the latch's feedback loop to stabilize during a change of state.

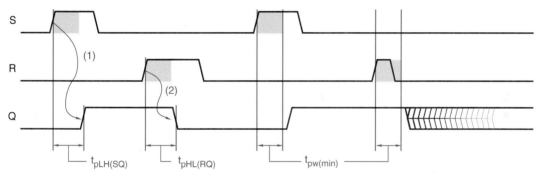

Figure 7-8 Timing parameters for an S-R latch.

7.2.2 $\overline{S}$-$\overline{R}$ Latch

An $\overline{S}$-$\overline{R}$ *latch* (read "S-bar-R-bar latch") with active-low set and reset inputs may be built from NAND gates as shown in Figure 7-9(a). In TTL and CMOS logic families, $\overline{S}$-$\overline{R}$ latches are used much more often than S-R latches because NAND gates are preferred over NOR gates.

   As shown by the function table, Figure 7-9(b), operation of the $\overline{S}$-$\overline{R}$ latch is similar to that of the S-R, with two major differences. First, $\overline{S}$ and $\overline{R}$ are active low, so the latch remembers its previous state when $\overline{S} = \overline{R} = 1$; the active-low inputs are clearly indicated in the symbols in (c). Second, when both $\overline{S}$ and $\overline{R}$ are asserted simultaneously, both latch outputs go to 1, not 0 as in the S-R latch. Except for these differences, operation of the $\overline{S}$-$\overline{R}$ is the same as the S-R, including timing and metastability considerations.

$\overline{S}$-$\overline{R}$ latch

Figure 7-9 $\overline{S}$-$\overline{R}$ latch: (a) circuit design using NAND gates; (b) function table; (c) logic symbol.

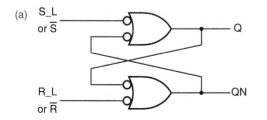

(b)

S_L	R_L	Q	QN
0	0	1	1
0	1	1	0
1	0	0	1
1	1	last Q	last QN

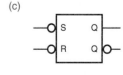

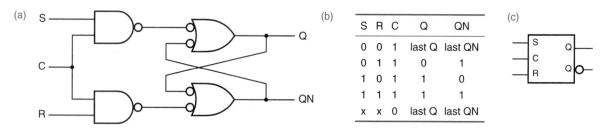

(a)   (b)

S	R	C	Q	QN
0	0	1	last Q	last QN
0	1	1	0	1
1	0	1	1	0
1	1	1	1	1
x	x	0	last Q	last QN

(c)

Figure 7-10 S-R latch with enable: (a) circuit using NAND gates; (b) function table; (c) logic symbol.

7.2.3 S-R Latch with Enable

S-R latch with enable

An S-R or $\overline{\text{S}}$-$\overline{\text{R}}$ latch is sensitive to its S and R inputs at all times. However, it may easily be modified to create a device that is sensitive to these inputs only when an enabling input C is asserted. Such an S-R *latch with enable* is shown in Figure 7-10. As shown by the function table, the circuit behaves like an S-R latch when C is 1, and retains its previous state when C is 0. The latch's behavior for a typical set of inputs is shown in Figure 7-11. If both S and R are 1 when C changes from 1 to 0, the circuit behaves like an S-R latch in which S and R are negated simultaneously—the next state is unpredictable and the output may become metastable.

7.2.4 D Latch

D latch

S-R latches are useful in control applications, where we often think in terms of setting a flag in response to some condition and resetting it when conditions change. So, we control the set and reset inputs somewhat independently. However, we often need latches simply to store bits of information—each bit is presented on a signal line, and we'd like to store it somewhere. A *D latch* may be used in such an application.

Figure 7-12 shows a D latch. Its logic diagram is recognizable as that of an S-R latch with enable, with an inverter added to generate S and R inputs from the

Figure 7-11 Typical operation of an S-R latch with enable.

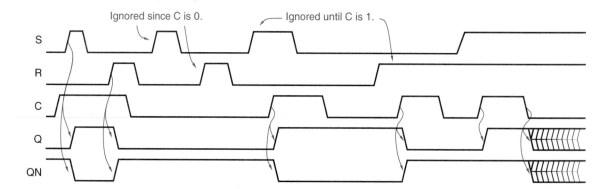

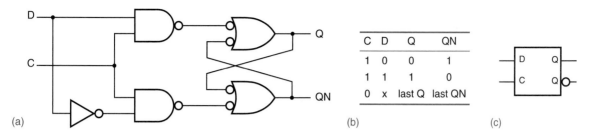

Figure 7-12 D latch: (a) circuit design using NAND gates; (b) function table; (c) logic symbol.

C	D	Q	QN
1	0	0	1
1	1	1	0
0	x	last Q	last QN

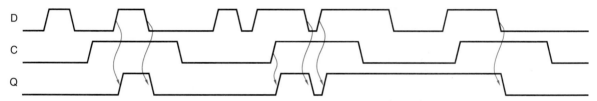

Figure 7-13 Functional behavior of a D latch for various inputs.

single D (data) input. This eliminates the troublesome situation in S-R latches, where S and R may be asserted simultaneously. The control input of a D latch, labeled C in (c), is sometimes named ENABLE, CLK, or G and is active low in some D-latch designs.

An example of a D latch's functional behavior is given in Figure 7-13. When the C input is asserted, the Q output follows the D input. In this situation, the latch is said to be "open" and the path from D input to Q output is "transparent"; the circuit is often called a *transparent latch* for this reason. When the C input is negated, the latch "closes"; the Q output retains its last value and no longer changes in response to D, as long as C remains negated.

transparent latch

More detailed timing behavior of the D latch is shown in Figure 7-14. Four different delay parameters are shown for signals that propagate from the C or D input to the Q output. For example, at transitions 1 and 4 the latch is initially

Figure 7-14 Timing parameters for a D latch.

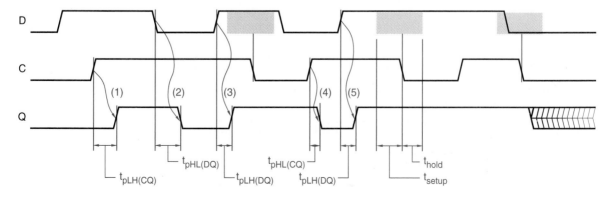

"closed" and the D input is the opposite of Q output, so that when C goes to 1 the latch "opens up" and the Q output changes after delay $t_{pLH(CQ)}$ or $t_{pHL(CQ)}$. At transitions 2 and 3 the C input is already 1 and the latch is already open, so that Q transparently follows the transitions on D with delay $t_{pHL(DQ)}$ and $t_{pLH(DQ)}$. Four more parameters specify the delay to the QN output, not shown.

Although the D latch eliminates the S = R = 1 problem of the S-R latch, it does not eliminate the metastability problem. As shown in Figure 7-14, there is a (shaded) window of time around the falling edge of C when the D input must not change. This window begins at time t_{setup} before the falling (latching) edge of C; t_{setup} is called the *setup time*. The window ends at time t_{hold} afterward; t_{hold} is called the *hold time*. If D changes at any time during the setup- and hold-time window, the output of the latch is unpredictable and may become metastable, as shown for the last latching edge in the figure.

7.2.5 Edge-Triggered D Flip-Flop

positive-edge-triggered
D flip-flop

master

slave

A *positive-edge-triggered D flip-flop* combines a pair of D latches, as shown in Figure 7-15, to create a circuit that samples its D input and changes its Q and QN outputs only at the rising edge of a controlling CLK signal. The first latch is called the *master*; it is open and follows the input when CLK is 0. When CLK goes to 1, the master latch is closed and its output is transferred to the second latch, called the *slave*. The slave latch is open all the while that CLK is 1, but changes only at the beginning of this interval, because the master is closed and unchanging during the rest of the interval.

dynamic-input
indicator

The triangle on the D flip-flop's CLK input indicates edge-triggered behavior and is called a *dynamic-input indicator*. Examples of the flip-flop's functional behavior for several input transitions are presented in Figure 7-16. The QM signal shown is the output of the master latch. Notice that QM changes only when CLK is 0. When CLK goes to 1, the current value of QM is transferred to Q, and QM is prevented from changing until CLK goes to 0 again.

Figure 7-17 shows more detailed timing behavior for the D flip-flop. All propagation delays are measured from the rising edge of CLK, since that's the only event that causes an output change. Different delays may be specified for LOW-to-HIGH and HIGH-to-LOW output changes.

Figure 7-15 Positive-edge-triggered D flip-flop: (a) circuit design using D latches; (b) function table; (c) logic symbol.

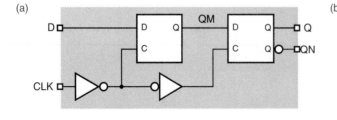

(b)

D	CLK	Q	QN
0	⌐_	0	1
1	_⌐	1	0
x	0	last Q	last QN
x	1	last Q	last QN

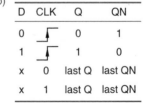

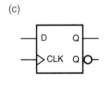

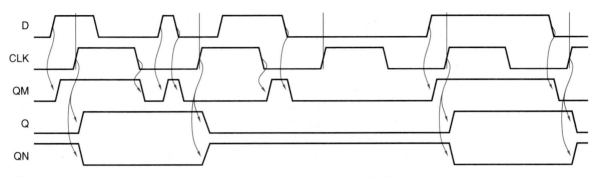

Figure 7-16 Functional behavior of a positive-edge-triggered D flip-flop.

Like a D latch, the edge-triggered D flip-flop has a setup and hold time window during which the D inputs must not change. This window occurs around the triggering edge of CLK, and is indicated by shaded color in Figure 7-17. If the setup and hold times are not met, the flip-flop output will usually go to a stable, though unpredictable, 0 or 1 state. In some cases, however, the output will oscillate or go to a metastable state halfway between 0 and 1, as shown at the second-to-last clock tick in the figure. If the flip-flop goes into the metastable state, it will return to a stable state on its own only after a probabilistic delay, as explained in Section 8.9. It can also be forced into a stable state by applying another triggering clock edge with a D input that meets the setup- and hold-time requirements, as shown at the last clock tick in the figure.

A *negative-edge-triggered* D *flip-flop* simply inverts the clock input, so that all the action takes place on the falling edge of CLK_L; by convention, a falling-edge trigger is considered to be active low. The flip-flop's function table and logic symbol are shown in Figure 7-18.

Some D flip-flops have *asynchronous inputs* that may be used to force the flip-flop to a particular state independent of the CLK and D inputs. These inputs, typically labeled PR (*preset*) and CLR (*clear*), behave like the set and reset

*negative-edge-triggered
D flip-flop*

asynchronous inputs
preset
clear

Figure 7-17 Timing behavior of a positive-edge-triggered D flip-flop.

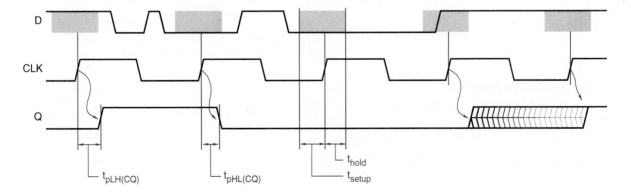

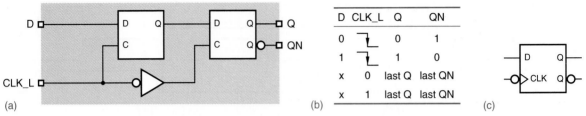

D	CLK_L	Q	QN
0	⬐	0	1
1	⬐	1	0
x	0	last Q	last QN
x	1	last Q	last QN

Figure 7-18 Negative-edge triggered D flip-flop: (a) circuit design using
D latches; (b) function table; (c) logic symbol.

inputs on an SR latch. The logic symbol and NAND circuit for an edge-triggered
D flip-flop with these inputs is shown in Figure 7-19. Although asynchronous
inputs are used by some logic designers to perform tricky sequential functions,
they are best reserved for initialization and testing purposes, to force a sequential
circuit into a known starting state; more on this when we discuss synchronous
design methodology in Section 8.7.

7.2.6 Edge-Triggered D Flip-Flop with Enable

enable input
clock-enable input

A commonly desired function in D flip-flops is the ability to hold the last value
stored, rather than load a new value, at the clock edge. This is accomplished by
adding an *enable input*, called EN or CE (*clock enable*). While the name "clock
enable" is descriptive, the extra input's function is not obtained by controlling

Figure 7-19 Positive-edge-triggered D flip-flop with preset and clear:
(a) logic symbol; (b) circuit design using NAND gates.

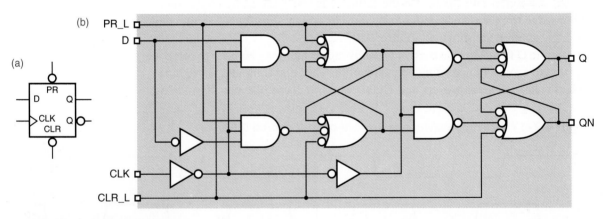

**TIME FOR A
COMMERCIAL**   Commercial TTL positive-edge-triggered D flip-flops do not use the master-slave
latch design of Figure 7-15 or Figure 7-19. Instead, flip-flops like the 74LS74 use the
six-gate design of Figure 7-20, which is smaller and faster. We'll show how to
formally analyze the next-state behavior of both designs in Section 7.9.

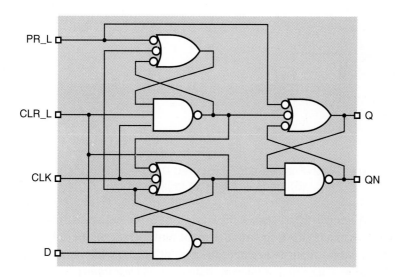

Figure 7-20
Commercial circuit for a positive-edge-triggered D flip-flop such as 74LS74.

the clock in any way whatsoever. Rather, as shown in Figure 7-21(a), a 2-input multiplexer controls the value applied to the internal flip-flop's D input. If EN is asserted, the external D input is selected; if EN is negated, the flip-flop's current output is used. The resulting function table is shown in (b). The flip-flop symbol is shown in (c); in some flip-flops, the enable input is active low, denoted by an inversion bubble on this input.

7.2.7 Scan Flip-Flop

An important flip-flop function for ASIC testing is so-called *scan capability*. *scan capability* The idea is to be able to drive the flip-flop's D input with an alternate source of data during device testing. When all of the flip-flops are put into testing mode, a test pattern can be "scanned in" to the ASIC using the flip-flops' alternate data inputs. After the test pattern is loaded, the flip-flops are put back into "normal" mode, and all of the flip-flops are clocked normally. After one or more clock ticks, the flip-flops are put back into test mode, and the test results are "scanned out."

Figure 7-21 Positive-edge-triggered D flip-flop with enable: (a) circuit design; (b) function table; (c) logic symbol.

(a)

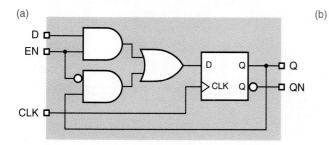

(b)

D	EN	CLK	Q	QN
0	1	⌐⌐	0	1
1	1	⌐⌐	1	0
x	0	⌐⌐	last Q	last QN
x	x	0	last Q	last QN
x	x	1	last Q	last QN

(c)

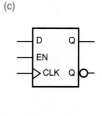

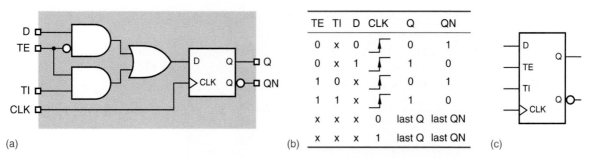

TE	TI	D	CLK	Q	QN
0	x	0	⌐_	0	1
0	x	1	⌐_	1	0
1	0	x	⌐_	0	1
1	1	x	⌐_	1	0
x	x	x	0	last Q	last QN
x	x	x	1	last Q	last QN

(a)   (b)   (c)

Figure 7-22 Positive-edge-triggered D flip-flop with scan: (a) circuit design; (b) function table; (c) logic symbol.

test-enable input, TE

test input, TI

scan chain

Figure 7-22(a) shows the design of a typical scan flip-flop. It is nothing more than a D flip-flop with a 2-input multiplexer on the D input. When the TE (*test enable*) input is negated, the circuit behaves like an ordinary D flip-flop. When TE is asserted, it takes its data from TI (*test input*) instead of from D. This functional behavior is shown in (b), and a symbol for the device is given in (c).

The extra inputs are used to connect all of an ASIC's flip-flops in a *scan chain* for testing purposes. Figure 7-23 is a simple example with four flip-flops in the scan chain. The TE inputs of all the flip-flops are connected to a global TE input, while each flip-flop's Q output is connected to another's TI input in serial (daisy-chain) fashion. The TI, TE, and TO (test output) connections are strictly for testing purposes; the additional logic connected to the D inputs and Q outputs needed to make the circuit do something useful is not shown.

To test the circuit, including the additional logic, the global TE input is asserted while *n* clock ticks occur and *n* test-vector bits are applied to the global TI input and are thereby scanned (shifted) into the *n* flip-flops; *n* equals 4 in Figure 7-23. Then TE is negated, and the circuit is allowed to run for one or more additional clock ticks. The new state of the circuit, represented by the new values in the *n* flip-flops, can be observed (scanned out) at TO by asserting TE while *n* more clock ticks occur. To make the testing process more efficient, another test vector can be scanned in while the previous result is being scanned out.

Figure 7-23 A scan chain with four flip-flops.

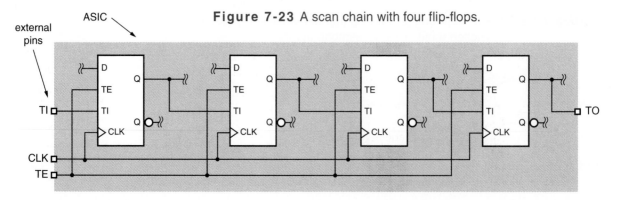

There are many different types of scan flip-flops, corresponding to different types of basic flip-flop functionality. For example, scan capability could be added to the D flip-flop with enable in Figure 7-21 by replacing its internal 2-input multiplexer with a 3-input one. At each clock tick the flip-flop would load D, TI, or its current state, depending on the values of EN and TE. Scan capability can also be added to other flip-flop types, such as J-K and T introduced later in this section.

*7.2.8 Master/Slave S-R Flip-Flop

We indicated earlier that S-R latches are useful in "control" applications, where we may have independent conditions for setting and resetting a control bit. If the control bit is supposed to be changed only at certain times with respect to a clock signal, then we need an S-R flip-flop that, like a D flip-flop, changes its outputs only on a certain edge of the clock signal. This subsection and the next two describe flip-flops that are useful for such applications.

If we substitute S-R latches for the D latches in the negative-edge-triggered D flip-flop of Figure 7-18(a), we get a *master/slave S-R flip-flop*, shown in Figure 7-24. Like a D flip-flop, the S-R flip-flop changes its outputs only at the falling edge of a control signal C. However, the new output value depends on input values not just at the falling edge, but during the entire interval in which C is 1 prior to the falling edge. As shown in Figure 7-25, a short pulse on S any time during this interval can set the master latch; likewise, a pulse on R can reset it. The value transferred to the flip-flop output on the falling edge of C depends on whether the master latch was last set or cleared while C was 1.

master/slave S-R flip-flop

Shown in Figure 7-24(c), the logic symbol for the master/slave S-R flip-flop does not use a dynamic-input indicator, because the flip-flop is not truly edge triggered. It is more like a latch that follows its input during the entire interval that C is 1 but changes its output to reflect the final latched value only when C goes to 0. In the symbol, a *postponed-output indicator* indicates that the output signal does not change until enable input C is negated. Flip-flops with this kind of behavior are sometimes called *pulse-triggered flip-flops*.

postponed-output indicator

pulse-triggered flip-flop

Figure 7-24 Master/slave S-R flip-flop: (a) circuit using S-R latches; (b) function table; (c) logic symbol.

(a)

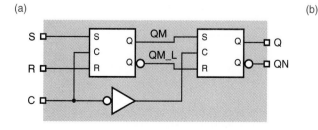

(b)

S	R	C	Q	QN
x	x	0	last Q	last QN
0	0	⎍	last Q	last QN
0	1	⎍	0	1
1	0	⎍	1	0
1	1	⎍	undef.	undef.

(c)

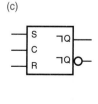

*Throughout this book, optional sections are marked with an asterisk.

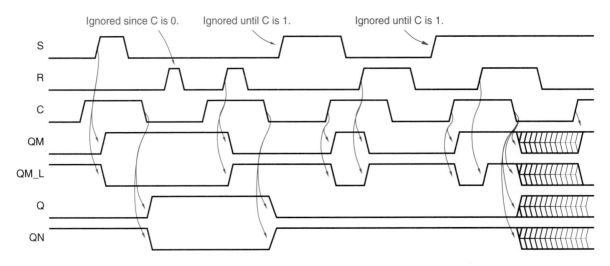

Figure 7-25 Internal and functional behavior of a master/slave S-R flip-flop.

The operation of the master/slave S-R flip-flop is unpredictable if both S and R are asserted at the falling edge of C. In this case, just before the falling edge, both the Q and QN outputs of the master latch are 1. When C goes to 0, the master latch's outputs change unpredictably and may even become metastable. At the same time, the slave latch opens up and propagates this garbage to the flip-flop output.

*7.2.9 Master/Slave J-K Flip-Flop

master/slave J-K flip-flop

The problem of what to do when S and R are asserted simultaneously is solved in a *master/slave J-K flip-flop*. The J and K inputs are analogous to S and R. However, as shown in Figure 7-26, asserting J asserts the master's S input only if the flip-flop's QN output is currently 1 (i.e., Q is 0), and asserting K asserts the master's R input only if Q is currently 1. Thus, if J and K are asserted simultaneously, the flip-flop goes to the opposite of its current state.

Figure 7-26 Master/slave J-K flip-flop: (a) circuit design using S-R latches; (b) function table; (c) logic symbol.

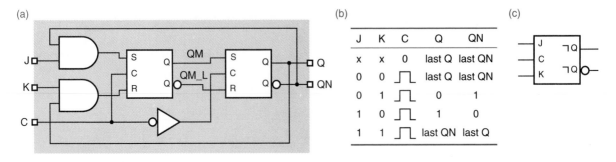

(a)

(b)

J	K	C	Q	QN
x	x	0	last Q	last QN
0	0	⎍	last Q	last QN
0	1	⎍	0	1
1	0	⎍	1	0
1	1	⎍	last QN	last Q

(c)

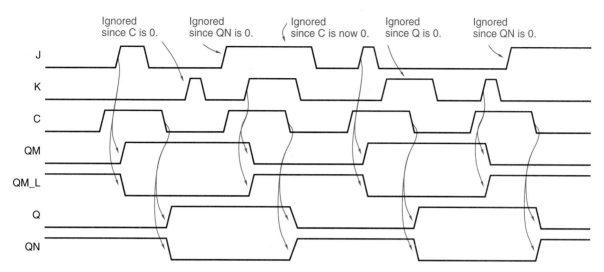

Figure 7-27 Internal and functional behavior of a master/slave J-K flip-flop.

Figure 7-27 shows the functional behavior of a J-K master/slave flip-flop for a typical set of inputs. Note that the J and K inputs need not be asserted at the end of the triggering pulse for the flip-flop output to change at that time. In fact, because of the gating on the master latch's S and R inputs, it is possible for the flip-flop output to change to 1 even though K and not J is asserted at the end of the triggering pulse. This behavior, known as *1s catching*, is illustrated in the second-to-last triggering pulse in the figure. An analogous behavior known as *0s catching* is illustrated in the last triggering pulse. Because of this behavior, the J and K inputs of a J-K master/slave flip-flop must be held valid during the entire interval that C is 1.

1s catching

0s catching

7.2.10 Edge-Triggered J-K Flip-Flop

The problem of 1s and 0s catching is solved in an *edge-triggered J-K flip-flop*, whose functional equivalent is shown in Figure 7-28. Using an edge-triggered D flip-flop internally, the edge-triggered J-K flip-flop samples its inputs at the

edge-triggered J-K flip-flop

Figure 7-28 Edge-triggered J-K flip-flop: (a) equivalent function using an edge-triggered D flip-flop; (b) function table; (c) logic symbol.

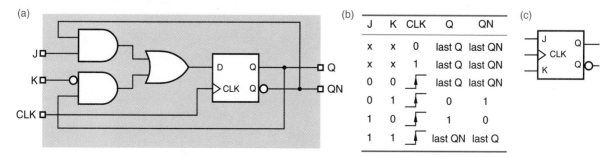

J	K	CLK	Q	QN
x	x	0	last Q	last QN
x	x	1	last Q	last QN
0	0	↑	last Q	last QN
0	1	↑	0	1
1	0	↑	1	0
1	1	↑	last QN	last Q

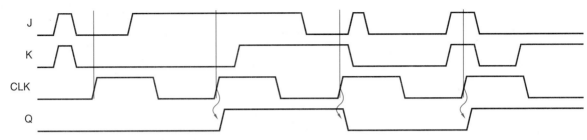

Figure 7-29 Functional behavior of a positive-edge-triggered J-K flip-flop.

rising edge of the clock and produces its next output according to the "charac-teristic equation" $Q^* = J \cdot Q' + K' \cdot Q$ (see Section 7.3.3).

Typical functional behavior of an edge-triggered J-K flip-flop is shown in Figure 7-29. Like the D input of an edge-triggered D flip-flop, the J and K inputs of a J-K flip-flop must meet published setup- and hold-time specifications with respect to the triggering clock edge for proper operation.

Because they eliminate the problems of 1s and 0s catching and of simulta-neously asserting both control inputs, edge-triggered J-K flip-flops have largely obsoleted the older pulse-triggered types. The *74x109* is a TTL positive-edge-triggered J-$\overline{\text{K}}$ flip-flop with an active-low K input (named $\overline{\text{K}}$ or K_L).

74x109

ANOTHER COMMERCIAL (FLIP-FLOP, THAT IS) The internal design of the 74LS109 is very similar to that of the 74LS74, which we showed in Figure 7-20. As shown in Figure 7-30, the '109 simply replaces the bottom-left gate of the '74, which realizes the characteristic equation $Q^* = D$, with an AND-OR structure that realizes the J-$\overline{\text{K}}$ characteristic equation, $Q^* = J \cdot Q' + K\_L \cdot Q$.

Figure 7-30
Internal logic diagram for the 74LS109 positive-edge-triggered J-$\overline{\text{K}}$ flip-flop.

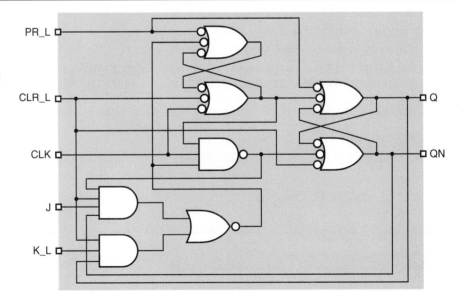

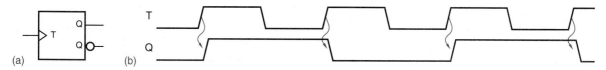

Figure 7-31 Positive-edge-triggered T flip-flop: (a) logic symbol; (b) functional behavior.

The most common application of J-K flip-flops is in clocked synchronous state machines. As we'll explain in Section 7.4.5, the next-state logic for J-K flip-flops is sometimes simpler than for D flip-flops. However, most state machines are still designed using D flip-flops, because the design methodology is a bit simpler and because most sequential programmable logic devices contain D, not J-K, flip-flops. Therefore, we'll give most of our attention to D flip-flops.

7.2.11 T Flip-Flop

A *T (toggle) flip-flop* changes state on every tick of the clock. Figure 7-31 shows the symbol and illustrates the behavior of a positive-edge-triggered T flip-flop. Notice that the signal on the flip-flop's Q output has precisely half the frequency of the T input. Figure 7-32 shows how to obtain a T flip-flop from a D or J-K flip-flop. T flip-flops are most often used in counters and frequency dividers, as we'll show in Section 8.4.

T flip-flop

In many applications of T flip-flops, the flip-flop need not be toggled on every clock tick. Such applications can use a *T flip-flop with enable*. As shown in Figure 7-33, the flip-flop changes state at the triggering edge of the clock only if the enable signal EN is asserted. Like the D, J, and K inputs on other edge-triggered flip-flops, the EN input must meet specified setup and hold times with respect to the triggering clock edge. The circuits of Figure 7-32 are easily modified to provide an EN input, as shown in Figure 7-34.

T flip-flop with enable

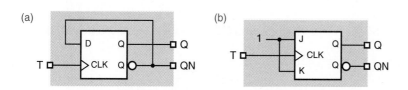

(a)   (b)

Figure 7-32
Possible circuit designs for a T flip-flop: (a) using a D flip-flop; (b) using a J-K flip-flop.

Figure 7-33 Positive-edge-triggered T flip-flop with enable: (a) logic symbol; (b) functional behavior.

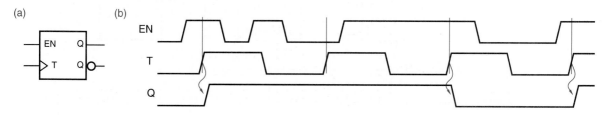

Figure 7-34
Possible circuits for a
T flip-flop with enable:
(a) using a D flip-flop;
(b) using a J-K flip-flop.

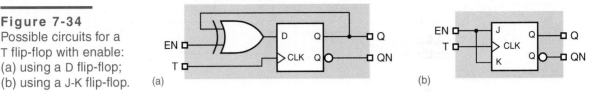

(a)   (b)

7.3 Clocked Synchronous State-Machine Analysis

Although latches and flip-flops, the basic building blocks of sequential circuits,
are themselves feedback sequential circuits that can be formally analyzed, we'll
first study the operation of *clocked synchronous state machines*, since they are
the easiest to understand. "State machine" is a generic name given to these
sequential circuits; "clocked" refers to the fact that their storage elements (flip-
flops) employ a clock input; and "synchronous" means that all of the flip-flops
use the same clock signal. Such a state machine changes state only when a triggering edge or "tick" occurs on the clock signal.

*clocked synchronous
state machine*

7.3.1 State-Machine Structure

Figure 7-35 shows the general structure of a clocked synchronous state machine.
The *state memory* is a set of *n* flip-flops that store the current state of the
machine, and it has 2^n distinct states. The flip-flops are all connected to a common clock signal that causes them to change state at each *tick* of the clock. What
constitutes a tick depends on the flip-flop type (edge triggered, pulse triggered,
etc.). For the positive-edge-triggered D and J-K flip-flops considered in this
section, a tick is the rising edge of the clock signal.

state memory

tick

The next state of the state machine in Figure 7-35 is determined by the
next-state logic F as a function of the current state and input. The *output logic G*

next-state logic
output logic

Figure 7-35 Clocked synchronous state-machine structure (Mealy machine).

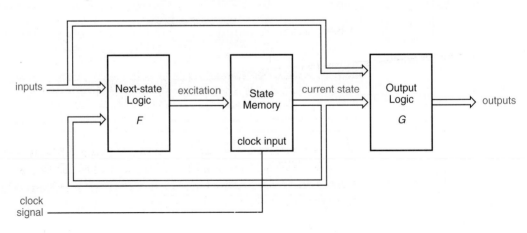

determines the output as a function of the current state and input. Both *F* and *G* are strictly combinational logic circuits. We can write

$$\text{Next state} \ = \ F(\text{current state, input})$$

$$\text{Output} \ = \ G(\text{current state, input})$$

State machines may use positive-edge-triggered D flip-flops for their state memory, in which case a tick occurs at each rising edge of the clock signal. It is also possible for the state memory to use negative-edge-triggered D flip-flops, D latches, or J-K flip-flops. However, inasmuch as most state machines are designed nowadays using programmable logic devices with positive-edge-triggered D flip-flops, that's what we'll concentrate on.

7.3.2 Output Logic

A sequential circuit whose output depends on both state and input as shown in Figure 7-35 is called a *Mealy machine*. In some sequential circuits the output depends on the state alone:

$$\text{Output} \ = \ G(\text{current state})$$

Such a circuit is called a *Moore machine*, and its general structure is shown in Figure 7-36.

Obviously, the only difference between the two state-machine models is in how outputs are generated. In practice, many state machines must be categorized as Mealy machines, because they have one or more *Mealy-type outputs* that depend on input as well as state. However, many of these same machines also have one or more *Moore-type outputs* that depend only on state.

In the design of high-speed circuits, it is often necessary to ensure that state-machine outputs are available as early as possible and do not change during each clock period. One way to get this behavior is to encode the state so that the state variables themselves serve as outputs. We call this an *output-coded state assignment*; it produces a Moore machine in which the output logic of Figure 7-36 is null, consisting of just wires.

output-coded state assignment

Figure 7-36 Clocked synchronous state-machine structure (Moore machine).

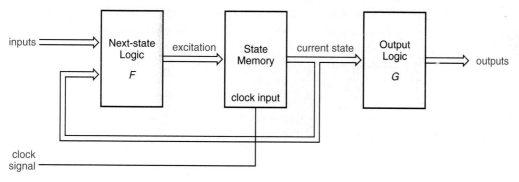

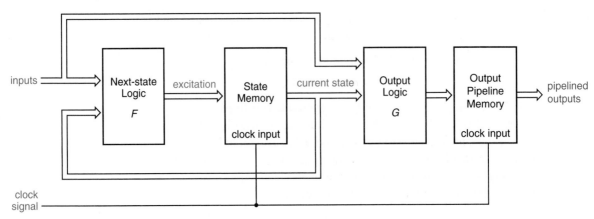

Figure 7-37 Mealy machine with pipelined outputs.

pipelined outputs

Another approach is to design the state machine so that the outputs during one clock period depend on the state and inputs during the *previous* clock period. We call these *pipelined outputs*, and they are obtained by attaching another stage of memory (flip-flops) to a Mealy machine's outputs, as in Figure 7-37.

With appropriate circuit or drawing manipulations, you can map one state-machine model into another. For example, you could declare the flip-flops that produce pipelined outputs from a Mealy machine to be part of its state memory and thereby obtain a Moore machine with an output-coded state assignment.

The exact classification of a state machine into one style or another is not so important. What's important is how you think about output structure and how it satisfies your overall design objectives, including timing and flexibility. For example, pipelined outputs are great for timing, but you can use them only in situations where you can figure out the desired next output value one clock period in advance. In any given application you may use different styles for different output signals. For example, we'll see in Section 7.11.5 that different statements can be used to specify different output styles in ABEL.

7.3.3 Characteristic Equations

characteristic equation

* suffix

The functional behavior of a latch or flip-flop can be described formally by a *characteristic equation* that specifies the flip-flop's next state as a function of its current state and inputs.

The characteristic equations of the flip-flops in Section 7.2 are listed in Table 7-1. By convention, the * suffix in Q* means "the next value of Q." Notice that the characteristic equation does not describe detailed timing behavior of the device (latching vs. edge-triggered, etc.), only the functional response to the control inputs. This simplified description is useful in the analysis of state machines, as we'll soon show.

Device Type	Characteristic Equation
S-R latch	$Q* = S + R' \cdot Q$
D latch	$Q* = D$
Edge-triggered D flip-flop	$Q* = D$
D flip-flop with enable	$Q* = EN \cdot D + EN' \cdot Q$
Master/slave S-R flip-flop	$Q* = S + R' \cdot Q$
Master/slave J-K flip-flop	$Q* = J \cdot Q' + K' \cdot Q$
Edge-triggered J-K flip-flop	$Q* = J \cdot Q' + K' \cdot Q$
T flip-flop	$Q* = Q'$
T flip-flop with enable	$Q* = EN \cdot Q' + EN' \cdot Q$

Table 7-1
Latch and flip-flop characteristic equations.

7.3.4 Analysis of State Machines with D Flip-Flops

Consider the formal definition of a state machine that we gave previously:

$$\text{Next state} = F(\text{current state, input})$$

$$\text{Output} = G(\text{current state, input})$$

Recalling our notion that "state" embodies all we need to know about the past history of the circuit, the first equation tells us that what we next need to know can be determined from what we currently know and the current input. The second equation tells us that the current output can be determined from the same information. The goal of sequential circuit analysis is to determine the next-state and output functions so that the behavior of a circuit can be predicted.

The analysis of a clocked synchronous state machine has three basic steps:

1. Determine the next-state and output functions F and G.

2. Use F and G to construct a *state/output table* that completely specifies the next state and output of the circuit for every possible combination of current state and input.   *state/output table*

3. (Optional) Draw a *state diagram* that presents the information from the previous step in graphical form.   *state diagram*

Figure 7-38 shows a simple state machine with two positive-edge-triggered D flip-flops. To determine the next-state function F, we must first consider the behavior of the state memory. At the rising edge of the clock signal, each D flip-flop samples its D input and transfers this value to its Q output; the characteristic equation of a D flip-flop is $Q* = D$. Therefore, to determine the next value of Q (i.e., Q^*), we must first determine the current value of D.

In Figure 7-38 there are two D flip-flops, and we have named the signals on their outputs Q0 and Q1. These two outputs are the state variables; their value is

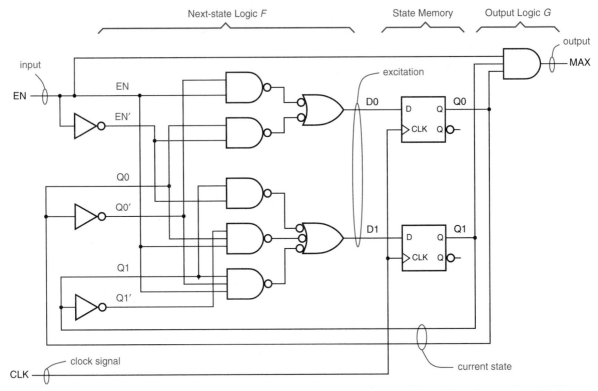

Figure 7-38 Clocked synchronous state machine using positive-edge-triggered D flip-flops.

excitation

excitation equation

the current state of the machine. We have named the signals on the corresponding D inputs D0 and D1. These signals provide the *excitation* for the D flip-flops at each clock tick. Logic equations that express the excitation signals as functions of the current state and input are called *excitation equations* and can be derived from the circuit diagram:

$$D0 \ = \ Q0 \cdot EN' + Q0' \cdot EN$$

$$D1 \ = \ Q1 \cdot EN' + Q1' \cdot Q0 \cdot EN + Q1 \cdot Q0' \cdot EN$$

* suffix

By convention, the next value of a state variable after a clock tick is denoted by appending a star to the state-variable name, for example, Q0* or Q1*. Using the characteristic equation of D flip-flops, Q* = D, we can describe the next-state function of the example machine with equations for the next value of the state variables:

$$Q0* \ = \ D0$$

$$Q1* \ = \ D1$$

Substituting the excitation equations for D0 and D1, we can write

$$Q0* \ = \ Q0 \cdot EN' + Q0' \cdot EN$$

$$Q1* \ = \ Q1 \cdot EN' + Q1' \cdot Q0 \cdot EN + Q1 \cdot Q0' \cdot EN$$

(a)

Q1 Q0	EN 0	EN 1
00	00	01
01	01	10
10	10	11
11	11	00
	Q1* Q0*	

(b)

S	EN 0	EN 1
A	A	B
B	B	C
C	C	D
D	D	A
	S*	

(c)

S	EN 0	EN 1
A	A, 0	B, 0
B	B, 0	C, 0
C	C, 0	D, 0
D	D, 0	A, 1
	S*, MAX	

Table 7-2
Transition, state, and state/output tables for the state machine in Figure 7-38.

These equations, which express the next value of the state variables as a function of current state and input, are called *transition equations*.

transition equation

For each combination of current state and input value, the transition equations predict the next state. Each state is described by two bits, the current values of Q0 and Q1: (Q1 Q0) = 00, 01, 10, or 11. [The reason for "arbitrarily" picking the order (Q1 Q0) instead of (Q0 Q1) will become apparent shortly.] For each state, our example machine has just two possible input values, EN = 0 or EN = 1, so there are a total of 8 state/input combinations. (In general, a machine with s state bits and i inputs has 2^{s+i} state/input combinations.)

Table 7-2(a) shows a *transition table* that is created by evaluating the transition equations for every possible state/input combination. Traditionally, a transition table lists the states along the left and the input combinations along the top of the table, as shown in the example.

transition table

The function of our example machine is apparent from its transition table—it is a 2-bit binary counter with an enable input EN. When EN = 0, the machine maintains its current count, but when EN = 1, the count advances by 1 at each clock tick, rolling over to 00 when it reaches a maximum value of 11.

If we wish, we may assign alphanumeric *state names* to each state. The simplest naming is 00 = A, 01 = B, 10 = C, and 11 = D. Substituting the state names for combinations of Q1 and Q0 (and Q1* and Q0*) in Table 7-2(a) produces the *state table* in (b). Here "S" denotes the current state and "S*" denotes the next state of the machine. A state table is usually easier to understand than a transition table, because in complex machines we can use state names that have meaning. However, a state table contains less information than a transition table because it does not indicate the binary values assumed by the state variables in each named state.

state names

state table

Once a state table is produced, we have only the output logic of the machine left to analyze. In the example machine there is only a single output signal, and it is a function of both current state and input (this is a Mealy machine). So we can write a single *output equation*:

$$MAX = Q1 \cdot Q0 \cdot EN$$

output equation

The output behavior predicted by this equation can be combined with the next-state information to produce a *state/output table* as shown in Table 7-2(c).

state/output table

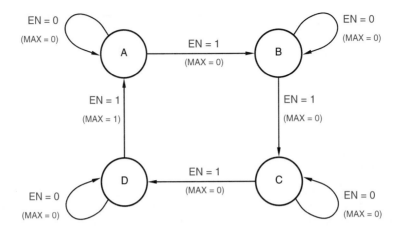

Figure 7-39
State diagram
corresponding to the
state machine of
Table 7-2.

State/output tables for Moore machines are slightly simpler. For example, in the circuit of Figure 7-38 suppose we removed the EN signal from the AND gate that produces the MAX output, producing a Moore-type output MAXS. Then MAXS is a function of the state only, and the state/output table can list MAXS in a single column, independent of the input values. This is shown in Table 7-3.

state diagram   A *state diagram* presents the information from the state/output table in a
node   graphical format. It has one circle (or *node*) for each state and an arrow (or
directed arc   *directed arc*) for each transition. Figure 7-39 shows the state diagram for our example state machine. The letter inside each circle is a state name. Each arrow leaving a given state points to the next state for a given input combination; it also shows the output value produced in the given state for that input combination.

The state diagram for a Moore machine can be somewhat simpler. In this case, the output values can be shown inside each state circle, since they are

Table 7-3
State/output table for
a Moore machine.

S	EN		MAXS
	0	**1**	
A	A	B	0
B	B	C	0
C	C	D	0
D	D	A	1
	S*		

A CLARIFICATION   The state-diagram notation for output values in Mealy machines is a little misleading. You should remember that the listed output value is produced continuously when the machine is in the indicated state and has the indicated input, not just during the transition to the next state.

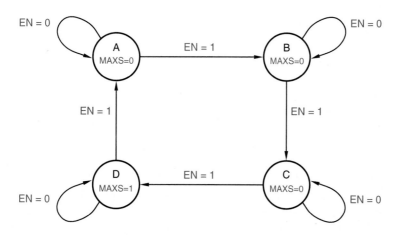

Figure 7-40
State diagram
corresponding to the
state machine of
Table 7-3.

functions of state only. The state diagram for a Moore machine using this convention is shown in Figure 7-40.

The original logic diagram of our example state machine, Figure 7-38, was laid out to match our conceptual model of a Mealy machine. However, nothing requires us to group the next-state logic, state memory, and output logic in this way. Figure 7-41 shows another logic diagram for the same state machine. To analyze this circuit, the designer (or analyzer, in this case) can still extract the required information from the diagram as drawn. The only circuit difference in

Figure 7-41 Redrawn logic diagram for a clocked synchronous state machine.

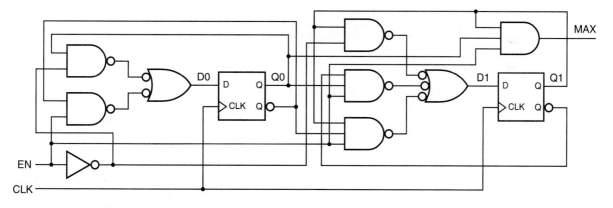

**LITTLE ARROWS,
LITTLE ARROWS
EVERYWHERE**   Since there is only one input in our example machine, there are only two possible input combinations, and two arrows leaving each state. In a machine with n inputs, we would have 2^n arrows leaving each state. This is messy if n is large. Later, in Figure 7-44, we'll describe a convention whereby a state needn't have one arrow leaving it for each input combination, only one arrow for each different next state.

SUGGESTIVE
DRAWINGS Using the transition, state, and output tables, we can construct a timing diagram that
shows the behavior of a state machine for any desired starting state and input
sequence. For example, Figure 7-42 shows the behavior of our example machine
with a starting state of 00 (A) and a particular pattern on the EN input.

Notice that the value of the EN input affects the next state only at the rising
edge of the CLOCK input; that is, the counter counts only if EN = 1 at the rising edge
of CLOCK. On the other hand, since MAX is a Mealy-type output, its value is affected
by EN at all times. If we also provide a Moore-type output MAXS as suggested in the
text, its value depends only on state as shown in the figure.

The timing diagram is drawn in a way that shows changes in the MAX and
MAXS outputs occurring slightly later than the state and input changes that cause
them, reflecting the combinational-logic delay of the output circuits. Naturally, the
drawings are merely suggestive; precise timing is normally indicated by a timing
table of the type suggested in Section 5.2.1.

the new diagram is that we have used the flip-flops' QN outputs (which are
normally the complement of Q) to save a couple of inverters.

In summary, the detailed steps for analyzing a clocked synchronous state
machine are as follows:

excitation equations   1. Determine the excitation equations for the flip-flop control inputs.

transition equations   2. Substitute the excitation equations into the flip-flop characteristic equa-
                             tions to obtain transition equations.

transition table   3. Use the transition equations to construct a transition table.

output equations   4. Determine the output equations.

Figure 7-42 Timing diagram for example state machine.

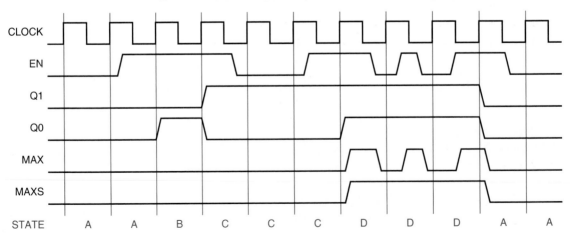

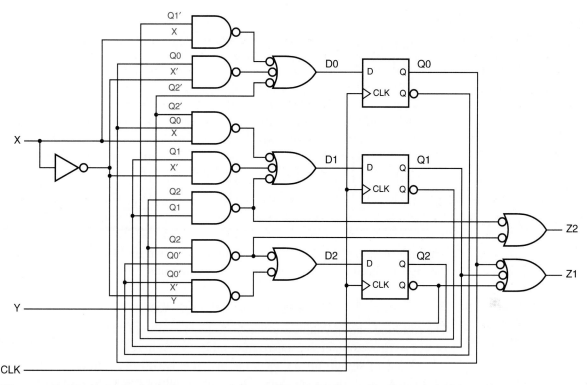

Figure 7-43 A clocked synchronous state machine with three flip-flops and eight states.

5. Add output values to the transition table for each state (Moore) or state/ input combination (Mealy) to create a transition/output table.   *transition/output table*

6. Name the states and substitute state names for state-variable combinations in the transition/output table to obtain a state/output table.   *state names* *state/output table*

7. (Optional) Draw a state diagram corresponding to the state/output table.   *state diagram*

We'll go through this complete sequence of steps to analyze another clocked synchronous state machine, shown in Figure 7-43. Reading the logic diagram, we find that the excitation equations are as follows:

$$DO = Q1' \cdot X + Q0 \cdot X' + Q2$$
$$D1 = Q2' \cdot Q0 \cdot X + Q1 \cdot X' + Q2 \cdot Q1$$
$$D2 = Q2 \cdot Q0' + Q0' \cdot X' \cdot Y$$

Substituting into the characteristic equation for D flip-flops, we obtain the transition equations:

$$Q0* = Q1' \cdot X + Q0 \cdot X' + Q2$$
$$Q1* = Q2' \cdot Q0 \cdot X + Q1 \cdot X' + Q2 \cdot Q1$$
$$Q2* = Q2 \cdot Q0' + Q0' \cdot X' \cdot Y$$

Table 7-4
Transition/output
and state/output
tables for the
state machine
in Figure 7-43.

(a)

		X Y				
Q2 Q1 Q0	00	01	10	11	Z1 Z2	
000	000	100	001	001	10	
001	001	001	011	011	10	
010	010	110	000	000	10	
011	011	011	010	010	00	
100	101	101	101	101	11	
101	001	001	001	001	10	
110	111	111	111	111	11	
111	011	011	011	011	11	
		Q2* Q1* Q0*				

(b)

		X Y			
S	00	01	10	11	Z1 Z2
A	A	E	B	B	10
B	B	B	D	D	10
C	C	G	A	A	10
D	D	D	C	C	00
E	F	F	F	F	11
F	B	B	B	B	10
G	H	H	H	H	11
H	D	D	D	D	11
		S*			

A transition table based on these equations is shown in Table 7-4(a). Reading the logic diagram, we can write two output equations:

$$Z1 = Q2 + Q1' + Q0'$$
$$Z2 = Q2 \cdot Q1 + Q2 \cdot Q0'$$

The resulting output values are shown in the last column of (a). Assigning state names A–H, we obtain the state/output table shown in (b).

A state diagram for the example machine is shown in Figure 7-44. Since our example is a Moore machine, the output values are written with each state. Each arc is labeled with a *transition expression*; a transition is taken for input combinations for which the transition expression is 1. Transitions labeled "1" are always taken.

transition expression

Figure 7-44 State diagram corresponding to Table 7-4.

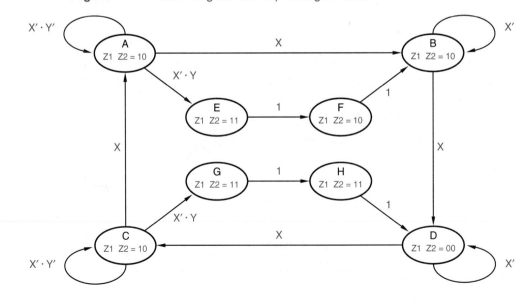

The transition expressions on arcs leaving a particular state must be mutually exclusive and all inclusive, as explained below:

- No two transition expressions can equal 1 for the same input combination,   *mutual exclusion*
since a machine can't have two next states for one input combination.

- For every possible input combination, some transition expression must   *all inclusion*
equal 1, so that all next states are defined.

Starting with the state table, a transition expression for a particular current state and next state can be written as a sum of minterms for the input combinations that cause that transition. If desired, the expression can then be minimized to give the information in a more compact form. Transition expressions are most useful in the *design* of state machines, where the expressions may be developed from the word description of the problem, as we'll show in Section 7.5.

*7.3.5 Analysis of State Machines with J-K Flip-Flops

Clocked synchronous state machines built from J-K flip-flops can also be analyzed by the basic procedure in the preceding subsection. The only difference is that there are two excitation equations for each flip-flop—one for J and the other for K. To obtain the transition equations, both of these must be substituted into the J-K's characteristic equation, $Q* = J \cdot Q' + K' \cdot Q$.

Figure 7-45 is an example state machine using J-K flip-flops. Reading the logic diagram, we can derive the following excitation equations:

$$
\begin{aligned}
J0 &= X \cdot Y' \\
K0 &= X \cdot Y' + Y \cdot Q1 \\
J1 &= X \cdot Q0 + Y \\
K1 &= Y \cdot Q0' + X \cdot Y' \cdot Q0
\end{aligned}
$$

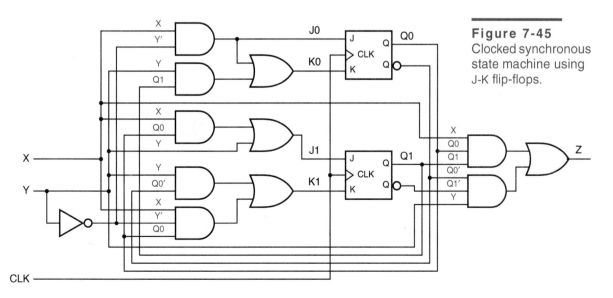

Figure 7-45
Clocked synchronous
state machine using
J-K flip-flops.

Table 7-5
Transition/output and state/output tables for the state machine in Figure 7-45.

(a)

	XY			
Q1 Q0	**00**	**01**	**10**	**11**
00	00, 0	10, 1	01, 0	10, 1
01	01, 0	11, 0	10, 0	11, 0
10	10, 0	00, 0	11, 0	00, 0
11	11, 0	10, 0	00, 1	10, 1
	Q1* Q0*, Z			

(b)

	XY			
S	**00**	**01**	**10**	**11**
A	A, 0	C, 1	B, 0	C, 1
B	B, 0	D, 0	C, 0	D, 0
C	C, 0	A, 0	D, 0	A, 0
D	D, 0	C, 0	A, 1	C, 1
	S*, Z			

Substituting into the characteristic equation for J-K flip-flops, we obtain the transition equations:

$$Q0* = J0 \cdot Q0' + K0' \cdot Q0$$
$$= X \cdot Y' \cdot Q0' + (X \cdot Y' + Y \cdot Q1)' \cdot Q0$$
$$= X \cdot Y' \cdot Q0' + X' \cdot Y' \cdot Q0 + X' \cdot Q1' \cdot Q0 + Y \cdot Q1' \cdot Q0$$
$$Q1* = J1 \cdot Q1' + K1' \cdot Q1$$
$$= (X \cdot Q0 + Y) \cdot Q1' + (Y \cdot Q0' + X \cdot Y' \cdot Q0)' \cdot Q1$$
$$= X \cdot Q1' \cdot Q0 + Y \cdot Q1' + X' \cdot Y' \cdot Q1 + Y' \cdot Q1 \cdot Q0' + X' \cdot Q1 \cdot Q0 + Y \cdot Q1 \cdot Q0$$

A transition table based on these equations is shown in Table 7-5(a). Reading the logic diagram, we can write the output equation:

$$Z = X \cdot Q1 \cdot Q0 + Y \cdot Q1' \cdot Q0'$$

The resulting output values are shown in each column of (a) along with the next state. Assigning state names A–D, we obtain the state/output table shown in (b). A corresponding state diagram that uses transition expressions is shown in Figure 7-46.

Figure 7-46
State diagram corresponding to the state machine of Table 7-5.

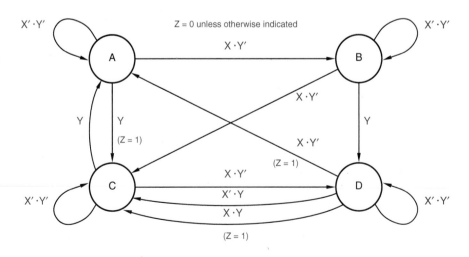

7.4 Clocked Synchronous State-Machine Design

The steps for designing a clocked synchronous state machine, starting from a word description or specification, are just about the reverse of the analysis steps that we used in the preceding section:

1. Construct a state/output table corresponding to the word description or specification, using mnemonic names for the states. (It's also possible to start with a state diagram; this method is discussed in Section 7.5.)   *state/output table*

2. (Optional) Minimize the number of states in the state/output table.   *state minimization*

3. Choose a set of state variables and assign state-variable combinations to the named states.   *state assignment*

4. Substitute the state-variable combinations into the state/output table to create a transition/output table that shows the desired next state-variable combination and output for each state/input combination.   *transition/output table*

5. Choose a flip-flop type (e.g., D or J-K) for the state memory. In most cases you'll already have a choice in mind at the outset of the design, but this step is your last chance to change your mind.

6. Construct an excitation table that shows the excitation values required to obtain the desired next state for each state/input combination.   *excitation table*

7. Derive excitation equations from the excitation table.   *excitation equations*

8. Derive output equations from the transition/output table.   *output equations*

9. Draw a logic diagram that shows the state-variable storage elements and realizes the required excitation and output equations. (Or realize the equations directly in a programmable logic device.)   *logic diagram*

In this section we'll describe each of these basic steps in state-machine design. Step 1 is the most important, since it is here that the designer really *designs*, going through the creative process of translating a (perhaps ambiguous) English-language description of the state machine into a formal tabular description. Step 2 is hardly ever performed by experienced digital designers, but designers bring much of their experience to bear in step 3.   *design*

Once the first three steps are completed, all of the remaining steps can be completed by "turning the crank," that is, by following a well-defined synthesis procedure. Steps 4 and 6–9 are the most tedious, but they are easily automated. For example, when you design a state machine that will be realized in a programmable logic device, you can use an ABEL compiler to do the cranking, as shown in Section 7.11.2. Still, it's important for you to understand the details of the synthesis procedure, both to give you an appreciation of the compiler's function and to give you a chance of figuring out what's really going on when the compiler produces unexpected results. Therefore, all nine steps of the state-machine design procedure are discussed in the remainder of this section.

STATE-TABLE
DESIGN AS A KIND
OF PROGRAMMING

Designing a state table (or equivalently, a state diagram) is a creative process that is like writing a computer program in many ways:

- You start with a fairly precise description of inputs and outputs, but a possibly ambiguous description of the desired relationship between them, and usually no clue about how to actually obtain the desired outputs from the inputs.

- During the design you may have to identify and choose among different ways of doing things, sometimes using common sense, and sometimes arbitrarily.

- You may have to identify and handle special cases that weren't included in the original description.

- You will probably have to keep track of several ideas in your head during the design process.

- Since the design process is not an algorithm, there's no guarantee that you can complete the state table or program using a finite number of states or lines of code. However, unless you work for the government, you must try to do so.

- When you finally run the state machine or program, it will do exactly what you told it to do—no more, no less.

- There's no guarantee that the thing will work the first time; you may have to debug and iterate on the whole process.

Although state-table design is a challenge, there's no need to be intimidated. If you've made it this far in your education, then you've probably written a few programs that worked, and you can become just as good at designing state tables.

7.4.1 State-Table Design Example

There are several different ways to describe a state machine's state table. Later, we'll see how ABEL and VHDL can specify state tables indirectly. In this section, however, we deal only with state tables that are specified directly, in the same tabular format that we used in the previous section for analysis.

We'll present the state-table design process as well as the synthesis procedure in later subsections, using the simple design problem below:

Design a clocked synchronous state machine with two inputs, A and B, and a single output Z that is 1 if:

- A had the same value at each of the two previous clock ticks, *or*

- B has been 1 since the last time that the first condition was true.

Otherwise, the output should be 0.

If the meaning of this specification isn't crystal clear to you at this point, don't worry. Part of your job as a designer is to convert such a specification into a state table that is absolutely unambiguous; even if it doesn't match what was originally intended, it at least forms a basis for further discussion and refinement.

REALIZING RELIABLE RESET

For proper system operation, the hardware design of a state machine should ensure that it enters a known initial state on power-up, such as the INIT state in our design example. Most systems have a RESET signal that is asserted during power-up.

The RESET signal is typically generated by an analog circuit. Such a reset circuit typically detects a voltage (say, 4.5 V) close to the power supply's full voltage, and follows that with a delay (say, 100 ms) to ensure that all components (including oscillators) have had time to stabilize before it "unresets" the system. The Texas Instruments TL7705 is such an analog reset IC; it has an internal 4.5-V reference for the detector and uses an external resistor and capacitor to determine the "unreset" time constant.

If a state machine is built using discrete flip-flops with asynchronous preset and clear inputs, the RESET signal can be applied to these inputs to force the machine into the desired initial state. If preset and clear inputs are not available, or if reset must be synchronous (as in systems using high-speed microprocessors), then the RESET signal may be used as another input to the state machine, with all of the next-state entries going to the desired initial state when RESET is asserted.

As an additional "hint" or requirement, state-table design problems often include timing diagrams that show the state machine's expected behavior for one or more sequences of inputs. Such a timing diagram is unlikely to specify unambiguously the machine's behavior for all possible sequences of inputs, but, again, it's a good starting point for discussion and a benchmark against which proposed designs can be checked. Figure 7-47 is such a timing diagram for our example state-table design problem.

The first step in the state-table design is to construct a template. From the word description, we know that our example is a Moore machine—its output depends only on the current state, that is, what happened in previous clock periods. Thus, as shown in Figure 7-48(a), we provide one next-state column for each possible input combination and a single column for the output values. The order in which the input combinations are written doesn't affect this part of the

Figure 7-47 Timing diagram for example state machine.

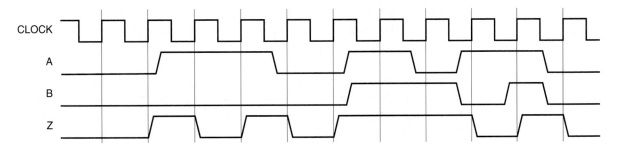

(a)

Meaning	S	00	01	11	10	Z
				A B		
Initial state	INIT					0
	. . .					
	. . .					
	. . .					
				S*		

(b)

Meaning	S	00	01	11	10	Z
				A B		
Initial state	INIT	A0	A0	A1	A1	0
Got a 0 on A	A0					0
Got a 1 on A	A1					0
				S*		

(c)

Meaning	S	00	01	11	10	Z
				A B		
Initial state	INIT	A0	A0	A1	A1	0
Got a 0 on A	A0	OK	OK	A1	A1	0
Got a 1 on A	A1					0
Got two equal A inputs	OK					1
				S*		

(d)

Meaning	S	00	01	11	10	Z
				A B		
Initial state	INIT	A0	A0	A1	A1	0
Got a 0 on A	A0	OK	OK	A1	A1	0
Got a 1 on A	A1	A0	A0	OK	OK	0
Got two equal A inputs	OK					1
				S*		

Figure 7-48 Evolution of a state table.

process, but we've written them in Karnaugh-map order to simplify the derivation of excitation equations later. In a Mealy machine we would omit the output column and write the output values along with the next-state values under each input combination. The leftmost column is simply an English-language reminder of the meaning of each state or the "history" associated with it.

The word description isn't specific about what happens when this machine is first started, so we'll just have to improvise. We'll assume that when power is first applied to the system, the machine enters an *initial state*, called INIT in this example. We write the name of the initial state (INIT) in the first row and leave room for enough rows (states) to complete the design. We can also fill in the value of Z for the INIT state; common sense says it should be 0 because there were *no* inputs beforehand.

Next, we must fill in the next-state entries for the INIT row. The Z output can't be 1 until we've seen at least two inputs on A, so we'll provide two states, A0 and A1, that "remember" the value of A on the previous clock tick, as shown in Figure 7-48(b). In both of these states Z is 0, since we haven't satisfied the conditions for a 1 output yet. The precise meaning of state A0 is "Got A = 0 on the previous tick, A ≠ 0 on the tick before that, and B ≠ 1 at some time since the previous pair of equal A inputs." State A1 is defined similarly.

At this point we know that our state machine has at least three states, and we have created two more blank rows to fill in. Hmmmm, this isn't such a good trend! In order to fill in the next-state entries for *one* state (INIT), we had to create *two* new states A0 and A1. If we kept going this way, we could end up with 65,535 states by bedtime! Instead, we should be on the lookout for existing states

initial state

(a)

Meaning	S	A B				Z
		00	01	11	10	
Initial state	INIT	A0	A0	A1	A1	0
Got a 0 on A	A0	OK	OK	A1	A1	0
Got a 1 on A	A1	A0	A0	OK	OK	0
Got two equal A inputs	OK	?	OK	OK	?	1
				S*		

(b)

Meaning	S	A B				Z
		00	01	11	10	
Initial state	INIT	A0	A0	A1	A1	0
Got a 0 on A	A0	OK0	OK0	A1	A1	0
Got a 1 on A	A1	A0	A0	OK1	OK1	0
Two equal, A=0 last	OK0					1
Two equal, A=1 last	OK1					1
				S*		

(c)

Meaning	S	A B				Z
		00	01	11	10	
Initial state	INIT	A0	A0	A1	A1	0
Got a 0 on A	A0	OK0	OK0	A1	A1	0
Got a 1 on A	A1	A0	A0	OK1	OK1	0
Two equal, A=0 last	OK0	OK0	OK0	OK1	A1	1
Two equal, A=1 last	OK1					1
				S*		

(d)

Meaning	S	A B				Z
		00	01	11	10	
Initial state	INIT	A0	A0	A1	A1	0
Got a 0 on A	A0	OK0	OK0	A1	A1	0
Got a 1 on A	A1	A0	A0	OK1	OK1	0
Two equal, A=0 last	OK0	OK0	OK0	OK1	A1	1
Two equal, A=1 last	OK1	A0	OK0	OK1	OK1	1
				S*		

Figure 7-49 Continued evolution of a state table.

that have the same meaning as new ones that we might otherwise create. Let's see how it goes.

In state A0, we know that input A was 0 at the previous clock tick. Therefore, if A is 0 again, we go to a new state OK with Z = 1, as shown in Figure 7-48(c). If A is 1, then we don't have two equal inputs in a row, so we go to state A1 to remember that we just got a 1. Likewise in state A1, shown in (d), we go to OK if we get a second 1 input in a row, or to A0 if we get a 0.

Once we get into the OK state, the machine description tells us we can stay there as long as B = 1, irrespective of the A input, as shown in Figure 7-49(a). If B = 0, we have to look for two 1s or two 0s in a row on A again. However, we've got a little problem in this case. The current A input may or may not be the second equal input in a row, so we may still be "OK" or we may have to go back to A0 or A1. We defined the OK state too broadly—it doesn't "remember" enough to tell us which way to go.

The problem is solved in Figure 7-49(b) by splitting OK into two states, OK0 and OK1, that "remember" the previous A input. All of the next states for OK0 and OK1 can be selected from existing states, as shown in (c) and (d). For example, if we get A = 0 in OK0, we can just stay in OK0; we don't have to create a new state that "remembers" three 0s in a row, because the machine's description doesn't require us to distinguish that case. Thus, we have achieved "closure" of the state table, which now describes a *finite*-state machine. As a sanity check, Figure 7-50 repeats the timing diagram of Figure 7-47, listing the states that should be visited according to our final state table.

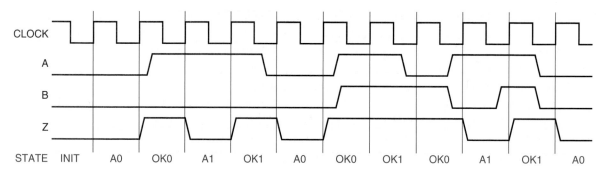

Figure 7-50 Timing diagram and state sequence for example state machine.

7.4.2 State Minimization

Figure 7-49(d) is a "minimal" state table for our original word description, in the sense that it contains the fewest possible states. However, Figure 7-51 shows other state tables, with more states, that also do the job. Formal procedures can be used to minimize the number of states in such tables.

equivalent states

The basic idea of formal minimization procedures is to identify *equivalent states*, where two states are equivalent if it is impossible to distinguish them by observing only the current and future *outputs* of the machine (and *not* the internal state variables). A pair of equivalent states can be replaced by a single state.

Two states S1 and S2 are equivalent if two conditions are true. First, S1 and S2 must produce the same values at the state-machine output(s); in a Mealy machine, this must be true for all input combinations. Second, for each input combination, S1 and S2 must have either the same next state or equivalent next states.

Thus, a formal state-minimization procedure shows that states OK00 and OKA0 in Figure 7-51(a) are equivalent because they produce the same output and their next-state entries are identical. Since the states are equivalent, state

Figure 7-51 Nonminimal state tables equivalent to Figure 7-49(d).

(a)

Meaning	S	A B 00	01	11	10	Z
Initial state	INIT	A0	A0	A1	A1	0
Got a 0 on A	A0	OK00	OK00	A1	A1	0
Got a 1 on A	A1	A0	A0	OK11	OK11	0
Got 00 on A	OK00	OK00	OK00	OKA1	A1	1
Got 11 on A	OK11	A0	OKA0	OK11	OK11	1
OK, got a 0 on A	OKA0	OK00	OK00	OKA1	A1	1
OK, got a 1 on A	OKA1	A0	OKA0	OK11	OK11	1
		S*				

(b)

Meaning	S	A B 00	01	11	10	Z
Initial state	INIT	A0	A0	A1	A1	0
Got a 0 on A	A0	OK00	OK00	A1	A1	0
Got a 1 on A	A1	A0	A0	OK11	OK11	0
Got 00 on A	OK00	OK00	OK00	A001	A1	1
Got 11 on A	OK11	A0	A110	OK11	OK11	1
Got 001 on A, B=1	A001	A0	AE10	OK11	OK11	1
Got 110 on A, B=1	A110	OK00	OK00	AE01	A1	1
Got bb...10 on A, B=1	AE10	OK00	OK00	AE01	A1	1
Got bb...01 on A, B=1	AE01	A0	AE10	OK11	OK11	1
			S*			

OK00 may be eliminated and its occurrences in the table replaced by OKA0, or vice versa. Likewise, states OK11 and OKA1 are equivalent.

To minimize the state table in Figure 7-51(b), a formal procedure must use a bit of circular reasoning. States OK00, A110, and AE10 all produce the same output and have almost identical next-state entries, so they might be equivalent. They are equivalent only if A001 and AE01 are equivalent. Similarly, OK11, A001, and AE01 are equivalent only if A110 and AE10 are equivalent. In other words, the states in the first set are equivalent if the states in the second set are, and vice versa. So, let's just go ahead and say they're equivalent.

IS THIS REALLY ALL NECESSARY?	Details of formal state-minimization procedures are discussed in advanced textbooks, cited in the References. However, these procedures are seldom used by most digital designers. By carefully matching state meanings to the requirements of the problem, experienced digital designers produce state tables for small problems with a minimal or near-minimal number of states, without using a formal minimization procedure. Also, there are situations where *increasing* the number of states may simplify the design or reduce its cost, so even an automated state-minimization procedure doesn't necessarily help. A designer can do more to simplify a state machine during the state-assignment phase of the design, discussed in the next subsection.

7.4.3 State Assignment

The next step in the design process is to determine how many binary variables are required to represent the states in the state table, and to assign a specific combination to each named state. We'll call the binary combination assigned to a particular state a *coded state*. The *total number of states* in a machine with n flip-flops is 2^n, so the number of flip-flops needed to code s states is $\lceil \log_2 s \rceil$, the smallest integer greater than or equal to $\log_2 s$.

coded state

total number of states

For reference, the state/output table of our example machine is repeated in Table 7-6. It has five states, so it requires three flip-flops. Of course, three flip-flops provide a total of eight states, so there will be $8 - 5 = 3$ *unused states*. We'll discuss alternatives for handling the unused states at the end of this subsection. Right now, we have to deal with lots of choices for the five coded states.

unused states

INITIAL VERSUS IDLE STATES	The example state machine in this subsection visits its initial state only during reset. Many machines are designed instead with an "idle" state that is entered both at reset and whenever the machine has nothing in particular to do.

Table 7-6
State and output table
for example problem.

S	00	01	11	10	Z
INIT	A0	A0	A1	A1	0
A0	OK0	OK0	A1	A1	0
A1	A0	A0	OK1	OK1	0
OK0	OK0	OK0	OK1	A1	1
OK1	A0	OK0	OK1	OK1	1
			S*		

Table header spans columns 00, 01, 11, 10 under **A B**.

The simplest assignment of s coded states to 2^n possible states is to use the first s binary integers in binary counting order, as shown in the first assignment column of Table 7-7. However, the simplest state assignment does not always lead to the simplest excitation equations, output equations, and resulting logic circuit. In fact, the state assignment often has a major effect on circuit cost, and it may interact with other factors, such as the choice of storage elements (e.g., D vs. J-K flip-flops) and the realization approach for excitation and output logic (e.g., a sum of products, a product of sums, or an ad hoc design).

So, how do we choose the best state assignment for a given problem? In general, the only formal way to find the *best* assignment is to try *all* the assignments. That's too much work, even for students. Instead, most digital designers rely on experience and several practical guidelines for making reasonable state assignments:

- Choose an initial coded state into which the machine can easily be forced at reset (00...00 or 11...11 in typical circuits).
- Minimize the number of state variables that change on each transition.

CAUTION: MATH

The number of different ways to choose m coded states out of a set of n possible states is given by a *binomial coefficient*, denoted $\binom{n}{m}$, whose value is $\dfrac{n!}{m! \cdot (n-m)!}$. (We used binomial coefficients previously in Section 2.10 in the context of decimal coding.) In our example, there are $\binom{8}{5}$ different ways to choose five coded states out of eight possible states, and 5! ways to assign the five named states to each different choice. So there are $\dfrac{8!}{5! \cdot 3!} \cdot 5!$ or 6,720 different ways to assign the five states of our example machine to combinations of three binary state variables. We don't have time to look at all of them.

State Name	Assignment			
	Simplest Q1–Q3	Decomposed Q1–Q3	One-hot Q1–Q5	Almost One-hot Q1–Q4
INIT	000	000	00001	0000
A0	001	100	00010	0001
A1	010	101	00100	0010
OK0	011	110	01000	0100
OK1	100	111	10000	1000

Table 7-7
Possible state assignments for the state machine in Table 7-6.

- Maximize the number of state variables that don't change in a group of related states (i.e., a group of states in which most of transitions stay in the group).

- Exploit symmetries in the problem specification and the corresponding symmetries in the state table. That is, suppose that one state or group of states means almost the same thing as another. Once an assignment has been established for the first, a similar assignment, differing only in one bit, should be used for the second.

- If there are unused states (i.e., if $s < 2^n$ where $n = \lceil \log_2 s \rceil$), then choose the "best" of the available state-variable combinations to achieve the foregoing goals. That is, don't limit the choice of coded states to the first s n-bit integers.

- Decompose the set of state variables into individual bits or fields, where each bit or field has a well-defined meaning with respect to the input effects or output behavior of the machine.

- Consider using more than the minimum number of state variables to make a decomposed assignment possible.

Some of these ideas are incorporated in the "decomposed" state assignment in Table 7-7. As before, the initial state is 000, which is easy to force either asynchronously (applying the RESET signal to the flip-flop CLR inputs) or synchronously (by ANDing RESET' with all of the D flip-flop inputs). After this point, the assignment takes advantage of the fact that there are only four states in addition to INIT, which is a fairly "special" state that is never reentered once the machine gets going. Therefore, Q1 can be used to indicate whether or not the machine is in the INIT state, and Q2 and Q3 can be used to distinguish among the four non-INIT states.

The non-INIT states in the "decomposed" column of Table 7-7 appear to have been assigned in binary counting order, but that's just a coincidence. State bits Q2 and Q3 actually have individual meanings in the context of the state machine's inputs and output. Q3 gives the previous value of A, and Q2 indicates that the conditions for a 1 output are satisfied in the current state. By decomposing the state-bit meanings in this way, we can expect the next-state and output logic to be simpler than in a "random" assignment of Q2,Q3 combinations to the non-INIT states. We'll continue the state-machine design based on this assignment in later subsections.

one-hot assignment

Another useful state assignment, one that can be adapted to any state machine, is the *one-hot assignment* shown in Table 7-7. This assignment uses more than the minimum number of state variables—it uses one bit per state. In addition to being simple, a one-hot assignment has the advantage of usually leading to small excitation equations, since each flip-flop must be set to 1 for transitions into only one state. An obvious disadvantage of a one-hot assignment, especially for machines with many states, is that it requires (many) more than the minimum number of flip-flops. However, the one-hot encoding is ideal for a machine with s states that is required to have a set of 1-out-of-s coded outputs indicating its current state. The one-hot-coded flip-flop outputs can be used directly for this purpose, with no additional combinational output logic.

The last column of Table 7-7 is an "almost one-hot assignment" that uses the "no-hot" combination for the initial state. This makes a lot of sense for two reasons: it's easy to initialize most storage devices to the all-0s state, and the initial state in this machine is never revisited once the machine gets going. Completing the state-machine design using this state assignment is considered in Exercises 7.37 and 7.40.

unused states

We promised earlier to consider the disposition of *unused states* when the number of states available with n flip-flops, 2^n, is greater than the number of states required, s. There are two approaches that make sense, depending on the application requirements:

- *Minimal risk.* This approach assumes that it is possible for the state machine somehow to get into one of the unused (or "illegal") states, perhaps because of a hardware failure, an unexpected input, or a design error. Therefore, all of the unused state-variable combinations are identified and explicit next-state entries are made so that, for any input combination, the unused states go to the "initial" state, the "idle" state, or some other "safe" state. This is an automatic consequence of some design methodologies if the initial state is coded 00. . . 00.

- *Minimal cost.* This approach assumes that the machine will never enter an unused state. Therefore, in the transition and excitation tables, the next-state entries of the unused states can be marked as "don't-cares." In most cases this simplifies the excitation logic. However, the machine's behavior if it ever does enter an unused state may be pretty weird.

We'll look at both of these approaches as we complete the design of our example state machine.

7.4.4 Synthesis Using D Flip-Flops

Once we've assigned coded states to the named states of a machine, the rest of the design process is pretty much "turning the crank." In fact, in Section 7.11.2 we'll describe software tools that can turn the crank for you. Just so that you'll appreciate those tools, however, we'll go through the process by hand in this subsection.

Coded states are substituted for named states in the (possibly minimized) state table to obtain a *transition table*. The transition table shows the next coded state for each combination of current coded state and input. Table 7-8 shows the transition and output table that is obtained from the example state machine of Table 7-6 on page 570 using the "decomposed" assignment of Table 7-7 on page 571.

transition table

The next step is to write an *excitation table* that shows, for each combination of coded state and input, the flip-flop excitation input values needed to make the machine go to the desired next coded state. This structure and content of this table depend on the type of flip-flops that are used (D, J-K, T, etc.). We *usually* have a particular flip-flop type in mind at the beginning of a design—and we *certainly* do in this subsection, given its title. In fact, most state-machine designs nowadays use D flip-flops, because of their availability in both discrete packages and programmable logic devices, and because of their ease of use (compare with J-K flip-flops in the next subsection).

excitation table

Of all flip-flop types, a D flip-flop has the simplest characteristic equation, $Q* = D$. Each D flip-flop in a state machine has a single excitation input, D, and the excitation table must show the value required at each flip-flop's D input for each coded-state/input combination. Table 7-9 shows the excitation table for our example problem. Since $D = Q*$, the excitation table is identical to the transition table, except for labeling of its entries. Thus, with D flip-flops, you don't really need to write a separate excitation table; you can just call the first table a *transition/excitation table*.

transition/excitation table

The excitation table is like a truth table for three combinational logic functions (D1, D2, D3) of five variables (A, B, Q1, Q2, Q3). Accordingly, we can

Q1 Q2 Q3	AB				Z
	00	01	11	10	
000	100	100	101	101	0
100	110	110	101	101	0
101	100	100	111	111	0
110	110	110	111	101	1
111	100	110	111	111	1
	Q1* Q2* Q3*				

Table 7-8
Transition and output table for example problem.

Table 7-9
Excitation and output table for Table 7-8 using D flip-flops.

Q1 Q2 Q3	AB 00	01	11	10	Z
000	100	100	101	101	0
100	110	110	101	101	0
101	100	100	111	111	0
110	110	110	111	101	1
111	100	110	111	111	1
	D1 D2 D3				

excitation maps

design circuits to realize these functions using any of the combinational design methods at our disposal. In particular, we can transfer the information in the excitation table to Karnaugh maps, which we may call *excitation maps*, and find a minimal sum-of-products or product-of-sums expression for each function.

5-variable Karnaugh map

Excitation maps for our example state machine are shown in Figure 7-52. Each function, such as D1, has five variables and therefore uses a *5-variable Karnaugh map*. A 5-variable map is drawn as a pair of 4-variable maps, where cells in the same position in the two maps are considered to be adjacent. These maps are a bit unwieldy, but if you want to design by hand any but the most trivial state machines, you're going to get stuck with 5-variable maps and worse. At least we had the foresight to label the input combinations of the original state table in Karnaugh-map order, which makes it easier to transfer information to the maps in this step. However, note that the *states* were not assigned in Karnaugh-map order; in particular, the rows for states 110 and 111 are in the opposite order in the map as in the excitation table.

Figure 7-52
Excitation maps for D1, D2, and D3 assuming that unused states go to state 000.

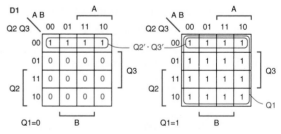

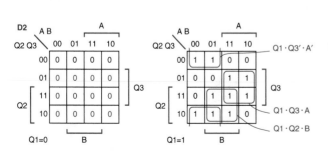

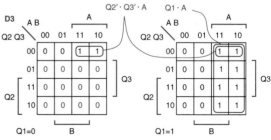

It is in this step, transferring the excitation table to excitation maps, that we discover why the excitation table is not quite a truth table—it does not specify functional values for *all* input combinations. In particular, the next-state information for the unused states, 001, 010, and 011, is not specified. Here we must make a choice, discussed in the preceding subsection, between a minimal-risk and a minimal-cost strategy for handling the unused states. Figure 7-52 has taken the minimal-risk approach: The next state for each unused state and input combination is 000, the INIT state. The three rows of colored 0s in each Karnaugh map are the result of this choice. With the maps completely filled in, we can now obtain minimal sum-of-products expressions for the flip-flop excitation inputs:

$$D1 = Q1 + Q2' \cdot Q3'$$
$$D2 = Q1 \cdot Q3' \cdot A' + Q1 \cdot Q3 \cdot A + Q1 \cdot Q2 \cdot B$$
$$D3 = Q1 \cdot A + Q2' \cdot Q3' \cdot A$$

An output equation can easily be developed directly from the information in Table 7-9. The output equation is simpler than the excitation equations, because the output is a function of state only. We could use a Karnaugh map, but it's easy to find a minimal-risk output function algebraically, by writing it as the sum of the two coded states (110 and 111) in which Z is 1:

$$Z = Q1 \cdot Q2 \cdot Q3' + Q1 \cdot Q2 \cdot Q3$$
$$= Q1 \cdot Q2$$

At this point, we're just about done with the state-machine design. If the state machine is going to be built with discrete flip-flops and gates, then the final step is to draw a logic diagram. On the other hand, if we are using a programmable logic device, then we only have to enter the excitation and output equations into a computer file that specifies how to program the device, as an example shows in Section 7.11.1. Or, if we were thinking ahead, we specified the machine using a state-machine description language like ABEL in the first place (Section 7.11.2), and the computer did all the work in this subsection for us!

MINIMAL-COST SOLUTION

If we choose in our example to derive minimal-cost excitation equations, we write "don't-cares" in the next-state entries for the unused states. The colored d's in Figure 7-53 are the result of this choice. The excitation equations obtained from this map are somewhat simpler than before:

$$D1 = 1$$
$$D2 = Q1 \cdot Q3' \cdot A' + Q3 \cdot A + Q2 \cdot B$$
$$D3 = A$$

For a minimal-cost output function, the value of Z is a "don't-care" for the unused states. This leads to an even simpler output function, $Z = Q2$. The logic diagram for the minimal-cost solution is shown in Figure 7-54.

Figure 7-53
Excitation maps for D1, D2, and D3 assuming that next states of unused states are "don't-cares."

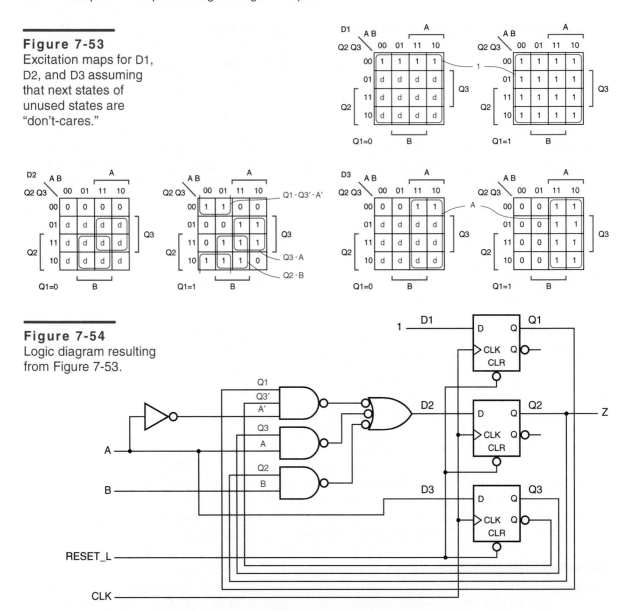

Figure 7-54
Logic diagram resulting from Figure 7-53.

*7.4.5 Synthesis Using J-K Flip-Flops

At one time, J-K flip-flops were popular for discrete SSI state-machine designs, since a J-K flip-flop embeds more functionality than a D flip-flop in the same size SSI package. By "more functionality" we mean that the combination of J and K inputs yields more possibilities for controlling the flip-flop than a single D input does. As a result, a state machine's excitation logic may be simpler using J-K flip-flops than using D flip-flops, which reduced package count when SSI gates were used for the excitation logic.

JUST FOR FUN While minimizing excitation logic was a big deal in the days of SSI-based design, the name of the game has changed with PLDs and ASICs. As you might guess from your knowledge of the AND-OR structure of combinational PLDs, the need to provide separate AND-OR arrays for the J and K inputs of a J-K flip-flop would be a distinct disadvantage in a sequential PLD.

In ASIC technologies, J-K flip-flops aren't so desirable either. For example, in LSI Logic Corp.'s LCA500K series of CMOS gate arrays, an FD1QP D flip-flop macrocell uses 7 "gate cells," while an FJK1QP J-K flip-flop macrocell uses 9 gate cells, over 25% more chip area. Therefore, a more cost-effective design usually results from sticking with D flip-flops and using the extra chip area for more complex excitation logic in just the cases where it's really needed.

Still, this subsection describes the J-K synthesis process "just for fun."

Up through the state-assignment step, the design procedure with J-K flip-flops is basically the same as with D flip-flops. The only difference is that a designer might select a slightly different state assignment, knowing the sort of behavior that can easily be obtained from J-K flip-flops (e.g., "toggling" by setting J and K to 1).

The big difference occurs in the derivation of an excitation table from the transition table. With D flip-flops, the two tables are identical; using the D's characteristic equation, $Q* = D$, we simply substitute $D = Q*$ for each entry. With J-K flip-flops, each entry in the excitation table has twice as many bits as in the transition table, since there are two excitation inputs per flip-flop.

A J-K flip-flop's characteristic equation, $Q* = J \cdot Q' + K' \cdot Q$, cannot be rearranged to obtain independent equations for J and K. Instead, the required values for J and K are expressed as functions of Q and Q* in a J-K *application* *J-K application table* table, Table 7-10. According to the first row, if Q is currently 0, all that is required to obtain 0 as the next value of Q is to set J to 0; the value of K doesn't matter. Similarly, according to the third row, if Q is currently 1, the next value of Q will be 0 if K is 1, regardless of J's value. Each desired transition can be obtained by either of two different combinations on the J and K inputs, so we get a "don't-care" entry in each row of the application table.

To obtain a J-K excitation table, the designer must look at both the current and desired next value of each state bit in the transition table and substitute the

Q	Q*	J	K
0	0	0	d
0	1	1	d
1	0	d	1
1	1	d	0

Table 7-10
Application table for
J-K flip-flops.

corresponding pair of J and K values from the application table. For the transition table in Table 7-8 on page 573, these substitutions produce the excitation table in Table 7-11. For example, in state 100 under input combination 00, Q1 is 1 and the required Q1* is 1; therefore, "d0" is entered for J1 K1. For the same state/input combination, Q2 is 0 and Q2* is 1, so "1d" is entered for J2 K2. Obviously, it takes quite a bit of patience and care to fill in the entire excitation table (a job best left to a computer).

As in the D synthesis example of the preceding subsection, the excitation table is *almost* a truth table for the excitation functions. This information is transferred to Karnaugh maps in Figure 7-55.

The excitation table does not specify next states for the unused states, so once again we must choose between the minimal-risk and minimal-cost approaches. The colored entries in the Karnaugh maps result from taking the minimal-risk approach.

Figure 7-55 Excitation maps for J1, K1, J2, K2, J3, and K3, assuming that unused states go to state 000.

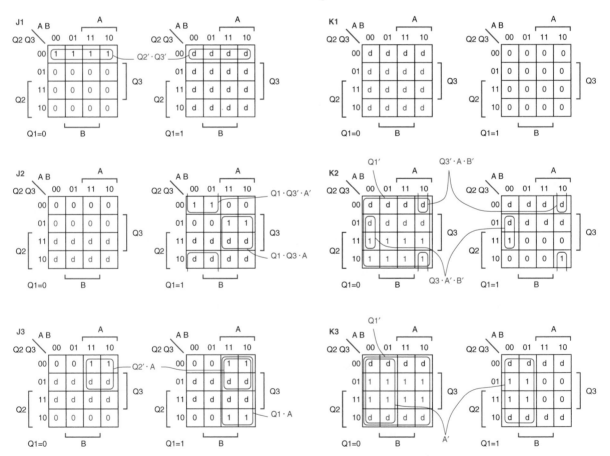

		A B			
Q1 Q2 Q3	00	01	11	10	Z
000	1d, 0d, 0d	1d, 0d, 0d	1d, 0d, 1d	1d, 0d, 1d	0
100	d0, 1d, 0d	d0, 1d, 0d	d0, 0d, 1d	d0, 0d, 1d	0
101	d0, 0d, d1	d0, 0d, d1	d0, 1d, d0	d0, 1d, d0	0
110	d0, d0, 0d	d0, d0, 0d	d0, d0, 1d	d0, d1, 1d	1
111	d0, d1, d1	d0, d0, d1	d0, d0, d0	d0, d0, d0	1
		J1 K1, J2 K2, J3 K3			

Table 7-11
Excitation and output table for the state machine of Table 7-8, using J-K flip-flops.

Note that even though the "safe" next state for unused states is 000, we didn't just put 0s in the corresponding map cells, as we were able to do in the D case. Instead, we still had to work with the application table to determine the proper combination of J and K needed to get $Q* = 0$ for each unused state entry, once again a tedious and error-prone process.

Using the maps in Figure 7-55, we can derive sum-of-products excitation equations:

$$J1 = Q2' \cdot Q3' \qquad\qquad K1 = 0$$
$$J2 = Q1 \cdot Q3' \cdot A' + Q1 \cdot Q3 \cdot A \qquad K2 = Q1' + Q3' \cdot A \cdot B' + Q3 \cdot A' \cdot B'$$
$$J3 = Q2' \cdot A + Q1 \cdot A \qquad\qquad K3 = Q1' + A'$$

These equations take two more gates to realize than do the preceding subsection's minimal-risk equations using D flip-flops, so J-K flip-flops didn't save us anything in this example, least of all design time.

MINIMAL-COST SOLUTION In the preceding design example, excitation maps for the minimal-cost approach would have been somewhat easier to construct, since we could have just put d's in all of the unused state entries. Sum-of-products excitation equations obtained from the minimal-cost maps (not shown) are as follows:

$$J1 = 1 \qquad\qquad K1 = 0$$
$$J2 = Q1 \cdot Q3' \cdot A' + Q3 \cdot A \qquad K2 = Q3' \cdot A \cdot B' + Q3 \cdot A' \cdot B'$$
$$J3 = A \qquad\qquad K3 = A'$$

The state encoding for the J-K circuit is the same as in the D circuit, so the output equation is the same, $Z = Q1 \cdot Q2$ for minimal risk, $Z = Q2$ for minimal cost.

A logic diagram corresponding to the minimal-cost equations is shown in Figure 7-56. This circuit has two more gates than the minimal-cost D circuit in Figure 7-54, so J-K flip-flops still didn't save us anything.

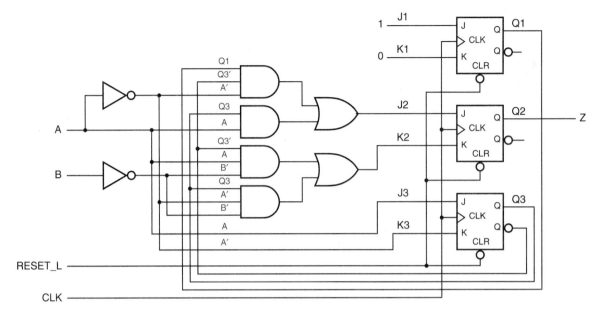

Figure 7-56 Logic diagram for example state machine using J-K flip-flops and minimal-cost excitation logic.

7.4.6 More Design Examples Using D Flip-Flops

We'll conclude this section with two more state-machine design examples using D flip-flops. The first example is a "1s-counting machine":

> Design a clocked synchronous state machine with two inputs, X and Y, and one output, Z. The output should be 1 if the number of 1 inputs on X and Y since reset is a multiple of 4, and 0 otherwise.

At first glance, you might think the machine needs an infinite number of states, since it counts 1 inputs over an arbitrarily long time. However, since the output indicates the number of inputs received *modulo 4*, four states are sufficient. We'll name them S0–S3, where S0 is the initial state and the total number of 1s received in S*i* is *i* modulo 4. Table 7-12 is the resulting state and output table.

Table 7-12
State and output table for 1s-counting machine.

				X Y		
Meaning	*S*	*00*	*01*	*11*	*10*	*Z*
Got zero 1s (modulo 4)	S0	S0	S1	S2	S1	1
Got one 1 (modulo 4)	S1	S1	S2	S3	S2	0
Got two 1s (modulo 4)	S2	S2	S3	S0	S3	0
Got three 1s (modulo 4)	S3	S3	S0	S1	S0	0
				S*		

Q1 Q2	XY				Z
	00	01	11	10	
00	00	01	11	01	1
01	01	11	10	11	0
11	11	10	00	10	0
10	10	00	01	00	0
	Q1* Q2* or D1 D2				

Table 7-13
Transition/excitation and output table for 1s-counting machine.

The 1s-counting machine can use two state variables to code its four states, with no unused states. In this case, there are only 4! possible assignments of coded states to named states. Still, we'll try only one of them. We'll assign coded states to the named states in Karnaugh-map order (00, 01, 11, 10) for two reasons: in this state table, it minimizes the number of state variables that change for most transitions, potentially simplifying the excitation equations; and it simplifies the mechanical transfer of information to excitation maps.

A transition/excitation table based on our chosen state assignment is shown in Table 7-13. Since we're using D flip-flops, the transition and excitation tables are the same. Corresponding Karnaugh maps for D1 and D2 are shown in Figure 7-57. Since there are no unused states, all of the information we need is in the excitation table; no choice is required between minimal-risk and minimal-cost approaches. The excitation equations can be read from the maps, and the output equation can be read directly from the transition/excitation table:

$$D1 = Q2 \cdot X' \cdot Y + Q1' \cdot X \cdot Y + Q1 \cdot X' \cdot Y' + Q2 \cdot X \cdot Y'$$
$$D2 = Q1' \cdot X' \cdot Y + Q1' \cdot X \cdot Y' + Q2 \cdot X' \cdot Y' + Q2' \cdot X \cdot Y$$
$$Z = Q1' \cdot Q2'$$

A logic diagram using D flip-flops and AND-OR or NAND-NAND excitation logic can be drawn from these equations.

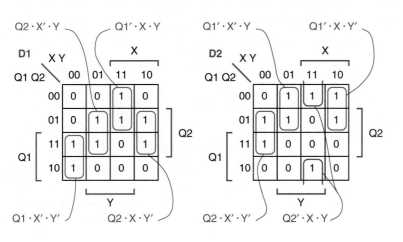

Figure 7-57
Excitation maps for D1 and D2 inputs in 1s-counting machine.

The second example is a "combination lock" state machine that activates an "unlock" output when a certain binary input sequence is received:

Design a clocked synchronous state machine with one input, X, and two outputs, UNLK and HINT. The UNLK output should be 1 if and only if X is 0 and the sequence of inputs received on X at the preceding seven clock ticks was 0110111. The HINT output should be 1 if and only if the current value of X is the correct one to move the machine closer to being in the "unlocked" state (with UNLK = 1).

It should be apparent from the word description that this is a Mealy machine. The UNLK output depends on both the past history of inputs and X's current value, and HINT depends on both the state and the current X (indeed, if the current X produces HINT = 0, then the clued-in user will want to change X before the clock tick).

A state and output table for the combination lock is presented in Table 7-14. In the initial state, A, we assume that we have received no inputs in the required sequence; we're looking for the first 0 in the sequence. Therefore, as long as we get 1 inputs, we stay in state A, and we move to state B when we receive a 0. In state B, we're looking for a 1. If we get it, we move on to C; if we don't, we can stay in B, since the 0 we just received might still turn out to be the first 0 in the required sequence. In each successive state, we move on to the next state if we get the correct input, and we go back to A or B if we get the wrong one. An exception occurs in state G; if we get the wrong input (a 0) there, the previous three inputs might still turn out to be the first three inputs of the required sequence, so we go back to state E instead of B. In state H, we've received the required sequence, so we set UNLK to 1 if X is 0. In each state, we set HINT to 1 for the value of X that moves us closer to state H.

Table 7-14
State and output table for combination-lock machine.

		X	
Meaning	*S*	*0*	*1*
Got zip	A	B, 01	A, 00
Got 0	B	B, 00	C, 01
Got 01	C	B, 00	D, 01
Got 011	D	E, 01	A, 00
Got 0110	E	B, 00	F, 01
Got 01101	F	B, 00	G, 01
Got 011011	G	E, 00	H, 01
Got 0110111	H	B, 11	A, 00
		S*, UNLK HINT	

Table 7-15
Transition/excitation table for combination-lock machine.

Q1 Q2 Q3	X 0	X 1
000	001, 01	000, 00
001	001, 00	010, 01
010	001, 00	011, 01
011	100, 01	000, 00
100	001, 00	101, 01
101	001, 00	110, 01
110	100, 00	111, 01
111	001, 11	000, 00
	Q1*Q2*Q3*, UNLK HINT	

The combination lock's eight states can be coded with three state variables, leaving no unused states. There are 8! state assignments to choose from. To keep things simple, we'll use the simplest, and assign the states in binary counting order, yielding the transition/excitation table in Table 7-15. Corresponding Karnaugh maps for D1, D2, and D3 are shown in Figure 7-58. The excitation equations can be read from the maps:

$$D1 = Q1 \cdot Q2' \cdot X + Q1' \cdot Q2 \cdot Q3 \cdot X' + Q1 \cdot Q2 \cdot Q3'$$
$$D2 = Q2' \cdot Q3 \cdot X + Q2 \cdot Q3' \cdot X$$
$$D3 = Q1 \cdot Q2' \cdot Q3' + Q1 \cdot Q3 \cdot X' + Q2' \cdot X' + Q3' \cdot Q1' \cdot X' + Q2 \cdot Q3' \cdot X$$

Figure 7-58 Excitation maps for D1, D2, and D3 in combination-lock machine.

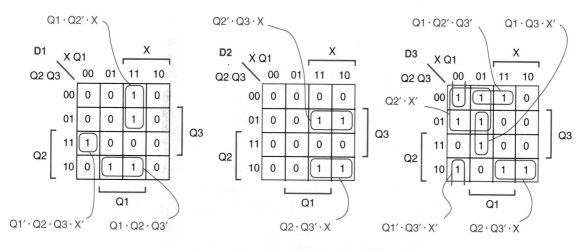

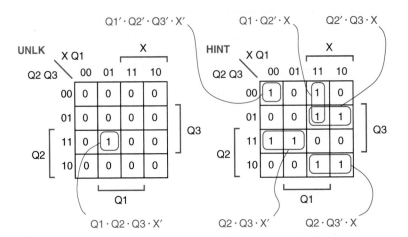

Figure 7-59
Karnaugh maps for
output functions
UNLK and HINT in
combination-lock
machine.

The output values are transferred from the transition/excitation and output table to another set of maps in Figure 7-59. The corresponding output equations are:

$$UNLK = Q1 \cdot Q2 \cdot Q3 \cdot X'$$

$$HINT = Q1' \cdot Q2' \cdot Q3' \cdot X' + Q1 \cdot Q2' \cdot X + Q2' \cdot Q3 \cdot X + Q2 \cdot Q3 \cdot X' + Q2 \cdot Q3' \cdot X$$

Note that some product terms are repeated in the excitation and output equations, yielding a slight savings in the cost of the AND-OR realization. If we went through the trouble of performing a formal multiple-output minimization of all five excitation and output functions, we could save two more gates (see Exercise 7.55).

7.5 Designing State Machines Using State Diagrams

Aside from planning the overall architecture of a digital system, designing state machines is probably the most creative task of a digital designer. Most people like to take a graphical approach to design—you've probably solved many problems just by doodling. For that reason, state diagrams are often used to design small- to medium-sized state machines. In this section we'll give examples of state-diagram design and describe a simple procedure for synthesizing circuits from the state diagrams. This procedure is the basis of the method used by CAD tools that can synthesize logic from graphical or even text-based "state diagrams."

Designing a state diagram is much like designing a state table, which, as we showed in Section 7.4.1, is much like writing a program. However, there is one fundamental difference between a state diagram and a state table, a difference that makes state-diagram design simpler but also more error prone:

- A state table is an exhaustive listing of the next states for each state/input combination. No ambiguity is possible.

- A state diagram contains a set of arcs labeled with transition expressions. Even when there are many inputs, only one transition expression is required per arc. However, when a state diagram is constructed, there is no guarantee that the transition expressions written on the arcs leaving a particular state cover all the input combinations exactly once.

In an improperly constructed (*ambiguous*) state diagram, the next state for some input combinations may be unspecified, which is generally undesirable, while multiple next states may be specified for others, which is just plain wrong. Thus, considerable care must be taken in the design of state diagrams; we'll give several examples.

ambiguous state diagram

Our first example is a state machine that controls the tail lights of a 1965 Ford Thunderbird, shown in Figure 7-60. There are three lights on each side, and for turns they operate in sequence to show the turning direction, as illustrated in Figure 7-61. The state machine has two input signals, LEFT and RIGHT, that carry the driver's request for a left turn or a right turn. It also has an emergency-flasher input, HAZ, that requests the tail lights to be operated in hazard mode—all six lights flashing on and off in unison. We also assume the existence of a free-running clock signal whose frequency equals the desired flashing rate for the lights.

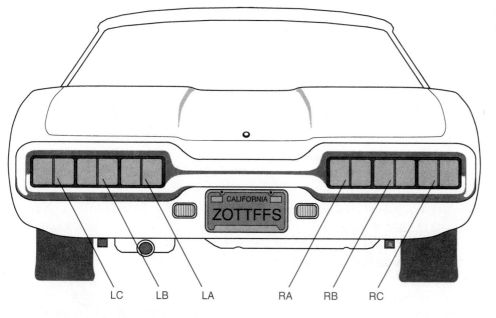

Figure 7-60
T-bird tail lights.

LC   LB   LA   RA   RB   RC

WHOSE REAR END?   Actually, Figure 7-60 looks more like the rear end of a Mercury Capri, which also had sequential tail lights.

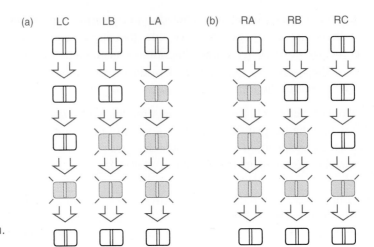

Figure 7-61
Flashing sequence for
T-bird tail lights:
(a) left turn; (b) right turn.

Figure 7-62
Initial state diagram
and output table for
T-bird tail lights.

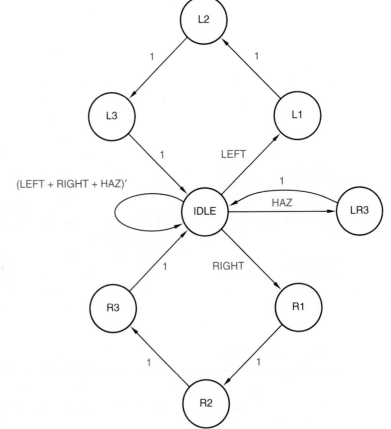

Output Table

State	LC	LB	LA	RA	RB	RC
IDLE	0	0	0	0	0	0
L1	0	0	1	0	0	0
L2	0	1	1	0	0	0
L3	1	1	1	0	0	0
R1	0	0	0	1	0	0
R2	0	0	0	1	1	0
R3	0	0	0	1	1	1
LR3	1	1	1	1	1	1

Given the foregoing requirements, we can design a clocked synchronous state machine to control the T-bird tail lights. We will design a Moore machine, so that the state alone determines which lights are on and which are off. For a left turn, the machine should cycle through four states in which the righthand lights are off and 0, 1, 2, or 3 of the lefthand lights are on. Likewise, for a right turn, it should cycle through four states in which the lefthand lights are off and 0, 1, 2, or 3 of the righthand lights are on. In hazard mode, only two states are required—all lights on and all lights off.

Figure 7-62 shows our first cut at a state diagram for the machine. A common IDLE state is defined in which all of the lights are off. When a left turn is requested, the machine goes through three states in which 1, 2, and 3 of the lefthand lights are on, and then back to IDLE; right turns work similarly. In the hazard mode, the machine cycles back and forth between the IDLE state and a state in which all six lights are on. Since there are so many outputs, we've included a separate output table rather than writing output values on the state diagram. Even without assigning coded states to the named states, we can write output equations from the output table, if we let each state name represent a logic expression that is 1 only in that state:

$$
\begin{aligned}
\mathsf{LA} &= \mathsf{L1} + \mathsf{L2} + \mathsf{L3} + \mathsf{LR3} & \mathsf{RA} &= \mathsf{R1} + \mathsf{R2} + \mathsf{R3} + \mathsf{LR3} \\
\mathsf{LB} &= \mathsf{L2} + \mathsf{L3} + \mathsf{LR3} & \mathsf{RB} &= \mathsf{R2} + \mathsf{R3} + \mathsf{LR3} \\
\mathsf{LC} &= \mathsf{L3} + \mathsf{LR3} & \mathsf{RC} &= \mathsf{R3} + \mathsf{LR3}
\end{aligned}
$$

There's one big problem with the state diagram of Figure 7-62—it doesn't properly handle multiple inputs asserted simultaneously. For example, what happens in the IDLE state if both LEFT and HAZ are asserted? According to the state diagram, the machine goes to two states, L1 and LR3, which is impossible. In reality, the machine would have only one next state, which could be L1, LR3, or a totally unrelated (and possibly unused) third state, depending on details of the state machine's realization (e.g., see Exercise 7.58).

The problem is fixed in Figure 7-63, where we have given the HAZ input priority. Also, we treat LEFT and RIGHT asserted simultaneously as a hazard request, since the driver is clearly confused and needs help.

The new state diagram is unambiguous because the transition expressions on the arcs leaving each state are mutually exclusive and all-inclusive. That is, for each state, no two expressions are 1 for the same input combination, and some expression is 1 for every input combination. This can be confirmed algebraically for this or any other state diagram by performing two steps:

1. *Mutual exclusion.* For each state, show that the logical product of each possible pair of transition expressions on arcs leaving that state is 0. If there are *n* arcs, then there are $n(n-1)/2$ logical products to evaluate.   *mutual exclusion*

2. *All inclusion.* For each state, show that the logical sum of the transition expressions on all arcs leaving that state is 1.   *all inclusion*

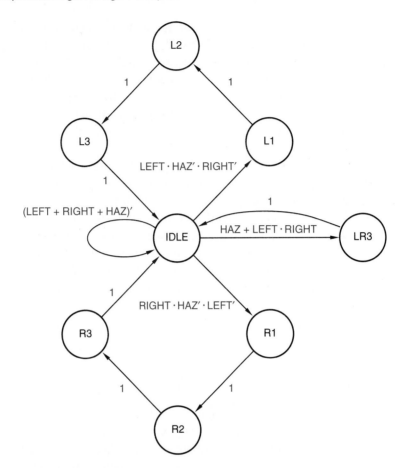

Figure 7-63
Corrected state
diagram for T-bird
tail lights.

If there are many transitions leaving each state, these steps, especially the first one, are very difficult to perform. However, typical state machines, even ones with lots of states and inputs, don't have many transitions leaving each state, since most designers can't dream up such complex machines in the first place. This is where the tradeoff between state-table and state-diagram design occurs. In state-table design, the foregoing steps are not required, because the structure of a state table guarantees mutual exclusion and all inclusion. But if there are a lot of inputs, the state table has *lots* of columns.

Verifying that a state diagram is unambiguous may be difficult in principle, but it's not too bad in practice for small state diagrams. In Figure 7-63, most of the states have a single arc with a transition expression of 1, so verification is trivial. Real work is needed only to verify the IDLE state, which has four transitions leaving it. This can be done on a sheet of scratch paper by listing the eight combinations of the three inputs and checking off the combinations covered by each transition expression. Each combination should have exactly one check. As another example, consider the state diagrams in Figures 7-44 and 7-46 on pages 560 and 562; both can be verified mentally.

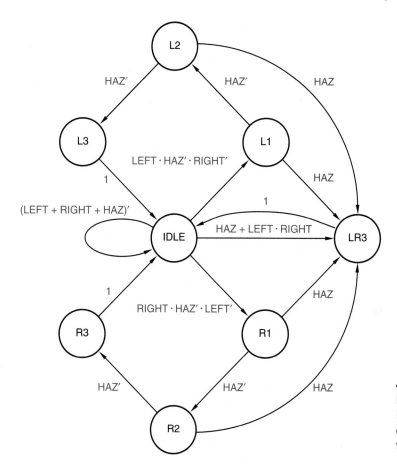

Figure 7-64
Enhanced state diagram for T-bird tail lights.

Returning to the T-bird tail-lights machine, we can now synthesize a circuit from the state diagram if we wish. However, if we want to change the machine's behavior, now is the time to do it, before we do all the work of synthesizing a circuit. In particular, notice that once a left- or right-turn cycle has begun, the state diagram in Figure 7-63 allows the cycle to run to completion, even if HAZ is asserted. While this may have a certain aesthetic appeal, it would be safer for the car's occupants to have the machine go into hazard mode as soon as possible. The state diagram is modified to provide this behavior in Figure 7-64.

Now we're finally ready to synthesize a circuit for the T-bird machine. The state diagram has eight states, so we'll need a minimum of three flip-flops to code the states. Obviously, many state assignments are possible (8! to be exact); we'll use the one in Table 7-16 for the following reasons:

1. An initial (idle) state of 000 is compatible with most flip-flops and registers, which are easily initialized to the 0 state.

2. Two state variables, Q1 and Q0, are used to "count" in Gray-code sequence for the left-turn cycle (IDLE→L1→L2→L3→IDLE). This minimizes the

Table 7-16
State assignment
for T-bird tail-lights
state machine.

State	Q2	Q1	Q0
IDLE	0	0	0
L1	0	0	1
L2	0	1	1
L3	0	1	0
R1	1	0	1
R2	1	1	1
R3	1	1	0
LR3	1	0	0

number of state-variable changes per state transition, which can often simplify the excitation logic.

3. Because of the symmetry in the state diagram, the same sequence on Q1 and Q0 is used to "count" during a right-turn cycle, while Q2 is used to distinguish between left and right.

4. The remaining state-variable combination is used for the LR3 state.

The next step is to write a sort of transition table. However, we must use a format different from the transition tables of Section 7.4.4, because the transitions in a state diagram are specified by expressions rather than by an exhaustive tabulation of next states. We'll call the new format a *transition list*, because it has one row for each transition or arc in the state diagram.

transition list

Table 7-17 is the transition list for the state diagram of Figure 7-64 and the state assignment of Table 7-16. Each row contains the current state, next state, and transition expression for one arc in the state diagram. Both the named and coded versions of the current state and next state are shown. The named states are useful for reference purposes, while the coded states are used to develop transition equations.

Once we have a transition list, the rest of the synthesis steps are pretty much "turning the crank." Synthesis procedures are described in Section 7.6. Although these procedures can be applied manually, they are usually embedded in a CAD software package; thus, Section 7.6 can help you understand what's going on (or going wrong) in your favorite CAD package.

We also encountered one "turn-the-crank" step in this section—finding the ambiguities in state diagrams. Even though the procedure we discussed can be easily automated, few if any CAD programs perform this step in this way. For example, one "state diagram entry" tool silently removes duplicated transitions and goes to the state coded "00...00" for missing transitions, without warning the user. Thus, in most design environments, the designer is responsible for writing a state-machine description that is unambiguous. The state-machine description languages at the end of this chapter provide a good way to do this.

S	Q2	Q1	Q0	Transition Expression	S*	Q2*	Q1*	Q0*
IDLE	0	0	0	(LEFT + RIGHT + HAZ)′	IDLE	0	0	0
IDLE	0	0	0	LEFT · HAZ′ · RIGHT′	L1	0	0	1
IDLE	0	0	0	HAZ + LEFT · RIGHT	LR3	1	0	0
IDLE	0	0	0	RIGHT · HAZ′ · LEFT′	R1	1	0	1
L1	0	0	1	HAZ′	L2	0	1	1
L1	0	0	1	HAZ	LR3	1	0	0
L2	0	1	1	HAZ′	L3	0	1	0
L2	0	1	1	HAZ	LR3	1	0	0
L3	0	1	0	1	IDLE	0	0	0
R1	1	0	1	HAZ′	R2	1	1	1
R1	1	0	1	HAZ	LR3	1	0	0
R2	1	1	1	HAZ′	R3	1	1	0
R2	1	1	1	HAZ	LR3	1	0	0
R3	1	1	0	1	IDLE	0	0	0
LR3	1	0	0	1	IDLE	0	0	0

Table 7-17
Transition list for
T-bird tail-lights
state machine.

*7.6 State-Machine Synthesis Using Transition Lists

Once a machine's state diagram has been designed and a state assignment has been made, the creative part of the design process is pretty much over. The rest of the synthesis procedure can be carried out by CAD programs.

As we showed in the preceding section, a transition list can be constructed from a machine's state diagram and state assignment. This section shows how to synthesize a state machine from its transition list. It also delves into some of the options and nuances of state-machine design using transition lists. Although this material is useful for synthesizing machines by hand, its main purpose is to help you understand the internal operation and the external quirks of CAD programs and languages that deal with state machines.

*7.6.1 Transition Equations

The first step in synthesizing a state machine from a transition list is to develop a set of transition equations that define each next-state variable V* in terms of the current state and input. The transition list can be viewed as a sort of hybrid

* This section and all of its subsections are optional.

truth table in which the state-variable combinations for the current state are listed explicitly and input combinations are listed algebraically. Reading down a V∗ column in a transition list, we find a sequence of 0s and 1s, indicating the value of V∗ for various (if we've done it right, all) state/input combinations.

A transition equation for a next-state variable V∗ can be written using a sort of hybrid canonical sum:

$$V* = \sum_{\text{transition-list rows where } V* = 1} (\text{transition p-term})$$

transition p-term

That is, the transition equation has one "transition p-term" for each row of the transition list that contains a 1 in the V∗ column. A row's *transition p-term* is the product of the current state's minterm and the transition expression.

Based on the transition list in Table 7-17, the transition equation for Q2∗ in the T-bird machine can be written as the sum of eight p-terms:

$$\begin{aligned}
Q2* = \ & Q2' \cdot Q1' \cdot Q0' \cdot (HAZ + LEFT \cdot RIGHT) \\
& + Q2' \cdot Q1' \cdot Q0' \cdot (RIGHT \cdot HAZ' \cdot LEFT') \\
& + Q2' \cdot Q1' \cdot Q0 \cdot (HAZ) \\
& + Q2' \cdot Q1 \cdot Q0 \cdot (HAZ) \\
& + Q2 \cdot Q1' \cdot Q0 \cdot (HAZ') \\
& + Q2 \cdot Q1' \cdot Q0 \cdot (HAZ) \\
& + Q2 \cdot Q1 \cdot Q0 \cdot (HAZ') \\
& + Q2 \cdot Q1 \cdot Q0 \cdot (HAZ)
\end{aligned}$$

Some straightforward algebraic manipulations lead to a simplified transition equation that combines the first two, second two, and last four p-terms above:

$$\begin{aligned}
Q2* = \ & Q2' \cdot Q1' \cdot Q0' \cdot (HAZ + RIGHT) \\
& + Q2' \cdot Q0 \cdot (HAZ) \\
& + Q2 \cdot Q0
\end{aligned}$$

Transition equations for Q1∗ and Q0∗ may be obtained in a similar manner:

$$\begin{aligned}
Q1* = \ & Q2' \cdot Q1' \cdot Q0 \cdot (HAZ') \\
& + Q2' \cdot Q1 \cdot Q0 \cdot (HAZ') \\
& + Q2 \cdot Q1' \cdot Q0 \cdot (HAZ') \\
& + Q2 \cdot Q1 \cdot Q0 \cdot (HAZ') \\
= \ & Q0 \cdot HAZ' \\
Q0* = \ & Q2' \cdot Q1' \cdot Q0' \cdot (LEFT \cdot HAZ' \cdot RIGHT') \\
& + Q2' \cdot Q1' \cdot Q0' \cdot (RIGHT \cdot HAZ' \cdot LEFT') \\
& + Q2' \cdot Q1' \cdot Q0 \cdot (HAZ') \\
& + Q2 \cdot Q1' \cdot Q0 \cdot (HAZ') \\
= \ & Q2' \cdot Q1' \cdot HAZ' \cdot (LEFT \oplus RIGHT) + Q1' \cdot Q0 \cdot HAZ'
\end{aligned}$$

Except for Q1∗, there's no guarantee that the transition equations above are in any sense minimal—in fact, the expressions for Q2∗ and Q0∗ aren't even in standard sum-of-products or product-of-sums form. The simplified equations, or the original unsimplified ones, merely provide an unambiguous starting point for whatever combinational design method you might choose to synthesize the excitation logic for the state machine—ad hoc, NAND-NAND, MSI-based, or whatever. In a PLD-based design, you could simply plug the equations into an ABEL program and let the compiler calculate the minimal sum-of-products expressions for the PLD's AND-OR array.

*7.6.2 Excitation Equations

While we're on the subject of excitation logic, note that so far we have derived only *transition equations*, not *excitation equations*. However, if we use D flip-flops as the memory elements in our state machines, then the excitation equations are trivial to derive from the transition equations, since the characteristic equation of a D flip-flop is $Q* = D$. Therefore, if the transition equation for a state variable $Qi*$ is

$$Qi* \;=\; \text{expression}$$

then the excitation equation for the corresponding D flip-flop input is

$$Di \;=\; \text{expression}$$

Efficient excitation equations for other flip-flop types, especially J-K, are not so easy to derive (see Exercise 7.63). For that reason, the vast majority of discrete, PLD-based, and ASIC-based state-machine designs employ D flip-flops.

*7.6.3 Variations on the Scheme

There are other ways to obtain transition and excitation equations from a transition list. If the column for a particular next-state variable contains fewer 0s than 1s, it may be advantageous to write that variable's transition equation in terms of the 0s in its column. That is, we write

$$V*' \;=\; \sum_{\text{transition-list rows where } V* = 0} (\text{transition p-term})$$

That is, $V*'$ is 1 for all of the p-terms for which $V*$ is 0. Thus, a transition equation for $Q2*'$ may be written as the sum of seven p-terms:

$$
\begin{aligned}
Q2*' \;=\; & Q2' \cdot Q1' \cdot Q0' \cdot ((\text{LEFT} + \text{RIGHT} + \text{HAZ})') \\
& + Q2' \cdot Q1' \cdot Q0' \cdot (\text{LEFT} \cdot \text{HAZ}' \cdot \text{RIGHT}') \\
& + Q2' \cdot Q1' \cdot Q0 \cdot (\text{HAZ}') \\
& + Q2' \cdot Q1 \cdot Q0 \cdot (\text{HAZ}') \\
& + Q2' \cdot Q1 \cdot Q0' \cdot (1) \\
& + Q2 \cdot Q1 \cdot Q0' \cdot (1) \\
& + Q2 \cdot Q1' \cdot Q0' \cdot (1) \\
\;=\; & Q2' \cdot Q1' \cdot Q0' \cdot \text{HAZ}' \cdot \text{RIGHT}' + Q2' \cdot Q0 \cdot \text{HAZ}' + Q1 \cdot Q0' + Q2 \cdot Q0'
\end{aligned}
$$

To obtain an equation for Q2∗, we simply complement both sides of the reduced equation.

To obtain an expression for a next-state variable V∗ directly, using the 0s in the transition list, we can complement the righthand side of the general V∗′ equation using DeMorgan's theorem, obtaining a sort of hybrid canonical product:

$$V* = \prod_{\text{transition-list rows where } V* = 0} (\text{transition s-term})$$

transition s-term

Here, a row's *transition s-term* is the sum of the maxterm for the current state and the complement of the transition expression. If the transition expression is a simple product term, then its complement is a sum, and the transition equation expresses V∗ in product-of-sums form.

*7.6.4 Realizing the State Machine

Once you have the excitation equations for a state machine, all you're left with is a multiple-output combinational logic design problem. Of course, there are many ways to realize combinational logic from equations, but the easiest way is just to type them into an ABEL or VHDL program and use the compiler to synthesize a PLD, FPGA, or ASIC realization.

Combinational PLDs, such as the PAL16L8 and GAL16V8 that we studied in Section 5.3, can be used to realize excitation equations up to a certain number of inputs, outputs, and product terms. Better yet, in Section 8.3 we'll introduce sequential PLDs that include D flip-flops on the same chip with the combinational AND-OR array. For a given number of PLD input and output pins, these sequential PLDs can realize larger state machines than their combinational counterparts, because the excitation signals never have to go off the chip. In Section 9.1.3, we'll show how to realize the T-bird tail-lights machine in a sequential PLD.

*7.7 Another State-Machine Design Example

This section gives one more example of state-machine design using a state diagram. The example provides a basis for further discussion of a few topics: unused states, output-coded state assignments, and "don't-care" state codings.

*7.7.1 The Guessing Game

Our final state-machine example is a "guessing game" that can be built as an amusing lab project:

Design a clocked synchronous state machine with four inputs, G1–G4, that are connected to pushbuttons. The machine has four outputs, L1–L4, connected to lamps or LEDs located near the like-numbered pushbuttons. There is also an ERR output connected to a red lamp. In normal operation the L1–L4 outputs display a 1-out-of-4 pattern. At each clock tick, the pattern is rotated by one position; the clock frequency is about 4 Hz.

Guesses are made by pressing a pushbutton, which asserts an input Gi. When any Gi input is asserted, the ERR output is asserted if the "wrong" pushbutton was pressed, that is, if the Gi input detected at the clock tick does not have the same number as the lamp output that was asserted before the clock tick. Once a guess has been made, play stops and the ERR output maintains the same value for one or more clock ticks until the Gi input is negated, then play resumes.

Clearly, we will have to provide four states, one for each position of the rotating pattern, and we'll need at least one state to indicate that play has stopped. A possible state diagram is shown in Figure 7-65. The machine cycles through states S1–S4 as long as no Gi input is asserted, and it goes to the STOP state when a guess is made. Each Li output is asserted in the like-numbered state.

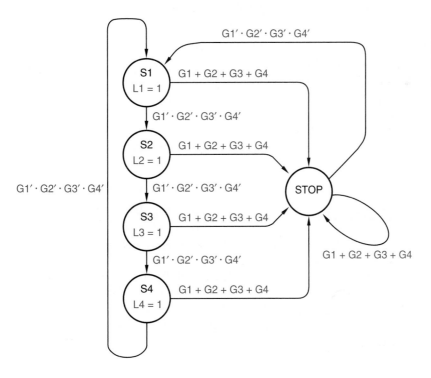

Figure 7-65
First try at a state diagram for the guessing game.

The only problem with this state diagram is that it doesn't "remember" in the STOP state whether the guess was correct, so it has no way to control the ERR output. This problem is fixed in Figure 7-66, which has two "stopped" states, SOK and SERR. On an incorrect guess, the machine goes to SERR, where ERR is asserted; otherwise, it goes to SOK. Although the machine's word description doesn't require it, the state diagram is designed to go to SERR even if the user tries to fool it by pressing two or more pushbuttons simultaneously, or by changing guesses while stopped.

A transition list corresponding to the state diagram is shown in Table 7-18, using a simple 3-bit binary state encoding with Gray-code order for the S1–S4 cycle. Transition equations for Q1* and Q0* can be obtained from the table as follows:

$$
\begin{aligned}
Q1* &= Q2' \cdot Q1' \cdot Q0 \cdot (G1' \cdot G2' \cdot G3' \cdot G4') \\
&+ Q2' \cdot Q1 \ \cdot Q0 \cdot (G1' \cdot G2' \cdot G3' \cdot G4') \\
&= Q2' \ \cdot Q0 \cdot G1' \cdot G2' \cdot G3' \cdot G4' \\
Q0* &= Q2' \cdot Q1' \cdot Q0' \cdot (G1' \cdot G2' \cdot G3' \cdot G4') \\
&+ Q2' \cdot Q1' \cdot Q0' \cdot (G2 + G3 + G4) \\
&+ Q2' \cdot Q1' \cdot Q0 \ \cdot (G1' \cdot G2' \cdot G3' \cdot G4') \\
&+ Q2' \cdot Q1' \cdot Q0 \ \cdot (G1 + G3 + G4) \\
&+ Q2' \cdot Q1 \ \cdot Q0 \ \cdot (G1 + G2 + G4) \\
&+ Q2' \cdot Q1 \ \cdot Q0' \cdot (G1 + G2 + G3) \\
&+ Q2 \ \cdot Q1' \cdot Q0 \ \cdot (G1 + G2 + G3 + G4)
\end{aligned}
$$

Using a logic-minimization program, the Q0* expression can be reduced to 11 product terms in two-level sum-of-products form. An expression for Q2* is best formulated in terms of the 0s in the Q2* column of the transition list:

$$
\begin{aligned}
Q2*' &= Q2' \cdot Q1' \cdot Q0' \cdot (G1' \cdot G2' \cdot G3' \cdot G4') \\
&+ Q2' \cdot Q1' \cdot Q0 \ \cdot (G1' \cdot G2' \cdot G3' \cdot G4') \\
&+ Q2' \cdot Q1 \ \cdot Q0 \ \cdot (G1' \cdot G2' \cdot G3' \cdot G4') \\
&+ Q2' \cdot Q1 \ \cdot Q0' \cdot (G1' \cdot G2' \cdot G3' \cdot G4') \\
&+ Q2 \ \cdot Q1' \cdot Q0' \cdot (G1' \cdot G2' \cdot G3' \cdot G4') \\
&+ Q2 \ \cdot Q1' \cdot Q0 \ \cdot (G1' \cdot G2' \cdot G3' \cdot G4') \\
&= (Q2' + Q1') \cdot (G1' \cdot G2' \cdot G3' \cdot G4')
\end{aligned}
$$

The last five columns of Table 7-18 show output values. Thus, output equations can be developed in much the same way as transition equations. However, since this example is a Moore machine, outputs are independent of the transition expressions; only one row of the transition list must be considered for each current state. The output equations are

$$L1 = Q2' \cdot Q1' \cdot Q0' \qquad L3 = Q2' \cdot Q1 \cdot Q0 \qquad ERR = Q2 \cdot Q1' \cdot Q0$$
$$L2 = Q2' \cdot Q1' \cdot Q0 \qquad L4 = Q2' \cdot Q1 \cdot Q0'$$

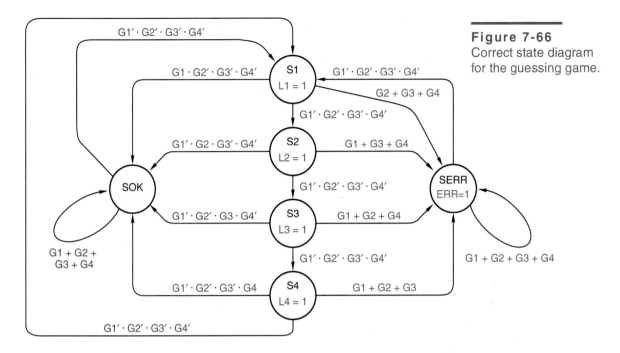

Figure 7-66
Correct state diagram
for the guessing game.

Table 7-18 Transition list for guessing-game machine.

Current State					Next State				Output				
S	Q2	Q1	Q0	Transition Expression	S*	Q2*	Q1*	Q0*	L1	L2	L3	L4	ERR
S1	0	0	0	G1′ · G2′ · G3′ · G4′	S2	0	0	1	1	0	0	0	0
S1	0	0	0	G1 · G2′ · G3′ · G4′	SOK	1	0	0	1	0	0	0	0
S1	0	0	0	G2 + G3 + G4	SERR	1	0	1	1	0	0	0	0
S2	0	0	1	G1′ · G2′ · G3′ · G4′	S3	0	1	1	0	1	0	0	0
S2	0	0	1	G1′ · G2 · G3′ · G4′	SOK	1	0	0	0	1	0	0	0
S2	0	0	1	G1 + G3 + G4	SERR	1	0	1	0	1	0	0	0
S3	0	1	1	G1′ · G2′ · G3′ · G4′	S4	0	1	0	0	0	1	0	0
S3	0	1	1	G1′ · G2′ · G3 · G4′	SOK	1	0	0	0	0	1	0	0
S3	0	1	1	G1 + G2 + G4	SERR	1	0	1	0	0	1	0	0
S4	0	1	0	G1′ · G2′ · G3′ · G4′	S1	0	0	0	0	0	0	1	0
S4	0	1	0	G1′ · G2′ · G3′ · G4	SOK	1	0	0	0	0	0	1	0
S4	0	1	0	G1 + G2 + G3	SERR	1	0	1	0	0	0	1	0
SOK	1	0	0	G1 + G2 + G3 + G4	SOK	1	0	0	0	0	0	0	0
SOK	1	0	0	G1′ · G2′ · G3′ · G4′	S1	0	0	0	0	0	0	0	0
SERR	1	0	1	G1 + G2 + G3 + G4	SERR	1	0	1	0	0	0	0	1
SERR	1	0	1	G1′ · G2′ · G3′ · G4′	S1	0	0	0	0	0	0	0	1

*7.7.2 Unused States

Our state diagram for the guessing game has six states, but the actual state machine, built from three flip-flops, has eight. By omitting the unused states from the transition list, we treated them as "don't-cares" in a very limited sense:

- When we wrote equations for Q1* and Q0*, we formed a sum of transition p-terms for state/input combinations that had an explicit 1 in the corresponding columns of the transition list. Although we didn't consider the unused states, our procedure implicitly treated them as if they had 0s in the Q1* and Q0* columns.

- Conversely, we wrote the Q2*′ equation as a sum of transition p-terms for state/input combinations that had an explicit 0 in the corresponding columns of the transition list. Unused states were implicitly treated as if they had 1s in the Q2* column.

As a consequence of these choices, all of the unused states in the guessing-game machine have a coded next state of 100 for all input combinations. That's safe, acceptable behavior should the machine stray into an unused state, since 100 is the coding for one of the normal states (SOK).

To treat the unused states as true "don't-cares," we would have to allow them to go to any next state under any input combination. This is simple in principle but may be difficult in practice.

At the end of Section 7.4.4 we showed how to handle unused states as "don't-cares" in the Karnaugh-map method for developing transition/excitation equations. Unfortunately, for all but the smallest problems, Karnaugh maps are unwieldy. Commercially available logic-minimization programs can easily handle larger problems, but many of them don't handle "don't-cares" or require the designer to insert special code to handle them. In ABEL state machines, don't-care next states can be handled fairly easily using the @DCSET directive, as we discuss in the box on page 632. In VHDL the process is a bit unwieldy.

*7.7.3 Output-Coded State Assignment

Let's look at another realization of the guessing-game machine. The machine's outputs are a function of state only; furthermore, a *different* output combination is produced in each named state. Therefore, we can use the outputs as state variables and assign each named state to the required output combination. This sort of *output-coded state assignment* can sometimes result in excitation equations that are simpler than the set of excitation and output equations obtained with a state assignment using a minimum number of state variables.

output-coded state assignment

Table 7-19 is the guessing-game transition list that results from an output-coded state assignment. Each transition/excitation equation has very few transition p-terms because the transition list has so few 1s in the next-state columns:

Table 7-19 Transition list for guessing-game machine using outputs as state variables.

S	L1	L2	L3	L4	ERR	Transition Expression	S*	L1*	L2*	L3*	L4*	ERR*
		Current State							Next State			
S1	1	0	0	0	0	$G1' \cdot G2' \cdot G3' \cdot G4'$	S2	0	1	0	0	0
S1	1	0	0	0	0	$G1 \cdot G2' \cdot G3' \cdot G4'$	SOK	0	0	0	0	0
S1	1	0	0	0	0	$G2 + G3 + G4$	SERR	0	0	0	0	1
S2	0	1	0	0	0	$G1' \cdot G2' \cdot G3' \cdot G4'$	S3	0	0	1	0	0
S2	0	1	0	0	0	$G1' \cdot G2 \cdot G3' \cdot G4'$	SOK	0	0	0	0	0
S2	0	1	0	0	0	$G1 + G3 + G4$	SERR	0	0	0	0	1
S3	0	0	1	0	0	$G1' \cdot G2' \cdot G3' \cdot G4'$	S4	0	0	0	1	0
S3	0	0	1	0	0	$G1' \cdot G2' \cdot G3 \cdot G4'$	SOK	0	0	0	0	0
S3	0	0	1	0	0	$G1 + G2 + G4$	SERR	0	0	0	0	1
S4	0	0	0	1	0	$G1' \cdot G2' \cdot G3' \cdot G4'$	S1	1	0	0	0	0
S4	0	0	0	1	0	$G1' \cdot G2' \cdot G3' \cdot G4$	SOK	0	0	0	0	0
S4	0	0	0	1	0	$G1 + G2 + G3$	SERR	0	0	0	0	1
SOK	0	0	0	0	0	$G1 + G2 + G3 + G4$	SOK	0	0	0	0	0
SOK	0	0	0	0	0	$G1' \cdot G2' \cdot G3' \cdot G4'$	S1	1	0	0	0	0
SERR	0	0	0	0	1	$G1 + G2 + G3 + G4$	SERR	0	0	0	0	1
SERR	0	0	0	0	1	$G1' \cdot G2' \cdot G3' \cdot G4'$	S1	1	0	0	0	0

$$
\begin{aligned}
L1* = {}& L1' \cdot L2' \cdot L3' \cdot L4 \cdot ERR' \cdot (G1' \cdot G2' \cdot G3' \cdot G4') \\
& + L1' \cdot L2' \cdot L3' \cdot L4' \cdot ERR' \cdot (G1' \cdot G2' \cdot G3' \cdot G4') \\
& + L1' \cdot L2' \cdot L3' \cdot L4' \cdot ERR \cdot (G1' \cdot G2' \cdot G3' \cdot G4') \\
L2* = {}& L1 \cdot L2' \cdot L3' \cdot L4' \cdot ERR' \cdot (G1' \cdot G2' \cdot G3' \cdot G4') \\
L3* = {}& L1' \cdot L2 \cdot L3' \cdot L4' \cdot ERR' \cdot (G1' \cdot G2' \cdot G3' \cdot G4') \\
L4* = {}& L1' \cdot L2' \cdot L3 \cdot L4' \cdot ERR' \cdot (G1' \cdot G2' \cdot G3' \cdot G4') \\
ERR* = {}& L1 \cdot L2' \cdot L3' \cdot L4' \cdot ERR' \cdot (G2 + G3 + G4) \\
& + L1' \cdot L2 \cdot L3' \cdot L4' \cdot ERR' \cdot (G1 + G3 + G4) \\
& + L1' \cdot L2' \cdot L3 \cdot L4' \cdot ERR' \cdot (G1 + G2 + G4) \\
& + L1' \cdot L2' \cdot L3' \cdot L4 \cdot ERR' \cdot (G1 + G2 + G3) \\
& + L1' \cdot L2' \cdot L3' \cdot L4' \cdot ERR \cdot (G1 + G2 + G3 + G4)
\end{aligned}
$$

There are no output equations, of course. The ERR* equation above is the worst in the group, requiring 16 terms to express in either minimal sum-of-products or product-of-sums form.

As a group, the equations developed above have just about the same complexity as the transition and output equations that we developed from Table 7-18. Even though the output-coded assignment does not produce a simpler set of equations in this example, it can still save cost in a PLD-based design, since fewer PLD macrocells or outputs are needed overall.

*7.7.4 "Don't-Care" State Codings

Out of the 32 possible coded states using five variables, only six are used in Table 7-19. The rest of the states are unused and have a next state of 00000 if the machine is built using the equations in the preceding subsection. Another possible disposition for unused states, one that we haven't discussed before, is obtained by careful use of "don't-cares" in the assignment of coded states to current states.

Table 7-20 shows one such state assignment for the guessing-game machine, derived from the output-coded state assignment of the preceding subsection. In this example, every possible combination of current-state variables corresponds to one of the coded states (e.g., 10111 = S1, 00101 = S3). However, *next states* are coded using the same unique combinations as in the preceding subsection. Table 7-21 shows the resulting transition list.

In this approach, each unused current state behaves like a nearby "normal" state; Figure 7-67 illustrates the concept. The machine is well behaved and goes to a "normal state" if it inadvertently enters an unused state. Yet the approach still allows some simplification of the excitation and output logic by introducing

Table 7-20
Current-state assignment for the guessing-game machine using don't-cares.

State	L1	L2	L3	L4	ERR
S1	1	x	x	x	x
S2	0	1	x	x	x
S3	0	0	1	x	x
S4	0	0	0	1	x
SOK	0	0	0	0	0
SERR	0	0	0	0	1

Figure 7-67
State assignment using don't-cares for current states.

Current coded states   Next coded states

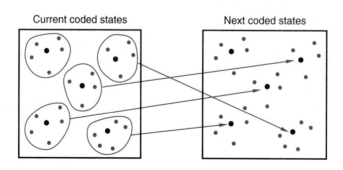

Table 7-21 Transition list for guessing-game machine using don't-care state codings.

Current State							Next State					
S	L1	L2	L3	L4	ERR	Transition Expression	S*	L1*	L2*	L3*	L4*	ERR*
S	1	x	x	x	x	$G1' \cdot G2' \cdot G3' \cdot G4'$	S2	0	1	0	0	0
S	1	x	x	x	x	$G1 \cdot G2' \cdot G3' \cdot G4'$	SOK	0	0	0	0	0
S	1	x	x	x	x	$G2 + G3 + G4$	SERR	0	0	0	0	1
S2	0	1	x	x	x	$G1' \cdot G2' \cdot G3' \cdot G4'$	S3	0	0	1	0	0
S2	0	1	x	x	x	$G1' \cdot G2 \cdot G3' \cdot G4'$	SOK	0	0	0	0	0
S2	0	1	x	x	x	$G1 + G3 + G4$	SERR	0	0	0	0	1
S3	0	0	1	x	x	$G1' \cdot G2' \cdot G3' \cdot G4'$	S4	0	0	0	1	0
S3	0	0	1	x	x	$G1' \cdot G2' \cdot G3 \cdot G4'$	SOK	0	0	0	0	0
S3	0	0	1	x	x	$G1 + G2 + G4$	SERR	0	0	0	0	1
S4	0	0	0	1	x	$G1' \cdot G2' \cdot G3' \cdot G4'$	S1	1	0	0	0	0
S4	0	0	0	1	x	$G1' \cdot G2' \cdot G3' \cdot G4$	SOK	0	0	0	0	0
S4	0	0	0	1	x	$G1 + G2 + G3$	SERR	0	0	0	0	1
SOK	0	0	0	0	0	$G1 + G2 + G3 + G4$	SOK	0	0	0	0	0
SOK	0	0	0	0	0	$G1' \cdot G2' \cdot G3' \cdot G4'$	S1	1	0	0	0	0
SERR	0	0	0	0	1	$G1 + G2 + G3 + G4$	SERR	0	0	0	0	1
SERR	0	0	0	0	1	$G1' \cdot G2' \cdot G3' \cdot G4'$	S1	1	0	0	0	0

don't-cares in the transition list. When a row's transition p-term is written, current-state variables that are don't-cares in that row are omitted; for example,

$$
\begin{aligned}
\text{ERR*} = \ & L1 \cdot (G2 + G3 + G4) \\
& + L1' \cdot L2 \cdot (G1 + G3 + G4) \\
& + L1' \cdot L2' \cdot L3 \cdot (G1 + G2 + G4) \\
& + L1' \cdot L2' \cdot L3' \cdot L4 \cdot (G1 + G2 + G3) \\
& + L1' \cdot L2' \cdot L3' \cdot L4' \cdot \text{ERR} \cdot (G1 + G2 + G3 + G4)
\end{aligned}
$$

Compared with the ERR* equation in the preceding subsection, the one above still requires 16 terms to express as a sum of products. However, it requires only five terms in minimal product-of-sums form, which makes its complement more suitable for realization in a PLD.

*7.8 Decomposing State Machines

Just like large procedures or subroutines in a programming language, large state machines are difficult to conceptualize, design, and debug. Therefore, when faced with a large state-machine problem, digital designers often look for opportunities to solve it with a collection of smaller state machines.

state-machine
decomposition

There's a well-developed theory of *state-machine decomposition* that you can use to analyze any given, monolithic state machine to determine whether it can be realized as a collection of smaller ones. However, decomposition theory is not too useful for designers who want to avoid designing large state machines in the first place. Rather, a practical designer tries to cast the original design problem into a natural, hierarchical structure, so that the uses and functions of submachines are obvious, making it unnecessary ever to write a state table for the equivalent monolithic machine.

main machine
submachines

The simplest and most commonly used type of decomposition is illustrated in Figure 7-68. A *main machine* provides the primary inputs and outputs and executes the top-level control algorithm. *Submachines* perform low-level steps under the control of the main machine and may optionally handle some of the primary inputs and outputs.

Perhaps the most commonly used submachine is a counter. The main machine starts the counter when it wishes to stay in a particular main state for n clock ticks, and the counter asserts a DONE signal when n ticks have occurred. The main machine is designed to wait in the same state until DONE is asserted. This adds an extra output and input to the main machine (START and DONE), but it saves $n - 1$ states.

An example decomposed state machine designed along these lines is based on the guessing game of Section 7.7. The original guessing game is easy to win after a minute of practice because the lamps cycle at a very consistent rate of

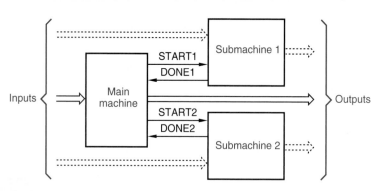

Figure 7-68
A typical, hierarchical state-machine structure.

| **A REALLY BAD JOKE** | Note that the title of this section has nothing to do with the "buried flip-flops" found in some PLDs. |

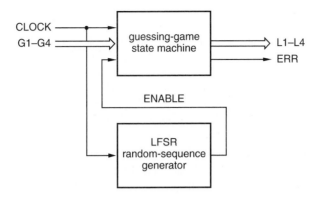

Figure 7-69
Block diagram of
guessing game with
random delay.

4 Hz. To make the game more challenging, we can double or triple the clock speed but allow the lamps to stay in each state for a random length of time. Then the user truly must guess whether a given lamp will stay on long enough for the corresponding pushbutton to be pressed.

A block diagram for the enhanced guessing game is shown in Figure 7-69. The main machine is basically the same as before, except that it advances from one lamp state to the next only if the enable input EN is asserted, as shown by the state diagram in Figure 7-70. The enable input is driven by the output of a pseudorandom sequence generator, a linear feedback shift register (LFSR).

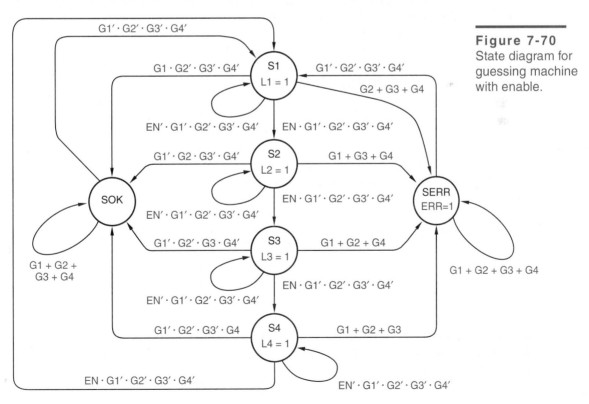

Figure 7-70
State diagram for
guessing machine
with enable.

A SHIFTY CIRCUIT   LFSR circuits are described in Section 8.5.8. In Figure 7-69, the low-order bit of an n-bit LFSR counter is used as the enable signal. Thus, the length of time that a particular lamp stays on depends on the counting sequence of the LFSR.

In the best case for the user, the LFSR contains 10...00; in this case the lamp is on for $n-1$ clock ticks, because it takes that long for the single 1 to shift into the low-order bit position. In the worst case, the LFSR contains 11...11, and shifting occurs for n consecutive clock ticks. At other times, shifting stops for a time determined by the number of consecutive 0s starting with the low-order bit of the LFSR.

All of these cases are quite unpredictable, unless the user has memorized the shifting cycle (2^n-1 states) or is very fast at Galois-field arithmetic. Obviously, a large value of n (≥ 16) provides the most fun.

Another obvious candidate for decomposition is a state machine that performs binary multiplication using the shift-and-add algorithm, or binary division using the shift-and-subtract algorithm. To perform an n-bit operation, these algorithms require an initialization step, n computation steps, and possible cleanup steps. The main machine for such an algorithm contains states for initialization, generic computation, and cleanup steps, and a modulo-n counter can be used as a submachine to control the number of generic computation steps executed.

*7.9 Feedback Sequential Circuits

The simple bistable and the various latches and flip-flops that we studied earlier in this chapter are all feedback sequential circuits. Each has one or more feedback loops that, ignoring their behavior during state transitions, store a 0 or a 1 at all times. The feedback loops are memory elements, and the circuits' behavior depends on both the current inputs and the values stored in the loops.

*7.9.1 Analysis

fundamental-mode circuit

Feedback sequential circuits are the most common example of *fundamental-mode circuits*. In such circuits, inputs are not normally allowed to change simultaneously. The analysis procedure assumes that inputs change one at a time, allowing enough time between successive changes for the circuit to settle into a stable internal state. This differs from clocked circuits, in which multiple inputs can change at almost arbitrary times without affecting the state, and all input values are sampled and state changes occur with respect to a clock signal.

*This section and all of its subsections are optional.

Like clocked synchronous state machines, feedback sequential circuits may be structured as Mealy or Moore circuits, as shown in Figure 7-71. A circuit with n feedback loops has n binary state variables and 2^n states.

To analyze a feedback sequential circuit, we must break the feedback loops in Figure 7-71 so that the next value stored in each loop can be predicted as a function of the circuit inputs and the current value stored in all loops. Figure 7-72 shows how to do this for the NAND circuit for a D latch, which has only one feedback loop. We conceptually break the loop by inserting a fictional buffer in the loop as shown. The output of the buffer, named Y, is the single state variable for this example.

Let us assume that the propagation delay of the fictional buffer is 10 ns (but any nonzero number will do) and that all of the other circuit components have zero delay. If we know the circuit's current state (Y) and inputs (D and C), then we can predict the value Y will have in 10 ns. The next value of Y, denoted Y∗, is a combinational function of the current state and inputs. Thus, reading the circuit diagram, we can write an *excitation equation* for Y∗:

excitation equation

$$Y* \; = \; (C \cdot D) + (C \cdot D' + Y')'$$
$$= \; C \cdot D + C' \cdot Y + D \cdot Y$$

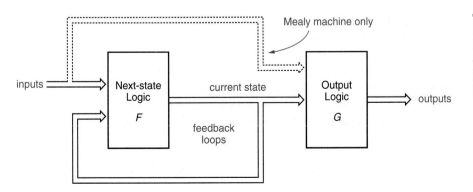

Figure 7-71
Feedback sequential
circuit structure for
Mealy and Moore
machines.

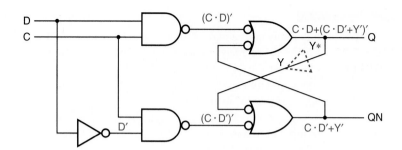

Figure 7-72
Feedback analysis
of a D latch.

transition table

Now the state of the feedback loop (and the circuit) can be written as a function of the current state and input, and enumerated by a *transition table* as shown in Figure 7-73. Each cell in the transition table shows the fictional-buffer output value that will occur 10 ns (or whatever delay you've assumed) after the corresponding state and input combination occurs.

Figure 7-73
Transition table for the
D latch in Figure 7-72.

Y			C D	
	00	01	11	10
0	0	0	1	0
1	1	1	1	0
		Y*		

A transition table has one row for each possible combination of the state variables, so a circuit with n feedback loops has 2^n rows in its transition table. The table has one column for each possible input combination, so a circuit with m inputs has 2^m columns in its transition table.

By definition, a fundamental-mode circuit such as a feedback sequential circuit does not have a clock to tell it when to sample its inputs. Instead, we can imagine that the circuit is evaluating its current state and input *continuously* (or every 10 ns, if you prefer). As the result of each evaluation, it goes to a next state predicted by the transition table. Most of the time, the next state is the same as the current state; this is the essence of fundamental-mode operation. We make some definitions below that will help us study this behavior in more detail.

JUST ONE LOOP   The way the circuit in Figure 7-72 is drawn, it may look like there are two feedback loops. However, once we make one break as shown, there are no more loops. That is, each signal can be written as a combinational function of the other signals, not including itself.

S	C D 00	01	11	10
S0	(S0)	(S0)	S1	(S0)
S1	(S1)	(S1)	(S1)	S0
		S*		

Figure 7-74
State table for the D latch in Figure 7-72. showing stable total states.

In a fundamental-mode circuit, a *total state* is a particular combination of *internal state* (the values stored in the feedback loops) and *input state* (the current value of the circuit inputs). A *stable total state* is a combination of internal state and input state such that the next internal state predicted by the transition table is the same as the current internal state. If the next internal state is different, then the combination is an *unstable total state*. We have rewritten the transition table for the D latch in Figure 7-74 as a *state table*, giving the names S0 and S1 to the states and drawing a circle around the stable total states.

total state
internal state
input state
stable total state

unstable total state
state table

To complete the analysis of the circuit, we must also determine how the outputs behave as functions of the internal state and inputs. There are two outputs and hence two *output equations*:

output equation

$$Q = C \cdot D + C' \cdot Y + D \cdot Y$$
$$QN = C \cdot D' + Y'$$

Note that Q and QN are *outputs*, not state variables. Even though the circuit has two outputs, which can theoretically take on four combinations, it has only one state variable Y, and hence only two states.

The output values predicted by the Q and QN equations can be incorporated in a combined state and output table that completely describes the operation of the circuit, as shown in Figure 7-75. Although Q and QN are normally complementary, it is possible for them to have the same value (1) momentarily, during the transition from S0 to S1 under the C D = 11 column of the table.

We can now predict the behavior of the circuit from the transition and output table. First of all, notice that we have written the column labels in our state tables in "Karnaugh map" order, so that only a single input bit changes

S	C D 00	01	11	10
S0	(S0), 01	(S0), 01	S1, 11	(S0), 01
S1	(S1), 10	(S1), 10	(S1), 10	S0, 01
		S*, Q QN		

Figure 7-75
State and output table for the D latch.

	C D			
S	00	01	11	10
S0	(S0), 01	(S0), 01	S1, 11	(S0), 01
S1	(S1), 10	(S1), 10	(S1), 10	S0, 01

S*, Q QN

Figure 7-76
Analysis of the D latch
for a few transitions.

between adjacent columns of the table. This layout helps our analysis because we assume that only one input changes at a time and that the circuit always reaches a stable total state before another input changes.

At any time, the circuit is in a particular internal state and a particular input is applied to it; we called this combination the total state of the circuit. Let us start at the stable total state "S0/00" (S = S0, C D = 00), as shown in Figure 7-76. Now suppose that we change D to 1. The total state moves to one cell to the right; we have a new stable total state, S0/01. The D input is different, but the internal state and output are the same as before. Next, let us change C to 1. The total state moves one cell to the right to S0/11, which is unstable. The next-state entry in this cell sends the circuit to internal state S1, so the total state moves down one cell, to S1/11. Examining the next-state entry in the new cell, we find that we have reached a stable total state. We can trace the behavior of the circuit for any desired sequence of single input changes in this way.

Now we can revisit the question of simultaneous input changes. Even though "almost simultaneous" input changes may occur in practice, we must assume that nothing happens simultaneously in order to analyze the behavior of sequential circuits. The impossibility of simultaneous events is supported by the varying delays of circuit components themselves, which depend on voltage, temperature, and fabrication parameters. What this tells us is that a set of n inputs that appear to us to change "simultaneously" may actually change in any of $n!$ different orders from the point of view of the circuit operation.

For example, consider the operation of the D latch as shown in Figure 7-77. Let us assume that it starts in stable total state S1/11. Now suppose that C and D

Figure 7-77
Multiple input changes
with the D latch.

	C D			
S	00	01	11	10
S0	(S0), 01	(S0), 01	S1, 11	(S0), 01
S1	(S1), 10	(S1), 10	(S1), 10	S0, 01

S*, Q QN

are both "simultaneously" set to 0. In reality, the circuit behaves as if one or the other input went to 0 first. Suppose that C changes first. Then the sequence of two left-pointing arrows in the table tells us that the circuit goes to stable total state S1/00. However, if D changes first, then the other sequence of arrows tells us that the circuit goes to stable total state S0/00. So the final state of the circuit is unpredictable, a clue that the feedback loop may actually become metastable if we set C and D to 0 simultaneously. The time span over which this view of simultaneity is relevant is the setup- and hold-time window of the D latch.

Simultaneous input changes don't always cause unpredictable behavior. However, we must analyze the effects of all possible orderings of signal changes to determine this; if all orderings give the same result, then the circuit output is predictable. For example, consider the behavior of the D latch starting in total state S0/00 with C and D simultaneously changing from 0 to 1; it always ends up in total state S1/11.

*7.9.2 Analyzing Circuits with Multiple Feedback Loops

In circuits with multiple feedback loops, we must break all of the loops, creating one fictional buffer and state variable for each loop that we break. There are many possible ways, which mathematicians call *cut sets*, to break the loops in a given circuit, so how do we know which one is best? The answer is that any *minimal cut set*—a cut set with a minimum number of cuts—is fine. Mathematicians can give you an algorithm for finding a minimal cut set, but as a digital designer working on small circuits, you can just eyeball the circuit to find one.

cut set

minimal cut set

Different cut sets for a circuit lead to different excitation equations, transition tables, and state/output tables. However, the stable total states derived from one minimal cut set correspond one-to-one to the stable total states derived from any other minimal cut set for the same circuit. That is, state/output tables derived from different minimal cut sets display the same input/output behavior, with only the names and coding of the states changed.

If you use more than the minimal number of cuts to analyze a feedback sequential circuit, the resulting state/output table will still describe the circuit correctly. However, it will use 2^m times as many states as necessary, where m is the number of extra cuts. Formal state-minimization procedures can be used to reduce this larger table to the proper size, but it's a much better idea to select a minimal cut set in the first place.

A good example of a sequential circuit with multiple feedback loops is the commercial circuit design for a positive edge-triggered TTL D flip-flop that we showed in Figure 7-20. The circuit is redrawn in simplified form in Figure 7-78, assuming that the original circuit's PR_L and CLR_L inputs are never asserted, and also showing fictional buffers to break the three feedback loops. These three loops give rise to eight states, compared with the minimum of four states used by the two-loop design in Figure 7-19. We'll address this curious difference later.

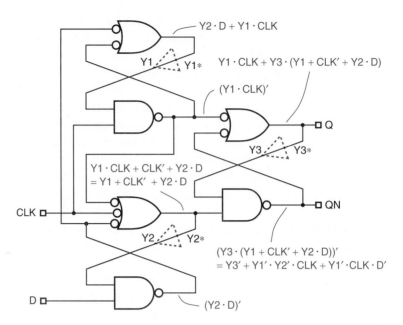

Figure 7-78
Simplified positive
edge-triggered
D flip-flop for analysis.

The following excitation and output equations can be derived from the logic diagram in Figure 7-78:

$$Y1* = Y2 \cdot D + Y1 \cdot CLK$$
$$Y2* = Y1 + CLK' + Y2 \cdot D$$
$$Y3* = Y1 \cdot CLK + Y1 \cdot Y3 + Y3 \cdot CLK' + Y2 \cdot Y3 \cdot D$$
$$Q = Y1 \cdot CLK + Y1 \cdot Y3 + Y3 \cdot CLK' + Y2 \cdot Y3 \cdot D$$
$$QN = Y3' + Y1' \cdot Y2' \cdot CLK + Y1' \cdot CLK \cdot D'$$

Figure 7-79
Transition table
for the D flip-flop
in Figure 7-78.

| Y1 Y2 Y3 | CLK D | | | |
	00	01	11	10
000	010	010	(000)	(000)
001	011	011	000	000
010	(010)	110	110	000
011	(011)	111	111	000
100	010	010	111	111
101	011	011	111	111
110	010	(110)	111	111
111	011	(111)	(111)	(111)
	Y1* Y2* Y3*			

	CLK D			
Y1 Y2 Y3	00	01	11	10
000	010	010	(000)	(000)
001	011	011	000	000
010	(010)	110	110	000
011	(011)	111	111	000
		Y1* Y2* Y3*		

Figure 7-80
Portion of the D flip-flop transition table showing a noncritical race.

The corresponding transition table is shown in Figure 7-79, with the stable total states circled. Before going further, we must introduce the concept of "races."

*7.9.3 Races

In a feedback sequential circuit, a *race* is said to occur when multiple internal variables change state as a result of a single input changing state. In the example of Figure 7-79, a race occurs in stable total state 011/00 when CLK is changed from 0 to 1. The table indicates that the next internal state is 000, a 2-variable change from 011.

race

As we've discussed previously, logic signals never really change "simultaneously." Thus, the internal state may make the change $011 \rightarrow 000$ as either $011 \rightarrow 001 \rightarrow 000$ or $011 \rightarrow 010 \rightarrow 000$. Figure 7-80 indicates that the example circuit, starting in total state 011/00, should go to total state 000/10 when CLK changes from 0 to 1. However, it may temporarily visit total state 001/10 or 010/10 along the way. That's OK, because the next internal state for both of these temporary states is 000; therefore, even in the temporary states, the excitation logic continues to drive the feedback loops toward the same stable total state, 000/10. Since the final state does not depend on the order in which the state variables change, this is called a *noncritical race*.

noncritical race

Now suppose that the next-state entry for total state 010/10 is changed to 110, as shown in Figure 7-81, and consider the case that we just analyzed. Starting in stable total state 011/00 and changing CLK to 1, the circuit may end up in internal state 000 or 111, depending on the order and speed of the internal variable changes. This is called a *critical race*.

critical race

WATCH OUT FOR CRITICAL RACES!   When you design a feedback-based sequential circuit, you must ensure that its transition table does not contain any critical races. Otherwise, the circuit may operate unpredictably, with the next state for racy transitions depending on factors like temperature, voltage, and the phase of the moon.

	CLK D			
Y1 Y2 Y3	00	01	11	10
000	010	010	(000)	(000)
001	011	011	000	000
010	(010)	110	110	110
011	(011)	111	111	000
100	010	010	111	111
101	011	011	111	111
110	010	(110)	111	111
111	011	(111)	(111)	(111)
		Y1* Y2* Y3*		

Figure 7-81
A transition table
containing a critical race.

*7.9.4 State Tables and Flow Tables

Analysis of the real transition table, Figure 7-79, for our example D flip-flip circuit, shows that it does not have any critical races; in fact, it has no races except the noncritical one we identified earlier. Once we've determined this fact, we no longer need to refer to state variables. Instead, we can name the state-variable combinations and determine the output values for each state/input combination to obtain a state/output table such as Figure 7-82.

Figure 7-82
State/output table
for the D flip-flop in
Figure 7-78.

			CLK D	
S	00	01	11	10
S0	S2 , 01	S2 , 01	(S0) , 01	(S0) , 01
S1	S3 , 10	S3 , 10	S0 , 10	S0 , 10
S2	(S2) , 01	S6 , 01	S6 , 01	S0 , 01
S3	(S3) , 10	S7 , 10	S7 , 10	S0 , 01
S4	S2 , 01	S2 , 01	S7 , 11	S7 , 11
S5	S3 , 10	S3 , 10	S7 , 10	S7 , 10
S6	S2 , 01	(S6) , 01	S7 , 11	S7 , 11
S7	S3 , 10	(S7) , 10	(S7) , 10	(S7) , 10
		S* , Q QN		

S	00	01	11	10
		CLK D		
S0	S2 , 01	S6 , 01	(S0) , 01	(S0) , 01
S2	(S2) , 01	S6 , 01	— , —	S0 , 10
S3	(S3) , 10	S7 , 10	— , —	S0 , 01
S6	S2 , 01	(S6) , 01	S7 , 11	— , —
S7	S3 , 10	(S7) , 10	(S7) , 10	(S7) , 10

S* , Q QN

Figure 7-83
Flow and output table
for the D flip-flop in
Figure 7-78.

The state table shows that for some single input changes, the circuit takes multiple "hops" to get to a new stable total state. For example, in state S0/11, an input change to 01 sends the circuit first to state S2 and then to stable total state S6/01. A *flow table* eliminates multiple hops and shows only the ultimate desti- *flow table* nation for each transition. The flow table also eliminates the rows for unused internal states—ones that are stable for no input combination—and eliminates the next-state entries for total states that cannot be reached from a stable total state as the result of a single input change. Using these rules, Figure 7-83 shows the flow table for our D flip-flop example.

The flip-flop's edge-triggered behavior can be observed in the series of state transitions shown in Figure 7-84. Let us assume that the flip-flop starts in internal state S0/10. That is, the flip-flop is storing a 0 (since Q = 0), CLK is 1, and D is 0. Now suppose that D changes to 1; the flow table shows that we move one cell to the left, still a stable total state with the same output value. We can change D between 0 and 1 as much as we want, and just bounce back and forth between these two cells. However, once we change CLK to 0, we move to

Figure 7-84
Flow and output table
showing the D flip-flop's
edge-triggered behavior.

S	00	01	11	10
		CLK D		
S0	S2 , 01	S6 , 01	(S0) , 01	(S0) , 01
S2	(S2) , 01	S6 , 01	— , —	S0 , 10
S3	(S3) , 10	S7 , 10	— , —	S0 , 01
S6	S2 , 01	(S6) , 01	S7 , 11	— , —
S7	S3 , 10	(S7) , 10	(S7) , 10	(S7) , 10

S* , Q QN

	CLK D			
S	00	01	11	10
SB	(SB) , 01	S6 , 01	(SB) , 01	(SB) , 01
S3	(S3) , 10	S7 , 10	— , —	SB , 01
S6	SB , 01	(S6) , 01	S7 , 11	— , —
S7	S3 , 10	(S7) , 10	(S7) , 10	(S7) , 10
	S* , Q QN			

Figure 7-85
Reduced flow and output table for a positive edge-triggered D flip-flop.

internal state S2 or S6, depending on whether D was 0 or 1 at the time; but still the output is unchanged. Once again, we can change D between 0 and 1 as much as we want, this time bouncing between S2 and S6 without changing the output.

The moment of truth finally comes when CLK changes to 1. Depending on whether we are in S2 or S6, we go back to S0 (leaving Q at 0) or to S7 (setting Q to 1). Similar behavior involving S3 and S7 can be observed on a rising clock edge that causes Q to change from 1 to 0.

In Figure 7-19 we showed a circuit for a positive edge-triggered D flip-flop with only two feedback loops and hence four states. The circuit that we just analyzed has three loops and eight states. Even after eliminating unused states, the flow table has five states. However, a formal state-minimization procedure can be used to show that states S0 and S2 are "compatible," so that they can be merged into a single state SB that handles the transitions for both original states, as shown in Figure 7-85. Thus, the job really could have been done by a four-state circuit. In fact, in Exercise 7.66 you'll show that the circuit in Figure 7-19 does the job specified by the reduced flow table.

*7.9.5 CMOS D Flip-Flop Analysis

CMOS flip-flops typically use transmission gates in their feedback loops. For example, Figure 7-86 shows the circuit design of the "FD1Q" positive-edge-triggered D flip-flop in LSI Logic's LCA500K series of CMOS gate arrays. Such a flip-flop can be analyzed in the same way as a purely logic-gate-based design, once you recognize the feedback loops. Figure 7-86 has two feedback loops, each of which has a pair of transmission gates in a mux-like configuration controlled by CLK and CLK', yielding the following loop equations:

$$Y1* = CLK' \cdot D' + CLK \cdot Y1$$
$$Y2* = CLK \cdot Y1' + CLK' \cdot Y2$$

Except for the double inversion of the data as it goes from D to Y2* (once in the Y1* equation and again in the Y2* equation), these equations are very reminiscent of the master/slave-latch structure of the D flip-flop in Figure 7-15.

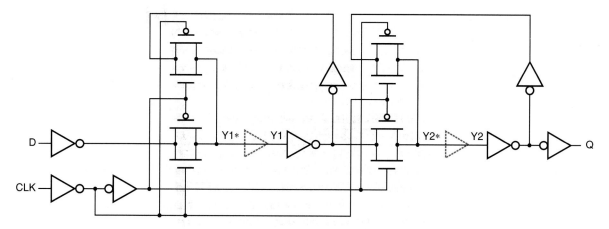

Figure 7-86 Positive edge-triggered CMOS D flip-flop for analysis.

Completing the formal analysis of the circuit is left as an exercise (7.73). Note, however, that since there are just two feedback loops, the resulting state and flow tables will have the minimum of just four states.

FEEDBACK CIRCUIT DESIGN

The feedback sequential circuits that we've analyzed in this section exhibit quite reasonable behavior, since, after all, they are latch and flip-flop circuits that have been used for years. However, if we throw together a "random" collection of gates and feedback loops, we won't necessarily get "reasonable" sequential circuit behavior. In a few rare cases we may not get a sequential circuit at all (see Exercise 7.67), and in many cases the circuit may be unstable for some or all input combinations (see Exercise 7.71). Thus, the design of feedback sequential circuits continues to be something of a black art and is practiced only by a small fraction of digital designers. Still, the next section introduces basic concepts that help you do simple designs.

---

*7.10 Feedback Sequential-Circuit Design

It's sometimes useful to design a small feedback sequential circuit, such as a specialized latch or a pulse catcher; this section will show you how. It's even possible that you might go on to be an IC designer and be responsible for designing high-performance latches and flip-flops from scratch. This section will serve as an introduction to the basic concepts you'll need, but you'll still need considerably more study, experience, and finesse to do it right.

*7.10.1 Latches

Although the design of feedback sequential circuits is generally a hard problem, some circuits can be designed pretty easily. Any circuit with one feedback loop

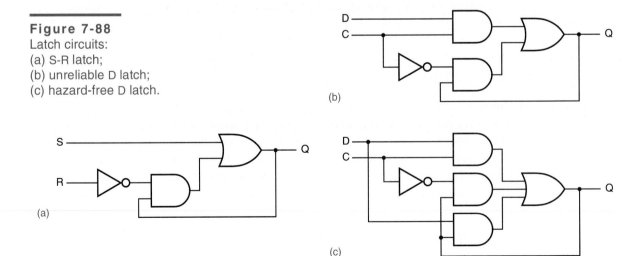

Figure 7-87
General structure
of a latch.

is just a variation of an S-R or D latch. It has the general structure shown in Figure 7-87 and an excitation equation with the following format:

$$Q* = (\text{forcing term}) + (\text{holding term}) \cdot Q$$

For example, the excitation equations for S-R and D latches are

$$Q* = S + R' \cdot Q$$
$$Q* = C \cdot D + C' \cdot Q$$

Corresponding circuits are shown in Figure 7-88(a) and (b).

hazard-free excitation logic

In general, *the excitation logic in a feedback sequential circuit must be hazard free;* we'll demonstrate this fact by way of an example. Figure 7-89(a) is a Karnaugh map for the D-latch excitation circuit of Figure 7-88(b). The map exhibits a static-1 hazard when D and Q are 1 and C is changing. Unfortunately, the latch's feedback loop may not hold its value if a hazard-induced glitch occurs. For example, consider what happens when D and Q are 1 and C changes from 1 to 0; the circuit should latch a 1. However, unless the inverter is very fast, the output of the top AND gate goes to 0 before the output of the bottom one goes to 1, the OR-gate output goes to 0, and the feedback loop stores a 0.

Figure 7-88
Latch circuits:
(a) S-R latch;
(b) unreliable D latch;
(c) hazard-free D latch.

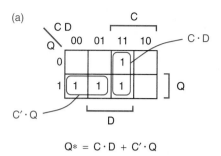

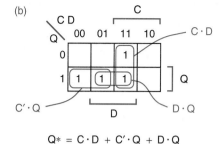

Figure 7-89
Karnaugh maps for
D-latch excitation
functions: (a) original,
containing a static-1
hazard; (b) hazard
eliminated.

Hazards can be eliminated using the methods described in Section 4.5. In the D latch, we simply include the consensus term in the excitation equation:

$$Q* = C \cdot D + C' \cdot Q + D \cdot Q$$

Figure 7-88(c) shows the corresponding hazard-free, correct D-latch circuit.

Now, suppose we need a specialized "D" latch with three data inputs, D1–D3, that stores a 1 only if D1–D3 = 010. We can convert this word description into an excitation equation that mimics the equation for a simple D latch:

$$Q* = C \cdot (D1' \cdot D2 \cdot D3') + C' \cdot Q$$

Eliminating hazards, we get

$$Q* = C \cdot D1' \cdot D2 \cdot D3' + C' \cdot Q + D1' \cdot D2 \cdot D3' \cdot Q$$

The hazard-free excitation equation can be realized with discrete gates or in a PLD, as we'll show in Section 8.2.6.

*7.10.2 Designing Fundamental-Mode Flow Table

To design feedback sequential circuits more complex than latches, we must first convert the word description into a flow table. Once we have a flow table, we can turn the crank (with some effort) to obtain a circuit.

When we construct the flow table for a feedback sequential circuit, we give each state a meaning in the context of the problem, much as we did in the design

---

PRODUCT-TERM EXPLOSION   In some cases, the need to cover hazards can cause an explosion in the number of product terms in a two-level realization of the excitation logic. For example, suppose we need a specialized latch with two control inputs, C1 and C2, and three data inputs as before. The latch is to be "open" only if both control inputs are 1, and is to store a 1 if any data input is 1. The minimal excitation equation is

$$Q* = C1 \cdot C2 \cdot (D1 + D2 + D3) + (C1 \cdot C2)' \cdot Q$$
$$= C1 \cdot C2 \cdot D1 + C1 \cdot C2 \cdot D2 + C1 \cdot C2 \cdot D3 + C1' \cdot Q + C2' \cdot Q$$

However, it takes six consensus terms to eliminate hazards (see Exercise 7.76).

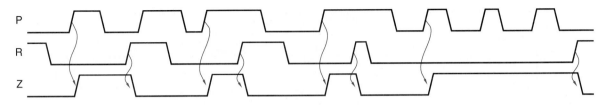

Figure 7-90 Typical functional behavior of a pulse-catching circuit.

primitive flow table

of clocked state machines. However, it's easier to get confused when constructing the flow table for a feedback sequential circuit, because not every total state is stable. Therefore, the recommended procedure is to construct a *primitive flow table*—one that has only one stable total state in each row. Since there is only one stable state per row, the output may be shown as a function of state only.

In a primitive flow table, each state has a more precise "meaning" than it might otherwise have, and the table's structure clearly displays the underlying fundamental-mode assumption: inputs change one at a time, with enough time between changes for the circuit to settle into a new stable state. A primitive flow table usually has extra states, but we can "turn the crank" later to minimize the number of states, once we have a flow table that we believe to be correct.

We'll use the following problem, a "pulse-catching" circuit, to demonstrate flow-table design:

> Design a feedback sequential circuit with two inputs, P (pulse) and R (reset), and a single output Z that is normally 0. The output should be set to 1 whenever a 0-to-1 transition occurs on P, and should be reset to 0 whenever R is 1. Typical functional behavior is shown in Figure 7-90.

Figure 7-91 is a primitive flow table for the pulse catcher. Let's walk through how this table was developed.

We assume that the pulse catcher is initially idle, with P and R both 0; this is the IDLE state, with Z = 0. In this state, if reset occurs (R = 1), we could probably stay in the same state, but since this is supposed to be a *primitive* flow table, we create a new state, RES1, so as not to have two stable total states in the same row. On the other hand, if a pulse occurs (P = 1) in the IDLE state, we definitely want to go to a different state, which we've named PLS1, since we've caught a pulse and we must set the output to 1. Input combination 11 is not considered in the IDLE state, because of the fundamental-mode assumption that only one input changes at a time; we assume the circuit always makes it to another stable state before input combination 11 can occur.

Next, we fill in the next-state entries for the newly created RES1 state. If reset goes away, we can go back to the IDLE state. If a pulse occurs, we must remain in a "reset" state, since, according to the timing diagram, a 0-to-1 transition that occurs while R is 1 is ignored. Again, to keep the flow table in primitive form, we must create a new state for this case, RES2.

Meaning	S	P R				Z
		00	01	11	10	
Idle, waiting for pulse	IDLE	(IDLE)	RES1	—	PLS1	0
Reset, no pulse	RES1	IDLE	(RES1)	RES2	—	0
Got pulse, output on	PLS1	PLS2	—	RES2	(PLS1)	1
Reset, got pulse	RES2	—	RES1	(RES2)	PLSN	0
Pulse gone, output on	PLS2	(PLS2)	RES1	—	PLS1	1
Got pulse, but output off	PLSN	IDLE	—	RES2	(PLSN)	0
				S*		

Figure 7-91 Primitive flow table for pulse-catching circuit.

Now that we have one stable total state in each column, we may be able to go to existing states for more transitions, instead of always defining new states. Sure enough, starting in stable total state PLS1/10, for R = 1 we can go to RES2, which fits the requirement of producing a 0 output. On the other hand, where should we go for P = 0? IDLE is a stable total state in the 00 column, but it produces the wrong output value. In PLS1, we've gotten a pulse and haven't seen a reset, so if the pulse goes away, we should go to a state that still has Z = 1. Thus, we must create a new state PLS2 for this case.

In RES2, we can safely go to RES1 if the pulse goes away. However, we've got to be careful if reset goes away, as shown in the timing diagram. Since we've already passed the pulse's 0-to-1 transition, we can't go to the PLS1 state, since that would give us a 1 output. Instead, we create a new state PLSN with a 0 output.

Finally, we can complete the next-state entries for PLS2 and PLSN without creating any new states. Notice that, starting in PLS2, we bounce back and forth between PLS2 and PLS1 and maintain a continuous 1 output if we get a series of pulses without an intervening reset input.

*7.10.3 Flow-Table Minimization

As we mentioned earlier, a primitive flow table usually has more states than required. However, there exists a formal procedure, discussed in the References, for minimizing the number of states in a flow table. This procedure is often complicated by the existence of don't-care entries in the flow table.

Fortunately, our example flow table is small and simple enough to minimize by inspection. States IDLE and RES1 produce the same output, and they have the same next-state entry for input combinations where they are both specified. Therefore, they are compatible and may be replaced by a single state

	P R				
S	00	01	11	10	Z
IDLE	(IDLE)	(IDLE)	RES	PLS	0
PLS	(PLS)	IDLE	RES	(PLS)	1
RES	IDLE	IDLE	(RES)	(RES)	0
			S*		

Figure 7-92
Reduced flow table for pulse-catching circuit.

(IDLE) in a reduced flow table. The same can be said for states PLS1 and PLS2 (replaced by PLS) and for RES2 and PLSN (replaced by RES). The resulting reduced flow table, which has only three states, is shown in Figure 7-92.

*7.10.4 Race-Free State Assignment

The next somewhat creative (read "difficult") step in feedback sequential circuit design is to find a race-free assignment of coded states to the named states in the reduced flow table. Recall from Section 7.9.3 that a race occurs when multiple internal variables change state as a result of a single input change. A feedback-based sequential circuit must not contain any critical races; otherwise, the circuit may operate unpredictably. As we'll see, eliminating races often necessitates *increasing* the number of states in the circuit.

A circuit's potential for having races in its transition table can be analyzed *state adjacency* by means of a *state adjacency diagram* for its flow table. The adjacency diagram *diagram* is a simplified state diagram that omits self-loops and does not show the direction of other transitions (A→B is drawn the same as B→A) or the input combinations that cause them. Figure 7-93 is an example fundamental-mode flow table and Figure 7-94(a) is the corresponding adjacency diagram.

adjacent states Two states are said to be *adjacent* if there is an arc between them in the state adjacency diagram. For race-free transitions, adjacent coded states must differ in only one bit. If two states A and B are adjacent, it doesn't matter whether

Figure 7-93
Example flow table for the state-assignment problem.

	X Y			
S	00	01	11	10
A	(A)	B	(A)	B
B	(B)	(B)	D	(B)
C	(C)	A	A	(C)
D	(D)	B	(D)	C
		S*		

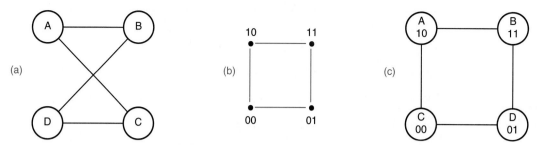

Figure 7-94 State-assignment example: (a) adjacency diagram; (b) a 2-cube; (c) one of eight possible race-free state assignments.

the original flow table had transitions from A to B, from B to A, or both. Any one of these transitions is a race if A and B differ in more than one variable. That's why we don't need to show the direction of transitions in an adjacency diagram.

The problem of finding a race-free assignment of states to n state variables is equivalent to the problem of mapping the nodes and arcs of the adjacency diagram onto the nodes and arcs of an n-cube. In Figure 7-94, the problem is to map the adjacency diagram (a) onto a 2-cube (b). You can visually identify eight ways to do this (four rotations times two flips), one of which produces the state assignment shown in (c).

Figure 7-95(a) is an adjacency diagram for our pulse-catching circuit, based on the reduced flow table in Figure 7-92. Clearly, there's no way to map this "triangle" of states onto a 2-cube. At this point, we can only go back and modify the original flow table. In particular, the flow table tells us the destination state that we *eventually* must reach for each transition, but it doesn't prevent us from going through other states on the way. As shown in Figure 7-96, we can create a new state RESA and make the transition from PLS to RES by going through RESA. The modified state table has the new adjacency diagram shown in Figure 7-95(b), which has many race-free assignments possible. A transition table based on the assignment in (c) is shown in Figure 7-98. Note that the PLS→RESA→RES transition will be slower than the other transitions in the

Figure 7-95 Adjacency diagrams for the pulse catcher: (a) using original flow table; (b) after adding a state; (c) showing one of eight possible race-free state assignments.

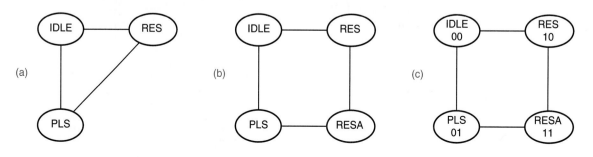

		P R			
S	00	01	11	10	Z
IDLE	(IDLE)	(IDLE)	RES	PLS	0
PLS	(PLS)	IDLE	RESA	(PLS)	1
RESA	—	—	RES	—	—
RES	IDLE	IDLE	(RES)	(RES)	0
			S*		

Figure 7-96
State table allowing a race-free assignment for the pulse-catching circuit.

original flow table because it requires two internal state changes, with two propagation delays through the feedback loops.

Even though we added a state in the previous example, we still got by with just two state variables. However, we may sometimes have to add one or more state variables to make a race-free assignment. Figure 7-97(a) shows the worst possible adjacency diagram for four states—every state is adjacent to every other state. Clearly, this adjacency diagram cannot be mapped onto a 2-cube. However, there is a race-free assignment of states to a 3-cube, shown in (b), where each state in the original flow table is represented by two equivalent states in the final state table. Both states in a pair, such as A1 and A2, are equivalent and produce the same output. Each state is adjacent to one of the states in every other pair, so a race-free transition may be selected for each next-state entry.

HANDLING THE GENERAL ASSIGNMENT CASE   In the general case of a flow table with 2^n rows, it can be shown that a race-free assignment can be obtained using $2n - 1$ state variables (see References). However, there aren't many applications for fundamental-mode circuits with more than a few states, so the general case is of little more than academic interest.

Figure 7-97
A worst-case scenario: (a) 4-state adjacency diagram; (b) assignment using pairs of equivalent states.

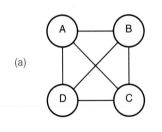

(a)

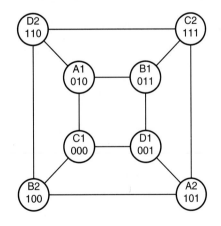

(b)

		P R			
Y1 Y2	00	01	11	10	Z
00	(00)	(00)	10	01	0
01	(01)	00	11	(01)	1
11	—	—	10	—	–
10	00	00	(10)	(10)	0
		Y1*	Y2*		

Figure 7-98
Race-free transition
table for the pulse-
catching circuit.

*7.10.5 Excitation Equations

Once we have a race-free transition table for a circuit, we can just "turn the crank" to obtain excitation equations for the feedback loops. Figure 7-99 shows Karnaugh maps derived from Figure 7-98, the pulse catcher's transition table. Notice that "don't-care" next-state and output entries give rise to corresponding entries in the maps, simplifying the excitation and output logic. The resulting minimal sum-of-products excitation and output equations are as follows:

$$Y1* = P \cdot R + P \cdot Y1$$
$$Y2* = Y2 \cdot R' + Y1' \cdot Y2 \cdot P + Y1' \cdot P \cdot R'$$
$$Z = Y2$$

Recall that the excitation logic in a feedback sequential circuit must be hazard free. The sum-of-products expressions we derived happen to be hazard free as well as minimal. The logic diagram of Figure 7-100 uses these expressions to build the pulse-catching circuit.

Figure 7-99 Karnaugh maps for pulse-catcher excitation and output logic.

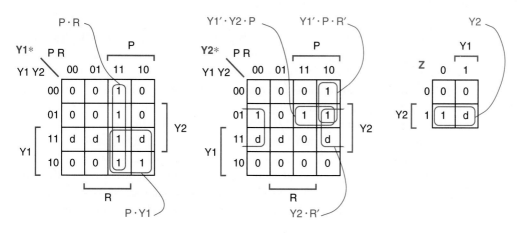

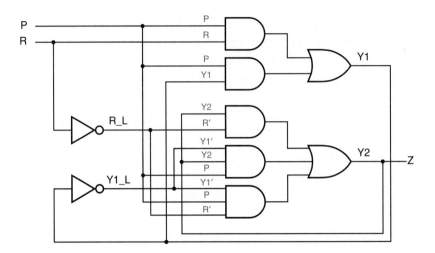

Figure 7-100
Pulse-catching
circuit.

*7.10.6 Essential Hazards

After all this effort, you'd think that we'd have a pulse-catching circuit that would operate reliably all of the time. Unfortunately, we're not quite there yet. A fundamental-mode circuit must generally satisfy five requirements for proper operation:

1. Only one input signal may change at a time, with a minimum bound between successive input changes.

2. There must be a maximum propagation delay through the excitation logic and feedback paths; this maximum must be less than the time between successive input changes.

3. The state assignment (transition table) must be free of critical races.

4. The excitation logic must be hazard free.

5. The minimum propagation delay through the excitation logic and feedback paths must be greater than the maximum timing skew through the "input logic."

Without the first requirement, it would be impossible to satisfy the major premise of fundamental-mode operation—that the circuit has time to settle into a stable total state between successive input changes. The second requirement says that the excitation logic is fast enough to do just that. The third requirement ensures that the proper state changes are made, even if the excitation circuits for different state variables have different delays. The fourth requirement ensures that state variables that aren't supposed to change on a particular transition don't.

The last requirement deals with subtle timing-dependent errors that can occur in fundamental-mode circuits, even ones that satisfy the first four requirements. An *essential hazard* is the possibility of a circuit's going to an

essential hazard

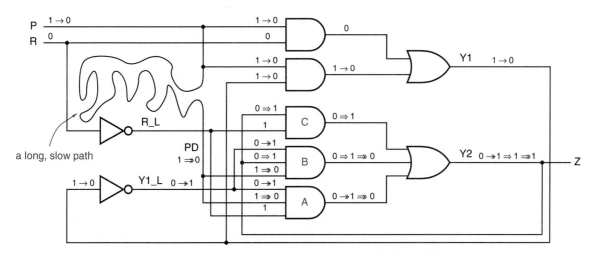

Figure 7-101 Physical conditions in pulse-catching circuit for exhibiting an essential hazard.

erroneous next state as the result of a single input change; the error occurs if the input change is not seen by all of the excitation circuits before the resulting state-variable transition(s) propagate back to the inputs of the excitation circuits. In a world where "faster is better" is the usual rule, a designer may sometimes have to *slow down* excitation logic to mask these hazards.

Essential hazards are best explained in terms of an example, our pulse-catching circuit. Suppose we built our circuit on a PCB or a chip, and we (or, more likely, our CAD system) inadvertently connected input signal P through a long, slow path at the point shown in Figure 7-101. Let's assume that this delay is longer than the propagation delay of the AND-OR excitation logic.

Now consider what can happen if P R = 10, the circuit is in internal state 10, and P changes from 1 to 0. According to the transition table, repeated in Figure 7-102, the circuit should go to internal state 00, and that's that. But let's look at the actual operation of the circuit, as traced in Figure 7-101:

- (Changes shown with "→") The first thing that happens after P changes is that Y1 changes from 1 to 0. Now the circuit is in internal state 00.

- (Changes shown with "→↠") Y1_L changes from 0 to 1. The change in Y1_L at AND gate A causes its output to go to 1, which in turn forces Y2 to 1. Whoops, now the circuit is in internal state 01.

- (Changes shown with "⇒") The change in Y2 at AND gates B and C causes their outputs to go to 1, reinforcing the 1 output at Y2. All this time, we've been waiting for the 1-to-0 change in P to appear at point PD.

- (Changes shown with "⇒↠") Finally, PD changes from 1 to 0, forcing the outputs of AND gates A and B to 0. However, AND gate C still has a 1 output, and the circuit remains in state 01—*the wrong state*.

Y1 Y2	P R 00	01	11	10	Z
00	00	00	10	01	0
01	01	00	11	01	1
11	—	—	10	—	—
10	00	00	10	10	0

Y1* Y2*

Figure 7-102
Transition table for the pulse-catching circuit, exhibiting an essential hazard.

The only way to avoid this erroneous behavior in general is to ensure that changes in P arrive at the inputs of all the excitation circuits before any changes in state variables do. Thus, the inevitable difference in input arrival times, called *timing skew*, must be less than the propagation delay of the excitation circuits and feedback loops. This timing requirement can generally be satisfied only by careful design *at the electrical circuit level*.

timing skew

In the example circuit, it would appear that the hazard is easily masked, even by non-electrical engineers, since the designer need only ensure that a straight wire has shorter propagation delay than an AND-OR structure—easy in most technologies.

Still, many feedback sequential circuits, such as the TTL edge-triggered D flip-flop in Figure 7-19, have essential hazards in which the input skew paths include inverters. In such cases, the input inverters must be guaranteed to be faster than the excitation logic; that's not so trivial in either board-level or IC design. For example, if the excitation circuit in Figure 7-101 were physically built using AND-OR-INVERT gates, the delay from input changes to Y1_L could be very short indeed, as short as the delay through a single inverter.

Essential hazards can be found in most but not all fundamental-mode circuits. There's an easy rule for detecting them; in fact, this is the definition of "essential hazard" in some texts:.

- A fundamental-mode flow table contains an essential hazard for a stable total state S and an input variable X if, starting in state S, the stable total state reached after three successive transitions in X is different from the stable total state reached after one transition in X.

THESE HAZARDS ARE, WELL, ESSENTIAL!   Essential hazards are called "essential" because they are inherent in the flow table for a particular sequential function and will appear in any circuit realization of that function. They can be masked only by controlling the delays in the circuit. Compare with static hazards in combinational logic, where we could eliminate hazards by adding consensus terms to a logic expression.

Thus, the essential hazard in the pulse catcher is detected by the arrows in Figure 7-102, starting in internal state 10 with P R = 10.

A fundamental-mode circuit must have at least three states to have an essential hazard, so latches don't have them. On the other hand, all flip-flops (circuits that sample inputs on a clock edge) do.

*7.10.7 Summary

In summary, you use the following steps to design a feedback sequential circuit:

1. Construct a primitive flow table from the circuit's word description.

2. Minimize the number of states in the flow table.

3. Find a race-free assignment of coded states to named states, adding auxiliary states or splitting states as required.

4. Construct the transition table.

5. Construct excitation maps and find a hazard-free realization of the excitation equations.

6. Check for essential hazards. Modify the circuit if necessary to ensure that minimum excitation and feedback delays are greater than maximum inverter or other input-logic delays.

7. Draw the logic diagram.

Also note that some circuits routinely violate the basic fundamental-mode assumption that inputs change one at a time. For example, in a positive-edge-triggered D flip-flop, the D input may change at the same time that CLK changes from 1 to 0, and the flip-flop still operates properly. The same thing certainly cannot be said at the 0-to-1 transition of CLK. Such situations require analysis of the transition table and circuit on a case-by-case basis if proper operation in "special cases" is to be guaranteed.

A FINAL QUESTION	Given the difficulty of designing fundamental-mode circuits that work properly, let alone ones that are fast or compact, how did anyone ever come up with the 6-gate, 8-state, commercial D flip-flop design in Figure 7-20? Don't ask me, I don't know!

7.11 ABEL Sequential-Circuit Design Features

7.11.1 Registered Outputs

ABEL has several features that support the design of sequential circuits. As we'll show in Section 8.3, most PLD outputs can be configured by the user to be *registered outputs* that provide a D flip-flop following the AND-OR logic, as in Figure 7-103. To configure one or more outputs to be registered, an ABEL

registered output

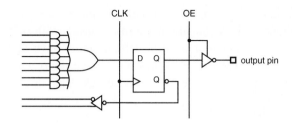

Figure 7-103
PLD registered output.

reg

program's pin declarations normally must contain an `istype` clause using the keyword "`reg`" (rather than "`com`") for each registered output. Table 7-22 is an example program that has three registered outputs and two combinational outputs.

As suggested by Figure 7-103, a registered output has at least two other attributes associated with it. The three-state buffer driving the output pin has an output-enable input OE, and the flip-flop itself has a clock input CLK. As shown in Table 7-22, the signals that drive these inputs are specified in the equations section of the program. Each input signal is specified as the corresponding main output signal name followed by an attribute suffix, `.CLK` or `.OE`. Some PLDs have flip-flops with additional controllable inputs. For example, asynchronous reset and preset inputs have attribute suffixes `.AR` and `.AP`; synchronous reset and preset inputs use suffixes `.SR` and `.SP`. And some PLDs provide flip-flop types other than D; their inputs are specified with suffixes like `.J` and `.K`.

.CLK
.OE
.AR
.AP
.SR
.SP

Within the `equations` section of the ABEL program, the logic values for registered outputs are established using the *clocked assignment operator, :=*. When the PLD is compiled, the expression on the righthand side will be applied to the D input of the output flip-flop. All of the same rules as for combinational outputs apply to output polarity control, on-set generation, don't-cares, and so on. In Table 7-22, the state bits Q1–Q3 are registered outputs, so they use clocked assignment, "`:=`". The UNLK and HINT signals are Mealy outputs, combinational functions of current state and input, so they use unclocked assignment, "`=`". A machine with pipelined outputs (Figure 7-37 on page 552), would instead use clocked assignment for such outputs.

clocked assignment operator, :=

ABEL's truth-table syntax (Table 4-16 on page 257) can also be used with registered outputs. The only difference is that the "`->`" operator between input and output items is changed to "`:>`".

clocked truth-table operator, :>

```
module CombLock
Title 'Combination-Lock State Machine'

" Input and Outputs
X, CLOCK        pin;
UNLK, HINT      pin istype 'com';
Q1, Q2, Q3      pin istype 'reg';

Q = [Q1..Q3];

Equations

Q.CLK = CLOCK; Q.OE = 1;

" State variables
Q1 := Q1 & !Q2 & X  #  !Q1 & Q2 & Q3 & !X  #  Q1 & Q2 & !Q3;
Q2 := !Q2 & Q3 & X  #  Q2 & !Q3 & X;
Q3 := Q1 & !Q2 & !Q3  #  Q1 & Q3 & !X  #  !Q2 & !X
      #  !Q1 & !Q3 & !X  #  Q2 & !Q3 & X;

" Mealy outputs
UNLK = Q1 & Q2 & Q3 & !X;
HINT = !Q1 & !Q2 & !Q3 & !X  #  Q1 & !Q2 & X  #  !Q2 & Q3 & X
       #  Q2 & Q3 & !X  #  Q2 & !Q3 & X;

end CombLock
```

Table 7-22
ABEL program using
registered outputs.

You can also design feedback sequential circuits in ABEL without using any of the language's sequential-circuit features. For example, in Section 8.2.6 we show how to specify latches using ABEL.

7.11.2 State Diagrams

The state-machine example in the previous subsection is just a transcription of the combination-lock machine that we synthesized by hand in Section 7.4.6 beginning on page 582. However, most PLD programming languages have a notation for defining, documenting, and synthesizing state machines directly, without ever writing a state, transition, or excitation table or deriving excitation equations by hand. Such a notation is called a *state-machine description language*. In ABEL, this notation is called a "state diagram," and the ABEL compiler does all the work of generating excitation equations that realize the specified machine.

state-machine description language

In ABEL, the keyword `state_diagram` indicates the beginning of a state-machine definition. Table 7-23 shows the textual structure of an ABEL "state diagram." Here *state-vector* is an ABEL set that lists the state variables of the machine. If there are n variables in the set, then the machine has 2^n possible states corresponding to the 2^n different assignments of constant values to

state_diagram

state-vector

Table 7-23
Structure of a "state diagram" in ABEL.

```
state_diagram state-vector
state state-value 1 : transition statement;
state state-value 2 : transition statement;
    . . .
state state-value 2^n : transition statement;
```

variables in the set. States are usually given symbolic names in an ABEL program; this makes it easy to try different assignments simply by changing the constant definitions.

state

The on-set for each state variable is created according to the information in the "state diagram." If a state variable also appears on the lefthand side of an equation in the `equations` section, the effects are cumulative (see the box on page 739 for a useful application of this behavior). The keyword `state` indicates that the next states and current outputs for a particular current state are about to be defined; a *state-value* is a constant that defines state-variable values for the current state. A *transition statement* defines the possible next states for the current state.

state-value
transition statement

GOTO statement
IF statement

ABEL has two commonly used transition statements. The *GOTO statement* unconditionally specifies the next state, for example "GOTO INIT". The *IF statement* defines the possible next states as a function of an arbitrary logic expressions. (There's also a seldom-used `CASE` statement that we don't cover.)

Table 7-24 shows the syntax of the ABEL IF statement. Here *TrueState* and *FalseState* are state values that the machine will go to if *LogicExpression* is true or false, respectively. These statements can be nested: *FalseState* can itself be another IF statement, and *TrueState* can be an IF statement if it is enclosed in braces. When multiple next states are possible, a nested IF-THEN-ELSE structure eliminates the ambiguities that can occur in hand-drawn state diagrams, where the transition conditions leaving a state can overlap (Section 7.5).

Our first example using ABEL's "state diagram" capability is based on our first state-machine design example from Section 7.4.1 on page 564. A state table for this machine was developed in Figure 7-49 on page 567. It is adapted to ABEL in Table 7-25. Several characteristics of this program should be noted:

- The definition of QSTATE uses three variables to encode state.
- The definitions of INIT–XTRA3 determine the individual state encodings.
- IF-THEN-ELSE statements are nested. A particular next state may appear in multiple places in one set of nested IF-THEN-ELSE clauses (e.g., see states OK0 and OK1).

Table 7-24
Structure of an ABEL IF statement.

```
IF LogicExpression THEN
    TrueState;
ELSE
    FalseState;
```

```
module SMEX1
title 'PLD Version of Example State Machine'

" Input and output pins
CLOCK, RESET_L, A, B              pin;
Q1..Q3                           pin istype 'reg';
Z                                pin istype 'com';

" Definitions
QSTATE = [Q1,Q2,Q3];             " State variables
INIT  = [ 0, 0, 0];
A0    = [ 0, 0, 1];
A1    = [ 0, 1, 0];
OK0   = [ 0, 1, 1];
OK1   = [ 1, 0, 0];
XTRA1 = [ 1, 0, 1];
XTRA2 = [ 1, 1, 0];
XTRA3 = [ 1, 1, 1];
RESET = !RESET_L;

state_diagram QSTATE

state INIT: IF RESET THEN INIT
            ELSE IF (A==0) THEN A0
            ELSE A1;

state A0:   IF RESET THEN INIT
            ELSE IF (A==0) THEN OK0
            ELSE A1;

state A1:   IF RESET THEN INIT
            ELSE IF (A==0) THEN A0
            ELSE OK1;

state OK0:  IF RESET THEN INIT
            ELSE IF (B==1)&(A==0) THEN OK0
            ELSE IF (B==1)&(A==1) THEN OK1
            ELSE IF (A==0) THEN OK0
            ELSE IF (A==1) THEN A1;

state OK1:  IF RESET THEN INIT
            ELSE IF (B==1)&(A==0) THEN OK0
            ELSE IF (B==1)&(A==1) THEN OK1
            ELSE IF (A==0) THEN A0
            ELSE IF (A==1) THEN OK1;

state XTRA1: GOTO INIT;
state XTRA2: GOTO INIT;
state XTRA3: GOTO INIT;

equations

QSTATE.CLK = CLOCK; QSTATE.OE = 1;
Z = (QSTATE == OK0) # (QSTATE == OK1);

END SMEX1
```

Table 7-25
An example of ABEL's state-diagram notation.

Table 7-26
Reduced equations
for SMEX1 PLD.

```
Q1 := (!Q2.FB & !Q3.FB & RESET_L
    # Q1.FB & RESET_L);
Q1.C = (CLOCK);
Q1.OE = (1);

Q2 := (Q1.FB & !Q3.FB & RESET_L & !A
    # Q1.FB & Q3.FB & RESET_L & A
    # Q1.FB & Q2.FB & RESET_L & B);
Q2.C = (CLOCK);
Q2.OE = (1);

Q3 := (!Q2.FB & !Q3.FB & RESET_L & A
    # Q1.FB & RESET_L & A);
Q3.C = (CLOCK);
Q3.OE = (1);

Z = (Q2 & Q1);
```

- Expressions like "(B==1)*(A==0)" were used instead of equivalents like "B*!A" only because the former are a bit more readable.

- The first IF statement in each of states INIT–OK1 ensures that the machine goes to the INIT state if RESET is asserted.

- Next-state equations are given for XTRA1–XTRA3 to ensure that the machine goes to a "safe" state if it somehow gets into an unused state.

- The single equation in the "equations" section of the program determines the behavior of the Moore-type output.

Table 7-26 shows the resulting excitation and output equations produced by the ABEL compiler (the reverse-polarity equations are not shown). Notice the use of variable names like "Q1.FB" in the righthand sides of the equations. Here, the ".FB" attribute suffix refers to the "feedback" signal into the AND-OR array coming from the flip-flop's Q output. This is done to make it clear that the signal is coming from the flip-flop, not from the corresponding PLD output pin, which can be selected in some complex PLDs. As shown in Figure 7-104, ABEL

USE IT OR ELSE

ABEL's IF-THEN-ELSE structure eliminates the transition ambiguity that can occur in state diagrams. However, the ELSE clause of an IF statement is optional. If it is omitted, the next state for some input combinations will be unspecified. Usually this is not the designer's intention.

Nevertheless, if you can guarantee that the unspecified input combinations will never occur, you may be able to reduce the size of the transition logic. If the @DCSET directive is given, the ABEL compiler treats the transition outputs for the unspecified state/input combinations as "don't-cares." In addition, it treats *all* transitions out of unused states as "don't-cares."

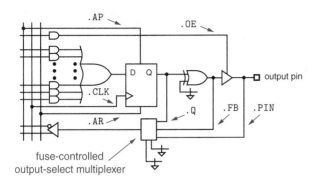

Figure 7-104
Output selection
capability in a
complex PLD.

actually allows you to select among three possible values on the righthand side
of an equation using an attribute suffix on the signal name:

.Q The actual flip-flop output pin before any programmable inversion.   *.Q*

.FB A value equal to the value that the output pin would have if enabled.   *.FB*

.PIN The actual signal at the PLD output pin. This signal is floating or driven   *.PIN*
      by another device if the three-state driver is not enabled.

Obviously, the .PIN value should not be used in a state-machine excitation
equation, since it is not guaranteed always to equal the state variable.

Despite the use of "high-level language," the program's author still had to
refer to the original, hand-constructed state table in Figure 7-49 to come up with
ABEL version in Table 7-25. A different approach is shown in Table 7-27. This
program was developed directly from the word description of the state machine,
which is repeated below:

Design a clocked synchronous state machine with two inputs, A and B, and
a single output Z that is 1 if:

— A had the same value at each of the two previous clock ticks, *or*
— B has been 1 since the last time that the first condition was true.

Otherwise, the output should be 0.

A key idea in the new approach is to remove the last value of A from the
state definitions, and instead to have a separate flip-flop that keeps track of it
(LASTA). Then only two non-INIT states must be defined: LOOKING ("still
looking for a match") and OK ("got a match or B has been 1 since last match").
The Z output is a simple combinational decode of the OK state.

PHANTOM **(OF THE)** **OPERAND**	Real CPLDs typically have only a two-input output-select multiplexer and omit the .FB input shown in Figure 7-104. When an equation calls for a signal with the .FB attribute, the ABEL compiler uses the corresponding .Q signal and simply adjusts it with the appropriate inversion (or not).

Table 7-27
A more "natural" ABEL program for the example state machine.

```
module SMEX2
title 'Alternate Version of Example State Machine'

" Input and output pins
CLOCK, RESET_L, A, B           pin;
LASTA, Q1, Q2                  pin istype 'reg';
Z                              pin istype 'com';

" Definitions
QSTATE  = [Q1,Q2];             " State variables
INIT    = [ 0, 0];             " State encodings
LOOKING = [ 0, 1];
OK      = [ 1, 0];
XTRA    = [ 1, 1];
RESET = !RESET_L;

state_diagram QSTATE

state INIT:    IF RESET THEN INIT ELSE LOOKING;

state LOOKING: IF RESET THEN INIT
               ELSE IF (A == LASTA) THEN OK
               ELSE LOOKING;

state OK:      IF RESET THEN INIT
               ELSE IF B THEN OK
               ELSE IF (A == LASTA) THEN OK
               ELSE LOOKING;

state XTRA:    GOTO INIT;

equations
LASTA.CLK = CLOCK; QSTATE.CLK = CLOCK; QSTATE.OE = 1;

LASTA := A;
Z = (QSTATE == OK);

END SMEX2
```

*7.11.3 External State Memory

In some situations, the state memory of a PLD-based state machine may be kept in flip-flops external to the PLD. ABEL provides a special version of the state_diagram statement to handle this situation:

state_diagram *current-state-variables* -> *next-state variables*

current-state-variables
next-state-variables

Here *current-state-variables* is an ABEL set that lists the input signals which represent the current state of the machine, and *next-state-variables* is a set that lists the corresponding output signals which are the excitation for external D flip-flops holding the state of the machine, for example,

state_diagram [CURQ1, CURQ2] -> [NEXTQ1, NEXTQ2]

```
state_diagram state-variables
state state-value 1 :
        optional equation ;
        optional equation ;
        . . .
        transition statement ;
state state-value 2 :
        optional equation ;
        optional equation ;
        . . .
        transition statement ;
. . .
state state-value 2ⁿ :
        optional equation ;
        optional equation ;
        . . .
        transition statement ;
```

Table 7-28
Structure of an ABEL
state diagram with
Moore outputs
defined.

*7.11.4 Specifying Moore Outputs

The output Z in our example state machine is a Moore output, a function of state only, and we defined this output in Tables 7-25 and 7-27 using an appropriate equation in the equations section of the program. Alternatively, ABEL allows Moore outputs to be specified along with the state definitions themselves. The transition statement in a state definition may be preceded by one or more optional equations, as shown in Table 7-28. To use this capability with the machine in Table 7-27, for example, we would eliminate the Z equation in the equations section and rewrite the state diagram as shown in Table 7-29.

As in other ABEL equations, when a variable such as Z appears on the left-hand side of multiple equations, the righthand sides are OR'ed together to form the final result (as discussed in Section 4.6.3). Also notice that Z is still specified

```
state_diagram QSTATE

state INIT:      Z = 0;
                 IF RESET THEN INIT ELSE LOOKING;
state LOOKING:   Z = 0;
                 IF RESET THEN INIT
                 ELSE IF (A == LASTA) THEN OK
                 ELSE LOOKING;
state OK:        Z = 1;
                 IF RESET THEN INIT
                 ELSE IF B THEN OK
                 ELSE IF (A == LASTA) THEN OK
                 ELSE LOOKING;
state XTRA:      Z = 0;
                 GOTO INIT;
```

Table 7-29
State machine with
embedded Moore
output definitions.

as a combinational, not a registered, output. If Z were a registered output, the desired output value would occur one clock tick after the machine visited the corresponding state.

*7.11.5 Specifying Mealy and Pipelined Outputs with WITH

Some state-machine outputs are functions of the inputs as well as state. In Section 7.3.2 we called them Mealy outputs or pipelined outputs, depending on whether they occurred immediately upon an input change or only after a clock edge. ABEL's *WITH statement* provides a way to specify these outputs side by side with the next states, rather than separately in the equations section of the program.

WITH statement

As shown in Table 7-30, the syntax of the WITH statement is very simple. Any next-state value which is part of a transition statement can be followed by the keyword WITH and a bracketed list of equations that are "executed" for the specified transition. Formally, let "E" an excitation expression that is true only when the specified transition is to be taken. Then for each equation in the WITH's bracketed list, the righthand side is AND'ed with E and assigned to the lefthand side. The equations can use either unclocked or clocked assignment to create Mealy or pipelined outputs, respectively.

Table 7-30
Structure of ABEL
WITH statement.

```
next-state WITH {
    equation;
    equation;
    . . .
}
```

We developed an example "combination lock" state machine with Mealy outputs in Table 7-14 on page 582. The same state machine is specified by the ABEL program in Table 7-31, using WITH statements for the Mealy outputs. Note that closing brackets take the place of the semicolons that normally end the transition statements for the states.

Based on the combination lock's word description, it is not possible to realize UNLK and HINT as pipelined outputs, since they depend on the current value of X. However, if we redefine UNLK to be asserted for the entire "unlocked" state, and HINT to be the actual recommended next value of X, we can create a new machine with pipelined outputs, as shown in Table 7-32. Notice that we used the clocked assignment operator for the outputs. More importantly, notice that the values of UNLK and HINT are different than in the Mealy example, since they have to "look ahead" one clock tick.

Because of "lookahead," pipelined outputs can be more difficult than Mealy outputs to design and understand. In the example above, we even had to modify the problem statement to accommodate them. The advantage of

```
module SMEX4
title 'Combination-Lock State Machine'

" Input and output pins
CLOCK, X                        pin;
Q1..Q3                          pin istype 'reg';
UNLK, HINT                      pin istype 'com';

" Definitions
S         = [Q1,Q2,Q3];         " State variables
ZIP       = [ 0, 0, 0];         " State encodings
X0        = [ 0, 0, 1];
X01       = [ 0, 1, 0];
X011      = [ 0, 1, 1];
X0110     = [ 1, 0, 0];
X01101    = [ 1, 0, 1];
X011011   = [ 1, 1, 0];
X0110111  = [ 1, 1, 1];

state_diagram S

state ZIP:      IF X==0 THEN X0     WITH {UNLK = 0; HINT = 1}
                ELSE ZIP           WITH {UNLK = 0; HINT = 0}

state X0:       IF X==0 THEN X0     WITH {UNLK = 0; HINT = 0}
                ELSE X01           WITH {UNLK = 0; HINT = 1}

state X01:      IF X==0 THEN X0     WITH {UNLK = 0; HINT = 0}
                ELSE X011          WITH {UNLK = 0; HINT = 1}

state X011:     IF X==0 THEN X0110 WITH {UNLK = 0; HINT = 1}
                ELSE ZIP           WITH {UNLK = 0; HINT = 0}

state X0110:    IF X==0 THEN X0     WITH {UNLK = 0; HINT = 0}
                ELSE X01101        WITH {UNLK = 0; HINT = 1}

state X01101:   IF X==0 THEN X0     WITH {UNLK = 0; HINT = 0}
                ELSE X011011       WITH {UNLK = 0; HINT = 1}

state X011011:  IF X==0 THEN X0110 WITH {UNLK = 0; HINT = 0}
                ELSE X0110111      WITH {UNLK = 0; HINT = 1}

state X0110111: IF X==0 THEN X0     WITH {UNLK = 1; HINT = 1}
                ELSE ZIP           WITH {UNLK = 0; HINT = 0}

equations
S.CLK = CLOCK;

END SMEX4
```

Table 7-31
State machine with embedded Mealy output definitions.

pipelined outputs is that, since they are connected directly to register outputs, they are valid a few gate-delays sooner after a state change than Moore or Mealy outputs, which normally include additional combinational logic. In the combination-lock example, it's probably not that important to open your lock or see your hint a few nanoseconds earlier. However, shaving off a few gate delays can be quite important in high-speed applications.

Table 7-32
State machine with embedded pipelined output definitions.

```
module SMEX5
title 'Combination-Lock State Machine'

" Input and output pins
CLOCK, X                        pin;
Q1..Q3, UNLK, HINT              pin istype 'reg';

" Definitions
S        = [Q1,Q2,Q3];          " State variables
ZIP      = [ 0, 0, 0];          " State encodings
X0       = [ 0, 0, 1];
X01      = [ 0, 1, 0];
X011     = [ 0, 1, 1];
X0110    = [ 1, 0, 0];
X01101   = [ 1, 0, 1];
X011011  = [ 1, 1, 0];
X0110111 = [ 1, 1, 1];

state_diagram S

    state ZIP:       IF X==0 THEN X0    WITH {UNLK := 0; HINT := 1}
                     ELSE ZIP           WITH {UNLK := 0; HINT := 0}

    state X0:        IF X==0 THEN X0    WITH {UNLK := 0; HINT := 1}
                     ELSE X01           WITH {UNLK := 0; HINT := 1}

    state X01:       IF X==0 THEN X0    WITH {UNLK := 0; HINT := 1}
                     ELSE X011          WITH {UNLK := 0; HINT := 0}

    state X011:      IF X==0 THEN X0110 WITH {UNLK := 0; HINT := 1}
                     ELSE ZIP           WITH {UNLK := 0; HINT := 0}

    state X0110:     IF X==0 THEN X0    WITH {UNLK := 0; HINT := 1}
                     ELSE X01101        WITH {UNLK := 0; HINT := 1}

    state X01101:    IF X==0 THEN X0    WITH {UNLK := 0; HINT := 1}
                     ELSE X011011       WITH {UNLK := 0; HINT := 1}

    state X011011:   IF X==0 THEN X0110 WITH {UNLK := 0; HINT := 1}
                     ELSE X0110111      WITH {UNLK := 1; HINT := 0}

    state X0110111:  IF X==0 THEN X0    WITH {UNLK := 0; HINT := 1}
                     ELSE ZIP           WITH {UNLK := 0; HINT := 0}

equations
S.CLK = CLOCK; UNLK.CLK = CLOCK; HINT.CLK = CLOCK;

END SMEX5
```

7.11.6 Test Vectors

Test vectors for sequential circuits in ABEL have the same uses and limitations as test vectors for combinational circuits, as described in Section 4.6.7. One important addition to their syntax is the use of the constant ".C." to denote a clock edge, $0 \rightarrow 1 \rightarrow 0$. Thus, Table 7-33 is an ABEL program, with test vectors, for a simple 8-bit register with a clock-enable input. A variety of vectors are used to test loading and holding different input values.

.C., clock edge

```
module REG8EN
title '8-bit register with clock enable'

" Input and output pins
CLK, EN, D1..D8        pin;
Q1..Q8                 pin istype 'reg';

" Sets
D = [D1..D8];
Q = [Q1..Q8];

equations

Q.CLK = CLK;

WHEN EN == 1 THEN Q := D ELSE Q := Q;

test_vectors ([CLK, EN,  D  ] -> [ Q  ])
             [.C.,   1, ^h00] -> [^h00]; " 0s in every bit
             [.C.,   0, ^hFF] -> [^h00]; " Hold capability, EN=0
             [.C.,   1, ^hFF] -> [^hFF]; " 1s in every bit
             [.C.,   0, ^h00] -> [^hFF]; " Hold capability
             [.C.,   1, ^h55] -> [^h55]; " Adjacent bits shorted
             [.C.,   0, ^hAA] -> [^h55]; " Hold capability
             [.C.,   1, ^hAA] -> [^hAA]; " Adjacent bits shorted
             [.C.,   1, ^h55] -> [^h55]; " Load with quick setup
             [.C.,   1, ^hAA] -> [^hAA]; " Again
END REG8EN
```

Table 7-33
ABEL program with test vectors for a simple 8-bit register.

   A typical approach to testing state machines is to write vectors that cause the machine not only to visit every state, but also to exercise every transition from every state. A key difference and challenge compared to combinational-circuit test vectors is that the vectors must first drive the machine into the desired state before testing a transition, and then come back again for each different transition from that state.

   Thus, Table 7-34 shows test vectors for the state machine in Table 7-27. It's important to understand that, unlike combinational vectors, these vectors work only if applied in exactly the order they are written. Notice that the vectors were written to be independent of the state encoding. As a result, they don't have to be modified if the state encoding is changed.

   We encounter another challenge if we attempt to create test vectors for the combination-lock state machine of Table 7-31 on page 637. This machine has a major problem when it comes to testing—it has no reset input. Its starting state at power-up may be different in PLD devices and technologies—the individual flip-flops may be all set, all reset, or all in random states. In the machine's actual application, we didn't necessarily need a reset input, but for testing purposes we somehow have to get to a known starting state.

Table 7-34 Test vectors for the state machine in Table 7-27.

```
test_vectors
([RESET_L, CLOCK, A, B] -> [QSTATE , LASTA, Z])
  [   0  ,  .C. , 0, 0] -> [INIT   ,  0  , 0]; " Check -->INIT (RESET)
  [   0  ,  .C. , 1, 0] -> [INIT   ,  1  , 0]; "  and LASTA flip-flop
  [   1  ,  .C. , 0, 0] -> [LOOKING,  0  , 0]; " Come out of initialization
  [   0  ,  .C. , 0, 0] -> [INIT   ,  0  , 0]; " Check LOOKING-->INIT (RESET)
  [   1  ,  .C. , 0, 0] -> [LOOKING,  0  , 0]; " Come out of initialization
  [   1  ,  .C. , 1, 0] -> [LOOKING,  1  , 0]; " --> LOOKING since 0!=1
  [   1  ,  .C. , 1, 0] -> [OK     ,  1  , 1]; " --> OK since 1==1
  [   0  ,  .C. , 0, 0] -> [INIT   ,  0  , 0]; " Check OK-->INIT (RESET)
  [   1  ,  .C. , 0, 0] -> [LOOKING,  0  , 0]; " Go back towards OK ...
  [   1  ,  .C. , 0, 0] -> [OK     ,  0  , 1]; " --> OK since 0==0
  [   1  ,  .C. , 1, 1] -> [OK     ,  1  , 1]; " --> OK since B, even though 1!=0
  [   1  ,  .C. , 1, 0] -> [OK     ,  1  , 1]; " --> OK since 1==1
  [   1  ,  .C. , 0, 0] -> [LOOKING,  0  , 0]; " --> LOOKING since 0!=1
```

synchronizing sequence

Luckily, the combination-lock machine has a *synchronizing sequence*—a fixed sequence of one or more input values that will always drive it to a certain known state. In particular, starting from any state, if we apply X=1 to the machine for four ticks, we will always be in state ZIP by the fourth tick. This is the approach taken by the first four vectors in Table 7-35. Until we get to the known state, we indicate the next state on the righthand side of the vector as being "don't care," so the simulator or the physical device tester will not flag a random state as an error.

Once we get going, we encounter something else that's new—Mealy outputs that must be tested. As shown by the fourth and fifth vectors, we don't have to transition the clock in every test vector. Instead, we can keep the clock fixed at 0, where the last transition left it, and observe the Mealy output values produced by the two input values of X. Then we can test the next state transition.

For the state transitions, we list the expected next state but we show the output values as don't-cares. For a correct test vector, the outputs must show the values attained *after* the transition, a function of the *next* state. Although it's possible to figure them out and include them, the complexity is enough to give you a headache, and they will be tested by the next CLOCK=0 vectors anyway.

Creating test vectors for a state machine by hand is a painstaking process, and no matter how careful you are, there's no guarantee that you've tested all its functions and potential hardware faults. For example, the vectors in Table 7-34

SYNCHRONIZING SEQUENCES AND RESET INPUTS

We lucked out with the combination lock; not all state machines have synchronizing sequences. This is why most state machines are designed with a reset input, which in effect allows a synchronizing sequence of length one.

Table 7-35 Test vectors for the combination-lock state machine of Table 7-31.

```
test_vectors
([CLOCK, X] -> [   S    , UNLK, HINT])
 [ .C. , 1] -> [.X.    , .X. , .X. ]; " Since no reset input, apply
 [ .C. , 1] -> [.X.    , .X. , .X. ]; "   a 'synchronizing sequence'
 [ .C. , 1] -> [.X.    , .X. , .X. ]; "   to reach a known starting
 [ .C. , 1] -> [ZIP    , .X. , .X. ]; "   state
 [ 0   , 0] -> [ZIP    , 0   , 1   ]; " Test Mealy outputs for both
 [ 0   , 1] -> [ZIP    , 0   , 0   ]; "   values of X
 [ .C. , 1] -> [ZIP    , .X. , .X. ]; " Test ZIP-->ZIP (X==1)
 [ .C. , 0] -> [X0     , .X. , .X. ]; "   and ZIP-->X0 (X==0)
 [ 0   , 0] -> [X0     , 0   , 0   ]; " Test Mealy outputs for both
 [ 0   , 1] -> [X0     , 0   , 1   ]; "   values of X
 [ .C. , 0] -> [X0     , .X. , .X. ]; " Test X0-->X0 (X==0)
 [ .C. , 1] -> [X01    , .X. , .X. ]; "   and X0-->X01 (X==1)
 [ 0   , 0] -> [X01    , 0   , 0   ]; " Test Mealy outputs for both
 [ 0   , 1] -> [X01    , 0   , 1   ]; "   values of X
 [ .C. , 0] -> [X0     , .X. , .X. ]; " Test X01-->X0 (X==0)
 [ .C. , 1] -> [X01    , .X. , .X. ]; " Get back to X01
 [ .C. , 1] -> [X011   , .X. , .X. ]; " Test X01-->X011 (X==1)
```

do not test (A LASTA) = 10 in state LOOKING, or (A B LASTA) = 100 in state OK. Thus, generating a complete set of test vectors for fault-detection purposes is a process best left to an automatic test-generation program. In Table 7-35 we petered out after writing vectors for the first few states; completing the test vectors is left as an exercise (7.92). Still, on the functional testing side, writing a few vectors to exercise the machine's most basic functions can weed out obvious design errors early in the process. More subtle design errors are best detected by a thorough system-level simulation.

7.12 VHDL Sequential-Circuit Design Features

Most of the VHDL features that are needed to support sequential-circuit design, in particular, processes, were already introduced in Section 4.7 and were used in the VHDL sections in Chapter 5. This section introduces just a couple more features and gives simple examples of how they are used. Larger examples appear in the VHDL sections of Chapter 8.

7.12.1 Feedback Sequential Circuits

A VHDL process and the simulator's event-list mechanism for tracking signal changes form the fundamental basis for handling feedback sequential circuits in VHDL. Remember that feedback sequential circuits may change state in response to input changes, and these state changes are manifested by changes propagating in a feedback loop until the feedback loop stabilizes. In simulation,

Table 7-36
Dataflow VHDL for
an S-R latch.

```
library IEEE;
use IEEE.std_logic_1164.all;

entity Vsrlatch is
  port (S, R: in STD_LOGIC;
        Q, QN: buffer STD_LOGIC );
end Vsrlatch;

architecture Vsrlatch_arch of Vsrlatch is
begin
  QN <= S nor Q;
  Q  <= R nor QN;
end Vsrlatch_arch;
```

this is manifested by the simulator putting signal changes on the event list and scheduling processes to rerun in "delta time" and propagate these signal changes until no more signal changes are scheduled.

Table 7-36 is a VHDL program for an S-R latch. The architecture contains two concurrent assignment statements, each of which gives rise to a process, as discussed in Section 4.7.9. These processes interact to create the simple latching behavior of an S-R latch.

The VHDL simulation is faithful enough to handle the case where both S and R are asserted simultaneously. The most interesting result in simulation occurs if you negate S and R simultaneously. Recall from the box on page 536 that a real S-R latch may oscillate or go into a metastable state in this situation. The simulation will potentially loop forever as each execution of one assignment statement triggers another execution of the other. After some number of repetitions, a well-designed simulator will discover the problem—delta time keeps advancing while simulated time does not—and halt the simulation.

**WHAT DO 'U'
WANT?**

It would be nice if the S-R-latch model in Table 7-36 produced a 'U' output whenever S and R were negated simultaneously, but it's not *that* good. However, the language is powerful enough that experienced VHDL designers can easily write a model with that behavior. Such a model would make use of VHDL's time-modeling facilities, which we haven't discussed, to model the latch's "recovery time" (see box on page 537) and force a 'U' output if a second input change occurred too soon. It's even possible to model a maximum assumed metastability resolution time in this way.

Note that if a circuit has the possibility of entering a metastable state, there's no guarantee that the simulation will detect it, especially in larger designs. The best way to avoid metastability problems in a system design is to clearly identify and protect its asynchronous boundaries, as discussed in Section 8.9.

```
library IEEE;
use IEEE.std_logic_1164.all;

entity VposDff is
  port (CLK, CLR, D: in STD_LOGIC;
        Q, QN: out STD_LOGIC );
end VposDff;

architecture VposDff_arch of VposDff is
begin
  process (CLK, CLR)
  begin
    if CLR='1' then Q <= '0'; QN <= '1';
    elsif CLK'event and CLK='1' then Q <= D; QN <= not D;
    end if;
  end process;
end VposDff_arch;
```

Table 7-37
Behavioral VHDL
for a positive-edge-
triggered D flip-flop.

7.12.2 Clocked Circuits

In practice, the majority of digital designs that are modeled using VHDL are clocked, synchronous systems using edge-triggered flip-flops. In addition to what we've already learned about VHDL, there's just one more feature needed to describe edge-triggered behavior. The *event attribute* can be attached to a signal name to yield a value of type boolean that is true if an event on the signal caused the encompassing process to run in the current simulation cycle, and false otherwise.

event attribute

Using the event attribute, we can model the behavior of a positive-edge-triggered D flip-flop with asynchronous clear as shown in Table 7-37. Here, the asynchronous clear input CLR overrides any behavior on the clock input CLK and is therefore checked first, in the "if" clause. If CLR is negated, then the "elsif" clause is checked, and its statements are executed on the rising edge of CLK. Note that "CLK'event" is true for any change on CLK, and "CLK='1'" is checked to limit triggering to just the rising edge of CLK. There are many other ways to construct processes or statements with edge-triggered behavior; Table 7-38 shows two more ways to describe a D flip-flop (without a CLR input).

In the test bench for a clocked circuit, one of things you need to do is to generate a system clock signal. This can be done quite easily with a loop inside a process, as shown in Table 7-39 for a 100-MHz clock with a 60% duty cycle.

```
process
  wait until CLK'event and CLK='1';
  Q <= D;
end process;

Q <= D when CLK'event and CLK='1' else Q;
```

Table 7-38
Two more ways to
describe a positive-
edge-triggered
D flip-flop.

SYNTHESIS STUFF

You may be wondering, how does a synthesis tool convert the edge-triggered behavior described in Table 7-37 or 7-38 into an efficient flip-flop realization? Most tools recognize only a few predetermined ways of expressing edge-triggered behavior and map those into predetermined flip-flop components in the target technology.

The Synopsis synthesis engine used in the Xilinx Foundation Series 1.5 software recognizes the "CLK'event and CLK='1'" expression that we use in this book. Even with that as a given, VHDL has many different ways of expressing the same functionality, as we showed in Table 7-38. Peter Ashenden, author of *The Designer's Guide to VHDL* (Morgan Kaufmann, 1996), ran these statements and one other, with some modification, through several different synthesis tools. Only one of them was able to synthesize three out of the four forms; most could handle only two. So, you need to follow the method that is prescribed by the tool you use.

Table 7-39
Clock process within a test bench.

```
architecture TB_arch of TB is
signal MCLK: STD_LOGIC;
signal ... -- Declare other input and output signals

process -- Clock generator
begin
  MCLK <= 1;  -- Start at 1 at time 0
  loop
    MCLK <= 0 after 6 ns;
    MCLK <= 1 after 4 ns;
  end loop;
end process;

process -- Generate the rest of the input stimuli, check outputs
begin
  ...
end;
```

References

The problem of metastability has been around for a long time. Greek philosophers wrote about the problem of indecision thousands of years ago. A group of modern philosophers named Devo sang about metastability in the title song of their *Freedom of Choice* album. And the U.S. Congress still can't decide how to "save" Social Security.

Scan capability was first deployed in latches, not flip-flops, in IBM IC designs decades ago. Edward J. McCluskey has a very good discussion of this and other scan methods in *Logic Design Principles* (Prentice Hall, 1986).

Most ASICs and MSI-, PLD- and FPGA-based designs use the sequential-circuit types described in this chapter. However, there are other types that are used in both older discrete designs (going all the way back to vacuum-tube logic) and as well as in modern, custom VLSI designs.

For example, clocked synchronous state machines are a special case of a more general class of *pulse-mode circuits*. Such circuits have one or more *pulse inputs* such that (a) only one pulse occurs at a time; (b) nonpulse inputs are stable when a pulse occurs; (c) only pulses can cause state changes; and (d) a pulse causes at most one state change. In clocked synchronous state machines, the clock is the single pulse input, and a "pulse" is the triggering edge of the clock. However, it is also possible to build circuits with multiple pulse inputs, and it is possible to use storage elements other than the familiar edge-triggered flip-flops. These possibilities are discussed very thoroughly in McCluskey's *Logic Design Principles* (Prentice Hall, 1986).

pulse-mode circuit
pulse input

A particularly important type of pulse-mode circuit that is discussed by McCluskey and others is the *two-phase latch machine*. The rationale for a two-phase clocking approach in VLSI circuits is discussed by Carver Mead and Lynn Conway in *Introduction to VLSI Systems* (Addison-Wesley, 1980). These machines essentially eliminate the essential hazards present in edge-triggered flip-flops by using pairs of latches that are enabled by nonoverlapping clocks.

two-phase latch machine

Methods for reducing both completely and incompletely specified state tables are described in advanced logic design texts, including McCluskey's 1986 book. A more mathematical discussion of these methods and other "theoretical" topics in sequential machine design appears in *Switching and Finite Automata Theory*, second edition, by Zvi Kohavi (McGraw-Hill, 1978).

As we showed in this chapter, improperly constructed state diagrams may yield an ambiguous description of next-state behavior. The "IF-THEN-ELSE" structures in HDLs like ABEL and VHDL can eliminate these ambiguities, but they were not the first to do so. *Algorithmic-state-machine (ASM)* notation, a flowchart-like equivalent of nested IF-THEN-ELSE statements, has been around for over 25 years. So-called *ASM charts* were pioneered at Hewlett-Packard Laboratories by Thomas E. Osborne and were further developed by Osborne's colleague Christopher R. Clare in a book, *Designing Logic Systems Using State Machines* (McGraw-Hill, 1973). Design and synthesis methods using ASM charts subsequently found a home in many digital design texts, including *The Art of Digital Design* by F. P. Prosser and D. E. Winkel (Prentice-Hall, 1987, second edition) and *Digital Design* by M. Morris Mano (Prentice-Hall, 1984), as well as in the first two editions of the book you're reading.

algorithmic state machine (ASM)
ASM chart

Another notation for describing state machines is an extension of "traditional" state-diagram notation called the mnemonic documented state (MDS) diagram; it was developed by William I. Fletcher in *An Engineering Approach to Digital Design* (Prentice-Hall, 1980). ASM charts and MDS diagrams have now been largely replaced by HDLs and their compilers.

Many CAD environments for digital design include a graphical state-diagram entry tool. Unfortunately, these typically support only traditional state diagrams, making it very easy for a designer to create an ambiguous description of next-state behavior. As a result, my personal recommendation is that you stay away from state-diagram editors and instead use an HDL to describe your state machines.

We mentioned the importance of synchronizing sequences in connection with state-machine test vectors. There's actually a very well developed but almost forgotten theory and practice of synchronizing sequences and somewhat less powerful "homing experiments," described by Frederick C. Hennie in *Finite-State Models for Logical Machines* (Wiley, 1968). Unless you've got this old classic on your bookshelf and know how to apply its teachings, please just remember to provide a reset input in every state machine that you design!

Drill Problems

7.1   Give three examples of metastability that occur in everyday life, other than ones discussed in this chapter.

7.2   Sketch the outputs of an S-R latch of the type shown in Figure 7-5 for the input waveforms shown in Figure X7.2. Assume that input and output rise and fall times are zero, that the propagation delay of a NOR gate is 10 ns, and that each time division below is 10 ns.

Figure X7.2

7.3   Repeat Drill 7.2 using the input waveforms shown in Figure X7.3. Although you may find the result unbelievable, this behavior can actually occur in real devices whose transition times are short compared to their propagation delay.

Figure X7.3

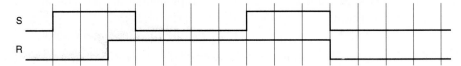

7.4   Figure 7-34 showed how to build a T flip-flop with enable using a D flip-flop and combinational logic. Show how to build a D flip-flop using a T flip-flop with enable and combinational logic.

7.5   Show how to build a J-K flip-flop using a T flip-flop with enable and combinational logic.

7.6   Show how to build an S-R latch using a single 74x74 positive-edge-triggered D flip-flop and *no* other components.

7.7   Show how to build a flip-flop equivalent to the 74x109 positive-edge-triggered J-$\overline{K}$ flip-flop using a 74x74 positive-edge-triggered D flip-flop and one or more gates from a 74x00 package.

7.8   Show how to build a flip-flop equivalent to the 74x74 positive-edge-triggered D flip-flop using a 74x109 positive-edge-triggered J-$\overline{K}$ flip-flop and *no* other components.

7.9   Analyze the clocked synchronous state machine in Figure X7.9. Write excitation equations, excitation/transition table, and state/output table (use state names A–D for Q1 Q2 = 00–11).

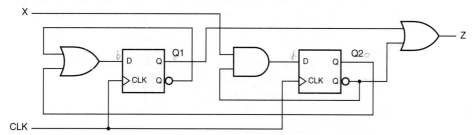

Figure X7.9

7.10   Repeat Drill 7.9, swapping AND and OR gates in the logic diagram. Is the new state/output table the "dual" of the original one? Explain.

7.11   Draw a state diagram for the state machine described by Table 7-6.

7.12   Draw a state diagram for the state machine described by Table 7-12.

7.13   Draw a state diagram for the state machine described by Table 7-14.

7.14   Construct a state and output table equivalent to the state diagram in Figure X7.14. Note that the diagram is drawn with the convention that the state does not change except for input conditions that are explicitly shown.

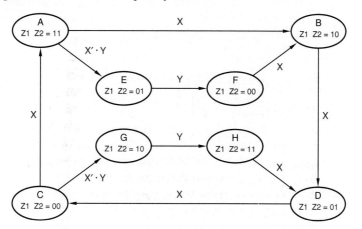

Figure X7.14

7.15   Analyze the clocked synchronous state machine in Figure X7.15. Write excitation equations, excitation/transition table, and state table (use state names A–H for Q2 Q1 Q0 = 000–111).

Figure X7.15

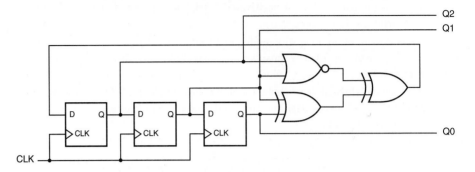

7.16   Analyze the clocked synchronous state machine in Figure X7.16. Write excitation equations, excitation/transition table, and state/output table (use state names A–H for Q1 Q2 Q3 = 000–111).

Figure X7.16

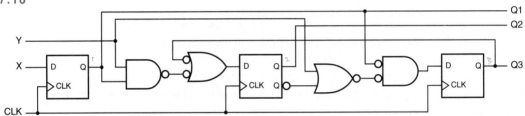

7.17   Analyze the clocked synchronous state machine in Figure X7.17. Write excitation equations, transition equations, transition table, and state/output table (use state names A–D for Q1 Q2 = 00–11). Draw a state diagram, and draw a timing diagram for CLK, X, Q1, and Q2 for 10 clock ticks, assuming that the machine starts in state 00 and X is continuously 1.

Figure X7.17

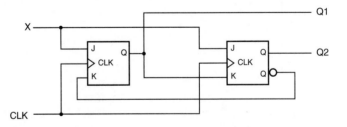

7.18   Analyze the clocked synchronous state machine in Figure X7.18. Write excitation equations, transition equations, transition table, and state/output table (use state names A–D for Q1 Q0 = 00–11). Draw a state diagram, and draw a timing diagram for CLK, EN, Q1, and Q0 for 10 clock ticks, assuming that the machine starts in state 00 and EN is continuously 1.

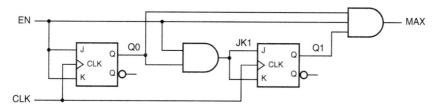

Figure X7.18

7.19   Analyze the clocked synchronous state machine in Figure X7.19. Write excitation equations, excitation/transition table, and state/output table (use state names A–D for Q1 Q2 = 00–11).

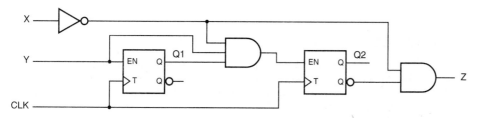

Figure X7.19

7.20   All of the state diagrams in Figure X7.20 are ambiguous. List all of the ambiguities in these state diagrams. (*Hint:* Use Karnaugh maps where necessary to find uncovered and double-covered input combinations.)

Figure X7.20

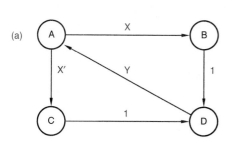

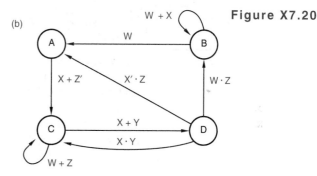

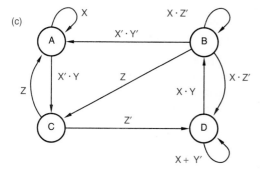

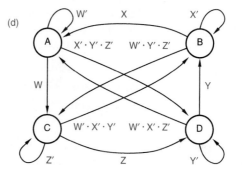

7.21 Synthesize a circuit for the state diagram of Figure 7-64 using six variables to encode the state, where the LA–LC and RA–RC outputs equal the state variables themselves. Write a transition list, a transition equation for each state variable as a sum of p-terms, and simplified transition/excitation equations for a realization using D flip-flops. Draw a circuit diagram using SSI and MSI components.

7.22 Starting with the transition list in Table 7-18, find a minimal sum-of-products expression for Q2*, assuming that the next states for the unused states are true don't-cares.

7.23 Modify the state diagram of Figure 7-64 so that the machine goes into hazard mode immediately if LEFT and RIGHT are asserted simultaneously during a turn. Write the corresponding transition list.

Exercises

7.24 Explain how metastability occurs in a D latch when the setup and hold times are not met, analyzing the behavior of the feedback loop inside the latch.

7.25 What is the minimum setup time of a pulse-triggered flip-flop such as a master/slave J-K or S-R flip-flop? (*Hint:* It depends on certain characteristics of the clock.)

7.26 Describe a situation, other than the metastable state, in which the Q and QN outputs of a 74x74 edge-triggered D flip-flop may be noncomplementary for an arbitrarily long time.

7.27 Compare the circuit in Figure X7.27 with the D latch in Figure 7-12. Prove that the circuits function identically. In what way is Figure X7.27, which is used in some commercial D latches, better?

Figure X7.27

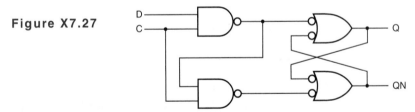

7.28 Suppose that a clocked synchronous state machine with the structure of Figure 7-35 is designed using D latches with active-high C inputs as storage elements. For proper next-state operation, what relationships must be satisfied among the following timing parameters?

t_{Fmin}, t_{Fmax} Minimum and maximum propagation delay of the next-state logic.

t_{CQmin}, t_{CQmax} Minimum and maximum clock-to-output delay of a D latch.

t_{DQmin}, t_{DQmax} Minimum and maximum data-to-output delay of a D latch.

t_{setup}, t_{hold} Setup and hold times of a D latch.

t_H, t_L Clock HIGH and LOW times.

7.29 Redesign the state machine in Drill 7.9 using just three inverting gates—NAND or NOR—and no inverters.

7.30   Draw a state diagram for a clocked synchronous state machine with two inputs, INIT and X, and one Moore-type output Z. As long as INIT is asserted, Z is continuously 0. Once INIT is negated, Z should remain 0 until X has been 0 for two successive ticks and 1 for two successive ticks, regardless of the order of occurrence. Then Z should go to 1 and remain 1 until INIT is asserted again. Your state diagram should be neatly drawn and planar (no crossed lines). (*Hint:* No more than ten states are required.)

7.31   Repeat Exercise 7.30, but write the state diagram in ABEL.

7.32   Design a clocked synchronous state machine that checks a serial data line for even parity. The circuit should have two inputs, SYNC and DATA, in addition to CLOCK, and one Moore-type output, ERROR. Devise a state/output table that does the job using just four states and include a description of each state's meaning in the table. Choose a 2-bit state assignment, write transition and excitation equations, and draw the logic diagram. Your circuit may use D flip-flops, J-K flip-flops, or one of each.

7.33   Repeat Exercise 7.32, but do the design using ABEL and a GAL16V8 PLD.

7.34   Design a clocked synchronous state machine with the state/output table shown in Table X7.34, using D flip-flops. Use two state variables, Q1 Q2, with the state assignment $A = 00$, $B = 01$, $C = 11$, $D = 10$.

Table X7.34

	X		
S	0	1	Z
A	B	D	0
B	C	B	0
C	B	A	1
D	B	C	0
	S*		

7.35   Repeat Exercise 7.34 using J-K flip-flops.

7.36   Write a new transition table and derive minimal-cost excitation and output equations for the state table in Table 7-6 using the "simplest" state assignment in Table 7-7 and D flip-flops. Compare the cost of your excitation and output logic (when realized with a two-level AND-OR circuit) with the circuit in Figure 7-54.

7.37   Repeat Exercise 7.36 using the "almost one-hot" state assignment in Table 7-7.

7.38   Suppose that the state machine in Figure 7-54 is to be built using 74LS74 D flip-flops. What signals should be applied to the flip-flop preset and clear inputs?

7.39   Write new transition and excitation tables and derive minimal-cost excitation and output equations for the state table in Table 7-6 using the "simplest" state assignment in Table 7-7 and J-K flip-flops. Compare the cost of your excitation and output logic (when realized with a two-level AND-OR circuit) with the circuit in Figure 7-56.

7.40   Repeat Exercise 7.39 using the "almost one-hot" state assignment in Table 7-7.

7.41   Construct an application table similar to Table 7-10 for each of the following flip-flop types: (a) S-R; (b) T with enable; (c) D with enable. Discuss the unique problem that you encounter when trying to make the most efficient use of don't-cares with one of these flip-flops.

7.42   Construct a new excitation table and derive minimal-cost excitation and output equations for the state machine of Table 7-8 using T flip-flops with enable inputs (Figure 7-33). Compare the cost of your excitation and output logic (when realized with a two-level AND-OR circuit) with the circuit in Figure 7-54.

7.43   Determine the full 8-state table of the circuit in Figure 7-54. Use the names U1, U2, and U3 for the unused states (001, 010, and 011). Draw a state diagram and explain the behavior of the unused states.

7.44   Repeat Exercise 7.43 for the circuit of Figure 7-56.

7.45   Write a transition table for the nonminimal state table in Figure 7-51(a) that results from assigning the states in binary counting order, INIT–OKA1 = 000–110. Write corresponding excitation equations for D flip-flops, assuming a minimal-cost disposition of the unused state 111. Compare the cost of your equations with the minimal-cost equations for the minimal state table presented in the text.

7.46   Write the application table for a T flip-flop with enable.

7.47   In many applications, the outputs produced by a state machine during or shortly after reset are irrelevant, as long as the machine begins to behave correctly a short time after the reset signal is removed. If this idea is applied to Table 7-6, the INIT state can be removed and only two state variables are needed to code the remaining four states. Redesign the state machine using this idea. Write a new state table, transition table, excitation table for D flip-flops, minimal-cost excitation and output equations, and logic diagram. Compare the cost of the new circuit with that of Figure 7-54.

7.48   Repeat Exercise 7.47 using J-K flip-flops, and use Figure 7-56 to compare cost.

7.49   Redesign the 1s-counting machine of Table 7-12, assigning the states in binary counting order (S0–S3 = 00, 01, 10, 11). Compare the cost of the resulting sum-of-products excitation equations with the ones derived in the text.

7.50   Repeat Exercise 7.49 using J-K flip-flops.

7.51   Repeat Exercise 7.49 using T flip-flops with enable.

7.52   Redesign the 1s-counting machine of Table 7-12 as an ABEL state diagram. Try to find a state assignment that minimizes the total number of product terms, assuming that you can use either polarity of output equations. How many different state assignments must you examine?

7.53   Redesign the combination-lock machine of Table 7-14, assigning coded states in Gray-code order (A–H = 000, 001, 011, 010, 110, 111, 101, 100). Compare the cost of the resulting sum-of-products excitation equations with the ones derived in the text.

7.54   Find a 3-bit state assignment for the combination-lock machine of Table 7-14 that results in less costly excitation equations than the ones derived in the text. (*Hint:* Use the fact that inputs 1–3 are the same as inputs 4–6 in the required input sequence.)

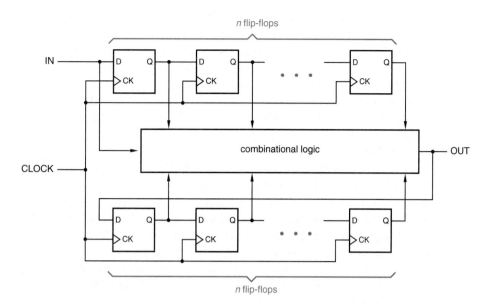

Figure X7.56

7.55   What changes would be made to the excitation and output equations for the combination-lock machine in Section 7.4.6 as the result of performing a formal multiple-output minimization procedure (Section 4.3.8) on the five functions? You need not construct 31 product maps and go through the whole procedure; you should be able to "eyeball" the excitation and output maps in Section 7.4.6 to see what savings are possible.

7.56   The output of a *finite-memory machine* is completely determined by its current input and its inputs and outputs during the previous n clock ticks, where n is a finite, bounded integer. Any machine that can be realized as shown in Figure X7.56 is a finite-memory machine. Note that a finite-*state* machine need not be a finite-*memory* machine; for example, a modulo-n counter with an enable input and a "MAX" output has only n states, but its output may depend on the value of the enable input at every clock tick since initialization. Show how to realize the combination-lock machine of Table 7-14 as a finite-memory machine.

finite-memory machine

7.57   Synthesize a circuit for the ambiguous state diagram in Figure 7-62. Use the state assignment in Table 7-16. Write a transition list, a transition equation for each state variable as a sum of p-terms, and simplified transition/excitation equations for a realization using D flip-flops. Determine the actual next state of the circuit, starting from the IDLE state, for each of the following input combinations on (LEFT, RIGHT, HAZ): (1,0,1), (0,1,1), (1,1,0), (1,1,1). Comment on the machine's behavior in these cases.

7.58   Suppose that for a state SA and an input combination I, an ambiguous state diagram indicates that there are two next states, SB and SC. The actual next state SD for this transition depends on the state machine's realization. If the state machine is synthesized using the method ($V* = \Sigma$ p-terms where $V* = 1$) to obtain transition/excitation equations for D flip-flops, what is the relationship between the coded states for SB, SC, and SD? Explain.

7.59   Repeat Exercise 7.58, assuming that the machine is synthesized using the method $(V* ' = \Sigma \text{ p-terms where } V* = 0)$.

7.60   Suppose that for a state SA and an input combination I, an ambiguous state diagram does not define a next state. The actual next state SD for this transition depends on the state machine's realization. Suppose that the state machine is synthesized using the method $(V* = \Sigma \text{ p-terms where } V* = 1)$ to obtain transition/excitation equations for D flip-flops. What coded state is SD? Explain.

7.61   Repeat Exercise 7.60, assuming that the machine is synthesized using the method $(V* ' = \Sigma \text{ p-terms where } V* = 0)$.

7.62   Given the transition equations for a clocked synchronous state machine that is to be built using master/slave S-R flip-flops, how can the excitation equations for the S and R inputs be derived? (*Hint:* Show that any transition equation, $Qi* = expr$, can be written in the form $Qi* = Qi \cdot expr1 + Qi' \cdot expr2$, and see where that leads.)

7.63   Repeat Exercise 7.62 for J-K flip-flops. How can the "don't-cares" that are possible in a J-K design be specified?

7.64   Draw a logic diagram for the output logic of the guessing-game machine in Table 7-18 using a single 74x139 dual 2-to-4 decoder. (*Hint:* Use active-low outputs.)

7.65   What does the personalized license plate in Figure 7-60 stand for? (*Hint:* It's the author's old plate, a computer engineer's version of OTTFFSS.)

7.66   Analyze the feedback sequential circuit in Figure 7-19, assuming that the PR_L and CLR_L inputs are always 1. Derive excitation equations, construct a transition table, and analyze the transition table for critical and noncritical races. Name the states, and write a state/output table and a flow/output table. Show that the flow table performs the same function as Figure 7-85.

7.67   Draw the logic diagram for a circuit that has one feedback loop but is *not* a sequential circuit. That is, the circuit's output should be a function of its current input only. In order to prove your case, break the loop and analyze the circuit as if it were a feedback sequential circuit, and demonstrate that the outputs for each input combination do not depend on the "state."

BUT flop
NBUT gate

7.68   A *BUT flop* may be constructed from an NBUT gate as shown in Figure X7.68. (An *NBUT gate* is simply a BUT gate with inverted outputs; see Exercise 5.31 for the definition of a BUT gate.) Analyze the BUT flop as a feedback sequential circuit and obtain excitation equations, transition table, and flow table. Is this circuit good for anything, or is it a flop?

7.69   Repeat Exercise 7.68 for the BUT flop in Figure X7.69.

Figure X7.68   **Figure X7.69**

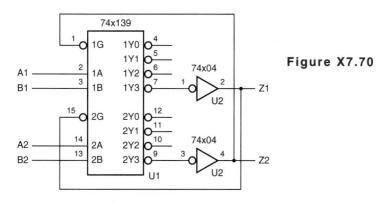

Figure X7.70

7.70 A clever student designed the circuit in Figure X7.70 to create a BUT gate. But the circuit didn't always work correctly. Analyze the circuit and explain why.

7.71 Analyze the feedback sequential circuit in Figure X7.71. Break the feedback loops, write excitation equations, and construct a transition and output table, showing the stable total states. What application might this circuit have?

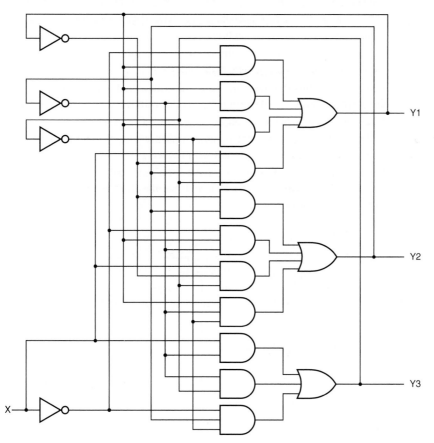

Figure X7.71

7.72   Show that a 4-bit ones'-complement adder with end-around carry is a feedback sequential circuit.

7.73   Complete the analysis of the positive-edge-triggered D flip-flop in Figure 7-86, including transition/output, state/output, and flow/output tables. Show that its behavior is equivalent to that of the D flip-flop in Figure 7-78.

7.74   We claimed in Section 7.10.1 that all single-loop feedback sequential circuits have an excitation equation of the form

$$Q* = (\text{forcing term}) + (\text{holding term}) \cdot Q$$

Why aren't there any practical circuits whose excitation equation substitutes Q' for Q above?

7.75   Simulate the latch circuit of Figure 7-88(b) under the conditions described in the text on page 616, either using a simulator in unit-delay mode or by hand assuming that each gate has a delay of 1 ns. Does the circuit behave as claimed in the text? Replace the inverter in the circuit with three inverters, repeat the simulation, and explain the results. What would you expect to happen in the real circuit?

7.76   Design a latch with two control inputs, C1 and C2, and three data inputs, D1, D2, and D3. The latch is to be "open" only if both control inputs are 1, and it is to store a 1 if any of the data inputs is 1. Use hazard-free two-level sum-of-products circuits for the excitation functions.

7.77   Repeat Exercise 7.76, but minimize the number of gates required; the excitation circuits may have multiple levels of logic.

7.78   Redraw the timing diagram in Figure 7-90, showing the internal state variables of the pulse-catching circuit of Figure 7-100, assuming that it starts in state 00.

7.79   The general solution for obtaining a race-free state assignment of 2^n states using 2^{n-1} state variables yields the adjacency diagram shown in Figure X7.79 for the $n = 2$ case. Compare this diagram with Figure 7-97. Which is better, and why?

7.80   Design a fundamental-mode flow table for a pulse-catching circuit similar to the one described in Section 7.10.2, except that the circuit should detect both 0-to-1 and 1-to-0 transitions on P.

Figure X7.79

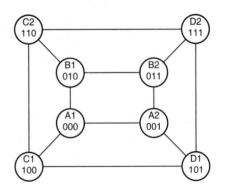

	X Y				
S	00	01	11	10	**Table X7.85**
A	B	C	—	(A)	
B	(B)	E	—	B	
C	F	(C)	—	E	
D	(D)	F	—	B	
E	D	(E)	—	(E)	
F	(F)	(F)	—	A	
		S*			

7.81   Design a fundamental-mode flow table for a double-edge-triggered D flip-flop, one that samples its inputs and changes its outputs on both edges of the clock signal.

7.82   Design a fundamental-mode flow table for a circuit with two inputs, EN and CLKIN, and a single output, CLKOUT, with the following behavior. A clock period is defined to be the interval between successive rising edges of CLKIN. If EN is asserted during an entire given clock period, then CLKOUT should be "on" during the next clock period; that is, it should be identical to CLKIN. If EN is negated during an entire given clock period, then CLKOUT should be "off" (constant 1) during the next clock period. If EN is both asserted and negated during a given clock period, then CLKOUT should be on in the next period if it had been off, and off if it had been on. After writing the fundamental-mode flow table, reduce it by combining "compatible" states if possible.

7.83   Design a circuit that meets the specifications of Exercise 7.82 using edge-triggered D flip-flops (74x74) or J-K flip-flops (74x109) and NAND and NOR gates without feedback loops. Give a complete circuit diagram and word description of how your circuit achieves the desired behavior.

7.84   Which of the circuits of the two preceding exercises is (are) subject to meta-stability, and under what conditions?

7.85   For the flow table in Table X7.85, find an assignment of state variables that avoids all critical races. You may add additional states as necessary, but use as few state variables as possible. Assign the all-0s combination to state A. Draw an adjacency diagram for the original flow table, and write the modified flow table and another adjacency diagram to support your final state-variable assignment.

7.86   Prove that the fundamental-mode flow table of any flip-flop that samples input(s) and change(s) outputs on the rising edge only of a clock signal CLK contains an essential hazard.

7.87   Locate the essential hazard(s) in the flow table for a positive-edge-triggered D flip-flop, Figure 7-85.

7.88   Identify the essential hazards, if any, in the flow table developed in Exercise 7.81.

7.89   Identify the essential hazards, if any, in the flow table developed in Exercise 7.82.

7.90   Build a verbal flip-flop—a logical word puzzle that can be answered correctly in either of two ways depending on state. How might such a device be adapted to the political arena?

7.91   Modify the ABEL program in Table 7-27 to use an output-coded state assignment, thereby reducing the total number of PLD outputs required by one.

7.92   Finish writing the test vectors, started in Table 7-35, for the combination-lock state machine of Table 7-31. The complete set of vectors should test all of the state transitions and all of the output values for every state and input combination.

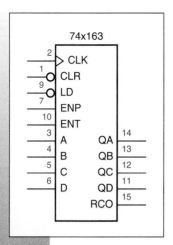

Sequential Logic Design Practices

T he purpose of this chapter is to familiarize you with the most commonly used and dependable sequential-circuit design methods. Therefore, we will emphasize *synchronous systems*—that is, systems in which all flip-flops are clocked by the same common clock signal. Although all the world does not march to the tick of a common clock, within the confines of a digital system or subsystem we can make it so. When we interconnect digital systems or subsystems that use different clocks, we can usually identify a limited number of asynchronous signals that need special treatment, as we'll show later.

We begin this chapter with a quick summary of sequential-circuit documentation standards. After revisiting the most basic building blocks of sequential-circuit design—latches and flip-flops—we describe some of the most flexible building blocks—sequential PLDs. Next we show how counters and shift registers are realized in both MSI devices and PLDs and show some of their applications. Finally, we show how these elements come together in synchronous systems and how the inevitable asynchronous inputs are handled.

8.1 Sequential-Circuit Documentation Standards

8.1.1 General Requirements

Basic documentation standards in areas like signal naming, logic symbols, and schematic layout, which we introduced in Chapter 5, apply to digital systems as a whole and therefore to sequential circuits in particular. We highlight the following ideas, however, for system elements that are specifically "sequential":

- *State-machine layout.* Within a logic diagram, a collection of flip-flops and combinational logic that forms a state machine should be drawn together in a logical format on the same page, so the fact that it is a state machine is obvious. (You shouldn't have to flip pages to find the feedback path!)

- *Cascaded elements.* In a similar way, registers, counters, and shift registers that use multiple ICs should have the ICs grouped together in the schematic so that the cascading structure is obvious.

- *Flip-flops.* The symbols for individual sequential-circuit elements, especially flip-flops, should rigorously follow the appropriate drawing standards, so that the type, function, and clocking behavior of the elements are clear.

- *State-machine descriptions.* State machines should be described by state tables, state diagrams, transition lists, or text files in a state-machine description language such as ABEL or VHDL.

- *Timing diagrams.* The documentation package for sequential circuits should include timing diagrams that show the general timing assumptions and timing behavior of the circuit.

- *Timing specifications.* A sequential circuit should be accompanied by a specification of the timing requirements for proper internal operation (e.g., maximum clock frequency), as well as the requirements for any externally supplied inputs (e.g., setup- and hold-time requirements with respect to the system clock, minimum pulse widths, etc.).

8.1.2 Logic Symbols

We introduced traditional symbols for flip-flops in Section 7.2. Flip-flops are always drawn as rectangular-shaped symbols and follow the same general guidelines as other rectangular-shaped symbols—inputs on the left, outputs on the right, bubbles for active levels, and so on. In addition, some specific guidelines apply to flip-flop symbols:

- A dynamic indicator is placed on edge-triggered clock inputs.

- A postponed-output indicator is placed on master/slave outputs that change at the end interval during which the clock is asserted.

- Asynchronous preset and clear inputs may be shown at the top and bottom of a flip-flop symbol—preset at the top and clear at the bottom.

The logic symbols for larger-scale sequential elements, such as the counters and shift register described later in this chapter, are generally drawn with all inputs, including presets and clears, on the left, and all outputs on the right. Bidirectional signals may be drawn on the left or the right, whichever is convenient.

Like individual flip-flops, larger-scale sequential elements use a dynamic indicator to indicate edge-triggered clock inputs. In "traditional" symbols, the names of the inputs and outputs give a clue of their function, but they are sometimes ambiguous. For example, two elements described later in this chapter, the 74x161 and 74x163 4-bit counters, have exactly the same traditional symbol, even though the behavior of their CLR inputs is completely different.

IEEE STANDARD SYMBOLS IEEE standard symbols have a rich set of notation that can provide an unambiguous definition of every signal's function. A guide to IEEE symbols, including all of the sequential elements in this chapter, can be found on the web at www.ddpp.com.

8.1.3 State-Machine Descriptions

So far we have dealt with six different representations of state machines:

- Word descriptions
- State tables
- State diagrams
- Transition lists
- ABEL programs
- VHDL programs

You might think that having all these different ways to represent state machines is a problem—too much for you to learn! Well, they're not all that difficult to learn, but there *is* a subtle problem here.

Consider a similar problem in programming, where high-level "pseudocode" or perhaps a flowchart might be used to document how a program works. The pseudocode may express the programmer's intentions very well, but errors, misinterpretations, and typos can occur when it is translated into real code. In any creative process, inconsistencies can occur when there are multiple representations of how things work.

The same kind of inconsistencies can occur in state-machine design. A logic designer may document a machine's desired behavior with a 100%-correct hand-drawn state diagram, but you can make mistakes translating the diagram into a program, and there are *lots* of opportunities to mess up if you have to "turn the crank" manually to translate the state diagram into a state table, transition table, excitation equations, and logic diagram.

inconsistent state-machine representations

The solution to this problem is similar to the one adopted by programmers who write self-documenting code using a high-level language. The key is to select a representation that is both expressive of the designer's intentions *and* translatable into a physical realization using an error-free, automated process. (You don't hear many programmers screaming "Compiler bug!" when their programs don't work the first time.)

The best solution (for now, at least) is to write state-machine "programs" directly in a high-level state-machine description language like ABEL or VHDL and to avoid alternate representations, other than general, summary word descriptions. Languages like ABEL and VHDL are easily readable and allow automatic conversion of the description into a PLD-, FPGA-, or ASIC-based realization. Some CAD tools allow state machines to be specified and synthesized using state diagrams, or even using sample timing diagrams, but these can often lead to ambiguities and unanticipated results. Thus, we'll use ABEL/VHDL approach exclusively for the rest of this book.

8.1.4 Timing Diagrams and Specifications

We showed many examples of timing diagrams in Chapters 5 and 7. In the design of synchronous systems, most timing diagrams show the relationship between the clock and various input, output, and internal signals.

Figure 8-1 shows a fairly typical timing diagram that specifies the requirements and characteristics of input and output signals in a synchronous circuit. The first line shows the system clock and its nominal timing parameters. The remaining lines show a range of delays for other signals.

For example, the second line shows that flip-flops change their outputs at some time between the rising edge of CLOCK and time t_{ffpd} afterward. External circuits that sample these signals should not do so while they are changing. The timing diagram is drawn as if the minimum value of t_{ffpd} were zero; a complete

Figure 8-1
A detailed timing diagram showing propagation delays and setup and hold times with respect to the clock.

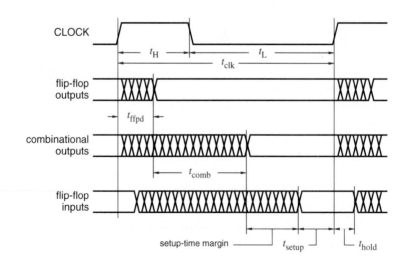

documentation package would include a timing table indicating the actual minimum, typical, and maximum values of t_{ffpd} and all other timing parameters.

The third line of the timing diagram shows the additional time, t_{comb}, required for the flip-flop output changes to propagate through combinational logic elements, such as flip-flop excitation logic. The excitation inputs of flip-flops and other clocked devices require a setup time of t_{setup}, as shown in the fourth line. For proper circuit operation we must have $t_{clk} - t_{ffpd} - t_{comb} > t_{setup}$.

Timing margins indicate how much "worse than worst-case" the individual components of a circuit can be without causing the circuit to fail. Well-designed systems have positive, nonzero timing margins to allow for unexpected circumstances (marginal components, brownouts, engineering errors, etc.) and clock skew (Section 8.8.1).

timing margin

The value $t_{clk} - t_{ffpd(max)} - t_{comb(max)} - t_{setup}$ is called the *setup-time margin;* if this is negative, the circuit won't work. Note that *maximum* propagation delays are used to calculate setup-time margin. Another timing margin involves the hold-time requirement t_{hold}; the sum of the *minimum* values of t_{ffpd} and t_{comb} must be greater than t_{hold}, and the *hold-time margin* is $t_{ffpd(min)} + t_{comb(min)} - t_{hold}$.

setup-time margin

hold-time margin

The timing diagram in Figure 8-1 does not show the timing differences between different flip-flop inputs or combinational-logic signals, even though such differences exist in most circuits. For example, one flip-flop's Q output may be connected directly to another flip-flop's D input, so that t_{comb} for that path is zero, while another's may go the ripple-carry path of a 32-bit adder before reaching a flip-flop input. When proper synchronous design methodology is used, these relative timings are not critical, since none of these signals affect the state of the circuit until a clock edge occurs. You merely have to find the longest delay path in one clock period to determine whether the circuit will work. However, you may have to analyze several different paths in order to find the worst-case one.

Another, perhaps more common, type of timing diagram shows only functional behavior and is not concerned with actual delay amounts; an example is shown in Figure 8-2. Here, the clock is "perfect." Whether to show signal changes as vertical or slanted lines is strictly a matter of personal taste in this and all other timing diagrams, unless rise and fall times must be explicitly indicated. Clock transitions are shown as vertical lines in this and other figures in keeping with the idea that the clock is a "perfect" reference signal.

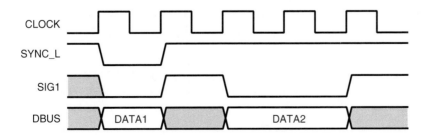

Figure 8-2
Functional timing of a synchronous circuit.

NOTHING'S PERFECT   In reality, there's no such thing as a perfect clock signal. One imperfection that most designers of high-speed digital circuits have to deal with is "clock skew." As we show in Section 8.8.1, a given clock edge arrives at different circuit inputs at different times because of differences in wiring delays, loading, and other effects.

Another imperfection, a bit beyond the scope of this text, is "clock jitter." A 10-MHz clock does not have a period of exactly 100 ns on every cycle—it may be 100.05 ns in one cycle and 99.95 ns in the next. This is not a big deal in such a slow circuit, but in a 500-MHz circuit the same 0.1 ns of jitter eats up 5% of the 2-ns timing budget. And the jitter in some clock sources is even higher!

The other signals in Figure 8-2 may be flip-flop outputs, combinational outputs, or flip-flop inputs. Shading is used to indicate "don't-care" signal values; crosshatching as in Figure 8-1 could be used instead. All of the signals are shown to change immediately after the clock edge. In reality, the outputs change sometime later, and inputs may change just barely before the next clock edge. However, "lining up" everything on the clock edge allows the timing diagram to display more clearly which functions are performed during each clock period. Signals that are lined up with the clock are simply understood to change sometime *after* the clock edge, with timing that meets the setup- and hold-time requirements of the circuit. Many timing diagrams of this type appear in this chapter.

Table 8-1 shows manufacturer's timing parameters for commonly used flip-flops, registers, and latches in CMOS and TTL. "Typical" values are for

Table 8-1 Propagation delay in ns of selected CMOS flip-flops, registers, and latches.

Part	Parameter	74HCT		74AHCT		74FCT		74LS	
		Typ.	Max.	Typ.	Max.	Min.	Max.	Typ.	Max.
'74	t_{pd}, CLK↑ to Q or $\overline{Q}$	35	44	6.3	10			25	40
	t_{pd}, $\overline{PR}$↓ or $\overline{CLR}$↓ to Q or $\overline{Q}$	40	50	8.1	13			25	40
	t_s, D to CLK↑	12	15		5				20
	t_h, D from CLK↑	3	3		0				5
	t_{rec}, CLK↑ from $\overline{PR}$↑ or $\overline{CLR}$↑	6	8		3.5				
	t_w, CLK low or high	18	23		5				25
	t_w, $\overline{PR}$ or $\overline{CLR}$ low	16	20		5				25
'174	t_{pd}, CLK↑ to Q	40	50	6.3	10			21	30
	t_{pd}, $\overline{CLR}$↓ to Q	44	55	8.1	13			23	35
	t_s, D to CLK↑	16	20		5				20
	t_h, D from CLK↑	5	5		0				5
	t_{rec}, CLK↑ from $\overline{CLR}$↑	12	15		3.5				25
	t_w, CLK low or high	20	25		5				20
	t_w, $\overline{CLR}$ low	25	31		5				20

Table 8-1 (continued) Propagation delay in ns of selected CMOS flip-flops, registers, and latches.

Part	Parameter	74HCT Typ.	74HCT Max.	74AHCT Typ.	74AHCT Max.	74FCT Min.	74FCT Max.	74LS Typ.	74LS Max.
'175	t_{pd}, CLK↑ to Q or $\overline{Q}$	33	41					21	30
	t_{pd}, $\overline{CLR}$↓ to Q or $\overline{Q}$	35	44					23	35
	t_s, D to CLK↑	20	25						20
	t_h, D from CLK↑	5	5						5
	t_{rec}, CLK↑ from $\overline{CLR}$↑	5	5						25
	t_w, CLK low or high	20	25						20
	t_w, $\overline{CLR}$ low	20	25						20
'273	t_{pd}, CLK↑ to Q	30	38	6.8	11	2	7.2	18	27
	t_{pd}, $\overline{CLR}$↓ to Q	32	40	8.5	12.6	2	7.2	18	27
	t_s, D to CLK↑	12	15		5	2			20
	t_h, D from CLK↑	3	3		0		1.5		5
	t_{rec}, CLK↑ from $\overline{CLR}$↑	10	13		2.5	2			25
	t_w, CLK low or high	20	25		6.5		4		20
	t_w, $\overline{CLR}$ low	12	15		6		5		20
'373	t_{pd}, C↑ to Q	35	44	8.5	14.5	2	8.5	24	36
	t_{pd}, D to Q	32	40	5.9	10.5	1.5	5.2	18	27
	t_s, D to C↓	10	13		1.5		2		0
	t_h, D from C↓	5	5		3.5		1.5		10
	t_{pHZ}, $\overline{OE}$ to Q	35	44		12	1.5	6.5	12	20
	t_{pLZ}, $\overline{OE}$ to Q	35	44		12	1.5	6.5	16	25
	t_{pZH}, $\overline{OE}$ to Q	35	44		13.5	1.5	5.5	16	28
	t_{pZL}, $\overline{OE}$ to Q	35	44		13.5	1.5	5.5	22	36
	t_w, C low or high	16	20		6.5		5		15
'374	t_{pd}, CLK↑ to Q	33	41	6.4	11.5	2	6.5	22	34
	t_s, D to CLK↑	12	15		2.5		2		20
	t_h, D from CLK↑	5	5		2.5		1.5		0
	t_{pHZ}, $\overline{OE}$ to Q	28	35		12	1.5	6.5		18
	t_{pLZ}, $\overline{OE}$ to Q	28	35		12	1.5	6.5		24
	t_{pZH}, $\overline{OE}$ to Q	30	38		12.5	1.5	5.5		28
	t_{pZL}, $\overline{OE}$ to Q	30	38		12.5	1.5	5.5		36
	t_w, CLK low or high	16	20		6.5		5		15
'377	t_{pd}, CLK↑ to Q	38	48			2	7.2	18	27
	t_s, D to CLK↑	12	15				2		20
	t_h, D from CLK↑	3	3				1.5		5
	t_s, $\overline{EN}$ to CLK↑	12	15				2		25
	t_h, $\overline{EN}$ from CLK↑	5	5				1.5		5
	t_w, CLK low or high	20	25				6		20

devices operating at 25°C, but, depending on the logic family, they could be for a typical part and nominal power-supply voltage, or they could be for a worst-case part at worst-case supply voltage. "Maximum" values are generally valid over the commercial operating range of voltage and temperature, except TTL values, which are specified at 25°C. Also note that the "maximum" values of t_s, t_h, t_{rec}, or t_w are the maximum values of the *minimum* setup time, hold time, recovery time, or pulse width that the specified part will exhibit.

Different manufacturers may use slightly different definitions for the same timing parameters, and they may specify different numbers for the same part. A given manufacturer may even use different definitions for different families or part numbers in the same family. Thus, all of the specifications in Table 8-1 are merely representative; for exact numbers *and* their definitions, you must consult the data sheet for the particular part and manufacturer.

8.2 Latches and Flip-Flops

8.2.1 SSI Latches and Flip-Flops

Several different types of discrete latches and flip-flops are available as SSI parts. These devices are sometimes used in the design of state machines and "unstructured" sequential circuits that don't fall into the categories of shift registers, counters, and other sequential MSI functions presented later in this chapter. However, SSI latches and flip-flops have been eliminated to a large extent in modern designs as their functions are embedded in PLDs and FPGAs. Nevertheless, a handful of these discrete building blocks still appear in many digital systems, so it's important to be familiar with them.

74x375

Figure 8-3 shows the pinouts for several SSI sequential devices. The only latch in the figure is the *74x375,* which contains four D latches, similar in function to the "generic" D latches described in Section 7.2.4. Because of pin limitations, the latches are arranged in pairs with a common C control line for each pair.

74x74

Among the devices in Figure 8-3, the most important is the *74x74,* which contains two independent positive-edge-triggered D flip-flops with preset and clear inputs. We described the functional operation, timing, and internal structure of edge-triggered D flip-flops in general, and the 74x74 in particular, in Section 7.2.5. Besides the use of the 74x74 in "random" sequential circuits, fast

74F74
74ACT74

versions of the part, such as the *74F74* and *74ACT74,* find application in synchronizers for asynchronous input signals, as discussed in Section 8.9.

74x109

The *74x109* is a positive-edge-triggered J-$\overline{\text{K}}$ flip-flop with an active-low K input (named $\overline{\text{K}}$ or K_L). We discussed the internal structure of the '109 in

74x112

Section 7.2.10. Another J-K flip-flop is the *74x112,* which has an active-low clock input.

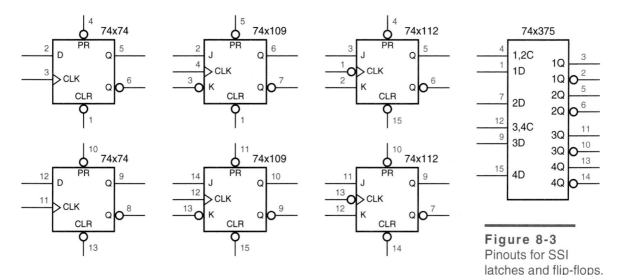

Figure 8-3
Pinouts for SSI
latches and flip-flops.

*8.2.2 Switch Debouncing

A common application of simple bistables and latches is switch debouncing. We're all familiar with electrical switches from experience with lights, garbage disposals, and other appliances. Switches connected to sources of constant logic 0 and 1 are often used in digital systems to supply "user inputs." However, in digital logic applications we must consider another aspect of switch operation, the time dimension. A simple make or break operation, which occurs instantly as far as we slow-moving humans are concerned, actually has several phases that are discernible by high-speed digital logic.

Figure 8-4(a) shows how a single-pole, single-throw (SPST) switch might be used to generate a single logic input. A pull-up resistor provides a logic-1 value when the switch is opened, and the switch contact is tied to ground to provide a logic-1 value when the switch is pushed.

As shown in (b), it takes a while after a push for the wiper to hit the bottom contact. Once it hits, it doesn't stay there for long; it bounces a few times before finally settling. The result is that several transitions are seen on the SW_L and DSW logic signals for each single switch push. This behavior is called *contact bounce*. Typical switches bounce for 10–20 ms, a very long time compared to the switching speeds of logic gates.

contact bounce

Contact bounce may or may not be a problem, depending on the switch application. For example, some computers have configuration information specified by small switches, called *DIP switches* because they are the same size as a dual in-line package (DIP). Since DIP switches are normally changed only when the computer is inactive, there's no problem. Contact bounce *is* a problem

DIP switch

* Throughout this book, optional sections are marked with an asterisk.

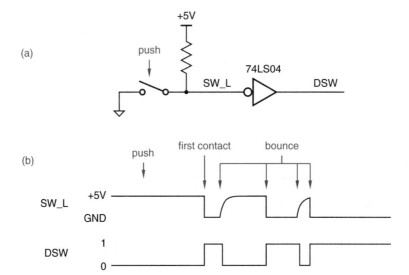

Figure 8-4
Switch input without debouncing.

debounce

if a switch is being used to count or signal some event (e.g., laps in a race). Then we must provide a circuit (or, in microprocessor-based systems, software) to *debounce* the switch—to provide just one signal change or pulse for each external event.

*8.2.3 The Simplest Switch Debouncer

Switch debouncing is a good application for the simplest sequential circuit, the bistable element of Section 7.1, which can be used as shown in Figure 8-5. This circuit uses a single-pole, double-throw (SPDT) switch. The switch contacts and wiper have a "break before make" behavior, so the wiper terminal is "floating" at some time halfway through the switch depression.

Before the button is pushed, the top contact holds SW at 0 V, a valid logic 0, and the top inverter produces a logic 1 on SW_L and on the bottom contact. When the button is pushed and contact is broken, feedback in the bistable holds SW at V_{OL} (≤ 0.5 V for LS-TTL), still a valid logic 0.

Next, when the wiper hits the bottom contact, the circuit operates quite unconventionally for a moment. The top inverter in the bistable is trying to maintain a logic 1 on the SW_L signal; the top transistor in its totem-pole output is "on" and connecting SW_L through a small resistance to +5 V. Suddenly, the switch contact makes a metallic connection of SW_L to ground, 0.0 V. Not surprisingly, the switch contact wins.

A short time later (30 ns for the 74LS04), the forced logic 0 on SW_L propagates through the two inverters of the bistable, so that the top inverter gives up its vain attempt to drive a 1, and instead drives a logic 0 onto SW_L. At this point, the top inverter output is no longer shorted to ground, and feedback in the bistable maintains the logic 0 on SW_L even if the wiper bounces off the bottom contact, as it does. (It does not bounce far enough to touch the top contact again.)

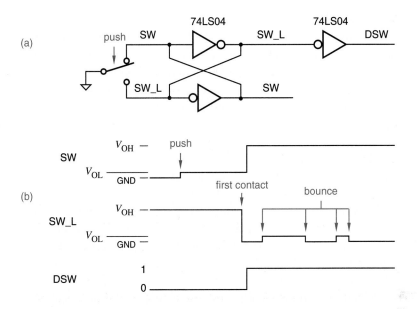

(a)

(b)

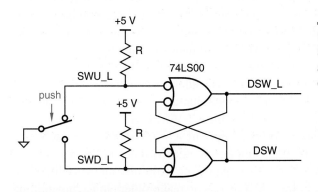

Figure 8-5
Switch input using
a bistable for
debouncing

Advantages of this circuit compared to other debouncing approaches are that it has a low chip count (one-third of a 74LS04), no pull-up resistors are required, and both polarities of the input signal (active-high and active-low) are produced. In situations where momentarily shorting gate outputs must be avoided, a similar circuit can be designed using an $\overline{S}$-$\overline{R}$ latch and pull-up resistors, as suggested in Figure 8-6.

Figure 8-6
Switch input using
an $\overline{S}$-$\overline{R}$ latch for
debouncing.

**WHERE WIMPY
WORKS WELL**

The circuit in Figure 8-5, while elegant, should not be used with high-speed CMOS devices, like the 74ACT04, whose outputs are capable of sourcing large amounts of current in the HIGH state. While shorting such outputs to ground momentarily will not cause any damage, it will generate a noise pulse on power and ground signals that may trigger improper operation of the circuit elsewhere. The debouncing circuit in the figure works well with wimpy logic families like HCT and LS-TTL.

*8.2.4 Bus Holder Circuit

In Sections 3.7.3 and 5.6 we described three-state outputs and how they are tied together to create three-state buses. At any time, at most one output can drive the bus; sometimes, no output is driving the bus, and the bus is "floating." When high-speed CMOS inputs are connected to a bus that is left floating for a long time (in the fastest circuits, more than a clock tick or two), bad things can happen. In particular, noise, crosstalk, and other effects can drive the high-impedance floating bus signals to a voltage level near the CMOS devices' input switching threshold, which in turn allows excessive current to flow in the device outputs. For this reason, it is desirable and customary to provide pull-up resistors that quickly pull a floating bus to a valid HIGH logic level.

Pull-up resistors aren't all goodness—they cost money and they occupy valuable printed-circuit-board real estate. A big problem they have in very high-speed circuits is the choice of resistance value. If the resistance is too high, when a bus goes from LOW to floating, the transition from LOW to pulled-up (HIGH) will be slow due to the high RC time constant, and input levels may spend too much time near the switching threshold. If the pull-up resistance is too low, devices trying to pull the bus LOW will have to sink too much current.

bus holder circuit

The solution to this problem is to eliminate pull-up resistors in favor of an active *bus holder circuit*, shown in Figure 8-7. This is nothing but a bistable with a resistor R in one leg of the feedback loop. The bus holder's INOUT signal is connected to the three-state bus line which is to be held. When the three-state output currently driving the line LOW or HIGH changes to floating, the bus holder's righthand inverter holds the line in its current state. When a three-state output tries to change the line from LOW to HIGH or vice versa, it must source or sink a small amount of additional current through R to overcome the bus holder. This additional current flow persists only for the short time that it takes for the bistable to flip into its other stable state.

The choice of the value of R in the bus holder is a compromise between low override current (high R) and good noise immunity on the held bus line (low R). A typical example, bus holder circuits in the 3.3-V CMOS LVC family specify a maximum override current of 500 μA, implying $R \approx 3.3 / 0.0005 = 6.6$ KΩ.

Bus holder circuits are often built into another MSI device, such as an octal CMOS bus driver or transceiver. They require no extra pins and very little chip area, so they are essentially free. And there's no real problem in having multiple (n) bus holders on the same signal line, as long as the bus drivers can provide n times the override current for a few nanoseconds during switching. Note that bus holders normally are not effective on buses that have TTL inputs attached to them (see Exercise 8.25).

Figure 8-7
Bus holder circuit.

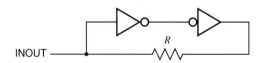

INOUT

8.2.5 Multibit Registers and Latches

A collection of two or more D flip-flops with a common clock input is called a *register*. Registers are often used to store a collection of related bits, such as a byte of data in a computer. However, a single register can also be used to store unrelated bits of data or control information; the only real constraint is that all of the bits are stored using the same clock signal.

register

Figure 8-8 shows the logic diagram and logic symbol for a commonly used MSI register, the *74x175*. The 74x175 contains four edge-triggered D flip-flops with a common clock and asynchronous clear inputs. It provides both active-high and active-low outputs at the external pins of the device.

74x175

The individual flip-flops in a '175 are negative-edge triggered, as indicated by the inversion bubbles on their CLK inputs. However, the circuit also contains an inverter that makes the flip-flops positive-edge triggered with respect to the device's external CLK input pin. The common, active-low, clear signal (CLR_L) is connected to the asynchronous clear inputs of all four flip-flops. Both CLK and CLR_L are buffered before fanning out to the four flip-flops, so that a device driving one of these inputs sees only one unit load instead of four. This is especially important if a common clock or clear signal must drive many such registers.

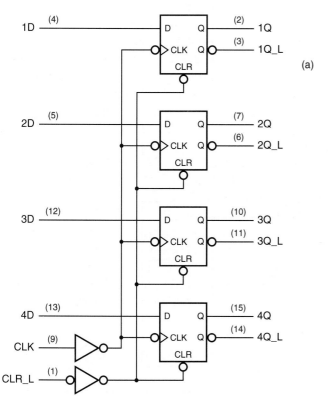

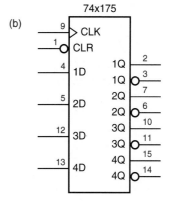

Figure 8-8
The 74x175 4-bit register:
(a) logic diagram, including pin numbers for a standard 16-pin dual in-line package;
(b) traditional logic symbol.

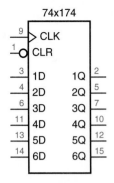

Figure 8-9
Logic symbol for the
74x174 6-bit register.

The logic symbol for the *74x174,* 6-bit register is shown in Figure 8-9. The internal structure of this device is similar to the 74x175's, except that it eliminates the active-low outputs and provides two more flip-flops instead.

Many digital systems, including computers, telecommunications devices, and stereo equipment, process information 8, 16, or 32 bits at a time; as a result, ICs that handle 8 bits are very popular. One such MSI IC is the *74x374* octal edge-triggered D flip-flop, also known simply as an 8-bit register. (Once again, "octal" means that the device has eight sections.)

As shown in Figure 8-10, the 74x374 contains eight edge-triggered D flip-flops that all sample their inputs and change their outputs on the rising edge of a common CLK input. Each flip-flop output drives a three-state buffer that in turn drives an active-high output. All of the three-state buffers are enabled by a

Figure 8-10
The 74x374 8-bit register:
(a) logic diagram, including pin
numbers for a standard 20-pin
dual in-line package;
(b) traditional logic symbol.

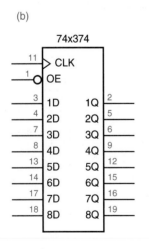

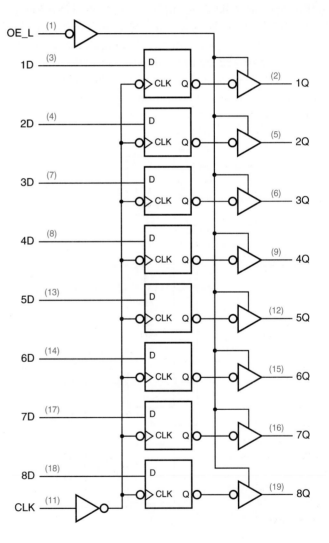

common, active-low OE_L (output enable) input. As in the other registers that we've studied, the control inputs (CLK and OE_L) are buffered so that they present only one unit load to a device that drives them.

One variation of the 74x374 is the *74x373,* whose symbol is shown in Figure 8-11. The '373 uses D latches instead of edge-triggered flip-flops. Therefore, its outputs follow the corresponding inputs whenever C is asserted, and they latch the last input values when C is negated. Another variation is the *74x273,* shown in Figure 8-12. This octal register has non-three-state outputs and no OE_L input; instead it uses pin 1 for an asynchronous clear input CLR_L.

The *74x377,* whose symbol is shown in Figure 8-13(a), is an edge-triggered register like the '374, but it does not have three-state outputs. Instead, pin 1 is used as an active-low clock-enable input EN_L. If EN_L is asserted (LOW) at the rising edge of the clock, then the flip-flops are loaded from the data inputs; otherwise, they retain their present values, as shown logically in (b).

Figure 8-11
Logic symbol for the
74x373 8-bit latch.

Figure 8-12
Logic symbol for the
74x273 8-bit register.

Figure 8-13 The 74x377 8-bit register with gated clock:
(a) logic symbol; (b) logical behavior of one bit.

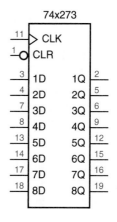

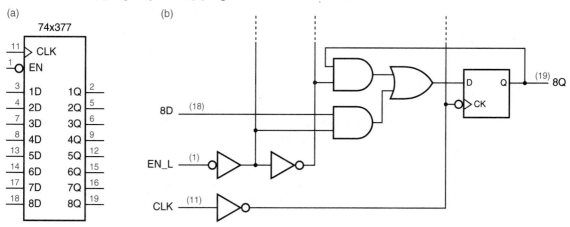

High pin-count surface-mount packaging supports even wider registers, drivers, and transceivers. Most common are 16-bit devices, but there are also devices with 18 bits (for byte parity) and 32 bits. Also, the larger packages can offer more control functions, such as clear, clock enable, multiple output enables, and even a choice of latching vs. registered behavior all in one device.

8.2.6 Registers and Latches in ABEL and PLDs

As we showed in Section 7.11, registers are very easy to specify in ABEL. For example, Table 7-33 on page 639 showed an ABEL program for an 8-bit register with enable. Obviously, ABEL allows the functions performed at the D inputs of register to be customized in almost any way desired, limited only by the number of inputs and product terms in the targeted PLD. We describe sequential PLDs in Section 8.3.

With most sequential PLDs, few if any customizations can be applied to a register's clock input (e.g, polarity choice) or to the asynchronous inputs (e.g., different preset conditions for different bits). However, ABEL does provide appropriate syntax to apply these customizations in devices that support them, as described in Section 7.11.1.

Very few PLDs have latches built in; edge-triggered registers are much more common and generally more useful. However, you can also synthesize a latch using combinational logic and feedback. For example, the excitation equation for an S-R latch is

$$Q* = S + R' \cdot Q$$

Thus, you could build an S-R latch using one combinational output of a PLD, using the ABEL equation "Q = S # !R & Q." Furthermore, the S and R signals above could be replaced with more complex logic functions of the PLD's inputs, limited only by the availability of product terms (seven per output in a 16V8C or 16L8) to realize the final excitation equation. The feedback loop can be created only when Q is assigned to a bidirectional pin (in a 16V8C or 16L8, pins IO2–IO7, not O1 or O8). Also, the output pin must be continuously output-enabled; otherwise, the feedback loop would be broken and the latch's state lost.

Probably the handiest latch to build out of a combinational PLD is a D latch. The basic excitation equation for a D latch is

$$Q* = C \cdot D + C' \cdot Q$$

However, we showed in Section 7.10.1 that this equation contains a static hazard, and the corresponding circuit does not latch data reliably. To build a reliable D latch, we must include a consensus term in the excitation equation:

$$Q* = C \cdot D + C' \cdot Q + D \cdot Q$$

The D input in this equation may be replaced with a more complicated expression, but the equation's structure remains the same:

$$Q* = C \cdot expression + C' \cdot Q + expression \cdot Q$$

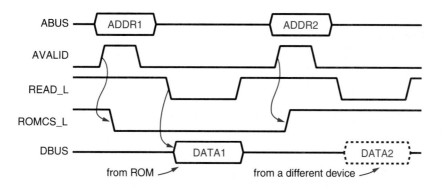

Figure 8-14
Timing diagram for a
microprocessor read
operation.

It is also possible to use a more complex expression for the C input, as we
showed in Section 7.10.1. In any case, it is very important for the consensus term
to be included in the PLD realization. The compiler can work against you in this
case, since its minimization step will find that the consensus term is redundant
and remove it.

Some versions of the ABEL compiler let you prevent elimination of
consensus terms by including a keyword "retain" in the property list of the
istype declaration for any output which is not to be minimized. In other
versions, your only choice is to turn off minimization for the entire design.

retain property

Probably the most common use of a PLD-based latch is to simultaneously
decode and latch addresses in order to select memory and I/O devices in microprocessor systems. Figure 8-14 is a timing diagram for this function in a typical
system. The microprocessor selects a device and a location within the device by
placing an address on its address bus (ABUS) and asserting an "address valid"
signal (AVALID). A short time later, it asserts a read signal (READ_L), and the
selected device responds by placing data on the data bus (DBUS).

Notice that the address does not stay valid on ABUS for the entire operation. The microprocessor bus protocol expects the address to be latched using
AVALID as an enable, then decoded, as shown in Figure 8-15. The decoder
selects different devices to be enabled or "chip-selected" according to the high-
order bits of the address (the 12 high-order bits in this example). The low-order
bits are used to address individual locations of a selected device.

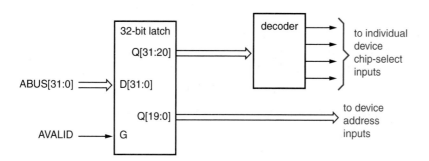

Figure 8-15
Microprocessor
address latching and
decoding circuit.

WHY A LATCH?   The microprocessor bus protocol in Figure 8-14 raises several questions:

- Why not keep the address valid on ABUS for the entire operation? In a real system using this protocol, the functions of ABUS and DBUS are combined (multiplexed) onto one three-state bus to save pins and wires.

- Why not use AVALID as the clock input to a positive-edge-triggered register to capture the address? There isn't enough setup time; in a real system, the address may first be valid at or slightly after the rising edge of AVALID.

- OK, so why not use AVALID to clock a negative-edge-triggered register? This works, but the latched outputs are available sooner; valid values on ABUS flow through a latch immediately, without waiting for the falling clock edge. This relaxes the access-time requirements of memories and other devices driven by the latched outputs.

Using a PLD, the latching and decoding functions for the high-order bits can be combined into a single device, yielding the block diagram in Figure 8-16. Compared with Figure 8-15, the "latching decoder" saves devices and pins, and it may produce a valid chip-select output more quickly (see Exercise 8.9).

Table 8-2 is an ABEL program for the latching decoder. Since it operates on only the high-order bits ABUS[31..20], it can decode addresses only in 1-Mbyte or larger chunks (2^{20} = 1 M). A read-only memory (ROM) is located in the highest 1-Mbyte chunk, addresses 0xfff00000–0xffffffff, and is selected by ROMCS. Three 16-Mbyte banks of read/write memory (RAM) are located at lower addresses, starting at addresses 0x00000000, 0x00100000, and 0x00200000, respectively. Notice how don't-cares are used in the definitions of the RAM bank address ranges to decode a chunk larger than 1 Mbyte. Other approaches to these definitions are also possible (e.g., see Exercise 8.1).

The equations in Table 8-2 for the chip-select outputs follow the D-latch template that we gave on page 674. The expressions that select a device, such as "ABUS==ROM," each generate a single product term, and each equation generates three product terms. Notice the use of the "retain" property in the pin declarations to prevent the compiler from optimizing away the consensus terms.

Figure 8-16
Using a combined
address latching and
decoding circuit.

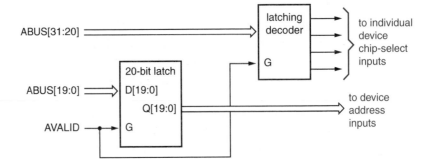

```
module latchdec
title 'Latching Microprocessor Address Decoder'

" Inputs
AVALID, ABUS31..ABUS20              pin;
" Latched and decoded outputs
ROMCS, RAMCS0, RAMCS1, RAMCS2       pin istype 'com,retain';

ABUS = [ABUS31..ABUS20];
ROM = ^hFFF;
RAMBANK0 = [0,0,0,0,0,0,0,0,.X.,.X.,.X.,.X.];
RAMBANK1 = [0,0,0,0,0,0,0,1,.X.,.X.,.X.,.X.];
RAMBANK2 = [0,0,0,0,0,0,1,0,.X.,.X.,.X.,.X.];

equations

ROMCS   = AVALID & (ABUS==ROM) # !AVALID & ROMCS
        # (ABUS==ROM) & ROMCS;
RAMCS0 = AVALID & (ABUS==RAMBANK0) # !AVALID & RAMCS0
        # (ABUS==RAMBANK0) & RAMCS0;
RAMCS1 = AVALID & (ABUS==RAMBANK1) # !AVALID & RAMCS1
        # (ABUS==RAMBANK1) & RAMCS1;
RAMCS2 = AVALID & (ABUS==RAMBANK2) # !AVALID & RAMCS2
        # (ABUS==RAMBANK2) & RAMCS2;

end latchdec
```

Table 8-2
ABEL program
for a latching
address decoder.

After seeing how easy it is to build S-R and D latches using combinational PLDs, you might be tempted to go further and try to build an edge-triggered D flip-flop. Although this is possible, it is expensive because an edge-triggered flip-flop has four internal states and thus two feedback loops, consuming two PLD outputs. Furthermore, the setup and hold times and propagation delays of such a flip-flop would be quite poor compared to those of a discrete flip-flop in the same technology. Finally, as we discussed in Section 7.10.6, the flow tables of all edge-triggered flip-flops contain essential hazards, which can be masked only by controlling path delays, difficult in a PLD-based design.

8.2.7 Registers and Latches in VHDL

Register and latch circuits can be specified using structural VHDL. For example, Table 8-3 is a structural VHDL program corresponding to the D latch circuit of Figure 7-12 on page 539. However, writing structural programs is not really our motivation for using VHDL; our goal is to use behavioral programs to model the operation of circuits more intuitively.

Table 8-4 is a process-based behavioral architecture for the D latch that requires, in effect, just one line of code to describe the latch's behavior. Note that the VHDL compiler "infers" a latch from this description—since the code

Table 8-3 VHDL structural program for the D latch in Figure 7-12.

```
library IEEE;
use IEEE.std_logic_1164.all;

entity Vdlatch is
  port (D, C: in STD_LOGIC;
        Q, QN: buffer STD_LOGIC );
end Vdlatch;

architecture Vdlatch_s of Vdlatch is
  signal DN, SN, RN: STD_LOGIC;
  component inv port (I: in STD_LOGIC; O: out STD_LOGIC ); end component;
  component nand2b port (I0, I1: in STD_LOGIC; O: buffer STD_LOGIC ); end component;
begin
  U1: inv port map (D,DN);
  U2: nand2b port map (D,C,SN);
  U3: nand2b port map (C,DN,RN);
  U4: nand2b port map (SN,QN,Q);
  U5: nand2b port map (Q,RN,QN);
end Vdlatch_s;
```

Table 8-4 VHDL behavioral architecture for a D latch.

```
architecture Vdlatch_b of Vdlatch is
begin
process(C, D, Q)
  begin
    if (C='1') then Q <= D; end if;
    QN <= not Q;
  end process;
end Vdlatch_b;
```

inferred latch

doesn't say what to do if C is not 1, the compiler creates an *inferred latch* to retain the value of Q between process invocations. In general, a VHDL compiler infers a latch for a signal that is assigned a value in an if or case statement if not all input combinations are accounted for.

event attribute

In order to describe edge-triggered behavior of flip-flops, we need to use one of VHDL's predefined signal attributes, the *event* attribute. If "SIG" is a signal name, then the construction "SIG'event" returns the value true at any delta time when SIG changes from one value to another, and false otherwise.

Using the event attribute, we can model a positive-edge triggered flip-flop as shown in Table 8-6. In the IF statement, "CLK'event" returns true on any clock edge, and "CLK='1'" ensures that D is assigned to Q only on a *rising* edge. Note that the process sensitivity list includes only CLK; changes on D at other times are not relevant in this functional model.

BUFFS 'N' STUFF   Note that in Table 8-3 we defined the type of Q and QN to be buffer rather than out, since these signals are used as inputs as well as outputs in the architecture definition. Then we had to define a special 2-input NAND gate nand2b with output type buffer to avoid having a type mismatch (out vs. buffer) in the component instantiations for U4 and U5. Alternatively, we could have used internal signals to get around the problem, as shown in Table 8-5. As you know by now, VHDL has many different ways to express the same thing.

Table 8-5 Alternative VHDL structural program for the D latch in Figure 7-12.

```
library IEEE;
use IEEE.std_logic_1164.all;

entity Vdlatch is
  port (D, C: in STD_LOGIC;
        Q, QN: out STD_LOGIC );
end Vdlatch;

architecture Vdlatch_s2 of Vdlatch is
  signal DN, SN, RN, IQ, IQN: STD_LOGIC;
  component inv port (I: in STD_LOGIC; O: out STD_LOGIC ); end component;
  component nand2 port (I0, I1: in STD_LOGIC; O: out STD_LOGIC ); end component;
begin
  U1: inv port map (D,DN);
  U2: nand2 port map (D,C,SN);
  U3: nand2 port map (C,DN,RN);
  U4: nand2 port map (SN,IQN,IQ);
  U5: nand2 port map (IQ,RN,IQN);
  Q <= IQ; QN <= IQN;
end Vdlatch_s2;
```

Table 8-6 VHDL behavioral model of an edge-triggered D flip-flop.

```
library IEEE;
use IEEE.std_logic_1164.all;

entity Vdff is
  port (D, CLK: in STD_LOGIC;
        Q: out STD_LOGIC );
end Vdff;

architecture Vdff_b of Vdff is
begin
process(CLK)
  begin
    if (CLK'event and CLK='1') then Q <= D; end if;
  end process;
end Vdff_b;
```

Table 8-7 VHDL model of a 74x74-like D flip-flop with preset and clear.

```
library IEEE;
use IEEE.std_logic_1164.all;

entity Vdff74 is
  port (D, CLK, PR_L, CLR_L: in STD_LOGIC;
          Q, QN: out STD_LOGIC );
end Vdff74;

architecture Vdff74_b of Vdff74 is
signal PR, CLR: STD_LOGIC;
begin
process(CLR_L, CLR, PR_L, PR, CLK)
  begin
    PR <= not PR_L; CLR <= not CLR_L;
    if (CLR and PR) = '1' then Q <= '0'; QN <= '0';
    elsif CLR = '1' then Q <= '0'; QN <= '1';
    elsif PR = '1' then Q <= '1'; QN <= '0';
    elsif (CLK'event and CLK='1') then Q <= D; QN <= not D;
    end if;
  end process;
end Vdff74_b;
```

Table 8-8 VHDL model of a 16-bit register with many features.

```
library IEEE;
use IEEE.std_logic_1164.all;

entity Vreg16 is
  port (CLK, CLKEN, OE_L, CLR_L: in STD_LOGIC;
        D: in STD_LOGIC_VECTOR(1 to 16);       -- Input bus
        Q: out STD_ULOGIC_VECTOR (1 to 16) ); -- Output bus (three-state)
end Vreg16;

architecture Vreg16 of Vreg16 is
signal CLR, OE: STD_LOGIC;  -- active-high versions of signals
signal IQ: STD_LOGIC_VECTOR(1 to 16); -- internal Q signals
begin
process(CLK, CLR_L, CLR, OE_L, OE, IQ)
  begin
    CLR <= not CLR_L;  OE <= not OE_L;
    if (CLR = '1') then IQ <= (others => '0');
    elsif (CLK'event and CLK='1') then
      if (CLKEN='1') then IQ <= D; end if;
    end if;
    if OE = '1' then Q <= To_StdULogicVector(IQ);
    else Q <= (others => 'Z'); end if;
  end process;
end Vreg16;
```

The D-flip-flop model can be augmented to include asynchronous inputs and a complemented output as in the 74x74 discrete flip-flop, as shown in Table 8-7. This more detailed functional model shows the non-complementary behavior of the Q and QN outputs when preset and clear are asserted simultaneously. However, it does not include timing behavior such as propagation delay and setup and hold times, which are beyond the scope of the VHDL coverage in this book.

Larger registers can be modeled, of course, by defining the data inputs and outputs to be vectors, and additional functions can also be included. For example, Table 8-8 models a 16-bit register with three-state outputs and clock-enable, output-enable, and clear inputs. An internal signal vector IQ is used to hold the flip-flop outputs, and three-state outputs are defined and enabled as in Section 5.6.4.

SYNTHESIS RESTRICTIONS   In Table 8-8, the first `elsif` statement theoretically could have included all of the conditions needed to assign D to IQ. That is, it could have read "`elsif (CLK'event) and (CLK='1') and (CLKEN='1') then ...`" instead of using a nested `if` statement to check CLKEN. However, it was written as shown for a very pragmatic reason.

Only a subset of the VHDL language can be synthesized by the VHDL compiler that was used to prepare this chapter; this is true of any VHDL compiler today. In particular, use of the "event" attribute is limited to the form shown in the example, and a few others, for detecting simple edge-triggered behavior. This gets mapped into edge-triggered D flip-flops during synthesis. The nested IF statement that checks CLKEN in the example leads to the synthesis of multiplexer logic on the D inputs of these flip-flops.

8.3 Sequential PLDs

8.3.1 Bipolar Sequential PLDs

The *PAL16R8*, shown in Figure 8-17, is representative of the first generation of   *PAL16R8*
sequential PLDs, which used bipolar (TTL) technology. This device has eight primary inputs, eight outputs, and common clock and output-enable inputs, and it fits in a 20-pin package.

The PAL16R8's AND-OR array is exactly the same as the one found in the PAL16L8 combinational PLD. However, the PAL16R8 has edge-triggered D flip-flops between the AND-OR array and its eight outputs, O1–O8. All of the flip-flops are connected to a common clock input, CLK, and change state on the rising edge of the clock. Each flip-flop drives an output pin through a 3-state buffer; the buffers have a common output-enable signal, OE_L. Notice that, like the combinational output pins of a PAL16L8, the registered output pins of the PAL16R8 contain the complement of the signal produced by the AND-OR array.

The possible inputs to the PAL16R8's AND-OR array are eight primary inputs (I1–I8) and the eight D flip-flop outputs. The connection from the D flip-flop outputs into the AND-OR array makes it easy to design shift registers, counters, and general state machines. Unlike the PAL16L8's combinational outputs, the PAL16R8's D-flip-flop outputs are available to the AND-OR array whether or not the O1–O8 three-state drivers are enabled. Thus, the internal flip-flops can go to a next state that is a function of the current state even when the O1–O8 outputs are disabled.

Many applications require combinational as well as sequential PLD outputs. The manufacturers of bipolar PLDs addressed this need by providing a few variants of the PAL16R8 that omitted the D flip-flops on some output pins, and instead provided input and output capability identical to that of the PAL16L8's bidirectional pins. For example, Figure 8-18 is the logic diagram of the *PAL16R6*, which has only six registered outputs. Two pins, IO1 and IO8, are bidirectional, serving both as inputs and as combinational outputs with separate 3-state enables, just like the PAL16L8's bidirectional pins. Thus, the possible inputs to the PAL16R6's AND-OR array are the eight primary inputs (I1–I8), the six D-flip-flop outputs, and the two bidirectional pins (IO1, IO8).

Table 8-9 shows eight standard bipolar PLDs with differing numbers and types of inputs and outputs. All of the PAL16xx parts in the table use the same AND-OR array, where each output has eight AND gates, each with 16 variables and their complements as possible inputs. The PAL20xx parts use a similar

PAL16R6

PAL16L8
PAL16R4
PAL16R8
PAL20L8
PAL20R4

▮ **Table 8-9** Characteristics of standard bipolar PLDs.

| Part number | Package pins | AND-gate inputs | Inputs to AND array | | |
			Primary inputs	Bidirectional combinational outputs	Registered outputs	Combinational outputs
PAL16L8	20	16	10	6	0	2
PAL16R4	20	16	8	4	4	0
PAL16R6	20	16	8	2	6	0
PAL16R8	20	16	8	0	8	0
PAL20L8	24	20	14	6	0	2
PAL20R4	24	20	12	4	4	0
PAL20R6	24	20	12	2	6	0
PAL20R8	24	20	12	0	8	0

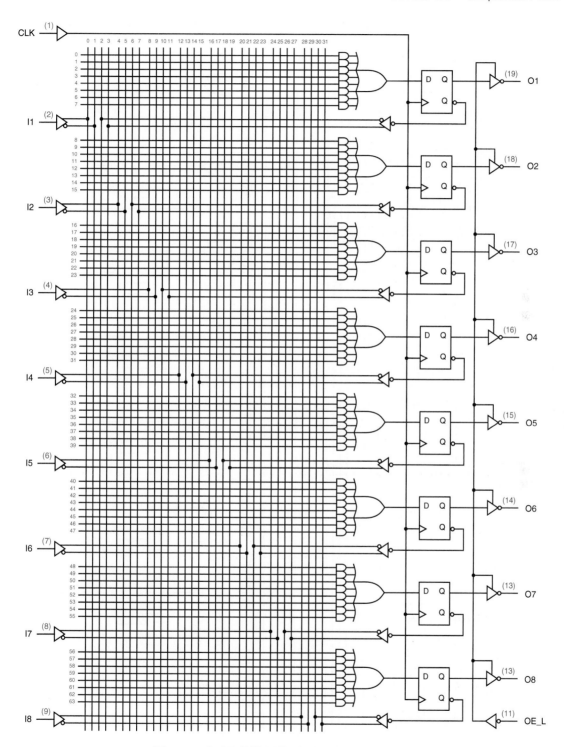

Figure 8-17 PAL16R8 logic diagram.

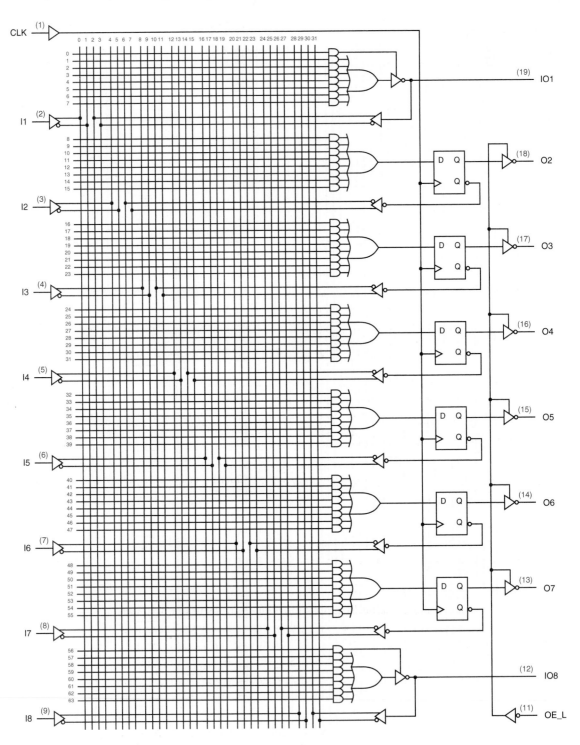

Figure 8-18 PAL16R6 logic diagram.

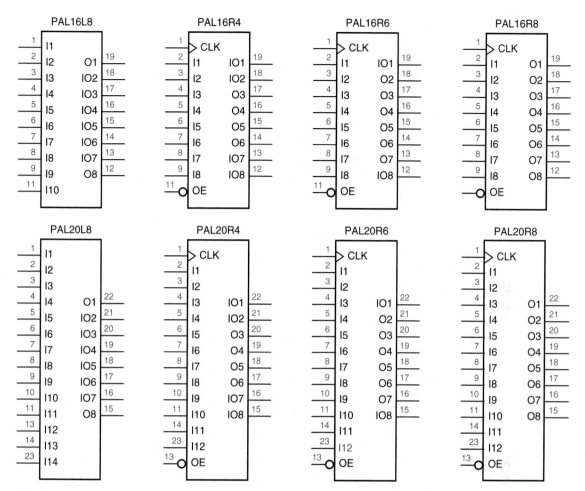

Figure 8-19 Logic symbols for bipolar combinational and sequential PLDs.

AND-OR array with 20 variables and their complements as possible inputs. Figure 8-19 shows logic symbols for all of the PLDs in the table.

PAL20R6
PAL20R8

8.3.2 Sequential GAL Devices

The GAL16V8 electrically erasable PLD was introduced in Section 5.3.3. Two "architecture-control" fuses are used to select among three basic configurations of this device. Section 5.3.3 described the *16V8C* ("complex") configuration, shown in Figure 5-27 on page 345, a structure similar to that of a bipolar combinational PAL device, the PAL16L8. The *16V8S* ("simple") configuration provides a slightly different combinational logic capability (see box on page 687).

16V8C

16V8S

    The third configuration, called the *16V8R*, allows a flip-flop to be provided on any or all of the outputs. Figure 8-20 shows the structure of the device when

16V8R

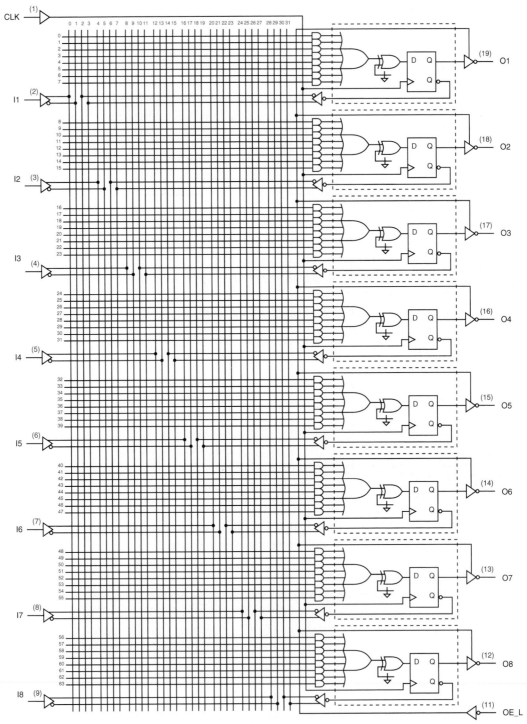

Figure 8-20 Logic diagram for the 16V8 in the "registered" configuration.

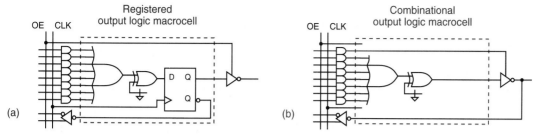

Figure 8-21 Output logic macrocells for the 16V8R: (a) registered; (b) combinational.

flip-flops are provided on all outputs. Notice that all of the flip-flops are controlled by a common clock signal on pin 1, as in the bipolar devices of the preceding subsection. Likewise, all of the output buffers are controlled by a common output-enable signal on pin 11.

The circuitry inside each dotted box in Figure 8-20 is called an *output logic macrocell*. The 16V8R is much more flexible than a PAL16R8 because each macrocell may be individually configured to bypass the flip-flop—that is, to produce a combinational output. Figure 8-21 shows the two macrocell configurations that are possible in the 16V8R; (a) is registered and (b) is combinational. Thus, it is possible to program the device to have any set of registered and combinational outputs, up to eight total.

output logic macrocell

The *20V8* is similar to the 16V8 but comes in a 24-pin package with four extra input-only pins. Each AND gate in the 20V8 has 20 inputs, hence the "20" in "20V8."

20V8

THE "SIMPLE" 16V8S
The "simple" 16V8S configuration of the GAL16V8 is not often used, because its capabilities are mostly a subset of the 16V8C's. Instead of an AND term, the 16V8S uses one fuse per output to control whether the output buffers are enabled. That is, each output pin may be programmed either to be always enabled or to be always disabled (except pins 15 and 16, which are always enabled). All of the output pins (except 15 and 16) are available as inputs to the AND array regardless of whether the output buffer is enabled.

The only advantage of a 16V8S compared to a 16V8C is that it has eight, not seven, AND terms as inputs to the OR gate on each output. The 16V8S architecture was designed mainly for emulation of certain now-obsolete bipolar PAL devices, some of which either had eight product terms per output or had inputs on pins 12 and 19, which are not inputs in the 16V8C configuration. With appropriate programming, the 16V8S can be used as a pin-for-pin compatible replacement for these devices, which included the PAL10H8, PAL12H6, PAL14H4, PAL16H2, PAL10L8, PAL12L6, PAL14L4, and PAL16L2.

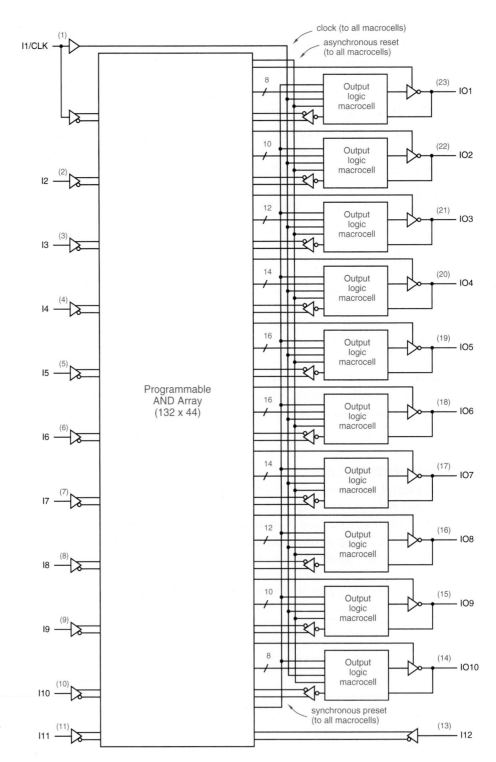

Figure 8-22
Logic diagram for
the 22V10.

The *22V10*, whose basic structure is shown in Figure 8-22, also comes in a 24-pin package but is somewhat more flexible than the 20V8. The 22V10 does not have "architecture control" bits like the 16V8's and 20V8's, but it can realize any function that is realizable with a 20V8, and more. It has more product terms, two more general-purpose inputs, and better output-enable control than the 20V8. Key differences are summarized below:

- Each output logic macrocell is configurable to have a register or not, as in the 20V8R architecture. However, the macrocells are different from the 16V8's and 20V8's, as shown in Figure 8-23.

- A single product term controls the output buffer, regardless of whether the registered or the combinational configuration is selected for a macrocell.

- Every output has at least eight product terms available, regardless of whether the registered or the combinational configuration is selected. Even more product terms are available on the inner pins, with 16 available on each of the two innermost pins. ("Innermost" is with respect to the right-hand side of the Figure 8-22, which also matches the arrangement of these pins on a 24-pin dual-inline package.)

- The clock signal on pin 1 is also available as a combinational input to any product term.

- A single product term is available to generate a global, asynchronous reset signal that resets all internal flip-flops to 0.

- A single product term is available to generate a global, synchronous preset signal that sets all internal flip-flops to 1 on the rising edge of the clock.

- Like the 16V8 and 20V8, the 22V10 has programmable output polarity. However, in the registered configuration, the polarity change is made at the output, rather than the input, of the D flip-flop. This affects the details of programming when the polarity is changed but does not affect the overall capability of the 22V10 (i.e., whether a given function can be realized). In fact, the difference in polarity-change location is transparent when you use a PLD programming language such as ABEL.

Figure 8-23 Output logic macrocells for the 22V10: (a) registered; (b) combinational.

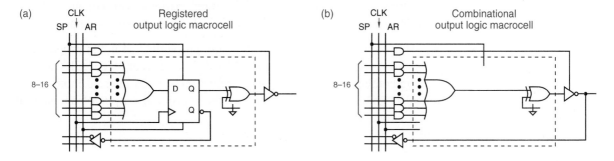

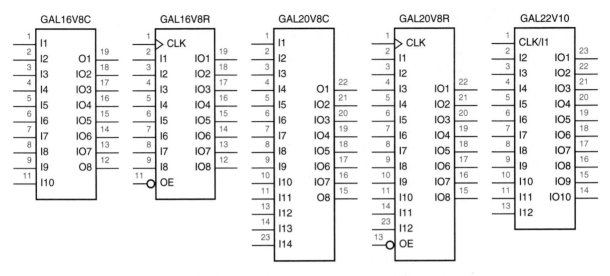

Figure 8-24 Logic symbols for popular GAL devices.

For most of the 1990s, the 16V8, 20V8, and 22V10 were the most popular and cost-effective PLDs (but see the box on page 693). Figure 8-24 shows generic logic symbols for these three devices. Most of the examples in the rest of this chapter can fit into the smallest of the three devices, the 16V8.

PALS? GALS? Lattice Semiconductor introduced GAL devices including the GAL16V8 and GAL20V8 in the mid-1980s. Advanced Micro Devices later followed up with a pin-compatible device which they call the PALCE16V8 ("C" is for CMOS, "E" is for erasable). Several other manufacturers make differently numbered but compatible devices as well. Rather than get caught up in the details of different manufacturers' names, in this chapter we usually refer to commonly used GAL devices with their generic names, 16V8, 20V8, and 22V10.

8.3.3 PLD Timing Specifications

Several timing parameters are specified for combinational and sequential PLDs. The most important ones are illustrated in Figure 8-25 and are explained below:

t_{PD}

feedback input

t_{PD} This parameter applies to combinational outputs. It is the propagation delay from a primary input pin, bidirectional pin, or "feedback" input to the combinational output. A *feedback input* is an internal input of the AND-OR array that is driven by the registered output of an internal macrocell.

t_{CO} This parameter applies to registered outputs. It is the propagation delay from the rising edge of CLK to a primary output. $\quad t_{CO}$

t_{CF} This parameter also applies to registered outputs. It is the propagation delay from the rising edge of CLK to a macrocell's registered output that connects back to a feedback input. If specified, t_{CF} is normally less than t_{CO}. However, some manufacturers do not specify t_{CF}, in which case you must assume that $t_{CF} = t_{CO}$. $\quad t_{CF}$

t_{SU} This parameter applies to primary, bidirectional, and feedback inputs that propagate to the D inputs of flip-flops. It is the setup time during which the input signal must be stable before the rising edge of CLK. $\quad t_{SU}$

t_{H} This parameter also applies to signals that propagate to the D inputs of flip-flops. It is the hold time during which the input signal must be stable after the rising edge of CLK. $\quad t_{H}$

f_{max} This parameter applies to clocked operation. It is the highest frequency at which the PLD can operate reliably and is the reciprocal of the minimum clock period. Two versions of this parameter can be derived from the previous specifications, depending on whether the device is operating with external feedback or internal feedback. $\quad f_{max}$

External feedback refers to a circuit in which a registered PLD output is connected to the input of another registered PLD with similar timing; for proper operation, the sum of t_{CO} for the first PLD and t_{SU} for the second must not exceed the clock period. $\quad$ *external feedback*

Internal feedback refers to a circuit in which a registered PLD output is fed back to a register in the same PLD; in this case, the sum of t_{CF} and t_{SU} must not exceed the clock period. $\quad$ *internal feedback*

Each of the PLDs that we described in previous sections is available in several different speed grades. The speed grade is usually indicated by a suffix on the part number, such as "16V8-10"; the suffix usually refers to the t_{PD}

Figure 8-25 PLD timing parameters.

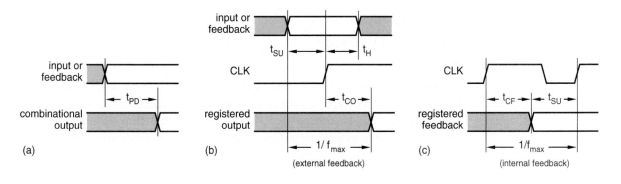

Table 8-10 Timing specifications, in nanoseconds, of popular bipolar and CMOS PLDs.

Part numbers	Suffix	t_{PD}	t_{CO}	t_{CF}	t_{SU}	t_H
PAL16L8, PAL16Rx, PAL20L8, PAL20Rx	-5	5	4	–	4.5	0
PAL16L8, PAL16Rx, PAL20L8, PAL20Rx	-7	7.5	6.5	–	7	0
PAL16L8, PAL16Rx, PAL20L8, PAL20Rx	-10	10	8	–	10	0
PAL16L8, PAL16Rx, PAL20L8, PAL20Rx	B	15	12	–	15	0
PAL16L8, PAL16Rx, PAL20L8, PAL20Rx	B-2	25	15	–	25	0
PAL16L8, PAL16Rx, PAL20L8, PAL20Rx	A	25	15	–	25	0
PALCE16V8, PALCE20V8	-5	5	4	–	3	0
GAL16V8, GAL20V8	-7	7.5	5	3	5	0
GAL16V8, GAL20V8	-10	10	7.5	6	7.5	0
GAL16V8, GAL20V8	-15	15	10	8	12	0
GAL16V8, GAL20V8	-25	25	12	10	15	0
PALCE22V10	-5	5	4	–	3	0
PALCE22V10	-7	7.5	4.5	–	4.5	0
GAL22V10	-10	10	7	2.5	7	0
GAL22V10	-15	15	8	2.5	10	0
GAL22V10	-25	25	15	13	15	0

specification, in nanoseconds. Table 8-10 shows the timing of several popular bipolar and CMOS PLDs. Note that only the t_{PD} column applies to the combinational outputs of a device, while the last four columns apply to registered outputs. All of the timing specifications are worst-case numbers over the commercial operating range.

When sequential PLDs are used in applications with critical timing, it's important to remember that they normally have longer setup times than discrete edge-triggered registers in the same technology, owing to the delay of the AND-OR array on each D input. Conversely, under typical conditions, a PLD actually has a negative hold-time requirement because of the delay through AND-OR array. However, you can't count on it having a negative hold time—the worst-case specification is normally zero.

8.4 Counters

The name *counter* is generally used for any clocked sequential circuit whose state diagram contains a single cycle, as in Figure 8-26. The *modulus* of a counter is the number of states in the cycle. A counter with m states is called a *modulo-m counter* or, sometimes, a *divide-by-m counter*. A counter with a non-power-of-2 modulus has extra states that are not used in normal operation.

counter
modulus

modulo-m counter
divide-by-m counter

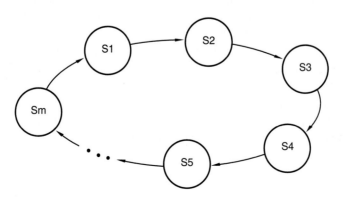

Figure 8-26
General structure of a counter's state diagram—a single cycle.

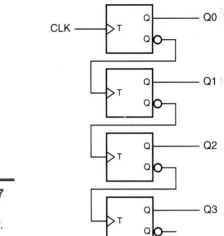

Figure 8-27
A 4-bit binary
ripple counter.

n-bit binary counter

Probably the most commonly used counter type is an *n-bit binary counter*. Such a counter has n flip-flops and has 2^n states, which are visited in the sequence 0, 1, 2, ... , $2^n - 1$, 0, 1, ... Each of these states is encoded as the corresponding n-bit binary integer.

8.4.1 Ripple Counters

An n-bit binary counter can be constructed with just n flip-flops and no other components, for any value of n. Figure 8-27 shows such a counter for $n = 4$. Recall that a T flip-flop changes state (toggles) on every rising edge of its clock input. Thus, each bit of the counter toggles if and only if the immediately preceding bit changes from 1 to 0. This corresponds to a normal binary counting sequence—when a particular bit changes from 1 to 0, it generates a carry to the next most significant bit. The counter is called a *ripple counter* because the carry information ripples from the less significant bits to the more significant bits, one bit at a time.

ripple counter

8.4.2 Synchronous Counters

Although a ripple counter requires fewer components than any other type of binary counter, it does so at a price—it is slower than any other type of binary counter. In the worst case, when the most significant bit must change, the output is not valid until time $n \cdot t_{TQ}$ after the rising edge of CLK, where t_{TQ} is the propagation delay from input to output of a T flip-flop.

synchronous counter

A *synchronous counter* connects all of its flip-flop clock inputs to the same common CLK signal, so that all of the flip-flop outputs change at the same time, after only t_{TQ} ns of delay. As shown in Figure 8-28, this requires the use of T flip-flops with enable inputs; the output toggles on the rising edge of T if and only if EN is asserted. Combinational logic on the EN inputs determines which, if any, flip-flops toggle on each rising edge of T.

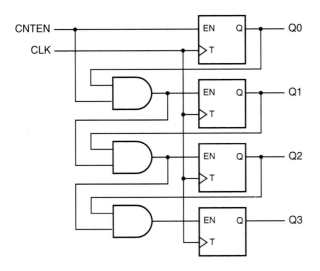

Figure 8-28
A synchronous 4-bit
binary counter with
serial enable logic.

As shown in Figure 8-28, it is also possible to provide a master count-enable signal CNTEN. Each T flip-flop toggles if and only if CNTEN is asserted and all of the lower-order counter bits are 1. Like the binary ripple counter, a synchronous n-bit binary counter can be built with a fixed amount of logic per bit—in this case, a T flip-flop with enable and a 2-input AND gate.

The counter structure in Figure 8-28 is sometimes called a *synchronous serial counter* because the combinational enable signals propagate serially from the least significant to the most significant bits. If the clock period is too short, there may not be enough time for a change in the counter's LSB to propagate to the MSB. This problem is eliminated in Figure 8-29 by driving each EN input with a dedicated AND gate, just a single level of logic. Called a *synchronous parallel counter,* this is the fastest binary counter structure.

synchronous serial counter

synchronous parallel counter

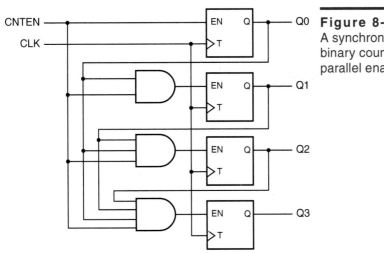

Figure 8-29
A synchronous 4-bit
binary counter with
parallel enable logic

8.4.3 MSI Counters and Applications

74x163

The most popular MSI counter is the *74x163,* a synchronous 4-bit binary counter with active-low load and clear inputs, with the traditional logic symbol shown in Figure 8-30. Its function is summarized by the state table in Table 8-11, and its internal logic diagram is shown in Figure 8-31.

The '163 uses D flip-flops rather than T flip-flops internally to facilitate the load and clear functions. Each D input is driven by a 2-input multiplexer consisting of an OR gate and two AND gates. The multiplexer output is 0 if the CLR_L input is asserted. Otherwise, the top AND gate passes the data input (A, B, C, or D) to the output if LD_L is asserted. If neither CLR_L nor LD_L is asserted, the bottom AND gate passes the output of an XNOR gate to the multiplexer output.

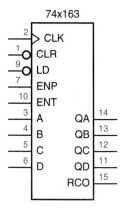

Figure 8-30
Traditional logic symbol for the 74x163.

Table 8-11 State table for a 74x163 4-bit binary counter.

Inputs				Current State				Next State			
CLR_L	LD_L	ENT	ENP	QD	QC	QB	QA	QD*	QC*	QB*	QA*
0	x	x	x	x	x	x	x	0	0	0	0
1	0	x	x	x	x	x	x	D	C	B	A
1	1	0	x	x	x	x	x	QD	QC	QB	QA
1	1	x	0	x	x	x	x	QD	QC	QB	QA
1	1	1	1	0	0	0	0	0	0	0	1
1	1	1	1	0	0	0	1	0	0	1	0
1	1	1	1	0	0	1	0	0	0	1	1
1	1	1	1	0	0	1	1	0	1	0	0
1	1	1	1	0	1	0	0	0	1	0	1
1	1	1	1	0	1	0	1	0	1	1	0
1	1	1	1	0	1	1	0	0	1	1	1
1	1	1	1	0	1	1	1	1	0	0	0
1	1	1	1	1	0	0	0	1	0	0	1
1	1	1	1	1	0	0	1	1	0	1	0
1	1	1	1	1	0	1	0	1	0	1	1
1	1	1	1	1	0	1	1	1	1	0	0
1	1	1	1	1	1	0	0	1	1	0	1
1	1	1	1	1	1	0	1	1	1	1	0
1	1	1	1	1	1	1	0	1	1	1	1
1	1	1	1	1	1	1	1	0	0	0	0

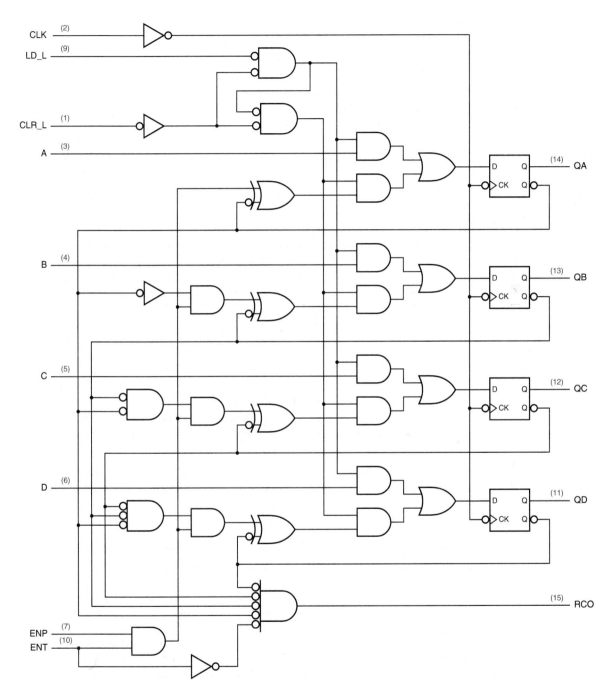

Figure 8-31 Logic diagram for the 74x163 synchronous 4-bit binary counter, including pin numbers for a standard 16-pin dual in-line package.

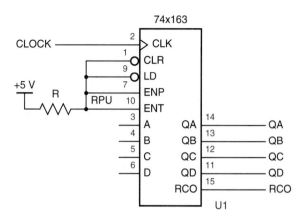

Figure 8-32
Connections for the 74x163 to operate in a free-running mode.

The XNOR gates perform the counting function in the '163. One input of each XNOR is the corresponding count bit (QA, QB, QC, or QD); the other input is 1, which complements the count bit, if and only if both enables ENP and ENT are asserted and all of the lower-order count bits are 1. The RCO ("ripple carry out") signal indicates a carry from the most significant bit position and is 1 when all of the count bits are 1 and ENT is asserted.

free-running counter

Even though most MSI counters have enable inputs, they are often used in a *free-running* mode in which they are enabled continuously. Figure 8-32 shows the connections to make a '163 operate in this way, and Figure 8-33 shows the resulting output waveforms. Notice that, starting with QA, each signal has half the frequency of the preceding one. Thus, a free-running '163 can be used as a divide-by-2, -4, -8, or -16 counter, by ignoring any unnecessary high-order output bits.

Note that the '163 is fully synchronous; that is, its outputs change only on the rising edge of CLK. Some applications need an asynchronous clear function,

Figure 8-33 Clock and output waveforms for a free-running divide-by-16 counter.

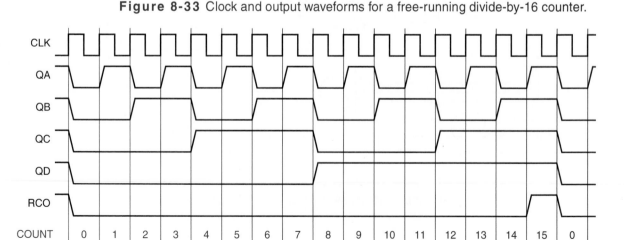

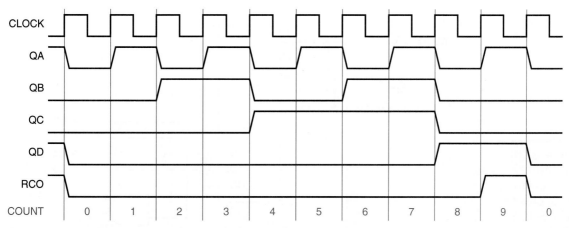

Figure 8-34 Clock and output waveforms for a free-running divide-by-10 counter.

as provided by the *74x161*. The '161 has the same pinout as the '163, but its CLR_L input is connected to the asynchronous clear inputs of its flip-flops.

The *74x160* and *74x162* are more variations with the same pinouts and general functions as the '161 and '163, except that the counting sequence is modified to go to state 0 after state 9. In other words, these are modulo-10 counters, sometimes called *decade counters*. Figure 8-34 shows the output waveforms for a free-running '160 or '162. Notice that although the QD and QC outputs have one-tenth of the CLK frequency, they do not have a 50% duty cycle, and the QC output, with one-fifth of the input frequency, does not have a constant duty cycle. We'll show the design of a divide-by-10 counter with a 50% duty-cycle output later in this subsection.

Although the '163 is a modulo-16 counter, it can be made to count in a modulus less than 16 by using the CLR_L or LD_L input to shorten the normal counting sequence. For example, Figure 8-35 shows one way of using the '163 as a modulo-11 counter. The RCO output, which detects state 15, is used to force

74x161

74x160
74x162

decade counter

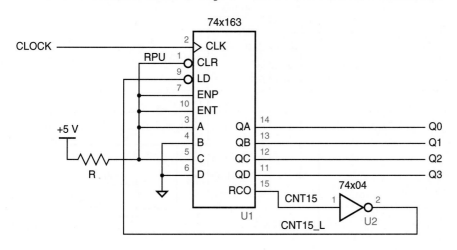

Figure 8-35
Using the 74x163 as a modulo-11 counter with the counting sequence 5, 6, …, 15, 5, 6, ….

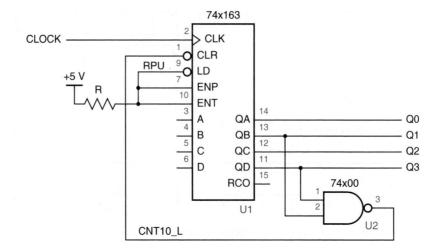

Figure 8-36
Using the 74x163 as a modulo-11 counter with the counting sequence 0, 1, 2, ..., 10, 0, 1, ...

the next state to 5, so that the circuit will count from 5 to 15 and then start at 5 again, for a total of 11 states per counting cycle.

A different approach for modulo-11 counting with the '163 is shown in Figure 8-36. This circuit uses a NAND gate to detect state 10 and force the next state to 0. Notice that only a 2-input gate is used to detect state 10 (binary 1010). Although a 4-input gate would normally be used to detect the condition CNT10 $= Q3 \cdot Q2' \cdot Q1 \cdot Q0'$, the 2-input gate takes advantage of the fact that no other state in the normal counting sequence of 0–10 has $Q3 = 1$ and $Q1 = 1$. In general, to detect state N in a binary counter that counts from 0 to N, we need to AND only the state bits that are 1 in the binary encoding of N.

There are many other ways to make a modulo-11 counter using a '163. The choice of approach—one of the preceding or a combination of them (as in Exercise 8.36)—depends on the application. As another example, in Section 2.10 we promised to show you how to build a circuit that counts in the

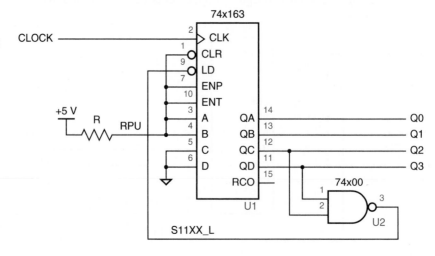

Figure 8-37
A 74x163 used as an excess-3 decimal counter.

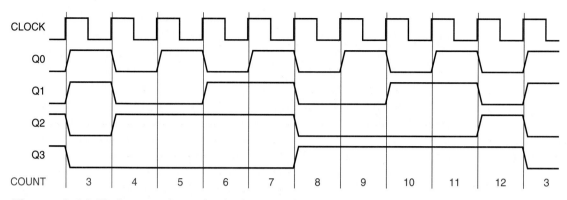

Figure 8-38 Timing waveforms for the '163 used as an excess-3 decimal counter.

excess-3 decimal code, shown in Table 2-9 on page 49. Figure 8-37 shows the connections for a '163 to count in the excess-3 sequence. A NAND gate detects state 1100 and forces 0011 to be loaded as the next state. Figure 8-38 shows the resulting timing waveforms. Notice that the Q3 output has a 50% duty cycle, which may be desirable for some applications.

A binary counter with a modulus greater than 16 can be built by cascading 74x163s. Figure 8-39 shows the general connections for such a counter. The CLK, CLR_L, and LD_L inputs of all the '163s are connected in parallel, so that all of them count or are cleared or loaded at the same time. A master count-enable (CNTEN) signal is connected to the low-order '163. The RCO4 output is asserted if and only if the low-order '163 is in state 15 *and* CNTEN is asserted; RCO4 is connected to the enable inputs of the high-order '163. Thus, both the carry information and the master count-enable ripple from the output of one

Figure 8-39 General cascading connections for 74x163-based counters.

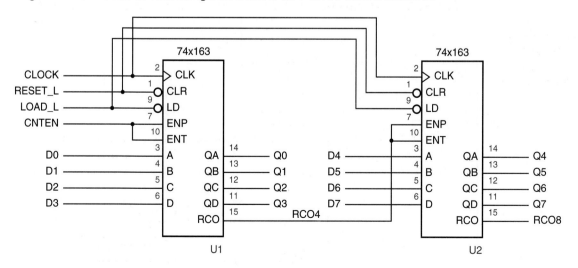

4-bit counter stage to the next. Like the synchronous serial counter of Figure 8-28, this scheme can be extended to build a counter with any desired number of bits; the maximum counting speed is limited by the propagation delay of the ripple carry signal through all of the stages (but see Exercise 8.38).

Even experienced digital designers are sometimes confused about the difference between the ENP and ENT enable inputs of the '163 and similar counters, since both must be asserted for the counter to count. However, a glance at the 163's internal logic diagram, Figure 8-31 on page 697, shows the difference quite clearly—ENT goes to the ripple carry output as well. In many applications this distinction is important.

For example, Figure 8-40 shows an application that uses two '163s as a modulo-193 counter that counts from 63 to 255. The MAXCNT output detects state 255 and stops the counter until GO_L is asserted. When GO_L is asserted, the counter is reloaded with 63 and counts up to 255 again. (Note that the value of GO_L is relevant only when the counter is in state 255.) To keep the counter

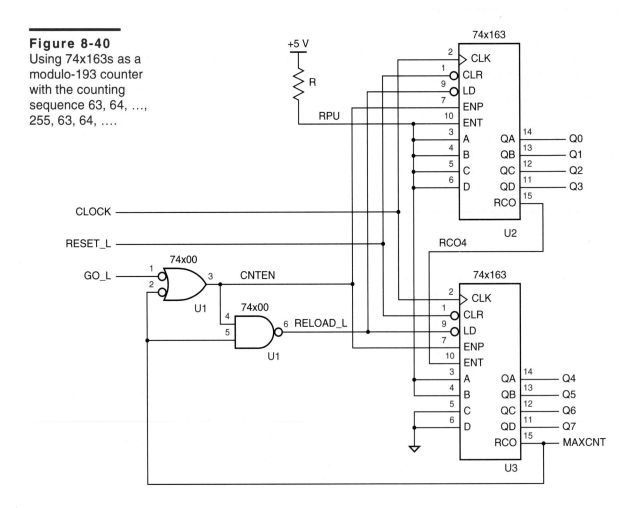

Figure 8-40
Using 74x163s as a
modulo-193 counter
with the counting
sequence 63, 64, ...,
255, 63, 64, ...

stopped, MAXCNT must be asserted in state 255 even while the counter is stopped. Therefore, the low-order counter's ENT input is always asserted, its RCO output is connected to the high-order ENT input, and MAXCNT detects state 255 even if CNTEN is not asserted (compare with the behavior of RCO8 in Figure 8-39). To enable counting, CNTEN is connected to the ENP inputs in parallel. A NAND gate asserts RELOAD_L to go back to state 63 only if GO_L is asserted and the counter is in state 255.

Another counter with functions similar to 74x163's is the *74x169,* whose logic symbol is shown in Figure 8-41. One difference in the '169 is that its carry output and enable inputs are active low. More importantly, the '169 is an *up/down counter;* it counts in ascending or descending binary order depending on the value of an input signal, UP/DN. The '169 counts up when UP/DN is 1 and down when UP/DN is 0.

74x169

up/down counter

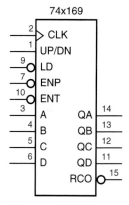

Figure 8-41
Logic symbol for the 74x169 up/down counter.

8.4.4 Decoding Binary-Counter States

A binary counter may be combined with a decoder to obtain a set of 1-out-of-*m*-coded signals, where one signal is asserted in each counter state. This is useful when counters are used to control a set of devices, where a different device is enabled in each counter state. In this approach, each output of the decoder enables a different device.

Figure 8-42 shows how a 74x163 wired as a modulo-8 counter can be combined with a 74x138 3-to-8 decoder to provide eight signals, each one representing a counter state. Figure 8-43 shows typical timing for this circuit. Each decoder output is asserted during a corresponding clock period.

Notice that the decoder outputs may contain "glitches" on state transitions where two or more counter bits change, even though the '163 outputs are glitch free and the '138 does not have any static hazards. In a synchronous counter like the '163, the outputs don't change at exactly the same time. More important,

decoding glitches

Figure 8-42
A modulo-8 binary counter and decoder.

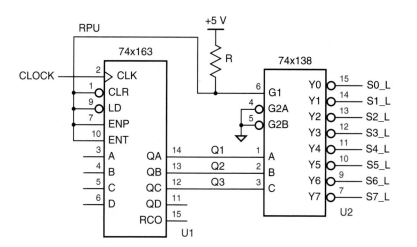

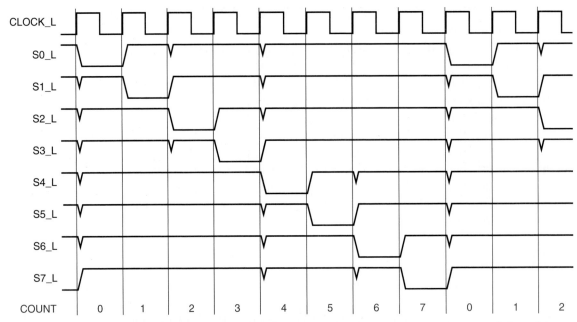

Figure 8-43 Timing diagram for a modulo-8 binary counter and decoder, showing decoding glitches.

function hazard

multiple signal paths in a decoder like the '138 have different delays; for example, the path from B to Y1_L is faster than the path from A to Y1_L. Thus, even if the input changes simultaneously from 011 to 100, the decoder may behave as if the input were temporarily 001, and the Y1_L output may have a glitch. In the present example, it can be shown that the glitches can occur in *any* realization of the binary decoder function; this problem is an example of a *function hazard.*

In most applications, the decoder output signals portrayed in Figure 8-43 would be used as control inputs to registers, counters, and other edge-triggered devices (e.g., EN_L in a 74x377, LD_L in a 74x163, or ENP_L in a 74x169). In such a case, the decoding glitches in the figure are not a problem, since they occur *after* the clock tick. They are long gone before the next tick comes along, when the decoder outputs are sampled by other edge-triggered devices. However, the glitches *would* be a problem if they were applied to something like the S_L or R_L inputs of an $\overline{\text{S}}$-$\overline{\text{R}}$ latch. Likewise, using such potentially glitchy signals as clocks for edge-triggered devices is a definite no-no.

If necessary, one way to "clean up" the glitches in Figure 8-43 is to connect the '138 outputs to another register that samples the stable decoded outputs on the next clock tick, as shown in Figure 8-44. Notice that the decoded outputs have been renamed to account for the 1-tick delay through the register. However, once you decide to pay for an 8-bit register, a less costly solution is to use an 8-bit "ring counter," which provides glitch-free decoded outputs directly, as we'll show in Section 8.5.6.

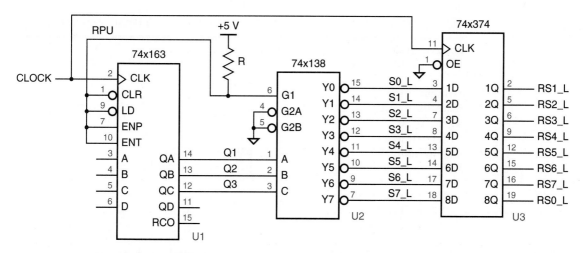

Figure 8-44 A modulo-8 binary counter and decoder with glitch-free outputs.

8.4.5 Counters in ABEL and PLDs

Binary counters are good candidates for ABEL- and PLD-based design, for several reasons:

- A large state machine can often be decomposed into two or more smaller state machines, where one of the smaller machines is a binary counter that keeps track of how long the other machine should stay in a particular state. This may simplify both the conceptual design and the circuit design of the machine.

- Many applications require almost-binary-modulus counters with special requirements for initialization, state detection, or state skipping. For example, a counter in an elevator controller may skip state 13. Instead of using an off-the-shelf binary counter and extra logic for the special requirements, a designer can specify exactly the required functions in an ABEL program.

- Most standard MSI counters have only 4 bits, while a single 24-pin PLD can be used to create a binary counter with up to 10 bits.

The most popular MSI counter is the 74x163 4-bit binary counter, shown in Figure 8-31 on page 697. A glance at this figure shows that the excitation logic for this counter isn't exactly simple, especially considering its use of XNOR gates. Nevertheless, ABEL provides a very simple way of defining counter behavior, which we describe next.

Recall that ABEL uses the "+" symbol to specify integer addition. When two sets are "added" with this operator, each is interpreted as a binary number; the rightmost set element corresponds to the least significant bit of the number. Thus, the function of a 74x163 can be specified by the ABEL program in Table 8-12. When the counter is enabled, 1 is added to the current state.

Table 8-12 ABEL program for a 74x163-like 4-bit binary counter.

```
module Z74X163
title '4-bit Binary Counter'

" Input pins
CLK, LD_L, CLR_L, ENP, ENT      pin;
A, B, C, D                      pin;
" Output pins
QA, QB, QC, QD                  pin istype 'reg';
RCO                             pin istype 'com';

" Set definitions
INPUT = [D, C, B, A];
COUNT = [QD, QC, QB, QA];

LD = !LD_L; CLR = !CLR_L;       " Active-level conversions

equations

COUNT.CLK = CLK;

COUNT := !CLR & ( LD & INPUT
                # !LD & (ENT & ENP) & (COUNT + 1)
                # !LD & !(ENT & ENP) & COUNT);

RCO = (COUNT == [1,1,1,1]) & ENT;

end Z74X163
```

Table 8-13 MInimized equations for the 4-bit binary counter in Table 8-12.

```
QA := (CLR_L & LD_L & ENT & ENP & !QA        QD := (CLR_L & LD_L & ENT & ENP
    # CLR_L & LD_L & !ENP & QA                            & !QD & QC & QB & QA
    # CLR_L & LD_L & !ENT & QA                # CLR_L & !LD_L & D
    # CLR_L & !LD_L & A);                     # CLR_L & LD_L & QD & !QB
                                              # CLR_L & LD_L & QD & !QC
                                              # CLR_L & LD_L & !ENP & QD
QB := (CLR_L & LD_L & ENT & ENP & !QB & QA    # CLR_L & LD_L & !ENT & QD
    # CLR_L & LD_L & QB & !QA                 # CLR_L & LD_L & QD & !QA);
    # CLR_L & LD_L & !ENP & QB
    # CLR_L & LD_L & !ENT & QB             RCO = (ENT & QD & QC & QB & QA);
    # CLR_L & !LD_L & B);

QC := (CLR_L & LD_L & ENT & ENP & !QC & QB & QA
    # CLR_L & LD_L & QC & !QA
    # CLR_L & LD_L & QC & !QB
    # CLR_L & LD_L & !ENP & QC
    # CLR_L & LD_L & !ENT & QC
    # CLR_L & !LD_L & C);
```

Table 8-13 shows the minimized logic equations that ABEL generates for the 4-bit counter. Notice that each more significant output bit requires one more product term. As a result, the size of counters that can be realized in a 16V8 or even a 20V8 is generally limited to five or six bits. Other devices, including the X-series PLDs and some CPLDs, contain an XOR structure that can realize larger counters without increasing product-term requirements.

Designing a specialized counting sequence in ABEL is much simpler than adapting a standard binary counter. For example, the ABEL program in Table 8-12 can be adapted to count in excess-3 sequence (Figure 8-38 on page 701) by changing the equations as follows:

```
COUNT := !CLR & ( LD & INPUT
        # !LD & (ENT & ENP) &
            ((COUNT==12) & 3) # ((COUNT!=12) & (COUNT + 1))
        # !LD & !(ENT & ENP) & COUNT);

RCO = (COUNT == 12) & ENT;
```

PLDs can be cascaded to obtain wider counters by providing each counter stage with a carry output that indicates when it is about to roll over. There are two basic approaches to generating the carry output:

- *Combinational.* The carry equation indicates that the counter is enabled and is currently in its last state before rollover. For a 5-bit binary up counter, we have

  combinational carry output

  ```
  COUT = CNTEN & Q4 & Q3 & Q2 & Q1 & Q0;
  ```

  Since CNTEN is included, this approach allows carries to be rippled through cascaded counters by connecting each COUT to the next CNTEN.

- *Registered.* The carry equation indicates that the counter is about to enter its last state before rollover. Thus, at the next clock tick, the counter enters this last state and the carry output is asserted. For a 5-bit binary up counter with load and clear inputs, we have

  registered carry output

  ```
  COUT := !CLR & !LD & CNTEN
          & Q4 & Q3 & Q2 & Q1 & !Q0
       # !CLR * !LD * !CNTEN
          & Q4 & Q3 & Q2 & Q1 & Q0
       # !CLR & LD
          & D4 & D3 & D2 & D1 & D0;
  ```

The second approach has the advantage of producing COUT with less delay than the combinational approach. However, external gates are now required between stages, since the CNTEN signal for each stage should be the logical AND of the master count-enable signal and the COUT outputs of all lower-order counters. These external gates can be avoided if the higher-order counters have multiple enable inputs.

8.4.6 Counters in VHDL

Like ABEL, VHDL allows counters to be specified fairly easily. The biggest challenge in VHDL, with its strong type checking, is to get all of the signal types defined correctly and consistently.

Table 8-14 is a VHDL program for a 74x163-like binary counter. Notice that the program uses the IEEE.std_logic_arith.all library, which includes the UNSIGNED type, as we described in Section 5.9.6 on page 427. This library includes definitions of "+" and "−" operators that perform unsigned addition and subtraction on UNSIGNED operands. The counter program declares the counter input and output as UNSIGNED vectors and uses "+" to increment the counter value as required.

In the program, we defined an internal signal IQ to hold the counter value. We could have used Q directly, but then we'd have to declare its port type as buffer rather than out. Also, we could have defined the type of ports D and Q to be STD_LOGIC_VECTOR, but then we would have to perform type conversions inside the body of the process (see Exercise 8.51).

Table 8-14 VHDL program for a 74x163-like 4-bit binary counter.

```
library IEEE;
use IEEE.std_logic_1164.all;
use IEEE.std_logic_arith.all;

entity V74x163 is
    port ( CLK, CLR_L, LD_L, ENP, ENT: in STD_LOGIC;
            D: in UNSIGNED (3 downto 0);
            Q: out UNSIGNED (3 downto 0);
            RCO: out STD_LOGIC );
end V74x163;

architecture V74x163_arch of V74x163 is
signal IQ: UNSIGNED (3 downto 0);
begin
process (CLK, ENT, IQ)
  begin
    if (CLK'event and CLK='1') then
      if CLR_L='0' then IQ <= (others => '0');
      elsif LD_L='0' then IQ <= D;
      elsif (ENT and ENP)='1' then IQ <= IQ + 1;
      end if;
    end if;
    if (IQ=15) and (ENT='1') then RCO <= '1';
    else RCO <= '0';
    end if;
    Q <= IQ;
  end process;
end V74x163_arch;
```

Table 8-15 VHDL architecture for counting in excess-3 order.

```
architecture V74xs3_arch of V74x163 is
signal IQ: UNSIGNED (3 downto 0);
begin
process (CLK, ENT, IQ)
  begin
    if CLK'event and CLK='1' then
      if CLR_L='0' then IQ <= (others => '0');
      elsif LD_L='0' then IQ <= D;
      elsif (ENT and ENP)='1' and (IQ=12) then IQ <= ('0','0','1','1');
      elsif (ENT and ENP)='1' then IQ <= IQ + 1;
      end if;
    end if;
    if (IQ=12) and (ENT='1') then RCO <= '1';
    else RCO <= '0';
    end if;
    Q <= IQ;
  end process;
end V74xs3_arch;
```

As in ABEL, specialized counting sequences can be specified very easily using behavioral VHDL code. For example, Table 8-15 modifies the 74x163-like counter to count in excess-3 sequence (3, ... , 12, 3, ...).

Unfortunately, some VHDL synthesis engines do not synthesize counters particularly well. In particular, they tend to synthesize the counting step using a binary adder with the counter value and a constant 1 as operands. This approach requires much more combinational logic than what we've shown for discrete counters; it is particularly wasteful in CPLDs and FPGAs containing T flip-flops, XOR gates, or other structures optimized for counters. In this case, a useful alternative is to write structural VHDL that is targeted to the cells available in a particular CPLD, FPGA, or ASIC technology.

For example, we can construct one bit-cell for a 74x163-like counter using the circuit in Figure 8-45. This circuit is designed to use serial propagation for the carry bits, so the same circuit can be used at any stage of an arbitrarily large counter, subject to fanout constraints on the common signals that drive all of the stages. The signals in the bit-cell have the following definitions:

CLK (common) The clock input for all stages.

LDNOCLR (common) Asserted if the counter's LD input is asserted and CLR is negated.

NOCLRORLD (common) Asserted if the counter's CLR and LD inputs are both negated.

CNTENP (common) Asserted if the counter's ENP input is asserted.

Di (per cell) Load data input for cell *i*.

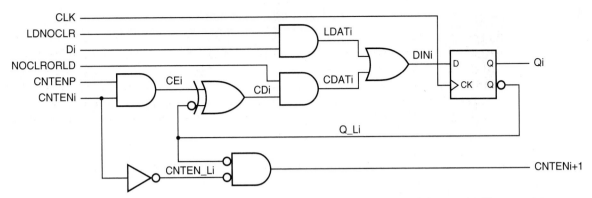

Figure 8-45 One bit-cell of a synchronous serial, 74x163-like counter.

CNTENi   (per cell) Serial count enable input for cell *i*.

CNTENi+1   (per cell) Serial count enable output for cell *i*.

Qi   (per cell) Counter output for cell *i*.

Table 8-16 is a VHDL program corresponding to the bit-cell in the figure. In the program, the **D** flip-flop component Vdffqqn is assumed to be already defined; it is similar to the D flip-flop in Table 8-6 with the addition of a **QN** (complemented) output. In an FPGA or ASIC design, a flip-flop component type would be chosen from the manufacturer's standard cell library.

Table 8-16 VHDL program for counter cell of Figure 8-45.

```
library IEEE;
use IEEE.std_logic_1164.all;

entity syncsercell is
  port( CLK, LDNOCLR, NOCLRORLD, CNTENP, D, CNTEN: in STD_LOGIC;
        CNTENO, Q: out STD_LOGIC );
end syncsercell;

architecture syncsercell_arch of syncsercell is
component Vdffqqn
  port( CLK, D: in STD_LOGIC;
        Q, QN: out STD_LOGIC );
end component;
signal LDAT, CDAT, DIN, Q_L: STD_LOGIC;
begin
  LDAT <= LDNOCLR and D;
  CDAT <= NOCLRORLD and ((CNTENP and CNTEN) xor not Q_L);
  DIN  <= LDAT or CDAT;
  CNTENO <= (not Q_L) and CNTEN;
  U1: Vdffqqn port map (CLK, DIN, Q, Q_L);
end syncsercell_arch;
```

> **A MATTER OF STYLE**   Note that Table 8-16 uses a combination of dataflow and structural VHDL styles. It could have been written completely structurally, for example using an ASIC manufacturer's gate component definitions, to guarantee that the synthesized circuit conforms exactly to Figure 8-45. However, most synthesis engines can do a good job of picking the best gate realization for the simple signal assignments used here.

Table 8-17 shows how to create an 8-bit synchronous serial counter using the cell defined previously. The first two assignments in the architecture body synthesize the common LDNOCLR and NOCLRORLD signals. The next two statements handle boundary conditions for the serial count-enable chain. Finally, the generate statement (introduced on page 453) instantiates eight 1-bit counter cells and hooks up the count-enable chain as required.

It should be clear that a larger or smaller counter can be created simply by changing a few definitions in the program. You can put VHDL's generic statement to good use here to allow you to change the counter's size with a one-line change (see Exercise 8.53).

Table 8-17 VHDL program for an 8-bit 74x163-like synchronous serial counter.

```
library IEEE;
use IEEE.std_logic_1164.all;

entity V74x163s is
  port( CLK, CLR_L, LD_L, ENP, ENT: in STD_LOGIC;
        D: in STD_LOGIC_VECTOR (7 downto 0);
        Q: out STD_LOGIC_VECTOR (7 downto 0);
        RCO: out STD_LOGIC );
end V74x163s;

architecture V74x163s_arch of V74x163s is
component syncsercell
  port( CLK, LDNOCLR, NOCLRORLD, CNTENP, D, CNTEN: in STD_LOGIC;
        CNTENO, Q: out STD_LOGIC );
end component;
signal LDNOCLR, NOCLRORLD: STD_LOGIC; -- common signals
signal SCNTEN: STD_LOGIC_VECTOR (8 downto 0); -- serial count-enable inputs
begin
  LDNOCLR <= (not LD_L) and CLR_L; -- create common load and clear controls
  NOCLRORLD <= LD_L and CLR_L;
  SCNTEN(0) <= ENT; -- serial count-enable into the first stage
  RCO <= SCNTEN(8);  -- RCO is equivalent to final count-enable output
  g1: for i in 0 to 7 generate     -- generate the eight syncsercell stages
      U1: syncsercell port map ( CLK, LDNOCLR, NOCLRORLD, ENP, D(i), SCNTEN(i),
                                 SCNTEN(i+1), Q(i));
  end generate;
end V74x163s_arch;
```

8.5 Shift Registers

8.5.1 Shift-Register Structure

shift register

serial input
serial output

serial-in, parallel-out
shift register

serial-to-parallel
conversion

parallel-in, serial-out
shift register

parallel-to-serial
conversion

parallel-in, parallel-out
shift register

A *shift register* is an *n*-bit register with a provision for shifting its stored data by one bit position at each tick of the clock. Figure 8-46 shows the structure of a serial-in, serial-out shift register. The *serial input,* SERIN, specifies a new bit to be shifted into one end at each clock tick. This bit appears at the *serial output,* SEROUT, after *n* clock ticks, and is lost one tick later. Thus, an *n*-bit serial-in, serial-out shift register can be used to delay a signal by *n* clock ticks.

A *serial-in, parallel-out shift register,* shown in Figure 8-47, has outputs for all of its stored bits, making them available to other circuits. Such a shift register can be used to perform *serial-to-parallel conversion,* as explained later in this section.

Conversely, it is possible to build a *parallel-in, serial-out shift register.* Figure 8-48 shows the general structure of such a device. At each clock tick the register either loads new data from inputs 1D–ND or it shifts its current contents, depending on the value of the LOAD/SHIFT control input (which could be named LOAD or SHIFT_L). Internally, the device uses a 2-input multiplexer on each flip-flop's D input to select between the two cases. A parallel-in, serial-out shift register can be used to perform *parallel-to-serial conversion*, as explained later in this section.

By providing outputs for all of the stored bits in a parallel-in shift register, we obtain the p*arallel-in, parallel-out shift register* shown in Figure 8-49. Such a device is general enough to be used in any of the applications of the previous shift registers.

Figure 8-46 Structure of a serial-in, serial-out shift register.

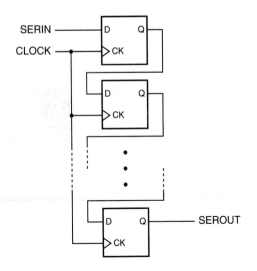

Figure 8-47 Structure of a serial-in, parallel-out shift register.

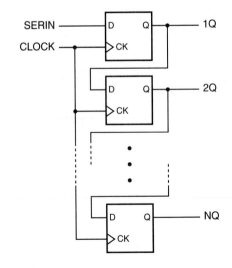

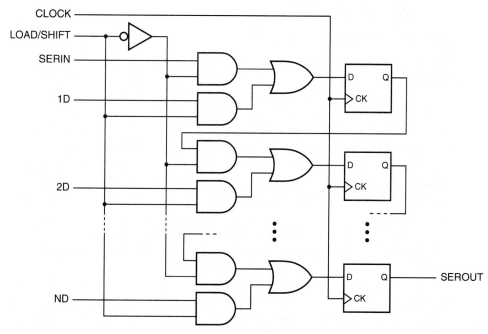

Figure 8-48 Structure of a parallel-in, serial-out shift register.

Figure 8-49 Structure of a parallel-in, parallel-out shift register.

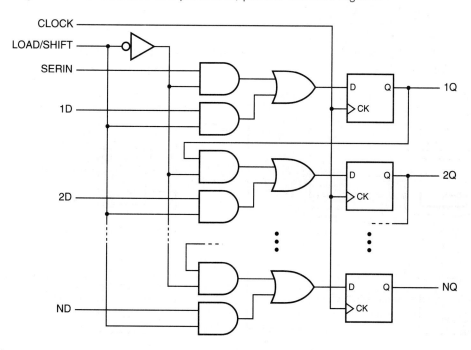

8.5.2 MSI Shift Registers

74x164

Figure 8-50 shows logic symbols for three popular MSI 8-bit shift registers. The *74x164* is a serial-in, parallel-out device with an asynchronous clear input (CLR_L). It has two serial inputs that are ANDed internally. That is, both SERA and SERB must be 1 for a 1 to be shifted into the first bit of the register.

74x166

The *74x166* is a parallel-in, serial-out shift register, also with an asynchronous clear input. This device shifts when SH/LD is 1 and loads new data otherwise. The '166 has an unusual clocking arrangement called a "gated clock" (see also Section 8.8.2); it has two clock inputs that are connected to the internal flip-flops as shown in Figure 8-50(c). The designers of the '166 intended for CLK to be connected to a free-running system clock, and for CLKINH to be asserted to inhibit CLK, so that neither shifting nor loading occurs on the next clock tick, and the current register contents are held. However, for this to work, CLKINH must be changed only when CLK is 1; otherwise, undesired clock edges occur on the internal flip-flops. A much safer way of obtaining a "hold" function is employed in the next devices that we discuss.

74x194

unidirectional shift register

bidirectional shift register

The *74x194* is an MSI 4-bit bidirectional, parallel-in, parallel-out shift register. Its logic diagram is shown in Figure 8-51. The shift registers that we've studied previously are called *unidirectional shift registers* because they shift in only one direction. The '194 is a *bidirectional shift register* because its contents may be shifted in either of two directions, depending on a control input. The two

Figure 8-50
Traditional logic symbols for MSI shift registers:
(a) 74x164 8-bit serial-in, parallel-out shift register;
(b) 74x166 8-bit parallel-in, serial-out shift register;
(c) equivalent circuit for 74x166 clock inputs;
(d) 74x194 universal shift register.

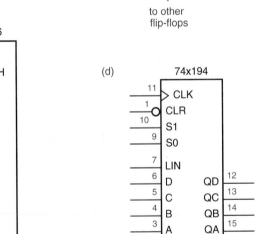

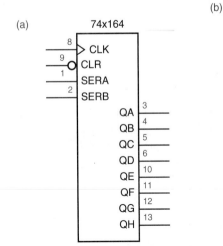

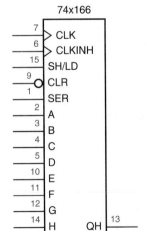

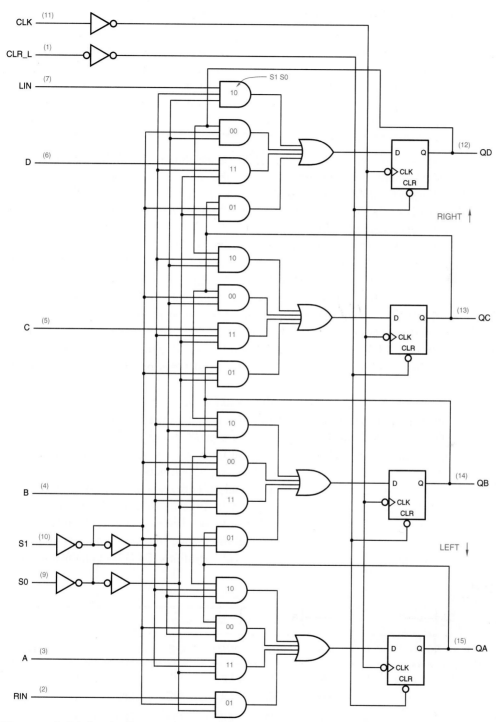

Figure 8-51 Logic diagram for the 74x194 4-bit universal shift register, including pin numbers for a standard 16-pin dual in-line package.

Table 8-18
Function table for the
74x194 4-bit universal
shift register.

Function	Inputs		Next state			
	S1	S0	QA*	QB*	QC*	QD*
Hold	0	0	QA	QB	QC	QD
Shift right	0	1	RIN	QA	QB	QC
Shift left	1	0	QB	QC	QD	LIN
Load	1	1	A	B	C	D

left
right

directions are called "left" and "right," even though the logic diagram and the logic symbol aren't necessarily drawn that way. In the '194, *left* means "in the direction from QD to QA," and *right* means "in the direction from QA to QD." Our logic diagram and symbol for the '194 are consistent with these names if you rotate them 90° clockwise.

Table 8-18 is a function table for the 74x194. This function table is highly compressed, since it does not contain columns for most of the inputs (A–D, RIN, LIN) or the current state QA–QD. Still, by expressing each next-state value as a function of these implicit variables, it completely defines the operation of the '194 for all 2^{12} possible combinations of current state and input, and it sure beats a 4,096-row table!

Note that the '194's LIN (left-in) input is conceptually located on the "righthand" side of the chip, but it is the serial input for *left* shifts. Likewise, RIN is the serial input for right shifts.

The '194 is sometimes called a *universal* shift register because it can be made to function like any of the less general shift-register types that we've discussed (e.g., unidirectional; serial-in, parallel-out; parallel-in, serial-out). In fact, many of our design examples in the next few subsections contain '194s configured to use just a subset of their available functions.

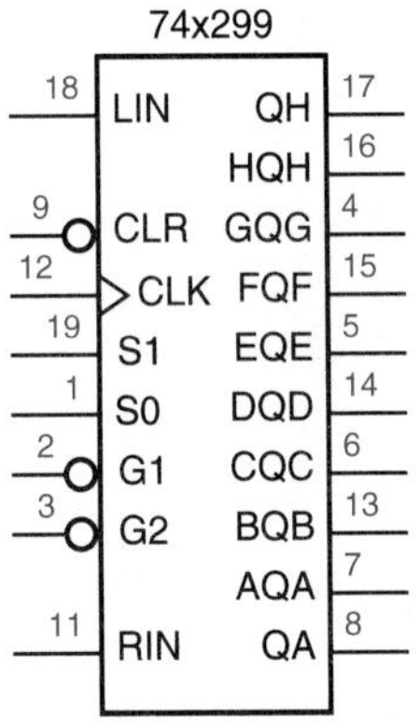

The *74x299* is an 8-bit universal shift register in a 20-pin package; its symbol and logic diagram are given in Figures 8-52 and 8-53. The '299's functions and function table are similar to the '194's, as shown in Table 8-19. To save pins, the '299 uses bidirectional three-state lines for input and output, as shown in the logic diagram. During load operations (S1 S0 = 11), the three-state

Figure 8-52
Traditional logic
symbol for
the 74x299.

Table 8-19 Function table for a 74x299 8-bit universal shift register.

Function	Inputs		Next state							
	S1	S0	QA*	QB*	QC*	QD*	QE*	QF*	QG*	QH*
Hold	0	0	QA	QB	QC	QD	QE	QF	QG	QH
Shift right	0	1	RIN	QA	QB	QC	QD	QE	QF	QG
Shift left	1	0	QB	QC	QD	QE	QF	QG	QH	LIN
Load	1	1	AQA	BQB	CQC	DQD	EQE	FQF	GQG	HQH

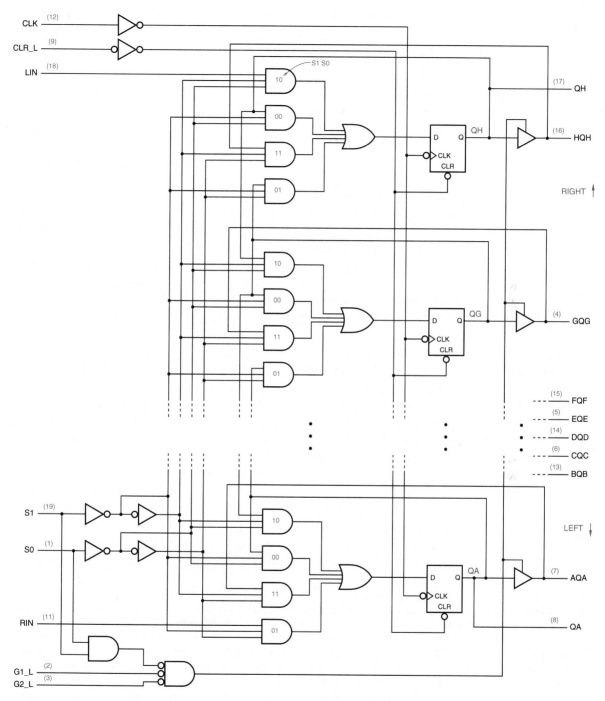

Figure 8-53 Logic diagram for the 74x299 8-bit universal shift register, including pin numbers for a standard 20-pin dual in-line package.

drivers are disabled and data is loaded through the AQA–HQH pins. At other times the stored bits are driven onto these same pins if G1_L and G2_L are asserted. The leftmost and rightmost stored bits are available at all times on separate output-only pins, QA and QH.

8.5.3 The World's Biggest Shift-Register Application

The most common application of shift registers is to convert parallel data into serial format for transmission or storage, and to convert serial data back to parallel format for processing or display (see Section 2.16.1). The most common example of serial data transmission, one that you almost certainly take part in every day, is in *digital telephony.*

digital telephony

CO

For years, TPCs (The Phone Companies) have been installing digital switching equipment in their central offices (*COs*). Most home phones have a two-wire analog connection to the central office. However, an analog-to-digital converter samples the analog voice signal 8,000 times per second (once every 125 μs) when it enters the CO and produces a corresponding sequence of 8,000 8-bit bytes representing the sign and amplitude of the analog signal at each sampling point. Subsequently, your voice is transmitted digitally on 64-Kbps *serial channels* throughout the phone network, until it is converted back to an analog signal by a digital-to-analog converter at the far-end CO.

serial channel

The 64-Kbps bandwidth required by a single digital voice signal is far *less* than can be obtained on a single digital signal line or switched by digital ICs. Therefore most digital telephone equipment *multiplexes* many 64-Kbps channels onto a single wire, saving both wires and digital ICs for switching. In the next subsection, we show how 32 channels can be processed by a handful of MSI chips; and these chips could be easily integrated into a single CPLD. This is a classic example of a *space/time tradeoff* in digital design—by running the chips faster, you can accomplish a larger task with fewer chips. Indeed, this is the main reason that the telephone network has "gone digital."

multiplex

space/time tradeoff

I STILL DON'T KNOW

ISDN (Integrated Services Digital Network) technology was developed in the late 1980s to extend full-duplex 144-Kbps serial digital channels to home phones. The idea was to carry two 64-Kbps voice conversations plus a 16-Kbps control channel on a single pair of wires, thus increasing the capacity of installed wiring.

In the first edition of this book, we noted that delays in ISDN deployment had led some people in the industry to rename it "Imaginary Services Delivered Nowhere." In the mid-1990s, ISDN finally took off in the U.S., but it was deployed not so much to carry voice as to provide "high-speed" connections to the Internet.

Unfortunately for TPCs, the growth in ISDN was cut short first by the deployment of inexpensive 56-Kbps analog modems and later by the growing availability of very high-speed connections (160 Kbps to 2 Mbps or higher) using newer DSL (Digital Subscriber Line) and cable-modem technologies.

A 74x166 parallel-in shift register performs the parallel-to-serial conversion. Bit 0 of the parallel data (D0–D7) is connected to the '166 input closest to the SDATA output, so bits are transmitted serially in the order 0 through 7.

During bit 7 of each timeslot, the BIT7_L signal is asserted, which causes the '166 to be loaded with D0–D7. The value of D0–D7 is irrelevant except during the setup- and hold-time window around the clock edge on which the '166 is loaded, as shown by shading in the timing diagram. This leaves open the possibility that the parallel data bus could be used for other things at other times (see Exercise 8.54).

A destination module can convert the serial data back into parallel format using the circuit of Figure 8-57. A modulo-256 counter built from a pair of '163s is used to reconstruct the timeslot and bit numbers. Although SYNC is asserted during state 255 of the counter on the source module, SYNC loads the destination module's counter with 0 so that both counters go to 0 on the same clock edge. The counter's high-order bits (timeslot number) are not used in the figure, but they may be used by other circuits in the destination module to identify the byte from a particular timeslot on the parallel data bus (PD0–PD7).

Figure 8-58 shows detailed timing for the serial-to-parallel conversion circuit. A complete received byte is available at parallel output of the 74x164 shift register during the clock period following the reception of the last bit (7) of the byte. The parallel data in this example is *double-buffered*—once it is fully received, it is transferred into a 74x377 register, where it is available on PD0-PD7 for eight full clock periods until the next byte is fully received. The BIT0_L signal enables the '377 to be loaded at the proper time. Additional registers and decoding could be provided to load the byte from each timeslot into a different register, making each byte available for 125 μs (see Section 8.57).

double-buffered data

Figure 8-58 Timing diagram for serial-to-parallel conversion.

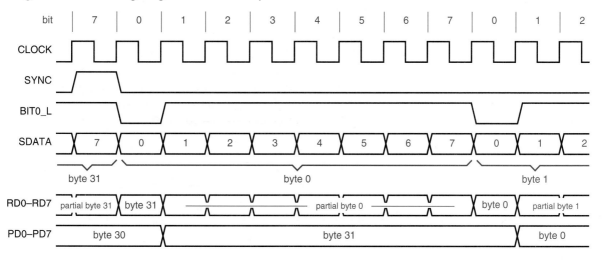

Once the received data is in parallel format, it can easily be stored or modified by other digital circuits; we'll give examples in Section 10.1.6. In digital telephony, the received parallel data is converted back into an analog voltage that is filtered and transmitted to an earpiece or speaker for 125 μs, until the next voice sample arrives.

8.5.5 Shift-Register Counters

shift-register counter

Serial/parallel conversion is a "data" application, but shift registers have "non-data" applications as well. A shift register can be combined with combinational logic to form a state machine whose state diagram is cyclic. Such a circuit is called a *shift-register counter.* Unlike a binary counter, a shift-register counter does not count in an ascending or descending binary sequence, but it is useful in many "control" applications nonetheless.

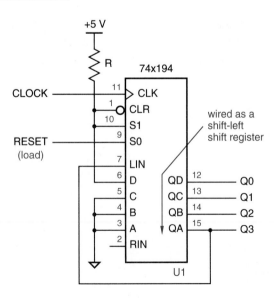

Figure 8-59
Simplest design for a 4-bit, 4-state ring counters with a single circulating 1.

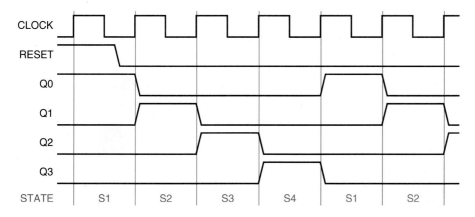

Figure 8-60
Timing diagram for a
4-bit ring counter.

8.5.6 Ring Counters

The simplest shift-register counter uses an n-bit shift register to obtain a counter with n states, and is called a *ring counter*. Figure 8-59 is the logic diagram for a 4-bit ring counter. The 74x194 universal shift register is wired so that it normally performs a left shift. However, when RESET is asserted, it loads 0001 (refer to the '194's function table, Table 8-18 on page 716). Once RESET is negated, the '194 shifts left on each clock tick. The LIN serial input is connected to the "leftmost" output, so the next states are 0010, 0100, 1000, 0001, 0010, ... Thus, the counter visits four unique states before repeating. A timing diagram is shown in Figure 8-60. In general, an n-bit ring counter visits n states in a cycle.

ring counter

    The ring counter in Figure 8-59 has one major problem—it is not robust. If its single 1 output is lost due to a temporary hardware problem (e.g., noise), the counter goes to state 0000 and stays there forever. Likewise, if an extra 1 output is set (i.e., state 0101 is created), the counter will go through an incorrect cycle of states and stay in that cycle forever. These problems are quite evident if we draw the *complete* state diagram for the counter circuit, which has 16 states. As shown in Figure 8-61, there are 12 states that are not part of the normal counting cycle. If the counter somehow gets off the normal cycle, it stays off it.

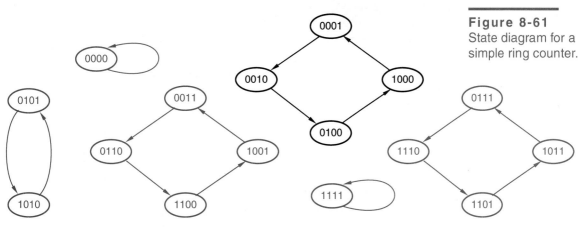

Figure 8-61
State diagram for a
simple ring counter.

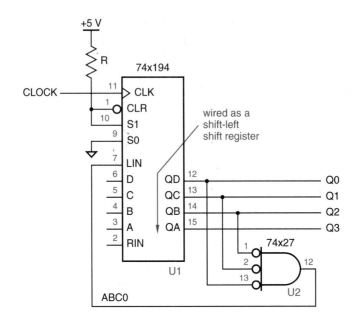

Figure 8-62
Self-correcting
4-bit, 4-state ring
counter with a single
circulating 1.

self-correcting counter

A *self-correcting counter* is designed so that all abnormal states have transitions leading to normal states. Self-correcting counters are desirable for the same reason that we use a minimal-risk approach to state assignment in Section 7.4.3: If something unexpected happens, a counter or state machine should go to a "safe" state.

self-correcting ring counter

A *self-correcting ring counter* circuit is shown in Figure 8-62. The circuit uses a NOR gate to shift a 1 into LIN only when the three least significant bits are 0. This results in the state diagram in Figure 8-63; all abnormal states lead back

Figure 8-63
State diagram for a
self-correcting ring
counter.

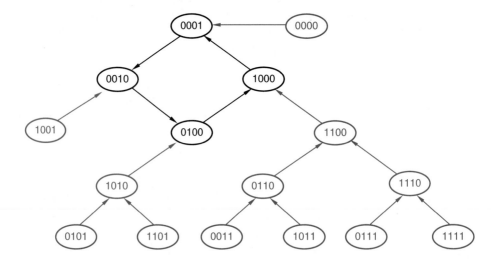

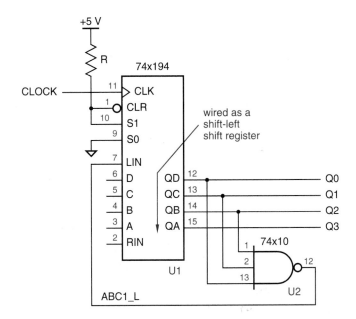

Figure 8-64
Self-correcting
4-bit, 4-state ring
counter with a single
circulating 0.

into the normal cycle. Notice that, in this circuit, an explicit RESET signal is not necessarily required. Regardless of the initial state of the shift register on power-up, it reaches state 0001 within four clock ticks. Therefore, an explicit reset signal is required only if it is necessary to ensure that the counter starts up synchronously with other devices in the system or to provide a known starting point in simulation.

In the general case, an n-bit self-correcting ring counter uses an $(n-1)$-input NOR gate and corrects an abnormal state within $n-1$ clock ticks.

In CMOS and TTL logic families, wide NAND gates are generally easier to come by than NORs, so it may be more convenient to design a self-correcting ring counter as shown in Figure 8-64. States in this counter's normal cycle have a single circulating 0.

The major appeal of a ring counter for control applications is that its states appear in 1-out-of-n decoded form directly on the flip-flop outputs. That is, exactly one flip-flop output is asserted in each state. Furthermore, these outputs are "glitch free"; compare with the binary counter and decoder approach of Figure 8-42 on page 703.

*8.5.7 Johnson Counters

An n-bit shift register with the complement of the serial output fed back into the serial input is a counter with $2n$ states and is called a *twisted-ring*, *Moebius*, or *Johnson counter*. Figure 8-65 is the basic circuit for a Johnson counter and Figure 8-66 is its timing diagram. The normal states of this counter are listed in

twisted-ring counter
Moebius counter
Johnson counter

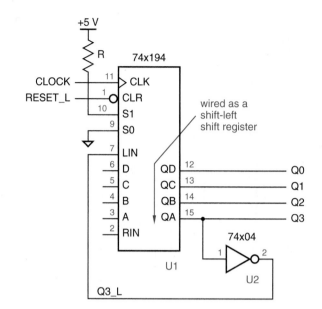

Figure 8-65
Basic 8-bit, 8-state
Johnson counter.

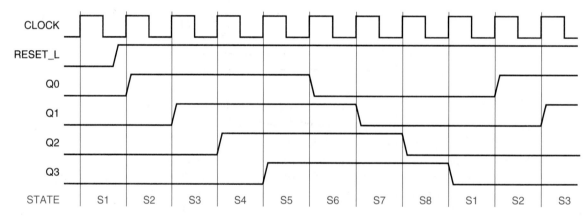

Figure 8-66 Timing diagram for a 4-bit Johnson counter.

Table 8-20
States of a 4-bit
Johnson counter.

State Name	Q3	Q2	Q1	Q0	Decoding
S1	0	0	0	0	$Q3' \cdot Q0'$
S2	0	0	0	1	$Q1' \cdot Q0$
S3	0	0	1	1	$Q2' \cdot Q1$
S4	0	1	1	1	$Q3' \cdot Q2$
S5	1	1	1	1	$Q3 \cdot Q0$
S6	1	1	1	0	$Q1 \cdot Q0'$
S7	1	1	0	0	$Q2 \cdot Q1'$
S8	1	0	0	0	$Q3 \cdot Q2'$

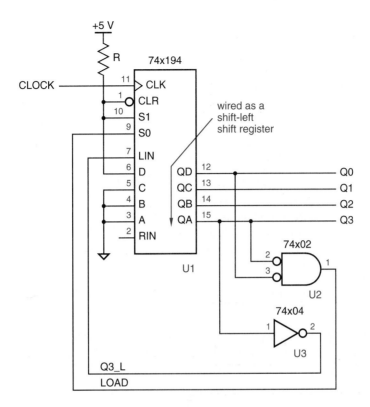

Figure 8-67
Self-correcting 4-bit,
8-state Johnson
counter.

Table 8-20. If both the true and complemented outputs of each flip-flop are available, each normal state of the counter can be decoded with a 2-input AND or NAND gate, as shown in the table. The decoded outputs are glitch free.

An n-bit Johnson counter has $2^n - 2n$ abnormal states and is therefore subject to the same robustness problems as a ring counter. A 4-bit *self-correcting Johnson counter* can be designed as shown in Figure 8-67. This circuit loads 0001 as the next state whenever the current state is 0xx0. A similar circuit using a single 2-input NOR gate can perform correction for a Johnson counter with any number of bits. The correction circuit must load 00...01 as the next state whenever the current state is 0x...x0.

self-correcting Johnson counter

THE SELF-CORRECTION CIRCUIT IS ITSELF CORRECT!

We can prove that the Johnson-counter self-correction circuit corrects any abnormal state as follows. An abnormal state can always be written in the form x...x10x...x, since the only states that can't be written in this form are normal states (00...00, 11...11, 01...1, 0...01...1, and 0...01). Therefore, within $n - 2$ clock ticks, the shift register will contain 10x...x. One tick later it will contain 0x...x0, and one tick after that the normal state 00...01 will be loaded.

*maximum-length
 sequence generator*

finite fields

*maximum-length
 sequence*

*8.5.8 Linear Feedback Shift-Register Counters

The n-bit shift register counters that we've shown so far have far less than the maximum of 2^n normal states. An n-bit *linear feedback shift-register (LFSR) counter* can have $2^n - 1$ states, almost the maximum. Such a counter is often called a *maximum-length sequence generator.*

The design of LFSR counters is based on the theory of *finite fields,* which was developed by French mathematician Évariste Galois (1811–1832) shortly before he was killed in a duel with a political opponent. The operation of an LFSR counter corresponds to operations in a finite field with 2^n elements.

Figure 8-68 shows the structure of an n-bit LFSR counter. The shift register's serial input is connected to the sum modulo 2 of a certain set of output bits. These feedback connections determine the state sequence of the counter. By convention, outputs are always numbered and shifted in the direction shown.

Using finite-field theory, it can be shown that for any value of n, there exists at least one feedback equation that makes the counter go through all $2^n - 1$ nonzero states before repeating. This is called a *maximum-length sequence.*

Table 8-21 lists feedback equations that result in maximum-length sequences for selected values of n. For each value of n greater than 3, there are many other feedback equations that result in maximum-length sequences, all different.

An LFSR counter designed according to Figure 8-68 can never cycle through all 2^n possible states. Regardless of the connection pattern, the next state for the all-0s state is the same—all 0s.

Figure 8-68 General structure of a linear feedback shift-register counter.

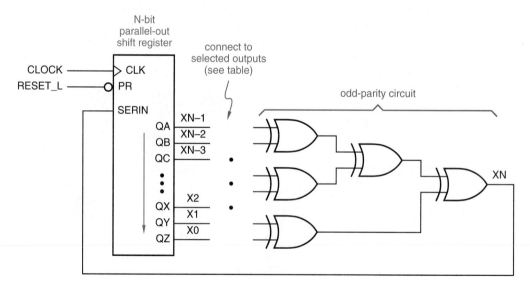

n	Feedback Equation
2	$X2 = X1 \oplus X0$
3	$X3 = X1 \oplus X0$
4	$X4 = X1 \oplus X0$
5	$X5 = X2 \oplus X0$
6	$X6 = X1 \oplus X0$
7	$X7 = X3 \oplus X0$
8	$X8 = X4 \oplus X3 \oplus X2 \oplus X0$
12	$X12 = X6 \oplus X4 \oplus X1 \oplus X0$
16	$X16 = X5 \oplus X4 \oplus X3 \oplus X0$
20	$X20 = X3 \oplus X0$
24	$X24 = X7 \oplus X2 \oplus X1 \oplus X0$
28	$X28 = X3 \oplus X0$
32	$X32 = X22 \oplus X2 \oplus X1 \oplus X0$

Table 8-21
Feedback equations
for linear feedback
shift-register counters.

WORKING IN THE FIELD

A finite field has a finite number of elements and two operators, addition and multiplication, that satisfy certain properties. An example of a finite field with P elements, where P is any prime, is the set of integers modulo P. The operators in this field are addition and multiplication modulo P.

According to finite-field theory, if you start with a nonzero element E and repeatedly multiply by a "primitive" element α, after $P-2$ steps you will generate the rest of the field's nonzero elements in the field before getting back to E. It turns out that in a field with P elements, any integer in the range $2, \ldots, P-1$ is primitive. You can try this yourself using $P = 7$ and $\alpha = 2$, for example. The elements of the field are $0, 1, \ldots, 6$, and the operations are addition and subtraction modulo 7.

The paragraph above gives the basic idea behind maximum-length sequence generators. However, to apply them to a digital circuit, you need a field with 2^n elements, where n is the number of bits required by the application. On one hand, we're in luck, because Galois proved that there exist finite fields with P^n elements for any integer n, as long as P is prime, including $P = 2$. On the other hand, we're out of luck, because when $n > 1$, the operators in fields with P^n (including 2^n) elements are quite different from ordinary integer addition and multiplication. Also, primitive elements are harder to find.

If you enjoy math, as I do, you'd probably be fascinated by the finite-field theory that leads to the LFSR circuits for maximum-length sequence generators and other applications; see the References. Otherwise, you can confidently follow the "cookbook" approach in this section.

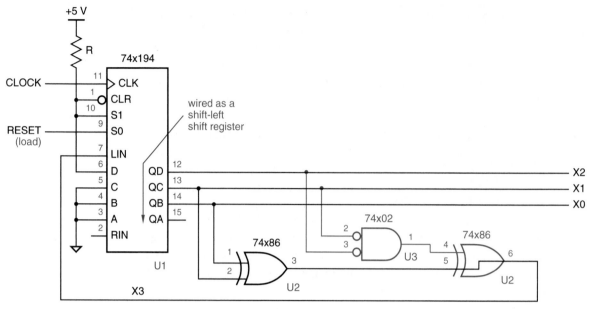

Figure 8-69 A 3-bit LFSR counter; modifications to include the all-0s state are shown in color.

The logic diagram for a 3-bit LFSR counter is shown in Figure 8-69. The state sequence for this counter is shown in the first three columns of Table 8-22. Starting in any nonzero state, 100 in the table, the counter visits seven states before returning to the starting state.

An LFSR counter can be modified to have 2^n states, including the all-0s state, as shown in color for the 3-bit counter in Figure 8-69. The resulting state sequence is given in the last three columns of Table 8-22. In an n-bit LFSR

Table 8-22
State sequences for the 3-bit LFSR counter in Figure 8-69.

Original Sequence			Modified Sequence		
X2	**X1**	**X0**	**X2**	**X1**	**X0**
1	0	0	1	0	0
0	1	0	0	1	0
1	0	1	1	0	1
1	1	0	1	1	0
1	1	1	1	1	1
0	1	1	0	1	1
0	0	1	0	0	1
1	0	0	0	0	0
.	.	.	1	0	0
.	.	.	.	.	.

counter, an extra XOR gate and an $n - 1$ input NOR gate connected to all shift-register outputs except X0 accomplishes the same thing.

The states of an LFSR counter are not visited in binary counting order. However, LFSR counters are typically used in applications where this characteristic is an advantage. A major application of LFSR counters is in generating test inputs for logic circuits. In most cases, the "pseudorandom" counting sequence of an LFSR counter is more likely than a binary counting sequence to detect errors. LFSRs are also used in the encoding and decoding circuits for certain error-detecting and error-correcting codes, including CRC codes, which we introduced in Section 2.15.4.

In data communications, LFSR counters are often used to "scramble" and "descramble" the data patterns transmitted by high-speed modems and network interfaces, including 100-Mbps Ethernet. This is done by XORing the LFSR's output with the user data stream. Even when the user data stream contains a long run of 0s or 1s, combining it with the LFSR's pseudorandom output improves the DC balance of the transmitted signal and creates a rich set of transitions that allows clocking information to be recovered more easily at the receiver.

8.5.9 Shift Registers in ABEL and PLDs

General-purpose shift registers can be specified quite easily in ABEL and fit nicely into typical sequential PLDs. For example, Figure 8-70 and Table 8-23 show how to realize a function similar to that of a 74x194 universal shift register using a 16V8. Notice that one of the I/O pins of the 16V8, pin 12, is used as an input.

The 16V8 realization of the '194 differs from the real thing in just one way—in the function of the CLR_L input. In the real '194, CLR_L is an asynchronous input, while in the 16V8 it is sampled along with other inputs at the rising edge of CLK.

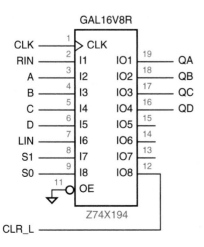

Figure 8-70
PLD realizations of a 74x194-like universal shift register with synchronous clear.

Table 8-23 ABEL program for a 4-bit universal shift register.

```
module Z74x194
title '4-bit Universal Shift Register'
Z74X194 device 'P16V8R';

" Input and output pins
CLK, RIN, A, B, C, D, LIN                pin 1, 2, 3, 4, 5, 6, 7;
S1, S0, CLR_L                            pin 8, 9, 12;
QA, QB, QC, QD                           pin 19, 18, 17, 16 istype 'reg';

" Active-level translation
CLR = !CLR_L;

" Set definitions
INPUT   = [ A,  B,  C, D ];
LEFTIN  = [ QB, QC, QD, LIN];
RIGHTIN = [RIN, QA, QB, QC ];
OUT     = [ QA, QB, QC, QD ];

CTRL  = [S1,S0];
HOLD  = (CTRL == [0,0]);
RIGHT = (CTRL == [0,1]);
LEFT  = (CTRL == [1,0]);
LOAD  = (CTRL == [1,1]);

equations
OUT.CLK = CLK;

OUT := !CLR & (
          HOLD & OUT
        # RIGHT & RIGHTIN
        # LEFT & LEFTIN
        # LOAD & INPUT);

end Z74x194
```

If you really need to provide an asynchronous clear input, you can use the 22V10, which provides a single product line to control the reset inputs of all of its flip-flops. This requires only a few changes in the original program (see Exercise 8.69).

The flexibility of ABEL can be used to create shift-register circuits with more or different functionality. For example, Table 8-24 defines an 8-bit shift register that can be cleared, loaded with a single 1 in any bit position, shifted left, shifted right, or held. The operation to be performed at each clock tick is specified by a 4-bit operation code, OP[3:0]. Despite the large number of "WHEN" cases, the circuit can be synthesized with only five product terms per output.

Table 8-24 ABEL program for a multi-function shift register.

```
module shifty
title '8-bit shift register with decoded load'

" Inputs and Outputs
CLK, OP3..OP0           pin;
Q7..Q0                  pin istype 'reg';

" Definitions

Q = [Q7..Q0];
OP = [OP3..OP0];

HOLD = (OP == 0);
CLEAR = (OP == 1);
LEFT = (OP == 2);
RIGHT = (OP == 3);
NOP = (OP >= 4) & (OP < 8);
LOADQ0 = (OP == 8);
LOADQ1 = (OP == 9);
LOADQ2 = (OP == 10);
LOADQ3 = (OP == 11);
LOADQ4 = (OP == 12);
LOADQ5 = (OP == 13);
LOADQ6 = (OP == 14);
LOADQ7 = (OP == 15);

Equations

Q.CLK = CLK;

WHEN HOLD THEN Q := Q;
ELSE WHEN CLEAR THEN Q := 0;
ELSE WHEN LEFT THEN Q := [Q6..Q0, Q7];
ELSE WHEN RIGHT THEN Q := [Q0, Q7..Q1];
ELSE WHEN LOADQ0 THEN Q := 1;
ELSE WHEN LOADQ1 THEN Q := 2;
ELSE WHEN LOADQ2 THEN Q := 4;
ELSE WHEN LOADQ3 THEN Q := 8;
ELSE WHEN LOADQ4 THEN Q := 16;
ELSE WHEN LOADQ5 THEN Q := 32;
ELSE WHEN LOADQ6 THEN Q := 64;
ELSE WHEN LOADQ7 THEN Q := 128;
ELSE Q := Q;

end shifty
```

Table 8-25 Program for an 8-bit ring counter.

```
module Ring8
title '8-bit Ring Counter'

" Inputs and Outputs
MCLK, CNTEN, RESTART                    pin;
S0..S7                                  pin istype 'reg';

equations

[S0..S7].CLK = MCLK;

S0 := CNTEN & !S0 & !S1 & !S2 & !S3 & !S4 & !S5 & !S6   " Self-sync
    # !CNTEN & S0                                       " Hold
    # RESTART;                                          " Start with one 1
[S1..S7] := !RESTART & ( !CNTEN & [S1..S7]     " Shift
                        # CNTEN & [S0..S6] );  " Hold
end Ring8
```

ABEL can be used readily to specify shift register counters of the various types that we introduced in previous subsections. For example, Table 8-25 is the program for an 8-bit ring counter. We've used the extra capability of the PLD to add two functions not present in our previous MSI designs: counting occurs only if CNTEN is asserted, and the next state is forced to S0 if RESTART is asserted.

Ring counters are often used to generate multiphase clocks or enable signals in digital systems, and the requirements in different systems are many and varied. The ability to reprogram the counter's behavior easily is a distinct advantage of an HDL-based design.

Figure 8-71 shows a set of clock or enable signals that might be required in a digital system with six distinct phases of operation. Each phase lasts for two ticks of a master clock signal, MCLK, during which the corresponding active-low phase-enable signal Pi_L is asserted. We can obtain this sort of timing from a ring counter if we provide an extra flip-flop to count the two ticks of each phase, so that a shift occurs on the *second* tick of each phase.

The timing generator can be built with a few inputs and outputs of a PLD. Three control inputs are provided, with the following behavior:

RESET   When this input is asserted, no outputs are asserted. The counter always goes to the first tick of phase 1 after RESET is negated.

RUN   When asserted, this input allows the counter to advance to the second tick of the current phase, or to the first tick of the next phase; otherwise, the current tick of the current phase is extended.

RESTART   Asserting this input causes the counter to go back to the first tick of phase 1, even if RUN is not asserted.

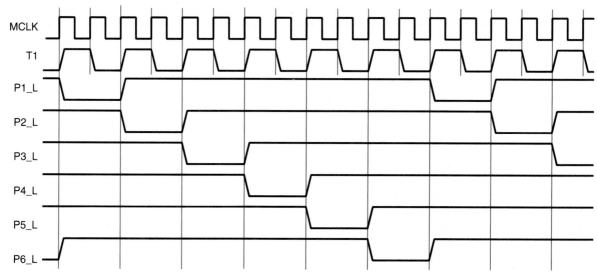

Figure 8-71 Six-phase timing waveforms required in a certain digital system.

Table 8-26 Program for a six-phase waveform generator.

```
module TIMEGEN6
title 'Six-phase Master Timing Generator'

" Input and Output pins
MCLK, RESET, RUN, RESTART                pin;
T1, P1_L, P2_L, P3_L, P4_L, P5_L, P6_L   pin istype 'reg';

" State definitions
PHASES = [P1_L, P2_L, P3_L, P4_L, P5_L, P6_L];
NEXTPH = [P6_L, P1_L, P2_L, P3_L, P4_L, P5_L];
SRESET = [1, 1, 1, 1, 1, 1];
P1 =    [0, 1, 1, 1, 1, 1];

equations
T1.CLK = MCLK; PHASES.CLK = MCLK;

WHEN RESET THEN {T1 := 1; PHASES := SRESET;}
ELSE WHEN (PHASES==SRESET) # RESTART THEN {T1 := 1; PHASES := P1;}
ELSE WHEN RUN & T1 THEN {T1 := 0; PHASES := PHASES;}
ELSE WHEN RUN & !T1 THEN {T1 := 1; PHASES := NEXTPH;}
ELSE {T1 := T1; PHASES := PHASES;}

end TIMEGEN6
```

Table 8-26 is a program that creates the required behavior. Notice the use of sets to specify the ring counter's behavior very concisely, with the RESET, RESTART, and RUN having the specified behavior in any counter phase.

Table 8-27 Alternate program for the waveform generator.

```
module TIMEGN6A
title 'Six-phase Master Timing Generator'

" Input and Output pins
MCLK, RESET, RUN, RESTART                      pin;
T1, P1_L, P2_L, P3_L, P4_L, P5_L, P6_L      pin istype 'reg';

" State definitions
TSTATE = [T1, P1_L, P2_L, P3_L, P4_L, P5_L, P6_L];
SRESET = [1, 1, 1, 1, 1, 1, 1];
P1F =    [1, 0, 1, 1, 1, 1, 1];
P1S =    [0, 0, 1, 1, 1, 1, 1];
P2F =    [1, 1, 0, 1, 1, 1, 1];
P2S =    [0, 1, 0, 1, 1, 1, 1];
P3F =    [1, 1, 1, 0, 1, 1, 1];
P3S =    [0, 1, 1, 0, 1, 1, 1];
P4F =    [1, 1, 1, 1, 0, 1, 1];
P4S =    [0, 1, 1, 1, 0, 1, 1];
P5F =    [1, 1, 1, 1, 1, 0, 1];
P5S =    [0, 1, 1, 1, 1, 0, 1];
P6F =    [1, 1, 1, 1, 1, 1, 0];
P6S =    [0, 1, 1, 1, 1, 1, 0];

equations
TSTATE.CLK = MCLK;
WHEN RESET THEN TSTATE := SRESET;

state_diagram TSTATE

state SRESET: IF RESET THEN SRESET ELSE P1F;

state P1F: IF RESET THEN SRESET ELSE IF RESTART THEN P1F
            ELSE IF RUN THEN P1S ELSE P1F;

state P1S: IF RESET THEN SRESET ELSE IF RESTART THEN P1F
            ELSE IF RUN THEN P2F ELSE P1S;

state P2F: IF RESET THEN SRESET ELSE IF RESTART THEN P1F
            ELSE IF RUN THEN P2S ELSE P2F;

state P2S: IF RESET THEN SRESET ELSE IF RESTART THEN P1F
            ELSE IF RUN THEN P3F ELSE P2S;

state P3F: IF RESET THEN SRESET ELSE IF RESTART THEN P1F
            ELSE IF RUN THEN P3S ELSE P3F;

state P3S: IF RESET THEN SRESET ELSE IF RESTART THEN P1F
            ELSE IF RUN THEN P4F ELSE P3S;

state P4F: IF RESET THEN SRESET ELSE IF RESTART THEN P1F
            ELSE IF RUN THEN P4S ELSE P4F;

state P4S: IF RESET THEN SRESET ELSE IF RESTART THEN P1F
            ELSE IF RUN THEN P5F ELSE P4S;
```

Table 8-27 (continued) Alternate program for the waveform generator.

```
state P5F: IF RESET THEN SRESET ELSE IF RESTART THEN P1F
           ELSE IF RUN THEN P5S ELSE P5F;

state P5S: IF RESET THEN SRESET ELSE IF RESTART THEN P1F
           ELSE IF RUN THEN P6F ELSE P5S;

state P6F: IF RESET THEN SRESET ELSE IF RESTART THEN P1F
           ELSE IF RUN THEN P6S ELSE P6F;

state P6S: IF RESET THEN SRESET ELSE IF RESTART THEN P1F
           ELSE IF RUN THEN P1F ELSE P6S;


end TIMEGN6A
```

The same timing-generator behavior in Figure 8-71 can be specified using a state-machine design approach, as shown in Table 8-27. This ABEL program, though longer, generates the same external behavior during normal operation as the previous one, and from a certain point of view it may be easier to understand. However, its realization requires 8 to 20 AND terms per output, compared to only 3 to 5 per output for the original ring-counter version. This is a good example of how you can improve circuit efficiency and performance by adapting a standard, simple structure to a "custom" design problem, rather than grinding out a brute-force state machine.

RELIABLE RESET   Notice in Table 8-27 that TSTATE is assigned a value in the equations section of the program, as well as being used in the state_diagram section. This was done for a very specific purpose, to ensure that the program goes to the SRESET state from any undefined state, as explained below.

ABEL augments the on-set for an output each time the output appears on the lefthand side of an equation, as we explained for combinational outputs on page 254. In the case of registered outputs, ABEL also augments the on-set of each state variable in the state vector for each "state" definition in a state_diagram. For each state-variable output, all of the input combinations that cause that output to be 1 in each state are added to the output's on-set.

The state machine in Table 8-27 has 2^7 or 128 states in total, of which only 13 are explicitly defined and have a transition into SRESET. Nevertheless, the WHEN equation ensures that anytime that RESET is asserted, the machine goes to the SRESET state. This is true regardless of the state definitions in the state_diagram. When RESET is asserted, the all-1s state encoding of SRESET is, in effect, ORed with the next state, if any, specified by the state_diagram. This approach to reliable reset would not be possible if SRESET were encoded as all 0s, for example.

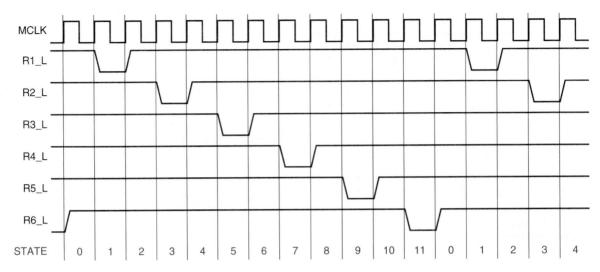

Figure 8-72 Modified timing waveforms for a digital system.

Now let's look at a variation of the previous timing waveforms that might be required in a different system. Figure 8-72 is similar to the previous waveforms, except that each phase output Ri is asserted for only one clock tick per phase. This change has a subtle but important effect on the design approach.

In the original design, we used a 6-bit ring counter and one auxiliary state bit T1 to keep track of the two states within each phase. With the new waveforms this is not possible. In the states between active-low pulses (STATE = 0, 2, 4, etc. in Figure 8-72) the phase outputs are all negated, so they can no longer be used to figure out which state should be visited next. Something else is needed to keep track of the state.

There are many different ways to solve this problem. One idea is to start with the original design in Table 8-26 but use the phase outputs P1_L, P2_L, and

Table 8-28 Additions to Table 8-26 for a modified six-phase waveform generator.

```
module TIMEG12K
...
R1_L, R2_L, R3_L, R4_L, R5_L, R6_L          pin istype 'com';
...
OUTPUTS = [R1_L, R2_L, R3_L, R4_L, R5_L, R6_L];

equations
...
!OUTPUTS = !PHASES & !T1;

end TIMEG12K
```

so on as internal states only. Then, each phase output Ri_L can be defined as a Moore-type combinational output that is asserted when the corresponding Pi_L is asserted *and* we are in the second tick of a phase. The additional ABEL code to support this first approach is shown in Table 8-28.

This first approach is easy, and it works just fine if the Pi_L signals are going to be used only as enables or other control inputs. However, it's a bad idea if these signals are going to be used as clocks, because they may have glitches, as we'll now explain. The Pi_L and T1 signals are all outputs from flip-flops clocked by the same master clock MCLK. Although these signals change at approximately the same time, their timing is never quite exact. One output may change sooner than another; this is called *output timing skew*. For example, suppose that on the transition from state 1 to 2 in Figure 8-71, P2_L goes LOW before T1 goes HIGH. In this case, a short glitch could appear on the R2_L output.

output timing skew

To get glitch-free outputs, we should design the circuit so that each phase output is the a registered output. One way to do this is to build a 12-bit ring counter and use only alternate outputs to yield the desired waveforms; an ABEL program using this approach is shown in Table 8-29.

Table 8-29 ABEL program for a modified six-phase waveform generator.

```
module TIMEG12
title 'Modified six-phase Master Timing Generator'

" Input and Output pins
MCLK, RESET, RUN, RESTART                    pin;
P1_L, P2_L, P3_L, P4_L, P5_L, P6_L           pin istype 'reg';
P1A, P2A, P3A, P4A, P5A, P6A                 pin istype 'reg';

" State definitions
PHASES = [P1A, P1_L, P2A, P2_L, P3A, P3_L, P4A, P4_L, P5A, P5_L, P6A, P6_L];
NEXTPH = [P6_L, P1A, P1_L, P2A, P2_L, P3A, P3_L, P4A, P4_L, P5A, P5_L, P6A];
SRESET = [1, 1, 1, 1, 1, 1, 1, 1, 1, 1, 1, 1];
P1 =    [0, 1, 1, 1, 1, 1, 1, 1, 1, 1, 1, 1];

equations

PHASES.CLK = MCLK;

WHEN RESET THEN PHASES := SRESET;
ELSE WHEN RESTART # (PHASES == SRESET) THEN PHASES := P1;
ELSE WHEN RUN THEN PHASES := NEXTPH;
ELSE PHASES := PHASES;

end TIMEG12
```

Table 8-30 Counter-based program for six-phase waveform generator.

```
module TIMEG12A
title 'Counter-based six-phase master timing generator'

" Input and Output pins
MCLK, RESET, RUN, RESTART              pin;
P1_L, P2_L, P3_L, P4_L, P5_L, P6_L    pin istype 'reg';
CNT3..CNT0                            pin istype 'reg';

" Definitions
CNT = [CNT3..CNT0];
P_L = [P1_L, P2_L, P3_L, P4_L, P5_L, P6_L];

equations

CNT.CLK = MCLK;  P_L.CLK = MCLK;

WHEN RESET THEN CNT := 15
ELSE WHEN RESTART THEN CNT := 0
ELSE WHEN (RUN & (CNT < 11)) THEN CNT := CNT + 1
ELSE WHEN RUN THEN CNT := 0
ELSE CNT := CNT;

P1_L := !(CNT == 0);
P2_L := !(CNT == 2);
P3_L := !(CNT == 4);
P4_L := !(CNT == 6);
P5_L := !(CNT == 8);
P6_L := !(CNT == 10);

end TIMEG12A
```

Still another approach is to recognize that, since the waveforms cycle through 12 states, we can build a modulo-12 binary counter and decode the states of that counter. An ABEL program using this approach is shown in Table 8-30. The states of the counter correspond to the "STATE" values shown in Figure 8-72. Since the phase outputs are registered, they are glitch free. Note that they are decoded one cycle early, to account for the one-tick decoding delay. Also, during reset, the counter is forced to state 15 rather than 0, so that the P1_L output is not asserted during reset.

8.5.10 Shift Registers in VHDL

Shift registers can be specified structurally or behaviorally in VHDL; we'll look at a few behavioral descriptions and applications. Table 8-31 is the function table for an 8-bit shift register with an extended set of functions. In addition to the hold, load, and shift functions of the 74x194 and 74x299, it performs circular

Table 8-31 Function table for an extended-function 8-bit shift register.

Function	Inputs			Next state							
	S2	S1	S0	Q7*	Q6*	Q5*	Q4*	Q3*	Q2*	Q1*	Q0*
Hold	0	0	0	Q7	Q6	Q5	Q4	Q3	Q2	Q1	Q0
Load	0	0	1	D7	D6	D5	D4	D3	D2	D1	D0
Shift right	0	1	0	RIN	Q7	Q6	Q5	Q4	Q3	Q2	Q1
Shift left	0	1	1	Q6	Q5	Q4	Q3	Q2	Q1	Q0	LIN
Shift circular right	1	0	0	Q0	Q7	Q6	Q5	Q4	Q3	Q2	Q1
Shift circular left	1	0	1	Q6	Q5	Q4	Q3	Q2	Q1	Q0	Q7
Shift arithmetic right	1	1	0	Q7	Q7	Q6	Q5	Q4	Q3	Q2	Q1
Shift arithmetic left	1	1	1	Q6	Q5	Q4	Q3	Q2	Q1	Q0	0

and arithmetic shift operations. In the *circular shift* operations, the bit that "falls off" one end during a shift is fed back into the other end. In the *arithmetic shift* operations, the edge input is set up for multiplication or division by 2; for a left shift, the right input is 0, and for a right shift, the leftmost (sign) bit is replicated.

circular shift
arithmetic shift

A behavioral VHDL program for the extended-function shift register is shown in Table 8-32. As in previous examples, we define a process and use the event attribute on the CLK signal to obtain the desired edge-triggered behavior. Several other features of this program are worth noting:

- An internal signal, IQ, is used for what eventually becomes the Q output, so it can be both read and written by process statements. Alternatively, we could have defined the Q output as type "buffer."

- The CLR input is asynchronous; because it's in the process sensitivity list, it is tested whenever it changes. And the IF statement is structured so that CLR takes precedence over any other condition.

- A CASE statement is used to define the operation of the shift register for the eight possible values of the select inputs S(2 downto 0).

- In the CASE statement, the "when others" case is required to prevent the compiler from complaining about approximately 2^{32} uncovered cases!

- The "null" statement indicates that no action is taken in certain cases. In case 1, note that no action is required; the default is for a signal to hold its value unless otherwise stated.

- In most of the cases the concatenation operator "&" is used to construct an 8-bit array from a 7-bit subset of IQ and one other bit.

Table 8-32 VHDL program for an extended-function 8-bit shift register.

```
library IEEE;
use IEEE.std_logic_1164.all;

entity Vshftreg is
    port (
        CLK, CLR, RIN, LIN: in STD_LOGIC;
        S: in STD_LOGIC_VECTOR (2 downto 0); -- function select
        D: in STD_LOGIC_VECTOR (7 downto 0); -- data in
        Q: out STD_LOGIC_VECTOR (7 downto 0) -- data out
    );
end Vshftreg;

architecture Vshftreg_arch of Vshftreg is
signal IQ: STD_LOGIC_VECTOR (7 downto 0);
begin
process (CLK, CLR, IQ)
  begin
    if (CLR='1') then IQ <= (others=>'0'); -- Asynchronous clear
    elsif (CLK'event and CLK='1') then
      case CONV_INTEGER(S) is
        when 0 => null;                           -- Hold
        when 1 => IQ <= D;                        -- Load
        when 2 => IQ <= RIN & IQ(7 downto 1);    -- Shift right
        when 3 => IQ <= IQ(6 downto 0) & LIN;    -- Shift left
        when 4 => IQ <= IQ(0) & IQ(7 downto 1);  -- Shift circular right
        when 5 => IQ <= IQ(6 downto 0) & IQ(7);  -- Shift circular left
        when 6 => IQ <= IQ(7) & IQ(7 downto 1);  -- Shift arithmetic right
        when 7 => IQ <= IQ(6 downto 0) & '0';    -- Shift arithmetic left
        when others => null;
      end case;
    end if;
    Q <= IQ;
  end process;
end Vshftreg_arch;
```

- Because of VHDL's strong requirements for type matching, we used the CONV_INTEGER function that we defined in Section 4.7.4 to convert the STD_LOGIC_VECTOR select input S to an integer in the CASE statement. Alternatively, we could have written each case label as a STD_LOGIC_VECTOR (e.g., ('0','1','1') instead of integer 3).

One application of shift registers is in ring counters, as in the six-phase waveform generator that we described on page 736 with the waveforms in Figure 8-71. A VHDL program that provides the corresponding behavior is

Table 8-33 VHDL program for a six-phase waveform generator.

```
library IEEE;
use IEEE.std_logic_1164.all;

entity Vtimegn6 is
    port (
        MCLK, RESET, RUN, RESTART: in STD_LOGIC; -- clock, control inputs
        P_L: out STD_LOGIC_VECTOR (1 to 6)       -- active-low phase outputs
    );
end Vtimegn6;

architecture Vtimegn6_arch of Vtimegn6 is
signal IP: STD_LOGIC_VECTOR (1 to 6); -- internal active-high phase signals
signal T1: STD_LOGIC;                 -- first tick within phase
begin
process (MCLK, IP)
  begin
    if (MCLK'event and MCLK='1') then
      if (RESET='1') then
        T1 <= '1'; IP <= ('0','0','0','0','0','0');
      elsif ((IP=('0','0','0','0','0','0')) or (RESTART='1')) then
        T1 <= '1'; IP <= ('1','0','0','0','0','0');
      elsif (RUN='1') then
        T1 <= not T1;
        if (T1='0') then IP <= IP(6) & IP(1 to 5); end if;
      end if;
    end if;
    P_L <= not IP;
  end process;
end Vtimegn6_arch;
```

shown in Table 8-33. As in the previous VHDL example, an internal active-high signal vector, IP, is used for reading and writing what eventually becomes the circuit's output; this internal signal is conveniently inverted in the last statement to obtain the required active-low output signal vector. The rest of the program is straightforward, but notice that it has three levels of nested IF statements.

A possible modification to the preceding application is to produce output waveforms that are asserted only during the second tick of each two-tick phase; such waveforms were shown in Figure 8-72 on page 740. One way to do this is to create a 12-bit ring counter and use only alternate outputs. In the VHDL realization only the six phase outputs, P_L(1 to 6), would appear in the entity definition. The additional six signals, which we name NEXTP(1 to 6), are local to the architecture definition. Figure 8-73 shows the relationship of these signals for shift-register operation, and Table 8-34 is the VHDL program.

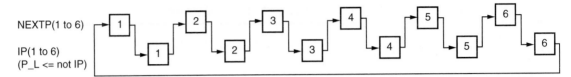

Figure 8-73 Shifting sequence for waveform generator 12-bit ring counter.

As in the previous program, a 6-bit active-high signal, IP, is declared in the architecture body and used for reading and writing what eventually becomes the circuit's active-low output, P_L. The additional 6-bit signal, NEXTP, holds the remaining 6 bits of state. Constants IDLE and FIRSTP are used to improve the program's readability.

■ Table 8-34 VHDL program for a modified six-phase waveform generator.

```
library IEEE;
use IEEE.std_logic_1164.all;

entity Vtimeg12 is
    port (
        MCLK, RESET, RUN, RESTART: in STD_LOGIC; -- clock, control inputs
        P_L: out STD_LOGIC_VECTOR (1 to 6)       -- active-low phase outputs
    );
end Vtimeg12;

architecture Vtimeg12_arch of Vtimeg12 is
signal IP, NEXTP: STD_LOGIC_VECTOR (1 to 6); -- internal active-high phase signals
begin
process (MCLK, IP, NEXTP)
  variable TEMP: STD_LOGIC_VECTOR (1 to 6);   -- temporary for signal shift
  constant IDLE: STD_LOGIC_VECTOR (1 to 6) := ('0','0','0','0','0','0');
  constant FIRSTP: STD_LOGIC_VECTOR (1 to 6) := ('1','0','0','0','0','0');
  begin
    if (MCLK'event and MCLK='1') then
      if (RESET='1') then IP <= IDLE;  NEXTP <= IDLE;
      elsif (RESTART='1') or (IP=IDLE and NEXTP=IDLE) then IP <= IDLE;  NEXTP <= FIRSTP;
      elsif (RUN='1') then
        if (IP=IDLE) and (NEXTP=IDLE) then NEXTP <= FIRSTP;
        else TEMP := IP;  IP <= NEXTP;  NEXTP <= TEMP(6) & TEMP(1 to 5);
        end if;
      end if;
    end if;
    P_L <= not IP;
  end process;
end Vtimeg12_arch;
```

Notice that a 6-bit variable, TEMP, is used just as a temporary place to hold the old value of IP when shifting occurs—IP is loaded with NEXTP, and NEXTP is loaded with the shifted, old value of IP. Because the assignment statements in a process are executed *sequentially*, we couldn't get away with just writing "IP <= NEXTP; NEXTP <= IP(6) & IP(1 to 5);". If we did that, then NEXTP would pick up the *new* value of IP, not the old. Notice also that since TEMP is a local variable, not a signal, its value is not visible outside the process. Also, a process invocation never uses the value of TEMP from a previous invocation. Therefore, the VHDL compiler need not synthesize any flip-flops to hold TEMP's value.

*8.6 Iterative versus Sequential Circuits

We introduced iterative circuits in Section 5.9.2. The function of an n-module iterative circuit can be performed by a sequential circuit that uses just one copy of the module but requires n steps (clock ticks) to obtain the result. This is an excellent example of a space/time trade-off in digital design.

As shown in Figure 8-74, flip-flops are used in the sequential-circuit version to store the cascading outputs at the end of each step; the flip-flop outputs are used as the cascading inputs at the beginning of the next step. The flip-flops must be initialized to the boundary-input values before the first clock tick, and they contain the boundary-output values after the nth tick.

Since an iterative circuit is a combinational circuit, all of its primary and boundary inputs may be applied simultaneously, and its primary and boundary outputs are all available after a combinational delay. In the sequential-circuit version, the primary inputs must be delivered sequentially, one per clock tick, and the primary outputs are produced with similar timing. Therefore, serial-out shift registers are often used to provide the inputs, and serial-in shift registers are used to collect the outputs. For this reason, the sequential-circuit version of an "iterative widget" is often called a "serial widget."

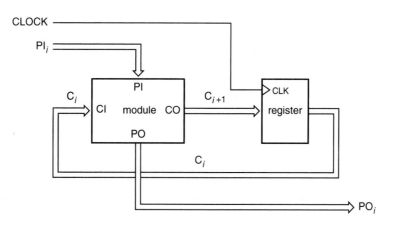

Figure 8-74
General structure of the sequential-circuit version of an iterative circuit.

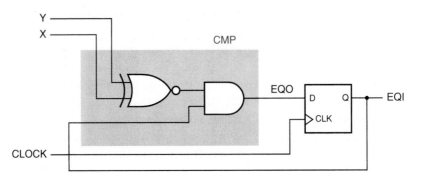

Figure 8-75
Simplified serial
comparator circuit.

serial comparator

For example, Figure 8-75 shows the basic design for a *serial comparator* circuit. The shaded block is identical to the module used in the iterative comparator of Figure 5-80 on page 423. The circuit is drawn in more detail using SSI chips in Figure 8-76. In addition, we have provided a synchronous reset input that, when asserted, forces the initial value of the cascading flip-flop to 1 at the next clock tick. The initial value of the cascading flip-flop corresponds to the boundary input in the iterative comparator.

With the serial comparator, an *n*-bit comparison requires *n* + 1 clock ticks. RESET_L is asserted at the first clock tick. RESET_L is negated and data bits are applied at the next *n* ticks. The EQI output gives the comparison result during the clock period following the last tick. A timing diagram for two successive 4-bit comparisons is shown in Figure 8-77. The spikes in the EQO waveform indicate the time when the combinational outputs are settling in response to new X and Y input values.

Figure 8-76
Detailed serial
comparator circuit.

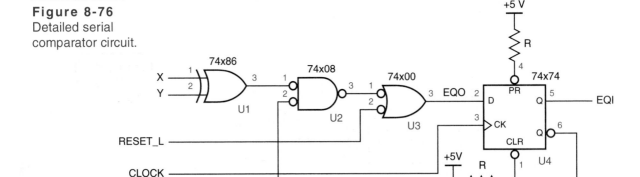

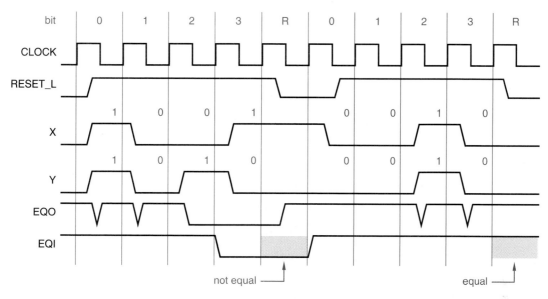

Figure 8-77 Timing diagram for serial comparator circuit.

A *serial binary adder* circuit for addends of any length can be constructed from a full adder and a D flip-flop, as shown in Figure 8-78. The flip-flop, which stores the carry between successive bits of the addition, is cleared to 0 at reset. Addend bits are presented serially on the A and B inputs, starting with the LSB, and sum bits appear on S in the same order.

serial binary adder

Because of the large size and high cost of digital logic circuits in the early days, many computers and calculators used serial adders and other serial versions of iterative circuits to perform arithmetic operations. Even though these arithmetic circuits aren't used much today, they are an instructive reminder of the space/time trade-offs that are possible in digital design.

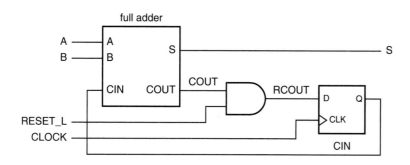

Figure 8-78
Serial binary adder
circuit.

8.7 Synchronous Design Methodology

synchronous system

In a *synchronous system*, all flip-flops are clocked by the same common clock signal, and preset and clear inputs are not used, except for system initialization. Although all the world does not march to the tick of a common clock, within the confines of a digital system or subsystem we can make it so. When we interconnect digital systems or subsystems that use different clocks, we can usually identify a limited number of asynchronous signals that need special treatment, as we'll show in Section 8.8.3.

Races and hazards are not a problem in synchronous systems, for two reasons. First, the only fundamental-mode circuits that might be subject to races or essential hazards are predesigned elements, such as discrete flip-flops or ASIC cells, that are guaranteed by the manufacturer to work properly. Second, even though the combinational circuits that drive flip-flop control inputs may contain static or dynamic or function hazards, these hazards have no effect, since the control inputs are sampled only *after* the hazard-induced glitches have had a chance to settle out.

Aside from designing the functional behavior of each state machine, the designer of a practical synchronous system or subsystem must perform just three well-defined tasks to ensure reliable system operation:

1. Minimize and determine the amount of clock skew in the system, as discussed in Section 8.8.1.

2. Ensure that flip-flops have positive setup- and hold-time margins, including an allowance for clock skew, as described in Section 8.1.4.

3. Identify asynchronous inputs, synchronize them with the clock, and ensure that the synchronizers have an adequately low probability of failure, as described in Sections 8.8.3 and 8.9.

Before we get into these issues, in this section we'll look at a general model for synchronous system structure and an example.

8.7.1 Synchronous System Structure

The sequential-circuit design examples that we gave in Chapter 7 were mostly individual state machines with a small number of states. If a sequential circuit has more than a few flip-flops, then it's not desirable (and often not possible) to treat the circuit as a single, monolithic state machine, because the number of states would be too large to handle.

data unit
control unit

Fortunately, most digital systems or subsystems can be partitioned into two or more parts. Whether the system processes numbers, digitized voice signals, or a stream of spark-plug pulses, a certain part of the system, which we'll call the *data unit,* can be viewed as storing, routing, combining, and generally processing "data." Another part, which we'll call the *control unit,* can be viewed as starting and stopping actions in the data unit, testing conditions, and deciding what to do next according to circumstances. In general, only the control unit must be designed as a state machine. The data unit and its components are typically handled at a higher level of abstraction, such as:

- *Registers*. A collection of flip-flops is loaded in parallel with many bits of "data," which can then be used or retrieved together.
- *Specialized functions*. These include multibit counters and shift registers, which increment or shift their contents on command.
- *Read/write memory*. Individual latches or flip-flops in a collection of the same can be written or read out.

The first two topics above were discussed earlier in this chapter, and the last is discussed in Chapter 10.

Figure 8-79 is a general block diagram of a system with a control unit and a data unit. We have also included explicit blocks for input and output, but we could have just as easily absorbed these functions into the data unit itself. The control unit is a state machine whose inputs include *command inputs* that indicate how the machine is to function, and *condition inputs* provided by the data unit. The command inputs may be supplied by another subsystem or by a user to set the general operating mode of the control state machine (RUN/HALT, NORMAL/TURBO, etc.), while the condition inputs allow the control state-machine unit to change its behavior as required by circumstances in the data unit (ZERO_DETECT, MEMORY_FULL, etc.).

  command input
  condition input

A key characteristic of the structure in Figure 8-79 is that the control, data, input, and output units all use the same common clock. Figure 8-80 illustrates the operations of the control and data units during a typical clock cycle:

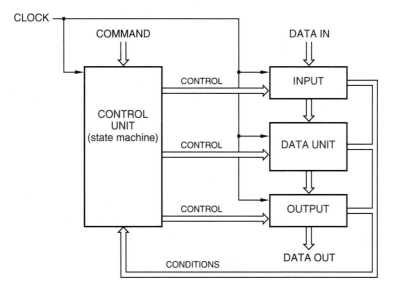

Figure 8-79
Synchronous system structure.

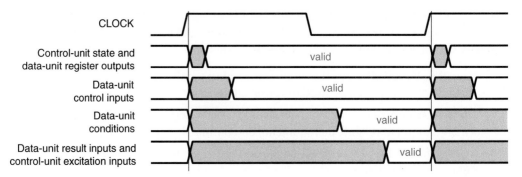

Figure 8-80 Operations during one clock cycle in a synchronous system.

1. Shortly after the beginning of the clock period, the control-unit state and the data-unit register outputs are valid.

2. Next, after a combinational logic delay, Moore-type outputs of the control-unit state machine become valid. These signals are control *inputs* to the data unit. They determine what data-unit functions are performed in the rest of the clock period—for example, selecting memory addresses, multiplexer paths, and arithmetic operations.

3. Near the end of the clock period, data-unit condition outputs such as zero- or overflow-detect are valid and are made available to the control unit.

4. At the end of the clock period, just before the setup-time window begins, the next-state logic of the control-unit state machine has determined the next state based on the current state and command and condition inputs. At about the same time, computational results in the data unit are available to be loaded into data-unit registers.

5. After the clock edge, the whole cycle may repeat.

Data-unit control inputs, which are control-unit state-machine outputs, may be of the Moore, Mealy, or pipelined Mealy type; timing for the Moore type was shown in Figure 8-80. Moore-type and pipelined-Mealy-type outputs control the data unit's actions strictly according to the current state and past inputs, which do not depend on *current* conditions in the data unit. In contrast, Mealy-type outputs may select different actions in the data unit according to *current* conditions in the data unit. This increases flexibility, but typically also

PIPELINED MEALY OUTPUTS	Some state machines have pipelined Mealy outputs, discussed in Section 7.3.2. In Figure 8-80, pipelined Mealy outputs would typically be valid early in the cycle, at the same time as control-unit state outputs. Early validity of these outputs, compared to Moore outputs that must go through a combinational logic delay, may allow the entire system to operate at a faster clock rate.

increases the minimum clock period for correct system operation, since the delay path may be much longer. Also, Mealy-type outputs must not create feedback loops. For example, a signal that adds 1 to an adder's input if the adder output is nonzero causes an oscillation if the adder output is -1.

8.7.2 A Synchronous System Design Example

To give you an overview of several elements of synchronous system design, this subsection presents a representative example of a synchronous system. The example is a *shift-and-add multiplier* for unsigned integers using the algorithm of Section 2.8. Its data unit uses standard combinational and sequential building blocks, and its control unit is described by a state diagram

shift-and-add multiplier

Figure 8-81 illustrates data-unit registers and functions that are used to perform an 8-bit multiplication:

MPY/LPROD   A shift register that initially stores the multiplier, and accumulates the low-order bits of the product as the algorithm is executed.

HPROD   A register that is initially cleared, and accumulates the high-order bits of the product as the algorithm is executed.

MCND   A register that stores the multiplicand throughout the algorithm.

F   A combinational function equal to the 9-bit sum of HPROD and MCND if the low-order bit of MPY/LPROD is 1, and equal to HPROD (extended to 9 bits) otherwise.

The MPY/LPROD shift register serves a dual purpose, holding both yet-to-be-tested multiplier bits (on the right) and unchanging product bits (on the left) as the algorithm is executed. At each step it shifts right one bit, discarding the multiplier bit that was just tested, moving the next multiplier bit to be tested to the rightmost position, and loading into the leftmost position one more product bit that will not change for the rest of the algorithm.

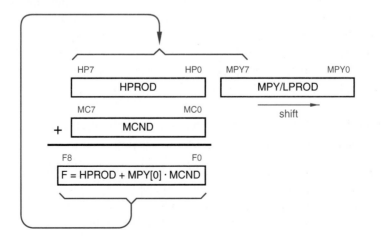

Figure 8-81
Registers and functions used by the shift-and-add multiplication algorithm.

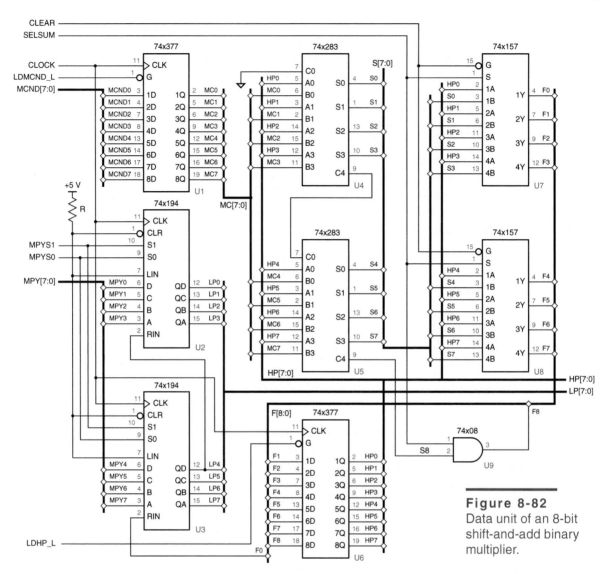

Figure 8-82
Data unit of an 8-bit shift-and-add binary multiplier.

Figure 8-82 is an MSI design for the data unit. The multiplier, MPY[7:0], and the multiplicand, MCND[7:0], are loaded into two registers before a multiplication begins. When the multiplication is completed, the product appears on HP[7:0] and LP[7:0]. The data unit uses the following control signals:

LDMCND_L When asserted, enables the multiplicand register U1 to be loaded.

LDHP_L When asserted, enables the HPROD register U6 to be loaded.

MPYS[1:0] When 11, these signals enable the MPY/LPROD register U2 and U3 to be loaded at the next clock tick. They are set to 01 during the multiplication operation to enable the register to shift right, and they are 00 at other times to preserve the register's contents.

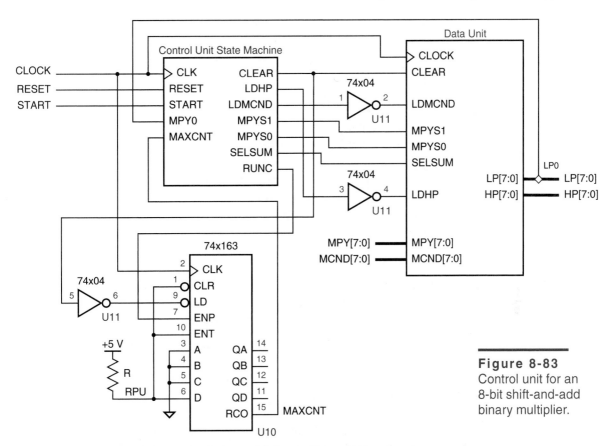

Figure 8-83
Control unit for an
8-bit shift-and-add
binary multiplier.

SELSUM   When this is asserted, the multiplexers U7 and U8 select the output
of the adders U4 and U5, which is the sum of HPROD and the
multiplicand MC. Otherwise, they select HPROD directly.

CLEAR   When asserted, the output of multiplexers U7 and U8 is zero.

The multiplier uses a control unit, shown along with the data-unit block in
Figure 8-83, to initialize the data unit and step through a multiplication. The
control unit is decomposed into a counter (U10) and a state machine with the
state diagram shown in Figure 8-84.

The state machine has the following inputs and outputs:

RESET   A reset input that is asserted at power-up.

START   An external command input that starts a multiplication.

MPY0   A condition input from the data unit, the next multiplier bit to test.

CLEAR   A control output that zeroes the multiplexer output and initializes
the counter.

LDMCND   A control output that enables the MCND register to be loaded.

LDHP   A control output that enables the HPROD register to be loaded.

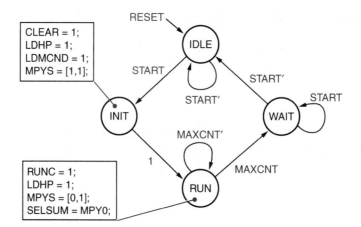

Figure 8-84
State diagram for the
control state machine
for a shift-and-add
binary multiplier.

RUNC   A control output that enables the counter to count.

MPYS[1:0]   Control outputs for **MPY/LPROD** shifting and loading.

SELSUM   A control output that selects between the shifted adder output or
shifted **HPROD** to be loaded back into **HPROD**.

The state diagram can be converted into a corresponding state machine
using any of a variety of methods, from turn-the-crank (a.k.a. hand-crafted)
design to automatic synthesis using a corresponding ABEL or VHDL description. The state machine has mostly Moore-type outputs; **SELSUM** is a Mealy-
type output. Two boxes in the state diagram list outputs that are asserted in the
INIT and **RUN** states; all outputs are negated at other times. The machine is
designed so that asserting **RESET** in any state takes it to the **IDLE** state.

After the **START** signal is asserted, a multiplication begins in the **INIT**
state. In this state, the counter is initialized to 1000_2, the multiplier and multiplicand are loaded into their respective registers, and **HPROD** is cleared. The **RUN**
state is entered next, and the counter is enabled to count. The state machine stays
in the **RUN** state for eight clock ticks, to execute the eight steps of the 8-bit shift-
and-add algorithm. During the eighth tick, the counter is in state 1111_2, so
MAXCNT is asserted and the state machine goes to the **WAIT** state. The machine
waits there until **START** is negated, to prevent a multiplication from restarting
until **START** is asserted once again.

The design details of the data and control units are interesting, but the most
important thing to see in this example is that all of the sequential circuit elements
for both data and control are edge-triggered flip-flops clocked by the same
common **CLOCK** signal. Thus, its timing is consistent with the model in
Figure 8-80, and the designer need not be concerned about races, hazards, and
asynchronous operations. Unless the state machine realization is very slow, the
overall circuit's maximum clock speed will be limited mainly by the propagation
delays through the data unit.

8.8 Impediments to Synchronous Design

Although the synchronous approach is the most straightforward and reliable method of digital system design, a few nasty realities can get in the way. We'll discuss them in this section.

8.8.1 Clock Skew

Synchronous systems using edge-triggered flip-flops work properly only if all flip-flops see the triggering clock edge at the same time. Figure 8-85 shows what can happen otherwise. Here, two flip-flops are theoretically clocked by the same signal, but the clock signal seen by FF2 is delayed by a significant amount relative to FF1's clock. This difference between arrival times of the clock at different devices is called *clock skew*.

clock skew

We've named the delayed clock in Figure 8-85(a) "CLOCKD." If FF1's propagation delay from CLOCK to Q1 is short, and if the physical connection of Q1 to FF2 is short, then the change in Q1 caused by a CLOCK edge may actually reach FF2 *before* the corresponding CLOCKD edge. In this case, FF2 may go to an incorrect next state determined by the *next* state of FF1 instead of the current state, as shown in (b). If the change in Q1 arrives at FF2 only slightly early relative to CLOCKD, then FF2's hold-time specification may be violated, in which case FF2 may become metastable and produce an unpredictable output.

If Figure 8-85 reminds you of the essential hazard shown in Figure 7-101, you're on to something. The clock-skew problem may be viewed simply as a manifestation of the essential hazards that exist in all edge-triggered devices.

We can determine quantitatively whether clock skew is a problem in a given system by defining t_{skew} to be the amount of clock skew and using the other timing parameters defined in Figure 8-1. For proper operation, we need

$$t_{ffpd(min)} + t_{comb(min)} - t_{hold} - t_{skew(max)} > 0$$

In other words, clock skew subtracts from the hold-time margin that we defined in Section 8.1.4.

Figure 8-85 Example of clock skew.

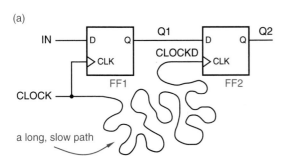

(a)

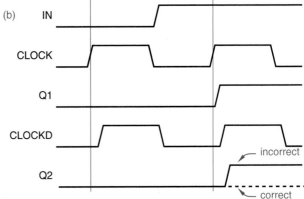

(b)

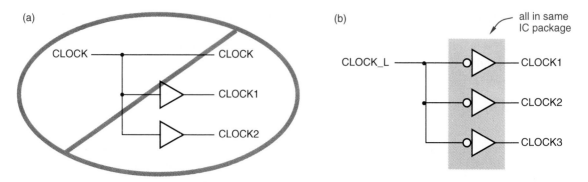

Figure 8-86 Buffering the clock: (a) excessive clock skew; (b) controllable clock skew.

Viewed in isolation, the example in Figure 8-85 may seem a bit extreme. After all, why would a designer provide a short connection path for data and a long one for the clock, when they could just run side by side? There are several ways this can happen; some are mistakes, while others are unavoidable.

In a large system, a single clock signal may not have adequate fanout to drive all of the devices with clock inputs, so it may be necessary to provide two or more copies of the clock signal. The buffering method of Figure 8-86(a) obviously produces excessive clock skew, since CLOCK1 and CLOCK2 are delayed through an extra buffer compared to CLOCK.

A recommended buffering method is shown in Figure 8-86(b). All of the clock signals go through identical buffers and thus have roughly equal delays. Ideally, all the buffers should be part of the same IC package, so that they all have similar delay characteristics and are operating at identical temperature and power-supply voltage. Some manufacturers build special buffers for just this sort of application and specify the worst-case delay variation between buffers in the same package, which can be as low as a few tenths of a nanosecond.

Even the method in Figure 8-86(b) may produce excessive clock skew if one clock signal is loaded much more heavily than the other; transitions on the more heavily loaded clock appear to occur later because of increases in output-transistor switching delay and signal rise and fall times. Therefore, a careful designer tries to balance the loads on multiple clocks, looking at both DC load (fanout) and AC load (wiring and input capacitance).

Another bad situation can occur when signals on a PCB or in an ASIC are routed automatically by CAD software. Figure 8-87 shows a PCB or ASIC with many flip-flops and larger-scale elements, all clocked with a common CLOCK signal. The CAD software has laid out CLOCK in a serpentine path that winds its way past all the clocked devices. Other signals are routed point-to-point between an output and a small number of inputs, so their paths are shorter. To make matters worse, in an ASIC some types of "wire" may be slower than others (polysilicon vs. metal in CMOS technology). As a result, a CLOCK edge may indeed arrive at FF2 quite a bit later than the data change that it produces on Q1.

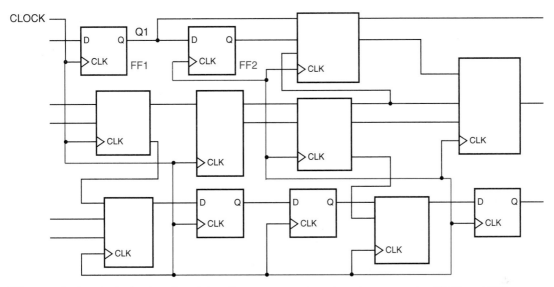

Figure 8-87 A clock-signal path leading to excessive skew in a complex PCB or ASIC.

One way to minimize this sort of problem is to arrange for **CLOCK** to be distributed in a tree-like structure using the fastest type of wire, as illustrated in Figure 8-88. Usually, such a "clock tree" must be laid out by hand or using a specialized CAD tool. Even then, in a complex design it may not be possible to guarantee that clock edges arrive everywhere before the earliest data change. A CAD timing analysis program is typically used to detect these problems, which

Figure 8-88 Clock-signal routing to minimize skew.

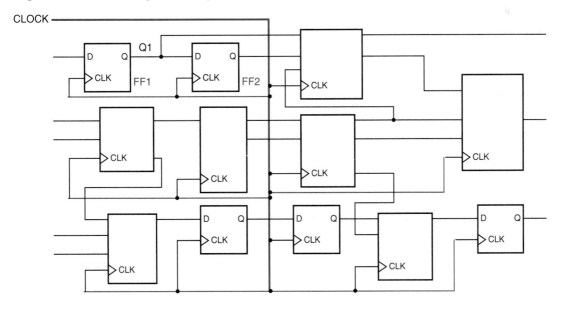

generally can be remedied only by inserting extra delay (e.g., pairs of inverters) in the too-fast data paths.

Although synchronous design methodology simplifies the conceptual operation of large systems, we see that clock skew can be a major problem when edge-triggered flip-flops are used as the storage elements. To control this prob- *two-phase latch design* lem, many high-performance systems and VLSI chips use a *two-phase latch design,* discussed in the References. Such designs effectively split each edge-triggered D flip-flop into its two component latches, controlling them with two nonoverlapping clock phases. The nonoverlap between the phases accommodates clock skew.

8.8.2 Gating the Clock

Most of the sequential MSI parts that we introduced in this chapter have synchronous function-enable inputs. That is, their enable inputs are sampled on the clock edge, along with the data. The first example that we showed was the 74x377 register with synchronous load-enable input; other parts included the 74x163 counter and 74x194 shift register with synchronous load-enable, count-enable, and shift-enable inputs. Nevertheless, many MSI parts, FPGA macros, and ASIC cells do not have synchronous function-enable inputs; for example, the 74x374 8-bit register has three-state outputs but no load-enable input.

So, what can a designer do if an application requires an 8-bit register with both a load-enable input *and* three-state outputs? One solution is to use a 74x377 to get the load-enable, and follow it with a 74x241 three-state buffer. However, this increases both cost and delay. Another approach is to use a larger, more expensive part, the 74x823, which provides both required functions as well as an asynchronous CLR_L input. A riskier alternative sometimes employed by designers who don't know any better is to use a '374, but to suppress its clock *gating the clock* input when it's not supposed to be loaded. This is called *gating the clock.*

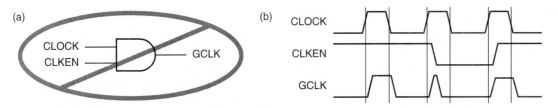

Figure 8-89 How not to gate the clock: (a) simple-minded circuit; (b) timing diagram.

Figure 8-89 illustrates an obvious but wrong approach to gating the clock. A signal CLKEN is asserted to enable the clock and is simply ANDed with the clock to produce the gated clock GCLK. This approach has two problems:

1. If CLKEN is a state-machine output or other signal produced by a register clocked by CLOCK, then CLKEN changes some time *after* CLOCK has already gone HIGH. As shown in (b), this produces glitches on GCLK and false clocking of the registers controlled by GCLK.

2. Even if CLKEN is somehow produced well in advance of CLOCK's rising edge (e.g., using a register clocked with the *falling* edge of CLOCK, an especially nasty kludge), the AND-gate delay gives GCLK excessive clock skew, which causes more problems all around.

A method of gating the clock that generates only minimal clock skew is shown in Figure 8-90. Here, both an ungated clock and several gated clocks are generated from the same active-low master clock signal. Gates in the same IC package are used to minimize the possible differences in their delays. The CLKEN signal may change arbitrarily whenever CLOCK_L is LOW, which is when CLOCK is HIGH. That's just fine; a CLKEN signal is typically produced by a state machine whose outputs change right after CLOCK goes HIGH.

The approach of Figure 8-90 is acceptable in a particular application only if the clock skew that it creates is acceptable. Furthermore, note that CLKEN

Figure 8-90 An acceptable way to gate the clock: (a) circuit; (b) timing diagram.

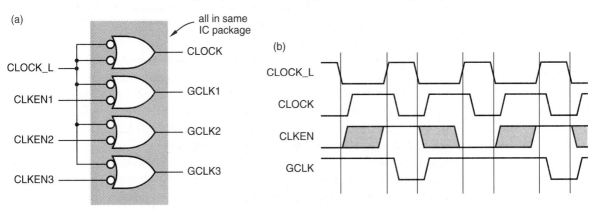

must be stable during the entire time that CLOCK_L is HIGH (CLOCK is LOW). Thus, the timing margins in this approach are sensitive to the clock's duty cycle, especially if CLKEN suffers significant combinational-logic delay (t_{comb}) from the triggering clock edge. A truly synchronous function-enable input, such as the 74x377's load-enable input in Figure 8-13, can be changed at almost any time during the entire clock period, up until a setup time before the triggering edge.

8.8.3 Asynchronous Inputs

Even though it is theoretically possible to build a computer system that is fully synchronous, you couldn't do much with it, unless you could synchronize your keystrokes with a 500-MHz clock. Digital systems of all types inevitably must deal with *asynchronous input signals* that are not synchronized with the system clock.

asynchronous input signal

Asynchronous inputs are often requests for service (e.g., interrupts in a computer) or status flags (e.g., a resource has become available). Such inputs normally change slowly compared to the system clock frequency, and they need not be recognized at a particular clock tick. If a transition is missed at one clock tick, it can always be detected at the next one. The transition rates of asynchronous signals may range from less than one per second (the keystrokes of a slow typist) to 100 MHz or more (access requests for a 500-MHz multiprocessor system's shared memory).

synchronizer

Ignoring the problem of metastability, it is easy to build a *synchronizer,* a circuit that samples an asynchronous input and produces an output that meets the setup and hold times required in a synchronous system. As shown in Figure 8-91, a D flip-flop samples the asynchronous input at each tick of the system clock and produces a synchronous output that is valid during the next clock period.

Figure 8-91
A single, simple synchronizer:
(a) logic diagram;
(b) timing.

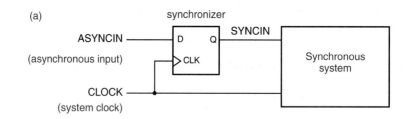

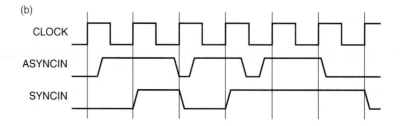

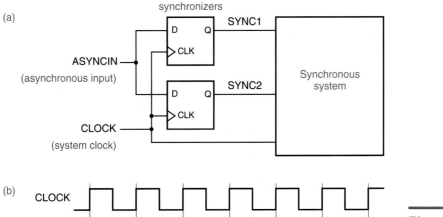

(a)

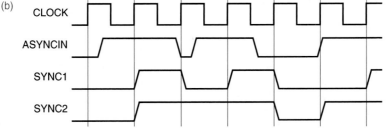

(b)

Figure 8-92
Two synchronizers
for the same
asynchronous input:
(a) logic diagram;
(b) possible timing.

It is essential for asynchronous inputs to be synchronized at only *one place* in a system; Figure 8-92 shows what can happen otherwise. Because of physical delays in the circuit, the two flip-flops will not see the clock and input signals at precisely the same time. Therefore, when asynchronous input transitions occur near the clock edge, there is a small window of time during which one flip-flop may sample the input as 1 and the other may sample it as 0. This inconsistent result may cause improper system operation, as one part of the system responds as if the input were 1, and another part responds as if it were 0.

Combinational logic may hide the fact that there are two synchronizers, as shown in Figure 8-93. Since different paths through the combinational logic will inevitably have different delays, the likelihood of an inconsistent result is even

Figure 8-93 An asynchronous input driving two synchronizers through combinational logic.

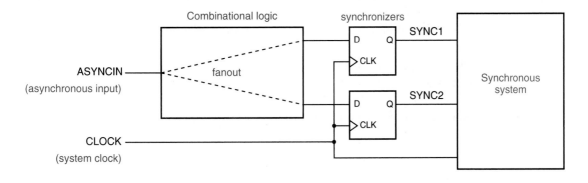

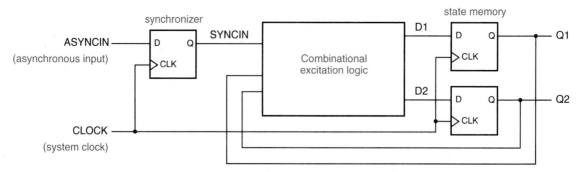

Figure 8-94 An asynchronous state-machine input coupled through a single synchronizer.

greater. This situation is especially common when asynchronous signals are used as inputs to state machines, since the excitation logic for two or more state variables may depend on the asynchronous input. The proper way to use an asynchronous signal as a state-machine input is shown in Figure 8-94. All of the excitation logic sees the same synchronized input signal, SYNCIN.

WHO CARES?	As you probably know, even the synchronizers in Figures 8-91 and 8-94 sometimes fail. The reason is that the setup and hold times of the synchronizing flip-flop are sometimes violated, because the asynchronous input can change at any time. "Well, who cares?" you may say. "If the D input changes near the clock edge, then the flip-flop will either see the change this time or miss it and pick it up next time; either way is good enough for me!" The problem is, there is a third possibility, discussed in the next section.

8.9 Synchronizer Failure and Metastability

We showed in Section 7.1 that when the setup and hold times of a flip-flop are not met, the flip-flop may go into a third, *metastable* state halfway between 0 and 1. Worse, the length of time it may stay in this state before falling back into a legitimate 0 or 1 state is theoretically unbounded. When other gates and flip-flops are presented with a metastable input signal, some may interpret it as a 0 and others as a 1, creating the sort of inconsistent behavior that we showed in Figure 8-92. Or the other gates and flip-flops may produce metastable outputs themselves (after all, they are now operating in the *linear* part of their operating range). Luckily, the probability of a flip-flop output remaining in the metastable state decreases exponentially with time, though never all the way to zero.

8.9.1 Synchronizer Failure

Synchronizer failure is said to occur if a system uses a synchronizer output while the output is still in the metastable state. The way to avoid synchronizer failure is to ensure that the system waits "long enough" before using a synchronizer's output, "long enough" so that the mean time between synchronizer failures is several orders of magnitude longer than the designer's expected length of employment.

synchronizer failure

Metastability is more than an academic problem. More than a few experienced designers of high-speed digital systems have built (and released to production) circuits that suffer from intermittent synchronizer failures. In fact, the initial versions of several commercial ICs are said to have suffered from metastability problems, for example, the AMD 9513 system timing controller, the AMD 9519 interrupt controller, the Zilog Z-80 Serial I/O interface, the Intel 8048 single-chip microcomputer, and the AMD 29000 RISC microprocessor. It makes you wonder, are the designers of these parts still employed?

There are two ways to get a flip-flop out of the metastable state:

1. Force the flip-flop into a valid logic state using input signals that meet the published specifications for minimum pulse width, setup time, and so on.

2. Wait "long enough," so the flip-flop comes out of metastability on its own.

Inexperienced designers often attempt to get around metastability in other ways, and they are usually unsuccessful. Figure 8-95 shows an attempt by a designer who thinks that since metastability is an "analog" problem, it must have an "analog" solution. After all, Schmitt-trigger inputs and capacitors can normally be used to clean up noisy signals. However, rather than eliminate metastability, this circuit enhances it—with the "right" components, the circuit will oscillate forever, once it is excited by negating S_L and R_L simultaneously. (*Confession:* It was the author who tried this over 20 years ago!) Exercises 8.97 and 8.94 give examples of valiant but also failed attempts to eliminate metastability. These examples should give you the sense that synchronizer problems can be very subtle, so you must be careful. The only way to make synchronizers reliable is to wait long enough for metastable outputs to resolve. We answer the question "How long is 'long enough'?" later in this section.

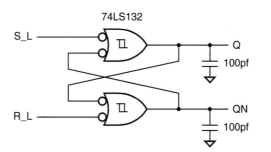

Figure 8-95
A failed attempt to build a metastable-proof $\overline{S}$-$\overline{R}$ flip-flop.

8.9.2 Metastability Resolution Time

t_r
metastability resolution time

If the setup and hold times of a D flip-flop are met, the flip-flop output settles to a new value within time t_{pd} after the clock edge. If they are violated, the flip-flop output may be metastable for an arbitrary length of time. In a particular system design, we use the parameter t_r, called the *metastability resolution time,* to denote the maximum time that the output can remain metastable without causing synchronizer (and system) failure.

For example, consider the state machine in Figure 8-94 on page 764. The available metastability resolution time is

$$t_r = t_{clk} - t_{comb} - t_{setup}$$

t_{clk}
t_{comb}
t_{setup}

where t_{clk} is the clock period, t_{comb} is the propagation delay of the combinational excitation logic, and t_{setup} is the setup time of the flip-flops used in the state memory.

8.9.3 Reliable Synchronizer Design

The most reliable synchronizer is one that allows the maximum amount of time for metastability resolution. However, in the design of a digital system, we seldom have the luxury of *slowing down* the clock to make the system work more reliably. Instead, we are usually asked to *speed up* the clock to get higher performance from the system. As a result, we often need synchronizers that work reliably with very short clock periods. We'll present several such designs, and show how to predict their reliability.

We showed previously that a state machine with an asynchronous input, built as illustrated in Figure 8-94, has $t_r = t_{clk} - t_{comb} - t_{setup}$. In order to maximize t_r for a given clock period, we should minimize t_{comb} and t_{setup}. The value of t_{setup} depends on the type of flip-flops used in the state memory; in general, faster flip-flops have shorter setup times. The minimum value of t_{comb} is zero and is achieved by the synchronizer design of Figure 8-96, whose operation we explain next.

Inputs to flip-flop FF1 are asynchronous with the clock and may violate the flip-flop's setup and hold times. When this happens, the META output may become metastable and remain in that state for an arbitrary time. However, we

Figure 8-96 Recommended synchronizer design.

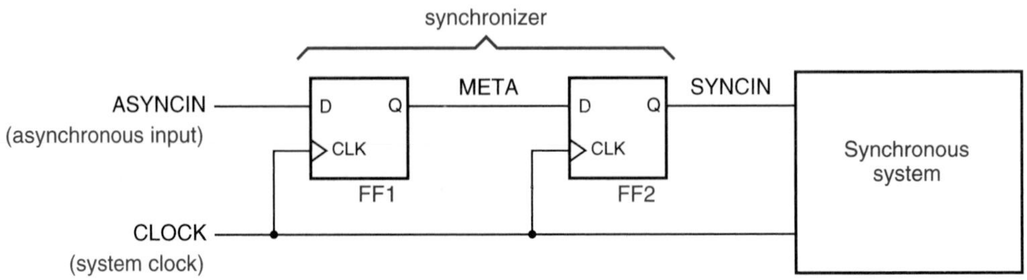

assume that the maximum duration of metastability after the clock edge is t_r. (We show how to calculate the probability that our assumption is correct in the next subsection.) As long as the clock period is greater than t_r plus the FF2's setup time, SYNCIN becomes a synchronized copy of the asynchronous input on the next clock tick without ever becoming metastable itself. The SYNCIN signal is distributed as required to the rest of the system.

8.9.4 Analysis of Metastable Timing

Figure 8-97 shows the flip-flop timing parameters that are relevant to our analysis of metastability timing. The published setup and hold times of a flip-flop with respect to its clock edge are denoted by t_s and t_h, and they bracket an interval called the *decision window*, when the flip-flop samples its input and decides to change its output if necessary. As long as the D input changes outside the decision window, as in (a), the manufacturer guarantees that the output will change and settle to a valid logic state before time t_{pd}. If D changes inside the decision window, as in (b), metastability may occur and persist until time t_r.

decision window

Theoretical research suggests, and experimental research has confirmed, that when asynchronous inputs change during the decision window, the duration of metastable outputs is governed by an exponential formula:

$$\text{MTBF}(t_r) = \frac{\exp(t_r/\tau)}{T_o \cdot f \cdot a}$$

Figure 8-97 Timing parameters for metastability analysis: (a) normal flip-flop operation; (b) metastable behavior.

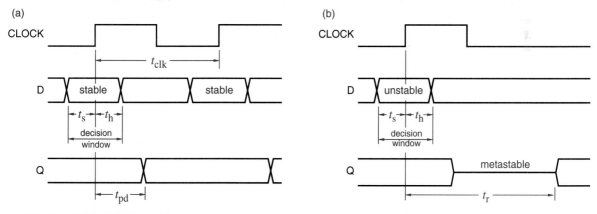

DETAILS, DETAILS   In our analysis of the synchronizer in Figure 8-96 we do not allow metastability, even briefly, on the output of FF2, because we assume that the system has been designed with zero timing margins. If the system can in fact tolerate some increase in FF2's propagation delay, the MTBF will be somewhat better than predicted.

f

a

T_o

τ

Here MTBF(t_r) is the mean time between synchronizer failures, where a failure occurs if metastability persists beyond time t_r after a clock edge, where $t_r \geq t_{pd}$. This MTBF depends on f, the frequency of the flip-flop clock; a, the number of asynchronous input changes per second applied to the flip-flop; and T_o and τ, constants that depend on the electrical characteristics of the flip-flop. For a typical 74LS74, $T_o \approx 0.4$ s and $\tau \approx 1.5$ ns.

Now suppose that we build a microprocessor system with a 10-MHz clock and use the circuit of Figure 8-96 to synchronize an asynchronous input. If ASYNCIN changes during the decision window of FF1, the output META may become metastable until time t_r. If META is still metastable at the beginning of the decision window for FF2, then the synchronizer fails, because FF2 may have a metastable output; system operation is unpredictable in that case.

Let us assume that the D flip-flops in Figure 8-96 are 74LS74s. The setup time t_s of a 74LS74 is 20 ns, and the clock period in our example microprocessor system is 100 ns, so t_r for synchronizer failure is 80 ns. If the asynchronous input changes 100,000 times per second, then the synchronizer MTBF is

$$\text{MTBF(80 ns)} = \frac{\exp(80/1.5)}{0.4 \cdot 10^7 \cdot 10^5} = 3.6 \cdot 10^{11}\,\text{s}$$

That's not bad, about 100 centuries between failures! Of course, if we're lucky enough to sell 10,000 copies of our system, one of them will fail in this way every year. But, no matter, let us consider a more serious problem.

Suppose we upgrade our system to use a faster microprocessor chip with a clock speed of 16 MHz. We may have to replace some components in our system to operate at the higher speed, but 74LS74s are still perfectly good at 16 MHz. Or are they? With a clock period of 62.5 ns, the new synchronizer MTBF is

$$\text{MTBF(42.5 ns)} = \frac{\exp(42.5/1.5)}{0.4 \cdot 1.6 \cdot 10^7 \cdot 10^5} = 3.1\,\text{s!}$$

UNDERSTANDING
A* AND *F

Although a flip-flop output can go metastable *only* if D changes during the decision window, the MTBF formula does not explicitly specify how many such input changes occur. Instead, it specifies the total number of asynchronous input changes per second, a, and assumes that asynchronous input changes are uniformly distributed over the clock period. Therefore, the fraction of input changes that actually occur during the decision window is "built in" to the clock-frequency parameter f—as f increases, the fraction goes up.

If the system design is such that input changes might be clustered in the decision window rather than being uniformly distributed (as when synchronizing a slow input with a fixed but unknown phase difference from the system clock), then a useful rule of thumb is to use a frequency equal to the reciprocal of the decision window (based on published setup and hold times), times a safety margin of 10.

The only saving grace of this synchronizer at 16 MHz is that it's so lousy, we're likely to discover the problem in the engineering lab before the product ships! Thank goodness the MTBF wasn't one year.

8.9.5 Better Synchronizers

Given the poor performance of the 74LS74 as a synchronizer at moderate clock speeds, we have a couple of alternatives for building more reliable synchronizers. The simplest solution, which works for most design requirements, is simply to use a flip-flop from a faster technology. Nowadays much faster technologies are available for flip-flops, whether discrete or embedded in PLDs, FPGAs, or ASICs.

Based on published data discussed in the References, Table 8-35 lists the metastability parameters for several common logic families and devices. These numbers are very much circuit-design and IC-process dependent. Thus, unlike guaranteed logic signal levels and timing parameters, published metastability numbers can vary dramatically among different manufacturers of the same part and must be used conservatively. For example, one manufacturer's 74F74 may give acceptable metastability performance in a design while another's may not.

Table 8-35 Metastability parameters for some common devices.

Reference	Device	τ (ns)	T_o (s)	t_r (ns)
Chaney (1983)	74LS74	1.50	$4.0 \cdot 10^{-1}$	77.71
Chaney (1983)	74S74	1.70	$1.0 \cdot 10^{-6}$	66.14
Chaney (1983)	74S174	1.20	$5.0 \cdot 10^{-6}$	48.62
Chaney (1983)	74S374	0.91	$4.0 \cdot 10^{-4}$	40.86
Chaney (1983)	74F74	0.40	$2.0 \cdot 10^{-4}$	17.68
TI (1997)	74LSxx	1.35	$4.8 \cdot 10^{-3}$	63.97
TI (1997)	74Sxx	2.80	$1.3 \cdot 10^{-9}$	90.33
TI (1997)	74ALSxx	1.00	$8.7 \cdot 10^{-6}$	41.07
TI (1997)	74ASxx	0.25	$1.4 \cdot 10^{3}$	14.99
TI (1997)	74Fxx	0.11	$1.9 \cdot 10^{8}$	7.90
TI (1997)	74HCxx	1.82	$1.5 \cdot 10^{-6}$	71.55
Cypress (1997)	PALC16R8-25	0.52	$9.5 \cdot 10^{-12}$	14.22*
Cypress (1997)	PALC22V10B-20	0.26	$5.6 \cdot 10^{-11}$	7.57*
Cypress (1997)	PALCE22V10-7	0.19	$1.3 \cdot 10^{-13}$	4.38*
Xilinx (1997)	7300-series CPLD	0.29	$1.0 \cdot 10^{-15}$	5.27*
Xilinx (1997)	9500-series CPLD	0.17	$9.6 \cdot 10^{-18}$	2.30*

Note that different authors and manufacturers may specify metastability parameters differently. For example, author Chaney and manufacturer Texas Instruments measure the metastability resolution time t_r from the triggering clock edge, as in our previous subsection. On the other hand, manufacturers Cypress and Xilinx define t_r as the *additional* delay beyond a normal clock-to-output delay time t_{pd}.

The last column in the table gives a somewhat arbitrarily chosen figure of merit for each device. It is the metastability resolution time t_r required to obtain an MTBF of 1000 years when operating a synchronizer with a clock frequency of 25 MHz and with 100,000 asynchronous input changes per second. For the Cypress and Xilinx devices, their parameter values yield a value of t_r, marked with an asterisk, consistent with their own definition as introduced above.

As you can see, the 74LS74 is one of the worst devices in the table. If we replace FF1 in the 16-MHz microprocessor system of the preceding subsection with a 74ALS74, we get

$$\text{MTBF(42.5 ns)} \; = \; \frac{\exp(42.5/0.87)}{12.5 \cdot 10^{-3} \cdot 1.6 \cdot 10^7 \cdot 10^5} \; = \; 8.2 \cdot 10^{10}\,\text{s}$$

If you're satisfied with a synchronizer MTBF of about 25 centuries per system shipped, you can stop here. However, if FF2 is also replaced with a 74ALS74, the MTBF gets better, since the 'ALS74 has a shorter setup time than the 'LS74, only 10 ns. With the 'ALS74, the MTBF is about 100,000 times better:

$$\text{MTBF(52.5 ns)} \; = \; \frac{\exp(52.5/0.87)}{12.5 \cdot 10^{-3} \cdot 2 \cdot 10^7 \cdot 10^5} \; = \; 8.1 \cdot 10^{15}\,\text{s}$$

Even if we ship a million systems containing this circuit, we (or our heirs) will see a synchronizer failure only once in 240 years. Now that's job security!

Actually, the margins above aren't as large as they might seem. (How large does 240 years seem *to you*?) Most of the numbers given in Table 8-35 are *averages* and are seldom specified, let alone guaranteed, by the device manufacturer. Furthermore, calculated MTBFs are extremely sensitive to the value of τ, which in turn may depend on temperature, voltage, and the phase of the moon. So the operation of a given flip-flop in an actual system may be much worse (or much better) than predicted by our table.

For example, consider what happens if we increase the clock in our 16-MHz system by just 25%, to 20 MHz. Your natural inclination might be to think that metastability will get 25% worse, or maybe 250% worse, just to be conservative. But, if you run the numbers, you'll find that the MTBF using 'ALS74s for both FF1 and FF2 goes down from $8.1 \cdot 10^{15}$ s to just $3.7 \cdot 10^9$ s, over a million times worse! The new MTBF of about 118 years is fine for one system, but if you ship a million of them, one will fail every hour. You've just gone from generations of job security to corporate goat!

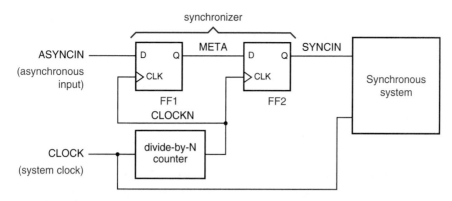

Figure 8-98
Multiple-cycle
synchronizer.

8.9.6 Other Synchronizer Designs

We promised to describe a couple of ways to build more reliable synchronizers. The first way was to use faster flip-flops, that is, to reduce the value of τ in the MTBF equation. Having said that, the second way is obvious—to *increase* the value of t_r in the MTBF equation.

For a given system clock, the best value we can obtain for t_r using the circuit of Figure 8-96 is t_{clk}, if FF2 has a setup time of 0. However, we can get values of t_r on the order of $n \cdot t_{clk}$ by using the *multiple-cycle synchronizer* circuit of Figure 8-98. Here we divide the system clock by n to obtain a slower synchronizer clock and longer $t_r = (n \cdot t_{clk}) - t_{setup}$. Usually a value of $n = 2$ or $n = 3$ gives adequate synchronizer reliability.

multiple-cycle synchronizer

In the figure, note that the edges of CLOCKN will lag the edges of CLOCK because CLOCKN comes from the Q output of a counter flip-flop that is clocked by CLOCK. This means that SYNCIN, in turn, will be delayed or skewed relative to other signals in the synchronous system that come directly from flip-flops clocked by CLOCK. If SYNCIN goes through additional combinational logic in the synchronous system before reaching its flip-flop inputs, their setup time may be inadequate. If this is the case, the solution in Figure 8-99 can be used. Here, SYNCIN is reclocked by CLOCK using FF3 to produce DSYNCIN, which will

Figure 8-99 Multiple-cycle synchronizer with deskewing.

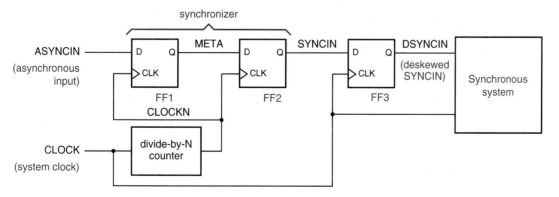

have the same timing as other flip-flop outputs in the synchronous system. Of course, the delay from CLOCK to CLOCKN must still be short enough that SYNCIN meets the setup time requirement of FF3.

In an n-cycle synchronizer, the larger the value of n, the longer it takes for an asynchronous input change to be seen by the synchronous system. This is simply a price that must be paid for reliable system operation. In typical microprocessor systems, most asynchronous inputs are for external events— interrupts, DMA requests, and so on—that need not be recognized very quickly, relative to synchronizer delays. In the time-critical area of main memory access, experienced designers use the processor clock to run the memory subsystem too, if possible. This eliminates the need for synchronizers and provides the fastest possible system operation.

At higher frequencies, the feasibility of the multiple-cycle synchronizer design shown in Figure 8-98 tends to be limited by clock skew. For this reason, *cascaded synchronizers*   rather than use a divide-by-n synchronizer clock, some designers use *cascaded synchronizers*. This design approach simply uses a cascade (shift register) of n flip-flops, all clocked with the high-speed system clock. This approach is shown in Figure 8-100.

With cascaded synchronizers, the idea is that metastability will be resolved with some probability by the first flip-flop, and failing that, with an equal probability by each successive flip-flop in the cascade. So the overall probability of failure is on the order of the nth power of the failure probability of a single-flip-flop synchronizer at the system clock frequency. While this is partially true, the MTBF of the cascade is poorer than that of a multiple-cycle synchronizer with the same delay ($n \cdot t_{clk}$). With the cascade, the flip-flop setup time t_{setup} must be subtracted from t_r, the available metastability resolution time, n times, but in a multiple-cycle design, it is subtracted only once.

PLDs that contain internal flip-flops can be used in synchronizer designs, where the two flip-flops in Figure 8-96 on page 766 are simply included in the PLD. This is very convenient in most applications, because it eliminates the need for external, discrete flip-flops. However, a PLD-based synchronizer typically has a poorer MTBF than a discrete circuit built with the same or similar

Figure 8-100 Cascaded synchronizer.

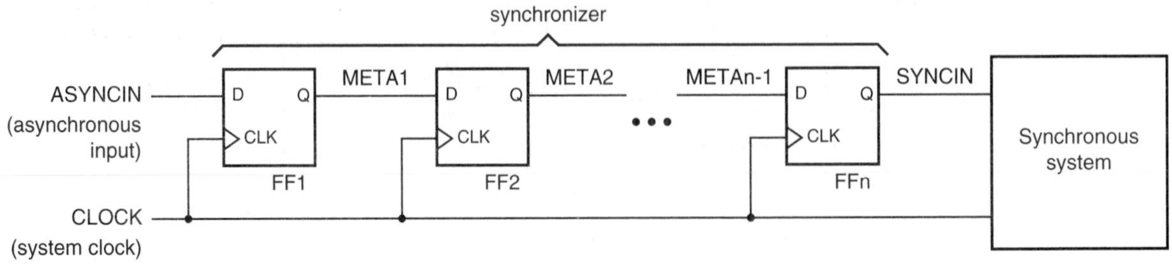

technology. This happens because each flip-flop in a PLD has a combinational logic array on its D input that increases its setup time and thereby reduces the amount of time t_r available for metastability resolution during a given system clock period t_{clk}. To maximize t_r without using special components, FF2 in Figure 8-96 should be a short-setup-time discrete flip-flop.

8.9.7 Metastable-Hardened Flip-Flops

In the late 1980s, Texas Instruments and other manufacturers developed SSI and MSI flip-flops that are specifically designed for board-level synchronizer applications. For example, the *74AS4374* was similar to the 74AS374, except *74AS4374* that each individual flip-flop was replaced with a pair of flip-flops, as shown in Figure 8-101. Each pair of flip-flops could be used as a synchronizer of the type shown in Figure 8-96, so eight asynchronous inputs could be synchronized with one 74AS4374.

The internal design of the 'AS4374 was improved to reduce τ and T_0 compared to other 74AS flip-flops, but the biggest improvement in the 'AS4374 was a greatly reduced t_{setup}. Because the entire synchronizer of Figure 8-96 is built on a single chip, there are no input or output buffers between FF1 and FF2, and t_{setup} for FF2 is only 0.5 ns. Compared to a conventional 74AS flip-flop with a 5 ns t_{setup}, and assuming that $\tau = 0.40$ ns, this improves the MTBF by a factor of exp(4.5/.40), or about 77,000.

In recent years, the move toward faster CMOS technologies and higher integration has largely obsoleted specialized parts like the 'AS4374. As you can see from the last few rows in Table 8-35 on page 769, fast PLDs and CPLDs are available with values of τ that rival the fastest discrete devices while offering the convenience of integrating synchronization with many other functions. Still, the approach used by 'AS4374 is worth emulating in FPGA and ASIC designs. That is, whenever you have control over the layout of a synchronizer circuit, it pays to locate FF1 and FF2 as close as possible to each other and to connect them with the fastest available wires, in order to maximize the setup time available for FF2.

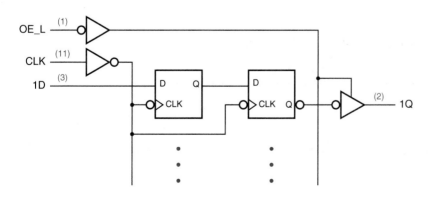

Figure 8-101
Logic diagram for the 74AS4374 octal dual-rank D flip-flop.

8.9.8 Synchronizing High-Speed Data Transfers

A very common problem in computer systems is synchronizing external data transfers with the computer system clock. A simple example is the interface between personal computer's network interface card and a 100-Mbps Ethernet link. The interface card may be connected to a PCI bus, which has a 33.33-MHz clock. Even though the Ethernet speed is an approximate multiple of the bus speed, the signal received on the Ethernet link is generated by another computer whose transmit clock is not synchronized with the receive clock in any way. Yet the interface must still deliver data reliably to the PCI bus.

Figure 8-102 shows the problem. NRZ serial data RDATA is received from the Ethernet at 100 Mbps. The digital phase-locked loop (DPLL) recovers a 100-MHz clock signal RCLK which is centered on the 100-Mbps data stream and

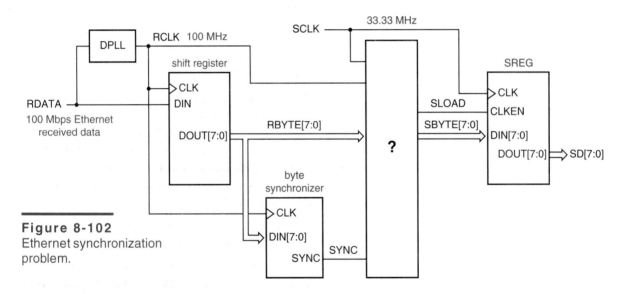

Figure 8-102
Ethernet synchronization problem.

ONE NIBBLE
AT A TIME

The explanation of 100-Mbps Ethernet reception above is oversimplified, but sufficient for discussing the synchronization problem. In reality, the received data rate is 125 Mbps, where each 4 bits of user data is encoded as a 5-bit symbol using a so-called 4B5B code. By using only 16 out of 32 possible 5-bit codewords, the 4B5B code guarantees that regardless of the user data pattern, the bit stream on the wire will have a sufficient number of transitions to allow clock recovery. Also, the 4B5B code includes a special code that is transmitted periodically to allow nibble (4-bit) and byte synchronization to be accomplished very easily.

As a result of nibble synchronization, a typical 100-Mbps Ethernet interface does not see an unsynchronized 100-MHz stream of bits. Instead, it sees an unsynchronized 25-MHz stream of nibbles. So, the details of a real 100-Mbps Ethernet synchronizer are different, but the same principles apply.

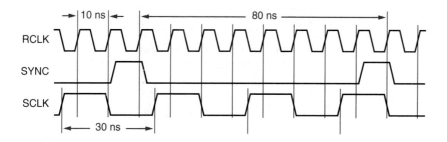

Figure 8-103
Ethernet link and
system clock timing.

allows data to be clocked bit-by-bit into an 8-bit shift register. At the same time, a byte synchronization circuit searches for special patterns in the received data stream that indicate byte boundaries. When it detects one of these, it asserts the SYNC signal and does so on every eighth subsequent RCLK tick, so that SYNC is asserted whenever the shift register contains an aligned 8-bit byte. The rest of the system is clocked by a 33.33 MHz clock SCLK. We need to transfer each aligned byte RBYTE[7:0] into a register SREG in SCLK's domain. How can we do it?

Figure 8-103 shows some of the timing. We immediately see that the byte-aligned signal, SYNC, is asserted for only 10 ns per byte. We have no hope of consistently detecting this signal with the asynchronous SCLK, whose period is a much longer 30 ns.

The strategy almost universally followed in this kind of situation is to transfer the aligned data first into a holding register HREG in the *receive* clock (RCLK) domain. This gives us a lot more time to sort things out—80 ns in this case. Thus, the "?" box in Figure 8-102 can be replaced by Figure 8-104, which shows HREG and a box marked "SCTRL." The job of SCTRL is to assert SLOAD during exactly one 30-ns SCLK period, so that the output of HREG is valid and stable for the setup and hold times of register SREG in the SCLK domain. SLOAD also serves as a "new-data available" signal for the rest of the interface, indicating that a new data byte will appear on SBYTE[7:0] during the

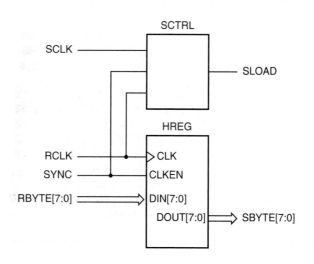

Figure 8-104
Byte holding register
and control.

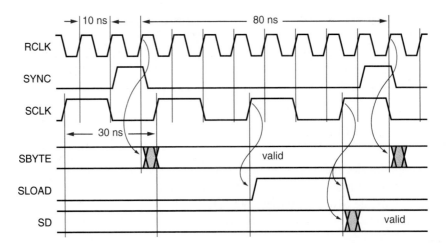

Figure 8-105
Timing for SBYTE
and possible timing
for SLOAD.

next SCLK period. Figure 8-105 shows possible timing for SLOAD based on this approach and the previous timing diagram.

Figure 8-106 is a circuit that can generate SLOAD with the desired timing. The idea is to use SYNC to set an S-R latch as a new byte becomes available. The output of this latch, NEWBYTE, is sampled by FF1 in the SCLK domain. Since NEWBYTE is not synchronized with SCLK, FF1's output SM may be metastable, but it is not used by FF2 until the next clock tick, 30 ns later. Assuming that the AND gate is reasonably fast, this gives plenty of metastability resolution time. FF2's output is the SLOAD signal. The AND gate ensures that SLOAD is only one SCLK period wide; if SLOAD is already 1, it can't be set to 1 on the next tick. This gives time for the S-R latch to be reset by SLOAD in preparation for the next byte.

A timing diagram for the overall circuit with "typical" timing is shown in Figure 8-107. Since SCLK is asynchronous to RCLK, it can have an arbitrary relationship with RCLK and SYNC. In the figure, we've shown a case where the next SCLK rising edge occurs well after NEWBYTE is set. Although the figure shows a window in which SM and SM1 could be metastable in the general case, metastability doesn't actually happen when the timing is as drawn. Later, we'll show what can happen if the SCLK edge occurs when NEWBYTE is changing.

We should make several notes about the circuit in Figure 8-106. First, the SYNC signal must be glitch free, since it controls the S input of a latch, and it must be wide enough to meet the minimum pulse-width requirement of the latch.

Figure 8-106
SCTRL circuit for
generating SLOAD.

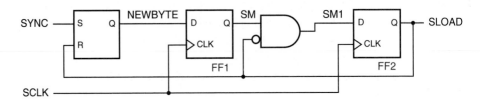

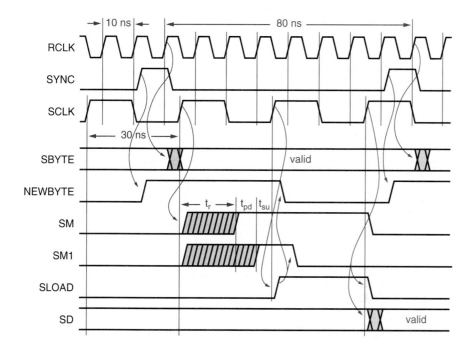

Figure 8-107
Timing for the SCTRL
circuit in Figure 8-106.

Since the latch is set on the leading edge of SYNC, we actually cheated a little; NEWBYTE may be asserted a little *before* a new byte is actually available in HREG. This is OK, because we know that it takes two SCLK periods from when NEWBYTE is sampled until SREG is loaded. In fact, we might have cheated even more if an earlier version of SYNC had been available (see Exercise 8.95).

Assuming that t_{su} is the setup time of a D flip-flop and t_{pd} is the propagation delay of the AND gate in Figure 8-106, the available metastability resolution time t_r is one SCLK period, 30 ns, minus $t_{su} + t_{pd}$, as shown in Figure 8-107. The timing diagram also shows why we can't use SM directly as the reset signal for the S-R latch. Since SM can be metastable, it could wreak havoc. For example, it could be semi-HIGH long enough to reset the latch but then fall back to LOW; in that case, SLOAD would not get set and we would miss a byte. By using instead the output of the synchronizer (SLOAD) both for the latch reset and for the load signal in the SCLK domain, we ensure that the new byte is detected and handled consistently in both clock domains.

The timing that we showed in Figure 8-107 is nominal, but we also have to analyze what happens if SCLK has a different phase relationship with RCLK and SYNC than what is shown. You should be able to convince yourself that if the SCLK edge occurs earlier, so that it samples NEWBYTE just as it's going HIGH, everything still works as before, and the data transfer just finishes a little sooner. The more interesting case is when SCLK occurs later, so that it just misses NEWBYTE as it's going HIGH and catches it one SCLK period later. This timing is shown in Figure 8-108.

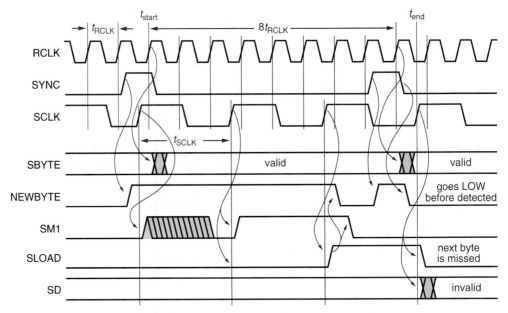

Figure 8-108 Maximum-delay timing for SCTRL circuit.

In the timing diagram, we have shown NEWBYTE going high around the same time as the SCLK edge—less than FF1's t_{su} before the edge. Thus, FF1 may not see NEWBYTE as HIGH or its output may become metastable, and it does not solidly capture NEWBYTE until one SCLK period later. Two SCLK periods after that, we get the SCLK edge that loads SBYTE into SREG.

This timing scenario is bad news, because by the time the load occurs, SBYTE is already changing to the *next* received byte. In addition, SLOAD happens to be asserted during and a little bit after the SYNC pulse for this next received byte. Thus, the latch has both S and R asserted simultaneously. If they are removed simultaneously, the latch output may become metastable. Or, as we've shown in the timing diagram, if NEWBYTE (R) is negated last, then the latch is left in the reset state, and this next received byte is never detected and loaded into the SCLK domain.

Thus, we need to analyze the maximum-delay timing case carefully to determine if a synchronizer will work properly. Figure 8-108 shows a starting reference point t_{start} for the SCTRL circuit, namely the RCLK edge on which a byte is loaded into HREG, at end of SYNC pulse. The ending reference point t_{end} is the SCLK edge on which SBYTE is loaded into SREG. The maximum delay between these two reference points, which we'll call t_{maxd}, is the sum of the following components:

$-t_{RCLK}$ Minus one RCLK period, the delay from t_{start} back to the edge on which SYNC was asserted. This number is negative because SYNC is asserted one clock tick before the tick that actually loads HREG.

t_{CQ} One flip-flop CLK-to-Q maximum delay. Assuming that SYNC is a direct flip-flop output in the RCLK domain, this is delay from the RCLK edge until SYNC is asserted.

t_{SQ} Maximum delay from S to Q in the S-R latch in Figure 8-106. This is the delay for NEWBYTE to be asserted.

t_{su} Setup time of FF1 in Figure 8-106. NEWBYTE must be asserted at or before the setup time to guarantee detection.

t_{SCLK} One SCLK period. Since RCLK and SCLK are asynchronous, there may be a delay of up to one SCLK period before the next SCLK edge comes along to sample NEWBYTE.

t_{SCLK} After NEWBYTE is detected by FF1, SLOAD is asserted on the next SCLK tick.

t_{SCLK} After SLOAD is asserted, SBYTE is loaded into SREG on the next SCLK tick.

Thus, $t_{maxd} = 3t_{SCLK} + t_{CQ} + t_{SQ} + t_{su} - t_{RCLK}$. A few other parameters must be defined to complete the analysis:

t_h The hold time of SREG.

$t_{CQ(min)}$ The minimum CLK-to-Q delay of HREG, conservatively assumed to be 0.

t_{rec} The recovery time of the S-R latch, the minimum time allowed between negating S and negating R (see box on page 537).

To be loaded successfully into SREG, SBYTE must be remain valid until at least time $t_{end} + t_h$. The point at which SBYTE changes and becomes invalid is 8 RCLK periods after t_{start}, plus $t_{CQ(min)}$. Thus, for proper circuit operation we must have

$$t_{end} + t_h \leq t_{start} + 8t_{RCLK}$$

For the maximum-delay case, we substitute $t_{end} = t_{start} + t_{maxd}$ into this relation and subtract t_{start} from both sides to obtain

$$t_{maxd} + t_h \leq 8t_{RCLK}$$

Substituting the value of t_{maxd} and rearranging, we obtain

$$3t_{SCLK} + t_{CQ} + t_{SQ} + t_{su} + t_h \leq 9t_{RCLK} \tag{8-1}$$

as the requirement for correct circuit operation. Too bad. Even if we assume very short component delays (t_{CQ}, t_{SQ}, t_{su}, t_h), we know that $3t_{SCLK}$ (90 ns) plus anything is going to be more than $9t_{RCLK}$ (also 90 ns). So this design will never work properly in the maximum-delay case.

Even if the load-delay analysis gave a good result, we would still have to consider the requirements for proper operation of the SCTRL circuit itself. In particular, we must ensure that when the SYNC pulse for the next byte occurs, it is not negated until time t_{rec} after SLOAD for the previous byte was negated. So, another condition for proper operation is

$$t_{end} + t_{CQ} + t_{rec} \leq t_{start} + 8t_{RCLK} + t_{CQ(min)}$$

Substituting and simplifying as before, we get another requirement that isn't met by our design:

$$3t_{SCLK} + 2t_{CQ} + t_{SQ} + t_{su} + t_{rec} \leq 9t_{RCLK} \qquad (8\text{-}2)$$

There are several ways that we can modify our design to satisfy the worst-case timing requirements. Early in our discussion, we noted that we "cheated" by asserting SYNC one RCLK period before the data in HREG is valid, and that we actually might get away with asserting SYNC even sooner. Doing this can help us meet the maximum-delay requirement, because it reduces the "$8t_{RCLK}$" term on the righthand side of the relations. For example, if we asserted SYNC two RCLK periods earlier, we would reduce this term to "$6t_{RCLK}$". However, there's no free lunch; we can't assert SYNC arbitrarily early. We must also consider a *minimum-delay* case, to ensure that the new byte is actually available in HREG when SBYTE is loaded into SREG. The minimum delay t_{maxd} between t_{start} and t_{end} is the sum of the following components:

$-nt_{RCLK}$ Minus n RCLK periods, the delay from t_{start} back to the edge on which SYNC was asserted. In the original design, $n = 1$.

$t_{CQ(min)}$ This is the minimum delay from the RCLK edge until SYNC is asserted, conservatively assumed to be 0.

t_{SQ} This is the delay for NEWBYTE to be asserted, again assumed to be 0.

$-t_{h}$ Minus the hold time of FF1 in Figure 8-106. NEWBYTE might be asserted at the end of the hold time and still be detected.

$0t_{SCLK}$ Zero times the SCLK period. We might get "lucky" and have the SCLK edge come along just as the hold time of FF1 is ending.

t_{SCLK} A one-SCLK-period delay to asserting SLOAD, as before.

t_{SCLK} A one-SCLK-period delay to loading SBYTE into SREG, as before.

In other words, $t_{mind} = 2t_{SCLK} - t_{h} - nt_{RCLK}$.

For this case, we must ensure that the new byte has propagated to the output of HREG when the setup time window of SREG begins, so we must have

$$t_{end} - t_{su} \geq t_{start} + t_{co},$$

where t_{co} is the maximum clock-to-output delay of HREG. Substituting $t_{end} = t_{start} + t_{mind}$ and subtracting t_{start} from both sides, we get

$$t_{mind} - t_{su} \geq t_{co}.$$

Substituting the value of t_{mind} and rearranging, we get the final requirement,

$$2t_{SCLK} - t_{h} - t_{su} - t_{co} \geq nt_{RCLK} \qquad (8\text{-}3)$$

If, for example, t_h, t_{su}, and t_{co} are 10 ns each, the maximum value of n is 3; we can't generate SYNC more than two clock ticks before its original position in Figure 8-108. This may or may not be enough to solve the maximum-delay problem, depending on other delay values; this is explored for a particular set of components in Exercise 8.95.

Moving the SYNC pulse earlier may not give enough delay improvement or may not be an available option in some systems. An alternative solution that can always be made to work is to increase the time between successive data transfers from one clock domain to the other. We can always do this because we can always transfer more bits per synchronization. In the Ethernet-interface example, we could collect 16 bits at a time in the RCLK domain and transfer 16 bits at a time to the SCLK domain. This changes the previously stated $8t_{RCLK}$ terms to $16t_{RCLK}$, providing a lot more margin for the maximum-delay timing requirements. Once 16 bits have been transferred into the SCLK domain, we can still break them into two 8-bit chunks if we need to process the data a byte at a time.

It may also be possible to improve performance by modifying the design of the SCTRL circuit. Figure 8-109 shows a version where SLOAD is generated directly by the flip-flop that samples NEWBYTE. In this way, SLOAD appears one SCLK period sooner than in our original SCTRL circuit. Also, the S-R latch is cleared sooner. This circuit works only if a couple of key assumptions are true:

1. A reduced metastability resolution time for FF1 is acceptable, equal to the time that SCLK is HIGH. Metastability must be resolved before SCLK goes LOW, because that's when the S-R latch gets cleared if SLOAD is HIGH.

2. The setup time of SREG's CLKEN input (Figure 8-102) is less than or equal to the time that SCLK is LOW. Under the previous assumption, the SLOAD signal applied to CLKEN might be metastable until SCLK goes LOW.

3. The time that SCLK is LOW is long enough to generate a reset pulse on RNEW that meets the minimum pulse-width requirement of the S-R latch.

Note that these behaviors makes proper circuit operation dependent on the duty cycle of SCLK. If SCLK is relatively slow and its duty cycle is close to 50%, this circuit generally works fine. But if SCLK is too fast or has a very small, very large, or unpredictable duty cycle, the original circuit approach must be used.

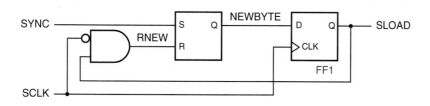

Figure 8-109
Half-clock-period SCTRL circuit for generating SLOAD.

All of these synchronization schemes require the clock frequencies to be within a certain range of each other for proper circuit operation. This must be considered for testing, where the clocks are usually run slower, and for upgrades, where one or both clocks may run faster. For example, in the Ethernet interface example, we wouldn't change the frequency of standard 100-Mbps Ethernet, but we might upgrade the PCI bus from 33 to 66 MHz.

The problems caused by clock-frequency changes can be subtle. To get a better handle on what can go wrong, it's useful to consider how a synchronizer works (or doesn't work!) if one clock frequency changes by a factor of 10 or more.

For example what happens to the synchronizer timing in Figure 8-107 if we change RCLK from 100 MHz to 10 MHz? At first glance, it would seem that all is well, since a byte now arrives only once every 800 ns, giving much more time for the byte to be transferred into the SCLK domain. Certainly, Eq. (8-1) on page 779 and Eq. (8-2) on page 780 are satisfied with much more margin. However, Eq. (8-3) is no longer satisfied, unless we reduce n to zero! This could be accomplished by generating SYNC one RCLK tick later than is shown in Figure 8-107.

But even with this change, there's *still* a problem. Figure 8-110 shows the new timing, including the later SYNC pulse. The problem is that the SYNC pulse is now 100 ns long. As before, NEWBYTE (the output of the S-R latch in

Figure 8-110 Synchronizer timing with slow (10-MHz) RCLK.

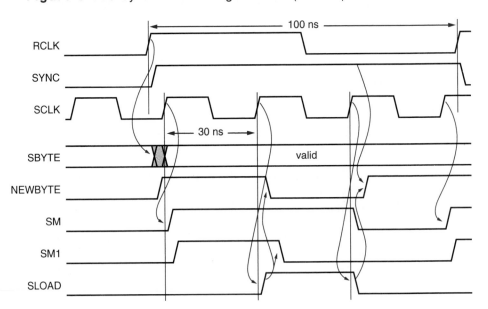

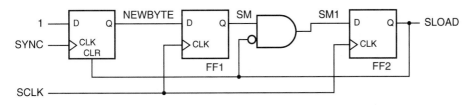

Figure 8-111
Synchronizer with
edge-triggered
SYNC detection.

Figure 8-106 on page 776) is set by SYNC and is cleared by SLOAD. The problem is that when SLOAD goes away, SYNC is still asserted, as shown in the new timing diagram. Thus, the new byte will be detected and transferred twice!

The solution to the problem is to detect only the leading edge of SYNC, so that the circuit is not sensitive to the length of the SYNC pulse. A common way of doing this is to replace the S-R latch with an edge-triggered D flip-flop, as shown in Figure 8-111. The leading edge of SYNC sets the flip-flop, while SLOAD is used as an asynchronous clear as before.

The circuit in Figure 8-111 solves the slow-RCLK problem, but it also changes the derivation of Eqns. 8-1 through 8-3 and may make timing more constrained in some areas (see Exercise 8.96). Another disadvantage that this circuit cannot be realized in a typical PLD, which has all flip-flops controlled by the same clock; instead, a discrete flip-flop must be used to detect SYNC.

After reading almost ten pages to analyze just one "simple" example, you should have a strong appreciation of the difficulty of correct synchronization-circuit design. Several guidelines can help you:

- Minimize the number of different clock domains in a system.
- Clearly identify all clock boundaries and provide clearly identified synchronizers at those boundaries.
- Provide sufficient metastability resolution time for each synchronizer so that synchronizer failure is rare, much more unlikely than other hardware failures.
- Analyze synchronizer behavior over a range of timing scenarios, including faster and slower clocks that might be applied as a result of system testing or upgrades.
- Simulate system behavior over a wide range of timing scenarios as well.

The last guideline above is a catch-all for modern digital designers, who usually rely on sophisticated, high-speed logic simulators to find their bugs. But it's not a substitute for following the first four guidelines. Ignoring them can lead to problems that cannot be detected by a typical, small number of simulation scenarios. Of all digital circuits, synchronizers are the ones for which it's most important to be "correct by design"!

References

Probably the first comprehensive set of examples of sequential MSI parts and applications appeared in *The TTL Applications Handbook*, edited by Peter Alfke and Ib Larsen (Fairchild Semiconductor, 1973). This highly practical and informative book was invaluable to this author and to many others who developed digital design curricula in the 1970s.

Another book that emphasizes design with larger-scale combinational and sequential logic functions is Thomas R. Blakeslee's *Digital Design with Standard MSI and LSI,* second edition, (Wiley, 1979). Blakeslee's coverage of the concept of space/time trade-offs is excellent, and he also one of the first to introduce the microprocessor as "a universal logic circuit."

Moving quickly from the almost forgotten to the yet-to-be discovered, manufacturers' Web sites are an excellent source of information on digital design practices. For example, a comprehensive discussion of bus-hold circuits can be found in Texas Instruments' "Implications of Slow or Floating CMOS Inputs" (publ. SCBA004B, December 1997), available on TI's web site at `www.ti.com`. Another discussion appears in Fairchild Semiconductor's "Designing with Bushold" (*sic*, publ. AN-5006, April 1999), at `www.fairchildsemi.com`.

Announcements and data sheets for all kinds of new, improved MSI and larger parts can also be found on the Web. Following a certain automobile manufacturer's proclamation in the 60s and then again in the late 90s that "wider is better," logic manufacturers have also introduced "wide-bus" registers, drivers, and transceivers that cram 16, 18, or even 32 bits of function into a high-pin-count surface-mount package. Descriptions of many such parts can be found at the Texas Instruments Web site (search for "widebus"). Other sites with a variety of logic data sheets, application notes, and other information include Motorola (`www.mot.com`), Fairchild Semiconductor (`www.fairchildsemi.com`), and Philips Semiconductor (`www.philipslogic.com`).

The field of logic design is fast moving, so much so that sometimes I wish that I wrote fiction, so that I wouldn't have to revise the "practices" discussions in this book every few years. Luckily for me, this book does cover some unchanging theoretical topics (a.k.a. "principles"), and this chapter is no exception. Logic hazards have been known since at least the 1950s, and function hazards were discussed by Edward J. McCluskey in *Logic Design Principles* (Prentice Hall, 1986). Galois fields were invented centuries ago, and their applications to error-correcting codes, as well as to the LFSR counters of this chapter, are described in introductory books on coding theory, including *Error-Control Techniques for Digital Communication* by A. M. Michelson and A. H. Levesque (Wiley-Interscience, 1985). A mathematical theory of state-machine decomposition has been studied for years; Zvi Kohavi devotes a chapter to the topic in his classic book *Switching and Finite Automata Theory*, second edition (McGraw-Hill, 1978). But let us now return to the less esoteric.

As we mentioned in Section 8.4.5, some PLDs and CPLDs contain XOR structures that allow large counters to be designed without a large number of product terms. This requires a somewhat deeper understanding of counter excitation equations, as described in Section 10.5 of the second edition of this book. This material can also be found on the web at www.ddpp.com.

The general topics of clock skew and multiphase clocking are discussed in McCluskey's *Logic Design Principles*, while an illuminating discussion of these topics as applied to VLSI circuits can be found in *Introduction to VLSI Systems* by Carver Mead and Lynn Conway (Addison-Wesley, 1980). Mead and Conway also provide an introduction to the important topic of *self-timed systems* that eliminate the system clock, allowing each logic element to proceed at its own rate. To give credit where credit is due, we note that all of the Mead and Conway material on system timing, including an outstanding discussion of metastability, appears in their Chapter 7 written by Charles L. Seitz.

self-timed systems

Thomas J. Chaney spent decades studying and reporting on the metastability problem. One of his more important works, "Measured Flip-Flop Responses to Marginal Triggering" (*IEEE Trans. Comput.*, Vol. C-32, No. 12, December 1983, pp. 1207–1209), reports some of the results that we showed in Table 8-35.

For the mathematically inclined, Lindsay Kleeman and Antonio Cantoni have written "On the Unavoidability of Metastable Behavior in Digital Systems" (*IEEE Trans. Comput.*, Vol. C-36, No. 1, January 1987, pp. 109–112); the title says it all. The same authors posed the question, "Can Redundancy and Masking Improve the Performance of Synchronizers?" (*IEEE Trans. Comput.*, Vol. C-35, No. 7, July 1986, pp. 643–646). Their answer in that paper was "No," but a response from a reviewer caused them to change their minds to "Maybe." Obviously, they've gone metastable themselves! Having two authors and a reviewer didn't improved their performance, so the obvious answer to their original question is "No"! In any case, Kleeman and Antonio's papers provide a good set of pointers to mainstream scholarly references on metastability.

The most comprehensive set of early references on metastability (not including Greek philosophers or Devo) is Martin Bolton's "A Guided Tour of 35 Years of Metastability Research" (*Proc. Wescon 1987*, Session 16, "Everything You Might Be Afraid to Know about Metastability," Wescon Session Records, www.wescon.com, 8110 Airport Blvd., Los Angeles, CA 90045).

In recent years, as system clock speeds have steadily increased, many IC manufacturers have become much more conscientious about measuring and publishing the metastability characteristics of their devices, often on the Web. Texas Instruments (www.ti.com) provides a very good discussion including test circuits and measured parameters for ten different logic families in "Metastable Response in 5-V Logic Circuits" by Eilhard Haseloff (TI pub. SDYA006, 1997). Cypress Semiconductor (www.cypress.com) publishes an application note, "Are Your PLDs Metastable?" (1997) that is an excellent reference including some history (going back to 1952!), an analog circuit analysis of metastability, test and measurement circuits, and metastability parameters for Cypress PLDs. Another recent note is "Metastability Considerations" from Xilinx Corporation (www.xilinx.com, publ. XAPP077, 1997), which gives measured parameters for their XC7300 and XC9500 families of CPLDs. Of particular interest is the clever circuit and methodology that allows them to count metastable events *inside* the device, even though metastable waveforms are not observable on external pins.

Most digital design textbooks now give good coverage to metastability, prompted by the existence of metastability in real circuits and perhaps also by competition—since 1990, the textbook you're reading has been promoting the topic by introducing metastability in its earliest coverage of sequential circuits. On the analog side of the house, Howard Johnson and Martin Graham provide a nice introduction and a description of how to observe metastable states in their *High-Speed Digital Design: A Handbook of Black Magic* (Prentice Hall, 1993).

Drill Problems

8.1 Suppose that in Table 8-2, the second RAM bank (RAMCS1) is decoded instead using the expression ((ABUS >= 0x010) & (ABUS < 0x020)). Does this yield the same results as the original expression, (ABUS == RAMBANK0)? Explain.

8.2 Determine the number of fuses in each of the PAL devices in Table 8-9.

8.3 How many fuses are contained in the 16V8 as described in the text? (The commercial device has additional fuses, not described in the text, for user-specific information and security.)

8.4 How many fuses are contained in the 22V10 as described in the text? (The commercial device has additional fuses, not described in the text, for user-specific information and security.)

8.5 Determine f_{max} with external feedback for all of the devices in Table 8-10.

8.6 Determine f_{max} with internal feedback for all of the GAL devices in Table 8-10.

8.7 Write an ABEL program for a 16V8 that gives it exactly the same functionality as a 74x374.

8.8 Write an ABEL program for a GAL16V8 or GAL20V8 that gives it exactly the same functionality as a 74x377.

8.9 Compare the propagation delays from AVALID to a chip-select output for the two decoding approaches shown in Figures 8-15 and 8-16. Assume that 74FCT373 latches and 10-ns 16V8C devices are used in both designs. Repeat for the delay from ABUS to a chip-select output.

8.10 What would happen if you replaced the edge-triggered D flip-flops in Figure 7-38 with D latches?

8.11 Modify the ABEL program in Table 8-12 to perform the function of a 74X162 decade counter.

8.12 Modify the ABEL program in Table 8-12 to perform the function of a 74X169 up/down counter. Does it still fit in a 16V8?

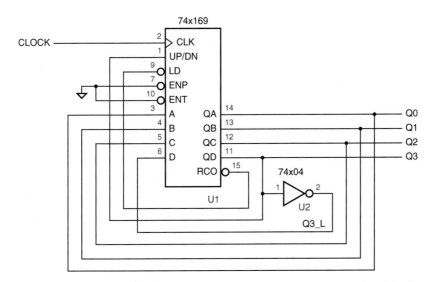

Figure X8.13

8.13   What is the counting sequence of the circuit shown in Figure X8.13?

8.14   What is the behavior of the counter circuit of Figure 8-36 if it is built using a 74x161 instead of a 74x163?

8.15   What is the behavior of the counter circuit of Figure 8-36 if the bottom input of U2 is connected to Q0 instead of Q1?

8.16   A 74x163 counter is hooked up with inputs ENP, ENT, A, and D always HIGH, inputs B and C always LOW, input LD_L = (QB · QC)', and input CLR_L = (QC · QD)'. The CLK input is hooked up to a free-running clock signal. Draw a logic diagram for this circuit. Assuming that the counter starts in state 0000, write the output sequence on QD QC QB QA for the next 15 clock ticks.

8.17   Determine the widths of the glitches shown in Figure 8-43 on the Y2_L output of a 74x138 decoder, assuming that the '138 is internally structured as shown in Figure 5-37(a) on page 358, and that each internal gate has a delay of 10 ns.

8.18   Starting with state 0001, write the sequence of states for a 4-bit LFSR counter designed according to Figure 8-68 and Table 8-21.

8.19   Calculate the minimum clock period of the data unit in Figure 8-82. Use the maximum propagation delays given in Table 5-3 for LS-TTL combinational parts. Unless you can get the real numbers from a TTL data book, assume that all registers require a 10-ns minimum setup time on inputs and have a 20-ns propagation delay from clock to outputs. Indicate any assumptions you've made about delays in the control unit.

8.20   Calculate the MTBF of a synchronizer built according to Figure 8-96 using 74F74s, assuming a clock frequency of 25 MHz and an asynchronous transition rate of 1 MHz. Assume the setup time of an 'F74 is 5 ns and the hold time is zero.

8.21   Calculate the MTBF of the synchronizer shown in Figure X8.21, assuming a clock frequency of 25 MHz and an asynchronous transition rate of 1 MHz. Assume that the setup time t_{setup} and the propagation delay t_{pd} from clock to Q or QN in a 74ALS74 are both 10 ns.

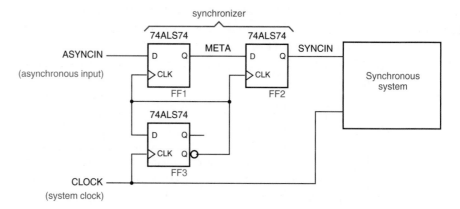

Figure X8.21

Exercises

8.22   What does the TTL Data Book have to say about momentarily shorting the outputs of a gate to ground as we do in the switch debounce circuit of Figure 8-5?

8.23   Investigate the behavior of the switch debounce circuit of Figure 8-5 if 74HCT04 inverters are used; repeat for 74AC04 inverters.

8.24   Suppose you are asked to design a circuit that produces a debounced logic input from an SPST (single-pole, *single*-throw) switch. What inherent problem are you faced with?

8.25   Explain why CMOS bus holder circuits don't work well on three-state buses with TTL devices attached. (*Hint:* Consider TTL input characteristics.)

8.26   Write a single VHDL program that combines the address latch and latching decoder of Figure 8-16 and Table 8-2. Use the signal name LA[19:0] for the latched address outputs.

8.27   Design a 4-bit ripple counter using four D flip-flops and no other components.

8.28   What is the maximum propagation delay from clock to output for the 4-bit ripple counter of Exercise 8.27 using 74HCT flip-flops? Repeat, using 74AHCT and 74LS74 flip-flops.

8.29   Design a 4-bit ripple *down* counter using four D flip-flops and no other components.

8.30   What limits the maximum counting speed of a ripple counter, if you don't insist on being able to read the counter value at all times?

8.31   Based on the design approach in Exercise 8.27 and the answer to Exercise 8.30, what is the maximum counting speed (frequency) for a 4-bit ripple counter using 74HCT flip-flops? Repeat, using 74AHCT and 74LS74 flip-flops.

8.32   Write a formula for the maximum clock frequency of the synchronous serial binary counter circuit in Figure 8-28. In your formula, let t_{TQ} denote the propagation delay from T to Q in a T flip-flop, t_{setup} the setup time of the EN input to the rising edge of T, and t_{AND} the delay of an AND gate.

8.33   Repeat Exercise 8.32 for the synchronous parallel binary counter circuit shown in Figure 8-29, and compare results.

8.34   Repeat Exercise 8.32 for an n-bit synchronous serial binary counter.

8.35   Repeat Exercise 8.32 for an n-bit synchronous parallel binary counter. Beyond what value of n is your formula no longer valid?

8.36   Using a 74x163 4-bit binary counter, design a modulo-11 counter circuit with the counting sequence 3, 4, 5, …, 12, 13, 3, 4, ….

8.37   Look up the internal logic diagram for a 74x162 synchronous decade counter in a data book, and write its state table in the style of Table 8-11, including its counting behavior in the normally unused states 10–15.

8.38   Devise a cascading scheme for 74x163s, analogous to the synchronous parallel counter structure of Figure 8-29, such that the maximum counting speed is the same for any counter with up to 36 bits (nine '163s). Determine the maximum counting speed using worst-case delays from a manufacturer's data book for the '163s and any SSI components used for cascading.

8.39   Design a modulo-129 counter using two 74x163s and a single inverter.

8.40   Write an ABEL program for an 8-bit module-N counter with load input using a PAL22v10, where the value of N is specified by a constant N in the program.

8.41   Design a clocked synchronous circuit with four inputs, N3, N2, N1, and N0, that represent an integer N in the range 0–15. The circuit has a single output Z that is asserted for exactly N clock ticks during any 16-tick interval (assuming that N is held constant during the interval of observation). (*Hints:* Use combinational logic with a 74x163 set up as a free-running divide-by-16 counter. The ticks in which Z is asserted should be spaced as evenly as possible, that is, every second tick when $N = 8$, every fourth when $N = 4$, and so on.)

8.42   Modify the circuit of Exercise 8.41 so that Z produces N *transitions* in each 16-tick interval. The resulting circuit is called a *binary rate multiplier* and was once sold as a TTL MSI part, the 7497. (*Hint:* Gate the clock with the level output of the previous circuit.)

binary rate multiplier

8.43   Repeat Exercises 8.41 and 8.42 using an 8-bit input N7..N0, and realize the circuit using an ABEL program for a single PAL22V10.

8.44   Repeat Exercises 8.41 and 8.42 using an 8-bit input N7..N0, and describe the desigbn using a behavioral VHDL program.

8.45   A digital designer (the author!) was asked at the last minute to add new functionality to a PCB that had room for just one more 16-pin MSI IC. The PCB already had a 16-MHz clock signal, MCLK, and a spare microprocessor-controlled select signal, SEL. The designer was asked to provide a new clock signal, UCLK, whose frequency would be 8 MHz or 4 MHz depending on the value of SEL. To make things worse, the PCB had no spare SSI gates, and UCLK was required to have a 50% duty cycle at both frequencies. It took the designer about five minutes to come up with a circuit. Now it's your turn to do the same. (*Hint:* The designer had long considered the 74x163 to be the fundamental building block of tricky sequential-circuit design.)

8.46   Design a modulo-16 counter, using one 74x169 and at most one SSI package, with the following counting sequence: 7, 6, 5, 4, 3, 2, 1, 0, 8, 9, 10, 11, 12, 13, 14, 15, 7, ….

8.47   Write an ABEL program for an 8-bit counter that realizes a counting sequence similar to the one in Exercise 8.46.

8.48   Design a binary up/down counter for the elevator controller in a 20-story building, using a single 16V8. The counter should have enable and up/down control inputs. It should stick at state 1 when counting down, stick at state 21 when counting up, and skip state 13 in either mode. Draw a logic diagram and write ABEL equations for your design.

8.49   Repeat the preceding exercise using VHDL.

8.50   Write a VHDL program for an n-bit counter that realizes a counting sequence similar to the one in Exercise 8.46. Write the program in such a way that the size of the counter can be changed by changing the value of a single constant N.

8.51   Modify the VHDL program in Table 8-14 so that the type of ports D and Q is STD_LOGIC_VECTOR, including conversion functions as required.

8.52   Modify the program in Table 8-16 to use structural VHDL, so it conforms exactly to the circuit in Figure 8-45, including the signal names shown in the figure. Define and use any of the following entities that don't already exist in your VHDL library: AND2, INV, NOR2, OR2, XNOR2, Vdffqqn.

8.53   Modify the program in Table 8-17 to use VHDL's generic statement, so that the counter size can be changed using the generic definition.

8.54   Design a parallel-to-serial conversion circuit with eight 2.048-Mbps, 32-channel serial links and a single 2.048-MHz, 8-bit, parallel data bus that carries 256 bytes per frame. Each serial link should have the frame format defined in Figure 8-55. Each serial data line SDATAi should have its own sync signal SYNCi; the sync pulses should be staggered so that SYNCi + 1 has a pulse one tick after SYNCi.

8.55   Show the timing of the parallel bus and the serial links, and write a table or formula that shows which parallel-bus timeslots are transmitted on which serial links and timeslots. Draw a logic diagram for the circuit using MSI parts from this chapter; you may abbreviate repeated elements (e.g., shift registers), showing only the unique connections to each one.

8.56   Repeat Exercise 8.54, assuming that all serial data lines must reference their data to a single, common SYNC signal. How many more chips does this design require?

8.57   Show how to enhance the serial-to-parallel circuit of Exercise 8-57 so that the byte received in each timeslot is stored in its own register for 125 μs, until the next byte from that timeslot is received. Draw the counter and decoding logic for 32 timeslots in detail, as well as the parallel data registers and connections for timeslots 31, 0, and 1. Also draw a timing diagram in the style of Figure 8-58 that shows the decoding and data signals associated with timeslots 31, 0, and 1.

8.58   Suppose you are asked to design a serial computer, one that moves and processes data one bit at a time. The first decision you must make is which bit to transmit and process first, the LSB or the MSB. Which would you choose, and why?

8.59   Design an 8-bit self-correcting ring counter whose states are 11111110, 11111101, ..., 01111111, using only two SSI/MSI packages.

8.60   Design two different 2-bit, 4-state counters, where each design uses just a single 74x74 package (two edge-triggered D flip-flops) and no other gates.

8.61   Design a 4-bit Johnson counter and decoding for all eight states using just four flip-flops and eight gates. Your counter need not be self-correcting.

8.62   Prove that an even number of shift-register outputs must be connected to the odd-parity circuit in an n-bit LFSR counter if it generates a maximum-length sequence. (Note that this is a necessary but not a sufficient requirement. Also, although Table 8-21 is consistent with what you're supposed to prove, simply quoting the table is not a proof!)

8.63   Prove that X0 must appear on the righthand side of any LFSR feedback equation that generates a maximum-length sequence. (*Note:* Assume the LFSR bit ordering and shift direction are as given in the text; that is, the LFSR counter shifts right, toward the X0 stage.)

8.64   Suppose that an n-bit LFSR counter is designed according to Figure 8-68 and Table 8-21. Prove that if the odd-parity circuit is changed to an even-parity circuit, the resulting circuit is a counter that visits $2^n - 1$ states, including all of the states except $11...11$.

8.65   Find a feedback equation for a 3-bit LFSR counter, other than the one given in Table 8-21, that produces a maximum-length sequence.

8.66   Given an n-bit LFSR counter that generates a maximum-length sequence ($2^n - 1$ states), prove that an extra XOR gate and an $n - 1$ input NOR gate connected as suggested in Figure 8-69 produce a counter with 2^n states.

8.67   Prove that a sequence of 2^n states is still obtained if a NAND gate is substituted for a NOR above, but that the state sequence is different.

8.68   Design an iterative circuit for checking the parity of a 16-bit data word with a single even-parity bit. Does the order of bit transmission matter?

8.69   Modify the shift-register program in Table 8-23 to provide an asynchronous clear input using a 22V10.

8.70   Write an ABEL program that provides the same functionality as a 74x299 shift register. Show how to fit this function into a single 22V10, or explain why it does not fit.

8.71   Determine the number of product terms required for each output of the RING8 PLD in Table 8-25. Does it fit in a 16R8 or 16V8R?

8.72   In what situations do the ABEL programs in Tables 8-26 and 8-27 give different operational results?

8.73   Modify the ABEL program in Table 8-26 so that the phases are always at least two clock ticks long, even if RESTART is asserted at the beginning of a phase. RESET should still take effect immediately.

8.74   Repeat the preceding exercise for the program in Table 8-27.

8.75   Suppose the timing generator of Table 8-26 is used to control a dynamic memory system, such that all six phases must be completed to read or write the memory. If the timing generator is reset during a write operation without completing all six phases, the memory contents may be corrupted. Modify the equations in Table 8-26 to avoid this problem.

8.76 A student proposed to create the timing waveforms of Figure 8-72 by starting with the ABEL program in Table 8-27 and changing the encoding of each of states P1F, P2F, ... , P6F so that the corresponding phase output is 1 instead of 0, so that the phase output is 0 only during the second tick of each phase, as required. Is this a good approach? Comment on the results produced by the ABEL compiler when you try this.

8.77 The output waveforms produced by the ABEL programs in Tables 8-29 and 8-30 are not identical when the RESTART and RUN inputs are changed. Explain the reason for this, and then modify the program in Table 8-30 so that its behavior matches that of Table 8-29.

8.78 The ABEL ring-counter implementation in Table 8-26 is not self-synchronizing. For example, describe what happens if the outputs [P1_L..P6_L] are initially all 0, and the RUN input is asserted without ever asserting RESET or RESTART. What other starting states exhibit this kind of non-self-synchronizing behavior? Modify the program so that it *is* self-synchronizing.

8.79 Repeat the preceding exercise for the VHDL ring-counter implementation in Table 8-33.

8.80 Design an iterative circuit with one input B_i per stage and two boundary outputs X and Y such that X = 1 if at least two B_i inputs are 1 and Y = 1 if at least two *consecutive* B_i inputs are 1.

8.81 Design a combination-lock machine according to the state table of Table 7-14 on page 582 using a single 74x163 counter and combinational logic for the LD_L, CLR_L, and A–D inputs of the '163. Use counter values 0–7 for states A–H.

8.82 Write an ABEL program corresponding to the state diagram in Figure 8-84 for the multiplier control unit.

8.83 Write a VHDL program corresponding to the state diagram in Figure 8-84 for the multiplier control unit.

8.84 Write a VHDL program that performs with the same inputs, outputs, and functions as the multiplier data unit in Figure 8-82.

8.85 Write a VHDL program that combines the programs in the two preceding exercises to form a complete 8-bit shift-and-add multiplier.

8.86 The text stated that the designer need not worry about any timing problems in the synchronous design of Figure 8-83. Actually, there is one slight worry. Look at the timing specifications for the 74x377 and discuss.

8.87 Determine the minimum clock period for the shift-and-add multiplier circuit in Figure 8-83, assuming that the state machine is realized in a single GAL16V8-20 and that the MSI parts are all 74LS TTL. Use worst-case timing information from the tables in this book. For the '194, t_{pd} from clock to any output is 26 ns, and t_s is 20 ns for serial and parallel data inputs and 30 ns for mode-control inputs.

8.88 Design a data unit and a control-unit state machine for multiplying 8-bit two's-complement numbers using the algorithm discussed in Section 2.8.

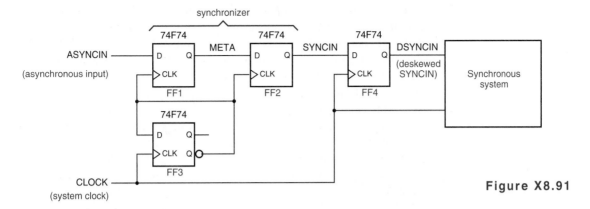

Figure X8.91

8.89 Design a data unit and control-unit state machine for dividing 8-bit unsigned numbers using the shift-and-subtract algorithm discussed in Section 2.9.

8.90 Suppose that the SYNCIN signal in Drill 8.21 is connected to a combinational circuit in the synchronous system, which in turn drives the D inputs of 74ALS74 flip-flops that are clocked by CLOCK. What is the maximum allowable propagation delay of the combinational logic?

8.91 The circuit in Figure X8.91 includes a deskewing flip-flop so that the synchronized output from the multiple-cycle synchronizer is available as soon as possible after the edge of CLOCK. Ignoring metastability considerations, what is the maximum frequency of CLOCK? Assume that for a 74F74, $t_{\text{setup}} = 5$ ns and $t_{\text{pd}} = 7$ ns.

8.92 Using the maximum clock frequency determined in Exercise 8.91, and assuming an asynchronous transition rate of 4 MHz, determine the synchronizer's MTBF.

8.93 Determine the MTBF of the synchronizer in Figure X8.91, assuming an asynchronous transition rate of 4 MHz and a clock frequency of 40 MHz, which is less than the maximum determined in Figure X8.91. In this situation, "synchronizer failure" really occurs only if DSYNCIN is metastable. In other words, SYNCIN may be allowed to be metastable for a short time, as long as it doesn't affect DSYNCIN. This yields a better MTBF.

8.94 Look up U.S. patent number 4,999,528, "Metastable-proof flip-flop," and describe why it doesn't always work as advertised. (*Hints:* Patents can be found at www.patents.ibm.com. There's enough information in this patent's abstract to figure out how the circuit can fail.)

8.95 In the synchronization circuit of Figures 8-102, 8-104, and 8-106, you can reduce the delay of the transfer of a byte from the RCLK domain to the SCLK domain if you use an earlier version of the SYNC pulse to start the synchronizer. Assuming that you can generate SYNC during any bit of the received byte, which bit should you use to minimize the delay? Also determine whether your solution satisfies the maximum-delay requirements for the circuit. Assume that all the components have 74AHCT timing and that the S-R latch is built from a pair of cross-coupled NOR gates, and show a detailed timing analysis for your answers.

8.96   Instead of using a latch in the synchronization control circuit of Figure 8-106, some applications use an edge-triggered D flip-flop as shown in Figure 8-111. Derive the maximum-delay and minimum-delay requirements for this circuit, corresponding to Eqs. (8-1) through (8-3), and discuss whether this approach eases or worsens the delay requirements.

8.97   A famous digital designer devised the circuit shown in Figure X8.97(a), which is supposed to eliminate metastability within one period of a system clock. Circuit M is a memoryless analog voltage detector whose output is 1 if Q is in the metastable state, 0 otherwise. The circuit designer's idea was that if line Q is detected to be in the metastable state when CLOCK goes low, the NAND gate will clear the D flip-flop, which in turn eliminates the metastable output, causing a 0 output from circuit M and thus negating the CLR input of the flip-flop. The circuits are all fast enough that this all happens well before CLOCK goes high again; the expected waveforms are shown in Figure X8.97(b).

Unfortunately, the synchronizer still failed occasionally, and the famous digital designer is now designing pockets for blue jeans. Explain, in detail, how it failed, including a timing diagram.

Figure X8.97

(a)

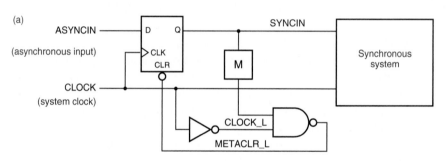

(b)

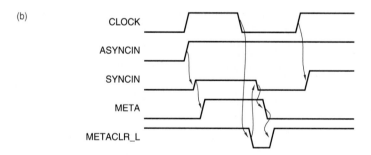

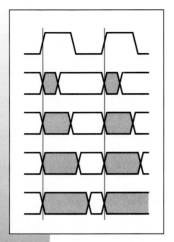

Sequential-Circuit Design Examples

J ust about every real digital system is a sequential circuit—all it takes is one feedback loop, latch, or flip-flop to make a circuit's present behavior depend on its past inputs. However, if we were to design or analyze a typical digital system as a single sequential circuit, we would end up with an enormous state table. For example, strictly speaking, a PC with only 16 Mbytes of main memory is a sequential circuit with well over $2^{128,000,000}$ states!

We noted in previous chapters that we typically deal with digital systems in smaller chunks, for example, partitioning them into data paths, registers, and control units (Section 8.7.1). In fact, a typical system has multiple functional units with well-defined interfaces and connections between them (as supported by VHDL, Section 4.7.2), and each functional unit may contain a hierarchy with several layers of abstraction (as supported by both VHDL and typical schematic drawing packages [Section 5.1.6]). Thus, we can deal with sequential circuits in much smaller chunks.

After all of this build-up, I have to admit that the design of complex, hierarchical digital systems is beyond the scope of this text. However, the heart of any system or any of its subsystems is typically a state machine or other sequential circuit, and that's something we *can* study here. So, this chapter will try to reinforce your understanding of sequential circuit and state-machine design by presenting several examples.

Early digital designers and many designers through the 1980s wrote out state tables by hand and built corresponding circuits using the synthesis methods that we described in Section 7.4. However, hardly anyone does that anymore. Nowadays, most state tables are specified with a hardware description language (HDL) such as ABEL, VHDL, or Verilog. The HDL compiler then performs the equivalent of our synthesis methods and realizes the specified machine in a PLD, CPLD, FPGA, ASIC, or other target technology.

This chapter gives state-machine and other sequential-circuit design examples in two different HDLs. In the first section we give examples in ABEL, and we target the resulting machines to small PLDs. Some of these examples illustrate the partitioning decisions that are needed when an entire circuit does not fit into a single component. In the second section we give examples in VHDL, which can be targeted to just about any technology.

Like Chapter 6, this chapter has as prerequisites only the chapters that precede it. Its two sections are written to be pretty much independent of each other, and the rest of the book is written so that you can read this chapter now or skip it and come back later.

9.1 Design Examples Using ABEL and PLDs

In Section 7.4 we designed several simple state machines by translating their word descriptions into a state table, choosing a state assignment, and finally synthesizing a corresponding circuit. We repeated one of these examples using ABEL and a PLD in Table 7-27 on page 634 and another in Table 7-31 on page 637. These designs were much easier to do using ABEL and a PLD, for two reasons:

- You don't have to be too concerned about the complexity of the resulting excitation equations, as long as they fit in the PLD.

- You may be able to take advantage of ABEL language features to make the design easier to express and understand.

Before looking at more examples, let's examine the timing behavior and packaging considerations for state machines that are built from PLDs.

9.1.1 Timing and Packaging of PLD-Based State Machines

Figure 9-1 shows how a generic PLD with both combinational and registered outputs might be used in a state-machine application. Timing parameters t_{CO} and t_{PD} were explained in Section 8.3.3; t_{CO} is the flip-flop clock-to-output delay and t_{PD} is the delay through the AND-OR array.

State variables are assigned to registered outputs, of course, and are stable at time t_{CO} after the rising edge of CLOCK. Mealy-type outputs are assigned to combinational outputs and are stable at time t_{PD} after any input change that

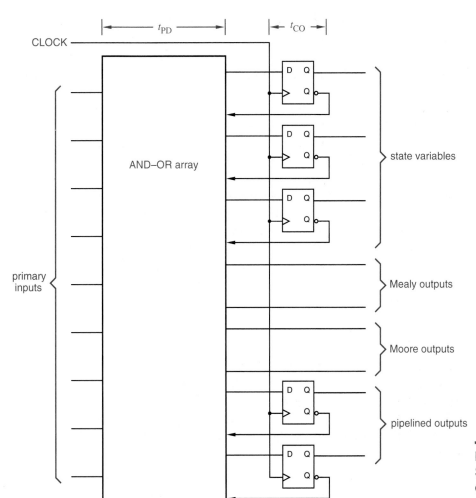

Figure 9-1
Structure and timing
of a PLD used as a
state machine.

affects them. Mealy-type outputs may also change after a state change, in which
case they become stable at time $t_{CO} + t_{PD}$ after the rising edge of CLOCK.

A Moore-type output is, by definition, a combinational logic function of
the current state, so it is also stable at time $t_{CO} + t_{PD}$ after the CLOCK. Thus,
Moore outputs may not offer any speed advantage over Mealy outputs in a PLD
realization. For faster propagation delay, we defined and used pipelined outputs
in Sections 7.3.2 and 7.11.5. In a PLD realization, these outputs come directly
from a flip-flop output and thus have a delay of only t_{CO} from the clock. Besides
having a shorter delay, they are also guaranteed to be glitch free, important in
some applications.

PLD-based state-machine designs are often limited by the number of input
and output pins available in a single PLD. According to the model of Figure 9-1,

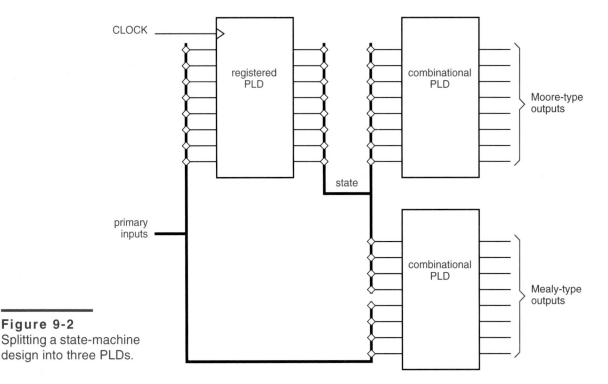

Figure 9-2
Splitting a state-machine
design into three PLDs.

one PLD output is required for each state variable and for each Mealy- or Moore-type output. For example, the T-bird tail-lights machine of Section 7.5 starting on page 585 requires three registered outputs for the state variables and six combinational outputs for the lamp outputs, too many for most of the PLDs that we've described, except for the 22V10.

On the other hand, an output-coded state assignment (Section 7.3.2) usually requires a smaller total number of PLD outputs. Using an output-coded state assignment, the T-bird tail-lights machine can be built in a single 16V8, as we'll show in Section 9.1.3.

Like any state variable or Moore-type output, an output-coded state variable is stable at time t_{CO} after the clock edge. Thus, an output-coded state assignment improves speed as well as packaging efficiency. In T-bird tail lights, turning on the emergency flashers 10 ns sooner doesn't matter, but in a high-performance digital system, an extra 10 ns of delay could make the difference between a maximum system clock rate of 100 MHz and a maximum of only 50 MHz.

If a design *must* be split into two or more PLDs, the best split in terms of both design simplicity and packaging efficiency is often the one shown in Figure 9-2. A single sequential PLD is used to create the required next-state behavior, and combinational PLDs are used to create both Mealy- and Moore-type outputs from the state variables and inputs.

RELIEF FOR A SPLITTING HEADACHE

Modern software tools for PLD-based system design can eliminate some of the trial and error that might otherwise be associated with fitting a design into a set of PLDs. To use this capability, the designer enters the equations and state-machine descriptions for the design, without necessarily specifying the actual devices and pinouts that should be used to realize the design. A software tool called a *partitioner* attempts to split the design into the smallest possible number of devices from a given family, while minimizing the number of pins used for interdevice communication. Partitioning can be fully automatic, partially user controlled, or fully user controlled.

Larger devices—CPLDs and FPGAs—often have internal, architectural constraints that may create headaches in the absence of expert software assistance. It appears to the designer, based on input, output, and total combinational logic and flip-flop requirements, that a design will fit in a single CPLD or FPGA. However, the design must still be split among multiple PLDs or logic blocks inside the larger device, where each block has only limited functionality.

For example, an output that requires many product terms may have to steal some from physically adjacent outputs. This may in turn affect whether adjacent pins can be used as inputs or outputs, and how many product terms are available to them. It may also affect the ability to interconnect signals between nearby blocks and the worst-case delay of these signals. All of these constraints can be tracked by *fitter* software that uses a heuristic approach to find the best split of functions among the blocks within a single CPLD or FPGA.

In many design environments, the partitioner and fitter software work together and interactively with the designer to find an acceptable realization using a set of PLDs, CPLDs, and FPGAs.

9.1.2 A Few Simple Machines

In Section 7.4 we designed several simple state machines using traditional methods. We presented an ABEL- and PLD-based design for the first of these in Table 7-25 on page 631, and then we showed how we could make the machine's operation more understandable by making better use of ABEL features in Table 7-27 on page 634.

Our second example was a "1s-counting machine" with the following specification:

> Design a clocked synchronous state machine with two inputs, X and Y, and one output, Z. The output should be 1 if the number of 1 inputs on X and Y since reset is a multiple of 4, and 0 otherwise.

We developed a state table for this machine in Table 7-12 on page 580. However, we can express the machine's function much more easily in ABEL, as shown in Table 9-1. Notice that for this machine we were better off not using ABEL's "state diagram" syntax. We could express the machine's function more naturally

Table 9-1
ABEL program for
ones-counting
state machine.

```
module onesctsm
title 'Ones-counting State Machine'
ONESCTSM device 'P16V8R';

" Inputs and outputs
CLOCK, RESET, X, Y      pin 1, 2, 3, 4;
Z                       pin 13 istype 'com';
COUNT1..COUNT0          pin 14, 15 istype 'reg';

" Sets
COUNT = [COUNT1..COUNT0];

equations
COUNT.CLK = CLOCK;
WHEN RESET THEN COUNT := 0;
ELSE WHEN X & Y THEN COUNT := COUNT + 2;
ELSE WHEN X # Y THEN COUNT := COUNT + 1;
ELSE COUNT := COUNT;

Z = (COUNT == 0);

end onesctsm
```

using a nested WHEN statement and the built-in addition operation. The first WHEN clause forces the machine to its initial state and count of 0 upon reset, and the succeeding clauses increase the count by 2, 1, or 0 as required when the machine is not reset. Note that ABEL "throws away" the carry bit in addition, which is equivalent to performing addition modulo-4. The machine easily fits into a GAL16V8 device. It has four states, because there are two flip-flops in its realization.

Another example from Section 7.4 is a combination-lock state machine (below we omit the HINT output in the original specification):

Design a clocked synchronous state machine with one input, X, and one output, UNLK. The UNLK output should be 1 if and only if X is 0 and the sequence of inputs received on X at the preceding seven clock ticks was 0110111.

RESETTING BAD HABITS   In Chapter 7 we started the bad habit of designing state tables without including an explicit reset input. There was a reason for this—each additional input potentially doubles the amount work we would have had to do to synthesize the state machine using manual methods, and there was little to be gained pedagogically.

Now that we're doing language-based state-machine design using automated tools, we should get into the habit of always providing an explicit reset input that sends the machine back to a known state. It's a requirement in real designs!

We developed a state table for this machine in Table 7-14 on page 582, and we wrote an equivalent ABEL program in Table 7-31 on page 637. However, once again we can take a different approach that is easier to understand. For this example, we note that the output of the machine at any time is completely determined by its inputs over the preceding eight clock ticks. Thus, we can use a so-called "finite-memory" approach to design this machine, where we explicitly keep track of the past seven inputs and then form the output as a combinational function of these inputs.

The ABEL program in Table 9-2 uses the finite-memory approach. It is written using sets to make modifications easy, for example, changing the combination. However, note that the HINT output would be just as difficult to provide in this version of the machine as in the original (see Exercise 9.2).

Table 9-2 Finite-memory program for combination-lock state machine.

```
module comblckf
title 'Combination-Lock State Machine'
"COMBLCKF device 'P16V8R';

" Input and output pins
CLOCK, RESET, X           pin 1, 2, 3;
X1..X7                    pin 12..18 istype 'reg';
UNLK                      pin 19;

" Sets
XHISTORY = [X7..X1];
SHIFTX   = [X6..X1, X];

equations

XHISTORY.CLK = CLOCK;
XHISTORY := !RESET & SHIFTX;

UNLK = !RESET & (X == 0) & (XHISTORY == [0,1,1,0,1,1,1]);

END comblckf
```

FINITE-MEMORY DESIGN	The finite-memory design approach to state-machine design can be generalized. Such a machine's outputs are completely determined by its current input and outputs during the previous *n* clock ticks, where *n* is a finite, bounded integer. Any machine that can be realized as shown in Figure 9-3 is a finite-memory machine. Note that a finite-*state* machine need not be a finite-*memory* machine. For example, the ones-counting machine in Table 9-1 has only four states but is not a finite-memory machine, its output depends on every value of X and Y since reset.

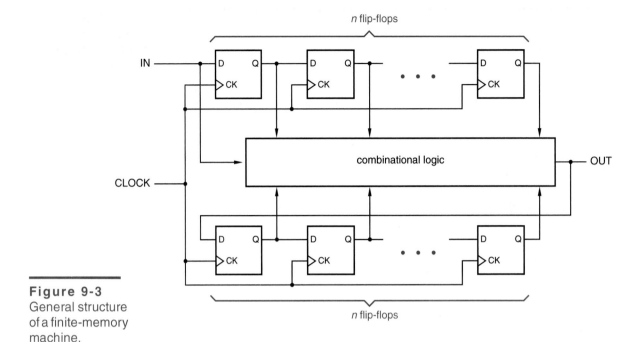

Figure 9-3
General structure
of a finite-memory
machine.

9.1.3 T-Bird Tail Lights

We described and designed a "T-bird tail-lights" state machine in Section 7.5. Table 9-3 is an equivalent ABEL "state diagram" for the T-bird tail-lights machine. There is a close correspondence between this program and the state diagram of Figure 7-64 on page 589 using the state assignment of Table 7-16 on page 590. Except for the added RESET input, the program produces exactly the same reduced equations as the explicit transition equations resulting from the transition list, which we worked out by hand in Section 7.6 on page 592.

The program in Table 9-3 handles only the state variables of the tail-lights machine. The output logic requires six combinational outputs, but only five more outputs are available in the 16V8 specified in the program. A second PLD could be used to decode the states, using the kind of partitioning that we showed in Figure 9-2. Alternatively, a larger PLD, such as the 22V10, could provide enough outputs for a single-PLD design.

An even better approach is to recognize that the output values of the tail-lights machine are different in each state, so we can also use an output-coded state assignment. This requires only six registered outputs and no combinational outputs of a 16V8, as shown in Figure 9-4. Only the device, pin, and state definitions in the previous ABEL program must be changed, as shown in Table 9-4. The six resulting excitation equations each use four product terms.

```
module tbirdsd
title 'State Machine for T-Bird Tail Lights'
TBIRDSD device 'P16V8R';

" Input and output pins
CLOCK, LEFT, RIGHT, HAZ, RESET    pin 1, 2, 3, 4, 5;
Q0, Q1, Q2                        pin 14, 15, 16 istype 'reg';

" Definitions
QSTATE = [Q2,Q1,Q0];             " State variables
IDLE   = [ 0, 0, 0];             " States
L1     = [ 0, 0, 1];
L2     = [ 0, 1, 1];
L3     = [ 0, 1, 0];
R1     = [ 1, 0, 1];
R2     = [ 1, 1, 1];
R3     = [ 1, 1, 0];
LR3    = [ 1, 0, 0];

equations
QSTATE.CLK = CLOCK;

state_diagram QSTATE
state IDLE: IF RESET THEN IDLE
            ELSE IF (HAZ # LEFT & RIGHT) THEN LR3
            ELSE IF LEFT THEN L1 ELSE IF RIGHT THEN R1
            ELSE IDLE;
state L1:   IF RESET THEN IDLE ELSE IF HAZ THEN LR3 ELSE L2;
state L2:   IF RESET THEN IDLE ELSE IF HAZ THEN LR3 ELSE L3;
state L3:   GOTO IDLE;
state R1:   IF RESET THEN IDLE ELSE IF HAZ THEN LR3 ELSE R2;
state R2:   IF RESET THEN IDLE ELSE IF HAZ THEN LR3 ELSE R3;
state R3:   GOTO IDLE;
state LR3:  GOTO IDLE;

end tbirdsd
```

Table 9-3
ABEL program for the T-bird tail-lights machine.

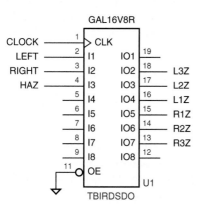

Figure 9-4
A single-PLD design for T-bird tail lights.

Table 9-4
Output-coded state
assignment for the
T-bird tail-lights
machine.

```
module tbirdsdo
title 'Output-Coded T-Bird Tail Lights State Machine'
TBIRDSDO device 'P16V8R';

" Input and output pins
CLOCK, LEFT, RIGHT, HAZ, RESET      pin 1, 2, 3, 4, 5;
L3Z, L2Z, L1Z, R1Z, R2Z, R3Z        pin 18..13 istype 'reg';

" Definitions
QSTATE = [L3Z,L2Z,L1Z,R1Z,R2Z,R3Z];  " State variables
IDLE  = [ 0,  0,  0,  0,  0,  0];  " States
L3    = [ 1,  1,  1,  0,  0,  0];
L2    = [ 0,  1,  1,  0,  0,  0];
L1    = [ 0,  0,  1,  0,  0,  0];
R1    = [ 0,  0,  0,  1,  0,  0];
R2    = [ 0,  0,  0,  1,  1,  0];
R3    = [ 0,  0,  0,  1,  1,  1];
LR3   = [ 1,  1,  1,  1,  1,  1];
```

9.1.4 The Guessing Game

A "guessing game" machine was defined in Section 7.7.1 starting on page 594, with the following description:

> Design a clocked synchronous state machine with four inputs, G1–G4, that are connected to pushbuttons. The machine has four outputs, L1–L4, connected to lamps or LEDs located near the like-numbered pushbuttons. There is also an ERR output connected to a red lamp. In normal operation, the L1–L4 outputs display a 1-out-of-4 pattern. At each clock tick, the pattern is rotated by one position; the clock frequency is about 4 Hz.
>
> Guesses are made by pressing a pushbutton, which asserts an input Gi. When any Gi input is asserted, the ERR output is asserted if the "wrong" pushbutton was pressed, that is, if the Gi input detected at the clock tick does not have the same number as the lamp output that was asserted before the clock tick. Once a guess has been made, play stops and the ERR output maintains the same value for one or more clock ticks until the Gi input is negated, then play resumes.

As we discussed in Section 7.7.1, the machine requires six states—four in which a corresponding lamp is on, and two for when play is stopped after either a good or a bad pushbutton push. An ABEL program for the guessing game is shown in Table 9-5. Two enhancements were made to improve the testability and robustness of the machine—a RESET input that forces the game to a known starting state, and the two unused states have explicit transitions to the starting state.

The guessing-game machine uses the same state assignments as the original version in Section 7.7.1. Using these assignments, the ABEL compiler

```
module ggame
Title 'Guessing-Game State Machine'
GGAME device 'P16V8R';

" Inputs and outputs
CLOCK, RESET, G1..G4          pin 1, 2, 3..6;
L1..L4, ERR                   pin 12..15, 19 istype 'com';
Q2..Q0                        pin 16..18 istype 'reg';

" Sets
G = [G1..G4];
L = [L1..L4];

" States
QSTATE = [Q2,Q1,Q0];
S1     = [ 0, 0, 0];
S2     = [ 0, 0, 1];
S3     = [ 0, 1, 1];
S4     = [ 0, 1, 0];
SOK    = [ 1, 0, 0];
SERR   = [ 1, 0, 1];
EXTRA1 = [ 1, 1, 0];
EXTRA2 = [ 1, 1, 1];

state_diagram QSTATE

state S1:  IF RESET THEN SOK ELSE IF G2 # G3 # G4 THEN SERR
             ELSE IF G1 THEN SOK ELSE S2;

state S2:  IF RESET THEN SOK ELSE IF G1 # G3 # G4 THEN SERR
             ELSE IF G2 THEN SOK ELSE S3;

state S3:  IF RESET THEN SOK ELSE IF G1 # G2 # G4 THEN SERR
             ELSE IF G3 THEN SOK ELSE S4;

state S4:  IF RESET THEN SOK ELSE IF G1 # G2 # G3 THEN SERR
             ELSE IF G4 THEN SOK ELSE S1;

state SOK:  IF RESET THEN SOK
              ELSE IF G1 # G2 # G3 # G4 THEN SOK ELSE S1;

state SERR:  IF RESET THEN SOK
               ELSE IF G1 # G2 # G3 # G4 THEN SERR ELSE S1;

state EXTRA1:  GOTO SOK;
state EXTRA2:  GOTO SOK;

equations

QSTATE.CLK = CLOCK;

L1  = (QSTATE == S1);
L2  = (QSTATE == S2);
L3  = (QSTATE == S3);
L4  = (QSTATE == S4);
ERR = (QSTATE == SERR);

end ggame
```

Table 9-5
ABEL program for
the guessing-game
machine.

Table 9-6
Product-term usage
in the guessing-game
state-machine PLD.

P-Terms	Fan-in	Fan-out	Type	Name
1/3	3	1	Pin	L1
1/3	3	1	Pin	L2
1/3	3	1	Pin	L3
1/3	3	1	Pin	L4
1/3	3	1	Pin	ERR
6/2	7	1	Pin	Q2.REG
1/7	7	1	Pin	Q1.REG
11/8	8	1	Pin	Q0.REG

23/32

Best P-Term Total: 16

Total Pins: 14

Average P-Term/Output: 2

cranks out minimized equations with the number of product terms shown in Table 9-6. The Q0 output just barely fits in a GAL16V8 (eight product terms). If we needed to save terms, the way in which we've written the program allows us to try alternate state assignments (see Exercise 9.4).

A more productive alternative might be to try an output-coded state assignment. We can use one state/output bit per lamp (L1..L4), and use one more bit (ERR) to distinguish between the SOK and SERR states when the lamps are all off. This allows us to drop the equations for L1..L4 and ERR from Table 9-5. The new assignment is shown in Table 9-7. With this assignment, L1 uses two product terms and L2..L4 use only one product term each. Unfortunately, the ERR output blows up into 16 product terms.

Table 9-7
ABEL definitions for
the guessing-game
machine with an
output-coded state
assignment.

```
module ggameoc
Title 'Guessing-Game State Machine'
"GGAMEOC device 'P16V8R';

" Inputs and outputs
CLOCK, RESET, G1..G4            pin 1, 2, 3..6;
L1..L4, ERR                     pin 12..15, 18 istype 'reg';

" States
QSTATE = [L1,L2,L3,L4,ERR];
S1      = [ 1, 0, 0, 0,  0];
S2      = [ 0, 1, 0, 0,  0];
S3      = [ 0, 0, 1, 0,  0];
S4      = [ 0, 0, 0, 1,  0];
SOK     = [ 0, 0, 0, 0,  0];
SERR    = [ 0, 0, 0, 0,  1];

. . .
```

Part of our problem with this particular output-coded assignment is that we're not taking full advantage of its properties. Notice that it is basically a "one-hot" encoding, but the state definitions in Table 9-7 require all five state bits to be decoded for each state. An alternate version of the coding using "don't-cares" is shown in Table 9-8.

In the new version, we are assuming that the state bits never take on any combination of values other than the ones we originally defined in Table 9-7. Thus, for example, if we see that state bit L1 is 1, the machine must be in state S1 regardless of the values of any other state bits. Therefore, we can set these bits to "don't care" in S1's definition in Table 9-8. ABEL will set each X to 0 when encoding a next state, but will treat each X as a "don't-care" when decoding the current state. Thus, we must take *extreme* care to ensure that decoded states are in fact mutually exclusive, that is, that no legitimate next state matches two or more different state definitions. Otherwise, the compiled results will not have the expected behavior.

The reduced equations that result from the output coding in Table 9-8 use three product terms for L1, one each for L2..L4, and only seven for ERR. So the

```
X = .X.;
QSTATE = [L1,L2,L3,L4,ERR];
S1     = [ 1, X, X, X,   X];
S2     = [ X, 1, X, X,   X];
S3     = [ X, X, 1, X,   X];
S4     = [ X, X, X, 1,   X];
SOK    = [ 0, 0, 0, 0,   0];
SERR   = [ X, X, X, X,   1];
```

Table 9-8
Output coding for the guessing-game machine using "don't cares."

DON'T-CARE, HOW IT WORKS   To understand how the don't-cares work in a state encoding, you must first understand how ABEL creates equations internally from state diagrams. Within a given state S, each transition statement (IF-THEN-ELSE or GOTO) causes the on-sets of certain state variables to be augmented according to the transition condition. The transition condition is an expression that must be true to "go to" that target, including being in state S. For example, all of the conditions specified in a state such as S1 in Table 9-5 are implicitly ANDed with the expression "QSTATE==S1". Because of the way S1 is defined using don't-cares, this equality check generates only a single literal (L1) instead of an AND term, leading to further simplification later.

For each target state in a transition statement, the on-sets of only the state variables that are 1 in that state are augmented according to the transition condition. Thus, when a coded state such as S1 in Table 9-8 appears as a target in any transition statement, only the on-set of L1 is augmented. This explains why the actual coding of state S1 as a target is 100000.

change was worthwhile. However, we must remember that the new machine is different from the one in Table 9-7. Consider what happens if the machine ever gets into an unspecified state. In the original machine with fully specified output coding, there are no next-states for the $2^5 - 6 = 26$ unspecified states, so the state machine will always go to the state coded 00000 (SOK) from unspecified states. In the new machine, "unspecified" states aren't really unspecified; for example, the state coded 11111 actually matches five coded states, S1–S4 and SERR. The next state will actually be the "OR" of next-states for the matching coded states. (Read the box on the previous page to understand why these outcomes occur.) Again, you need to be careful.

RESETTING EXPECTATIONS

Reading the guessing-game program in Table 9-5, you would expect that the RESET input would force the machine to the SOK state, and it does. However, the moment that you have unspecified or partially coded states as in Tables 9-7 or 9-8, don't take anything for granted.

Referring to the box on the previous page, remember that transition statements in ABEL state machines augment the on-sets of state variables. If a particular, unused state combination does not match any of the states for which transition statements were written, then no on-sets will be augmented. Thus, the only transition from that state will be to the state with the all-0s coding.

For this reason, it is useful to code the reset state or a "safe" state as all 0s. If this is not possible, but the all-0s state is still unused, you can explicitly provide a transition from the all-0s state to a desired safe state.

9.1.5 Reinventing Traffic-Light Controllers

Our final example is from the world of cars and traffic. Traffic-light controllers in California, especially in the fair city of Sunnyvale, are carefully designed to *maximize* the waiting time of cars at intersections. An infrequently used intersection (one that would have no more than a "yield" sign if it were in Chicago) has the sensors and signals shown in Figure 9-5. The state machine that controls the traffic signals uses a 1-Hz clock and a timer and has four inputs:

NSCAR   Asserted when a car on the north-south road is over either sensor on either side of the intersection.

EWCAR   Asserted when a car on the east-west road is over either sensor on either side of the intersection.

TMLONG   Asserted if more than five minutes has elapsed since the timer started; remains asserted until the timer is reset.

TMSHORT   Asserted if more than five seconds has elapsed since the timer started; remains asserted until the timer is reset.

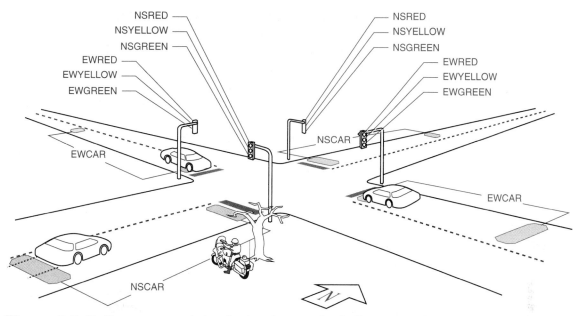

Figure 9-5 Traffic sensors and signals at an intersection in Sunnyvale, California.

The state machine has seven outputs:

NSRED, NSYELLOW, NSGREEN   Control the north-south lights.

EWRED, EWYELLOW, EWGREEN   Control the east-west lights.

TMRESET   When asserted, resets the timer and negates TMSHORT and TMLONG. The timer starts timing when TMRESET is negated.

A typical, municipally approved algorithm for controlling the traffic lights is embedded in the ABEL program of Table 9-9. This algorithm produces two frequently seen behaviors of "smart" traffic lights. At night, when traffic is light, it holds a car stopped at the light for up to five minutes, unless a car approaches on the cross street, in which case it stops the cross traffic and lets the waiting car go. (The "early warning" sensor is far enough back to change the lights before the approaching car reaches the intersection.) During the day, when traffic is heavy and there are always cars waiting in both directions, it cycles the lights every five seconds, thus minimizing the utilization of the intersection and maximizing everyone's waiting time, thereby creating a public outcry for more taxes to fix the problem.

The equations for the TMRESET output are worth noting. This output is asserted during the "double-red" states, NSDELAY and EWDELAY, to reset the timer in preparation for the next green cycle. The desired output signal could be generated on a combinational output pin by decoding these two states, but we have chosen instead to generate it on a registered output pin by decoding the *predecessors* of these two states.

■ **Table 9-9** Sunnyvale traffic-lights program.

```
module svaletl
title 'State Machine for Sunnyvale, CA, Traffic Lights'
SVALETL device 'P16V8R';

" Input and output pins
CLOCK, !OE                        pin 1, 11;
NSCAR, EWCAR, TMSHORT, TMLONG     pin 2, 3, 8, 9;
Q0, Q1, Q2, TMRESET_L             pin 17, 16, 15, 14 istype 'reg';

" Definitions
LSTATE  = [Q2,Q1,Q0];            " State variables
NSGO    = [ 0, 0, 0];            " States
NSWAIT  = [ 0, 0, 1];
NSWAIT2 = [ 0, 1, 1];
NSDELAY = [ 0, 1, 0];
EWGO    = [ 1, 1, 0];
EWWAIT  = [ 1, 1, 1];
EWWAIT2 = [ 1, 0, 1];
EWDELAY = [ 1, 0, 0];

state_diagram LSTATE
state NSGO:                        " North-south green
   IF (!TMSHORT) THEN NSGO         " Minimum green is 5 seconds.
   ELSE IF (TMLONG) THEN NSWAIT    " Maximum green is 5 minutes.
   ELSE IF (EWCAR & !NSCAR)        " If E-W car is waiting and no one
          THEN NSGO                "  is coming N-S, make E-W wait!
   ELSE IF (EWCAR & NSCAR)         " Cars coming in both directions?
          THEN NSWAIT              " Thrash!
   ELSE IF (!NSCAR)                " Nobody coming N-S and not timed out?
          THEN NSGO                " Keep N-S green.
   ELSE NSWAIT;                    " Else let E-W have it.

state NSWAIT: GOTO NSWAIT2;        " Yellow light is on for two ticks for safety.
state NSWAIT2: GOTO NSDELAY;       " (Drivers go 70 mph to catch this turkey green!)
state NSDELAY: GOTO EWGO;          " Red in both directions for added safety.

state EWGO: " East-west green; states defined analogous to N-S
   IF (!TMSHORT) THEN EWGO
   ELSE IF (TMLONG) THEN EWWAIT
   ELSE IF (NSCAR & !EWCAR) THEN EWGO
   ELSE IF (NSCAR & EWCAR) THEN EWWAIT
   ELSE IF (!EWCAR) THEN EWGO ELSE EWWAIT;

state EWWAIT: GOTO EWWAIT2;
state EWWAIT2: GOTO EWDELAY;
state EWDELAY: GOTO NSGO;

equations

LSTATE.CLK = CLOCK;  TMRESET_L.CLK = CLOCK;
!TMRESET_L := (LSTATE == NSWAIT2)  " Reset the timer when going into
           + (LSTATE == EWWAIT2);  "   state NSDELAY or state EWDELAY.
end svaletl
```

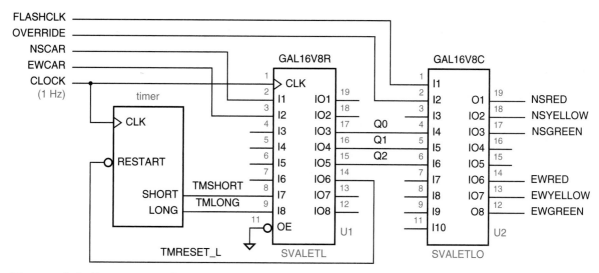

Figure 9-6 Sunnyvale traffic-light controller using two PLDs.

Table 9-10 Output logic for Sunnyvale traffic lights.

```
module svaletlo
title 'Output logic for Sunnyvale, CA, Traffic Lights'
"SVALETLO device 'P16V8C';

" Input pins
FLASHCLK, OVERRIDE, Q0, Q1, Q2   pin 1, 2, 4, 5, 6;

" Output pins
NSRED, NSYELLOW, NSGREEN          pin 19, 18, 17 istype 'com';
EWRED, EWYELLOW, EWGREEN          pin 14, 13, 12 istype 'com';

" Definitions (same as in state machine SVALETL)
...

equations

NSRED = !OVERRIDE & (LSTATE != NSGO) & (LSTATE != NSWAIT) & (LSTATE != NSWAIT2)
     # OVERRIDE & FLASHCLK;
NSYELLOW = !OVERRIDE & ((LSTATE == NSWAIT) # (LSTATE == NSWAIT2));
NSGREEN  = !OVERRIDE & (LSTATE == NSGO);

EWRED = !OVERRIDE & (LSTATE != EWGO) & (LSTATE != EWWAIT) & (LSTATE != EWWAIT2)
     # OVERRIDE & FLASHCLK;
EWYELLOW = !OVERRIDE & ((LSTATE == EWWAIT) # (LSTATE == EWWAIT2));
EWGREEN  = !OVERRIDE & (LSTATE == EWGO);

end svaletlo
```

The ABEL program in Table 9-9 defines only the state variables and one registered Moore output for the traffic controller. Six more Moore outputs are needed for the lights, more than remain on the 16V8. Therefore, a separate combinational PLD is used for these outputs, yielding the complete design shown in Figure 9-6 on the preceding page. An ABEL program for the output PLD is given in Table 9-10. We've taken this opportunity to add an OVERRIDE input to the controller. This input may be asserted by the police to disable the controller and put the signals into a flashing-red mode (at a rate determined by FLASHCLK), allowing them to manually clear up the traffic snarls created by this wonderful invention.

A traffic-light state machine including output logic can be built in a single 16V8, shown in Figure 9-7, if we choose an output-coded state assignment. Only the definitions in the original program of Table 9-9 must be changed, as shown in Table 9-11. This PLD does not include the OVERRIDE input and mode, which is left as an exercise (9.7).

Table 9-11 Definitions for Sunnyvale traffic-lights machine with output-coded state assignment.

```
module svaletlb
title 'Output-Coded State Machine for Sunnyvale Traffic Lights'
"SVALETLB device 'P16V8R';

" Input and output pins
CLOCK, !OE                            pin 1, 11;
NSCAR, EWCAR, TMSHORT, TMLONG         pin 2, 3, 8, 9;
NSRED, NSYELLOW, NSGREEN              pin 19, 18, 17 istype 'reg';
EWRED, EWYELLOW, EWGREEN              pin 16, 15, 14 istype 'reg';
TMRESET_L, XTRA                       pin 13, 12 istype 'reg';

" Definitions
LSTATE  = [NSRED,NSYELLOW,NSGREEN,EWRED,EWYELLOW,EWGREEN,XTRA]; " State vars
NSGO    = [   0,      0,      1,     1,      0,       0,    0]; " States
NSWAIT  = [   0,      1,      0,     1,      0,       0,    0];
NSWAIT2 = [   0,      1,      0,     1,      0,       0,    1];
NSDELAY = [   1,      0,      0,     1,      0,       0,    0];
EWGO    = [   1,      0,      0,     0,      0,       1,    0];
EWWAIT  = [   1,      0,      0,     0,      1,       0,    0];
EWWAIT2 = [   1,      0,      0,     0,      1,       0,    1];
EWDELAY = [   1,      0,      0,     1,      0,       0,    1];
```

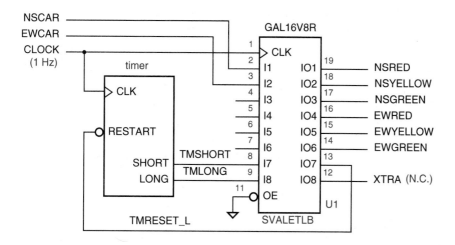

Figure 9-7
Traffic-light state machine using output-coded state assignment in a single PLD.

9.2 Design Examples Using VHDL

As we explained Section 7.12, the basic VHDL language features that we introduced way back in Section 4.7, including processes, are just about all that is needed to model sequential-circuit behavior. Unlike ABEL, VHDL does not provide any special language elements for modeling state machines. Instead, most programmers use a combination of existing "standard" features—most notably enumerated types and case statements—to write state-machine descriptions. We'll use this method in the examples in this section.

9.2.1 A Few Simple Machines

In Section 7.4.1 we illustrated the state-table design process using the simple design problem below:

Design a clocked synchronous state machine with two inputs, A and B, and a single output Z that is 1 if:

  − A had the same value at each of the two previous clock ticks, *or*
  − B has been 1 since the last time that the first condition was true.

Otherwise, the output should be 0.

In an HDL-based design environment, there are many possible ways of writing a program that meets the stated requirements. We'll look at several.

The first approach is to construct a state and output table by hand and then manually convert it into a corresponding program. Since we already developed such a state table in Section 7.4.1, why not use it? We've written it again in Table 9-12, and we've written a corresponding VHDL program in Table 9-13.

Table 9-12
State and output table for the example state machine.

S	00	01	11	10	Z
			A B		
INIT	A0	A0	A1	A1	0
A0	OK0	OK0	A1	A1	0
A1	A0	A0	OK1	OK1	0
OK0	OK0	OK0	OK1	A1	1
OK1	A0	OK0	OK1	OK1	1
			S*		

Table 9-13
VHDL program for state-machine example.

```
library IEEE;
use IEEE.std_logic_1164.all;

entity smexamp is
  port ( CLOCK, A, B: in STD_LOGIC;
         Z: out STD_LOGIC );
end;

architecture smexamp_arch of smexamp is
type Sreg_type is (INIT, A0, A1, OK0, OK1);
signal Sreg: Sreg_type;
begin

  process (CLOCK) -- state-machine states and transitions
  begin
    if CLOCK'event and CLOCK = '1' then
      case Sreg is
        when INIT => if    A='0' then Sreg <= A0;
                     elsif A='1' then Sreg <= A1;    end if;
        when A0 =>   if    A='0' then Sreg <= OK0;
                     elsif A='1' then Sreg <= A1;  end if;
        when A1 =>   if    A='0' then Sreg <= A0;
                     elsif A='1' then Sreg <= OK1;  end if;
        when OK0 =>  if    A='0' then Sreg <= OK0;
                     elsif A='1' and B='0' then Sreg <= A1;
                     elsif A='1' and B='1' then Sreg <= OK1;  end if;
        when OK1 =>  if    A='0' and B='0' then Sreg <= A0;
                     elsif A='0' and B='1' then Sreg <= OK0;
                     elsif A='1' then Sreg <= OK1;            end if;
        when others => Sreg <= INIT;
      end case;
    end if;
  end process;

  with Sreg select  -- output values based on state
    Z <= '0' when INIT | A0 | A1,
         '1' when OK0 | OK1,
         '0' when others;

end smexamp_arch;
```

As usual, the VHDL entity declaration specifies only inputs and outputs—CLOCK, A, B, and Z in this example. The architecture definition specifies the state machine's internal operation. The first thing it does is to create an enumerated type, Sreg_type, whose values are identifiers corresponding to the state names. Then it declares a signal, Sreg, which will be used to hold the machine's current state. Because of the way Sreg is used later, it will map into an edge-triggered register in synthesis.

The statement part of the architecture has two concurrent statements—a process and a selected-assignment statement. The process is sensitive only to CLOCK and establishes all of the state transitions, which occur on the rising edge of CLOCK. Within the process, an "if" statement checks for the rising edge, and a case statement enumerates the transitions for each state.

The case statement has six cases, corresponding to the five explicitly defined states and a catch-all for other states. For robustness, the "others" case sends the machine back to the INIT state. In each case we've used a nested "if" statement to explicit cover all combinations of inputs A and B. However, it's not strictly necessary to include the combinations where there is no transition from the current state; Sreg will hold its current value if no assignment is made on a particular process invocation.

The selected-assignment statement at the end of Table 9-13 handles the machine's single Moore output, Z, which is set to a value as a function of the current state. It would be easy to define Mealy outputs within this statement as well. That is, Z could be a function of the inputs as well as the current state. Since the "with" statement is a concurrent state, any input changes will affect the Z output as soon as they occur.

We really should have included a reset input in Table 9-13 (see box on page 808). A RESET signal is easily accommodated by modifying the entity declaration and adding a clause to the "if" statement in the architecture definition. If RESET is asserted, the machine should go to the INIT state, otherwise the case statement should be executed. Depending on whether we check RESET before or after the clock-edge check, we can create either an asynchronous or a synchronous reset behavior (see Exercise 9.10).

What about the state assignment problem? Table 9-13 gives no information on how state-variable combinations are to be assigned to the named states, or even how many binary state variables are needed in the first place!

A synthesis tool is free to associate any integer values or binary combinations it likes with the identifiers in an enumerated type, but a typical tool will assign integers in the order the state names are listed, starting with 0. It will then use the smallest possible number of bits to encode those integers, that is, $\lceil \log_2 s \rceil$ bits for s states. Thus, the program in Table 9-13 will typically synthesize with the same, "simplest" state assignment that we chose in the original example in Table 7-7 on page 571. However, VHDL supports a couple of ways that we can force a different assignment.

Table 9-14
Using an attribute to force an enumeration encoding.

```
library IEEE;
use IEEE.std_logic_1164.all;
library SYNOPSYS;
use SYNOPSYS.attributes.all;
...
architecture smexampe_arch of smexamp is
type Sreg_type is (INIT, A0, A1, OK0, OK1);
attribute enum_encoding of Sreg_type: type is
           "0000 0001 0010 0100 1000";
signal Sreg: Sreg_type;
...
```

One way to force an assignment is to use VHDL's "attribute" statement as shown in Table 9-14. Here, "enum_encoding" is a user-defined attribute whose value is a string that specifies the enumeration encoding to be used by the synthesis tool. The VHDL language processor ignores this value, but passes the attribute name and its value to the synthesis tool. The attribute "enum_encoding" is defined and known by most synthesis tools, including tools from Synopsys, Inc. Notice that a Synopsys "attributes" package must be "used" by the program; this is necessary for the VHDL compiler to recognize "enum_encoding" as a legitimate user-defined attribute. By the way, the state assignment that we've specified in the program is equivalent to the "almost one-hot" coding in Table 7-7 on page 571.

Another way to force an assignment, without relying on external packages or synthesis attributes, is to define the state register more explicitly using standard logic data types. This approach is shown in Table 9-15. Here, Sreg is defined as a 4-bit STD_LOGIC_VECTOR, and constants are defined to allow the states to be referenced by name elsewhere in the program. No other changes in the program are required.

Going back to our original VHDL program in Table 9-13, one more interesting change is possible. As written, the program defines a conventional Moore-type state machine with the structure shown in Figure 9-8(a). What

Table 9-15
Using standard logic and constants to specify a state encoding.

```
library IEEE;
use IEEE.std_logic_1164.all;
...
architecture smexampc_arch of smexamp is
subtype Sreg_type is STD_LOGIC_VECTOR (1 to 4);
constant INIT: Sreg_type := "0000";
constant A0  : Sreg_type := "0001";
constant A1  : Sreg_type := "0010";
constant OK0 : Sreg_type := "0100";
constant OK1 : Sreg_type := "1000";
signal Sreg: Sreg_type;
...
```

happens if we convert the output logic's selected-assignment statement into a case statement and move it into the state-transition process? By doing this, we create a machine that will most likely be synthesized with the structure shown in (b). This is essentially a Mealy machine with pipelined outputs whose behavior is indistinguishable from that of the original machine, except for timing. We've reduced the propagation delay from CLOCK to Z by producing Z directly on a register output, but we've also increased the setup-time requirements of A and B to CLOCK, because of the extra propagation delay through the output logic to the D input of the output register.

Figure 9-8 State-machine structures implied by VHDL programs: (a) Moore machine with combinational output logic; (b) pipelined Mealy machine with output register.

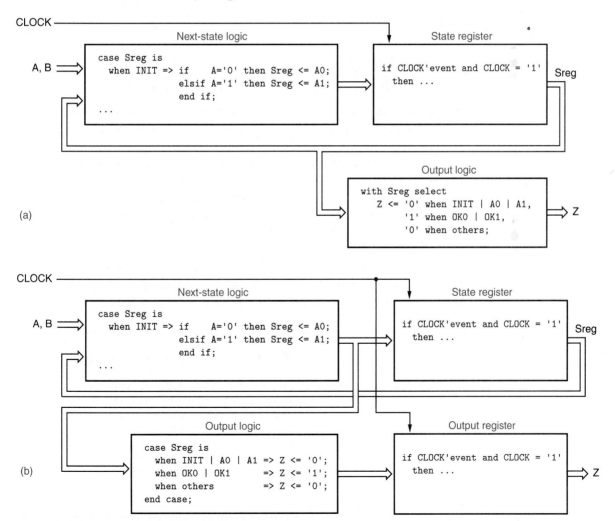

TRICKY TIMING   When we write a VHDL architecture corresponding to Figure 9-8(b), it is very important to add `Sreg` to the sensitivity list of the process. Notice that the output-logic `case` statement determines the value of Z as a function of `Sreg`. Throughout the first execution of the process after a rising clock edge, `Sreg` contains the old state value. This is true because `Sreg` is a signal, not a variable. As explained in Section 4.7.9, signals that are changed in a process do not acquire their new values until at least one delta delay *after* the process has executed. By putting `Sreg` in the sensitivity list, we ensure that the process is executed a second time, so the final value of Z is based on the new value of `Sreg`.

All of the solutions to the example state-machine design problem that we've shown so far rely on the state table that we originally constructed by hand in Section 7.4.1. However, it is possible to write a VHDL program directly, without writing out a state table by hand.

Based on the original problem statement on page 813, the key simplifying idea is to remove the last value of A from the state definitions, and instead to have a separate register that keeps track of it (LASTA). Then only two non-INIT states

Table 9-16
Simplified state machine for VHDL example problem.

```
architecture smexampa_arch of smexamp is
type Sreg_type is (INIT, LOOKING, OK);
signal Sreg: Sreg_type;
signal lastA: STD_LOGIC;
begin
  process (CLOCK) -- state-machine states and transitions
  begin
    if CLOCK'event and CLOCK = '1' then
      lastA <= A;
      case Sreg is
        when INIT    =>                       Sreg <= LOOKING;
        when LOOKING => if    A=lastA then Sreg <= OK;
                        else              Sreg <= LOOKING;
                        end if;
        when OK      => if    B='1'   then Sreg <= OK;
                        elsif A=lastA then Sreg <= OK;
                        else              Sreg <= LOOKING;
                        end if;
        when others  =>                       Sreg <= INIT;
      end case;
    end if;
  end process;

  with Sreg select  -- output values based on state
    Z <= '1' when OK,
         '0' when others;

end smexampa_arch;
```

must be defined: LOOKING ("still looking for a match") and OK ("got a match or B has been 1 since last match"). A VHDL architecture based on this approach is shown in Table 9-16. In the CLOCK-driven process, the first assignment statement creates the LASTA register, and the case statement creates the 3-state machine. At the end of the program, the Z output is defined as a simple combinational decode of the OK state.

Another simple state-machine example is a "ones-counting machine" with the following specification:

> Design a clocked synchronous state machine with two inputs, X and Y, and one output, Z. The output should be 1 if the number of 1 inputs on X and Y since reset is a multiple of 4, and 0 otherwise.

We developed a state table for this machine in Table 7-12 on page 580. However, we can make use of the counting capabilities in the IEEE std_logic_arith package to write a VHDL program for this problem directly.

Table 9-17 shows our solution. As always, there are many different ways to solve the problem, and we have picked a way that illustrates several different language features. Within the architecture, we declare a subtype COUNTER which is a 2-bit UNSIGNED value. We then declare a signal COUNT of this type to hold the ones count, and a constant ZERO of the same type for initializing and checking the value of COUNT.

Table 9-17
VHDL program for a ones-counting machine.

```
library IEEE;
use IEEE.std_logic_1164.all;
use IEEE.std_logic_arith.all;

entity Vonescnt is
  port ( CLOCK, RESET, X, Y: in STD_LOGIC;
         Z: out STD_LOGIC );
end;

architecture Vonescnt_arch of Vonescnt is
subtype COUNTER is UNSIGNED (1 downto 0);
signal COUNT: COUNTER;
constant ZERO: COUNTER := "00";
begin

process (CLOCK)
  begin
    if CLOCK'event and CLOCK = '1' then
      if RESET = '1' then COUNT <= ZERO;
      else COUNT <= COUNT + ('0', X) + ('0', X);
      end if;
    end if;
  end process;

Z <= '1' when COUNT = ZERO else '0';

end Vonescnt_arch;
```

Table 9-18
Alternative VHDL
process for ones-
counting machine.

```
process (CLOCK)
variable ONES: STD_LOGIC_VECTOR (1 to 2);
begin
  if CLOCK'event and CLOCK = '1' then
    ONES := (X, Y);
    if RESET = '1' then COUNT <= ZERO;
    else case ONES is
            when "01" | "10" => COUNT <= COUNT + "01";
            when "11"        => COUNT <= COUNT + "10";
            when others      => null;
         end case;
    end if;
  end if;
end process;
```

Within the process, we use the usual method to check for a rising edge on
CLOCK. The "if" clause performs a synchronous reset when required, and the
"else" clause elegantly adds 0, 1 or 2 to COUNT depending on the values of X
and Y. Recall that an expression such as "('0', X)" is an array literal; here we
get an array of two STD_LOGIC elements, '0' and the current value of X. The
type of this literal is compatible with UNSIGNED, since the number and type of
elements are the same, so they can be combined using the "+" operation defined
in the std_logic_arith package. Outside the process, the concurrent signal-
assignment statement sets the Moore output Z to 1 when COUNT is zero.

For synthesis purposes, the "if" statement and assignment to COUNT in
Table 9-17 don't necessarily yield a compact or speedy circuit. With a simple-
minded synthesis tool, it could yield two 2-bit adders connected in series.
Another approach is shown in Table 9-18. An intelligent tool may be able to
synthesize a more compact incrementer for each of the two additions. In any
case, formulating the choices in a case statement allows the two adders or
incrementers to operate in parallel, and a multiplexer can be used to select one of
their outputs according to the choices.

A final example for this subsection is the combination-lock state machine
from Section 7.4 (below we omit the HINT output in the original specification):

Design a clocked synchronous state machine with one input, X, and one
output, UNLK. The UNLK output should be 1 if and only if X is 0 and the
sequence of inputs received on X at the preceding seven clock ticks was
0110111.

```
library IEEE;
use IEEE.std_logic_1164.all;

entity Vcomblck is
  port ( CLOCK, RESET, X: in STD_LOGIC;
         UNLK: out STD_LOGIC );
end;

architecture Vcomblck_arch of Vcomblck is
signal XHISTORY: STD_LOGIC_VECTOR (7 downto 1);
constant COMBINATION: STD_LOGIC_VECTOR (7 downto 1) := "0110111";
begin

  process (CLOCK)
  begin
    if CLOCK'event and CLOCK = '1' then
      if RESET = '1' then XHISTORY <= "0000000";
      else XHISTORY <= XHISTORY(6 downto 1) & X;
      end if;
    end if;
  end process;

  UNLK <= '1' when (XHISTORY=COMBINATION) and (X='0') else '0';

end Vcomblck_arch;
```

Table 9-19
VHDL program for finite-memory design of combination-lock state machine.

We developed a state table for this machine in Table 7-14 on page 582. But once again we can take a different approach that is easier to understand. Here, we note that the output of the machine at any time is completely determined by its inputs over the preceding eight clock ticks. Thus, we can use the so-called "finite-memory" approach to design this machine (see box on page 801). With this approach, we explicitly keep track of the past seven inputs and then form the output as a combinational function of these inputs.

The VHDL program in Table 9-19 is a finite-memory design. Within the architecture, the process merely keeps track of the last seven values of X using what's essentially a shift register, bit 7 being the oldest value of X. (Recall that the "&" operator in VHDL is array concatenation.) Outside of the process, the concurrent signal-assignment statement sets the Mealy output UNLK to 1 when X is 0 and the 7-bit history matches the combination.

9.2.2 T-Bird Tail Lights

We described and designed a "T-bird tail-lights" state machine in Section 7.5. Table 9-20 is an equivalent VHDL program for the T-bird tail-lights machine. The state transitions in this machine are defined exactly the same as in the state diagram of Figure 7-64 on page 589. The machine uses an output-coded state assignment, taking advantage of the fact that the tail-light output values are different in each state.

■ **Table 9-20** VHDL program for the T-bird tail-lights machine.

```vhdl
entity Vtbird is
  port ( CLOCK, RESET, LEFT, RIGHT, HAZ: in STD_LOGIC;
         LIGHTS: buffer STD_LOGIC_VECTOR (1 to 6) );
end;

architecture Vtbird_arch of Vtbird is
constant IDLE: STD_LOGIC_VECTOR (1 to 6) := "000000";
constant L3  : STD_LOGIC_VECTOR (1 to 6) := "111000";
constant L2  : STD_LOGIC_VECTOR (1 to 6) := "110000";
constant L1  : STD_LOGIC_VECTOR (1 to 6) := "100000";
constant R1  : STD_LOGIC_VECTOR (1 to 6) := "000001";
constant R2  : STD_LOGIC_VECTOR (1 to 6) := "000011";
constant R3  : STD_LOGIC_VECTOR (1 to 6) := "000111";
constant LR3 : STD_LOGIC_VECTOR (1 to 6) := "111111";
begin
  process (CLOCK)
  begin
    if CLOCK'event and CLOCK = '1' then
      if RESET = '1' then LIGHTS <= IDLE; else
        case LIGHTS is
          when IDLE => if HAZ='1' or (LEFT='1' and RIGHT='1') then LIGHTS <= LR3;
                       elsif LEFT='1'                         then LIGHTS <= L1;
                       elsif RIGHT='1'                        then LIGHTS <= R1;
                       else                                        LIGHTS <= IDLE;
                       end if;
          when L1  => if HAZ='1' then LIGHTS <= LR3; else LIGHTS <= L2; end if;
          when L2  => if HAZ='1' then LIGHTS <= LR3; else LIGHTS <= L3; end if;
          when L3  => LIGHTS <= IDLE;
          when R1  => if HAZ='1' then LIGHTS <= LR3; else LIGHTS <= R2; end if;
          when R2  => if HAZ='1' then LIGHTS <= LR3; else LIGHTS <= R3; end if;
          when R3  => LIGHTS <= IDLE;
          when LR3 => LIGHTS <= IDLE;
          when others => null;
        end case;
      end if;
    end if;
  end process;
end Vtbird_arch;
```

IDLE MUSINGS   In VHDL state machines, it's not necessary to make an explicit assignment of a next state if it's the same state that the machine is already in. In the execution of a process, a VHDL signal keeps its value if no assignment is made to it. Thus, in Table 9-20, the final "`else`" clause could be omitted in the IDLE state, with no effect on the machine's behavior.

Separately, the robustness of the state machine in Table 9-20 could be improved by replacing the "`null`" statement in the "`when others`" case with a transition to the IDLE state.

9.2.3 The Guessing Game

A "guessing-game" machine was defined in Section 7.7.1 starting on page 594, with the following description:

Design a clocked synchronous state machine with four inputs, G1–G4, that are connected to pushbuttons. The machine has four outputs, L1–L4, connected to lamps or LEDs located near the like-numbered pushbuttons. There is also an ERR output connected to a red lamp. In normal operation, the L1–L4 outputs display a 1-out-of-4 pattern. At each clock tick, the pattern is rotated by one position; the clock frequency is about 4 Hz.

Guesses are made by pressing a pushbutton, which asserts an input Gi. When any Gi input is asserted, the ERR output is asserted if the "wrong" pushbutton was pressed, that is, if the Gi input detected at the clock tick does not have the same number as the lamp output that was asserted before the clock tick. Once a guess has been made, play stops and the ERR output maintains the same value for one or more clock ticks until the Gi input is negated, then play resumes.

As we discussed in Section 7.7.1, the machine requires six states—four in which a corresponding lamp is on, and two for when play is stopped after either a good or a bad pushbutton push. A VHDL program for the guessing game is shown in Table 9-21. This version also includes a RESET input that forces the game to a known starting state.

The program is pretty much a straightforward translation of the original state diagram in Figure 7-66 on page 597. Perhaps its only noteworthy feature is in the "SOK | SERR" case. Since the next-state transitions for these two states are identical (either go to S1 or stay in the current state), they can be handled in one case. However, this tricky style of saving typing isn't particularly desirable from the point of view of state-machine documentation or maintainability. In the author's case, the trick's primary benefit was to help fit the program on one book page!

Table 9-21 VHDL program for the guessing-game machine.

```
library IEEE;
use IEEE.std_logic_1164.all;

entity Vggame is
  port ( CLOCK, RESET, G1, G2, G3, G4: in  STD_LOGIC;
         L1, L2, L3, L4, ERR:             out STD_LOGIC );
end;

architecture Vggame_arch of Vggame is
type Sreg_type is (S1, S2, S3, S4, SOK, SERR);
signal Sreg: Sreg_type;
begin

  process (CLOCK)
  begin
    if CLOCK'event and CLOCK = '1' then
      if RESET = '1' then Sreg <= SOK; else
        case Sreg is
          when S1 => if    G2='1' or G3='1' or G4='1' then Sreg <= SERR;
                     elsif G1='1'                      then Sreg <= SOK;
                     else                                   Sreg <= S2;
                     end if;
          when S2 => if    G1='1' or G3='1' or G4='1' then Sreg <= SERR;
                     elsif G1='1'                      then Sreg <= SOK;
                     else                                   Sreg <= S3;
                     end if;
          when S3 => if    G1='1' or G2='1' or G4='1' then Sreg <= SERR;
                     elsif G1='1'                      then Sreg <= SOK;
                     else                                   Sreg <= S4;
                     end if;
          when S4 => if    G1='1' or G2='1' or G3='1' then Sreg <= SERR;
                     elsif G1='1'                      then Sreg <= SOK;
                     else                                   Sreg <= S1;
                     end if;
          when SOK | SERR => if G1='0' and G2='0' and G3='0' and G4='0'
                             then Sreg <= S1; end if;
          when others => Sreg <= S1;
        end case;
      end if;
    end if;
  end process;

  L1  <= '1' when Sreg = S1   else '0';
  L2  <= '1' when Sreg = S2   else '0';
  L3  <= '1' when Sreg = S3   else '0';
  L4  <= '1' when Sreg = S4   else '0';
  ERR <= '1' when Sreg = SERR else '0';

end Vggame_arch;
```

The program in Table 9-21 does not specify a state assignment; a typical synthesis engine will use three bits for Sreg and assign the six states in order to binary combinations 000–101. For this state machine, it is also possible to use an output coded state assignment, using just the lamp and error output signals that are already required. VHDL does not provide a convenient mechanism for grouping together the entity's existing output signals and using them for state, but we can still achieve the desired effect with the changes shown in Table 9-22. Here we used a comment to document the correspondence between outputs and the bits of the new, 5-bit Sreg, and we changed each of the output assignment statements to pick off the appropriate bit instead of fully decoding the state.

Table 9-22
VHDL architecture for guessing game using output-coded state assignment.

```
architecture Vggameoc_arch of Vggame is
signal Sreg: STD_LOGIC_VECTOR (1 to 5);
-- bit positions of output-coded assignment: L1, L2, L3, L4, ERR
constant S1:   STD_LOGIC_VECTOR (1 to 5) := "10000";
constant S2:   STD_LOGIC_VECTOR (1 to 5) := "01000";
constant S3:   STD_LOGIC_VECTOR (1 to 5) := "00100";
constant S4:   STD_LOGIC_VECTOR (1 to 5) := "00010";
constant SERR: STD_LOGIC_VECTOR (1 to 5) := "00001";
constant SOK:  STD_LOGIC_VECTOR (1 to 5) := "00000";
begin

  process (CLOCK)
  ...              (no change to process)
  end process;

  L1  <= Sreg(1);
  L2  <= Sreg(2);
  L3  <= Sreg(3);
  L4  <= Sreg(4);
  ERR <= Sreg(5);

end Vggameoc_arch;
```

9.2.4 Reinventing Traffic-Light Controllers

If you read the ABEL example in Section 9.1.5, then you've already heard me rant about the horrible traffic-light controllers in Sunnyvale, California. They really *do* seem to be carefully designed to *maximize* the waiting time of cars at intersections. In this section we'll design a traffic-light controller with distinctly Sunnyvale-like behavior.

An infrequently used intersection (one that would have no more than a "yield" sign if it were in Chicago) has the sensors and signals shown in

Table 9-23 VHDL program for Sunnyvale traffic-light controller.

```vhdl
library IEEE;
use IEEE.std_logic_1164.all;

entity Vsvale is
  port ( CLOCK, RESET, NSCAR, EWCAR, TMSHORT, TMLONG: in  STD_LOGIC;
         OVERRIDE, FLASHCLK:                          in  STD_LOGIC;
         NSRED, NSYELLOW, NSGREEN:                    out STD_LOGIC;
         EWRED, EWYELLOW, EWGREEN, TMRESET:           out STD_LOGIC );
end;

architecture Vsvale_arch of Vsvale is
type Sreg_type is (NSGO, NSWAIT, NSWAIT2, NSDELAY,
                   EWGO, EWWAIT, EWWAIT2, EWDELAY);
signal Sreg: Sreg_type;
begin

process (CLOCK)
begin
  if CLOCK'event and CLOCK = '1' then
    if RESET = '1' then Sreg <= NSDELAY; else
      case Sreg is
        when NSGO =>                                           -- North-south green.
          if    TMSHORT='0' then Sreg <= NSGO;                 -- Minimum 5 seconds.
          elsif TMLONG='1'  then Sreg <= NSWAIT;               -- Maximum 5 minutes.
          elsif EWCAR='1' and NSCAR='0' then Sreg <= NSGO;     -- Make EW car wait.
          elsif EWCAR='1' and NSCAR='1' then Sreg <= NSWAIT;   -- Thrash if cars both ways.
          elsif EWCAR='0' and NSCAR='1' then Sreg <= NSWAIT;   -- New NS car? Make it stop!
          else                            Sreg <= NSGO;        -- No one coming, no change.
          end if;
        when NSWAIT  => Sreg <= NSWAIT2;                       -- Yellow light,
        when NSWAIT2 => Sreg <= NSDELAY;                       --   two ticks for safety.
        when NSDELAY => Sreg <= EWGO;                          -- Red both ways for safety.
        when EWGO =>                                           -- East-west green.
          if    TMSHORT='0' then Sreg <= EWGO;                 -- Same behavior as above.
          elsif TMLONG='1'  then Sreg <= EWWAIT;
          elsif NSCAR='1' and EWCAR='0' then Sreg <= EWGO;
          elsif NSCAR='1' and EWCAR='1' then Sreg <= EWWAIT;
          elsif NSCAR='0' and EWCAR='1' then Sreg <= EWWAIT;
          else                            Sreg <= EWGO;
          end if;
        when EWWAIT  => Sreg <= EWWAIT2;
        when EWWAIT2 => Sreg <= EWDELAY;
        when EWDELAY => Sreg <= NSGO;
        when others  => Sreg <= NSDELAY;                       -- "Reset" state.
      end case;
    end if;
  end if;
end process;
```

Figure 9-5 on page 809. The state machine that controls the traffic signals uses a 1-Hz clock and a timer and has four inputs:

NSCAR   Asserted when a car on the north-south road is over either sensor on either side of the intersection.

EWCAR   Asserted when a car on the east-west road is over either sensor on either side of the intersection.

TMLONG   Asserted if more than five minutes has elapsed since the timer started; remains asserted until the timer is reset.

TMSHORT   Asserted if more than five seconds has elapsed since the timer started; remains asserted until the timer is reset.

The state machine has seven outputs:

NSRED, NSYELLOW, NSGREEN   Control the north-south lights.

EWRED, EWYELLOW, EWGREEN   Control the east-west lights.

TMRESET   When asserted, resets the timer and negates TMSHORT and TMLONG. The timer starts timing when TMRESET is negated.

A typical, municipally approved algorithm for controlling the traffic lights is embedded in the VHDL program of Table 9-23. This algorithm produces two frequently seen behaviors of "smart" traffic lights. At night, when traffic is light, it holds a car stopped at the light for up to five minutes, unless a car approaches on the cross street, in which case it stops the cross traffic and lets the waiting car go. (The "early warning" sensor is far enough back to change the lights before the approaching car reaches the intersection.) During the day, when traffic is heavy and there are always cars waiting in both directions, it cycles the lights every five seconds, thus minimizing the utilization of the intersection and maximizing everyone's waiting time, thereby creating a public outcry for more taxes to fix the problem.

Table 9-23 (continued) VHDL program for Sunnyvale traffic-light controller.

```
TMRESET   <= '1' when Sreg=NSWAIT2 or Sreg=EWWAIT2 else '0';
NSRED     <= FLASHCLK when OVERRIDE='1' else
             '1' when Sreg/=NSGO and Sreg/=NSWAIT and Sreg/=NSWAIT2 else '0';
NSYELLOW  <= '0' when OVERRIDE='1' else
             '1' when Sreg=NSWAIT or Sreg=NSWAIT2 else '0';
NSGREEN   <= '0' when OVERRIDE='1' else '1' when Sreg=NSGO else '0';
EWRED     <= FLASHCLK when OVERRIDE='1' else
             '1' when Sreg/=EWGO and Sreg/=EWWAIT and Sreg/=EWWAIT2 else '0';
EWYELLOW  <= '0' when OVERRIDE='1' else
             '1' when Sreg=EWWAIT or Sreg=EWWAIT2 else '0';
EWGREEN   <= '0' when OVERRIDE='1' else '1' when Sreg=EWGO else '0';

end Vsvale_arch;
```

While writing the program, we took the opportunity to add two inputs that weren't in the original specification. The OVERRIDE input may be asserted by the police to disable the controller and put the signals into a flashing-red mode at a rate determined by the FLASHCLK input. This allows them to manually clear up the traffic snarls created by this wonderful invention.

Table 9-24 Definitions for Sunnyvale traffic-lights machine with output-coded state assignment.

```
library IEEE;
use IEEE.std_logic_1164.all;

entity Vsvale is
  port ( CLOCK, RESET, NSCAR, EWCAR, TMSHORT, TMLONG: in  STD_LOGIC;
         OVERRIDE, FLASHCLK:                          in  STD_LOGIC;
         NSRED, NSYELLOW, NSGREEN:                    out STD_LOGIC;
         EWRED, EWYELLOW, EWGREEN, TMRESET:           out STD_LOGIC );
end;

architecture Vsvaleoc_arch of Vsvale is
signal Sreg: STD_LOGIC_VECTOR (1 to 7);
-- bit positions of output-coded assignment:  (1) NSRED, (2) NSYELLOW, (3) NSGREEN,
--                                 (4) EWRED, (5) EWYELLOW, (6) EWGREEN, (7) EXTRA
constant NSGO:    STD_LOGIC_VECTOR (1 to 7) := "0011000";
constant NSWAIT:  STD_LOGIC_VECTOR (1 to 7) := "0101000";
constant NSWAIT2: STD_LOGIC_VECTOR (1 to 7) := "0101001";
constant NSDELAY: STD_LOGIC_VECTOR (1 to 7) := "1001000";
constant EWGO:    STD_LOGIC_VECTOR (1 to 7) := "1000010";
constant EWWAIT:  STD_LOGIC_VECTOR (1 to 7) := "1000100";
constant EWWAIT2: STD_LOGIC_VECTOR (1 to 7) := "1000101";
constant EWDELAY: STD_LOGIC_VECTOR (1 to 7) := "1001001";

begin

process (CLOCK)
...             (no change to process)
end process;

TMRESET  <= '1' when Sreg=NSWAIT2 or Sreg=EWWAIT2 else '0';
NSRED    <= Sreg(1);
NSYELLOW <= Sreg(2);
NSGREEN  <= Sreg(3);
EWRED    <= Sreg(4);
EWYELLOW <= Sreg(5);
EWGREEN  <= Sreg(6);

end Vsvaleoc_arch;
```

Like most of our other examples, Table 9-23 does not give a specific state assignment. And like many of our other examples, this state machine works well with an output-coded state assignment. Many of the states can be identified by a unique combination of light-output values. But there are three pairs of states that are not distinguishable by looking at the lights alone: (NSWAIT, NSWAIT2), (EWWAIT, EWWAIT2), and (NSDELAY, EQDELAY). We can handle these by adding one more state variable, "EXTRA", that has different values for the two states in each pair. This idea is realized in the modified program in Table 9-24.

Exercises

9.1   Write an ABEL program for the state machine described in Exercise 7.30.

9.2   Modify the ABEL program of Table 9-2 to include the HINT output from the original state-machine specification in Section 7.4.

9.3   Redesign the T-bird tail-lights machine of Section 9.1.3 to include parking-light and brake-light functions. When the BRAKE input is asserted, all of the lights should go on immediately, and stay on until BRAKE is negated, independent of any other function. When the PARK input is asserted, each lamp is turned on at 50% brightness at all times when it would otherwise be off. This is achieved by driving the lamp with a 100-Hz signal DIMCLK with a 50% duty cycle. Draw a logic diagram for the circuit using one or two PLDs, write an ABEL program for each PLD, and write a short description of how your system works.

9.4   Find a 3-bit state assignment for the guessing-game machine in Table 9-5 that reduces the maximum number of product terms per output to 7. Can you do even better?

9.5   The operation of the guessing game in Section 9.1.4 is very predictable; it's easy for a player to learn the rate at which the lights change and always hit the button at the right time. The game is more fun if the rate of change is more variable.

Modify the ABEL state machine in Table 9-5 so that in states S1–S4, the machine advances only if a new input, SEN, is asserted. (SEN is intended to be hooked up to a pseudorandom bit-stream generator.) Both correct and incorrect button pushes should be recognized whether or not SEN is asserted. Determine whether your modified design still fits in a 16V8.

9.6   In connection with the preceding exercise, write an ABEL program for an 8-bit LFSR using a single 16V8, such that one of its outputs can be used as a pseudorandom bit-stream generator. After how many clock ticks does the bit sequence repeat? What is the maximum number of 0s that occur in a row? of 1s?

9.7   Add an OVERRIDE input to the traffic-lights state machine of Figure 9-7, still using just a single 16V8. When OVERRIDE is asserted, the red lights should flash on and off, changing once per second. Write a complete ABEL program for your machine.

9.8   Modify the behavior of the ABEL traffic-light-controller machine in Table 9-9 to have more reasonable behavior, the kind you'd like to see for traffic lights in your own home town.

9.9　　Write a VHDL program for the state machine described in Exercise 7.30.

9.10　Show how to modify Table 9-13 to provide an asynchronous RESET input that forces the state machine to the INIT state. Repeat for a synchronous version that forces the state machine to the INIT state if RESET is asserted on the rising clock edge.

9.11　Write and test a VHDL program corresponding to Figure 9-8(b). Using an available synthesis tool, determine whether the circuit resulting from this version differs from the one resulting from Figure 9-8(a), and how.

9.12　Write a VHDL program for the ones-counting state machine as described by the state table in Table 7-12.

9.13　Write a VHDL program for the combination-lock state machine as described by the state table in Table 7-14.

9.14　Modify the VHDL program of Table 9-19 to include the HINT output from the original state-machine specification in Section 7.4.

9.15　Repeat Exercise 9.3 using VHDL, assuming you are targeting your design to a single CPLD or FPGA.

9.16　Modify the VHDL guessing-game program of Table 9-21 according to the idea in Exercise 9.5. Add another process to the program to provide a pseudorandom bit-stream generator according to Exercise 9.6.

9.17　Modify the behavior of the VHDL traffic-light-controller machine in Table 9-23 to have more reasonable behavior, the kind you'd like to see for traffic lights in your own home town.

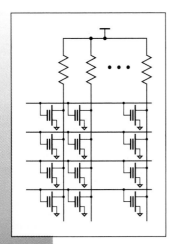

Memory, CPLDs, and FPGAs

A ny sequential circuit has memory of a sort, since each flip-flop or latch stores one bit of information. However, we usually reserve the word "memory" to refer to bits that are stored in a structured way, usually as a two-dimensional array in which one row of bits is accessed at a time.

This chapter describes several different types of memory organizations and commercially available memory chips. The same kinds of memory may be embedded into larger VLSI chips, where they are combined with other circuits to perform a useful function.

The applications of memory are many and varied. In a microprocessor's central processing unit (CPU), a "read-only memory" may be used to define the primitive steps that are performed to execute instructions in the CPU's instruction set. Alongside the CPU, a fast "static memory" may serve as a cache to hold recently used instructions and data. And a microprocessor's main-memory subsystem may contain hundreds of millions of bits in "dynamic memory" that store complete operating systems, programs, and data.

Applications of memory are not limited to microprocessors or even to purely digital systems. For example, equipment in the public telephone system uses read-only memories to perform certain transformations on digitized voice signals, and fast "static memories" as a "switching fabric" to route digitized voice between subscribers. Many portable audio compact-disc players "read ahead" and store a few seconds of audio in a "dynamic memory" so that the unit can keep playing even if it is physically jarred (this requires over 1.4 million bits per second of stored audio). And there are many examples of modern audio/visual equipment that use memories to temporarily store digitized signals for enhancement through digital signal processing.

This chapter begins with a discussion of read-only memory and its applications and then describes two different types of read/write memory. Optional subsections discuss the internal structure of the different memory types.

The last two sections in this chapter discuss CPLDs and FPGAs. These devices are akin to memories in that they are large, regular structures that can be used in a variety of applications. By enabling very fast development of customized logic functions, they have become essential building blocks in modern digital design.

10.1 Read-Only Memory

read-only memory (ROM)

address input

data output

A *read-only memory (ROM)* is a combinational circuit with n inputs and b outputs, as shown in Figure 10-1. The inputs are called *address inputs* and are traditionally named A0, A1, ... , An−1. The outputs are called *data outputs* and are typically named D0, D1, ... , Db−1.

A ROM "stores" the truth table of an n-input, b-output combinational logic function. For example, Table 10-1 is the truth table of a 3-input, 4-output combinational function; it could be stored in a $2^3 \times 4$ (8×4) ROM. Neglecting propagation delays, a ROM's data outputs at all times equal the output bits in the truth-table row selected by the address inputs.

Since a ROM is a combinational circuit, you would be correct to say that it's not really a memory at all. In terms of digital circuit operation, you can treat

Figure 10-1
Basic structure of a $2^n \times b$ ROM.

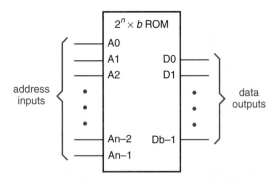

MEMORY—NOT!   Most types of read-only memory are not really memory in the strictest sense of the word, because they are combinational, not sequential, circuits. They are called "memory" because of the organizational paradigm that describes their function.

Inputs			Outputs			
A2	A1	A0	D3	D2	D1	D0
0	0	0	1	1	1	0
0	0	1	1	1	0	1
0	1	0	1	0	1	1
0	1	1	0	1	1	1
1	0	0	0	0	0	1
1	0	1	0	0	1	0
1	1	0	0	1	0	0
1	1	1	1	0	0	0

Table 10-1
Truth table for a 3-input, 4-output combinational logic function.

a ROM like any other combinational logic element. However, you can also think of information as being "stored" in the ROM when it is manufactured or programmed (we'll say more about how this is done in Section 10.1.4).

Although we think of ROM as being a type of memory, it has an important difference from most other types of integrated-circuit memory. ROM is *non-volatile memory*; that is, its contents are preserved even if no power is applied.

nonvolatile memory

10.1.1 Using ROMs for "Random" Combinational Logic Functions

Table 10-1 is actually the truth table of a 2-to-4 decoder with an output-polarity control, a function that can be built with discrete gates as shown in Figure 10-2. Thus, we have two different ways to build the decoder—with discrete gates, or with an 8×4 ROM that contains the truth table, as shown in Figure 10-3.

The assignment pattern of decoder inputs and outputs to ROM inputs and outputs in Figure 10-3 is a consequence of the way that the truth table in Table 10-1 is constructed. Thus, the physical ROM realization of the decoder is

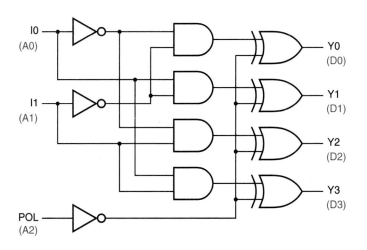

Figure 10-2
A 2-to-4 decoder with output-polarity control.

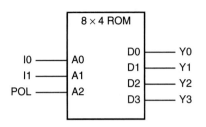

Figure 10-3
Connections to build the
2-to-4 decoder using an 8×4
ROM that stores Table 10-1.

not unique. That is, we could write the rows or columns of the truth table in a different order and use a physically different ROM to perform the same logic function, simply by assigning the decoder signals to different ROM inputs and outputs. Another way to look at this is that we can rename the individual address inputs and data outputs of the ROM.

For example, swapping the bits in the D0 and D3 columns of Table 10-1 would give us the truth table for a physically different ROM. However, the new ROM could still be used to build the 2-to-4 decoder simply by swapping the Y0 and Y3 labels in Figure 10-3. Likewise, if we shuffled the data rows of the truth table as shown in Table 10-2, we would get another different ROM, but it could still be used as the 2-to-4 decoder with a rearrangement of the address inputs, A0 = POL, A1 = I0, A2 = I1.

When constructing a ROM to store a given truth table, input and output signals reading from right to left in the truth table are normally assigned to ROM address inputs and data outputs with ascending labels. Each address or data combination may then be read as a corresponding binary integer with the bits numbered in the "natural" way. A data file is typically used to specify the truth table to be stored in the ROM when it is manufactured or programmed. The data file usually gives the address and data values as hexadecimal numbers. For example, a data file may specify Table 10-2 by saying that ROM addresses 0–7 should store the values E, 1, D, 2, B, 4, 7, 8.

Table 10-2
Truth table with data
rows shuffled.

Inputs			Outputs			
A2	A1	A0	D3	D2	D1	D0
0	0	0	1	1	1	0
0	0	1	0	0	0	1
0	1	0	1	1	0	1
0	1	1	0	0	1	0
1	0	0	1	0	1	1
1	0	1	0	1	0	0
1	1	0	0	1	1	1
1	1	1	1	0	0	0

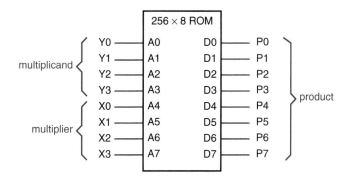

Figure 10-4
Connections to
perform a 4×4
unsigned binary
multiplication using
a 256×8 ROM.

Another simple example of a function that can be built with ROM is 4×4 unsigned binary multiplication. We showed an ABEL program for this function in Section 5.11.2 and found that the number of product terms required (36) was too high to obtain with just one pass through a conventional PLD's AND-OR array. Alternatively, we can realize the function with one pass through a $2^8 \times 8$ (256×8) ROM with the connections shown in Figure 10-4.

A ROM's contents are normally specified by a file that contains one entry for every address in the ROM. For example, Table 10-3 is a hexadecimal listing of the 4×4 multiplier ROM contents. Each row gives a starting address in the ROM and specifies the 8-bit data values stored at 16 successive addresses.

The nice thing about ROM-based design is that you can usually write a simple program in a high-level language to calculate what should be stored in the ROM. For example, it took only a few minutes to write a C program, shown in Table 10-4, that generated the contents of Table 10-3.

Table 10-3 Hexadecimal text file specifying the contents of a 4×4 multiplier ROM.

```
00:  00 00 00 00 00 00 00 00 00 00 00 00 00 00 00 00
10:  00 01 02 03 04 05 06 07 08 09 0A 0B 0C 0D 0E 0F
20:  00 02 04 06 08 0A 0C 0E 10 12 14 16 18 1A 1C 1E
30:  00 03 06 09 0C 0F 12 15 18 1B 1E 21 24 27 2A 2D
40:  00 04 08 0C 10 14 18 1C 20 24 28 2C 30 34 38 3C
50:  00 05 0A 0F 14 19 1E 23 28 2D 32 37 3C 41 46 4B
60:  00 06 0C 12 18 1E 24 2A 30 36 3C 42 48 4E 54 5A
70:  00 07 0E 15 1C 23 2A 31 38 3F 46 4D 54 5B 62 69
80:  00 08 10 18 20 28 30 38 40 48 50 58 60 68 70 78
90:  00 09 12 1B 24 2D 36 3F 48 51 5A 63 6C 75 7E 87
A0:  00 0A 14 1E 28 32 3C 46 50 5A 64 6E 78 82 8C 96
B0:  00 0B 16 21 2C 37 42 4D 58 63 6E 79 84 8F 9A A5
C0:  00 0C 18 24 30 3C 48 54 60 6C 78 84 90 9C A8 B4
D0:  00 0D 1A 27 34 41 4E 5B 68 75 82 8F 9C A9 B6 C3
E0:  00 0E 1C 2A 38 46 54 62 70 7E 8C 9A A8 B6 C4 D2
F0:  00 0F 1E 2D 3C 4B 5A 69 78 87 96 A5 B4 C3 D2 E1
```

Table 10-4 Program to generate the text file specifying
the contents of a 4 × 4 multiplier ROM.

```
#include <stdio.h>

/* Procedure to print d as a hex digit. */
void PrintHexDigit(int d)
{
    if (d<10) printf("%c", '0'+d);
    else printf("%c", 'A'+d-10);
}

/* Procedure to print i as two hex digits.  */
void PrintHex2(int i)
{
    PrintHexDigit((i / 16) % 16);
    PrintHexDigit(i % 16);
}

void main()
{
  int x, y;

  for (x=0; x<=15; x++) {
    PrintHex2(x*16); printf(":");
    for (y=0; y<=15; y++) {
      printf(" ");
      PrintHex2(x*y);
    }
    printf("\n");
  }
}
```

*10.1.2 Internal ROM Structure

The mechanism used by ROMs to "store" information varies with different ROM technologies. In most ROMs, the presence or absence of a diode or transistor distinguishes between a 0 and a 1.

Figure 10-5 is the schematic of a primitive 8 × 4 ROM that you could build yourself using an MSI decoder and a handful of diodes. The address inputs select one of the decoder outputs to be asserted. Each decoder output is called a
word line because it selects one row or word of the table stored in the ROM. The figure shows the situation with A2–A0 = 101 and ROW5_L asserted.

word line

bit line

Each vertical line in Figure 10-5 is called a *bit line* because it corresponds to one output bit of the ROM. An asserted word line pulls a bit line LOW if a diode is connected between the word line and the bit line. There is only one

*Throughout this book, optional sections are marked with an asterisk.

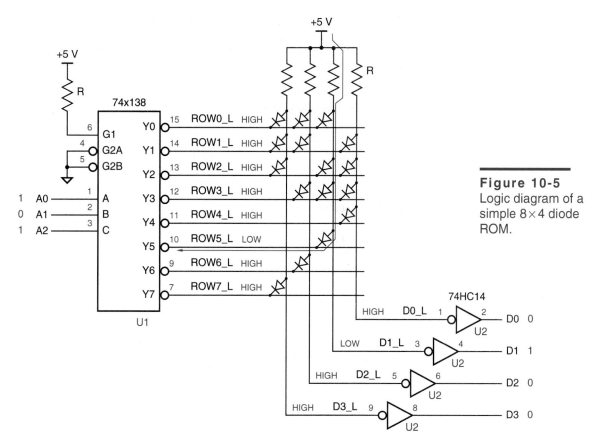

Figure 10-5
Logic diagram of a simple 8×4 diode ROM.

diode in row 5, and the corresponding bit line (D1_L) is pulled LOW. The bit lines are buffered through inverters to produce the D3–D0 ROM outputs, 0010 for the case shown.

In the ROM circuit of Figure 10-5, each intersection between a word line and a bit line corresponds to one bit of "memory." If a diode is present at the intersection, a 1 is stored; otherwise, a 0 is stored. If you were to build this circuit in the lab, you would "program" the memory by inserting and removing diodes at each intersection. Primitive though it may seem, owners of the DEC PDP-11 minicomputer (circa 1970) made use of similar technology in the M792 32×16 "bootstrap ROM module." The module was shipped with 512 diodes soldered in

DETAILS, DETAILS	Inverters with CMOS input thresholds are used in Figure 10-5 to improve the circuit's noise margins, because the 0.7-volt drop across the diodes creates a LOW level on the bit lines that is not really so low. Fortunately, with the 74HC04, a LOW level is anything less than 1.35 V. Of course, you'd never build this circuit except in an academic lab—it's easier just to buy and program a commercial ROM chip.

SNEAK PATHS   The ROM circuit of Figure 10-5 must use diodes, not direct connections, at each location where a 1 is to be stored. Figure 10-6 shows what happens if just a few diodes—such as the ones in row 3—are replaced with direct connections. Suppose the address inputs are 101; then ROW5_L is asserted, and only D1_L is supposed to be pulled LOW to create an output of 0010. However, the direct connections allow current to flow along *sneak paths*, so that bit lines D2_L and D0_L are also pulled LOW, and the (incorrect) output is 0111. When diodes are used, sneak paths are blocked by reverse-biased diodes, and correct outputs are obtained.

place, and owners programmed it by clipping out the diode at each location where a 0 was to be stored.

The diode pattern shown in Figure 10-5 corresponds to the 2-to-4-decoder truth table of Table 10-1. This doesn't seem very efficient—we used a 3-to-8 decoder and a bunch of diodes to build the ROM version of a 2-to-4 decoder. We could have used a subset of the 3-to-8 decoder directly! However, we'll show a more efficient ROM structure and some more useful design examples later.

Figure 10-6
Sneak paths
in a ROM.

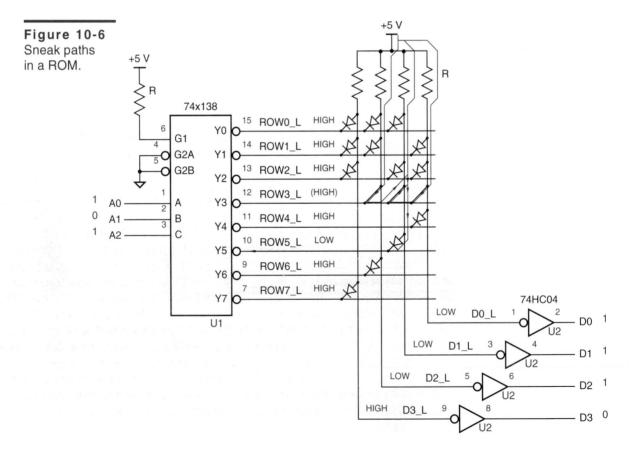

*10.1.3 Two-Dimensional Decoding

Suppose you wanted to build a 128×1 ROM using the kind of structure described in the preceding subsection. Have you ever thought about what it would take to build a 7-to-128 decoder? Try 128 7-input NAND gates to begin with, and add 14 buffers and inverters with a fanout of 64 each! ROMs with a million bits or more are available commercially; trust me, they do not contain 20-to-1,048,576 decoders. Instead, a different structure, called *two-dimensional decoding*, is used to reduce the decoder size to something on the order of the square root of the number of addresses.

two-dimensional decoding

The basic idea in two-dimensional decoding is to arrange the ROM cells in an array that is as close as possible to square. For example, Figure 10-7 shows a possible internal structure for a 128×1 ROM. The three high-order address bits, A6–A4, are used to select a row. Each row stores 16 bits starting at address (A6, A5, A4, 0, 0, 0, 0). When an address is applied to the ROM, all 16 bits in the selected row are "read out" in parallel on the bit lines. A 16-input multiplexer selects the desired data bit based on the low-order address bits.

Figure 10-7 Internal structure of a 128×1 ROM using two-dimensional decoding.

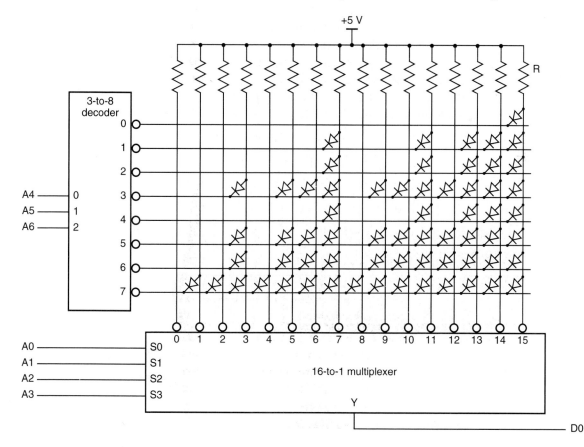

TRANSISTORS AS ROM ELEMENTS

MOS ROMs actually use a transistor instead of a diode at each location where a bit is to be stored; Figure 10-8 shows the basic idea. The row decoder has active-high outputs. When a row line is asserted, the NMOS transistors in that row are turned on, which pulls the corresponding bit lines low. A similar idea can be used in ROMs built with bipolar transistors.

By the way, the diode pattern in Figure 10-7 was not chosen at random. It performs a very useful 7-input combinational logic function that would require 35 4-input AND gates to build as a minimal two-level AND-OR circuit (see Exercise 10.7). The ROM version of this function would actually save a fair amount of engineering effort and space in a board-level design.

Two-dimensional decoding allows a 128×1 ROM to be built with a 3-to-8 decoder and a 16-input multiplexer (whose complexity is comparable to that of a 4-to-16 decoder). A $1M \times 1$ ROM could be built with a 10-to-1024 decoder and a 1024-input multiplexer—not easy, but a lot simpler than the one-dimensional alternative.

Besides reducing decoding complexity, two-dimensional decoding has one other benefit—it leads to a chip whose physical dimensions are close to square, important for chip fabrication and packaging. A chip with a $1M \times 1$ physical array would be *very* long and skinny and could not be built economically.

In ROMs with multiple data outputs, the storage arrays corresponding to each data output may be made narrower in order to achieve an overall chip layout that is closer to square. For example, Figure 10-9 shows the possible layout of a $32K \times 8$ ROM.

Figure 10-8
MOS transistors as storage elements in a ROM.

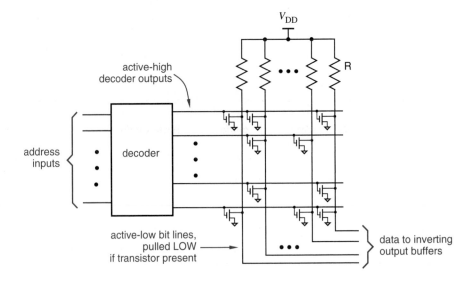

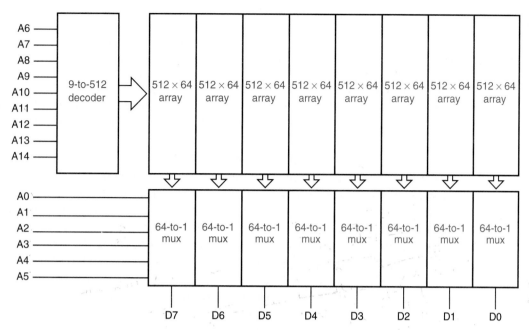

Figure 10-9 Possible layout of a 32K × 8 ROM.

10.1.4 Commercial ROM Types

Unless you visit computer museums, it's impossible nowadays to find ROM modules built with discrete diodes. A modern ROM is fabricated as a single IC chip; one that stores four megabits can be purchased for under $5. Various methods are used to "program" the information stored in a ROM, as discussed below and summarized in Table 10-5.

Most of the early integrated-circuit ROMs were *mask-programmable ROMs* (or, simply, *mask ROMs*). A mask ROM is programmed by the pattern of connections and no-connections in one of the *masks* used in the IC manufacturing process. To program or write information into the ROM, the customer gives

mask-programmable ROM

mask ROM

mask

Table 10-5 Commercial ROM types.

Type	Technology	Read cycle	Write cycle	Comments
Mask ROM	NMOS, CMOS	10–200 ns	4 weeks	Write once; low power
Mask ROM	Bipolar	< 100 ns	4 weeks	Write once; high power; low density
PROM	Bipolar	< 100 ns	10–50 μs/byte	Write once; high power; no mask charge
EPROM	NMOS, CMOS	25–200 ns	10–50 μs/byte	Reusable; low power; no mask charge
EEPROM	NMOS	50–200 ns	10–50 μs/byte	10,000–100,000 writes/location limit

the manufacturer a listing of the desired ROM contents, using a floppy disk or other transfer medium. The manufacturer uses this information to create one or more customized masks to manufacture ROMs with the required pattern. ROM manufacturers impose a *mask charge* of several thousand dollars for the "customized" aspects of mask-ROM production. Because of mask charges and the four-week delay typically required to obtain programmed chips, mask ROMs are normally used today only in very high-volume applications. For low-volume applications there are more cost-effective choices, discussed next.

mask charge

A *programmable read-only memory (PROM)* is similar to a mask ROM, except that the customer may store data values (i.e., "program the PROM") in just a few minutes using a *PROM programmer*. A PROM chip is manufactured with all of its diodes or transistors "connected." This corresponds to having all bits at a particular value, typically 1. The PROM programmer can be used to set desired bits to the opposite value. In bipolar PROMs, this is done by vaporizing tiny *fusible links* inside the PROM corresponding to each bit. A link is vaporized by selecting it using the PROM's address and data lines, and then applying a high-voltage pulse (10–30 V) to the device through a special input pin.

programmable read-only memory (PROM)

PROM programmer

fusible link

Early bipolar PROMs had reliability problems. Sometimes the stored bits changed because of incompletely vaporized links that would "grow back," and sometimes intermittent failures occurred because of floating shrapnel inside the IC package. However, these problems were worked out, and reliable fusible-link technology is used nowadays not only in bipolar PROMs, but also in the bipolar PLD circuits that we described in Section 5.3.4.

An *erasable programmable read-only memory (EPROM)* is programmable like a PROM, but it can also be "erased" to the all-1s state by exposing it to ultraviolet light. No, the light does not cause fuses to grow back! Rather, EPROMs use a different technology, called "floating-gate MOS."

erasable programmable read-only memory (EPROM)

As shown in Figure 10-10, an EPROM has a *floating-gate MOS transistor* at every bit location. Each transistor has two gates. The "floating" gate is not connected and is surrounded by extremely high-impedance insulating material. To program an EPROM, the programmer applies a high voltage to the non-floating gate at each bit location where a 0 is to be stored. This causes a temporary breakdown in the insulating material and allows a negative charge to accumulate on the floating gate. When the high voltage is removed, the negative charge remains. During subsequent read operations, the negative charge prevents the MOS transistor from turning on when it is selected.

floating-gate MOS transistor

EPROM manufacturers "guarantee" that a properly programmed bit will retain 70% of its charge for at least 10 years, even if the part is stored at 125°C, so EPROMs definitely fall into the category of "nonvolatile memory." However, they can also be erased. The insulating material surrounding the floating gate becomes slightly conductive if it is exposed to ultraviolet light with a certain wavelength. Thus, EPROMs can be erased by exposing the chips to ultraviolet

EPROM erasing

V_{DD}

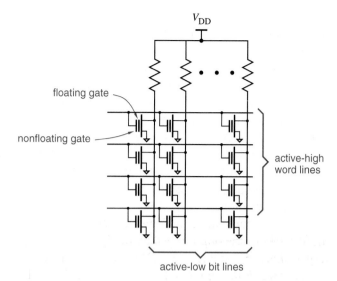

floating gate

nonfloating gate

active-high
word lines

active-low bit lines

Figure 10-10
Storage matrix in
an EPROM using
floating-gate MOS
transistors.

light, typically for 5–20 minutes. An EPROM chip is normally housed in a package with a quartz lid through which the chip may be exposed to the erasing light.

Probably the most common application of EPROMs is to store programs in microprocessor systems. EPROMs are typically used during program development, where the program or other information in the EPROM must be repeatedly changed during debugging. However, ROMs and PROMs usually cost less than EPROMs of similar capacity. Therefore, once a program is finalized, a ROM or PROM may be used in production to save cost. Most of today's PROMs are actually EPROMS housed in inexpensive packages without quartz lids; these are sometimes called *one-time programmable (OTP) ROMs*.

one-time programmable (OTP) ROM

An *electrically erasable programmable read-only memory (EEPROM)* is like an EPROM, except that individual stored bits may be erased electrically. The floating gates in an EEPROM are surrounded by a much thinner insulating layer and can be erased by applying a voltage of the opposite polarity as the charging voltage to the nonfloating gate. Large EEPROMs (1 Mbit or larger) allow erasing only in fixed-size blocks, typically 128–512 Kbits (16–64 Kbytes) at a time. These memories are usually called *flash EPROMs* or *flash memories*, because erasing occurs "in a flash."

electrically erasable programmable read-only memory (EEPROM)

flash EPROM
flash memory

As indicated in Table 10-5, programming or "writing" an EEPROM location takes much longer than reading it, so an EEPROM is no substitute for the read/write memories discussed later in this chapter. Also, because the insulating layer is so thin, it can be worn out by repeated programming operations. As a result, EEPROMs can be reprogrammed only a limited number of times, as few as 10,000 times per location. Therefore, EEPROMs are typically used for storing data that must be preserved when the equipment is not powered, but that doesn't change very often, such as the default configuration data for a computer.

Popular ROMs for microprocessors and other moderate-speed ROM applications include the *2764*, *27128*, *27256*, and *27512* EPROMs, whose logic symbols are shown in Figure 10-11. The figure also indicates pins that must be connected to a constant signal during normal read-only operation. Inputs labeled V_{CC} must be connected to +5 volts, inputs labeled V_{IH} must be connected to a valid HIGH logic signal, and "N.C." means "no connection." Different input configurations are used to program and test the device; the pin labeled VPP is used to apply the programming voltage.

Larger ROMs are available with more bits and in some cases wider data outputs of 16 or 32 bits. In 1999, one of the most impressive was the AMD 29LV640 flash memory, with 64 Mbits in 256 32-Kbyte sectors. Much smaller ROMs are also produced with 3-bit serial interfaces for specialized applications, such as downloading the programming information into FPGAs.

Multiple flash memories are often packaged into a single credit-card-size module for applications requiring large amounts of nonvolatile storage. The most common application of these modules is in digital cameras, where storing a single high-resolution image may require as much as 4 Mbytes. In 1999, the largest flash card was sold by industry leader SanDisk Corporation and contained 192 Mbytes (1536 Mbits) of memory.

Figure 10-11 Logic symbols for standard EPROMs in 28-pin dual in-line packages.

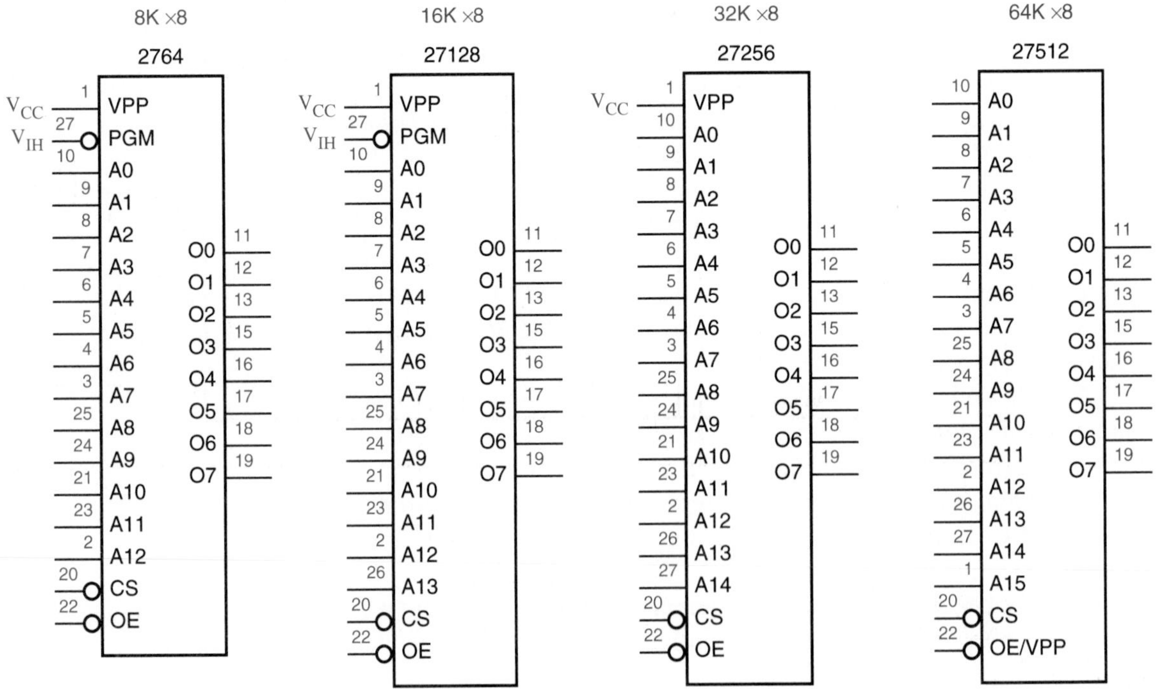

10.1.5 ROM Control Inputs and Timing

The outputs of a ROM must often be connected to a three-state bus, where different devices may drive the bus at different times. Therefore, most commercial ROM chips have three-state data outputs and an *output-enable (OE) input* that must be asserted to enable the outputs.

output-enable (OE) input

Many ROM applications, especially program storage, have multiple ROMs connected to a bus, where only one ROM drives the bus at a time. Most ROMs have a *chip-select (CS) input* to simplify the design of such systems. In addition to OE, a ROM's CS input must be asserted to enable the three-state outputs.

chip-select (CS) input

Figure 10-12 shows how the OE and CS inputs could be used when connecting four 32K × 8 ROMs to a microprocessor system that requires 128 Kbytes of ROM. The microprocessor has an 8-bit data bus and a 20-bit address bus, for a maximum address space of 1 Mbyte (2^{20} bytes). The ROM is supposed to be located in the highest 128K of the address space. To obtain this behavior, a NAND gate is used to produce the HIMEM_L signal, which is asserted when the address bus contains an address in the highest 128K (A19–A17 = 111). A 74x139 decoder then selects one of the four 32K × 8 ROMs. The selected ROM drives

Figure 10-12 Address decoding and ROM enabling in a microprocessor system.

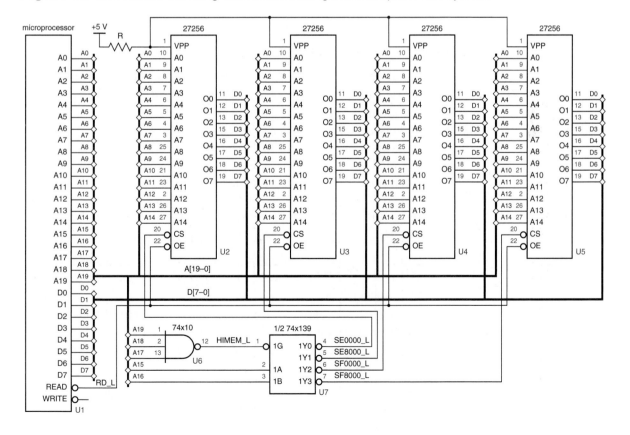

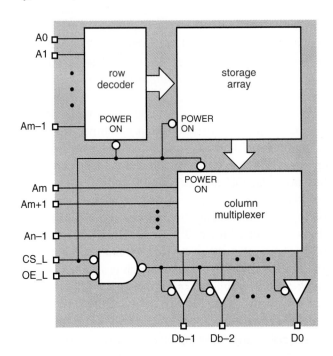

Figure 10-13
Internal ROM
structure, showing
use of control inputs.

the data bus only when the microprocessor requests a read operation by asserting READ, which is connected to all of the OE inputs.

As we've described it so far, a CS input is no more than a second output-enable input that is ANDed with OE to enable the three-state outputs. However, in many ROMs, CS also serves as a *power-down input*. When CS is negated, power is removed from the ROM's internal decoders, drivers, and multiplexers. In this *standby mode* of operation, a typical ROM consumes less than 10% of the power it uses in *active mode* with CS asserted. In Figure 10-12, at most one ROM is selected at any time, so the total power consumption of all the ROM chips is much closer to that of one chip than four.

Figure 10-13 shows how CS and OE inputs are used inside a typical ROM. Figure 10-14 shows typical ROM timing, including the following parameters:

power-down input

standby mode
active mode

t_{AA}
*access time from
  address*

t_{AA} *Access time from address.* The access time from address of a ROM is the propagation delay from stable address inputs to valid data outputs. When designers talk about "a 100-ns ROM," they are usually referring to this parameter.

t_{ACS}
*access time from chip
  select*

t_{ACS} *Access time from chip select.* The access time from chip select of a ROM is the propagation delay from the time CS is asserted until the data outputs are valid. In some chips, this is longer than the access time from address, because the chip takes a little while to "power up." In others, this time is shorter because CS controls only output enabling.

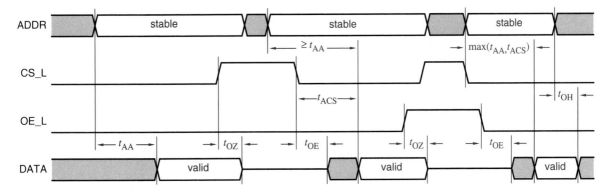

Figure 10-14 ROM timing.

t_{OE} *Output-enable time.* This parameter is usually much shorter than access time. The output-enable time of a ROM is the propagation delay from OE and CS both asserted until the three-state output drivers have left the Hi-Z state. Depending on whether the address inputs have been stable long enough, the output data may or may not be valid at that point.

t_{OE}
output-enable time

t_{OZ} *Output-disable time.* The output-disable time of a ROM is the propagation delay from the time OE or CS is negated until the three-state output drivers have entered the Hi-Z state.

t_{OZ}
output-disable time

t_{OH} *Output-hold time.* The output-hold time of a ROM is the length of time that the outputs remain valid after a change in the address inputs, or after OE_L or CS_L is negated.

t_{OH}
output-hold time

As with other components, the manufacturer specifies maximum and, sometimes, typical values for all timing parameters. Usually, minimum values are also specified for t_{OE} and t_{OH}. The minimum value of t_{OH} is usually specified to be 0; that is, the minimum combinational-logic delay through the ROM is 0.

NOT ALL INPUTS ARE CREATED EQUAL   Given the decoding and multiplexing structure in Figure 10-13, you might think that the address access time is shorter from some address inputs than it is from others. In a given application, if some address-input signals were delayed relative to the others, you could recover this delay by connecting the slower signals to the "faster" ROM inputs. After all, any input signal can be connected to any ROM address input if you're willing to rearrange the ROM contents accordingly.

However, ROM manufacturers don't specify which, if any, inputs are faster. In fact, the internal electrical characteristics of most ROMs may be such that the difference is not enough to bother with.

UNBELIEVABLY FAST ROMS   The actual delay through a ROM is never 0, of course, but it can easily be just slightly less than what you need to meet a nonzero hold-time requirement elsewhere in a design. So it's best to assume that t_{OH} for a ROM is 0, unless you really know what you're doing.

10.1.6 ROM Applications

As we mentioned earlier, the most common application of ROMs is for program storage in microprocessor systems. However, in many applications a ROM can provide a low-cost realization of a complex or "random" combinational logic function. In this section we'll give a couple of examples of ROM-based circuits that are used in the digital telephone network.

When an analog voice signal enters a typical digital telephone system, it is sampled 8,000 times per second and converted into a sequence of 8-bit bytes representing the analog signal at each sampling point. In Section 8.5.3 we showed how these digital voice samples could be converted between parallel and serial formats, but we did not describe the coding of the 8-bit bytes themselves. We'll do that now, and then show how ROM-based circuits can easily deal with this highly encoded information.

The simplest 8-bit encoding of the sign and amplitude of an analog signal would be an 8-bit integer in the two's-complement or signed-magnitude system; this is called a *linear encoding*. However, an 8-bit linear encoding yields a *dynamic range*—the ratio between range of representable numbers and the smallest representable difference—of only 2^8 or 256. For you audiophiles, this corresponds to a dynamic range in signal power of 20 log 256 or about 48 dB. By comparison, compact audio discs use a 16-bit linear encoding with a theoretical dynamic range of 20 log 2^{16} or about 96 dB.

linear encoding
dynamic range

Instead of a linear encoding, the North American telephone network uses an 8-bit *companded encoding* called μ-*law PCM* (pulse-code modulation). (The European network uses a different 8-bit companding formula called *A-law*.) Figure 10-15 shows the format of an 8-bit coded byte, a sort of floating-point representation containing sign (*S*), exponent (*E*), and mantissa (*M*) fields. The analog value *V* represented by a byte in this format is given by the formula

companded encoding
μ-law PCM
A-law PCM

$$V = (1-2S) \cdot [(2^E) \cdot (2M + 33) - 33]$$

Figure 10-15
Format of a μ-law PCM byte.

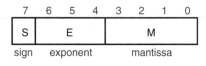

MONEY FOR NOTHING. . . With µ-law PCM you don't get something for nothing, of course. In the companded encoding, the difference between successive coded values is greater at high amplitudes than it is at low amplitudes, so the encodings of large analog signals suffer from more quantizing distortion (caused by the difference between the sampled analog value and the nearest available coded representation). Expressed as a percentage of peak signal amplitude, though, quantizing distortion is roughly constant across the entire range of representable values.

An analog signal represented in this format can range from $-8159 \cdot k$ to $+8159 \cdot k$, where k is an arbitrary scale factor. The range of signals is $2 \cdot 8159$ and the smallest difference that can be represented is only 2 (when $E = 0$), so the dynamic range is $20 \log 8159$ or about 78 dB, quite an improvement over an 8-bit linear code.

Now let's look at some ROM applications involving µ-law coded voice signals. Believe it or not, in many types of phone connections your voice is purposely attenuated by a few decibels to make things work better. In an analog phone network, attenuation is performed by a simple, passive analog circuit, but not so in the digital world. Given a µ-law PCM byte, a *digital attenuator* must produce a different PCM byte that represents the original analog signal multiplied by a specified attenuation factor.

digital attenuator

One way to build a digital attenuator is shown in Figure 10-16. The input byte is applied to a µ-law decoder that expands the byte according to the formula given earlier to produce a 14-bit signed-magnitude integer. This 14-bit linear value is then multiplied by a 14-bit binary fraction corresponding to the desired attenuation amount. The fractional bits of the product are discarded, and the result is reencoded into a new 8-bit µ-law PCM byte. Each block in the figure could be built with perhaps a dozen MSI chips or a CPLD or FPGA.

Figure 10-17 is the logic diagram for a digital attenuator that uses a single, inexpensive 8K × 8 ROM instead. This ROM can apply any of 32 different attenuation factors to a µ-law input byte—it simply stores 32 different attenuation tables. The high-order address bits select a table, and the low-order address bits select an entry. Each entry is a precomputed µ-law byte corresponding to the

Figure 10-16 Block diagram of a digital attenuator using discrete components.

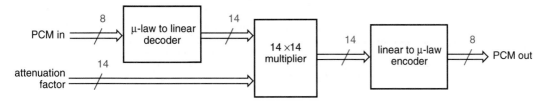

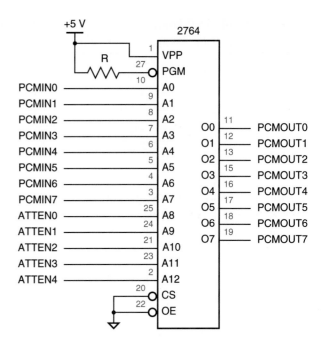

Figure 10-17
A digital attenuator.

given attenuation code and input byte. Table 10-6 is a C program that generates the ROM contents. The details of the `UlawToLinear` and `LinearToUlaw` functions are left as an exercise.

Another ROM application in digital telephony is a digital conference circuit. In the analog telephone network, it's fairly easy to make a conference connection between three or more parties. Just connect the analog phone wires together and you get an analog summing junction, so each person hears everyone else (it's not quite that easy to do it right, but you get the idea). In the digital network, of course, chaos would result from just shorting together digital output signals. Instead, a digital conference circuit must include a digital *adder* that produces output samples corresponding to sums of the input samples.

. . . AND THE
CHECK'S FOR
FREE

The digital attenuator is another good example of the many cost-effective space/time trade-offs that became possible when the telephone network "went digital." A single analog voice signal produces an 8-bit μ-law PCM sample once every 125 μs, but the digital attenuator ROM can produce a valid output in a few hundred nanoseconds or less. Thus, in a digital telephone system, a single ROM chip can attenuate hundreds of digital voice streams, where hundreds of analog attenuator networks would have been required in the old days.

■ **Table 10-6** Program to generate the contents of an 8K × 8, 32-position
attenuator ROM for μ-law coded bytes.

```c
#include <stdio.h>
#include <math.h>

extern int UlawToLinear(int in);
extern int LinearToUlaw(int x);

void main()
{
  int i, j, position;
  int pcmIN, linearOUT, pcmOUT;
  double atten, attenDB, fpcmOUT;

  for (position=0; position<=31; position++) {  /* Make 32 256-byte tables. */
    printf("%i attenuation (dB): ");  /* Get amount in dB from designer,       */
    scanf("%f\n", attenDB);    /*  negative for attenuation, positive for gain. */
    atten = exp(log(10)*attenDB/10); /* Convert to fraction. */
    for (i=0; i<=15; i++) {  /* Construct output file in rows of 16. */
      printf("%04x:", position*256 + i*16);
      for (j=0; j<=15; j++) {
        pcmIN = i*16 + j;
        fpcmOUT = atten * UlawToLinear(pcmIN);
        if (fpcmOUT >=0) linearOUT = floor(fpcmOUT + 0.5);  /* Rounding */
        else linearOUT = ceil(fpcmOUT - 0.5);
        pcmOUT = LinearToUlaw(linearOUT);
        printf(" %2x", pcmOUT);
      }
      printf("\n");
    }
  }
}
```

You know how to build a binary adder for two 8-bit operands, but binary
adders cannot process μ-law PCM bytes directly. Instead, the 8-bit PCM bytes
must be converted into 14-bit linear format, then added, and then reencoded. As
in the digital attenuator circuit of Figure 10-16, an MSI design for this function
would require many chips. Alternatively, the function could be performed by a
single 64K × 8 ROM, as shown in Figure 10-18. The ROM has 16 address
inputs, accommodating two 8-bit μ-law PCM operands. For each pair of operand
values, the corresponding ROM address contains the precomputed μ-law PCM
sum. Table 10-7 is a C program that can be used to generate the ROM contents.

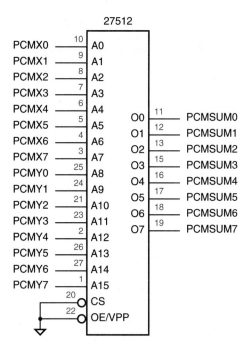

Figure 10-18
An adder circuit for
μ-law coded bytes.

These examples illustrate the many advantages of building complex combinational functions with ROMs. We usually consider a function to be "complex" if we know that a gate-level circuit for the function will be a real pain to design. However, most "complex" functions, like the examples in this subsection, have fairly straightforward word descriptions. Such a description can generally be translated into a computer program that "figures out" what the function's outputs should be for every possible input combination, so the function can be built simply by dropping its truth table into a ROM.

In addition to ease and speed of design, a ROM-based circuit has other important advantages:

- For a moderately complex function, a ROM-based circuit is usually faster than a circuit using multiple SSI/MSI devices and PLDs, and often faster than an FPGA or custom LSI chip in a comparable technology.

- The program that generates the ROM contents can easily be structured to handle unusual or undefined cases that would require additional hardware in any other design. For example, the adder program in Table 10-7 easily handles out-of-range sums. (Also see Exercise 10.22.)

- A ROM's function is easily modified just by changing the stored pattern, usually without changing any external connections. For example, the PCM attenuator and adder ROMs in this subsection can be changed to use 8-bit A-law PCM, the standard digital voice coding in Europe.

Table 10-7 Program to generate the contents of a 64K × 8 adder ROM
for μ-law coded bytes.

```
#include <stdio.h>
#include <math.h>
#define MINLINEAR -8159
#define MAXLINEAR 8159

void main()
{
  int i, j, linearSum;
  int pcmINX, pcmINY;

  for (pcmINY=0; pcmINY<=255; pcmINY++) {  /* For all Y samples... */
    for (i=0; i<=15; i++) {  /* Construct output file in rows of 16. */
      printf("%04x:", pcmINY*256 + i*16);
      for (j=0; j<=15; j++) {  /* For all X samples... */
        pcmINX = i*16 + j;
        linearSum = UlawToLinear(pcmINX) + UlawToLinear(pcmINY);
        /* The next two lines perform "clipping" on overflow. */
        if (linearSum < MINLINEAR) linearSum = MINLINEAR;
        if (linearSum > MAXLINEAR) linearSum = MAXLINEAR;
        printf(" %02x", LinearToUlaw(linearSum));
      }
      printf("\n");
    }
  }
}
```

- The prices of ROMs and other structured logic devices are always dropping, making them more economical, and their densities are always increasing, expanding the scope of problems that can be solved with a single chip.

There are a few disadvantages of ROM-based circuits, too:

- For simple to moderately complex functions, a ROM-based circuit may cost more, consume more power, or run slower than a circuit using a few SSI/MSI devices and PLDs or a small FPGA.

- For functions with more than 20 inputs, a ROM-based circuit is impractical because of the limit on ROM sizes that are available. For example, you wouldn't want to build a 16-bit adder in ROM—it would require billions and billions of bits.

10.2 Read/Write Memory

read/write memory (RWM)

random-access memory (RAM)

The name *read/write memory (RWM)* is given to memory arrays in which we can store and retrieve information at any time. Most of the RWMs used in digital systems nowadays are *random-access memories (RAMs)*, which means that the time it takes to read or write a bit of memory is independent of the bit's location in the RAM. From this point of view, ROMs are also random-access memories, but the name "RAM" is generally used only for read/write random-access memories.

static RAM (SRAM)

dynamic RAM (DRAM)

In a *static RAM (SRAM)* ("S-ram"), once a word is written at a location, it remains stored as long as power is applied to the chip, unless the same location is written again. In a *dynamic RAM (DRAM)* ("D-ram"), the data stored at each location must be refreshed periodically by reading it and then writing it back again, or else it disappears. We'll discuss both types in this section.

volatile memory

nonvolatile memory

Most RAMs lose their memory when power is removed; they are a form of *volatile memory.* Some RAMs retain their memory even when power is removed; they are called *nonvolatile memory.* Examples of nonvolatile RAMs are old-style magnetic core memories and modern CMOS static memories in an extra-large package that includes a lithium battery with a 10-year lifetime. Recently, nonvolatile *ferroelectric RAMs* have been introduced; these devices combine magnetic and electronic elements on a single IC chip that retains its state even when power is not applied, just like the old-style core memories.

ferroelectric RAM

10.3 Static RAM

10.3.1 Static-RAM Inputs and Outputs

Like a ROM, a RAM has address and control inputs and data outputs, but it also has data inputs. The inputs and outputs of a simple $2^n \times b$-bit static RAM are shown in Figure 10-19. The control inputs are comparable to those of a ROM, with the addition of a *write-enable (WE) input.* When WE is asserted, the data inputs are written into the selected memory location.

write-enable (WE) input

The memory locations in a static RAM behave like D latches, rather than edge-triggered D flip-flops. This means that whenever the WE input is asserted, the latches for the selected memory location are "open" (or "transparent"), and input data flows into and through the latch. The actual value stored is whatever is present when the latch closes.

Static RAM normally has just two defined access operations:

Read   An address is placed on the address inputs while CS and OE are asserted. The latch outputs for the selected memory location are delivered to DOUT.

Write An address is placed on the address inputs and a data word is placed on DIN; then CS and WE are asserted. The latches in the selected memory location open, and the input word is stored.

SERIAL-ACCESS MEMORY

Random-access memory can be contrasted with *serial-access memory*, where some location is immediately accessible at any time, but other locations take additional steps to access.

Some early computers used electromechanical serial-access memory devices, such as delay lines and rotating drums. Instructions and data were stored in a rotating medium with only one location under the "read/write head" at any time. To access a random location, the machine would have to wait until the constant rotation brought that location under the head.

In the 1970s, electronic equivalents of serial-access rotating memories were developed, including memories based on charge-coupled devices (CCDs) and others that used magnetic bubbles. Both types of devices were roughly equivalent to very large serial-in, serial-out shift registers with their serial output connected back into the serial input. This connection point was the logical equivalent of a hard disk's "read/write head." To read a particular location, you would clock the shift register until the desired bit appeared at the serial output, and to write the location, you would substitute the desired new value at the serial input.

Although they offered higher density (more bits) than DRAMs at the time they were developed, CCD and magnetic-bubble memories never gained much commercial acceptance. One reason for this was the enormous inconvenience of serial access. Another was that they were never more than a couple years ahead of DRAMs in the densities that they could achieve.

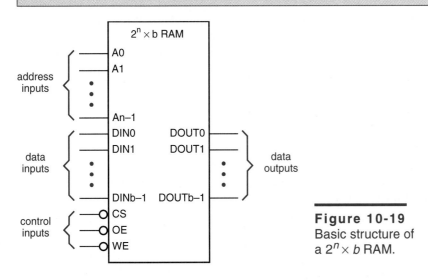

Figure 10-19
Basic structure of a $2^n \times b$ RAM.

A certain amount of care is needed when accessing SRAM, because it is possible to inadvertently "clobber" one or more locations while writing to a selected one, if the SRAM's timing requirements are not met. The following subsection gives details on SRAM internal structure to show why this is so, and the next one explains the actual timing behavior and requirements.

10.3.2 Static-RAM Internal Structure

SRAM cell

Each bit of memory (or *SRAM cell*) in a static RAM has the same functional behavior as the circuit in Figure 10-20. The storage device in each cell is a D latch. When a cell's SEL_L input is asserted, the stored data is placed on the cell's output, which is connected to a bit line. When both SEL_L and WR_L are asserted, the latch is open and a new data bit is stored.

SRAM cells are combined in an array with additional control logic to form a complete static RAM, as shown in Figure 10-21 for an 8 × 4 SRAM. As in a simple ROM, a decoder on the address lines selects a particular row of the SRAM to be accessed at any time.

Although Figure 10-21 is a somewhat simplified model of internal SRAM structure, it accurately portrays several important aspects of SRAM behavior:

- During read operations, the output data is a combinational function of the address inputs, as in a ROM. No harm is done by changing the address lines while the output data bus is enabled. The access time for read operations is specified from the time that the last address input becomes stable.

- During write operations, the input data is stored in *latches*. This means that the data must meet certain setup and hold times with respect to the *trailing* edge of the latch enable signal. That is, the input data at a latch's D input need not be stable at the moment WR_L is asserted internally; it must only be stable a certain time before WR_L is negated.

- During write operations, the address inputs *must* be stable for a certain setup time before WR_L is asserted internally and for a hold time after WR_L is negated. Otherwise, data may be "sprayed" all over the array because of the glitches that may appear on the SEL_L lines when the address inputs of the decoder are changed.

write cycle

- Internally, WR_L is asserted only when both CS_L and WE_L are asserted. Therefore, a *write cycle* begins when both CS_L and WE_L are asserted and ends when either is negated. Setup and hold times for address and data are specified with respect to these events.

Figure 10-20 Functional behavior of a static-RAM cell.

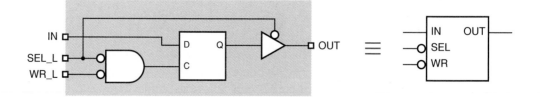

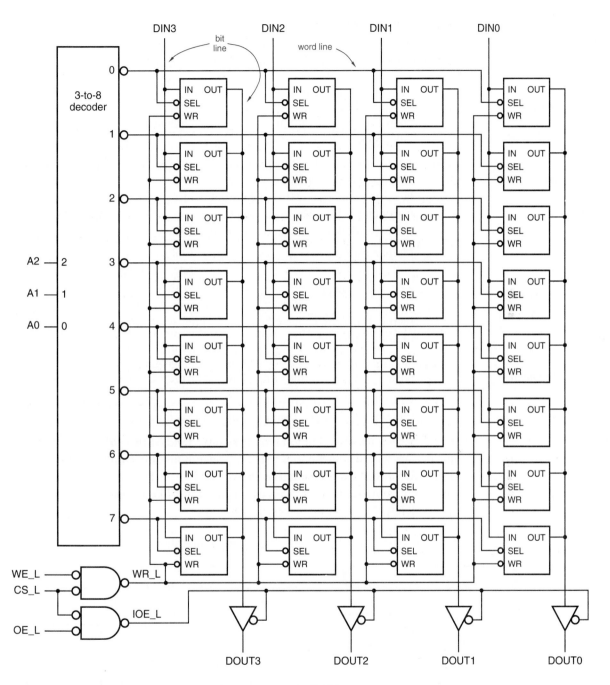

Figure 10-21 Internal structure of an 8×4 static RAM.

10.3.3 Static-RAM Timing

Figure 10-22 shows the timing parameters that are typically specified for read operations in a static RAM; they are described below:

t_{AA}
access time from address

t_{AA} *Access time from address.* Assuming that OE and CS are already asserted, or will be soon enough not to make a difference, this is how long it takes to get stable output data after a change in address. When designers talk about a "70-ns SRAM," they're usually referring to this number.

t_{ACS}
access time from chip select

t_{ACS} *Access time from chip select.* Assuming that the address and OE are already stable, or will be soon enough not to make a difference, this is how long it takes to get stable output data after CS is asserted. Often this parameter is identical to t_{AA}, but sometimes it's longer in SRAMs with a "power-down" mode and shorter in SRAMs without one.

t_{OE}
output-enable time

t_{OE} *Output-enable time.* This is how long it takes for the three-state output buffers to leave the high-impedance state when OE and CS are both asserted. This parameter is normally less than t_{ACS}, so it is possible for the RAM to start accessing data internally before OE is asserted; this feature is used to achieve fast access times while avoiding "bus fighting" in many applications.

t_{OZ}
output-disable time

t_{OZ} *Output-disable time.* This is how long it takes for the three-state output buffers to enter the high-impedance state after OE_L or CS_L is negated.

t_{OH}
output-hold time

t_{OH} *Output-hold time.* This parameter specifies how long the output data remains valid after a change in the address inputs.

If you've been paying attention, you may have noticed that the timing diagram and timing parameters for SRAM read operations are identical to what we discussed for ROM read operations in Section 10.1.5. That's the way it is; when they're not being written, SRAMs can be used just like ROMs. The same is not true for DRAMs, as we'll see later.

Figure 10-22 Timing parameters for read operations in a static RAM.

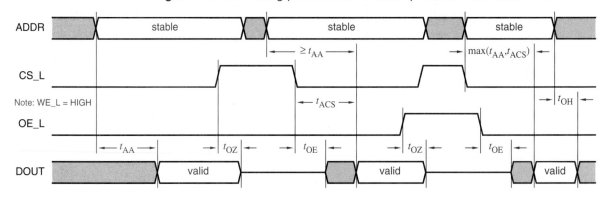

Timing parameters for write operations are shown in Figure 10-23 and are described below:

t_{AS} *Address setup time before write.* All of the address inputs must be stable at this time before both CS and WE are asserted. Otherwise, the data stored at unpredictable locations may be corrupted.

t_{AS}
address setup time

t_{ACS} *Address hold time after write.* Analogous to t_{AS}, all address inputs must be held stable until this time after CS or WE is negated.

t_{AH}
address hold time

t_{CSW} *Chip-select setup before end of write.* CS must be asserted at least this long before the end of the write cycle in order to select a cell.

t_{CSW}
chip-select setup

t_{WP} *Write-pulse width.* WE must be asserted at least this long to reliably latch data into the selected cell.

t_{WP}
write-pulse width

t_{DS} *Data setup time before end of write.* All of the data inputs must be stable at this time before the write cycle ends. Otherwise, the data may not be latched.

t_{DS}
data setup time

t_{DH} Data hold time after end of write. Analogous to t_{DS}, all data inputs must be held stable until this time after the write cycle ends.

t_{DH}
data hold time

Manufacturers of SRAMs specify two write-cycle types, *WE-controlled* and *CS-controlled*, as shown in the figure. The only difference between these cycles is whether WE or CS is the last to be asserted and the first to be negated when enabling the SRAM's internal write operation.

WE-controlled write
CS-controlled write

The write-timing requirements of SRAMs could be relaxed somewhat if, instead of using latches, the cells contained edge-triggered D flip-flops with a common clock input and enable inputs tied controlled by SEL and WR. However, this just isn't done, because it would at least double the chip area of each cell, since a D flip-flop is built from two latches. Thus, the logic designer is left to reconcile the SRAM's latch-type timing with the edge-triggered register and state-machine timing used elsewhere in a system.

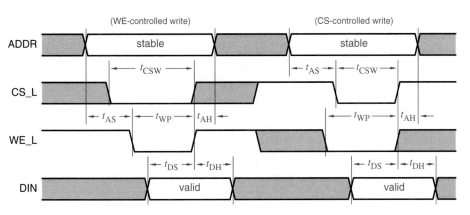

Figure 10-23
Timing parameters for write operations in a static RAM.

A large SRAM does not contain a physical array whose dimensions equal the logical dimensions of the memory. As in a ROM, the SRAM cells are laid out in an almost square array, and an entire row is read internally during read operations. For example, the layout of a 32K × 8 SRAM chip might be very similar to that of a 32K × 8 ROM shown in Figure 10-9 on page 841. During read operations, column multiplexers pass the required data bits to the output data bus, as specified by a subset of the address bits (A5–A0 in the ROM example). For write operations, the write-enable circuitry is designed so that only one column in each subarray is enabled, as determined by the same subset of the address bits.

10.3.4 Standard Static RAMs

Static RAMs are available in many sizes and speeds. In 1999, the largest conventional SRAMs used CMOS technology, were organized as 256K × 16, 512K × 8, or 1M × 4 bits (4 Mbits), and had access times as fast as 10 ns. The fastest TTL/CMOS compatible SRAMs also used CMOS technology and contained only 1K × 4 bits (4 Kbits) but had access times of 2.7 ns!

Part numbers for SRAM components are not standardized among manufacturers, although the components themselves often have identical pinouts and are interchangeable. Figure 10-24 shows part numbers and pinouts for Hitachi

Figure 10-24
Logic symbols for standard SRAMs in 28- and 32-pin DIP packages.

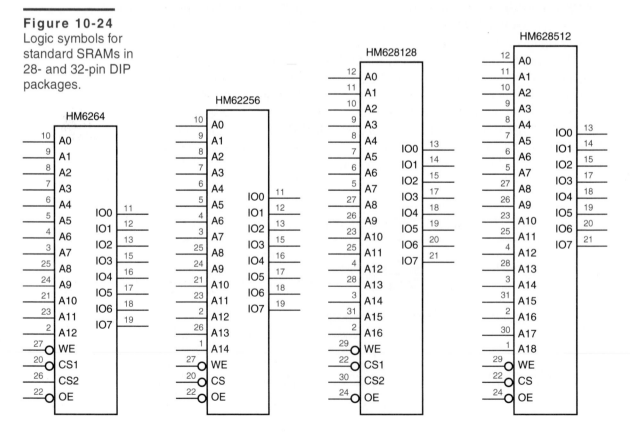

SRAMs, ranging in size from standard 8K × 8 to 512K × 8, that you might use in a digital design lab project; they all use a 5-V power supply and come in DIP packages. These devices are also available in more compact packages, and newer SRAM devices are typically available only in compact packages and use lower power-supply voltages, such as 3.3 V or 2.5 V.

The first two SRAMs in the figure use 28-pin DIP packages and have pinouts that are patterned after the like-size EPROMs in Figure 10-11. The *HM6264* 8K × 8 SRAM is similar to a 2764 EPROM, except that it has a write-enable input and two chip-select inputs instead of one; CS1_L and CS2 both must be asserted to enable the chip. The *HM62256* 32K × 8 SRAM is similar to a 27256 EPROM, with the addition of a write-enable input. Both chips are available from many manufacturers, often with different part numbers, and have access times in the range of 15–150 ns. The last two SRAMs in the figure use 32-pin DIP packages. The *HM628128* is a 128K × 8 SRAM and the *HM628512* is a 512K × 8 SRAM

The SRAMs in Figure 10-24 have bidirectional data buses—that is, they use the same data pins for both reading and writing. This necessitates a slight change in their internal control logic, shown in Figure 10-25. The output buffer is automatically disabled whenever WE_L is asserted, even if OE_L is asserted. However, the timing parameters and requirements for read and write operations are practically identical to what we described in the preceding subsection.

A common use of SRAMs is to store data in small microprocessor systems, often in "embedded" applications (telephones, toasters, electric bumpers, etc.). General-purpose computers more often use DRAMs, discussed in Section 10-33, because their density is greater and their cost per bit lower.

Very fast SRAMs are often used in the "cache" memories of high-performance computers to store frequently used instructions and data. In fact, the "need for speed" in PC cache-memory applications led to the development and widespread deployment of a faster, clocked SRAM interface. The standard SRAMs in this subsection are now called *asynchronous SRAMs* to distinguish them from the new style discussed next.

HM6264

HM62256

HM628128

asynchronous SRAM

Figure 10-25 Output-buffer control in an SRAM with a bidirectional data bus.

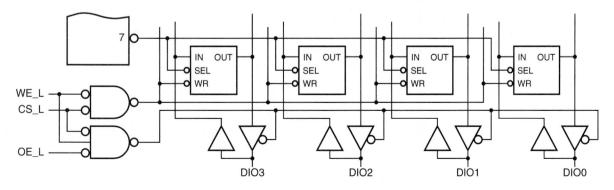

*10.3.5 Synchronous SRAM

synchronous SRAM (SSRAM)

A new variety of SRAM, called a *synchronous SRAM (SSRAM)* ("S-S-ram"), still uses latches internally but has a clocked interface for control, address, and data. As shown in Figure 10-26, internal edge-triggered registers AREG and CREG are placed on the signal paths for address and control. As a result, an operation that is set up before the rising edge of the clock is performed internally during a subsequent clock period. Register INREG captures the input data for write operations and, depending on whether the device has "pipelined" or "flow-through" outputs, register OUTREG is or is not provided to hold the output data from a read operation.

late-write SSRAM with flow-through outputs

The first variety of SSRAM to be introduced was the *late-write SSRAM with flow-through outputs.* For a read operation, shown in Figure 10-27(a), the control and address inputs are sampled at the rising edge of the clock, and the

Figure 10-26
Internal structure of a synchronous SRAM.

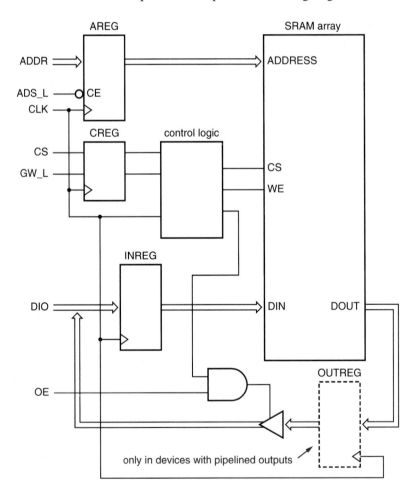

only in devices with pipelined outputs

internal address register AREG is loaded only if ADS_L is asserted. During the next clock period, the internal SRAM array is accessed and read data is delivered to the device's DIO data-bus pins. The device also supports a burst mode, in which data at sequence of addresses is read. In this mode AREG behaves as a counter, eliminating the need to apply a new address at each cycle. (The control signals that support burst mode are not shown in Figures 10-26 or 10-27.)

For a write operation, shown in (b), the write data is stored temporarily in an on-chip register INREG, which is sampled one clock tick *after* the address register is loaded. Therefore, ADS_L must be inhibited for at least one tick after loading the address, so that the address in AREG is still valid when the write takes place. The write takes place during the clock period following the edge on which the "global write" control signal GW_L is asserted. As with reading, the device has a burst mode where a sequence of addresses can be written without supplying a new address.

Note that the "late-write" protocol makes it impossible to write to two different, nonsequential addresses in successive clock periods; the SRAM array is idle for one clock period between writes (except in burst mode). From the point of view of internal chip capabilities, this behavior is not necessary. However, the late-write protocol was designed this way to match the bus protocols of microprocessors that use these SSRAMs in their cache subsystems.

A *late-write SSRAM with pipelined outputs* is like the previous version, except that a register OUTREG is placed between the SRAM array output and the device output for read operations. As shown in Figure 10-28, this delays the read output data at the device pins until the beginning of the next clock period,

late-write SSRAM with pipelined outputs

Figure 10-27 Timing behavior for a late-write SSRAM with flow-through outputs: (a) read operations; (b) write operations.

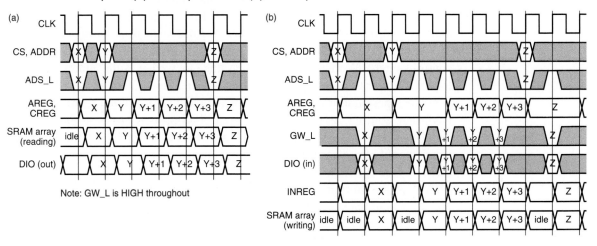

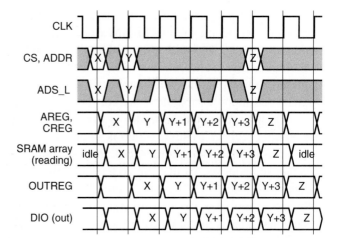

Figure 10-28
Read-timing behavior
for a late-write
SSRAM with
pipelined outputs.

but it also provides the benefit that the data is now valid for almost the entire clock period. The write cycle behaves the same as with flow-through outputs. Compared to flow-through outputs, pipelined outputs provide much better setup time for the device receiving the read data, and therefore may allow operation at higher clock frequencies.

As we showed in Figure 10-26, conventional SSRAMs share the same pins for both input data and output data. During a given clock period the data I/O pins can be used for reading or writing but not both. If you study the pattern of data-bus and SRAM-array use in both styles of late-write SSRAM, you'll find cases where it's not possible to initiate a read one clock cycle after initiating a write or vice versa, due to resource conflict (see Exercise 10.32). Thus, late-write *turn-around penalty* SSRAMs suffer a *turn-around penalty*, a clock period in which the internal SRAM array must be idle when a read is followed by a write or vice versa.

zero-bus-turn-around The turn-around penalty is eliminated in so called *zero-bus-turn-around (ZBT) SSRAM* *(ZBT) SSRAMs*. The timing for a *ZBT SSRAM with flow-through outputs* is shown in Figure 10-29. The type of operation (read or write) is selected by a control signal R/$\overline{W}$ that sampled at the same clock edge as the address. Regardless of whether the operation is a read or a write, the DIO bus is used during the next clock period to transfer the read or write data. As a result, there is no data-bus-usage conflict, as long as OE is controlled properly to avoid bus-fighting between successive cycles. However, if a write is followed by a read, both operations would like to use the SRAM array during the same clock period. To avoid this resource conflict, the write operation is deferred until the next available SRAM cycle. This opportunity occurs when either another write operation or no operation is initiated on the address and control lines.

Although a ZBT SSRAM can use access the internal SRAM array on every clock cycle, this performance improvement is not without a price. While a write operation is pending, the write address and related information must be stored in another register, WAREG, since AREG is reused by other operations; this costs

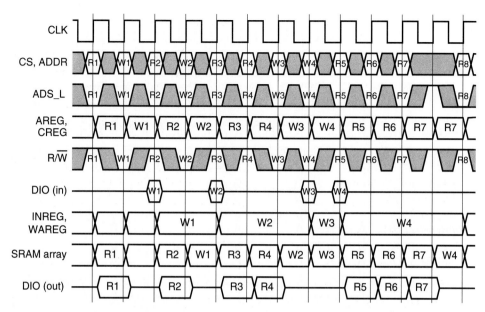

Figure 10-29 Timing behavior for a ZBT SSRAM with flow-through outputs.

chip area. More significantly for some applications, a write operation may be deferred indefinitely if it is immediately followed by a continuous series of read operations. This anomaly may require tricky controller design to detect the case where one of these read operations attempts to access the address that was just written, since the value stored in the SRAM array is "stale"!

A *ZBT SSRAM with pipelined outputs* adds **OUTREG** to the read data path but is otherwise similar to the previous device. In this device, both reads and writes use the **DIO** bus during the second clock period following the clock edge in which the operation was initiated. As in the previous device, writes to the internal SRAM array are deferred until an available cycle, so that reads can take precedence. Timing is shown in Figure 10-30. As implied by the timing, two levels of internal registers are needed for write address and data, since up to two writes may be deferred while a sequence of reads occurs.

Synchronous SRAM access protocols are very beneficial in high-speed systems. For example, address, control, and write inputs can be applied with more-or-less conventional setup and hold times with respect to the system clock, and read data on pipelined output pins is available for almost a complete clock cycle. Very importantly, the designer does not have to worry about the tricky circuits and timing paths that are otherwise needed to enable conventional SRAM latch-style operation.

SSRAMs were available in 1999 with clock frequencies as high as 166 MHz. Note that among the four styles of SSRAM that we described, no single one is the "best." The best SSRAM is the one that best fits the bus protocol and other requirements of the system in which it is used.

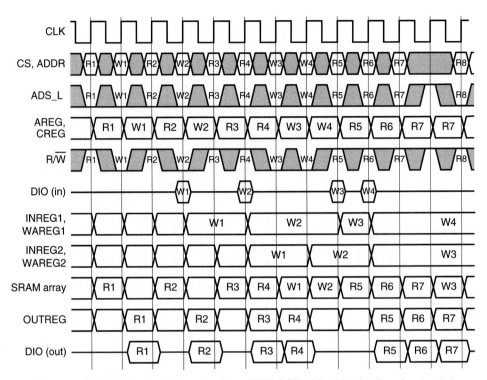

Figure 10-30 Timing behavior for a ZBT SSRAM with pipelined outputs.

10.4 Dynamic RAM

The basic memory cell in an SRAM, a *D latch*, requires four gates in a discrete design, and four to six transistors in a custom-designed SRAM LSI chip. In order to build RAMs with higher density (more bits per chip), chip designers invented memory cells that use as little as one transistor per bit.

10.4.1 Dynamic-RAM Structure

dynamic RAM (DRAM)

It is not possible to build a bistable element with just one transistor. Instead, the memory cells in a *dynamic RAM (DRAM)* store information on a tiny capacitor accessed through a MOS transistor. Figure 10-31 shows the storage cell for one bit of a DRAM, which is accessed by setting the word line to a HIGH voltage. To store a 1, a HIGH voltage is placed on the bit line, which charges the capacitor through the "on" transistor. To store a 0, LOW voltage on the bit line discharges the capacitor.

precharge

To read a DRAM cell, the bit line is first *precharged* to a voltage halfway between HIGH and LOW, and then the word line is set HIGH. Depending on whether the capacitor voltage is HIGH or LOW, the precharged bit line is pulled slightly higher or slightly lower. A *sense amplifier* detects this small change and recovers a 1 or 0 accordingly. Note that reading a cell destroys the original

sense amplifier

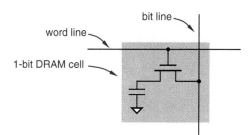

Figure 10-31
Storage cell for one bit
in a DRAM.

voltage stored on the capacitor, so that the recovered data must be written back into the cell after reading.

The capacitor in a DRAM cell has a very small capacitance, but the MOS transistor that accesses it has a very high impedance. Therefore, it takes a relatively long time (many milliseconds) for a HIGH voltage to discharge to the point that it looks more like a LOW voltage. In the meantime, the capacitor stores one bit of information.

Naturally, using a computer would be no fun if you had to reboot every few milliseconds because its memory contents disappeared (the typical behavior of Windows notwithstanding). Therefore, DRAM-based memory systems use *refresh cycles* to update every memory cell periodically, once every four milliseconds in early DRAMs. This involves sequentially reading the somewhat degraded contents of each cell into a D latch and writing back a nice solid LOW or HIGH value from the latch. Figure 10-32 illustrates the electrical state of a cell after a write and a sequence of refresh operations.

refresh cycle

The first DRAMs, introduced in the early 1970s, contained only 1,024 bits, but modern DRAMs are available containing 256 *megabits* or more. If you had to refresh every cell, one at a time, in four milliseconds, you'd have a problem— that works out to far less than 1 ns per cell, and includes no time for useful read and write operations. Fortunately, as we'll show, DRAMs are organized using two-dimensional arrays, and a single operation refreshes an entire row of the array. Early DRAM arrays had 256 rows, requiring 256 refresh operations every four milliseconds, or one about every 15.6 μsec. Newer arrays have 4096 rows but need to be refreshed only once every 64 msec, which still works out to one row per 15.6 μsec. A refresh operation typically takes less than 100 ns, so the DRAM is available for useful read and write operations over 99% of the time.

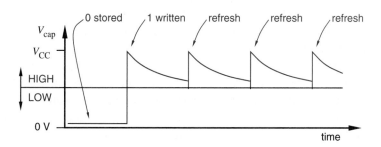

Figure 10-32
Voltage stored in a
DRAM cell after
writing and refresh
operations.

Figure 10-33
Internal structure of
a 64K × 1 DRAM.

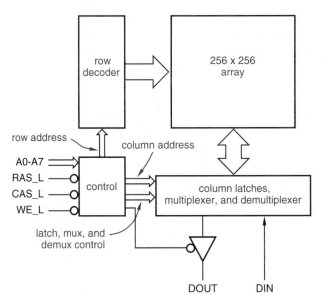

*multiplexed address
inputs*

*RAS_L (row address
strobe)*

*CAS_L (column address
strobe)*

Figure 10-33 is a block diagram of the internal structure of a 64K × 1 DRAM. The logical array has 64K × 1 bits, but the physical array is square, containing 256 × 256 bits. Although the memory has 64K locations, the chip has only eight *multiplexed address inputs*. A complete 16-bit address is presented to the chip in two steps controlled by two signals—*RAS_L (row address strobe)* and *CAS_L (column address strobe)*. Multiplexing the address inputs saves pins, important for compact design of memory systems, and also fits quite naturally with the two-step DRAM access methods that we'll describe shortly.

Larger DRAMs have larger arrays and often have multiple arrays. One advantage of multiple arrays is to ease the electrical and physical design problems that would occur with an extremely large array. But even more important is the parallelism that can occur with multiple arrays. As we'll see in the next subsection, DRAM operation is much more complicated than SRAM operation. Taking advantage of the multiple arrays in larger, high-speed DRAMs, a modern DRAM controller can perform several operations in parallel—for example, completing a write operation in one array while initiating a read operation in another. This increases the effective throughput of the memory.

10.4.2 Dynamic-RAM Timing

There are many different timing scenarios for different DRAM types and operations. In this section we'll describe the most common cycles for conventional DRAM and relate them to the internal structure of the device. The most striking aspect of conventional DRAM timing is that there is no clock. Instead, DRAM operations are initiated and completed on both the rising and falling edges of RAS_L and CAS_L. This is very trick timing indeed, but the industry has somehow lived with it for over 25 years!

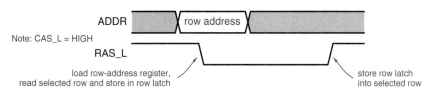

Figure 10-34
DRAM RAS-only
refresh-cycle timing.

The timing for a *RAS-only refresh cycle* is shown in Figure 10-34. This *RAS-only refresh cycle*
cycle is used to refresh a row of memory without actually reading or writing any
data at the external pins of the DRAM chip. The cycle begins when a row
address is applied to the multiplexed address inputs (eight bits in the case of a
64K × 1 DRAM) and RAS_L is asserted. The DRAM stores the row address in
an internal *row-address register* on the falling edge of RAS_L and reads the *row-address register*
selected row of the memory array into an on-chip *row latch*. When RAS_L is *row latch*
negated, the contents of the row are written back from the row latch. To refresh
an entire 64K × 1 DRAM, the system designer must ensure that 256 such cycles,
using all 256 possible row addresses, are executed every four milliseconds. An
external 8-bit counter may be used to generate the row addresses, and a timer is
used to initiate a refresh cycle once every 15.6 μs.

A *read cycle*, shown in Figure 10-35, begins like a refresh cycle, where a *read cycle*
selected row is read into the row latch. Next, a column address is applied to the
multiplexed address inputs and is stored in an on-chip *column-address register* *column-address register*
on the falling edge of CAS_L. The column address is used to select one bit of the

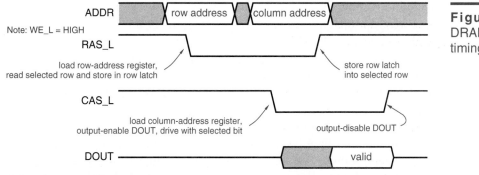

Figure 10-35
DRAM read-cycle
timing.

TRICKY TIMING   You may have noticed that the valid intervals of row and column addresses in
Figure 10-35 are skewed to the right with respect to the falling edges of RAS_L and
CAS_L, which load the addresses into on-chip edge-triggered registers. DRAMs
typically have a short setup-time requirement for the address inputs (often 0 ns) but
a relatively large, positive hold-time requirement (7–20 ns). This can cause some
problems in the design of an otherwise fully synchronous system that uses a single
clock and flip-flops with zero hold times, and is often solved by distasteful methods
involving tapped delay lines or using the opposite edge of the clock. This is one of
the reasons that many designers consider DRAM system design a tricky, black art.

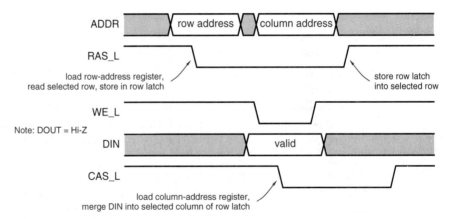

Figure 10-36
DRAM write-cycle timing.

just-read row, which is made available on the DRAM's DOUT pin. A three-state pin, DOUT is output enabled as long as CAS_L is asserted. In the meantime, the entire row is written back into the array as soon as RAS_L is negated.

write cycle
WE_L (write enable)

A *write cycle*, shown in Figure 10-36, also begins like a refresh or read cycle. However, *WE_L (write enable)* must be asserted before CAS_L is asserted to select a write cycle. One thing this accomplishes is to disable DOUT for the rest of the cycle, even though CAS_L will be asserted subsequently. Once the selected row is read into the row latch, WE_L also forces the input bit on DIN to be merged into the row latch, in the bit position selected by the column address. When the row is subsequently written back into the array on the rising edge of RAS_L, it contains a new value in the selected column.

Several other types of cycles, not shown in the figure, are available in typical DRAMs:

CAS-before-RAS
refresh cycle

- A *CAS-before-RAS refresh cycle* performs a refresh operation without requiring a row address to be supplied by an external counter. Instead, the DRAM chip uses an internal row-address counter. If CAS_L is asserted before RAS_L is asserted, the DRAM refreshes the row selected by the counter and then increments the counter. This simplifies DRAM system design, eliminating the external refresh counter and reducing from three (row, column, refresh) to two the number of multiplexed sources that must drive the DRAM address inputs.

read-modify-write cycle

- A *read-modify-write cycle* begins like a read cycle and allows data to be read on DOUT when CAS_L is asserted. However, WE_L may be asserted subsequently to write new data back at the same location.

page-mode read cycle

- A *page-mode read cycle* allows up to an entire row ("page") of data to be read without repeating the entire RAS-CAS cycle. Since the entire row is already stored in the row latch, this cycle simply requires CAS_L to be pulsed LOW multiple times while RAS_L remains asserted continuously. A new column address is provided and a new bit becomes available on DOUT on each falling edge of CAS_L. This cycle provides much faster memory access during a sequence of reads to consecutive or otherwise nearby addresses; such sequences often occur in microprocessors when they're fetching instructions or filling caches.

- Similarly, a *page-mode write cycle* allows multiple bits of a row to be written with a single RAS_L and multiple CAS_L cycles. *page-mode write cycle*

Two other multiple-access modes, *static-column mode* and *nibble mode*, were provided at one time in certain DRAM chips, but these became obsolete with the marketplace dominance of so-called *extended-data-out (EDO) DRAMs*. In these devices, CAS_L no longer controls the output enable during read cycles; instead, a separate OE_L control input is provided. This is important for high-speed page-mode read cycles, as it provides valid output data for a longer period of time. Instead turning off between successive CAS_L pulses, the output is left on continuously (as controlled by OE_L), and page-mode read data is valid from the beginning of one CAS_L pulse to the beginning of the next.

static-column mode
nibble mode
extended-data-out
  (EDO) DRAM

10.4.3 Synchronous DRAMs

The RAS/CAS edge-based access protocol of conventional DRAMs is not only tricky, it's difficult to make it go fast and still meet timing margins when interfacing with the rest of a system. Thus, the early 1990s saw the development of *synchronous DRAMs (SDRAMs)* that use a more conventional clocked interface, and by the late 1990s these devices dominated the PC memory market.

synchronous DRAM
  (SDRAM)

Synchronous DRAMs retain the multiplexed addressing style of conventional DRAMs—row and column addresses are delivered in two steps. However, an SDRAM's control signals as well as its address inputs are sampled only on the rising edge of a common clock signal (CLK) with a frequency as high as 133 MHz. In addition, an SDRAM has a clock-enable signal (CKE) so that inputs are ignored if CKE is not asserted. For write operations, data is sampled on the clock edge, and for reads, output data is delivered on the clock edge.

Just like a conventional DRAM, an SDRAM requires multiple internal steps to perform an operation, and this means multiple clock ticks externally. Internally, SDRAMs have multiple DRAM banks, typically four, and they can have multiple operations pending simultaneously.

At each clock tick, an SDRAM's control signals RAS_L, CAS_L, and WE_L are interpreted as a "command word," as opposed to having individual significance. At the same time, the high-order address bits are interpreted as a "bank select" to indicate which bank a command applies to. For example, an intelligent SDRAM controller can use four ticks to initiate read operations in four different banks, and then come back and read the results one per clock tick as they complete.

Internal timing in an SDRAM is derived from the clock input CLK. Typically, a RAS signal to the internal array will be asserted immediately after the clock tick when a read or write command is given. To meet internal timing requirements, the chip generates an internal CAS signal at a later clock tick. How many ticks later is a function of the CLK frequency and the speed grade of the SDRAM chip itself. To accommodate differing requirements, the RAS-to-CAS delay, called the CAS latency, is programmable. This and several other important operating parameters must be downloaded into an SDRAM at initialization. Downloading is fairly easy; the "load parameters" command is recognized when RAS_L, CAS_L, and WE_L control signals are all asserted simultaneously, and the parameters themselves are transferred on the address lines.

10.5 Complex Programmable Logic Devices

Since their introduction years ago, programmable logic devices such as the 16V8 and 22V10 have been very flexible workhorses of digital design. As IC technology advanced, there was naturally great interest in creating larger PLD architectures to take advantage of increased chip density. The question is, why didn't manufacturers just scale the existing architectures?

For example, if DRAM densities increased by a factor of 64 over the last 10 years, why couldn't manufacturers scale the 16V8 to create a "128V64"? Such a device would have 64 input pins, 64 I/O pins, and some number of 128-variable product terms (say, 8) for each of its 128 logic macrocells. It could combine the functions of a larger collection of 16V8s and offer terrific performance and flexibility in using any input in any output function.

Or could it? Yes, a 128V64 would be very flexible, but it would not have very good performance. Unlike the 16V8, which has 32 inputs (16 signals and their complements) per AND term, this device would have 256. Due to capacitive effects, leakage currents, and so on, such a large wired-AND structure would be at least eight times slower than the 16V8's AND array.

Figure 10-37
General CPLD architecture.

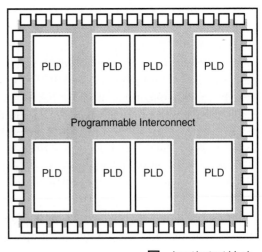

☐ = input/output block

Worse from a manufacturer's point of view, a 128V64 would not make very cost-effective use of chip area. It would use about 64 times the chip area of a 16V8, but provide the same number of inputs and outputs as only eight 16V8s. That is, for n times as much logic (in terms of inputs, outputs and AND terms), the 128V64 uses n^2 as much chip area. In terms of efficiency of silicon use, a clever designer would be better off partitioning a desired function into eight 16V8s, if such a partitioning were possible.

This is where the idea of complex programmable logic devices (CPLDs) came from. As shown in Figure 10-37, a CPLD is just a collection of individual PLDs on a single chip, accompanied by a programmable interconnection structure that allows the PLDs to be hooked up to each other on-chip in the same way that a clever designer might do with discrete PLDs off-chip. Here, the chip area for n times as much logic is only n times the area of a single PLD plus the area of the programmable interconnect structure.

Different manufacturers have taken many different approaches to the general architecture shown in the figure. Areas in which they differ include the individual PLDs (AND array and macrocells), the input/output blocks, and the programmable interconnect. We'll discuss each of these areas in the rest of this section, using the Xilinx 9500-series CPLD architecture as a representative example.

10.5.1 Xilinx XC9500 CPLD Family

The Xilinx XC9500 series is a family of CPLDs with a similar architecture but varying numbers of external input/output (I/O) pins and internal PLDs (which Xilinx calls function blocks—FBs). As we'll see later, each internal PLD has 36 inputs and 18 macrocells and outputs and might be called a "36V18." As shown in Table 10-8, devices in the family are named according to the number of macrocells they contain. The smallest has 2 FBs and 36 macrocells, and the largest has 16 FBs and 288 macrocells.

Another important feature of this and most CPLD families is that a given chip, such as the XC95108, is available in several different packages. This is important not only to accommodate different manufacturing practices but also to provide some choice and potential savings in the number of external I/O pins provided. In most applications, it is not necessary for all internal signals of a state machine or subsystem to be visible to and used by the rest of the system.

So, even though the XC95108 has 108 internal macrocells, the outputs of at most 69 of them can be connected externally in the 84-pin-PLCC version of the device. In fact, many of the 69 I/O pins would typically be used for inputs, in which case even fewer outputs would be visible externally. That's OK; the remaining macrocell outputs are still quite usable internally, since they can be hooked up internally through the CPLD's programmable interconnect. Macrocells whose outputs are usable only internally are sometimes called *buried macrocells*.

buried macrocell

Table 10-8 Function blocks and external I/O pins in Xilinx 9500-series CPLDs.

	Part Number					
	XC9536	*XC9572*	*XC95108*	*XC95144*	*XC95216*	*XC95288*
FBs / macrocells	2 / 36	4 / 72	6 / 108	8 / 144	12 / 216	16 / 288
Package	*Device I/O Pins*					
44-pin VQFP	34					
44-pin PLCC	34	34				
48-pin CSP	34					
84-pin PLCC		69	69			
100-pin TQFP		72	81	81		
100-pin PQFP		72	81	81		
160-pin PQFP			108	133	133	
208-pin HQFP					166	168
352-pin BGA					166	192

Another important dimension of Table 10-8 is the horizontal one. All but two of the packages support at least two different devices in the same package and, as it turns out, with compatible pinouts. This is a life-saver in making last-minute design changes. For example, suppose you were to target a design to an XC9572 in an 84-pin PLCC. You might find that the 69 I/O pins of the device are quite adequate. You would like to use the XC9572 for its low cost, but you might be a little nervous if your initial design used 68 out of the 72 available internal macrocells (I know that I would be!). Based on Table 10-8, you can sleep well knowing that if bug fixes or design-specification changes necessitate a more complex design internally, you can always move up to the XC95108 in the same package and pick up another 36 macrocells.

Figure 10-38 is a block diagram of the internal architecture of a typical XC9500-family CPLD. Each external I/O pin can be used as an input, an output, or a bidirectional pin according to the device's programming, as discussed later. The pins at the bottom of the figure can also be used for special purposes. Any of three pins can be used as "global clocks" (GCK); as we'll see later, each macrocell can be programmed to use a selected clock input. One pin can be used as a "global set/reset" (GSR); again, each macrocell can programmably use this signal as an asynchronous preset or clear. Finally, two or four pins (depending on device) can be used as "global three-state controls" (GTS); one of these signals can be selected in each macrocell to output-enable the corresponding output driver when the macrocell's output is hooked up to an external I/O pin.

Only four FBs are shown in the figure, but the XC9500 architecture scales to accommodate 16 FBs in the XC95288. Regardless of the specific family

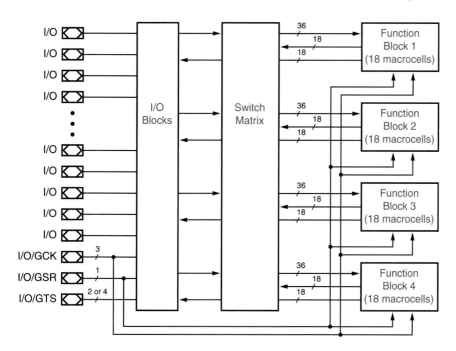

Figure 10-38
Architecture of Xilinx
9500-family CPLDs.

member, each FB programmably receives 36 signals from the switch matrix. The inputs to the switch matrix are the 18 macrocell outputs from each of the FBs and the external inputs from the I/O pins. We'll say more about how the switch matrix hooks things up in Section 10.5.4.

Each FB also has 18 outputs that run "under" the switch matrix as drawn in Figure 10-38 and connect to the I/O blocks. These are merely the output-enable signals for the I/O-block output drivers; they're used when the FB macrocell's output is hooked up to an external I/O pin.

10.5.2 Function-Block Architecture

The basic structure of an XC9500 FB is shown in Figure 10-39. The programmable AND array has just 90 product terms. Compared to 16V8- and 22V10-style PLDs, the XC9500 and most CPLD macrocells have fewer AND terms per macrocell—where the 16V8 has 8 and the 22V10 has 8–16, the XC9500 has only 5. However, this is not all that bad, because of *product-term allocation*. The

product-term allocation

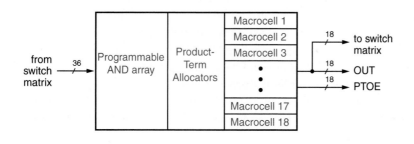

Figure 10-39
Architecture of
function block (FB).

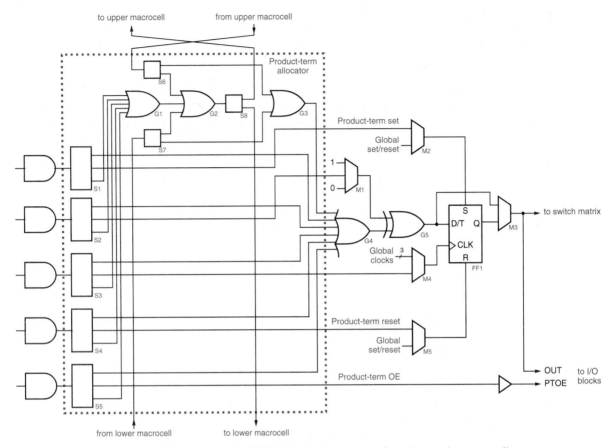

Figure 10-40 XC9500 product-term allocator and macrocell.

product-term allocator   XC9500 and other CPLDs have *product-term allocators* that allow a macrocell's unused product terms to be used by other nearby macrocells in the same FB.

Figure 10-40 is a logic diagram of the XC9500 product-term allocator and macrocell. In this figure, the rectangular boxes labeled S1–S8 are programmable signal-steering elements that connect their input to one of their two or three outputs. The trapezoidal boxes labeled M1–M5 are programmable multiplexers that connect one of their two to four inputs to their output.

The five AND gates associated with the macrocell appear on the lefthand side of the figure. Each one is connected to a signal-steering box whose top output connects the product term to the macrocell's main OR gate G4. Considering just this, only five product terms are available per macrocell. However, the top, sixth input of G4 connects to another OR gate G3 that receives product terms from the macrocells above and below the current one.

Any of the macrocell's product terms that are not otherwise used can be steered through S1–S5 to be combined in an OR gate G1 whose output can eventually be steered to the macrocell above or below by S8. Before steering,

ONE WAY   In a given XC9500 macrocell, you wouldn't normally steer daisy-chained product terms back in the direction from which they came. For example, if product terms are arriving at S6 from above, we can use them locally by having S6 steer them to G3, or S6 can steer them to G2. In the latter case, S8 should steer the output of G2 down to the lower macrocell; there's no point in steering the product terms back up to the upper macrocell. In fact, if S6 and S8 in this macrocell were steering product terms up, and S7 and S8 the macrocell above were steering product terms down, we would have a nasty loop.

these product terms may be combined with product terms from below or above through S6, S7, and G2. Thus, product terms can be "daisy-chained" through successive macrocells to create larger sums of products. In principle, all 90 product terms in the FB could be combined and steered to one macrocell, although that would leave 17 out of the FB's 18 macrocells with no product terms at all.

Besides depriving other macrocells of product terms, daisy-chaining terms has an additional price. A small additional delay is incurred for each "hop" made by steered product terms. This delay can be minimized by careful allocation of product-term-hungry macrocells so they are adjacent to macrocells with low product-term requirements. For example, a macrocell can make use of 13 product terms with only one extra hop delay if it is positioned between two macrocells that use only one product term each.

The third, middle choice for each of the steering boxes S1–S5 is to use the product term for a "special function." The special functions are flip-flop clock, set, and reset; XOR control; and output enable. Most of these special functions are not normally used.

Getting closer to the heart of the macrocell, OR gate G4 forms a sum-of-products expression using all selected product terms and feeds it into XOR gate G5. The other input of G5 can be 0, 1, or a product term, as selected by multiplexer M1. Setting this input to 1 inverts G4's sum-of-products expression, so the macrocell can be configured to use either polarity of minimized logic equations. Setting this input to a product term is useful in the design of counters. The product term is arranged to be 1 when the lower-order counter bits are 1 and counting is enabled, and the output of G4 is arranged to be the current value of the counter bit; thus, the counter bit is complemented as required.

The macrocell's flip-flop FF1 can be programmed to behave either as a D flip-flop or as a T flip-flop with enable; the latter is also useful in some styles of counter realization. Multiplexer M4 selects the flip-flop's clock input from one of four sources—the CPLD's three global clock inputs or a product term. This last choice is a no-no in synchronous design methodologies, except in well-defined synchronization applications, as in Figure 8-111 on page 783.

The flip-flop also has asynchronous set and reset inputs with input sources controlled by multiplexers M2 and M5. In most applications, set or reset would be connected to the CPLD's global set/reset input and would be used only at system initialization. However, these inputs can also be used to access an S-R latch in which the CLK input is not used, or, if you're very careful, you can use CLK and S or R in synchronization applications as in Figure 8-111.

A final multiplexer M3 selects either the flip-flop output or its data input to be used as the macrocell output, OUT. This signal is sent to the switch matrix, where it can be used by any other macrocell. It is also sent to the I/O blocks, along with a product term selected by S5 that can be used as the output-enable signal PTOE if needed.

10.5.3 Input/Output-Block Architecture

I/O block (IOB)

The structure of the XC9500 *I/O block (IOB)* is shown in Figure 10-41. There are seven, count them, seven choices of output-enable signals for the three-state driver buffer. It can be always on, always off, controlled by the product term PTOE from the corresponding macrocell, or controlled by any of up to four global output enables. The global output enables are selectable as active-high or active-low versions of the external GTS pins.

The XC9500's IOB is a good example of an important trend in CPLD and FPGA I/O architectures—providing many "analog" controls in addition to "logic" ones like output enables. Three different analog controls are provided:

- *Slew-rate control.* The rise and fall time of the output signals can be set to be fast or slow. The fast setting provides the fastest possible propagation delay, while the slow setting helps to control transmission-line ringing and system noise at the expense of a small additional delay.

- *Pull-up resistor.* When enabled, the pull-up resistor prevents output pins from floating as the CPLD is powered up. This is useful if the outputs are used to drive active-low enable inputs of other logic that is not supposed to be enabled during power up.

- *User-programmable ground.* This feature actually reallocates an I/O pin to be a ground pin, not a signal pin at all. This is useful in high-speed, high-slew-rate applications. Extra ground pins are needed to handle the high dynamic currents that flow when multiple outputs switch simultaneously.

In addition to these features, the XC9500 family provides compatibility with both 5-V and 3.3-V external devices. The input buffer and the internal logic run from a 5-V power supply (V_{CCINT}). Depending on the operating voltage of external devices, the output driver uses either a 5-V or a 3.3-V supply (V_{CCIO}). Notice that the pull-up resistor pulls to the I/O supply voltage, V_{CCIO}. Diodes $D1$ and $D2$ are used to clamp voltages above V_{CCINT} or below ground that can occur due to transmission-line ringing.

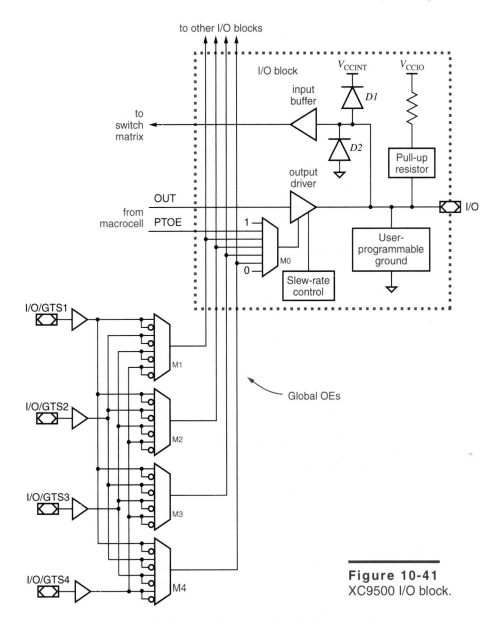

Figure 10-41
XC9500 I/O block.

10.5.4 Switch Matrix

Theoretically, a CPLD's programmable interconnect should allow any internal PLD output or external input pin to be connected to any internal PLD input. Likewise, it should allow any internal PLD output to connect to any external output pin. But if you think about this, you'll see that we're back to the same n^2 problem that we would have in building a 128V64.

The case of a typical Xilinx XC9500 family member, the XC95108, is shown in Figure 10-42. There are 108 internal macrocell outputs and 108 external-pin inputs, a total of 216 signals which should be connected as inputs to the switch matrix. Since the XC95108 has 6 FBs with 36 inputs each, the switch matrix should have (coincidentally) 216 outputs, each of which is a 216-input multiplexer driving one input of an FB's AND array.

A switch matrix such as the one shown in the figure can be built in a chip as a rectangular structure, with a column for each input, a row for each output, and a pass transistor (or transmission gate) at each crosspoint to control whether a given input is connected to a given output. But this is still a big structure—216 rows and 216 columns in the example.

With today's high-density IC technology, the problem is not so much size but speed. Having a large number of transistors connected to each row or column makes for high capacitance, which makes for slow speed. Therefore, CPLD manufacturers look for ways to reduce the size of the switch matrix.

For example, we can observe in Figure 10-42 that not every switch-matrix input must be connectable to every output. We only need every input to be connectable to *some* input of each FB, since any FB input can connect to any AND gate within the FB's AND array.

Then again, the requirement is not quite as simple as we just stated. Suppose that the switch-matrix output for FB input i ($0 \leq i \leq 35$) is created by a simple 8-input multiplexer whose inputs are switch-matrix inputs $8i$ through $8i + 7$. There are many interconnection patterns that cannot be accommodated by such a switch matrix. For example, connecting switch-matrix inputs 0–35 to the same FB would not be possible. As soon as switch-matrix input 0 is connected to FB input 0, switch-matrix inputs 1–7 are blocked.

Thus, the switch-matrix connectivity requirement must be stated more broadly: For each FB, any combination of switch-matrix inputs must be connectable to some combination of the FB's inputs.

Figure 10-42
XC95108 switch-matrix requirements.

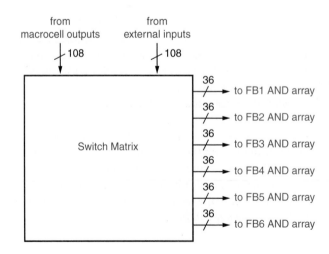

A typical CPLD switch matrix is a compromise between the minimal multiplexer scheme and a full, nonblocking crosspoint array. With anything less than a nonblocking crosspoint array, the problem of allocating input-output connections in the switch matrix is nontrivial. For each different CPLD-based design, a set of switch-matrix connections must be found by "fitter" software provided to the designer by the CPLD's manufacturer.

Finding a complete set of connections through a sparse switch matrix is one of those *NP*-complete problems that you hear about in computer science. In lay terms, that means that for some designs, the fitter software may have to run a lot longer than you care to wait to find out whether there is a solution. If the switch matrix has too few crosspoints, as in our minimal multiplexer example, even the best fitter software running "forever" will not be able to find a complete set of connections for some designs.

Thus, the design of a CPLD's switch matrix is a compromise between chip performance (speed, area, cost) and fitter-software capabilities. The fitter software usually determines not only the final connections through the switch matrix, but also the assignment of CPLD inputs and outputs to FBs, macrocells, and external pins, and of "buried logic" to FBs and macrocells. These assignments interact in turn with both the switch-matrix-connection and product-term allocation problems. The solutions to these problems are the "secret sauce" of CPLD chip and software design, and are not typically disclosed by CPLD manufacturers.

PIN LOCKING   Another important issue in CPLD chip and software design is *pin locking*. In most CPLD applications, it's OK to let the fitter software pick any pins that it likes for the device's external input and output signals. However, once the design is complete and a PCB has been fabricated, the designer would like to "lock down" the pin assignments so they remain the same even if small (or large!) changes are made to the design for bug-fixing purposes. This spares everyone the time, expense, and hassle of reworking or redesigning and refabricating the PCB.

"Locked" pin assignments are typically specified in a file that is read by the fitter software. With early CPLDs and FPGAs, locking down pins before making a small change did not guarantee success—the fitter would throw up its hands and complain that it was too constrained. If you unlocked the pins, the fitter could find a new allocation that worked, but it might be completely scrambled from the original.

These problems were not necessarily the fault of the fitter software; the CPLDs and FPGAs simply did not have rich enough internal connectivity to support frequent design changes under the constraint of pin locking. The device manufacturers have learned from this experience and improved their internal device architectures to accommodate frequent design changes. For example, some devices contain an "output switch matrix" that guarantees that any internal macrocell signal can be connected to any external I/O pin.

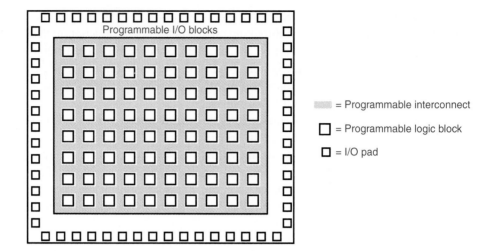

= Programmable interconnect

= Programmable logic block

= I/O pad

Figure 10-43
General FPGA chip
architecture.

10.6 Field-Programmable Gate Arrays

A field-programmable gate array (FPGA) is kind of like a CPLD turned inside-out. As shown in Figure 10-43, the logic is broken into a large number of programmable logic blocks that are individually smaller than a PLD. They are distributed across the entire chip in a sea of programmable interconnections, and the entire array is surrounded by programmable I/O blocks. An FPGA's programmable logic block is less capable than a typical PLD, but an FPGA chip contains a lot more logic blocks than a CPLD of the same die size has PLDs.

Xilinx, Inc. invented FPGAs, and in this section we'll use one of their popular families, the XC4000E, to illustrate FPGA architecture.

10.6.1 Xilinx XC4000 FPGA Family

configurable logic block (CLB)

The programmable logic blocks in the Xilinx XC4000E family of FPGAs are called *configurable logic blocks (CLBs)*. The smallest part, the XC3003E, contains a 10×10 array of CLBs, and the largest, the XC4025E, contains a 32×32 array for a total of 1,024 CLBs. Xilinx also created extended XC4000EX and XC4000XL families, based on the XC4000E family, that have additional resources and features not discussed here. The largest member of the extended families, the XC4085XL, has 3,136 CLBs. Table 10-9 from Xilinx shows the family members that were available in 1999.

Like a CPLD family, the XC4000 family spans a range of device sizes and input/output capabilities. The table column labeled "Max. User I/O" refers to the maximum number of input/output blocks that are provided on-chip. However, XC4000 devices are available in a variety of packages, and not all of the I/Os are brought out to external pins of the smaller packages. As with CPLDs, the FPGA user has the opportunity to migrate a design from smaller to larger devices in the same package, or from a smaller package to a larger one.

Table 10-9 Resources in Xilinx XC4000-series FPGAs.

Device	CLB Matrix	Total CLBs	Max. User I/O	Flip-Flops	Max. RAM bits (no logic)	Max. Gates (no RAM)	Typical Gate Range (Logic and RAM)
XC4002XL	8×8	64	64	256	2,048	1,600	1,000–3,000
XC4003E	10×10	100	80	360	3,200	3,000	2,000–5,000
XC4005E/XL	14×14	196	112	616	6,272	5,000	3,000–9,000
XC4006E	16×16	256	128	768	8,192	6,000	4,000–12,000
XC4008E	18×18	324	144	936	10,368	8,000	7,000–15,000
XC4010E/XL	20×20	400	160	1,120	12,800	10,000	7,000–20,000
XC4013E/XL	24×24	576	192	1.536	18,432	13,000	10,000–30,000
XC4020E/XL	28×28	784	224	2,016	25,088	20,000	13,000–40,000
XC4025E	32×32	1,024	256	2,560	32,768	25,000	15,000–45,000
XC4028EX/XL	32×32	1,024	256	2,560	32,768	28,000	18,000–50,000
XC4036EX/XL	36×36	1,296	288	3,168	41,472	36,000	22,000–65,000
XC4044XL	40×40	1,600	320	3,840	51,200	44,000	27,000–80,000
XC4052XL	44×44	1,936	352	4,576	61,952	52,000	33,000–100,000
XC4062XL	48×48	2,304	384	5,376	73,728	62,000	40,000–130,000
XC4085XL	56×56	3,136	448	7,168	100,352	85,000	55,000–180,000

The "Flip-Flops" column counts all of the flip-flops in the device, two per CLB and two per I/O block, as we'll see later. Only a fraction of the available flip-flops are used in a typical design, but this number is a figure of merit that designers look at when roughly sizing an FPGA for an design. The "Max. RAM Bits" column is another such figure of merit. As we'll see, instead of being used for logic, each CLB can be configured as a small SRAM storing up to 32 bits.

The "Max. Gates" number is fuzzy. For the XC4002XL, the table claims that each CLB can perform the same function as about 25 gates in a discrete design. The XC4003E is even better, at 30 gates per CLB. Is this reasonable? And just what is a "gate"? Do XORs or wide NAND gates count as one gate? You can decide for yourself later, after you know more about the CLB architecture. At the same time, you can decide whether this column was created by engineering or by marketing people.

The last column of the table must certainly have been written by marketing folks, since the high end of each "typical" gate range is higher than the maximum in the column before it! Actually, there *is* a reasonable explanation for this seeming inconsistency. This column assumes that 20–30% of the CLBs are used for memory rather than logic, at 32 bits of SRAM per CLB. The minimum

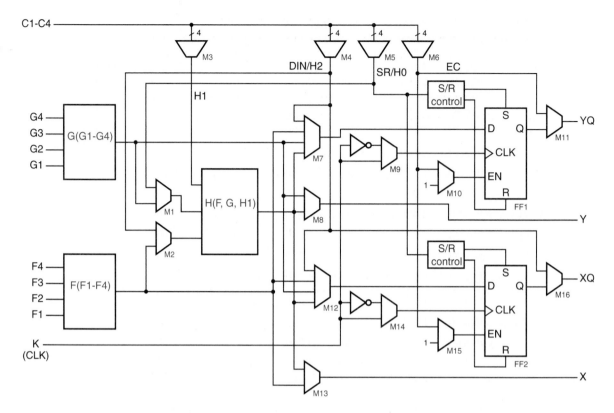

Figure 10-44 XC4000 configurable logic block.

gate-level implementation of an SRAM cell is a D latch using four gates (Figure X7.27 on page 650), so we can count each CLB as 128 gates when it's used as SRAM. The number in the last column, therefore, is the number of gates used for logic *plus* the number of equivalent gates used for SRAM.

10.6.2 Configurable Logic Block

Since an FPGA can have lots and lots of CLBs, it's important that we understand them first! Figure 10-44 shows the internal structure of an XC4000-series CLB.

The CLBs most important programmable elements are the logic-function generators F, G, and H. Both F and G can perform any combinational logic function of their four inputs, and H can perform any combinational logic function of its three inputs.

How do F, G, and H work? Given the task of building a "universal" function generator for 4-input logic functions, what kind of gate-level circuit would you come up with? Think about it, and we'll come back to it later.

As in our CPLD discussion, the trapezoidal boxes in Figure 10-44 represent programmable multiplexers. Notice that the outputs of F and G as well as additional CLB inputs can be directed to inputs of H by multiplexers M1–M3, so it's possible to realize some functions of more than four inputs. A taxonomy of the functions that can be realized by F, G, and H in a single CLB is given below:

- Any function of up to four variables, plus any second function of up to four unrelated variables, plus any third function of up to three unrelated variables.
- Any single function of five variables (see Exercise 10.34).
- Any function of four variables, plus some second functions of six unrelated variables.
- Some functions of up to nine variables, including parity checking and a cascadable equality checker for two 4-bit inputs (see Exercises 10.35 and 10.36).

With appropriate programming of multiplexers M7–M8 and M12–M13, the outputs of the function generators can be directed to CLB outputs X and Y, or they can be captured in edge-triggered D flip-flops FF1 and FF2. The flip-flops can use the rising or falling edge of a common clock input, K, as selected by multiplexers M9 and M14. They can also make use of a clock-enable signal, EC, selected by M10 and M15. The sources of EC and three other internal signals are selected from a set of four miscellaneous inputs C1–C4 by multiplexers M3–M6 at the top of the CLB.

The XQ and YQ outputs of the CLB carry the flip-flop outputs out of the CLB. If a flip-flop is not used in the CLB, multiplexer M11 or M16 can select XQ or YQ to be a "bypass output" that is simply a copy of a CLB input selected by M4 or M6.

The block labeled "S/R control" from each flip-flop determines whether the flip-flop is set or reset at configuration. It also determines whether the flip-flop responds to a global set/reset signal (not shown) or to the CLB's SR signal selected by multiplexer M5.

Wow, that's a lot of programmability! Naturally, the configuration of CLBs within an XC4000 part, whether it has 3,136 CLBs or only 64, is not carried out by hand. The manufacturer provides a fitter tool that allocates, configures, and connects CLBs to match a higher-level design description written in ABEL, VHDL, Verilog, or schematic form.

Let's come back to our question of how to build a universal function generator for 4-input logic functions. It's a hard problem if you think about it at the gate level, but pretty easy if you think about it from another point of view. Any 4-input logic function can be described by its truth table, which has 16 rows. Suppose we store the truth table in a 16-word by 1-bit-wide memory. When we apply the function's four input bits to the memory's address lines, its data output is the value of the function for that input combination.

This is exactly the approach that was taken by the FPGA designers at Xilinx. Function generators F and G are actually just very compact and fast 16×1 SRAMs, and H is an 8×1 SRAM. When a CLB is used to perform logic, the truth tables of logic functions F, G, and H are loaded into SRAM at configuration time from an external read-only memory. The programmable multiplexers in Figure 10-44 are also controlled by "SRAM," actually individual D latches that are also loaded at configuration time. This programming is done for all of the CLBs in the FPGA.

Besides convenience, using memory to store the truth tables has another important benefit. Any XC4000 CLB can be configured at start-up to be used as memory rather than logic. Several different modes can be configured:

- *Two 16×1 SRAMs.* F and G are used as SRAMs with independent address and write-data inputs. They share a common write-enable input, however.

- *One 32×1 SRAM.* The same four address bits are used for F and G, and a fifth address bit is applied to the H function generator and the write-enable circuitry to select between F and G, the upper and lower halves of the memory.

- *Asynchronous or synchronous.* For write operations, the SRAMS above can be configured to have "normal" asynchronous latching behavior, or they can be configured to occur on a designated edge of the K clock signal.

- *One 16×1 dual-port SRAM.* The two sets of address inputs are used to independently read and write different locations in the same SRAM. Only synchronous write operations are supported in this mode.

In these modes, function inputs F1–F4 and G1–G4 supply address, other CLB inputs H0–H2 provide data inputs and the write-enable signal, and data outputs are provided on the F and G generator outputs and can be captured in flip-flops FF1 and FF2 or exit at CLB outputs X and Y.

10.6.3 Input/Output Block

I/O block (IOB)

The structure of the XC4000 *I/O block (IOB)* is shown in Figure 10-45. An I/O pin can be used for input or output or both.

The XC4000 IOB has more "logic" controls than its cousin in the XC9500 CPLD. In particular, its input and output paths contain edge-triggered D flip-flops selectable by multiplexers M5–M7. Placing input and output flip-flops "up close" to the device I/O pins is especially useful in FPGAs. On output, relatively long delays from internal CLB flip-flop outputs to the IOBs can make it difficult to connect to external synchronous systems at very high clock rates. On input,

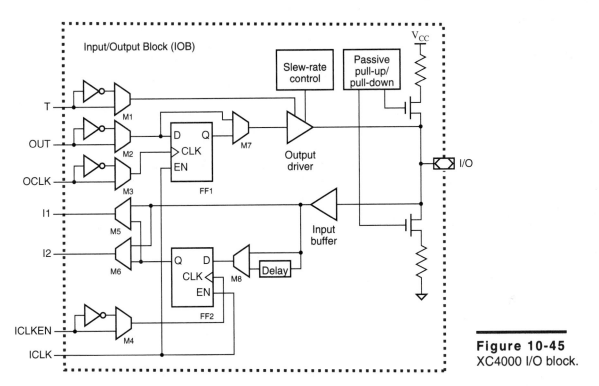

Figure 10-45
XC4000 I/O block.

long delays from the I/O pins to CLB flip-flop inputs can make it difficult to meet external system setup and hold times if external inputs are clocked directly into a CLB flip-flop without being captured first by a flip-flop at the IOB pin. Of course, using IOB flip-flops is possible only if the FPGA's external interface specifications allow "pipelining" of inputs and outputs.

For pipelined inputs, the XC4000 IOB actually goes one step further by providing a delay element, selectable by M8, in series with the D input of input flip-flop FF2. The effect of this element is to delay the D input relative to the FPGA's internal copies of the system clock, guaranteeing that the input will have a zero hold-time requirement with respect to the external system clock. This benefit comes at the expense of increased setup time, of course.

The IOB's other logic controls are selectable polarity, using multiplexers M1–M4, for its four inputs that come from the CLB array via the programmable interconnect. These inputs, OUT, T, OCLK, and ICLKEN, are the output bit, its three-state enable, the output clock, and the input clock enable, respectively.

Like the XC9500 IOB, the XC4000 IOB also has analog controls. The output driver's slew rate is programmable, and a pull-up or pull-down resistor may be connected to the I/O pin.

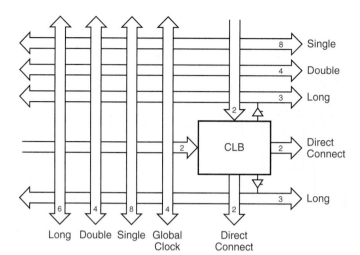

Figure 10-46
XC4000 general
interconnect
structure.

10.6.4 Programmable Interconnect

Well, we saved the best for last. The XC4000 programmable interconnect architecture is fascinating example of a structure that provides rich, symmetric connectivity in a small silicon area.

As we showed in Figure 10-43 on page 882, each CLB in an FPGA is embedded in the interconnect structure, which is really just wires with programmable connections to them. Figure 10-46 gives a little more detail of the XC4000's connection scheme. Wires are not really "owned" by any one CLB, but an XC4000 CLB array is created by tiling the chip with exactly the structure shown in the figure. For example, 100 copies of this figure make the 10×10 CLB array of an XC4003.

The number in each arrow indicates the number of wires in that signal path. Thus, you can see that a CLB has two wires (outputs) each going to the CLBs immediately below and to the right of it. It also connects to three groups of wires above, one below, and four to its left. Signals on these wires can flow in either direction.

The four wires in the group labeled "Global Clock" are optimized for use as clock inputs to the CLBs, providing short delay and minimal skew. The two "Singles" groups are optimized for flexible connectivity between adjacent blocks, without the small number and unidirectional limitation of wires in the "Direct Connect" groups.

It's possible to connect a CLB to another that's more than one hop away using "Single" wires, but they have to go through a programmable switch for each hop, which adds delay. Wires in the "Doubles" groups travel past two CLBs before hitting a switch, so they provide shorter delays for longer connections.

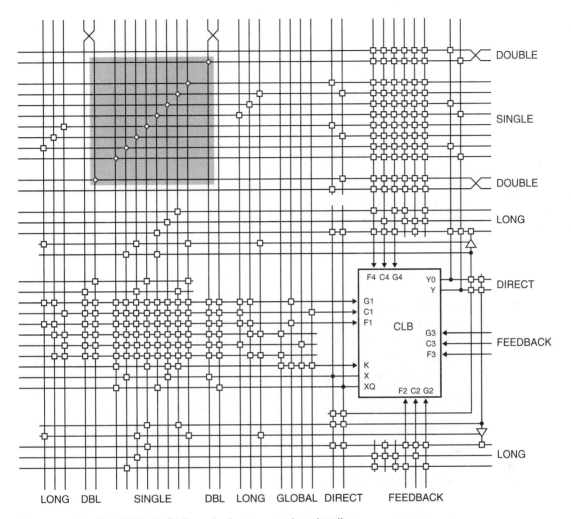

Figure 10-47 XC4000 CLB and wire connection details.

For really long connections, the "Long" groups don't go through any programmable switches at all; instead, they travel all the way across or down a row or column and are driven by three-state drivers near the CLB.

Figure 10-47 shows a CLB and wires in a lot more detail. The small squares are programmable connections—a horizontal wire is connected or not to a vertical one, depending on the state of the programming bit (again, in a latch) for that switch. Additional, specialized programmable interconnect is provided at the edges of the CLB array for connections to the IOBs.

In the figure, the shaded area in color is called a *programmable switch matrix (PSM)*. The PSM is shown in more detail in Figure 10-48. Each of the

programmable switch matrix (PSM)

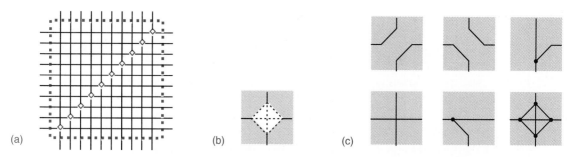

Figure 10-48 XC4000 programmable connections: (a) programmable switch matrix (PSM); (b) programmable switch element (PSE); (c) a few possible connections.

programmable switch element (PSE)

diamonds in (a) is a *programmable switch element (PSE)* that can connect any line to any other, as shown in (b). With four lines, there are 6 possible pairwise connections as shown in (b), and the PSE has a transmission gate for each one of them. Some, none, or all of the transmission gates in a PSE may be enabled—again, by configuration bits stored in latches. Thus, many different connection patterns are possible, as shown in (c).

The PSM is essential for hooking things up in the wiring structure of Figure 10-47. By enabling and disabling connections, PSEs extend or isolate wire segments in the "Single" and "Double" groups. More importantly, the PSM allows signals to "turn the corner" by connecting a horizontal wire to a vertical one. Without this, CLBs would not be able to connect to others in a different row or column of the array.

While the PSM is essential, using it has a price—signals incur a small delay each time they hop through a PSE. Therefore, high-quality FPGA fitter software searches for not just any CLB placement and wire connections that work. The "placement and routing" tool spends a lot of time trying to optimize device performance by finding a placement that allows short connections, and then routing the connections themselves.

Like CPLDs, FPGAs are judged by the flexibility of their architectures and the consistency of the results obtained from a fitter after small design changes are made. There's nothing more frustrating than making a small change to a large design and finding that it no longer meets timing requirements. Thus, FPGA manufacturers have learned to provide "extra" resources in their architectures to help ensure consistent results.

GOOD PRACTICE   Placement and routing is actually a pretty well understood problem, because it is the major part of the "back end" of any custom chip design. Thus, the same kind of tools and the same tool vendors are involved with placement and routing for both FPGAs and ASICS. So, you might like to consider any FPGA experience that you get to be good practice for ASIC design!

References

Manufacturers of ROMs and RAMs publish data books with their device specifications for their parts, and most publish individual data sheets and application notes on their web sites. Information on legacy ultraviolet-erasable EPROMs can be found at Fairchild Semiconductor's site, `www.fairchildsemi.com`. Advanced Micro Device's site, `www.amd.com`, has comprehensive information on more recent EPROMs and flash memory.

Web sites that list all kinds of SRAMs and DRAMs, old and new, include Hitachi Semiconductor (`semiconductor.hitachi.com`) and NEC Electronics (`www.necel.com`). The latest and greatest SSRAMs are offered by companies like Integrated Device Technology (`www.idt.com`), Motorola Semiconductor (`www.mot.com`), and Micron Technology (`www.micron.com`).

Industry publications such as *Electronic Systems* (`www.estd.com`) and *Electronic Design* carry tutorial and update articles from time to time that describe the capabilities and applications of the latest memory chips; just go to the Web sites and search for "SDRAM" to get started. New DRAM architectures include Rambus DRAM (RDRAM), enhanced synchronous DRAM (ESDRAM), SyncLink DRAM (SLDRAM), and double-data-rate synchronous DRAM (DDR-DRAM).

In addition to the memories discussed in this chapter, several types of "specialty" memory devices have widespread use. Probably the most common are *first-in, first-out (FIFO) memories*; these are typically used to transfer data from one processor or clock domain to another. The web sites of IDT, Motorola, and Texas Instruments (`www.ti.com`) are good sources of information on FIFOs. Although it might seem that FIFOs magically solve the metastability problems that occur when crossing clock domains, they do not. For example, see "FIFO Memories: Solution to Reduce FIFO Metastability" by Tom Jackson (Texas Instruments publ. SCAA011A, March 1996).

first-in, first-out (FIFO) memories

Another type of specialty memory is the dual-port memory, which has two independent sets of address, data, and control lines and allows independent operations to be performed on both ports simultaneously. Integrated Device Technology is a leading source for these devices; in addition to data sheets, their Web site has an excellent set of application notes on the devices.

Several manufacturers introduced different CPLD and FPGA architectures at about the same time. While Xilinx (`www.xilinx.com`) was and is the leader in FPGAs, the CPLD field is much more crowded. The first two CPLD families were introduced in the early 1990s by Altera Corporation (`www.altera.com`) and Advanced Micro Devices. AMD's CPLD product line is now owned by former competitor Lattice Semiconductor (`www.latticesemi.com`). Cypress Semiconductor (`www.cypress.com`) is another important player in the CPLD market. All of these companies have excellent resources available on their Web sites.

Drill Problems

10.1 Determine the ROM size needed to realize the combinational logic function in each of the following figures: 4-39(b), 5-39, 5-77, 6-1, and 6-6.

10.2 Determine the ROM size needed to realize the combinational logic function performed by each of the following MSI parts: 74x49, 74x139, 74x153, 74x257, 74x381, 74x682.

10.3 Draw a logic symbol for and determine the size of a ROM that realizes a combinational logic function that can perform the function of either a 74x381 or a 74x382 depending on the value of a single MODE input.

10.4 Draw a logic symbol for and determine the size of a ROM that realizes an 8×8 combinational multiplier.

10.5 Show how to design a 2M × 8 SRAM using HM628512 SRAMs and a combinational MSI part as building blocks.

Exercises

10.6 Our discussion of ROM sneak paths in connection with Figure 10-6 claimed that with A2–A0 = 101, bit lines D2_L and D0_L are pulled LOW through the direct connections. That's not really correct, unless the 74x138 is replaced with a decoder with open-collector outputs. Explain.

10.7 Describe the logic function of seven variables that is performed by the 128×1 ROM in Figure 10-7. Starting with the ROM pattern, one way to describe the logic function is to write the corresponding truth table and canonical sum. However, since the canonical sum has 64 7-variable product terms, you might want to look for a simple but precise word description of the function.

10.8 For the two-output logic function of Figure 4-39(b), compare the number of diodes or transistors needed in the AND-OR, AND, or storage array for PLA, PAL, and ROM realizations.

10.9 Show how to double the number of different attenuation amounts that can be selected in the digital attenuator of Figure 10-17 without increasing the size of the ROM.

10.10 Write the C functions UlawToLinear and LinearToUlaw that are required in Table 10-6. In LinearToUlaw, you should select the μ-law byte whose theoretical value is closest to the linear input. (*Hint:* The most efficient approach builds, at program initialization, a 256-entry table that is used by both functions.) Both functions should report out-of-range inputs.

10.11 Modify the C program in Table 10-6 to perform clipping in cases where the designer has specified a particular attenuation amount to be a gain, and multiplying an input value by the attenuation factor produces an out-of-range result.

10.12 Write a C program to generate the contents of a 256×8 ROM that converts from 8-bit binary to 8-bit Gray code. (*Hint:* Your program should embody the second Gray-code construction method in Section 2.11.)

10.13 Write a C program to generate the contents of a 256×8 ROM that converts from 8-bit Gray code to 8-bit binary. (*Hints:* You may use the results of Exercise 10.12. It doesn't matter if your program is slow.)

10.14 A certain communication system has been designed to transmit ASCII characters serially on a medium that requires an average signal level of zero, so a 5-out-of-10 code is used to code the data. Each 7-bit ASCII input character is transmitted as a 10-bit word with five 0s and five 1s. Write a C program to generate the contents of a 128×10 ROM that converts ASCII characters into coded words.

10.15 The receiving end of the system in Exercise 10.14 must convert each 10-bit coded word back into a 7-bit ASCII character. Write a C program to generate the contents of a $1K \times 8$ ROM that converts coded words onto ASCII characters. The extra output bit should be an "error" flag that indicates when a noncode word has been received.

10.16 How many ROM bits would be required to build a 16-bit adder/subtracter with mode control, carry input, carry output, and two's-complement overflow output? Be more specific than "billions and billions," and explain your answer.

10.17 Repeat Exercise 10.16 assuming that you may use *two* ROMs, so that the delay for a 16-bit addition or subtraction is twice the delay through one ROM. Assume that the two ROMs must be identical in both size and programming. Try to minimize the total number of ROM bits required, and sketch the resulting adder/subtracter circuit. Can the number of ROM bits be further reduced if the two ROMs are allowed to be different?

10.18 Show how, using additional SSI/MSI parts, a 2764 can be used as a $64K \times 1$ ROM. What is the access time of the $64K \times 1$ ROM?

10.19 Show how, using additional SSI/MSI parts, a 2764 can be used as a $2K \times 32$ ROM. You may assume the existence of a free-running clock signal whose period is slightly longer than the access time of the 2764. What is the access time of the $2K \times 32$ ROM?

10.20 Show how to build the μ-law adder circuit of Figure 10-18 with a $32K \times 8$ ROM and two XOR gates. Also, write a C program to generate the ROM contents.

10.21 Determine the ROM size needed to build the fixed-point to floating-point encoder of Figure 6-3. Draw a logic diagram using a commercially available ROM.

10.22 Write a C program to generate the ROM contents for Exercise 10.21. Unlike the original MSI solution, your C program should perform rounding; that is, for each fixed-point number it should generate the nearest possible floating-point number.

10.23 Draw a complete logic diagram for a ROM-based circuit that performs combinational multiplication of a pair of 8-bit unsigned or two's-complement integers. Signed versus unsigned operation should selected by a single input, SIGNED. You may use any of the commercial ROMs in Figure 10-11, as well as discrete gates.

10.24 Write and test a C program that generates the contents of the ROM(s) in Exercise 10.23.

10.25 Write a C program that generates a $256K \times 4$ ROM that computes the next move in a Tic-Tac-Toe game, using the input and output encodings of Section 6.2.7. Your program must be smart enough to pick a winning move whenever possible.

10.26 Repeat Exercise 10.25 using a $32K \times 4$ ROM. To accomplish this, the board state must be encoded in only 15 bits. Explain your coding algorithm, and write C functions to translate a cell number in either direction between your encoding and the encoding of Section 6.2.7.

10.27 For each of the timing parameters defined in Section 10.3.3, determine whether its value for the $128K \times 8$ SRAM you designed in Drill 10.5 is the same as the HM628512's. If it is different, indicate the new value. Use the worst-case values in the 74FCT column of Table 5-3 to determine the delays for the MSI part.

10.28 Using an HM6264 $8K \times 8$ SRAM, a handful of MSI parts, and a PLD as building blocks, design an $8K \times 8$ late-write SSRAM with flow-through outputs.

10.29 Using an HM6264 $8K \times 8$ SRAM, a handful of MSI parts, and a PLD as building blocks, design an $8K \times 8$ ZBT SSRAM with pipelined outputs.

10.30 Define the relevant timing parameters for the synchronous SRAM as defined in Exercise 10.28.

10.31 Calculate the values of the timing parameters you defined in Exercise 10.30 for your solution to Exercise 10.28.

10.32 In the style of Figure 10-27, draw the timing diagram for a late-write SSRAM with flow-through outputs for a series of interleaved reads and writes in the pattern R-R-W-W-R-W-R-W. Run the individual cycles as close together as possible, but be sure to account for resource conflicts that prevent back-to-back cycles. What is the average utilization of the SRAM array if the SSRAM is presented with a continuous streams of R-W-R-W-R-W requests?

10.33 Repeat the preceding exercise for a late-write SSRAM with pipelined outputs.

10.34 Using one of the theorems in Section 4.1, prove that the Xilinx XC4000-series CLB can realize any logic function of five variables.

10.35 Show how to use the function generators in an XC4000-series CLB to realize a 9-bit even parity function.

10.36 Show how to use the function generators in an XC4000-series CLB to realize an equality checker for two 4-bit operands. You should be able to cascade your design to check $4n$-bit operands with just n CLBs.

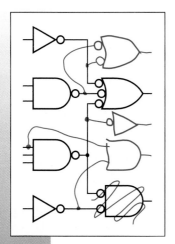

Additional Real-World Topics

T his chapter contains a collection of "real-world" topics that are certain to interest you at some point if you are (or become) a working digital designer. We barely scratch the surface of each of these topic areas, so be sure to consult the References in areas that are of particular interest.

11.1 Computer-Aided Design Tools

"If it wasn't hard, they wouldn't call it hardware."

Many digital designers with twenty years of experience consider this statement to be indisputable. Yet more and more, digital design is being carried out using *software*, and it's getting easier as a result.

The terms *computer-aided design (CAD)* and *computer-aided engineering (CAE)* are used to refer to software tools that aid the development of circuits, systems, and many other things. "CAD" is the more general term and applies to tools both inside and outside the electronics area, including architectural and mechanical design tools, for example. Within electronics, "CAD" often refers to *physical-design* tools, such as IC- and PCB-layout programs. "CAE" is used more often to refer to *conceptual-design* tools, such as schematic editors, circuit simulators, and PLD compilers. However, a lot of people in electronics (including the author) tend to use the two terms interchangeably. In this section, we'll discuss some CAD/CAE tools used by digital designers.

IS HARDWARE NOW EASY-WARE?	Since more and more hardware design and debugging is being carried out using software tools, is it really getting easier? Not necessarily. In the author's experience, the increasing use of CAD means that instead of spending time fighting with soldering irons and test clips, many designers can now spend their time fighting buggy programs running in a buggy software environment.

11.1.1 Hardware Description Languages

In previous decades, most logic design was performed graphically, using block diagrams and schematics. However, the rise of synthesizable HDLs coupled with programmable logic devices and very-large-scale ASIC technology in the 1990s has radically changed the way that the largest digital designs are done.

In traditional software design, high-level programming languages like C, C++, and Java have raised the level of abstraction so that programmers can design larger, more complex systems with some sacrifice in performance compared to hand-tuned assembly-language programs. The situation for hardware design is similar. The circuit produced by a VHDL or Verilog synthesis tool may not be as small or fast as one designed and tweaked by hand by an experienced designer, but in the right hands these tools can support much larger system designs. This is, of course, a requirement if we're ever to take advantage of the millions of gates offered by the most advanced CPLD, FPGA, and ASIC technologies.

The use of HDLs has taken off with the availability of synthesis tools, but *register-transfer language* non-synthesizable languages have been around for a while. Most prominent are *register-transfer languages*, which have been used for decades to describe the operation of synchronous systems. Such a language combines the control-flow notation of a state-machine description language with a means for defining and operating on multibit registers. Register-transfer languages have been especially useful in computer design, where individual machine-language instructions are defined as a sequence of more primitive steps involving loading, storing, combining, and testing registers.

11.1.2 Schematic Capture

Except in homework assignments, your first step in designing a circuit usually is to convince someone that your proposed approach is the right one. That means preparing block diagrams and presentation slides, and discussing your ideas *design review* with managers and peers in a preliminary *design review*. Once your project has been approved, you can safely start on the "fun stuff," that is, drawing the schematic.

Schematics used to be drawn by hand, but most are now prepared using *schematic editor* *schematic editors,* CAD programs that run on engineering workstations. The *schematic capture* process of creating a schematic on a computer is often called *schematic capture*.

WHAT'S A WORKSTATION?	An "engineering workstation" nowadays can be anything from a $500 PC to a $50,000 parallel processor. The "canonical" engineering workstation in terms of cost (what a typical company will spend per engineer) and performance (among the best) is a $5,000 Sun workstation, whatever model you get for that price this year.

(Huh? How did it get loose?) This term is used because the schematic editor captures more information than just a drawing. Most information in the schematic has a "type" associated with it, so that selected information can be retrieved automatically as required later in the design process.

In order to simplify both entry and retrieval of information in a schematic, a schematic editor typically provides *fields* for each information type, either hidden or drawn automatically nearby the associated schematic element. Some typical fields and their uses are described below.

fields

- *Component type.* By specifying a component type (e.g., resistor, capacitor, 74FCT374, GAL16V8), the designer can call out a predrawn symbol from a *component library.* Some components have user-selectable alternate symbols (e.g., DeMorgan equivalents for gates). The component type is used both for manufacturing documentation and for simulation.

component type

component library

- *Component value.* Most analog components have a value that must be specified by this field (e.g., 2.7 kΩ). Additional distinguishing information may appear in a "rating" or "tolerance" field (e.g., 1/4 W, 1%). Digital components may give a speed rating here (e.g., -10 for a 10-ns PLD).

component value

- *Approved-part number.* Using all of the distinguishing information from the previous fields, the CAD software may automatically select a part number from the company's "approved parts list" (e.g., 126-10117-0272 for a 2.7-kΩ, 1/4-W, 1% metal-film resistor), or a warning if no such part is available. (No, in large companies you can't use just any part in the catalog; all parts must be approved through a *parts-qualification* process.)

approved-part number

parts qualification

- *Reference designator.* This alphanumeric label distinguishes among multiple instances of similar parts and is used just about everywhere.

reference designator

- *Component location.* Using a coordinate system, this field may indicate the physical position of the component on an assembled PCB, simplifying the task of finding it during debugging. Of course, the final component location generally is not known during the initial circuit design. However, CAE software with *back-annotation* capability can insert this information into the schematic when the PCB layout is completed.

component location

back annotation

- *Pin numbers.* These specify the assignment of the signal pins on each component and are usually prespecified in the symbol that is called up from the component library. However, for SSI and other parts that have

pin number

multiple identical sections in a single package, pin numbers (and even reference designators) might not be specified until the PCB layout is completed; this is another case where back annotation is handy.

wire type

- *Wire type.* Usually the only wire types are "ordinary" and "bus," where a bus is just a set of ordinary wires that are grouped for drawing purposes. However, other wire types, such as extra-wide PCB traces for analog signals that carry high current, may be indicated in some systems; such information is passed to the PCB-layout program.

signal name

- *Signal name.* The user may (and should) associate a name with each signal. The name is especially useful in simulation and debugging, but it does not affect the physical PCB layout.

connection flag

- *Connection flag.* This drawing element indicates the connection of one signal line to another one on the same or a different page. Software can ensure that all outgoing connections are matched by incoming ones. In a hierarchical schematic, software can use the connection flags to generate signal names in a "logic symbol" that represents the entire page.

Once you've "captured" a circuit's schematic in a CAD system, there are lots of things you can do with it besides printing it. At least two documents can be generated for manufacturing purposes:

parts list

- *Parts list.* This is a list of components and their reference designators.

net
net list

- *Net list.* A *net* is a set of pins that are all connected to the same electrical node or signal. A *net list* specifies all such connections that are required by the schematic, typically using an alphabetically sorted list of signal names. For each signal name, it lists the device pins (identified by reference designator and pin number) where that signal is connected.

  Alternatively, the net list may be sorted by reference designator and pin number, and show the signal name connected to each pin; sometimes this

pin list

  is called a *pin list*.

The parts list and the net list are the main inputs to the PCB-layout process. The designer may also use the parts list to estimate the cost and reliability of the circuit, and the manufacturing department obviously can use it to order the parts.

11.1.3 Timing Drawings and Specifications

Block diagrams and schematics are not the only drawings used in digital system design. Timing diagrams are an essential part of almost any documentation package.

Larger systems are decomposed into smaller subsystems that communicate through well-defined interfaces. The definition of an interface typically contains not only signal names and functions, but also specifications of expected timing behavior. The starting point is typically a specification of maximum and sometimes minimum clock frequencies. Inputs to a subsystem have their setup- and hold-time requirements with respect to a clock edge specified. Outputs of a subsystem have their minimum and maximum delays from a clock edge specified.

Programs such as TimingDesigner (`www.chronology.com`) can automate the tedious task of creating complicated timing diagrams and specifications, where changes in one timing parameter may propagate to many others.

11.1.4 Circuit Analysis and Simulation

The component library of a sophisticated CAD system contains more than just a symbol for each component. For ICs, the library may contain a *component model* describing the device's logical and electrical operation. In fact, HDLs like VHDL and Verilog were originally invented strictly for the purpose of modeling the logical behavior of components and their interconnection in systems. Using these models, logical and timing errors can be found in simulation.

component model

As a minimum, the model for an IC indicates whether each pin is an input or output. With just this information, a *design-rule checker* can detect some of the more common "stupid mistakes" in a design, such as shorted outputs and floating inputs. With the addition of input loading and output driving characteristics for each pin, the program can also determine whether any output's fanout capability is exceeded.

design-rule checker

The next step is *timing verification*. Even without a detailed model of an IC's logical behavior, the component library can provide a worst-case delay value for each input-to-output path, and setup and hold times for clocked devices. Using this information, a *timing verifier* can find the worst-case delay paths in the overall circuit, so the designer can determine whether the timing margins are adequate.

timing verification

timing verifier

Finally, the library may contain a detailed model of each component's logical behavior, in which case *simulation* may be used to predict the overall circuit's behavior for any given sequence of inputs. The designer provides an input sequence, and a *simulator* program determines how the circuit would respond to that sequence. The simulator output is usually displayed graphically, in the form of timing waveforms that the designer could see on an oscilloscope or logic analyzer if the same inputs were applied to a real circuit. In such an environment, it is possible to debug an entire circuit without "breadboarding" it, and to assemble a PCB that works on the very first try.

simulation

simulator

Simulators use different models for different types of components. Analog simulators like SPICE may use mathematical models with just one parameter for a resistor and with anywhere from a few to dozens of parameters for a bipolar transistor. Based on these models and the circuit's structure, the simulator develops and solves equations to determine the circuit's behavior.

Digital simulators normally use one of the two following models for each logic element:

behavioral model

- *Behavioral model.* In such a model, the software "understands" the basic function of the logic element. For example, an AND gate is understood to produce a 1 output with delay t_{pHL} after all inputs are 1, and a 0 output with delay t_{pHL} after any input is 0.

structural model

- *Structural model.* Larger elements may be modeled by connecting groups of smaller elements that have behavioral models.

We saw that VHDL supports both types of model. For example, Table 5-15 on page 368 gave a behavioral model for a 74x138 3-to-8 decoder. By using one of VHDL's clauses ("after" and other mechanisms not covered in this book), we could specify the delay from input to output. Alternatively, Table 5-14 gave a structural model of a 2-to-4 decoder as a set of gates connected according to the internal logic diagram of Figure 5-32 on page 352. Assuming that timing information already exists for the individual gates in the structure (inv and and3), the larger structural model can predict circuit timing more accurately, since it can follow the actual circuit path taken by each different input transition.

software model

Both of the foregoing models are called *software models* because the chip's operation is simulated entirely by software. Some chips, including microprocessors and other LSI parts, require huge software models. In many cases, the chip manufacturers don't provide such models (they're considered to be proprietary information), and the models would take months or years for a user to develop. To get around this problem, innovative CAD companies have developed a third type of model:

physical model
hardware model

- *Physical model*, sometimes called a *hardware model*. A real, working copy of the chip is connected to the computer that's running the simulation. The simulator applies inputs to the real chip and observes its outputs, which are then used as inputs to the other models in the simulation.

uninitialized sequential circuits

A digital designer's input to a logic simulator is a schematic and a sequence of input vectors to be applied to the circuit in simulated time. The first output that the designer usually sees is "xxxxx," meaning that the simulator can't figure out what the output will be. This usually occurs because the circuit contains sequential circuits (flip-flops and latches) that are not initialized. To obtain meaningful simulator output, the designer must ensure that all circuits are explicitly reset to a known value, even if reset is not required for correct operation in the real system. For example, the self-correcting ring counters in Figures 8-62 and 8-64 on page 726 cannot be simulated, because the simulator has no way of knowing their starting state.

simulation inefficiency

Once the simulated circuit has been initialized, the designer's ability to probe its operation is limited only by the simulator's speed. Simulation of complex circuits can be very slow, compared to the speed of the real thing. Depending on the level of detail being simulated, the simulation time can be 10^3 to 10^8 times longer than the interval being simulated. A strictly functional

simulation, in which all components are assumed to have zero delay, is fastest. A much more realistic simulation, one that calculates and considers a worst-case range of delays for all signal paths, is the slowest.

11.1.5 PCB Layout

Printed-circuit-board layout can be carried out by a program or by a person (a PCB designer) or by a combination of the two. PCB designers are usually people who like to solve puzzles. Their job is to fit all of the components in the schematic onto a board of a given size, and then to hook up all the connections as required by the schematic—and usually to do so with the smallest possible PCB size and number of interconnection layers.

PCB design can be done automatically by software, but the results aren't always so good. Both the PCB designer and the circuit designer normally must guide the initial component-placement steps, taking into account many practical concerns, including the following:

- *Mechanical constraints.* Certain connectors, indicators, and other components may have to be placed at a particular location on the PCB because of a predetermined mechanical requirement.

  mechanical constraints

- *Critical signal paths.* An automatic placement program can place components in a way that minimizes the average length of interconnections, but designers usually have a more intuitive feel for what areas of the circuit and which specific signals have the most critical requirements. In fact, the circuit designer normally provides the PCB designer with a *floorplan,* a suggested placement for groups of components that reflects the designer's ideas about which things need to be close to each other.

  critical signal paths

  floorplan

- *Thermal concerns.* Some areas of the PCB may have better cooling than others when the board is installed in a system, and certainly some components run hotter than others. The designers must ensure that the operating temperature range of the components is not exceeded.

  thermal concerns

  The temperature anywhere on a well-designed PCB in a properly cooled system typically will not exceed 10°C above ambient, but a power-hungry chip (such as a microprocessor) placed in an area with no air flow could create a temperature rise of 30°C or more. If the ambient temperature were 40°C, the actual temperature at the microprocessor would be 70°C, which is at the limit of the commercial operating range for most chips.

- *Radio-frequency emissions.* Electronic equipment radiates radio-frequency energy as a side effect. Government agencies in the U.S. and other countries specify the maximum permissible level of emissions; in the U.S. the limits are spelled out in *FCC Part 15* regulations.

  radio-frequency emissions

  FCC Part 15

  A circuit's emissions levels usually depend not only on component performance and circuit configuration, but also on the physical layout of the

PCB. During layout, special attention must therefore be paid to component placement, grounding, isolation, and decoupling, in order to minimize unwanted (and illegal!) radio-frequency emissions.

stupid mistakes

- *Stupid mistakes*. Perhaps the most important step in PCB design, especially when using new or unfamiliar components, is checking for stupid mistakes. Almost every digital designer, even successful ones, can tell you stories of laying out a PCB using the mirror image of a new part's pinout, putting mounting holes or connectors in the wrong place, using the wrong version of an IC package, using the wrong sex of a connector, leaving out the inner connection layers when the board is fabricated, and so on. Remember, "Computers don't make mistakes, humans do!"

routing

Once component placement is done, all of the signals must be hooked up; this is called *routing*. This phase of PCB layout is most easily automated; in fact, a decent "auto-router" can route a moderate-sized PCB overnight. Still, design engineers may have to give special attention to the routing of critical-path traces and to other issues such as placement of signal terminations and test points, which are discussed later in this chapter.

11.2 Design for Testability

When you buy a new piece of equipment, you expect it to work properly "right out of the box." However, even with modern automated equipment, it's impossible for an equipment manufacturer to guarantee that every unit produced will be perfect. Some units don't work because they contain individual components that are faulty, or because they were assembled incorrectly or sloppily, or because they were damaged in handling. Most manufacturers prefer to discover these problems in the factory, rather than get a bad reputation when their customers receive a faulty product. This is the purpose of *testing*.

testing
go/no-go test

The most basic test, a *go/no-go test*, yields just one bit of information—is the system 100% functional or not? If the answer is yes, the system can be shipped. If not, the next action depends on the size of the system. The appropriate response for a digital watch would be to throw it out. If many watches are turning up faulty, it's more efficient to repair what must be a consistent problem in the assembly process or an individual component than to try to salvage individual units.

diagnostic test

On the other hand, you wouldn't want to toss out a $10,000 computer that fails to boot. Instead, a more detailed *diagnostic test* may be run to locate the particular subsystem that is failing. Depending on cost, the failed subsystem may be either repaired or replaced. A typical digital subsystem is a PCB whose assembled cost with components ranges from $50 to $1,000. The repair/replace decision is an economic trade-off between the cost of the assembled PCB and the estimated time to locate and repair the failure.

This is where *design for testability (DFT)* comes in. "DFT" is a general term applied to design methods that lead to more thorough and less costly testing. Many benefits ensue from designing a system or subsystem so that failures are easy to detect and locate:

design for testability (DFT)

- The outcome of a go/no-go test is more believable. If fewer systems with hidden faults are shipped, fewer customers get upset, which yields obvious economic as well as psychological benefits.

- Diagnostic tests run faster and produce more accurate results. This reduces the cost of salvaging a subsystem that fails the go/no-go test, making it possible to manufacture more systems at lower cost.

- Both go/no-go and diagnostic tests require less test-engineering time to develop.

- Although the savings in test-engineering time may be offset by added design-engineering effort to include DFT, any increase in overall product development cost usually can be offset by decreased manufacturing cost.

11.2.1 Testing

Digital circuits are tested by applying *test vectors* which consist of inputs combinations and expected output combinations. A circuit "passes" if its outputs match what's expected. In the worst case, an n-input combinational circuit requires 2^n test vectors. However, if we know something about the circuit's physical realization and make some assumptions about the type of failures that can occur, the number of vectors required to test the circuit fully can be greatly reduced. The most common assumption is that failures are *single stuck-at faults,* that is, that they can be modeled as a single input or output signal stuck at logic 0 or logic 1. Under this assumption, an 8-input NAND gate, which might otherwise require 256 test vectors, can be fully tested with just nine—11111111, 01111111, 10111111, …, 11111110.

test vector

single stuck-at fault

Under the single-fault assumption, it's easy to come up with test vectors for individual logic elements. However, the problem in practice is *applying* the test vectors to logic elements that are buried deep in a circuit, and *seeing* the results. For example, suppose that a circuit has a dozen combinational and sequential logic elements between its primary inputs and the inputs of an 8-input NAND gate that we want to test. It's not at all obvious what primary-input vector, or sequence of primary-input vectors, must be applied to generate the test vector 11111111 at the NAND-gate inputs. Furthermore, it's not obvious what else might be required to propagate the NAND gate's output to a primary output of the circuit.

Sophisticated *test-generation programs* deal with this complexity and try to create a *complete test set* for a circuit, that is, a sequence of test patterns that fully tests each logic element in the circuit. However, the computation required can be huge, and it's quite often just not possible to generate a complete test set.

test-generation program

complete test set

controllability

observability

test points

DFT methods attempt to simplify test-pattern generation by enhancing the "controllability" and "observability" of logic elements in a circuit. In a circuit with good *controllability*, it's easy to produce any desired values on the internal signals of the circuit by applying an appropriate test-vector input combination to the primary inputs. Similarly, good *observability* means that any internal signal can be easily propagated to a primary output for comparison with an expected value by the application of an appropriate primary-input combination. The most common method of improving controllability and observability is to add *test points,* additional primary inputs and outputs that are used during testing.

11.2.2 Bed-of-Nails and In-Circuit Testing

test fixture
nail
bed of nails
automatic tester
in-circuit testing

overdrive

In a digital circuit that is fabricated on a single PCB, the "ultimate" in observability is obtained by using every pin of every IC as a test point. This is achieved by building a special *test fixture* that matches the layout of the PCB and contains a spring-loaded pin (*nail*) at each IC-pin position. The PCB is placed on this *bed of nails,* and the nails are connected to an *automatic tester* that can monitor each pin as required by a test program.

Going one step further, *in-circuit testing* also achieves the "ultimate" in controllability. This method not only monitors the signals on the bed of nails for observability, but also connects each nail to a very low impedance driving circuit in the tester. In this way, the tester can override (or *overdrive*) whatever circuit on the PCB normally drives each signal, and directly generate any desired test vector on the internal signals of the PCB. Overdriving an opposing gate output causes excessive current flow in both the tester and the opposing gate, but the tester gets away with it for short periods (milliseconds).

To test an 8-input NAND gate, an in-circuit tester needs only to provide the nine test vectors mentioned previously, and can ignore whatever values the rest of the circuit is trying to drive onto the eight input pins. And the NAND-gate output can be observed directly on the output pin, of course. With in-circuit testing, each logic element can be tested in isolation from the others.

HITTING THE NAIL ON THE PIN

The rise of high-density surface-mount packaging has made bed-of-nails testing considerably more difficult than it was with thru-hole components like DIPs. Since components may be mounted on both sides of the PCB, a special test fixture called a clam shell may be needed to connect nails to both sides of the PCB.

Furthermore, the pins of many surface-mount devices are so small and their spacing is so tight (25 mils or less), that it may be impossible to reliably land a test pin on them. In such cases, the PCB designer may have to explicitly provide test pads, extra copper patches that are large enough for a test pin to contact (e.g., 50 mils in diameter). A separate test pad must be provided for each signal that does not connect to a larger (e.g., 62-mil through-hole) component pad somewhere on the PCB.

Although in-circuit testing greatly enhances the controllability and observability of a PCB-based circuit, logic designers must still follow a few DFT guidelines for the approach to be effective. Some of these are listed below.

- *Initialization.* It must be possible to initialize all sequential circuit elements to a known state.

Since the preset and clear input pins of registers and flip-flops are available to an in-circuit tester, you would think that this is no problem. However, Figure 11-1(a) shows a classic case of a circuit (a Gray-code counter) that cannot be initialized, since the flip-flops go to an unpredictable state when PR and CLR are asserted simultaneously. The correct way to handle the preset and clear inputs is shown in (b).

- *Clock generation.* The tester must be able to provide its own clock signal *without* overdriving the on-board clock signals.

Testers usually must override an on-board clock, for several reasons: The speed at which they can apply test vectors is limited; they must allow extra time for overdriven signals to settle; and sometimes they must stop the clock. However, overdriving the clock is a no-no. An overdriven signal may "ring" and make several transitions between LOW and HIGH before finally settling to the level that the tester wants. On a clock signal, such transitions can create unwanted state changes.

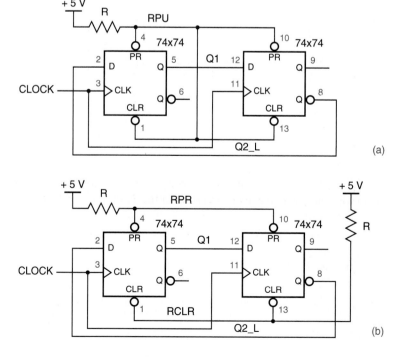

Figure 11-1
Flip-flops with pull-up resistors for unused inputs: (a) untestable; (b) testable.

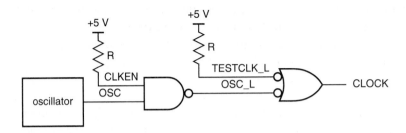

Figure 11-2
Clock driver circuit
that allows a tester to
cleanly override the
system clock.

Figure 11-2 shows a recommended clock driver circuit. To inject its own clock, the tester pulls CLKEN LOW and inserts its clock on TESTCLK_L. Since the tester is not overdriving any gate outputs, the resulting CLOCK signal is clean. In general, any normally glitch-free signal that is used as a clock input or other asynchronous input must not be overdriven by the tester, and would have to be treated in a way similar to Figure 11-2. This is another reason why synchronous design with a single clock is so desirable.

- *Grounded inputs.* In general, ground should not be used as a constant-0 logic source.

The in-circuit tester can overdrive *most* signals, but it can't overdrive ground. Therefore, signals that require a constant-0 input during normal operation should be tied to ground through a resistor, which allows the tester to set them to 1 as required during testing. For example, consider what would happen if we used a GAL16V8 PLD to generate a customized set of clock waveforms from a master clock, as we did in the ABEL program of Table 8-26 on page 737. If not for testing considerations, we could connect the 16V8's pin 11, its active-low global output-enable pin, directly to ground. However, it should be connected through a resistor, so that the tester can float the PLD outputs and drive the clock signals P1_L–P6_L itself. Although the tester could theoretically overdrive these signals, it shouldn't if they are used as clocks.

- *Bus drivers.* In general, it should be possible to disable the drivers for wide buses so that the tester can drive the bus without having to overdrive all the signals on the bus.

That is, it should be possible to output-disable all of the three-state drivers on a bus, so that the tester drives a "floating" bus. This reduces electrical stress both on the tester and on multibit drivers (e.g., 74x244) that might otherwise be overheated and damaged by having all of their outputs overdriven simultaneously.

11.2.3 Scan Methods
In-circuit testing works fine, up to a point. It doesn't do much good for custom VLSI chips and ASICs, because the internal signals simply aren't accessible. Even in board-level circuits, high-density packaging technologies such as surface mounting greatly increase the difficulty of providing a test point for

every signal on a PCB. As a result, an increasing number of designs are using "scan methods" to provide controllability and observability.

A *scan method* attempts to control and observe the internal signals of a circuit using only a small number of test points. A *scan-path method* considers any digital circuit to be a collection of flip-flops or other storage elements interconnected by combinational logic, and is concerned with controlling and observing the state of the storage elements. It does this by providing two operating modes: a normal mode, and a *scan mode* in which all of the storage elements are reorganized into a giant shift register. In scan mode, the state of the circuit's *n* storage elements can be read out by *n* shifts (*observability*), and a new state can be loaded at the same time (*controllability*).

Figure 11-3 shows a circuit designed using a scan-path method. Each storage element in this circuit is a scan flip-flop (Section 7.2.7) that can be loaded from one of two sources. The test enable (TE) input selects the source—normal data (D) or test data (T). The T inputs are daisy-chained to create the scan path shown in color. By asserting ENSCAN for 11 clock ticks, a tester can read out the current state of the flip-flops and load a new state. The test engineer is left with the job of deriving test sets for the individual combinational logic blocks, which can be fully controlled and observed using the scan path and the primary inputs and outputs.

Scan-path design is used most often in custom VLSI and ASIC design, because of the impossibility of providing a large number of conventional test points. However, the two-port flip-flops used in scan-path design do increase

scan method
scan-path method

scan mode

Figure 11-3 Circuit containing a scan path, shown in color.

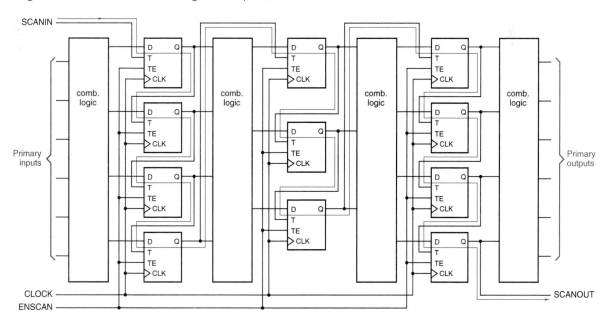

chip area. For example, in LSI Logic Corp.'s LCA500K series of CMOS gate arrays, an **FD1QP D** flip-flop macrocell uses 7 "gate cells," while an **FD1SQP D** scan flip-flop macrocell uses 9 gate cells, almost a 30% increase in silicon area. However, the overall increase in chip area is much less, since flip-flops are only a fraction of the chip, and large "regular" memory structures (e.g., RAM) may be tested by other means. In any case, the improvement in testability may actually *reduce* the cost of the packaged chip when the cost of testing is considered. For large ASIC designs with rich, complicated control structures, scan-path design should be considered a requirement.

11.3 Estimating Digital System Reliability

reliability

Qualitatively, the *reliability* of a digital system is the likelihood that it works correctly when you need it. Marketing and sales people like to say that the systems they sell have "high reliability," meaning that the systems are "pretty likely to keep working." However, savvy customers ask questions that require more concrete answers, such as, "If I buy 100 of these systems, how many will fail in a year?" To provide these answers, digital design engineers are often responsible for calculating the reliability of the systems they design, and in any case they should be aware of the factors that affect reliability.

Quantitatively, reliability is expressed as a mathematical function of time:

$R(t)$ = Probability that the system still works correctly at time t

Reliability is a real number between 0 and 1; that is, at any time $0 \leq R(t) \leq 1$. We assume that $R(t)$ is a monotonically decreasing function; that is, failures are permanent and we do not consider the effects of repair. Figure 11-4 is a typical reliability function.

The foregoing definition of reliability assumes that you know the mathematical definition of probability. If you don't, reliability and the equivalent probability are easiest to define in terms of an experiment. Suppose that we were to build and operate N identical copies of the system in question. Let $W_N(t)$ denote the number of them that would still be working at time t. Then,

$$R(t) = \lim_{n \to \infty} W_N(t)/N$$

Figure 11-4
Typical reliability
function for a system.

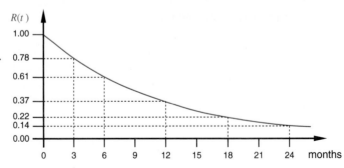

That is, if we build *lots* of systems, $R(t)$ is the fraction of them that are still working at time t. When we talk about the reliability of a single system, we are simply using our experience with a large population to estimate our chances with a single unit.

It would be very expensive to compute $R(t)$ if the only way to do it was to perform an experiment—to build and monitor N copies of the system. Worse, for any t, we wouldn't know the value of $R(t)$ until time t had elapsed in real time. Thus, to answer the customer's question posed earlier, we'd have to build a bunch of systems and wait a year; by then, our potential customer would have purchased something else.

Instead, we can estimate the reliability of a system by combining reliability information for its individual components, using a simple mathematical model. The reliability of mature components (e.g., 74FCT CMOS chips) may be known and published based on actual experimental evidence, while the reliability of new components (e.g., a Sexium microprocessor) may be estimated or extrapolated from experience with something similar. In any case, a component's reliability is typically described by a single number, the "failure rate" described next.

11.3.1 Failure Rates

The *failure rate* is the number of failures that occur per unit time in a component or system. In mathematical formulas, failure rate is usually denoted by the Greek letter λ. Since failures occur infrequently in electronic equipment, the failure rate is measured or estimated using many identical copies of a component or system. For example, if we operate 10,000 microprocessor chips for 1,000 hours, and eight of them fail, we would say that the failure rate is

failure rate

λ

$$\lambda = \frac{8 \text{ failures}}{10^4 \text{ chips} \cdot 10^3 \text{ hours}} = (8 \cdot 10^{-7} \text{ failures/hour})/\text{chip}$$

That is, the failure rate of a single chip is $8 \cdot 10^{-7}$ failures/hour.

The actual process of estimating the reliability of a batch of chips is not nearly as simple as we've just portrayed it; for more information, see the References. However, we can use individual component failure rates, derived by whatever means, in a straightforward mathematical model to predict overall system reliability, as we'll show later in this section.

Since the failure rates for typical electronic components are so small, there are several scaled units that are commonly used for expressing them: percent failures per 10^3 hours, failures per 10^6 hours, and failures per 10^9 hours. The last unit is called a *FIT*:

FIT

$$1 \text{ FIT} = 1 \text{ failure}/(10^9 \text{ hours})$$

In our earlier example, we would say that $\lambda_{\text{microprocessor}} = 800$ FITs.

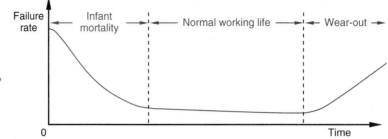

Figure 11-5
The "bathtub curve" for electronic-component failure rates.

The failure rate of a typical electronic component is a function of time. As shown in Figure 11-5, a typical component has a high failure rate during its early life, during which most manufacturing defects make themselves visible; failures during this period are called *infant mortality*. Infant mortality is why manufacturers of high-quality equipment perform *burn-in*—operating the equipment for 8 to 168 hours before shipping it to customers. With burn-in, most infant mortality occurs at the factory rather than at the customer's premises. Even without a thorough burn-in period, the seemingly stingy 90-day warranty offered by most electronic equipment manufacturers does in fact cover most of the failures that occur in the first few *years* of the equipment's operation (but if it fails on the 91st day, that's tough!). This is quite different from the situation with an automobile or other piece of mechanical equipment, where "wear and tear" increases the failure rate as a function of time.

infant mortality
burn-in

Once an electronic component has successfully passed the burn-in phase, its failure rate can be expected to be pretty much constant. Depending on the component, there may be *wear-out* mechanisms that occur late in the component's life, increasing its failure rate. In the old days, vacuum tubes often wore out after a few thousand hours because their filaments deteriorated from thermal stress. Nowadays, most electronic equipment reaches obsolescence before its solid-state components start to experience wear-out failures. For example, even though it's been over 25 years since the widespread use of EPROMs began, many of which were guaranteed to store data for only 10 years, we haven't seen a rash of equipment failures caused by their bits leaking away. (Do you know anyone with a 10-year-old PC or VCR?)

wear-out

Thus, in practice, the infant-mortality and wear-out phases of electronic-component lifetime are ignored, and reliability is calculated on the assumption that failure rate is constant during the normal working life of electronic equipment. This assumption, which says that a failure is equally likely at any time in a component's working life, allows us to use a simplified mathematical model to predict system reliability, as we'll show later in this section.

There *are* some other factors that can affect component failure rates, including temperature, humidity, shock, vibration, and power cycling. The most significant of these for ICs is temperature. Many IC failure mechanisms involve chemical reactions between the chip and some kind of contaminant, and these are accelerated by higher temperatures. Likewise, electrically overstressing a transistor, which heats it up too much and eventually destroys it, is worse if the device temperature is high to begin with. Both theoretical and empirical evidence support the following widely used rule of thumb:

- An IC's failure rate roughly doubles for every 10°C rise in temperature.

This rule is true to a greater or lesser degree for most other electronic parts.

Note that the temperature of interest in the foregoing rule is the *internal* temperature of the IC, not the ambient temperature of the surrounding air. A power-hogging component in a system without forced-air cooling may have an internal temperature as much as 40–50°C higher than ambient. A well-placed fan may reduce the temperature rise to 10–20°C, reducing the component's failure rate by perhaps a factor of 10.

11.3.2 Reliability and MTBF

For components with a constant failure rate λ, it can be shown that reliability is an exponential function of time:

$$R(t) = e^{-\lambda t}$$

The reliability curve in Figure 11-4 is such a function; it happens to use the value $\lambda = 1$ failure/ year.

Another measure of the reliability of a component or system is the *mean time between failures (MTBF)*, the average time that it takes for a component to fail. For components with a constant failure rate λ, it can be shown that MTBF is simply the reciprocal of λ:

$$\text{MTBF} = 1/\lambda$$

mean time between failures (MTBF)

11.3.3 System Reliability

Suppose that we build a system with m components, each with a different failure rate, $\lambda_1, \lambda_2, \ldots, \lambda_m$. Let us assume that for the system to operate properly, *all* of its components must operate properly. Basic probability theory says that the system reliability is then given by the formula

$$R_{\text{sys}}(t) = R_1(t) \cdot R_2(t) \cdot \cdots \cdot R_m(t)$$
$$= e^{-\lambda_1 t} \cdot e^{-\lambda_{2(1)} t} \cdot \cdots \cdot e^{-\lambda_m t}$$
$$= e^{-(\lambda_1 + \lambda_2 + \cdots + \lambda_m)t}$$
$$= e^{-\lambda_{\text{sys}} t}$$

where

$$\lambda_{\text{sys}} = \lambda_1 + \lambda_2 + \cdots + \lambda_m$$

Thus, system reliability is also an exponential function, using a composite failure rate λ_{sys} that is the sum of the individual component failure rates.

The constant-failure-rate assumption makes it very easy to determine the reliability of a system—simply add the failure rates of the individual components to get the system failure rate. Individual component failure rates can be obtained from manufacturers, reliability handbooks, or company standards.

For example, a portion of one company's standard is listed in Table 11-1. Since failure rates are just estimates, this company simplifies the designer's job by considering only broad categories of components, rather than giving "exact" failure rates for each component. Other companies use more detailed lists, and some CAD systems maintain a component failure-rate database, so that a circuit's composite failure rate can be calculated automatically from its parts list.

Suppose that we were to build a single-board system with a VLSI microprocessor, 16 memory and other LSI ICs, 2 SSI ICs, 4 MSI ICs, 10 resistors, 24 decoupling capacitors, and a 50-pin connector. Using the numbers in Table 11-1, we can calculate the composite failure rate,

$$\lambda_{sys} = 1000 + 16 \cdot 250 + 2 \cdot 90 + 4 \cdot 160 + 10 \cdot 10 + 24 \cdot 15 + 50 \cdot 10 + 500 \text{ FITs}$$
$$= 7280 \text{ failures} / 10^9 \text{ hours}$$

The MTBF of our single-board system is $1/\lambda$, or about 15.6 years. Yes, small circuits really can be that reliable. Of course, if we include a typical power supply, with a failure rate of over 10,000 FITs, the MTBF is more than halved.

Table 11-1 Typical component failure rates at 55°C.

Component	Failure Rate (FITs)
SSI IC	90
MSI IC	160
LSI IC	250
VLSI microprocessor	500
Resistor	10
Decoupling capacitor	15
Connector (per pin)	10
Printed-circuit board	1000

11.4 Transmission Lines, Reflections, and Termination

Nothing happens instantly, especially where digital circuits are concerned. In particular, consider the fact that the "speed-of-light" propagation delay of electrical signals in *wire* is on the order of 1.5–2 ns per foot (the exact delay depends on characteristics of the wire). When wire delays are similar to the transition times of the signals that they carry, we must treat wires not as zero-delay, perfect conductors, but as the "transmission lines" that they really are.

Transmission-line behavior includes signal changes that would not be predicted by a "DC" analysis of circuit operation. The most significant changes occur during an interval of approximately $2T$ after an output changes state, where T is the delay from the output to the far end of the wire that it drives.

Transmission-line behavior usually doesn't affect the logical operation of devices whose signal transition times and propagation delays are much longer than $2T$. Thus, transmission-line behavior typically is considered only for 74LS connections over a couple feet in length, for 74AS, 74AC and 74ACT connections over a foot, and for 74FCT and ECL connections over 3–6 inches. Naturally, you can sometimes get away with longer connections depending on the application, but there are also situations where transmission-line theory must be applied to even shorter connections. Even some VLSI chips achieve high enough speeds that transmission-line behavior must be considered in the internal chip design.

11.4.1 Basic Transmission-Line Theory

Two conductors in parallel constitute the simplest *transmission line*. Consider a pair of conductors with infinite length, as shown in Figure 11-6(a). If we instantaneously put a voltage source across the pair, a certain current flows to create a voltage wave that travels along the pair. The ratio of voltage to current, V_{out}/I_{out}, depends on the physical characteristics of the conductors and is called the *characteristic impedance Z_0* of the conductor pair.

transmission line

characteristic impedance Z_0

Figure 11-6 Transmission lines: (a) with infinite length; (b) with finite length, terminated with characteristic impedance.

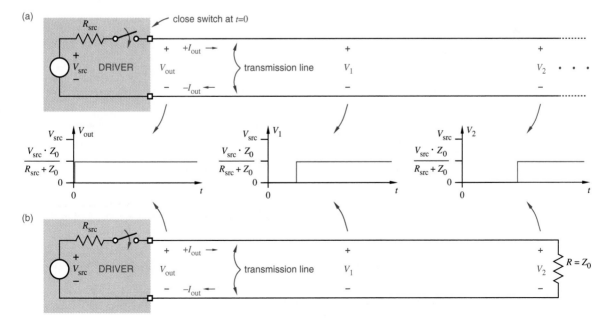

The magnitude of V_{out} is determined by viewing the series combination of R_{src} and Z_0 as a voltage divider, so that

$$V_{out} = V_{src} \cdot \frac{Z_0}{R_{src} + Z_0}$$

Of course, we don't have any infinitely long conductors. However, suppose that we have a 5-foot-long pair of conductors, and we place a resistance equal to Z_0 across the far end. If we instantaneously place a voltage source across the pair as shown in (b), the same current flows forever as in the infinite-length case. Thus, when a line is terminated by its characteristic impedance, we needn't consider transmission-line effects any further.

The situation is different for a transmission line of finite length that is not terminated in its characteristic impedance. An extreme case, in which the far end is short-circuited, is shown in Figure 11-7(a). For simplicity, we assume that $R_{src} = Z_0$ in this example. Initially, all the driver sees is the line's characteristic impedance, and a voltage wave with amplitude $V_{src}/2$ happily propagates down the line. However, when the wave hits the far end at time T, it sees the short-circuit. In order to satisfy Kirchhoff's laws, a voltage wave of the *opposite*

Figure 11-7 Transmission lines: (a) short-circuited; (b) open-circuited.

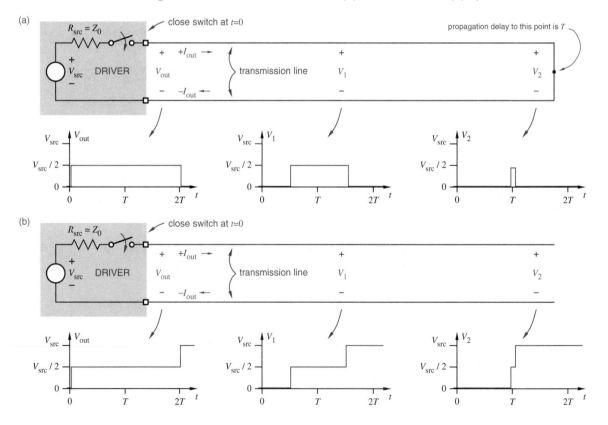

polarity propagates back down the line, canceling the original wave. The far end has *reflected* the original wave, and the driver sees the short-circuit at time $2T$.

Another extreme case, in which the far end is open-circuited, is shown in (b). Everything starts out as before. However, when the initial voltage wave hits the far end, the current has nowhere to go, and so a voltage wave of the *same* polarity propagates back up the line, adding to the original voltage. When the reflected wave reaches the driver at time $2T$, the voltage everywhere is V_{src}, and nothing more happens.

In the general case, the amplitude of the wave reflected at the end of a transmission line is determined by the *reflection coefficient, ρ* (rho). The value of ρ depends on Z_0 and Z_{term}, the *termination impedance* that appears at the end of the line:

$$\rho = \frac{Z_{term} - Z_0}{Z_{term} + Z_0}$$

When a voltage wave with amplitude V_{wave} hits the end of a transmission line, a wave with amplitude $\rho \cdot V_{wave}$ is reflected. Note that three simple cases match our previous discussion:

$Z_{term} = Z_0$ When a transmission line is terminated in its characteristic imped-
              ance, the reflection coefficient is 0.

$Z_{term} = 0$ The reflection coefficient of a short-circuited line is -1, producing a
            reflection of equal magnitude and opposite polarity.

$Z_{term} = \infty$ The reflection coefficient of an open-circuited line is $+1$, producing a
             reflection of equal magnitude and the same polarity.

If the source impedance (R_{src}) does not equal Z_0, then reflections occur at the near end of the line as well as at the far end. Each end of the line has its own value of ρ. The *principle of superposition* applies, so that voltage at any point on the line and instant in time is the sum of that point's initial condition and all waves that have passed it so far. The end of a transmission line is said to be *matched* if it is terminated with its characteristic impedance.

Figure 11-8 shows the behavior of a transmission line that is not matched at either end. (Note that the driver's ideal voltage source V_{src} is considered to have a resistance of 0 Ω for this analysis.) Reflections occur at both ends of the line, with smaller and smaller waves reflecting back and forth. The voltage everywhere on the line asymptotically approaches $0.9V_{src}$, the value that would be predicted by a "DC" analysis of the circuit using Ohm's law.

11.4.2 Logic-Signal Interconnections as Transmission Lines

So, how does transmission-line theory affect logic signals? Let's consider the case of a 74HC CMOS output driving a CMOS input at the end of a transmission line consisting of a signal line and ground, as shown in Figure 11-9. An "on"

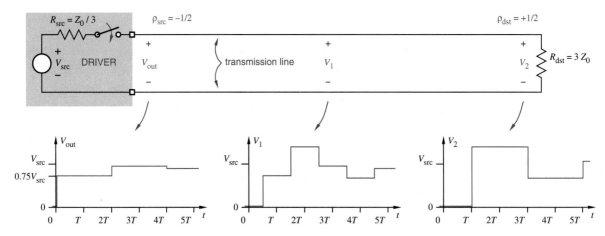

Figure 11-8 A transmission line that is not matched at either end.

p-channel or *n*-channel output transistor in the HCT family has resistance of roughly 100–200 Ω; for simplicity let's assume 150 Ω in either the HIGH or the LOW state. The characteristic impedance of a typical PCB trace with respect to ground is on the order of 100–150 Ω; let's assume 150 Ω for convenience, so the reflection coefficient on the driving end of the line is 0. A typical CMOS input has an impedance over 1 MΩ, so the reflection coefficient on the receiving end is effectively +1.

In the figure, we consider the case of a LOW-to-HIGH transition. When the CMOS driver switches from LOW to HIGH, the 5 V source in the driver sees the 150-Ω resistance of the driver in series with the 150-Ω Z_0 of the line, so a 2.5-V wave propagates down the line. After time T, this wave reaches the receiving gate U2 on the far end and is reflected. After time $2T$, the reflected wave reaches the sending end and is absorbed without a reflection, because $\rho = 0$ at that end.

Figure 11-9 Reflections on a logic signal line changing from LOW to HIGH.

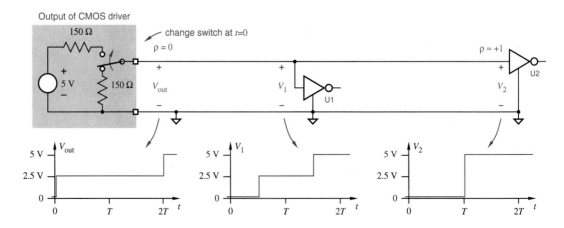

Everything works fine as far as receiving gate U2 is concerned—it sees an instantaneous transition from 0 V to 5 V at time T after the driver switches. However, consider the waveform seen by another receiving gate U1 positioned halfway between the driver and U2. As shown in the figure, U1 sees an input of only 2 V for an interval T. A receiving gate positioned closer to the driver would see this input voltage even longer. This is a problem, because 2.5 V is right at the switching threshold for 5-V CMOS inputs. If this input voltage were maintained for long, U2's output could oscillate or produce a nonlogic voltage.

Any logic-signal connection, no matter how long or how short, is a transmission line. However, the foregoing analysis is quite idealized. In practice, transmission-line behavior causes no problem if T is somewhat shorter than the transition times of logic signals or the propagation delays of logic gates. In such a case, the reflections typically settle out before the receivers on the line have a chance to notice them.

A somewhat nastier case occurs when a high-performance output switches from HIGH to LOW. For example, as shown in Figure 11-10, in the LOW state an FCT CMOS output looks roughly like a 10-Ω resistor to ground. Thus, the reflection coefficient on the sending end is about −0.88. When the output is first switched LOW, it sees a voltage divider consisting of the 10-Ω resistor in series with the line impedance of 150 Ω. Since the line was originally at 5 V, the new voltage at the output of the divider must be

$$V_{out} = \frac{10\Omega}{150\Omega + 10\Omega} \cdot 5\ V = 0.31\ V$$

Therefore, a voltage wave $(0.31 - 5.0) = -4.69$ V is propagated down the line. When this wave hits the receiving end at time T, it produces a reflection of equal sign and magnitude (since $\rho = 1$). Thus, the voltage at the receiving end is now $V_2 = 5.0 - 4.69 - 4.69 = -4.38$ V, a negative voltage! This is called *undershoot*.   *undershoot*

Figure 11-10 Reflections on a logic signal line changing from HIGH to LOW.

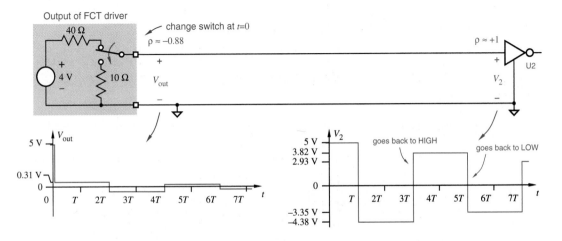

When the -4.69 V reflection gets back to the sending end at time $2T$, yet another reflection occurs. This time, since ρ is negative, the reflected wave has a positive amplitude, $-0.88 \cdot -4.69 = +4.10$ V. The output voltage is now modified by the sum of the incoming reflection and the new outgoing reflection: $V_{out} = 0.31 - 4.49 + 4.10 = -0.08$ V. This is no problem. However, when the $+4.10$ V reflection reaches the receiving end at time $3T$, the ensuing positive reflection yields $V_2 = -4.38 + 4.10 + 4.10 = 3.82$ V, back to positive again!

As suggested by the figure, the reflections continue, with the voltage at both the sending and receiving ends asymptotically approaching 0 V, the value one would predict with a "DC" analysis of the circuit. This oscillating pattern is

ringing

called *ringing*.

The large magnitude of the ringing on the receiving end may be a problem, since V_2 does not settle to a range that is less than the LOW-state input voltage (0.8 V) until time $27T$. Thus, the effective propagation delay of the circuit has been increased by many times the wire delay. Worse, if the signal is a clock, extra edges will be detected on the ringing transitions, falsely clocking the flip-flops at the receiving end.

Once again, the transmission line behavior that we've described causes no problems if T is much shorter than the transition times of logic signals or the propagation delays of logic gates. In addition, TTL and many CMOS input

clamping diodes

structures include *clamping diodes,* the normally reverse-biased diodes from each input to ground that we showed in Figure 3-75 on page 157. Since these diodes change the receiver's input impedance from very high to very low for negative voltages, they limit the negative excursion at time T to about 1 V. This reduces the reflection back to the sending end, which in turn reduces the excursion at time $3T$ to less than 1 V. The inputs of some devices also have diodes connected to V_{CC} to handle the overshoot that occurs when low-impedance drivers make a LOW-to-HIGH transition.

11.4.3 Logic-Signal Terminations

Reflections can be eliminated by terminating transmission lines in their characteristic impedance. Two methods are commonly used.

Thévenin termination

In a *Thévenin termination*, a pair of resistors is placed at the end of the line as shown in Figure 11-11(a). This termination has the Thévenin equivalent circuit shown in (b). Several factors are considered when choosing the resistor values:

1. The Thévenin equivalent resistance R_{Thev} should equal or be close to the line's characteristic impedance, Z_0. The Thévenin resistance is the parallel combination of $R1$ and $R2$, that is, $(R1 \cdot R2) / (R1 + R2)$.

2. The Thévenin equivalent voltage V_{Thev} may be chosen to optimize the current sinking and sourcing requirements of the driving gate. Here, V_{Thev} equals $V_{CC} \cdot R2 / (R1 + R2)$.

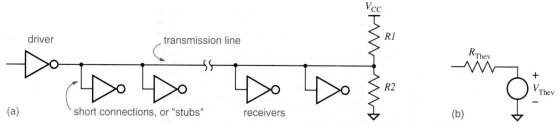

Figure 11-11 Thévenin termination: (a) circuit; (b) equivalent circuit for termination.

- For example, if the driving gate can source and sink equal amounts of current in the HIGH and LOW states (as in standard CMOS, not TTL compatible), then V_{Thev} can be halfway between V_{OL} and V_{OH}.
- On the other hand, if the driving gate can sink more current than it can source (as in TTL and TTL-compatible CMOS), then a higher value of V_{Thev} is chosen to lessen the driver's current sourcing requirements in the HIGH state, at the expense of increasing its current sinking requirements in the LOW state.

3. For three-state buses, the Thévenin equivalent voltage V_{Thev} may be chosen to yield a valid logic level when the bus is not being driven. It is especially important in this case to choose a voltage that is not close to the switching threshold of the receivers, lest they draw excessive current or oscillate.

4. The final resistor values are chosen to match standard component values (e.g., 150, 220, 270, 330, 390, 470 ohms).

For example, a "standard" termination in some TTL applications has $R1 = 220\,\Omega$ and $R2 = 330\,\Omega$, yielding $R_{Thev} = 132\,\Omega$ and $V_{Thev} = 3.0$ V. In the LOW state, a driver must sink $(3.0\text{ V})/(132\,\Omega) = 22.7$ mA, and in the HIGH state, no current is needed to maintain a valid logic level.

With a well-matched Thévenin termination at the end of a transmission line, little or no reflection will occur. The downside of this termination is that it consumes DC power at all times, and relatively high driver currents are required. A *source termination* (sometimes called a *series termination*) overcomes these problems.

source termination
series termination

As shown in Figure 11-12, a resistance equal to Z_0 minus R_d, the typical output impedance of the driver, is placed in series with the output, physically close to the output itself. Timing waveforms for this termination are illustrated in

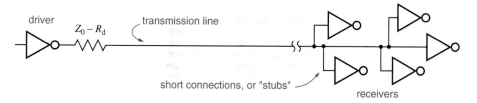

Figure 11-12
Source termination.

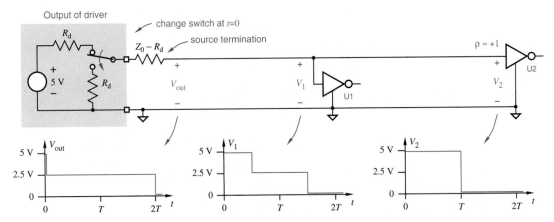

Figure 11-13 Reflections on a source-terminated line changing from LOW to HIGH.

Figure 11-13. On a transition from V_{CC} to 0 (or vice versa), the driver sees a series resistance of $2Z_0$—the series resistor plus the line itself. Half of the voltage change is developed across the series resistor and the other half across the line itself. Thus, after a one-way transmission delay T, the far-end receiver U2 sees a voltage change of $V_{CC}/2$. Assuming a reflection coefficient of 1, this full value is immediately reflected back toward the driver, bringing the voltage at U2 all the way up to V_{CC}. When this reflection reaches the driver, it is absorbed in the matched series resistance Z_0, and the line is stable.

Source termination works well for drivers whose output impedances in the LOW and HIGH states are close to equal for practical purposes (e.g., CMOS). Typical designs use series resistors of 15–40 Ω for 50–100-Ω transmission lines.

Source termination is undesirable if any receiving gates, such as U1 in the figure, are positioned only partway down the transmission line. On a transition, these "middle" gates see an initial transition of only $V_{CC}/2$; they see the full transition only after the reflection from the far end goes by. Since $V_{CC}/2$ is likely to be near the switching threshold of the receiving gates, this is not a good place to be for very long. Therefore, source termination is normally used only when a PCB is physically laid out in a way similar to Figure 11-12, with a relatively long line driving one load or a few that are clustered close together.

References

CAD tools are just tools, and you won't find any scholarly references on how to use them. However, *IEEE Design & Test of Computers* (*D&T*) is an excellent ongoing source of information about CAD tools. The best source of information for a particular CAD tool is, naturally, the supplier's documentation. When a particular tool is used *very* widely, an independent tutorial guide may appear; a good example is *SPICE: A Guide to Circuit Simulation and Analysis Using Pspice*® by Paul W. Tuinenga (Prentice Hall, 1988).

Register-transfer languages are discussed in most computer design books, including *Computer Architecture and Organization,* second edition, by John P. Hayes (McGraw-Hill, 1988). *IEEE Computer* magazine had a special issue on hardware description languages in February 1985, and *D&T* had a special issue on VHDL in April 1986.

The authoritative treatment of design for testability in the context of logic design is in Edward J. McCluskey's *Logic Design Principles* (Prentice Hall, 1986). A thorough treatment of the subject from the point of view of manufacturing and test engineers is *Digital Test Engineering* by J. Max Cortner (Wiley, 1987). Once again, *D&T* is an excellent ongoing information source. In the August 1989 special issue on built-in self-test methods, an article by Kenneth P. Parker, "The Impact of Boundary Scan on Board Test," gives a very practical perspective on the growing importance of scan methods in board-level design.

The required mathematical background and general methods for analyzing system reliability can be found in a handy little book by J. C. Cluley, *Electronic Equipment Reliability* (Halsted Press/Wiley, 1974). To see how ICs actually fail, and how their failure rates can be predicted in less than a lifetime (yours or an IC's), you should read *Failure Mechanisms in Semiconductor Devices* by E. A. Amerasekera and D. S. Campbell (Wiley, 1987).

An excellent introduction to transmission-line theory and practice from the point of view of a digital designer can be found in Thomas R. Blakeslee's *Digital Design with Standard MSI and LSI,* second edition (Wiley, 1979).

Highly recommended reading and an authoritative treatment of the analog aspects of digital design that I love to harp on is *High-Speed Digital Design* by Howard W. Johnson and Martin Graham (Prentice Hall, 1993). Although the book's subtitle is *A Handbook of Black Magic*, the authors go a long way toward demystifying the analog behavior that all high-speed digital system designers must cope with.

INDEX

Note: Page numbers for defining references are given in bold color.

XILINX STUDENT EDITION

Access the Xilinx Design Series Home Page for useful information and help

www.xilinx.com/programs/xds1.htm

This page will always contain the latest version of this information.

This CD-ROM set is designed for use with Prentice Hall's "Xilinx Design Series" educational products. This software is designed for first time installation of the student version of Xilinx Foundation Series Express F1.5 software, as well as to upgrade Xilinx Foundation Series F1.3 and F1.4 software to version F1.5 and to provide other electronic materials that may be useful. To properly execute the VHDL and Verilog synthesis software module-Synopsys FPGA Express ™ module of the Xilinx Foundation Series-software a "license.dat" file is required. If you are upgrading from 1.3 or 1.4 please access: http://www.xilinx.com/programs/xsefaq1.htm for information on making a simple modification to your existing license file. If this is a new installation then using the serial number from the coupon which is included in your Xilinx Design Series package from Prentice Hall (ISBN 0130205869), you can obtain a "license.dat" file from the Xilinx Design Series Home Page. Purchase of the Prentice Hall package is required to obtain a license file.

Basic installation instructions are printed here. Complete instructions are on the Design Environment CD-ROM disk 1 of 2 at D:\readme.wri. These instructions are complete at press time. Updated versions of these instructions may be found at

www.xilinx.com/programs/xds1.htm

Section I—Xilinx Foundation Series Express F1.5

Section II—Other tools
- a. (Optional) Xilinx CORE Generator 1.5
- b. (Optional) XESS Tools for XS40 or XS95 boards
- c. (Optional) Aldec's Active-VHDL 3.1 simulator (Student Version)
- d. (Optional) SimuCad's SILOSIII Verilog simulator (Demo Version)

Section III – License file setup

Section IV - (Optional) Adobe Acrobat and datasheets

To upgrade from F1.3 or F1.4 to 1.5, start at Section I A.

To install on a new computer, start at Section I B.

Section I—Xilinx Foundation Series Express F1.5

Section IA—Uninstall F1.3 or F1.4

1. Uninstall the Foundation Project Manager (Design Environment): Click on Start, Settings, Control Panel, Add/Remove Programs

2. Uninstall the Foundation Design Implementation Tools.

3. If uninstall does not work, then delete everything including and below C:\Xilinx directory (assuming that was the directory under which you installed your F1.3 or F1.4 Design Implementation Tools).

Section I.B—Install F1.5

1. Insert and install Disk 1 of 2, Design Environment.

   Begin installation: Click on Start, Run and select D:\ setup.exe When "Xilinx Foundation F1.5 Setup" window appears, click "Next" and then follow the prompts.

   Your CD KEY is: $\boxed{\text{FXAA2743390}}$ Do not lose this number!

2. If you are upgrading go to http://www.xilinx.com/programs/xds1.htm for full instructions.

   Note: You may wish to deselect XC4000E and SpartanXL devices as this will save you over 50MB of disk space. You can run setup.exe again to add these devices later.

3. Insert and install Disk 2 of 2, Documentation.

   Begin installation: Click on Start, Run and select D:\ setup.exe

   When "Xilinx Foundation F1.5 Documentation Setup" window appears, click "Next" and then follow the prompts. The CD KEY is not required for this step.

4. To save disk space, you can deselect the "Multimedia Foundation Demo" in the Select Products to Install dialog box.

5. Under "Registry Settings Options" do not change any options.

6. When finished, reboot your computer to allow Registry settings to take place.

Section II—Other tools files for F1.5

There are other tools available on disk 2.

In general, the setup is similar to what you have just done with the foundation tools. Click on the Start button, then RUN and move to the directory for the tool that you wish to install. Double click on the "setup.exe" file and follow the prompts.

The labs in the Practical Xilinx Designer lab book may be found on disk 2 in D:\XESS\ XLabs\… There is a subdirectory for each lab.

Information on system requirements, are on the Design Environment Disk CD-ROM disk 1 of 2 at D:\readme.wri as well as on the software packaging

The complete installation guide can be found on disk 2 at D:\Finstall\fndinst.pdf. Acrobat reader must be installed to read this file. Acrobat reader can be installed from the ACROREAD directory on disk 2.

Section III—License file for F1.5

For version F1.5, the "license.dat" file is required to run the Synopsys FPGA Express VHDL and Verilog synthesis tools.

If you are upgrading an existing Xilinx Foundation Student Series F1.3 or F1.4 installation up to F1.5, you do not need to make any changes to your system. Your existing "license.dat" is sufficient. If using the Express tool see the next paragraph.

If you are installing the Xilinx Foundation Series Express F1.5 software on a new computer, or you just did not obtain a "license.dat" file previously, then go to the web to obtain your license file and read the instructions at: http://www.xilinx.com/programs/xds1.htm

Section IV—(Optional) Adobe Acrobat and datasheets

There are several documents included in Adobe Acrobat format on this CD-ROM set for your reference. For more comprehensive documentation on the Xilinx Foundation Series Express F1.5 software and installation, the commercial F1.5 Quick Start Guide is available at D:\installdoc\fndinst.pdf.

These are the key documents that are included on "Disk 2 of 2, Foundation Documentation":

D:\Applinx\Databook.pdf   Xilinx 1998 Databook
D:\Applinx\coregen.pdf   Xilinx Core Solutions Databook

Other key documents which can be found using the Dynatext Document viewer in the Xilinx Book directory:
Foundations Series Quickstart Guide
Foundation Series User Guide
Verilog Reference Guide
VHDL Reference Guide

If you need to install the Acrobat reader, run D:\ACRORED\WIN32\Ar32e30.exe and follow the instructions to install the Acrobat Reader software.

In addition, there are several lab projects available at the Xilinx University Program home page under "Presentation Materials and Lab Files".

See **www.xilinx.com/programs/univ.htm**

Minimum PC Requirements:

Pentium® Processor recommended   Windows '95 or NT 4.0

48MB RAM recommended 350MB (plus 100MB swap space) free hard drive space before installation.

Depending on options you select, the F1.5 Foundation Series Express "Base" installation will consume 250 to 500 Mbytes of hard drive space. Does not include Active-HDL or Silos III.

Please note that all support for the Xilinx Student Edition software is web based. Access: **http://www.xilinx.com/programs/univ.htm** for troubleshooting and other information.

XILINIX END USER LICENSE

Attention: By opening this case, you are consenting to be bound by and are becoming a party to this agreement. If you do not agree to all the terms of this agreement, return the unopened case, along with the entire package withwithj proof of payment to authorized dealer where you took delivery within ten days; and get a full refund.

1. License. XILINX, Inc. ("XILINX") hereby grants you a nonexclusive license to use the application, demonstration, and system software included on this disk, diskette, tape or CD ROM, and related documentation (the "Software") solely by the number of simultaneous users for which you have paid XILINX a license fee, and solely for your use in developing designs for XILINX Programmable Logic devices or internal business purposes. You own the media on which the Software is recorded, but XILINX and its licensors retain title to the Software and to any patents, copyrights, trade secrets and other intellectual property rights therein. This License allows you to use the Software on a single computer. In addition, you may make up to the number of the copies of the Software, if used on separate computers, or permit up to the number of simultaneous users to use the Software, if used in a network environment, as permitted in a separate written agreement between you and XILINX, and make one copy of the Software in machine-readable form for backup purposes only. You must reproduce on each copy of the Software the copyright and any other proprietary legends that were on the original copy of the Software. You may also transfer the Software, including any backup copy of the Software you may have made, the related documentation, and a copy of this License to another party provided the other party reads and agrees to accept the terms and conditions of this License prior to your transfer of the Software to the other party, and provided that you retain no copies of the Software yourself.

2. Restrictions. The Software contains copyrighted material, trade secrets, and other proprietary information. In order to protect them you may not decompile, reverse engineer, disassemble, or otherwise reduce the Software to a human-perceivable form. You may not modify or prepare derivative works of the Software in whole or in part. You may not publish any data or information that compares the performance of the Software with software created or distributed by others.

3. Termination. This License is effective until terminated. You may terminate this License at any time by destroying the Software and all copies thereof. This License will terminate immediately without notice from XILINX if you fail to comply with any provision of this License. Upon termination you must destroy the Software and all copies thereof.

4. Governmental Use. The Software is commercial computer software developed exclusively at Xilinx's expense. Accordingly, pursuant to the Federal Acquisition Regulations (FAR) Section 12.212 and Defense FAR Supplement Section 227.2702, use, duplication and disclosure of the Software by or for the Government is subject to the restrictions set forth in this License Agreement. Manufacturer is XILINX, INC., 2100 Logic Drive, San Jose, California 95124.

5. Limited Warranty and Disclaimer. XILINX warrants that, for a period of ninety (90) days from the date of delivery to you of the Software as evidenced by a copy of your receipt, the media on which the Software is furnished will, under normal use, be free from defects in material and workmanship. XILINX's and its Licensors' and Distributor's entire liability to you and your exclusive remedy under this warranty will be for XILINX, at its option, after return of the defective Software media, to either replace such media or to refund the purchase price paid therefor and terminate this Agreement. EXCEPT FOR THE ABOVE EXPRESS LIMITED WARRANTY, THE SOFTWARE IS PROVIDED TO YOU "AS IS". XILINX AND ITS LICENSORS AND DISTRIBUTORS MAKE AND YOU RECEIVE NO OTHER WARRANTIES OR CONDITIONS, EXPRESS, IMPLIED, STATUTORY OR OTHERWISE, AND XILINX SPECIFICALLY DISCLAIMS ANY IMPLIED WARRANTIES OF MERCHANTABILITY, NONINFRINGEMENT, OR FITNESS FOR A PARTICULAR PURPOSE. XILINX does not warrant that the functions contained in the Software will meet your requirements, or that the operation of the Software will be uninterrupted or error free, or that defects in the Software will be corrected. Furthermore, XILINX does not warrant or make any representations regarding use or the results of the use of the Software in terms of correctness, accuracy, reliability or otherwise.

6. Limitation of Liability. IN NO EVENT WILL XILINX OR ITS LICENSORS OR DISTRIBUTORS BE LIABLE FOR ANY LOSS OF DATA, LOST PROFITS, COST OF PROCUREMENT OF SUBSTITUTE GOODS OR SERVICES, OR FOR ANY SPECIAL, INCIDENTAL, CONSEQUENTIAL OR INDIRECT DAMAGES ARISING FROM THE USE OR OPERATION OF THE SOFTWARE OR ACCOMPANYING DOCUMENTATION, HOWEVER CAUSED AND ON ANY THEORY OF LIABILITY. THIS LIMITATION WILL APPLY EVEN IF XILINX HAS BEEN ADVISED OF THE POSSIBILITY OF SUCH DAMAGE. THIS LIMITATION SHALL APPLY NOTWITHSTANDING THE FAILURE OF THE ESSENTIAL PURPOSE OF ANY LIMITED REMEDIES HEREIN.

7. Export Restriction. You agree that you will not export or reexport the Software, reference images or accompanying documentation in any form without the appropriate United States and foreign government licenses. Your failure to comply with this provision is a material breach of this Agreement.

8. Third Party Beneficiary. You understand that portions of the Software and related documentation have been licensed to XILINX from third parties and that such third parties are intended third party beneficiaries of the provisions of this license Agreement.

9. General. This License shall be governed by the laws of the State of California, and without reference to conflict of laws principles. If for any reason a court of competent jurisdiction finds any provision of this License, or portion thereof, to be unenforceable, that provision of the License shall be enforced to the maximum extent permissible so as to effect the intent of the parties, and the remainder of this License shall continue in full force and effect. This License constitutes the entire agreement between the parties with respect to the use of this Software and related documentation, and supersedes all prior or contemporaneous understandings or agreements, written or oral, regarding such subject matter.

INTRODUCTION TO

HUMAN DISEASE

THIRD EDITION

Thomas H. Kent, MD
Professor of Pathology
University of Iowa
Iowa City, Iowa

Michael Noel Hart, MD
Professor of Pathology
University of Iowa
Iowa City, Iowa

APPLETON & LANGE
Norwalk, Connecticut

0-8385-4347-2

93 94 95 96 97 / 10 9 8 7 6 5 4 3 2 1

Prentice Hall International (UK) Limited, *London*
Prentice Hall of Australia Pty. Limited, *Sydney*
Prentice Hall Canada, Inc., *Toronto*
Prentice Hall Hispanoamericana, S.A., *Mexico*
Prentice Hall of India Private Limited, *New Delhi*
Prentice Hall of Japan, Inc., *Tokyo*
Simon & Schuster Asia Pte. Ltd., *Singapore*
Editora Prentice Hall do Brasil Ltda., *Rio de Janeiro*
Prentice Hall, *Englewood Cliffs, New Jersey*

Library of Congress Cataloging-in-Publication Data

Kent, Thomas H. (Thomas Hugh), 1934–
    Introduction to human disease / Thomas H. Kent, Michael Noel Hart.
  —3rd ed.
    p. cm.
    Includes index.
    ISBN 0-8385-4347-2
    1. Pathology.   2. Allied health personnel. I. Hart, Michael
Noel, 1938– . II. Title.
    [DNLM: 1. Pathology. QZ 4 K37i]
  RB111.K45 1992
  616.07—dc20
  DNLM/DLC
  for Library of Congress
                                          92-11057
                                          CIP

Acquisitions Editor: Cheryl Mehalik
Production Editor: Sheilah Holmes
Designer: Janice Barsevich

PRINTED IN THE UNITED STATES OF AMERICA

Contents

Preface

Introduction to Human Disease is a textbook of pathology for health science students other than medical students. We have noticed, however, that medical students use it to obtain basic concepts before reading the more detailed and encyclopedic texts designed for them. We have also noticed that our family members use it to get types of information desired by lay people. This broad usage pleases us because it was our intent to provide a text that covers all aspects of human disease with minimal requirements for prerequisite knowledge.

We are gratified by the continued use of the first and second editions by instructors who teach pathology courses to a wide variety of health science students, including dental, pharmacy, physician assistant, pathology assistant, dental assistant, medical technology, radiology technology, nursing, and graduate students. We feel that all health science workers have a need for a common vocabulary and broad understanding of human disease. For this reason, we define terms as clearly and specifically as possible and describe all of the common and important diseases of humans, including mental illnesses.

The third edition contains considerable new material, with emphasis on some important viral infections (hepatitis, papilloma virus infections, human immunodeficiency virus infection), more detailed classification of lung cancer, aspects of cancer diagnosis including fine needle biopsy, several lesions of the gastrointestinal tract commonly found at endoscopy, and dietary excesses. Some of the other topics that were rewritten and expanded include aging, amyloidosis, several types of cancer, diseases with known gene defects, diagnosis and treatment of eye diseases, causes and frequency of venereal disease, and atopic allergies.

This text is divided into four sections. Section I provides the most fundamental vocabulary and concepts, a broad analysis of the most common and significant diseases, and a discussion of the tools and process of diagnosis.

Section II provides a framework for the basic types of human disease: reactions to injury, neoplasia, genetically determined disease, and intrauterine injury.

Section III approaches disease from the viewpoint that we most commonly look at it, by organ system. Within each chapter, we review the most important anatomy and physiology, provide an overview of the most frequent and important diseases encountered, discuss diagnostic techniques (symptoms, signs, laboratory tests, and radiological and clinical procedures), profile the diseases, and discuss the consequences of failure of the organ to function.

Section IV presents diseases that tend to affect multiple organs and within each group share causative mechanisms. Topics included are infections, immune reactions, external injury by physical and chemical agents encompassing environmental diseases, and disorders caused by nutritional deprivations and excesses. We have found that these chapters are easier to learn after the diseases of the organs have been studied. They can, however, be inserted earlier in a course without any prerequisites other than Sections I and II.

Each chapter contains review questions. These address the most important aspects of the chapter. They can be considered the learning objectives. We have not included any references. We recommend any of the standard textbooks of pathology or internal medicine for additional reading. These texts contain an abundance of citations for the student who wishes to pursue any topic in greater depth.

ACKNOWLEDGMENTS

We would like to thank the following individuals who reviewed chapters for the third edition: Robert S. Bar, Jo Benda, Kent Bottles, Thomas A. Farrell, Michael Finkelstein, John D. Kemp, Charles F. Lynch, Frank A. Mitros, Steven A. Moore, Jeffrey C. Murray, John D. Olson, Hal B. Richerson, Robert L. Schelper, Thomas J. Waldschmidt, and George Winokur.

William E. Erkonen reviewed the radiologic illustrations and accompanying text. He provided several new radiographic images to add to those provided by Kenneth D. Dolan for the second edition.

Joel Carl made several new charts. Most of the line drawings made by Jo Ann Reynolds for the first edition were retained.

Section

I

An Overview

The purpose of this section is to give you (1) the general vocabulary used to discuss and classify disease, (2) a feeling for the general frequency and significance of disease so that you can separate the forest from the trees, and (3) a perspective of the resources commonly used in diagnosis so that you can bridge the gap between their scientific basis as learned in the basic sciences and their practical application in the care of patients.

Introduction to Pathology

PATHOLOGY

The term *pathology* has several meanings. In the broadest sense, pathology is the study of disease. All health scientists are lifelong students of pathology because, in one way or another, all are interested in altering the course of disease through better understanding of its nature.

A course in pathology, such as the one you are undertaking, provides a concentrated study of the nature of disease and lays the foundation for further study of disease within specific disciplines.

Pathology is also a name applied to one of the disciplines of medicine, one that deals with analysis of body fluids and tissues for diagnostic purposes and with teaching and research relating to

TABLE 1–1. ROLES OF A PATHOLOGIST

Role	Subject
Experimental pathology	Research
Academic pathology	Teaching, research, anatomic and/or clinical pathology
Anatomic pathology	Morphologic examinations
Autopsy pathology	Postmortem study of the body
Surgical pathology	Biopsies and resected tissues
Cytopathology	Individual cells removed by scraping or washing
Clinical pathology	Laboratory tests
Chemistry	Chemical analysis
Microbiology	Microorganisms
Hematology	Blood and bone marrow, blood clotting
Blood banking	Blood transfusion services
Immunopathology	Antigen and antibody detection

fundamental aspects of disease. The roles of pathologists are outlined in Table 1–1.

Pathologists usually practice laboratory medicine or study basic aspects of disease within a department of pathology associated with a hospital and/or medical school. Sometimes pathologists subdivide themselves on the basis of special interests into experimental pathologists, anatomic pathologists, or clinical pathologists. Experimental pathologists are basic scientists who spend the majority of their time investigating the causes and mechanisms of disease. Anatomic pathologists perform autopsies, examine all tissues removed from live patients (surgical pathology), and examine cell preparations used to screen for cancer cells (cytology or cytopathology). Clinical pathologists analyze various specimens removed from patients, such as blood, urine, feces, spinal fluid, or sputum, for chemical substances, microorganisms, antigens and antibodies, blood cells, and coagulation factors. Most pathologists in community medical centers practice both anatomic and clinical pathology. In health science teaching centers, pathologists are likely to combine teaching and experimental pathology with an interest in one or more of the special areas of anatomic or clinical pathology and are often referred to as *academic pathologists.*

DISEASE

Disease is a structural or functional change within the body judged to be abnormal. Minor changes that are of no importance may be judged to be variations of the normal state rather than disease. Pathology not only includes the study of basic structural and functional changes associated with a disease, but also includes the study of causes that lead to the structural and functional changes and the manifestations that result from them. Furthermore, pathology is concerned with the sequence of events that leads from cause to structural and functional abnormalities and finally to manifestations. This sequence is referred to as the *pathogenesis* of disease. The term *etiology* means the study of causes. It is commonly misused as a synonym for cause.

Diagnosis is the process of assigning a name to a patient's condition. The name applied (the noun) is also called the *diagnosis.* If possible, a diagnosis should be the name of the disease that the patient has, e.g., multiple sclerosis. Assigning this name implies that the illness will follow a course similar to that of other patients with the same disease. A diagnosis is a generalization (oversimplification) used for convenience of communication and thinking. Sometimes the findings cannot be expressed in terms of a disease, e.g., paralysis of unknown cause. In such cases, the clinical problem is used as the diagnosis until the patient's disease becomes evident. Some clusters of findings commonly encountered with more than one disease have been named (assigned a diagnostic term). These clusters of findings are called *syndromes.* For example, leakage of protein into the urine, low serum protein, and edema are a common set of findings in long-standing diseases of the renal glomerulus. This constellation is called the nephrotic syndrome because it is not one disease, but a set of findings common to several diseases. Therefore, the patient may be diagnosed as having the nephrotic syndrome until the specific disease is known.

CHART 1–1. PATHOLOGY: THE STUDY OF DISEASE

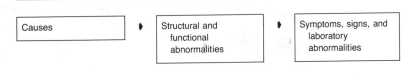

MANIFESTATIONS OF DISEASE

We will use the term *manifestations* to refer to the data that can be gathered about an individual patient, namely, symptoms, signs, and laboratory abnormalities (Table 1–2). *Symptoms* are evidence of disease perceived by the patient, such as pain, a lump, or diarrhea. The written description of symptoms in the patient's record is re-

TABLE 1–2. MANIFESTATIONS OF DISEASE

Type of Manifestation	Nature of Data	Name for Collection of Results
Symptoms	Patient's perceptions	History
Signs	Examiner's observations	Physical examination
Laboratory abnormalities	Results of tests and special procedures	Laboratory findings

ferred to as the *history*. *Signs* are physical observations made by the person who examines the patient, such as tenderness, a mass, or abnormal heart sounds. Signs are recorded under *physical examination*. *Laboratory abnormalities* in the broad sense refers to the observations made by tests or special procedures, such as x-rays, blood counts, or biopsies.

STRUCTURAL DISEASES

Structural diseases are those diseases characterized by structural changes within the body as the most basic abnormality. They are also referred to as *organic diseases*. Structural changes are called *lesions* and may be biochemical or morphologic (visible). The term *lesion* is most often used in reference to morphologic change whether it be at the gross (naked eye), microscopic, or electron microscopic level. Three broad categories suffice to classify most structural diseases (Table 1–3), although some diseases fall into more than one category and some are difficult to classify.

Genetic and developmental diseases are defined for the purposes of this book as diseases that are caused by abnormalities in the genetic makeup of the individual (genes or chromosomes) or abnormalities due to changes in utero (during embryonic and fetal development). The range of abnormalities in this category is very broad, extending from deformities present at birth (congenital anomalies) to biochemical changes caused by genes but influenced by environment so they appear later in life, e.g., diabetes mellitus. Due to overlap in classification schemes, many genetic and developmental diseases can also be classified as injuries, inflammations, proliferations, or even neoplasms.

Acquired injuries and inflammatory diseases are diseases due to internal or external forces or agents that destroy cells or intercellular substances, deposit abnormal substances (foreign bodies or ma-

TABLE 1–3. MAJOR DISEASE CATEGORIES

Genetic and developmental diseases
Acquired injuries and inflammatory diseases
Hyperplasias and neoplasms

terials produced by the body), or cause the body to injure itself by means of the inflammatory process. External agents of injury include physical and chemical substances and microbes. The major internal mechanisms of injury are vascular insufficiency, immunologic reactions, and metabolic disturbances. The direct effects of injury are referred to as *necrosis* if cells are killed in the injured area and *sublethal cell injury* if the injured cells are capable of recovery. Sublethal cell injury may also be called *degeneration,* but this term has many other vague connotations. The few types of necrosis have myriads of causes. Acute sublethal cell injury tends to appear similar regardless of cause or cells involved, but chronic sublethal cell injury has many variations. There are two general reactions to injury: inflammation and repair. *Inflammation* is a vascular and cellular reaction that attempts to localize the injury, destroy the offending agent, and remove damaged cells and other materials. *Repair* is the replacement of damaged tissue by new tissue of the same type and/or by fibrous connective tissue. Inflammation is a stereotyped response with several important variations. Unlike necrosis and inflammation, repair is greatly influenced by the type of tissue or organ that has been injured.

Hyperplasias and neoplasms is a category used to describe diseases characterized by increases in cell populations. Repair also may involve increases in cell populations, but the purpose is obviously to replace that which has been lost. In hyperplasia and neoplasia, the cell increase is beyond normal. *Hyperplasia* is a proliferative reaction to a prolonged external stimulus and will usually regress when the stimulus is removed. *Neoplasia* is presumed to result from a genetic change producing a single population of new (neoplastic) cells, which can proliferate beyond the degree allowed by the mechanisms that normally govern cell proliferation. Neoplasms are divided into two groups, benign and malignant, based on whether they will remain localized (benign) or will continue to grow and spread (malignant). *Cancer* is synonymous with malignant neoplasm. The situation is made more complicated by certain types of hyperplasias that slowly evolve, presumably through a series of genetic changes induced by external agents, into malignant neoplasms.

TABLE 1–4. EXAMPLES OF VARYING EFFECTS OF STRUCTURAL AND FUNCTIONAL DISEASES

Disease	Type of Disease	Nature of Manifestations
Tension headaches	Functional (muscle spasm)	Functional (pain)
Benign tumor of the breast that produces a mass	Structural (tumor)	Structural (mass)
Exogenous obesity caused by craving for food	Functional (hunger)	Structural (obesity)
Cancer of the esophagus that prevents eating	Structural (cancer)	Functional (inability to eat)

FUNCTIONAL DISEASES

Functional diseases are those diseases in which the onset begins without the presence of any lesions (biochemical or morphologic). The basic change is a physiologic or functional change and is referred to as a *pathophysiologic change*. Some long-standing functional diseases may, however, lead to secondary structural changes. The examples in Table 1–4 illustrate how either organic or functional diseases can have manifestations that are either structural or functional in nature.

Many mental illnesses are considered functional disorders, although some may have a genetic or other organic basis. The more mental illnesses are investigated, the more it is appreciated that there is likely to be an organic basis (on a biochemical level) to many of them. The most common functional disorders are tension headache and functional bowel syndrome, disorders that probably are due to unconscious stimulation of the autonomic nervous system.

CAUSES OF DISEASE

Diseases are initiated by injury, which may be either external or internal in origin. Agents acting from without are termed *exogenous*, those acting from within are referred to as *endogenous*.

External causes of disease are divided into physical, chemical, and microbiologic (Table 1–5). Direct physical injury by an object is called *trauma*. Other physical agents causing disease include heat and cold, electricity, atmospheric pressure changes, and radiation (electromagnetic and particulate). Chemical injuries are generally subdivided by the manner of injury into poisoning (accidental, homicidal, or suicidal) and drug reactions (toxic effects of prescription or proprietary drugs taken to treat disease). Microbiologic inju-

TABLE 1-5. EXTERNAL CAUSES OF DISEASE

Physical injury	Microbiologic injury
Trauma	Bacteria
Heat–cold	Fungi
Electricity	Rickettsia
Pressure	Viruses
Ionizing radiations	Protozoa
Chemical injury	Helminths
Poisoning	
Drug reactions	

ries are usually classified by the type of offending organism (bacteria, fungi, rickettsia, viruses, protozoa, and helminths) and are called *infections*.

Internal causes of disease fall into three large categories (Table 1–6). Vascular diseases may involve obstruction of blood supply to an organ or tissue, bleeding (hemorrhage), or altered blood flow such as occurs with heart failure. Immunologic diseases are those caused by aberrations of the immune system. Failure of the immune system to work when it is needed results in immunodeficiency disease. Overreaction or unwanted reactions of the immune system cause allergic (hypersensitivity) diseases. Metabolic diseases encompass a wide variety of biochemical disorders that may be primarily genetically determined or secondary effects of acquired disease. Metabolic diseases are most commonly categorized as abnormalities primarily involving lipids, carbohydrates, proteins, minerals, vitamins, and fluid. Some large categories of disease cannot be classified according to internal or external causes because the cause is not known, e.g., most neoplasms. Diseases of unknown cause are termed *idiopathic*. Adverse reactions resulting from treatment by a health specialist produce *iatrogenic* disease. *Nosocomial* diseases are those acquired from a hospital environment.

TABLE 1-6. INTERNAL CAUSES OF DISEASE

Vascular	Metabolic
Obstruction	Abnormal metabolism or deficiency of:
Bleeding	Lipid
Deranged flow	Carbohydrate
Immunologic	Protein
Immune deficiency	Mineral
Allergy	Vitamins
	Fluids

THE CARE OF PATIENTS

The "workup" of a patient encompasses three major steps: (1) History taking involves talking to the patient (or relatives) to ascertain the patient's symptoms and reviewing any other past or present medical problems that might relate to them; (2) physical examination involves systematic looking, feeling, listening, and sometimes even smelling the accessible parts of the body for signs of illness; and (3) when needed, laboratory tests and radiologic and clinical procedures are ordered to detect chemical and physiologic abnormalities which aid in establishing a diagnosis.

When the workup is completed, a diagnosis is made. As we have seen, this may be a disease, a prominent manifestation, or a syndrome. The diagnosis applied to a patient's illness plays the major role in selection of treatment. It also serves to indicate the likelihood of a favorable or unfavorable outcome. The expected outcome of a disease is called its *prognosis*.

The application of treatment to a patient is an attempt to alter the natural course of the patient's disease whether it be prevention of death, lessening the destructive effects, shortening the illness, or reducing symptoms. Follow-up of patients is important to determine whether the desired outcome has occurred or whether treatment has failed to control the disease or whether complications have developed. *Complications* are secondary problems known to occur with a disease in some but not all instances.

Diagnosis of specific diseases is not only useful to the patient for determining treatment and prognosis, but is also useful to future patients. It is through collection of data by disease category that our knowledge of prognosis, effectiveness of treatment, and frequency of complications is derived. Sometimes these data also further our understanding of the cause of a disease. This is particularly true when the distribution of the disease gives clues to possible causative factors. Major breakthroughs in our understanding of the causes of disease, however, tend to come from experimentation and/or technical advances that allowed finer definition of the body's structure and function.

The application of the process of patient care described above is limited by the availability of resources, by the nature of disease, and our ability to understand disease. The greatest effects on health care have come from knowledge of cause leading to preventative measures, particularly through improved nutrition, control of infectious diseases, and avoidance of toxic substances. New knowledge of this type is difficult and expensive to obtain with only occasional major breakthroughs. Application lags behind current knowledge and ability; for example, underdeveloped countries have, for educational and economic reasons, poor prevention of infectious disease and developed countries fail, for social reasons, to prevent the effects of tobacco and alcohol. In the United States, a great deal of effort is being directed toward the effects and complications of disease. For example, cancer is attacked with complex surgical, radiologic, and chemical techniques; surgical techniques are used to circumvent arteries obstructed by atherosclerosis; and organs that have failed are replaced by transplants from other individuals.

Economic considerations play an important role in the care of patients. The cost of diagnosis and treatment must be weighed against possible benefits. The concentration of serious disease in the elderly complicates cost/benefit decisions because of their already limited life expectancy. The cost of diagnosis may be small or large. If the cost is large, the benefits of each of the possible diagnoses must be weighed against the cost. Only through knowledge of disease can cost/benefit decisions be made about individual patients. In many cases the knowledge needed is highly technical and specific to the diseases under consideration. However, it is important to have a broad foundation of knowledge about the more common diseases, such as is presented in this book, before becoming enmeshed in the details of one's own area of specialization.

REVIEW QUESTIONS

1. What is pathology?
2. What do pathologists do?
3. What categories of information are useful to describe a disease?
4. What is the major difference between a functional and an organic (structural) disease?
5. What are the fundamental characteristics of the three major forms of organic disease?
6. What are the six major types of causes of disease? Which causes are external and which are internal in origin?
7. What do the following terms mean?
    Pathogenesis
    Etiology
    Diagnosis
    Syndrome
    History
    Symptom
    Sign
    Lesion
    Congenital anomaly

Necrosis
Inflammation
Repair
Hyperplasia
Neoplasia
Benign
Malignant
Cancer
Pathophysiology
Trauma
Infection
Idiopathic
Iatrogenic
Nosocomial
Prognosis
Complications

Most Frequent and Significant Diseases

CAUSES OF DEATH (MORTALITY)

CAUSES OF DISABILITY (MORBIDITY)

MEASURES OF MORTALITY AND MORBIDITY

FREQUENCY OF ACUTE DISEASES

FREQUENCY OF CHRONIC DISEASES

DISEASE FREQUENCY RELATED TO VISITS TO FAMILY PHYSICIANS

AGING

REVIEW QUESTIONS

CAUSES OF DEATH (MORTALITY)

Diseases causing death are described in terms of mortality rate (number per 100,000 population dying per year). These statistics are compiled by government agencies from death certificates, which must be filled out by a physician at the time of death. Approximately 74 percent of deaths in the United States are accounted for by the five most frequent causes (Fig. 2–1).

Heart disease accounts for nearly four out of every ten deaths in the United States. The vast majority of these deaths are caused by atherosclerosis—a degenerative disease of arteries that over a course of many years obstructs the coronary arteries by the development of lipid-rich thickenings in the arterial lining. These areas

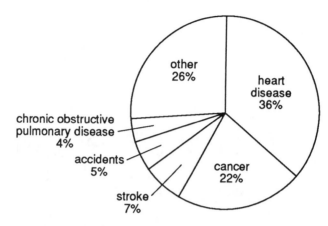

Figure 2–1. Major causes of death. *(Data from Vital Statistics of the United States, 1987.)*

of thickening are called *atherosclerotic plaques*. At some point, the blood supply suddenly becomes inadequate, and as a result a portion of the heart muscle is killed. The area of dead muscle is called a *myocardial infarct*. Myocardial infarcts are more frequent and occur at an earlier age in men than in women.

Cancers (malignant neoplasms) are the second leading cause of death in the United States accounting for one out of every five deaths. Cancer of the sex organs (breast, uterus, ovary, prostate), gastrointestinal tract, and lung account for 60 percent of all cancer deaths. The frequency of cancer increases dramatically between ages 50 and 80.

Stroke (cerebrovascular disease) is the third leading cause of death in the United States, accounting for 1 out of every 11 deaths. *Stroke* is the common name for cerebral vascular accident, which is caused by injury to an area of the brain, usually resulting from vascular obstruction or bleeding into the brain. Strokes, like heart disease, are most often due to obstruction of arteries resulting from atherosclerosis. Atherosclerosis of the arteries leading to the brain leads to death of a portion of the brain called a *cerebral infarct*. Strokes caused by bleeding usually result from high blood pressure (hypertension) or rupture of an aneurysm (a dilated outpouching of an artery). Strokes are most common in the elderly and commonly present with weakness of one side of the body or difficulty with speech.

Trauma caused by accidents is the fourth leading cause of death in the United States, accounting for 1 out of every 18 deaths. Automobile accidents are the most common cause of traumatic death. It is estimated that 50 percent of drivers responsible for automobile accidents are under the influence of alcohol. The young and elderly are most commonly affected by accidents.

Chronic obstructive pulmonary disease is the fifth leading cause of death, with most of these deaths due to emphysema caused by cigarette smoking.

CAUSES OF DISABILITY (MORBIDITY)

Another way diseases may be significant is in terms of disability. Disabilities are health problems that interfere with a person's normal physical, mental, or emotional functions. Health problems are what motivate the patient to seek health care. The frequency of disability within a population is called *morbidity.* The term *prognosis* refers to the outcome of a disease. Prognosis includes both morbidity and mortality estimates.

MEASURES OF MORTALITY AND MORBIDITY

The frequency of disease can be measured at a given point in time or over a period of time. Whichever method is used, the observations should be made on a predetermined population so that the results will not be biased. Examples of predetermined populations are all of the people living in one state, or all of the men in one state, or all of the women in the state aged 50 to 59.

Mortality rate is a measure of the number of people dying in a given time period. It is usually expressed as the number dying per 100,000 population per year. *Incidence* is a measure of the number of newly diagnosed patients in a given time period, usually a year. Persons who had the disease diagnosed before the year began are not counted. Incidence is a useful measure for those diseases of short duration where the persons with the disease either get well soon or die. Such diseases are easy to count over a year's time.

Prevalence refers to the number of persons with a disease at any one point in time. It is more difficult to measure because the population must be surveyed at one time, and it may be difficult to decide whether a person has the disease or not. For example, does a person who recently had a cancer removed still have cancer? One cannot always tell. Prevalence is best used as a measure of long-standing diseases that are neither cured nor lead to death within a few years.

Other measures derived from mortality rate, incidence, and prevalence data are often useful for comparative purposes. The relative incidence of a group of conditions, e.g., various cancers of a particular organ, can be expressed as a percentage of the total. Survival rate is the percentage of people with a particular condition that live for a given period of time after diagnosis; e.g., the 5-year survival rate of breast cancer is the percentage of all women with breast cancer who were alive 5 years after diagnosis, regardless of whether they still have cancer or not. The age adjusted or relative survival rate adjusts the rate for those who might have died from other causes based on their age.

CHART 2–1. MEASURES OF MORTALITY AND MORBIDITY

$$\text{Mortality rate} = \frac{\text{number of persons dying per year}}{\text{number of persons in population}}$$

$$\text{Incidence} = \frac{\text{newly diagnosed cases per year}}{\text{number of persons in population}}$$

$$\text{Prevalence} = \frac{\text{number of persons with a disease at a given point in time}}{\text{number of persons in population}}$$

FREQUENCY OF ACUTE DISEASES

Acute diseases are those that last a short period of time, usually a few days to a few weeks. Any health condition that causes the patient to seek medical consultation or to miss work or school or prevents normal daily activities is classified as a disease (illness). More than half of acute diseases are respiratory illnesses, mostly

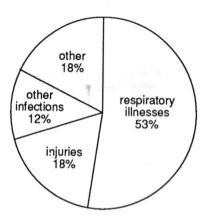

Figure 2–2. Frequency of acute diseases. *(Data from Department of Health, Education and Welfare, Health United States 1975, Publication No. (HRA) 76-1232.)*

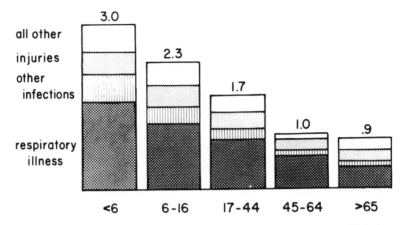

Figure 2–3. Number of acute illnesses per person per year by age group. *(Data from Department of Health, Education, and Welfare, Health United States 1975, Publication No. (HRA) 76-1232.)*

acute viral diseases (Fig. 2–2). Approximately one-sixth of acute diseases result from injuries of various types. The other one-third of acute diseases are about equally divided among nonrespiratory infections and a variety of noninfectious, noninjury type diseases. Acute conditions are best measured in terms of the number of diagnosed cases per population per year (incidence).

The frequency of acute illnesses decreases with age (Fig. 2–3). For example, in one survey, children under 6 years of age had three acute illnesses per year compared to one per year for persons over age 44.

FREQUENCY OF CHRONIC DISEASES

Chronic diseases are those that last for a long time, often for the patient's lifetime. Unlike acute disease, the frequency of chronic disease dramatically increases with age. Prevalence is a good measure of chronic disease, because it indicates the proportion of the population that has the condition at any one time. The most prevalent chronic disease is periodontal disease (inflammation of the gums), which affects about one-third of persons between 45 and 65 and about one-half of persons over 65. Patients often do not report periodontal disease as a disabling condition, so it is not listed in the comparative data given below. Mental diseases are also quite prevalent but are not accurately reported, because criteria for diagnosis vary and social stigma inhibit reporting.

The five leading causes of chronic disease reported in a na-

TABLE 2–1. PREVALENCE OF CHRONIC DISEASES (PERCENT OF POPULATION AT AGES INDICATED HAVING THE CONDITION)

Condition	All Ages	<17	17–44	45–64	≥65
Arthritis	9	—	4	20	38
Hearing impairment	7	1	4	11	29
Hypertension	6	—	4	13	20
Heart disease	5	1	2	9	20
Visual impairment	5	1	2	9	20
Chronic bronchitis	3	4	2	4	4
Asthma	3	3	3	3	4
Diabetes	2	—	1	4	8
Back ailments	—	—	5	7	7
Hernia	—	—	—	3	6
Peptic ulcer	—	—	—	3	3

(Data from Department of Health, Education and Welfare, Health United States 1975, Publication No. (HRA) 76–1232.)

tional survey (Table 2–1) are all associated with aging (arthritis, hearing impairment, hypertension, heart disease, and visual impairment). Note that arthritis is the most prevalent chronic disease and occurs predominantly in the elderly. Diabetes mellitus, back ailments, abdominal hernias, and peptic ulcers are other common chronic diseases that increase in frequency with age. Two chronic conditions that are about equally frequent in the young and old are chronic bronchitis and asthma.

DISEASE FREQUENCY RELATED TO VISITS TO FAMILY PHYSICIANS

Common acute diseases encountered in a family practice office are listed in Table 2–2 in approximate order of frequency. Upper respiratory infections are by far the most common.

TABLE 2–2. COMMON ACUTE DISEASES

Upper respiratory infections, including pharyngitis and tonsilitis
Superficial injuries
Middle ear infections (otitis media)
Acute bronchitis
Anxiety states
Strains and sprains
Urinary tract infections
Acute gastroenteritis
Pneumonia

(Based on data presented in Baker C: What's different about family medicine. J Med Educ 49:229, 1974; Marsh GN, et al.: Anglo-American contrasts in general practice. Br Med J 1:1321, 1976.)

TABLE 2–3. COMMON CHRONIC DISEASES AND CONDITIONS

Hypertension
Pregnancy
Diabetes mellitus
Depression
Low back problems
Degenerative arthritis
Small benign skin lesion, such as warts, nevi, and keratoses
Heart disease
Asthma
Iron-deficiency anemia

(Based on data presented in Baker C: What's different about family medicine. J Med Educ 49:229, 1974; Marsh GN, et al: Anglo-American contrasts in general practice. Br Med J 1:1321, 1976.)

Common chronic diseases or conditions encountered in a family practice office are listed in Table 2–3 in approximate order of frequency. Acute conditions outnumber chronic conditions. The management of hypertension has become the most frequent reason for office visits among patients with chronic disease.

AGING

Humans, like other species, have a finite lifespan. This lifespan is determined by the process of aging, a normal process affecting all individuals that is progressive and irreversible. The human lifespan is about 85 years for females and a few years less for males. The lifespan is that age at which 50 percent of the population will die from the effects of aging.

The aging process begins at physical maturity (age 17) but does not have a marked influence on death rate until much later in life. The mechanism of death from aging is increased susceptibility to disease. For example, aged individuals will die from respiratory infections, accidents, or exposure to cold that would not have killed them at a younger age. Although medical care can protect an elderly person against a potentially fatal disease such as influenza, increased susceptibility to disease may soon be expressed by death from another cause.

There is no simple explanation of aging. Both genetic and acquired factors are involved. From a genetic standpoint, each species has a maximum lifespan that is attained by a few individuals. The maximum lifespan of about 110 years for humans has not changed during the time that the average life expectancy has risen from the 20 to 85 years of age. Some families or inbred groups of people tend to live longer than others, perhaps due to a lesser tendency to develop certain diseases. On the other hand, individuals with the rare genetic disease called *progeria* show the changes of aging at a very young age. From an experimental standpoint, cells in culture will undergo a limited number of cell divisions before dying out. The exact significance of this observation is unknown, although cell metabolism and cell turnover rates do decrease somewhat with age.

Evidence for the accumulation of acquired injuries as the cause of aging relate to alterations in cells and intercellular substances. Mutations, damage to DNA, and cell death due to radiation or the formation of free radicals may account for decreased cell turnover and decreased cell function that is observed with aging. Cross-linking of collagen and other long-lived connective tissue substances makes them more rigid and less functional leading to changes such as degenerative arthritis, wrinkling of the skin, and cataracts.

Diseases that increase in frequency with age can be roughly divided into those that are age dependent and those that are age related. Age-dependent diseases occur to some extent in all individuals with time. Examples include senile degeneration of the skin, degenerative arthritis, osteoporosis, presbyopia, endocrine changes associated with atrophy of ovaries and testes, and hyperplasia of the prostate. Age-related diseases are not part of the aging process itself, since all individuals are not affected. Examples include many types of cancer, actinic and seborrheic keratoses, atherosclerosis, hypertension, Alzheimer's disease, diverticulosis of the colon, and cataracts. Still another group of diseases accumulate in frequency with age, although their onset is not age related. Examples include chronic lung disease from smoking and gallstones.

There is considerable variation in the degree to which various body systems are affected by aging. Decrease in immunity is not overtly obvious, but is expressed by an increased susceptibility to and severity of infections in the elderly and by an increased rate of development of cancer, due to a failure to reject cancer cells as foreign. Manifestations of aging in the endocrine system include impaired glucose tolerance that predisposes to diabetes mellitus, decreased thyroid function, and decreased gonadal function leading to osteoporosis and other changes. Neuronal loss is part of the aging process, but its most severe form (Alzheimer's disease) may have a specific cause. The musculoskeletal system and skin are particularly affected by aging as can be readily seen in aged individuals. Decrease in function of the heart, lungs, liver, and kidneys can be demonstrated, but the direct changes of aging have relatively little effect compared to the major diseases of these organs, some of which are age related. The intestines show little morphologic evidence of aging, but decreased motility is frequently manifested by constipation and diverticulosis in the elderly.

The aims of a good health care system are to alleviate disability and prolong life. The evidence is clear that lifespan is finite and determined by the aging process. Some individuals reach the species limit; others die earlier due to acute or chronic disease. Figure 2–4 compares the survival of populations having primitive and modern medical care with the survival curve of an ideal health care society where all deaths are due to aging. In the society with primitive health care, there is high infant mortality, uncontrolled serious infections, and high death rate from accidents. The improvement in survival in a society with modern medical care, represented by the shaded area A, is due to good maternal–infant health care, vaccination against serious infections, and high recovery rate from infections and accidents. The shaded area B represents further gains that could be made toward an ideal health care system where all deaths would be due to the consequences of aging.

In the United States, the diseases that are responsible for most of these "early" deaths (shaded area B) are heart disease, stroke, and cancer. There is also a distinct potential for preventing the early mortality by eliminating poor maternal–infant care, which occurs in selected portions of our society.

Heart disease and stroke deaths are in part due to aging itself, but there is also an environmental factor (presumably diet) that causes accelerated atherosclerosis in the United States as compared to societies with a low-cholesterol diet such as in Japan. Japanese, like Americans, die from the effects of aging on the vascular system, but they rarely die at an early age from myocardial infarcts. This evidence suggests that some gain toward the ideal health curve could be made by a change in diet. The gain that might be expected from preventing cancer is less than the gain that might be expected from retarding the progress of atherosclerosis. This is

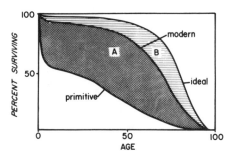

Figure 2–4. Survival curves in societies with primitive, modern, and ideal health care. See text for explanation of A and B. *(Modified from Hayflick M: The cell biology of human aging. N Engl J Med 295:1302, 1976.)*

so because cancer causes fewer deaths than atherosclerosis, and most cancer deaths occur near the time that death would occur from aging itself, regardless of the nature of the final insult.

REVIEW QUESTIONS

1. What are the leading causes of death? What percentage of people die from each?
2. How are incidence, prevalence, and mortality rates defined? Can you calculate them if given appropriate data?
3. What is the relative frequency of the leading causes of acute disease and how does this relate to age?
4. What are the most common chronic diseases? Which ones are age related?
5. Which acute and chronic diseases most frequently bring patients to their family physician?
6. How does aging relate to the causes of death?
7. What factors determine the limits of life expectancy?
8. What is the meaning of the following terms?
   Morbidity
   Prognosis
   Acute
   Chronic
   Aging
   Age-dependent disease
   Age-related disease

Diagnostic Resources

APPROACH TO PATIENT CARE

Symptomatic Disease

The most common approach to disease is to wait for the patient to seek help because of symptoms of disease. The health practitioner, presented with a sick patient, proceeds in a systematic fashion to help the patient by the steps outlined in Table 3–1. This process may take minutes or weeks, depending on the complexity of the disease. Most diagnoses can be made from the history with physical examination and laboratory tests providing confirmatory evidence.

TABLE 3–1. STEPS IN CARE OF SYMPTOMATIC DISEASE

1. Gather facts:
   History
   Physical examination
   Laboratory tests
2. Interpret the facts and attach a summarizing label (diagnosis)
3. Treat the patient, if feasible
4. Follow up on results of treatment

Asymptomatic Disease

Another approach to disease is to try to discover the disease in its asymptomatic stage, that is, before the patient notices it. Early diagnosis allows treatment to be started at an early stage of the disease. This approach can be applied to the individual through regular checkups such as regular dental appointments, well-baby examinations, and periodic physical examinations. This approach is also applied to defined populations through procedures such as tuberculosis testing of school children, school physicals, and army physicals. The attempt to discover disease in its early stage is called *screening*. The goal of screening is to either cure a disease by catching it early (e.g., cancer) or to begin treatment early to delay the progression of the disease (e.g., hypertension). It is of limited value to screen for diseases that cannot be treated or to spend large sums of money to find rare diseases.

Potential Disease

A third and obviously most desirable approach to disease is the prevention of its occurrence. The discipline that deals with prevention of disease is called *preventive medicine*. Most of the classic infectious diseases that formerly killed a significant proportion of the population (smallpox, plague, typhoid fever, typhus, measles, diphtheria, whooping cough, tuberculosis) now fall into the category of potential diseases that are preventable by immunization and good sanitation. Other outstanding examples of preventable diseases are dental caries and periodontal disease, which have been greatly reduced by use of fluorides and preventive dentistry. We also know that reduction in alcoholism would prevent deaths from liver disease and accidents. Reduction in smoking would prevent many cancers (particularly lung cancer) and chronic obstructive pulmonary disease from developing. It is also likely that alteration of the types of lipids in the diet would lead to reduced atherosclerosis.

TABLE 3–2. SCREENING TESTS AND PROCEDURES

Test or Procedure	Purpose
Cervical (Pap) smear	Early detection of uterine cancer
Blood count	Detection of anemia
Urinalysis	Detection of urinary infection or kidney disease
Fecal occult blood test	Detection of hidden bleeding from colon cancer
Serum lipids (especially cholesterol)	Detection of tendency to have atherosclerosis
Serology	Detection of syphilis prior to pregnancy
Dental x-rays	Finding caries (cavities)
Chest x-ray	Detection of tuberculosis or lung cancer
Mammography	Detection of breast cancer
Visual acuity tests	Detection of visual problems in preschool children and automobile license applicants
Audiograms	Detection of hearing problems in school children
Tuberculin skin test	Detection of children who might have tuberculosis
Electrocardiogram	Detection of asymptomatic myocardial infarcts
Sigmoidoscopy	Detection of colon cancer

SCREENING

The techniques of screening for asymptomatic disease include history, physical examination, and laboratory tests. Screening by history involves a checklist of symptoms that may suggest further investigation. Examples of screening by physical examination include dental examination for caries, palpation of breasts for lumps, and listening to a baby's heart to detect murmurs. Some of the more important and widely used screening tests and procedures are listed in Table 3–2 along with their purpose.

DIAGNOSTIC TESTS AND PROCEDURES

Test is used here to refer to an analysis performed on a specimen removed from a patient. A *procedure* involves doing some manipulation of the patient beyond that usually done during physical examination. Some procedures are done to obtain specimens for a test. Most tests are performed by or supervised by a pathologist. Procedures are performed by various types of physicians, including radiologists.

Clinical Procedures

Primary health care practitioners may perform some common or simple tests and procedures themselves. For example, laboratory tests such as urinalysis, blood counts, and throat cultures may be done is a physician's office. Manipulative procedures such as sigmoidoscopy may be done by the primary care physician, or the patient may be referred to a specialist for more complex procedures. One function of the various medical specialists is the performance of specific manipulative procedures to detect disease in hidden areas of the body. For example, the urologist looks in the bladder with a cystoscope and the gastroenterologist looks into the stomach with an endoscope. Other procedures will be mentioned in the chapters that deal with diseases of the various organs.

Radiologic Procedures

Radiology is the discipline of medicine performed by radiologists and includes diagnostic radiology, nuclear medicine, and radiation therapy. Some of the more common radiologic procedures used for diagnosis are listed in Table 3–3.

Diagnostic radiology involves x-ray and other imaging procedures such as ultrasound and magnetic resonance imaging that are used to locate and describe morphologic lesions in a living patient.

TABLE 3–3. RADIOLOGIC PROCEDURES

Test or Procedure	Lesions Commonly Detected
Dental x-rays	Caries
Sinus films	Sinusitis
Chest x-ray	Any lesion that replaces normally air-filled lungs
Upper gastrointestinal series	Defects or tumors of the esophagus, stomach, and upper small intestine
Barium enema	Tumors, ulcers, and diverticula of the colon
Gallbladder series	Gallstones
Intravenous urogram	Decrease in kidney function, obstruction of the urinary tract
Bone films	Any lesion that destroys bone; fractures
Myelogram	Obstruction of the space surrounding the spinal cord
Arteriogram	Obstruction or displacement of arteries
Computerized tomography	Tumors and other lesions
Ultrasound	Gallstones, cysts, twin pregnancy
Magnetic resonance	Tumors and other lesions
Nuclear isotope scans	Tumors, altered tissue uptake of specific substances

X-ray procedures are dependent upon differing absorption properties encountered in the x-ray path to produce images of the tissue. In conventional x-ray techniques, the net amount of x-radiation that passes through the body exposes film to produce a *roentgenogram* (colloquially called an x-ray). Radiodense material, such as barium, or radiolucent material, such as air, can be introduced into body passageways to provide contrast with tissue. Alternatively, the x-rays that pass through the body may be viewed with a fluoroscope, an instrument that uses a fluorescent plate to detect x-rays. A roentgenogram is a static image; a fluoroscopic image is dynamic and allows the radiologist to watch movements such as barium as it passes down the esophagus. A chest roentgenogram of a patient with advanced tuberculosis (Fig. 3–1) shows radiodensity in the diseased left lung and radiopacity of the air-filled right lung.

Computerized tomography is a sophisticated x-ray technique in which the x-ray absorption patterns through planes of tissue are analyzed and recorded by computer for each point in the plane. The computerized images can be translated into printed images or viewed on a screen. Because computerized tomography looks at a single plane at a time, lesions can be more sharply defined and precisely localized. As with conventional x-ray techniques, contrast materials can be used to outline hollow organs. Lesions that are hidden deep in the body such as the kidney cancer with spread

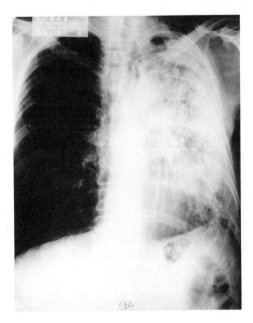

Figure 3–1. Chest roentgenogram, tuberculosis of left lung.

to the liver (Fig. 3–2) are well delineated by computerized tomography.

Magnetic resonance imaging is similar to computerized tomography in its use of a computer to record tissue characteristics in tissue planes, but differs in that it does not use x-rays. The image is produced by displacing protons in atomic nuclei with radiofrequency signals while the body is surrounded by a strong magnet. The affected protons release a similar radiofrequency signal that can be evaluated by computer to produce images of a section through the body. Different physical characteristics of protons among the elements allow production of two types of images. T1 images give a strong signal for lipid and T2 images give a strong signal for water. Because of cost, the use of magnetic resonance is limited to major medical centers. It is very useful for locating lesions, especially neoplasms. The anatomic detail provided is illustrated in Figure 3–3.

Ultrasound examination measures the reflection of high frequency sound waves as they pass through body tissues. The greatest contrast is provided by interfaces of soft tissues and liquids;

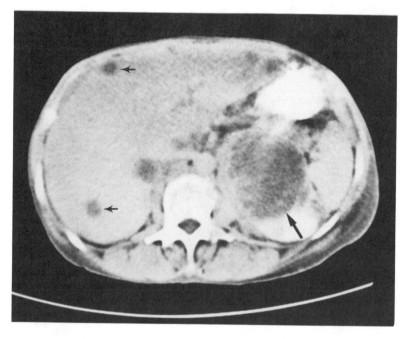

Figure 3–2. Abdominal cross-section by computerized tomography, cancer of left kidney (*large arrows*) with metastases to liver (*small arrows*).

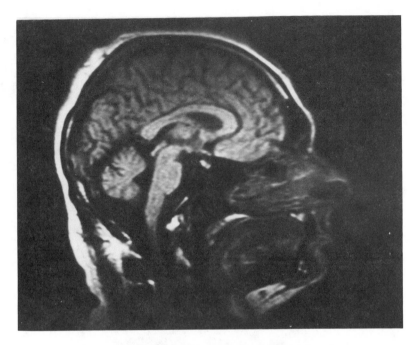

Figure 3–3. Normal head by magnetic resonance imaging.

therefore, this technique has its greatest usefulness in studying cystic structures such as gallbladder, kidney, and gravid uterus. It is the procedure of choice for detecting gallstones. In pregnancy, it can be used without risk of radiation to the fetus. Twins and ectopic pregnancies are easily detected (Fig. 3.4).

The subspecialty of nuclear medicine involves the injection of various radioactive materials into the bloodstream and subsequently determining their degree of localization within tissue. The body is scanned externally for radioactivity and the results recorded as a *nuclear isotope scan*. Areas of decreased concentration within an organ suggest a space-occupying lesion such as a neoplasm (negative image); some neoplasms exhibit an increased uptake of isotope (positive image) (Fig. 3–5). The functional activity of an organ can also be evaluated; for example, the amount of radioactive iodine taken up by the thyroid gland reflects thyroid function.

Radiation therapy is the branch of radiology involved in treatment of cancer and other conditions with x-rays and gamma rays. Some cancers may be cured by radiation; others are treated to slow the progress of the disease and delay complications. When cure is not expected, the therapy is referred to as *palliative*.

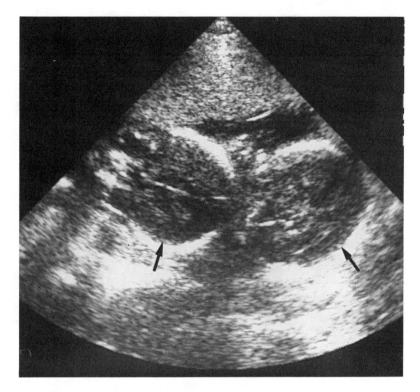

Figure 3–4. Ultrasound of abdomen of pregnant woman, twins (*arrows*).

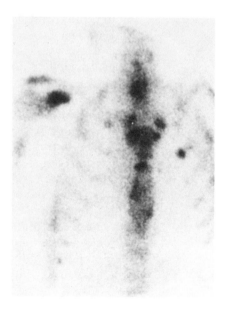

Figure 3–5. Nuclear isotope bone scan demonstrating spread of a breast cancer to bone (*dark areas*).

Anatomic Pathology Tests and Procedures

Surgical pathology involves the diagnosis of lesions in pieces of tissue removed from a patient. Diagnosis is based on gross (naked-eye) and microscopic examination and interpretation by a pathologist. *Biopsy* is the procedure for obtaining small specimens. Partial (incisional) biopsy specimens include only part of the lesion and are done primarily for diagnosis. Needle biopsy, which involves the insertion of a needle into a solid organ and aspiration of a core of tissue, is widely used for the diagnosis of liver, kidney, and prostate disease. Excisional biopsy specimens include the entirety of a small lesion and are done for both diagnosis and treatment. The removal of large specimens in the operating room is called *resection*. Resected specimens are usually removed primarily for treatment purposes. Biopsy specimens ordinarily require 1 to 2 days for preparation of microscopic slides by histotechnicians and then microscopic examination and diagnosis by a pathologist. When more rapid diagnosis is needed, a frozen section can be prepared in a few minutes for interpretation by the pathologist.

All tissues removed from a patient are examined by a pathologist to insure accuracy of diagnosis. Gross examination is sufficient for some specimens such as teeth, placentas, varicose veins, and tonsils from children. All other specimens are examined microscopically.

Cytology specimens contain cells sloughed or scraped from body surfaces. These are used primarily to detect cancer cells, and the majority of cytology specimens are from the uterine cervix (Pap smears). Ordinarily the stained smears are examined by a cytotechnologist, and any abnormalities are interpreted by a pathologist. Any body fluid, such as urine, sputum, cerebrospinal fluid, and pleural fluid, can be used for cytologic study. Fine-needle aspiration is a technique that uses a small caliber needle to obtain aspirated tissue for cytologic examination. Fine-needle aspiration is faster and less expensive than open biopsy for the diagnosis of certain cancers, but may not always provide a specimen that is adequate for diagnosis. The vast majority of cytologic examinations are done to detect cancer.

An autopsy is the postmortem examination of a body. Organs of the neck, chest, abdomen, and cranium are ordinarily examined to make a final evaluation of the nature and extent of disease and to determine the probable cause of death. Biochemical, microbiologic, and immunologic tests can be performed if needed. The autopsy is performed by a pathologist with the help of a trained technician or a mortician. An autopsy requires advanced permission of the next of kin. In the case of unexpected death or suspicion of homicide, the medical examiner or coroner may authorize an autopsy.

Clinical Pathology Tests and Procedures

Clinical pathology or *laboratory medicine* is the branch of pathology that performs laboratory tests on tissues and fluids other than those already described under surgical pathology and cytology. The clinical pathology laboratory is subdivided into sections of chemistry, hematology, microbiology, immunopathology, and blood bank. Biochemical tests on blood account for the largest number of tests done. Biochemical tests may be used to evaluate organ function or to detect relatively specific abnormalities. A complete blood count (CBC) is the most common hematologic test and consists of measurement of hemoglobin, counting of white and red blood cells, and microscopic evaluation for morphologic changes in the blood cells. Special hematologic tests are available for evaluation of various types of anemia and evaluation of blood coagulation. Blood banking or transfusion medicine involves the procurement, typing, testing, processing, storage, and administering of blood components as well as evaluation of adverse reactions to transfusion therapy. The processing of blood is analogous to the processing of milk in that the various components are removed for specialized use. Immunopathology involves the detection of antigens and antibodies in blood and tissue and the study of lymphocytes. Immunologic techniques are used to detect a wide variety of diseases including immunodeficiencies, allergic (hypersensitivity) diseases, and certain cancers. Bacterial culture is the most common test performed in the microbiology laboratory because bacteria are easily grown and can be tested for their sensitivity to antibiotics. The ease and means of detection of other types of microbes is variable.

Forensic Pathology

Forensic pathologists are pathologists specially trained to investigate accidental and criminal deaths. There are relatively few forensic pathologists in the United States, and most work in large metropolitan areas. In most communities, the investigation of accidental, sudden, or suspected criminal deaths is carried out by the county medical examiner or coroner, who may be any physician appointed to the office. When an autopsy is needed, the county medical examiner calls upon a general pathologist to perform the autopsy. In unusual cases, a forensic pathologist may be brought in to help.

Public Health Laboratories

Public health laboratories are established by governments to help in the control of communicable diseases. For this reason, they perform many microbiologic tests, such as blood tests for syphilis, rabies tests, and viral cultures to identify epidemics of virus infection. Water testing is another important aspect of community health. State health laboratories may serve as a reference laboratory and provide a link with the National Communicable Disease Laboratory in Atlanta, Georgia. Certain highly contagious diseases require reporting to state authorities.

REVIEW QUESTIONS

1. What strategies do health care practitioners use to deal with symptomatic disease, asymptomatic disease, and potential disease?
2. Which diseases are commonly detected by screening? What screening test or procedure is used for each?
3. What types of diagnostic tests and procedures are performed or directed by each of the following?
   Primary care physician
   Medical specialist
   Radiologist
   Pathologist
   Forensic pathologist
   Public health laboratory
4. What is the meaning of the following terms?
   Test
   Procedure
   Incisional biopsy
   Excisional biopsy
   Biopsy specimen
   Resected specimen
   Needle biopsy
   Fine-needle aspiration

Section

II

Basic Disease Processes

The purpose of this section is to give you a clear picture of the fundamental mechanisms of disease. The knowledge learned here will apply to almost every disease presented throughout the rest of the book. We have chosen to present this material rather concisely so that we can spend more time on individual diseases. We recommend that you study these chapters very thoroughly, review them as you progress through the book, and consult other texts to broaden your understanding of these topics.

Chapter

4

Injury, Inflammation, and Repair

REVIEW OF STRUCTURE AND FUNCTION

The body is made up of cells and intercellular substances that are capable of undergoing dynamic change to carry out body functions, including self-renewal. One or several of the more than 100 cell types, along with appropriate intercellular substances, comprise a tissue. The term *tissue* is usually used to refer to a functional grouping of cells and intercellular substances at the microscopic level. An *organ* is one or more tissues arranged into a tissue mass that carries out a major body function. For example, liver tissue forms one massive organ called the liver, whereas, loose connective tissue is a general type of tissue that may be a part of many organs. The cells that carry out the main function of an organ, and are usually most abundant and often unique to the organ, are called parenchymal cells. For example, hepatocytes, renal tubular epithelial cells, cardiac muscle cells, and osteocytes are the parenchymal cells of their respective organs (liver, kidney, heart, and bone).

Typical components of a cell are illustrated in Figure 4–1. The nucleus is surrounded by a nuclear membrane and contains loosely arranged chromatin, which stains with basic dyes such as hematoxylin because of its high content of deoxyribonucleic acid (DNA). During cell division, the chromatin aggregates into discrete strands, or chromosomes, which replicate and separate to form two daughter cells. The nucleus is vital to the cell because the genetic code in its DNA is the ultimate regulator of cell function. Histologically, changes in the nucleus are used as indicators of cell death.

The cytoplasm of a cell consists of the cell membrane, cytoplasmic organelles, and a soluble phase called the cytosol. The cell membrane protects the cell from physical injury and selectively regulates the entrance and exit of various ions and nutrients, such as amino acids and sugars. Movement of water to and from the cell is dependent upon movement of ions and nutrients. When the

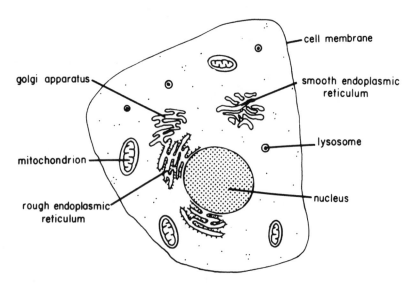

Figure 4–1. Components of a cell.

regulation of water movement fails, the cell may swell or shrink. Cytoplasmic organelles include structures such as mitochondria, rough and smooth endoplasmic reticulum, Golgi apparatus, and lysosomes. Mitochondria are complex, membranous structures that generate energy for use by the cell. Injuries that interfere with energy production often cause the mitochondria to swell and later condense. The endoplasmic reticulum is a tortuous set of membranes. Rough endoplasmic reticulum is lined by small basophilic granules called *ribosomes* because of their high content of ribonucleic acid. Proteins produced under the enzymatic control of ribosomes are carried along the rough endoplasmic reticulum to the Golgi apparatus, where they are stored for secretion. The smooth endoplasmic reticulum also serves to transport materials through the cell and is the site of production of many biochemical substances other than proteins. Lysosomes are membrane-bound packets of digestive enzymes. Lysosomes may coalesce to surround and digest foreign substances that have been engulfed (phagocytosed) by the cell. Worn out or injured parts of the cytoplasm may also be digested by lysosomes, a process known as autophagocytosis.

A simple classification of tissue components is shown in Table 4–1. The two most varied classes of cells are epithelial cells and connective tissue cells. The distinction between these two classes of cells is very important in pathology because they react quite differently in disease situations. Epithelial cells work with each other in tight clusters to carry out specialized functions, such as protection of body surfaces, secretion of specific products, and special metabolic functions. Injury interferes with their specialized function and causes them to revert to a more primitive stage for

TABLE 4–1. STRUCTURAL ELEMENTS OF TISSUES

Cells
   Epithelium
   Connective tissue cells
      Fixed: Fibrocytes, chondrocytes,
           osteocytes, endothelial cells
      Motile: Blood cells
   Muscle cells
   Nervous tissues cells
Intercellular substances
   Basement membranes
   Ground substance
   Collagen
   Elastin
   Cartilage
   Bone

purposes of reproduction to replace cells that have been killed. Connective tissue cells are more loosely arranged and are involved in general support functions, such as providing physical support and promoting the appropriate movement of fluids and nutrients. An example of the arrangement of the various cell types into a tissue is shown in Figure 4–2.

White blood cells (leukocytes) are very mobile nonepithelial cells that are specialized to aid in attacking foreign substances. Each type has a characteristic morphologic appearance (Fig. 4–3). *Neutrophils*, also called *polymorphonuclear leukocytes* or *polys*, and monocytes can engulf (phagocytose) and digest foreign materials such as bacteria. When monocytes leave the bloodstream and enter tissue they are called *macrophages* or *histiocytes*. Lymphocytes direct the attack against persistent foreign materials by remembering the chemical structure of the foreign materials. Lymphocytes release substances (lymphotoxins) that kill cells in the area of the foreign material and other substances that attract macrophages to the area. The macrophages, in turn, can phagocytose both the foreign material and dead cells to prevent further spread of the foreign substance. Some lymphocytes also transform into plasma cells to produce antibodies. Antibodies attach to the unique chemical structure of the foreign substance (antigen) to aid in neutralizing or destroying the foreign substance. The other two types of white blood cells, basophils and eosinophils, are much less abundant. Both are involved in allergic reactions and are discussed in Chapter 27.

The connective tissue structure of the body involves physical

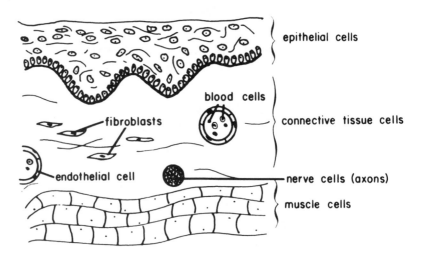

epithelial cells

blood cells

fibroblasts

connective tissue cells

endothelial cell

nerve cells (axons)

muscle cells

Figure 4–2. Comparative features of major cell types in the wall of the oral cavity.

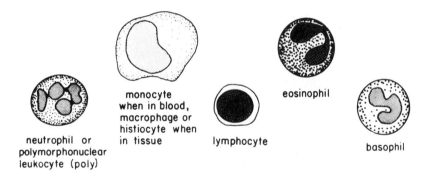

Figure 4–3. Types of white blood cells.

support and transportation. Gross tissues for physical support include bone, cartilage, tendons, fascia, and other fibrous tissues. At the microscopic level, connective tissues contain ground substance, which allows passage of fluid and nutrients. Basement membranes surround epithelial clumps and allow passage of fluid and nutrients to and from the epithelial cells. Collagen is the most abundant component of connective tissues. The amount of collagen relates to the strength and fibrous nature of the connective tissue. Thus, loose connective tissue contains little collagen and dense connective tissue contains much collagen. Vessels course through the supporting connective tissue to allow fluids to be carried close to epithelial cells and other active tissues such as muscle.

Regulation of tissue fluids is a function of movement of water in and out of small blood vessels (capillaries and venules) and uptake of fluid by lymphatics. Fluid and nutrients pass out of the small blood vessels and diffuse through the ground substance and basement membranes so that exchange with cells can occur. Reverse movement can occur back into blood vessels or into lymphatic vessels. The most important factors in this regulatory process are the pressure in the vessels, the osmotic pressure difference between tissue and blood due to the relative amount of large protein molecules present, and the size of the pores between endothelial cells in the lining of small vessels. This regulatory process is important in situations of injury, because larger amounts of fluid and nutrients are needed to react to the injury.

Physiologic replacement of cells is a normal process that closely relates to the repair of injuries to be discussed later in this chapter. Certain body cells wear out rapidly and are continually being replaced. These include blood cells and cells lining body surfaces such as skin and intestinal mucosa. Most glandular epithelial cells and cells that form the supportive connective tissue undergo very slow replacement but are capable of more rapid replacement if necessary. Other types of cells, such as cardiac muscle cells and neurons, cannot be replaced.

EVENTS FOLLOWING INJURY

The events following injury are a continuum of necrosis, inflammation, and repair. Necrosis is the death of cells or tissue in a localized area of the body due to injury, either exogenous or endogenous. *Necrosis* consists of the direct effects of the injury plus the changes that occur in cells after they die. Mild forms of injury may produce sublethal cell injury without necrosis, changes that may be referred to as *degeneration*. We will avoid use of the word degeneration and use the term necrosis with the realization that lethal and sublethal cellular changes occur together in varying proportions. *Inflammation* is the vascular and cellular response to necrosis or sublethal cell injury and is the body's mechanism of limiting the spread of injury and removing necrotic debris. *Repair* refers to the body's attempt to replace dead cells, whether by regeneration of the original tissue or replacement by connective tissue.

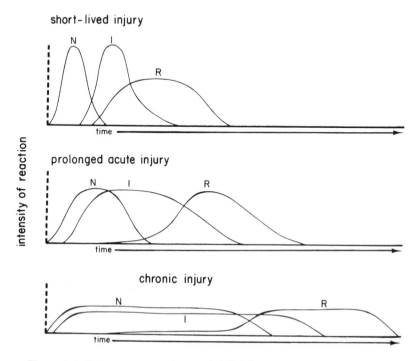

Figure 4–4. Time sequences of necrosis (N), inflammation (I), and repair (R).

The relative intensity of necrosis, inflammation, and repair depends on the magnitude of the injury, the duration of injury, and, to some extent, on the location within the body and the nature of the injury. Some examples of how necrosis, inflammation, and repair vary with intensity and duration of injury are shown in Figure 4–4. In general, inflammation begins shortly after the effects of cell injury are evident. Repair is usually not well established until necrosis ceases, although in very chronic injuries, all three processes are likely to occur together. Inflammation itself, when intense, is a cause of necrosis. The body, in a sense, sacrifices some of its own tissue to isolate an injurious agent.

INJURY

The critical difference between sublethal (reversible) cell injury and necrosis is whether the cell can recover or is already dead. Certain changes in the nucleus, as seen in microscopic sections, indicate cell death. Nuclear changes may include condensation of the nucleus (*pyknosis*), fragmentation of the nucleus (*karyorrhexis*), and lysis or fading of the nucleus (*karyolysis*). These nuclear changes take a number of hours to develop, so cells that have died and are fixed in formaldehyde may not show these changes. For example, a person who has a myocardial infarct (heart attack) may die within 12 hours and not show these histologic changes of necrosis. If the patient had lived, some of the myocardial cells would have developed the histologic changes of necrosis whereas others may have developed reversible changes and then recovered. Reversible cell injury is characterized by preservation of the nucleus and variable changes in the cytoplasm such as swelling or condensation. These histologic changes reflect biochemical changes in the cell. There is no exact biochemical end point that determines cell death, but depletion in the cell's energy system and alteration of cell membrane permeability are important events in cell death. Once the nucleus is destroyed or the cell membrane disrupted, the cell cannot recover. After cell death, enzymes released from the cell's own lysosomes begin to digest the remains of the cell. Other events associated with necrosis include influx of calcium, dissolution of ribosomes, clumping of DNA followed by its enzymatic digestion, and finally rupture of the cell membrane.

Acute changes resulting from sudden injury to a cell or tissue lead to cell death or recovery within a short time. Necrosis is often associated with acute injury. Chronic injury (mild continuous injury) leads to cumulative effects on the cells and the tissues. Necrosis is not usually prominent and may be of such a low degree as to be imperceptible in histologic sections.

Acute Injury and Necrosis

The effects of acute injury have their most prominent effect on cells, because cells are more susceptible to injury than are the non-cellular connective tissue elements. Cells may be injured by any of the exogenous and endogenous causes discussed in Chapter 1.

Lack of oxygen (anoxia) or reduced oxygen (hypoxia) is one of the most common causes of acute injury and necrosis. Cells are vulnerable to hypoxia in proportion to their oxygen requirements; thus metabolically active cells are selectively vulnerable. Selective vulnerability is well illustrated by cases of systemic anoxia from such causes as carbon monoxide poisoning, blood loss, or suffocation. In these situations, neurons in the brain and the kidneys' tubular epithelial cells are more vulnerable to necrosis than other types of cells. Localized hypoxia due to poor blood flow is called *ischemia*. When severe, ischemia leads to necrosis of the cells in the area of the deranged blood supply. An area of ischemic necrosis is called an *infarct*. Infarcts are most commonly due to obstruction of arteries, especially arteries that supply a defined segment of tissue. As noted in Chapter 2, atherosclerotic plaques that obstruct coronary arteries and lead to myocardial infarcts are responsible for a high percentage of all deaths. Atherosclerotic obstruction is also important in producing infarcts of the brain, legs, kidneys, and other sites.

Thrombi and emboli are also important causes of ischemic necrosis (infarcts). A *thrombus* is a blood clot that forms during life in a blood vessel due to activation of the coagulation mechanism. It is composed of layers of fibrin and entrapped blood cells. An *embolus* is any particulate object that travels in the bloodstream from one site to another and is most commonly a thrombus, but may be composed of other substances such as bone marrow, fat, air, or cancer tissue. Thrombotic emboli most commonly originate in the leg veins with spread through the vena cava and right heart to the pulmonary arteries or from the left side of the heart with spread through the aorta to various organs such as brain, legs, kidneys, spleen, and intestines. Bone marrow and fat emboli occur from trauma to bones; when severe, the fat globules pass through the pulmonary vessels and gain access to the systemic circulation where they may cause obstruction of the small vessels of the brain.

Trauma, infection, and hypersensitivity are other common causes of acute injury and necrosis. Trauma disrupts cells by direct physical force; the effects are dependent on the site injured and nature of the force applied. Although there are many types of infections and the degree of injury varies widely, most of the damage is produced by the body's own inflammatory reaction to the invading microorganism. The basic mechanisms of this process will be discussed; a much more detailed account of the reactions to the major groups of offending organisms and specific pathogens can be found in Chapter 26. Immunologic mechanisms are an important part of the inflammatory reaction and also contribute to the damage produced by inflammation, but the immune damage is usually less than the potential damage that could be inflicted by the offending agent. When an immunologic or presumed immunologic reaction occurs only in certain individuals (those that are sensitive), the reaction is called a hypersensitivity reaction or *allergy*. Poison ivy, hay fever, hives, and contact dermatitis are common examples. Sometimes the body's immune system reacts to its own tissues (*autoimmune reaction*) producing destructive diseases such as rheumatoid arthritis, lupus erythematosus, and thyroiditis. Chapter 25 details the mechanisms and classification of immunologic disease. It would be helpful at this time to examine the first portions of Chapters 26 and 27, which describe the common types of physical and chemical injuries.

The morphology of reversible cell injury and necrosis is a continuum (Fig. 4–5). If the injury is mild and functional changes following the injury go away in a few hours, we can anticipate that the cells involved underwent sublethal changes with return to normal. If functional changes persist, it is likely that at least some cells underwent necrosis; recovery will then depend upon regeneration, a process discussed later in the chapter. Study of biopsy specimens or tissues removed at autopsy allows the pathologist the opportunity to evaluate the extent of injury. Changes in reversibly damaged cells will be limited to the cytoplasm; necrotic cells will have both cytoplasmic and nuclear changes. Typically, the early change is cytoplasmic swelling producing enlarged cells with pale cytoplasm. Later the cytoplasm may be shrunken and more

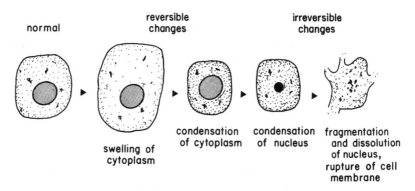

Figure 4–5. Cellular changes leading to cell death.

densely eosinophilic than normal. The development of nuclear changes—pyknosis, karyorrhexis, or karyolysis—indicates progression to necrosis. Thus, when reversible changes predominate, a tissue will be enlarged; when necrosis predominates, a tissue will be of normal size or shrunken.

Necrotic tissue takes on different gross and microscopic appearances depending on circumstances; recognition of these differences allows prediction of the cause of the necrosis. *Coagulation necrosis* is most commonly caused by anoxia, whether it be generalized or ischemic. The involved tissue undergoes slow disintegration with denaturation of cellular protein by the dead cells' own lysosomal enzymes. In many tissues, the coagulation process evolves slowly over a number of days producing the characteristic preservation of cell and tissue outlines until the later stages of the process. The pathologist can recognize the coagulation necrosis of an infarct by its pale yellow color and solid but soft texture. The location, size, and shape of the infarct will be dependent on the area supplied by the blocked artery.

Liquefaction necrosis is most commonly caused by certain types of bacteria, known as pyogenic bacteria. Pyogenic bacteria attract neutrophils into the area and the enzymes released by the neutrophils liquefy the dead tissue producing a thick, creamy mixture of dead tissue and neutrophils called *pus* or *purulent exudate.* When cut into, an area of liquefaction necrosis will exude pus and leave a hole in the tissue.

Caseous necrosis is most commonly caused by *Mycobacterium tuberculosis* (the bacteria that causes tuberculosis) or by certain types of fungi. Caseous necrosis looks different because it is necrosis of diseased tissue. The causative organisms are attacked by large numbers of lipid-containing macrophages. Necrosis of these macrophages produces a solid, amorphous, cheesy mass. Microscopically, a confluent mass of red cytoplasm with scattered nuclear dust remains. The alert pathologist, however, will use special staining techniques to demonstrate the causative organisms in the caseous material and sometimes the organisms can be cultured.

Enzymatic fat necrosis, which occurs in the pancreas and surrounding adipose tissue due to leakage of that organ's digestive enzymes, produces chalky, yellow-white nodules somewhat resembling caseous necrosis. The location, however, is limited to the pancreas and surroundings.

Gangrenous necrosis (gangrene) is coagulation necrosis with superimposed decomposition by saphrophytic bacteria. It is similar to postmortem decomposition except that only a portion of the body is dead. In *gas gangrene,* however, the organisms causing the gangrene include a strain of bacteria of the genus *Clostridium* that produces gas and a necrotizing toxin. The toxin of gas gangrene can spread to normal tissue and produce lethal effects.

Chronic Injury: Atrophy and Accumulations

Chronic injury may produce a decrease in tissue size (atrophy) or accumulation of material within cells or between cells. Atrophy may be due to a decrease in the size of cells or a decrease in the number of cells or both. A gradual loss of cells is the most common mechanism. Some of the types of atrophy will be briefly discussed.

Senile atrophy is caused by aging. As persons age, their tissues often become smaller and decrease in functional capacity, presumably as a natural part of the aging process. For example, the brains of elderly people become smaller, while decreased memory and slowed thought processes provide some evidence of decreased cellular function.

Disuse atrophy occurs when the cells unable to carry on their normal function undergo atrophy. For example, when an arm or leg is placed in a cast, the muscle cells become smaller and show a decreased ability to contract.

Pressure atrophy results from steady pressure on tissue, such as might be produced by the mass of an expanding tumor. Bedsores are another common example. They occur in chronically bedridden patients due to continued external pressure on the skin.

Endocrine atrophy is due to decreased hormonal stimulation. Certain organs are maintained in a functional state by the action of hormones upon them. Insufficient hormone results in atrophy in the respective organ. For example, the decrease in estrogen and progesterone at the time of menopause results in atrophy of the breasts and the uterus.

Chronic injuries associated with accumulation of substances are quite different from atrophy. Many times, cells will slowly accumulate their own metabolic products or exogenous materials, with resultant decrease in cell function over a period of time. The storage of these materials may even result in an enlarged cell, albeit with decreased function. These types of chronic cell or tissue degeneration are classified according to the type of material accumulated.

Accumulation of lipid within cells is called *fatty change* or *fatty metamorphosis*. Fatty change should be distinguished from adiposity. In adiposity, there is an increased storage of fat in fat cells; in fatty change, fat droplets appear as an abnormality in parenchymal cells. Fatty change may be either acute or chronic, and characteristically occurs in cells involved in fat metabolism, especially the liver. The liver takes in lipid in the form of triglycerides (from dietary absorption) and free fatty acids (from adipose tissue stores or absorption). Triglycerides and free fatty acids are metabolized in the liver to lipoprotein, a much more soluble form of lipid that can be exported for use by other tissues. Droplets of triglyceride may form in hepatocytes due to decreased production of lipoprotein or increased uptake of lipid from the blood. Causes of fatty liver include mobilization of more fat than the liver can handle such as occurs in diabetes mellitus, excess dietary intake as in obesity, chemical injury as in alcoholism and carbon tetrachloride poisoning, and acute starvation where there is depletion of the proteins needed to form lipoproteins. In chronic alcoholism, the liver may become more than twice normal size due to the accumulation of fat in hepatocytes. In diabetes mellitus, there is decreased uptake of fat in adipose tissue and increased accumulation in the liver. Alcoholism is the most common cause of clinically significant fatty liver in affluent societies.

Glycogen storage is an example of accumulation of carbohydrate. It occurs in rare genetic conditions where specific enzymes for glycogen breakdown are missing. The glycogen accumulates in various organs and eventually causes malfunction.

Accumulations of excess protein become compacted, producing a dense, homogeneous, eosinophilic deposit called *hyalin*. Excess collagen and compacted fibrin clots are the most common causes of hyalinization. Amyloid is a hyalin deposit of protein that has a crystalline chemical structure, which polarizes light and stains with the dye Congo red. Certain small proteins leak from the blood and crystallize to form extracellular deposits of amyloid. Examples of proteins that can form amyloid include immunoglobulin light chains derived from abnormal proliferations of plasma cells, serum amyloid-associated protein produced by the liver in prolonged chronic inflammations, and beta protein deposited in the brain in Alzheimer's disease. Amyloid deposits develop very slowly, affect organ function late in the course of disease, and are not reversible.

Accumulations of minerals and pigments include calcification, hemosiderosis, and brown atrophy. In some situations, the deposition of minerals or pigments is associated with obvious tissue injury, but in others it is difficult to prove that excessive accumulation of a given pigment or mineral is deleterious to that tissue. Calcification is of two types. Excessive blood calcium, which may result from certain metabolic disorders, leads to calcium accumulation in tissues, especially those that excrete acid from the body such as renal tubules, lung, and gastric mucosa. This is termed *metastatic calcification.* As noted above, dying cells take on calcium. When this calcium remains as a deposit in the area of necrosis, it is known as *dystrophic calcification.* Most dystrophic calcification causes no problem in itself, but since calcium is radiopaque, it allows the radiologist to spot areas of disease. For example, foci of caseous necrosis in the lung caused by tuberculosis will remain evident to the radiologist for years or a breast cancer with focal calcification due to areas of necrosis may be discovered by mammography.

Hemosiderosis and *hemochromatosis* are terms applied to excessive iron accumulation in tissues; the former term implies iron accumulation in tissues; the latter term implies a more serious condition associated with tissue damage. Normally, the absorption of iron from the intestines is carefully regulated so that the body has enough for production of red blood cells but not too much. Excessive iron may be introduced into the body by blood transfusion or excessive absorption may occur due to genetic causes, dietary overload, or increased need produced by hemolytic anemias (anemias caused by excessive destruction of red blood cells). The excess iron in the form of ferritin combines with protein to form hemosiderin, a brown pigment that accumulates in cells, especially macrophages and hepatocytes. Hemochromatosis is usually due to a genetic defect in regulation of iron uptake and its damaging effects are most felt in the liver and pancreatic islets with resultant cirrhosis of the liver and diabetes mellitus. Periodic withdrawal of blood lowers body iron stores and reduces the progress of this disease.

Brown atrophy is an old term applied to the brown color of the heart and the liver that develops with aging from the accumulation of *lipofuscin* pigment in myocardial fibers and hepatocytes. This poorly defined pigment, composed of lipid, carbohydrate, and protein, is the residue of lysosomal digestion of cellular debris and has no clinical significance other than being a marker for aging or increased cellular damage.

CHART 4–1. EFFECTS OF INJURY

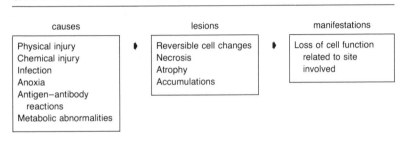

causes	lesions	manifestations

Physical injury
Chemical injury
Infection
Anoxia
Antigen–antibody
   reactions
Metabolic abnormalities
	Reversible cell changes
	Necrosis
	Atrophy
	Accumulations
		Loss of cell function
		   related to site
		   involved

INFLAMMATION

Inflammation is the protective response that the body mounts in response to injury. The term is appropriate in that an inflamed lesion, such as a burn, is red, hot, and painful like fire. As we have seen, necrosis reflects the destructive effects of injury to cells. Inflammation focuses on materials—fluid, chemicals, and cells— brought to the injured area and how these materials limit the injury and remove necrotic debris. The process of inflammation involves very complex chemical and, to a lesser extent, neural mechanisms that serve to turn the "fire fighters" on quickly and mobilize more reserves, but also to turn the process off so that the cellular and chemical fire fighters do not destroy any more normal tissue than is necessary to control the spread of injury.

The nature of the inflammatory response is stereotyped; the degree and duration vary depending on the cause and time course of the injury. The stereotyped response is described under acute inflammation and the variations produced by prolonged injury and certain agents are described under chronic inflammation. Although inflammation is described as a protective response, it also has its damaging effects. The potential for drugs to modify the inflammatory response has stimulated continued research to unravel its complex biochemical control mechanisms.

Acute Inflammation

Acute inflammation is usually described in terms of the vascular and cellular response. The vascular response consists of an increase in blood flow to the inflamed area and increased vascular permeability that allows leakage of water, electrolytes, and serum proteins into the tissue spaces. The cellular response refers to the movement of leukocytes, predominantly neutrophils and monocytes, from the blood into the tissue.

These events produce the *cardinal signs of inflammation:* redness, swelling, heat, pain, and loss of function. The increased blood flow in dilated vessels is called *hyperemia* and causes *redness.* The leakage of fluid into the tissue is called *edema* and causes *swelling.* If the lesion is on the body surface, the increased blood in the area causes *heat* because the temperature of the skin is normally less than that of the blood. *Pain* results from the pressure of the swelling and the action of kinins on nerve endings. *Loss of function* results from the attempt to protect the painful, swollen lesion.

The effects of inflammation will be to destroy or limit the spread of the causative agent and to clean up the debris in preparation for repair. In simple injuries, such as a burn, a cut, or a chemical injury where the chemical has been diluted away, the causative

agent is no longer a threat and the inflammatory reacti
proportional to the amount of tissue damage. Tissue dan
will incite a mild inflammatory reaction, enough to brii
cytes to digest and remove the debris from the dead (_ ana
increase lymph flow to carry away fluid from the lesion. Both neutrophils and macrophages engulf particulate matter, a process
called *phagocytosis.*

Phagocytes (neutrophils and macrophages) play a key role in
the inflammatory process. They move from their normal central
location in the bloodstream to the periphery as the vessel dilates
and the bloodstream slows—a process called *margination.* The leukocytes then stick to the endothelial cells (*adhesion*) due to complementary molecules on the leukocytes and endothelial cells that are
activated by various chemical mediators of the inflammatory process. Once chemically stuck to the endothelial cell, leukocytes
crawl between endothelial cell into the tissue (*emigration*). Neutrophils *migrate* faster from the vessel to the injured site, arriving
within minutes and accumulating over hours. Macrophages are
slower moving and peak later than neutrophils. Neutrophils may
die soon after arrival at the injured site to liberate their powerful
digestive enzymes or they may phagocytose and digest cellular
debris and foreign material before dying.

Neutrophils are particularly important in certain types of bacterial infections such as those caused by *Staphylococcus aureus,*
various streptococci (including pneumococcus), gonococcus, men-
ingococcus, coliform bacteria, anaerobic bacteria from the intestines, and other organisms. These organisms are particularly important because many are part of the normal flora of the skin,
mouth, respiratory tract, and intestines; they are ever ready to
cause infection whenever host defense mechanisms break down at
these sites and are responsible for a large number of infections.
Neutrophils engulf bacteria like they do with other particulate matter, but they also can recognize and move toward certain unknown
chemicals contained in bacteria by the process of chemotaxis.
Chemotaxis is the movement of white blood cells in response to a
chemical gradient; it may be positive or negative. Bacteria have
evolved mechanisms of resisting phagocytosis such as the thick
polysaccharide capsule of some strains of pneumococcus (*Streptococcus pneumoniae*); the host can counter by producing antibodies
that attach to the capsule of the pneumococcus and are easily recognized by phagocytes. Such phagocytosis-promoting antibodies
are called *opsonins.* Opsonins come into play with organisms that
have been previously encountered or when antibodies have been
artificially induced by immunization such as is sometimes done in
patients who are particularly susceptible to pneumococcal pneu-

monia. The brisk neutrophil reaction to these bacteria often results in the death of many neutrophils and much tissue breakdown to produce pus; for this reason the organisms are referred to as pyogenic (pus forming).

Macrophages arrive later and are hardier than neutrophils; they carry the major load in cleaning up the inflammatory debris including the dead neutrophils. The relative numbers of neutrophils and macrophages will depend on the amount and nature of the dead tissue and whether highly chemotactic foreign substances, such as pyogenic bacteria, are present. A staphylococcal infection will have lots of neutrophils; injured adipose tissue from a blow will have mostly macrophages gobbling up the spilled lipid. Macrophages also predominate in reactions to large inert foreign particles such as talc or suture material. They surround the foreign material and often form multinucleated giant cells and remain for a long time.

The role of noncellular elements in the inflammatory focus is more difficult to visualize. The increased fluidity of the lesion facilitates movement of cells and chemicals and also promotes increased lymph flow to carry fluid debris away from the area. It may also serve to dilute offending agents such as toxins and antigens. Fibrinogen is a soluble blood protein that may leak into the inflamed site and be converted to a stringy polymer, fibrin. This process involves several enzymes and is activated by exposure to tissue. The formation of fibrin serves as a barrier to the spread of injury; for example, the scab formed over a scrape of the skin is composed largely of fibrin and serves to keep bacteria out and fluid in. Because fibrinogen is a very large protein, it leaks into tissue only when the increase in vascular permeability is severe; even then some of the fibrin that is formed is lysed by the enzyme fibrinolysin. These control mechanisms prevent the formation of fibrin in mild injuries when it is not needed.

Patients who lack the ability to mount the acute inflammatory reaction succumb to infections that are easily warded off by a normal person, so the reaction is obviously very important. The acute inflammatory reaction can also be very damaging, so its control is important. It must be activated quickly and turned off when no longer needed. The control mechanisms are complex. We will outline them here mentioning the major elements involved. Much more detail is given in standard pathology textbooks.

The inflammatory reaction is initiated by local factors in the injured tissue. Stimulation of small nerve endings causes arteriolar dilatation, but this reaction is not an essential or prominent event. The release of histamine from mast cells is the most important initial event. Mast cells are scattered throughout the connective tissue of the body and release histamine when injured. Histamine diffuses from the injured site to cause vasodilatation and increased permeability of small venules. The venules leak plasma proteins, particularly albumen, drawing water into the tissue by osmotic pressure. Mast cells become depleted and the released histamine is diluted and inactivated, so the inflammatory reaction cannot be sustained by histamine. Parenthetically, it should be noted that histamine can be released from mast cells by two types of immunologic reactions: atopic allergy and immune complex reactions (see Chapter 27 for details). If the injury involves pyogenic bacteria, another important initiating factor is the chemotactic factor released from these organisms.

Once initiated, the inflammatory reaction can be quick and greatly amplified by chemicals circulating in the blood. Three chemical systems are involved: the kinin system, the complement system, and the coagulation system. In each case, an inactive protein precursor is activated by a series of enzymatic steps, with products of the reaction itself acting as catalysts to further speed the reaction. To counter the dangers of accidental triggering of these reactions, there are inhibitors, enzymes that destroy products of the reactions, and the dilutional effect of the bloodstream.

When kinins leak through the venule made permeable by histamine, they are activated to become bradykinin. Bradykinin itself causes increased vascular permeability and is thought to be the major factor in sustaining the flow of fluid and chemicals to the inflammatory site by a self-perpetuating reaction. Bradykinin also acts on nerve endings to cause pain. At some point, bradykinin will be deactivated faster than it forms and the vascular response will gradually subside.

When fibrinogen leaks through the permeable vessels along with other blood coagulation factors, the coagulation (clotting) system is activated to polymerize fibrinogen to fibrin. The role of fibrin has already been mentioned. Both the kinin and coagulation systems are initiated by a tissue factor known as the Hageman factor or Factor XII.

The splitting of complement into several active factors is initiated by complexes of antigen and antibody (see Chapter 27 for details) and by an alternate pathway by bacterial endotoxins and some normal tissue proteins. Complement fragment C5a is an important mediator of chemotaxis and fragments C3a and C5a have components known as *anaphylatoxins* that cause increased vascular permeability by stimulating release of histamine from mast cells and blood platelets.

The exact role of each of these reactions in a particular inflammatory lesion is difficult to evaluate and we have not discussed many of the chemical intermediates that have varying degrees of inflammatory and catalytic activity. It is more important to appreciate that the blood plays a sophisticated role in the regulation of the inflammatory reaction.

In addition to the chemicals from the plasma, neutrophils bring products that help amplify and sustain the reaction. Neutrophil enzymes can activate the complement and kinin systems, but perhaps more importantly they provide essential products for synthesis of prostaglandins.

Prostaglandins and leukotrienes are metabolites of arachidonic acid that are produced locally by cells and act as short-range hormones. The chemistry of these compounds is complex and the number of compounds produced and their effects are confusing. Suffice it to say that prostaglandins and leukotrienes are produced in response to inflammation and act locally to sustain the reaction. They are involved in producing vasoconstriction, vasodilatation, increased vascular permeability, chemotaxis, and fever. The anti-inflammatory action of aspirin and indomethacin, at least in part, appears to be due to inhibition of prostaglandin synthesis.

We have already noted some important variations in the inflammatory process. Reactions with lots of neutrophils cause tissue destruction but are important in containing pyogenic bacteria and antigen–antibody complexes. Macrophages will be prominent when there is lots of dead tissue to remove or foreign substances to surround or engulf. Edema will predominate when lots of histamine is released as in atopic allergy and immune complex reactions. Fibrin will form a protective barrier on injured surfaces. More variations will be described under chronic inflammation.

CHART 4–2. ACUTE INFLAMMATION

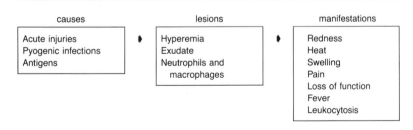

causes	lesions	manifestations
Acute injuries Pyogenic infections Antigens	Hyperemia Exudate Neutrophils and macrophages	Redness Heat Swelling Pain Loss of function Fever Leukocytosis

Chronic Inflammation

Chronic means persistent for a long time. In that sense, chronic inflammation may result from acute inflammation that persists because the cause is not completely eliminated or it may be associated with a cause that never was acute but is continuing at a low level for a long time.

The term *chronic inflammation* is also used as a label for the histologic picture typically associated with prolonged inflammation. As will be noted later, some chronic inflammations have a more specific appearance (e.g., granulomatous inflammation) and some clinically acute inflammations mimic chronic inflammation histologically. Let us first describe the typical appearance of chronic inflammation and then deal with the variations and their pathogenesis.

Since the injury in chronic inflammation is usually low grade, edema and hyperemia are less pronounced than in acute inflammation and few or no neutrophils are present. The area is infiltrated predominantly with lymphocytes, plasma cells, and less conspicuously with macrophages (Fig. 4–6). Plasma cells are often prominent and easily recognized. They are derived from B lymphocytes in the tissue and their primary function is to produce antibodies. Presumably the antibodies produced by the plasma cells attach to foreign material in the area and allow the neutrophils and macrophages to phagocytose this material. The lymphocytes, which are mostly nucleus with a small rim of cytoplasm, play a much larger role than their innocuous appearance suggests. We know that different types of lymphocytes can perform various functions. They can recognize foreign materials, kill host cells in the area of foreign antigens to isolate the foreign substance, transform into plasma cells to produce antibody, and direct the traffic of other inflammatory cells, especially macrophages. However, histologically, we cannot tell which lymphocytes are doing what and why. Macrophages may play the same role as they do in acute inflammation (phagocytosis and digestion of debris), but they may also become directly cytotoxic to host cells under certain conditions.

Another hallmark of chronic inflammation, regardless of type, is the laying down of new fibrous tissue in the area. Whenever there is tissue injury in the presence of chronic inflammation, there is a fibrous tissue proliferation that tends to wall-off the injured area and provide strength to the defective area. By judging the extent and the age of the new fibrous tissue, one can estimate the chronicity of the inflammation. The process of fibrous repair is discussed in the section on Repair.

Grossly, chronic inflammation has the same features as acute inflammation—edema, redness, heat, pain, and loss of function—but these features are much less pronounced and more variable in their intensity. Contraction of the developing fibrous tissue may distort the lesion and surrounding tissue and give the lesion a variegated, firm, glistening, gray appearance.

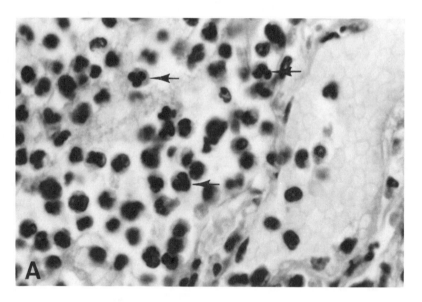

Figure 4–6. A. Microscopic appearance of acute inflammation characterized by presence of neutrophils (*arrows*).

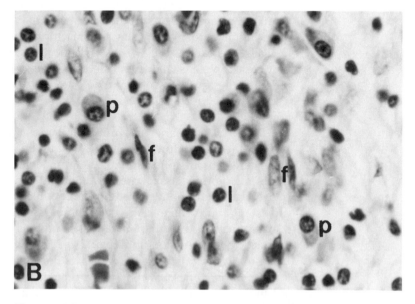

Figure 4–6. B. Microscopic appearance of chronic inflammation characterized by the presence of lymphocytes (l), plasma cells (p), and fibroblasts (f).

The pathogenesis of chronic inflammation involves persistence of the causative agent and a host reaction that is predominantly immunologic in nature. The immune response, which may be of one or several types, produces a more varied picture than the acute inflammatory response. A discussion of several patterns of chronic inflammatory responses follows.

Chronic inflammation due to persistent acute inflammation is usually caused by pyogenic bacteria. Foreign bodies and necrotic tissue provide a haven for these organisms to proliferate and cause continuing foci of acute reaction. The superimposed chronic inflammation is characterized by many plasma cells, which are producing antibodies to help fight the festering bacterial infection, and by fibrous tissue, which is attempting to wall-off the area. The reaction to a splinter is typical of this situation.

If the splinter (foreign body) is removed, the acute inflammatory reaction will eliminate the bacteria and digest the small amount of necrotic tissue. But, if the splinter is not removed, bacteria will often lurk in the foreign body and continue the inflammatory process into a chronic stage. Large amounts of necrotic tissue produce an even greater effect. Necrotic tissue is an ideal culture media for bacteria and is not accessible to the body's vascular transport system that delivers inflammatory cells and antibiotics. Consequently, the offending bacteria will gain the upper hand until the necrotic tissue is removed, either surgically or more gradually by the inflammatory reaction. In these types of chronic inflammation, there will usually be a focus of acute inflammation near the source of infection surrounded by a zone of chronic inflammation. Such lesions are sometimes called subacute inflammation or combined acute and chronic inflammation. Therapy should be directed at the removal of the cause; antibiotics are often of limited help because they cannot reach the offending organisms.

Another pattern of chronic inflammation is the persistence of low-grade injury without an initial acute phase. Agents include microorganisms, antigens, and, less frequently, chemicals. The variable patterns produced by these diverse agents and the several immunologic mechanisms involved tend to produce specific diseases rather than the stereotyped, nonspecific reaction of acute and persistent acute inflammation. Syphilis, hay fever, contact dermatitis, and viral infections will serve as examples to illustrate the diversity of chronic inflammation. Many others can be found in Chapters 26 to 29. The common denominator of these chronic inflammatory reactions is that they employ T lymphocytes to attack the offending agent (cellular immunity) and/or B lymphocytes to produce antibody to it (humoral immunity). Consequently, lymphocytes and plasma cells predominate in the lesions.

The spirochete of syphilis (a bacterium) enters the skin and proliferates with little or no acute inflammation. Ten to 90 days later, the initial lesion at the site of entry develops as a chronic inflammatory lesion. It is slightly red, swollen, and firm with little or no pain and soon ulcerates. Histologically, there are many plasma cells, macrophages, fibroblasts, and new capillaries—hallmarks of chronic inflammation. Soon after the lesion develops, the body is loaded with antibody to the organism and the lesion gradually resolves, presumably because the antibodies allow the macrophages to remove the organisms. This is an example of an organism that produces little tissue damage itself and is not chemotactic to neutrophils. Damage to tissue is associated with development of the immune response, but the mechanism of injury is poorly understood. Killing the organisms with penicillin stops the process; failure to do so may be followed much later by serious chronic inflammatory lesions of the aorta or central nervous system.

In hay fever, the patient has a predisposed sensitivity to airborne allergens such as ragweed pollen, molds, or house dust. Immunoglobulin E (IgE) antibodies to the offending agent attach themselves to mast cells, and, when they encounter the antigen, cause the mast cells to release the contents of their cytoplasmic granules, including histamine and other substances, which in turn increase vascular permeability and secretion of mucus. Although the mechanism of this reaction is similar to the one occurring in acute inflammation, the prolonged exposure to the allergen causes it to be chronic. Antibody production is associated with an increase in plasma cells in the nasal mucosa and an infiltrate of eosinophils is also typical of inflammations involving IgE antibodies. Eosinophils are thought to degrade substances produced by certain types of allergic reactions. Therapy can be directed toward avoidance of exposure to the antigen, toward desensitizing the immune response, or toward suppressing the inflammatory reaction with antihistamines and vasoconstrictor drugs.

Contact dermatitis, another type of allergy in which the individual's hypersensitivity to an environmental antigen is involved, may be acute or chronic, depending on the dose of antigen and duration of exposure. Prolonged, low-grade exposure might occur with sensitivity to a cosmetic or metal in a watch band, thus producing a chronic, slightly edematous, mildly red lesion that may itch. It is characterized histologically by a lymphocytic infiltrate, spotty necrosis of epidermal cells, and a mild, prolonged chronic inflammatory reaction to the necrotic cells. In this situation, the antigen entering the epidermis is recognized by a small population of T lymphocytes. Upon encountering the antigen, these sensitized T lymphocytes recruit cytotoxic lymphocytes or macrophages that kill the host cells containing the antigen, either directly or by elaboration of toxic proteins called lymphotoxins. By sacrificing the host cells, the spread of antigen is limited. In contact dermatitis, the immune system of the susceptible person is fooled into recognizing an antigen that is harmless to nonallergic individuals.

The delayed hypersensitivity reaction is an important mechanism involved in the production of many chronic diseases. It is associated with a variety of allergies as illustrated by contact dermatitis, in protection against many microorganisms, in several types of autoimmune diseases, and in rejection of tissues transplanted from one individual to another. All of these situations are associated with lymphocytic infiltration and killing of host cells, producing the histologic picture of chronic inflammation. Lymphocytes usually predominate over plasma cells.

Viral infections often elicit cellular immune reactions of the delayed hypersensitivity type. These can develop abruptly, producing a clinically acute illness with a chronic inflammatory reaction histologically. For example, in many viral infections the organisms grow intracellularly in selected cells in the body. During the incubation period (10 to 14 days), some of the antigen from the virus is complexed with the surface of the infected cell. T lymphocytes recognize this antigen complex, become sensitized to it, and then recruit cytotoxic lymphocytes and macrophages to destroy the infected cells, thus eliminating the necessary environment for the virus. The cost of this protective reaction is the necrosis of host cells with release of tissue products that produce fever and other systemic manifestations. The onset of illness occurs suddenly as sensitized T lymphocytes reach the infected cells; recovery is gradual as injured host cells recover or are replaced. It should be noted that delayed hypersensitivity is not the only mechanism in viral-induced injury, and delayed hypersensitivity is involved in a variety of other types of infections.

Both humoral and cellular immune reactions can be accelerated by previous immunization. Antibodies are important in preventing antigens from entering the body and in helping neutrophils and macrophages destroy them immediately after entry. When sensitized lymphocytes are already present, they can destroy an infected site in a day or two; the organisms do not have time to proliferate and spread. Smallpox has been eradicated from the world by making sure that all exposed persons were immunized. At the opposite extreme, suppression of the immune response is desirable in autoimmune diseases such as rheumatoid arthritis and with organ transplants where the effects are undesirable.

The above examples are only a sample of the diversity of chronic inflammatory diseases that will be expanded upon throughout the book. A knowledge of the agent and its distribution and dose, along with the type of immunologic reaction induced by the agent, serve to more easily understand the disease produced by the agent and to direct therapy and future research. The pathogenesis of many chronic inflammatory diseases is still incompletely understood. As you progress through the book, refer to Chapter 25 for more information on immunologic mechanisms.

CHART 4–3. CHRONIC INFLAMMATION

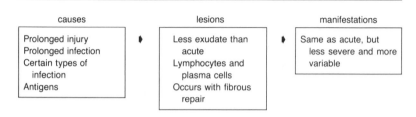

causes	lesions	manifestations

Prolonged injury
Prolonged infection
Certain types of
  infection
Antigens
	Less exudate than
	  acute
	Lymphocytes and
	  plasma cells
	Occurs with fibrous
	  repair
		Same as acute, but
		less severe and more
		variable

Granulomatous Inflammation

Granulomatous inflammation is a specific type of chronic inflammation characterized by focal collections of closely packed, plump macrophages (Fig. 4–7). Granulomatous inflammation occurs in response to certain indigestible organisms and other foreign materials and often involves an element of cell-mediated immunity to the foreign material. In granulomatous inflammation, T lymphocytes become sensitized to the offending agent and recruit large numbers of macrophages that engulf the antigenic agent. These macrophages are called *epithelioid cells* because their abundant cytoplasm and close approximation to each other in aggregates make them resemble epithelial cells.

The macrophages in a granuloma often form a few giant cells, which makes them easily recognizable. Granulomas often become large enough to produce grossly visible pale, yellow nodules and in tuberculosis and fungal infections may become quite large and undergo caseous necrosis. Small granulomas heal by fibrosis. Large caseous granulomas are walled off by a fibrous rim and calcify.

The classic cause of granulomatous inflammation is tuberculosis. Fortunately, it has been brought under control in the United States, but it is still a very significant problem in many areas of the world. See Chapter 12 for more details on tuberculosis. Other microorganisms, particularly the fungi causing histoplasmosis, coccidioidomycosis, blastomycosis, and cryptococcosis, produce granulomatous disease that is very similar to tuberculosis.

Another relatively common granulomatous disease is sar-

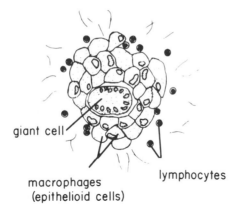

giant cell

macrophages
(epithelioid cells)

lymphocytes

Figure 4–7. A granuloma.

coidosis, a disease of unknown cause characterized by widespread noncaseous granulomas. It produces a mild to moderately debilitating illness often involving lungs and lymph nodes as well as many other organs. It is more common in young adult women, in blacks, and in southern United States. Most patients are asymptomatic or have a mild illness that disappears in a few years. A few die from progressive granulomatous involvement of the lungs or complications of the lesions.

Foreign body granulomas are less stereotyped than the "tuberculoid" or "sarcoid" granulomas described above. They result from indigestible "foreign" material being surrounded by epithelioid cells and giant cells. Common causes are suture material, splinters, talc, mineral oil inhaled into the lung, crystalline cholesterol deposits derived from blood or bile, and large microorganisms such as helminths. The "foreign" body is usually quite evident within the granuloma.

The term granuloma has been used in different ways. The best use is for the characteristic histologic appearance, which suggests that it would be profitable to look for a specific cause. Round lesions seen on chest x-rays are often termed granulomas because most of these are due to old healed tuberculosis or fungal infection.

CHART 4–4. GRANULOMATOUS INFLAMMATION

causes	lesions	manifestations
Tuberculosis Fungal infections Foreign bodies Sarcoidosis	Focal collections of plump macrophages and giant cells Often multiple foci Caseous necrosis with some causes	Nonspecific, may be none Positive tests for causative organisms Tissue destruction which may affect organ function

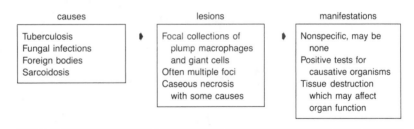

Exudates and Transudates

Exudates should be distinguished from transudates. A *transudate* is a collection of fluid in tissue or in a body space due to increased hydrostatic or decreased osmotic pressure in the vascular system without loss of protein into the tissue. Thus transudates are watery with low protein content. *Exudates* are the result of increased osmotic pressure in the tissue due to high protein content and are caused by inflammation or obstruction of lymphatic flow. Thus a swelling with cloudy or protein rich fluid is caused by inflammation or lymphatic obstruction, whereas a swelling with thin watery fluid might be caused by heart failure with its increased venous pressure or by depleted serum proteins. Exudates tend to be more localized than transudates because most inflammations are localized whereas the effects of increased hydrostatic pressure or depleted serum proteins are usually generalized.

Serous exudate contains fluid as well as small amounts of protein and often implies a lesser degree of damage. For example, the fluid content of blisters that follow skin burns is serous exudate.

A *fibrinous exudate* is an exudate composed of large amounts of fibrinogen from the blood that is polymerized to form fibrin. For example, in bacterial pneumonia fibrinous exudate forms a mesh that helps trap the bacteria; on a skin wound, dried fibrinous exudate forms a scab.

Purulent exudate (also called *pus*) is an exudate that is loaded with live and dead leukocytes, mostly neutrophils. An inflammatory reaction with much purulent exudate is called *suppurative inflammation*. A localized collection of pus is an *abscess*. When pus fills a body cavity such as the pleural cavity, the term *empyema* is used.

Gross Inflammatory Lesions

Lesions with relatively specific gross appearance include abscess, cellulitis, and ulcer. An *abscess* is a localized, usually spherical lesion containing liquified dead tissue and neutrophils (purulent exudate) (Fig. 4–8). Pyogenic bacteria are the most typical cause of abscesses because they liberate chemotactic factors and proliferate to produce an exuberant acute inflammatory reaction. The host is induced to destroy and liquify its own tissue to limit the spread of the offending agent. Abscesses are typically caused by bacteria of the skin (staphylococci), oral cavity (streptococci including pneumococci, anaerobic bacteria), and lower intestinal tract (coliform bacteria and anaerobic bacteria). The bacteria, which enter the tissue and proliferate due to obstruction of ducts and glands, tissue injury, or foreign bodies, are opportunists. Examples of abscesses include the *boil* or *furuncle* usually due to an obstructed skin appendage or foreign body, a *paronychia* due to purulent infection around

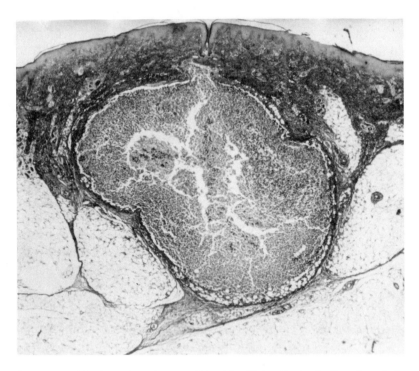

Figure 4–8. Histology of an abscess of the skin with localized collection of purulent exudate.

a fingernail, and pimples of acne due to greasy secretions with obstruction and infection of sebaceous glands associated with onset of sexual maturity. Abscesses also occur in areas where there is change from one tissue type to another such as around the nares, teeth, and anus. The combination of foreign material, necrotic tissue, and bacteria trapped in wounds or in operative incisions is particularly likely to produce an abscess. In more serious breakdowns of the body defense mechanisms, such as perforation of the intestines, large areas of necrosis, and infarcts in organs open to the bacterial environment, such as lung, intestine, or legs, anaerobic organisms of intestinal, oral, or soil origin often cause abscesses.

The typical small abscess is red, hot, swollen, and quite painful. When the abscess reaches a "head," the center is liquified and fluctuant and the edge is beginning to wall off. Puncture of the abscess causes an outpouring of pus, relief of pain, and more rapid healing. If punctured before this stage, the abscess may be spread. Larger abscesses are more irregular and may spread in tissue spaces and cause extensive damage. If the host wins, an abscess will be walled off and replaced by fibrous tissue after drainage or resolution (resorption) of the purulent exudate. If uncontained, an abscess may enlarge, spread, and kill the host. For example, acute appendicitis with abscess formation can lead to death; early appendectomy prevents such an outcome.

Cellulitis refers to a spreading acute inflammatory process. This type of inflammation is commonly seen with streptococcal bacterial infections and is due to the body's inability to confine the organism. Cellulitis is seen in the skin and subcutaneous tissue and is characterized by nonlocalized edema and redness.

An *ulcer* is a locally excavated area of skin or mucous membrane secondary to an injury and the subsequent inflammation. Ulcers are commonly seen in the stomach and duodenum secondary to local injury by acid from the stomach. Bedsores, resulting from pressure atrophy, are another example of ulcers.

REPAIR

Regeneration

The body's two basic methods of repair following tissue destruction are *regeneration* and *fibrous connective tissue repair* (scarring or fibrosis). Regeneration is replacement of the destroyed tissue by cells similar to those previously present, i.e., the parenchymal cells of the organ are regenerated. For example, the epidermal surface of a cut is replaced by epidermis, fractured bone is united by bone, or scattered dead liver cells are replaced by new liver cells. In fibrous connective tissue repair, tissue previously present is replaced by fibrous tissue (scar). For example, the dermal edges of a cut are united by scar, a bone fracture that is not properly united is healed by scar tissue, extensively damaged liver may be replaced by fibrous tissue. Many tissue injuries heal in part by regeneration and in part by fibrosis.

Regeneration is the most desirable form of repair, because normal function is often restored. As a prerequisite to regeneration, cells next to those that have died must be able to multiply. For example, neurons and cardiac muscle fibers cannot undergo cell division in adults; therefore, these cells cannot regenerate. Tissues that are continuously replacing their cells under normal circumstances also have great capacity for regeneration. Examples of such tissues include the epidermis, the mucosal lining of the intestinal tract, and bone marrow. Bone marrow can replace itself when only a few cells survive an injury. Epidermis and intestinal mucosa can repair defects up to several centimeters in diameter through the process of regeneration (Fig. 4–9). Most of the tissues of the body

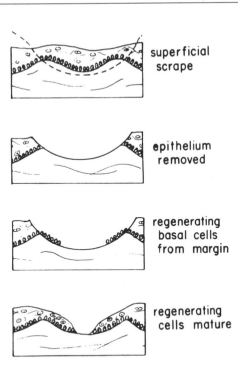

superficial
scrape

epithelium
removed

regenerating
basal cells
from margin

regenerating
cells mature

normal
structure
restored

Figure 4–9. Complete regeneration of lost surface epithelium.

normally undergo cell replacement at a slow rate and are intermediate in their ability to regenerate. Regeneration can usually occur in parenchymal organs if the architectural framework is not destroyed. Tissues particularly noted for their regenerative capacity include liver, renal tubules, and bone. More complex structures, such as lung and renal glomeruli, do not regenerate.

Regeneration is particularly important when there has been widespread damage to a vital organ. When there is generalized hypoxia, renal tubular cells, hepatocytes, and neurons are most susceptible to necrosis because of their high metabolic requirements. Typical causes of generalized hypoxia include shock, cardiac arrest followed by successful resuscitation, and carbon monoxide poisoning. Renal and hepatic function is likely to be restored because the few surviving cells can regenerate and restore normal structure. Brain function cannot be restored when neurons die because they cannot regenerate. Bone marrow is an organ that may be wiped out by drug therapy for cancer and other causes; it has tremendous capacity to regenerate from only a few cells and, in some cases, may be restored by cells from another individual by bone marrow transplantation.

Fibrous Connective Tissue Repair (Scarring or Fibrosis)

Fibrosis can occur in any tissue and produces the same result regardless of site, namely the formation of a dense, tough mass of collagen called a *scar.* Unlike regeneration, replacement by fibrous tissue does not restore the original function. The purpose of fibrosis is to provide a strong bridge between normal tissue and the damaged area.

The process of fibrous repair is also called *organization* and consists of a granulation tissue stage and a scar formation stage. *Granulation tissue* consists of capillaries and fibroblasts. The process, which begins when the injury has been stabilized, is initiated by the ingrowth of new capillaries and fibroblasts into the injured area. The capillaries bring blood to provide the nutrition for the repair process. Capillaries also carry away liquid remains of dead tissue and particulate material removed by macrophages. The removal process is called *resolution.* The fibroblasts proliferate rapidly

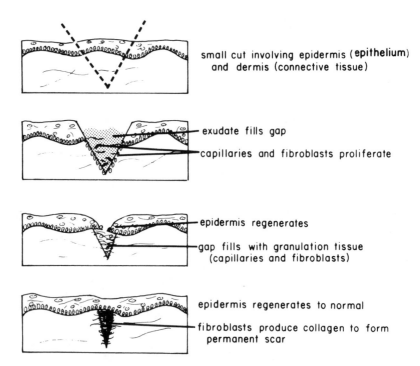

small cut involving epidermis (**epithelium**) and dermis (connective tissue)

exudate fills gap

capillaries and fibroblasts proliferate

epidermis regenerates

gap fills with granulation tissue (capillaries and fibroblasts)

epidermis regenerates to normal

fibroblasts produce collagen to form permanent scar

Figure 4–10. Fibrous connective tissue repair and epithelial regeneration in a skin cut.

and then initiate the stage of scar formation by laying down collagen. Initially, there are small amounts of loose collagen within the mass of capillaries and fibroblasts. With time, more collagen is formed and the number of capillaries and fibroblasts decreases. The final stage, which takes weeks to months, involves shrinking and condensation of the fibrous scar (Fig. 4–10).

Wound Repair

The process of repair of wounds is artificially separated into repair by primary union and secondary union, depending on whether the wound edges are placed together or left separated. The best example of repair by primary union is that which follows a clean surgical incision of the skin in which there is minimal tissue damage and the edges of the wound are closely approximated by tape or sutures. In this example, the narrow space between the two wound edges fills with a small amount of serum, which quickly dries and clots forming a scab. Within 1 to 2 days, the narrow zone of acute inflammation at the wound edges has lessened and new capillaries begin to bridge the gap between the wound edges. By this time, the epithelium has already grown across the surface of this gap. Within a few more days, fibroblasts grow across the subepithelial portions of the wound gap and begin to deposit collagen, the collagen eventually contracts, pulling the wound edges together and giving them strength. Although this incision may appear well healed by about 2 weeks, it may take a month or more for the strength of the scar tissue to equal that of the original tissue.

Repair by secondary union utilizes the same basic process as primary union, except that there is greater injury with consequent greater tissue damage and more inflammation to resolve (Fig. 4–11). To fill the void left by tissue damage, there is a tremendous proliferation of capillaries and fibroblasts, which actually start growing after the injury is just a few days old and acute inflammation may still be intense. After a week or more, the wound will be filled with this tissue, composed largely of capillaries, fibroblasts, and variable numbers of residual acute inflammatory cells (neutrophils) with some chronic inflammatory cells (lymphocytes). This tissue is friable and red with oozing blood and is called *granulation tissue*. It is what is seen when one picks a scab off a skin wound. Granulation tissue eventually is replaced as more and more collagen is deposited by fibroblasts. Fibroblasts and collagen have inherent contractile properties, which aid in shrinking a wound and drawing the edges together. It may take a long time for a wound that heals by secondary union to achieve strength approximating that of the normal tissue. If a skin wound is very large, the epithelium may never completely bridge the wound, and skin may need to be

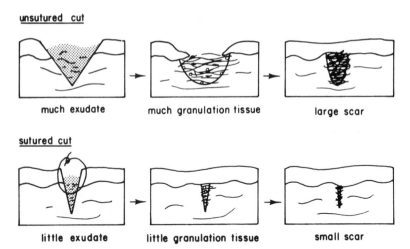

unsutured cut

much exudate   much granulation tissue   large scar

sutured cut

little exudate   little granulation tissue   small scar

Figure 4–11. Repair of an unsutured and sutured skin cut leading to secondary and primary union, respectively.

grafted to the wound site from another area of the body. Transplanted skin usually grows quite readily in such a situation, because the underlying granulation tissue is so rich in capillaries. One of the greatest impediments to healing and repair of a wound is the amount of dead tissue and foreign material (dirt, bacteria, shrapnel, etc.) present. It might take the body's inflammatory cells many months to phagocytize a large amount of dead tissue and foreign material, and the presence of bacteria in a wound may produce necrotic tissue and inflammatory cells as fast as they are removed. For this reason, the medical care of a large wound should always include thorough cleaning and *debridement* (removal of foreign material and necrotic tissue).

Inflammation and subsequent repair of tissue is a very dynamic process that is influenced by numerous modifying factors. The following factors that may detract from the body's ability to most effectively deal with an injury:

1. Virulence of the infective organisms; for example, staphylococcus is more capable of destroying tissue than alpha streptococcus.
2. Advancing age; elderly people heal more slowly than younger people for various reasons.
3. Poor nutrition; protein and vitamin C are needed to produce collagen.
4. Diabetes; small blood vessels are abnormal in diabetics and consequently they do not deliver materials to the tissues optimally.

5. Steroid therapy; steroids inhibit the inflammatory response by preventing vascular permeability, hindering cellular digestion of debris, and blocking antigen–antibody reactions. Steroids can be very useful in those situations where inhibition of inflammation is desired; for example, steroids are used to slow the destructive inflammation of rheumatoid arthritis.

CHART 4–5. REPAIR

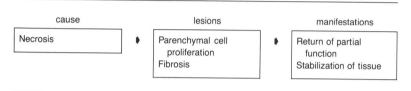

cause	lesions	manifestations
Necrosis	Parenchymal cell proliferation Fibrosis	Return of partial function Stabilization of tissue

REVIEW QUESTIONS

1. How does the duration of an injury affect the interrelationship between injury, inflammation, and repair?
2. What are the major causes, lesions, and manifestations of necrosis, inflammation, and repair?
3. What functional and histologic changes indicate cell death?
4. How do the various types of necrosis, atrophy, and accumulations differ in appearance and cause?
5. What distinguishes metastatic from dystrophic calcification?
6. How do acute inflammation, chronic inflammation, and granulomatous inflammation differ in cause, histologic appearance, and manifestations?
7. What is the sequence of events in a typical acute inflammatory reaction and how does this relate to the local and systemic signs and laboratory findings?
8. How do chemical mediators control the acute inflammatory response?
9. How do serous, fibrinous, and purulent exudates differ in composition and gross appearance?
10. How do abscesses, cellulitis, and ulcers differ in gross appearance and location within the body?
11. What is the usual cause of an abscess? What type of inflammatory cell is most characteristic of an abscess?
12. What determines whether repair will be by regeneration or fibrous connective tissue repair? What is the sequence of each and how do the outcomes differ?

13. What is the difference in sequence and outcomes of repair of an open versus a closed (sutured) wound?
14. What conditions inhibit wound healing?
15. What do the following terms mean?
    Pyknosis
    Karyorrhexis
    Karyolysis
    Hypoxia
    Anoxia
    Ischemia
    Infarct
    Thrombus
    Embolus
    Allergy
    Autoimmune reaction
    Gangrene
    Fatty metamorphosis
    Amyloid
    Hyalin
    Hyperemia
    Edema
    Chemotaxis
    Phagocytosis
    Anaphylatoxins
    Transudate
    Exudate
    Pus
    Suppuration
    Pyogenic
    Empyema
    Resolution
    Granulation tissue
    Organization
    Debridement

Hyperplasias and Neoplasms

REVIEW OF STRUCTURE AND FUNCTION

The cells of the body are derived from one cell, the fertilized ovum. During embryonic development successive cell divisions lead to more and more specialized cell types and the less differentiated embryonic cells disappear. The cells of the developed organism can be divided into germ cells (normally confined to the gonads) and somatic cells. Somatic cells are classified under four major categories: epithelial cells, connective tissue cells, muscle cells, and nervous tissue cells.

Epithelial cells are generally those that arise from the embryonic ectoderm and endoderm to form the lining of body spaces and surfaces and the various glands. Surface lining cells are of squamous, transitional, or columnar types. Stratified squamous epithelium forms a tough protective barrier, often with a layer of keratin on the surface. Stratified squamous epithelium lines the skin, mouth, pharynx, larynx, esophagus, and anus. Transitional epithelium is also multilayered but lacks keratin. Transitional epithelium is confined to the urinary tract including renal pelvis, ureter, bladder, and urethra. Columnar epithelium is usually composed of one layer of tall cells, which often are mucus secreting. The surface columnar epithelium is often in continuity with underlying glands; its pathologic reactions are similar to those of glandular epithelium, hence, it is often referred to as glandular epithelium. Columnar epithelium lines the nose, trachea, bronchi, stomach, small intestine, colon, and many of the ducts leading to glands such as the bile ducts and breast ducts.

Epithelial cells comprising the various epithelial organs may be arranged as glands (acini), tubules, or cords. Glandular organs include breast, salivary glands, thyroid, and pancreas. The kidney is an example of an organ composed predominantly of tubules. The liver, adrenal, and pituitary are arranged in cords or sheets, with blood sinusoids between the sheets of cells.

Connective tissue cells, which are mostly derived from mesoderm, are recognized by their lack of close approximation with other cells and by the substances they produce. Fibroblasts are associated with collagen, chondrocytes with cartilage, osteocytes with bone, and endothelial cells with blood vessels. White blood cells are recognized by their round appearance and distinctive nuclei.

Muscle cells are also derived from mesoderm but resemble epithelial cells in their close approximation to each other. They differ from epithelium by their elongated fiber-like structure and abundant contractile cytoplasm.

Nervous tissue cells are derived from ectoderm and include neurons and their supporting cells. Neurons have very long processes (axons), which carry electrical impulses. The supporting cells in the brain and spinal cord are glial cells (astrocytes and oligodendroglia). Supporting cells in peripheral nerves are Schwann cells.

The tendency of cells to undergo hyperplasia and neoplasia is roughly related to their involvement in physiologic replacement. Surface epithelial cells undergo continuous replacement, and this process of replacement can be accelerated by mild injury. Glandular epithelial cells are more stable but can proliferate following injury. Connective tissue cells also have great capacity to proliferate. Blood cells are replaced continuously. Fixed connective tissue cells are replaced more slowly except when stimulated by injury. Heart muscle cells cannot be replaced, and skeletal muscle cells have very limited replacement capacity. Smooth muscle cells, especially those in small blood vessels, can proliferate. Of the nervous tissue cells, only the neurons and oligodendroglia are incapable of replacing themselves following injury.

The proliferative capacity of cells relates to the process of differentiation. Differentiation is the process of maturation from a nonspecific cell type to a specialized cell. For example, the stem cell in the bone marrow is a poorly differentiated cell whose main function is to divide and produce daughter cells. The daughter cell passes through several intermediate stages of differentiation until it becomes a mature, differentiated white or red blood cell. The stages of differentiation are also well illustrated in the epidermis— the basal cells are poorly differentiated and concerned primarily with cell division; the surface cells are differentiated with formation of keratin to protect the surface.

BASIC DEFINITIONS

Hyperplasias and neoplasms are both characterized by proliferation of cells that increase tissue mass. They differ on the basis of cause and growth potential.

Hyperplasias are exaggerated responses to various stimuli or presumed stimuli (a nonexaggerated response to injury is called

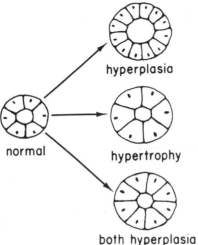

hyperplasia

normal

hypertrophy

both hyperplasia and hypertrophy

Figure 5–1. Hyperplasia and hypertrophy of a gland.

repair). Hyperplasia may recede if the stimulus is removed, provided permanent structural changes have not occurred. Hyperplasia typically involves tissues that undergo physiologic replacement but may involve stable tissues.

Neoplasms are growths that proliferate with varying degrees of independence from normal cellular control mechanisms. Neoplasms are presumed to arise by mutation or altered genetic control, they behave as if they were an independent parasitic organism struggling to overwhelm the host. Neoplasms may arise from embryonic cells, germinal cells, or mature somatic cells. Like hyperplasias, the majority of neoplasms arise from cells that undergo physiologic replacement. Therefore, hyperplasias and neoplasias cannot be distinguished on the basis of cell type alone.

Hyperplasias may be caused by a wide variety of stimuli, such as a remote response to inflammation (e.g., lymph node hyperplasia, bone marrow hyperplasia), hormone excess or hormone deficiency (e.g., thyroid goiter), chronic irritation (e.g., skin callus), or unknown factors (e.g., hyperplasia of the prostate gland in elderly men). Hypertrophy should be distinguished from hyperplasia; both frequently occur together. *Hypertrophy* refers to an increase in cell size, whereas *hyperplasia* refers to increase in cell numbers (Fig. 5–1). The term *hypertrophy* is best applied to muscles, as the muscle fiber (cell) enlarges as a response to increased work load rather than undergoing hyperplasia.

Most neoplasms represent an overgrowth of a single cell type; however, neoplasms of germ cells and embryonic cells may contain more than one cell type. A germ cell neoplasm that produces tissue representing more than one of the three germ cell layers (ectoderm, mesoderm, endoderm) is called a *teratoma*. Neoplasms thought to be derived from embryonic cells are termed *embryonal*.

Neoplasms are classified on the basis of the cell type they resemble and from which they presumably arise. Almost all cell types in the body can undergo neoplasia, but some do so much more frequently than others. The growth potential for each type of neoplasm is best determined by analysis of patients who have had similar neoplasms.

Most neoplasms can be classified as benign or malignant (a few are intermediate). *Benign neoplasms* are localized, single masses of cells that remain localized at their site of origin and limited in their growth (although some become very large). *Malignant neoplasms* are defined by their potential to invade and metastasize at some point in their life history. *Invasion* refers to direct extension of neoplastic cells into surrounding tissue without regard to tissue boundaries. *Metastasis* means transplantation of cells to a new site.

The neoplastic cells must be transported through vascular channels or body spaces and must grow at the new site. Metastasis is the single most important feature of malignant neoplasms, although some types can kill by local invasion alone. The term *cancer* is a synonym for *malignant neoplasm*. The term *tumor* classically meant swelling or mass, but now is more commonly used as a synonym for neoplasm.

SIGNIFICANCE OF HYPERPLASIAS AND NEOPLASMS

Hyperplasias and neoplasms commonly produce masses that are discovered by direct vision, palpation, radiographic imaging, or by presumption from the effects of the mass on organ function. Once discovered, the nature of a mass must be determined to be inflammation, hyperplasia, or neoplasia because the treatment for these processes is radically different. Subclassification within each category allows for even more rational treatment based on past experience with similar lesions.

With inflammation and hyperplasia, therapy focuses on the causative stimulus. With some forms of hyperplasia, the potential development of cancer is also a concern. With a benign neoplasm, therapeutic concern is limited to accurate diagnosis and removal of the lesion. With a malignant neoplasm, therapy is based on an estimate of the possibility for complete destruction of the neoplasm. Curative therapy is the attempt to remove all of the cancer, whether by surgical operation, radiation, or administration of drugs. Palliative therapy, whether by operative removal, radiation, or by administration of drugs, attempts to control the effects of the cancer rather than to cure it. Overall, about 50 percent of patients with malignant neoplasms die of the cancer (this does not include skin cancers).

An example will illustrate the differences in approach to hyperplasias, benign neoplasms, and malignant neoplasms. A woman discovers a lump in her breast. The mass is removed by excisional biopsy. If the pathologist classifies the lesion as a benign neoplasm, nothing further need be done and the patient is cured. If the lesion is a malignant neoplasm and judged to be potentially curable, some combination of surgical and radiation therapy will be used to remove or kill all of the cancer cells. If the neoplasm has spread beyond the breast and regional lymph nodes, palliative therapy may be used. If the pathologist diagnoses the breast lump as hyperplasia, the situation is more complex. It is unlikely that the stimulus can be removed (presumed to be the patient's own hormonal variations). If the pathologist judges the hyperplasia to be a potential site of malignancy (precancerous), the surgeon is faced with the choice of whether to remove the breast(s) to prevent cancer or to see the patient at regular intervals to detect the development of a cancer in its early stage. If the pathologist judges the lesion to be simple hyperplasia, it is still likely that further lumps will develop, because hyperplasia, unlike neoplasia, tends to be multifocal or diffuse. When another lump develops, the same diagnostic procedure must be repeated.

CLASSIFICATION OF NEOPLASMS

Most neoplasms are named by the cell or tissue they resemble plus a suffix or word indicating neoplasm. For a benign neoplasm, the suffix *-oma* is added after the name of the tissue, for example, *fibroma*. For a malignant neoplasm, either the term *carcinoma* or *sarcoma* is added to the name of the tissue. If the malignant neoplasm is from epithelium, *carcinoma* is used; if from nonepithelial tissue, *sarcoma* is used (Fig. 5–2). A few additional types of neoplasm of nonepithelial or ambiguous origin are named separately. Benign neoplasms, which follow standard nomenclature, are listed in Table 5–1.

Unfortunately there are several glaring exceptions to this naming system. A *hematoma* is a collection of blood, not a neoplasm. A *lymphoma* is a malignant neoplasm of lymph node cells (there is no benign counterpart). A *melanoma* is a malignant neoplasm of melanocytes (the benign counterpart is called a *nevus*). *Glioma* is used to refer to all neoplasms of the supporting cells of the brain (glial cells). Gliomas represents an example of neoplasms that are ambiguous in terms of the usual classification. They are benign in that they usually do not metastasize and malignant in that most kill

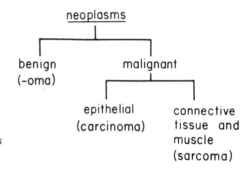

Figure 5–2. Major categories and suffixes for neoplasms.

TABLE 5–1. NAMES OF BENIGN NEOPLASMS

Cell or Tissue of Origin	Name
Squamous epithelium	Squamous papilloma
Glandular or surface columnar epithelium	Adenoma
Fibrous tissue	Fibroma
Adipose tissue	Lipoma
Cartilage	Chondroma
Bone	Osteoma
Blood vessels	Hemangioma
Smooth muscle	Leiomyoma
Nerve sheath	Neurilemoma

the patient. Biologically they represent a spectrum from slow growth to rapid growth.

The naming system for the most common types of malignant neoplasms is given in Table 5–2. Two exceptions worth noting include the common use of *hepatoma* to mean liver-cell carcinoma and *hypernephroma* to mean renal tubular cell carcinoma.

TABLE 5–2. NAMES OF MALIGNANT NEOPLASMS

Cell or Tissue of Origin	Name
Epithelium	
Site not specified	Carcinoma
Squamous epithelium of skin	Squamous cell carcinoma of skin
Basal cells of epithelium of skin	Basal cell carcinoma (unique to skin)
Colonic mucosa	Adenocarcinoma of colon
Breast glands	Adenocarcinoma of breast
Bronchial epithelium of lung	Bronchogenic carcinoma
Prostatic glands	Adenocarcinoma of prostate
Bladder mucosa	Transitional cell carcinoma of bladder
Endometrium	Adenocarcinoma of endometrium
Cervix	Squamous cell carcinoma of cervix
Stomach mucosa	Adenocarcinoma of stomach
Pancreatic ducts	Adenocarcinoma of pancreas
Connective tissue and muscle	
Site not specified	Sarcoma
Lymphoid tissue	Malignant lymphoma
Bone marrow	Leukemia
Plasma cells in bone marrow	Multiple myeloma
Cartilage	Chondrosarcoma
Bone	Osteosarcoma
Fibrous tissue	Fibrosarcoma
Smooth muscle	Leiomyosarcoma
Other	
Site not specified	Malignant neoplasm
Glial cells	Glioma
Melanocytes	Malignant melanoma
Germ cells	Teratocarcinoma

MORPHOLOGIC FEATURES

Hyperplasias and Hypertrophies

Hyperplasias represent a very diverse group of conditions about which it is difficult to generalize. Diffuse organ hyperplasias are characterized by an increase in the normal cellular elements of the organ. Common examples include lymph node hyperplasia with increased lymphocytes and macrophages in response to an inflammatory condition, physiologic breast hyperplasia with increased number and size of glands in response to pregnancy, nodular goiter (enlarged thyroid) with enlarged thyroid acini in response to prolonged iodine deficiency, and hyperplasia of the prostate with enlarged glands and fibromuscular stroma in response to uncertain stimuli in elderly men. These types of hyperplasia are grossly diffuse (they lack the discreteness of neoplasms) and, histologically, some cells appear normal but may be somewhat distorted in their arrangement and increased in density. Sometimes degeneration and fibrosis within a hyperplastic organ produces a nodularity that may diagnostically require differentiation from cancer.

Hyperplasias of epithelial surface cells are particularly important as sites of cancer development in some organs. Surface hyperplasias produce slightly raised lesions due to piling up of cells. Where they can be visualized grossly (skin, oral cavity, respiratory tract, cervix), these lesions appear more opaque than surrounding surface. In mucous membranes, opaque white lesions are called *leukoplakias*. Surface hyperplasias are easily confused grossly with inflammation or early neoplastic lesions. Microscopically, the hyperplastic nature will be evident, but the pathologist will need to judge whether it is a simple innocuous hyperplasia or whether it is premalignant hyperplasia. A premalignant lesion is one in which there is an increased likelihood of cancer as compared to adjacent normal tissue. If premalignant, the relative potential for malignancy must be estimated, based on analysis of similar lesions in other patients. As a general rule, this judgment will be based on the atypical appearance of the cells (atypia) in the hyperplastic surface epithelium. This atypical change includes increased size and staining of the nucleus, irregular shape of the nucleus, and decreased amount and maturity of the cytoplasm.

A change closely associated with hyperplasia is metaplasia. *Metaplasia* is a change from one type of tissue to another. Most commonly this involves change from columnar epithelium to stratified squamous epithelium (*squamous metaplasia*). In addition, the metaplastic tissue is often hyperplastic because both occur together in response to mild injury.

Hypertrophy is organ enlargement at the gross level and increased cell size at the microscopic level. It is usually a diffuse process. The best examples of hypertrophy are those of muscle such as the enlarged skeletal muscles of weight lifters, the enlarged cardiac muscle that occurs with high blood pressure, and the enlarged uterine smooth muscle in pregnancy.

CHART 5–1. HYPERPLASIA

causes	lesion	manifestations
Excessive stimulation of an organ Chronic irritation of surface epithelium	Organ or tissue enlargement due to increase in number of cells	Organ enlargement Surface thickening Sometimes alteration of organ function

Benign Neoplasms

Benign neoplasms are relatively easy to recognize grossly and microscopically, because they produce a single mass that is discrete from surrounding tissue. When originating on a body surface, benign neoplasms extend outwardly, producing a polyp (Fig. 5–3). A *polyp* is any abnormal protrusion from a mucosal surface. Polypoid growth is characteristic of benign neoplasms on surfaces; however, polypoid growth sometimes occurs with other types of lesions, such as inflammations, hyperplasias, and malignant neoplasms.

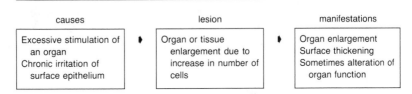

benign neoplasm originating in a tissue

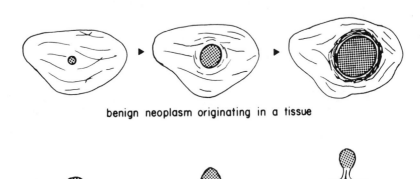

benign neoplasm originating on a surface

Figure 5–3. Comparison of a benign neoplasm in solid tissue such as breast with a benign neoplasm developing in an organ with mucosal surface such as colon. The breast neoplasm is encapsulated; the colon neoplasm is polypoid.

Benign neoplasms that originate within solid organs or connective tissue usually compress tissue around them to form a fibrous rim (capsule) (Figs. 5–3, 5–5A). Because of the fibrous rim, benign tumors are easily separated from surrounding tissue during removal. This can be a distinguishing feature from malignant neoplasms. Histologically, benign neoplasms very closely resemble their cells of origin. Minimal to moderate degrees of cellular atypia are encountered with various types of benign neoplasms. The normal architectural arrangement of tissue or organ is lost within the neoplasm. The most important criteria are the discreteness of the lesion and the uniform, relatively mature appearance of the cells.

Malignant Neoplasms

Early lesions are hardest to diagnose but are most important. Diagnostic differentiation from premalignant surface hyperplasia may be difficult. Two criteria are required for the diagnosis of a malignant neoplasm—establishing that the cells are neoplastic and demonstrating invasion. The cells are judged to be neoplastic by microscopic criteria. Proof would require genetic study which is impractical or impossible. Microscopic criteria for identification of malignant cells include uniformity of cell population, formation of a mass with replacement of the normal components of the organ, and, to a variable extent, the degree of cellular atypia. Atypical hyperplasias may have similar degrees of atypia to some neo-

normal
cell

benign
neoplastic
cell

malignant
cell

Figure 5–4. Comparison of cells and glands from normal tissue, benign neoplasm, and malignant neoplasm.

normal
gland
tissue

benign
neoplastic
gland
tissue

malignant
gland
tissue

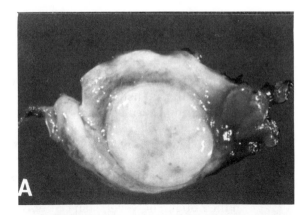

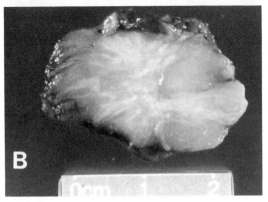

Figure 5–5. Excisional biopsy specimens of breast neoplasms. **A.** Spherical, encapsulated benign neoplasm (fibroadenoma). **B.** Invasive, crab-like streaks of a breast cancer (adenocarcinoma).

plasms, but they fail to form a mass or replace normal landmarks. Very severe atypia is nearly diagnostic of cancer, but many cancers have a relatively mild degree of atypia, which is not diagnostic in itself (Fig. 5–4).

Invasion and metastasis are the principal criteria used to distinguish benign and malignant neoplasms. Metastasis is the most definitive criterion but is not a desirable means of diagnosing cancer because it occurs late in the evolution of a malignant neoplasm. Proof of metastasis requires biopsy to demonstrate that the apparent metastatic lesion contains neoplastic cells rather than inflammatory cells, because infections can also "metastasize."

Local invasiveness is a more commonly used criterion for cancer than is metastasis (Fig. 5–5B). As mentioned earlier, invasion is characterized by infiltration of cancer cells with poor respect for tissue boundaries. Invasiveness is recognized grossly by irregularity of tumor tissue margins, failure of the tumor to separate from surrounding tissue during removal, and, when advanced, direct spread beyond the organ of origin.

Additional features that are helpful but not entirely specific in separating malignant from benign neoplasms include hardness (many malignant neoplasms provoke proliferation of fibrous tissue with scarring), necrosis (soft yellow areas resulting from the cancer outgrowing its blood supply), ulceration (necrotic tissue of surface cancers sloughs, leaving an ulcer defect), and multilobulation (uncommon in benign neoplasms). Malignant neoplasms are usually larger than benign neoplasms, but this is a very unreliable generalization. Both benign and malignant neoplasms are usually solitary in their origin; however, skin cancers and a few benign neoplasms are characteristically multiple (nevi, leiomyomas of the uterus, neurofibromas, and sometimes adenomas of the colon).

CHART 5–2. NEOPLASIA

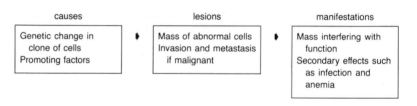

causes	lesions	manifestations
Genetic change in clone of cells Promoting factors	Mass of abnormal cells Invasion and metastasis if malignant	Mass interfering with function Secondary effects such as infection and anemia

LIFE HISTORY AND SPREAD OF CANCER

Cancers have a long life history, most of which occurs before there is any lesion that can be called a cancer. We know of the early events through experimentation and inference rather than through direct observations of each type of cancer that occurs in humans.

Genetic alteration is the basis for development of cancer, but, of all the mutations that occur in cells, very few lead to cancer. Further, most cells with cancer potential do not express that potential. A cell must undergo an alteration or series of alterations called *initiation* to acquire the autonomous growth potential referred to as cancer. Agents that so alter cells are called *carcinogens*. Carcinogens may be physical, chemical, or biologic agents. Once a carcinogen has altered a cell, another agent called a *promoter* furthers the expression of the cancer in that cell and its progeny. Some carcinogens act as both initiators and promoters but many others serve only one function or the other. The mechanism of action of the carcinogen is direct or indirect at the level of the cell's deoxyribonucleic acid (DNA). The net result is replication of DNA and consequent cell division outside of the normal control mechanisms.

Transformation is the process by which cells with cancer potential gradually establish themselves and give rise to a cancer. The cancer cells must proliferate rapidly enough to maintain and expand the size of their population. Further, they must either influence the surrounding tissue from which they originated to supply their nutritional and metabolic needs, or they must adapt themselves nutritionally and metabolically. Antagonism between the body and these aberrant cells is continuous. If the body gains the upper hand, the aberrant population may disappear. If the aberrant cells gain the upper hand, they can become established.

These colonies of aberrant cells are not necessarily neoplasms. If they have permanently and irreversibly established themselves as an expanding population in the body, then they are considered neoplasms. If they are not firmly established and there remains a potential for their spontaneous disappearance (regression), they are said to be preneoplastic. If the preneoplastic states are clinically discoverable, surgical removal of the aberrant cell populations can be undertaken before a cancer actually develops. Unfortunately, it is not possible to detect preneoplastic phases for most types of neoplasms. There are a few cancer types where preneoplastic tissue may manifest itself in a detectable manner. Squamous epithelium may go through a slow series of changes, including simple epithelial hyperplasia, atypical hyperplasia (often called *dysplasia*), and finally carcinoma in situ. With *carcinoma in situ*, the transformed cells are confined to the epithelial layer and have not yet manifested themselves as frank neoplastic cells by invading the surrounding tissue (Fig. 5–6). The in situ stage can be micro-

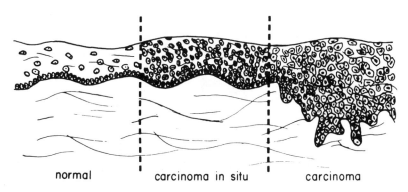

normal   carcinoma in situ   carcinoma

Figure 5–6. Invasion of the basement membrane differentiates carcinoma from carcinoma in situ.

scopically detected in the uterine cervix, oral cavity, and larynx. Cancer can be prevented by removal of either the dysplastic lesion or the in situ lesion.

There are a number of lesions from which cancers arise in noticeable frequency that are not in themselves definable as carcinoma in situ. These predisposing lesions are definable by statistical association with the development of cancer. Examples are a hyperplastic lesion of the gums due to ill-fitting dentures or atypical hyperplasia and squamous metaplasia of bronchi in a chronic smoker. In some sites, particularly epithelial surfaces, chronic inflammation may be predisposing for the development of neoplasms.

Local invasion is a critical step in the development of overt cancer. For epithelial cancers (carcinomas), this is the point where malignant cells break through the basement membrane that separates epithelium from connective tissue. Once this break occurs, the cancer can grow and spread to destroy tissue, produce a mass, and further spread via the vascular system and body passageways. Local growth may be slow or rapid. Some cancers kill mainly by local destruction. Examples include blockage of ureters by local spread of carcinoma of the cervix, obstruction of the common bile duct by carcinoma of the pancreas, and obstruction of the esophagus by esophageal carcinoma.

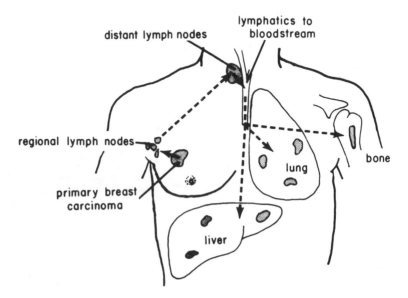

Figure 5–7. Typical pattern of metastasis from a breast cancer. For carcinomas, metastasis via lymphatics usually precedes metastasis via the blood.

Regional and distant spread through lymphatics is an important step in the life history of most carcinomas. Sarcomas less commonly spread by this route. Lymph nodes can filter out the cancer cells for a while, but eventually the cancer cells spread beyond the nodes and to the bloodstream for wide dissemination. Absence of lymph node involvement is a favorable stage for surgical cure. Some patients with regional spread (involved lymph nodes) can also be cured. Those with distant spread rarely survive. Sarcomas usually spread directly in the bloodstream and are filtered out first in the lungs, where they may grow and produce metastatic nodules. The spread of breast carcinoma is illustrated in Figure 5–7.

GRADING AND STAGING

The terms *grade* and *stage* refer to the histologic differentiation and degree of spread of cancers, respectively. *Differentiation* refers to the degree of resemblance of the cancer to its tissue of origin. Thus, cancers are commonly graded as well-differentiated, moderately differentiated, poorly differentiated, or undifferentiated. Poorly differentiated neoplasms exhibit more cellular atypia and less structural similarity to the tissue of origin than do well-differentiated neoplasms.

Staging reflects the degree of local invasion and regional and distant metastasis. Rules for staging have been established for each of the common cancer types. These rules for staging involve evaluation for in situ change, localization to the organ of origin, direct spread beyond the organ, lymph node metastasis, and distant metastasis. Both grading and staging are used to predict likelihood of cure, but staging usually is the better predictor.

REVIEW QUESTIONS

1. What is the difference in pathogenesis of hyperplasias and neoplasms? What is the significance of this difference? How does hypertrophy differ from hyperplasia?
2. How do hyperplasias, benign neoplasms, and malignant neoplasms differ in behavior, morphology, and treatment?
3. How does the appearance of hyperplasias and neoplasms differ from inflammation?
4. What conventions are used to name the various types of benign and malignant neoplasms? What are the exceptions to these conventions?

5. Which organs or tissues commonly undergo hyperplasia? Which organs characteristically undergo hypertrophy rather than hyperplasia?
6. What is the significance of surface epithelial hyperplasia?
7. What criteria are used to diagnose cancer?
8. What are the temporal relationships among mutation, transformation, premalignant lesions, in situ cancer, local invasion, metastasis, and death?
9. How do carcinomas and sarcomas differ in their spread?
10. What do *grading* and *staging* refer to and why are they used?
11. What do the following terms mean?

   Tumor
   Metaplasia
   Squamous metaplasia
   Polyp
   Initiation
   Promotion
   Carcinogen
   Predisposing condition
   Dysplasia
   Carcinoma in situ
   Differentiation

Cancer

Chapter 5 covered the basic concepts of cell proliferative disorders, separating hyperplasias from neoplasms and benign from malignant neoplasms. The emphasis in this chapter is on malignant neoplasms (cancers)—their frequency and significance, etiology, manifestations, natural history, and treatment. This chapter is an overview; much more will be said about specific cancers in Section III (Major Organ-Related Diseases).

FREQUENCY AND SIGNIFICANCE

Cancer is often fatal and ranks as the second leading cause of death in the United States. However, the term *cancer* encompasses a large number of specific types of malignant neoplasms that must be considered individually, because their behavior, their treatment, and probably their causes vary considerably. The prognosis of a cancer depends upon the natural history of that type of cancer, the extent of spread at the time of discovery, and efficacy of existing therapy for that particular type of cancer. In general, the incidence of malignant tumors is about twice the mortality rate. Stated differently, the overall survival rate of cancer is approximately 50 percent. There is a great variability in the behavior of different cancers. Some types, such as carcinoma of the pancreas, almost always kill the patient, whereas others, such as skin carcinoma, rarely kill the patient.

Three important variables relating to cancer frequency and significance are site of development, sex, and age. Cancers developing from epithelium (carcinomas) outnumber cancers from non-epithelial cells (sarcomas, leukemias, and lymphomas) by 6 to 1. The most common cancers of humans are actually basal and squamous cell carcinomas of the skin, comprising about 40 percent of all cancers and 99 percent of all skin cancers. Despite their frequency, they are very seldom fatal because they are readily visualized, grow slowly, metastasize only rarely, and can be completely excised. In contrast, malignant melanoma comprises only 1 percent of all skin malignancies but is fatal in about 20 percent of patients. Thus, a large proportion of the mortality associated with cancers of the skin is accounted for by the least common type. In addition to being the most common cancers, basal and squamous cell carcinomas are the only common cancers that are frequently treated in a physician's office and thus escape hospital statistics. Thus, these

TABLE 6–1. FREQUENCY AND SURVIVAL OF COMMON CANCERS IN MALES

	Relative Incidence (%)	Percent of Cancer Deaths (%)	Relative 5-Year Survival (%)
Lung	19	34	13
Prostate	22	12	72
Colon	14	11	54
Bladder	10	5	77
Leukemias/Lymphomas	7	8	33–74
	72	70	

TABLE 6–2. FREQUENCY AND SURVIVAL OF COMMON CANCERS IN FEMALES

	Relative Incidence (%)	Percent of Cancer Deaths (%)	Relative 5-Year Survival Rate (%)
Breast	32	18	76
Colon	14	13	54
Cervix/Endometrium	8	4	68/86
Lung	11	21	13
Leukemias/Lymphomas	6	6	33–74
	71	62	

two common skin cancers are often omitted from overall collections of cancer statistics.

Aside from skin cancers, carcinomas of the lung, colon, breast, prostate, and uterus are the most common types. The five most common types in males and females are given in Tables 6–1 and 6–2.

Lung cancer is responsible for over one-third of cancer deaths in males and nearly one-fifth of cancer deaths in females. Treatment is relatively ineffective (13% 5-year survival), but prevention of most cases (by cessation of smoking) is possible. Treatment of colon cancer (surgical removal) cures more than 50 percent of patients, and, although we know that a dietary factor is the likely cause, we have no practical way of preventing its occurrence at the present time. Breast and uterine cancers are often detected early and are quite accessible to surgical and radiation therapy, but their high frequency still accounts for a large number of cancer deaths. Prostate cancer is a disease of elderly men, so its impact is lessened by the fact that these patients are likely to die of other causes.

TABLE 6–3. AGE PEAK OF COMMON CARCINOMAS

Site	Peak Age
Cervix	40s and 50s
Breast	40s and 50s
Endometrium	50s
Ovary	50s to 70s
Lung	60s
Bladder	70s
Colon	70s
Pancreas	70s
Stomach	70s
Prostate	70s

TABLE 6–4. CANCER TYPES CHARACTERISTICALLY OCCURRING IN PERSONS UNDER AGE 30

0–10 years
  Acute lymphocytic leukemia
  Neuroblastoma
  Wilm's tumor of kidney
  Retinoblastoma
  Medulloblastoma
10–20 years
  Osteogenic sarcoma
20–30 years
  Hodgkin's disease
  Adenocarcinoma of thyroid
  Testicular cancers

Leukemias and lymphomas have variable survival rates, depending on the specific type. Aggressive radiation and chemotherapy of leukemias and lymphomas consume a disproportionate share of medical resources and are associated with more complications than those cancers treated by surgical therapy alone.

In general, cancer is much more common in older persons; this is particularly true for carcinomas. Of the ten most common carcinomas, most have a peak frequency in the seventies (Table 6–3). Cancers of the breast and female genital tract tend to occur in midlife. Cancer of the lung also presents somewhat earlier than carcinomas in general, with the peak number of new cases occurring in the sixties. Of the many other less common types of cancer, most have a characteristic age incidence. Although cancer is much less frequent in younger people, some types occur predominantly in the young (Table 6–4). All of the common cancers of young people listed in this table are nonepithelial, except for adenocarcinoma of the thyroid, a carcinoma most common in young adult women.

ETIOLOGY

Cancers arise in a variety of circumstances, which suggests that more than one factor is involved in the development of any neoplasm. Experimental studies suggest that cancers go through progressive changes before becoming clinically evident. Some even become more aggressive after being temporarily inhibited by cancer therapy. As indicated in Chapter 5, neoplasms are usually considered to originate from a single, genetically altered cell (initiation). This cell, in turn, divides to form a discrete population of altered cells (a clone). Further alterations may occur and give rise to new clones. Agents causing transformation of mutant cells into cancer may differ from those causing the initial mutation. If the new clone has a higher rate of cell division than the old clone, the new will outgrow the old. As a general rule, the more a cell deviates in character from analogous normal cells (i.e., the more poorly differentiated it becomes), the more rapidly it grows. The more poorly differentiated cells are usually more invasive and more likely to metastasize.

Prevention of cancer will be dependent upon discovery of initiating agents that cause the original genetic change or the circumstances that cause malignant transformation of mutated cells. At the present time, three initiating factors can be identified— radiation, oncogenic viruses, and chemical carcinogens. Their roles as carcinogens will be briefly discussed.

The recognition of a high incidence of scrotal skin cancer in London chimney sweeps led to the discovery that repeated occupational exposure to coal tar was carcinogenic. Later, through experimentation, methylcholanthrene was identified as a specific polycyclic hydrocarbon that could cause cancer. Since then, a large number of man-made compounds have been discovered that are potentially carcinogenic. Chemical carcinogens also occur in nature. One of the most potent carcinogens, known as aflatoxin, is a product of fungi that contaminates peanuts, corn, and other human and animal foods.

The mechanism of chemical carcinogenesis is complex. Generally, carcinogens must be metabolized by cells to an active metabolite, which, in turn, interacts with deoxyribonucleic acid (DNA), ribonucleic acid (RNA), or cell proteins. The interaction with DNA is potentially mutagenic. However, the cell has an enzyme system that actively repairs DNA defects such as those caused by carcinogens. It may be that the establishment of a carcinogenic effect depends on failure to repair DNA. The appearance of tumors takes months to years after exposure to a carcinogen (latent period). What happens in this intervening time is not well known. The latent period can be reduced by increasing the frequency of exposure and size of the dose of carcinogen. Other factors modifying the effects of chemical carcinogens include age, sex, diet, genetic factors, and immune deficiencies. Certain chemicals that do not cause cancer when given alone (promoters), enhance the chance of neoplastic conversion when applied to tissues after exposure to carcinogens.

There are many examples of human cancer associated with recognizable exposure to carcinogenic chemicals. Many of the examples involve occupational chemical exposure, resulting in increased incidence of a specific cancer in a definable population. Compounds implicated include certain dyes, vinyl chloride, alkylating agents, and asbestos. Hormones may increase the incidence of certain neoplasms by mechanisms that are not understood. For example, administration of diethylstilbesterol to pregnant women has been found to correlate with increased incidence of an otherwise rare vaginal adenocarcinoma in their female offspring. Based on present knowledge, the cancer-inducing chemical agent(s) affecting the largest number of people is/are contained in cigarette smoke. Smokers have a 10- to 50-fold greater chance of developing bronchogenic carcinoma than nonsmokers, and the risk can be significantly correlated with the number of packs smoked per day multiplied by the number of years an individual has smoked (pack-years). Squamous cell carcinomas of the oral region, larynx, and esophagus and transitional cell carcinomas of the urinary bladder are also significantly associated with cigarette smoking.

Ultraviolet radiation, x-radiation, and gamma radiation are all carcinogenic, with the effect dependent on dose, duration, and the portion of the body exposed. As with chemical carcinogens, a latent period of years to decades intervenes between exposure and the appearance of neoplasms. Just how radiation initiates cancer is not known, although we do know that radiation produces localized breaks in DNA strands.

Examples of radiation-induced cancer include thyroid carcinoma in children, a neoplasm that was rarely encountered before the advent of therapeutic radiation. Now 75 percent of children with this cancer have a history of radiation of the neck for conditions that are no longer treated in this manner. Survivors of the Hiroshima and Nagasaki atomic blasts who were exposed to whole body gamma radiation have an increased incidence of leukemias and thyroid carcinomas. Radiologists not uncommonly developed skin cancers before the benefits of protective shielding were realized. People having extended exposure to the sun (farmers, mariners, and sunbathers) experience a high incidence of basal or squamous cell skin carcinomas due largely to the ultraviolet component of sunlight.

Both RNA- and DNA-containing viruses have oncogenic (tumor-producing) potential. In the DNA groups, members of the papova, herpes, pox, and adenovirus groups have been found to produce tumors in animals. Tumor-producing RNA viruses are called oncornaviruses (onco + RNA) or retroviruses. In cells infected by DNA oncogenic viruses, the viral genome is incorporated in host cell genome and is expressed with it. This insertion of viral genome occurs in only a few cells of an infected population. In cells infected by RNA oncogenic viruses, the virus carries an enzyme, reverse transcriptase, that makes a copy of DNA complementary to the virus' RNA, and this DNA becomes inserted into the host genome.

Viral causation of neoplasms has been demonstrated in numerous animal cancers, but few strong cases of demonstrable viral oncogenesis in humans can be found. Papovaviruses cause squamous papillomas in humans including the common wart (verruca vulgaris), venereal warts (condyloma acuminata), and laryngeal papillomas. Several viruses have been associated with malignant neoplasms in humans, but the exact role of the virus in causing these cancers has not been established and most people with these viruses do not get cancer. Epstein-Barr virus, the cause of infectious mononucleosis, has been associated with Burkitt's lymphoma, a neoplasm most prevalent in Africa, and with undifferentiated nasopharyngeal carcinoma. Specific strains of human papilloma virus play an important role in the etiology of carcinoma of the uterine cervix. Hepatitis B virus has been associated with hepatocarcinoma.

LOCAL AND SYSTEMIC MANIFESTATIONS

Local manifestations include mass, pain, obstruction, hemorrhage, pathologic fracture, and infection. Systemic manifestations include infection, anemia, cachexia, and hormone production. Any one of these manifestations is present in a minority of patients with cancer; in fact, many cancers are asymptomatic until late in their course.

Mass

If a cancer becomes very large or is situated in a location that is readily visible or palpable, it may be first noted as a mass. Otherwise, one or more of the complications listed below may become evident before a mass is noted.

Pain

Cancer produces pain by local destruction of tissue, with invasion of nerves, by obstructing hollow organs such as the intestine, and by causing inflammation, which, in turn, is associated with pain. In advanced cases, surgical interruption of nerve tracts is sometimes needed to relieve the discomfort. Many cancers are nonpainful.

Obstruction

Body passageways may be obstructed by tumors growing within their lumens or by external compression by a mass. Symptoms depend on the passages involved. Examples of internal obstruction include the obstruction of a bronchus by lung cancer or bowel obstruction by colonic carcinoma. An example of external obstruction is the blockage of the common bile duct by a carcinoma of the pancreas.

Hemorrhage

Cancers on internal or external surfaces may ulcerate and bleed, leading to either acute or chronic blood loss. Inapparent blood in feces (occult blood) may be detected by a simple chemical test.

Pathologic Fracture

Primary bone cancer or metastatic cancers may invade and locally destroy bone, weakening it so that fracture may occur with minimal injury. Cancers of lung, breast, and prostate are especially likely to metastasize to bone. Multiple myeloma, a neoplastic proliferation of plasma cells, is a primary cancer of bone marrow that commonly destroys adjacent bone. A third mechanism by which a neoplasm can cause pathologic fracture is by induction of osteoporosis. Osteoporosis may be caused by inactivity or by rare tumors secreting high levels of corticosteroids or ACTH.

Infection

Infection is a common and often disastrous complication of neoplasia. A variety of overlapping mechanisms produce infections in patients with cancer. At the local level, erosion or ulceration of tumors of epithelial surfaces allows entry of microorganisms. Impaired host responses to infection can arise from a number of causes. Leukopenia may result from extensive replacement of bone marrow in leukemias and lymphomas or in other metastatic cancers. In leukemias and lymphomas, the circulating neoplastic leukocytes are deficient in phagocytic activity. Inadequate nutrition and other factors in the terminally debilitated cancer patient may be associated with a decline in the general capacities of the immune system. Chemotherapy of cancer has a side effect of suppressing the bone marrow's production of leukocytes. Because lymphocyte quantity and quality are often deficient in neoplastic disease, cell-mediated immunity may be particularly impaired. Immune deficiency is manifested by unusual causal agents of infections, such as fungi, protozoa, and viruses, and bacteria that are ordinarily not pathogens.

Anemia

Anemia is one of the most frequent manifestations of malignant neoplasms. Major mechanisms include decreased erythropoiesis resulting from the effects of chemotherapeutic drugs or radiation on bone marrow, bone marrow replacement by neoplastic cells, and blood loss caused by ulceration of the cancer.

Cachexia

Cachexia is the generalized wasting that occurs in the terminal cancer patient. It is probably caused by a combination of anorexia (loss of appetite), nutritional problems, and demands of the rapidly growing neoplasm.

Hormone Production

A few neoplasms secrete hormones that lead to effects associated with excess hormone. Benign neoplasms, particularly those of the endocrine glands, are most commonly involved (see Chapter 25 for specific examples). Sometimes, malignant endocrine gland neoplasms or other neoplasms such as small cell carcinoma of the lung produce hormones in sufficient quantity to cause clinical effects.

DIAGNOSIS

The diagnosis of cancer is based on investigation of the cause of a patient's symptoms or on screening. The diagnosis is confirmed by biopsy, or sometimes by blood smear or cytology. The astute physician will recognize that a patient's symptoms may be due to cancer and will do the appropriate physical examination or order appropriate laboratory tests, radiologic procedures, and endoscopic examinations. Before carrying out irreversible treatment procedures that always entail some risk, a final biopsy report should be in hand, because noncancerous lesions often mimic cancer clinically.

Case finding for cancer is often done, especially in the elderly, when the patient is seen for some other disease. Asymptomatic cancer of the breast or prostate may be detected by palpation, and the skin can be inspected for malignant melanoma and carcinoma. The most common routine screening tests for cancer are cytology of the uterine cervix, testing feces for occult blood, mammography, and chest x-ray. The screening procedures produce a lot of false-positive results that require follow-up tests and biopsies.

NATURAL HISTORY

Each malignant neoplasm possesses its own expected behavior pattern. For example, it is known from past experience that a given malignant neoplasm will likely metastasize to certain organs and will likely take the life of a patient within a certain length of time. Thus, we speak in terms of 5- or 10-year survival for neoplasms, meaning that a certain percentage of patients with that particular neoplasm will still be alive (with or without disease) 5 or 10 years after initial diagnosis. This prognostication of tumor behavior is far from exact. One person with a malignant metastatic neoplasm may die within months of diagnosis, whereas another patient with the same type of neoplasm may live for several years.

Infection is the most common cause of death in cancer patients. Commonly, as the terminally ill patient with a malignant neoplasm becomes more and more immobilized and bedridden, the lungs fail to remove secretions, allowing bacteria to proliferate and cause pneumonia. Infections of the genitourinary tract are also common in terminally ill patients. Whatever the initial site of the infection, many patients eventually develop bacteremia, with spread of organisms to other organs. The predisposing factors for terminal infections are the previously mentioned immune and white cell deficiencies, immobilization, obstruction of body passageways, and general debilitation manifested by cachexia.

Cachexia itself seems to be the cause of death in many cancer patients. Cachexia may be the result of loss of appetite or of a mass obstructing the gastrointestinal tract in some instances. In other instances, there is no logical explanation for cachexia. In some cases, the increased metabolic needs of a neoplasm can explain cachexia, but at other times, patients with very little neoplastic mass become cachectic.

Metabolic and endocrine effects of neoplasm and hemorrhage are occasional causes of death of cancer patients. Often, a single immediate cause of death in a patient with terminal cancer is not identifiable.

TREATMENT

The basic modality of cancer treatment is surgical removal. Even in those cases where total removal of a malignant neoplasm cannot be accomplished, some tumor tissue must be biopsied to establish a tissue (histologic) diagnosis. If a tissue diagnosis is not established for a given cancer, then treatment such as chemotherapy or radiation therapy would be based on less than optimum knowledge of the type of tumor present; sometimes, the diagnosis of cancer may be erroneous. The amount of the cancer removed at operation depends upon its accessibility to surgical removal and its amenability to other modes of therapy. For example, a surgeon may remove very little of a cerebellar medulloblastoma because surgical exposure of the tumor is difficult and there is a risk of destroying adjacent brain structures. In addition, the surgeon knows that this type of cancer responds readily to radiation therapy. Conversely, when a cancer is discovered in the large bowel, total removal is anticipated because the tumor is usually accessible to the surgeon in its entirety. In the latter example, even if metastasis has already occurred, removal of the primary tumor may prevent bowel obstruction.

Radiation therapy is employed in the treatment of localized neoplastic masses that are not surgically accessible (inoperable) or in situations where surgery would be impractical or deleterious to the patient. Radiation therapy is also used following surgery to treat residual neoplasm. Another use of radiation therapy is in the treatment of lymph nodes in the expected metastatic pathway of a particular type of neoplasm. For example, carcinoma of the medial aspect of the breast would be expected to metastasize via the internal mammary lymphatic channels; consequently, these lymph node channels are often irradiated prophylactically following mastectomy. Radiation therapy is generally most effective with tumors composed of rapidly dividing cells. Only a limited number of cancer types can be cured by radiation. For many types of cancers, radiation therapy is palliative; that is, it temporarily abates the progress of a cancer without curing it.

Chemotherapy employs a wide variety of powerful metabolic inhibitors and other cell-killing chemicals that are often used in various combinations for treatment of different cancers. The rationale behind the use of all chemotherapeutic agents is that they will kill or inhibit the growth of neoplastic cells to a greater degree than that of the body's normal cells. However, in some instances not enough differential response exists, and consequently, it is understandable that serious side effects from chemotherapy can occur, most of which are based on the deleterious effects on rapidly metabolizing normal body tissues such as bone marrow. Chemotherapy has been most successful for certain types of leukemias and lymphomas and for choriocarcinoma in females.

Hormonal therapy may cause significant regression of some cancers of the breast and prostate gland and prolong life. The response may be induced by administering hormone or by removing hormone-producing organs such as testes, ovaries, adrenal glands, or pituitary gland. Breast cancers that have estrogen receptors on their cell surfaces (about two-thirds) often regress when estrogen levels are lowered by oophorectomy, adrenalectomy, or hypophysectomy. Similarly, some prostate cancers regress when testosterone levels are lowered by castration or by administration of an opposing hormone, estrogen. Hormonal therapy is most used as palliative therapy for metastatic cancer.

REVIEW QUESTIONS

1. What are the most common carcinomas? How do they differ in frequency, sex ratios, and survival rate?
2. Which carcinomas occur predominantly in the middle aged? in the elderly?
3. Which cancers occur predominantly in each of the first three decades of life?
4. What types of agents have been implicated in the development of cancers?
5. What are six local manifestations of cancer? How does cancer cause these manifestations?
6. What is the pathogenesis of the three major systemic manifestations of cancer?
7. What are the forms of cancer therapy? What are the advantages and disadvantages of each?

Chapter

7

Genetic and Developmental Diseases

REVIEW OF STRUCTURE AND FUNCTION
Chromosomal Duplication and Fertilization · Genetic Inheritance · Embryonic Development · Fetal Development · Perinatal Period · Infancy · Childhood · Adolescence

DEFINITIONS

FREQUENCY AND SIGNIFICANCE OF DEVELOPMENTAL ABNORMALITIES
  Monogenetic (Single-Gene) Diseases
  Polygenetic (Multigene) Disorders
  Chromosomal Diseases
  Embryonic Anomalies
  Fetal Diseases
  Perinatal Diseases
  Diseases of Infancy
  Diseases of Childhood and Adolescence

SYMPTOMS, SIGNS, AND TESTS
  Genetic Diseases
  Chromosomal Diseases
  Embryonic Anomalies
  Fetal Diseases

EXAMPLES OF SPECIFIC DISEASES
  Genetic Diseases
    Diabetes Mellitus · Gout · Sickle-Cell Anemia · Hemophilia · Phenylketonuria · Muscular Dystrophy · Cystic Fibrosis
  Chromosomal Diseases
    Down's Syndrome · Klinefelter's Syndrome · Turner's Syndrome · Fragile X-Syndrome
  Embryonic Anomalies
    Tetralogy of Fallot · Agenesis of One Kidney · Meckel's Diverticulum · Meningomyelocele
  Fetal Diseases
    Erythroblastosis Fetalis · Congenital Syphilis · Fetal Alcohol Syndrome

REVIEW QUESTIONS

REVIEW OF STRUCTURE AND FUNCTION

The development of a mature individual takes approximately 18 years and can be divided into the following stages: fertilization, embryonic period, fetal period, perinatal period, infancy, childhood, and adolescence.

Fertilization involves the uniting of a sperm and ovum, with each contributing 23 chromosomes to the newly created human organism. These chromosomes contain all of the genetic information needed to control the succeeding stages of development. Because many human diseases result from abnormalities in the genetic makeup as a result of events occurring before fertilization, the events occurring before fertilization will be briefly reviewed.

All of the cells of a normal mature individual have 46 chromosomes. These chromosomes may duplicate themselves and divide to form two daughter cells each with 46 chromosomes. This process is called *mitosis* and can occur in most of the cells in the body (Fig. 7–1). The germ cells that develop into sperm and ova undergo a different type of cell division called *meiosis*. In meiosis, only one chromosome from each pair is passed on to each gamete (sperm or ovum) (Fig. 7–1). Thus, each gamete has only 23 chromosomes.

Because each pair of chromosomes is unique with an orderly arrangement of its subunits (genes), abnormalities of separation of the chromosome pairs or breakage of chromosomes during meiosis will be reflected in the chromosomal makeup of a new individual at the time of fertilization. Abnormalities in chromosome number

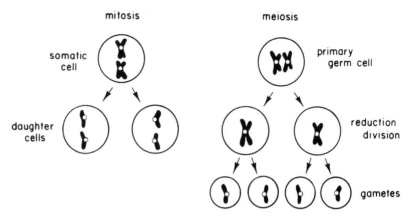

Figure 7–1. Comparison of mitosis in somatic cells with meiosis in germ cells. A homologous pair of chromosomes that have already duplicated but are still connected by a centromere are illustrated at the top. In mitosis, the centromere is duplicated and each daughter cell receives one of each of the duplicated chromosomes from each pair. In meiosis, the paired chromosomes align opposite each other and are separated at the first (reduction) division. The second division sends only one chromosome duplicate from the original pair to each gamete.

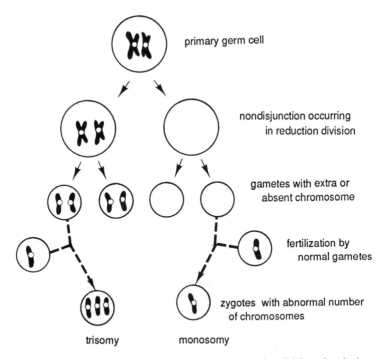

primary germ cell

nondisjunction occurring
in reduction division

gametes with extra or
absent chromosome

fertilization by
normal gametes

zygotes with abnormal number
of chromosomes

trisomy   monosomy

Figure 7–2. The results of nondisjunction at reduction division of meiosis.

result from nondisjunction, a process in which two chromosomes go to one gamete and none to another (Fig. 7–2). When such gametes result in fertilization, the zygote formed has three homologous chromosomes (trisomy) or only one (monosomy). Abnormalities in chromosome structure result from breakage of chromosomes during the process of cell division. Resulting fragments may pass to the wrong daughter cells, resulting in duplication or deletion of parts of a chromosome, or fragments may exchange places with fragments from nonhomologous chromosomes, resulting in translocations (Fig. 7–3). Occasionally, nondisjunction, duplication, deletion, or translocation occur in the first or second cell division after fertilization, resulting in an individual with cells of differing chromosomal makeup. This is called *mosaicism*. Because of the vast amount of genetic information carried by each chromosome, the new organism may be severely affected by abnormalities in chromosomes.

Chromosomes are made of deoxyribonucleic acid (DNA) molecules arranged into specific sequences. These sequences define the subunits of a chromosome known as *genes*. One or more genes are responsible for determining a genetic trait. An abnormality in a gene may result in the genetic trait being expressed in an abnormal way. Each individual harbors a few abnormal genes that may cause disease and many genes that cause nonharmful traits such as red

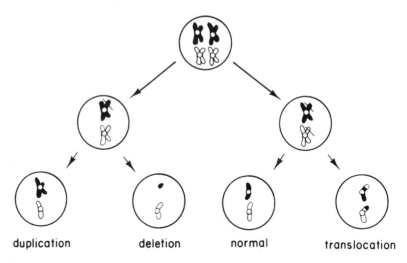

Figure 7–3. Illustration of how breaks in chromosomes can result in duplication, deletion, and translocation.

hair. There are two ways an individual can acquire an abnormal gene. When a gene is altered chemically, it is called *mutation*. When a mutation occurs during meiosis, the resulting sperm or ovum will pass the abnormal gene to the new individual. This abnormal gene will be present in all the cells of the new individual and will be passed to subsequent generations by the germ cells. Thus, abnormal genes can arise by mutation of germ cells or by passage from previous generations. It should also be noted that mutations can occur in nongerm cells (somatic cells), but the effects are limited because of the limited number of new cells resulting from any given somatic cell. As we have noted in previous chapters, neoplasms can result from mutations in somatic cells.

Not all abnormal genes produce an abnormal genetic trait. Each genetic trait is influenced by two genes, one on each of the paired chromosomes. There is one exception to this rule; males have some unpaired genes on the sex chromosomes (X and Y). One of the genes of a pair may be dominant over the other, that is, it will determine the genetic trait that the individual expresses.

A recessive gene will be expressed only if it is paired with another recessive gene. Thus, recessive genes for abnormal traits can only be expressed if both genes are abnormal or if they are located on the X chromosome of a male, where there is no second gene for the trait.

Sometimes abnormal dominant genes or pairs of abnormal recessive genes fail to cause expression of the abnormal trait. This is called *nonpenetrance*. If the abnormal trait is expressed to a variable extent in different individuals, it is called *variable expressivity*. To complicate matters even further, the expression of some genetic traits is dependent on multiple factors such as other genes and environmental factors including time. It is difficult to determine the genetic makeup (genotype) of an individual except for completely expressed dominant traits that become evident early in life. The term *phenotype* refers to those traits that are manifest. Most of the abnormal genes in the population are not manifest. Abnormal genes are only discovered by inference, by establishing which family members have the disease and deducing which ones carried the abnormal genes.

Embryonic development occurs in the first 8 weeks after fertilization. This developmental sequence, which transforms a single fertilized ovum to an individual with body structure and organs, may be influenced by abnormal genes or by environmental factors such as nutrition, infection, chemicals, and radiation exposure.

Fetal development occurs from the eighth week after fertilization to birth. The fetal period is primarily one of growth and maturation of organs and tissues and results in an individual who can survive and grow without the direct support of a maternal blood supply. The lungs, kidneys, and liver require the longest time to reach this stage.

The perinatal period is defined as the period from 2 weeks before birth to 4 weeks after birth. The major events of this period are survival from the trauma of birth and adjustment to external life. The lungs and vascular system must adjust to breathing and rerouting of oxygenated blood. Kidney and liver function increase at this time to meet new demands.

The period of infancy is defined as 1 month to 2 years of age. Major events during this period include growth, development of motor and intellectual functions, and development of immunologic defense against foreign substances.

The period of childhood covers the period from infancy to puberty and is primarily one of growth and refinement of motor and intellectual functions.

Adolescence is the final developmental period before the process of aging begins. Sexual maturation is the major event that distinguishes this period.

DEFINITIONS

Abnormalities of development may be due to altered genetic structure or environmental effects or a combination of the two. The interaction of genetic and environmental factors is complex, and for a given disease may not be well understood. We will use the term developmental abnormality in a broad sense to refer to diseases that affect normal maturation.

A genetic disease is an established disease caused by an abnormal gene. Simply having an abnormal gene is not a disease. Although abnormal genes are acquired at the time of fertilization, most genetic diseases do not develop until some time after birth if they develop at all. Single-gene defects encompass the classic genetic diseases in which the abnormal gene can be traced through family trees. Multiple-gene or complex gene defects involve more than one abnormal gene and sometimes environmental factors for their expression. Inheritance patterns are not clear-cut, but there is some tendency for the disease to occur in families. Diabetes mellitus is an example of a complex genetic disease.

Chromosomal diseases are also genetic in nature but will be treated separately in this text because they differ considerably from other genetic diseases. Chromosomal abnormalities are defined by visible misplacement of chromosomes. Chromosomes can be visualized by a process known as *karyotyping*. This involves culture of cells, inducing the cells to divide, arresting the dividing cells during mitosis, and squashing and staining the cells so that the chromosomes can be seen under a microscope (Fig. 7–4). Sex chromosome numbers can be evaluated by a simpler test known as a *buccal smear*. X-chromatin bodies (Barr bodies) can be seen on the periphery of cell nuclei with ordinary stains when more than one X chromosome is present (Fig. 7–5). Thus, Barr bodies are absent in normal males. The Y chromosome (F body) can be seen using a fluorescent staining technique. Chromosomal diseases often result

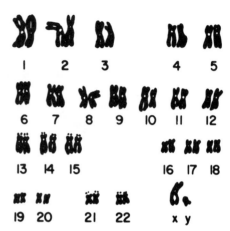

Figure 7–4. A normal male karyotype.

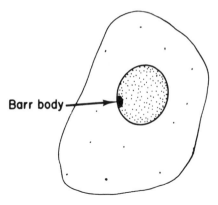

Figure 7–5. Squamous epithelial cell from a female with a Barr body at the nuclear membrane.

in abnormalities so severe as to preclude reproduction and further passage of the abnormality.

Abnormalities developing between the time of fertilization and birth are divided into two types, based on whether they are initiated during the embryonic period (first 8 weeks) or the fetal period (the final 32 weeks to birth). Changes occurring in the embryonic period produce gross malformations of body or organ structure and are referred to as *embryonic* or *congenital anomalies*. Diseases of the fetal period are less common and are more similar to diseases of later life in that they produce destructive lesions.

Congenital means present at birth. Embryonic anomalies and fetal diseases are present at birth but may go undetected until adult life (e.g., two ureters on one side). Genetic and chromosomal diseases may also be congenital; however, many genetic diseases are not manifest until sometime after birth and therefore are not congenital. Familial diseases are diseases in which several family members have the same disease. Familial diseases are not necessarily genetic in origin (e.g., common colds tend to be familial but not genetic). We generally tend to think of genetic diseases when we talk of familial disease because of genes being passed from generation to generation. However, many individuals with genetic diseases do not have a family history of the disease because of the recessive nature of many of these diseases or because the disease is caused by a new germ cell mutation. Also, some embryonic anomalies are caused by abnormal genes and may be familial (e.g., polycystic kidneys, cleft lip and palate). In the latter situation, the disease can be classified as both congenital and familial. Most genetic diseases are not congenital. Although the abnormal gene may be present, the disease usually does not develop by the time of birth and may never develop.

FREQUENCY AND SIGNIFICANCE OF DEVELOPMENTAL ABNORMALITIES

Monogenetic (Single-Gene) Diseases

These diseases are inherited by one of four mechanisms: autosomal recessive, autosomal dominant, sex-linked recessive, sex-linked dominant. Autosomal diseases are much more common than sex-linked diseases, and recessive diseases are much more common than dominant diseases. Many of the thousands of types of genetic diseases are incompletely penetrant (not all persons with the appropriate genes will have the disease) and many are variably expressed (not all persons will have the disease to the same severity). Most genetic diseases are uncommon or rare; yet, when taken in aggregate, most families will be influenced by some type of multifactorial genetic disease and may have members with single-gene diseases. Dominant genes are easily recognized when completely penetrant, because the presence of the disease identifies those individuals with the gene, and the line of inheritance can be followed through each generation (Fig. 7–6).

Recessive disorders usually appear sporadically. When they occur, about one-fourth of siblings are involved (both parents are carriers, so the chance of a child getting both abnormal genes is $1/2 \times 1/2$). The family history for others with the disease will usually be negative unless the recessive gene has a very high frequency in the population or there was marriage among related individuals (Fig. 7–7).

Sex-linked recessive disorders will appear in every other generation. Females with the abnormal gene on the X chromosome will not have the disease, because they have an opposing normal gene on the other X chromosome; males with the gene will have

Single-Gene Diseases

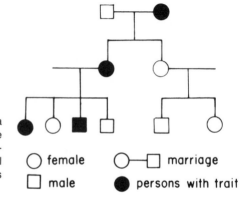

Figure 7–6. Family tree for a dominantly inherited trait. The abnormality appears in the parent of each involved individual unless it arises by mutation or is nonpenetrant.

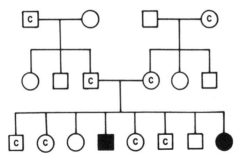

Figure 7–7. Family tree for a recessively inherited trait. The trait appears in one-fourth of offspring when both parents carry the recessive gene. The trait is often absent in previous generations, although carriers (represented by the letter c) are likely to have been present but undetected.

the disease, because the Y chromosome has no gene to oppose the abnormal one on the X chromosome (Fig. 7–8).

Recessive disorders are very difficult to control, because the abnormal genes can be perpetuated through many generations before two carriers mate. Many dominant diseases will arise as the result of mutations and, if severe, will disappear as involved individuals fail to reproduce. Factors favoring the perpetuation of dominant diseases include late onset, mild disease, nonpenetrance, and variable expressivity.

Polygenetic (Multigene) Disorders

These disorders are difficult to define, because inheritance patterns are not clear and because similar disorders may be caused by nongenetic mechanisms. Diabetes mellitus is a good example. Diabetes tends to occur in families but not in a predictable manner. Some cases of diabetes are due to destruction of the pancreas, a nongenetic cause. Some congenital anomalies, such as cleft lip and congenital heart disease, are established as genetic by a definite tendency to occur in families without other explanation for the familial tendency. In addition to the multigene disorders that are reasonably well established as genetic, there is a much broader category of multifactorial disease that may have abnormal genes as one of the

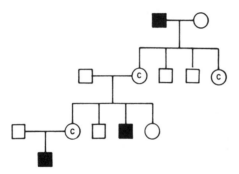

Figure 7–8. Family tree for a sex-linked recessive trait. The trait only appears in males. Uncles or grandfathers are likely to be affected. The abnormal gene may be passed through carrier females for generations.

factors. Such multifactorial diseases probably include many cases of atherosclerotic heart disease, hypertension, and cancer.

Chromosomal Diseases

Most diseases of autosomal chromosomes are so severe as to be incompatible with embryonic development. These account for one-third of the spontaneous abortions that occur during the first third of pregnancy. The other two-thirds of early spontaneous abortions are of unknown cause. Of the types of chromosomal diseases compatible with survival to birth, Down's syndrome (trisomy 21 or mongolism) is most common. Other autosomal chromosomal disorders compatible with survival to birth usually produce severely deformed infants who die within a few months. Abnormalities of sex chromosome number produce relatively mild abnormalities that often are not detected until puberty, when secondary sex characteristics are found to be abnormal.

The overall frequency of genetic disease is difficult to establish because of delayed onset of some types and the variable effects of environmental factors. An estimate of the frequency of genetic disease in children is given in Table 7–1.

Embryonic Anomalies

Approximately 2 percent of newborns are said to have significant congenital anomalies. The causes of these anomalies can be roughly estimated as follows: 65 percent unknown, 20 percent genetic, 5 percent chromosomal, and 10 percent environmental. The genetic causes can be classified as monogenetic (abnormal gene or pair of genes), polygenetic (effects of several genes), or multifactional (combined effect of genes and environmental factors). The chromosomal causes have been discussed previously. Of the environmental causes, radiation is quite capable of causing anomalies but care with diagnostic and therapeutic use of radiation makes this a rare cause. Rubella virus is an important cause of congenital anomalies, but its effects have been reduced by vaccination. Other

TABLE 7–1. INCIDENCE OF GENETIC DISEASES AS PERCENT OF LIVE BIRTHS

Genetic Disease	Incidence (%)
Chromosomal abnormalities	0.4
Single-gene defects	2.0
Multigene disorders	2.6
Total	5.0

(Data from The National Health Education Committee, The Killers and Cripplers—Facts on Major Diseases in the U.S. Today. New York: D. McKay, 1976, p. 127.)

infectious agents, such as cytomegalovirus, herpes virus, and toxoplasma, usually produce their effects during fetal development. Maternal metabolic disorders, such as iodine deficiency and diabetes, can cause anomalies. Drugs, environmental chemicals, and alcohol account for about half of the known nongenetic causes of developmental defects. Thalidomide is a drug that produced many babies with short arms before its effects were discovered. Other drugs, such as anticonvulsants, are less predictable in their effects and more important for maternal health. The effects of environmental chemicals has received much attention, but the effects are difficult to evaluate unless there is a dramatic increase in the frequency of anomalies above the background level. Alcohol abuse is a preventable cause of growth and mental retardation in infants.

Among the various types of anomalies, congenital heart defects are common and most important. Many can be successfully repaired surgically. Abnormalities of the kidney and urinary tract are also quite common; they are often hidden, in which case they are discovered later in life or at autopsy. The gastrointestinal tract harbors many different anomalies, many of which are treatable. Anomalies of the central nervous system are often serious and nonrepairable. Approximately 15 percent of embryonic anomalies are multiple, involving several organs.

Fetal Diseases

Fetal diseases are less frequent than other types of developmental disorders. The most important is erythroblastosis fetalis, which is caused by the mother producing antibodies to the fetus' red blood cells. This disease is a concern whenever the mother has Rh negative red blood cells and the fetus has Rh positive red blood cells. All women should be tested for their Rh status during their first pregnancy, because erythroblastosis fetalis can be prevented or controlled (see below). Syphilis in the fetus is uncommon, but it can be completely prevented if the mother is treated before the middle of pregnancy. Therefore, all pregnant women should be tested for syphilis early in pregnancy to prevent congenital syphilis. Fetal rubella can also be prevented by insuring that the mother is immunized prior to becoming pregnant. Alcohol and other drugs are important causes of damage to the fetus.

Perinatal Diseases

Prematurity and cerebral palsy are two major categories of disease caused during the perinatal period. Approximately 10 percent of babies are born prematurely, as indicated by subnormal birth weight. Major problems resulting from prematurity include brain damage and development of hyaline membrane disease due to immaturity of the lungs. If there is no brain damage and the infant survives, a normal life may be expected. Brain damage in the perinatal period often results in cerebral palsy, a condition characterized by abnormal muscular coordination with or without mental retardation.

Diseases of Infancy

Major problems of infancy that relate specifically to this developmental stage include infections due to lack of previous experience with infectious agents and nutritional problems. Antibodies acquired from the mother tend to protect the infant for about 6 months, after this the infant must experience the infection or be immunized in order to develop antibodies against infectious agents. A common nutritional problem is the failure to get enough iron. Babies fed on milk and fruit, without sources of iron such as cereals, will develop iron-deficiency anemia beginning around 6 months of age, when iron stores acquired from the mother are used up.

Diseases of Childhood and Adolescence

Diseases of these periods rarely relate to developmental problems. They differ from diseases of older age groups only in relative frequency. Accidents and infections are more common, whereas degenerations and neoplasms (except for a few types) are uncommon.

SYMPTOMS, SIGNS, AND TESTS

Genetic Diseases

Each specific disease has its own highly individual manifestations, which vary in time of onset and nature of the lesions. Some genetic diseases are present at birth (congenital), but most appear later. Some genetic diseases present as inflammations, degenerations, or growth disturbances. For example, cystic fibrosis presents with pancreatitis and pneumonia, Friedreich's ataxia presents with degeneration of nerve cells, and retinoblastoma presents as a malignant neoplasm of the eye in children. A large number of genetic disorders produce specific recognizable biochemical defects. These are called inborn errors of metabolism and therefore may be also classified as metabolic diseases.

Biochemical defects can be subdivided into defects in structural protein and defects in enzymes. An example of structural protein defect is sickle-cell anemia, in which the hemoglobin of red blood cells is defective, resulting in premature death of red blood cells. Enzyme defects are subdivided on the basis of the type of substance affected into disorders of carbohydrate, protein, lipid, and mineral metabolism. An example of an inborn error of carbohydrate metabolism is glycogen storage disease, in which glucose can be converted into glycogen for storage, but one of the enzymes needed to convert the glycogen back to glucose is missing. This results in excessive storage of glycogen and enlargement of organs in which glycogen is stored. Biochemical tests are available for some inborn errors of metabolism, but many are diagnosed mainly by clinical findings. A few genetic diseases can be diagnosed before birth by testing of the amniotic fluid (Fig. 7–9). Several biochemical diseases, including sickle-cell anemia, phenylketonuria, galactosemic, and hypothyroidism, are routinely tested for at birth, using a small blood sample obtained from the head.

Chromosomal Diseases

These diseases are suspected and often diagnosed by physical examination. Karyotyping should be used to substantiate the diagnosis. Karyotyping is performed only at large referral centers, but is often needed for accurate diagnosis and counseling.

Embryonic Anomalies

These gross anomalies are detected by physical examination, radiologic procedures, or exploratory surgery.

Fetal Diseases

The method of diagnosis depends on the nature of the disease. Infections are diagnosed by culture or measurement of antibodies (e.g., serology for syphilis). Immunologic diseases are diagnosed by measuring antigen and antibodies (e.g., Rh factor in fetal and

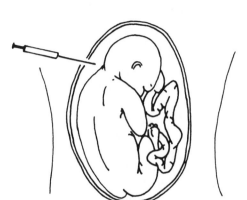

Figure 7–9. Amniocentesis to obtain fluid and cells for the detection of some genetic, chromosomal, and fetal diseases.

maternal red blood cells). Hydrocephalus, polycystic kidneys, and severe bone disease can be detected by ultrasound.

EXAMPLES OF SPECIFIC DISEASES

Genetic Diseases

Relatively common and important examples of genetic diseases include diabetes mellitus, gout, sickle-cell anemia, hemophilia, phenylketonuria, muscular dystrophy, and cystic fibrosis.

Diabetes mellitus is probably the most important genetic disease, because it is common (2 percent of the population), requires treatment (diet and drugs), and has many complications. It is discussed in Chapter 25.

Gout is about one-tenth as common as diabetes (2 per 1000 persons). It is due to an unknown biochemical defect that causes overproduction of uric acid. Uric acid is a normal breakdown product of purines (used to build DNA) and is excreted by the kidney. When the blood level of uric acid is high, crystals sometimes form in joints, leading to the acute inflammation of gout. If untreated, deposits of urate crystals build up over a period of years in cartilage of the joints and the ear. These deposits are called tophi. Acute and chronic changes in joints are referred to as acute and chronic gout, respectively. For some reason, gout affects men, not women, and is a disease of older men. Three types of drug treatment are available: allopurinol inhibits the body's production of uric acid, probenecid promotes excretion of uric acid by the kidneys (and thereby may promote formation of uric acid stones in the kidney), and colchicine inhibits the acute inflammatory reaction in joints to relieve pain temporarily.

The gene for sickle-cell anemia causes a structural defect in the hemoglobin molecule. When two such genes are present (homozygous), the blood cells of such individuals often are sickle-shaped. Under stressful circumstances, such as a respiratory infection, the red cells are destroyed, producing severe anemia (so-called sickle-cell crisis). Over a period of years, the disease is fatal. This is a recessive disorder prevalent among blacks. Heterozygotes can be identified because their red blood cells show a slight tendency to sickle.

The two types of hemophilia (classic hemophilia and Christmas disease) are similar, sex-linked disorders. Hemophilia is an example of a rare genetic disease (2 per 100,000 persons) that commands a large amount of medical care in the treatment of severe bleeding and long-term prevention of bleeding. A specific enzyme in the blood-clotting mechanism (factor VIII or factor IX) is deficient, leading to spontaneous bleeding into joints or uncontrollable bleeding when cut. The missing factor can be replaced on a temporary basis, but at great expense.

Phenylketonuria is caused by an absence of phenylalanine hydroxylase, which results in accumulation of phenylketones. Over a period of months, brain damage results. Diagnosis can be made by testing newborns for phenylketones in the blood. This is an example of screening to detect a rare disease.

The most common type of muscular dystrophy (Duchenne) occurs in males because it is a sex-linked recessive disorder. Skeletal muscles degenerate over a period of years, leading to death in childhood. The genetic defect presumably involves the structure of muscle fibers. Diagnosis of muscular dystrophy requires expertise in differentiation from other genetic and acquired musculoskeletal disorders.

Cystic fibrosis is an autosomal recessive disorder characterized by thick secretions that lead to obstruction of body passageways. The pancreas is particularly affected, because blockage of ducts leads to decreased delivery of pancreatic enzymes to the intestine and thus to poor digestion of food and marked weight loss. In the lungs, the obstruction of bronchi predisposes to bouts of pneumonia caused by the trapped bacteria. Prevention of pneumonia may prolong life to adulthood. Analysis of chloride secretion in sweat provides a diagnostic test (sweat chloride test). The nature of the biochemical defect is unknown.

The genes that cause sickle-cell disease, phenylketonuria, and Duchenne's muscular dystrophy have been identified using DNA techniques, resulting in improved testing for carriers and prenatal diagnosis, and providing hope for advances in treatment. Techniques in DNA analysis are being applied to many other diseases, including cystic fibrosis.

CHART 7–1. GENETIC DISEASES

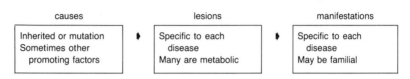

causes	lesions	manifestations
Inherited or mutation Sometimes other promoting factors	Specific to each disease Many are metabolic	Specific to each disease May be familial

Chromosomal Diseases

The most common types of chromosomal disorders include increases (trisomy) or decreases (monosomy) in the normal complement of autosomes or in the sex chromosomes, combining of chromatin material, or the deletion or duplication of parts of one or more arms of the involved chromosome. Examples include Down's syndrome, Klinefelter's syndrome, Turner's syndrome, and the fragile X syndrome.

Down's syndrome (trisomy 21) occurs in 1 of every 1000 newborns and is the most common genetic cause of mental retardation. It is a chromosomal disorder associated with mental deficiency and other minor common dysmorphic characteristics. With appropriate surgical and medical treatment of anatomic (cardiac septal defects), physiologic (anemia), and neoplastic (leukemia) disorders, as well as the provision of adequate educational, vocational, and social opportunities, the life expectancy and quality of life for persons with Down's syndrome have markedly improved.

Klinefelter's syndrome occurs in males with one Y and two X chromosomes. It is usually not diagnosed until puberty, when lack of male sexual development, breast enlargement, and chubby appearance may be detected. One in every 1000 males is affected.

Turner's syndrome occurs in females with only one X chromosome who fail to show normal female secondary sex development at puberty. They are short in stature with a broad neck. The condition is less common than Klinefelter's syndrome.

Fragile X-syndrome produces mild to moderate mental retardation and is the second most common genetic cause of mental retardation. A fragile site on an X chromosome is transmitted from generation to generation but is expressed only in males. Unlike most chromosomal diseases, this condition is familial with a sex-linked recessive pattern of inheritance.

CHART 7–2. CHROMOSOMAL ABNORMALITIES

cause	lesion	manifestations
Defective or missing chromosomes	Gross changes in body structure or sex characteristics	Observable changes in body structure Abnormal chromosomes Mental retardation

Embryonic Anomalies

Tetralogy of Fallot, agenesis of one kidney, Meckel's diverticulum, and meningomyelocele are representative examples from the four most commonly involved organs.

Tetralogy of Fallot is one of the more common serious anomalies of the heart in which a defect between the two chambers of the heart and misplacement of the aorta allow blood to flow from the right side of the heart to the left side without being oxygenated by the lungs. The infants are often cyanotic (blue) from lack of oxygen in their red blood cells and have a loud heart murmur. The defects can be repaired in some cases.

One kidney may be small (hypoplastic) or missing (agenesis). This only presents a problem if the other kidney becomes diseased and needs to be removed. Otherwise this anomaly may go undetected.

Meckel's diverticulum is an outpouching of the distal small intestine and is a remnant of an embryonic connection between the intestine and yolk sac. It occurs in 2 percent of the population and is usually harmless.

Meningomyelocele is an outpouching (-*cele*) of meninges (covering of spinal cord) and spinal cord (-*myelo*-) through a defect in the bony structure of lumbar vertebrae. The brainstem often has an associated defect, with the brain being pushed into the foramen magnum (hole at base of skull). Compression of the brainstem may block the flow of cerebrospinal fluid, resulting in hydrocephalus (enlargement of brain and head due to increased fluid in the ventricles of the brain). Surgical repair of the meningomyelocele will prevent secondary infection of the protruding mass but will neither correct the damage done to nerves nor the hydrocephalus.

CHART 7–3. EMBRYONIC ANOMALIES

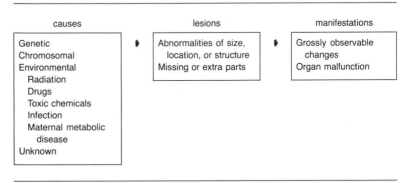

causes	lesions	manifestations

Genetic
Chromosomal
Environmental
  Radiation
  Drugs
  Toxic chemicals
  Infection
  Maternal metabolic
    disease
Unknown
	Abnormalities of size,
	location, or structure
	Missing or extra parts
		Grossly observable
		changes
		Organ malfunction

Fetal Diseases

Examples include erythroblastosis fetalis and congenital infections.

Erythroblastosis fetalis occurs when maternal antibodies destroy fetal red blood cells. In the latter part of pregnancy or at the time of delivery, some Rh+ fetal red blood cells leak across the placenta and enter the maternal circulation. The mother forms antibodies to the Rh factor. This rarely causes any problem with the first pregnancy. With a second or later pregnancy involving an Rh positive fetus, maternal serum antibodies may cross the placenta and destroy the fetal red blood cells. This is called *erythroblastosis fetalis*. The severity varies greatly. The fetus may go into heart failure because of the anemia produced by destruction of red blood cells and be stillborn, or the problem may not develop until after birth, when massive destruction of blood cells may produce jaundice. The bilirubin pigment responsible for the jaundice is a breakdown product of red blood cells. Bilirubin in high concentration can produce brain damage and mental retardation. It is now possible to limit maternal antibody production caused by leakage of fetal red cells at the time of delivery. Gamma globulin containing Rh positive antibodies is given to the mother just after delivery. These antibodies tie up the Rh positive antigen on the fetal cells that have leaked into the maternal circulation, thus preventing the mother's immune system from recognizing them as foreign and producing antibodies to them. If this procedure is followed for each pregnancy involving an Rh negative mother and Rh positive fetus, the incidence of erythroblastosis is reduced.

Syphilis is a disease that may go unrecognized, especially in females, where the primary lesion is hidden in the vagina. The spirochetes (organisms causing syphilis) may circulate in the blood for months without being noticed. During the second half of pregnancy, the spirochetes are capable of crossing the placenta and causing syphilis in the fetus. The severity of the disease in the fetus varies greatly; it may cause death of the fetus and stillbirth or may only be discovered years later as mental retardation or deformity of bones. The disease can be prevented by giving the mother penicillin during the first half of pregnancy. Maternal syphilis is easily diagnosed by testing for antibodies to the spirochetes (serology). For this reason, a serologic test for syphilis is recommended at the time of the first prenatal visit.

Congenital syphilis is rare in the United States because we understand the disease and know how to treat it. Toxoplasmosis and cytomegalovirus infections can be transmitted transplacentally and can cause acute illnesses or destructive brain lesions. Herpes simplex infection can be transmitted during birth and produce a generalized fatal infection.

Heavy alcohol intake, especially during early pregnancy, produces the *fetal alcohol syndrome*, which is characterized by growth retardation, abnormal facial appearance, septal defects in the heart, and mental retardation due to malformation of the brain.

CHART 7–4. FETAL DISEASES

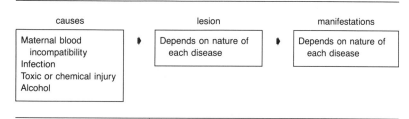

causes	lesion	manifestations
Maternal blood incompatibility Infection Toxic or chemical injury Alcohol	Depends on nature of each disease	Depends on nature of each disease

REVIEW QUESTIONS

1. What are the causes of genetic diseases, chromosomal abnormalities, and congenital anomalies? At what stage of development do these diseases become manifest?
2. What is the value of karyotyping?
3. What are the differences among congenital diseases, familial diseases, and genetic diseases?
4. What are the differences in inheritance patterns among multigene genetic disease, variably expressed inherited disease, and the four types of single-gene inheritance?
5. Approximately what percentage of births are associated with single-gene abnormalities, recognized multigene abnormalities, chromosomal abnormalities, and embryonic anomalies? What is the relative frequency of the various inheritance patterns for single-gene abnormalities? What organs are most frequently involved by embryonic anomalies?
6. What types of diseases are common in the perinatal period, in infancy, and in childhood?
7. How are genetic diseases, chromosomal abnormalities, and embryonic congenital anomalies likely to differ in their appearance? Illustrate these differences by citing specific diseases.
8. What do the following terms mean?
    Trisomy
    Monosomy
    Duplication
    Deletion
    Translocation
    Mosaicism
    Nonpenetrant
    Variable expressivity
    Inborn error of metabolism

Section III

Major Organ-Related Diseases

The purpose of this section is to survey the diseases of humans as they cause problems in the various organs. Most diseases present with a problem that can be associated with an organ; most laypersons can tell you the organ that is causing their problem. To deal with these problems, it is necessary to know which are most likely, how they can be diagnosed, and the specific features of the possible diagnoses. With the exception of Chapter 22 (Mental Illness), each chapter is organized for your convenience according to the following format:

- *Review of Structure and Function*
- *Most Frequent and Serious Problems*
- *Symptoms, Signs, and Tests*
- *Specific Diseases*
  Genetic/Developmental Diseases
  Inflammatory/Degenerative Diseases
  Hyperplastic/Neoplastic Diseases
- *Organ Failure*

Degenerative diseases is used here as a nonspecific classification to include all acquired non-neoplastic conditions that are presumably the result of exogenous or endogenous injuries.

Heart

REVIEW OF STRUCTURE AND FUNCTION

The heart functions as two pumps working synchronously to move blood to all sites in the body (Figs. 8–1 and 8–2). The right side of the heart receives poorly oxygenated blood from the vena cava and pumps it under relatively low pressure to the lungs where it is oxygenated. The left side of the heart receives the well-oxygenated blood from the pulmonary veins and pumps it at high pressure to all areas of the body. The right heart is less muscular than the left heart due to the lower pressure on the right side and minimal

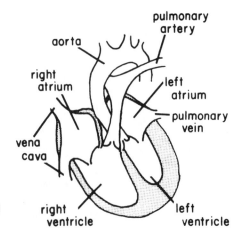

Figure 8–1. Great vessels and chambers of the heart.

pulmonary vascular resistance. The muscles of each side of the heart are capable of undergoing hypertrophy to pump blood at higher pressure under disease conditions, such as when blood has to be pumped through a narrowed valve. Within each side of the heart, there is a receiving chamber (atrium) and a major pumping chamber (ventricle). Delicate fibrous valves open and close efficiently with a quick snap in response to pressure changes, producing the normal heart sounds. Valves control the inflow of blood in the ventricle and prevent back flow. If the valves fail to close completely, blood may be pumped by the ventricle back into the atrium during contraction (systole), or blood may flow back into the ventricle from the major artery during relaxation (diastole). In either case, the ventricle would have to pump the blood more than once, with resultant loss of efficiency of the heart. If the valves fail to open completely, more pressure is needed to force the blood

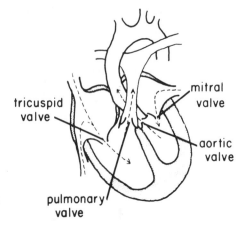

Figure 8–2. Blood flow and valves of the heart.

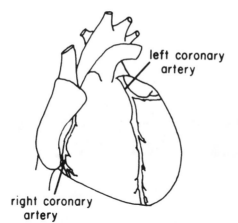

Figure 8–3. Coronary arteries.

through the narrowed valves. To some extent, the heart can gain this needed increase in force by hypertrophy of its muscles. Atrial muscle enlarges when the atrioventricular valves (tricuspid or mitral) are narrowed and fail to open completely, and ventricular muscle enlarges when the pulmonary or aortic valves are narrowed and fail to open completely.

Cardiac muscle must have a generous supply of oxygenated blood to provide the fuel for its high energy needs. Blood is supplied to the muscle via two medium-sized arteries, the right and left coronary arteries, which have their origin from the aorta immediately above the aortic valve (Fig. 8–3). In addition to the rich blood supply, an electrical pulse is needed to initiate each rhythmical contraction of the heart. This pulse is generated spontaneously within a pacemaker focus called the sinoatrial node and is conducted to the ventricles by the arterioventricular node and specialized bundles (Fig. 8–4). The pulse may be disturbed by disease

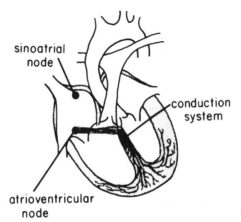

Figure 8–4. Conduction system.

in the area of the node, by metabolic changes in the blood, such as changes in serum potassium level, and by a number of drugs.

MOST FREQUENT AND SERIOUS PROBLEMS

Major diseases of the heart result from ischemic injury to cardiac muscle, damaged heart valves, and increased work load for the pumps produced by altered blood flow or high blood pressure in pulmonary or systemic vascular circuits.

Compromise of the blood supply to the heart (ischemia) caused by narrowing of the main coronary arteries from atherosclerosis is by far the most common cause of death in the United States and thus is the most common cause of cardiac deaths (Fig. 8–5). Atherosclerosis leads to cardiac muscle dysfunction through myocardial infarction and/or abnormal heart rhythms (arrhythmias). Coronary atherosclerosis not only is common as a cause of death, it is also common as a cause of disability, which may last for weeks to years prior to death. The major forms of disability are angina pectoris and heart failure.

Hypertensive heart disease, the second most common heart problem, is underestimated in the mortality figures, because it often occurs in conjunction with coronary atherosclerotic heart disease. Hypertensive heart disease is not a primary disease of the heart, but rather it is caused by high blood pressure in the systemic

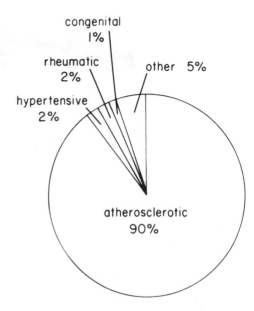

Figure 8–5. Heart diseases, percent as cause of death. *(Data from Bureau of Census, Statistical Abstracts, 1976.)*

arteries causing an increased work load for the heart. Congenital heart disease and rheumatic endocarditis are less common, but are particularly important as causes of valve deformity. Many types of congenital heart disease also produce abnormalities in the pathway of blood flow leading to poor oxygenation of blood and increased work load on the heart. Chronic lung disease can cause hypertension in the pulmonary arteries and lead to heart failure. Heart failure secondary to lung disease is called *cor pulmonale*. The frequency of cor pulmonale as a cause of death is not known, because these patients are usually recorded as having died from lung disease.

Most persons succumbing to congenital heart disease die in the first year of life (Fig. 8–6). Those that survive have less severe anomalies or undergo surgical correction of their defect and have variable survival times extending to old age. Rheumatic and hypertensive heart diseases have a prolonged course, so that most deaths occur in older people. Atherosclerotic heart disease has a shorter course (Fig. 8–7), and the mortality rate rises rapidly after age 30 and continues to increase at all older age levels (Fig. 8–6).

Sudden death resulting from heart disease is most often associated with one of several complications of coronary atherosclerosis but may also be associated with other types of heart disease. Sudden death due to a "heart attack" generally is the result of ventricular fibrillation, probably due to a sudden episode of myocardial ischemia. Ventricular fibrillation is an uncoordinated, ineffective,

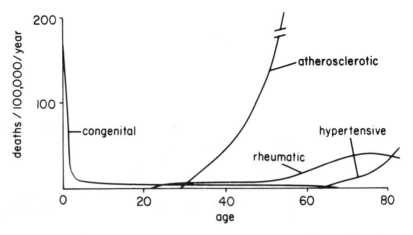

Figure 8–6. Mortality rate by age for various types of heart disease. *(Data from Monthly Vital Stat Rep 25:16, 1977; Moriyama IM, Krueger DE, Stamler J: Cardiovascular Diseases in the United States. Cambridge, Harvard University Press, 1971, p. 263.)*

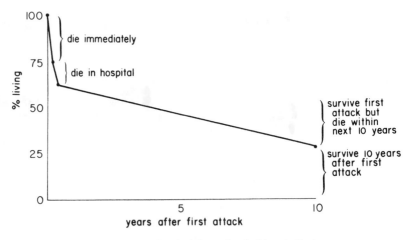

Figure 8–7. Survival time after first heart attack.

weak contraction of ventricular muscle due to spontaneous generation of impulses within the muscle rather than coordinated electrical stimulation through the conduction system. Overdosage with
certain cardiac drugs may also produce ventricular fibrillation.
Rupture of the heart is not common but does produce sudden
death, because the pericardial sac fills with blood and prevents
adequate pumping of blood out of the heart. Rupture may be due
to the softening of a large myocardial infarct or trauma (car accident, stabbing, gunshot wound). Marked narrowing of the aortic
valve (aortic stenosis) also is associated with sudden death.

SYMPTOMS, SIGNS, AND TESTS

Sudden severe pain is a symptom of many but not all myocardial
infarcts. The pain of myocardial infarction is persistent and may be
a squeezing chest pain or may be pain referred to the shoulder,
arm, neck, or jaw. The term *angina pectoris* refers to transient chest
pain brought on by exercise or emotional stress and relieved by rest
or vasodilator drugs. Angina pectoris is due to transient hypoxia to
the heart with spasm of atherosclerotic coronary arteries. Congestive heart failure is a cluster of symptoms, signs, and laboratory
measurements that indicates inadequate pumping of the heart
relative to the demands of the peripheral tissues. The resultant
buildup of pressure in the pulmonary or systemic veins leads to
venous distention and edema.

Important aspects of physical examination, which relate to
possible heart disease, include listening for murmurs, estimating

heart size, measuring blood pressure, and checking for signs of heart failure. Heart murmurs are abnormal sounds of the heart usually heard with the aid of a stethoscope. The process of listening through a stethoscope is called *auscultation*. Murmurs are caused by abnormal flow of blood through the heart valves or major vessels. A heart murmur may be functionally unimportant or may indicate serious underlying disease.

Arterial blood pressure is usually referred to simply as the *blood pressure* and is measured using a sphygmomanometer. A sphygmomanometer is a cuff with an attached pressure gauge that can be applied to an extremity and inflated above arterial pressure. By listening over an artery distal to the cuff, one can hear the pulse when the pressure drops below the systolic pressure. The pulse sound disappears when the blood pressure drops below the diastolic pressure.

Venous blood pressure, which is used as an indicator of congestive heart failure, is estimated on physical examination by the degree of distention of neck veins and, when necessary, is measured accurately through a catheter placed intravenously. The manifestations of heart failure are discussed at the end of this chapter in the section on Organ Failure.

Procedures and laboratory tests used in the evaluation of heart disease include chest x-ray, electrocardiogram (ECG), serum enzyme levels, and cardiac catheterization. Chest x-ray is used to evaluate the size and shape of the heart. Enlargement of the heart usually reflects muscle hypertrophy or dilatation of the chambers.

Electrocardiography is carried out by the placement of a series of electrodes at various standard locations on the body surface. From electrical activity at these sites, the electrical activity of the heart can be inferred. ECG is most often used in the diagnosis of myocardial infarcts, either recent or old. Results may be normal, nonspecific, or relatively specific and usually require interpretation by an experienced physician. ECG is also used to define rhythm disturbances (arrythmias) and to monitor the course of a disease or the effect of therapy. An ECG taken after exercise is used to detect evidence of ischemia due to atherosclerotic narrowing of the coronary arteries.

Enzymes released from dying muscle can be measured in the blood and are helpful in determining if and when an acute myocardial infarct has occurred. Creatinine phosphokinase (CK), aspartate aminotransferase (AST), and lactic dehydrogenase (LDH) have peak elevations at different times following onset of a myocardial infarct (Fig. 8–8).

Cardiac catheterization is used for more extensive evaluation of serious cardiac problems. Catheters are tubes that can be

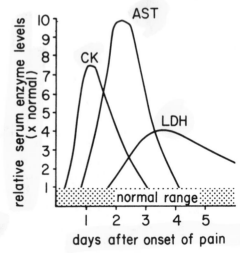

Know

Figure 8–8. Patterns of serum enzyme elevations following a myocardial infarct.

threaded into the vessels and heart to measure pressure and to allow injection of radiopaque dyes. X-rays taken following injection of radiopaque dyes into the vascular system are called *angiograms.* Abnormalities in the rate and route of blood flow in various parts of the heart and great vessels can be detected through cardiac catheterization. The angiograms taken during catheterization reveal the anatomy of the heart chambers and blood vessels. Radiopaque dyes can even be injected directly into the coronary arteries to outline atherosclerotic plaques. Newer, noninvasive techniques for evaluating heart disease include echocardiography, which utilizes ultrasound to define anatomic structures, and radioisotope techniques, which are used to evaluate the rate of flow in the ventricles and coronary arteries.

SPECIFIC DISEASES

Genetic/Developmental Diseases

Developmental abnormalities of the heart are almost all embryonic anomalies, which collectively are called *congenital heart disease.* Of all the organs, the heart and great vessels are the most common sites of congenital defects, and the defects are likely to have serious consequences. Early diagnosis and surgical correction of the defect alter the life expectancy in many cases. Ventricular septal defects account for 30 percent of congenital heart diseases. The variety of other types each comprise less than 10 percent. A few classic types are briefly presented below to illustrate the variety of blood flow disturbances of congenital disease of the heart and great vessels.

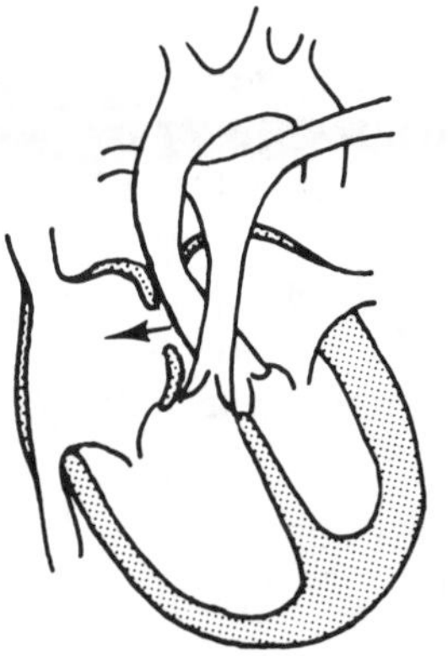

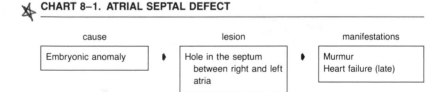

Figure 8–9. Atrial septal defect.

Atrial Septal Defect

Know

During the fetal period, there is a hole (called *foramen ovale*) in the septum between the right and left atria that usually closes shortly after birth. If this septal defect remains open and is large (Fig. 8–9), blood may flow from the left atrium to the right atrium causing increased work load on the right side of the heart.

CHART 8–1. ATRIAL SEPTAL DEFECT

cause		lesion		manifestations
Embryonic anomaly	▶	Hole in the septum between right and left atria	▶	Murmur Heart failure (late)

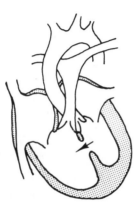

Figure 8–10. Ventricular septal defect.

Ventricular Septal Defect

A hole in the interventricular septum (Fig. 8–10) is more serious than an atrial septal defect, because there is a greater pressure difference between the two ventricles than between the two atria. Some blood flows from the left ventricle (high pressure) to the right ventricle (low pressure), forcing both ventricles to pump the same blood more than once. Over a number of years, the added work load will lead to left and/or right heart failure.

CHART 8–2. VENTRICULAR SEPTAL DEFECT

cause	lesion	manifestations
Embryonic anomaly	Hole in the septum between right and left ventricles	Murmur Heart failure

Tetralogy of Fallot

This is the most common cause of cyanotic congenital heart disease. *Cyanosis* refers to the blue color of the skin caused by poor oxygenation of blood. In cyanotic heart disease, the poor oxygenation is caused by shunting of blood from the right heart to the systemic circulation, so that some blood fails to reach the lung to receive oxygen. The four changes of tetralogy (Fig. 8–11) are (1) transposition of the aorta so that both ventricles empty into the aorta, (2) a stenosis (narrowing) of the pulmonary outflow tract,

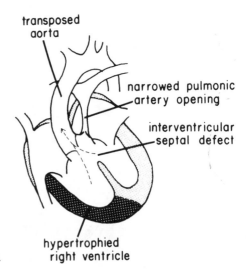

Figure 8–11. Tetralogy of Fallot.

(3) a ventricular septal defect allowing the blood that cannot get through the pulmonary valve to pass into the systemic circulation, and (4) right ventricular hypertrophy caused by the increased force required of the right ventricle. Because only a fraction of the blood will be oxygenated, the child usually will be cyanotic. Life expectancy is variable, depending on the severity of the defects. Surgical correction of the abnormalities may allow prolonged survival.

CHART 8–3. TETRALOGY OF FALLOT

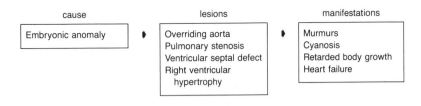

cause	lesions	manifestations
Embryonic anomaly	Overriding aorta Pulmonary stenosis Ventricular septal defect Right ventricular hypertrophy	Murmurs Cyanosis Retarded body growth Heart failure

Coarctation of Aorta

Coarctation is a narrow fibrous constriction in the descending thoracic aorta (Fig. 8–12). Proximal to the coarctation the blood pressure is usually elevated, and distal to the coarctation it is decreased. Diminution of the femoral artery pulse in a child may lead one to suspect the presence of this defect, and a marked difference in blood pressure in the arm versus the leg strengthens the suspicion. Surgical correction usually is curative if the disease is diagnosed early. The disease may easily go undetected, however, and, if so, the high blood pressure proximal to the coarctation eventually leads to left ventricular hypertrophy and heart failure.

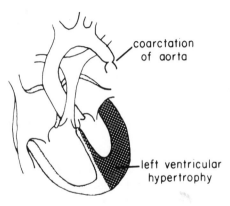

coarctation
of aorta

left ventricular
hypertrophy

Figure 8–12. Coarctation of aorta.

CHART 8–4. COARCTATION OF AORTA

cause		lesion		manifestations
Embryonic anomaly	▸	Stenosis of descending thoracic aorta	▸	Decreased femoral pulse Hypertension in upper extremity Murmur Heart failure (late)

Inflammatory/Degenerative Diseases

Degenerative and inflammatory diseases of the heart are the most important group of diseases in the United States. They are of diverse causes, but the outcome is similar in that they cripple this vital organ. Hypertensive heart disease is not really a primary heart disease but is discussed here because of its importance and similarity to other heart disease. See Chapter 9 for further discussion of hypertension.

Coronary Artery Atherosclerosis (Arteriosclerotic Heart Disease)

The intimal plaques that build up with age in the right and left coronary arteries produce the most significant disease in the United States. The etiology and epidemiology of atherosclerosis are discussed in Chapter 9.

We are concerned here with how these small plaques may kill or cripple an individual. Simplistically, we can say that the plaques narrow the lumens of the coronary arteries, thus reducing blood flow to heart muscle and to the conduction system of the heart (Fig. 8–13). In practice, the correlation between the number and size of plaques and significant damage to the heart is not so simple, because the patterns of circulation vary considerably depending on anastomoses (connections) between right and left coronary arteries. For example, a single appropriately placed plaque may kill one person at a young age, whereas on rare occasions another person may gradually develop complete occlusion of both coronary arteries and survive to old age. Increased vascularization of the myocardium due to development of anastomoses occurs with recovery from a myocardial infarct and with aging.

Symptoms of inadequate blood flow in the coronary arteries may be acute, chronic, or both. In either case, the coronary arteries have one or more areas of significant narrowing by atherosclerotic plaque. An area of marginally adequate blood flow may suddenly become inadequate when there is an increased need for oxygen by the myocardium due to exercise or emotional stress or when the artery is blocked due to thrombosis (formation of blood clot), vaso-

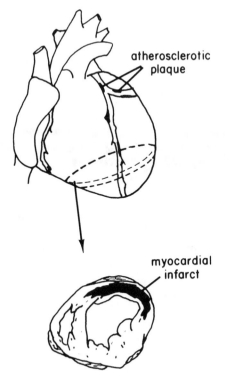

atherosclerotic
plaque

myocardial
infarct

Figure 8–13. Coronary artery atherosclerosis and myocardial infarct.

constriction (spasm), or a dissection of blood into the plaque (plaque hemorrhage) with resultant pushing of the plaque into the lumen of the coronary artery. The roles of thrombosis in coronary artery obstruction has been debated; currently, it is thought to be an important factor in most cases even though it is now always present at autopsy in fatal cases.

The results of sudden insufficiency of coronary artery blood flow are arrhythmia, infarct, or both. Persons who die suddenly of "heart attacks" have an arrhythmia with a failure of the electrical impulse that times the pump; persons who survive the attack usually develop a myocardial infarct. Arrhythmias are usually reversible, so emergency resuscitation is practical; infarcts can heal.

Chronic insufficiency of coronary artery blood flow causes pain accentuated by exertion (angina pectoris) because of the relative ischemia produced by the increased need of the myocardium for oxygen. Vasodilator drugs usually relieve the pain.

The degree of coronary artery obstruction by atherosclerotic plaque may be evaluated directly and indirectly. Radiopaque dyes injected into the coronary arteries to produce arteriograms are used to evaluate areas of obstruction; however, they are not always accurate in predicting the degree of functional obstruction. Electrocardiograms taken during exercise may demonstrate electrical changes associated with ischemia. With the development of techniques for bypassing the coronary arteries with vein grafts, the evaluation of the degree of coronary atherosclerosis has become more important. Current evidence, however, suggests that bypass is most useful with severe angina and in some patients in the early stage of infarction. It should be noted that the viewing of collapsed atherosclerotic arteries at autopsy is not a very accurate measure of their functional state during life.

When an area of myocardium dies, the lesion is called a *myocardial infarct* and the process is called *myocardial infarction*. The major site of involvement is the left ventricle, although some infarcts extend into the right ventricle. Once an area of muscle is killed by lack of oxygen, the following sequence of changes occurs:

1. For 12 to 18 hours, the heart appears normal both grossly and microscopically.
2. During days 1 to 5, there is progressive softening and disintegration of the dead muscle fibers and an inflammatory reaction at the edge of the necrotic area.
3. Healing begins at day 5 with the ingrowth of fibroblasts and capillaries (granulation tissue), which replace the dead muscle and form collagen.
4. By 2 weeks, sufficient collagen has been laid down to give new strength to the softened area, and the collagen continues to accumulate and contract for weeks to months. The end result is a dense tough scar not unlike that which can occur in the skin following a cut.

Complications can develop at any stage of the process of myocardial infarction. Arrhythmias, such as ventricular fibrillation, may occur. Ventricular fibrillation produces sudden death except in some cases where medical care is immediately available. Artificial maintenance of heart contraction and breathing (cardiopulmonary resuscitation) may be successful. Another possible complication is heart failure due to loss of function of a large area of myocardium. Heart failure may cause venous congestion in the lungs with seepage of fluid into the pulmonary air spaces (pulmonary edema). When more severe, there may be an inability to maintain the arterial blood pressure (cardiogenic shock). Heart failure in the first few days after a myocardial infarct is more common in persons whose heart has been weakened by previous myocardial infarcts.

Following a myocardial infarct, the area of damage may extend due to further inadequacy of the blood supply. A major reason for requiring bed rest is to prevent intolerable demand on the heart through exercise. Extension of the infarct may tip the patient into heart failure or cause a fatal arrhythmia. If sufficient muscle is lost, by single or multiple infarcts, the remaining muscle hypertrophies in response to the increased demand placed on it. Thus, myocardial infarcts are one cause of cardiac enlargement. Myocardial hypertrophy is a reflection of the heart's inadequacy to function normally, and eventually this will be expressed by heart failure. Hypertrophy is a late complication of myocardial infarct, because it requires time to develop.

Three less common complications of large infarcts are rupture, ventricular aneurysm, and endocardial thrombus formation. Myocardial rupture occurs when blood dissects through necrotic myocardium into the pericardial space with resultant compression of the heart and sudden death. Rupture occurs at the time of maximum softening (about 5 to 7 days) of a large infarct. Aneurysms are sac-like outpouchings of the heart or vessels. A ventricular aneurysm is an outward bulging of the scar of a large, healed left ventricular infarct. The aneurysm tends to make the left ventricle an ineffective pump because the blood is not effectively moved out of the aneurysm. Mural thrombosis is the formation of a thrombus over the inner endocardial wall of an infarct. Sometimes material breaks loose from a mural thrombus, producing free-floating material in the systemic arterial circulation (arterial emboli) that can lodge at distant sites producing infarcts. Brain, intestines, and extremities are most commonly affected. For example, a patient recovering from a myocardial infarct may suddenly develop a stroke because thrombotic material has embolized from the heart to cerebral vessels causing an infarct of the brain.

Symptoms and signs of myocardial infarcts are quite variable. Most patients have no symptoms prior to their first myocardial infarct, although a few may experience angina pectoris. Angina pectoris is more frequent in persons who have had one or more previous myocardial infarcts and have a marginally adequate blood supply to the heart. Although pain is the major symptom of myocardial infarction, infarcts may be asymptomatic and heal without the patient being aware of them. Thus, some patients may first present with heart failure due to advanced disease.

CHART 8–5. CORONARY ARTERY ATHEROSCLEROSIS

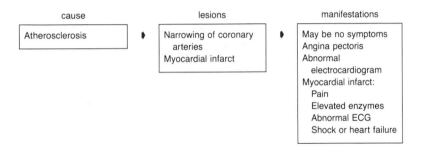

cause	lesions	manifestations

Atherosclerosis	Narrowing of coronary arteries
	Myocardial infarct
		May be no symptoms
		Angina pectoris
		Abnormal electrocardiogram
		Myocardial infarct:
		  Pain
		  Elevated enzymes
		  Abnormal ECG
		  Shock or heart failure

Rheumatic Heart Disease

Rheumatic fever is a hypersensitivity disorder occurring in a small percentage of persons following a streptococcal infection, usually pharyngitis. Apparently, the protein of group A hemolytic streptococci is similar to the proteins in the heart and other connective tissues of these susceptible individuals, so that the antibodies that develop against the streptococci, not only attack the bacteria but also the host tissue (especially the heart and joints). The inflammatory reaction produced by this hypersensitivity is usually mild or asymptomatic but may produce a full-blown illness called *rheumatic fever* characterized by myocarditis and arthritis. The illness is not usually serious in the acute stages and may gradually resolve without apparent residual effects. The sequelae of one or more episodes of rheumatic fever cause inflammation of the heart valves (valvulitis) rather than the myocardium and lead to scarring and deformity of the valves (Fig. 8–14).

At variable times following a known or presumed acute illness, chronic damage to heart valves may become evident. The functional valve damage is produced by narrowing of valve opening (stenosis) and/or failure of the valve to close completely (valvular insufficiency). Fibrosis of the mitral valve, the valve most commonly involved, usually leads to left heart failure. The left heart failure is due either to backup of blood into the lungs caused by stenosis or to regurgitation of blood back through the insufficient valve. Stenosis and insufficiency of the aortic valve, the other commonly affected valve, cause left ventricular hypertrophy and eventually heart failure. Surgical replacement of damaged valves may prolong life.

In addition to heart failure, the valves may become secondarily infected, producing infective endocarditis (discussion follows).

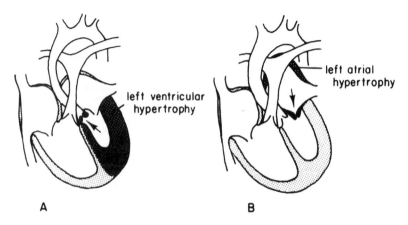

A. left ventricular hypertrophy

B. left atrial hypertrophy

Figure 8–14. A. Aortic stenosis. **B.** Mitral stenosis.

Thus, rheumatic fever itself is less serious than its potential sequelae, rheumatic valvulitis and infective endocarditis. The pathogenesis of rheumatic heart disease is summarized in Table 8–1.

To prevent the long chain of events of rheumatic heart disease, pharyngitis caused by group A hemolytic streptococci should be treated with antibiotics to minimize the development of antibodies to the streptococcal antigen. Accurate diagnosis of group A hemolytic streptococci by culture and prolonged treatment (10 days) is effective in preventing rheumatic fever.

CHART 8–6. RHEUMATIC HEART DISEASE

cause	lesions	manifestations
Hypersensitivity reaction to antigen of group A hemolytic streptococci	Diffuse myocarditis (early) Fibrous thickening of heart valves (late)	Fever, arthritis, cardiac abnormalities (early) Murmur, heart failure (late)

Infective Endocarditis

Infective endocarditis, in contrast to rheumatic valvulitis, is caused by organisms living on the heart valves and producing an inflammatory reaction. Bacteria are the most common cause, hence the name *bacterial endocarditis*. Most cases occur on previously dam-

TABLE 8–1. PATHOGENESIS OF RHEUMATIC HEART DISEASE

Sequence of Steps	Causes	Lesions	Manifestations
Ordinary strep throat	Throat infection due to group A streptococcus	Pharyngitis	Sore throat
Rheumatic fever	Antibodies to group A streptococcus cross-react with human tissues	Carditis, arthritis	Fever, joint pains, murmurs, may be no symptoms
Recurrent rheumatic carditis	Further throat infections with group A streptococci	Recurrent carditis	Usually none
Latent rheumatic valvulitis	Long latent period (years), presumed continuation of previous injury	Scarring of heart valves	Murmurs, heart failure (late)
Rheumatic valvulitis complicated by subacute bacterial endocarditis	Transient bacteremia with organisms normally present in mouth (usually alpha hemolytic streptococci)	Infective endocarditis	Fever, leukocytosis, bacteremia, emboli, heart failure

aged valves, most often caused by rheumatic heart disease but also by congenital valve deformities. The bacteria often enter the bloodstream through the mouth and stick to the damaged valve to initiate the infection. Alpha hemolytic streptococcus (*Streptococcus viridans*), a normal resident of the throat, is the most common offending organism. Because alpha hemolytic streptococci are very weak pathogens, they produce a smoldering infection on the heart valve (subacute bacterial endocarditis), which gradually destroys the valve and leads to death either by heart failure or by damage from emboli dislodged from the valve. In contrast, when a more virulent organism such as *Staphylococcus aureus* lodges on a heart valve, an acute infection (acute bacterial endocarditis) ensues, and organisms dislodged from the valve travel to other organs producing multiple abscesses and rapid death.

Many cases of infective endocarditis can be prevented by preventing rheumatic fever, because rheumatic valvulitis is the most common predisposing factor to the development of infective endocarditis. Those persons who already have damaged heart valves can be protected from the bacteremia resulting from manipulative procedures, such as dental extractions, by antibiotics. Once developed, early diagnosis of infective endocarditis may result in treatment that will reduce the amount of damage, but overall the disease is very serious.

CHART 8–7. INFECTIVE ENDOCARDITIS

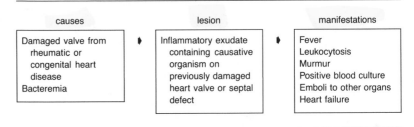

causes	lesion	manifestations
Damaged valve from rheumatic or congenital heart disease Bacteremia	Inflammatory exudate containing causative organism on previously damaged heart valve or septal defect	Fever Leukocytosis Murmur Positive blood culture Emboli to other organs Heart failure

Hypertensive Heart Disease

High blood pressure (hypertension) will lead to an increased work load on the heart, which leads to cardiac hypertrophy and eventually heart failure (Fig. 8–15). Hypertension also accelerates the development of atherosclerotic plaques, so it is often combined with coronary atherosclerotic heart disease. Hypertension within the systemic arterial blood circuit is quite common and increases in frequency with age. It is usually of unknown cause, hence the name *idiopathic* or *essential hypertension*. However, every patient

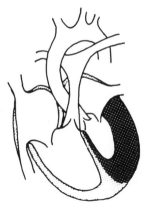

Figure 8–15. Effect of systemic hypertension—left ventricular hypertrophy.

should be medically evaluated for causes that are curable, such as coarctation of aorta, narrowing of renal artery, which leads to increased release of hormones causing hypertension, or tumors of the adrenal gland, which secrete hormones that elevate blood pressure.

CHART 8–8. HYPERTENSIVE HEART DISEASE

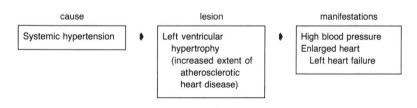

cause	lesion	manifestations
Systemic hypertension	Left ventricular hypertrophy (increased extent of atherosclerotic heart disease)	High blood pressure Enlarged heart Left heart failure

Cor Pulmonale

Hypertension can also occur in the pulmonary arterial system due to long-standing diffuse pulmonary disease or failure of the left heart. Pulmonary hypertension with right heart failure caused by chronic lung disease is called *cor pulmonale*. Pulmonary hypertension leads to right ventricular hypertrophy and eventually to right heart failure (Fig. 8–16).

Hyperplastic/Neoplastic Diseases

Hypertrophy has already been discussed. All neoplasms of heart, whether benign, malignant, or metastatic, are rare.

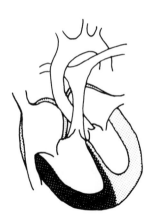

Figure 8-16. Effect of pulmonary hypertension—right ventricular hypertrophy.

ORGAN FAILURE

Inadequacy of cardiac pumping action may take two forms—cardiogenic shock and congestive heart failure. Shock is inadequate perfusion of tissues with oxygen due to low blood pressure. Congestion is distention of veins due to increased pressure within the veins. Shock, whether due to heart damage or other conditions such as blood loss, is a serious acute condition. If blood pressure is not restored quickly, the patient will die. Congestion, on the other hand, usually develops gradually and is less life threatening.

Cardiogenic shock is usually the result of extensive myocardial infarction and rapidly leads to death in the majority of patients. Drugs given to constrict the vascular bed and improve the force of contraction of the heart are sometimes effective in controlling the shock.

Congestive heart failure means the inability of the heart to pump enough blood to meet the demands of peripheral tissue. When this occurs, there is increased pressure in the pulmonary or systemic veins, or both, resulting in "backing up" of blood with consequent transudation of fluid into the pulmonary alveoli and tissues supplied by the systemic circulation. Congestive heart failure, like cardiogenic shock, is most commonly a sequela of myocardial infarcts but also occurs in the later stages of other forms of heart disease.

Congestive heart failure is divided into left and right heart failure, although both commonly occur together. Left heart failure involves increased venous pressure from the left side of the heart into the lungs and is usually manifest by shortness of breath (dyspnea) due to transudation of fluid into pulmonary alveoli (pulmonary edema). Right heart failure causes enlargement of the liver and spleen (hepatosplenomegaly) due to congestion; edema, par-

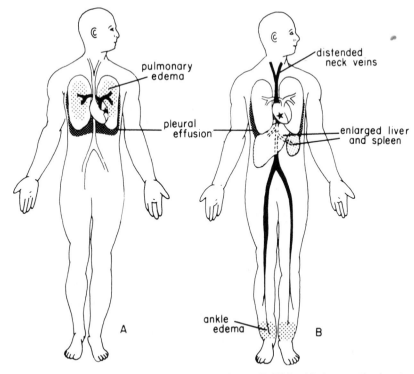

Figure 8–17. A. Left-sided congestive heart failure. **B.** Right-sided congestive heart failure.

ticularly noticeable in the ankles because of their dependent position; and distention of neck veins due to increased venous pressure (Fig. 8–17). Left heart failure may lead to right heart failure because increased pressure is transmitted through the pulmonary circuit to the right heart.

Severe chronic heart disease is often expressed as congestive heart failure. The severity and duration of heart failure varies. At its worst, it leads to death. In its mild forms, it may be controllable for many years with rest and drugs, such as digitalis, which improve myocardial contractal force, and diuretics, which promote renal excretion of edema fluid.

REVIEW QUESTIONS

1. What are the major causes of heart disease? What is the relative frequency of each type?
2. What is the pathogenesis of three causes of sudden death from heart disease?

3. How are each of the following tests or procedures helpful in evaluation of a patient with heart disease?
   Blood pressure
   Electrocardiogram
   Serum enzyme levels
   Chest x-ray
   Cardiac catheterization
4. What are the causes, lesions, and major manifestations of each of the specific heart diseases discussed in the chapter?
5. What are the steps involved in the evolution of a myocardial infarct?
6. What is the pathogenesis of the various complications of myocardial infarcts?
7. What is the pathogenesis of each stage of rheumatic heart disease and its complications?
8. Which two types of heart disease predispose to infective endocarditis? What are the consequences of infective endocarditis?
9. How is hypertension related to atherosclerosis?
10. How do cardiogenic shock and congestive heart failure differ in terms of acuteness, seriousness, cause, and manifestations?
11. What do the following terms mean?
    Ventricular fibrillation
    Cardiac arrhythmia
    Angina pectoris
    Heart murmur
    Myocardial infarct
    Ventricular aneurysm
    Mural thrombosis
    Rheumatic fever
    Rheumatic valvulitis
    Cardiogenic shock
    Congestive heart failure

Vascular System

REVIEW OF STRUCTURE AND FUNCTION

The vascular system actually comprises two systems—the blood vascular system and the lymphatic vascular system. The blood vascular system is a continuous-flow system transporting blood to and from the heart via arteries, capillaries, and veins. Because the blood does not directly surround tissue cells, the vascular system must allow for exchange of gases, nutrients, and metabolic wastes across its walls. This exchange takes place primarily in the smallest component of the blood vascular system, the capillaries and small venules where blood flow is slowest and where the walls are the thinnest.

The arteries and veins have three concentric layers—intima, media, and adventitia (Fig. 9–1). The intima is lined by broad flat endothelial cells with an underlying basement membrane and a few connective tissue cells for support. The endothelial cells in the smaller vessels control the exchange between blood and tissue. The media is composed of elastic tissue and smooth muscle. Either sympathetic stimulation or circulating adrenergic hormones can produce contraction of the muscle layer, which decreases the caliber of vessels. Cholinergic hormones and, in some vessels, parasympathetic stimulation cause vasodilation. The adventitia is composed of supporting connective tissue cells.

The arterial side of the vascular system has intraluminal pressures of about 100 mm Hg throughout most of its length. In accordance with this high pressure, arteries are thick-walled muscular structures. Major determinants of pressure are the amount of blood flowing through the vessels and the resistance of the vessels to this flow. Therefore, either an increased amount of blood or decreased vessel caliber due to contraction will elevate the blood pressure. Normally, pressure is controlled mainly by the degree of dilatation or constriction of arteries and arterioles (very small arteries). The amount of blood flow is a function of the rate of contraction of the heart and the volume of blood pumped with each contraction.

The venous side of the vascular system has a mean pressure of less than 30 mm Hg. Accordingly, veins have thinner muscular walls than arteries and larger lumens to carry blood at a slower rate under less pressure. The function of veins is to return blood to the heart from the tissue capillary beds. The force that pushes blood in the veins back to the heart is the pressure that is left after the

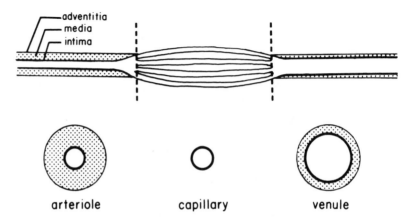

Figure 9–1. Comparison of arterioles, capillaries, and venules.

passage of blood through arteries, arterioles, and capillaries. This force is only adequate to return blood to the heart when the body is in a horizontal position. Unless compensated for, blood in head and upper parts of the body would gravitate to the abdomen and lower extremities in a person standing upright. Among the compensatory mechanisms to prevent this are venomotor reflexes that constrict the lower body veins, the massaging action of skeletal muscles, and action of respiratory movements on the great veins in the chest. In addition, veins contain valves, which aid in the vertical flow of blood against gravity.

Obstruction of blood flow in veins by such mechanisms as compression, thrombosis, or heart failure elevates venous and capillary blood pressures, causing accumulation of blood in these highly distensible structures. Distention of veins and capillaries due to increased venous pressure is called *congestion*. Continued congestion results in leakage of fluids through capillary endothelium into tissues. This accumulation of fluid is called *edema*, and it results in a disturbance of tissue–capillary exchange of oxygen, nutrients, and metabolites.

The portal vein is an exceptional vein that collects blood from most of the intestinal tract and spleen and carries it to the liver, where nutrients and metabolic products are removed before it is returned to the heart via the hepatic vein.

The capillaries are small vessels that connect arteries and veins and are essentially composed of an intimal layer. Capillaries comprise by far the largest volume of the vascular system, although most are in a state of collapse at any given time. Capillaries are extremely adaptable in being able to open in response to increased blood flow needs and in being able to proliferate to help repair an injured area. Capillary flow is controlled by contraction or relaxation of small arteries (arterioles) and possibly by small veins (venules).

In some organs, such as liver, spleen, adrenal gland, and pituitary gland, the capillaries take the form of sinusoids. Sinusoids are dilated capillaries that are more continuously open to blood flow than ordinary capillaries. Sinusoids are associated with specialized functions, such as filtering out particulate matter, including damaged blood cells in the liver and spleen, and hormone regulation in the adrenal and pituitary glands. The most structurally complex capillaries are found in the renal glomerulus, where filtration of waste products into the urine is the specialized function.

The lymph vascular system (lymphatic system) is unique in that it has no vessels flowing into the system. Lymphatic vessels originate as blind-ended, thin-walled channels that extend to lymph nodes and beyond, becoming larger, and eventually emptying into the thoracic duct. The thoracic duct flows into the superior vena cava, thus emptying lymph into the venous blood. To a lesser extent, lymphatic vessels may empty into other veins. Lymphatic vessels function as a low-pressure drainage system, much like the tiles draining a farmer's field, as they return some of the excess tissue fluid back into the venous system. Also, lymphocytes may reenter the circulation from tissue via this route.

MOST FREQUENT AND SERIOUS PROBLEMS

Atherosclerosis is a disease characterized by lipid, calcium, and fibrous deposits in the intima of large- and medium-sized arteries and is by far the most significant disease in the United States in terms of death and morbidity. We have already described its effects on the heart, which account for the most common cause of death. The arteries to the brain are the second most important site of atherosclerosis. Cerebral infarcts resulting from atherosclerosis of cerebral vessels are the third most common cause of death in the United States. Atherosclerosis of the arteries of the legs and of the main artery to the intestines (superior mesenteric artery) less commonly leads to arterial obstruction and infarction, but when it occurs, the effects are quite serious.

The formation of thrombi (blood clots formed within vessels) in the deep veins of the legs is a relatively common and sometimes serious problem that occurs particularly in bedridden patients. The most important complication of thrombi occurs when they break away from the vessel wall to form emboli. The emboli are then carried through the veins to the right side of the heart, where they are pumped into the branches of the pulmonary arteries, producing obstruction of blood flow to portions of the lungs.

Varicosities are permanently dilated venous channels. They most commonly occur in the legs (varicose veins) and about the anus (hemorrhoids). Varicose veins may cause discomfort, disfigurement, edema, ulceration of the lower extremities, and predispose to the formation of thrombi. Hemorrhoids may be asymptomatic or may cause itching, pain on defecation, bright red bleeding, or sudden severe pain if thrombosis occurs.

High blood pressure (hypertension) is one of the most important conditions for the health care team to understand, because its recognition and treatment can prevent or delay complications for a large number of people. Most cases of hypertension are idiopathic (called *essential hypertension*). The effects of hypertension have a latent period of many years and take the form of accelerated atherosclerosis leading to myocardial and cerebral infarcts, heart failure, cerebral hemorrhage, and occasionally renal failure.

SYMPTOMS, SIGNS, AND TESTS

The manifestations of arterial obstruction due to atherosclerosis, inflammation, thrombosis, embolism, or external compression are either infarcts or more gradual loss of tissue due to insufficient blood supply (ischemic atrophy). Either infarcts or ischemic atrophy will result in loss of function in the organs supplied by the affected vessel. Pain may also occur.

The integrity of arteries can be directly evaluated by arteriography, a procedure in which radiopaque dye is injected into arteries and successive x-ray films are taken to show the caliber of the vessels and distribution of blood flow. For example, a cerebral angiogram, made by injecting radiopaque dye into the carotid artery, may demonstrate an aneurysm of the circle of Willis or displacement of cerebral vessels by a tumor. Commonly, a carotid angiogram will demonstrate atherosclerotic narrowing or occlusion of the carotid artery, a significant predisposing factor to strokes (Fig. 9–2).

Functional disease of arteries caused by high or low blood pressure has entirely different symptoms than arterial occlusion.

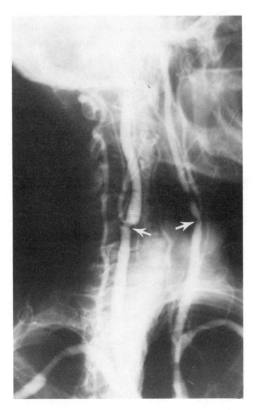

Figure 9–2. An arteriogram of a carotid artery with areas of narrowing (*arrows*).

High blood pressure (hypertension) is usually asymptomatic, but it may be manifested by headaches and dizziness. Low blood pressure (hypotension), if severe, will be manifested by symptoms of shock, which include faintness, cold skin due to vasoconstriction, and reduced blood flow to the brain and the kidneys.

The functional status of the arterial system is usually evaluated by measuring the arterial blood pressure with a sphygmomanometer and by determining the fullness of the pulse using a finger placed over a pulsating artery.

The status of the venous system can be estimated by inspection or measured using specialized instruments. Distention of neck veins when the patient is in an upright position is an obvious sign of increased pressure in the venous system. More subtle increases can be measured by inserting a pressure gauge into an arm vein. Thrombosis of deep leg veins causes cyanotic congestion, edema, and sometimes pain. Ultrasound techniques measuring the Doppler effect provide a sophisticated method of evaluating venous flow at deep sites, including deep vein thrombosis of the legs. Incompetent valves in superficial leg veins lead to distended, tortuous, protruding, rope-like varicosities that are obvious on inspection. Likewise, hemorrhoidal and esophageal varicosities are evaluated by inspection of the anus and by endoscopy, respectively.

Generalized disease of small vessels, such as occurs with hypertension and diabetes mellitus, may be manifested by decreased visual acuity, and can be evaluated by looking at the eyegrounds (back of the eye) with an ophthalmoscope (funduscopic examination). Small vessels are easily seen here, because they lie immediately beneath the highly translucent retina.

SPECIFIC DISEASES

Genetic/Developmental Diseases

Congenital variations in the vascular system are common but rarely cause problems except for congenital heart diseases and anomalies of the great vessels.

Angiomas

Hemangiomas are local proliferations of capillaries that may be present at birth. They are common in the skin, where they vary from small red dots to large cosmetically distracting "port wine stains" (Fig. 9–3). Because the dilated vessels fill with blood, they are red to blue and blanch under the pressure of a finger as the blood is pushed out of them. Hemangiomas of the skin rarely cause any problem, but they are occasionally removed for cosmetic rea-

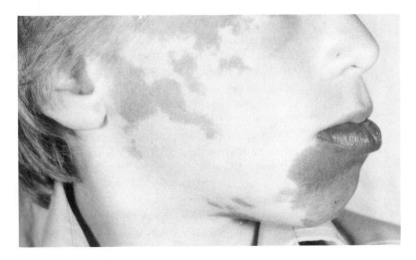

Figure 9-3. Hemangiomas of the face (port wine stains).

sons. Lymphangiomas, dilated masses of lymphatics, are much less common. When they occur in the neck of an infant (cystic hygroma), they may be frightening in appearance, but they usually regress with age. Hemangiomas and lymphangiomas may be considered as congenital anomalies when their growth is commensurate with that of the patient's body. If their growth is disproportionately great, they are considered to be neoplasms of vessels.

CHART 9-1. HEMANGIOMAS

cause	lesion	manifestations
Congenital anomalies	Excessive capillary proliferation, often of the skin	Localized redness of the skin, which blanches with pressure Rarely bleed

Inflammatory/Degenerative Diseases

Atherosclerosis and hypertension will be discussed first and in more detail than the other conditions. The term *atherosclerosis* should be distinguished from the term *arteriosclerosis*. *Arteriosclerosis* means hardening of the arteries and is the general term used to include thickening of the intima and media of small arterioles (arteriolosclerosis or arteriolarsclerosis), calcification of the media of large arteries (medial sclerosis), and plaque formation in the intima of large arteries (atherosclerosis) (Fig. 9–4). Arterio-losclerosis occurs in response to increased blood pressure and thus is associated with hypertension. It is seen most commonly in the renal arterioles and may contribute to decreased renal function. In fact, decreased blood flow to the kidney is known to cause hypertension, so arteriolosclerosis may be contributing to a vicious cycle of events.

Medical calcification occurs in the medium-sized to large arteries of the thyroid, genital tract, and extremities with aging and is not considered to be a cause of significant disease.

Atherosclerosis

Atherosclerosis is a degenerative condition, with prevalence and severity increasing with advancing age. The basic lesion of atherosclerosis is a fibrofatty deposit in the intima of blood vessels, particularly in the major muscular arteries (Fig. 9–5). This deposit is called an *atheroma* or, more commonly, a *plaque,* and its continued growth will eventually occlude an artery, leading to an infarct in the territorial distribution of that artery. Deposits occur in large- and medium-sized arteries, particularly those of the aorta, heart, brain, kidneys, legs, and intestines. The early lesions may be found in children dying from unrelated diseases, and most persons over the age of 30 will have grossly visible atherosclerotic lesions in their larger arteries. Complications of atherosclerosis (decreased blood flow, thrombosis, or embolism) begin to appear in persons in their thirties and increase with each year of age thereafter. The increase in frequency of complications in women is usually delayed

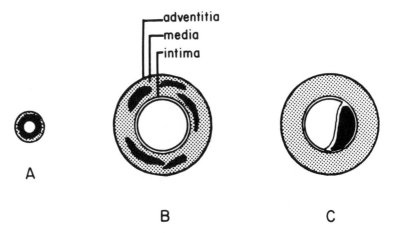

Figure 9–4. Types of arteriosclerosis: **A.** Arteriolosclerosis. **B.** Medial calcification. **C.** Atherosclerosis.

until after menopause, presumably due to a protective effect of female hormones. The low-pressure pulmonary arteries are not usually involved.

An atherosclerotic lesion starts with the accumulation of smooth muscle cells and macrophages in the vessel intima. Lipid accumulates in these cells giving them a foamy appearance histologically and grossly imparting a yellow color to the plaque. These small fatty plaques occur in all populations. In populations with a high prevalence of atherosclerotic vascular disease, the plaques gradually enlarge over years and become progressively more fibrous (sclerotic). The sclerosis presumably occurs in response to the low-grade degeneration and necrosis of cells in the lesion. A white-yellow, shaggy, crusted appearance to the developing fibrous plaque is due to crystallization of the cholesterol that is accumulating in the plaque and to dystrophic calcification.

Eventually, atherosclerotic plaques may cause harm in one of several ways. The lumen of the blood vessel may be obstructed to produce ischemia distally. Ulceration of the surface of the plaque invites thrombus formation there, which may in turn lead to occlusion or embolization. The plaque may produce damage to the vessel wall so that it balloons under pressure to form an aneurysm, a complication most often found in the abdominal aorta (Fig. 9–6) and iliac arteries.

Atherosclerosis in the coronary arteries tends to produce occlusion, superimposed thrombosis, or sometimes ulceration with hemorrhage into a plaque. Plaques in the carotid arteries tend to produce occlusion (see Fig. 9–2). Atherosclerosis of the abdominal aorta, the most common site for atherosclerosis, is usually asymptomatic even when extensive (Fig. 9–7). Complications of aortic atherosclerosis occur if a tributary is blocked by plaque, if embolization occurs from the surface of the plaque, or if the aorta bulges to form an aneurysm. Atherosclerosis in the arteries of the lower extremity may cause an infarct of the foot (gangrene) (Fig. 9–8) or produce more gradual ischemic changes in the leg. The latter are associated with ischemic atrophy of muscle and pain that is aggravated by walking and relieved by rest (*intermittent claudication*).

It is not understood at present what initiates and perpetuates the atherosclerotic process. The more prevalent theories propose initial damage followed by lipid infiltration of the intima, possibly augmented by platelet aggregation and microclotting. Smooth muscle cells with phagocytic properties, called *myointimal cells*, and macrophages phagocytose the lipoprotein, cholesterol, and other lipid to form the early atherosclerotic lesion. Collagenous and elastic connective tissue is formed in response to the lipid and con-

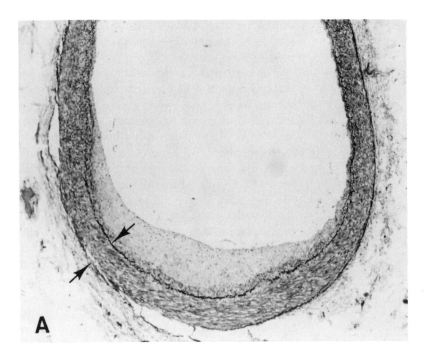

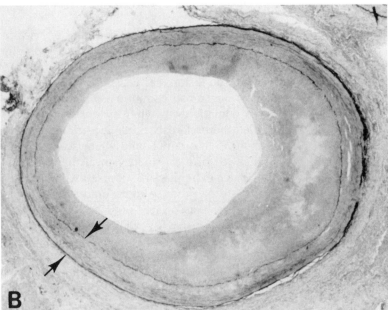

Figure 9–5. Progressive stages of narrowing of the lumen of coronary artery by atherosclerotic plaque (**A, B, C**). The internal and external elastic lamina define the media of the artery (*arrows*). In **A,** the early intimal thickening at the bottom of the photograph is contrasted with the normal intima. In **B** the plaque is mostly fibrous tissue, with patchy areas of lipid deposit (*lighter areas*).

tributes to the bulk of the developing lesion. Lipid deposition is clearly the factor that perpetuates the evolution of the lesion. But what causes the lipid deposition? Much attention has been paid to the role of injury to the endothelial surface as an initiating factor. This would explain the location of plaques near bifurcations of arteries where the pressure of the arterial stream might cause injury and the increased frequency of plaques in hypertensive patients where the injury might be more severe. Another interesting observation is that the myointimal cells can be derived from one clone of proliferating cells, suggesting that they might be tiny hyperplasias or even neoplasias. In summary, serum lipid abnormalities promote the development of atherosclerosis, but local factors, which are poorly understood, are necessary for plaque development. The complexity and graded severity of both serum and local factors makes isolation of specific causes difficult.

All people have some degree of atherosclerosis, yet one individual may die of a myocardial infarct from an atherosclerotic, occluded coronary artery at the age of 35, whereas another individual may die of cancer at the age of 90 with only mild atherosclerosis. The recognition of this variation in individual susceptibility has led to the elucidation of risk factors, those conditions

C

Figure 9–5. Continued. As seen in **B,** the plaque is mostly fibrous tissue, with patchy areas of lipid deposit (lighter areas).

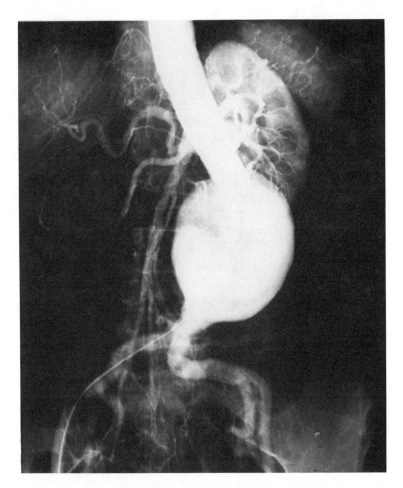

Figure 9–6. Arteriogram of atherosclerotic aneurysm of aorta at typical location between the renal and iliac arteries. Note the catheter in the right iliac artery that was used to inject the radiopaque dye.

that when present will render an individual more susceptible to atherosclerosis. The more important of these risk factors are:

1. *Age.* In general, the older a person, the more atherosclerosis present.
2. *Sex.* Men have a much higher incidence of serious atherosclerosis. In women, the frequency of atherosclerosis increases after menopause and approaches that of men in the elderly.
3. *Family.* Some families have a much higher incidence of atherosclerosis and resulting diseases than other families. In

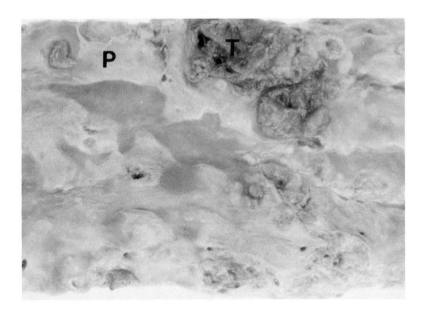

Figure 9–7. Atherosclerosis of the aorta. Note plaques (P) and fibrin thrombi (T) on the surface of plaques.

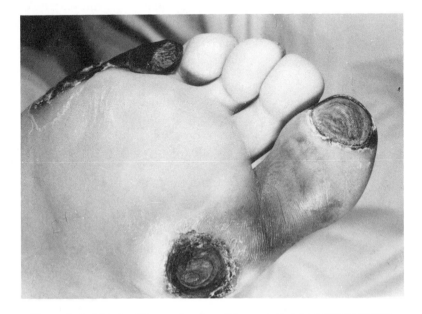

Figure 9–8. Infarcts of the great and small toes resulting from atherosclerosis.

some instances, this may be related to a common diet, but in other cases, there is a clear genetic influence such as familial hypercholesterolemia.

4. *Diet.* People with obesity and/or high caloric intake of carbohydrates and lipids have a greater incidence of atherosclerosis.

5. *Hypertension.* The greater the hypertension, the greater the risk of atherosclerosis. The reason is not known, but it may be related to trauma to the arterial intima, as veins very seldom develop significant atherosclerosis. On the other hand, when a vein is transplanted to a position where it carries arterial blood, it may develop atherosclerosis very rapidly. Thus, atherosclerosis may develop in veins that are utilized as coronary artery bypass grafts.

6. *Cigarette smoking.* The death rate from coronary atherosclerosis in heavy smokers may double that of nonsmokers.

7. *Stress.* Emotional stress is often blamed for an increased incidence of atherosclerotic coronary disease, but this is not well substantiated.

8. *Carbohydrate and lipid derangements.* Familial elevations of serum lipids, diabetes mellitus, and, to a much lesser extent, hypothyroidism are predisposing factors to atherosclerosis. Serum cholesterol values over 240 mg/dl are generally considered to carry an increased risk, although this figure is relative and risk can only be crudely estimated in any individual by taking all risk factors into consideration. Further, serum cholesterol levels represent the total of several forms of cholesterol in the serum and some forms are more strongly associated with the development of atherosclerosis than others. Lipoproteins, the major carriers of cholesterol in the serum, have been classified on the basis of their density and electrophoretic migration. High serum levels of low-density lipoproteins are associated with more severe atherosclerosis. Hyperlipoproteinemic states have been classified on the basis of serum levels of lipoproteins, cholesterol, and triglycerides. The classification of hyperlipidemias is not very exact, except perhaps for some familial types that carry a high risk for atherosclerosis at an early age. Diabetics have more atherosclerosis on the average than nondiabetics, but in spite of this many diabetics, especially those with later onset of the disease, live a normal lifespan. Atherosclerosis tends to be more severe in insulin-dependent diabetes of long duration.

Of all these risk factors, the major treatable ones are hypertension, smoking, and hyperlipoproteinemias. Treatment of atherosclerosis involves taking care of complications as they arise and preventing further ones from developing. An enormous portion of our health care resources is utilized in managing myocardial infarcts and angina pectoris, inserting coronary artery bypass grafts, caring for stroke victims, removing plaques from carotid arteries, bypassing occlusions and aneurysms of the aorta and leg arteries with grafts, and amputating gangrenous legs.

Surgical therapy for atherosclerosis has become very important and dramatic over the last several decades. Abdominal aortic aneurysms are bypassed by using a tube of biologically inert material. Ultrasound techniques allow diagnosis and measurement of asymptomatic aneurysms so that they can be treated after they reach a critical size and before they fatally rupture. Operations for carotid and coronary artery atherosclerosis are reserved for symptomatic patients, and then only after careful evaluation of potential risks and benefits. Carotid artery plaques, which occur at the bifurcation of internal and external carotid arteries and reduce cerebral blood flow, are removed by opening the artery and shelling out the plaque, a process called *endarterectomy*. Atherosclerotic lesions in coronary arteries may be bypassed using a portion of the patient's own vein (usually saphenous) carefully stitched into the ascending aorta proximally and into the coronary artery distal to the occlusion. Techniques have been developed to insert balloon catheters into coronary arteries that can be expanded to break and crush plaques without subjecting the patient to an operation.

In recent years, more attention to preventing or delaying the development of atherosclerosis has produced some hopeful results. Preventive measures focus on reducing risk factors. Early diagnosis and treatment of hypertension clearly has a beneficial effect. Cessation of smoking, even by those who have smoked for many years, reduces the frequency of atherosclerotic complications. Smoking appears to exert its effects both by affecting the development of plaques and by causing vasopasms, which increase the likelihood of complications. Dietary preventative measures attempt to reduce serum cholesterol in individuals at high risk and in the population as a whole by altering eating patterns. In an individual, serum cholesterol can be lowered by dietary manipulation, but the effect is limited by genetic factors and the outcome depends on a combination of risk factors. In the United States there has been some drop in the incidence of heart disease and this has been attributed to reduced cholesterol intake, but cause and effect are hard to prove. Some populations in the world have very little atherosclerosis. When people from these populations emigrate to high-risk areas, they develop atherosclerosis. This provides a very compelling argument for attempts to prevent atherosclerosis among high-risk populations.

CHART 9–2. ATHEROSCLEROSIS

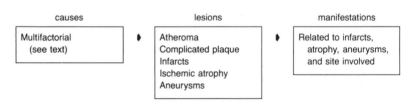

Hypertension

Hypertension means high blood pressure. The upper limits of normal blood pressure are generally considered to be 90 mm Hg diastolic and 140 mm Hg systolic. Transient hypertension, such as is caused by anxiety or physical activity needs to be distinguished from hypertensive disease in which there is sustained hypertension leading to gradual development of cardiovascular lesions. The systolic pressure is created by the force of the heart's contraction as it expels blood into the aorta. Large arteries expand to dampen the force of the pulse. Loss of the ability to expand, as occurs in the aorta with arteriosclerosis, produces systolic hypertension. Thus, elderly persons with atherosclerosis tend to have high systolic blood pressure. Systolic hypertension also may be produced by increased cardiac output, such as occurs with exercise, hyperthyroidism, or fever. The diastolic pressure represents the minimum pressure within arteries and is created by arterial relaxation, as during cardiac diastole. Injury from prolonged hypertension correlates best with the diastolic pressure, because eventually the small, rigid peripheral arteries and arterioles will not allow adequate tissue delivery of blood during diastole.

Essential hypertension refers to primary or idiopathic hypertensive disease. It is the most common form, accounting for over 90 percent of persons with hypertensive disease. *Secondary hypertension refers to hypertensive disease in which a cause is evident.*

Although an exact cause of essential hypertension cannot be identified, it is clearly a multifactorial disease involving both genetic and environmental factors. The genetic factor is suggested by the tendency for essential hypertension to be familial and by the high prevalence among American blacks. The genetic factor probably involves multiple genes and does not produce clear lines of inheritance. Among the many possible environmental factors, a high salt diet has been identified as a factor by animal experiments and by some epidemiologic studies. Psychologic stimuli appear to aggravate hypertension in some individuals. It is possible that many cases of essential hypertension are perpetuated by the hypertensive effects on the kidney vesels, which results in increased renin secretion, which, in turn, sustains the hypertension. This mechanism is described below.

Causes of secondary hypertension include renal disease, endocrine disorders, brain disorders, coarctation of the aorta, arteriovenous fistulas, and toxemia of pregnancy. Three types of renal disorders that produce hypertension are (1) chronic destructive renal disease, such as chronic glomerulonephritis, chronic pyelonephritis, or polycystic kidneys; (2) acute renal disease, such as glomerulonephritis; and (3) narrowing of the main renal artery due to atherosclerosis or smooth muscle hypertrophy. Chronic destructive renal disease is the most common cause of secondary hypertension. Partial occlusion of the renal artery is diagnosed by arteriograms and is often surgically treatable. Unilateral renal artery stenosis causes the affected kidney to secrete increased amounts of renin, an enzyme that activates angiotensin. Angiotensin causes hypertension by contracting small arteries as well as by causing release of aldosterone, which, in turn, aggravates the hypertension because of its salt-retaining effects. Renin may also be important as a cause of hypertension in acute or chronic renal parenchymal disease, although it is less commonly found in elevated levels in the blood under these circumstances.

Endocrine causes of hypertension usually relate to increased secretion of adrenal hormones. Corticosteroids and aldosterone from the adrenal cortex cause salt retention and often lead to hypertension. Epinephrine and norepinephrine are produced by a rare neoplasm of the adrenal medulla called a *pheochromocytoma* and cause hypertension by constricting blood vessels. Endocrine causes of hypertension are often treatable by removal of a benign neoplasm or hyperplastic gland.

Increased intracranial pressure is often associated with hypertension. Coarctation of the aorta produces hypertension proximal to the coarctation by obstruction of the aorta and is treatable by surgical removal of the coarcted segment. Arteriovenous fistulas are connections between arteries and veins that cause more rapid drainage of blood from arteries, producing a low diastolic pressure and a greater cardiac output leading to systolic hypertension. Toxemia of pregnancy is a cause of hypertension in pregnant women. Toxemia of pregnancy is discussed in Chapter 17.

Hypertension takes years to produce its damaging effects. The basic lesion consists of thickening and rigidity of small arteries and arterioles due to the prolonged increase in pressure, especially in the kidney. Hypertrophy of the left ventricle due to the increased pumping force required of the heart is an almost invariable finding. Acceleration of the development of atherosclerosis also occurs commonly in hypertensive patients. Most hypertensive patients die from heart disease due to heart failure. This is the result of the increased work imposed on the left ventricle as well as the damage to the myocardium from the concomitant coronary artery atherosclerosis. The second most common cause of death in hypertensives is from strokes as a result of bleeding from the damaged small arteries in the brain.

Small arteries are best visualized clinically in the retina, where they show thickening and focal narrowing. In fact, the degree of arteriolar involvement can be assessed by ophthalmoscopic examination of the retina. Retinal hemorrhages may lead to blindness. The renal involvement is occasionally severe enough to produce renal failure. At autopsy, the damage to arterioles can be seen microscopically in the kidneys. Rarely, the course of hypertensive disease may become accelerated, with severe renal and retinal vascular disease and very high blood pressure, often in the range of 300/150 mm Hg or more, and manifested by severe headaches and convulsions. This is termed *malignant hypertension* and leads to rapid death within months if not controlled. Severe thickening of arteries and arterioles with obliteration of the lumens are seen in patients who die of malignant hypertension (Fig. 9–9).

Most hypertensive disease is asymptomatic; thus, it must be diagnosed by screening. When symptoms develop, they are usually complications of the lesions described above. Heart lesions produce myocardial infarcts or heart failure; brain lesions produce strokes, headaches, and sometimes dizziness or light-headedness; retinal lesions produce blind spots; and renal lesions produce proteinuria. Severe epistaxis (nose bleeds), hemoptysis (coughing up blood), and metrorrhagia (uterine bleeding at an abnormal time) may occur due to the effects of hypertension on small vessels.

Prognosis depends on many factors, such as the levels of blood pressure, variability in the natural course of the disease, occurrence with other potentiating disease, such as atherosclerosis, presence of a treatable cause, stage at time of diagnosis, ability to alter environmental factors, such as stress and diet, and effectiveness of drug therapy in reducing the blood pressure. Untreated hypertensive disease might be expected to run a course of 20 years or so, with the first 15 years being asymptomatic. Although it is difficult to measure the effects of treatment in a predominantly elderly group of patients, there is no doubt that treatment adds years to the life of an average hypertensive patient.

The strategy of approach to hypertensive disease is threefold: (1) make the diagnosis early through screening, (2) undergo thorough medical investigation to discover the 5 to 10 percent who have secondary hypertension that may be curable, and (3) reduce

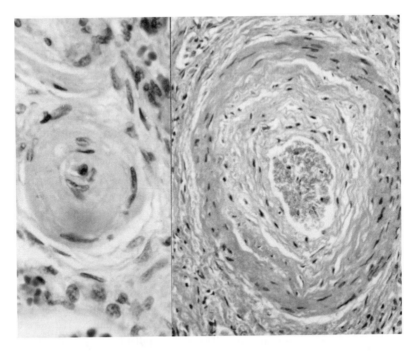

Figure 9–9. Changes in arteriole (*left*) and small artery (*right*) associated with severe hypertension. The arteriole is hyalinized, with obliteration of the lumen. The small artery is greatly thickened by concentric layers of connective tissue and smooth muscle. The narrowed lumen contains red blood cells.

the blood pressure through environmental manipulation and drug therapy. The environmental manipulation refers to proper diet and exercise as well as attaining psychologic well-being. Drug therapy requires that the patient be under continuous medical surveillance. As mentioned in Chapter 2, carrying out these strategies accounts for hypertension being the most common chronic disease in terms of visits to family physicians.

CHART 9–3. HYPERTENSION

causes		lesions		manifestations

Unknown (most)
Chronic renal disease
Acute
  glomerulonephritis
Renal artery stenosis
Pheochromocytoma
Corticosteroids
Coarctation of aorta
Toxemia of pregnancy
	▶	Arteriolosclerosis
		Left ventricular
		  hypertrophy
		Accelerated
		  atherosclerosis
			▶	Elevated blood
				  pressure
				Complications related
				  to myocardial infarct,
				  heart failure, cerebral
				  hemorrhage, cerebral
				  infarct, renal failure

Vasculitis

There are several uncommon, relatively low-grade and sporadic, noninfectious inflammatory diseases that affect arteries and sometimes veins. Their exact causes are usually not evident, but in most instances they are probably autoimmune in nature. Vasculitis produces significant diagnostic problems, because its variable distribution produces unpredictable and often serious effects. Collections of acute and/or chronic inflammatory cells are found in the walls of and scattered around the vessels.

The most common type of vasculitis is systemic lupus erythematosus, which is noted for a butterfly-shaped rash over the cheeks, effusions (collections of watery fluid) in joints and other body cavities such as the pericardial sac and pleural space. Muscle weakness may occur, and eventual involvement of the kidney vessels may lead to renal failure. Serum antibodies to cell nuclei provide the antinuclear antibody test.

Polyarteritis nodosa is another type of vasculitis that consists of nodular inflammatory thickenings of medium-sized arteries, particularly those of the kidneys, intestines, and skeletal muscles. Inflammation of these arteries (Fig. 9–10) may lead to luminal occlusion, resulting in small- to medium-sized infarcts in many organs.

CHART 9–4. VASCULITIS

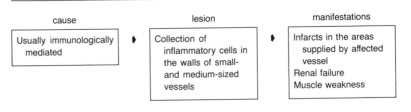

cause		lesion		manifestations
Usually immunologically mediated	▸	Collection of inflammatory cells in the walls of small- and medium-sized vessels	▸	Infarcts in the areas supplied by affected vessel Renal failure Muscle weakness

Functional Vascular Disease

Various disease states exist in which arterioles and venules constrict secondarily to local influences such as precipitation of intraluminal proteins, chemicals applied to the adventitia, or disturbance of autonomic function. One of these states is called Raynaud's phenomenon. In this condition, the small vessels of the extremities (usually hands) constrict markedly, leaving the extremity cold and blue. Raynaud's phenomenon occurs with some collagen–vascular diseases, such as lupus erythematosus, but at other times the reason for its occurrence is unknown.

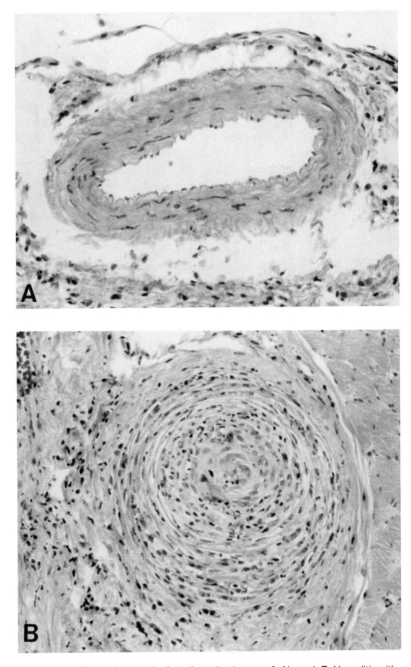

Figure 9–10. Photomicrograph of medium-sized artery. **A.** Normal. **B.** Vasculitis with infiltration of inflammatory cells and obliteration of vascular lumen.

Thrombophlebitis

Literally, thrombophlebitis means thrombosis in an inflamed vein. The term *phlebothrombosis* may be more appropriate, because the veins are not usually inflamed. Thrombophlebitis occurs most frequently in the deep veins of the legs or pelvis, where thrombosis is the important finding. Thrombophlebitis may also occur in arm veins as a complication of intravenous catheters that are left in place for several days. Sluggish and turbulent blood flow is probably more important in causing thrombosis than is inflammation. If the thrombi are dislodged from the vein wall, they will become emboli and will be carried to the lung (Fig. 9–11), where they may produce varying pulmonary symptoms and sometimes death.

Pulmonary emboli are common in hospitalized patients, particularly following surgical operations and childbirth, presumably because lack of exercise causes stasis of blood in leg veins and blood platelet concentration is increased. The prevention of thrombi in leg veins is the reason for early ambulation following

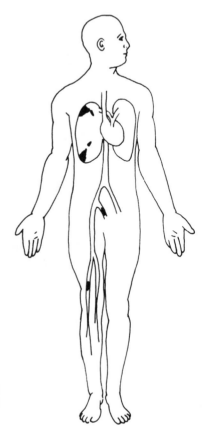

Figure 9–11. Thrombi in leg and pelvic veins leading to pulmonary embolism and infarcts.

operation or childbirth. Most venous thrombi and even pulmonary emboli are undiagnosed; this is known because they are found frequently at autopsy when they were clinically unsuspected. Local symptoms of thrombophlebitis may include leg pain and tenderness, but these symptoms are unreliable diagnostic indicators. Accurate diagnosis depends upon techniques involving injection of radiopaque dyes or use of ultrasound to detect venous occlusions. Treatment consists of anticoagulant drugs and may involve surgical ligation of the veins to the legs to prevent pulmonary emboli. With time, the thrombi are either dissolved or replaced by scar.

CHART 9–5. THROMBOPHLEBITIS

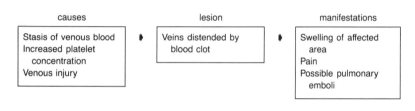

causes	lesion	manifestations

Stasis of venous blood
Increased platelet
  concentration
Venous injury
	Veins distended by
	blood clot
		Swelling of affected
		  area
		Pain
		Possible pulmonary
		  emboli

Varicose Veins

Varicose veins are dilated tortuous veins. Three important sites of involvement are legs, anus, and lower esophagus. Esophageal varices are caused by increased pressure in the portal vein, usually due to cirrhosis of the liver, and may readily bleed, with a fatal outcome. Hemorrhoids are enlarged veins of the lower rectum and anus, usually without definable cause, although they are common following childbirth due to external pressure and consequent venous distention from pregnancy. Esophageal varices and hemorrhoids are discussed in Chapters 14 and 15.

Varicose veins of the lower extremities commonly involve the greater saphenous vein, a superficial vein just beneath the skin surface of the medial side of the leg. Normally, valves in the leg veins aid in the return of blood upward to the heart by preventing backflow once the blood has been forced upward by muscular activity. Varicose veins have incompetent valves, so hydrostatic pressure is transmitted through the entire course of the vein. The increased internal pressure leads to dilated, tortuous, and visible veins, and standing for long periods of time may further lead to edema, and eventually, ulceration of the skin from the pressure of the edema fluid. Because there are many alternate venous routes for blood to return to the heart, these superficial varicosities can be removed by a surgeon when symptoms indicate the need.

CHART 9–6. VARICOSE VEINS

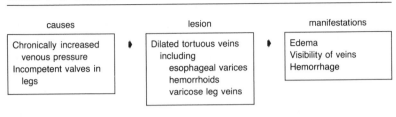

causes		lesion		manifestations
Chronically increased venous pressure Incompetent valves in legs	●	Dilated tortuous veins including esophageal varices hemorrhoids varicose leg veins	●	Edema Visibility of veins Hemorrhage

Hyperplastic/Neoplastic Diseases

Except for hemangiomas, discussed earlier, benign and malignant tumors of vessels are rare.

ORGAN FAILURE

Failure of the cardiovascular system to maintain adequate blood pressure is called *shock*, and it is clinically manifest by decreased blood pressure, increased heart rate, decreased urine output, and altered states of consciousness. The two important determinants of arterial blood pressure are the amount of blood pumped by the heart and the resistance of arteries, which is determined by muscular and elastic tension in their walls, as well as by the amount of blood in the vascular system. The types of shock relate to failure of these functions (Table 9–1). Cardiogenic shock, which was discussed in the previous chapter, results when the heart cannot pump enough blood under sufficient arterial pressure to maintain oxygenation of tissues. Hemorrhagic shock results when bleeding leads to an insufficient amount of blood in the vascular system to maintain arterial pressure. Neurogenic shock results when small vessels dilate resulting in an insufficient amount of blood for the remainder of the vascular system. Syncope (fainting) is most often due to neurogenic shock and often occurs in response to emotional situations.

TABLE 9–1. TYPES AND CAUSES OF SHOCK

Type	Cause
Cardiogenic	Inadequate pumping by the heart
Hemorrhagic	Loss of blood
Neurogenic	Inappropriate dilation of small vessels, resulting in pooling of blood

In neurogenic shock, pooling of blood in the lower body results in brain ischemia. After approximately 10 seconds of ischemia, neurons cease to function properly and the person loses consciousness. Syncope is relieved by lowering the position of the head so that less pressure is needed to get blood to the head. In the more severe forms of shock (cardiogenic, hemorrhagic), the kidney, along with the brain, is a sensitive indicator of the severity of the shock. Urine output will be low if the kidney has been damaged by the decreased circulation of blood. Hemorrhagic shock may be treated by expanding the vascular compartment with plasma or whole blood. Severe shock is often treated with drugs (pressor agents) that contract small vessels, thereby increasing the resistance of arteries and elevating the blood pressure.

Failure of the systemic and pulmonary venous systems to return blood to the heart results in congestion and edema due to the increased hydrostatic pressure in venules and capillaries. Failure of the portal venous system has an additional effect, namely, the creation of anastomotic channels between the lower-pressure portal system and the systemic venous system. These anastomoses occur at many sites in the abdomen, but most importantly in the submucosa of the distal esophagus, where the dilated venous channels are prone to rupture and produce fatal bleeding into the gastrointestinal tract. Failure of the systemic and pulmonary venous systems most commonly results from right and left heart failure, respectively. Failure of the portral venous system most commonly results from cirrhosis of the liver, because the fibrous scarring in the liver acts as an external compressor of the portal veins.

Failure of the capillaries to regulate fluid exchange between blood and tissue results in edema. Three factors determine the ability of capillaries and venules to control fluid exchange: (1) the hydrostatic pressure transmitted from the venules, (2) the osmotic pressure determined by the amount of solutes (particularly protein) in blood and tissue fluid, and (3) the integrity of endothelial cells. As previously mentioned, a generalized increase in hydrostatic pressure leading to edema is a result of heart failure. Altered osmotic pressure leading to edema may result from decreased serum proteins or increased electrolyte (salt) content of tissues. Decreased serum proteins result from a variety of disease conditions, including poor absorption of dietary protein due to malabsorption, chronic liver disease with low production of albumin, and injury to renal glomeruli with loss of protein. Increased salt retention in the tissues occurs from increased dietary intake or decreased renal excretion of salt. Disruption of capillary endothelial cell integrity or function leading to edema is usually a localized accompaniment of the inflammatory reaction caused in part by the release of vasodilatory substances such as histamine.

Edema that is generalized throughout the body and associated with accumulation of fluid in the chest cavity (hydrothorax) and abdominal cavity (ascites) is called *anasarca*. Anasarca is particularly likely when serum proteins are very low or when low serum protein level is associated with salt retention and heart failure. The edema fluid produced by increased hydrostatic pressure in the blood or decreased osmotic pressure in the blood has a low protein content and is called a *transudate*. The edema fluid associated with endothelial injury, as occurs with inflammation, has a high protein content due to the leaky endothelium and thus is an *exudate*.

Failure of the lymphatic system may result from permanent and extensive obstruction to the many lymphatic channels of a particular region of the body. The result is a chronic edema due to failure to drain excess tissue fluid. This type of chronic lymphatic edema, called *elephantiasis*, produces marked enlargement of the affected region, which has a firm doughy consistency. Chronic lymphedema may be secondary to obstruction of lymph nodes by cancer or to surgical removal of lymph nodes with resultant backup of lymphatic fluid, a possible complication of radical mastectomy in which axillary lymph nodes are removed. In the tropics, elephantiasis is caused by a helminthic worm infection called *filariasis*. The worms cause fibrosis and blockage of lymphatic vessels.

REVIEW QUESTIONS

1. What organs are most frequently affected by atherosclerosis?
2. How are each of the following tests utilized in the diagnosis of vascular disease?
   Sphygmomanometry
   Angiography
   Ultrasound
   Funduscopic examination
   Venous pressure measurement
   Antinuclear antibody test
3. What are the causes, lesions, and major manifestations of each of the specific vascular diseases discussed in this chapter?
4. What are the three different types of arteriosclerosis and which type is the most significant? Why?
5. What are the steps in the development of atherosclerosis?
6. What are four ways in which atherosclerotic deposits can cause complications?
7. What are eight important risk factors in atherosclerosis?

8. What are the major sites of atherosclerosis and the potential manifestations of involvement in each of these sites?
9. How does essential hypertension differ from secondary hypertension in terms of cause and treatment?
10. What are the major effects of hypertension?
11. How does malignant hypertension differ from benign hypertension in terms of lesions and manifestations?
12. What is the relationship between hypertension and atherosclerosis?
13. What are the most common sites of involvement of each of the common types of vasculitis?
14. How do the manifestations and complications of varicose veins differ from those of thrombophlebitis?
15. What do fainting and hemorrhage have in common?
16. How do venous and arterial failure differ in terms of causal mechanisms, manifestations, and complications?
17. What are the major mechanisms causing localized and/or generalized edema?
18. What do the following terms mean?
    Thrombus
    Embolus
    Cystic hygroma
    Arteriosclerosis
    Arteriolosclerosis
    Medial sclerosis
    Atherosclerosis
    Intermittent claudication
    Aneurysm
    Raynaud's phenomenon
    Phlebothrombosis
    Pulmonary emboli
    Esophageal varices
    Hemorrhoids
    Hemorrhagic shock
    Syncope
    Hydrostatic pressure
    Osmotic pressure
    Anasarca
    Elephantiasis

Hematopoietic System

REVIEW OF STRUCTURE AND FUNCTION
Blood Cells · Lymphoid Tissue · Mononuclear Phagocytic System · Bone Marrow · Production and Destruction of Red Blood Cells

MOST FREQUENT AND SERIOUS PROBLEMS
Anemia · Leukocytosis · Leukemia, Lymphoma, and Multiple Myeloma

SYMPTOMS, SIGNS, AND TESTS
Lymphadenopathy, Splenomegaly, Hepatomegaly · Petechiae · Hematocrit · Hemoglobin · Red Blood Cell Count · Mean Corpuscular Volume · Mean Corpuscular Hemoglobin Concentration · White Blood Cell Count · White Blood Cell Differential Count · Platelet Count · Reticulocyte Count · Bone Marrow Examination · Lymph Node Biopsy

SPECIFIC DISEASES
  Genetic/Developmental Diseases
  Inflammatory/Degenerative Diseases
    Anemias in General
    Blood-Loss Anemias
    Hemolytic Anemias
    Anemias with Decreased Red Blood Cell Production
    Disorders of White Blood Cells
    Disorders of Platelets
    Infections
  Hyperplastic/Neoplastic Diseases
    Polycythemia
    Leukemias
    Lymphomas
    Multiple Myeloma

ORGAN FAILURE

REVIEW QUESTIONS

REVIEW OF STRUCTURE AND FUNCTION

The major functional components of the hematopoietic system are blood, bone marrow, lymphoid tissues, mononuclear phagocytic system, and immune system. Unlike other systems, these components are located within several organs and have overlapping functions. The major organs where these functional components reside are bone marrow, blood vessels, spleen, lymph nodes, and thymus. We will concentrate on the blood and its cellular components and the bone marrow with brief mention of lymphoid tissues and the mononuclear phagocytic system. The immune system is covered in Chapter 27.

The blood consists of plasma and cells. The plasma alone will clot due to the conversion of fibrinogen to fibrin. If the fibrin is removed, the remaining fluid is called *serum*. The cells include red blood cells, which will fall to the bottom of a tube of blood upon standing, and the platelets and white blood cells, which will form a thin white layer between the serum and red blood cells called the *buffy coat*.

The cells of the hematopoietic system are produced at one site (bone marrow) and live their mature lives at other sites (blood, lymphoid, and mononuclear phagocytic tissues). In the bone marrow, common precursor cells produce offsprings that mature along any one of several pathways to produce erythrocytes, platelets, granulocytes, lymphocytes, and monocytes.

Mature erythrocytes live their 120-day life span in the blood carrying out the highly specialized function of oxygen transport; when senile they are removed in the spleen and other mononuclear phagocytic tissues and their chemical constituents are returned to the body pool.

Red blood cells (erythrocytes) are specialized cells that have no nucleus and whose hemoglobin-filled cytoplasm is shaped like a biconcave disk. The principal function of red blood cells is oxygen transport. The amount of oxygen that can be transported by the blood is determined by the number of red cells in circulation and the amount of hemoglobin in them.

Millions of old red blood cells are removed from the circulation each hour by mononuclear phagocytes in the spleen, liver, and, to a lesser extent, other sites. The most important product of the breakdown process is the iron portion of the hemoglobin molecule, which must be stored by the mononuclear phagocytic system for later production of new red blood cells. Proteins from the degraded red cells (and white cells) are returned to the body's protein pool. The major product of red blood cell breakdown, which requires excretion, is bilirubin. Bilirubin is derived from the noniron-containing portion of the heme molecule. Bilirubin is carried by the blood to the liver, conjugated by liver cells, and excreted into the intestine through the bile duct. Serum bilirubin levels may rise if there is increased breakdown of red blood cells, if the liver is diseased, or if the bile duct is obstructed.

Platelets are actually fragments of a bone marrow cell, the megakaryocyte. The megakaryocyte remains in the bone marrow but its cytoplasmic fragments enter the blood where they are ready to participate in the blood clotting system when needed. Platelets are short lived and must be replaced continuously. The role of platelets is discussed in Chapter 11.

Granulocytes also live their short lives (half a day for neutrophils) in the blood where they are ever ready to participate in an inflammatory reaction; they are removed in the same manner as red blood cells.

After leaving the bone marrow, lymphocytes undergo further maturation. Some differentiate in the thymus to become T lymphocytes and become involved in cell-mediated immunity. Others differentiate in other lymphoid tissues to become B lymphocytes that are capable of further transforming into plasma cells for antibody production. Non-B, non-T lymphocytes have still other functions (see Chapter 27 for details).

Monocytes are the most widespread of the bone marrow derived cells. Some circulate in the blood ready to participate in an inflammatory reaction; others reside in tissues, particularly the sinusoids of liver, spleen, lymph nodes, and bone marrow, but also in practically every tissue of the body. In tissues they are referred to by many names: macrophages, histiocytes, reticuloendothelial cells, and Kupffer cells (in liver). Tissue macrophages as a whole, which carry out the scavenger function of removing debris including foreign materials and the body's own dead cells, are referred to as the *mononuclear phagocytic system* (previously called the *reticuloendothelial system*). Macrophages also play a major role in cellular immune responses.

The bone marrow consists of specialized connective tissue through which flows many capillaries. Filling the tissue are immature forms of the various blood cells to intermediate forms to mature forms (Fig. 10–1). Red blood cell intermediate forms are called rubricytes or normoblasts up until the time that the nucleus is extruded to form a red blood cell. Immature red blood cells retain

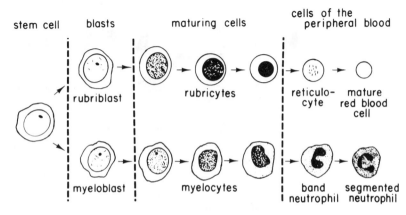

Figure 10–1. Maturation sequence of red blood cells and polymorphonuclear leukocytes.

basophilic material in their cytoplasm and are called reticulocytes. An increase in reticulocytes in the blood is an indication of early release from the bone marrow and thus suggests accelerated red blood cell production. Red cells proliferate in clusters recognized by the small, round, dense nucleus of normoblasts in the cluster. Granulocytes mature from myeloblast to myelocyte to mature granulocytes. The band, or stab, neutrophil occurs at a stage just before the lobes of the nucleus become separated by a thin strand. When found in increased numbers in the blood (more than 6 percent), band neutrophils indicate increased production and early release of neutrophils from the bone marrow. The characteristic neutrophilic, basophilic, or eosinophilic granules are acquired in the later stages of myelocyte differentiation. Granulocytic cells are spread diffusely throughout the bone marrow; their normal concentration is about four times that of red blood precursor cells. Monocytes develop from monoblasts and are mixed in with the granulocytes.

The bone marrow serves as a storage site for blood cells, which can be released when needed. The bone marrow can increase its production of blood cells in response to increased demand. Under normal circumstances erythrocytes live 120 days and neutrophils a half day. Erythropoietin is a hormone, released from the kidney, that stimulates erythropoiesis. When there are too few erythrocytes in circulation, more hormone is released to accelerate production of new cells. The production of neutrophils is thought to be mediated by a hormonal substance released from damaged tissue. Mechanisms that control production of other bone marrow elements are poorly understood.

MOST FREQUENT AND SERIOUS PROBLEMS

Overall, the most common clinical problem relating to the hematopoietic system is anemia. Anemia is a decrease in the circulating red blood cell mass. Thus, it is a finding rather than a disease per se. It may be due to decreased production of red blood cells or to increased destruction or loss of red blood cells. The most common types of anemia are: (1) iron-deficiency anemia resulting from dietary deficiency, (2) iron-deficiency anemia resulting from chronic bleeding from the uterus or gastrointestinal tract, (3) anemia associated with various chronic diseases, and (4) vitamin B_{12}–deficiency anemia (pernicious anemia) and folic acid-deficiency anemia. Anemia may be serious in itself if severe; otherwise, it is often a clue to the discovery and treatment of its underlying cause.

Most disorders of white blood cells are secondary effects of

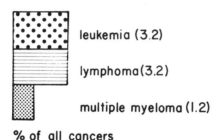

% of all cancers

Figure 10–2. Relative frequency of primary hematopoietic cancer.

other diseases rather than primary in the hematopoietic system. For example, most infections are associated with an increased need for white blood cells and thus cause leukocytosis. If the infection is severe or prolonged, myeloid and/or lymphoid hyperplasia results.

Primary cancers of the hematopoietic system comprise leukemias, lymphomas, and multiple myeloma. These cancers of white blood cells and their derivatives account for 7.6 percent of all cancers and most of the deaths from primary disease of the hematopoietic system (Fig. 10–2).

SYMPTOMS, SIGNS, AND TESTS

Most symptoms of hematopoietic system disease are nonspecific; they can be caused by diseases of other systems. The symptoms of anemia vary from no symptoms to heart failure. Heart failure occurs when there is an insufficient amount of blood to pump and, thus, insufficient oxygenation of the tissues of the body. Nonspecific symptoms of anemia include headache, easy fatigability, loss of appetite, heartburn, shortness of breath, edema of the ankles, and numbness and tingling sensations.

Physical examination may detect enlargement of lymph nodes (lymphadenopathy), spleen (splenomegaly), and liver (hepatomegaly), which can be due to a wide variety of hematopoietic and nonhematopoietic diseases. The occurrence of tiny hemorrhages in the skin (petechiae) is an important finding that suggests a decrease in the number of circulating platelets. Other types of hemorrhage, such as nosebleeds or ecchymoses (large areas of hemorrhage into the skin), may be associated with decreased platelet count or they may be associated with a coagulation disorder (discussed in Chapter 11). Pallor of the skin is found with severe anemia but is an unreliable sign for mild anemia.

Laboratory tests for hematopoietic disease include analysis of blood cells, biopsy of lymph nodes or bone marrow, and special tests for specific diseases. Analysis of blood cells is used for screening and diagnostic purposes. The most commonly used tests in this category are the hematocrit, hemoglobin, red blood cell count, white blood cell count, white blood cell differential count, red blood cell morphology, platelet count, and reticulocyte count.

The hematocrit, hemoglobin, red blood cell count, and blood smear are used to indicate the presence or absence of anemia, and to characterize it as microcytic, normocytic, or macrocytic and as normochromic or hypochromic. Anemia is defined as a decrease in circulating red cell mass as measured by a low hematocrit, hemoglobin, or red blood cell count. Hematocrit is the volume of the red blood cells as compared with other blood elements. When blood is centrifuged, the red cells form the sediment. The percentage of the blood occupied by the red blood cells is the hematocrit. The hemoglobin is a measurement of the amount of hemoglobin in grams per deciliter (g/dl). The red blood cell count, performed by counting the cells in a small chamber, is the number of red blood cells per cubic millimeter of blood.

The size and hemoglobin concentration of red blood cells can be visualized directly by placing a blood smear under a microscope or may be calculated from the hematocrit, hemoglobin, and red blood cell count. The blood smear is a useful, rapid means of classifying red blood cells as macrocytic, normocytic–normochromic, or microcytic–hypochromic (Fig. 10–3). Other changes in red blood cell morphology may also be present and may give more specific clues to the cause of the anemia. The mean corpuscular volume (MCV), or mean red blood cell size, is calculated by dividing the hematocrit by the red blood cell count. The mean corpuscular hemoglobin concentration (MCHC) is calculated by dividing the hemoglobin by the hematocrit. Table 10–1 gives the normal values for measurements of red blood cells and interpretations of high and low values.

The white blood cell count and differential white blood cell count are used to evaluate white blood cells. The white blood cell count involves the counting of cells in a chamber and calculating the number of cells per cubic millimeter. The differential count involves identifying consecutive white blood cells on a smear and calculating the percentage of each type present. Absolute counts can be calculated from the percentages of each cell type and the total count. Absolute counts are of more value than the relative percentages, because the numbers of specific types of leukocytes may vary independently of each other. Table 10–2 gives the normal values and names applied to increases and decreases.

Platelets (thrombocytes) are evaluated by the platelet count, expressed as thousands per cubic millimeter. Normal values are

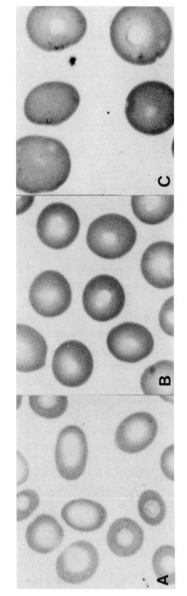

Figure 10–3. Classification of red blood cells by size and hemoglobin content as visualized on a blood smear. **A.** Hypochromic, microcytic red blood cells. **B.** Normochromic, normocytic red blood cells. **C.** Macrocytic red blood cells.

TABLE 10–1. MEASUREMENTS OF RED BLOOD CELLS

Test	Normal Values	Name of Low Value	Name of High Value
Hematocrit	Male: 40–54% Female: 37–49%	Anemia	Polycythemia
Hemoglobin	Male: 14.1–18.0 g/dl Female: 12.3–16.2 g/dl	Anemia	Polycythemia
Red blood cell count	Male: 4.7–6.1 million/mm³ Female: 4.2–5.6 million/mm³	Anemia	Polycythemia
MCV	82–97 cu μ	Microcytosis (microcytic anemia)	Macrocytosis (macrocytic anemia)
MCHC	32–36 g/dl	Hypochromia (hypochromic anemia)	Hyperchromia (rarely occurs)

between 150,000 and 400,000 per cubic millimeter. A platelet decrease is called *thrombocytopenia* and an increase *thrombocytosis*.

The reticulocyte count is a measure of the percentage of immature red blood cells in circulation. Values greater than 2 percent are referred to as *reticulocytosis* and indicate an increased rate of production and release of new red blood cells.

Biopsy of the bone marrow is performed by boring a needle into the iliac crest, vertebra, or sternum to obtain bone marrow tissue. The tissue can be smeared on a slide and stained or embedded in paraffin and sectioned. Bone marrow examination is used in the diagnosis of hematopoietic cancers and selected other hematopoietic diseases. Lymph node biopsy is most often used to evaluate the presence or absence of cancer, but it is also used to diagnose rare types of chronic infections.

Examples of special hematology tests include *hemoglobin electrophoresis* to evaluate genetic abnormalities in the hemoglobin molecule, *sickle-cell preparation* to induce the sickle-shaped red blood

TABLE 10–2. MEASUREMENTS OF WHITE BLOOD CELLS (WBCs)

Test	Normal Values	Name of High Value	Name of Low Value
WBC count	4,300–11,600/mm³	Leukocytosis	Leukopenia
WBC differential count			
Neutrophils	42–81%	Granulocytosis or neutrophilic leukocytosis	Granulocytopenia or neutropenia
Lymphocytes	10–47%	Lymphocytosis	Lymphopenia
Monocytes	0–10%	Monocytosis	Not applicable
Eosinophils	0–7%	Eosinophilia	Not applicable
Basophils	0–1%	Basophilia	Not applicable

cells associated with sickle-cell anemia, *red cell fragility test* to detect spherical red blood cells (spherocytes), which burst when exposed to hypotonic solutions. Iron concentrations may be measured in serum. Diseases that are the result of antibodies against red or white cells are evaluated by special immunohematology procedures.

SPECIFIC DISEASES

Genetic/Developmental Diseases

There are several important hereditary defects that cause anemia including sickle-cell disease, thalassemia, hereditary spherocytosis, and glucose-6-phosphatase deficiency. For clarity, they will be discussed in the following section with the other types of anemia. The much rarer genetic defects of white blood cells, which lead to death in childhood from repeated infections, are discussed in Chapter 26.

Inflammatory/Degenerative Diseases

This section deals with anemias, disorders of white blood cells, disorders of platelets, and certain inflammatory disorders that characteristically affect the mononuclear phagocytic and lymphoid tissues.

Many of these conditions are indirect results of injury, inflammation, and repair. Others, included in this section for convenience, are better classified as genetic disorders, metabolic disorders, or both.

Anemias in General

Anemia may be a finding in many diseases, including primary diseases of red blood cells and diseases that secondarily involve the hematopoietic system. A pathogenetic classification is given in Table 10–3. Anemia may be due either to removal of red blood cells at a rate that exceeds the replacement capacity of the bone marrow (blood-loss anemias and hemolytic anemias) or to decreased production of red blood cells by the bone marrow. Blood-loss anemia involves loss of blood from the vascular system, either externally to the body or internally. Hemolytic anemia involves destruction of red blood cells within the vascular or mononuclear phagocytic systems. When the mononuclear phagocytic system (particularly the spleen) removes red blood cells before their normal lifespan is up, it is called *extravascular hemolysis*. If the red blood cells are actually destroyed in the bloodstream, it is called *intravascular hemolysis*.

A laboratory-oriented approach to the classification of anemia

TABLE 10–3. CLASSIFICATION OF ANEMIA

Blood-loss anemias
  Acute blood-loss anemia
  Chronic blood-loss anemia
Hemolytic anemias
  Sickle-cell anemia
  Thalassemia
  Hereditary spherocytosis
  Glucose-6-phosphate dehydrogenase deficiency
  Immune hemolytic anemia
  Hypersplenism
  Microangiopathic hemolytic anemia
Anemias with decreased red blood cell production
  Deficiency anemia
    Iron deficiency
    Vitamin B_{12}–deficiency (pernicious anemia)
    Folic acid deficiency
  Anemia of chronic disease
  Myelophthisic anemia
  Aplastic anemia

is based on common laboratory findings that help separate the most frequently encountered types of anemia in an efficient manner. Evaluation of red blood cell size and hemoglobin concentration gives three major categories of anemia. Microcytic hypochromic anemia occurs with iron deficiency, with some cases of anemia of chronic disease, and with a few other rare diseases. Macrocytic (normochromic) anemia is usually due to vitamin B_{12} or folic acid deficiency. Normocytic normochromic anemia is a feature of most other types of anemia. The reticulocyte count helps separate anemias of decreased production (normal or low reticulocyte count) from those of increased destruction or loss (elevated reticulocyte count). The laboratory approach to classification must be applied with care, for there are some ambiguities. For example, chronic blood loss is not associated with an elevated reticulocyte count, because the mechanism by which chronic blood loss produces anemia is iron deficiency. Thus, chronic blood-loss anemia is both a blood-loss anemia and a decreased-production anemia. Another example is thalassemia. In thalassemia, there is both decreased production of red blood cells because of the deficient hemoglobin production and increased extravascular hemolysis because the resulting cells are defective and consequently more subject to removal.

The classification of anemia is obviously very important, because incorrect interpretation may result in failure to diagnose the underlying disease or apply the proper treatment. We suggest that you consider each type of anemia in terms of its pathogenetic mechanism, laboratory classification, and the context in which it is likely to occur.

Blood-Loss Anemias

Acute blood loss will produce anemia within a few hours because of hemodilation, a process that allows replacement of blood serum before the bone marrow can replace lost cells. The red blood cells remaining in the anemic blood will be normochromic and normocytic. Within a few days, an elevated reticulocyte count will herald the stepped-up production and release of new cells from the bone marrow. The bone marrow is capable of replacing a large amount of lost blood. For example, a blood donor who gives one pint of blood suffers no ill effects. Blood lost need not be replaced unless the amount is such that hemodynamic effects, such as shock, occur.

Chronic blood loss is the slow loss of small amounts of blood over a period of time. The most common causes are excessive menstrual bleeding and bleeding from the gastrointestinal tract. The bone marrow has plenty of capacity to replace this type of blood loss. Anemia occurs late in the course of chronic blood loss due to failure to recycle the iron from the lost red blood cells. Thus, chronic blood-loss anemia is a subcategory of iron-deficiency anemia. Chronic blood loss is discovered by taking a menstrual history or checking the feces for blood (occult fecal blood test). Other forms of chronic blood loss are usually obvious from the patient's history.

CHART 10–1. BLOOD-LOSS ANEMIA

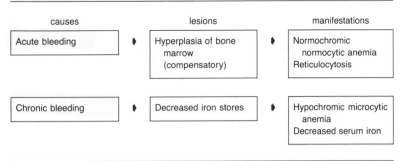

causes	lesions	manifestations
Acute bleeding	Hyperplasia of bone marrow (compensatory)	Normochromic normocytic anemia Reticulocytosis
Chronic bleeding	Decreased iron stores	Hypochromic microcytic anemia Decreased serum iron

Hemolytic Anemias

In hemolytic anemias, red blood cells are prematurely removed from the bloodstream (extravascular hemolysis) or destroyed within the bloodstream (intravascular hemolysis). Extravascular hemolysis is more common. Hemolysis may be caused by the bone marrow producing defective red blood cells that are destined to a short life or by events that affect normal cells after they are released from the bone marrow.

In hemolytic anemia, there is an increased production of bilirubin as macrophages degrade the dead red blood cells. The transportation of this pigment to the liver for excretion is often manifest by an elevation of serum bilirubin and by mild jaundice. The bone marrow, which is stimulated by the anemia to produce more erythrocytes, becomes hyperplastic and releases more immature erythrocytes than normal, causing a reticulocytosis. With intravascular hemolysis, there also will be free hemoglobin in the blood and urine and a decrease in serum haptoglobin, a serum protein that binds free hemoglobin. Patients with hemolytic anemia do not become depleted of iron because iron from the destroyed cells reenters the body's iron pool. In fact, increased absorption of iron and transfusion therapy often leads to hemosiderosis (excessive iron deposition) in these patients.

The mechanisms of hemolytic anemia include hereditary defects in red blood cells that decrease their lifespan, antibodies to red blood cells that cause their destruction or premature removal by the mononuclear phagocytic system, premature removal of red blood cells by the spleen due to chronic passive congestion (*hypersplenism*), and mechanical injury to red blood cells by rough surfaces in the bloodstream (*microangiopathic hemolytic anemia*). Important diseases that illustrate these mechanisms will be discussed.

Sickle-cell anemia is one of several genetic abnormalities of hemoglobin structure due to an altered sequence of amino acids in the globin molecule. The diseases produced by these genetic defects are called *hemoglobinopathies,* and the abnormal hemoglobin is designed by a letter or name of a place where it was first described. Sickle-cell anemia occurs in persons with two genes for hemoglobin S (the homozygous state). Persons with one hemoglobin S gene (the heterozygous state) are said to carry the sickle trait and can be identified by a positive sickle-cell preparation test (Fig. 10–4), but do not have anemia. Identification of the trait is quite useful for purposes of genetic counseling. Hemoglobin S is a genetic abnormality predominantly of blacks; about 10 percent are heterozygous carriers and about 1 percent are homozygous persons with the disease. The abnormal hemoglobin results in some red blood cells becoming sickle shaped under situations of low oxygen tension. The sickle cells are not only more susceptible to rupture and premature death, but they also tend to sludge and obstruct small blood vessels. Vascular obstructions are particularly common in the spleen and bone, where they produce multiple

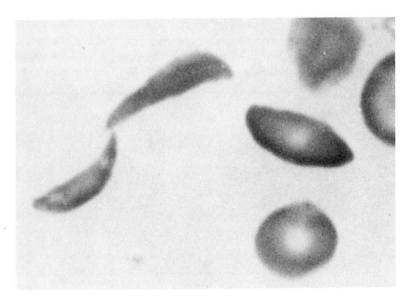

Figure 10–4. Sickled red blood cells.

small infarcts over a period of years. Patients with sickle-cell anemia often live reasonably well with their anemia except during periods of so-called crisis, when more cells become sickled, leading to abdominal and bone pain from small infarcts and jaundice from the increased breakdown of red blood cells. Leg ulcers are a common complication of long-standing disease. Most patients with sickle-cell anemia die by age 30 to 40.

CHART 10–2. SICKLE-CELL ANEMIA

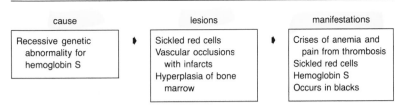

cause		lesions		manifestations
Recessive genetic abnormality for hemoglobin S	▶	Sickled red cells Vascular occlusions with infarcts Hyperplasia of bone marrow	▶	Crises of anemia and pain from thrombosis Sickled red cells Hemoglobin S Occurs in blacks

Thalassemia is a genetic defect affecting the rate of synthesis of normal hemoglobin (hemoglobin A) due to a deficient production of alpha or beta globin. There is a compensatory increase in a type of hemoglobin found in the fetus (hemoglobin F) or in a type of hemoglobin found normally in small amounts (hemoglobin A2).

Thalassemia is most common among persons of Mediterranean descent. Thalassemia major occurs in homozygous individuals with severe anemia developing in infancy and leads to death in childhood or adolescence. There is a decrease in production of red blood cells because of increased destruction of immature red blood cells in the bone marrow.

CHART 10–3. THALASSEMIA

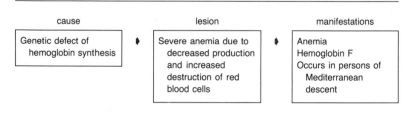

cause	lesion	manifestations
Genetic defect of hemoglobin synthesis	Severe anemia due to decreased production and increased destruction of red blood cells	Anemia Hemoglobin F Occurs in persons of Mediterranean descent

Hereditary spherocytosis is a genetic defect of the red blood cell membrane with an autosomal dominant inheritance pattern. The abnormal red blood cells are spherical rather than the normal flat, biconcave disks. As they filter through the spleen, they are more easily removed than normal cells. The anemia is usually mild and often not discovered until adulthood. The spleen is enlarged because it traps the abnormal red blood cells. Removal of the spleen usually is curative. This is an example of a dominant disease that is perpetuated because its relative mild nature and prolonged course allows persons with the disease to reach maturity and reproduce.

CHART 10–4. HEREDITARY SPHEROCYTOSIS

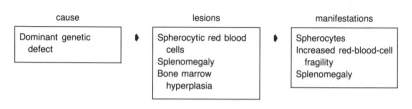

cause	lesions	manifestations
Dominant genetic defect	Spherocytic red blood cells Splenomegaly Bone marrow hyperplasia	Spherocytes Increased red-blood-cell fragility Splenomegaly

Glucose-6-phosphate dehydrogenase deficiency is a genetic enzyme defect that only becomes manifest when red blood cells are exposed to certain oxidant drugs such as antimalarial drugs, sulfas, nitrofurantoin, aspirin, and other analgesics. These patients have a mutant enzyme that becomes deficient in older red blood cells, allowing oxidants to damage the cell membrane and produce hemolysis. Discontinuance of the drug and normal replacement by young red blood cells end the hemolytic episode. The genetic abnormality is sex-linked and is present in 10 percent of blacks. Males are much more prone to develop anemia because most affected females are heterozygous carriers. The disease can be prevented by screening high-risk individuals (black males) for the defect and then avoiding exposure to oxidant drugs.

CHART 10–5. GLUCOSE-6-PHOSPHATE DEHYDROGENASE DEFICIENCY

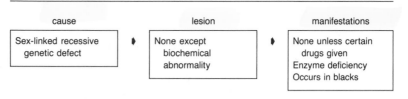

cause	lesion	manifestations
Sex-linked recessive genetic defect	None except biochemical abnormality	None unless certain drugs given Enzyme deficiency Occurs in blacks

Immune hemolytic anemia may be associated with antibodies that activate complement and lyse the red blood cell in the bloodstream or with antibodies that facilitate removal of the red cell by the spleen. Transfusion of blood with a major incompatibility between the donor's and recipient's ABO systems and severe cases of erythroblastosis fetalis are examples of causes of intravascular hemolysis. Certain drugs may induce antibodies to red blood cells and thus cause immune hemolytic anemia. In many cases of immune hemolytic anemia, the source of the antigen is the patient's own Rh factor; thus, the anemia is classified as autoimmune hemolytic anemia. The Coombs' test measures the presence of antibodies attached to the surface of red blood cells and therefore is the test used to detect immune hemolytic anemias.

Hypersplenism is most commonly caused by chronic passive congestion of the spleen, a condition where the venous pressure is increased due to obstruction of the portal venous system, usually due to cirrhosis of the liver. The venous congestion causes the spleen to remove more blood cells than normal, thus producing a type of extravascular hemolytic anemia. The condition is suspected in a patient with cirrhosis and enlarged spleen and who often has leukopenia and thrombocytopenia along with anemia.

Microangiopathic hemolytic anemia is caused by rough surfaces in the bloodstream as sometimes produced by prosthetic heart valves, rough atherosclerotic plaques, and disseminated intravascular thrombosis. The blood smear will show the fractured cells, called schiztocytes or helmet cells.

CHART 10–6. HEMOLYTIC ANEMIAS

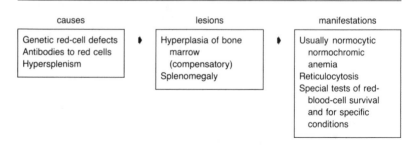

causes	lesions	manifestations

Genetic red-cell defects
Antibodies to red cells
Hypersplenism
	Hyperplasia of bone
	  marrow
	  (compensatory)
	Splenomegaly
		Usually normocytic
		  normochromic
		  anemia
		Reticulocytosis
		Special tests of red-
		  blood-cell survival
		  and for specific
		  conditions

Anemias with Decreased Red Blood Cell Production

The bone marrow may fail to put out enough red blood cells if it has an inadequate supply of nutrients to produce red blood cells, if its function is suppressed by the presence of chronic disease, or if the bone marrow tissue is insufficient in amount. The blood smear is quite helpful in initial screening because deficiency of iron produces a microcytic, hypochromic anemia and deficiency of either vitamin B_{12} or folic acid produces a macrocytic anemia. Most other types of anemias are normocytic and normochromic.

Iron-deficiency anemia is a common type of anemia. It may be due to loss of iron or inadequate intake of iron. Loss of iron most commonly results from chronic blood loss, as has been discussed. Inadequate intake of iron occurs in infants fed on a milk and fruit diet without meat or supplemental iron; in women during the menstrual years (a combination of inadequate intake and increased loss); during pregnancy, when iron must be provided to the fetus; and in chronic intestinal diseases associated with malabsorption of iron. In iron-deficiency anemia, the cells produced are smaller and paler than normal because there is less hemoglobin per cell (microcytic hypochromic anemia). The anemia is often mild and unrecognized by the patient. Because iron stores are depleted, the serum iron is decreased and the proteins that bind iron in the serum are increased (elevated serum iron-binding capacity). Administration of iron causes an elevation of the reticulocyte count in a few days and increases in hemoglobin after about 10 days.

Vitamin B_{12}–deficiency anemia is caused by failure to absorb vitamin B_{12} from the intestinal tract and not from deficiency of the diet. Dietary vitamin B_{12} (extrinsic factor) must combine with a protein produced in the gastric mucosa (intrinsic factor) and be carried to the distal small intestine before it can be absorbed and carried to the bone marrow or body storage sites. Atrophy of the gastric mucosa, occurring mostly in persons over 60, with resultant insufficiency of intrinsic factor, is the most common cause of vitamin B_{12}–deficiency. The disease produced is called *pernicious anemia*. Vitamin B_{12}–deficiency causes disordered synthesis of DNA, resulting in accumulation of large, abnormal red blood cell precursors (megaloblasts) in the bone marrow. The maturation of red blood cells is delayed, and the red cells released into the bloodstream are larger than normal (macrocytic). The macrocytic cells have a tendency to hemolyze more than normal cells, so the resulting anemia may have a hemolytic component as well as a production deficiency. The basic defect also affects other blood cells to a lesser degree. Pernicious anemia may be associated with permanent destruction of the spinal cord, which results in loss of coordination. Pernicious anemia is suspected when a macrocytic anemia with megaloblasts in the bone marrow is found, and it is confirmed by low serum levels of vitamin B_{12}. The Schilling test, which measures the degree of absorption of vitamin B_{12} from the intestine, is also a useful diagnostic test. Injections of vitamin B_{12} cure the anemia and are continued at regular intervals to prevent recurrence of the disease.

Folic acid deficiency also results in impaired DNA synthesis and produces a macrocytic anemia similar to pernicious anemia, except that the spinal cord degeneration does not occur. Folic acid deficiency results from inadequate diet such as is common in alcoholics, from increased need such as occurs in pregnancy, and from chronic intestinal diseases that produce malabsorption. Folic acid levels in the serum are used to distinguish folic acid deficiency from vitamin B_{12}–deficiency.

Anemia of chronic disease is an anemia of unknown cause that is diagnosed in patients with chronic disease after exclusion of other causes. It is the most common type of anemia and is unresponsive to therapy. Associated diseases include long-standing infections, cancer, chronic inflammatory diseases, such as rheumatoid arthritis, and chronic renal disease. Ten to 15 percent of hospitalized patients may exhibit this type of anemia. The cost of differentiating this untreatable type of anemia from other treatable forms is considerable. Differentiation from mild iron-deficiency anemia is a particular problem because both may be borderline microcytic with borderline serum iron levels and both may be present at the same time. The pathogenesis of the anemia is unclear, although there is suppression of red blood cell reproduction, reluctance of mononuclear phagocytic cells to release stored iron for production of new cells, and a mildly shortened red blood cell survival time. Diagnosis rests on a history of chronic disease in addition to laboratory studies that show normocytic, normochromic (or sometimes mildly microcytic, hypochromic) red blood cells, normal to slightly low serum iron and serum ferritin levels, and normal serum iron-binding capacity (elevated in iron deficiency). Also, in contrast to iron-deficiency anemia, mononuclear phagocytic cells in the bone marrow contain abundant iron.

Myelophthisic anemia refers to anemia caused by replacement of the bone marrow by diseased tissue such as cancer or fibrous tissue. Cancer is the most common cause. Leukemias, lymphomas, multiple myeloma, and metastatic carcinoma from lung, breast, or prostate can replace the bone marrow so that blood cells are produced in subnormal numbers. In addition to the anemia, there will be leukopenia and thrombocytopenia. Replacement of the bone marrow by fibrous tissue is called *myelofibrosis.* It may be the result of irradiation or drugs or of unknown cause. To compensate for bone marrow destruction, hematopoietic cells may take up residence in other sites, such as spleen, liver, and lymph nodes, a process called *myeloid metaplasia* or *extramedullary hematopoiesis.*

Aplastic anemia is an atrophy of the bone marrow, usually of unknown cause, but sometimes caused by chemical poisons such as benzene, drugs such as anticancer agents and chloramphenacol, and by radiation. As in myelophthisic anemia, all blood cell elements may be reduced. Bone marrow biopsy is useful in distinguishing myelophthisic anemia from aplastic anemia, the former displaying the disease replacing the bone marrow and the latter displaying only a hypocellular bone marrow.

CHART 10–7. ANEMIA WITH DECREASED PRODUCTION

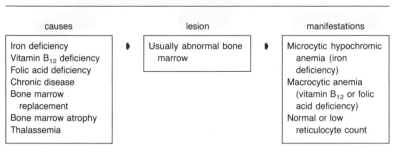

causes	lesion	manifestations

Iron deficiency
Vitamin B$_{12}$ deficiency
Folic acid deficiency
Chronic disease
Bone marrow
  replacement
Bone marrow atrophy
Thalassemia
	Usually abnormal bone
	  marrow
		Microcytic hypochromic
		  anemia (iron
		  deficiency)
		Macrocytic anemia
		  (vitamin B$_{12}$ or folic
		  acid deficiency)
		Normal or low
		  reticulocyte count

Disorders of White Blood Cells

Degenerative and inflammatory disorders involving white blood cells are almost always secondary to disease of some other system. Granulocytosis is characteristic of acute inflammation; lymphocytosis and monocytosis occur with some chronic inflammations; and eosinophilia is characteristic of parasitic infections and some types of allergy. Neutropenia and lymphopenia occasionally occur with some types of infections. White blood cell count may be decreased (leukopenia) by excessive removal with hypersplenism or insufficient production in association with myelophthisic or aplastic anemia.

Disorders of Platelets

Thrombocytopenia is more common and more significant than thrombocytosis. Mechanisms of thrombocytopenia include increased platelet destruction and decreased platelet production. Causes of increased destruction include antibodies to platelets, increased utilization of platelets such as occurs in some blood coagulation disorders, and hypersplenism. Occasionally, treatment with any of a variety of drugs may be associated with the development of antibodies to platelets. Causes of decreased production of platelets include myelophthisic and aplastic anemia.

   Idiopathic thrombocytopenic purpura (ITP) refers to thrombocytopenia without evident cause and is believed to be due to antibodies to platelets in most instances. It occurs as a short-lived disorder in children following infections. It also occurs in adults, especially young women, without a precipitating episode and with a prolonged course. In chronic cases, removal of the spleen often results in remission, because the spleen can no longer remove the antibody-coated platelets. Corticosteroid drugs may cause a temporary rise in platelets.

   Regardless of cause, a markedly depressed platelet count will be associated with bleeding from small blood vessels to produce petechiae. Thrombocytosis is associated with a few uncommon diseases and usually produces no ill effects.

CHART 10–8. THROMBOCYTOPENIA

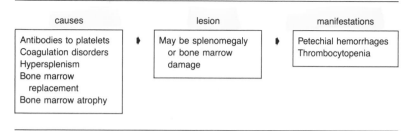

causes	lesion	manifestations
Antibodies to platelets Coagulation disorders Hypersplenism Bone marrow replacement Bone marrow atrophy	May be splenomegaly or bone marrow damage	Petechial hemorrhages Thrombocytopenia

Infections

As mentioned previously, infections frequently cause a secondary hyperplasia of myeloid, lymphoid, and mononuclear phagocytic tissues. A few types of infections, mostly chronic types, have their major effects in the hematopoietic system. Malaria is caused by small protozoa that live in red blood cells and cause episodes of red cell destruction manifested by fever and anemia. Infectious mononucleosis is caused by a virus that produces enlargement of lymphoid tissue, including lymph nodes, pharyngeal lymphoid tissue, and spleen. The disease runs a prolonged course, producing weakness, sore throat, and lymphadenopathy. Atypical reactive lymphocytes in the blood and a positive mono spot test for antibodies to the virus are the means of diagnosis.

Granulomatous diseases such as tuberculosis and systemic fungal infections have a strong tendency to localize in organs with much mononuclear phagocytic tissue, including lymph nodes, spleen, liver, and bone marrow. Sarcoidosis is an idiopathic granulomatous disease that also produces widespread granulomatous lesions in the mononuclear phagocytic tissues.

Hyperplastic/Neoplastic Diseases

The diseases called *leukemias* are white blood cell cancers characterized by extensive bone marrow replacement with neoplastic white blood cells. The condition *leukemia* is an elevated blood cell count due to leukemic cells, and is usually, but not always, present in patients with leukemia (the disease). Leukemias can be either granulocytic or lymphocytic. *Lymphomas* are also white blood cell cancers and are characterized by involvement of sites peripheral to the bone marrow, often with production of mass lesions (as opposed to diffuse involvement of bone marrow and leukemia), and are mostly of lymphocytic origin. *Multiple myeloma* is a cancer of plasma cells, usually arising in the bone marrow without producing leukemia. The effects of multiple myeloma are due to cancerous replacement of bone marrow and production of abnormal immunoglobulins that are deposited in other organs. Without treatment hematopoietic malignancies are uniformly fatal. Establishment of the curative potential of various drug regimens and/or radiation therapy and the introduction of bone marrow transplantation have made the classification of these diseases based on cell type much more important. Only a general outline of this complicated subject will be presented.

Polycythemia

Polycythemia is an increase in red blood cells due to persistent overproduction. It is a form of hyperplasia that may be primary (of unknown cause) or secondary. Primary polycythemia is called *polycythemia vera*. Secondary polycythemia is mediated by erythropoietin. The causes of excess erythropoietin production include (1) hypoxia due to chronic lung disease, cyanotic heart disease, or living at a high altitude and (2) excess production of erythropoietin by any one of several rare neoplasms.

Secondary polycythemia due to hypoxia is necessary to sustain oxygenation of tissue. Polycythemia leads to difficulty by increasing the viscosity of the blood, which may lead to thrombosis and hemorrhage. Polycythemia vera is much less common than secondary polycythemia and appears to be a primary proliferative disease of bone marrow. Other blood cells may be involved, and the disease may lead to myelofibrosis, and, in rare cases, leukemia develops.

CHART 10–9. POLYCYTHEMIA

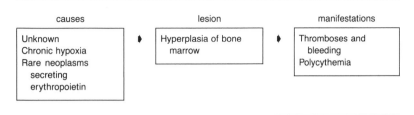

causes	lesion	manifestations
Unknown Chronic hypoxia Rare neoplasms secreting erythropoietin	Hyperplasia of bone marrow	Thromboses and bleeding Polycythemia

Leukemias

Leukemias comprise several types of malignant neoplasms of white blood cells that originate and spread diffusely in the bone marrow and usually produce high white blood cell counts. Leukemic cells often diffusely infiltrate other organs, such as spleen, liver, and lymph nodes.

Leukemias are usually classified by the type of white cell involved and the chronicity of the disease. The degree of differentiation of the leukemic cells relates closely to the likely duration; thus, acute leukemias have poorly differentiated cells and a rapid course, whereas chronic leukemias have well-differentiated cells and a slow course. Most leukemias involve either lymphocytes or granulocytes (mostly neutrophils). Monocytic leukemia is less common. Acute lymphocytic leukemia is the most common type of childhood leukemia and is rapidly fatal unless treated very aggressively with multiagent chemotherapy. Successful therapy may lead to long-term survival and cure. Chronic lymphocytic leukemia is a much different disease; it occurs in the elderly and runs a prolonged, indolent course with many patients dying from other causes before the leukemia has time to kill the patient. Both acute and chronic granulocytic (myelogenous) leukemia occur predominantly in adults and are less likely to be cured than leukemia in children.

Acute leukemias are composed of more primitive or less well-differentiated cell types; chronic leukemias are composed mainly of mature cells with some primitive blast forms. Acute leukemias have an abrupt onset with bleeding due to thrombocytopenia, anemia, fatigue, fever, and weight loss. The white blood count may be normal or elevated. In chronic leukemia, the symptoms, although similar, appear gradually and by the time they develop the white blood count may be very high and organs, such as spleen, liver, and lymph nodes, enlarged by leukemic infiltrates. Diagnosis is made by examination of the blood smear and bone marrow, and patients are often referred to major centers for treatment.

CHART 10–10. LEUKEMIAS

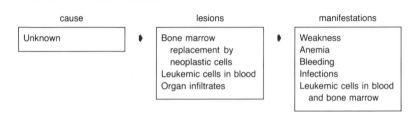

cause	lesions	manifestations
Unknown	Bone marrow replacement by neoplastic cells Leukemic cells in blood Organ infiltrates	Weakness Anemia Bleeding Infections Leukemic cells in blood and bone marrow

Lymphomas

Lymphomas comprise several types of malignant neoplasms of lymphocytes and histiocytes that originate in lymphoid tissues outside of the bone marrow, most often in lymph nodes. Lymphomas usually produce mass lesions, in contrast with leukemias in which disease is concentrated within the bone marrow. Further, lymphomas usually do not spill malignant cells into the bloodstream, i.e., they are aleukemic.

The classification of lymphomas is based on cell type and is very complex. Major categories include Hodgkin's disease and non-Hodgkin's lymphoma. Hodgkin's disease is characterized by a large, malignant cell with a multilobed nucleus containing prominent nucleoli that is known as a *Reed-Sternberg cell*. Unlike most malignancies, Hodgkin's disease contains many benign cells including lymphocytes, histiocytes, neutrophils, eosinophils, and fibroblasts; these additional cells are used for subclassification and their presence or absence relates to prognosis. Non-Hodgkin's lymphomas are classified on the basis of cell size (lymphocytic vs histiocytic), cell maturity (well differentiated vs poorly differentiated), immunologic markers (T cell, B cell, or neither), histologic pattern (nodular vs diffuse), and details of cell structure.

The ultimate purpose of trying to subclassify disease is to better estimate prognosis and provide optimal therapy; considerable progress has been made with regard to lymphomas. The survival from Hodgkin's disease, once considered to be zero, is greatly improved for all types with aggressive radiation therapy and/or chemotherapy, and survival is now expected in many patients with the less malignant types of disease. Nodular poorly differentiated lymphocytic lymphoma has a long median survival without therapy and is not cured by chemotherapy. In contrast, diffuse histiocytic lymphoma can be cured or life greatly prolonged by aggressive chemotherapy, but has a short survival without therapy.

Diagnosis and classification of a lymphoma is made by biopsy; patients are often referred to major centers for therapy.

CHART 10–11. LYMPHOMAS

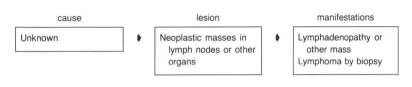

cause	lesion	manifestations
Unknown	Neoplastic masses in lymph nodes or other organs	Lymphadenopathy or other mass Lymphoma by biopsy

Multiple Myeloma

Multiple myeloma is a malignant neoplasm of plasma cells. For unknown reasons, it arises in the bone marrow and grows to replace bone marrow with localized destruction of surrounding bone. Another characteristic feature is the production of immunoglobulins, which can be detected in the blood and urine. Multiple myeloma is a disease of the middle-aged and elderly that presents with anemia, infection, multifocal destructive bone lesions, and sometimes renal failure from immunoglobulin precipitates in renal tubules. Although chemotherapy has prolonged survival, the ultimate prognosis is poor.

CHART 10–12. MULTIPLE MYELOMA

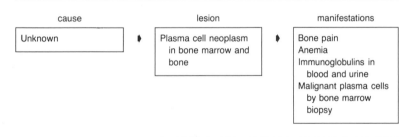

cause	lesion	manifestations
Unknown	Plasma cell neoplasm in bone marrow and bone	Bone pain Anemia Immunoglobulins in blood and urine Malignant plasma cells by bone marrow biopsy

ORGAN FAILURE

Bone marrow failure has been discussed above as myelophthisic anemia and aplastic anemia, conditions associated with replacement or atrophy of the bone marrow. The results of bone marrow failure are anemia, leukopenia, and thrombocytopenia. These effects, in turn, lead to increased likelihood of infection and bleeding. Failure of the lymphoid system is associated with immune deficiency (discussed in Chapter 26). Failure of the mononuclear phagocytic system is not clearly defined.

REVIEW QUESTIONS

1. What are the most common types of anemias and cancers of the hematopoietic system?
2. What are the manifestations of anemia?
3. What are petechiae and what do they suggest?
4. What are the names of the abnormalities defined by hematocrit, hemoglobin, red blood cell count, MCV, MCHC, and morphology of red blood cells as seen on a blood smear?
5. What abnormalities can be defined by the white blood cell count and white blood cell differential count?
6. What is the value of a bone marrow or lymph node biopsy?
7. What abnormality is detected by the following tests?
    Reticulocyte count
    Hemoglobin electrophoresis
    Sickle-cell preparation
    Red-cell fragility test
    Serum iron
    Coombs' test
    Mono spot test
8. What are the causes, lesions, and major manifestations of each of the diseases or groups of diseases discussed in this chapter?
9. What are the major categories of anemia as classified by pathogenesis and by laboratory findings?
10. What are the major causes of leukocytosis, leukopenia, and thrombocytopenia?
11. What are the major differences between leukemias and lymphomas?
12. What are the causes and effects of bone marrow failure?
13. What do the following terms mean?
    Reticuloendothelial system
    Mononuclear phagocytic system
    Intravascular hemolysis

Extravascular hemolysis
Pernicious anemia
Myelophthisic anemia
Aplastic anemia
Reed-Sternberg cell
Multiple myeloma

Bleeding and Clotting Disorders

Bleeding and clotting disorders are treated in this separate chapter because they involve both the vascular and hematopoietic systems and their classification does not fit into the scheme used for other chapters.

REVIEW OF STRUCTURE AND FUNCTION

Hemostasis is the process that prevents excessive bleeding following injury. The mechanisms of hemostasis involve a complex interaction of blood vessels, platelets, and chemical coagulation factors in plasma. Hemostasis is highly regulated by numerous activating and inhibiting mechanisms that allow the process to proceed rapidly but not excessively under normal conditions.

This chapter was coauthored by John Olson, M.D.

Blood vessels confine blood and allow it to circulate. The endothelial cells lining blood vessels normally prevent the activation of blood platelets and plasma coagulation factors, but when the endothelium is removed or injured, platelets and the coagulation mechanism are activated to prevent leakage of blood from the injured vessel. The activating mechanisms will be discussed. Blood vessels undergo spastic contraction in response to injury, thus aiding hemostasis by decreasing the blood flow to the injured area. Spasm of the smooth muscle of small arteries (arterioles) can occlude the vascular lumen and stop bleeding from an injured vessel in a few minutes. Shunting of blood to noninjured vessels provides collateral circulation for healing of the injured vessels. Spasm in medium and larger arteries may aid in hemostasis by decreasing blood flow but is often not sufficient to stop bleeding.

Platelets are cytoplasmic fragments of megakaryocytes. Normally 150,000 to 400,000 platelets per microliter circulate in the blood with a lifespan of 8 to 10 days each. If they are not needed for hemostasis, they are removed by the mononuclear phagocytic system as they become senescent.

The major functions of platelets at the site of an injury are to physically obstruct blood flow, to release chemicals that further the hemostatic process, and to facilitate clot retraction in the healing phase. Platelets will stick to any surface other than normal endothelium—a process called *adhesion*. Endothelial cells produce a potent inhibitor of these functions called *prostacyclin*, which is thought to prevent platelet activation under normal conditions. Minor injury to the endothelial layer with exposure of underlying extracellular connective tissue components will activate platelets and coagulation factors. Adhesion is accompanied by *aggregation* of platelets to each other to form a *platelet plug*. In small vessels, the plug itself may be sufficient to effect hemostasis. Aggregation is potentiated by collagen (exposed by the endothelial injury), by epinephrine (released from adrenal glands during stress and from the platelets themselves), by arachidonic acid (from platelets and other cells in the injured areas), and by thrombin (an enzyme in the coagulation mechanism).

Chemicals released from platelet granules as they adhere and aggregate facilitate all three components of the hemostatic mechanism and are essential to the coagulation mechanism. Vasoactive amines cause vascular smooth muscle contraction. Epinephrine and arachiodonic acid accelerate aggregation of additional platelets. Components of the platelet surface are a necessary ingredient of the plasma coagulation mechanism. The role of platelets in the repair process will be discussed later.

The process of converting plasma from a liquid to a solid is called *blood coagulation* or *blood clotting*. The solidification step in this process involves the conversion of fibrinogen, a large, soluble plasma protein produced by the liver, to form fibrin monomer. Fibrin monomer polymerizes to form fibrin, a stringy, strong, insoluble protein. As noted in Chapter 4, fibrinogen may leak into tissues during inflammation and coagulate to form a fibrinous exudate. When the process occurs in a blood vessel the product is called a *thrombus*. Thrombi can contain fibrin with entrapped red blood cells and platelets (Fig. 11–1).

Thrombosis, the process of thrombus formation, is a complex process involving tissue, platelet, and plasma coagulation factors. The formation of fibrin occurs through a series of enzymatic steps often referred to as the coagulation cascade because the product of each step catalyzes the next step, and often preceding steps so that the reaction can accelerate rapidly. The reaction can be slowed or stopped by inhibitors of the reaction. Factors involved in the reaction have been designated by Roman numerals, which are used interchangeably with their names. Some knowledge of this highly complex scheme is needed to understand and diagnose bleeding and clotting disorders.

The complex process of blood coagulation is presented in an abbreviated form (Fig. 11–2). As mentioned, the mechanism is designed to achieve the goal of converting fibrinogen to a stable fibrin clot and thereby converting the blood from a liquid to a solid. The conversion of fibrinogen to fibrin is dependent upon two active enzymes. The first is the enzyme thrombin that converts fibrinogen to fibrin monomer. The fibrin monomer then polymerizes and is reacted upon by a second enzyme called fibrin stabilizing factor (Factor XIII). Factor XIII converts the polymer of fibrin monomer to a stable fibrin clot. These active enzymes and the others of the coagulation mechanism are derived from inactive precursors. In the case of Factor XIII, the inactive precursor is converted to the active enzyme by thrombin. Thrombin also is derived from an inactive precursor called prothrombin (Factor II). The conversion of prothrombin to thrombin is dependent upon the active form of Factor X. Factor X with the cofactor, Factor V, in the presence of calcium ions and phospholipids convert prothrombin to thrombin. The cofactor, Factor V, also circulates in an inactive form and is activated by thrombin.

Factor X can be converted to its active form, Factor Xa, by two distinct mechanisms. In the first mechanism (traditionally called the extrinsic pathway), tissue juices, which contain tissue factor, combine with plasma Factor VII. This complex of tissue factor and Factor VII, in the presence of calcium ions, converts Factor X to Factor Xa. In the second mechanism (traditionally called the intrin-

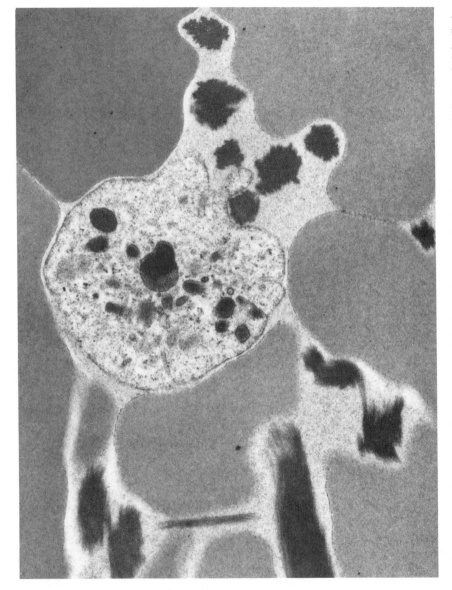

Figure 11–1. Fibrin fibers with an entrapped platelet and portions of several red blood cells (*larger dark bodies*).

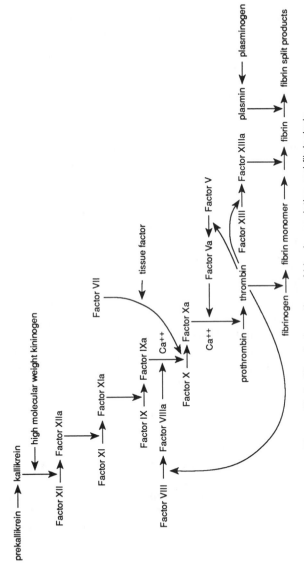

Figure 11–2. Chemical reactions of blood coagulation and fibrinolysis.

sic pathway), Factor X can be converted to Factor Xa by the activated form of plasma Factor IX. Plasma Factor IX also requires a cofactor (Factor VIIIa) and calcium ions to facilitate its activity. Factor VIII is also converted to its active form, Factor VIIIa, by thrombin. Factor IX is converted to Factor IXa by Factor XIa and Factor XI is converted to Factor XIa by the active form of Factor XII.

Factor XII along with prekallikrein and high molecular weight kininogen are part of a complex system called the contact activation system. Contact activating factors are of interest because their absence greatly prolongs the clotting time of blood in the test tube. These contact factors are of interest in explaining abnormal laboratory tests but of little consequence in the hemostatic mechanism in the patient, as patients who lack them do not bleed. The mechanism by which contact activation is initiated in vivo is unknown, however, exposure of charged molecules such as collagen may be important.

It is apparent that once a clot has formed, it is necessary to have a mechanism for the appropriate removal of the clot in the repair process. This removal is accomplished by the fibrinolytic system. The active enzyme plasmin is formed from plasminogen in the presence of a variety of activators. Tissue-derived plasminogen activator and urokinase appear to be the physiologic activators. Streptokinase, an enzyme derived from the bacteria streptococcus, is also used as a therapeutic agent to activate plasminogen. Plasmin acts upon fibrin to break down the insoluble fibrin clot into soluble fragments called fibrin-split products. Some large thrombi are not completely lysed, in which case they are replaced by granulation tissue and become scar tissue. A thrombus converted to scar tissue is called an *organized thrombus*.

Hemostasis is needed to prevent bleeding, but unregulated intravascular activation of hemostasis is not otherwise desirable. Several inhibitors of the coagulation mechanism have been identified to protect against the devastating consequences of unchecked coagulation. The most prominent of these inhibitors is named antithrombin III. Antithrombin III binds to and inactivates thrombin and the other activated enzymes in the coagulation mechanism. If the active enzymes were generated at a slow rate, or if free active enzymes were found in the circulation in small quantities, antithrombin III could bind to and inactivate them. In addition, another inhibitor called protein C, when activated in the presence of its cofactor (protein S), can inactivate the active form of Factor V and Factor VIII. These three molecules plus many others then are capable of slowing or stopping the coagulation process. Finally, the endothelium produces prostacycline, which is a potent inhibitor of platelet function, and provides a surface upon which other reactions can occur.

MOST FREQUENT AND SERIOUS PROBLEMS

Everyone experiences bleeding due to physical injury (trauma). Most of the time traumatic hemorrhage is controlled by the normal hemostatic mechanisms. Along with maintenance of an airway, control of bleeding is the major consideration in accident victims. Pressure on bleeding vessels can be used as an emergency measure, but access to medical facilities where the lost blood volume can be replaced and surgical control of internal bleeding can be effected saves many lives.

Bleeding may occur with many vascular diseases, but those that cause sudden rupture of a large artery are most life threatening. For example, fatal hemorrhage may occur from a ruptured aortic or cerebral aneurysm or from a bleeding peptic ulcer. Trauma and surgical operations increase the likelihood of significant hemorrhage in patients with preexisting vascular disease.

Spontaneous bleeding may be anticipated when the platelet count drops below 10,000 per microliter or when platelets are defective. Typical causes of platelet deficiency include bone marrow failure where production is deficient and peripheral destruction of platelets as seen with idiopathic thrombocytopenic purpura where antibodies cause premature removal of platelets. Patients with temporary platelet deficiency, such as might be caused by cancer chemotherapy, can be treated with platelet transfusions.

Hereditary coagulation disorders, such as hemophilia, are rare, but people with these disorders require considerable medical care over many years. Acquired coagulation disorders are typically associated with other diseases, with drugs, or are induced by anticoagulation therapy.

Thrombosis is usually secondary to some other condition. Deep vein thrombosis and pulmonary embolism are quite common and may occur in persons who are minimally ill such as after childbirth or an operation. Disseminated intravascular coagulation usually occurs in patients with significant underlying disease.

SYMPTOMS, SIGNS, AND TESTS

A patient with a tendency to bleed, whether known by a history of bleeding or by abnormal laboratory tests, is said to have a *hemorrhagic diathesis.*

Bleeding from decreased or deficient platelets causes bleeding from small vessels in the skin and internal surfaces such as nose, gastrointestinal tract, and urinary tract in the form of *petechiae* (small hemorrhages in the skin) (Fig. 11–3), nose bleeds, hematuria (blood in the urine), and occult fecal blood (positive test for blood

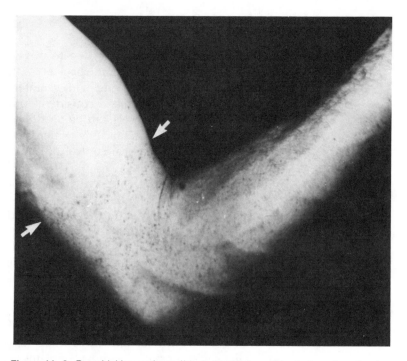

Figure 11–3. Petechial hemorrhage (*black dots*) induced by placing a tourniquet on the arm (*at the site of the arrow*) in a patient with platelet deficiency.

in the stool). Tissue bleeding, such as occurs with coagulation deficiencies, tends to occur at one site at a time with minor trauma and produces a localized collection of blood (*hematoma*). Typical sites are joints (Fig. 11–4), soft tissues, and brain.

A history of previous bleeding, such as after circumcision or tooth extraction, and a family history of bleeding may provide the clues to diagnosis of a hereditary coagulation defect. Such a history is particularly important in patients scheduled for an operation because the history may call attention to a problem when screening tests for bleeding disorders are normal.

Laboratory tests for bleeding disorders are used to determine the specific pathogenesis of the problem (vascular, platelet, or coagulation disorder), to monitor the process of the problem or the effect of therapy, and to preoperatively screen patients who are at high risk. Some of the more common tests will be discussed here.

The *platelet count* is simple and informative. It is performed on most hospitalized patients as a part of a complete blood count. Although the normal platelet count is 150,000 to 400,000 per microliter, spontaneous hemorrhage due to thrombocytopenia is rare if the count exceeds 10,000 per microliter. Abnormalities in platelet function will not be identified by this test.

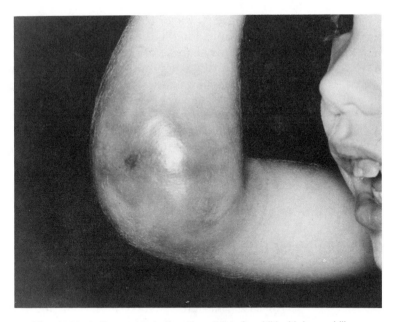

Figure 11–4. Hematoma in the elbow joint of a child with hemophilia.

Bleeding time is the time needed for bleeding to stop after a small skin incision at a standard venous pressure (obtained by putting a tourniquet on the arm). Normal hemostatic mechanisms stop the bleeding in 6 to 9 minutes. Bleeding will be prolonged with a low platelet count, with abnormalities of platelet function, and with diseases affecting small vessels. If the platelet count is normal and bleeding time prolonged, other special tests may be needed to distinguish between abnormal platelets and vascular disease.

The *activated partial thromboplastin time* tests for deficiencies in the coagulation mechanism except for Factor VII. A sample of the patient's blood is drawn into a container containing citrate. The citrate binds calcium, an essential ingredient in clotting, so that the sample can be taken to the laboratory without clotting. At the start of the test three ingredients are added: calcium, a surface contact material, such as finely ground glass, to activate the contact system, and phospholipid as a substitute for platelets. Normally fibrin will solidify the plasma in this test in about 30 seconds. The test is sensitive to defects in the intrinsic pathway, such as hemophilia, and less sensitive to prothrombin and fibrinogen deficiencies. Specific tests are available to test for each of the factors that might be deficient.

The *prothrombin time* is a more sensitive measure of prothrombin and fibrinogen deficiency, but also reflects adequacy of Factors

V, VII, and X. Factor VII activates Factor X when tissue factor and calcium are added to the citrated plasma (see Fig. 11–1). Normally fibrin forms in about 12 seconds in this test. The prothrombin time is quite sensitive to prothrombin, Factor VII, Factor V, and Factor X deficiencies and less sensitive to fibrinogen deficiency. The *fibrinogen assay* can be used to measure fibrinogen levels.

Measurement of *fibrin degradation products* reflects fibrinolysis in vivo (the amount of fibrin or fibrinogen being split by plasmin). The test is particularly useful in the diagnosis of disseminated intravascular coagulation, which will be discussed later.

SPECIFIC DISEASES

Disorders Characterized by Hemorrhage

Vascular Disorders

The most common vascular causes of hemorrhage are trauma and diseases that erode blood vessels. In these situations, the cause and effect are usually obvious and bleeding is controlled by normal hemostatic process or mechanically by the patient or physician. One should be aware that patients on anticoagulant therapy may bleed excessively from minor or major trauma. Some forms of vasculitis may be associated with petechial hemorrhage. They need to be distinguished from platelet disorders using the appropriate tests.

Bleeding from vascular disruption may be life threatening. Immediate consideration should be given to stop the bleeding and to replace the blood volume. If the bleeding site is accessible, pressure, suture, or a tourniquet can be employed. If the bleeding is internal, access to a hospital is critical. If sufficient blood is lost, plasma or a plasma expander should be given during and after transport to a hospital. Replacement of red blood cells usually is not needed because the patient has a considerable reserve capacity, but blood volume must be replaced immediately to prevent shock.

Platelet Disorders

Decreased numbers of platelets leads to petechial hemorrhages, hematuria, and gastrointestinal hemorrhage. The cause is either decreased production or increased destruction of platelets. Production problems are usually due to extensive bone marrow disease as was discussed in Chapter 10. A bone marrow examination will confirm decreased megakaryocytes; whereas, with increased platelet destruction megakaryocytes will be normal or increased.

There are three major mechanisms of increased platelet destruction. Antibodies to platelets, as occurs in idiopathic thrombocytopenic purpura (Chapter 10), cause their premature removal by the spleen. Diseases of the spleen can cause excessive removal of normal platelets as well as other blood cells—a condition called *hypersplenism*. Finally, disseminated intravascular coagulation can use up platelets faster than the bone marrow can produce new ones.

Disorders of platelet function may be hereditary or acquired. Hereditary disorders are rare but interesting as they help us understand platelet function. Patients with *von Willebrand's disease*, an autosomal dominant disease, lack a plasma factor (von Willebrand factor) that facilitates platelet adhesion. They also have a deficiency of Factor VIII (hemophiliac factor), so they may give positive tests for platelet function (bleeding time) and coagulation function (activated partial thromboplastin time). It is important to identify these patients because they may not manifest significant bleeding problems until operated upon. Other hereditary platelet disorders include a deficiency of platelet receptors for von Willebrand factor, a deficiency of platelet aggregation and clot retraction, and deficiency of the release of platelet components.

Acquired disorders of platelet function include renal failure and aspirin therapy. In severe renal failure (uremia), waste products accumulate in the blood that may interfere with the platelet's ability to support coagulation. A platelet exposed to aspirin will no longer be able to release its storage granules. This, obviously, does not usually cause a problem, as millions of people consume aspirin. Aspirin may potentiate an existing bleeding tendency, however. Other drugs also affect platelet function.

Coagulation Disorders

Coagulation disorders may be hereditary or acquired. Hereditary disorders are nature's experiments that have greatly enhanced our understanding of the role of the various coagulation factors.

Hemophilia is the most common of the hereditary coagulation disorders, and both forms are sex-linked recessive. Hemophilia A or classic hemophilia is a deficiency of Factor VIII and occurs in about 1 of 5000 males (females are rarely affected) and about 20 percent appear as new mutants (negative family history). This was a prevalent disease among European royalty because of inbreeding. Hemophilia B or Christmas disease is a deficiency of Factor IX and is about one-tenth as common as hemophilia A and somewhat less severe. Persons with hemophilia bleed into joints, muscle, and soft tissues and can bleed excessively from minor operations such as circumcision or tooth extraction. The diagnosis is based on the history of bleeding, screening tests that indicate a coagulation factor defect (normal bleeding time, platelets, and prothrombin time; prolonged activated partial thromboplastin time), and specific tests for Factors VIII and IX. Hemophiliacs suffer lifelong bleeding problems and typically become crippled by bleeding into joints, leading to degenerative arthritis. Factor VIII and IX are available commercially for lifelong replacement and emergency therapy, but are costly and can transmit other diseases (hepatitis and acquired immune deficiency syndrome).

Other hereditary coagulation disorders are autosomally transmitted, much rarer, usually less severe, and lack specific treatment. Fresh-frozen plasma may be used to provide the missing factor in emergencies or prior to an operation.

Acquired causes of bleeding due to coagulation factor deficiency are caused by inadequate production of coagulation factors, excessive destruction of coagulation factors, or drugs that inhibit coagulation. All of the plasma factors are produced in the liver (except for von Willebrand factor, which is a product of endothelial cells). Severe acute or chronic liver disease may be associated with bleeding because of deficiency of one or more of these factors. Some of the factors require vitamin K for their synthesis in the liver. Vitamin K is a fat-soluble vitamin produced by bacteria in the intestines, so reduction of the bacteria with antibiotics or malabsorption of fats can lead to a hemorrhagic diathesis due to vitamin K deficiency. Newborns, who lack intestinal flora, are given vitamin K parenterally to prevent *hemorrhagic disease of the newborn.*

Coagulation factors may be consumed rapidly by excessive coagulation, thus paradoxically superimposing a hemorrhagic diathesis on top of a coagulation process. This has led to the term *consumption coagulopathy,* which, in its most severe form, may become disseminated intravascular coagulation (discussion follows). Another form of destruction of coagulation factors occurs when antibodies to exogenous Factor VIII or Factor IX develop in hemophilia A or B, respectively.

Coumadin (one of several coumarin drugs) and heparin are used to prevent thrombosis, so they cause a hemorrhagic diathesis at therapeutic levels and active bleeding at toxic levels. Coumadin inhibits vitamin K utilization by liver cells, thus inhibiting production of coagulation factors. Coumadin takes several days to act and its effect is long-acting. Activity is measured by the prothrombin time and counteracted by high doses of vitamin K.

The inhibition of thrombin by antithrombin III is greatly accelerated by heparin. It also inhibits Factors XI, IX, and X, so the activated partial thromboplastin time can be used to monitor the degree of anticoagulation. Heparin is used for acute therapy in hospitals (it has to be given by injection). Its half-life is only about

90 minutes. Protamine is used to treat an overdose of heparin. Coumadin is more appropriate for long-term therapy and can be taken orally.

Disorders Characterized by Thrombosis

Vascular Disorders
Vascular changes that lead to thrombosis include acute and chronic injuries to vessel walls (including the endocardium) and turbulent blood flow. The most important of these conditions, including thrombi in leg veins, thrombi on atherosclerotic plaques, mural thrombi over a myocardial infarct, and atrial thrombi, are discussed in Chapters 8 and 9. Any vascular injury or site of turbulence can lead to thrombosis.

Platelet Disorders
Excessive platelets (thrombocytosis) is uncommon but can lead to thrombosis. When the platelet counts exceed 1,000,000 per microliter, there is an increased likelihood of thrombosis. Causes include idiopathic thrombocythemia, bone marrow hyperplasia secondary to loss or destruction of red blood cells, cancer, splenectomy, and inflammatory diseases. More common is the relative increase in platelets that follows trauma, operations, and childbirth, which, along with venous stasis caused by bed rest, predisposes to deep leg vein thromboses and subsequent pulmonary embolism.

Coagulation Disorders
Disseminated intravascular coagulation is a clinical syndrome caused by coagulation within the bloodstream. Multiple small thrombi may plug many small vessels throughout the body; however, bleeding often predominates due to consumption of coagulation factors and platelets. Coagulation tests reveal many defects and fibrin degradation products are elevated. Disseminated intravascular coagulation is caused by release of tissue thrombaoplastin or activation of the contact system. It is usually associated with serious underlying disease such as massive trauma, shock, septicemia, or cancer, but it also can occur as a complication of pregnancy due to tissue factors entering the maternal bloodstream from the placenta. Disseminated intravascular coagulation is often a terminal event; however, if there is reason to believe that the underlying cause can be treated, anticoagulation therapy may buy time for treatment of underlying cause of the coagulopathy.

Rare hereditary deficiencies of antithrombin III, protein C, or protein S cause spontaneous thrombosis due to lack of these inhibitors of coagulation. As with platelets, coagulation factors are increased after trauma, operations, and childbirth, thus contributing to the tendency to form thrombi under these conditions.

As mentioned earlier, heparin and vitamin K antagonists are used to prevent thrombosis. Long-term therapy with aspirin to prevent thrombosis in coronary and cerebral arteries is based on the known effects of aspirin on platelet function; however, the long-term effects of such therapy require more study. Attempts are also made to dissolve existing thrombi by using enzymes that convert plasminogen to plasmin such as tissue plasminogen activator, streptokinase, and urokinase. Catheters have been used to inject these enzymes into recently plugged coronary and pulmonary arteries.

REVIEW QUESTIONS

1. Thrombi, hematomas, and fibrinous exudates all contain fibrin. How do they differ?
2. What are the roles of vessels, platelets, and plasma coagulation factors in hemostasis?
3. How are bleeding disorders classified? What types are most common? Most life threatening?
4. What signs of hemorrhage would distinguish a patient with hemophilia from one with platelet deficiency?
5. What is the significance of an abnormal value for each of the following tests?
    Platelet count
    Bleeding time
    Activated partial thromboplastin time
    Prothrombin time
    Fibrin degradation products
6. What is the mechanism and possible treatment for bleeding in each of the following conditions?
    Trauma
    Vasculitis
    Idiopathic thrombocytopenic purpura
    Thrombocytopenia due to hypersplenism
    Hemophilia A
    Hemophilia B
    Severe liver disease
    Hemorrhagic disease of the newborn
    Disseminated intravascular coagulation
    Coumadin therapy
    Heparin therapy
7. How does atherosclerosis predispose to arterial thrombosis?

How does a major surgical operation predispose to venous thrombosis?

8. What are the causes, pathogenesis, and prognosis for intravascular thrombosis?
9. What treatments are used to prevent or dissolve thrombi?
10. What do the following terms mean?
    Hemostasis
    Fibrinolysis
    Organized thrombus
    Hemorrhagic diathesis
    Petechiae

Chapter 12

Lung

REVIEW OF STRUCTURE AND FUNCTION

The main function of the lungs is to transfer oxygen from the
atmosphere to the blood and carbon dioxide from the blood to the
atmosphere. The lungs also serve as a filter of air and blood, and,

like the kidney and liver, they are involved in the detoxification and excretion of certain toxins and normal metabolites.

To carry out its main function, air must be moved from the atmosphere to the terminal units of the lung, a process called *ventilation*, and gas must pass across tissue from air to blood and blood to air, a process called *gas exchange* or *perfusion*. Pathologic processes interfering with ventilation are referred to as *obstructive* and those interfering with gas exchange as *restrictive*.

The functional units for gas exchange are thin-walled, wide-mouthed sacs called *alveoli*. A cluster of alveoli with their associated respiratory bronchioles from an *acinus* (Fig. 12–1). Alveolar walls consist of capillaries, a scant amount of connective tissue, and epithelial lining cells called Type I and Type II pneumocytes. Type I pneumocytes have very thin cytoplasm, much like an endothelial cell, and comprise most of the inner lining of the alveolus. Type II pneumocytes secrete surfactant, a phospholipid that lowers surface tension, and proliferate when the lining is injured. Monocytes emigrate from the capillaries to the alveolar surface or lumen where they are known as alveolar macrophages and play their usual roles as scavengers of foreign particles. The capillaries bulge into the alveolus. The endothelium and epithelium share basement membranes in areas, providing a very short distance for gases to move across tissue.

The ventilatory pathway is tree-like with the trachea as a trunk and branching bronchi as limbs. Bronchi occupy only a small fraction of the lung space compared with the alveoli (Fig. 12–2). Movement of air in the tracheobronchial tree is brought about by muscular movement of the thorax and diaphragm and elastic recoil of the lungs. The bronchial muscle controls the size of the bronchi and thus the resistance to air flow.

Although the skin, mouth, and intestines are more exposed to gross environmental contamination, the lung is more vulnerable because the lining of the distal air sacs is designed for gas ex-

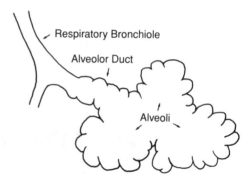

Respiratory Bronchiole

Alveolor Duct

Alveoli

Figure 12–1. Portion of a pulmonary acinus.

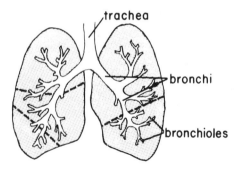

Figure 12–2. Gross anatomy of the trachea and lungs. Lobes are separated by dashed lines.

change, not protection. Proximal to the acinus, the bronchi and trachea are lined by columnar epithelial cells with cilia that propel solid material upward. This is aided by mucus secreted by surface goblet cells and mucosal mucous glands. When excessive material accumulates in the bronchi, the cough reflex helps expel it. Excessive particulate material in alveoli is countered by increased numbers of alveolar macrophages.

MOST FREQUENT AND SERIOUS PROBLEMS

The lung is the site of a wide variety of inflammatory and destructive diseases as well as being the most common primary site of lethal cancer. The majority of deaths from disease of the lung can be attributed to cigarette smoking, the major cause of emphysema and bronchogenic carcinoma.

Pneumonia means an inflammation of the lungs with consolidation and is most often used to describe infections of the lung. The term *pneumonitis* is used in a less specific sense to indicate inflammation. Bacterial pneumonias commonly occur terminally in debilitated persons dying from other causes. Antibiotics have greatly reduced the effects of primary bacterial pneumonias. Viral and mycoplasmal pneumonias are fairly common but rarely cause death.

Chronic bronchitis and emphysema cause a great deal of disability from chronic lung disease. Most cases are associated with smoking. *Pneumoconioses* are diseases of the lung caused by deposition and reaction to inhaled particles such as silica, asbestos, or cane fibers. They occur in miners and others exposed to particulate matter, and may cause destructive pulmonary fibrosis.

Asthma is a common nondestructive lung disease that has its greatest effect on children and young adults. It requires considerable medical attention; acute episodes may be severe enough to require emergency treatment.

In the United States, lung cancer causes over one-third of cancer deaths in men. In women, the frequency of lung cancer has increased rapidly and now causes more deaths than breast cancer.

SYMPTOMS, SIGNS, AND TESTS

Difficulty with breathing is called *dyspnea*, and lack of breathing is called *apnea*. *Orthopnea* refers to difficulty in breathing while reclining. A person who is not getting enough oxygen will attempt to breathe faster and may often become frightened. A person not getting rid of enough carbon dioxide will be somnolent (carbon dioxide narcosis) and may even be unaware of danger. Partial obstruction of major airways will produce noisy breathing such as wheezing.

Cough is due to irritation of bronchi or accumulation of fluid. Fluid coughed up from the lungs is *sputum*. *Hemoptysis*, the coughing of blood, suggests serious destructive disease such as cancer or tuberculosis.

Physical examination involves observation of the rate and type of breathing. Rapid breathing, *tachypnea*, occurs in response to increased oxygen need whether it be from lung disease or increased general body needs such as fever. *Cyanosis* is blue skin due to inadequate oxygen saturation of the blood. Difficulty with inhaling suggests bronchospasm, as in asthma, whereas difficulty in exhaling is typical of emphysema. *Percussion*, thumping on a finger placed against the chest, may reveal dull or low-pitched sounds suggestive of underlying pulmonary consolidation or pleural effusion. Listening to the chest with a *stethoscope*, a cupped diaphragm that, when placed on the chest wall, transmits sounds from the lung to the examiner's ear via air-filled tubes, may be even more revealing. Bubbly sounds, called *rales*, are created by turbulent air–fluid mixtures and indicate transudate or exudate in the lung as might be caused by heart failure or pneumonia. Diminished normal breath sounds suggest consolidation or pleural effusion.

The ultimate measure of pulmonary function is the oxygen and carbon dioxide content of arterial blood. These tests along with measurement of the capacity to move air in and out of the lung allow assessment of the severity of chronic lung disease. With ventilation defects, the capacity to move air in and out will be decreased and lungs will not empty as much as normal. With perfusion defects, blood gases will be abnormal with relatively normal ventilation. Low blood oxygen is *hypoxemia* and high blood carbon dioxide is *hypercapnia*.

The chest roentgenogram (x-ray) is the major diagnostic tool used to look for lung lesions. Once a lesion is found, it is often necessary to determine whether it is inflammatory or neoplastic. If inflammatory, is it due to infection or other cause? If neoplastic, is it primary or metastatic? To make these determinations, samples of lung need to be obtained for culture, cytologic evaluation, or histologic evaluation. Sputum can be used for culture and cytology. Biopsy of trachea, bronchi, and lung can be done at bronchoscopy (endoscopic examination of the trachea and bronchi). Fluid or tissue obtained from the pleural cavity or pleura with a needle can be used for culture, cytology, or histology. If these techniques fail, open lung biopsy may be necessary to establish the nature of the lung disease.

SPECIFIC DISEASES

Genetic/Developmental Diseases

Congenital anomalies of the lung, which are rare, include hypoplastic lobes, lobes with a separate blood supply, and multiple cysts. At birth, pulmonary function suddenly becomes critical, so it is not surprising to find a serious disease, hyaline membrane disease, occurring at this time. Cystic fibrosis is a genetic disease that has more delayed effects.

Respiratory Distress Syndrome of the Newborn (Hyaline Membrane Disease)

The importance of pulmonary alveolar surfactant is nowhere more evident than at birth. If sufficient surfactant is present, the alveoli will inflate and allow perfusion of gases, otherwise they will collapse, a condition called *atelectasis*. Surfactant is excreted into the amniotic fluid in utero where it can be measured in samples obtained by amniocentesis. Such samples are useful in determining when elective premature delivery is safe for the fetus. If surfactant is deficient, a proteinaceous precipitate rich in fibrin builds up in dilated bronchioles and alveolar ducts (Fig. 12–3). This eosinophilic membrane, although now known to be a nonspecific manifestation to epithelial injury, accounts for the name hyaline membrane disease. Oxygen therapy, which is essential in this situation, also causes injury to the lining cells leading to more hyaline membrane formation. Improved ventilation techniques and awareness of oxygen toxicity, have lowered the age and birth weight at which prematurely born infants can survive. Untreated infants with respiratory distress syndrome live 1 or 2 days. With treatment, they may survive long enough for the lungs to mature and recover to normal. Prolonged survival, however, may be associated with perma-

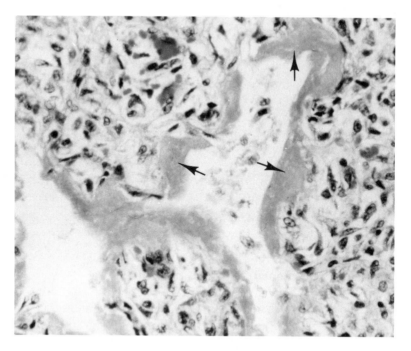

Figure 12–3. Microscopic view of hyaline membranes (*arrows*) in pulmonary alveoli.

nent lung damage in some infants and may unmask other complications of prematurity such as necrotizing enterocolitis. Prematurity is the most common underlying cause of neonatal mortality and hyaline membrane disease is the usual finding at autopsy.

CHART 12–1. RESPIRATORY DISTRESS SYNDROME

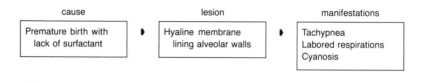

cause		lesion		manifestations
Premature birth with lack of surfactant	▸	Hyaline membrane lining alveolar walls	▸	Tachypnea Labored respirations Cyanosis

Cystic Fibrosis

This autosomal recessive disease occurs in about 1 of every 2000 whites with an estimated carrier rate of 1 in 20. It is rare in blacks and orientals. It is discussed here because most patients die of lung disease, and attention to medical care of the lung disease has raised the average life expectancy from childhood to early adulthood.

Although the genetic or biochemical nature of the defect is unknown, the pathologic effects are the result of plugging of various ducts and body passageways with thick mucus.

CHART 12–2. CYSTIC FIBROSIS

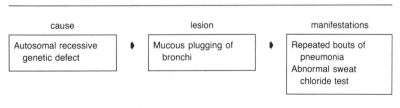

cause	lesion	manifestations
Autosomal recessive genetic defect	Mucous plugging of bronchi	Repeated bouts of pneumonia Abnormal sweat chloride test

Inflammatory/Degenerative Diseases

For this discussion we will group the diseases into four categories: infections, chronic obstructive pulmonary diseases, acute and chronic noninfectious interstitial diseases, and vascular diseases.

Infections

Bacteria are the principal causes of lobar pneumonia, bronchopneumonia, and abscesses. Tuberculosis is an entirely different type of bacterial infection. Several fungi cause pulmonary lesions similar to tuberculosis, whereas others act more like opportunistic bacteria. Viruses and mycoplasma produce an interstitial pneumonia. We have chosen to treat tuberculosis in some detail here, rather than in Chapter 26. The control of tuberculosis is a major medical triumph in the United States that we sometimes take for granted.

Lobar Pneumonia

Lobar pneumonia involves an area of lung (often a single lobe, hence the name) in a diffuse manner and is caused by *Streptococcus pneumoniae* (pneumococcus) in most cases. It used to be the most common form of pneumonia and involved relatively healthy as well as debilitated people. Most strains of the organism are very susceptible to antibiotics such as penicillin, and the disease responds dramatically to treatment. However, a few resistant strains have emerged, justifying the development of vaccines. The bacteria and resulting exudate spread through the pores in alveolar walls in a wave-like fashion, incapacitating the involved areas. This classic bacterial inflammatory disease can be diagnosed by the finding of respiratory distress, fever, leukocytosis, lobar density on chest x-ray (Figs. 12–4 and 12–5A), many gram-positive diplococci on

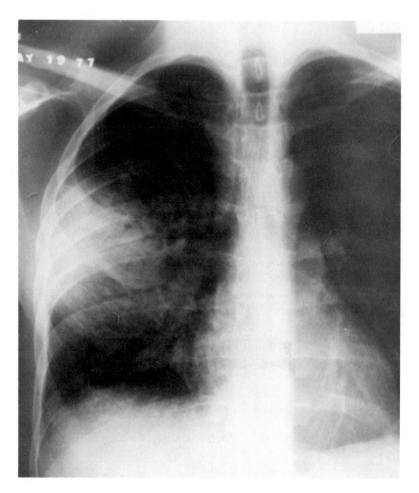

Figure 12–4. Lobar pneumonia by chest x-ray. Note radiodensity in right middle lobe caused by inflammatory exudate filling the alveoli.

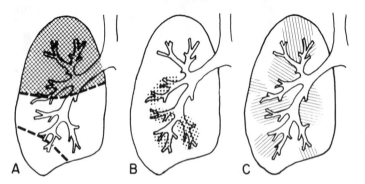

Figure 12–5. Comparison of lobar (**A**), broncho- (**B**), and interstitial (**C**) pneumonia. Note lobar, bronchial, and patchy distributions.

Gram's stain smear of sputum (coughed up from the lungs), and culture of *S. pneumoniae*. The organism itself is always around as an inconstant and variable component of the normal throat flora. There is great variability in the pathogenicity of various strains of pneumococci.

CHART 12–3. LOBAR PNEUMONIA

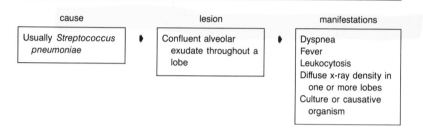

cause	lesion	manifestations
Usually *Streptococcus pneumoniae*	Confluent alveolar exudate throughout a lobe	Dyspnea Fever Leukocytosis Diffuse x-ray density in one or more lobes Culture or causative organism

Bronchopneumonia

In contrast to lobar pneumonia, bronchopneumonia has a patchy distribution throughout the lungs, with multiple foci of infection occurring at sites where clumps of bacteria and debris lodge in bronchi (Fig. 12–5B). The essential causative factor is obstruction of small bronchi by mucus or aspirated gastric contents, so that bacteria that are normally present in small numbers are trapped and multiply to produce foci of infection in the lung. Various types of bacteria and other nonbacterial organisms may be involved. The lesion consists of inflammatory exudate filling bronchi and alveoli. The extent and location depends on the underlying cause and source of causative organisms. If lung lesions, such as an obstructing cancer, viral bronchitis, or chronic lung disease of various types, are the predisposing cause, the bacteria causing the pneumonia are likely to be normal residents of the upper respiratory tract that were not removed normally. In debilitated patients and those with abdominal infections, gram-negative intestinal organisms, including anaerobic bacteria, are commonly the cause. Circulating bacteria in septic patients may seed the lung and cause multifocal pneumonia.

Bronchopneumonia is more difficult to diagnose and treat than lobar pneumonia, because multiple organisms may be involved, the pneumonia may be masked by underlying disease, and x-ray changes are less distinctive. Bronchopneumonia is a very significant cause of death in debilitated and bedridden patients with chronic diseases.

Aspiration pneumonia is a subtype of bronchopneumonia in which particulate material carrying bacteria gets into the lungs and gravitates to the lower, more dependent lobes. This occurs most often in postoperative and debilitated persons and is enhanced by lack of full expansion of the lungs. The postoperative patient wants neither to breath deeply nor to cough because it hurts, but both are important to prevent bronchopneumonia.

Legionnaire's disease is a bacterial pneumonia due to inhalation of *Legionella pneumophilia,* an organism that lives in water storage tanks and cooling systems. The confluent areas of pneumonia may be extensive enough to be fatal if not treated in time.

Pneumonia is one of the commonest causes of death in immunosuppressed patients. Immunosuppression is a consequence of protection of organ transplant recipients from transplant rejection and of chemo- and radiation therapy for cancer. The organisms, which are usually harmless to persons with normal defense mechanisms, include protozoa (*Pneumocystis carinii*), viruses (cytomegalovirus), fungi (aspergillus, mucor, cryptococcus), and bacteria (staphylococcus, gram-negative bacteria).

CHART 12–4. BRONCHOPNEUMONIA

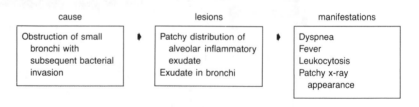

cause	lesions	manifestations
Obstruction of small bronchi with subsequent bacterial invasion	Patchy distribution of alveolar inflammatory exudate Exudate in bronchi	Dyspnea Fever Leukocytosis Patchy x-ray appearance

Abscess

Lung abscesses are localized areas of suppuration, which may develop as a complication of bacterial pneumonia. They may result from aspiration of a foreign body, obstruction of a bronchus by neoplasm, infection in an infarct, or from sepsis. Massive bacterial infection of the pleural cavity with outpouring of purulent exudate is called *empyema.* This serious condition usually results from underlying lung infection.

Interstitial Pneumonia

When the inflammatory reaction involves the walls of alveoli with relatively little exudation and relatively few polymorphonuclear leukocytes, the reaction is called *interstitial pneumonia* or *pneumonitis.* It is distributed throughout all lobes but tends to be more central than peripheral (Fig. 12–5C).

Interstitial pneumonias due to infection usually are associated with an upper respiratory infection and vary from mild and unnoticed by the patient to severe with cough, fever, and malaise. The term primary atypical pneumonia has been applied to acute infectious interstitial pneumonias. The most commonly identified agent (most often the agent is unidentified) is *Mycoplasma pneumoniae,* a bacteria-like organism. Viral causes include influenza A and B, respiratory syncytial virus, and rhinoviruses. The course of mycoplasmal pneumonia may be shortened by antibiotic therapy. Antibiotics may also be useful to treat bacterial complications of viral pneumonia.

CHART 12–5. INTERSTITIAL PNEUMONIA

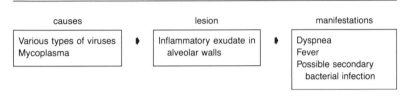

causes	lesion	manifestations
Various types of viruses Mycoplasma	Inflammatory exudate in alveolar walls	Dyspnea Fever Possible secondary bacterial infection

Tuberculosis

Although the disease is under control in the United States, it still occurs, often unexpectedly. Tuberculosis is the most significant lung disease in many areas of the world. The initial or primary lesion is usually asymptomatic. Later, usually years later, secondary lesions develop in some individuals, and these lesions gradually destroy the lung, often with the diseased individual being unaware of the severity of the illness. Such persons spread organisms into the environment and to other individuals.

Tuberculosis is caused by the bacteria *Mycobacterium tuberculosis* and some related atypical strains. The organism is spread from person to person by coughing. A cluster of organisms lodge at the periphery of the lung to set up a focal granulomatous reaction. A few organisms spread through lymphatic vessels to hilar lymph nodes, setting up another granulomatous focus. These foci usually undergo necrosis (caseous type), and they heal by fibrosis and calcification. They may remain indolent for years as small calcified nodules and are then known as the *primary,* or *Ghon complex* (Fig. 12–6A). If this primary infection is severe enough, a few organisms may be carried through the bloodstream, seeding the infection in other organs or new sites in the lung. These foci may lie dormant for months to years and then spring up as isolated tuberculous lesions in organs such as kidney, fallopian tubes,

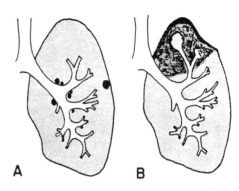

Figure 12–6. **A.** Ghon complex with solitary peripheral primary granuloma and secondary granulomas in hilar lymph nodes. **B.** Advanced pulmonary tuberculosis with cavities (connecting to bronchi), granulomas, and scarring of lung and pleura.

bone, or brain. Occasionally, especially in children, the spread from the lungs will be massive, with multiple small foci (miliary lesions) developing in many organs. Meningeal involvement usually leads to fatal outcome (tuberculous meningitis).

Much more common than the spread to other organs is the development of new lung lesions either through breakdown of the primary lesion or by way of reinfection from another individual. These secondary lung infections are different in terms of localization, amount of tissue destruction, and site in the lungs. Because of previous exposure, lymphocytes now recognize the tubercle bacilli as foreign, resulting in a more rapid reaction and greater amount of necrosis. The disease is usually found in the apices (top) of the lungs, because the organisms presumably grow better with the higher oxygen concentrations found in the upper lobes of the lung. The necrosis may involve the bronchial walls. This allows tissue to be sloughed out through the bronchi, leaving a cavity in the area of necrosis (Fig. 12–6B). This sloughed tissue may be coughed, thereby spreading the disease to other persons.

Healing by scar formation destroys the involved portion of lung. The balance between destruction and healing may go on for years. In days gone by, rest was the only treatment, and hospitals were filled with incapacitated infective persons. These patients were usually cachectic and appeared to be "consumed" by their disease, consequently, the term *consumption* became synonymous with *tuberculosis*. Presently, antituberculous drugs control the progression of disease in many persons, and many are able to have parts of their lungs removed, while the drugs prevent recurrence of the disease. Through massive screening efforts, the number of persons with tuberculosis has been greatly reduced, and advanced cavitary disease is rare.

There are two methods of screening, each with its advantages. Skin testing measures hypersensitivity—whether a person's lymphocytes have encountered tubercle bacilli before. Hypersensitive persons may be checked for active disease by a chest x-ray. Skin testing is best for large-scale screening of children, because it is less expensive and when positive tests are found they are more significant than in adults. The reason they are more significant is because children have had less time to acquire an innocuous primary infection, and contacts are more likely to be traceable. Chest x-rays are more definitive than skin tests in finding significant lesions and are less likely to uncover falsely positive cases. For example, an irregular apical lesion in an asymptomatic adult is quite likely to be tuberculosis. The definitive diagnosis is made by culture of sputum or concentrated gastric contents (the coughed-up lung material is normally swallowed). It takes several weeks for the organism to grow in culture. Patients with positive cultures are those who are infectious.

CHART 12–6. TUBERCULOSIS

cause	lesions	manifestations
Mycobacterium tuberculosis	Ghon complex Secondary apical lesions with granulomas, cavities, and fibrosis	Positive skin test Apical lung lesions by x-ray Coughing Dyspnea Cachexia if advanced

Fungal Diseases

Two primary fungal infections produce diseases much like tuberculosis. Histoplasmosis, which occurs predominantly in midwestern United States, and coccidioidomycosis, which occurs in southwestern United States, are noteworthy. Like tuberculosis, they are usually asymptomatic diseases but may disseminate throughout the lung to cause serious acute illnesses manifested by dyspnea, fever, and incapacitation. Organisms are inhaled after becoming airborne in dust or in dried bird droppings. The lung lesions are granulomatous inflammations but usually do not cavitate as does tuberculosis. Other fungi, such as candida, aspergillus, and mucor, are opportunistic pathogens that may produce pneumonia.

Atelectasis

This means collapse of lung and is not a disease but rather a finding. There are two causative mechanisms (Fig. 12–7): (1) external pressure on the lung, such as fluid in the pleural space or poor

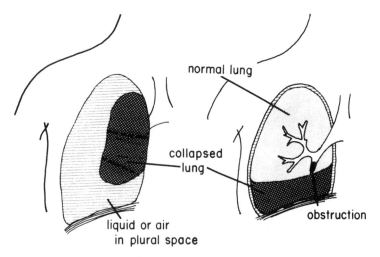

Figure 12–7. Atelectasis produced by external compression (**A**) and by bronchial obstruction (**B**).

expansion of the chest following surgical operation or (2) complete obstruction of a bronchus with reabsorption of the air in alveoli distal to the obstruction. In contrast to complete obstruction, partial bronchial obstruction usually produces an expanded segment of lung, because air gets in past the obstruction more easily than it gets out (ball-valve effect).

CHART 12–7. ATELECTASIS

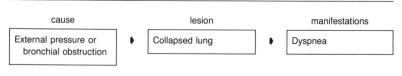

cause		lesion		manifestations
External pressure or bronchial obstruction	●	Collapsed lung	●	Dyspnea

Chronic Obstructive Pulmonary Disease (COPD)

This term and the acronym refer collectively to diseases characterized by chronic partial obstruction of airways and inadequate ventilation. COPD is more often used to indicate chronic bronchitis and/or emphysema, which have a similar etiology, than asthma and bronchiectasis, which are different diseases. Asthma may be a lifelong disease, but tends to be more severe in children and young adults. Bronchiectasis is really a complication of previous disease rather than a disease in itself. Chronic bronchitis and emphysema will be discussed together because they are overlapping conditions usually due to heavy cigarette smoking.

Asthma

Asthma is labored breathing due to spasm of bronchioles. It is characterized by episodic attacks characterized by forced inhalation due to the bronchospasm. The patient appears short of breath, produces wheezing sounds, and may be frightened. Bronchodilating drugs may relieve the symptoms or the patient may seek emergency medical help. The patient appears normal between episodes. There is little permanent damage to the lung unless the disease is complicated by repeated bacterial infections. The bronchial muscle may become hypertrophied and there is hyperplasia of mucous glands in the bronchi associated with increased mucous secretion. During recovery from an attack, coughing leads to expectoration of large amounts of mucus.

The etiology of asthma is complex, involving external allergens or irritants and increased reactivity of the bronchi. Some patients have obvious external allergens as the cause, such as those who become asthmatic during the hay fever season. Avoidance of ragweed pollen and/or desensitization may prevent the asthmatic attacks. Other agents that should be evaluated by history and/or skin testing include house dust, animal dander, foods, aspirin, chemicals, molds, and fungi. In many cases, a specific allergen or irritant cannot be identified; therefore, prevention and treatment are directed at reducing the reactivity of the bronchial muscles and mucous glands. Respiratory tract infections, such as the common cold, typically trigger nonallergic types of asthma. Cold, stress, and exercise may also trigger an asthmatic attack. A severe prolonged attack is called *status asthmaticus* and occasionally is fatal. Most patients can be controlled by avoiding precipitating factors, desensitization of those with atopic allergies (see Chapter 27 for more detail), and administration of bronchodilating agents and corticosteroids. This common disease requires considerable medical attention and may require monitoring of drug levels in patients receiving maintenance bronchodilators.

CHART 12–8. ASTHMA

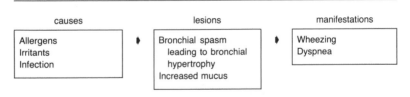

causes	lesions	manifestations

Allergens
Irritants
Infection
	Bronchial spasm
	  leading to bronchial
	  hypertrophy
	Increased mucus
		Wheezing
		Dyspnea

Bronchiectasis

Bronchiectasis is a dilatation of a bronchus or group of bronchi due to chronic inflammation (Fig. 12–8). Classically, it occurs in lower lobes or the right middle lobe following childhood infections complicated by pneumonia. The prevention of bacterial complications of childhood viral pneumonias and the success of measles vaccine have greatly reduced the incidence of this condition. Bronchiectasis can occur with a variety of chronic pulmonary diseases such as cystic fibrosis, tuberculosis, cancer, and obstruction of a bronchus by aspirated foreign body. Diagnosis is made by bronchogram (a roentgenogram of the lung after infusion of radiopaque dye into the bronchial tree). When the disease is isolated to one lobe, surgical removal of the diseased lobe ends the continuing inflammatory and destructive process.

CHART 12–9. BRONCHIECTASIS

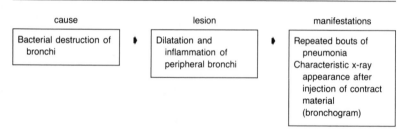

cause	lesion	manifestations
Bacterial destruction of bronchi	Dilatation and inflammation of peripheral bronchi	Repeated bouts of pneumonia Characteristic x-ray appearance after injection of contract material (bronchogram)

Chronic bronchitis has been arbitrarily defined to be a clinical term meaning persistent cough with sputum production for at least 3 months in at least 2 consecutive years. Thus chronic bronchitis is a symptom invented because of its high specificity (about 90%) for heavy smokers. *Emphysema* is defined as permanent enlargement

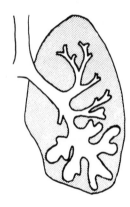

Figure 12–8. Dilated lower lobe bronchi of bronchiectasis.

of acinar air spaces due to destruction of their walls (Fig. 12–9). Thus emphysema is an anatomic abnormality that is usually observed at autopsy. Chronic bronchitis and emphysema are not really diagnoses, but rather nonspecific reactions to injury made more specific by a common cause (smoking). No wonder the acronym COPD has become popular! With these definitional problems out of the way, let us try to describe these conditions.

Chronic Bronchitis

Chronic bronchitis is correlated histologically with hyperplastic bronchial mucous glands and goblet cells. Some degree of obstruction may occur in small bronchioles due to mucus and mild inflammatory changes, but changes at autopsy are not striking. Chronic bronchitis is a reversible condition, as the symptoms and, presumably, the underlying lesions go away when smoking is stopped. Patients with chronic bronchitis tend to have frequent respiratory infections. In the late stages, they have hypoxia and hypercapnea associated physiologically with resistance to air flow in bronchioles.

Emphysema

Smokers gradually develop more and more emphysema so that there is a great overlap in those with chronic bronchitis and those with emphysema. In emphysema, the ventilation problem is mainly due to trapping of air so that oxygen poor, carbon dioxide rich air is not exchanged effectively with fresh air. This occurs because the walls of alveoli have been destroyed leaving large air spaces with little elastic recoil to force air out. The total size of the lungs increases, often with an expanded fixed chest wall (*barrel chest*) and depressed diaphragm. The lungs are more radiolucent on chest roentgenograms due to the increased air/tissue ratio. Retention of carbon dioxide stimulates the respiratory center to increase the rate of breathing. Eventually, the respiratory center becomes refractory to high carbon dioxide levels and low blood oxygen becomes the stimulus for breathing. At this end stage of the disease, oxygen therapy may kill a patient due to removal of the stimulus for breathing (low oxygen). Some patients with emphysema develop pulmonary hypertension leading to right heart failure (cor pulmonale). They are also more susceptible to pneumonia.

Several types of emphysema have been identified that likely have different causes. In centriacinar emphysema, the common type found in smokers, the respiratory bronchioles in the center of an acinus are dilated and the alveoli at the periphery of the acinus are spared. The upper lobes tend to be more affected. In advanced

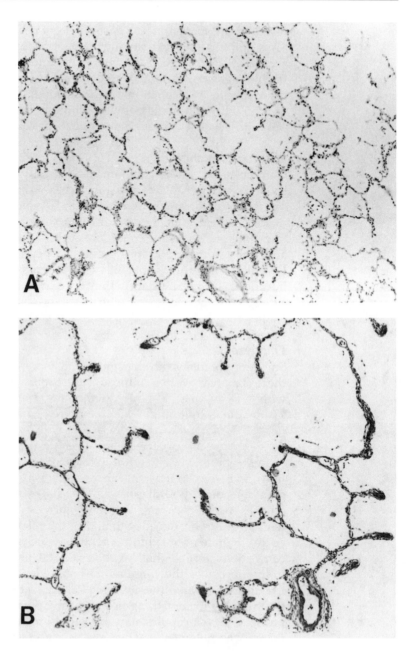

Figure 12–9. Normal lung (**A**) and emphysema (**B**) taken at the same magnification. Note that the alveoli are larger and fewer in number in emphysema.

stages grossly dilated blebs occur. Panacinar emphysema involves all of the alveoli and tends to be more severe in the lower lobes of the lung. This type of emphysema is of interest because it occurs in homozygous alpha-1-antitrypsin deficiency. Alpha-1-antitrypsin inactivates proteases (including elastase). Without the counter activity, elastase, activated by inflammation, can destroy elastic tissue in alveolar walls leading to emphysema. In patients with the deficiency, the rate of destruction is greatly enhanced by smoking. The evidence suggests that smoking produces injury that attracts neutrophils, and neutrophils release elastase that destroys elastic tissue in alveolar walls. The process occurs slowly and locally in respiratory bronchioles in smokers with normal alpha-1-antitrypsin deficiency, but is rapid and generalized in those with the deficiency. Other forms of emphysema include paraseptal emphysema that occurs in the outer part of the acinus, particularly in subpleural areas. This form may be the cause of spontaneous pneumothorax in young adults. Rupture of a bleb forces air into the pleural space, collapsing the lung, and leading to sudden dyspnea. If the lobe does not seal spontaneously, a tube may be placed in the pleural space to siphon off air until healing occurs. Irregular emphysema occurs with various chronic destructive diseases of the lung in association with scarring.

CHART 12–10. EMPHYSEMA

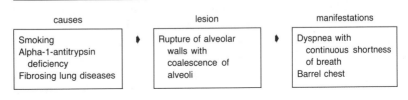

causes	lesion	manifestations
Smoking Alpha-1-antitrypsin deficiency Fibrosing lung diseases	Rupture of alveolar walls with coalescence of alveoli	Dyspnea with continuous shortness of breath Barrel chest

Acute Noninfectious Interstitial Disease

Diffuse acute and chronic injuries to alveolar walls by inhaled fumes and small particulate matter or by hypoxemia and blood-borne toxins generally cause restrictive pulmonary disease with inadequate perfusion of oxygen leading to dyspnea, tachypnea, and, when severe, cyanosis. The airway is not obstructed.

Acute alveolar injuries, which have been given many names depending on cause, are collectively referred to as *adult respiratory distress syndrome* or *diffuse alveolar damage*. The damage to pneumocytes produces edema and inflammatory cell infiltrate that enlarge the alveolar walls. Exudate into alveoli is scant compared to the purulent exudate of bacterial pneumonia. Fibrinous exudate and necrotic lining cells may form hyaline membranes in the alveoli.

Causes of diffuse alveolar damage include shock (sometimes called shock lung), oxygen toxicity, nitrogen dioxide, hypersensitivity reaction to volatile chemicals, and drugs. The outcome may be death, recovery, or progression to chronic interstitial lung disease.

Chronic Noninfectious Interstitial Disease

Chronic diffuse interstitial disease may be due to the same agents that cause acute interstitial disease. The most common known causes are pneumoconioses and sarcoidosis. Fibrosis of the alveolar walls leads to restrictive disease of variable severity. Complications include pulmonary hypertension leading to cor pulmonale and increased susceptibility to infection.

Pneumoconiosis refers collectively to environmentally induced pulmonary fibrosis resulting from inhalation of particulate matter. *Anthracosis* is carbon accumulation in the lungs and its draining lymph nodes. It is more severe in city dwellers, factory workers, and coal miners. Anthracosis is generally harmless because carbon by itself elicits very little fibrous reaction. On the other hand, coal miners may develop progressive fibrosis after many years in the mines. This is probably due to the fact that coal miners inhale silica as well as carbon, although the role of silica versus carbon versus infection in producing coal miners' pneumoconiosis is debated. *Silicosis* is caused by inhalation of crystalline forms of silica and occurs in miners and fabricators of silica products such as glass cutters, sand blasters, pottery workers, and foundry workers. Very small crystalline particles of silica are potent inducers of fibrosis. *Asbestosis,* a condition caused by the inhalation of fibrous silicates, is associated with pulmonary fibrosis, bronchogenic carcinoma, and mesothelioma of the pleura (a malignant neoplasm of pleural lining cells). The effects of asbestosis are somewhat unpredictable as many people harbor the fibers without obvious deleterious effect. Because asbestos has been so widely used as an insulating and fireproofing product, it will be decades before its effects subside. *Berylliosis* appears to be a hypersensitivity reaction (it occurs in only a fraction of exposed persons) and may produce acute diffuse injury or a chronic granulomatous reaction. Beryllium is a metal used in alloys because of its strength and heat resistance.

Sarcoidosis is a generalized noncaseating granulomatous disease of unknown cause and with varying clinical manifestations associated with involvement of various organs. In some individuals, pulmonary fibrosis is the dominant process and these patients have a worse prognosis than others with predominantly lymph node disease. Sarcoidosis is most common in young adults and is more common in women, blacks, and residents of southern United States. Immunologically, delayed hypersensitivity reactions are decreased or absent and antibody-mediated reactions are normal or enhanced.

Diffuse idiopathic pulmonary fibrosis occurs when no cause is found for progressive diffuse alveolar injury. The course is unpredictable but often fatal in months to a few years.

CHART 12–11. PNEUMOCONIOSES

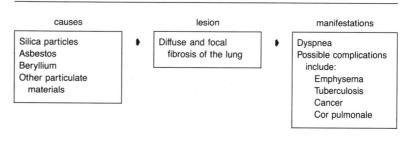

causes	lesion	manifestations

Silica particles
Asbestos
Beryllium
Other particulate
  materials
	Diffuse and focal
	  fibrosis of the lung
		Dyspnea
		Possible complications
		  include:
		    Emphysema
		    Tuberculosis
		    Cancer
		    Cor pulmonale

Vascular Conditions

Pulmonary congestion results from left heart failure with increased venous pressure in the pulmonary veins. When severe or prolonged, fluid passes through the alveolar walls into alveoli, producing pulmonary edema, and into the pleural space, producing hydrothorax. Both of these complications impair breathing. Right heart failure due to lung disease (cor pulmonale) may be caused by acute or chronic lung diseases that result in increased pulmonary blood pressure. For example, acute cor pulmonale may result from pulmonary emboli, and chronic cor pulmonale is associated with emphysema and fibrotic lung lesions. In both of these conditions, blood can only be forced into and through the lungs by increased right heart pressure.

Pulmonary embolism results from thrombi in the leg or pelvic veins being carried to the right heart and into the pulmonary arteries. Pulmonary emboli are sometimes apparent clinically but are much more often found at autopsy. Sometimes they play a role in causing death, but when small, they usually resolve without producing significant injury. Occasionally they cause infarcts.

Hyperplastic/Neoplastic Diseases

Cancer of the lung is one of the three most common nonskin cancers in the United States (along with colon and breast) and is the leading killer. The lung is also a leading site for metastatic cancer. Benign lung neoplasms are rare.

Primary lung cancers are divided into three types based on location and histology: bronchogenic carcinoma, bronchoalveolar carcinoma, and bronchial carcinoid.

Bronchogenic Carcinoma

By far the most common type of primary lung cancer is bronchogenic carcinoma, so named because it arises from large bronchi in the central or hilar region of the lung (Figs. 12–10, 12–11, 12–12). Smoking is the major contributing factor in most cases. Asbestos and radioactive ores are also factors in exposed individuals. The total exposure to cigarettes is estimated in terms of pack-years (the number of packs of cigarettes smoked per day times the number of years smoked). The number of pack-years correlates with the likelihood of developing lung cancer. Although more lung cancers occur in men, the relative frequency in women is rising.

Bronchogenic carcinomas are subclassified histologically into squamous cell carcinoma (most common), adenocarcinoma, large cell undifferentiated carcinoma, and small cell undifferentiated carcinoma (oat cell carcinoma). The first three subtypes are treated similarly. If there is no apparent spread by the time of diagnosis (30%), a surgical cure is possible by removing the lesion; otherwise, palliative therapy is indicated. Small cell carcinoma, which is identified by small cells that look like lymphocytes and contain neuroendocrine granules, almost always has spread beyond the point of surgical cure by the time it becomes manifest. Radiation and chemotherapy prolong life but do not effect a cure. Some small-cell carcinomas become manifest due to hormone production.

The behavior of bronchogenic carcinomas is variable. About 20 percent are localized at the time of diagnosis, and the overall cure rate is about 12 percent. Bronchogenic carcinomas spread in many directions, including local invasion of the lung, pleura, and mediastinum and metastasis to lymph nodes of the lung and neck.

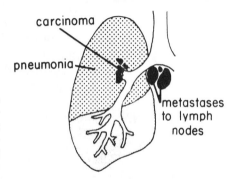

Figure 12–10. Bronchogenic carcinoma, arising in the right upper lobe bronchus with obstruction leading to pneumonia and metastasis to hilar lymph nodes producing a mediastinal mass.

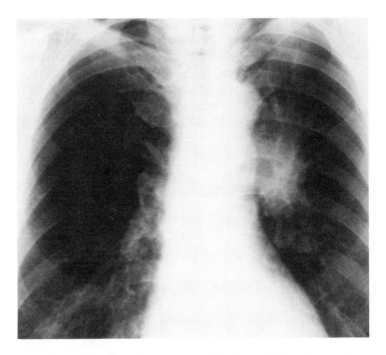

Figure 12-11. Bronchogenic carcinoma of the hilus of left lung by chest x-ray.

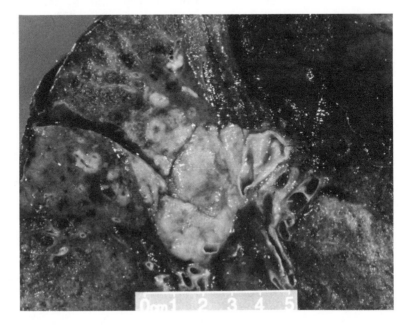

Figure 12-12. Bronchogenic carcinoma with pneumonia distal to the tumor.

Hoarseness may result from involvement of the recurrent laryngeal nerve. Lung cancer may also metastasize through the bloodstream to many organs. Metastasis to the brain and bone is notorious for producing clinical problems, whereas liver metastasis is frequently evident in advanced stages. Symptoms may include shortness of breath, weight loss, pneumonia secondary to bronchial obstruction, or coughing of blood (hemoptysis). The diagnosis may be suggested by x-ray but is usually made by sputum cytology and biopsy.

CHART 12–12. BRONCHOGENIC CARCINOMA

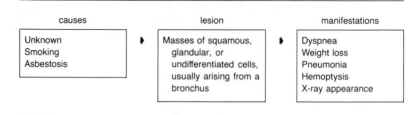

causes		lesion		manifestations
Unknown Smoking Asbestosis	▶	Masses of squamous, glandular, or undifferentiated cells, usually arising from a bronchus	▶	Dyspnea Weight loss Pneumonia Hemoptysis X-ray appearance

Bronchoalveolar Carcinoma

Adenocarcinomas that arise in the periphery of the lung, spread in alveoli, and are not related to smoking are called bronchoalveolar carcinomas. They are uncommon but have up to a 50 percent cure rate with resection. This small group of tumors must be differentiated from more peripherally located bronchogenic carcinomas and metastatic adenocarcinomas.

Bronchial Carcinoid

Carcinoid tumors are low-grade carcinomas that arise from neuroendocrine cells. This rare variety of lung cancer arises in the hilar region, has very characteristic clusters of neuroendocrine cells, and spreads very slowly. Most have not spread to lymph nodes by the time of diagnosis and, even if they have, progression may occur over many years.

Metastatic Cancer

The lung is a common site of metastatic cancer spread by the bloodstream. Clumps of tumor cells are carried through the bloodstream to the periphery of the lungs, and some grow to produce spherical tumor nodules. Because sarcomas preferentially spread via the bloodstream, the lung is a characteristic site of sarcoma metastases. Many advanced carcinomas also spread to the lung.

Other Lung Tumors

Benign neoplasms of the lung are rare. Most benign nodules in the lung are old healed granulomas from a primary tuberculous or fungal infection. Many patients with benign lung nodules undergo excisional biopsies, however, because the x-ray appearance of the nodules often resembles carcinoma. Occasional carcinomas arise in the periphery of the lung; these are adenocarcinomas and produce a single large mass, in contrast to the multiple nodules of metastatic carcinoma.

ORGAN FAILURE

Death is associated with failure of pulmonary, cardiac, or brain function. Failure of any one of these organs causes the others to fail. Pulmonary causes of death may be acute and chronic. Acute obstruction of the trachea and major bronchi may cause rapid death. Obstruction may be external (strangulation) or internal (aspiration of a large foreign body, massive hemorrhage). Acute restrictive disease may be caused by pulmonary edema or pneumonia. Inadequate expansion of the lung may be caused by pneumothorax, hydrothorax, multiple rib fractures, or muscular paralysis.

Chronic lung disease causes death by producing slowly worsening hypoxia, by cor pulmonale, or by superimposed pneumonia. Lung cancer may cause death by diverse mechanisms including infection, obstruction, and cachexia.

REVIEW QUESTIONS

1. How do obstructive and restrictive pulmonary disease differ in terms of location, causes, and manifestations?
2. Why is the impact of lung disease so great? Can anything be done about it?
3. How are lung diseases detected and diagnosed?
4. What are the causes, lesions, and major manifestations of each of the specific diseases listed in this chapter?
5. What is the cause and significance of respiratory distress syndrome of the newborn?
6. What are the similarities and differences among lobar, broncho-, and interstitial pneumonia?
7. Why is pulmonary tuberculosis such an important disease? How do primary and secondary forms differ?

8. What are the two mechanisms of atelectasis?
9. What do the terms chronic obstructive pulmonary disease, chronic bronchitis, and emphysema mean?
10. How does emphysema differ from chronic noninfectious interstitial disease in terms of lesions, cause, and physiologic effect?
11. What is asthma?
12. Why is bronchogenic carcinoma the most important cancer in the United States today?
13. By what mechanisms does lung disease cause death?
14. What do the following terms mean?

> Ventilation
> Perfusion
> Pneumonia
> Dyspnea
> Orthopnea
> Hemoptysis
> Tachypnea
> Rales
> Hypercapnia
> Hyaline membrane
> Legionnaire's disease
> Emphysema
> Mycoplasma
> Ghon complex
> Adult respiratory distress syndrome
> Pneumoconiosis
> Anthracosis
> Silicosis
> Asbestosis
> Berylliosis
> Sarcoidosis

Oral Region, Upper Respiratory Tract, and Ear

REVIEW OF STRUCTURE AND FUNCTION

The structure and function of these organs are generally known to most persons, and their importance is exemplified by the number of specialists involved in health care of diseases of these organs (dentists—including many subspecialists; otolaryngologists, otologists, plastic surgeons, speech pathologists). The oral cavity, including teeth, tongue, and walls of the mouth, serves in the mastication and swallowing of food, as a phonetic box for speech, and as a secondary pathway for breathing. The salivary glands (parotid, submandibular, lingual, and minor salivary glands of the oral mucosa) provide moisture to soften and, of lesser importance, to add carbohydrate-digesting enzymes to the food.

The nose is a passageway for breathing and also partially filters air. Its upper portion contains the sense organ for smell. The pharynx serves as a passageway for air, provides the musculature for swallowing, and contains the openings of the eustachian tubes, which serve as pressure outlets for the middle ears. The pharynx also contains abundant lymphoid tissue, which aids in the recognition of antigens such as foreign materials and microorganisms. The larynx is a major air passage to the lungs and contains the vocal cords.

The ear detects sound and also contains in its inner portion the vestibular apparatus, which is a sensory organ for body equilibrium.

MOST FREQUENT AND SERIOUS PROBLEMS

Diseases of these organs account for an enormous proportion of health care problems. The most common include dental caries, periodontal disease, and upper respiratory infections. Other common diseases include sinusitis, allergic rhinitis, otitis media, and deafness. Of the relatively few potentially fatal conditions, carcinoma of the mouth, pharynx, and larynx are most common.

SYMPTOMS, SIGNS, AND TESTS

Dental caries is either asymptomatic or causes pain. Symptoms of periodontal disease include loose teeth and bleeding gums. Most instances of dental caries and periodontal disease are detected by screening (routine dental examination). The propensity of the nasal mucosa to secrete mucoid exudate in response to mild stimuli is well known to all sufferers of hay fever and the common cold. Pain is the principal symptom of sinusitis, pharyngitis, and otitis media. Loss of hearing occurs with many types of ear diseases. Bleeding from the mouth may signal periodontal disease, an ulcerating neoplasm, or may often be the first manifestation of a systemic bleeding disorder.

Most diseases of the organs under consideration are evaluated by direct vision with or without the aid of specialized instruments such as mirrors, laryngoscopes, and otoscopes. X-rays are particularly helpful in detection of disease in teeth and nasal sinuses. Function tests for hearing employ audiometers. The most important laboratory test is the throat culture for beta hemolytic streptococci, because treatment of streptococcal pharyngitis prevents rheumatic fever (see Chapter 8 for further discussion). Other laboratory tests are of value under a limited number of circumstances.

SPECIFIC DISEASES

Genetic/Developmental Diseases

Various rare types of abnormalities occur in the oral and ear structures that may be either isolated or occur in conjunction with defects of other organs. Cleft lip and cleft palate are two of the more common abnormalities of this area. Various forms and degrees of improper dental development may occur. The most common is *malocclusion* (improper contact between upper and lower teeth). Bizarre abnormalities of facial structure may occur in children with severe mental retardation; fortunately, most of these are rare.

Cleft Lip and Cleft Palate

Cleft lip, a defect on either side of the midline of the upper lip (Fig. 13–1), occurs with or without cleft palate, a defect in the roof of the mouth connecting with the nasal cavity (Fig. 13–2). These

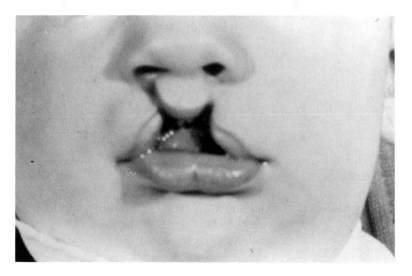

Figure 13–1. Cleft lip.

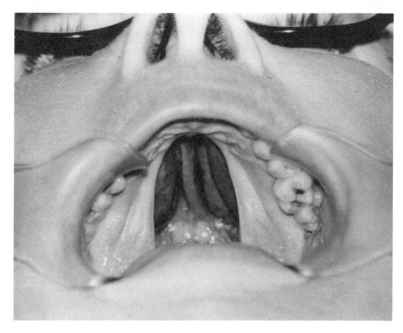

Figure 13–2. Cleft palate.

failures of late embryonic development usually can be corrected surgically to improve voice, mastication, and cosmetic appearance.

CHART 13–1. CLEFT LIP AND CLEFT PALATE

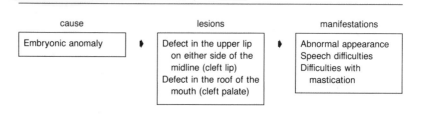

cause	lesions	manifestations
Embryonic anomaly	Defect in the upper lip on either side of the midline (cleft lip) Defect in the roof of the mouth (cleft palate)	Abnormal appearance Speech difficulties Difficulties with mastication

Inflammatory/Degenerative Diseases

Most health problems of the oral region, upper respiratory tract, and ear are acute or chronic infections. The most significant inflammatory diseases of these regions will be discussed followed by several less frequent but significant degenerative conditions of the ear.

Dental Caries

Caries is a microbial disease of the calcified portion of the teeth (enamel and dentin) characterized by demineralization. Caries is the disease that produces cavities in teeth. Disease activity is greatest in childhood so that caries is the predominant cause of tooth loss in persons under 35 years of age.

The cause of caries is complex, although bacteria are essential factors and a diet high in carbohydrate content is a contributing factor. Bacteria adhere to the tooth surface in the form of a tenacious nonvisible mass called *dental plaque*. The plaque can be made visible by staining with the red dye basic fuchsin. Daily removal of the plaque through tooth cleaning and reduction of sugar in the diet reduce the incidence of caries. Introduction of fluoride ion into the hydroxyapatite crystal of the dental enamel markedly decreases the susceptibility of teeth to formation of caries. Introduction of fluoride into drinking water and tooth paste has reduced the incidence of caries by more than 75 percent. Caries may erode into the central connective tissue core of a tooth (dental pulp), producing infection and necrosis of the pulp. Complications of pulpitis include tooth loss, periapical abscess, periapical granuloma, and periapical cyst. Abscess is the most acute and destructive complication. A periapical granuloma is not really a granuloma but rather a low-grade inflammation with abundant granulation tissue. Periapical cysts form from islands of odontogenic epithelium that are trapped in a periapical granuloma and proliferate to form a squamous-lined cyst. Root canal therapy can prevent tooth loss due to pulpitis.

CHART 13–2. DENTAL CARIES

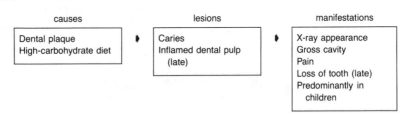

causes	lesions	manifestations
Dental plaque High-carbohydrate diet	Caries Inflamed dental pulp (late)	X-ray appearance Gross cavity Pain Loss of tooth (late) Predominantly in children

Periodontal Disease

Periodontal disease refers to a usually painless chronic low-grade inflammation of the supporting tissues of teeth. It is a disease of adults that increases in prevalence with age and is the major cause of tooth loss in adults. As with caries, dental plaque is the major etiologic factor. Accumulation of plaque on the tooth between tooth and gingiva produces a low-grade inflammation of the gingiva, which may be recognized by direct inspection as changes in color, texture, and amount of gingival tissue. Daily removal of plaque from between the teeth and gingiva with dental floss and brushing is the major means of preventing and reducing periodontal disease. If allowed to progress, the low-grade inflammation extends to the underlying alveolar bone and connective tissue, thus loosening the teeth and causing them to be extruded. Most adults have some degree of periodontal disease, but its severity varies greatly.

CHART 13–3. PERIODONTAL DISEASE

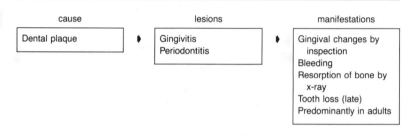

cause	lesions	manifestations
Dental plaque	Gingivitis Periodontitis	Gingival changes by inspection Bleeding Resorption of bone by x-ray Tooth loss (late) Predominantly in adults

Acute Necrotizing Ulcerative Gingivitis (Vincent's Infection, Trench Mouth)

This distinct form of acute, severe gingivitis occurs predominantly in young adults and is characterized by ulcerated and bleeding gums. It is caused by fusiform and spirochetal bacteria, normally present in the mouth and associated with poor oral hygiene or underlying disease that lowers host resistance.

Herpes Stomatitis (Cold Sores)

Herpes simplex type I is a very prevalent virus of humans. Following primary infection, herpes has the ability to lie dormant in nervous tissue until activated by stress, trauma, or immunosuppression. Its favorite site is the mucocutaneous border of the lips, gums, and hard palate, where it produces blisters followed by ulcerating and crusting inflammatory lesions when activated by sunburn or the fever of a common cold (Fig. 13–3). The lesion will recur at intervals throughout life. Genital herpes is caused by a different virus (type II).

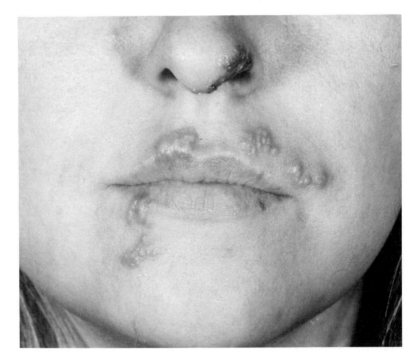

Figure 13–3. Herpes simplex cold sores.

Aphthous Stomatitis (Canker Sores)

Aphthous means spot, *stoma* means mouth, so aphthous stomatitis is a painful inflammatory spot in the mouth. These are very discrete, shallow, painful ulcers within the mouth that gradually heal in 7 to 10 days (Fig. 13–4). The lesions are similar to herpes stomatitis in that they are activated periodically, but differ from herpes in their discrete appearance and occur on nonkeratinized areas of the oral mucosa. The causative agent is unknown.

Upper Respiratory Infection (URI)

This term encompasses a wide spectrum of illnesses in terms of sites, causative organisms, and severity. Inflammation of nasal sinuses, lung, and conjunctiva are commonly associated with upper respiratory infections. Terms based on site include *rhinitis* (nose), *pharyngitis, tonsillitis,* and *laryngitis.* Terms based on cause include viral sore throat and streptococcal sore throat. Nonspecific terms include *common cold, sore throat, flu,* and *coryza. Coryza* is a descriptive term meaning profuse nasal discharge. Most upper respiratory infections are caused by viruses of various types that are not or-

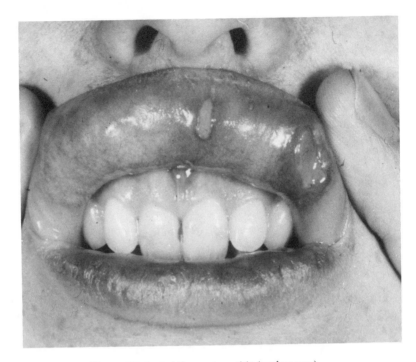

Figure 13–4. Aphthous stomatitis (canker sore).

dinarily identified. Rhinoviruses are the most common cause. A few cases are caused by group A beta hemolytic streptococci. It needs to be emphasized that the amount of redness and exudate in the throat and the white blood count are not reliable in distinguishing viral from bacterial upper respiratory infections. Both may be mild or severe, both may be associated with elevated white blood counts. Throat culture will detect beta hemolytic streptococcus and the numbers (many versus few) present will help distinguish persons with strep throat from persons who are merely carriers of the organism.

Upper respiratory infections are the most significant cause of work loss in adults, but complications are more common in children and in the elderly. In children, the abundant lymphoid tissue of the nasopharynx (adenoids) and pharyngeal tonsils enlarge in response to upper respiratory infections. Commonly, the enlargement of the adenoids surrounding openings of the eustachian tubes blocks these openings and predisposes to middle ear infection (otitis media). Repeated bouts of otitis media may be cause for removal of the adenoids. Enlarged tonsils may partially obstruct breathing and sometimes are removed for this reason. Pneumonia is a frequent complication of upper respiratory infections in both children and the elderly. Epidemics of upper respiratory infections increase the death rate of the elderly. Sometimes, upper respiratory infections predominantly involve the larynx, producing edema, partial obstruction, and difficult inspiration. In children, this produces a condition known as *croup*, a laryngeal spasm characterized by a loud, high-pitched, inspiratory sound, which is frightening to the patient because of inability to breathe and to bystanders because of the labored respiration. Usually the edema can be reduced with great relief. In adults, laryngitis is manifest by a rough quality to the voice (hoarseness).

Two other uncommon but serious respiratory infections are whooping cough and *Hemophilus influenzae* infection. These will be discussed in the chapter on infectious diseases (Chapter 26).

CHART 13–4. UPPER RESPIRATORY INFECTIONS

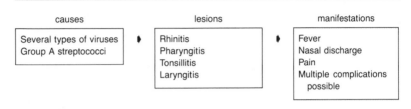

causes	lesions	manifestations
Several types of viruses Group A streptococci	Rhinitis Pharyngitis Tonsillitis Laryngitis	Fever Nasal discharge Pain Multiple complications possible

Sinusitis

Infection in the paranasal sinuses is a cause of pain and headache in some individuals. Obstruction of the opening to sinus cavities leads to fluid accumulation and infection (Fig. 13–5). Treatment that reduces the inflammatory swelling in the lining epithelium of the sinus and its opening may allow the exudate to drain and the infection to be brought under control. Sometimes the sinus opening must be surgically enlarged to allow drainage. Chronic infections with continuous drainage into the pharynx (postnasal drip) may lead to pulmonary infections.

Allergic Rhinitis (Hay Fever)

Hay fever is an outpouring of watery exudate from the nose due to mild inflammation, which is in turn caused by exposure to antigenic substances in the environment. Ragweed pollen is the most important seasonal allergen, although various other pollens may be the cause in some individuals. House dust and dander from pets are important causes of chronic allergic rhinitis. In some individuals, the boggy, swollen nasal mucosa may form polyps, which obstruct breathing and require surgical removal.

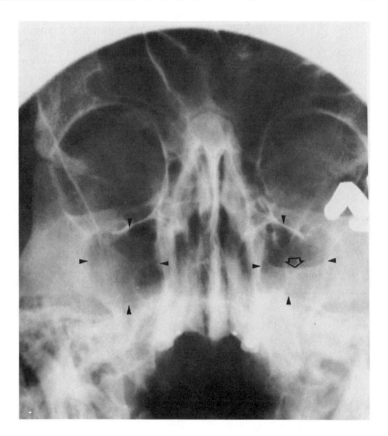

Figure 13–5. Facial x-ray showing paranasal sinuses (*arrow heads*) and air/fluid level (*open arrow*) in patient's left sinus due to chronic sinusitis and obstruction of the opening to the sinus.

Otitis Media

Otitis media is inflammation of the middle ear. It is most common in young children as a sequela of upper respiratory infections. Pharyngeal edema causes obstruction of the eustachian tube, with resultant spread of infection through the eustachian tube to the middle ear. In most instances, the otitis media is presumed to be caused by bacteria, although culture is usually not performed because of the inaccessible location of the middle ear behind the eardrum. Antibiotic therapy usually is curative, although some children are subject to repeated episodes of otitis media. Otitis media is suspected in a child who may be irritable and who can often localize the pain to one ear. Inspection through an otoscope reveals a swollen, inflamed eardrum (Fig. 13–6). Complications of otitis media include rupture of the eardrum, with drainage of pus and mastoiditis due to spread of the infection to surrounding mas-

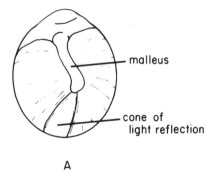

A

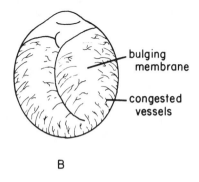

B

Figure 13–6. Eardrum: normal **(A)** and otitis media **(B)**. In otitis media, normal landmarks on the eardrum (visualization of the malleus through the eardrum and the light reflex) disappear and bulging of the membrane and vascular congestion become evident.

toid bone. The development of mastoiditis is particularly serious, because it may lead to permanent hearing loss or spread of infection into the brain. Chronic low-grade otitis media may lead to the formation of a mass of kerantinized tissue called a *cholesteatoma*.

CHART 13–5. OTITIS MEDIA

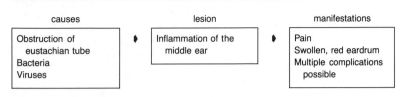

causes	lesion	manifestations
Obstruction of eustachian tube Bacteria Viruses	Inflammation of the middle ear	Pain Swollen, red eardrum Multiple complications possible

Meniere's Disease

Meniere's disease is a degenerative disease of unknown cause in which the vestibular apparatus becomes dilated. The symptoms of Meniere's disease are *tinnitus* (ringing in the ears), deafness, and episodic *vertigo* (illusion of movement).

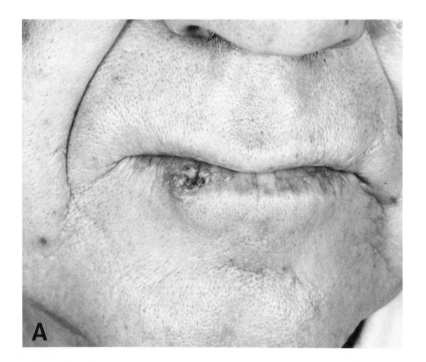

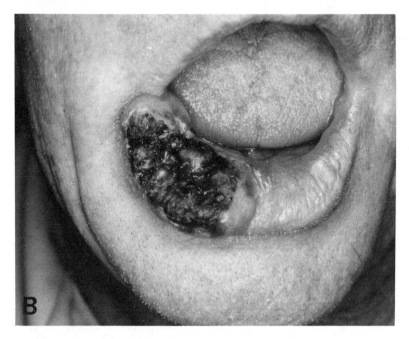

Figure 13–7. Squamous cell carcinoma of the lower lip; early lesion **(A)** and advanced lesion **(B)**.

Presbycusis

Hearing loss accompanying increasing age is called *presbycusis* and may be due to continued exposure to noise as well as vascular changes.

Otosclerosis

Otosclerosis is a fixation of the small bones in the middle ear, which conduct vibrations from the eardrum to the inner ear. Its cause is unknown, and it most commonly occurs in young women, resulting in gradual hearing loss. Operative mobilization of these tiny bones may improve sound conduction.

Hyperplastic/Neoplastic Diseases

Squamous Cell Carcinoma

Carcinoma of the lip usually occurs on the lower lip because exposure to sunlight is the most important etiologic factor. Smoking is the most important factor in the genesis of laryngeal cancer. Some poorly differentiated pharyngeal cancers are associated with the Epstein-Barr virus. Most importantly, mouth cancers are greatly potentiated by the combined use of alcohol and tobacco and appear to be related to the dose of each. The early precancerous lesions in alcohol/tobacco abusers appear as reddened areas on thin, nonkeratinized areas of squamous epithelium (floor of mouth, ventrolateral tongue, and soft palate). Chewing tobacco has a high association with the development of oral cancer.

Squamous cell carcinomas present as expanding, ulcerating, and encrusted lesions (Fig. 13–7). Most lip cancers (90%) are cured because the lesion can be detected early before lymph node metastasis has occurred and because resection or irradiation are relatively easy procedures. About two-thirds of laryngeal carcinomas are cured by resection, which is sometimes combined with irradiation. Carcinomas of the mouth and pharynx have the worst prognosis (about 50%) because lymph node metastasis is often present by the time of diagnosis.

Salivary Gland Tumors

Most salivary gland neoplasms arise in the parotid gland, and the majority are benign. The so-called *mixed tumor*, or pleomorphic adenoma, is the most common of several types.

ORGAN FAILURE

Loss of ability to masticate (chew) is associated with loss of teeth, fractures of the mandible, extensive cancer operations, and severe congenital abnormalities. Swallowing may be defective with paralysis from poliomyelitis or stroke. Obstruction of the airway may be due to allergic nasal polyps or enlarged lymphoid tissue in the pharynx. Impaction of a large piece of meat in the lower pharynx is an occasional cause of sudden death.

Any person with a detectable hearing impairment may be described as relatively deaf, although the essence of deafness as we generally use the word is the inability to hear and understand the spoken voice. The types of deafness are usually divided into *nerve deafness*, in which there is interruption of the cochlear nerve, and *conduction deafness*, in which there is disease of the middle ear or occlusion of the external auditory canal. This separation is somewhat artificial, as many conditions, such as infection, can damage several components of the ear.

The true incidence of deafness as well as the relative causes of deafness are difficult to assess. It is estimated that there is a prevalence rate of up to 16 cases of deafness per 1000 school children in which the hearing loss is severe enough to at least require a hearing aid. About half of these cases are due to genetic conditions, whereas in the other half deafness is acquired. Most of the acquired cases in the past have been due to rubella in which there is inner ear damage. The prevalence of deafness increases linearly with age to the point where over 60 percent of persons above the age of 65 have a major impairment of hearing. The most common cause of hearing loss in the elderly is usually ascribed to presbycusis; however, this is begging the question, as presbycusis simply means hearing loss in the elderly. It is generally agreed that continued noise exposure, degeneration of ossicles, trauma, and infection are the most common contributing factors to deafness in adults; however, the relative incidence of each of these factors is unknown. Other known causes of deafness in persons of all ages include otosclerosis, cholesteatoma, Meniere's disease, neoplasms, psychogenic causes, and toxins. Some of the drugs that are known to produce sensorineural hearing loss are aspirin, neomycin, quinine, and streptomycin.

REVIEW QUESTIONS

1. What are the most common diseases of the mouth, ear, and upper respiratory tract? How are they usually diagnosed?
2. Why is a throat culture such an important laboratory test?
3. What are the causes, lesions, and major manifestations of each of the specific diseases discussed in this chapter?
4. How does cleft lip differ from cleft palate?

5. What are the similarities and differences between dental caries and periodontal disease?
6. What are the major complications of upper respiratory infections?
7. How do squamous cell carcinomas of the lip, mouth, and larynx differ as to cause and prognosis?
8. What is the difference between nerve deafness and conduction deafness?
9. What do the following terms mean?
   Dental plaque
   Coryza
   Croup
   Mastoiditis
   Cholesteatoma
   Tinnitus
   Vertigo
   Presbycusis
   Mixed tumor

Alimentary Tract

Colonic Polyps
Carcinoma of the Colon

ORGAN FAILURE

REVIEW QUESTIONS

REVIEW OF STRUCTURE AND FUNCTION

The *digestive* or *gastrointestinal system* includes the *oral region* (mouth, salivary glands, pharynx), the *alimentary tract* (esophagus, stomach, small intestine, large intestine, including the vermiform appendix, and anus), and the *pancreaticobiliary tract* (liver, gallbladder, bile ducts, pancreas). The function of the alimentary tract is to digest masticated food, to absorb digestion products, and to excrete the residue along with certain waste products excreted by the liver through the bile duct. The segments of the alimentary tract are reviewed in Fig. 14–1.

The esophagus is a straight tube that serves to carry food from the pharynx to the stomach. The stomach is a distensible organ, particularly in its upper part, or body. The stomach body mucosal cells secrete hydrochloric acid and proteolytic enzymes, which aid in digestion. The mucosa of the lower part of the stomach, or antrum, is lined by mucous cells. The extreme distal end of the stomach is often called the *pylorus*. The small intestine is artificially divided into duodenum, jejunum, and ileum. Although some di-

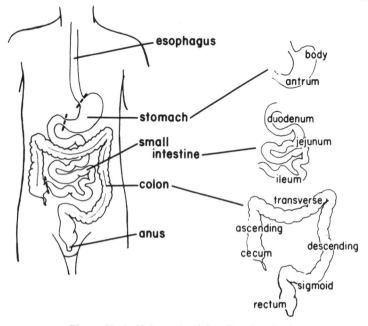

Figure 14–1. Major parts of the alimentary tract.

gestion occurs at the surface of the small intestinal mucosa, its major function is absorption. The large intestine, or colon, is also artificially divided into segments consisting of cecum, ascending colon, transverse colon, descending colon, sigmoid colon, and rectum. The vermiform appendix is a nonfunctional vestigial structure attached to the cecum. The colon is a storage reservoir for undigested food and also is a site of water absorption.

The alimentary tract has four layers—mucosa, submucosa, muscularis, and serosa (Fig 14–2A). The mucosal layer has three components—epithelial cells lining the surface, supporting connective tissue called the *lamina propria,* and a unique thin muscular layer called the *muscularis mucosae* (Fig. 14–2B). The structure of inner mucosal layer is varied to provide the specialized function of each level of the tract: the esophagus is lined by stratified squamous epithelium, which promotes rapid gliding of masticated food from the mouth to the stomach; the stomach is lined by a thick glandular mucosa, which provides mucus, acid, and proteolytic enzymes to help break up food; the small intestinal mucosa has a villous structure, which provides a large surface of cells for active absorption; and the large intestinal mucosa is lined by abundant mucus-secreting cells, which facilitate storage and evacuation of the residue. Beneath the mucosa is the submucosa, a layer unique to the alimentary tract, which provides structural support to the tract because of the abundant collagenous tissue present. The muscle layer contracts rhythmically to move materials through the alimentary tract. The serosal layer is a thin, smooth membrane present on the outer surface of those parts of the alimentary tract that lie within the abdominal cavity. It keeps the highly tortuous loops of bowel from becoming tangled. The serosa is continuous with the mesentery—the connective tissue attachment of the bowel that contains blood vessels, lymphatics, and nerves.

In order for the intestinal tract to carry out its job of digestion, absorption, and excretion, the contents of the lumen must be pro-

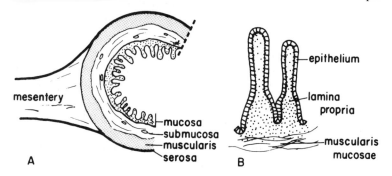

Figure 14–2. A. Layers of the wall of the alimentary tract. **B.** Components of the mucosal layer.

pelled along the tract at the appropriate rate, and reverse movement must be prevented. A wave of muscle contractions (peristalsis) carries a bolus of swallowed food down the esophagus, and a sphincter at the lower end of the esophagus prevents regurgitation. Contractions in the stomach emanate as waves from the body of the stomach and proceed toward the distal stomach (pylorus). They serve to mix the digesting food and push the partially digested contents into the duodenum. The greater the distention of the stomach, the stronger the contractions and the more rapid the emptying.

The muscle of the pylorus only partially closes the outlet to the stomach, so intestinal contents can regurgitate into the stomach, particularly when the small intestine is not emptying normally. Excessive distention of the stomach may result in vomiting, a process in which the esophageal sphincter opens to allow abdominal contractions to forcibly propel gastric and small intestinal contents upward through the esophagus. Under normal conditions, movement of lumenal contents of the small intestine is most rapid in the upper small intestine and decelerates distally. Contents gradually pass from the ileum to the cecum, and reverse movement is partially prevented by the ileocecal valve. The rate of movement in the colon depends on the amount of solid material that remains and the water content. Water is absorbed in the colon to make the contents solid, and the solid residue is moved to the left side of the colon and rectum. When the rectum becomes distended, an urge for voluntary relaxation of the anal sphincter to induce defecation occurs.

The digestive process begins in the mouth, where carbohydrate-splitting enzymes (amylases) from the salivary glands mix with food during mastication. In the stomach, proteolytic enzymes (pepsin) and hydrochloric acid are added to speed the digestive process. The greatest volume of digestive enzymes are added to the digesting mixture of food in the duodenum and originate from the pancreas. These include more amylases, more proteolytic enzymes (trypsins), and fat-splitting enzymes (lipases). In addition, bile salts, which are contained in bile secreted by the liver and stored in the gallbladder, are added to emulsify lipids into small water-soluble packets called *micelles*. The final phase of the digestive process occurs in the surface of small intestinal epithelial cells, where carbohydrate-splitting enzymes (disaccharidases) and protein-splitting enzymes (dipeptidases) are present. Complex endocrine and nervous mechanisms coordinate the timing of the secretion and storage of digestive enzymes, hydrochloric acid, and bile salts so that appropriate amounts are available when needed. For example, the sight of food may induce salivation and gastric secretions via nervous stimulation; distention of the gastric antrum causes release of gastrin, a hormone that stimulates acid production and gastric emptying; and emptying of food into the duodenum causes secretion of the hormones secretin and pancreozymin, which, in turn, cause the pancreas to secrete more fluid and enzymes and the gallbladder to empty bile into the duodenum.

Because many lesions of the alimentary tract are treated by surgical removal, it is important to know what parts can be removed and the physiologic consequences of removal. Short segments of large intestine, small intestine, and stomach and the entire vermiform appendix can be removed without significant functional impairment. Mild to moderate disability may be encountered with removal of all of the large intestine, large segments of small intestine, large segments or all of the stomach, and any part of the esophagus. Continuity of the alimentary tract with the pancreas and liver are essential for survival. Of the major operations that cause functional impairment, the two most common are removal of the distal two-thirds of the stomach for treatment of peptic ulcer and removal of the colon for inflammatory bowel disease.

MOST FREQUENT AND SERIOUS PROBLEMS

The most common problem confronting patients is functional alteration of movement through the alimentary tract. *Constipation* refers to infrequent and/or difficult evacuation of feces. *Diarrhea* refers to abnormally frequent and liquid stools. Constipation is frequently associated with a low-bulk diet and with aging. Diarrhea may be a manifestation of anxiety or due to organic disease. The most common functional alteration of the alimentary tract is called *functional bowel syndrome,* or *spastic colon.* The symptoms are variable and include constipation, diarrhea, and crampy abdominal pain. The functional bowel syndrome mimics many organic diseases and is the most common malady of the alimentary tract.

There are many organic diseases of the alimentary tract that cause temporary or prolonged disability but are rarely fatal. The most common of these is viral enteritis (intestinal flu). Manifestations include nausea, vomiting, and diarrhea. The course is self-limited and the cause is most often a virus. Acute overindulgence in alcoholic beverages produces an illness similar to viral enteritis, although muscle aches and pains and fever are less frequent and the duration is often shorter. Diverticulosis of the colon and esophageal hiatal hernia are common conditions in older people but only occasionally cause symptoms. Inguinal hernia is an outpouching of the abdominal cavity into the groin that may trap loops of bowel. Preventative repair of inguinal hernias is the most common abdominal operation. Acute appendicitis is the most frequent indication for operations on the alimentary tract itself. Diseases of the anus are very common and an important source of discomfort to patients. Mild anal problems include itching in children due to pinworms, simple itching in adults (pruritis ani), and hemorrhoids. More severe anal lesions include thrombosed hemorrhoids, fissures, abscesses, and pilonidal sinuses.

Common conditions that require more extensive health care include duodenal and gastric ulcers, carcinoma of the colon, polyps of the colon, and inflammatory bowel disease (ulcerative colitis and Crohn's disease). Of these lesions, carcinoma of the colon is fatal about 40 percent of the time, the others being infrequently fatal. Lesions of the alimentary tract with a very low survival rate include carcinoma of the esophagus, carcinoma of the stomach, esophageal varices, infarction of the bowel due to occlusion of the mesenteric artery, and perforation with generalized peritonitis regardless of cause.

SYMPTOMS, SIGNS, AND TESTS

Many manifestations of alimentary tract disease can be related to hemorrhage, altered motility, and perforation. Hemorrhage may be mild or severe and may originate from the upper or lower part of the tract. Severe hemorrhage from the upper alimentary tract (esophagus, stomach, duodenum) leads to *hematemesis* (vomiting of blood). Vomited blood is bright red if fresh or has the appearance of coffee grounds if it has been altered by acid in the stomach. Severe upper tract bleeding will also produce melena (black tarry stools) due to alteration of the blood as it passes down the tract. Severe bleeding from the lower alimentary tract produces bright red blood in the stool (*hematochezia*). Mild bleeding from the upper or lower tract will not be noticed by the patient or produce gross color changes in the feces, but can be detected by chemical tests of the feces for hemoglobin (occult fecal blood test). In any patient with severe gastrointestinal bleeding, it is most important to look for evidence of low blood volume so that transfusions can be given before shock develops. Mild, prolonged bleeding will lead to loss of iron and eventually to iron-deficiency anemia. Mild bleeding tends to be an early manifestation of gastrointestinal cancers, so fecal blood tests and blood counts constitute important routine screening tests for persons in the cancer age range (over 40).

Manifestations of altered motility may be due to obstruction or altered activity of the bowel. Obstruction of the alimentary tract may be partial or complete. Complete obstruction will lead to distention of the bowel proximal to the obstruction and then to vomiting. The distended segment will be muscularly active in its attempt to overcome the obstruction, producing a sloshing of the liquid contents with resultant increase in "bowel sounds." The two most common causes of intestinal obstruction are trapped bowel in an inguinal hernia and twisting of bowel around a fibrous adhesion resulting from a previous abdominal operation. Obstruction usually requires surgical operation for diagnosis and treatment.

Altered motility without obstruction may be responsible for vomiting, ileus, diarrhea, constipation, or dysphagia. Vomiting is a common nonspecific symptom that may be associated with generalized illness, central nervous system disease, or intestinal disease. *Ileus* refers to dilation of the intestines. *Adynamic ileus* refers to ileus with absent bowel sounds because the intestinal musculature is hypoactive; the bowel is not physically obstructed. Adynamic ileus occurs most commonly during the first few days after abdominal operations and is one of the reasons for intravenous feeding following operation. When ileus is present, gastrointestinal contents may be vomited and aspirated producing pneumonia.

Diarrhea and constipation are most commonly functional; however, organic diseases of the colon must be considered if these symptoms are severe or prolonged.

Dysphagia is difficulty in swallowing. It may be caused by a variety of esophageal lesions.

Perforation of the intestinal tract into the peritoneal cavity is life threatening because of the massive seeding of intestinal bacteria and ease of spread of the peritonitis within the cavity. Inflammation of the peritoneal lining causes pain, muscle contraction leading to a rigid abdomen, and adynamic ileus. Systemic manifestations include fever and leukocytosis. Causes of perforation include peptic ulcer, trauma to the abdomen, infarcts, and untreated appendicitis.

Pain is also an important manifestation of gastrointestinal disease. The location, duration, and time sequence of pain may help localize the cause. Sudden severe abdominal pain suggests the possibility of perforation or sudden twisting of bowel. Duodenal ulcers produce a chronic, nonradiating, burning, epigastric pain perceived as deep to the abdominal wall. Heartburn, a burning substernal discomfort, is a typical symptom of esophagitis. The initial manifestation of appendicitis is pain in the umbilicus, which shifts to the right lower quadrant of the abdomen as it becomes more severe.

Because there is a wide variety of lesions that can produce the major manifestations described above, it is usually necessary to search for specific lesions using radiologic and endoscopic procedures. The radiologist can outline the normal and abnormal anatomic features of most segments of the gastrointestinal tract by placing a radiopaque material (barium) in the lumen. *Upper GI series* refers to radiologic examination of esophagus, stomach, and upper small intestine. *Barium enema* refers to radiologic examination of the colon. *Endoscopic procedures* refers to the use of a tube or scope to look inside a body passageway. Most endoscopes are equipped with biopsy forceps to obtain tissue so that the pathologist can make a specific histologic diagnosis. *Sigmoidoscopy*, the most common endoscopic procedure, is used to visualize the rectum and lower sigmoid colon. Using highly developed instruments, the specialized gastroenterologist may also directly visualize and biopsy lesions at higher levels of the colon and at all levels of the upper gastrointestinal tract. *Upper gastrointestinal endoscopy* is used to visualize and biopsy lesions of the esophagus, stomach, and the first portion of the duodenum. *Colonoscopy* is used to visualize and biopsy lesions of the entire colon and distal ileum.

The measurement of acid in the stomach is called *gastric analysis*. Its use is limited except to demonstrate the absence of acid (achlorhydria) or rare patients with very high acid output. Laboratory tests on gastrointestinal contents, blood, and urine may be used to evaluate absorption from the alimentary tract by calculating oral intake and output from feces and urine or rise and fall of blood levels of the nutrient under study. Several types of tests for carbohydrate and fat may be used to detect malabsorption. Examination of feces for ova and parasites is used to find ameba and helminths in the gastrointestinal tract. Fecal culture is used to identify specific bacterial pathogens.

SPECIFIC DISEASES

Genetic/Developmental Diseases

Developmental abnormalities include a number of embryonic malformations that are usually surgically correctable unless they are associated with other more serious malformations or are associated with problems of prematurity. In addition to malformations, pyloric stenosis, a condition of uncertain cause, and Hirschsprung's disease, a disease of defective innervation, will be discussed.

Malformations

Meckel's diverticulum is an outpouching of the ileum (Fig. 14–3A) that results from a failure of the embryonic connection with the yolk sac to disappear. It occurs in 2 percent of the population and is usually discovered at autopsy. Occasionally, disease in a Meckel's diverticulum leads to hemorrhage, inflammation, or obstruction of the ileum.

Esophageal atresia is the absence of part of the esophagus so that the upper esophagus ends as a blind pouch (Fig. 14–3B). Usually there is a connection of the trachea to the distal esophagus (tracheoesophageal fistula).

Congenital diaphragmatic hernia is a hole in the diaphragm

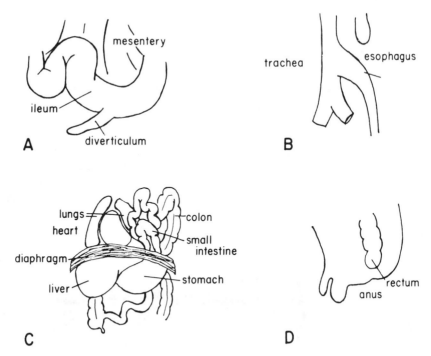

Figure 14–3. Four examples of embryonic anomalies of the alimentary tract. **A.** Meckel's diverticulum of the ileum. **B.** Tracheoesophageal fistula. **C.** Diaphragmatic hernia. **D.** Imperforate anus.

separating the abdomen from the thorax (Fig. 14–3C). Portions of the alimentary tract may herniate through the hole at birth or later to compress the lungs and impair breathing.

Imperforate anus is a failure of the anus to connect with the rectum (Fig. 14–3D). There may be associated openings (fistulae) from the rectum to the bladder, urethra, or vagina.

These malformations as well as others that either compromise

CHART 14–1. MALFORMATIONS

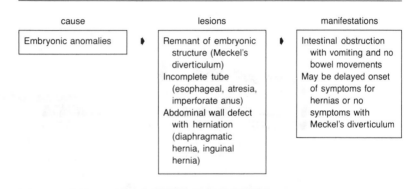

cause		lesions		manifestations
Embryonic anomalies	▶	Remnant of embryonic structure (Meckel's diverticulum) Incomplete tube (esophageal, atresia, imperforate anus) Abdominal wall defect with herniation (diaphragmatic hernia, inguinal hernia)	▶	Intestinal obstruction with vomiting and no bowel movements May be delayed onset of symptoms for hernias or no symptoms with Meckel's diverticulum

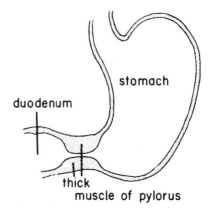

Figure 14–4. Pyloric stenosis.

breathing or obstruct the intestinal tract must be repaired surgically to prevent a fatal outcome.

Congenital Pyloric Stenosis

Congenital pyloric stenosis is a narrowing of the outlet of the distal stomach due to hypertrophy of the muscle (Fig. 14–4). The cause is unknown, and for some reason, symptoms of projectile vomiting after feeding do not begin until 2 to 4 weeks after birth. Other than Meckel's diverticulum, this is the most common developmental abnormality of the alimentary tract. It occurs almost exclusively in boys. A simple operation to split the thickened pyloric muscle effects a cure.

CHART 14–2. CONGENITAL PYLORIC STENOSIS

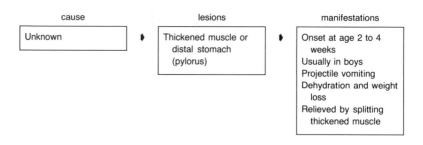

Hirschsprung's Disease

A lack of nerve (ganglion) cells in the rectum results in defective evacuation of feces from the rectum, with consequent massive distention of the colon with feces (megacolon) (Fig. 14–5). Hirschsprung's disease is suspected in children with chronic constipation and distended abdomen, diagnosed by rectal biopsy to demonstrate absence of ganglion cells, and treated by surgical removal of the aganglionic segment.

CHART 14–3. HIRSCHSPRUNG'S DISEASE

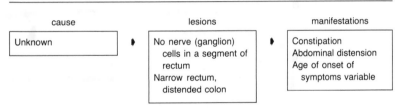

cause	lesions	manifestations
Unknown	No nerve (ganglion) cells in a segment of rectum Narrow rectum, distended colon	Constipation Abdominal distension Age of onset of symptoms variable

Inflammatory/Degenerative Diseases

The diseases discussed in this section illustrate the great variety of conditions that involve the alimentary tract. Infectious diarrheas and food poisoning are diseases acquired from the environment. Esophagitis, gastritis, and peptic ulcers are nagging, mostly chronic problems. The malabsorption syndrome, although relatively uncommon, illustrates the many ways that small intestinal function can be interfered with. Acute appendicitis is well known because it is frequent and strikes suddenly without warning. Pseudomembranous enterocolitis is an acute, life-threatening condition usually induced by antibiotic therapy. Ulcerative colitis and Crohn's disease, collectively known as inflammatory bowel disease, are chronic, usually lifelong conditions that require a lot of medical care. Colonic diverticulosis is an acquired structural defect with inflammatory complications. Inguinal hernia may be either congenital or acquired and is important as a cause of intestinal obstruction and infarction.

Infectious Diarrheas and Food Poisoning

Organisms from the environment may produce gastrointestinal disease in several ways. Most microorganisms that are ingested are destroyed by the acid in the stomach and are digested or pass

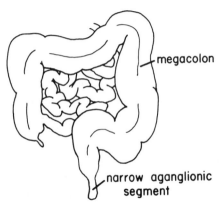

megacolon

narrow aganglionic segment

Figure 14–5. Hirschsprung's disease.

through the alimentary tract without causing harm. Highly virulent organisms ingested in small numbers can escape destruction in the stomach and invade the mucosa of the upper small intestine (viruses) or distal small intestine and colon (*ameba, shigella, campylobacter*). The intestines may also serve as a portal of entry for systemic disease (e.g., typhoid fever). Intestinal diseases caused by highly virulent organisms are typically spread by the fecal–oral route. Small amounts of fecal material contaminate much of man's immediate environment and, in many areas of the world, the water supply is also contaminated. Microorganisms that are less virulent and/or are more susceptible to destruction in the stomach must be ingested in larger numbers in order to get sufficient numbers into the intestines to cause disease. Intestinal disease caused by microorganisms of low or intermediate virulence (e.g., most salmonellae) are typically associated with food poisoning—sufficient numbers of organisms must proliferate in food and be ingested in order to overcome the host's defenses. Another form of food poisoning involves bacteria growing in food and producing a toxin. The toxin, not the bacteria, causes disease of the stomach and upper small intestine (staphylococcal food poisoning) or the toxin is absorbed and damages the nervous system (botulism).

Viral enteritis (intestinal flu) is the most common type of gastrointestinal infection and is caused by normal daily contact with infected individuals. Rotavirus (mostly in infants) and Norwalk agent are the most common causes, but a variety of other viruses may cause gastroenteritis. Viral enteritis is both seasonal (epidemic) and sporadic. Children are more commonly involved than adults. The virus grows in surface epithelial cells of the small intestine, leading in 1 to 2 days to nausea, vomiting, diarrhea, and malaise (a vague feeling of discomfort) that lasts from one to several days. Diagnosis is presumptive, because there are no routine tests for viral enteritis. Prevention is not practical, and treatment is not needed unless there is excessive fluid loss from vomiting and diarrhea.

Traveler's diarrhea refers to any acute diarrheal illness in travelers to foreign countries. Natives can obviously be affected by the same disease, but they are more likely to be immune or have a low-grade chronic involvement. Causes are multiple. Recent studies have implicated enterotoxin-producing strains of *Escherichia coli* as the most common cause. These strains are not easily separated from harmless strains of *E. coli* normally present in the colon. Enterotoxin-producing *E. coli* proliferate in the small intestine and liberate their enterotoxin, which in turn causes outpouring of fluid into the intestinal lumen. The affected person may be mildly inconvenienced by diarrhea and abdominal cramps or completely incapacitated for one to several days. Traveler's diarrhea is most frequent in places where the water supply is contaminated by sewage. Prevention consists of avoidance of nonsterile fluids and uncooked foods such as salads. Specific diagnosis is difficult.

Staphylococcal food poisoning and salmonella food infection are both caused by contamination of food. When several people who have eaten together become ill with vomiting and diarrhea, food poisoning should be suspected. If the time from exposure to illness is short (1 to 4 hours), it is likely that toxins produced by certain strains of staphylococci have caused the illness. The bacteria can contaminate unrefrigerated food and liberate their enterotoxins. The enterotoxins, when ingested with the food, injure the mucosal lining of the stomach and small intestine, producing an illness that usually lasts less than 1 day. Salmonella food infection occurs when large numbers of salmonella are ingested and reach the distal alimentary tract, where these bacteria are capable of invading the mucosa and causing illness. Because large numbers of organisms are required to induce the infection, growth of the organisms in food is usually required. The illness begins 1 to 2 days after ingesting contaminated food. The longer time from exposure to illness, as compared to staphylococcus food poisoning, is the result of the time needed for the bacteria to traverse the intestinal tract and initiate an infection in the mucosal lining. Outbreaks of food poisoning are investigated by public health authorities, who search for food handlers (carriers) who harbor the causative organism, thus contaminating food. Staphylococcal food poisoning can be prevented by keeping food refrigerated. Salmonella food infection can be prevented by keeping food refrigerated or by cooking contaminated food. Staphylococcal food poisoning is diagnosed by culturing the food; salmonella gastroenteritis is diagnosed by stool culture.

Shigellosis and amebiasis are infections of the colon caused by high virulent bacteria of the genus *Shigella* and the protozoan *Entameba histolytica,* respectively. The inflammation of the mucosal lining is often severe, leading to ulceration and dysentery (bloody diarrhea). Shigellosis has an acute onset and lasts for several days, whereas amebiasis has a gradual onset and may be chronic. These are classic diseases of overcrowding and poor sanitation such as occurs in mental institutions and war situations. The occurrence of dysentery calls for a search for shigella by culturing feces and for ameba by direct smear examination of feces. Of the infectious diarrheas discussed, shigellosis and amebiasis are the only ones that clearly require antimicrobial therapy to prevent complications and shorten the illness.

Campylobacter enteritis, due to bacteria of the genus *Campylobacter,* has been found to be among the more common causes of colitis with diarrhea or dysentery.

CHART 14-4. INFECTIOUS DIARRHEAS AND FOOD POISONING

	causes	lesions	manifestations
Viral enteritis	Viruses	Superficial damage to small-intestinal mucosa	Acute onset Diarrhea and vomiting Dehydration if severe Often epidemic 1- to 2-day duration
Traveler's diarrhea	Toxigenic *Escherichia coli* Many other possible agents	Often none	Acute onset Diarrhea Abdominal cramps Generalized malaise 1- to 2-day duration
Staphylococcal food poisoning	Toxin produced by staphylococci growing in food	Superficial damage to gastric and upper small-intestinal mucosa	Vomiting and diarrhea Acute onset 1 to 4 hours after eating food Usually several people involved 1-day duration
Salmonella food infection	Eating food contaminated by large numbers of salmonella	Mild inflammation of colonic mucosa	Acute onset 1 to 2 days after eating contaminated food Diarrhea and vomiting Malaise Duration of several days Resolves without treatment
Shigellosis	Bacterial infection by shigella	Severe inflammation of colonic mucosa with ulcers	Acute onset Bloody diarrhea Abdominal cramps Malaise Unsanitary environment Positive stool culture Lasts several days and occasionally fatal if untreated
Amebiasis	Protozoal infection by *Entameba histolytica*	Discrete colonic ulcers	Gradual onset Bloody diarrhea Unsanitary environment or recent travel to endemic areas Ameba seen on fecal smear Gradual recovery accelerated by treatment

Reflux Esophagitis

Reflux esophagitis is inflammation of the mucosa and submucosa of the lower esophagus due to reflux of acid gastric contents into the lower esophagus. It is caused by incompetence of the sphincter at the lower end of the esophagus. In many, but not all, cases there is an associated esophageal hiatal hernia. Hiatal hernia is a sliding (herniation) of the stomach through the hole (hiatus) in the diaphragm normally occupied by the esophagus (Fig. 14–6). Esophageal hiatal hernia is a common finding that increases in frequency with age. Most hiatal hernias are asymptomatic and not associated with an incompetent lower esophageal sphincter.

Reflux esophagitis may be a mild inflammation or may progress to ulceration. The ulcerative area of the esophagus may undergo metaplasia to a gastric-type or intestinal-type epithelium (*Barrett's esophagus*), and after many years adenocarcinomas may arise in the metaplastic epithelium. If persistent, ulceration leads to scarring, and contraction of the scar tissue causes stenosis (stricture) of the esophagus in some patients. The most common symptom of reflux esophagitis is heartburn, a burning sensation in the chest beneath the sternum. Bleeding may also occur from the ulceration. Stricture causes dysphagia. In severe cases, an operation to repair the incompetent sphincter may be carried out.

CHART 14–5. REFLUX ESOPHAGITIS

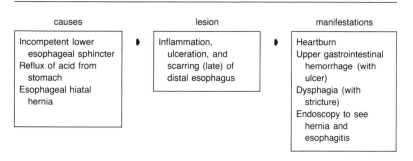

causes	lesion	manifestations
Incompetent lower esophageal sphincter Reflux of acid from stomach Esophageal hiatal hernia	Inflammation, ulceration, and scarring (late) of distal esophagus	Heartburn Upper gastrointestinal hemorrhage (with ulcer) Dysphagia (with stricture) Endoscopy to see hernia and esophagitis

Gastritis

Acute gastritis may be caused by injurious agents, such as aspirin, alcohol, and staphylococcal enterotoxin, that produce acute superficial injury to the mucosa that rapidly heals in a day or so and requires no specific treatment.

There are two major forms of chronic gastritis: fundal gastritis and helicobacter gastritis. Fundal gastritis involves the proximal stomach where the acid and pepsinogen-secreting cells are located.

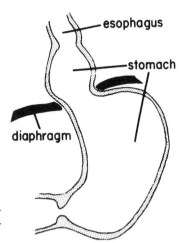

Figure 14–6. Esophageal hiatal hernia. Proximal stomach is herniated through the diaphragm into the chest.

Over many years, these secretory cells are destroyed and replaced by metaplastic gastric antral or intestinal epithelium. In man, extensive loss of acid-secreting cells leads to atrophic gastritis and achlorhydria (lack of hydrochloric acid) and to associated deficiency of intrinsic factor secretion, which is the cause of pernicious anemia. The cause of fundal gastritis is believed to be autoimmume. See Chapter 10 for more information on pernicious anemia.

Helicobacter gastritis is so prevalent that for most people it can almost be considered a normal condition associated with a commensal organism. _Helicobacter pylori_ is a small curved bacterium that lives in the mucous layer of the gastric epithelial surface and incites an active low-grade acute and chronic inflammatory reaction most prominent in the distal (antral) part of the stomach. The condition becomes more frequent with age and affects about one-fifth of adults in the United States and probably more in many areas of the world. Gastritis is usually asymptomatic, and so-called symptoms of gastritis (epigastric discomfort, nausea, bloating) do not correlate well with the presence of gastritis histologically. Helicobacter gastritis may play some role in the development of gastric and duodenal ulcers and gastric carcinoma. Elimination of helicobacter may tip the balance in favor of ulcer healing; however, the organism should not be considered the most important factor in causing the ulcer. The cancer relationship is mild and likely related to the regenerative metaplastic changes resulting from prolonged chronic gastritis. Helicobacter gastritis is usually diagnosed by seeing the organisms in biopsy specimens, but the organism can be cultured and urease production in gastric juice is presumptive evidence that the organism is present. Helicobacter can be reduced or eliminated with bismuth subsalicylate and amoxicillin therapy, but

treatment of all persons with this condition is imprac▮
necessary.

Peptic Ulcer

Peptic ulcers are chronic ulcers due to injury produceu ьy gastric
secretions containing hydrochloric acid and proteolytic enzymes.
They occur most commonly in the first part of the duodenum,
where the stomach empties its acid contents onto the more susceptible intestinal mucosa. Less frequently, ulcers occur in the stomach
(Fig. 14–7). In men, duodenal ulcers are approximately seven times
more common than gastric ulcers. In women, the ratio of duodenal
to gastric ulcer is two to one. A peptic ulcer consists of a sharply
punched out area of tissue necrosis that is covered by inflammatory
exudate with underlying granulation tissue and fibrosis.

Many peptic ulcers cause pain, but occasionally they are
asymptomatic until one of the more serious complications such as
massive bleeding, perforation, or obstruction occurs. Usually there
is not a direct explanation of why the ulcer developed, but sometimes severe stress or heavy intake of drugs, such as aspirin or
corticosteroids, are implicated as contributory factors. Although
acid is necessary to produce peptic ulcers, this relationship cannot
be quantitated in a particular individual. Although most duodenal
ulcers heal with medical therapy, recurrences are common. Some
patients require surgical therapy. Gastric ulcers are more likely to
recur than duodenal ulcers and may require surgical treatment.
Bed rest and abstinence from smoking are well-documented factors
that promote healing. Antacids to neutralize gastric acid and H_2-
histamine receptor antagonists to reduce histamine stimulation of
gastric acid production aid in ulcer healing. Elimination of helicobacter may be helpful (see gastritis above).

Some patients have recurrent or chronic duodenal ulcers,

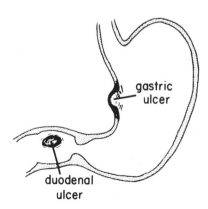

Figure 14–7. Peptic ulcer, typical location of duodenal and gastric ulcers.

which may lead to one of the four complications requiring an operation—severe hemorrhage, perforation, obstruction, or unrelieved pain.

Operations on the stomach are designed to reduce the effects of gastric acid by removing the gastric antrum, which secretes the acid-stimulating hormone gastrin, and by partial removal of the gastric body to remove a portion of the acid-secreting glands. In addition, vagotomy (cutting the vagus nerves) helps reduce the nervous stimulation to acid production.

CHART 14–6. PEPTIC ULCER

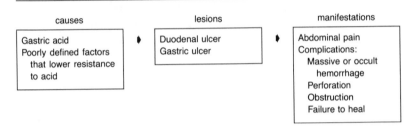

causes	lesions	manifestations

Gastric acid
Poorly defined factors
  that lower resistance
  to acid
	Duodenal ulcer
	Gastric ulcer
		Abdominal pain
		Complications:
		  Massive or occult
		    hemorrhage
		  Perforation
		  Obstruction
		  Failure to heal

Malabsorption Syndrome

Failure to digest and/or absorb food, leading to weight loss and steatorrhea (fat in the stool) is called the *malabsorption syndrome.* There are many diseases that cause the malabsorption syndrome. Specialized knowledge and procedures are required for their diagnosis. Severe malabsorption is rare but should be suspected when there is unexplained weight loss and the stools are large and bulky and very foul smelling (owing to bacterial action on unabsorbed fats). Celiac disease, one of the most common causes of severe malabsorption, is a sporadic, apparently genetic disease with onset at any age. In celiac disease, gluten, a protein found in grains (especially wheat), causes a mild prolonged injury to the small intestinal mucosa. This injury causes degeneration of surface epithelial cells, resulting in loss of villi and inflammatory cell infiltration of the lamina propria (Fig. 14–8). A gluten-free diet effects a remission of the disease. As a gluten-free diet is expensive and because there are many other causes of malabsorption and weight loss, a diagnosis of celiac disease should be firmly established by a specialist before treatment is started. Chronic pancreatitis causes malabsorption when enough pancreatic tissue is destroyed to deplete the production of digestive enzymes.

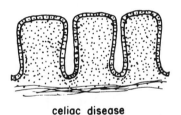

celiac disease

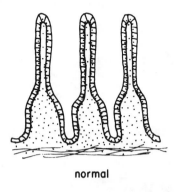

normal

Figure 14–8. Small intestinal mucosa with celiac disease compared with normal small intestine.

CHART 14–7. MALABSORPTION SYNDROME

causes		lesions		manifestations
Gluten sensitivity (celiac disease) Pancreatic insufficiency Altered continuity of gastrointestinal tract Many other causes	▶	Mucosal degeneration with atrophy and inflammation (celiac disease) Inflammation of the pancreas with atrophy Surgical alteration or removal of parts of alimentary tract	▶	Large, bulky, foul-smelling stools Weight loss Abnormalities of tests for fat and carbohydrate absorption May be abnormal intestinal biopsy (celiac disease)

Acute Appendicitis

The vermiform appendix is a vestigial outpouching of the proximal colon that has no known function but may become acutely inflamed at any time for no obvious reason. The bacteria causing the inflammation are those normally present in the colon. Sometimes a calcified mass of feces (fecalith) blocks the lumen of the appendix, thus predisposing to infection (Fig. 14–9). It has been learned through experience that surgical removal of the acutely inflamed

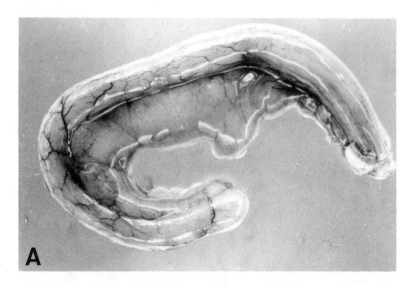

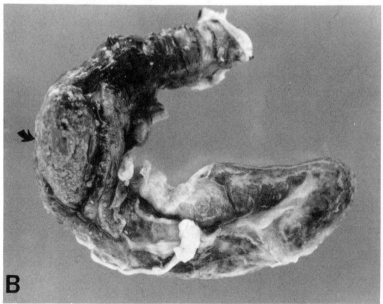

Figure 14–9. A. Normal appendix. **B.** Swollen, inflamed appendix with exudate on surface and fecalith (*arrow*) in lumen.

appendix is less risky than the complications that might develop from leaving it in. The complications include abscess formation, perforation leading to peritonitis, and spread of the infection to the liver. Appendicitis begins with pain, usually starting in the umbilical region and moving to the right lower quadrant of the abdomen.

The inflammatory irritation of the peritoneum causes tenderness to palpation, and the abdominal muscles become rigid. Appendicitis is the most common indication for laparotomy (surgically opening the abdominal cavity). However, diseases of adjacent organs, such as the right ovary and fallopian tube, distal ileum, and mesenteric lymph nodes, can closely mimic appendicitis. Appendicitis is prevented in those individuals who have their appendix removed during abdominal operations carried out for other reasons. This is called incidental appendectomy.

CHART 14–8. ACUTE APPENDICITIS

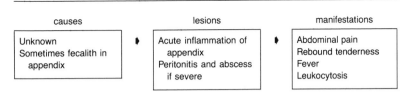

causes	lesions	manifestations
Unknown Sometimes fecalith in appendix	Acute inflammation of appendix Peritonitis and abscess if severe	Abdominal pain Rebound tenderness Fever Leukocytosis

Pseudomembranous Enterocolitis

The false membrane of pseudomembranous enterocolitis is caused by superficial mucosal injury with necrotic cells, mucus, fibrin, and acute inflammatory cells forming a sticky exudate on the mucosal surface of colon and/or small intestine. Although there are several causes that will induce this type of injury, the most clinically important is the production of a necrotizing toxin by *Clostridium difficile*. The typical situation for this disease occurs in patients on prolonged antibiotic therapy. The therapy suppresses the normal intestinal flora and allows the toxigenic *Clostridium difficile*, which is normally present in small numbers, to proliferate and produce its toxin. The rapidly developing diarrheal disease may be fatal if not suspected and treated with change in antibiotics and fluid replacement.

Inflammatory Bowel Disease

Inflammatory bowel disease refers collectively to two chronic inflammatory diseases of the distal intestinal tract called *ulcerative colitis* and *Crohn's disease*. Both diseases are of unknown cause and produce weight loss and variable degrees of diarrhea, hemorrhage, and abdominal pain. Owing to the similarity of presentation, the specific diagnosis may not be immediately obvious, hence the need for the more general term inflammatory bowel disease. Pathologically, ulcerative colitis is a diffusely distributed mucosal disease of the colon (Fig. 14–10A). Crohn's disease is a patchy disease involv-

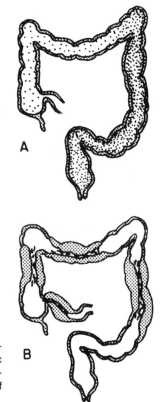

Figure 14–10. Inflammatory bowel disease. **A.** Ulcerative colitis with diffuse involvement of colonic mucosa. **B.** Crohn's disease with marked thickening of the wall of the ileum and patchy thickening of the wall of the colon.

ing all layers of the bowel wall; it occurs most commonly in the distal ileum but also often involves the colon in a patchy manner (Fig. 14–10B). About half of the cases have a granulomatous component to the chronic inflammatory reaction. Both diseases occur most commonly in young adults and disrupt their lives to a variable degree as the severity waxes and wanes. In ulcerative colitis, the risk of adenocarcinoma developing in the chronically inflamed and often dysplastic mucosa rises rapidly after 10 years. Total removal of the colon cures the disease and prevents cancer and the resulting permanent ileostomy is much preferable to the continuing risk of cancer. Complications of Crohn's disease resulting from the full wall inflammation and fibrosis include obstruction of the bowel and development of fistula tracts through the bowel wall and into adjacent tissues or organs. Perianal fistulas are particularly common. Resection of the diseased segment often relieves the complications, but recurrence may occur in previously normal segments. Drugs (azulfidine and corticosteroids) are used as anti-inflammatory agents with variable degrees of success. *Regional en-*

teritis and *granulomatous colitis* are synonyms for Crohn's disease of the small intestine and colon, respectively.

CHART 14–9. INFLAMMATORY BOWEL DISEASE

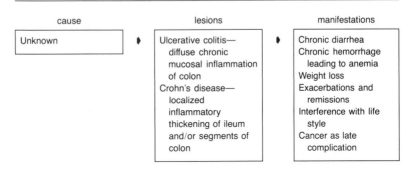

cause	lesions	manifestations
Unknown	Ulcerative colitis— diffuse chronic mucosal inflammation of colon Crohn's disease— localized inflammatory thickening of ileum and/or segments of colon	Chronic diarrhea Chronic hemorrhage leading to anemia Weight loss Exacerbations and remissions Interference with life style Cancer as late complication

Colonic Diverticulosis

Multiple outpouchings (diverticula) of the colon develop frequently with advancing age at points of weakness in the muscular wall of the colon, especially the sigmoid colon. They are readily diagnosed by barium enema (Fig. 14–11). Diverticulosis is usually

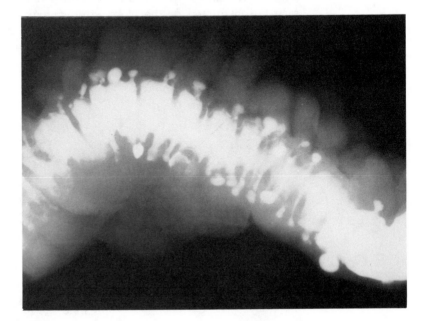

Figure 14–11. Colonic diverticulosis. X-ray of barium enema study shows barium-filled lumen and diverticula.

an asymptomatic anatomic alteration but occasionally a diverticulum becomes inflamed producing diverticulitis. The inflammation may subside or persist leading to operative removal of the diseased segment. More rarely the artery overlying a diverticulum is eroded producing massive rectal bleeding.

CHART 14–10. COLONIC DIVERTICULOSIS

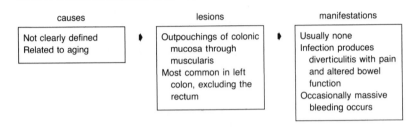

causes	lesions	manifestations
Not clearly defined Related to aging	Outpouchings of colonic mucosa through muscularis Most common in left colon, excluding the rectum	Usually none Infection produces diverticulitis with pain and altered bowel function Occasionally massive bleeding occurs

Inguinal Hernia

Inguinal hernia, commonly referred to as a *rupture,* is an outpouching of the outer peritoneal lining of the abdominal cavity into the groin (inguinal region). Loops of bowel can enter the pouch (Fig. 14–12), and if they become caught or twisted, the intestine may be obstructed or the blood supply may be cut off causing infarction of the intestine. Because obstruction and/or infarction of the intestine is life threatening, it is better to surgically repair the hernia before, rather than after, the bowel gets caught. Inguinal

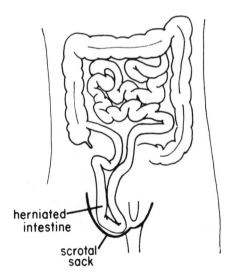

Figure 14–12. Inguinal hernia with entrapped loop of bowel.

herniated intestine

scrotal sack

hernias are much more common in men due to a congenital defect resulting from the peritoneum being carried into the inguinal region as the testes descend from the abdomen into the scrotum. Some inguinal hernias in older people are due to a weakness in the wall of the abdomen without a congenital hernia sac. As a person ages and as a result of pressure, the hernia sac may enlarge and allow entrapment of bowel. In spite of the fact that symptoms usually develop in the aged, repair of congenital hernia sacs in males is best carried out in infancy or childhood.

CHART 14–11. INGUINAL HERNIA

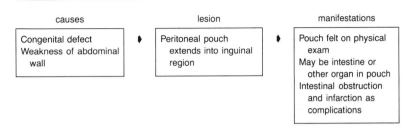

causes	lesion	manifestations
Congenital defect Weakness of abdominal wall	Peritoneal pouch extends into inguinal region	Pouch felt on physical exam May be intestine or other organ in pouch Intestinal obstruction and infarction as complications

Hyperplastic/Neoplastic Diseases

Alimentary tract cancers account for 20 percent of all cancers (skin cancers excluded). Colon cancer is by far the most common and also the most curable type. In addition, there are many benign tumors of the intestines, the most common occurring as polyps in the colon. There are many other types of neoplastic and non-neoplastic tumors of the alimentary tract. Most present with bleeding or obstruction and are diagnosed by the pathologist after removal of the tumor by the surgeon. Only carcinomas of the colon and stomach and polyps of the colon will be discussed here.

Carcinoma of the Stomach

Carcinoma of the stomach is less than half as common as carcinoma of the colon in the United States, although in some parts of the world it is the most common cancer. Dietary factors are suspected as the cause, but no specific agent has been proven. They arise from the gastric mucosa, spread into the gastric wall, and metastasize to regional lymph nodes, liver, and other distant sites (Fig. 14–13). Symptoms, such as loss of appetite, weight loss, pain, anemia, or abdominal mass, occur late, so that surgical cure is effected in only 15 percent of patients.

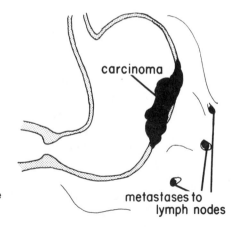

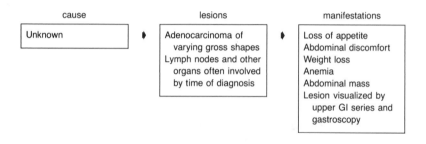

Figure 14–13. Carcinoma of the stomach.

CHART 14–12. CARCINOMA OF STOMACH

cause		lesions		manifestations
Unknown	▶	Adenocarcinoma of varying gross shapes Lymph nodes and other organs often involved by time of diagnosis	▶	Loss of appetite Abdominal discomfort Weight loss Anemia Abdominal mass Lesion visualized by upper GI series and gastroscopy

Colonic Polyps

A polyp is any protrusion from a mucosal surface and could represent an inflammatory lesion, a benign neoplasm, or a malignant neoplasm. Several types of polyps occur in the colon, the most common being hyperplastic polyps and two types of adenomas.

Hyperplastic polyps are small, slightly raised exaggerations of normal mucosal crypts. They are more common than adenomas and are totally innocuous. They are sometimes biopsied because they cannot always be distinguished from small adenomas.

The two types of adenomas (tubular and villous) are distinguished on the basis of their size, growth pattern, and malignant potential. *Tubular adenomas* (also called adenomatous polyps) (Fig. 14–14A) are small (usually under 2 centimeters), pedunculated (large head with narrow stalk), composed predominately of glands, and uncommonly contain cancer at the time of diagnosis.

Villous adenomas (Fig. 14–14B) are large (usually over 2 centimeters), grow as a raised broad-based mass, composed predominantly of villous type epithelium, and frequently (about 20% contain cancer at the time of diagnosis. Adenomatous polyps are

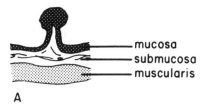

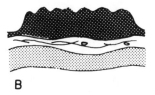

A

B

Figure 14–14. Two types of colonic polyps. **A.** A pedunculated adenomatous polyp. **B.** A sessile villous adenoma.

removed by excisional biopsy because they cannot be accurately separated from other types of polyps without histologic examination and because they are believed to be precancerous. If feasible, villous adenomas are removed by surgical resection, because they may already contain cancer and because they often recur if not completely removed.

CHART 14–13. COLONIC POLYPS

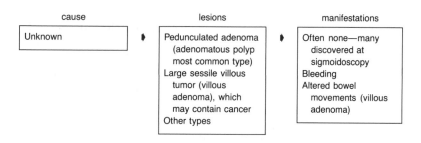

cause	lesions	manifestations
Unknown	Pedunculated adenoma (adenomatous polyp most common type) Large sessile villous tumor (villous adenoma), which may contain cancer Other types	Often none—many discovered at sigmoidoscopy Bleeding Altered bowel movements (villous adenoma)

Carcinoma of the Colon

Adenocarcinoma of the colon is the most common internal cancer. It usually develops spontaneously without evident cause, but it also occurs as a complication of familial adenomatous polyposis, ulcerative colitis, and villous adenoma. Early diagnosis is important, because the disease is potentially curable by surgical resection. The cancer arises in the mucosa and gradually increases in size as it grows through the colonic wall to the serosa and expands size as it grows through the colonic wall to the serosa and expands into the lumen (Fig. 14–15). At any time after it has invaded beyond the mucosa, it may spread through lymphatic channels to the

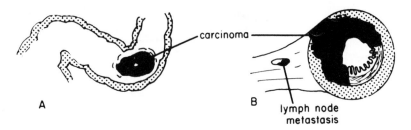

Figure 14–15. Carcinoma of the colon. **A.** Mucosal surface view. **B.** Cross-section with invasion through the muscle layer and metastasis to lymph nodes.

lymph nodes in the pericolonic fat. Cancer cells may gain access to the bloodstream by further spreading up the lymphatic system to the thoracic duct or by directly invading veins. Blood-borne metastases are most frequently found in the liver; most patients with advanced colonic cancer have a large nodular liver filled with metastatic cancer. If surgical removal is carried out before the lymph nodes are involved, the cure rate is high. Once the liver is involved, there is no chance for cure.

The main manifestations of colonic carcinomas are bleeding and altered bowel habits. Almost all colon cancers bleed due to ulceration of their surface. This may be detected by the patient noting gross blood in the stool, by the development of iron-deficiency anemia, or by screening the stool for occult blood. Because bleeding occurs from the surface, the cancer may still be in an early stage when bleeding is detected. Altered bowel habits resulting from colonic carcinoma may take the form of diarrhea, constipation, or narrow pencil-like stools. Development of colonic obstruction or pain implies more extensive tumor growth. Diagnosis is made in one of two ways—endoscopy (sigmoidoscopy or colonoscopy) or barium enema (Fig. 14–16).

CHART 14–14. CARCINOMA OF COLON

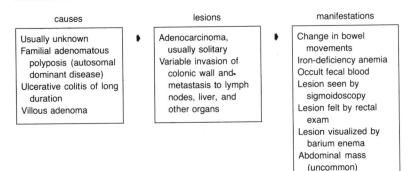

causes		lesions		manifestations

Usually unknown
Familial adenomatous
  polyposis (autosomal
  dominant disease)
Ulcerative colitis of long
  duration
Villous adenoma
	▸	Adenocarcinoma,
		  usually solitary
		Variable invasion of
		  colonic wall and
		  metastasis to lymph
		  nodes, liver, and
		  other organs
			▸	Change in bowel
				  movements
				Iron-deficiency anemia
				Occult fecal blood
				Lesion seen by
				  sigmoidoscopy
				Lesion felt by rectal
				  exam
				Lesion visualized by
				  barium enema
				Abdominal mass
				  (uncommon)

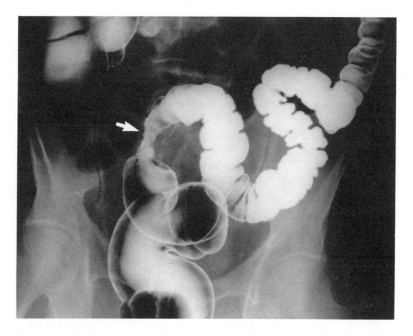

Figure 14–16. Barium enema illustrating a constricting colon cancer (*arrow*) with proximal dilated, air-filled colon due to obstruction of the lumen by the cancer.

ORGAN FAILURE

Acute failure of the absorption process can be tolerated for a number of days. Severe vomiting or diarrhea, however, may lead to fatal loss of fluids. Infants are most susceptible to fluid loss and much more frequently require intravenous fluid replacement for vomiting and diarrhea than do adults. The most striking acute organ failure occurs when the small intestine is massively damaged, such as with occlusion of the mesenteric artery by a thrombus, leading to infarction and death. The alimentary tract functions amazingly well in the face of chronic diseases unless obstruction develops. Ulcerative lesions of the intestinal tract are quite resistant to infection unless there is perforation into the peritoneal cavity.

REVIEW QUESTIONS

1. What is the functional bowel syndrome and why may it be difficult to distinguish from organic disease?
2. What are the common alimentary tract diseases that cause mild or short-lived disability and few deaths?

3. Which alimentary tract diseases cause serious prolonged illnesses or are frequently fatal?
4. What are the manifestations of bleeding, altered motility, and perforation of the alimentary tract?
5. How are radiologic and endoscopic procedures used to diagnose alimentary tract diseases?
6. What are the causes, lesions, and manifestations of each of the specific diseases discussed in the chapter?
7. How are most developmental abnormalities of the alimentary tract treated and what is the likely outcome?
8. How do the infectious diarrheas and food poisonings differ in terms of causative agents, likely situation for their occurrence, and timing of onset and recovery?
9. What conditions may mimic appendicitis?
10. What is the relationship between reflux esophagitis and esophageal hiatal hernia?
11. Why is carcinoma of the colon a more important medical problem than carcinoma of the stomach?
12. What do the following terms mean?

    Digestive or gastrointestinal system
    Alimentary tract
    Diarrhea
    Constipation
    Hematemesis
    Hematochezia
    Melena
    Ileus
    Adynamic ileus
    Upper GI series
    Barium enema
    Sigmoidoscopy
    Upper gastrointestinal endoscopy
    Gastric analysis
    Hernia
    Adenomatous polyp
    Villous adenoma

Liver, Gallbladder, and Pancreas

REVIEW OF STRUCTURE AND FUNCTION

The liver and the pancreas are glandular organs with excretory
ducts emptying into the second portion of the duodenum, usually
at a common site called the *papilla of Vater* (Fig. 15–1). The excretory
ducts of the liver are called *bile ducts*. The gallbladder is a storage
reservoir connected to the bile ducts by the cystic duct.

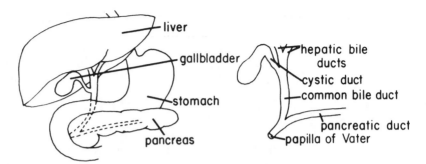

Figure 15–1. Major components of the liver, biliary tract, and pancreas.

The liver is the largest glandular organ in the body. Most of the blood from the abdominal organs is carried to the liver via the portal veins so that it can be filtered past the glandular cells of the liver (hepatocytes) before being returned to the heart via the hepatic vein. Because portal blood has little oxygen left after passing through the abdominal organs, the liver has a second source of blood, the hepatic artery, to provide oxygenated blood. The bulk of the liver is composed of hepatocytes, large epithelial cells capable of carrying out many metabolic functions. The hepatocytes are aligned in cords or plates with sinusoids between the plates to percolate the blood from the portal areas to the central vein (Fig. 15–2). Between the cell membranes of adjacent hepatocytes are tiny canaliculi that carry bile produced by the hepatocytes to the portal area, where they empty into epithelial-lined bile ducts. Thus, blood flows into the liver through the hepatic artery and portal vein and enters the sinusoids from the widely dispersed portal areas. In the sinusoids, waste products and nutrients are

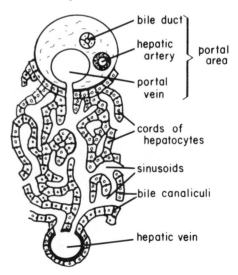

Figure 15–2. Basic histologic structure of the liver.

removed and metabolized by the hepatocytes. The metabolites may be returned to the blood, stored in the hepatocytes, or excreted into bile canaliculi to form bile. The bile is carried through the canaliculi to the portal areas, where it enters the bile ducts and later is emptied into the intestine after storage in the gallbladder.

The liver also contains many mononuclear phagocytic cells, which are present in the form of fixed macrophages called *Kupffer cells*. Kupffer cells line the sinusoids and are normally inconspicuous. They phagocytize particulate material from the blood.

Of the many metabolic functions of the liver, five will be briefly reviewed: (1) production of bile salts, (2) excretion of bilirubin, (3) metabolism of nitrogenous substances, (4) production of serum proteins, and (5) detoxification of drugs and poisons.

The liver's main contribution to the gastrointestinal tract function is bile salts. Bile salts are bipolar molecules derived from cholesterol that aggregate into spherical masses (micelles). Lipids are soluble in the inside of the micelle, whereas the outer surface is water soluble. Bile salts render cholesterol soluble in bile to prevent the formation of cholesterol crystals, and in the intestine, they dissolve other lipid molecules so that they can be absorbed.

The most notable of the liver's excretion products is bilirubin, a hemoglobin breakdown product, which is produced from worn out red blood cells. Approximately $1/120$ of a person's red blood cells die every day. The liver cannot remove all of the bilirubin from the blood immediately, so in the normal state of equilibrium between bilirubin production and its removal by the liver there is about 1 mg/dl of bilirubin in the blood. After the liver cells remove bilirubin from the blood, they conjugate it with glucuronide molecules to make it water soluble and excrete it into the bile. Many other substances are handled in a manner similar to that for bilirubin, but bilirubin is more important because of the large amount excreted and because it is yellow. When bilirubin accumulates in the blood, the patient's skin and eyes appear yellow (*jaundiced*), indicating illness. This may result from too many red cells dying, from failure of liver cells to remove and excrete bilirubin, or from blockage of bile ducts.

Another important metabolic product of hepatocytes is urea nitrogen. This is handled differently from bilirubin. The breakdown of dead cells produces nitrogenous products such as ammonia. Also, the breakdown of protein by bacteria in the intestines produces ammonia, which can be absorbed into the blood by the portal vein. The liver converts these nitrogenous products to urea nitrogen and returns the urea nitrogen to the bloodstream, because it cannot excrete urea nitrogen effectively in the bile. It remains for the kidney to excrete the urea nitrogen. From these considerations, it is easy to see that severe liver failure results in accumulation of ammonia in the blood and severe kidney failure results in accumulation of urea nitrogen in the blood.

Another notable chemical product of the liver is the production of serum proteins, particularly albumin and the specific globulins necessary for blood coagulation. Prothrombin is one of the specific globulins that is important for blood coagulation. The liver requires vitamin K to produce prothrombin; consequently, low prothrombin levels in the blood may be due to severe liver disease or vitamin K deficiency.

Many drugs and poisons are metabolized in the liver by way of conjugation with other compounds to increase their solubility or aid in their excretion. Thus, severe liver disease is likely to delay the excretion of drugs and enhance the effects of poisons.

The excretory ducts of the liver (bile ducts) form a long tree-like network with the trunk (common bile duct) emptying into the duodenum at the same point as the pancreatic duct. The gallbladder is an outpouching of the common bile duct that acts as a storage reservoir for bile. The gallbladder empties its contents into the duodenum at a time (after meals) when bile salts are needed for fat absorption. This reservoir function is not essential, as the gallbladder can be removed without loss of digestive function.

The pancreas is a long, narrow glandular organ lying horizontally in the midabdomen behind the peritoneum. Its tail stretches toward the spleen on the left, and its head nestles behind the proximal duodenum and distal stomach. The pancreatic duct runs the length of the pancreas and empties into the duodenum after joining the bile duct. The bulk of the pancreas is made up of glands (acini) that secrete digestive enzymes into the pancreatic duct. When activated by intestinal juices, these enzymes digest carbohydrate (amylase), fat (lipase), and protein (trypsins). Pancreatic enzymes are essential to life. When the pancreas is destroyed, it is not possible to give enough enzymes orally to restore normal function, so a state of chronic malnutrition ensues.

Scattered among the pancreatic glands are clusters of endocrine cells known as the *islets of Langerhans,* which produce insulin and other hormones. Removal of the entire pancreas or severe destruction will produce diabetes mellitus due to lack of insulin. Insulin therapy is then vital to life.

MOST FREQUENT AND SERIOUS PROBLEMS

The most common problem occurs in the gallbladder, namely, the formation of gallstones. Gallstones are usually asymptomatic but have the potential of producing serious complications such as obstruction of bile flow due to migration of stones into the common bile duct.

Viral hepatitis is probably the most common liver disease, particularly in younger people. It usually resolves completely but occasionally persists for many months or more rarely develops into a serious chronic inflammation leading to cirrhosis. *Cirrhosis* refers to fibrosis and nodular regeneration of the liver and is the characteristic lesion of serious chronic liver disease regardless of cause. Chronic alcoholism is the most common cause of cirrhosis, but only a fraction of chronic alcoholics develop cirrhosis. The most common neoplastic condition of the liver is metastatic carcinoma.

Inflammation of the pancreas is uncommon and usually mild; however, there are serious acute, often fatal forms and chronic forms that lead to destruction of the pancreas. The more severe acute and chronic forms are usually associated with alcoholism, whereas mild forms are often associated with gallstones. Carcinoma of the pancreas ranks sixth in frequency among carcinomas and is one of the most lethal of all cancers. Although insulin is produced in the pancreas, the relative lack of insulin (diabetes mellitus) is not usually related to any obvious pancreatic disease. Therefore, diabetes is treated as a metabolic disorder rather than a pancreatic disorder.

SYMPTOMS, SIGNS, AND TESTS

Jaundice is an obvious and often serious symptom or sign of liver disease. It may be caused by increased hemoglobin breakdown, liver disease, or bile duct obstruction. Nonspecific digestive disturbances may be associated with acute hepatitis and gallstones. Pain is a feature of acute cholecystitis (inflammation of the gallbladder), gallstones in the common bile duct, pancreatitis, and late stages of carcinoma of the pancreas. Enlargement of the liver (hepatomegaly) is particularly prominent in alcoholic fatty liver and metastatic cancer to the liver.

Laboratory tests are often used to diagnose the type of liver disease. They are somewhat inappropriately referred to as *liver function tests* because they are not usually specific for liver function. Serum tests related to liver disease usually include a routine screening battery of bilirubin, total protein, albumin, aspartate aminotransferase (AST), and alkaline phosphatase. The causes of elevated bilirubin are the same as the causes of jaundice. The serum bilirubin is at least twice normal when jaundice becomes apparent. Low levels of serum protein, particularly albumin, occur with severe chronic liver disease such as cirrhosis. Aspartate aminotransferase is elevated with liver necrosis (as in hepatitis), and alkaline phosphatase is usually elevated with bile duct obstruction (often before the bilirubin is elevated). Prothrombin time is a blood coagulation test that reflects serum levels of prothrombin. Prothrombin deficiency may be caused by very severe liver disease or vitamin K deficiency.

Biopsy of the liver can be performed with a needle inserted through the skin or at the time of surgical opening of the abdomen (laparotomy). Biopsy is the most reliable way of establishing the nature of the more serious liver diseases such as cirrhosis, chronic hepatitis, granulomas, and cancer.

Tests for gallbladder disease are designed to visualize gallstones. Twenty percent of gallstones are calcified and can be seen on an x-ray of the abdomen. Ultrasound is the preferred method of demonstrating gallstones because it is noninvasive, demonstrates most stones, and gives an immediate answer. Cholecystography involves swallowing a radiopaque dye and waiting for it to be absorbed, excreted by the liver, and concentrated by the gallbladder. Stones will then appear as holes on roentgenograms of the radiopaque gallbladder. If a stone blocks the cystic duct, no bile or radiopaque dye can enter the gallbladder; thus, the gallbladder is said to be nonfunctional.

Tests for pancreatic disease include measurement of serum amylase and lipase as indicators of acute or active pancreatic injury and tests for intestinal malabsorption as indicators of chronic damage with inadequate production of digestive enzymes.

SPECIFIC DISEASES

Genetic/Developmental Diseases

Neonatal Liver Disease
Jaundice at or shortly after birth may be due to increased breakdown of red blood cells, parenchymal liver disease, or atresia of bile ducts. Normally there is an increased breakdown of red blood cells at birth producing an elevated serum bilirubin. This is accentuated by prematurity, because the liver is less able to remove and excrete bilirubin, and by erythroblastosis fetalis, because of the presence of maternal antibodies to the neonate's red blood cells. Neonatal infections and rare genetic metabolic defects may produce parenchymal injury at this time. Absence of bile ducts (biliary atresia) usually leads to progressive damage to the liver and death within a year. Occasionally extrahepatic atresia can be repaired surgically. Recent evidence suggests that biliary atresia is due to fetal inflammatory disorders rather than an embryonic anomaly.

Cystic Fibrosis of the Pancreas

This recessively inherited autosomal disorder will occur in about one-fourth of siblings when both parents are carrying the gene that causes cystic fibrosis. The pancreatic ducts are filled with thick mucoid material, leading to cystic dilatation of the ducts and fibrosis of the parenchyma. This leads to malabsorption and weight loss of variable severity. As described under lung, bouts of pneumonia also occur, and the pancreatic or pulmonary deficiencies lead to death in childhood. Treatment is directed at preventing pneumonia and replacing pancreatic enzymes.

Inflammatory/Degenerative Diseases

The liver is involved to a variable extent by many systemic diseases producing nonspecific changes such as fatty change (fatty metamorphosis), chronic inflammation of portal areas, and enlargement of Kupffer cells. In this section, we will concentrate on the major primary diseases of the liver, gallbladder, and pancreas.

Viral Hepatitis

Several viruses may affect the liver as part of systemic viral infections and a few have their major effect on the liver. The diseases caused by the latter group are yellow fever, a very serious disease that occurs in the tropics, and viral hepatitis, a term applied to at least five viral infections that have a similar clinical illness and liver lesions. Viral antigens and/or antibodies have been identified for viral hepatitis A, B, C, and delta.

Patients with viral hepatitis feel ill with loss of appetite (anorexia) and distaste for cigarettes (if they smoke). Jaundice may be the first clue to the presence of liver disease, although not all patients become yellow. Physical examination reveals an enlarged tender liver and the urine appears dark. In the acute stage of the illness, serum AST is always strikingly elevated (over 1000 units versus a normal of less than 40), indicating liver cell necrosis. There is no specific treatment other than rest and a good diet. Unlike many other viral illnesses, the disease runs a prolonged or subacute course, with 85 percent of patients having recovered by 6 weeks. Although viral hepatitis causes considerable disability, it is the long-term complications and potential for spread of the disease that make it a major health problem.

Hepatitis A, caused by an RNA virus, is the most benign form. The virus is spread sporadically or in epidemics by fecal contamination of water and food. The incubation period (2 to 6 weeks) is the shortest of the three forms and the onset of symptoms is the most abrupt. Diagnosis is made by finding elevated IgM anti-hepatitis A antibodies in the serum a few days to 3 months after onset of illness. Elevated IgG anti-hepatitis A antibodies without

IgM antibodies indicates past illness and lifetime immunity to the disease. Hepatitis A is rarely fatal in the acute stage and does not lead to chronic hepatitis. The virus is shed in the feces for about 2 weeks before and 1 week after onset of illness; after that the patient is no longer a carrier of the disease.

In contrast, hepatitis B, caused by a DNA virus, is potentially a much more serious disease. The incubation period is long (1 to 6 months) and the onset of symptoms more gradual. Diagnosis is made by detecting serum antigens and antibodies, but is much more complex than for hepatitis A. The antigen in the surface coat of the virus is called hepatitis B surface antigen (HBsAg) and two antigens in the core of the virus are called core antigen and e antigen (HBcAg and HBeAg). The surface antigen is shed into the blood for several weeks before onset of illness and persists into the recovery period. Antibody to the surface antigen rises sometime after recovery and is a good indication of recovery and lifelong immunity. Presence of the e antigen parallels disease activity and its persistence suggests chronic disease. Anti-HBc and anti-HBe indicate current or recent infection and do not confer immunity. These markers have proved useful in sorting out the various clinical forms of hepatitis B.

The most severe form of hepatitis B is massive or fulminant hepatic necrosis, an uncommon but often fatal form that kills most of the liver's cells. Chronic forms of hepatitis B, which may or may not be associated with an acute stage, are divided into two types. Chronic persistent hepatitis is a mild illness with prolonged (more than 6 months) elevations of aspartate aminotransferase levels. It is not associated with permanent liver damage and eventually recovery occurs. Chronic active hepatitis is clinically similar but liver biopsy reveals beginning fibrosis and progression to cirrhosis is expected in many cases. Finally, the carrier state occurs in asymptomatic patients and in those with chronic hepatitis. The carrier state is diagnosed by elevated serum HBsAg and anti-HBc. The presence of anti-HBe suggests that the carrier has active disease.

Hepatitis B is a major health problem because about one of every 1000 persons is a carrier and the potential effects of the disease are serious. Hepatitis B virus may be spread by feces, urine, and body secretions, but more commonly it is spread by blood transfusion, contaminated breaks in the skin, or transplacentally. Drug addicts, homosexual men, institutionalized children, blood recipients, dentists, and selected other health care workers are particularly at risk. A vaccine prepared from surface antigen is available for persons at high risk for the disease.

Hepatitis C is caused by an RNA virus, has an incubation

period that overlaps hepatitis A and B (2 weeks to 6 months), and is spread in a fashion similar to hepatitis B. After immunologic markers were found for hepatitis B, hepatitis C became the cause of most cases of post-transfusion hepatitis. Now that C can be recognized in blood, the incidence of post-transfusion hepatitis will greatly diminish. Hepatitis C is even more likely than B to become chronic and can lead to cirrhosis.

Delta agent, a rare cause of hepatitis, requires the presence of hepatitis B (usually a chronic infection) in order to cause hepatitis. The frequency of fulminant or chronic hepatitis is greater with delta agent than with hepatitis B alone. Other forms of hepatitis may be recognized such as a rare epidemic form identified epidemiologically as hepatitis E.

CHART 15–1. VIRAL HEPATITIS

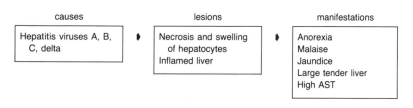

causes	lesions	manifestations
Hepatitis viruses A, B, C, delta	Necrosis and swelling of hepatocytes Inflamed liver	Anorexia Malaise Jaundice Large tender liver High AST

Chemical Injuries to the Liver

Many drugs and toxic chemicals have been associated with liver injury, but the degree of injury is usually mild, and the cause and effect relationship is difficult to prove. Some agents cause injury in selected individuals only (hypersensitivity), whereas other agents cause injury directly related to the dose of the injurious substance. Among the former type, the most important and serious is the reaction to a commonly used anesthetic agent, halothane. A very small percentage of patients anesthetized with halothane, particularly when used for a second operation, develop very severe massive necrosis of liver cells much like that occurring in acutely fatal cases of viral hepatitis. Isoniazide, a drug used for long-term treatment of tuberculosis, is also associated with occasional idiosyncratic cases of massive hepatic necrosis. Serious, although very uncommonly encountered, examples of direct toxic injury include ingestion or inhalation of carbon tetrachloride or chloroform and mushroom poisoning. Because the liver metabolizes many drugs and toxic chemicals, it is not surprising that patients with severe acute liver disease (hepatitis) or severe chronic liver disease (cirrhosis) may be more susceptible to drugs and toxins regardless of the site at which they produce injury.

CHART 15–2. CHEMICAL INJURIES TO THE LIVER

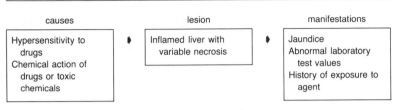

causes	lesion	manifestations
Hypersensitivity to drugs Chemical action of drugs or toxic chemicals	Inflamed liver with variable necrosis	Jaundice Abnormal laboratory test values History of exposure to agent

Alcoholic Liver Disease

Prolonged heavy ingestion of alcoholic beverages leads to acute and chronic liver disease in some individuals. Experimentally, mild histologic changes can be produced in persons ingesting "socially acceptable" amounts of alcohol. Three types of changes occur in the liver during the long time interval required for development of cirrhosis—fatty change, hepatitis, and fibrosis. *Fatty change* is reversible but is usually found to a variable extent at all stages of the disease. Fatty change appears to be the most universal effect of alcohol. Although more severe changes are also dose related, there appears to be some degree of individual susceptibility that is not adequately explained by our current knowledge. Acute cell damage occurs with bouts of heavy drinking and results in inflammation of the liver called *alcoholic hepatitis*. After repeated episodes of injury, fibrosis develops and gradually causes the disruption of the hepatic architecture. Regenerative nodules of hepatocytes and fibrosis constitute the end stage of the disease, called *cirrhosis* (discussion follows). Clinically, alcoholic liver disease is often asymptomatic until advanced. Advanced disease presents as cirrhosis. Some patients present with a large fatty liver and jaundice. Others develop bouts of delirium tremens, especially upon withdrawal from alcohol. *Delirium tremens* is characterized by tremors, combative behavior, and hallucinations (often the patient sees strange objects).

CHART 15–3. ALCOHOLIC LIVER DISEASE

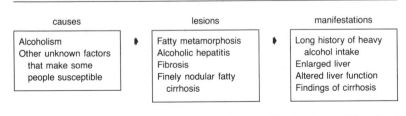

causes	lesions	manifestations
Alcoholism Other unknown factors that make some people susceptible	Fatty metamorphosis Alcoholic hepatitis Fibrosis Finely nodular fatty cirrhosis	Long history of heavy alcohol intake Enlarged liver Altered liver function Findings of cirrhosis

Cirrhosis

Cirrhosis is the term used to describe the end stage of most serious chronic types of liver disease. It is characterized by fibrosis and nodular regeneration of hepatocytes. Normally, the liver can easily replace damaged hepatocytes. If the delicate connective tissue framework is destroyed or replaced by bands of fibrous tissue, the replacement process results in clusters of regenerated hepatocytes to form grossly visible nodules.

Of the many causes of cirrhosis, alcoholism and chronic hepatitis account for a large portion of the cases, but in many patients with cirrhosis, the exact cause cannot be determined. Other types of cirrhosis, which are uncommon, include biliary cirrhosis, resulting from prolonged bile duct obstruction; pigmentary cirrhosis of hemochromatosis, resulting from massive storage of iron; Wilson's disease, with cirrhosis resulting from an inherited defect in copper metabolism; and cirrhosis associated with α-1-antitrypsin deficiency.

The morphology of the cirrhosis may be helpful in determining its cause. Most instances of alcoholic cirrhosis produce a finely nodular liver (Fig. 15–3A). Histologically, many hepatocytes are filled with fat. The finely nodular cirrhosis of alcoholism is sometimes called *Laennec's cirrhosis.* Idiopathic cirrhosis and cirrhosis associated with chronic hepatitis generally have larger regenerative nodules and broader and more irregular bands of fibrous tissue (Fig. 15–3B). This is called *postnecrotic cirrhosis.*

The development of cirrhosis takes months to years. It is usually asymptomatic until serious irreversible changes in the architecture of the liver lead to any of several possible complications, including portal hypertension, esophageal varices, ascites, and hepatic encephalopathy. These complications usually result from altered blood flow through the liver. Blood may flow through the fibrous septae with reduced flow through hepatic sinusoids, and thus, blood has reduced contact with hepatocytes. The altered flow also leads to elevated blood pressure in the portal vein (portal hypertension). Failure of blood to flow past liver cells may lead to accumulation of nitrogenous breakdown products (such as ammonia) in the blood. This is associated with hepatic encephalopathy characterized by depression of the central nervous system, with stupor or coma. The shunting of blood away from the hepatocytes also explains why drugs are poorly metabolized in cirrhotic persons. Portal hypertension leads to development of the collateral venous connections to the vena caval system. Most notably, the veins of the lower esophagus become dilated (esophageal varices) by increased collateral blood flow, and they may rupture to produce massive hemorrhage into the alimentary tract. Another

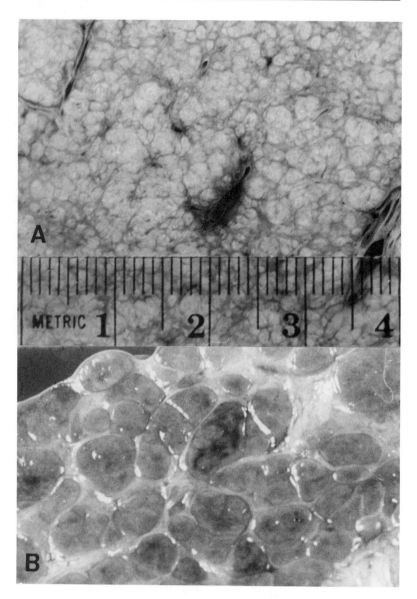

Figure 15–3. Comparison of finely nodular alcoholic cirrhosis **(A)** with the coarsely nodular irregular postnecrotic cirrhosis **(B)**.

common feature of portal hypertension is passive congestion of the spleen leading to splenomegaly. Occasionally, the splenomegaly of portal hypertension is associated with increased destruction of blood cells by the spleen leading to anemia, leukopenia, and/or thrombocytopenia (hypersplenism). Ascites also occurs. *Ascites* is the accumulation of massive amounts of watery fluid in the peritoneal cavity so that the abdomen becomes bloated. Its causation is complicated and involves low serum protein levels, increased back pressure in the portal vein and lymphatics of the liver, and retention of salt. Jaundice is a variable feature of cirrhosis and is caused by compression of bile ducts due to the fibrosis and to depressed function of the hepatocytes due to altered blood flow. Altered metabolism of sex hormones by the liver leads to estrogenic effects, including small dilated blood vessels in skin (spider angiomas) and breast enlargement in men (gynecomastia).

The average alcoholic cirrhotic patient comes to medical attention because of the development of one or more of the following: jaundice, ascites, bleeding esophageal varices, delirium tremens, or hepatic encephalopathy. If the cirrhosis is well advanced, the patient will die within a few years of encephalopathy, bleeding, or superimposed infection. If the diagnosis is made at an early stage of alcoholic liver disease, abstinence from alcohol can greatly prolong life for some individuals. Most nonalcoholic types of cirrhosis are progressive. Biopsy is the best way to evaluate the extent of the disease. Operations to relieve the portal hypertension and thus decrease the risk of esophageal varices are sometimes carried out.

CHART 15–4. CIRRHOSIS

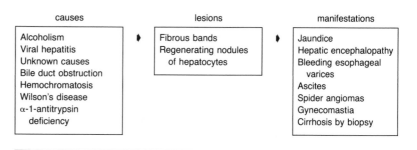

causes	lesions	manifestations

Alcoholism
Viral hepatitis
Unknown causes
Bile duct obstruction
Hemochromatosis
Wilson's disease
α-1-antitrypsin
  deficiency
	Fibrous bands
	Regenerating nodules
	  of hepatocytes
		Jaundice
		Hepatic encephalopathy
		Bleeding esophageal
		  varices
		Ascites
		Spider angiomas
		Gynecomastia
		Cirrhosis by biopsy

Cholecystitis and Gallstones

Bile is rich in cholesterol, which is barely held in solution by bile salts and phospholipids. In the gallbladder, bile is concentrated by absorption of its water content. If the cholesterol comes out of solution, it forms crystals, which, along with the bilirubin pigment and calcium in the bile, form stones (Fig. 15–4). The stones vary greatly in number, size, shape, and color. The reasons why some people develop gallstones and others do not is not clear, although women are much more prone to develop gallstones than men, and

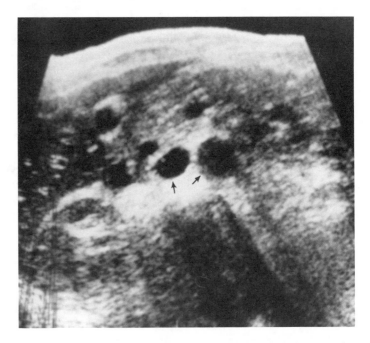

Figure 15–4. Ultrasound image of abdominal wall (*skin at top*) demonstrating gall-stones (*arrows*) in gallbladder (*light area*).

the likelihood of developing gallstones increases with age. Native Americans have a very high rate of gallstone formation.

The stones in the gallbladder are associated with a very low-grade inflammation, which we label *chronic cholecystitis.* Gallstones are important because of their complications. The two most common complications of gallstones are acute cholecystitis, which is painful and makes the patient quite ill, and migration of gallstones down the cystic duct into the common bile duct, obstructing its distal narrow end to produce jaundice. The main reason for removing the gallbladder (cholecystectomy) is to prevent or treat these two serious complications.

Uncomplicated chronic cholecystitis with gallstones is usually asymptomatic but may be associated with symptoms such as pain following meals, especially when fatty or spicy food is ingested. Cholecystectomy may relieve symptoms of uncomplicated cholecystitis. Symptoms attributed to cholecystitis are sometimes due to other causes and are not relieved by cholecystectomy.

The diagnosis of gallstones is usually established by radiologic procedures. Roentgenograms of the abdomen usually fail to reveal gallstones, because only 20 percent of patients have enough calcium in their stones to make them radiopaque. The classic method of demonstrating gallstones, called a cholecystogram, involves ingestion of a radiopaque dye that is absorbed by the liver and excreted into the bile. If the gallbladder is normal, the dye is concentrated along with bile. Radiolucent gallstones produce less dense filling defects. If the cystic duct is blocked by a stone, dye cannot enter into the gallbladder and it is said to be nonfunctional. A newer technique for demonstrating gallstones, ultrasound, involves the creation of an image of sound waves as they are deflected from the stones (Fig. 15–4). Ultrasound can be done rapidly and without ingestion of drugs or exposure to x-rays, but as with cholecystography, it produces some false-negative results.

CHART 15–5. CHOLECYSTITIS AND GALLSTONES

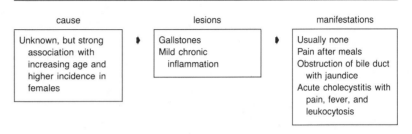

cause	lesions	manifestations
Unknown, but strong association with increasing age and higher incidence in females	Gallstones Mild chronic inflammation	Usually none Pain after meals Obstruction of bile duct with jaundice Acute cholecystitis with pain, fever, and leukocytosis

Pancreatitis

Inflammation of the pancreas differs from inflammation in other organs, because the powerful digestive enzymes produced by acinar cells may escape from the cells or ducts to digest the pancreas itself and surrounding adipose tissue. This is called *enzymatic necrosis*. The mechanisms by which pancreatic enzymes escape into tissue are not well understood, although obstruction of the pancreatic duct and stimulation of secretion, such as occurs following a heavy meal, are thought to be important factors.

Most cases of severe pancreatitis are associated with alcoholism. Acute hemorrhagic pancreatitis results in extensive necrosis and hemorrhage into the pancreas, leading to pain and shock. This form usually follows a heavy, alcohol-laden meal and is often fatal in spite of attempts to decrease pancreatic secretion by evacuating food from the stomach and administering drugs that block secretion. Severe chronic pancreatitis occurs in chronic alcoholics and is a slowly progressive disease much like cirrhosis. After many years of heavy drinking, these patients develop malabsorption due to replacement of pancreatic acini by fibrous tissue and diabetes mellitus due to destruction of the islets of Langerhans. Severe abdominal pain may occur with chronic pancreatitis. End-stage pancreatitis in the alcoholic is less common than cirrhosis, and the two conditions may or may not be present together.

Other diseases associated with pancreatitis include gallstones and trauma. Patients with gallstones frequently have episodes of mild pancreatitis, which may either be asymptomatic or associated with abdominal pain. These episodes cease when the gallbladder is removed. It is likely that small gallstones migrate from the gallbladder down the bile duct and temporarily block the pancreatic duct before passing into the duodenum. Obstruction of the pancreatic duct may be the cause of pancreatitis. This form of pancreatitis is much milder than that seen with alcoholism. Trauma to the pancreas, whether caused by accident or surgical operation on the pancreas, can also lead to pancreatitis due to release of enzymes.

The manifestations of pancreatitis usually include pain in the midabdomen, that may bore through to the back. Acute or active pancreatitis is associated with increased levels of amylase and lipase in the blood and amylase in the urine. In severe chronic pancreatitis, the acini may be destroyed, so that enzyme levels are no longer elevated. In chronic pancreatitis, there may be evidence of malabsorption or diabetes. Also, stimulation of the pancreas with secretin, a hormone that causes fluid secretion by the pancreas, produces subnormal amounts of pancreatic secretions in the duodenum (secretin test).

CHART 15–6. PANCREATITIS

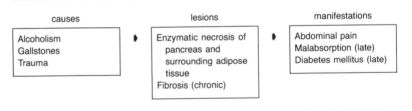

causes	lesions	manifestations
Alcoholism Gallstones Trauma	Enzymatic necrosis of pancreas and surrounding adipose tissue Fibrosis (chronic)	Abdominal pain Malabsorption (late) Diabetes mellitus (late)

Hyperplastic/Neoplastic Diseases

In the liver, metastases are much more common than primary cancer. Hepatocarcinoma is the most common primary tumor of the liver, and it most often occurs as a complication of nonalcoholic types of cirrhosis. Gallbladder cancer is rare, usually fatal, and is usually associated with gallstones. Cancer of the pancreas accounts for 3 percent of cancers and is almost always fatal.

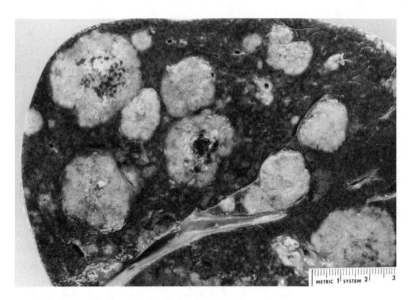

Figure 15–5. Liver metastases in a patient who presented with jaundice and an undiagnosed carcinoma of the colon.

Metastatic Cancer of the Liver

Abdominal cancers, such as carcinoma of the colon, stomach, and pancreas, characteristically metastasize to the liver, but metastases of other neoplasms, such as leukemias and lymphomas, are also common. A few patients present with liver metastases without symptoms referrable to the primary cancer site. They have large, nodular livers and may have jaundice (Fig. 15–5). Liver metastases are not curable.

CHART 15–7. METASTATIC CANCER TO LIVER

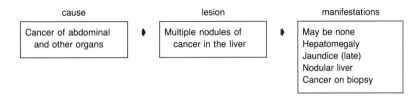

cause		lesion		manifestations
Cancer of abdominal and other organs	▶	Multiple nodules of cancer in the liver	▶	May be none Hepatomegaly Jaundice (late) Nodular liver Cancer on biopsy

Carcinoma of the Pancreas

Most cases of carcinoma of the pancreas develop in the head of the pancreas, producing jaundice due to obstruction of the common bile duct and pain due to involvement of nerves in surrounding tissues.

CHART 15–8. CARCINOMA OF PANCREAS

cause		lesion		manifestations
Unknown	▸	Adenocarcinoma of head or tail of pancreas	▸	Jaundice Weight loss Pain

ORGAN FAILURE

Acute liver failure results from massive death of hepatocytes and is associated with hepatic encephalopathy. Viral hepatitis and halothane reactions are the usual causes. Chronic liver failure is associated with cirrhosis. *Nonfunctioning gallbladder* is defined by a failure to take up radiopaque dyes designed to visualize the gallbladder. Stones wedged at the neck of the cystic duct are the usual cause. Acute pancreatic failure is not a recognized condition. Chronic pancreatic failure leads to the malabsorption syndrome if pancreatic acini are extensively destroyed and to diabetes mellitus if islets of Langerhans are extensively destroyed.

REVIEW QUESTIONS

1. What are the most frequent diseases of the liver, gallbladder, and pancreas? Which are likely to be fatal?
2. What are three mechanisms by which jaundice may be produced?
3. What tests are used to detect liver disease? How do abnormalities of the tests correlate with changes in the liver?
4. How is a liver biopsy performed and what is its value?
5. How are gallstones demonstrated?
6. What are the causes, lesions, and manifestations of the diseases discussed in this chapter?
7. What is the most common outcome of viral hepatitis? What other outcomes occur?
8. What are the effects of alcohol on the liver and pancreas?
9. What are the causes and effects of portal hypertension, esophageal varices, ascites, and hepatic encephalopathy?
10. How do gallstones produce problems?
11. What do the following terms mean?
    Jaundice
    Delirium tremens
    Laennec's cirrhosis
    Postnecrotic cirrhosis
    Nonfunctioning gallbladder
    Cholecystography

Kidney, Lower Urinary Tract, and Male Genital Organs

REVIEW OF STRUCTURE AND FUNCTION

Diseases of the male genital organs are discussed with diseases of the urinary tract because of the intimate anatomic relationship of the male genital organs and lower urinary tract (Fig. 16–1).

The kidneys are bilateral retroperitoneal organs that receive blood from the renal arteries and are drained by the renal veins. Urine formed by the kidney leaves through the ureter. The kidneys themselves consist of cortex, medulla, and pelvis (Fig. 16–2). The cortex contains all of the glomeruli and most of the tubules. The medulla contains specialized parts of the tubules (loops of Henle) and the collecting tubules. The pelvis is a space lined by transitional epithelium that transmits urine from the collecting tubules to the ureters.

The kidneys' function is to regulate the concentration of salt, water, and hydrogen ions in the body and to excrete waste products such as urea and creatinine or foreign substances such as drugs and their metabolites. To accomplish this, the kidneys receive 20 percent of the circulating blood each minute. The main renal arteries branch several times into smaller arteries and arterioles, which eventually supply the renal glomeruli of the organ's cortex.

The basic functional unit of the kidney is called a *nephron* and consists of glomerulus, tubules, and associated vessels (Fig. 16–3).

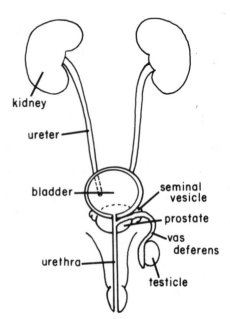

Figure 16–1. Components of urinary tract and male genital system.

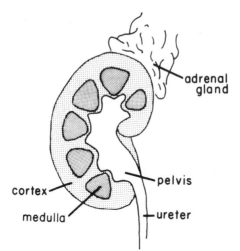

Figure 16–2. The kidney and adjacent structures.

Each glomerulus is a tuft of capillaries that are lined by endothelial cells. The entire tuft is covered with a layer of thin epithelial cells (Fig. 16–4). Separating the two cell layers is an important basement membrane, which participates in the filtration function of the glomerulus. Blood leaving the glomerulus passes into a capillary network that surrounds epithelial-lined tubules. The waste mate-

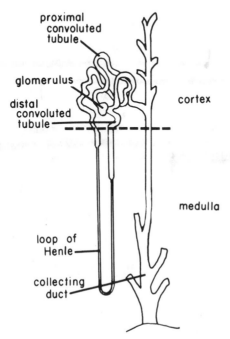

Figure 16–3. Components of a nephron.

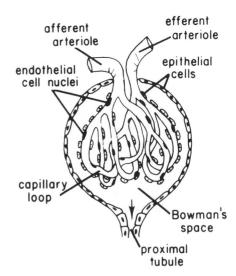

Figure 16–4. Components of a glomerulus.

rials that have been filtered through the glomeruli pass into these tubules. The tubules eventually merge to form collecting ducts, which carry urine to the renal pelvis. Throughout the length of the tubules there is constant exchange between the tubules and the capillary networks that surround them. Consequently, the urine that emerges from the kidney has a much different chemical composition than the initial filtrate emerging from the glomeruli. Many substances, such as glucose, chloride, and drugs, continually pass back and forth between the tubules and the capillaries that surround them. Passage of substances from the capillaries to the tubules is termed *secretion*, while the opposite, passage from the tubules to capillaries, is termed *resorption*. The final concentrations of any substance in the urine is partly determined by the amount of secretion versus resorption. One of the most important exchanges between renal tubular epithelium and surrounding capillaries takes place in the distal tubules and involves exchange of potassium for sodium. The kidney also contributes to the regulation of the body's acid–base balance by either secreting or absorbing hydrogen ion as required to control blood pH. The final composition of the urine is determined in the collecting tubules, where water is absorbed from urine back into the blood under the influence of antidiuretic hormone derived from the pituitary gland. The urine then passes through the pelvis of the kidneys and the ureters to the bladder. All the supporting tissue between the tubules and glomeruli is referred to as the interstitium of the kidney.

In the walls of arterioles supplying the glomerulus, there is a collection of cells called the *juxtaglomerular apparatus*. These cells secrete renin, an enzyme that acts on a protein secreted by the liver (renin substrate), converting it to angiotensin I. Angiotensin I is converted to angiotensin II by another enzyme (converting enzyme) produced by endothelial cells of the lung and kidney. Angiotensin II stimulates vasoconstriction (thereby elevating blood pressure) and enhances sodium and potassium absorption (by stimulating release of aldosterone from the adrenal cortex).

The lower urinary tract consists of the ureters, bladder, and urethra plus accessory glands of the tract. The tract is essentially similar in both male and female with regard to the ureters and the bladder. The ureters descend from the pelvis of the kidneys in the retroperitoneal tissue to the bladder. The function of the ureter is solely to transmit urine from the kidney to the bladder. No absorption of urine, or any of its contents, occurs along this muscular tubule. The bladder is a large muscular organ in which urine is collected from the ureters and passed down the urethra through the urethrovesicular outlet. The bladder outlet involves a complex intertwining of muscle fibers, which on contraction open the sphincter to allow discharge of bladder contents. The urethra of the female passes directly from the bladder through the urogenital diaphragm to the urethral meatus. In the male, however, the urethra consists of several additional segments that pass through the center of the prostate gland and extend on into the penis. In the prostate gland, the ejaculatory ducts penetrate through the substance of the prostate to connect the vas deferens to the urethra.

The testes consist basically of the seminiferous tubules where sperm are produced. Sperm are stored in the epididymis and vas deferens and are propelled along the vas deferens during ejaculation by muscular contraction. Sperm are also stored in the seminal vesicles, which lie along the vas deferens just posterior to the prostate.

The male prostate encircles the neck of the bladder in the retroperitoneal space. It is composed of three lobes and is a glandular organ, the ducts of which empty into the urethra as it passes through the prostate gland. Prostatic secretions comprise the major portion of seminal fluid and have a high antibacterial activity.

MOST FREQUENT AND SERIOUS PROBLEMS

Bacterial infections are the most frequent problems affecting the kidney and the lower urinary tract. Bladder infection (cystitis) is common in females, because bacteria from the perineum gain access to the bladder via the short female urethra. Prostatitis is fairly common in younger men. Prostatic enlargement is very common in older men and the consequent obstruction of the urethra predisposes to kidney as well as bladder infections. Renal calculi (stones) are not uncommon problems in adults. Diseases of the renal glomeruli are collectively of major importance, since they are the major cause of chronic renal failure. Transitional cell carcinoma of the bladder and adenocarcinoma of the prostate are relatively common in older individuals. Cancers of the kidneys and testes are uncommon.

SYMPTOMS, SIGNS, AND TESTS

Most patients with cystitis or urethritis experience *frequency of urination*, *dysuria* (painful urination), or *nocturia* (increased night time urination). In addition, the urine may be clouded by pus. Prostatitis may manifest simply as low back pain. Pyelonephritis may be asymptomatic or may present acutely with intense flank pain in addition to systemic signs of infection such as fever and leukocytosis. Diseases of the renal glomeruli may present with *hematuria* (blood in the urine), *proteinuria* (protein in the urine), or systemic signs such as edema and hypertension. Renal calculi characteristically present with intense, sharp *flank pains* that radiate toward the groin as the calculi migrate from the renal pelvis to the bladder. Acute necrosis of the renal tubules may result from either renal ischemia or the action of certain toxins and often is manifest by decreased output of urine (*oliguria*) or even complete absence of urine (*anuria*).

The nephrotic syndrome is a constellation of signs and laboratory abnormalities resulting from damage to the glomerular filtering apparatus. Proteinurias and hypoproteinemia are caused by excessive loss of protein through the defective glomerular basement membrane. Edema results from lowered plasma oncotic pressure due to protein loss. Hyperlipidemia is also present in the nephrotic syndrome for poorly understood reasons; it may be due to decreased lipid transport by proteins. The most common causes of the nephrotic syndrome are lipoid nephrosis, focal glomerulosclerosis, and membranous glomerulonephritis, which will be discussed later in this chapter.

Physical examination of the urinary tract and genital organs consists of inspection of the penis or vulva for signs of exudation or ulceration from venereal infections, palpation of the abdomen for tumors of the kidney or a distended bladder, and, in the male, palpation of the inguinal ring for hernia or undescended testis, palpation of the testes for tumors, and rectal examination to feel for enlargement of the prostate.

Urinalysis is the most important routine laboratory test performed, because it can detect the presence of many common urinary tract disorders, especially infections. A urinalysis normally includes tests for the specific gravity, pH, and presence of protein, sugar, blood, and ketones, as well as microscopic examination to check for the possible presence of red blood cells, white blood cells, bacteria, crystals, and *casts* from damaged renal tubules (Fig. 16–5).

Urine culture for bacteria may be performed when there are symptoms of urinary tract infection or increased amounts of white blood cells in the urine. Roughly, when there are over 100,000 bacteria per milliliter of urine, a diagnosis of urinary tract infection may be made. Lesser numbers suggest contamination of the specimen during collection. Bacterial antibiotic sensitivity studies help determine the most rational antibiotic therapy.

Renal function is most commonly evaluated by measuring levels of two substances in the blood that are excreted by filtration through the glomerulus, namely, *urea nitrogen* and *creatinine*. The normal blood levels of these substances are determined by an equilibrium between normal breakdown of nitrogenous compounds to produce urea nitrogen and normal muscle breakdown to produce creatinine, as well as the ability of the glomeruli to filter and the tubules to resorb these substances. Significant renal disease often raises the blood levels of these substances, because they are inade-

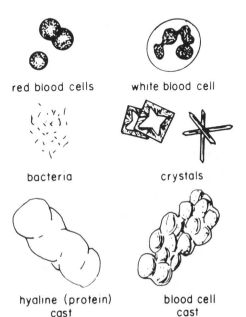

red blood cells   white blood cell

bacteria   crystals

hyaline (protein) cast   blood cell cast

Figure 16–5. Possible findings on microscopic examination of urine.

quately filtered in glomerular diseases or inadequately secreted and resorbed in tubular diseases.

Renal biopsies are often performed for the evaluation of glomerular disease by inserting a thin needle through the skin of the flank and extracting a core of renal tissue. Immunofluorescent techniques and electron microscopy, in addition to light microscopy, are usually performed on renal biopsies. A better understanding of glomerular diseases has been reached by this procedure.

Other examinations used in evaluating renal and urinary tract disease include *cystograms,* in which a radiopaque dye is introduced into the bladder by catheter and x-rays are taken to elucidate bladder morphology and function.

The intravenous urogram (IVU) is a commonly used test to look for gross structural changes in the kidneys and ureters. Radiopaque (iodinated) contrast material, injected intravenously, is filtered by the glomeruli into the tubules and concentrated in the renal pelvis and ureters. Distortion of the normal pattern of dye in the collecting system suggests a mass or cyst. Calculi in the renal pelvis appear as filling defects. Obstruction of the urinary tract dilates the pelvis and/or ureter.

Catheterization of the bladder refers to the process of placing a tube in the bladder, usually for the purpose of draining the bladder or collecting a urine sample. An indwelling catheter is one that is left in place for some time. *Cystoscopy* is the visualization of the bladder using a scope and is used to detect and biopsy bladder lesions. Seminal fluid examination is useful in detecting inflammatory cells and bacteria in cases of prostatitis and in evaluating sperm counts for determination of fertility.

SPECIFIC DISEASES

Genetic/Developmental Diseases

There are numerous developmental abnormalities of the genitourinary system, many of which occur in conjunction with anomalies of other organ systems. A few of the more common anomalies will be discussed.

Agenesis and Hypoplasia

Complete agenesis of the kidney occurs very infrequently and is obviously incompatible with life if bilateral. Various degrees of hypoplasia may also be rarely encountered, and the final outcome of these cases depends upon the degree of hypoplasia plus the chance of superimposed insults such as pyelonephritis.

Polycystic Kidneys

Adult polycystic disease is a fairly common (1 in 500 persons) autosomal dominant disease in which thin-walled cysts of various size cause massive bilateral renal enlargement (Fig. 16–6). The disease is usually not discovered until adulthood when hypertension or chronic renal failure occur. Although cysts occur in other organs, especially the liver, these are of little consequence, thus these patients are good candidates for renal dialysis and/or renal transplantation. *Infantile polycystic disease* is a rare, fatal autosomal recessive condition associated with bilateral renal cysts and hepatic fibrosis.

CHART 16–1. ADULT POLYCYSTIC DISEASE

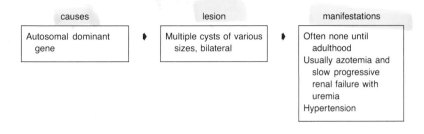

causes	lesion	manifestations
Autosomal dominant gene	Multiple cysts of various sizes, bilateral	Often none until adulthood Usually azotemia and slow progressive renal failure with uremia Hypertension

Dysplastic (Multicystic) Kidney

Malformation of embryonic development of nephrons with formation of cartilage and cysts may be unilateral or bilateral and is often associated with other anomalies and obstruction of the urinary

Figure 16–6. Adult polycystic kidney.

tract. Prognosis for this common noninherited condition depends on the amount of normal renal parenchyma and seriousness of the associated anomalies.

Cryptorchism

Cryptorchism is a failure of the testes to descend into the scrotum. It is not uncommon. Bilateral cryptorchism will cause sterility if not surgically corrected in childhood. Descent normally does not occur until after birth and absence of descent 6 months after birth is not unusual. If surgical descent must be tried, it is usually undertaken around 5 years of age.

CHART 16–2. CRYPTORCHISM

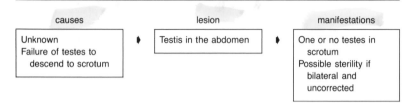

causes	lesion	manifestations
Unknown Failure of testes to descend to scrotum	Testis in the abdomen	One or no testes in scrotum Possible sterility if bilateral and uncorrected

Inflammatory/Degenerative Diseases

In this section, diseases of the glomeruli, infections of the kidney and lower urinary tract, interstitial nephritis, and kidney stones are considered. Acute degeneration and necrosis of renal tubules is discussed under Organ Failure.

Glomerular Diseases

Diseases of glomeruli may be divided into those that are primary diseases of the glomeruli and those that are secondary to systemic diseases such as diabetes mellitus or lupus erythematosus. *Nephritis* means inflammation of the kidney, and *glomerulonephritis* specifically designates inflammatory disease of the glomeruli. Many important primary diseases of glomeruli are inflammatory and are triggered by allergic (immune) injury of either of two types. In immune complex glomerulonephritis, antigen–antibody complexes form in the blood and are deposited on the basement membrane of the glomerulus as the glomerulus attempts to filter these complexes. In this context, the kidney is a passive recipient of these damaging antigen–antibody complexes. The second type is anti-basement membrane glomerulonephritis, in which antibodies are formed against the basement membrane itself, which acts as the antigen. The antibodies in this type of nephritis may be generated to react against a foreign protein (possibly a virus) that shares common antigenic properties with the glomerular basement membrane. This sharing of common antigenic properties is termed cross-reactivity. Both immune complex deposition and anti-basement membrane deposition damage the glomerular capillary structures, allowing excessive filtration of protein. In acute glomerulonephritis, red blood cells also leak through the glomerular structures, producing hematuria. One of the most common and important immune complex diseases is acute poststreptococcal glomerulonephritis.

Acute Poststreptococcal Glomerulonephritis

This disease is an immune complex glomerulonephritis that follows an antecedent infection with certain strains of group A streptococci anywhere in the body, usually a pharyngitis. One to 4 weeks following the initial streptococcal infection, antibodies are formed against the streptococcus antigens, and this antigen–antibody complex is deposited on the glomerular basement membrane (Fig. 16–7; Fig. 16–8). The subsequent inflammation of the basement membrane leads to the cardinal clinical signs and symptoms of hematuria and proteinuria due to incompetence of the glomerular filtering apparatus. Generalized edema occurs because

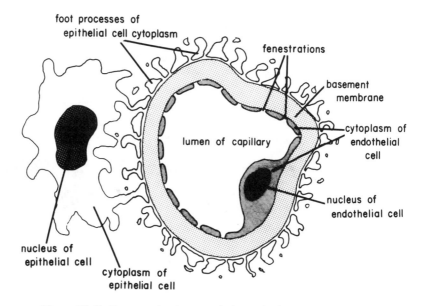

Figure 16–7. Components of a normal glomerulus by electron microscopy.

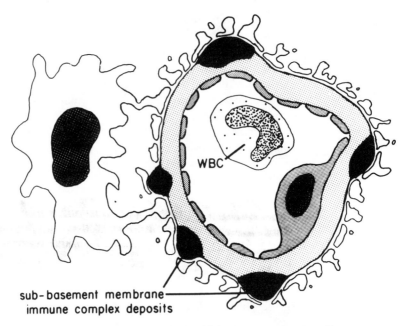

Figure 16–8. Glomerulus with immune complex deposits.

of loss of protein, and hypertension occurs because of stimulation of the juxtaglomerular apparatus. The diagnosis is made on the basis of hematuria, edema, and hypertension, plus a history of streptococcal infection. Other findings may include elevated urea nitrogen and creatinine in the blood, because these breakdown products cannot be excreted in normal amounts. The lesion is an inflamed glomerulus secondary to immune complex deposits. The recovery rate is approximately 95 percent in children and slightly lower in adults, the remainder in both groups develop progressive renal failure. Numerous drugs, as well as other infectious agents, may also cause immune complex glomerulonephritis.

CHART 16–3. ACUTE POSTSTREPTOCOCCAL GLOMERULONEPHRITIS

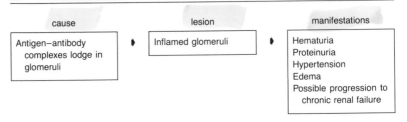

cause	lesion	manifestations
Antigen–antibody complexes lodge in glomeruli	Inflamed glomeruli	Hematuria Proteinuria Hypertension Edema Possible progression to chronic renal failure

Chronic Glomerulonephritis

Chronic Glomerulonephritis refers to a variety of prolonged, often progressive, renal diseases. They may be initiated by immune complex deposition, by anti-basement membrane antibodies, or by nonimmunologic degeneration of unknown cause. Many of these diseases do not occur in an acute phase as does poststreptococcal glomerulonephritis; rather, the inflammation or degeneration and consequent scarring proceed slowly and insidiously. Often the diagnosis is made only after the patient complains of fatigue or edema and subsequent urinalysis reveals proteinuria. At this time, the kidney may be so severely damaged that the ordinary clues as to the cause of the disease are lost. Consequently, many patients are said to have chronic glomerulonephritis without regard to cause.

Many patients, especially children, initially manifest chronic renal disease by the *nephrotic syndrome.* The nephrotic syndrome is not a disease per se but a complex of signs and symptoms (syndrome), which includes hypoproteinemia, proteinuria, hyperlipidemia, and edema. The syndrome may be associated with lipoid nephrosis, membranous glomerulonephritis, focal glomerulosclerosis, diabetes mellitus, amyloidosis, progressive poststreptococcal glomerulonephritis, and many other less common diseases. Of these, lipoid nephrosis and membranous glomerulonephritis are deserving of further description.

Lipoid nephrosis is the most common cause of the nephrotic syndrome in children and adults. It accounts for 80 percent of the nephrotic syndrome in children and 30 percent in adults. The lesion of lipoid nephrosis is damaged epithelial foot processes adjacent to the glomerular capillary membrane (Fig. 16–9) associated with leakage of protein into Bowman's space and thus into the urine. The damage can only be seen by electron microscopy. Further, there are no antibody deposits in the glomeruli of patients with lipoid nephrosis, and the cause as well as the pathogenesis of the disease are not known.

The second most common cause of the nephrotic syndrome in adults is *membranous glomerulonephritis.* Membranous glomerulonephritis is characterized by deposits of antigen–antibody embedded within the thickened glomerular basement membrane as observed in the electron microscope. These deposits disturb the normal permeability of the glomerular basement membrane, resulting in a large amount of protein, particularly albumin, passing from the vascular to the urinary space. In most cases, the nature of the antigen as well as the cause of the glomerulonephritis is unknown, and the disease is consequently termed *idiopathic membranous*

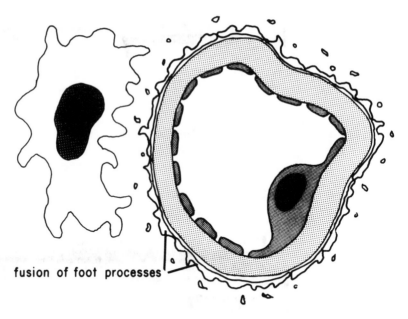

fusion of foot processes

Figure 16–9. Glomerulus in lipoid nephrosis with foot process fusion and no immune complex deposits.

glomerulonephritis. In a few instances, the antigen is known. Occasionally, the antigen is hepatitis virus. A bout of hepatitis may be associated with membranous glomerulonephritis with the nephrotic syndrome. In some instances, membranous glomerulonephritis may follow treatment with certain drugs such as gold salts, which are used to treat rheumatoid arthritis, or may be associated with certain other diseases such as syphilis.

Diabetes mellitus commonly causes chronic renal damage because of its effect on small blood vessels but only rarely results in the nephrotic syndrome.

Some patients with the nephrotic syndrome will recover spontaneously, while others appear to respond to steroid therapy. Some will progress to chronic renal failure and may be kept alive for variable periods by *dialysis*—the so-called artificial kidney. Renal transplantation is possible in some patients with chronic renal disease.

Pyelonephritis

The most important inflammatory kidney disease is pyelonephritis, an acute or chronic bacterial infection predominantly involving the renal tubules and most commonly caused by *Escherichia coli* or

CHART 16–4. CHRONIC GLOMERULONEPHRITIS

causes	lesion	manifestations
Immune complexes Anti-basement membrane antibodies Unknown causes	Inflamed and scarred glomeruli	May be none (early) Proteinuria Hematuria Hypertension Uremia (late)

other gram-negative bacteria such as proteus, pseudomonas, enterobacter, and klebsiella. The organisms may enter the kidney via the bloodstream or, most commonly, in retrograde fashion through the bladder and ureters. In the latter case, obstruction often plays an important role in the pathogenesis of the infection, because stagnation of urine consequent to urethral obstruction at any level favors the multiplication of bacteria and obviates their chances of being washed downstream. The patient with acute pyelonephritis will have more or less acute onset of flank pain as well as fever. Microscopically, the kidney will show variable amounts of acute inflammatory cells in the interstitial and tubular tissue, with relative sparing of the glomeruli. Later, the kidney will show scars in these areas. The patient's urine will have increased amounts of protein, casts, white blood cells, and bacteria. Often, the white cells will be so numerous as to constitute *pyuria* (pus in the urine). The casts are cylindrical protein deposits from damaged tubular epithelium (casts of the tubular lumens). The offending organism can usually be cultured from the urine. The treatment of pyelonephritis is antibiotic therapy and alleviation of the obstruction. In some patients, the disease may smolder for a long time in spite of antibiotic therapy. These are called *chronic pyelonephritis* and are often fostered by continued or intermittent obstructions of the urinary tract, although in some cases the reason for the perpetuation of the disease is not known.

CHART 16–5. PYELONEPHRITIS

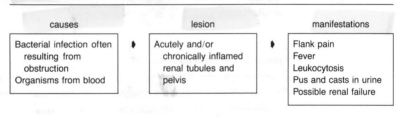

causes	lesion	manifestations

Bacterial infection often
  resulting from
  obstruction
Organisms from blood
	Acutely and/or
	  chronically inflamed
	  renal tubules and
	  pelvis
		Flank pain
		Fever
		Leukocytosis
		Pus and casts in urine
		Possible renal failure

Interstitial Nephritis

Chronic renal inflammations limited for the most part to the interstitium are encountered in patients who give a history of ingesting excessive amounts of analgesics, especially those containing phenacetin.

Cystitis

Infection of the bladder is called *cystitis* and is usually caused by the same organisms that cause pyelonephritis. It is one of the more common infections encountered. The symptoms of cystitis are dysuria (painful urination), frequency (frequent urination), and urgency (repeated or continuous urge to urinate). The urine findings are the same as in pyelonephritis, except that casts are not found in cystitis. Infectious microorganisms invade the urinary tract by two major pathways—up the urethra or, rarely, via the circulatory system. The urethral route is the one most plaguing to women. In the male, the long penile urethra plus the presence of antibacterial secretions from the prostate discourage this route. Fecal soiling of the urethral meatus and sexual activity are common factors allowing organisms to enter the urinary tract in females. Instrumentation such as catheterization or cystoscopy is commonly complicated by infection in both sexes.

CHART 16–6. CYSTITIS

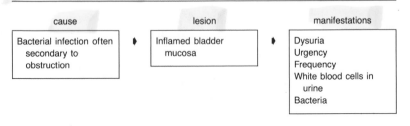

cause	lesion	manifestations
Bacterial infection often secondary to obstruction	Inflamed bladder mucosa	Dysuria Urgency Frequency White blood cells in urine Bacteria

Prostatitis

Prostatitis may be an acute or chronic infection and is often associated with considerable pain and discomfort. Leukocytes are encountered in the urine in prostatitis.

Kidney Stones (Calculi)

Crystallization of minerals in the urine to form hard, stone-like masses is common. Kidney stones are most often composed of calcium and various other substances excreted by the kidney such as oxalates. Stones containing calcium are visible on x-ray. Other stones, such as those composed of urates or cystine, are not visible by x-ray. Most commonly, small stones form in adults without clear-cut cause; however, low urine volume due to dehydration, chronic urinary tract infection, and prolonged bed rest with liberation of calcium from the bones can be precipitating factors. Less commonly, stones are caused by serious underlying diseases, including hyperparathyroidism and severe bone disease, which result in increased calcium in the urine. Gout and cystinosis are metabolic diseases that cause increased excretion of urates and cystine, respectively, leading to stone formation in some instances.

Passage of a small stone from the renal pelvis into the ureter produces sudden severe flank pain, which patients describe as worse than anything they have ever experienced. Immediate treatment consists of pain medications with the hope that the stone will pass down the urinary tract and be recovered in the urine. If the stone fails to dislodge from the ureter, a urologist may have to remove the stone using catheters passed through the urethra, bladder, and ureter to dislodge the impacted stone. A stone impacted in the ureter may lead to hydronephrosis (Figs. 16–10, 16–11). Occasionally, large stones are encountered that fill the renal pelvis. These are referred to as *staghorn calculi* because of their appearance.

CHART 16–7. KIDNEY STONES

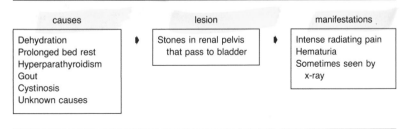

causes	lesion	manifestations
Dehydration Prolonged bed rest Hyperparathyroidism Gout Cystinosis Unknown causes	Stones in renal pelvis that pass to bladder	Intense radiating pain Hematuria Sometimes seen by x-ray

Hyperplastic/Neoplastic Diseases

Hyperplasia of the prostate and carcinoma of the prostate and bladder are common and very significant diseases of the elderly. Cancers of the testis and kidney are much less common, are of several types, and occur in younger individuals.

Hyperplasia of the Prostate

Enlargement of the prostate is caused by hyperplasia of the glandular parenchyma and its fibromuscular stroma in the periurethral area, probably due to relative hormonal imbalance in the elderly. As the bulk of this central tissue enlarges, the peripheral tissue

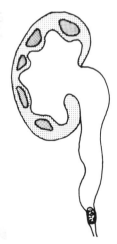

Figure 16–10. Renal stone lodged in ureter with dilated ureter and renal pelvis (hydronephrosis).

begins to atrophy. The hyperplastic process does not take place evenly throughout the affected regions but occurs as multiple nodules. The cause of prostatic hyperplasia is not known. The most popular theories involve imbalances in the blood androgen–estrogen ratios. The most common symptom of prostatic hyper-

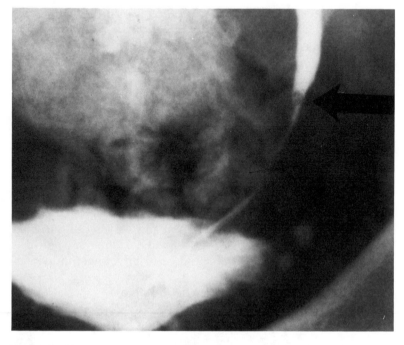

Figure 16–11. Intravenous urogram demonstrating a calculus lodged in the ureter (*arrow*) with dye-filled, dilated ureter above and dye-filled urinary bladder below.

plasia is difficulty in initiating and stopping urination. The major complication of prostatic enlargement is obstruction of the urinary tract at the outlet of the bladder (Fig. 16–12). Urinary tract infections frequently occur as a consequence of the obstruction. The usual treatment is surgical resection of the gland, and the most common surgical procedure is the *transurethral resection* (TUR) of excessive prostatic tissue. This operation entails the use of a special instrument inserted through the penis to cut away chips of prostatic tissue surrounding the urethra.

CHART 16–8. HYPERPLASIA OF PROSTATE

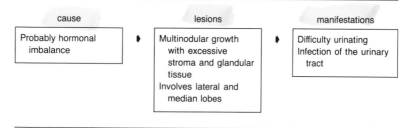

cause	lesions	manifestations
Probably hormonal imbalance	Multinodular growth with excessive stroma and glandular tissue Involves lateral and median lobes	Difficulty urinating Infection of the urinary tract

Adenocarcinoma of the Prostate

Cancer of the prostate is rare under the age of 50 but progressively more common thereafter, attaining a very high incidence in the elderly. Carcinoma of the prostate presents as a hard, irregular nodule in the gland, usually in the posterior lobe (Fig. 16–13). Its

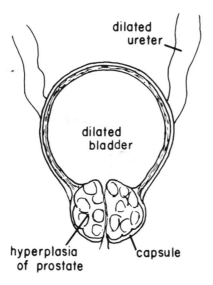

dilated ureter

dilated bladder

hyperplasia of prostate

capsule

Figure 16–12. Hyperplasia of the prostate with obstruction of the urethra.

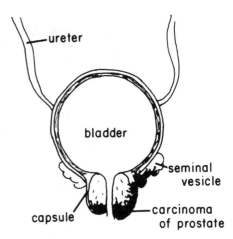

Figure 16–13. Carcinoma of the prostate with invasion of capsule and seminal vesicle.

presence can often be found by palpation of the prostate from the rectum. Low back pain and weight loss, x-ray appearance of pelvic bone lesions, and elevated serum acid and alkaline phosphatase all suggest metastatic dissemination. Occasionally, prostatic carcinomas produce urethral obstruction and may require transurethral resection to relieve the obstruction. Prostatic carcinoma preferentially metastasizes to bone. Carcinomas found while still contained within the prostate itself can sometimes be successfully treated by surgical excision. When metastatic spread has occurred, estrogen therapy with or without castration to remove the source of testosterone may be employed as palliative therapy. It is not uncommon for elderly men to live 10 years or more with prostatic cancer. On the other hand, in some patients the disease progresses rapidly, especially in younger ones with normal testosterone levels.

CHART 16–9. ADENOCARCINOMA OF PROSTATE

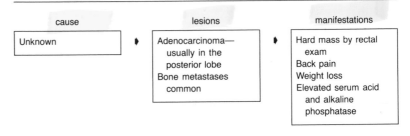

cause	lesions	manifestations
Unknown	Adenocarcinoma— usually in the posterior lobe Bone metastases common	Hard mass by rectal exam Back pain Weight loss Elevated serum acid and alkaline phosphatase

Transitional Cell Carcinoma of the Bladder

Carcinoma of the urinary bladder is the fifth most common cancer, occurs much more commonly in males, and has an age peak in the seventies. In the United States, cigarette smoking may be the most important etiologic factor. There is a high incidence among employees of industries that manufacture certain chemicals, notably aniline dyes. Transitional cell carcinomas usually present with painless hematuria. Diagnosis is made by cystoscopy and biopsy. Prognosis varies greatly depending on whether the tumor is superficial and well differentiated or extends into the muscularis propria and is poorly differentiated (Fig. 16–14). Superficial lesions can be removed by transurethral resection, but they tend to recur or be multiple.

CHART 16–10. TRANSITIONAL CELL CARCINOMA OF BLADDER

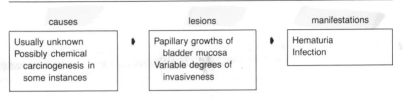

causes	lesions	manifestations
Usually unknown Possibly chemical carcinogenesis in some instances	Papillary growths of bladder mucosa Variable degrees of invasiveness	Hematuria Infection

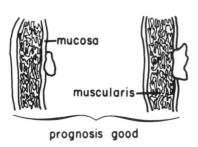

prognosis good

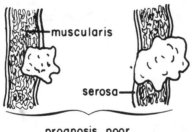

prognosis poor

Figure 16–14. Transitional cell carcinoma of the bladder with varying degrees of invasion.

Cancers of the Kidney

Cancer of the kidney is relatively uncommon. In children, malignant tumors of primitive renal elements occur and are called *Wilm's tumor.* Adenocarcinomas of renal epithelial origin comprise most of the renal carcinomas in adults. Both types of tumors often present with hematuria and/or a flank mass. Treatment consists of surgical removal of the affected kidney plus chemotherapy or radiation therapy if extension beyond the kidney has occurred.

Cancers of the Testis

Several types of cancers of the germ cells of the testis occur. The most common is called *seminoma.* These cancers are uncommon but tend to occur most often in young adult men. They are usually detected as a mass. Seminomas have the most favorable outcome; however, other types are often cured by aggressive chemotherapy and radiation therapy even in the presence of metastases.

ORGAN FAILURE

Acute Renal Failure

The common life-threatening lesion of the renal tubules is *acute tubular necrosis,* an acute degeneration and necrosis of renal epithelial cells in the proximal and/or distal convoluted tubules. The common mechanisms of acute tubular necrosis are ischemic and toxic injury. Ischemic necrosis results from shock, with diversion of blood flow away from the kidney and insufficient oxygen to keep the tubular cells alive. Toxic necrosis is caused by a variety of poisons, the most common of which are methyl alcohol, ethylene glycol (antifreeze), mushroom poisons, chloroform, carbon tetrachloride, several antibiotic agents, and many heavy metals, especially mercury. Necrosis of the renal tubular epithelium rapidly leads to oliguria and even anuria. Consequent water and potassium retention in the patient results in a life-threatening situation. The clinical course of such a patient is monitored by following levels of the blood urea nitrogen and creatinine. The higher the values, the worse the patient's condition. If the patient can be kept alive by careful monitoring of blood electrolytes and possible dialysis (artificial kidney) for about 2 weeks, then the renal epithelium begins to regenerate. This regeneration is attended by significant diuresis (excessive urinary output), because the immature epithelium cannot efficiently reabsorb water.

Chronic Renal Failure

Most renal failure is chronic and insidious. Often the first detectable evidence of compromised renal function is the retention of the nitrogenous breakdown products of protein metabolism, reflected by elevated serum levels of creatinine and urea nitrogen. The retention of nitrogenous wastes in the blood is called *azotemia*. Progressive renal failure will result in body water disturbances, electrolyte (sodium, potassium, chloride, calcium) imbalances, and retention of acids, with profound effects on the body, including slowed mental activity, muscle weakness, and anemia. The sum total of all these effects of advanced renal failure is called *uremia*. Chronic renal failure is most often the result of chronic glomerulonephritis but may occasionally be due to severe pyelonephritis, long-standing obstruction of the urinary tract, or severe vascular disease of the kidney (arteriosclerosis).

In acute renal failure, the patient may be dialyzed if there is reasonable hope that kidney function may return to a near normal state. With chronic renal failure, the patient may be dialyzed at regular intervals on a permanent basis, or, depending on the nature of the underlying disease, a kidney transplant may be considered.

REVIEW QUESTIONS

1. What are the most common diseases of the kidney, lower urinary tract, and male genital system? What is the major effect of each?
2. What is the significance of each of the following signs or symptoms?
     Hematuria
     Dysuria
     Nocturia
     Frequency
     Anuria
     Flank pain
     Pyuria
3. What are the following tests and procedures used for?
     Urinalysis
     Blood urea nitrogen
     Creatinine
     Intravenous urogram
     Cystogram
     Cystoscopy
     Renal biopsy
4. What are the causes, lesions, and major manifestations of each of the specific diseases discussed in this chapter?
5. How do immune complex glomerulonephritis and anti-basement membrane glomerulonephritis differ in terms of cause and lesion?

6. How do acute and chronic glomerulonephritis differ?
7. What is the usual cause and pathogenesis of pyelonephritis?
8. What are the common causes and predisposing factors of kidney stones?
9. How do prostatic hyperplasia and prostatic carcinoma differ in terms of causes, lesions, manifestations, and possible complications?
10. How do acute and chronic renal failure differ in terms of causes, lesions, and manifestations?
11. What do the following terms mean?
   Juxtaglomerular apparatus
   Casts
   Nephritis
   Calculi
   Transurethral resection
   Wilm's tumor
   Seminoma
   Oliguria
   Diuresis
   Azotemia
   Uremia

Female Genital Organs

REVIEW OF STRUCTURE AND FUNCTION

Female genital organs include the vulva (labia majora, labia minora, clitoris), vagina, uterus (cervix, body), fallopian tubes, and ovaries (Fig. 17–1). The vulva, vagina, and outer aspects of the cervix are lined by a protective stratified squamous epithelium.

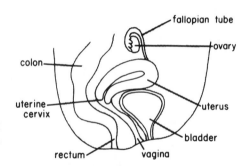

Figure 17–1. Female genital organs in relationship to colon and rectum.

Bartholin's glands at the outlet of the vagina produce a mucoid secretion.

The uterine cervix is the distal, narrow portion of the uterus that projects into the vagina. The vaginal surface of the cervix is covered by stratified squamous epithelium. At the os (opening), there is a transition to the columnar mucus-secreting epithelium of the endocervix. The glandular mucosa lining the uterine cavity (endometrium) is the site for implantation of a fertilized ovum. The body (fundus) of the uterus is a muscular organ with a glandular epithelial lining. The muscular wall stretches and hypertrophies during pregnancy, and its contraction is important in accomplishing childbirth.

The ovaries and endometrium function synchronously to provide opportunity for pregnancy to occur. The menstrual cycle is timed from the onset of bleeding because this date is easily determined. However, the critical events are more accurately timed from the day of ovulation as determined by a rise in basal body temperature. In the 14 days following ovulation, the endometrium, under the influence of progesterone secreted by the ovary, undergoes secretory changes leading to sloughing if implantation does not occur. During and after the 3 to 5 days of menstruation, the endometrium is in proliferative phase for about 14 days, but the length of this phase is less predictable than the secretory phase. If implantation occurs, the stromal cells of the secretory endometrium becomes large and plump producing decidualized endometrium or decidua. The complex endocrine control of the menstrual cycle and pregnancy involves hormones from the pituitary glands, adrenal glands, ovaries, and placenta. The details of these mechanisms and pathologic variations will be left to other sources.

MOST FREQUENT AND SERIOUS PROBLEMS

The most frequent health problems relating to the female genital system include birth control, sexual counseling, prenatal care and childbirth, menopausal symptoms, infections, and cancer screening.

Serious conditions associated with pregnancy include ectopic pregnancy, septic abortion, toxemia of pregnancy, hemorrhage, complications of delivery, and endometritis following delivery. Spontaneous abortion (miscarriage) and elective abortion are common.

Sexually transmitted (*venereal*) infections produce both acute and long-term problems in females. Gonorrhea is the most commonly reported communicable disease, but infections due to chlamydia, although less commonly reported, may be even more common. Gonorrhea and chlamydia infection produce urethritis and cervicitis. When they spread to the fallopian tubes they produce a more serious infection called *pelvic inflammatory disease*. Viral infections caused by human papilloma virus is an important factor in the genesis of carcinoma of the cervix. Herpes simplex commonly affects the cervix, vagina, and vulva.

After breast, the uterus and ovaries are the second most common site of cancer in females. Cytologic diagnosis of precancerous lesions has led to a decrease in carcinoma of the cervix so that it is now less common than endometrial carcinoma. Ovarian cancer is less common than uterine cancer, but much more likely to be fatal. Leiomyoma, a benign tumor, of the uterus is more common than cancer, but usually requires no treatment.

SYMPTOMS, SIGNS, AND TESTS

Major symptoms include bleeding, pain, vaginal discharge, and endocrine effects. Normal vaginal bleeding (menstruation), which occurs for 3 to 5 days at intervals of approximately 28 days from menarche to menopause (except during pregnancy) must be distinguished from abnormal uterine bleeding. Bleeding may be abnormal in amount, timing, or character. Hormonal changes are the most common cause of abnormal bleeding, but bleeding is also one of the principal symptoms of cancer. Several names have been coined to describe patterns of bleeding: menorrhagia (excessive menses), vaginal spotting (small amounts of blood not associated with menses), dysmenorrhea (painful menstruation), dysfunctional uterine bleeding (abnormal bleeding without causative lesion).

Cramping pain is common during menstruation (dysmenorrhea). Many women experience a sharp, one-sided abdominal pain at midcycle due to peritoneal irritation caused by rupture of an ovarian follicle at the time of ovulation. Causes of severe pain include ruptured ectopic pregnancy, acute pelvic inflammatory disease (salpingitis), and twisted or ruptured ovarian cysts. Pruritis (itching) is a form of pain that is commonly associated with atrophic changes in the vulvar skin in postmenopausal women and infections of the vulva and vagina in younger women.

Nonbloody vaginal discharge is associated with mild superficial infections, such as trichomonas and candida vaginitis. Symptoms related to endocrine changes are common prior to onset of menstruation (often called premenstrual syndrome) and at the time of menopause (hot flashes).

Signs of gynecologic disease are discovered by systematic examination of the female genitalia referred to as *pelvic examination*. This involves direct inspection of the vulva, examination of the vagina and cervix through a speculum (a device used to spread the vaginal wall), and bimanual palpation of the uterus, fallopian tubes, and ovaries. Bimanual examination is so named because one of the examiner's hands is placed on the abdomen and the fingers of the other hand are inserted into the vagina and the organs are palpated between the two hands. Bimanual rectal examination allows palpation of posterior uterine lesions.

The most common signs are pelvic mass and flat or raised lesions of the vulva and cervix. Pregnancy is the most common mass. Leiomyomas of the uterus are also common and the various ovarian tumors are usually discovered by palpation of a mass. Carcinomas of the vulva and cervix are visible by inspection. Condylomata acuminata are common venereal warts of the vulva, vagina, and cervix. The vulva may be involved by many skin diseases.

The most common test is the Papanicolaou smear (Pap smear) of the uterine cervix (Fig. 17–2). Pap smears at regular intervals allow detection of cervical cancer before it becomes invasive and can be credited with a sharp reduction in mortality from this disease. Blood counts are important for detection of iron-deficiency anemia, a common condition in women due to loss of iron in menstrual blood and transfer of iron to the fetus during pregnancy. Pregnancy tests involve the measurement of chorionic gonadotropin, a hormone, in urine or blood.

The female genital tract is rivaled only by skin in the frequency of use of biopsy as a diagnostic tool. Biopsy is used to define the nature of vulvar lesions, especially those with malignant potential. The cervix is biopsied whenever there is a visible lesion or a positive Pap smear. Staining of the cervix and examination at high magnification with a colposcope aids in selecting the most appropriate biopsy sites. Cone biopsy refers to the removal of a cone of tissue including the cervical os and endocervical lining for systematic histologic evaluation. Endometrial biopsy is essential for evaluation of endometrial lesions and to exclude endometrial carcinoma (cytology is much more reliable in detecting cervical

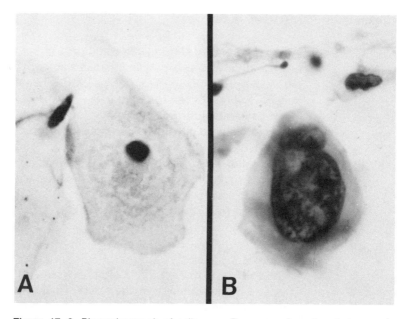

Figure 17–2. Photomicrograph of cells on a Pap smear from the uterine cervix. **A.** Normal surface squamous epithelial cell with small nucleus and abundant cytoplasm. **B.** Neoplastic squamous epithelial cell with large atypical nucleus and scant cytoplasm.

lesions). One procedure for endometrial biopsy is called dilation and curettage (D & C) and involves dilating the cervical os and scraping out tissue with a curette. Small samples of endometrium can be obtained without dilating the cervix.

Diseases of the fallopian tubes and ovaries may be evaluated by ultrasound, a noninvasive procedure, and by *laparoscopy,* a procedure involving the insertion of an endoscope into the peritoneal cavity through an incision at the umbilicus. These procedures are particularly useful in diagnosis of ectopic pregnancy, endometriosis, and pelvic inflammatory disease.

SPECIFIC DISEASES

Genetic/Developmental Diseases

Compared to other organ systems, congenital anomalies of the female genital organs are relatively uncommon. Various degrees of uterine duplication occur. Resection of a uterine septum that divides the endometrial cavity into two compartments may improve chances for a successful pregnancy. Cysts formed from embryonic remnants occur adjacent to the fallopian tubes and in the lateral wall of the cervix and vagina.

Inflammatory/Degenerative Diseases

Gonorrheal and Chlamydial Infections

Gonococcal and chlamydial organisms account for millions of cases of venereally transmitted disease in the United States each year (see Chapter 26 for discussion of these diseases). Females with asymptomatic urethritis and cervicitis are carriers. If the organisms spread to the fallopian tubes, acute or chronic salpingitis may occur. Acute salpingitis with abdominal pain and fever must be differentiated from acute appendicitis since the treatment for the latter is surgical. Acute salpingitis is treated with antibiotics. Resolution of acute salpingitis often leads to scarring of the fallopian tube. Scarring of the tube leads to two important complications: infertility and ectopic pregnancy.

Syphilis

The painless primary ulcer of syphilis (chancre) will go unnoticed if located in the vagina or on the cervix. The key to control of syphilis is finding and treating such asymptomatic carriers. (For further discussion of this disease see Chapter 26).

Herpes Infection

Herpes simplex viruses cause blistering lesions of the squamous epithelium of the vulva and cervix. Most genital herpes is caused by herpes simplex virus type II, but some infections are caused by type I virus. At birth the virus may rarely be spread to the newborn producing a fatal disseminated infection.

Condyloma Acuminatum

Several serotypes of human papilloma viruses can cause multiple squamous papillomas of the vulva, anus, vagina, and cervix. These lesions are similar to the common skin wart (verruca vulgaris) except that they are commonly venereally transmitted and they tend to occur on moist mucous membranes in both males and females. Lesions on the cervix are usually flat. The papilloma virus and condylomatous changes in the cervix have been statistically associated with the development of carcinoma of the cervix.

Superficial Vaginal Infections

Trichomonas vaginalis (Fig. 17–3A) is a protozoan that commonly produces a bubbly vaginal discharge and may cause small reddened areas in the vagina and cervix. Trichomoniasis does not cause any serious problems, is easily diagnosed by microscopic examination of a fresh drop of vaginal fluid, and can be treated with drugs. Candida (Fig. 17–3B) are fungi that are normally pres-

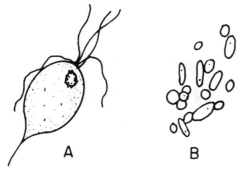

Figure 17–3. A. *Trichomonas vaginalis* organism. **B.** Yeast form of candida organisms.

ent in small numbers but can proliferate to produce a superficial vaginitis, especially during pregnancy, in association with oral contraceptive therapy, in diabetic women, and as a complication of antibiotic therapy. Candidiasis can be diagnosed by smear or culture of vaginal discharge. Treatment consists of antimicrobial agents and control of the underlying condition.

Hyperplastic/Neoplastic Diseases

There is a wide variety of neoplasms, hyperplasias, and cystic diseases of the female genital system. Carcinomas of the vulva, cervix, endometrium, and ovary account for the vast majority of cancers. Leiomyoma of uterus is important because of its frequency. Endometriosis, a condition of uncertain cause, represents ectopic endometrium that responds to hormonal stimulation. Collectively, ovarian tumors are fairly common. They present as masses and are diagnosed by a pathologist after removal.

Endometriosis

Endometriosis is the occurrence of endometrial tissue outside the uterus, usually in the ovary, but also on the peritoneal surface of adjacent organs. The origin of this tissue is unclear (metaplasia or ectopia). Bleeding may occur at the time of menstruation with the resultant hematoma producing severe dysmenorrhea. The hematoma may be walled off by fibrous tissue with the liquid contents forming a "chocolate" cyst. In addition to pain, endometriosis is often associated with infertility.

Leiomyoma of the Uterus

Solitary or multiple benign neoplasms of uterine smooth muscle are estrogen dependent, i.e., they develop during childbearing years and undergo atrophic changes after menopause. Atrophy with fibrous replacement accounts for the common synonym, *fibroid*. Leiomyoma may become quite large and distort the uterus

(Fig. 17–4). They are usually asymptomatic but may cause pelvic pressures, bleeding, or infertility.

CHART 17–1. LEIOMYOMA OF THE UTERUS

cause	lesions	manifestations
Unknown (dependent on estrogen for growth)	Benign neoplasm of smooth muscle Often multiple	Mass Bleeding Infertility

Carcinoma of the Endometrium

Adenocarcinoma of the endometrium is now the most common female genital cancer (excluding breast cancer). Prolonged estrogen stimulation is associated with increased frequency of endometrial carcinoma such as occurs in patients with breast cancer being treated with hormones, estrogen-secreting ovarian tumors, and obesity. Current types of oral contraceptive therapy do not predispose to endometrial cancer. Other conditions associated with endometrial cancer include diabetes, hypertension, and infertility. Endometrial carcinomas, which are often well differentiated, begin

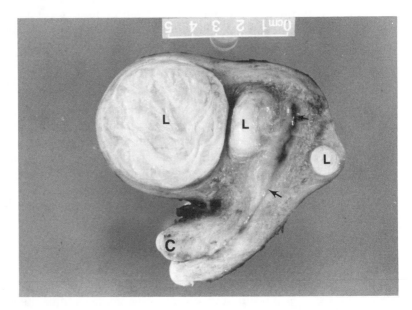

Figure 17–4. Uterus with three leiomyomas (L). Arrows indicate endometrial cavity. C is cervix.

in a normal or hyperplastic endometrium. They gradually grow into the wall of the uterus (Fig. 17–5), but usually bleed and are diagnosed before they metastasize.

Radiation therapy and surgical removal of the uterus and ovaries can be combined to give an overall survival rate of about 80 percent because most of the cancers are confined to the body of the uterus at the time of diagnosis. The high cure rate may in part be due to the fact that most of the cancers develop after menopause so that bleeding is recognized by the patient as abnormal.

CHART 17–2. CARCINOMA OF ENDOMETRIUM

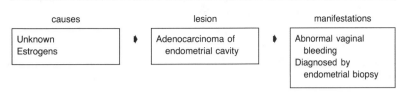

causes	lesion	manifestations
Unknown Estrogens	Adenocarcinoma of endometrial cavity	Abnormal vaginal bleeding Diagnosed by endometrial biopsy

Carcinoma of the Cervix

Squamous cell carcinomas of the cervix almost always arise near the os at the junction of ectocervix and endocervix. Fortunately, carcinoma of the cervix usually develops in stages over a number of years so that Pap smears at regular intervals should provide a diagnosis in time for a cure. The evidence is overwhelming that some sexually transmissible agent is involved in the pathogenesis of carcinoma of the cervix; women who have not had intercourse do not develop the cancer. The changes in the cervical epithelium that precede overt cancer with invasion represent a continuum, but

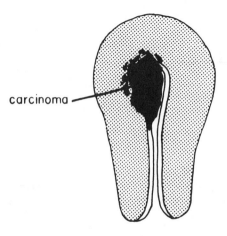

carcinoma

Figure 17–5. Carcinoma of the endometrium.

for convenience are graded as follows: cervical intraepithelial neo-plasia grade I (CIN-I or mild dysplasia), CIN-II (or moderate dysplasia), CIN-III (encompassing severe dysplasia and carcinoma-in-situ), and microinvasive squamous cell carcinoma. It is not clear exactly when the developing lesion becomes a cancer, but metastases cannot occur before microinvasion occurs and rarely occurs before a grossly visible cancer is present.

The grading of the developing lesion is done by cytology and biopsy and is subject to observer variability. CIN-I lesions are usually followed to see if they regress or become more severe. CIN-III lesions are removed by hysterectomy, cone biopsy, or laser cautery. CIN-II lesions are variably treated. Cone biopsy is used for treatment when the patient wishes to preserve childbearing potential. Invasive lesions are treated by surgical removal and/or radiation depending on the stage of the disease. Advanced carcinoma of the cervix (Fig. 17–6) usually kills by local invasion leading to obstruction of the ureters and renal failure.

CHART 17–3. CARCINOMA OF THE CERVIX

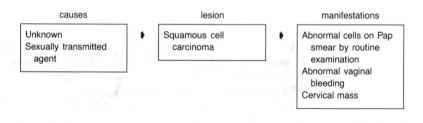

causes	lesion	manifestations
Unknown Sexually transmitted agent	Squamous cell carcinoma	Abnormal cells on Pap smear by routine examination Abnormal vaginal bleeding Cervical mass

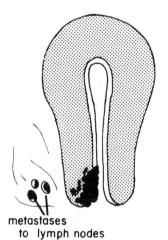

metastases
to lymph nodes

Figure 17–6. Carcinoma of uterine cervix.

Ovarian Tumors

The ovary has the potential to produce a greater variety of tumors than any other organ. These take the form of cysts and neoplasms of varying malignancy. Of the neoplasms, epithelial neoplasms derived from surface cells are most common. These epithelial neoplasms differentiate along three lines. Serous tumors form from simple ciliated columnar epithelium, mucinous tumors have cells containing a drop of mucus, and endometrioid tumors are glandular tumors resembling endometrium. The serous and mucinous tumors are usually cystic and are classified into three groups: cystadenoma (a benign epithelial-lined cyst), lesions with low malignant potential (cysts with more cellular lining and capacity to implant in the peritoneum, but limited growth potential), and cystadenocarcinoma (obviously invasive). The survival rate for the carcinomas is very low; thus ovarian cancer kills as many women as endometrial and cervical cancers combined. Tumors with low malignant potential are usually cured if localized to the ovary, but they are often large and may spill into the peritoneum where they have a tendency to recur. The endometrioid tumors are usually malignant.

Ovarian tumors derived from ovarian stroma or germ cells tend to be benign, although a few have malignant potential. The stromal tumors may secrete female or male hormones. The most common germ cell tumor is a benign teratoma, often called a *dermoid cyst*. Teratomas are composed of a variety of tissues including skin, various types of glands, cartilage, brain, and even teeth.

Non-neoplastic cysts are quite common and include follicular cysts (enlarged graafian follicles), corpus luteum cyst (enlarged corpus luteum), and chocolate cyst of endometriosis.

The diagnosis of ovarian tumors is based on pathologic examination after removal. Since most ovarian tumors are small and benign, professional judgment is required to decide which ones should be removed. Hemorrhage from a cystic tumor or twisting and infarction of the tumor may produce an acutely painful abdomen and require operative removal of the cyst or tumor.

Diseases of Pregnancy

The number of problems that can complicate pregnancy is so large that they are separately dealt with by the subspecialty of obstetrics. We will discuss a few important problems that arise at various times during and after pregnancy. Numerous other abnormal conditions, many of which occur at the time of labor and delivery, are beyond the scope of this text.

Spontaneous Abortion

At least one out of five pregnancies terminates spontaneously during the first third of pregnancy. These abortions usually result from a major defect in embryonic development due to abnormal chromosomes. The chromosome abnormality is usually confined to the individual sperm or ovum involved, so the next pregnancy is likely to be normal. The patient presents with cramps and bleeding, and the placental tissue either passes spontaneously or is found in the cervical os or vagina. Dilation and curettage may be needed to remove remaining placental tissue to prevent infection. The pathologist frequently finds only placental tissue without evidence of the embryo.

Ectopic Pregnancy

This results from implantation of the fertilized ovum before it reaches the endometrium. It most often occurs in a fallopian tube (Figs. 17–7, 17–8) but may occur in the abdominal cavity. As the placenta enlarges, blood vessels are likely to rupture and produce serious internal hemorrhage. The symptoms include pain from hemorrhage into the fallopian tube, vaginal bleeding, and light-headedness from blood loss. Laparoscopy leads to a specific diagnosis, and surgical removal of the ectopic pregnancy prevents fatal hemorrhage. Ectopic pregnancies are more likely to occur in fallopian tubes that have been partially obstructed by previous pelvic inflammatory disease.

CHART 17–4. ECTOPIC PREGNANCY

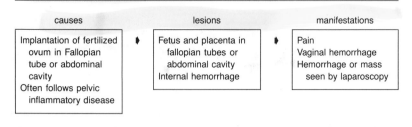

causes	lesions	manifestations
Implantation of fertilized ovum in Fallopian tube or abdominal cavity Often follows pelvic inflammatory disease	Fetus and placenta in fallopian tubes or abdominal cavity Internal hemorrhage	Pain Vaginal hemorrhage Hemorrhage or mass seen by laparoscopy

Septic Abortion

This refers to infection of the uterus superimposed on abortion and more commonly occurs when abortions are carried out by untrained persons using nonsterile technique. The infection may spread to the peritoneal cavity and cause death. Treatment consists of removal of the infected tissue along with antibiotics.

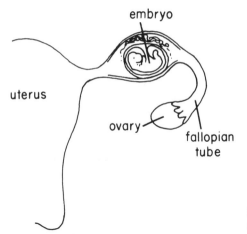

Figure 17–7. Ectopic pregnancy in fallopian tube.

Toxemia of Pregnancy

This common syndrome of unknown cause develops in the last third of pregnancy and is characterized by high blood pressure, edema, and proteinuria. Toxemia may progress to a severe, life-threatening stage with convulsions. Toxemia with convulsions is called *eclampsia*. Toxemia is most common in first pregnancies, in

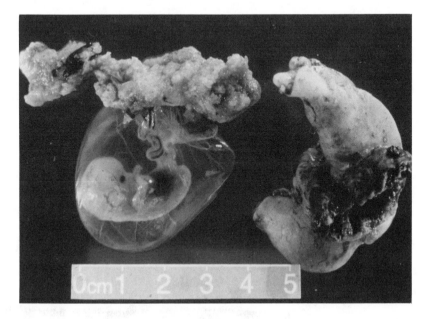

Figure 17–8. Ruptured ectopic pregnancy. Fallopian tube is on right, embryo on left.

pregnancies with multiple fetuses, and in patients with previous high blood pressure or renal disease. It has been postulated that release of thromboplastin (tissue factor) from the placenta causes a low-grade disseminated intravascular coagulation with thrombotic lesions in the renal glomeruli leading to protein loss and activation of the renin–angiotensin mechanism to produce high blood pressure. Medical measures may lessen the effects to some extent, but termination of pregnancy is the only cure. In making the decision to terminate pregnancy, the danger of eclampsia to the mother must be weighed against dangers of prematurity to the fetus. Amniocentesis with measurement of surfactant levels may be helpful in determining the maturity of the fetus. Toxemia may or may not recur in a subsequent pregnancy. Detection of toxemia is a major reason for frequent medical visits in the last third of pregnancy.

CHART 17–5. TOXEMIA OF PREGNANCY

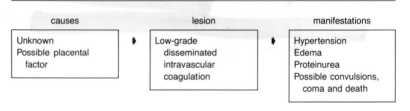

causes	lesion	manifestations
Unknown Possible placental factor	Low-grade disseminated intravascular coagulation	Hypertension Edema Proteinurea Possible convulsions, coma and death

Gestational Trophoblastic Neoplasms

Neoplastic placental tissue, which results from genetic material entirely derived from sperm, presents a spectrum of lesions whose behavior can be benign or malignant. Most are benign, grape-like masses of edematous placental villi that are noninvasive and are named *hydatidiform mole*. Some are locally invasive into the uterine wall (invasive mole). A malignant metastasizing variant, *choriocarcinoma*, occurs in about 2 percent of these neoplasms. Gestational trophoblastic neoplasms secrete chorionic gonadotropic that can be measured in blood or urine. This provides an effective diagnostic test and a test for determining when all of the neoplasm has been removed. Further, choriocarcinoma is one of the few cancers that can be cured by chemotherapy even when metastases are present.

ORGAN FAILURE

Fertility and infertility are both major health care problems. The majority of women desire temporary or permanent infertility at some time during the childbearing years. Temporary infertility can be produced by mechanical blockage of the cervix (diaphragm), spermicidal chemicals, mechanical prevention of implantation (intrauterine device), hormonal prevention of ovulation (oral contraceptives), and by hormonal prevention of implantation (the morning-after pill). Permanent infertility in the female can be accomplished by clamping or resecting a portion of the fallopian tubes or by hysterectomy. Occasionally tubes can be repaired to restore fertility. The condom is the major means available to produce temporary infertility in the male. Vasectomy provides permanent infertility. All of these techniques have desirable and undesirable features. Medical expertise and counseling are extensively used in order to reach optimum solutions.

Undesirable, involuntary infertility is also a common problem. Lesions that cause infertility in the female include polycystic ovaries, endometriosis, chronic salpingitis, leiomyomas, and congenital anomalies of the uterus. A low-grade endometritis of unknown cause may cause infertility and antibiotic therapy sometimes results in a successful pregnancy. Imbalance in the normal endocrine cycle may also lead to infertility. An infertility evaluation involves examination of male seminal fluid for number and quality of sperm, injection studies to determine the patency of the fallopian tubes, body temperature measurements to determine whether ovulation has occurred, and endometrial biopsy to evaluate the hormonal response of the endometrium, and search for evidence of endometritis. Sometimes the cause of infertility cannot be found.

REVIEW QUESTIONS

1. What are the most frequent health problems of the female genital system?
2. What are the typical causes of abnormal vaginal bleeding, pelvic pain, vaginal discharge, and pelvic mass?
3. How are each of the following tests or procedures helpful in evaluation of a patient with health problems referable to the female genital system?
   Pelvic examination
   Pap smear
   Cone biopsy
   Dilation and curettage
   Laparoscopy
4. What are the causes, lesions, and manifestations of each of the specific diseases discussed in this chapter?

5. What are the possible complications of the various infections of the female genital tract?
6. Why is cervical carcinoma preventable?
7. Why is ectopic pregnancy life threatening?
8. What are the methods of voluntary infertility and causes of involuntary infertility?
9. What do the following terms mean?
   Menarche
   Menopause
   Menorrhagia
   Dysmenorrhea
   Human chorionic gonadotropin
   Fibroid
   Dermoid cyst
   Follicular cyst
   Eclampsia
   Hydatidiform mole
   Choriocarcinoma

Breast

REVIEW OF STRUCTURE AND FUNCTION

The mature female breast is composed of eight to ten separate sets of glandular units, each of which empties into the nipple by means of an excretory duct (Fig. 18–1). The branching ducts of each unit connect with many glandular lobules. Within each glandular lobule, there are epithelial buds surrounded by loose connective tissue. Between glandular lobules, there is a variable amount of dense connective tissue and adipose tissue.

At birth, male and female breasts consist only of ducts. The ducts may be hypertrophied at birth, producing slight breast enlargement due to the influence of maternal hormones. Otherwise,

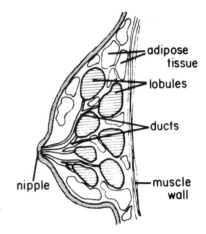

Figure 18–1. Structures of normal mature breast.

the breasts remain quiescent until puberty, when both male and female breasts undergo ductal hyperplasia. The reaction is of minimal extent and regresses in the male. Proliferation of ducts, glandular buds within lobules, and, to a great extent, connective tissues account for the enlargement of the female breast at puberty. The cyclic hormonal stimulation of the menstrual cycle results in mild hyperplasia and regression of glandular buds and edema of the connective tissue. In some women, this results in a feeling of fullness at the end of the menstrual cycle. High levels of the hormones estrogen, progesterone, and prolactin produced during pregnancy cause a striking change in the structure of the breast. The glandular buds become glands with luminal secretion. The glandular proliferation compresses connective tissue so that the entire breast appears to be composed of glands, as compared to the nonlactating breast, which appears to be predominantly connective tissue. Following the period of lactation, the physiologic glandular hyperplasia gradually regresses, although glands are somewhat more abundant than before the first pregnancy. After menopause, there is gradual atrophy of glands, so that in very old women the breasts consist mainly of ducts, dense connective tissue, and some adipose tissue.

MOST FREQUENT AND SERIOUS PROBLEMS

Carcinoma of the breast is by far the most important breast disease. About 1 of every 15 women will develop breast cancer and about 1 of every 25 will die from it. Other lesions of the breast are important because they may be causally related to breast cancer (epithelial hyperplasia) or need to be differentiated from it (fibro-

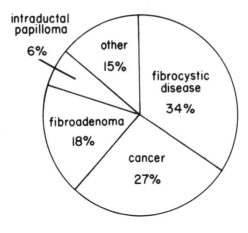

Figure 18–2. Relative frequency of breast lesions as diagnosed at the time of operation *(Data from Leis HP: The diagnosis of breast cancer. CA; Cancer Journal for Clinicians, 27:211, 1977.)*

adenoma, fat necrosis, fibrocystic disease). The relative frequency of breast diseases that produce masses and require biopsy for diagnosis is shown in Figure 18–2 and the age distribution at the time of biopsy is shown in Figure 18–3.

Infection of the breast (acute mastitis) is usually a complication of lactation. The extreme variation in normal size of the breasts is a problem for some women who undergo reduction mammoplasties (to remove tissue) or insertion of prostheses (to increase the apparent size).

SYMPTOMS, SIGNS, AND TESTS

Symptoms and signs of breast disease are similar as patient and physician make the same observations. The most important finding is a painless lump, which is usually discovered by the patient, particularly if she practices routine self-examination. The probability of the lump being cancer increases with age (Fig. 18–3), but

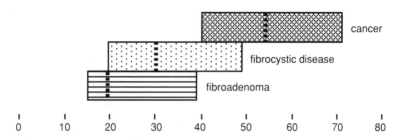

Figure 18–3. Age frequencies of major breast lesions, with mean represented by dashed line. *(Data from Leis HP: The diagnosis of breast cancer. CA; Cancer Journal for Clinicians, 27:212, 1977.)*

most lumps that are biopsied are not cancer (Fig. 18–2). Further, many lumps that are clinically judged to be benign are not biopsied. Other signs of breast cancer will be discussed later.

Secretion of milk unassociated with pregnancy is called *galactorrhea*. It may be associated with oral contraceptive therapy; otherwise, investigation for the presence of a prolactin-secreting pituitary tumor or other cause should be undertaken.

X-ray examination of the breast (*mammography*) is used to evaluate breast lumps and to screen patients for early breast cancer. Mammography may demonstrate masses (Fig. 18–4) and is particularly useful because cancer tends to have foci of dystrophic calcification that are easily seen radiologically.

Biopsy is the definitive test of breast disease. The decision to biopsy lies in the hands of the surgeon who must judge the likelihood of a mass being carcinoma. It is impractical to biopsy all breast masses and unwise to biopsy too few.

Measurement of estrogen and progesterone receptor sites on breast cancer cells is often performed on biopsy tissue to predict

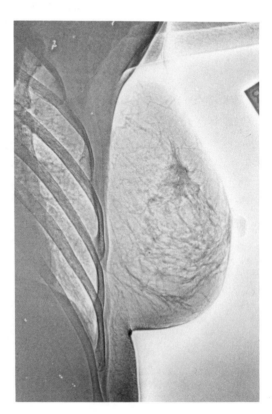

Figure 18–4. Mammogram demonstrating a small radiodense cancer in upper breast.

the likely response of the cancer to hormone manipulation therapy, if needed at a future time.

SPECIFIC DISEASES

Genetic/Developmental Diseases

Supernumerary Nipples
Extra nipples may be located anywhere along the milk line, which extends from clavical to midgroin.

Accessory Breast Tissue
Breast tissue may extend into the axilla. Enlargement of the tissue occasionally may be confused with metastatic carcinoma in axillary lymph nodes.

Inflammatory/Degenerative Diseases
Inflammatory conditions of the breast are not common. In addition to acute mastitis, there is an unusual form of chronic inflammation of unknown cause called *comedomastitis*, which affects one or more of the major ducts and occurs in women near the time of menopause. Trauma to the breast can produce localized *fat necrosis*, which becomes scarred and may produce a hard lump simulating carcinoma. Trauma and infection, however, do not cause cancer.

Acute Mastitis
Ordinary bacteria from the skin, such as *Staphylococcus aureus*, may gain access to the breast through the nipple and cause an acute inflammatory reaction. This is most likely to occur during lactation. The involved area becomes painful, red, hot, and swollen. Abscesses often form, with drainage of pus. If not severe or excessively painful, breast feeding can continue since the organisms likely originated from the baby's mouth anyway and decompression of the breast aids the healing process.

CHART 18–1. ACUTE MASTITIS

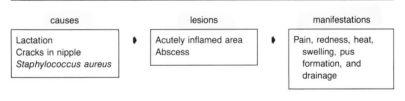

causes	lesions	manifestations
Lactation Cracks in nipple *Staphylococcus aureus*	Acutely inflamed area Abscess	Pain, redness, heat, swelling, pus formation, and drainage

Hyperplastic/Neoplastic Diseases

Fibroadenoma

This benign tumor is peculiar in that it is composed of both glandular and fibrous elements, which form a well-encapsulated nodule (Fig. 18–5). It occurs as a painless, movable lump, most commonly in women 15 to 35 years of age. Excisional biopsy results in diagnosis and cure.

CHART 18–2. FIBROADENOMA

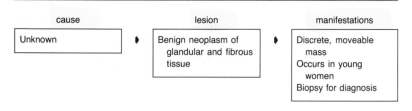

cause	lesion	manifestations
Unknown	Benign neoplasm of glandular and fibrous tissue	Discrete, moveable mass Occurs in young women Biopsy for diagnosis

Fibrocystic Disease

Fibrocystic disease is a general term used to encompass a variety of changes that occur in the breast. The most conspicuous changes are fibrosis and cystic dilation of ducts (Fig. 18–6). The most important changes, however, relate to hyperplasia of breast ducts because ductal hyperplasia may be a precursor of carcinoma. It is assumed that the stimuli for ductal hyperplasia also promote the development of cancer. The greater the extent of ductal hyperplasia and the more atypical the cells are, the more one is concerned about the future likelihood of cancer. Overall the association

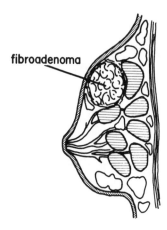

fibroadenoma

Figure 18–5. Fibroadenoma of breast.

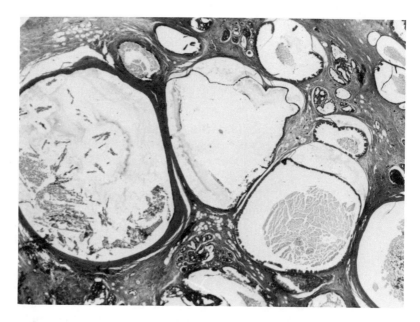

Figure 18–6. Fibrocystic disease, with many large cysts surrounded by fibrous tissue.

between fibrocystic disease and cancer is low. In examining biopsies of fibrocystic disease, pathologists try to identify those patients who are at higher risk for cancer. This is of aid to the surgeon in deciding to biopsy the breast again when another lump is found. Sometimes the degree of proliferative activity is so great that prophylactic mastectomy is warranted.

The cause of fibrocystic disease appears to be fluctuating estrogen stimulation of the breast. Combined oral contraceptive therapy with estrogens and progesterones appears to decrease the extent of this condition. Mild degrees of fibrocystic disease are so frequent and nonspecific histologically that it is difficult to interpret statistical findings of possible associations with cancer.

CHART 18–3. FIBROCYSTIC DISEASE

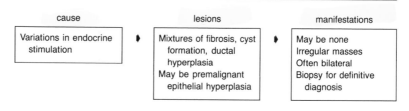

cause	lesions	manifestations
Variations in endocrine stimulation	Mixtures of fibrosis, cyst formation, ductal hyperplasia May be premalignant epithelial hyperplasia	May be none Irregular masses Often bilateral Biopsy for definitive diagnosis

Intraductal Papilloma

A localized hyperplasia within a large duct is called intraductal papilloma. A serous or bloody nipple discharge or a small mass call attention to these lesions. Removal is indicated because they cannot be distinguished clinically from carcinoma.

Carcinoma of the Breast

Breast carcinoma is the most common cancer in women in the United States, but not in all areas of the world. It increases in frequency after age 30, thus affecting many women in the prime of life. The 5-year survival rate is 75 percent. Carcinoma of the breast arises most often from the ductal epithelium and sometimes from the glandular epithelium of the lobules. Carcinoma may develop in any area of the breast. The most common site is the upper, outer quadrant of the breast.

Grossly, a breast cancer may be of any size. It usually can be felt on physical examination when it reaches 1 to 2 centimeters in diameter. When advanced, it may cause dimpling of the skin, changes in skin texture, retraction of the nipple, bloody discharge from the nipple, or even a protruding mass with ulceration (Fig. 18–7).

Microscopically, carcinomas of the breasts are composed of cells in a more or less glandular arrangement (adenocarcinoma) and surrounded by variable amounts of connective tissue. The neoplastic cells invade and replace normal breast tissue and often induce more fibrosis around the invading neoplastic cells. The his-

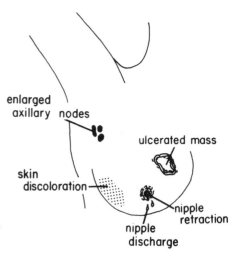

Figure 18–7. Signs produced by carcinoma of the breast.

tology varies from well-differentiated glands to undifferentiated masses of cancer cells. The natural history of untreated breast cancer includes metastasis to axillary lymph nodes and widespread metastasis to many organs. Occasionally, tumors are more selective in their site of metastasis, such as those that spread to bone, producing bone pain and sometimes fractures.

Extensive efforts have been made to subclassify breast cancers in hopes of better understanding their cause and behavior. Three subtypes will be briefly defined. Inflammatory carcinoma refers to a swollen, red appearance of the skin overlying a cancer and is caused by extensive permeation of lymphatic vessels; prognosis is poor. Medullary carcinomas are large, fairly discrete masses with lymphocytic reaction; prognosis is better than average. Paget's disease produces visible changes in the skin of the nipple and areola due to infiltration of the epidermis by cancer cells which have metastasized from an underlying cancer; prognosis is less than average.

Treatment of breast cancer is variable and subject to constant study to find the optimum methods. Nevertheless, the primary goal is to remove or destroy all of the cancer cells. Wide excision of the mass or mastectomy may be combined with radiation therapy. Resection of axillary lymph nodes provides information on the stage of the disease and may improve the cure rate when only a few of the nodes nearest to the breast are involved.

The progress of metastatic breast cancer can be delayed, but the long-term outlook is not favorable once the cancer has spread beyond the regional lymph nodes. Bone metastases, which are relatively common, often can be ameliorated by radiation therapy. Hormones affect the growth rate of many breast cancers. At the time the cancer is first removed, cancer cells can be evaluated for the presence of receptor sites for estrogen and progesterone. The majority of cancers with these receptors regress when the level of estrogen in the body is reduced by removal of ovaries, adrenals, or the pituitary gland. Such therapy is instituted after metastasis occurs in women with receptor-positive cancer cells.

The clinical course of breast cancer is quite variable. The majority are cured by removal of the primary lesions. Some cancers grow rapidly and are fatal within months, others are very slow growing and may recur years after removal of the primary lesion. Thus 5- and 10-year survival rates do not adequately reflect the overall mortality and morbidity of this disease. The variable survival rates and long duration make it difficult to evaluate the efficiency of various therapies.

CHART 18–4. CARCINOMA OF THE BREAST

causes		lesions		manifestations

Unknown
Fibrocystic disease
  mildly predisposing
	◆	Adenocarcinoma
		Metastases to regional
		  lymph nodes and
		  other organs
			◆	Mass
				Changes in overlying
				  skin
				Nipple discharge
				Enlarged axillary lymph
				  nodes
				Mammography
				Biopsy

Lesions of the Male Breast

Enlargement of the male breast due to proliferation of ducts and connective tissues is called *gynecomastia*. If bilateral, gynecomastia is usually due to estrogen simulation such as occurs with estrogen-secreting neoplasms, estrogenic drugs such as digitalis and marijuana, cirrhosis of the liver with decreased metabolism of the normally small amounts of estrogen produced by males, Klinefelter's syndrome, and aging with decreased countereffects of androgens. Unilateral gynecomastia typically occurs in younger men, is idiopathic, and is of no particular consequence. Pseudogynecomastia is an increase in adipose tissue in the region of the male breast, a common finding in the obese and elderly.

Carcinoma of the male breast accounts for 1 percent of all breast cancers. It is a disease of the elderly with a somewhat worse prognosis than cancer of the female breast.

CHART 18–5. GYNECOMASTIA

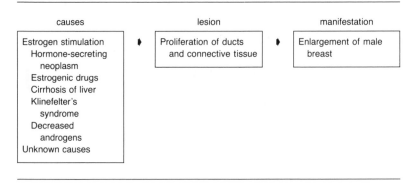

causes		lesion		manifestation

Estrogen stimulation
  Hormone-secreting
    neoplasm
  Estrogenic drugs
  Cirrhosis of liver
  Klinefelter's
    syndrome
  Decreased
    androgens
Unknown causes
	◆	Proliferation of ducts
		  and connective tissue
			◆	Enlargement of male
				  breast

REVIEW QUESTIONS

1. What is the most significant manifestation of breast disease and why?
2. How are breast masses detected and evaluated?
3. What are the causes, lesions, and major manifestations of the diseases discussed in the chapter?
4. Why is fibrocystic disease clinically important?
5. What steps are involved in the development, diagnosis, treatment, and outcome of carcinoma of the breast?
6. What do the following terms mean?
   Mammography
   Galactorrhea
   Comedomastitis
   Fat necrosis
   Gynecomastia

Skin

REVIEW OF STRUCTURE AND FUNCTION
Functions · Epidermis · Dermis

MOST FREQUENT AND SERIOUS PROBLEMS
Cuts · Abscesses · Acne · Nevi · Warts · Eczematous Dermatitis · Seborrheic Dermatitis · Seborrheic Keratosis · Actinic Keratosis · Psoriasis · Skin Cancer · Burns · Drug Reactions · Lesions of Systemic Diseases

SYMPTOMS, SIGNS, AND TESTS
Pruritus · Pain · Rash · Macule · Nodule · Papule · Pustule · Vesicle · Wheal (Hive) · Scale · Crust · Biopsy · Culture · Smear

SPECIFIC DISEASES
  Genetic/Developmental Diseases
    Hemangiomas
    Hair Disorders
    Disorders of Pigmentation
  Inflammatory/Degenerative Diseases
    Viral Exanthems
    Verruca (Warts)
    Acne
    Abscess
    Impetigo
    Syphilis
    Superficial Fungal Infections
    Dermatitis in General
    Contact Dermatitis
    Poison Ivy
    Atopic Dermatitis
    Seborrheic Dermatitis
    Urticaria (Hives)
    Psoriasis
  Hyperplastic/Neoplastic Diseases
    Senile (Solar) Degeneration and Actinic Keratosis
    Seborrheic Keratosis
    Melanocytic Nevus
    Malignant Melanoma
    Other Pigmented Lesions
    Basal Cell and Squamous Cell Carcinomas

ORGAN FAILURE

REVIEW QUESTIONS

REVIEW OF STRUCTURE AND FUNCTION

The skin functions as a barrier between the body and its external environment, protecting the body from injury by external forces and preventing excessive loss of body fluids. The skin also constitutes a major sense organ; cutaneous sensation of touch, temperature, pressure, and pain is essential for man to maintain orientation with his environment. In addition, the skin plays a vital role in regulation of body temperature, both by controlling the amount of blood brought near the surface for heat exchange and through the process of sweating, which lowers skin temperature through vaporization.

Histologically, as depicted in Figure 19–1, the skin consists of an outer covering of stratified squamous epithelium (epidermis); an underlying layer of fibrous connective tissue (dermis), which contains the hair follicles, sebaceous and sweat glands, blood vessels, and sensory nerves; and a deep layer of adipose tissue (subcutis or subcutaneous tissue).

The epidermis has three major morphologic divisions. The deepest, the basal layer, is a single layer of columnar germinative cells that gives rise to all other epidermal cells by mitotic division. The broad middle Malpighian zone consists of several layers of cells undergoing progressive maturation to become keratinized surface cells. As cells migrate upward through the Malpighian layer they become flattened, acquire keratin fibrils, and form rows parallel to the skin surface. The outermost layer consists of anuclear keratinized cells. The thickness of the keratin layer varies greatly in different body sites; for example, the keratin layer is very thick over the palm and fingers and thin over the skin of the forearm. Scattered throughout the basal layer are larger, pale cells

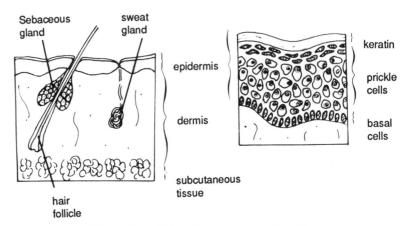

Figure 19–1. Histologic components of the skin.

which may contain brown granules. These cells, called *melanocytes,* produce brown melanin pigment, which helps protect the skin from sunlight damage and contributes to skin color. The concentration of melanocytes varies widely, with melanocytes being twice as numerous on the face as on the trunk.

The dermis consists of fibrous tissue intermixed with elastin fibers. The high collagen concentration provides great skin resistance to mechanical force and the elastin allows the skin to return to its normal form after mechanical deformation. When elastin fibers are destroyed with aging or disease, the skin becomes loose and wrinkled. A gel-like ground substance holds the dermal fibers together and permits the skin to change shape slowly, as with changes in body weight. *Eccrine glands* are present in the deep dermis all over the body and are responsible for the production of sweat in response to heat stress. *Apocrine glands* occur in a restricted distribution (axillae, pubis, perineum, nipple region, ear canal) and produce a sticky proteinaceous fluid in response to hormonal stimuli; apocrine glands are not functional until puberty. With the exception of palms, soles, and portions of the genitalia, the entire body surface is covered by hair; the hairs are produced by division of cells lining the hair follicle. Each follicle undergoes recurring cycles of hair growth, regression, and rest. Attached to each hair follicle is a *sebaceous gland,* which secretes lipid-rich sebum; the function of sebum in man is not known.

MOST FREQUENT AND SERIOUS PROBLEMS

Among the most common dermatologic problems that may prompt a person to seek medical attention are cuts, abscesses, acne, nevi (moles), warts (verruca vulgaris), eczematous dermatitis, seborrheic dermatitis (dandruff), and rashes of various types. Small, greasy, warty lesions (seborrheic keratoses) commonly occur on the face, trunk, and extremities of persons past middle age and are usually multiple. Older persons who have been exposed to sunlight for many years frequently have small precancerous lesions (actinic or solar keratoses) or early skin cancers on exposed surfaces. Psoriasis is a chronic, scaling skin disease, which sometimes is quite distressing to the patient.

Life-threatening skin conditions include extensive burns, severe drug reactions with sloughing of portions of the epidermis (exfoliative dermatitis), and malignant melanomas. Loss or destruction of large areas of epidermis is always potentially lethal because of the resulting loss of body fluids and because of the high risk of secondary infection, usually by pseudomonas.

Skin lesions may be important indicators of systemic disease. Cutaneous lesions, most commonly a red rash over both cheeks and the bridge of the nose, occur in most patients with systemic lupus erythematosus and often constitute the first sign of disease. Deposits of urate crystals frequently occur in patients with gout; these chalky deposits most commonly occur in the subcutis of the helix of the ear and over the elbows and digits of hands and feet. Well-circumscribed areas of brown atrophic skin and foci of dermal connective tissue degeneration are common on the lower extremities of diabetics. Small, soft, yellow papules (xanthomas) are common on the face and extremities of persons with consistently elevated plasma lipid levels (hyperlipidemia and uncontrolled diabetes) and may be the first sign of disease. Some degree of hyperpigmentation of skin occurs in 90 percent of pregnant women, with the most striking manifestation being mask-like pigmentation of the face. Freckles and nevi commonly appear darker during pregnancy.

SYMPTOMS, SIGNS, AND TESTS

Most skin diseases are not life threatening, but many are distressing to patients because of unsightly appearance, itching (*pruritus*), or pain. Unlike diseases of other body systems, skin diseases are readily visible both to the patient and the doctor. The mere discovery of a skin lesion may bring the patient to a doctor or he/she may seek medical attention because of pain or itching. Distress over the appearance of skin lesions and fear of cancer are perhaps the most common concerns of patients with skin disease.

Often, the gross appearance of the lesions and the history of their development is sufficient to make a diagnosis. The gross appearance is described in terms of size, location on the body, multiplicity, color, shape, and texture. Lesions that are flat are termed *macular* and those that are raised are called *papular*. Rashes are temporary eruptions on the skin that have a multitude of causes. General causes of rashes include systemic infections (measles, rubella, chicken pox, scarlet fever, secondary syphilis), hypersensitivity reactions to food or drugs, and local reactions to surface contact with topical drugs and allergenic or irritating substances. The following terms are commonly used to describe the gross appearance of skin lesions. Their appearance is depicted in Figure 19–2.

- *Macule*–Change in skin color; not raised or depressed
- *Nodule*–Knot or lump; elevated solid lesion
- *Papule*–Very small lump
- *Pustule or abscess*–Elevated skin lesion containing pus
- *Vesicle or blister*–Bubble-like swelling containing air or fluid
- *Wheal or hive*–Ridge-like reddened elevation caused by edema and congestion

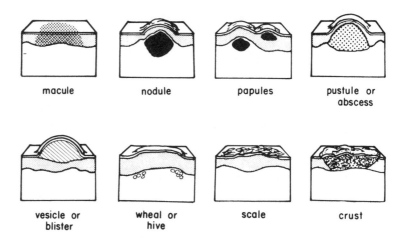

Figure 19–2. The appearance of various types of skin lesions.

- *Scale*–Flaky superficial material (keratin) that easily separates from the skin
- *Crust*–Hardened, adherent serum on skin surface over a lesion

Biopsies are used in evaluation of many chronic eruptions and nodular lesions. Excisional biopsy may be used in the evaluation of neoplastic lesions and often constitutes the sole treatment. Laboratory tests are only indicated under selected circumstances, such as culturing a purulent lesion for bacteria, preparing a smear when fungal infection is suspected, or ordering blood tests when systemic disease is suggested.

SPECIFIC DISEASES

Genetic/Developmental Diseases

Hemangiomas have been mentioned previously in Chapter 9. Major inherited disorders of hair distribution and pigmentation are included in this category, because they can be considered specific diseases. Acquired problems of hair distribution and pigmentation are possible manifestations of a wide variety of diseases and will not be specifically discussed.

Hemangiomas

These congenital proliferations of small blood vessels in the dermis present as elevated red or purple lesions. They increase slowly in size, usually commensurate with body growth. Even small hemangiomas, if located on the face, may be important for cosmetic reasons. Large hemangiomas *(port wine stains)* that cover the forehead and cheek on one side are genetically determined. Port wine stains that involve one side of the forehead may be associated with vascular lesions in the brain, seizure disorders, and mental retardation.

Hair Disorders

The distribution, color, and texture of scalp and body hair are genetically determined and influenced by hormones. Excessive body hair *(hirsutism)* may be particularly distressing to female patients. The early onset of baldness *(alopecia)* in men is often inherited as a dominant trait. It is not manifest in women, because its expression is influenced by male sex hormones.

Disorders of Pigmentation

Albinism is an uncommon autosomal recessive hereditary condition in which melanocytes are unable to produce normal amounts of melanin pigment. Patients lack normal pigmentation in skin, hair, and irides. The absence of normal cutaneous melanin renders the patient markedly sensitive to sunlight and predisposes to increased incidence of basal and squamous cell cancers. *Xeroderma pigmentosum* is a rare autosomal recessive condition characterized by intolerance of skin and eyes to sunlight, with development of skin cancers in childhood or early adult life. These patients lack enzymes necessary to repair damage to DNA caused by sunlight.

CHART 19–1. DISORDERS OF PIGMENTATION

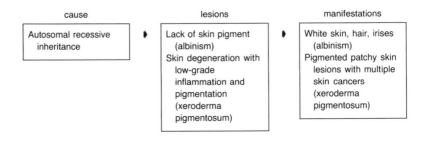

cause	lesions	manifestations
Autosomal recessive inheritance	Lack of skin pigment (albinism) Skin degeneration with low-grade inflammation and pigmentation (xeroderma pigmentosum)	White skin, hair, irises (albinism) Pigmented patchy skin lesions with multiple skin cancers (xeroderma pigmentosum)

Inflammatory/Degenerative Diseases

The first part of this section will deal with the various types of skin infections and the situations in which they occur. This will be followed by a discussion of the various forms of dermatitis. Although *dermatitis* literally means inflammation of the skin, the term is usually used in a narrower sense to describe patchy noninfectious inflammations that are usually chronic and often allergic in nature. Finally, acute edema (urticaria) and chronic scaling skin diseases will be discussed.

Viral Exanthems

Rashes of various viral diseases are most often seen in children, and their clinical appearance and distribution are the basis for diagnosis of measles, rubella, and chicken pox. See Chapter 26 on infectious diseases for further discussion.

Verruca (Warts)

Human papilloma viruses of several strains cause benign neoplastic proliferations of stratified squamous epithelium called squamous papillomas. Three different patterns of disease, presumably caused by different strains of papilloma virus, are verruca vulgaris, condyloma acuminatum, and juvenile laryngeal polyps.

Verruca vulgaris (common wart) occurs on exposed body parts, particularly the fingers and back of the hand, and is a raised, horny, dry lesion (Fig. 19–3A). Warts on the palms and soles (palmar and plantar warts) are flat or elevated dry lesions that may be very painful. Warts on the face and neck may form tiny finger-like projections (filiform warts). Verruca vulgaris are common in children but may occur at any age. Although warts eventually regress spontaneously, they may cause considerable pain and discomfort, especially when frequently traumatized. Various techniques are used to destroy the lesions, but eradication may be difficult due to repeated autoinoculation.

Condyloma acuminata occurs on anogenital mucous membranes and is spread by venereal contact, particularly in young adults (Fig. 19–3B). Juvenile laryngeal papillomas occur in selected patients and are strikingly prone to recur but eventually disappear in adulthood.

CHART 19–2. VERRUCA (WARTS)

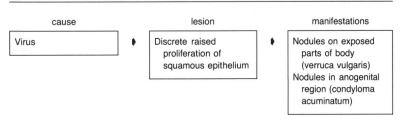

cause	lesion	manifestations
Virus	Discrete raised proliferation of squamous epithelium	Nodules on exposed parts of body (verruca vulgaris) Nodules in anogenital region (condyloma acuminatum)

Acne

Multiple recurrent crops of nodules and pustules on the face and upper back during puberty and early adulthood are called acne (Fig. 19–4). The basic lesion in acne is plugging of hair follicles and associated sebaceous gland ducts with lipid and keratin material, producing slightly raised white heads or blackheads (comedones). The inflammatory reaction of acne is thought to be due to irritating fatty acids liberated from the sebaceous material by *Propionibac-*

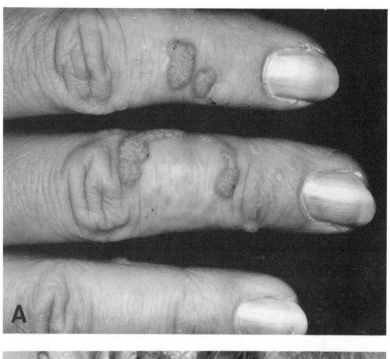

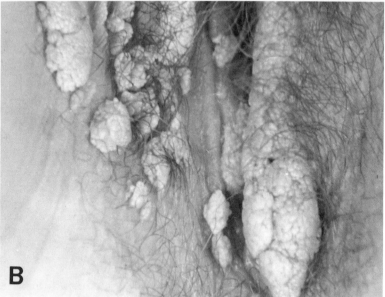

Figure 19–3. A. Verruca vulgaris on fingers. **B.** Condyloma acuminatum on vulva.

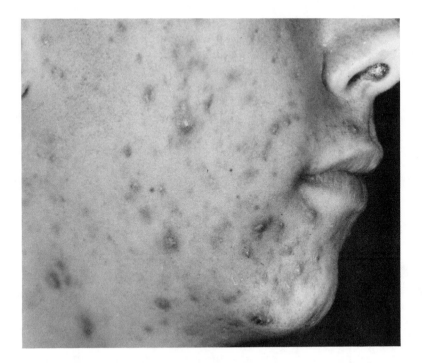

Figure 19–4. Acne.

terium acnes, a bacteria that is able to grow when it is trapped in the plugged pilosebaceous follicle. Other more virulent organisms may infect the lesions producing tiny abscesses. The severity of acne varies greatly among individuals, suggesting hereditary and hormonal influences on sebaceous glands. Acne may be aggravated by emotional stress and administration of adrenocorticosteroids. Severe cases lead to facial scarring, which may be disfiguring. Treatment is directed toward reduction of greasiness of facial skin to avoid plugging of sebaceous glands and, in severe cases, toward reduction of bacteria in the lesions.

CHART 19–3. ACNE

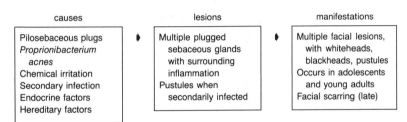

causes	lesions	manifestations
Pilosebaceous plugs *Proprionibacterium acnes* Chemical irritation Secondary infection Endocrine factors Hereditary factors	Multiple plugged sebaceous glands with surrounding inflammation Pustules when secondarily infected	Multiple facial lesions, with whiteheads, blackheads, pustules Occurs in adolescents and young adults Facial scarring (late)

Abscess

Skin abscesses when small and solitary are called *furuncles* or *boils*. *Staphylococcus aureus* is the most common offending organism in skin abscesses. Skin abscesses occur at sites of skin trauma, in obstructed skin appendages, and around embedded foreign material. Abscesses commonly occur in the folds around the nails *(paronychia)* (Fig. 19–5), around the nose, around splinters, in the hairy skin over the tip of the spine between the buttocks *(pilonidal abscess)*, around the anus, and on the back of the neck. Abscesses commonly develop in obstructed hair follicles and enlarge to form painful, elevated, red areas, which subsequently develop a pale, soft center that represents liquefaction necrosis of tissue with accumulation of purulent exudate. If uncomplicated, the furuncle is gradually walled off, ruptures to release the exudate, and then collapses to heal with a small scar. Healing may be accelerated by puncturing the abscess to release the exudate, but this should not be undertaken until liquefaction necrosis is well established and granulation tissue has formed to line the abscess cavity. Recurrent abscesses, such as pilonidal abscesses, may require an operation to remove the infection. Splinters and other foreign materials should be removed, preferably before an abscess develops.

A *carbuncle* is a group of confluent furuncles with associated

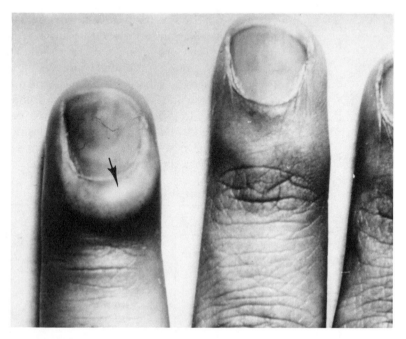

Figure 19–5. Paronychia, with swelling due to collection of purulent exudate around the nail (arrow).

connecting sinus tracts and multiple openings on the skin. This uncommon lesion usually occurs on the back of the neck. Diabetics are especially prone to develop carbuncles because of their reduced resistance to infection.

CHART 19–4. ABSCESS

causes	lesion	manifestations
Breaks in skin Foreign bodies *Staphylococcus aureus*	Abscess in dermis or subcutis furuncles paronychia pilonidal abscess carbuncle	Pain, swelling, heat, redness Purulent discharge

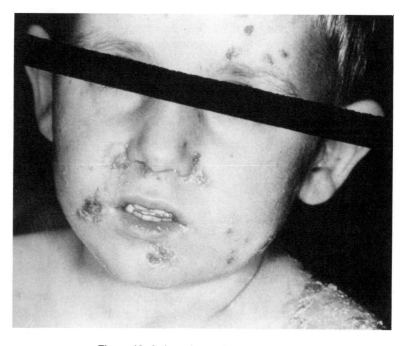

Figure 19–6. Impetigo on face and shoulder.

Impetigo is contagious and may spread rapidly among very young children in crowded conditions.

CHART 19–5. IMPETIGO

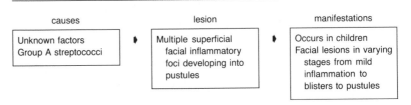

causes	lesion	manifestations
Unknown factors Group A streptococci	Multiple superficial facial inflammatory foci developing into pustules	Occurs in children Facial lesions in varying stages from mild inflammation to blisters to pustules

Syphilis

The primary lesion of syphilis, known as a *chancre*, occurs at the site of entry of the causative organism, *Treponema pallidum*, and takes about 3 weeks to develop. Common sites of chancres include the mucous membranes of the penis, vagina, cervix, anus, and mouth. The chancre is usually a single, painless, superficial papule with superficial central ulceration that heals spontaneously in 4 to 12 weeks. The chancre contains many spirochetes and is highly contagious even though serologic tests for syphilis may be negative at this stage. The secondary stage of syphilis may be associated with a variety of cutaneous manifestations, most commonly a faint rash on the trunk or extremities, often with involvement of the palms and soles. For further discussion of syphilis see Chapter 26.

Superficial Fungal Infections

In the superficial fungal infections, the fungi are located in the keratin layers of the skin, nail, or hair and can often be demonstrated in a smear prepared from skin scrapings. The dermis is not affected. The two main types of superficial fungal infection of the skin are *ringworm (tinea)* and *candidiasis*. Ringworm is not caused by a worm but by various fungal species. Regardless of the site affected, the lesions are typically itchy, brown or red scaly patches; blisters, cracks or localized thickenings may also occur. Some forms of ringworm are transmitted from person to person or animal to person; in other forms, the mode of acquisition is not known. The most common form of ringworm is athlete's foot *(tinea pedis)*, which affects the soles and interdigital spaces of the feet. Hereditary and immunologic factors, sweating, and type of foot covering all appear to play a part in the etiology of this infection. *Tinea cruris* (jock itch) produces symmetric involvement of the perineum, buttocks, and inner thighs. Ringworm of the scalp *(tinea capitis)* and beard *(tinea barbae)* produce breaking of hairs and focal hair loss. Ringworm infections of the nails produce opaque, discolored painless patches within the nail; in late stages of infection, the nail is deformed and eventually destroyed. Most ringworm infections are eradicated by treatment with topical agents.

Superficial infections with *Candida albicans* (candidiasis) may be focal or diffuse and produce red, itchy, scaly patches, often with blisters or pustules at the margin. Factors that predispose to superficial candida infection include diabetes mellitus, leukemia, and lymphoma; treatment with antibacterial or immunosuppressive drugs; and chronic immersion in water. Involvement of intertriginous and interdigital skin is likely in diabetics. Candidiasis of the perioral skin (perleche) produces red cracks and fissures at the corners of the mouth. People who chronically have their hands in water (bartenders, dishwashers) may develop candidiasis of the fingernails manifest by painless, red swelling of the skin around the nail and brown discoloration and ridging of the nail. Superficial candidiasis may be difficult to eradicate.

CHART 19–6. SUPERFICIAL FUNGAL INFECTIONS

causes	lesions	manifestation
Tinea Candida	Fungi in epidermis Thickening and sloughing of keratin layer	Crusting, oozing, or cracking skin lesions in characteristic areas

Dermatitis in General

Dermatitis is a generic, clinical term used to describe a wide variety of skin conditions all characterized by inflammation of the skin. *Eczema* is a term often used to describe dermatitis, particularly that of nonspecific type that evolves slowly from an acute to a chronic dermatitis. Dermatitis occurs in response to a wide variety of stimuli, including drugs, chemical allergens, ultraviolet radiation, local trauma, and various metabolic and immunologic disorders. Dermatitis accounts for about one-third of patients who consult a dermatologist. Acute dermatitis is red and oozing, with many small blisters and crusts (Fig. 19–7A); in later stages, the skin becomes reddened and thickened, with accentuation of normal skin lines (Fig. 19–7B). Severe pruritus may accompany any stage of disease. Thickening of the lesion is largely a reaction to prolonged scratching.

Histologic features in dermatitis are usually nonspecific, and diagnosis of the specific type of dermatitis depends primarily on the clinical evolution and distribution of the lesions.

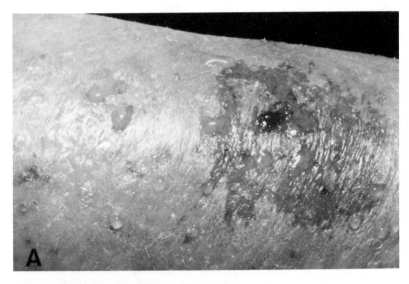

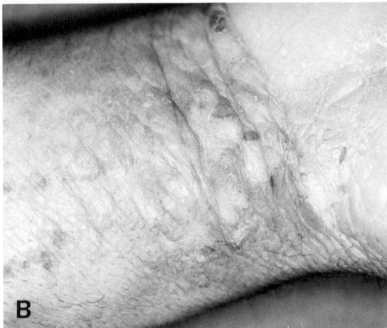

Figure 19–7. A. Acute stage of dermatitis. **B.** Chronic stage of dermatitis.

Ideally a classification of dermatitis would be based on cause. In many patients, however, cause cannot be determined, and some clinical types are of unknown cause. The types described in Table 19–1 represent a mixed classification based on cause and

TABLE 19–1. CLASSIFICATION AND CHARACTERISTICS OF DERMATITIS

Type	Morphologic Pattern	Location	Frequency	Severity
Contact	Eczema	Site of allergen contact	Most common	Distressing, recurrent
Atopic	Eczema	Begins on face, extremities late	Common	Disfiguring, impaired function
Seborrheic	Greasy, scaling	Scalp, face, trunk	Common	Widely variable
Light-induced	Eczema or rash	Exposed skin	Common	Usually mild
Exfoliative	Sloughing of superficial epidermis	Total body	Uncommon	May be fatal

morphology. Contact, atopic, and seborrheic dermatitis will be discussed below. Light-induced dermatitis is diagnosed by history of exposure to sunlight and location on exposed surfaces. Exfoliative dermatitis is the most severe and least common type. Drugs are the cause in many cases of dermatitis.

Contact Dermatitis

This is a type of eczematous dermatitis that occurs as a delayed hypersensitivity reaction to chemical allergens such as clothes, cosmetics, jewelry, and various metals. Since skin changes occur at the site of allergen contact, the location and configuration of the skin eruption often provides an important clue to the causative agent. A suspected substance can be placed on a patch on the back to test for reactivity *(patch test)*. A local reaction to the offending substance will occur under the patch after 24 to 28 hours. Avoidance of the causative allergen cures the eruption; avoidance of the allergen often entails a change in work or personal activities. Contact dermatitis may be subclassified by the causative agents. Poison ivy is a common example.

CHART 19–7. CONTACT DERMATITIS

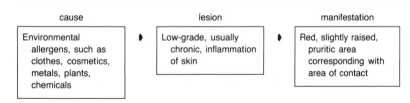

cause	lesion	manifestation
Environmental allergens, such as clothes, cosmetics, metals, plants, chemicals	Low-grade, usually chronic, inflammation of skin	Red, slightly raised, pruritic area corresponding with area of contact

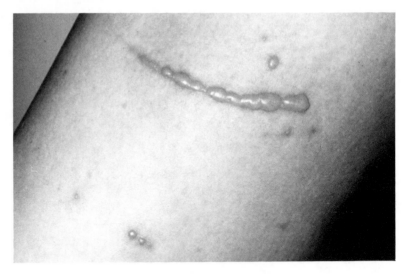

Figure 19–8. Poison ivy blisters.

Poison Ivy

Poison ivy is an example of allergic contact dermatitis. Vesicles, often linear, and redness of the skin develop at the site of contact with the oil of poison ivy leaves (Fig. 19–8). This reaction occurs in hypersensitive persons about 24 hours after contact. Oozing from the vesicles may spread the eruption to other body parts. When poison ivy is severe, it may be advisable to treat the patient with corticosteroids to reduce the inflammatory reaction. Poison oak plants and sumac bushes may produce a similar allergy.

CHART 19–8. POISON IVY

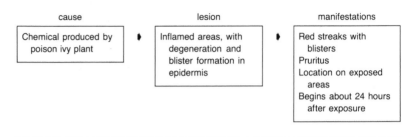

cause	lesion	manifestations
Chemical produced by poison ivy plant	Inflamed areas, with degeneration and blister formation in epidermis	Red streaks with blisters Pruritus Location on exposed areas Begins about 24 hours after exposure

Atopic Dermatitis

Atopic allergy occurs by a different mechanism from contact dermatitis (see Chapter 27 for discussion of mechanism). Descriptively, atopic dermatitis is a type of eczematous dermatitis that occurs in individuals having other manifestations of atopic allergy (asthma, hayfever, urticaria). Atopic dermatitis usually begins in infancy with eczema of the face and scalp, then recurs in children and adults with more generalized involvement of the trunk and extremities. This condition is characterized by severe itching. Vesicular lesions predominate in infants, while children and adults develop dry, thickened skin lesions. A specific offending substance is not identified, but exacerbations of this disease have been related to substances in the environment, temperature variation, and emotional stress, as well as by hereditary predisposition. Patch testing is not useful in atopic dermatitis.

CHART 19–9. ATOPIC DERMATITIS

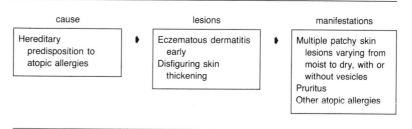

Seborrheic Dermatitis

This common form of dermatitis occurs in sites of greatest concentration of sebaceous glands—scalp, face, ears, neck, axillae, breasts, umbilicus, and anogenital regions. Seborrheic dermatitis usually begins in childhood as fine scaling of the scalp *(cradle cap)* and may continue throughout life with variable extension to other areas. The scaly lesions often resemble psoriasis but are usually yellow and greasy. In men, severe seborrheic dermatitis may be associated with a chronic pustular eruption in the beard area. The specific cause is unknown.

CHART 19–10. SEBORRHEIC DERMATITIS

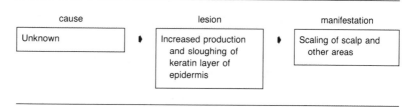

Urticaria (Hives)

Urticaria is an acute patchy eruption with raised edematous areas *(wheal)* surrounded by red margins *(flare)*. The lesions itch. The lesions seldom persist for more than 48 hours but may recur in successive episodes. Urticaria results from local release of histamine in the skin and is usually due to atopic type of allergic reaction commonly caused by an allergen in food such as chocolate or shellfish or by a drug such as penicillin.

CHART 19–11. URTICARIA

cause		lesion		manifestations
Histamine release due to allergens such as foods or drugs or unknown causes	▶	Edema of skin with vascular congestion at margin	▶	Large patchy areas of wheal and flare Pruritus

Psoriasis

Psoriasis is a chronic inflammatory disease of skin of unknown cause that varies greatly in severity and is characterized by thickened areas of skin with silver-colored scales (Fig. 19–9). Lesions are

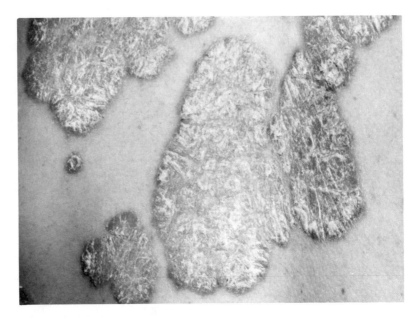

Figure 19–9. Well-demarcated, raised, red, scaly lesions characteristic of psoriasis.

most common on the knees and elbows. Proliferation of the epidermis is a prominent feature of this disease. In severe cases, various ointments and ultraviolet light are used to control the proliferative activity. Psoriasis may be quite distressing to the patient but is not life threatening.

CHART 19–12. PSORIASIS

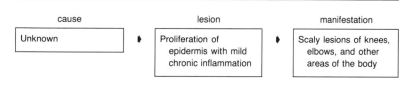

cause		lesion		manifestation
Unknown	‣	Proliferation of epidermis with mild chronic inflammation	‣	Scaly lesions of knees, elbows, and other areas of the body

Hyperplastic/Neoplastic Diseases

The skin is the most common site of neoplastic and preneoplastic change. Long exposure to sunlight in association with aging is important in the vast majority of these lesions. Although some of the lesions discussed are degenerative in nature, they are discussed here because they represent a spectrum of changes. The neoplastic lesions involve two separate cell lines—the squamous cells and the melanocytes of the epidermis. Melanocytes are epithelial-like cells that are derived embryologically from neural elements. The relative frequency of the three major types of cancer of the skin is shown in Figure 19–10.

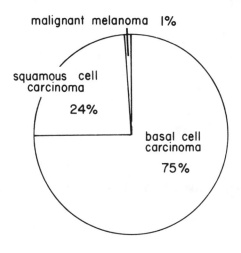

Figure 19–10. Relative frequency of skin cancers derived from squamous cells (basal and squamous cell carcinomas) and from melanocytes (malignant melanoma).

Senile (Solar) Degeneration and Actinic Keratosis

With age, a number of changes occur in exposed skin, predominantly as a result of years of exposure to sunlight. The degree of skin change is directly related to the duration and intensity of solar exposure. The most characteristic change is degeneration of dermal collagen and elastic fibers, which results in excessive wrinkling of the skin. Other skin changes attributed to solar degeneration include skin atrophy, areas of decreased pigmentation (*vitiligo*), areas of increased pigmentation (*lentigo*), collections of small blood vessels (*telangiectasia*), and keratoses. Thus, the face of an elderly farmer or seaman is likely to have areas of pigmentation and depigmentation as well as a ruddy complexion due to telangiectasia and keratoses (Fig. 19–11).

Actinic keratoses (also called senile or solar keratoses) occur as multiple, scaly lesions on sun-exposed skin of persons in or past middle life; they are more common in fair-skinned persons. The actinic keratosis is an area of atypical epidermal proliferation usually accompanied by thickening of the keratin layer (*hyperkeratosis*). Squamous cell carcinoma frequently arises in an actinic keratosis. About one-fourth of patients with actinic keratosis eventually develop squamous cell carcinoma.

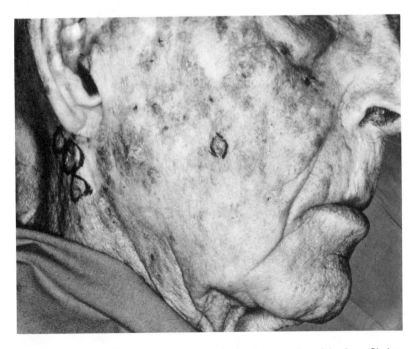

Figure 19–11. Multiple actinic keratoses and solar degeneration of the face. Circles are sites to be biopsied to distinguish actinic keratosis from early cancer.

CHART 19–13. SENILE DEGENERATION

causes		lesions		manifestations
Long exposure to sunlight Aging	▸	Degeneration of collagen and elastic fibers Telangiectases Increased or decreased pigmentation Actinic keratoses	▸	Wrinkling of skin of face and hands Pigmented or depigmented spots on exposed areas Ruddy complexion Small hyperkeratotic foci

Seborrheic Keratosis

Seborrheic keratoses are benign neoplastic proliferations of basal cells of the epidermis that occur on the trunk, face, and arms of persons in middle life and beyond. The seborrheic keratosis is a warty, brown, greasy lesion that appears to be tacked on to the skin and can often be easily scraped off (Fig. 19–12). The stimulus that triggers this proliferative response is not known. Seborrheic keratoses are not premalignant.

CHART 19–14. SEBORRHEIC KERATOSIS

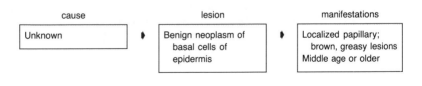

cause		lesion		manifestations
Unknown	▸	Benign neoplasm of basal cells of epidermis	▸	Localized papillary; brown, greasy lesions Middle age or older

Melanocytic Nevus

The common nevus (*melanocytic nevus, nevocellular nevus,* mole) results from a benign proliferation of melanocytes of the epidermis. At first the melanocytes or nevus cells proliferate at the junction of the epidermis and dermis (*junctional nevus*); later the cells migrate into the dermis and form clumps of nevus cells (*intradermal nevus*). When both components are present they are called *compound nevi.* Melanocytic nevi develop in most light-skinned individuals any time from childhood to early adulthood and evolve from junctional to intradermal nevi over a number of years. They may eventually disappear. They may be brown to black or even colorless, depending on the amount of melanin pigment produced. They vary from flat to pedunculated depending on the number of nevus cells present (Fig. 19–13).

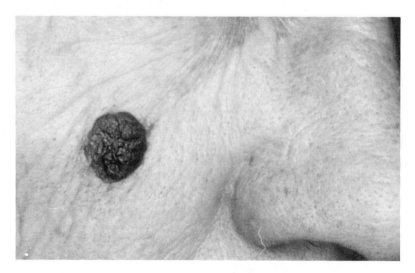

Figure 19–12. A large seborrheic keratosis of the face.

The vast majority of melanocytic nevi are harmless and are only occasionally removed for cosmetic reasons. It is estimated that about 30 percent of malignant melanomas arise from melanocytic nevi, but the chance of any one nevus becoming malignant is extremely small. The risk is considerably greater in large congenital nevi *(giant hairy nevi)*. A mole that increases in size, ulcerates, bleeds, or undergoes color changes should be evaluated by a physician for possible malignant melanoma.

CHART 19–15. MELANOCYTIC NEVUS

causes		lesions		manifestation
Congenital malformation ?Hereditary	▸	Clusters of melanocytes at dermal/epidermal junction and/or in dermis	▸	Small brown spots; may be flat, raised, or pedunculated

Malignant Melanoma

Skin cancers composed of melanocytes are much more dangerous than those derived from squamous cells *(basal cell* and *squamous cell carcinomas)*. Currently it is estimated that about 20 percent of patients with malignant melanoma die from metastases. This number would be much greater if many lesions were not removed at an early stage. Once metastases have occurred, malignant melanoma

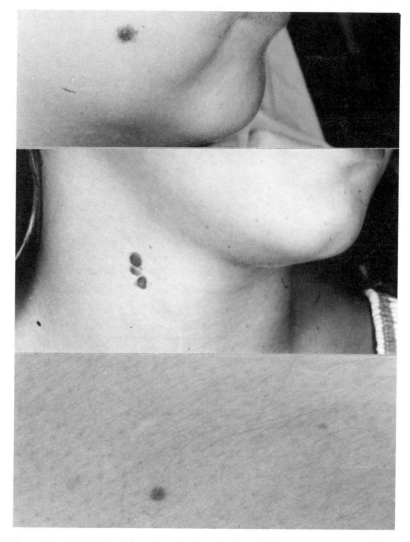

Figure 19–13. Melanocytic nevi varying from flat to raised and sharply demarcated to less discrete.

has a tendency to spread widely and rapidly to many organs in the body, but occasionally malignant cells will lie dormant for several years before producing obvious metastases.

Four forms of malignant melanoma are recognized based on clinical and histologic appearance and behavior of the lesions. *Superficial spreading melanoma* is characterized by spreading growth of malignant cells within the epidermis producing a flat to slightly raised lesion, often with central regression of the lesion and a variegated color pattern (Fig. 19–14A). If not removed, downward growth into the dermis and metastases will occur after months to years of superficial growth. *Lentigo maligna melanoma* also spreads radially in the epidermis and eventually invades the dermis; however, it develops on the exposed skin of the face of older persons and becomes very large because it grows for years before invading. Before it invades the dermis it is called lentigo maligna or Hutchinson's freckle. *Acral-lentiginous melanoma* is composed of more spindle-shaped melanocytes that tend to grow into the dermis without producing a nodule. They tend to occur on the hands and feet and mucous membranes such as the vulva. Acral-lentiginous melanoma is less likely to be discovered early and more dangerous than the preceding types. *Nodular melanoma* grows downward and expands outward to form a nodule (Fig. 19–14B). The likelihood of metastasis in nodular melanoma, as well as with other types, correlates with the vertical thickness of the tumor. Classification and staging of malignant melanomas are done by a pathologist after a wide excisional biopsy of the lesion is performed.

CHART 19–16. MALIGNANT MELANOMA

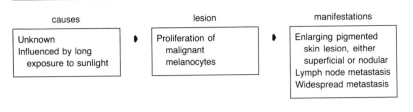

causes	lesion	manifestations
Unknown Influenced by long exposure to sunlight	Proliferation of malignant melanocytes	Enlarging pigmented skin lesion, either superficial or nodular Lymph node metastasis Widespread metastasis

Other Pigmented Lesions

Freckles are localized areas of hyperpigmentation that occur in response to sunlight. Freckles begin in childhood and usually increase in number with repeated exposure to sunlight. On the other hand, *lentigo senilis* (age spots) are localized areas of increased melanocytes in the epidermis or exposed areas of skin in persons past middle age. Actinic keratoses, seborrheic keratoses, and basal cell carcinomas may also be dark colored when they contain increased amounts of melanin.

Basal Cell and Squamous Cell Carcinomas

Skin cancer is the most common type of cancer in man and in most instances is related to exposure to sunlight. Skin cancers are often multiple and occur most frequently on the face and hands of older adults. There are two distinctive cellular types of skin cancer—

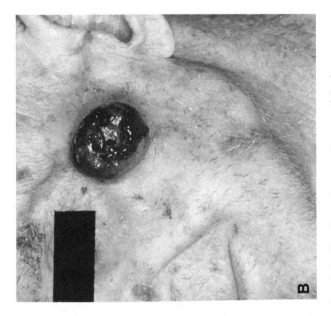

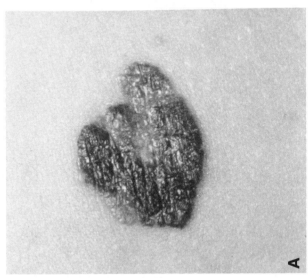

Figure 19–14. A. Superficial spreading melanoma of trunk. **B.** Nodular malignant melanoma of the face.

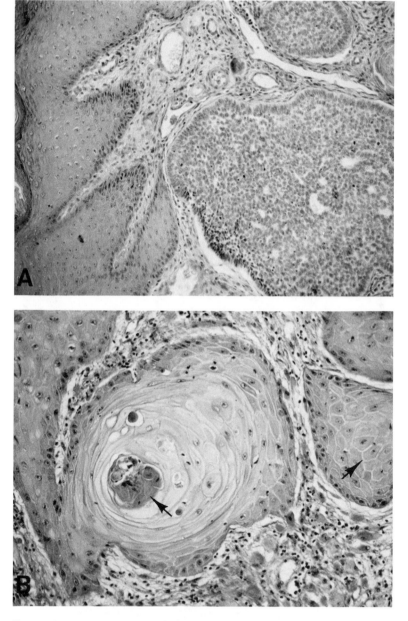

Figure 19–15. Histology of basal cell (**A**) and squamous cell (**B**) carcinomas. Compare with normal epidermis in **A** *(left).* The squamous cell carcinoma exhibits keratin formation *(left arrow)* and prickle cells with intercellular spaces *(right arrow).*

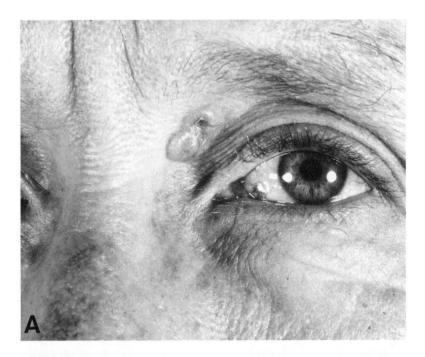

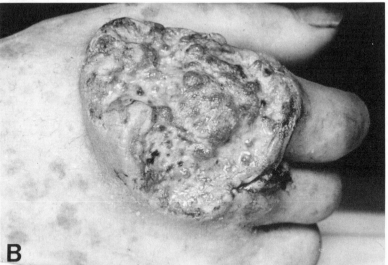

Figure 19–16. A. A basal cell carcinoma of the eyelid. **B.** A large, ulcerated squamous cell carcinoma of the hand.

basal cell carcinoma and squamous cell carcinoma. Basal cell carcinoma is more common (75% of skin cancers) and is made up of cells that resemble those of the basal cell layer of epidermis (Fig. 19–15A). The basal cell carcinoma usually presents as a slowly enlarging, raised lesion that does not ulcerate until advanced (Fig. 19–16A); if untreated, it will slowly enlarge and destroy normal tissue, but it does not metastasize. Basal cell carcinoma usually does not arise in any pre-existing skin lesion.

Squamous cell carcinoma is composed of clumps of cells in layers that resemble the layers of the epidermis (Fig. 19–15B) and has a slightly greater growth potential than basal cell carcinoma. Squamous cell carcinomas commonly arise in pre-existing skin lesions, such as actinic keratoses, burn scars, and chronic ulcers, and present as enlarging, often ulcerated lesions (Fig. 19–16B). If untreated, some skin carcinomas may metastasize, usually only to regional lymph nodes. Large, untreated skin cancers may cause death from secondary infection or, rarely, from hemorrhage.

Cancers can also arise from the skin appendages, blood vessels, and connective tissue, but these neoplasms are all rare.

CHART 19–17. BASAL CELL AND SQUAMOUS CELL CARCINOMAS

cause		lesion		manifestation
Long exposure to sunlight or x-rays	▸	Malignant proliferation of basal cells or differentiating squamous cells to form small nodular lesions	▸	Enlarging nodule on exposed areas of skin

ORGAN FAILURE

Lethal failure of skin function occurs with extensive, severe burns and rare cases of diffuse blistering diseases such as epidermolysis and pemphigus vulgaris. In these conditions, there may be inability to control fluid loss and to prevent infection.

REVIEW QUESTIONS

1. What kinds of skin diseases are most common?
2. Which skin conditions are likely to be fatal?
3. Which systemic diseases commonly have skin manifestations?

4. What do the following descriptive terms used to describe skin lesions mean?
    Macule
    Nodule
    Papule
    Pustule
    Vesicle
    Wheal
    Eczema
    Scale
    Crust
5. What are the causes, lesions, and manifestations of the specific diseases described in this chapter?
6. How do verruca vulgaris and condyloma acuminatum differ?
7. What are the various types of skin abscesses?
8. What infectious agents cause skin lesions and what is the nature of the lesions they cause?
9. What are the similarities among the various types of dermatitis?
10. What are the skin lesions associated with exposure to sunlight?
11. How do neoplasms of squamous cells and melanocytes differ in frequency and prognosis?
12. How can extensive skin lesions lead to a fatal outcome?

REVIEW OF STRUCTURE AND FUNCTION

The *globe* of the eye (eyeball) sits inside the bony orbit and is protected anteriorly by the eyelid. The extraocular muscles attach to its outer surface. The optic nerve together with blood vessels enter the globe posteriorly. The globe is composed of three layers (Fig. 20–1)—the outer *scleral layer,* a tough fibrous coating; the *choroid* or *uveal layer,* a pigmented layer of connective tissue

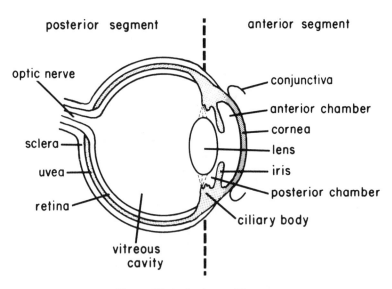

Figure 20–1. Anatomy of the eye.

through which nerves and blood vessels course; and the inner, *retinal layer.*

Anteriorly, the scleral layer is continuous with the cornea, the transparent structure through which light first passes into the globe. The *cornea* is the initial refracting surface of the eye. The exterior of the sclera is covered by the *conjunctiva,* a sheet of cells that reflect onto the inner surface of the eyelids at approximately 1 centimeter posterior to its origin at the corneoscleral junction. The *lens,* the suspending ligaments of the lens, the *iris,* and the muscular *ciliary body* are all located in the anterior portion of the eye. Contraction of the ciliary body controls the focal length of the lens. The iris controls the amount of light reaching the lens by varying the size of its aperture *(pupil).* Decrease in pupil size *(miosis)* is achieved by contraction of the circular smooth muscles in the iris, which are stimulated by autonomic nerve fibers from the third cranial nerve, the oculomotor.

Posteriorly, the retina is the site of the light- and color-sensing rod and cone neurons. Impulses generated in these cells by light are transmitted by a chain of neurons back through the optic nerve to the visual cortex of the cerebrum. The ciliary body and the lens and its ligaments mark the boundary between the anterior and posterior segments of the globe. Within the anterior segment is the anterior chamber, which is filled with a watery fluid, *aqueous humor.* Aqueous humor is secreted by the ciliary body and diffuses forward, around the lens, from the posterior to the anterior chamber. It drains out of the anterior chamber slowly through a series of tissue spaces at the periphery of the chamber. The posterior segment is filled with a clear gelatinous substance, the *vitreous humor.*

MOST FREQUENT AND SERIOUS PROBLEMS

The most common overall problem affecting the eyes is a decrease in visual acuity (indistinct focusing of the visual image on the retina) from refractive error. Approximately one-third of the population wears eyeglasses for correction of the main types of refractive error, which include myopia, hyperopia, presbyopia, and astigmatism. Trauma to various parts of the eye, especially to the cornea, are very common problems and can lead to significant pain and disability. Acute conjunctivitis, colloquially called *pink-eye,* may occur by itself or be associated with another bacterial or viral infection such as an upper respiratory infection. Infections in the adjacent sinuses can spread to the orbit resulting in serious complications. Strabismus is deviation of one or both eyes that cannot be overcome by the patient. Strabismus is relatively common and is due to several causes. Cataract is an opacity of the lens leading to decreased visual acuity. Cataracts may be the result of birth defect, infection, or trauma, but most commonly they arise de novo in elderly persons. Glaucoma is a disease caused by an increase in intraocular pressure, with resultant damage to the optic nerve and its fibers inside the eye. Afflictions of the small retinal vessels occur commonly in individuals with hypertension and diabetes. Tumors of the eye itself are relatively rare.

SYMPTOMS, SIGNS, AND TESTS

The most common manifestation of eye disease is decreased visual acuity. Decreased visual acuity may be a symptom when experienced by a person or a sign when manifested as a result of a visual acuity test. Decreased visual acuity may result from glaucoma, cataracts, and vascular diseases, as well as from myopia or hyperopia. Visual field defects (focal areas of blindness) also may be signs or symptoms and result from diseases affecting the optic pathways of the central nervous system or from diseases of the retina. Pain is often a manifestation of trauma, infection, or acutely increased intraocular pressure from closed-angle glaucoma. Blurring of vision accompanies various systemic diseases, especially those of toxic and metabolic origin. *Papilledema* (swelling of the head of the optic nerve as seen through the ophthalmoscope) may reflect an increase in intracranial pressure, usually due to an expanding intracranial mass. *Photophobia* (uncomfortable sensitivity to light) is commonly seen with infections inside and outside the eyeball. *Nystagmus* (flickering eye movements) may be caused by a variety of central nervous system lesions.

Clinical tests for eye disease include the utilization of eye charts for testing *visual acuity*. The person being examined reads letters or numbers from a chart at a distance of 20 feet. Normal visual acuity (20/20) means that the person can accurately read with either eye the smallest figures that are readable to a normal control population. A visual acuity of 20/40 means that the affected eye can accurately read at 20 feet what a normal eye can read at 40 feet. These charts are widely used as screening devices for school children and for persons applying for drivers' licenses.

Tests for mapping visual field defects are employed to detect area(s) of the retina or optic pathways affected by a particular disease process (Fig. 20–2). In administering *visual field tests*, the person being examined is asked to focus on a small stationary spot while a test spot is moved to different points of a circular map. In those areas where the person cannot see the test spot, he or she has a visual field defect.

The *ophthalmoscope* is a hand-held instrument used to examine the retina by utilizing lenses of various refractive powers to focus a beam of light on the retina. Visualization of the retina with the ophthalmoscope is referred to as a *funduscopic examination*. A *tonometer* is a small instrument that is placed directly on the eyeball to measure the intraocular pressure, which may be elevated in glaucoma. A *slit-lamp* is a binocular instrument that projects a beam of light into the eye and is used for detailed examination of the cornea, anterior chamber, iris, and lens, and is useful in the evalua-

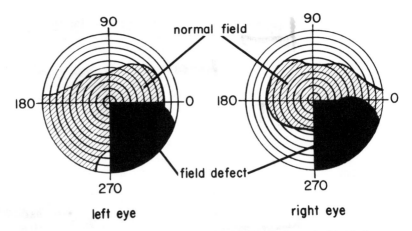

Figure 20–2. Visual field chart with visual field defects marked in black.

tion of glaucoma, trauma, and other conditions. Various types of lenses are utilized in the evaluation of myopia and hyperopia to determine the *refractive error* (degree by which the cornea, lens, and humor fail to focus light rays on the retina).

SPECIFIC DISEASES

Genetic/Developmental Diseases

Isolated congenital eye defects are rare, but genetic factors appear to predispose individuals to certain eye disorders such as myopia, glaucoma, strabismus, cataracts, and retinoblastoma (a malignant neoplasm). *Cyclopia,* in which a single, malformed eye is centrally situated, may occur in conjunction with severe brain malformation.

Inflammatory/Degenerative Diseases

Except for injuries and infections, the diseases discussed here represent a heterogenous group of development and acquired conditions. Although not discussed further, it should be noted that infections of the maxillary, ethmoid, and frontal sinuses can extend into the orbit and cause damage to the eye.

Trauma and Chemical Injury

The eye is susceptible to many different types of injuries, which include penetrating injuries, blunt injuries, and chemical injuries. The most common penetrating injury occurs when metal strikes metal. The manifestations of penetrating injuries are usually readily apparent and receive the immediate attention of an ophthalmic surgeon. The manifestations of blunt injuries are usually less apparent, but may be no less severe. When the eye is hit directly by a blunt object (e.g., a paddle ball), compression/decompression occurs, which can produce damage to several intraocular structures. Such damage can, in time, result in retinal detachment, intraocular hemorrhage, dislocation of the lens, and/or glaucoma. Of the different types of chemicals that can damage the eye, concentrated alkali solutions such as nitrogen fertilizers produce the most damage to the cornea and adjacent tissues. Complications of severe chemical burns include corneal ulcer and corneal perforation.

Conjunctivitis

Inflammation of the conjunctiva with vascular congestion producing a red or pink eye (Fig. 20–3) may be secondary to viral, chlamydial, bacterial, and fungal infections or allergy. Symptoms

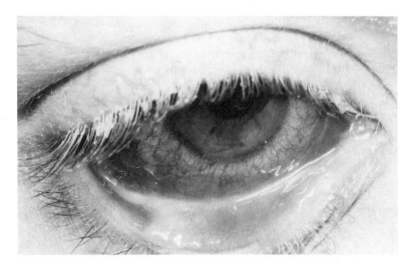

Figure 20–3. Conjunctivitis with congested conjunctival vessels and exudate on lower lid.

include irritation, blurred vision, and photophobia. Conjunctivitis most often occurs in association with an upper respiratory infection but may occur by itself. Measles often present with conjunctivitis. Conjunctivitis associated with hay fever may be very troublesome because of photophobia and itching. Viral and allergic types of conjunctivitis are prone to secondary bacterial infections; hence, the treatment of conjunctivitis often consists of administration of eye drops that contain antibiotics and topical antihistamines. Conjunctivitis caused by *trachoma* (a chlamydial disease) is one of the most common causes of blindness in many parts of the world because it results in corneal scarring. Inflammation of other parts of the eye are less common, but keratitis (corneal inflammation), uveitis, and retinitis can be very devastating to sight.

CHART 20–1. CONJUNCTIVITIS

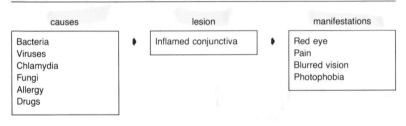

causes	lesion	manifestations
Bacteria Viruses Chlamydia Fungi Allergy Drugs	Inflamed conjunctiva	Red eye Pain Blurred vision Photophobia

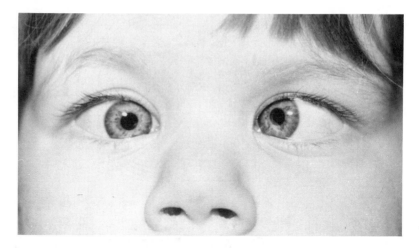

Figure 20–4. Strabismus.

Strabismus (Heterotropia)

Strabismus is an improper alignment of the visual axis (the line of vision), with one or both eyes at fault. The result is that each eye points in a different direction and the eyes cannot fixate on the same visual object simultaneously (Fig. 20–4). Strabismus is a common condition, most often manifest in childhood and often referred to as crossed eyes in layman's terms. Normally children attain alignment by 3 to 4 months of age. A variety of factors may cause strabismus—paralysis of an extraocular muscle, refractive errors, opacities of the cornea or lens, diseases of the retina, and diseases of the optic nerve or brain. Some types of strabismus may be corrected with lenses; others are corrected surgically by altering the insertions of extraocular muscles to realign the eyes. If not discovered and treated early, strabismus can cause irreversible loss of vision in one eye called *amblyopia* (lazy eye).

CHART 20–2. STRABISMUS

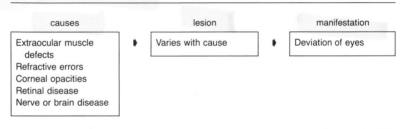

causes	lesion	manifestation
Extraocular muscle defects Refractive errors Corneal opacities Retinal disease Nerve or brain disease	Varies with cause	Deviation of eyes

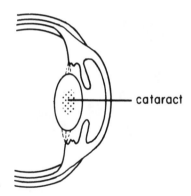

Figure 20–5. Cataract.

Cataract

Cataract is an opacification of the lens (Fig. 20–5) that most commonly occurs in older individuals *(senile cataract)*. Less commonly, cataracts may be secondary to trauma or may be accelerated by diabetes mellitus in poor control. Congenital cataracts are rare, and some types have a familial tendency. Most cataracts develop slowly, resulting in progressive diminution of vision. In most cases, treatment of cataracts consists of surgical extraction of the lens followed by placement of a prosthetic intraocular lens.

CHART 20–3. CATARACT

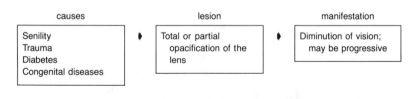

causes	lesion	manifestation
Senility Trauma Diabetes Congenital diseases	Total or partial opacification of the lens	Diminution of vision; may be progressive

Glaucoma

Glaucoma is a condition characterized by a rise in intraocular pressure sufficient to damage optic nerve fibers. The increase in intraocular pressure is almost always due to obstruction of the normal exit of anterior chamber fluid (aqueous humor) in the angle where the iris meets the corneal–scleral junction. Glaucoma is divided into open-angle, closed-angle, and rare congenital forms. In *open-angle glaucoma*, there are gross or microscopic abnormalities of the angle tissues. Open-angle glaucoma is the more common type and is chronic and insidious in onset. The patient is unaware of slowly progressive damage to the optic nerve with the consequent visual loss until the condition is far advanced.

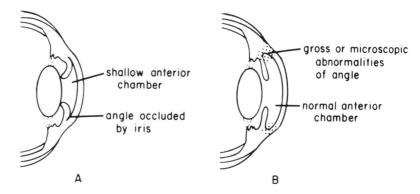

Figure 20–6. A. Closed-angle glaucoma. **B.** Open-angle glaucoma.

In *closed-angle glaucoma*, there is obstruction of the angle by the iris (Fig. 20–6). The adhesion of the iris to the angle structures may be reversible or, if inflammation and scarring has occurred, permanent. The symptoms of closed-angle glaucoma may be acute, with pain, nausea, vomiting, and a sudden decrease in vision.

Glaucoma is diagnosed by demonstration of increased intraocular pressure as measured with a tonometer and by documented defects in peripheral visual fields. It is treated by various adrenergic and anti-cholinergic drugs and by carbonic anhydrase inhibitors, which partially inhibit the formation and secretion of aqueous humor and, to some degree, increase outflow.

CHART 20–4. GLAUCOMA

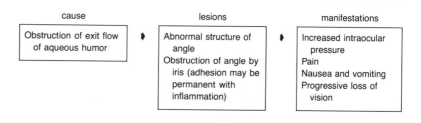

cause	lesions	manifestations
Obstruction of exit flow of aqueous humor	Abnormal structure of angle Obstruction of angle by iris (adhesion may be permanent with inflammation)	Increased intraocular pressure Pain Nausea and vomiting Progressive loss of vision

Hypertensive Retinopathy

Hypertension causes progressive arteriolosclerosis of the retinal vessels, eventually resulting in loss of vision because of compromised oxygen delivery to the retinal neurons. Hypertensive retinopathy is diagnosed by funduscopic examination, because the sclerosis of the retinal arteries is seen as characteristic alterations in

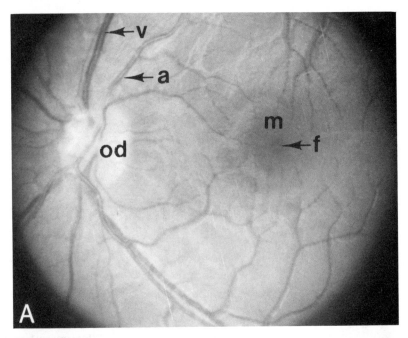

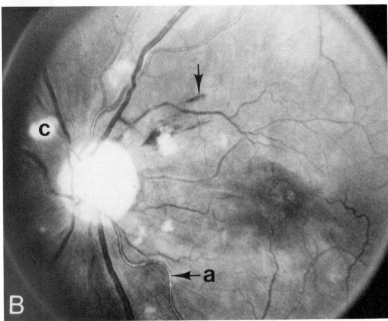

Figure 20–7. A. Normal retina, showing optic disk (od), macula (m) with fovea in center (f), artery (a), and vein (v). **B.** Hypertensive retinopathy, with flame-shaped hemorrhages *(arrow)*, narrow, tortuous artery (a), and cotton wool patches (c) caused by focal ischemia.

the light reflection from these vessels (Fig. 20–7). Control of the hypertension will usually result in some visual improvement.

CHART 20–5. HYPERTENSIVE RETINOPATHY

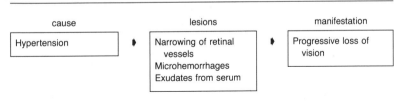

cause		lesions		manifestation
Hypertension	▸	Narrowing of retinal vessels Microhemorrhages Exudates from serum	▸	Progressive loss of vision

Diabetic Retinopathy

Long-standing diabetes mellitus leads to retinal vessel disease manifested by progressive arteriosclerosis similar to that which occurs with hypertension. In addition, diabetes is associated with the development of vascular abnormalities. Early vascular lesions consist of *capillary aneurysms* and *microhemorrhages,* the latter eventually resulting in visual loss (Fig. 20–8). The most advanced vascu-

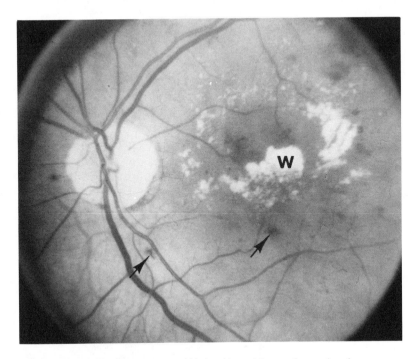

Figure 20–8. Diabetic retinopathy, with focal hemorrhages *(arrows)* and waxy exudates (W) forming a ring around the macula.

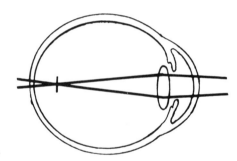

Figure 20–9. Myopic eye.

lar lesions consist of proliferation of newly formed abnormal vessels, which can cause retinal detachment. Treatment of retinal vessel hemorrhage may be effectively accomplished using a lasar beam to coagulate small areas of the retina by thermal energy.

CHART 20–6. DIABETIC RETINOPATHY

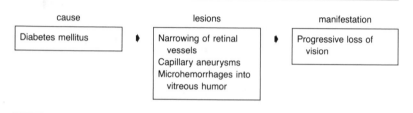

cause	lesions	manifestation
Diabetes mellitus	Narrowing of retinal vessels Capillary aneurysms Microhemorrhages into vitreous humor	Progressive loss of vision

Myopia (Nearsightedness)

Myopia is a condition of refractive error in which light entering the eye is focused at a point anterior to the retina (Fig. 20–9). This may be due either to abnormal curvature and refractive power of the cornea and lens, or it may be due to relative elongation of the eyeball. Myopia often develops in childhood for poorly understood reasons and progresses until early adulthood, when the condition stabilizes because of normal loss of elasticity of the eye structures. Myopia tends to be familial and is usually treated successfully with corrective lens. Severe myopia may be associated with multiple defects in intraocular structures, including the angle and retina.

CHART 20–7. MYOPIA

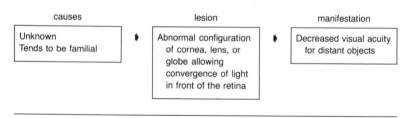

causes	lesion	manifestation
Unknown Tends to be familial	Abnormal configuration of cornea, lens, or globe allowing convergence of light in front of the retina	Decreased visual acuity for distant objects

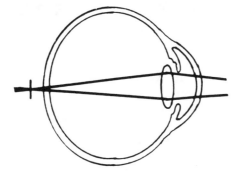

Figure 20–10. Hyperopic eye.

Hyperopia (Farsightedness)

This condition is different from myopia in that light entering the eye tends to focus at a point posterior to the retina, with resultant poor vision for near objects and better vision for far objects (Fig. 20–10). Hyperopia is the result of a relatively short eyeball or reduced refractive power of the cornea or lens. Children can often overcome an underlying mild hyperopic refractive error because of lens elasticity but often lose this ability as they grow older. The lens loses its elasticity and ability to adjust its focusing power *(accommodation)* with advancing age and is less able to focus light rays closer to the retina rather than behind it. Thus, a mild hyperopia can become manifest in later years and only then require corrective lens.

CHART 20–8. HYPEROPIA

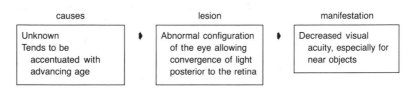

causes	lesion	manifestation
Unknown Tends to be accentuated with advancing age	Abnormal configuration of the eye allowing convergence of light posterior to the retina	Decreased visual acuity, especially for near objects

Presbyopia

Presbyopia is a refractive error produced by loss of elasticity of the lens in the elderly. It represents an inability to accommodate. Accommodation means the ability of the lens to change its shape in order to focus light rays from nearby objects onto the retina. Although the optical defect of presbyopia is similar to that of hyperopia inasmuch as light rays are focused behind the retina, the pathogenesis of the two conditions is different. Hyperopia is due to the relatively short length of the eyeball, whereas presbyopia results from the degenerative effects of aging on the lens. Presbyopia is corrected by reading glasses or bifocals.

Astigmatism

Astigmatism is uneven focusing of light entering the eye due to unequal curvature of the cornea or acquired irregularities in the corneal surface or lens. Astigmatism may occur by itself or may accompany myopia or hyperopia. Unless unusually severe, astigmatism is correctable with appropriate lenses.

CHART 20–9. ASTIGMATISM

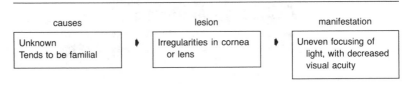

causes	lesion	manifestation
Unknown Tends to be familial	Irregularities in cornea or lens	Uneven focusing of light, with decreased visual acuity

Hyperplastic/Neoplastic Diseases

Retinoblastomas and melanomas comprise over 90 percent of all primary intraocular eye tumors. With rare exceptions, retinoblastoma occurs in children, while melanoma occurs in adults.

Retinoblastoma

This is a rare malignant tumor of primitive neurons, which are the precursors of the retinal ganglion cells. Retinoblastomas occur in children; they tend to be familial and have an autosomal dominant inheritance pattern with incomplete penetrance. Retinoblastomas are bilateral about 30 percent of the time. Treatment is removal of the eye *(enucleation)*, often combined with radiation.

CHART 20–10. RETINOBLASTOMA

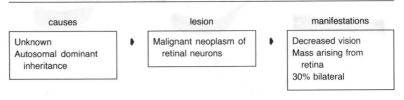

causes	lesion	manifestations
Unknown Autosomal dominant inheritance	Malignant neoplasm of retinal neurons	Decreased vision Mass arising from retina 30% bilateral

Malignant Melanoma

This is a malignant tumor arising from the choroid pigment-containing cells and is primarily a tumor of adults. Most ocular melanomas arise from the choroid layer, although they can arise from the iris or ciliary body. They carry a better prognosis than melanomas of the skin.

ORGAN FAILURE

Approximately 4 of every 1000 persons in the United States are legally blind. *Legal blindness* is defined as visual acuity of 20/200 or less in the better eye with best correction. Blindness may be caused by lesions of the cornea, lens, vitreous humor, retina, and optic nerve. Congenital blindness or blindness developing in infancy can be due to a wide variety of developmental, inflammatory, or traumatic conditions. Oxygen toxicity causes fibrosis of the retina, a condition called retrolental fibroplasia. Attention to the amount of oxygen given to newborns with respiratory distress syndrome has greatly reduced the incidence of this condition. The most common of the many causes of blindness in adults are glaucoma, diabetic retinopathy, macular degeneration, senile cataract, optic nerve atrophy, and retinitis pigmentosa. Degeneration of the macula with resultant loss of central vision may be due to several causes, the most common of which is an idiopathic condition called *senile macular degeneration* characterized by damage to retinal pigment epithelium with underlying vascular proliferation. Optic atrophy may be caused by occlusion of the small vessels supplying the optic nerve, by optic neuritis such as occurs in multiple sclerosis, and by masses that press on the optic nerve. Retinitis pigmentosa, usually a recessively inherited condition, results in progressive destruction of rods and cones beginning in childhood and leading to blindness.

REVIEW QUESTIONS

1. What symptom is common to most people wearing eyeglasses regardless of their underlying eye disorder?
2. How are each of the following instruments or procedures helpful in evaluation of eye problems?
   Visual acuity tests
   Visual field tests
   Funduscopic examination
   Tonometer
   Slit-lamp
3. What are the causes, lesions, and major manifestations of each of the specific diseases or disorders discussed in this chapter?
4. How do abnormalities in the cornea, lens, or retina lead to strabismus?

5. What is the pathogenesis of visual loss in glaucoma?
6. How does diabetic retinopathy differ from hypertensive retinopathy?
7. What is the pathogenesis of decreased visual acuity in myopia, hyperopia, presbyopia, and astigmatism?
8. What are the leading causes of blindness?
9. What do the following terms mean?

   Visual acuity
   Papilledema
   Photophobia
   Nystagmus
   20/20 vision
   Visual axis
   Amblyopia
   Capillary aneurysms
   Microhemorrhages
   Accommodation
   Enucleation
   Legal blindness

Bones and Joints

REVIEW OF STRUCTURE AND FUNCTION

The human body contains 206 bones, many of which have joints at
their ends to connect them to adjacent bones. The major bones and
joints are labeled on Figure 21–1.

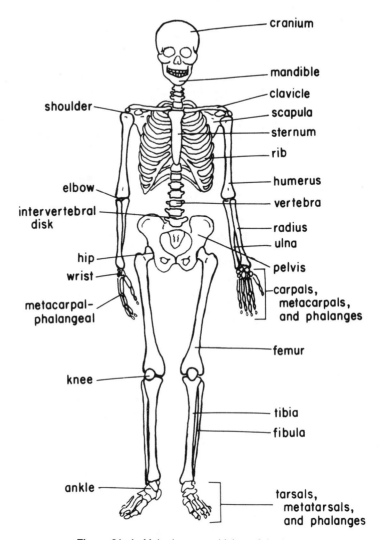

Figure 21–1. Major bones and joints of the body.

Bone provides a framework for the attachment of muscles, supports weight bearing, and protects internal organs from injury. It also plays a major role in calcium and phosphorus metabolism. Bone marrow, which may be considered part of bone, is discussed in Chapter 10. Joints allow movement of body parts and control the extent of movement.

The components of a typical long bone are shown in Figure 21–2. Between the epiphysis and metaphysis lies the epiphyseal plate, a layer of cartilage that undergoes growth and ossification until the bone has reached its mature length.

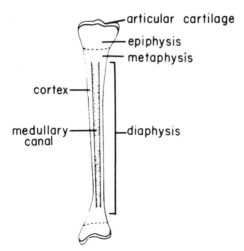

Figure 21–2. Structure of a long bone—the tibia.

Joints consist of the articular and adjacent surfaces of bone bridged externally by a fibrous capsule. The joint is lined by flattened mesothelial-like synovial lining cells and contains a small amount of serous fluid.

The knee joint (Fig. 21–3) is more complex, but particularly important because it is a frequent site of injury and disease. Unlike other joints, the knee joint contains fibrocartilagenous pads called *menisci* and contains a bone, the patella, within an overlying tendon. The spinal column (Fig. 21–4) has pads between vertebral bodies that consist of nucleus pulposis surrounded by fibrocartilage. The spinal canal, which houses the delicate spinal cord and emerging spinal nerves, lies posterior to the vertebral bodies and anterior to the more complex bony parts of the vertebrae. Like the

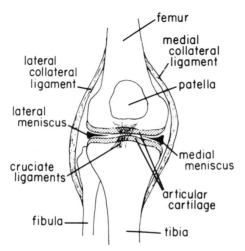

Figure 21–3. Structure of a large, complex joint—the knee.

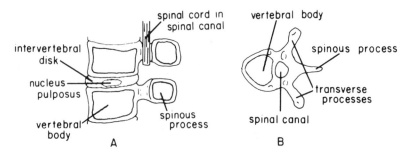

Figure 21–4. Structure of the spinal column. **A.** Sagittal plane. **B.** Transverse plane.

knee, the spine is particularly vulnerable to the effects of weight bearing.

The flat bones such as ribs, sternum, pelvis, and cranium serve to protect the thorax, abdomen, and brain. They also (along with the vertebrae) are major sites of hematopoiesis.

Bone, in spite of its solid consistency, is an active tissue that responds to physical stress, metabolic conditions, injury, and disease. Bone is formed by osteoblasts. They lay down a collagenous matrix called osteoid and promote deposition of hydroxyapatite crystals within the collagenous matrix to form bone. During development, osteoblasts may differentiate from primitive mesenchymal cells and form bone directly (intramembranous bone formation) or invade and replace a growing cartilagenous plate with bone (enchondral bone formation). Intramembranous bone formation typically occurs in flat bones and endochondral bone formation typically occurs in long tubular bones with a dense cortex.

In childhood, most of the increase in body height is due to enchondral bone formation in long bones. Disease or injury to epiphyseal plates will produce asymmetric body growth and deformity. Bone continues to remodel throughout adult life. When injured, such as by fracture, osteoblasts proliferate to form new bone. Chondrocytes may also produce cartilage at the site of injury and this may be invaded by osteoblasts and replaced by bone. Osteoblasts must be able to differentiate from more primitive cells in adult life, as metaplastic bone is seen in sites of injury remote from bone.

Osteoclasts, the cells responsible for bone destruction, are multinucleated cells seen at the edge of bony seams.

MOST FREQUENT AND SERIOUS PROBLEMS

By far the most frequent affliction of bone is fracture. Fractures may be caused by trauma alone or may be pathologic, meaning that the bone was already weakened by other disease such as metastatic cancer or osteoporosis. Sites commonly involved with fractures vary with age and sex. For example, traumatic fractures are most common in 20- to 40-year-old males and are common in the extremities. In children, the most common fractures are of the clavicle and the humerus. Elderly women are prone to fracture their hips or vertebrae due to osteoporosis.

If arthritis is considered to be primarily a disease of joints, then joint disease is much more prevalent than bone disease. This generalization also holds for traumatic injuries, as strains and sprains, which affect joint function, are more common than fractures of bone.

Strains and sprains are among the top ten causes of patients' seeking health care for acute disease, while low back problems and degenerative arthritis are among the top ten causes of patients seeking health care for chronic disease. The relative frequency of the three major forms of chronic arthritis is shown in Figure 21–5. The prevalence of these forms of arthritis is very high; 5 percent of the population suffers from degenerative arthritis, 2.5 percent from rheumatoid arthritis, and 0.5 percent from gouty arthritis.

Generalized loss of bone (*osteopenia*) may be due to a wide variety of causes such as osteoporosis, osteomalacia, rickets, hyperparathyroidism, and metastatic cancer. Osteoporosis in postmenopausal women is the most common of these conditions and the vertebral column is the most significant site of involvement by painful fractures.

Metastatic cancer is the most common malignancy of bone and often requires considerable health care because of associated pain and pathologic fracture. Of the several types of bone cancers, multiple myeloma is the most common in adults and osteosarcoma is the most common in young people, particularly adolescents.

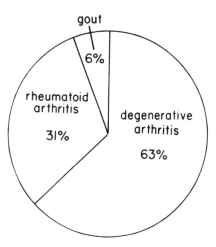

Figure 21–5. Relative frequency of major forms of chronic arthritis. *(Data from the National Health Education Committee: The Killers and Cripplers— Facts on Major Diseases in the US Today. New York, D. McKay, 1976.)*

SYMPTOMS, SIGNS, AND TESTS

The most common symptoms of bone and joint diseases are pain, decreased mobility, and deformity. Almost all fractures of bone are associated with pain due to disruption of sensory nerves. Usually, the fracture is obvious because of the attendant deformity, although in some fractures the bone is not displaced. The muscles surrounding a fracture site will undergo intense sustained contraction (spasm) in an attempt to protect the fractured area and this spasm will often cause additional pain.

Joint stiffness, decreased mobility, and varying degrees of pain are associated with chronic arthritis (rheumatoid and degenerative); whereas, the cardinal signs of inflammation (redness, heat, swelling, and pain) are associated with acute infectious arthritis.

Physical examination of patients with bone and joint disease involves careful attention to evaluation of joint mobility, gait, and neurologic examination (see Chapter 23 for description), as well as looking for deformity or masses.

X-ray films are the primary laboratory results used for evaluation of most bone and joint diseases. The metabolic activity of bone is evaluated by a battery of serum tests, including calcium, phosphorus, and alkaline phosphatase. Calcium and phosphorus are principal constituents of bone, and alkaline phosphatase is an enzyme present in the serum that is produced by the osteoblast and is elevated in many bone diseases involving proliferation of osteoblasts. The erythrocyte sedimentation rate may be used as an indicator of chronic inflammation and is useful in following patients with rheumatoid arthritis. Tests for rheumatoid factor (latex fixation titer) and serum uric acid levels may be useful aids in the diagnosis of rheumatoid arthritis and gout, respectively. Cultures are important in the diagnosis of acute arthritis and osteomyelitis. Biopsy is used to diagnose bone tumors.

SPECIFIC DISEASES

Genetic/Developmental Diseases

Developmental abnormalities fall into two broad groups. Selected structural defects include embryonic anomalies and localized deformities that arise during fetal development and childbirth. The most common types, many of which need early diagnosis and treatment to prevent permanent deformity, are discussed below. These include clubfoot, intoeing, congenital dislocation of the hip, and torticollis. The wide variety of other types include conditions such as missing limbs and extra digits. The second major group of developmental abnormalities are generalized genetic disorders. These are much less common than the localized types, and their treatment is symptomatic rather than curative. Three types are discussed briefly—achondroplasia, osteogenesis imperfecta, Marfan's syndrome.

Clubfoot (Talipes Equinovarus)

Clubfoot occurs in approximately 1 of every 1000 births. Clubfoot consists of a downward (equino) and inward (varus) turning of the foot (Fig. 21–6). It may occur bilaterally and is more frequent in males. Although the precise cause of clubfoot is not known, there is evidence that a genetic factor is involved. The deformity is evident at birth and can usually be corrected by placing the foot in casts for 2 to 6 months to gradually correct the deformity.

CHART 21–1. CLUBFOOT

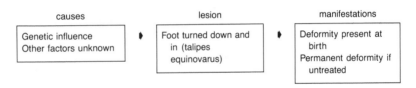

causes	lesion	manifestations
Genetic influence Other factors unknown	Foot turned down and in (talipes equinovarus)	Deformity present at birth Permanent deformity if untreated

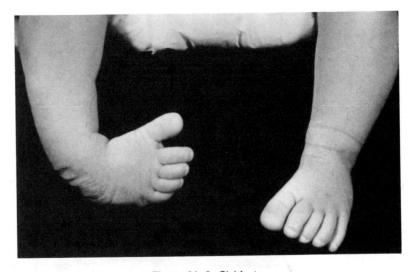

Figure 21–6. Clubfoot.

Intoeing

An inward turning of the toes is a very common finding in infants. Three common causes are femoral anteversion, tibial torsion, and metatarsus adductus. *Femoral anteversion* is a turning in of the femur. *Tibial torsion* is an internal twisting of the tibia. *Metatarsus adductus* is an inward curvature (varus deformity) of the forefoot. These conditions usually correct themselves. In some cases of metatarsus adductus, special braced shoes or casts may be needed to correct the deformity and prevent permanent intoeing (pigeon toes deformity). No specific cause is known for these conditions.

CHART 21–2. INTOEING

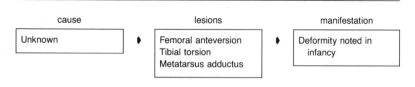

cause	lesions	manifestation
Unknown	Femoral anteversion Tibial torsion Metatarsus adductus	Deformity noted in infancy

Congenital Dislocation of the Hip

Congenital dislocation of the hip may be detected during infancy or childhood. It consists of a malformation of the acetabulum (hip socket) that allows displacement of the head of the femur in relation to the acetabulum (Fig. 21–7). The occurrence is influenced by heredity and much more frequently involves girls than boys. The most common clinical signs are shortening of the involved extremity, with abnormal folds of skin in the buttocks. Treatment with appropriate splinting, if done in time, prevents permanent deformity and crippling.

CHART 21–3. CONGENITAL DISLOCATION OF THE HIP

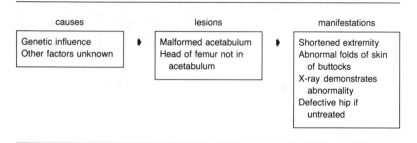

causes	lesions	manifestations
Genetic influence Other factors unknown	Malformed acetabulum Head of femur not in acetabulum	Shortened extremity Abnormal folds of skin of buttocks X-ray demonstrates abnormality Defective hip if untreated

Torticollis (Wry Neck)

Torticollis means turned neck and presents within the first 3 months of infancy as a neck pulled to one side. Although the cause is usually not clinically evident, it is believed to be due to injury to

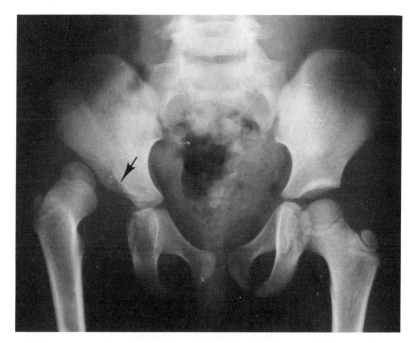

Figure 21–7. X-ray of congenital dislocation of the hip. Note irregularity of acetabulum on the involved side *(arrow).*

the sternocleidomastoid muscle during the fetal period by abnormal intrauterine positioning or injury at birth. Scarring and contraction of the muscle leads to the pulling of the neck to one side (Fig. 21–8). Treatment consists of manipulative stretching or surgical cutting of the muscle to prevent permanent facial deformity.

CHART 21–4. TORTICOLLIS

cause		lesion		manifestations
Presumed injury to sternocleidomastoid muscle	♦	Scarring of sternocleidomastoid muscle	♦	Neck turned and fixed to one side Onset in early infancy

Achondroplasia
Achondroplasia is a rare autosomal dominant disorder most often arising by mutation. The anatomic defect consists of poorly organized epiphyseal cartilage, which results in reduced growth of long bones and short stature (Fig. 21–9). A similar disorder occurs in basset hounds. Persons affected may die before birth or during infancy, or they may have a normal life expectancy.

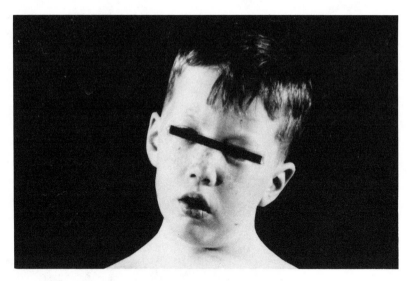

Figure 21–8. Torticollis.

CHART 21–5. ACHONDROPLASIA

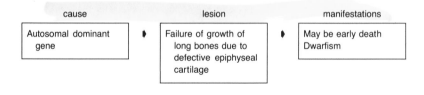

cause		lesion		manifestations
Autosomal dominant gene	▶	Failure of growth of long bones due to defective epiphyseal cartilage	▶	May be early death Dwarfism

Osteogenesis Imperfecta

This genetic condition associated with abnormal collagen formation usually follows an autosomal dominant inheritance pattern. Severe forms cause death in utero or soon after birth. The more common delayed form is characterized by thin bones that fracture with minimal trauma. The abnormal collagen production results in thin blue sclera, fractures of the bony ossicles of the ear leading to deafness, and deformed, hypoplastic teeth.

CHART 21–6. OSTEOGENESIS IMPERFECTA

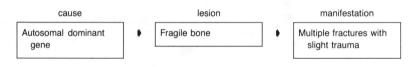

cause		lesion		manifestation
Autosomal dominant gene	▶	Fragile bone	▶	Multiple fractures with slight trauma

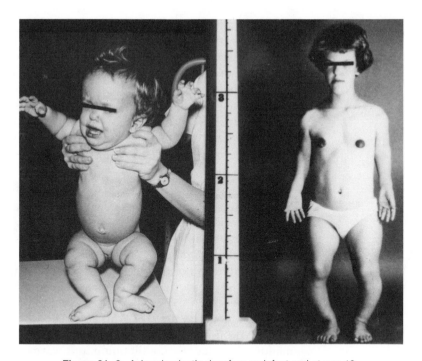

Figure 21-9. Achondroplastic dwarf as an infant and at age 13.

Marfan's Syndrome

This is another autosomal dominant disorder, which involves elastic tissue of bone, blood vessels, and other sites. Unlike the conditions discussed above, it may go undetected, because the changes are subtle and only certain complications prove to be harmful. Involved individuals are tall and slender with long narrow fingers and toes and asymmetrical skulls. About half have dislocation of the lens of the eye. Weakening of the media of the aorta due to the defective connective tissue leads to the most serious complication, rupture of the aorta with fatal hemorrhage. This is likely to occur suddenly during exercise.

CHART 21-7. MARFAN'S SYNDROME

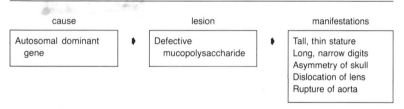

cause		lesion		manifestations
Autosomal dominant gene	▶	Defective mucopolysaccharide	▶	Tall, thin stature Long, narrow digits Asymmetry of skull Dislocation of lens Rupture of aorta

Inflammatory/Degenerative Diseases

Under this category, we will discuss the effects of trauma (strains, sprains, fractures), infection (acute arthritis, osteomyelitis), and chronic disorders of varied etiology. Low back pain and curvatures of the spine (scoliosis and kyphosis) are clinical syndromes with multiple causes. There are many other inflammatory and degenerative conditions of bones and joints, often made specific by their location and symptomatology, that are beyond the scope of this text.

Injuries to Joints and Muscles

An acute injury to a joint with tearing of the joint capsule and ligaments around the joint is called a *sprain*. Hemorrhage around or into the joint may also be present. Twisting of the ankle is a common cause of a sprain and usually involves rupture of a ligament on the lateral side of the foot. *Whiplash injury*, caused by sudden extension of the neck, is a sprain in which ligaments and other tissues supporting the cervical spine are torn.

Tearing of a muscle and/or its tendon due to excessive use and stretching is called a *strain*. Muscle strains (pulled muscles) produce disruption of the muscle with hemorrhage and mild inflammation. Athletes use conditioning and warm-up exercises to prevent strains.

Bones may be traumatically dislocated from their joint sockets. A partial dislocation is called a *subluxation*.

The knee joint is a common site of acute injury, especially in athletes. Tears may occur in the menisci, cruciate ligaments, and medial and lateral ligaments. Arthroscopic surgery is an endoscopic technique used to examine the inside of the knee joint and remove torn fragments of tissue. This technique has much less morbidity than open knee operations.

CHART 21–8. SPRAINS AND STRAINS

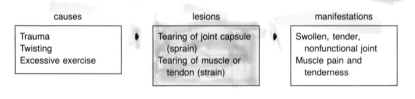

causes	lesions	manifestations
Trauma Twisting Excessive exercise	Tearing of joint capsule (sprain) Tearing of muscle or tendon (strain)	Swollen, tender, nonfunctional joint Muscle pain and tenderness

Low Back Pain

Owing to an upright posture, man is uniquely susceptible to low back pain. The problem of weight bearing thrust upon the lumbar spine is accentuated by obesity, weak abdominal muscles, poor posture, and sudden physical stresses. These factors, however, are more likely to cause problems when there is underlying disease of

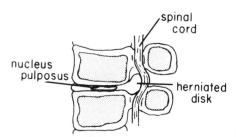

Figure 21–10. Hemiated intervertebral disc.

the spine. The evaluation of persistent low back pain, an extremely common problem, involves evaluation of the patient for underlying disease of the spine, which may or may not be found. Pain may be due to compression of nerves, muscle spasm (a protective mechanism), or sprained ligaments.

Generalized diseases that can cause or accentuate low back pain include degenerative arthritis, rheumatoid arthritis, ankylosing spondylitis, osteoporosis, and metastatic cancer. Diseases that are specific to the spine include herniated intervertebral disc and congenital fusion defects of vertebrae, especially the fifth lumbar vertebra. Neurologic examination to evaluate for nerve compression and roentgenograms of the lower spine will reveal most of these conditions, but one of the most common causes of back pain, herniated intervertebral disc, may require further study.

Herniated intervertebral disc, usually in the lumbar region, is a common and important cause of low back pain, because the herniated disc puts pressure on spinal nerves. The pain may radiate down the back of the leg, a condition called *sciatica* because the sciatic nerve is often involved. The soft central material (nucleus pulposus) is pushed out through the surrounding fibrocartilage (annulus fibrosus) and displaces the capsule to impinge on a spinal nerve (Fig. 21–10). Radiologic examination after injection of radiopaque dye into the spinal canal (myelogram) is used to demonstrate compression of the spinal cord and nerve roots (Figs. 21–11, 21–12). Operative removal of the disc may be required to relieve pain and prevent paralysis.

CHART 21–9. LOW BACK PAIN

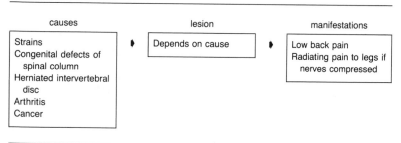

causes	lesion	manifestations
Strains Congenital defects of spinal column Herniated intervertebral disc Arthritis Cancer	Depends on cause	Low back pain Radiating pain to legs if nerves compressed

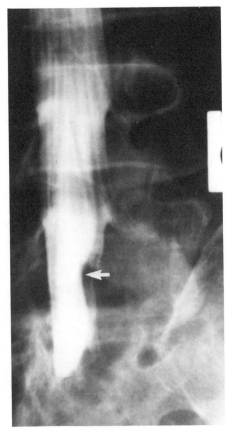

Figure 21–11. Myelogram demonstrating compression of the radiopaque dye in the spinal canal by a herniated intervertebral disc *(arrow).*

Scoliosis and Kyphosis

Scoliosis is abnormal lateral curvature of the spine, and kyphosis is an abnormal forward bending of the upper spine producing a hunched back (Fig. 21–13). By the age of 14, approximately 2 percent of people have some degree of scoliosis. Most of the time the cause is unknown. Known causes include congenital deformity, paralysis of one side of the body, and diseases involving the vertebrae. Kyphosis is most often due to arthritis of the spine but may be due to other spinal column diseases.

CHART 21–10. SCOLIOSIS AND KYPHOSIS

causes		lesions		manifestations
Unknown factors Congenital deformities Paralysis Arthritis Other diseases of spine	▶	Lateral deviation of spine (scoliosis) Hunchback (kyphosis)	▶	Observable deformity May impair activity if severe

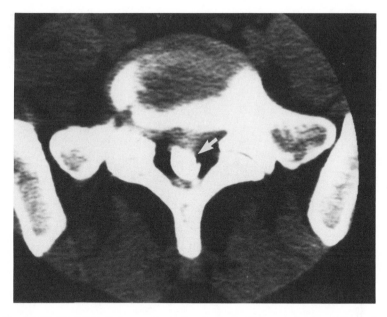

Figure 21–12. Magnetic resonance scan demonstrating compression of the radiopaque dye in the spinal canal *(arrow)* and failure of dye to fill around the nerve root on the right.

Fractures

A fracture is any disruption of the continuity of bone. Most fractures are caused by trauma. Spontaneous fractures or fractures resulting from slight trauma suggest the possibility that the fracture was caused by underlying disease of bone (pathologic fracture).

Many terms are used to describe the nature of the fracture (Fig. 21–14). *Incomplete fractures* produce cracks without separation of the bone; with *complete fractures,* the bone is separated into two or more parts (Fig. 21–15). *Comminuted fractures* are ones in which more than two fragments are produced. A *compression fracture* is one in which the bones are pushed together rather than apart. These commonly occur in vertebrae. *Open fractures* produce disruption of the skin; *closed fractures* do not. A stable fracture tends to maintain its position following fracture, an unstable one does not.

The sites of fractures vary in frequency with age and sex, because these factors are related to the likelihood of various types of injury and the possibility of underlying disease. For example, arm fractures are common in children due to pulling or falling; spine and hip fractures are common in elderly women because these weight-bearing bones are affected by osteoporosis. Bone has great power to heal, so that continuity can be accomplished in a few weeks and remodeling and return to normal strength can oc-

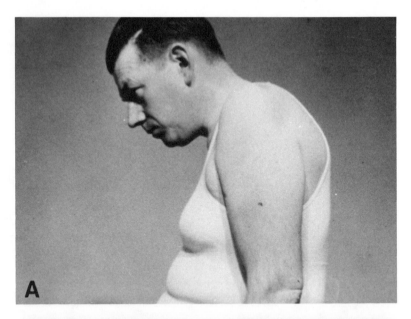

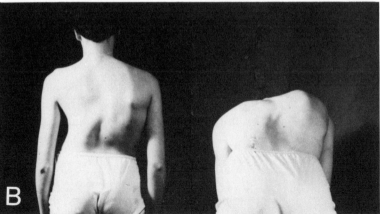

Figure 21–13. A. Kyphosis. **B.** Scoliosis, with winging of the scapula when bent forward.

cur in a few months. Fracture healing is much more rapid in the young than in the elderly.

The process of fracture healing involves the proliferation of osteoblasts from the fracture margins to form new cartilage and bone. The immature bone and cartilage is gradually remodeled into mature bone. Bone is usually produced in excess, but eventually, through the process of remodeling, the bone returns to normal structure.

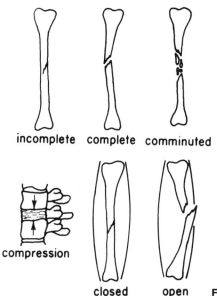

incomplete   complete   comminuted

compression

closed   open   **Figure 21–14.** Types of fractures.

There are several important factors that can prevent this normal healing sequence. The broken fragments must be close to each other or the ends will fail to unite (nonunion). The fracture must be a stable fracture or, if unstable, it must be artificially stabilized using splints, casts, traction (steady pulling by means of weights), or operatively inserted metal pins, screws, or plates. With fractures of the neck of the femur, the head of the femur is sometimes removed and replaced with a metal ball. Nonstabilized fractures also lead to nonunion. One of the worst complications of fracture is infection. Open fractures, particularly those in which the trauma

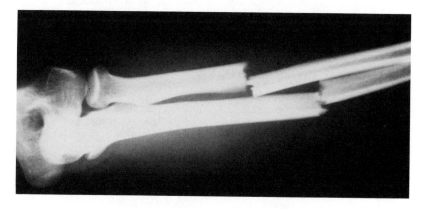

Figure 21–15. Complete fractures of radius and ulna.

drives dirt into the wound, and fractures that are artificially opened in the operative room to accomplish immobilization are subject to the possibility of infection. Cleaning of the wound, removal of dead tissue, and antibiotics are used to prevent infection. The consequences of an infected fracture are chronic osteomyelitis, nonunion, and eventual deformity. Another factor affecting healing is the extent and location of the fracture. Comminuted fractures heal well unless they are displaced, involve joints, or cannot be stabilized. Fractures in the middle of long and flat bones generally heal well, while fractures involving joints are likely to produce problems with joint mobility. Fractures that disrupt the blood supply to the bone will not heal well. Finally, pathologic fractures may not heal, depending on the nature of the disease causing them.

CHART 21–11. FRACTURES

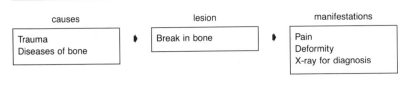

causes	lesion	manifestations
Trauma Diseases of bone	Break in bone	Pain Deformity X-ray for diagnosis

Acute Arthritis

Severe acute arthritis is usually caused by pyogenic bacteria such as staphylococcus. Acute arthritis is occasionally associated with gonorrhea, as the gonococcus organism gains access to the blood and preferentially locates in the joint space, causing a swollen, painful, red joint. Staphylococci may be spread to the joint through the blood but more often may cause arthritis when the joint is opened due to trauma, adjacent disease, or operation. Acute arthritis produces rapid destruction of the joint lining and is likely to lead to permanent destruction and bony ankylosis of the joint.

CHART 21–12. ACUTE ARTHRITIS

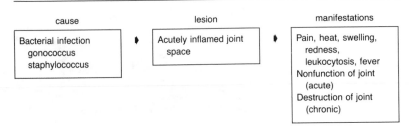

cause	lesion	manifestations
Bacterial infection gonococcus staphylococcus	Acutely inflamed joint space	Pain, heat, swelling, redness, leukocytosis, fever Nonfunction of joint (acute) Destruction of joint (chronic)

Osteomyelitis

Although *osteomyelitis* means inflammation of bone, the term is usually used in a more restricted sense to mean infection of bone. The route of infection is the basis of classification of the two major types of osteomyelitis. Hematogenous osteomyelitis involves spread of the causative organism through the blood to localize in one bone to set up a focus of infection. The site of entry of the organism is usually a skin infection (which may go unnoticed), the organism is usually *Staphylococcus aureus,* and the site of the osteomyelitis is usually the metaphysis of a long bone near but not involving the epiphysis. Children are most commonly affected. If untreated, the infection spreads, producing necrosis of the bone. The purulent infection may produce draining sinuses, and the necrotic bone must be removed, because antibiotics will not be effective against bacteria lurking in dead bone. The clinical findings are pain and other local and systemic signs of inflammation. By the time x-ray changes occur, the bone is necrotic and simple antibiotic treatment may not effect a cure. The outcome in advanced cases includes recurrence of the infection and bone deformity with crippling.

The other form of osteomyelitis, which is much more common, is called secondary osteomyelitis, because the infection spreads to bone secondarily from an adjacent site of infection or open wound. The most common causes are infected operative sites (most commonly involving repair of fractures), soft tissue infections adjacent to bone, and gangrene of the toes with ulceration and infection. Treatment consists of removal of dead tissue and bone, open drainage of the infected area, and antibiotics. Permanent damage is likely.

CHART 21–13. OSTEOMYELITIS

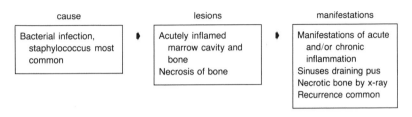

cause	lesions	manifestations
Bacterial infection, staphylococcus most common	Acutely inflamed marrow cavity and bone Necrosis of bone	Manifestations of acute and/or chronic inflammation Sinuses draining pus Necrotic bone by x-ray Recurrence common

Osteoporosis

Osteoporosis is a generalized quantitative decrease in bone with the remaining bone having a normal amount of mineral and matrix (Fig. 21–16). The osteopenia (quantitative loss of bone) is usually

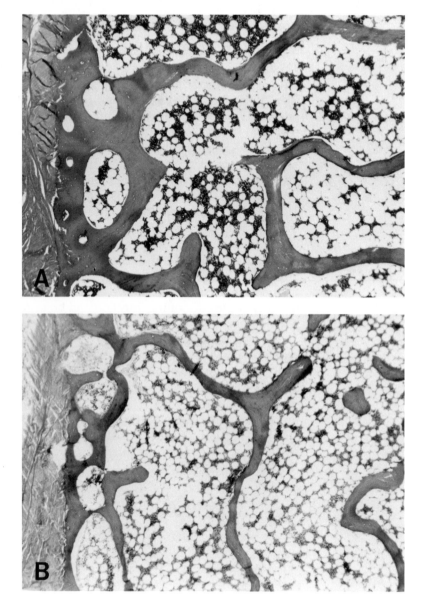

Figure 21–16. Histologic section of vertebra. **A.** Normal from 20-year-old. **B.** Osteoporosis from 80-year-old with narrow, widely separated bone spicules and thin cortex.

demonstrated on roentgenograms. The normality of mineral and matrix is suggested by finding normal serum levels of calcium, phosphorus, and alkaline phosphatase. Conditions that cause metabolic breakdown of bone, such as osteomalacia, hyperparathyroidism (see Chapter 26), and Paget's disease, typically have elevated alkaline phosphatase levels and altered calcium and phosphorus levels.

Thinning of bone occurs normally with aging, so the diagnosis of osteoporosis is relative. Most symptomatic osteoporosis occurs in postmenopausal women, but it may occur in elderly men, in patients on long-term corticosteroid therapy, and with prolonged bed rest. Persons with large bones are less affected than persons with small bones and thin trabecular bones (such as vertebrae) are more affected than dense cortical bones.

No single specific cause for osteoporosis has been identified; multiple factors are likely involved in its pathogenesis. Once developed, reversion back to normal bone thickness does not appear to be possible. Therefore, prevention of the development of the disease is most important and this must be done years (decades) before the disease is likely to become manifest. Therapies that have been directed at this goal include high calcium intake, vitamin D supplements, estrogens in menopausal women, testosterone in men with low levels, exercise, and various other drugs. However, in most instances, preventive treatment is not employed and complications are treated as they develop.

The most common complications of osteoporosis are compression fractures of vertebra and hip fractures. Vertebral disease is associated with pain and height reduction. Hip fracture, which occurs with minimal trauma in osteoporotic persons, is a severe acute illness that is often fatal or leads to permanent disability, particularly in frail elderly women.

CHART 21–14. OSTEOPOROSIS

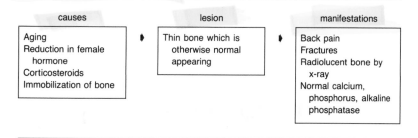

causes		lesion		manifestations

Aging
Reduction in female
  hormone
Corticosteroids
Immobilization of bone
	▶	Thin bone which is
		  otherwise normal
		  appearing
			▶	Back pain
				Fractures
				Radiolucent bone by
				  x-ray
				Normal calcium,
				  phosphorus, alkaline
				  phosphatase

Osteomalacia and Rickets

These are relatively rare conditions characterized by softening of bone. They differ from osteoporosis in that there is inadequate deposition of calcium and phosphorus, leaving an excess of the protein matrix of the bone. Osteomalacia is the adult form characterized by bone softening. Rickets is the childhood form with both softening and decreased growth of bones. The majority of cases are secondary to poor intake or poor utilization of vitamin D, with consequent improper deposition of calcium and phosphorus in bone. In children, untreated rickets leads to markedly deformed bones. Osteomalacia in adults may cause fractures. Serum levels of calcium, phosphorus, and alkaline phosphatase are abnormal.

CHART 21–15. OSTEOMALACIA AND RICKETS

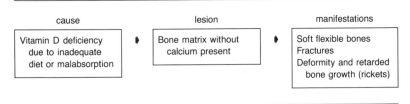

cause	lesion	manifestations
Vitamin D deficiency due to inadequate diet or malabsorption	Bone matrix without calcium present	Soft flexible bones Fractures Deformity and retarded bone growth (rickets)

Degenerative Arthritis

This very common disease is also called *degenerative joint disease* and *osteoarthritis*. It occurs most often in the middle aged to elderly and is manifest by joint stiffness and often pain. The lesion consists of wearing of the articular joint cartilage, with subsequent deformity of the cartilage and of the bone, resulting in stiffness and decreased motion. New growth of bone at the margins of the joint lead to so-called lipping (osteophytes), which further limits movement of the joint (Fig. 21–17B). Degenerative joint disease is more common in women and typically involves the weight-bearing joints and the distal finger joints (Fig. 21–18A). This pattern of involvement, plus a typical x-ray picture, help distinguish degenerative from rheumatoid arthritis. Degenerative arthritis is likely to develop with time in injured joints or joints subject to undue stress, such as might occur with congenital dislocation of the hip or a knee subject to football injuries.

CHART 21–16. DEGENERATIVE ARTHRITIS

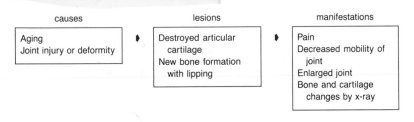

causes	lesions	manifestations
Aging Joint injury or deformity	Destroyed articular cartilage New bone formation with lipping	Pain Decreased mobility of joint Enlarged joint Bone and cartilage changes by x-ray

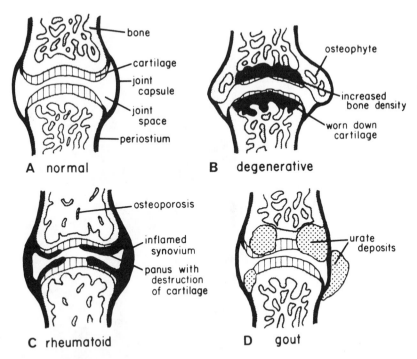

Figure 21–17. Sehematic representation of normal joint (**A**) and lesions of degenerative arthritis (**B**), rheumatoid arthritis (**C**), and gout (**D**).

Rheumatoid Arthritis

This is a chronic inflammatory disease that predominantly affects joints, but may produce lesions in other organs. The joint lesions, which produce pain, stiffness, and deformity, are due to inflammation of the synovial lining. As the inflammation extends on to the joint surface, it destroys cartilage and produces a layer of granulation tissue called a *pannus*. Eventually the entire joint surface may be destroyed and replaced by fibrous tissue. Fusion of joints is called *ankylosis*.

Rheumatoid arthritis is more common in women than men and varies in severity from mild joint stiffness to severe cases, showing distortion and ankylosis of many joints, with almost total loss of function. Usually the metacarpal–phalangeal joints are initially affected, followed by wrists, knees, and elbows. Involvement of the metacarpal–phalangeal joints leads to characteristic ulnar deviation of the fingers (Fig. 21–17B). Many patients with severe rheumatoid arthritis will also have chronic inflammation and vasculitis involving other organs such as the heart, muscle, lungs, skin, and possibly the eye.

Morphologically, the joint lesion consists of a low-grade

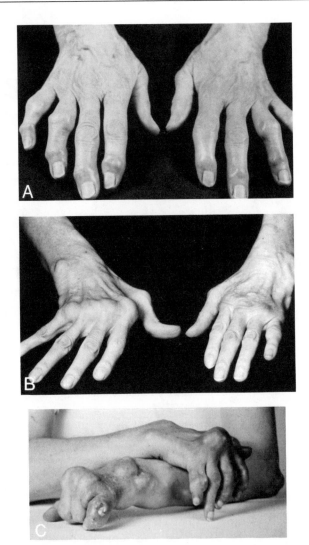

Figure 21–18. Hand changes in advanced cases of degenerative arthritis (**A**), rheumatoid arthritis (**B**), and gout (**C**). In degenerative arthritis, the distal interphalangeal joints show swelling due to formation of osteophytes. Rheumatoid arthritis is characterized by ulnar deformity and involvement of metacarpal–phalangeal joints. In gout, tophi produce irregular nodules about the joints. *(Reproduced from Clinical Slide Collection on the Rheumatic Diseases, produced by The Arthritis Foundation, New York, copyright 1972.)*

chronic inflammation of the synovial lining and joint surface, with destruction of the joint cartilage, fibrosis around the joint, and osteoporosis of the surrounding bone due to disuse (Fig. 21–18C).

The cause of rheumatoid arthritis is not known, but it is associated with an abnormal immunoglobulin M, which is directed against the body's normal immunoglobulin G (in effect, an antibody against an antibody). This abnormal antibody is called the rheumatoid factor and may be tested for in the laboratory as an aid to diagnosis. It is thought that this abnormal immunoglobulin complex is deposited on the synovial membrane, eliciting an inflammatory response that results in proliferation and thickening of the synovium. Severe rheumatoid arthritis is often treated with corticosteroids, because these drugs inhibit both the formation of immunoglobulin complexes and inflammation.

In recent years, surgeons have had much success in replacing severely ankylosed joints with artificial (prosthetic) joints. Knee and hip joints can be replaced as well as finger joints. Variants of rheumatoid arthritis include *juvenile arthritis,* in which the involvement occurs very early in life and may be extremely severe.

Ankylosing spondylitis is an inflammatory arthritis predominantly involving the spine and sacroiliac joints. Long considered a variant of rheumatoid arthritis because the lesions are histologically similar, it is now considered by many to be a separate disease because it occurs in young men, has a strong association with the inherited HLA-B27 antigen, and has a familial tendency. The deformity of the spine with stooped posture often leads to severe disability.

CHART 21–17. RHEUMATOID ARTHRITIS

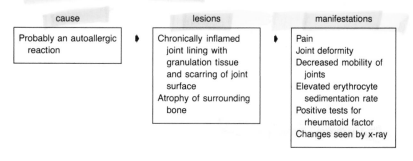

cause		lesions		manifestations
Probably an autoallergic reaction	▸	Chronically inflamed joint lining with granulation tissue and scarring of joint surface Atrophy of surrounding bone	▸	Pain Joint deformity Decreased mobility of joints Elevated erythrocyte sedimentation rate Positive tests for rheumatoid factor Changes seen by x-ray

Gout

This is an inherited disease in which there is abnormal metabolism of uric acid. The excess uric acid may be deposited in many tissues, particularly joints, and eventuates in painful arthritis. In the chronic stage, these urate deposits accumulate to form tophi (Figs. 21–17D, 21–18C). Gout occurs almost exclusively in men over 30 and clinically manifests by bouts of painful arthritis, particularly in the fingers, wrists, ankles, knees, and toes, especially the great toe. Treatment of gout consists of long-term administration of agents that promote the excretion of uric acid or inhibit its production. See Chapter 7 for further discussion.

CHART 21–18. GOUT

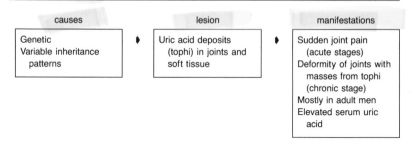

causes	lesion	manifestations

Genetic
Variable inheritance
  patterns
	Uric acid deposits
	  (tophi) in joints and
	  soft tissue
		Sudden joint pain
		  (acute stages)
		Deformity of joints with
		  masses from tophi
		  (chronic stage)
		Mostly in adult men
		Elevated serum uric
		  acid

Ganglion

A ganglion is a smooth cystic swelling that arises from joint capsules, most commonly on the wrist. They are often associated with continued trauma and may be painful, although they usually arise insidiously as a simple swelling. Surgical removal may be undertaken if the ganglion is bothersome to the patient.

CHART 21–19. GANGLION

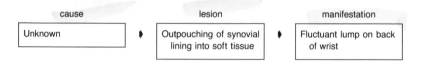

cause	lesion	manifestation

Unknown	Outpouching of synovial
	  lining into soft tissue
		Fluctuant lump on back
		  of wrist

Hyperplastic/Neoplastic Diseases

There are many types of neoplasms and non-neoplastic tumors of bone, cartilage, joints, and tendon sheaths. Most are rare and will not be discussed. The most common malignant tumors are metastatic cancers from sites such as breast, lung, prostate, and kidney, as well as multiple myeloma and lymphomas. Paget's disease of bone is a peculiar hyperplastic disease.

Paget's Disease of Bone

Paget's disease or osteitis deformans is a localized or multifocal enlargement of bone of unknown cause that affects about 2 percent of the population, typically in persons over 40 years of age. Initially the affected bone may be more porous, but there is a gradual haphazard bony proliferation leading to some deformity of the bone and occasionally to pathologic fracture. High serum alkaline phosphatase reflects the active bone remodeling, but there is no defect in calcium and phosphorous metabolism. Most patients are asymptomatic. Rarely, osteosarcoma may develop in the lesion.

Osteosarcoma

This is a malignant bone-forming tumor arising in bone, thus it is to be considered a neoplasm arising from osteoblasts. Osteosarcoma arises most commonly in the long bones, especially near the knees (Fig. 21–19), but it may be seen in other bones. Osteosarcoma occurs in children and in young adults and presents with pain or swelling. Most patients present with a bony mass and no evidence of metastasis, but occult metastases are probably present because amputation alone is followed by overt matastases and death in 80 percent of patients. This has led to aggressive radiation and chemotherapy at the time of removal of the mass and a cure rate of over 60 percent. Some cases occur in adult years after radiation of bone or in association with Paget's disease of bone.

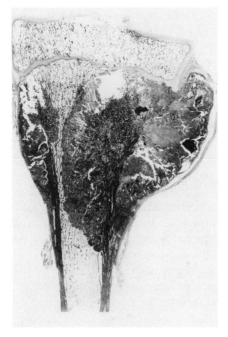

Figure 21–19. Large histologic section of a typical osteosarcoma of the distal femur involving metaphysis and spreading through the periosteum.

CHART 21–20. OSTEOSARCOMA

causes		lesion		manifestations
Unknown Radiation Paget's disease of bone	▶	Sarcoma with bone production	▶	Mass in bone Pathologic fracture X-ray appearance Biopsy Usually in long bones of teenagers Lung metastases

ORGAN FAILURE

The main function of the skeletal system is to maintain support and mobility for everyday activity. Injury or disuse of major joints and fractures of weight-supporting bones are likely to lead to considerable incapacity of movement. Extensive severe arthritis or widespread bone metastases may confine patients to a wheelchair or bed.

REVIEW QUESTIONS

1. What is the relative frequency of joint disease as compared to bone disease? How are they likely to differ in their consequences?
2. What are the most common musculoskeletal problems and which are most serious?
3. What are the common manifestations of bone and joint disease?
4. What are the symptoms, signs, and laboratory abnormalities of each of the specific diseases discussed in this chapter?
5. How do localized developmental abnormalities differ from generalized genetic abnormalities of the skeletal system in terms of frequency and likely outcome?
6. What is a herniated intervertebral disc and what are its effects?
7. How do fractures heal? What factors impair healing?
8. How do hematogenous and secondary osteomyelitis differ in terms of cause, persons affected, and outcome?
9. How do osteoporosis and osteomalacia differ in terms of cause, histology, and laboratory findings?
10. How do degenerative and rheumatoid arthritis differ in frequency, cause, morphology, and location?

11. How do metastatic cancer of bone and osteosarcoma differ in frequency, age, and outcome?
12. What do the following terms mean?
    Pathologic fracture
    Rheumatoid factor
    Talipes equinovarus
    Femoral anteversion
    Tibial torsion
    Metatarsus adductus
    Whiplash injury
    Ankylosis
    Sciatica
    Incomplete fracture
    Comminuted fracture
    Compression fracture
    Open fracture
    Unstable fracture
    Nonunion of a fracture
    Senile osteoporosis
    Juvenile arthritis
    Ankylosing spondylitis
    Paget's disease

Chapter 22

Skeletal Muscle

REVIEW OF STRUCTURE AND FUNCTION

Skeletal muscle is the largest organ in the body and utilizes about 10 percent of the body's oxygen in the resting state but as much as 80 percent or more with intense exercise. All skeletal muscles are separated into bundles called *fascicles*, which are enclosed in connective tissue. Fascicles are in turn composed of individual muscle fibers (cells), each of which courses the entire length of the muscle (Fig. 22–1). Muscle fibers are innervated by branches of axons from

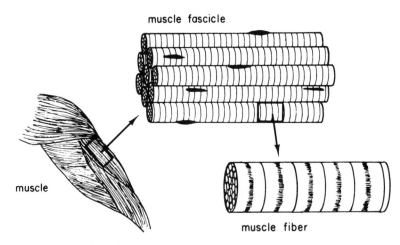

muscle fascicle

muscle

muscle fiber

Figure 22–1. Schematic representation of skeletal muscle, with a fascicle and an individual fiber. Each muscle fiber courses the entire length of the muscle. The alternating light and dark bands are called striations and can be seen by light microscopy.

anterior horn neurons in the spinal cord. In muscles that need fine discriminatory movements, such as the eye, one neuron may innervate as few as four to six muscle fibers; whereas in large muscles used for strength and weight bearing (gluteus or quadriceps), one neuron may innervate up to 2000 muscle fibers. Muscle maintains its tone by a complex system of nerves that wrap around special fibers (spindles) and send information regarding the degree of muscle contraction back to the spinal cord neurons. Chemically, muscle is composed primarily of two proteins called *actin* and *myosin*, which are filaments arranged alternately in parallel rows. These filaments slide back and forth beside each other, thereby performing the contraction and relaxation process. The overlapping of the filaments delineates the I and the A bands, which can be seen by light microscopy as striations (Fig. 22–1).

Muscle fibers are divided into types I and II on the basis of their histochemical reactions (Fig. 22–2). These types also correspond somewhat to physiologic properties. For example, type I fibers are more utilized for slow, sustained contractions, whereas type II fibers contract more quickly. Type II fibers respond to exercise by hypertrophy and to disuse by atrophy. In humans, the two fiber types are evenly distributed throughout all muscles, but in many animals an individual muscle may be composed entirely of one fiber type. For example, in domestic birds such as chickens, the dark muscles (leg, thigh) consist entirely of type I fibers, while the light muscles (wings, breast) are all type II fibers.

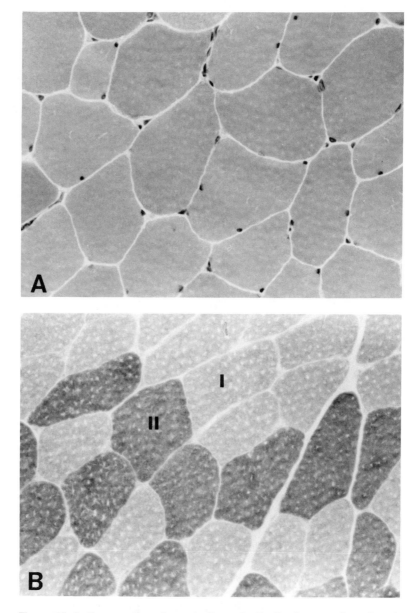

Figure 22–2. Cross-section of muscle fibers. **A.** Routine hematoxylin and eosin stained section. **B.** Stained with an enzyme stain (ATPase) to demonstrate type I and II fibers.

MOST FREQUENT AND SERIOUS PROBLEMS

Probably the most common affliction of muscle is simply weakness with or without muscle atrophy. This may occur as a part of a generalized disease such as cancer or any disease that results in prolonged immobilization. Muscular dystrophies are important muscle diseases because they are often inherited and, consequently, genetic counseling becomes an important aspect of disease prevention. The most frequently occurring and severe dystrophy is Duchenne muscular dystrophy. Whenever there is disease of the nervous system, there will also be associated muscle weakness. Therefore, victims of traumatic nerve injuries, strokes, and many other nervous system diseases will all display muscle weakness. Trauma may result in muscle weakness from either muscle destruction or nerve damage.

SYMPTOMS, SIGNS, AND TESTS

Weakness is the common denominator of muscle disease, whether it be primary disease of muscle *(myopathic)* or secondary to disease of the nervous system *(neurogenic)*. The pattern of weakness often affords the physician an important clue as to the type of muscle disease present. For example, muscular dystrophies most often involve proximal muscle groups, whereas diseases of nerves are more likely to result in atrophy and weakness of the more distal parts of the extremities. If weakness persists for any length of time, the muscle will become atrophic, irrespective of the cause of the disorder. Conversely, atrophy of muscle obviously will result in weakness. Pain is occasionally associated with muscle disease and when present often signifies muscle inflammation.

Although the presence of muscle disease is often obvious because of weakness and atrophy, many times the patient's symptoms may be nondescript or vague. In such cases, there are certain laboratory tests that aid in the diagnosis of muscle disease and at the same time help quantify the degree of muscle damage. *Aldolase* and *creatine kinase* are two enzymes normally involved in the metabolism of muscle that are present in the serum in increased quantities following many disorders that damage muscle. They are usually more elevated in myopathic than they are in neurogenic disorders, and the degree of elevation of these enzymes will roughly parallel the extent of muscle damage. Electromyography will also help to separate intrinsic muscle disorders from neurogenic muscle disorders and to quantify the extent of muscle damage. *Electromyography* is accomplished by inserting a needle

TABLE 22–1. COMPARISON OF NORMAL, MYOPATHIC, AND NEUROGENIC MUSCLE DISORDERS

	Normal	Myopathic	Neurogenic
Symptoms or signs	None	Proximal weakness, possible pain	Distal weakness, often in a nerve distribution; possible sensory loss
Electromyography	No spontaneous activity in muscle	Asynchronous spontaneous activity	Synchronous activity of small amplitude
Serum enzymes (creatine kinase and adolase)	Normal	Often markedly elevated	Mildly elevated or normal
Biopsy	Normal fiber configuration	Variable size of fibers; degenerative fibers; possible fibrosis	Atrophic fibers in small groups

into a muscle and recording the electrical activity. The most reliable means of separating myopathic from neurogenic causes of muscle disease is muscle biopsy, a procedure that is easily performed on most muscles. The major differences between neurogenic and myopathic muscle disorders are summarized in Table 22–1.

SPECIFIC DISEASES

Genetic/Developmental Diseases

There are numerous types of primary muscle degeneration collectively called *dystrophies*. Since many of the more common types of dystrophy are genetically determined, we will consider the entire group under developmental rather than under degenerative disorders, keeping in mind that the exact cause of some types of dystrophy is undetermined.

Dystrophy literally means poor nutrition and was originally applied to muscle disorders thought to be of simple cause and not due secondarily to disease of nerves. Today there are many diseases termed *dystrophic* that have various causes. Many are hereditary with onset at an early age. Others are definitely hereditary but do not become manifest until adult life. Still others do not appear to be hereditary at all. Dystrophies are classified according to the pattern of muscle involvement—group(s) of muscles affected—or according to the type of microscopic lesion. Different dystrophies initially show slightly different lesions, but eventually all will lead to muscle atrophy with replacement of muscle by adipose and

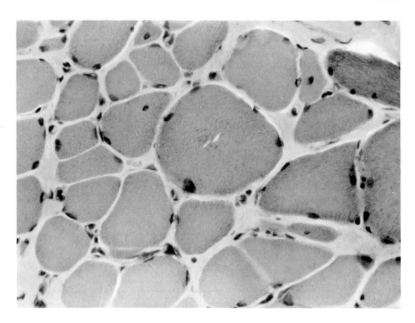

Figure 22–3. Photomicrograph of dystrophic muscle. Note the excessive amount of connective tissue, marked variation in fiber size, and large, rounded fibers. Central nuclei are present in some fibers.

fibrous tissue. Most of the dystrophies involve proximal muscles of the extremities and the pelvis and shoulder girdles in preference to distal muscles of the extremities. There is no known cure for any of the dystrophies, and all follow a variable but fairly predictable course. Therapy must be supportive. By light microscopy, most dystrophies show variable fiber size, fiber splitting, increased amounts of connective tissue, and large rounded fibers (Fig. 22–3). Other microscopic findings are more specific for particular types of dystrophy.

Duchenne Dystrophy

The most common and serious of the dystrophies is Duchenne dystrophy, which is inherited as a sex-linked recessive disorder and affects males within the first few years of life, with an expected lifespan of only 12 to 18 years. As in most myopathic conditions, the weakness is predominantly of the proximal muscle groups. Boys with Duchenne dystrophy characteristically develop "pseudohypertrophy" of the calves, in which the calves appear large and muscular but are actually replaced by adipose tissue (Fig. 22–4). The heart muscle may also be involved in this disease. The diagnosis is based upon typical age of occurrence, family history, and

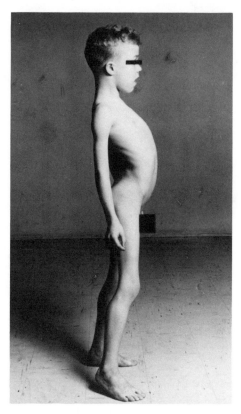

Figure 22–4. Child with Duchenne muscular dystrophy. Note the enlarged calves.

findings of intrinsic muscle disease (myopathic) by electromyography, enzyme tests, and muscle biopsy. The biopsy shows disruption and loss of both type I and type II muscle fibers with replacement by connective tissue.

CHART 22–1. DUCHENNE MUSCULAR DYSTROPHY

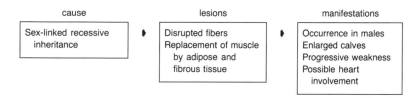

cause		lesions		manifestations
Sex-linked recessive inheritance	▸	Disrupted fibers Replacement of muscle by adipose and fibrous tissue	▸	Occurrence in males Enlarged calves Progressive weakness Possible heart involvement

Myotonic Dystrophy

This is an autosomal dominant disease associated with muscle weakness and a characteristic inability to release contraction *(myotonia)*. Patients with myotonic dystrophy will shake hands with

Figure 22–5. Patient with myotonic dystrophy. Note elongated face, frontal baldness, and lack of facial muscle tone.

someone and then are unable to let go. Patients with myotonic dystrophy also may have cataracts, frontal balding, heart disease, and gonadal atrophy.

Typical facial appearance is shown in Figure 22–5. Type I fibers are preferentially involved by atrophy, splitting, and encasement in fibrous tissue. In many fibers, nuclei migrate to the center of the fiber. The electromyographic findings are very characteristic in this disease. Patients may lead a long life but are often quite disabled.

CHART 22–2. MYOTONIC DYSTROPHY

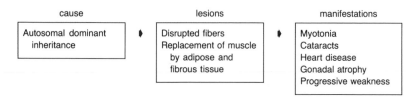

cause	lesions	manifestations
Autosomal dominant inheritance	Disrupted fibers Replacement of muscle by adipose and fibrous tissue	Myotonia Cataracts Heart disease Gonadal atrophy Progressive weakness

Other Dystrophies

Other, less common dystrophies named by site of involvement include limb–girdle dystrophy (limbs, pelvic and pectoral girdles), facioscapulohumeral (face, scapula, and humerus), and oculopharyngeal. More recently discovered dystrophies, which are often called *congenital myopathies,* are named according to histologic appearance of the muscle. These include nemaline myopathy, in which small thread-like rods are found in the muscle; central core disease, in which each fiber has a central, pale-staining area on cross-section; and central nuclear myopathy, in which the muscle fiber nuclei are in the center of the fiber instead of at the normal, peripheral position. These disorders are all manifest by variable degrees and types of weakness.

Inflammatory/Degenerative Diseases

Neuroscience practitioners commonly divide diseases of muscle into myopathic types, in which there is primary affliction of muscle, and neurogenic types, in which the muscle affliction is secondary to disease of the nerves that innervate the muscle. The muscular dystrophies are one group of the myopathies category. Other types of myopathic disorders include all those conditions in which muscle weakness follows another disease process, such as immunologic disease, vascular disease, neoplasia, or metabolic disease.

Autoimmune Diseases

Polymyositis is an autoimmune inflammatory disease that affects muscle in preference to other organs and is characterized by a lymphocytic infiltrate. A similar lymphocytic infiltration of muscle fibers is seen in other autoimmune diseases such as lupus erythematosus and rheumatoid arthritis. These diseases, collectively referred to as autoimmune inflammatory myopathies, produce muscle weakness, often accompanied by pain and tenderness, and elevation of serum creatine kinase. Autoimmune inflammatory myopathies can often be successfully treated with corticosteroids because these drugs are antilymphocytic in addition to being anti-inflammatory. Paradoxically, muscle weakness and atrophy can also result from corticosteroid therapy in susceptible persons.

Neurogenic Disorders

As alluded to previously, any affliction of peripheral nerve or spinal cord motor neuron will result in muscle weakness and wasting. Acceptable treatment demands the separation of nervous system

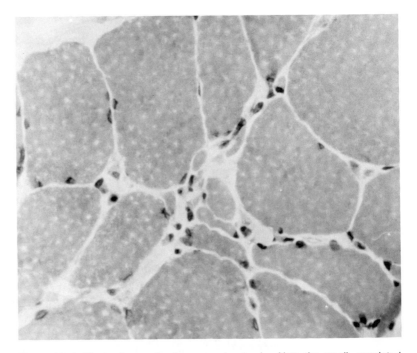

Figure 22–6. Photomicrograph of neurogenic atrophy. Note the small, angulated fibers in a group.

disease from that of primary muscle disease. Histologically, muscles that are atrophic secondary to lesions of the nervous system show a characteristic pattern of atrophy in which atrophic fibers have sharply angulated contours and occur in groups (Fig. 22–6). The group occurrence is due to the fact that adjacent fibers are all innervated by the same axon.

In addition to laboratory tests, helpful clues to the presence of neurogenic disease include the pattern of muscle involvement and the presence of sensory symptoms such as decreased pain. For example, muscle weakness in the distribution of a motor nerve or loss of sensation in the same area as the weakness would indicate a neurogenic disorder. In addition, disease of nerves tend to affect the more distal muscles in the extremities first. Treatment of neurogenic muscular diseases consists of dedicated physical therapy to prevent irreversible atrophy and fibrosis of muscle and contracture of joints.

The most common neurogenic disorder of muscle is that which is secondary to peripheral nerve injury. Other important primary neurogenic diseases that severely affect muscle are

Werdnig–Hoffman's disease in infants and *amyotrophic lateral sclerosis* in adults. In both diseases there is idiopathic degeneration of the motor neurons in the spinal cord, with consequent muscle wasting and eventual death due to respiratory failure.

CHART 22–3. NEUROGENIC DISORDERS

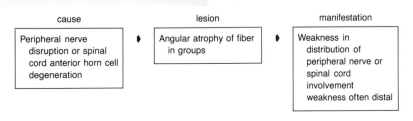

The course of the disease may be improved by use of anti-cholinesterase drugs. Normally cholinesterase degrades acetylcho-line; anticholinesterase drugs slow this process, thus making more acetylcholine available to initiate muscle contraction. If a thymoma

Myasthenia Gravis

Myasthenia gravis is characterized by a progressive decrease in muscle strength associated with activity and a return of strength after rest. It is a disease of the junction of nerve endings with muscle, so it does not fall clearly into either of the myopathic or neurogenic categories discussed above. In myasthenia gravis, anti-bodies against the acetylcholine receptor are present in the serum of most affected persons, resulting in degeneration of the receptor. Consequently, the nerve impulse, which is normally transmitted by acetylcholine, is ineffectively transferred to the muscle resulting in progressively weaker muscle contractions.

The initiating cause of the antibody production is not known; therefore, the disease can be considered an autoimmune disease. In many cases the disease is associated with a neoplasm of the thymus. The onset of myasthenia gravis is usually insidious. Al-most any muscle may be affected; however, half of the patients with this disease have *diplopia* (double vision) as a first symptom because of frequent involvement of extraocular muscles. My-asthenia gravis runs a course of years but is ultimately fatal in most cases due to slowly developing atrophy of muscles, especially those required for respiration.

is found by chest x-ray, its removal may be associated with clinical improvement.

CHART 22–4. MYASTHENIA GRAVIS

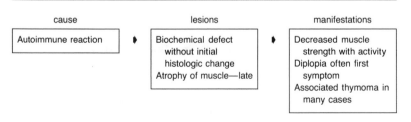

cause		lesions		manifestations
Autoimmune reaction	▶	Biochemical defect without initial histologic change Atrophy of muscle—late	▶	Decreased muscle strength with activity Diplopia often first symptom Associated thymoma in many cases

Metabolic Myopathies

Many metabolic conditions can result in muscular weakness. Some of these are inherited, such as the glycogen storage diseases, lipid metabolism disorders, and familial periodic paralysis. Other metabolic disorders are situational, such as myopathy due to excessive alcoholic intake or uremia. The inherited metabolic myopathies may surface in childhood or adulthood depending on the relative amounts of key enzymes in the patient's system. In familial periodic paralysis, the patient may become profoundly weak within a matter of minutes due to poorly understood fluctuations in serum potassium. These episodes can occur irregularly throughout a patient's life. Metabolic myopathies are diagnosed by history and selected tests. The muscle biopsy may be helpful in some, such as glycogen storage disease.

Hyperplastic/Neoplastic Diseases

Rhabdomyosarcoma is a rare primary malignant neoplasm of skeletal muscle. Metastases of carcinomas to skeletal muscle are uncommon; sarcomas will occasionally spread through skeletal muscle.

ORGAN FAILURE

A single muscle or group of muscles may fail *(paralysis)* because of focal dystrophic, traumatic, neurogenic, or inflammatory involvement. Simultaneous failure of most of a person's muscle mass may be acute or chronic. Acute paralysis of all muscles follows administration of curare-like drugs, which block the myoneural junction. These drugs are often used during operative procedures when complete muscle relaxation is required. Acute paralysis of muscle may also follow rapidly progressive peripheral nerve diseases in which there is generalized inflammation of nerves or nerve roots. Chronic muscle failure is the end result of any neurogenic or myopathic disorder and consists of replacement of muscle by fibrous and adipose tissue. Whether acute or chronic, muscle failure eventually affects the diaphragm and intercostal muscles, and the patient will succumb to respiratory paralysis.

REVIEW QUESTIONS

1. What are the two major divisions of muscle disease in terms of pathogenesis?
2. How are each of the following tests or procedures helpful in evaluation of a person with muscle disease?
    Serum aldolase and creatine kinase
    Electromyography
    Muscle biopsy
3. What are the causes, lesions, and major manifestations of each of the specific muscle diseases listed in this chapter?
4. How might the signs and symptoms of neurogenic muscle disorders differ from those of myopathic disorders?
5. Why is it important to separate dystrophic, inflammatory, and neurogenic causes of muscle weakness?
6. What do the following terms mean?
    Dystrophy
    Pseudohypertrophy of calves
    Myotonia
    Polymyositis
    Werdnig–Hoffman disease
    Amyotrophic lateral sclerosis
    Rhabdomyosarcoma
    Paralysis

Chapter 23

Central Nervous System

REVIEW OF STRUCTURE AND FUNCTION

The central nervous system consists of the brain and spinal cord (Fig. 23–1). The brain consists of the cerebrum, cerebellum, and

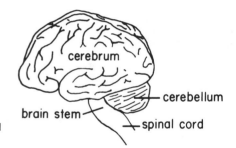

Figure 23–1. Parts of the central nervous system.

brain stem (midbrain, pons, and medulla). The cerebrum consists of an outer layer of gray matter (cortex), deep gray matter, and white matter (Fig. 23–2). The cortex is replete with neurons that are employed for intellectual (cognitive) functions as well as for sensory and motor functions above the vegetative level. The deep gray matter consists of groups of neurons such as the thalamus and basal ganglia that perform the same type of functions as the cortex, albeit at a much more primitive level.

The white matter is composed primarily of the axons and their myelin sheaths. Axons are long processes of neurons that connect with neurons in other parts of the brain and spinal cord. Axons of spinal cord neurons innervate skeletal muscles. Thus a voluntary thought generated from neurons of the cerebral cortex can control movement of skeletal muscles. Of course, some axons convey sensory impulses in the opposite direction, from the spinal cord to various parts of the brain. The cerebellum, which is situated in the posterior inferior aspect of the skull, is mainly responsible for coordination of motor functions. The brain stem is a relay between brain and spinal cord and is also a control center for heart rate, respiration rate, sleep and wakefulness, integration of eye movements, and other functions.

The brain is covered by meninges, which include an outer,

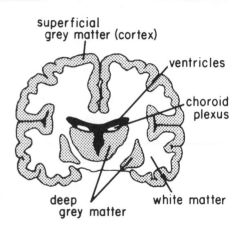

Figure 23–2. Coronal section of the brain.

tough membrane called the *dura*, which is next to the skull, and an inner lace-like membrane, the *pia–arachnoid*, which lies directly over the cortex. Meninges also form a continuous covering over the spinal cord.

Cerebrospinal fluid is utilized for metabolic exchange, as an excretory vehicle, and as a means to absorb pressure changes in the central nervous system. The cerebrospinal fluid is formed in the ventricles of the brain by secretion from the choroid plexus and by filtration through the ependyma. It flows from the three ventricles in the anterior part of the brain, through a narrow aqueduct, to the area of the brain stem (medulla). At this point, it passes out of the ventricular system and percolates between the layers of the pia–arachnoid membrane, bathing the brain and spinal cord before being absorbed back into veins.

Most of the blood flows to the brain through the two internal carotid arteries anteriorly and the paired vertebral arteries posteriorly. The carotid arteries supply blood to the bulk of the cerebrum, whereas the vertebral arteries supply blood only to the brain stem, cerebellum, and posterior part of the cerebrum (occiput). Although the vertebral arteries carry much less blood than the carotids, they supply the vital areas in the brain stem, including cranial nerve nuclei and control centers for respiration and consciousness. These vessels all interconnect at the base of the brain, forming the *circle of Willis*. Consequently, occlusion of one major artery to the brain may not necessarily result in deprivation of blood to its area of distribution, because blood can be "borrowed" from other arteries via the circle of Willis. The major arteries branch into smaller arteries as they course through the pia–arachnoid membrane and eventually penetrate the cortex and deeper structures. The anatomy and physiology of the capillaries in the brain differ from those in the rest of the body. The brain capillaries are constructed and function in such a manner as to prevent passage of many substances into the brain that can easily reach other body tissues. This selective exclusion of substances is termed the *blood–brain barrier*. For example, certain antibiotics will not pass the barrier and consequently are not useful in the treatment of brain infections. The actual site of the blood–brain barrier is the endothelium but the astrocytes lying just outside of the capillaries give signals to the endothelium, which help to govern the passage of molecules between the blood and the brain. The capillaries entering the choroid plexus are different from those in the brain parenchyma; thus, some drugs that cannot enter the brain parenchyma may enter the cerebrospinal fluid. However, this blood–cerebrospinal fluid barrier does not allow indiscriminate passage of all drugs.

Microscopically, the important cellular constituents of the brain stem and spinal cord are the *neurons, astrocytes,* and *oligo-*

dendroglia. Neurons are the large cells found in gray matter that conduct nervous impulses. Their efferent processes (axons) may extend for long distances in gray and white matter. Their short afferent processes (dendrites) connect to other neurons through synapses. Astrocytes with their spider-like processes provide structural support to the central nervous system. Astrocytes also regulate the blood–brain barrier and tissue electrolytes. When the brain is injured, astrocytes proliferate, much like fibroblasts, to form a glial scar composed of glial processes but lacking collagen. Oligodendroglia manufacture and maintain the myelin sheath that surrounds and protects axons and dendrites.

MOST FREQUENT AND SERIOUS PROBLEMS

The major diseases of the brain are cerebrovascular accidents (strokes), traumatic injuries, infections (meningitis, encephalitis, and abscess), and neoplasms. Strokes are the third leading cause of death in the United States. Developmental disorders and degenerative diseases, such as multiple sclerosis, Parkinson's disease, and senile dementia, are also significant. Headaches and epilepsy are very important in terms of prevalence and morbidity but may be manifestations of a variety of diseases.

SYMPTOMS, SIGNS, AND TESTS

The most common presenting symptoms of central nervous system disease are headache, diminution or loss of a motor function, sensory loss, seizures, and disturbances in intellectual or memory capabilities.

The neurologic examination of a patient presenting with one or more of these symptoms will include examination of the motor and sensory systems and testing of cognitive function. The motor system examination involves observation of the patient's gait, posture, and symmetry of muscle mass, as well as testing for muscle strength, coordination, and quality of reflexes. Abnormalities of any of these parameters could be due to lesions of the cerebrum, cerebellum, spinal cord, peripheral nerves, or muscle. Examination of the sensory system entails eliciting a careful history of abnormal sensations (*dysesthesias* and *paresthesias*) and testing for diminished or absent sensory perception on various areas of the body by means of pinprick or application of heat, cold, or vibration. Lesions causing abnormalities of sensation may be located in the peripheral nerves, spinal cord, or cerebral cortex. Testing of reflexes is an important part of a neurologic examination. A decreased reflex may indicate a lesion in the appropriate peripheral nerve, with resultant inability to either transmit the sensory impulse back to the spinal cord or to transmit the motor impulse out to the muscle. A hyperactive reflex, such as an exaggerated knee jerk, represents an intact nerve between the knee and the spinal cord, albeit without the modifying control of the reflex normally mediated by the central nervous system. Testing of cognitive (memory, intellect) functions of the cerebral cortex entails asking the patient to repeat special phrases and perform arithmetic tasks. Other aspects of the neurologic examination may include tests for the integrity of the cranial nerves, observations of abnormal movements, and specific tests for the ability to perform coordinated movements.

The neurologic examiner will attempt to categorize the findings as focal (referable to a specific area of nervous system involvement) or generalized (involving integrated functions of the whole brain). Examples of focal signs or symptoms are *hemiparesis* (weakness of one side of the body), localized areas of sensory deprivation, abnormalities of one or two cranial nerves, or localized headaches. Examples of generalized signs and symptoms are intellectual impairment, generalized headaches, stupor, or loss of consciousness (*coma*). One of the major causes of generalized signs and symptoms is increased intracranial pressure. Because the brain is enclosed in a rigid skull, an increase in volume anywhere within the cranial cavity will rapidly cause a generalized increase in pressure throughout the entire brain. This effect will follow the development of any mass lesion in the cranial cavity, such as a neoplasm, hematoma, abscess, or localized edema surrounding a lesion. The increased intracranial pressure may also be the result of generalized edema secondary to diffuse infection. The only major opening in the skull is the foramen magnum, and the substance of the brain tends to be pushed toward this foramen as a consequence of any increased intracranial pressure (Fig. 23–3). As the cerebrum is forced into the space where the cerebellum lies, the oculomotor nerve becomes pinched, resulting in pupillary constriction of the same side as the lesion. This affords the physician a valuable clue as to the side of the brain lesion. The downward and backward excursion of brain substance toward the foramen magnum will result in hemorrhage into the brain stem, with coma and rapid death due to involvement of respiratory and activating centers if the pressure is unrelieved. The treatment of increased intracranial pressure is removal of any space-occupying lesion. In addition, steroid drugs and osmotic agents, such as mannitol, may help relieve brain edema by drawing interstitial and intracellular fluid back into the vascular system.

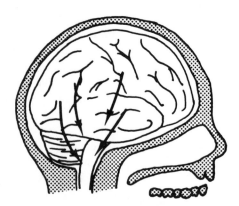

Figure 23–3. Forces resulting from increased intracranial pressure pushing the brain into the foramen magnum.

The most important laboratory examination utilized in the evaluation of central nervous system disease is the analysis of cerebrospinal fluid. Cerebrospinal fluid is usually obtained by inserting a needle into the lumbar pia–arachnoid space in a sitting or reclining patient. As the fluid is being withdrawn, the pressure is measured with a manometer to detect elevations in intracranial pressure. The fluid is then examined under the microscope for the presence of leukocytes, red blood cells, neoplastic cells, and microorganisms. Chemical determinations are made for protein and sugar. Serologic tests are utilized for the detection of syphilis and certain viral agents. If an infectious agent is suspected, the fluid is cultured.

Several radiologic procedures are used in evaluation of the patient with a neurologic lesion. Skull x-rays are used to detect fractures of the skull. A skull fracture connotes an injurious force of sufficient magnitude to also damage the underlying brain. A patient presenting with a localized lesion in the brain often undergoes angiography. Radiopaque dye is injected into the appropriate artery (most often carotid) and a simultaneous x-ray is taken to look for abnormal distribution or distortion of vessels in the region of a lesion such as a neoplasm, abscess, or hematoma. Angiograms are also utilized to demonstrate vessel occlusion in the patient with a cerebrovascular accident and to find the site of rupture of an intracranial aneurysm.

The *electroencephalogram* (EEG) is a device for evaluating electrical activity simultaneously in various areas of the brain. Normal neurons discharge electrically in certain known patterns. Abnormalities in patterns denote neuronal disturbance, which may be predicative of injury in specific areas. Patients with seizures may have violent focal disturbances in neuronal electrical activity, thereby localizing the site where the seizure originates. A damaged area in the brain may generate abnormal electrical activity by EEG even when the disturbance is not of sufficient magnitude to cause a clinical seizure. The patient with generalized signs and symptoms may show diffuse EEG abnormalities. The EEG is also used to determine if brain death has occurred in some patients who are in a deep coma.

Computerized tomography (CT scan) is used extensively to study the brain, ventricles, and subarachnoid spaces. Plain CT scans will allow evaluation of ventricular size, the presence of blood, or an infarct. Intravenous contrast material may be injected to detect a brain tumor that has sufficient vascular supply to become enhanced. Magnetic resonance (MR) produces even better images than a CT scan, but is not as available.

SPECIFIC DISEASES

Genetic/Developmental Diseases

Developmental abnormalities are more important in the brain than in any other single organ, with the possible exception of the heart. Persons with brain developmental abnormalities may live for many years with very little functional deficit or may be quite retarded and require constant nursing care. Developmental abnormalities of the central nervous system are usually divided into malformations and destructive brain lesions.

Malformations

Malformations are the result of deleterious forces acting upon the embryonic or fetal brain roughly within the first half of gestation. Malformations may be mediated genetically or may be the result of infection or hypoxic or traumatic insult to the brain. Further brain development following an insult early in gestation will result in abnormal brain structure (malformation). Individuals with brain malformations are often severely retarded mentally, unable to care for themselves, and, consequently, confined to hospitals. Other persons with brain malformations function at various levels in society. Down's syndrome (mongolism) is an example of a malformation caused by a chromosomal abnormality. Persons with Down's syndrome vary widely in intellectual capabilities. The structural abnormalities of the brain in Down's syndrome are not striking and consist of abnormal variation in brain shape and location of neurons.

One of the most common malformations is spina bifida, in which the posterior arches and spines of some vertebrae are absent. This defect is often discovered incidentally on x-rays, but if severe, a meningomyelocele will result. Meningomyelocele is a defect in the spinal column through which spinal cord and meninges protrude into the skin of the back. The cause of meningomyelocele is not known. It may result in severe paralysis of the legs but is compatible with life. Anencephaly is a severe malformation in which the entire forebrain is missing. Infants with anencephaly are stillborn or die soon after birth. Hydrocephalus may also result from a malformation that occludes the flow of cerebrospinal fluid. Hydrocephalus is discussed later in this chapter.

CHART 23–1. MALFORMATIONS

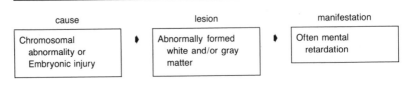

cause	lesion	manifestation
Chromosomal abnormality or Embryonic injury	Abnormally formed white and/or gray matter	Often mental retardation

Destructive Brain Lesions

These occur in the last half of gestational life or during the first 2 years after birth. Since the brain is reasonably well formed during the second half of gestation, injuries at this time result in destructive lesions with actual loss of brain substance in various areas. Most destructive lesions occur at the time of labor and delivery or in the neonatal period in premature infants and are due to anoxia from prolonged and difficult labor or to respiratory distress following delivery. Infections, especially meningitis, may also cause destructive brain lesions. Destructive brain lesions vary greatly in severity. Most patients have motor problems (weakness or incoordination), although one-third or more also are mentally retarded. Clinically, patients with destructive brain lesions are usually referred to as having *cerebral palsy,* which is defined as a nonprogressive condition manifested by motor retardation and sometimes accompanied by mental retardation. Malformations may also be a cause of cerebral palsy, but more often the mental retardation they cause will overshadow the motor retardation. External influences such as maternal diet, drugs, radiation, and toxins can adversely affect brain development. The severity of the defect will depend upon the stage of development at the time of insult.

CHART 23–2. DESTRUCTIVE BRAIN LESIONS

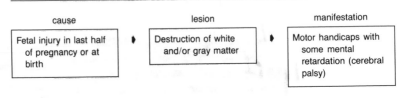

cause	lesion	manifestation
Fetal injury in last half of pregnancy or at birth	Destruction of white and/or gray matter	Motor handicaps with some mental retardation (cerebral palsy)

Inflammatory/Degenerative Diseases

Numerous infectious diseases involve the brain preferentially. Some of the more common processes and diseases will be discussed in this chapter, whereas, others, such as central nervous system syphilis, poliomyelitis, and HIV infection, are covered in the chapter on infectious diseases. The more common manifestations of degenerative diseases of the brain and spinal cord are also discussed in this section.

Meningitis

Meningitis means inflammation of the pia–arachnoid and is most often caused by bacteria. Meningitis most commonly occurs by itself but may be associated with other infections, such as pharyngitis or pneumonia. The onset is usually abrupt, and the major signs and symptoms are fever, headache, neck rigidity, and pain due to muscle spasm from nerve irritation. *Escherichia coli* is a common cause of meningitis in newborn infants, while *Hemophilus influenzae* commonly causes meningitis in small children.

Streptococcus pneumoniae and *Neisseria meningitidis* are more often the cause of meningitis in older children and adults. Neisseria meningitis is especially important, because it can occur in epidemics. Bacteria usually gain access to the brain and spinal cord via the blood. The diagnosis of acute meningitis is made on the basis of cerebrospinal fluid findings of neutrophilic leukocytes, decreased sugar, and the presence of organisms. Treatment consists of immediate antibiotic therapy. If treatment is not immediate or is inadequate, the presence of the bacteria and the leukocytes will result in alterations in the blood–brain barrier, leading to edema with consequent increased intracranial pressure and death of the patient. The lesion of acute meningitis is mainly that of purulent exudate in the subarachnoid space. Less commonly, bacteria will locate in the brain parenchyma rather than the meninges, forming an abscess that behaves as an expanding mass lesion and, if untreated, is almost always fatal.

Chronic meningitis may be caused by tuberculosis or several types of fungal organisms, the most common being *Cryptococcus neoformans*. The inflammatory cells in chronic meningitis are predominantly monocytes and lymphocytes rather than neutrophils, and the disease often smolders at the base of the brain for weeks to months, gradually affecting more and more cranial nerves at their point of exit from the brain. If a patient does not die from acute or chronic meningitis, there is always a danger of developing hydrocephalus from the obliteration of the subarachnoid space by fibrous tissue, with resultant failure to absorb cerebrospinal fluid.

CHART 23–3. MENINGITIS

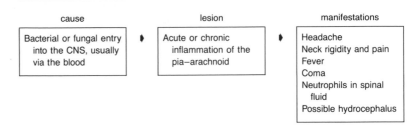

cause		lesion		manifestations
Bacterial or fungal entry into the CNS, usually via the blood	→	Acute or chronic inflammation of the pia–arachnoid	→	Headache Neck rigidity and pain Fever Coma Neutrophils in spinal fluid Possible hydrocephalus

Encephalitis

Encephalitis refers to a more or less diffuse inflammation of the whole brain. It is usually caused by viral infections. Bacterial, fungal, and protozoal infections usually affect the meninges or cause localized abscesses rather than encephalitis. Many viral encephalitides in the United States are mosquito-borne and occur in epidemics in the warm months of the year. Common types include St. Louis, equine, and Venezuelan encephalitis. Patients with any of these viral encephalitides usually present with generalized signs and symptoms of irritability, drowsiness, and headache. Specific diagnosis depends on culture and identification of the viral agent from the cerebrospinal fluid by serologic testing. In contrast to bacterial infections, viral encephalitis is usually accompanied by a cerebrospinal fluid lymphocytosis. There is no specific treatment for these diseases, and patients will either die, recover fully, or recover with variable neurologic deficit.

Herpes simplex virus type I also can cause an encephalitis. The same virus that causes oral blisters in susceptible persons may, on rare occasions, invade the brain and result in severe destruction of large areas of the brain, most often the temporal lobes.

CHART 23–4. ENCEPHALITIS

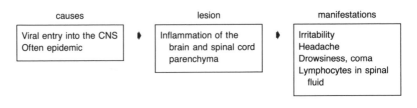

causes		lesion		manifestations
Viral entry into the CNS Often epidemic	→	Inflammation of the brain and spinal cord parenchyma	→	Irritability Headache Drowsiness, coma Lymphocytes in spinal fluid

Rabies

Although death from rabies occurs only rarely in the United States, it is such a feared disease that virtually every animal bite raises a concern for rabies. The reservoir of the virus is in wild animals, especially fox and skunk. These animals may transmit the disease to domestic pets, which in turn transmit the virus to humans via bites. The virus travels up the peripheral nerve to the brain and, once the brain is infected, death is virtually inevitable. The incubation period is proportional to the distance from the bite to the brain. After a bite by a domestic animal, the animal should be watched closely for 12 days. If the animal is still alive and well after 12 days, it is unlikely that it has rabies. However, if the animal shows signs of having rabies during the 12-day period, immunization of the victim should be initiated. If there is any suspicion by way of abnormal behavior or symptoms of rabies in the animal and the patient's bite is on the head or neck (close to the brain), immunization should start immediately, since it takes approximately 2 weeks after initiation of immunization to develop effective levels of antibody. Veterinarians are usually prophylactically immunized. If a person is bitten by a wild animal, especially one that displays abnormal behavior, every attempt should be made to locate the animal and submit it to an appropriate laboratory for rabies analysis. Rabies analysis consists of inoculating mice with brain tissue from the suspected animal plus a search for the viral inclusion bodies of rabies in the neurons of the suspected animal. These inclusion bodies are called *Negri bodies* and are found in the cytoplasm of neurons by light microscopy and by immunofluorescence, in which fluorescent-labeled antibodies are directed against the inclusions on tissue section and visualized with a microscope having an ultraviolet light source.

CHART 23–5. RABIES ENCEPHALITIS

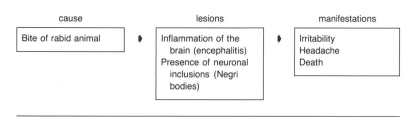

cause	lesions	manifestations
Bite of rabid animal	Inflammation of the brain (encephalitis) Presence of neuronal inclusions (Negri bodies)	Irritability Headache Death

Myelitis

Myelitis is an infection of the spinal cord. Poliomyelitis is a specific infection of the gray matter of the spinal cord. The poliomyelitis virus preferentially destroys the gray matter of the cord, killing the anterior horn motor neurons with resultant paralysis. Poliomyelitis is no longer the dreaded disease that it was prior to 1960 because of successful immunization programs.

Cerebrovascular Accident (Stroke)

A stroke is a sudden neurologic deficit caused by either vascular occlusion from thrombosis or embolism or from hemorrhage into the brain. The majority of cerebrovascular accidents are caused by emboli, which arise by separating from a thrombus in a large vessel such as the carotid artery (Fig. 23–4) or perhaps the heart. The embolus then travels distally, where it lodges in a brain vessel and results in an infarct. Most thrombi initially form because the endothelium of the vessel in which they arise has been damaged by atherosclerosis. Consequently, the common denominator of most cerebrovascular accidents is atherosclerotic vascular disease. This is why cerebrovascular accidents become increasingly prevalent in the elderly.

Whether from emboli or thrombi, vascular occlusions result in infarcts in the brain tissue supplied by the affected vessel. The damaged brain tissue loses function within minutes and becomes soft and necrotic within a few days. Later, tissue is lost from the area, leaving a cystic cavity (Figs. 23–5, 23–6).

The middle cerebral artery is the largest cerebral artery, and it is most often occluded by emboli, because it is a direct continuation of the carotid artery. Occlusion of the middle cerebral artery is important, because this artery supplies the part of the cortex controlling motor function. Involvement of the motor cortex will produce weakness (*paresis*) or paralysis on the opposite side of the body. If the dominant side of the brain (the side that primarily controls speech and motor function, usually the left side) is involved, the patient will also have *aphasia* (impaired language function).

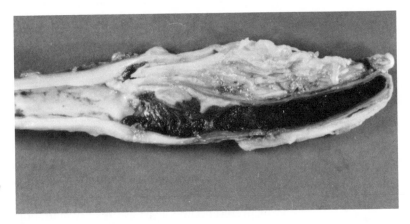

Figure 23–4. Laminated thrombus has filled the lumen of this carotid artery in a patient who died of a massive brain infarct.

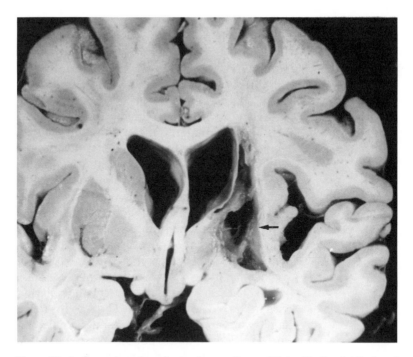

Figure 23–5. Coronal section of a cerebrum with an old infarct in the distribution of the middle cerebral artery with cystic changes.

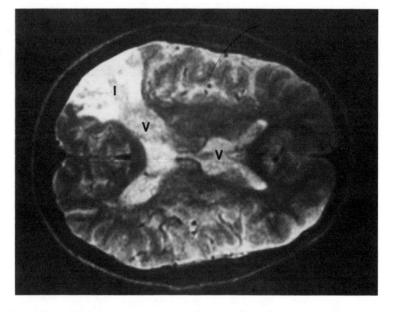

Figure 23–6. Magnetic resonance scan of large peripheral cystic infarct (I). Centrally, the fluid-filled ventricles can be seen (V).

Cerebrovascular accidents involving the vertebral arteries or their branches may also cause paralysis because of injury to motor fibers in the brain stem coursing between the brain and spinal cord. Large cerebrovascular accidents in the brain stem will usually kill the patient because of interruption of the nervous centers that control respiration.

Cerebrovascular accidents are also caused by rupture of vessels and bleeding into the brain (brain hemorrhage). The ruptured vessel has usually been weakened by arteriosclerosis, in a patient with hypertension. The signs and symptoms of a brain hemorrhage depend upon its location and size, but almost half of the patients with large brain hemorrhages will die within hours, because the accumulation of blood displaces adjacent tissue and rapidly elevates the intracranial pressure (Fig. 23–7).

A third important cause of stroke is ruptured saccular *aneurysm*. Saccular (berry) aneurysms occur predominantly in the vicinity of the circle of Willis, where vessels branch (Figs. 23–8, 23–9). They are saccular outpouchings of vessels due to deficiencies in the blood vessel wall. The reasons for development of these aneurysms are poorly understood. Saccular aneurysms are present in 2 to 5 percent of the population, but most do not rupture. When they do rupture, blood is spilled into the subarachnoid space and can be detected in the cerebrospinal fluid. Consequently, examination of the cerebrospinal fluid following a stroke may distinguish ruptured aneurysm from the other important causes of strokes.

Overall, about one-third of the patients with a cerebrovascular accident will die, one-third will be left with a serious neurologic deficit, and one-third will recover (Fig. 23–10). The individual's prognosis depends upon the amount of brain involved and the quality of supportive care received. Many persons sustain small infarcts that never result in neurologic deficit, because they occur in noncritical, so-called "silent" areas of the brain. Little can be

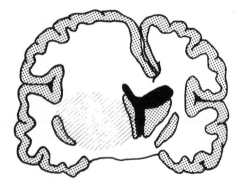

Figure 23–7. Area of hypertensive hemorrhage with displacement of adjacent brain tissues.

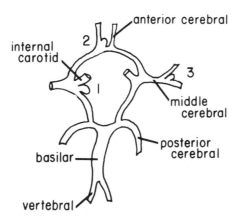

Figure 23–8. Circle of Willis and three most frequent locations of aneurysms.

done to reverse the damage done by a cerebrovascular accident, but much can be done with good physical therapy to rehabilitate the patient who has sustained a cerebrovascular accident.

When the brain is damaged by ischemia following a cerebrovascular accident, the neurons die within minutes to hours and are never replaced. The oligodendroglia are likewise very vulnerable to injury and readily die following ischemia. The astrocytes proliferate rapidly and repair the injury structurally by forming a

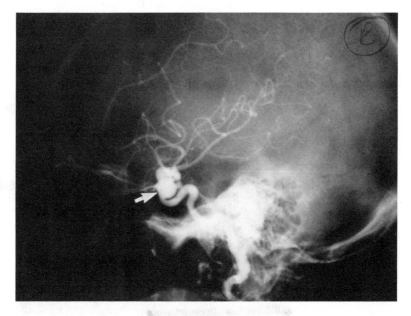

Figure 23–9. Carotid angiogram demonstrating radiopaque dye in a saccular aneurysm of the middle cerebral artery (*arrow*).

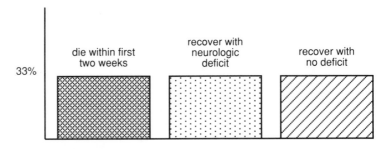

Figure 23–10. Prognosis for stroke patients.

scar (glial scar). The astrocytes and their processes are the central nervous system analog of fibroblasts and connective tissue. Monocytes enter from the blood after injury and aid in clearing away the debris.

CHART 23–6. CEREBROVASCULAR ACCIDENT

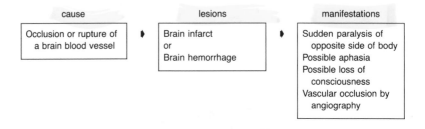

Trauma

The brain is especially vulnerable to injuries in modern times with high-speed transportation. The more common types of brain injuries include concussion, contusion, epidural hemorrhage, subdural hemorrhage, and penetrating injury. Patients who have sustained head trauma may have a concussion and recover completely, only to lapse into coma several hours later. The usual reason for late deterioration is that a subdural or epidural bleed developed immediately following the injury but did not affect the patient until a critical amount of blood accumulated. For this reason, any patient who has had sufficient head trauma to sustain a concussion should be watched closely for 12 to 24 hours.

Concussion

Concussion is a momentary loss of consciousness and loss of reflexes following head trauma, with amnesia for the traumatic event and complete recovery. No structural damage can be detected in the brain.

CHART 23–7. CONCUSSION

cause		lesion		manifestations
Cranial trauma	▶	None	▶	Momentary loss of consciousness Momentary loss of reflexes Amnesia of the event

Contusion

Contusions are bruises of the surface of the brain sustained at the time of traumatic impact. Contusions occurring on the same side of the brain as the trauma are termed *coup lesions,* whereas those on the opposite side are *contrecoup lesions.* If the head is in motion at the time of impact, the contrecoup lesion will often be larger than the coup lesion, because the force of the blow is magnified as it is transmitted to the opposite side. Contusions result in hemorrhages from small blood vessels in the brain. These hemorrhages cause further vessel occlusion and consequent edema, rendering the patient vulnerable to the sequelae of increased intracranial pressure.

CHART 23–8. CONTUSION

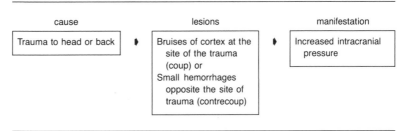

cause		lesions		manifestation
Trauma to head or back	▶	Bruises of cortex at the site of the trauma (coup) or Small hemorrhages opposite the site of trauma (contrecoup)	▶	Increased intracranial pressure

Epidural Hematoma

Epidural hemorrhage occurs between the dura and the skull (Fig. 23–11). It is associated with severe trauma in which the skull is usually fractured. Because an artery is ruptured (middle meningeal), the blood accumulates rapidly, and the patient will die within hours unless the hematoma is removed by an operation. Often the patient with an epidural hematoma sustains a concussion followed by a lucid interval prior to the onset of the signs of increased intracranial pressure.

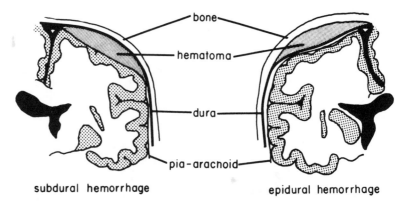

subdural hemorrhage   epidural hemorrhage

Figure 23–11. Comparative locations of epidural and subdural hematomas.

CHART 23–9. EPIDURAL HEMORRHAGE

cause		lesion		manifestation
Traumatic rupture of a dural artery	♦	Rapid accumulation of blood between the dura and skull	♦	Rapid increase in intracranial pressure

Subdural Hematoma

Subdural hematoma is a collection of blood beneath the dura (Fig. 23–11). It is a common sequelae of head injury and is due to rupture of veins on the dorsum of the brain. Because this bleeding is venous, the blood does not accumulate as rapidly as in an epidural hemorrhage and consequently is not quite as life threatening, although usually the blood must be removed by a surgical procedure to prevent compression of the underlying brain. Subdural hematomas are relatively more common in infants and elderly persons and often are discovered in patients who have no history of trauma.

CHART 23–10. SUBDURAL HEMORRHAGE

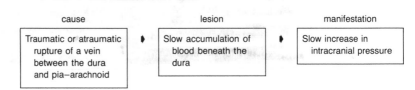

cause		lesion		manifestation
Traumatic or atraumatic rupture of a vein between the dura and pia–arachnoid	♦	Slow accumulation of blood beneath the dura	♦	Slow increase in intracranial pressure

Penetrating Injuries

Most penetrating injuries are from bullets. The damage to the brain from a bullet is proportional to the square of the velocity of the bullet; consequently, high-speed bullets do much more damage to the brain than low-speed bullets. Other dangers from penetrating injuries are the impaction of fractured bone splinters into the brain plus the strong likelihood of introducing infection into the open wound.

Penetrating and crushing injuries of the spinal cord are very common and if severe may result in complete paralysis of the body below the lesion. A very common form of spinal cord injury is herniation of the cushioning material (disc) between vertebrae. Herniation of discs is discussed in Chapter 21. Spinal cord injuries are often surgical emergencies.

Multiple Sclerosis

Multiple sclerosis (MS) is a common disease that affects many young and middle-aged adults throughout the world, primarily in northern hemispheric countries. There are at least 100,000 persons in the United States with multiple sclerosis at any one time, and there is a slightly higher incidence among women than men. The

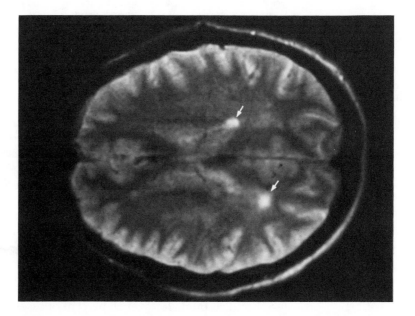

Figure 23–12. Magnetic resonance image of brain demonstrating two plaques (*arrows*) of multiple sclerosis in the white matter.

basic lesion is focal loss of the myelin sheath (demyelination), which appears to render axons incapable of properly transmitting a nervous impulse. Because this loss of myelin can occur anywhere in the brain or spinal cord, the symptoms may vary considerably from one patient to another. Visual impairment is usually present to some degree, because multiple sclerosis preferentially affects the optic nerves as well as the tissue surrounding the brain ventricles and the spinal cord. The cause of multiple sclerosis is not known, although it is strongly suspected to be a virus or an immunologic reaction to a virus. The disease span is usually 5 to 25 years, with the course of the disease alternately remitting and relapsing. The patient eventually becomes quite debilitated from muscle weakness. The diagnosis is made on the basis of clinical history and physical findings that support a multifocal neurologic deficit plus the findings of increased IgG protein in the cerebrospinal fluid. The lesions can be seen using magnetic resonance imaging (Fig. 23–12).

CHART 23–11. MULTIPLE SCLEROSIS

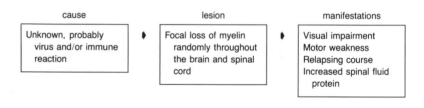

cause	lesion	manifestations
Unknown, probably virus and/or immune reaction	Focal loss of myelin randomly throughout the brain and spinal cord	Visual impairment Motor weakness Relapsing course Increased spinal fluid protein

Senile Dementia

Dementia means a decrease in cognitive function, usually accompanied by loss of memory for recent events. *Senile dementia* is a descriptive term for a condition of elderly persons who have poor memory for recent events, pick at their clothes, get lost easily, and are often irritable.

The degree of dementia in a patient is proportional to the loss of substance in the frontal lobes, the region of the brain that is associated with higher cognitive function. The loss of substance may be due to trauma or stroke (infarct) but it is more often secondary to generalized atrophy and degeneration of the neurons in the gray matter (Fig. 23–13). This latter condition is called *Alzheimer's disease* and accounts for over 60 percent of cases of chronic dementia. Alzheimer's disease is ordinarily a disease of the aged but occasionally affects persons in their forties. Characteristic silver-staining neuritic plaques and neurofibrillary tangles are found in

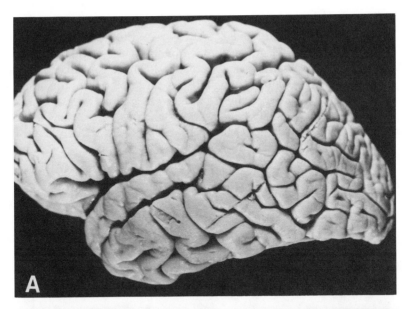

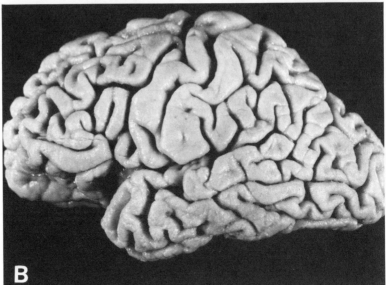

Figure 23–13. Atrophic brain of Alzheimer's disease (**B**) compared with normal (**A**).

the cerebral cortex and allow the neuropathologist to diagnose Alzheimer's disease. It is not known what causes neurons to degenerate resulting in Alzheimer's disease. The social impact of this disease is significant. It is estimated that currently there are over 600,000 persons with Alzheimer's disease in the United States and with an increasing population of elderly persons, there may be as many as 800,000 cases by the year 2000. A significant increase in the quantity and quality of resources necessary to care for these patients will be needed. These resources include more and better training of health care professionals in the home and nursing home care of these patients. Dementia may also accompany numerous metabolic conditions such as uremia or electrolyte and fluid imbalance. Many of these types of dementia are reversible.

CHART 23–12. SENILE DEMENTIA

cause	lesions	manifestations

Alzheimer's disease
Other organic brain
  diseases
	Brain atrophy
	Diffuse loss of neurons
	  in the cerebral cortex
		Poor recent memory
		Irritability
		Lapses in social
		  restraints

Parkinson's Disease

Parkinson's disease is caused by degeneration of certain portions of the extrapyramidal (involuntary) motor system, especially the substantia nigra nucleus in the midbrain. Parkinson's disease usually affects older people and results in tremors at rest, mask-like facial expression, and rigidity of skeletal muscles. A shuffling gait is characteristic. Many older people appear to have minor degrees of Parkinson's disease, and about 10 percent of persons with Parkinson's disease will also be demented. Treatment of Parkinson's disease with L-dopa results in some symptomatic relief in about half of the patients.

CHART 23–13. PARKINSON'S DISEASE

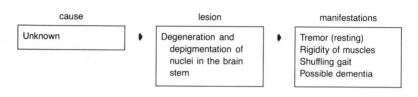

cause	lesion	manifestations

Unknown	Degeneration and
	  depigmentation of
	  nuclei in the brain
	  stem
		Tremor (resting)
		Rigidity of muscles
		Shuffling gait
		Possible dementia

Hydrocephalus

Hydrocephalus literally means water brain, and it may occur congenitally or may arise at any time after birth from a variety of causes. In hydrocephalic individuals, the ventricles enlarge (Fig. 23–14) secondarily to a block in the flow of cerebrospinal fluid at some

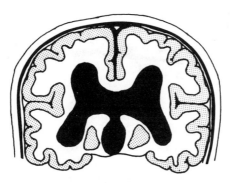

Figure 23–14. Expanded ventricles with hydrocephalus (compare with Fig. 23–3).

level. The most common type of congenital hydrocephalus is stenosis (closure) of the aqueduct between the third and fourth ventricles. As the ventricles expand with accumulated cerebrospinal fluid, the head may enlarge enormously. In older children and adults, the causes of hydrocephalus are more often tumors that block the flow of cerebrospinal fluid and meningeal scarring secondary to meningitis or hemorrhage, with consequent failure of cerebrospinal fluid to be reabsorbed into the venous system. In older children and adults, the head does not usually enlarge, because the skull is well formed; rather, the increased pressure from the accumulated fluid in the ventricles causes pressure atrophy of the surrounding white and gray tissue, resulting in mental deterioration. If the increased pressure is not relieved, the brain may herniate toward the foramen magnum. Pressure may be relieved by placement of a tube that acts as a shunt between the ventricles and the veins, heart, or peritoneal cavity.

CHART 23–14. HYDROCEPHALUS (CONGENITAL)

cause		lesion		manifestations
Usually from failure of the aqueduct of Sylvius to properly open	▸	Enlarged brain ventricles from the pressure of increased spinal fluid	▸	Enlarged head Mental and motor deterioration

CHART 23–15. HYDROCEPHALUS (ACQUIRED)

cause		lesion		manifestations
Scarring of the meninges or blockage of a ventricle	▸	Enlarged brain ventricles from the pressure of increased spinal fluid	▸	Normal-sized head Mental and motor deterioration

Epilepsy

This is a condition of recurrent seizures. Seizures are focal and/or generalized disturbances of neuronal electrical activity, which may be manifested by abnormal movements or sensations and loss of reflexes, memory, or consciousness. Epilepsy may follow recovery from trauma or central nervous system infections or may be induced by malformations or neoplasms of the brain. In most cases, persons with epilepsy have a lifelong history of seizures without known cause. Seizures may also be due to more acute conditions such as electrolyte imbalance, high fever, uremia, and eclampsia. Seizure activity in persons with epilepsy is usually controlled with barbiturates or phenytoin-type drugs.

CHART 23–16. EPILEPSY

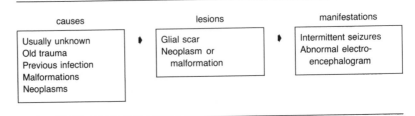

Hyperplastic/Neoplastic Diseases

Brain Neoplasms

Neoplasms of the brain and spinal cord are not as neatly separated into benign and malignant varieties as are tumors elsewhere. The reason is that small, slowly growing tumors of the brain that would be benign in other locations may readily disrupt vital functions in strategic locations such as the brain stem, killing the patient. Other benign brain tumors may occur in deep areas of the brain where the surgeon cannot gain access to them without destroying adjacent vital brain structures.

Brain tumors are the second most common neoplasms occurring in children (leukemias are first). Two-thirds of brain tumors in children occur in the posterior fossa, predominantly in the cerebellum. Conversely, two-thirds of brain tumors in adults occur in the anterior parts of the brain, in the white matter of the hemispheres.

The most common presenting signs and symptoms in patients with brain tumors are those of increased pressure because of the mass lesion, often with accompanying edema. Generalized symptoms such as headaches, vomiting, blurred vision, and seizures may all result from increased intracranial pressure, while focal signs may accompany brain tumors, depending upon where the tumor arises.

The treatment of brain tumors almost always entails an operation, because some of the tumor tissue needs to be examined in order to establish a histologic diagnosis. In general, if a tumor is of a slow-growing type, the surgeon will attempt to remove as much as possible. Whereas, if a tumor appears to be malignant, the surgeon will sample enough tissue to establish a definitive diagnosis and then further treat the patient with radiation and/or chemotherapy.

The most common brain tumors are those arising from astrocytes. Slower-growing types are called *astrocytomas*. The fast-growing malignant type is called *glioblastoma multiforme*. Glioblastomas are the most common brain tumor in adults and usually result in death within a year or two following diagnosis. As the tumors grow, they produce an expanding, poorly defined mass (Fig. 23–15). Almost no patients survive.

The second most common brain tumor is the *meningioma*, a benign neoplasm that arises from the dura and is slow growing

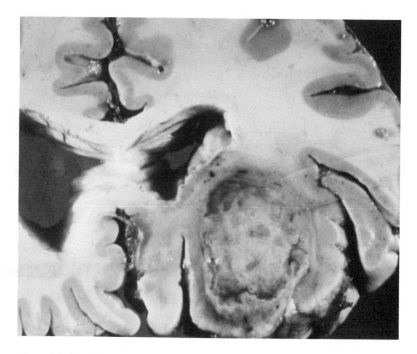

Figure 23–15. Glioblastoma multiforme, with mass infiltrating and displacing of surrounding structures.

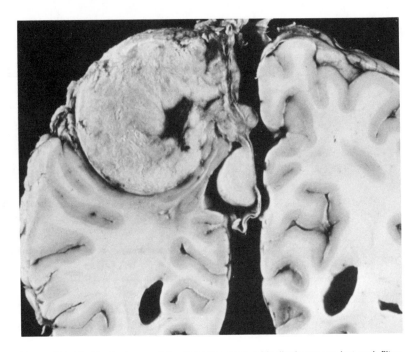

Figure 23–16. Meningioma arising from the dura, with displacement but no infiltration of the brain. *(Courtesy of Margaret Jones.)*

and well circumscribed without infiltration of surrounding brain (Fig. 23–16). Since most meningiomas arise from the dorsum (top) of the head, the surgeon can usually excise the entire tumor, and the patient's prognosis is good. If a meningioma arises at the base of the brain, however, the surrounding vital structures such as hypothalamus, brain stem, and blood vessels make surgical excision much more difficult, and the prognosis is correspondingly worse.

Pituitary tumors (adenomas) are also common and likewise are difficult to remove because of location. Some pituitary adenomas secrete growth hormone, resulting in gigantism or acromegaly, syndromes that are discussed further in Chapter 25. Brain tumors arising from oligodendroglia, ependymal cells, and neuronal cell lines occur but are less common. One of the most important brain tumors in children is the *medulloblastoma*, which arises in the cerebellum from primitive cells that are neuronal precursors. These tumors are malignant but do respond well to radiation therapy, and the patient may even be cured.

Metastatic tumors to the brain are common. They usually are removed only if there is just one focus of metastasis and the patient is in sufficiently good health to withstand the surgical procedure.

Tumors of the spinal cord are much less common than brain tumors.

CHART 23-17. BRAIN NEOPLASMS

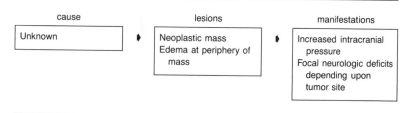

cause	lesions	manifestations
Unknown	Neoplastic mass Edema at periphery of mass	Increased intracranial pressure Focal neurologic deficits depending upon tumor site

ORGAN FAILURE

As the brain is a composite of numerous groups (nuclei) of neurons with partially related functions, any group of neurons may fail, resulting in a focal neurologic deficit. Generalized brain failure results from diffuse brain disease and places the patient in a vegetative state, incapable of performing basic mental or motor functions (*coma*). The patient in coma is alive because of the continuing function of the brain stem, which is often the last part of the brain affected by diffuse disease. Coma may be reversible if structural damage to the brain has not taken place. In many areas of the United States, a patient is considered legally dead if two successive electroencephalograms taken 24 hours apart both show complete absence of electrical activity in the brain, irrespective of the status of the heart. Interpretation of the death of the patient by demonstrating death of the brain is somewhat controversial and poses ethical as well as legal dilemmas. Other criteria for determining brain death should include body temperature and reflex responses determined by neurologic examination. Since barbiturates may cause decreased or absent electrical activity in the brain that is reversible, blood levels of these substances should also be measured before pronouncing brain death.

REVIEW QUESTIONS

1. What is the difference between a focal and a generalized neurologic deficit?
2. What are the causes of increased intracranial pressure? Why is increased intracranial pressure dangerous?

3. How is each of the following tests utilized in diagnosing central nervous system disease?
   Cerebrospinal fluid analysis
   Angiography
   Electroencephalogram
   Computerized tomography
   Magnetic resonance imaging

4. What are the causes, lesions, and manifestations of each specific disease listed in this chapter?

5. How does the pathogenesis and clinical expression of developmental brain malformations differ from developmental brain destructive lesions?

6. How does meningitis differ from encephalitis in terms of pathogenesis, location of lesion, and clinical expression?

7. When should rabies be suspected? How should it be diagnosed and treated?

8. What is the relationship between cerebrovascular accidents and atherosclerosis?

9. Which cells in the brain are most vulnerable to anoxia from cerebrovascular accidents?

10. Why should a patient who recovers from a brain concussion be closely watched?

11. What are the differences between epidural and subdural hematomas in terms of pathogenesis and development of symptoms?

12. How can senile dementia be distinguished from the dementia associated with Parkinson's disease?

13. How does congenital hydrocephalus differ from acquired hydrocephalus in terms of pathogenesis and clinical expression?

14. Why should almost all patients with brain tumors be operated on?

15. What do the following terms mean?
    Down's syndrome
    Anencephaly
    Meningomyelocele
    Spina bifida
    Cerebral palsy
    Aphasia
    Astrocytoma
    Glioblastoma multiforme
    Meningioma
    Medulloblastoma

Chapter 24

Mental Illness

NATURE OF MENTAL ILLNESS

Mental or psychiatric illnesses are those that affect behavior, emotion, and cognition. There is no precise definition of what constitutes mental health; rather, it is a matter of experienced judgment whether a person has normal intellectual capacity and is functioning normally as a member of society. The degree to which a person is emotionally impaired (i.e., whether or not the patient receives a psychiatric diagnosis) depends primarily on (1) the degree to which the symptoms interfere with his or her ability to function appropriately and to obtain gratification in life experiences and (2) the degree to which the symptoms disturb others. For example, many persons who might be diagnosed as having schizophrenia function well as semi-isolates, holding a regular job and maintaining a restricted social life.

Psychiatric disorders have traditionally been separated from diseases of the nervous system because most mental illnesses have not been shown to be associated with demonstrable lesions. However, many mental symptoms are known to occur as a result of neurologic, metabolic, or endocrinologic disorders, suggesting that there are biologic bases for mental illness. Mental illnesses that are caused by detectable brain lesions are called *organic mental disorders* or *organic brain syndromes.*

Health care specialists who deal primarily with mental and emotional illnesses include psychiatrists, psychologists, and social workers. Psychiatrists are physicians who are trained to diagnose and treat psychiatric illness. Psychologists usually hold advanced degrees in psychology and are concerned with the study of normal and abnormal behavior. Psychologists often play a role in diagnosis of subtle psychiatric illnesses, by using specific psychologic tests. Psychologists may also be involved in psychologic or behavioral aspects of treatment of patients. Social workers are especially concerned with the social and physical environments of patients with psychiatric illness, just as they are with patients having organic disease. Psychiatric illnesses are often first detected by primary care practitioners.

MOST FREQUENT AND SERIOUS PROBLEMS

Psychiatric illnesses collectively are a common and very serious problem. Each year at least 500,000 persons in the United States enter a mental hospital for the first time. At any given moment, approximately 1 out of every 500 persons is hospitalized under psychiatric care. Underscoring these high incidence and prevalence rates is the fact that in one recent year 6 billion doses of tranquilizers (1.2 million pounds) were produced. The most common diagnoses associated with first admissions and for chronic care, respectively, are shown in Figures 24–1 and 24–2.

Affective disorders are estimated to affect 7 percent of the population and are the most common severe psychiatric disorders. Affective disorders are a group of illnesses characterized by a disturbance of mood accompanied by a manic or depressive syndrome that is not due to any other physical or mental disorder.

Schizophrenia, another common severe psychiatric disorder, affects about 1 percent of the population. It usually begins between 15 and 24 years of age. At older ages the types of psychiatric

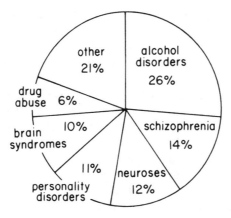

Figure 24–1. Relative incidence of mental illnesses based on first admissions to state or county mental hospitals. *(Data from Statistical Note 97, September 1973, Survey and Reports Section, National Institute of Mental Health, Chevy Chase, Maryland.)*

diseases vary considerably among males and females. Alcoholic disorders are probably the most common psychiatric problem in males of all ages over 24, followed by the affective disorders, schizophrenia, and personality disorders. In females of comparable ages, schizophrenia and anxiety syndromes are much more frequent than alcoholic and personality disorders. In middle-aged and older people, depression is a frequent and serious psychiatric disorder. In the elderly, permanent organic brain diseases, often collectively referred to as *chronic brain syndromes*, become more prevalent. Although chronic brain syndromes are associated with brain atrophy, the symptoms are usually behavioral in nature. Dementia is discussed in Chapter 23.

Suicide may be associated with acute or chronic environmental stress, with alcoholism, with emotional disorders (especially those

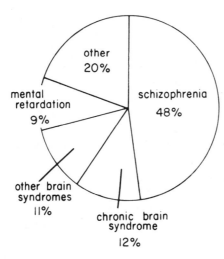

Figure 24–2. Relative prevalence of mental illness associated with long-term hospitalization. *(Data from Statistical Note 72, December 1972, Survey and Reports Section, National Institute of Mental Health, Chevy Chase, Maryland.)*

in which depression plays a major role), or with chronic illnesses (especially chronic lung diseases or illnesses in which chronic pain is a major factor). Suicide among patients with psychiatric illness is three to fifteen times more frequent than among the general population. The incidence of suicide varies considerably in different populations, being influenced by age, sex, population density, occupation, climate, and alcoholism. In the United States, suicide is the eleventh leading cause of death overall, but the third leading cause of death in the 18 to 30 age group. There are eight attempts for every suicide committed.

Psychiatric emergencies in addition to suicide attempts include acute toxic psychoses (delirium), alcohol and other drug withdrawal, manic excitement, and acute anxiety or panic attacks. Acute toxic psychoses may be secondary to intoxication with or withdrawal from drugs or alcohol, infection, metabolic disorders, or trauma.

SYMPTOMS, SIGNS, AND TESTS

Mental illnesses are manifested by a wide range of symptoms, subjective feelings, and behaviors, the majority of which are not specifically pathologic when viewed in isolation. Because the symptoms are nonspecific and because many are known to be associated with physical illnesses as well as with organic brain syndromes, a complete physical examination and screening laboratory tests are an essential part of a psychiatric evaluation. X-rays of the chest and skull, electroencephalogram (EEG), and psychological assessment, such as neuropsychological testing and personality inventory, are usually a part of a thorough psychiatric evaluation.

As in physical illnesses, a detailed history is essential and is usually obtained from family members as well as the patient. The mental status examination, a specific component of the psychiatric evaluation, is of great value in differentiating between organic and nonorganic mental diseases. The mental status examination includes evaluation of (1) orientation to time, place, and person; (2) memory, both recent and remote; (3) intellectual functions, including general fund of information and arithmetic ability; (4) judgment; (5) mood and affect, (6) speech pattern; and (7) deliriums and hallucinations.

Table 24–1 classifies and defines many of the manifestations of mental illness. It should be remembered that many symptoms and signs are experienced by normal people. When they are disproportionately intense or prolonged in relation to their stimulus or when they significantly interfere with the individual's ability to function in and gain gratification from his or her environment, they may

TABLE 24–1. TYPES OF MANIFESTATIONS OF PSYCHIATRIC ILLNESS

I. Disturbances of consciousness
   Confusion—lack of orientation of time, place, or person
   Delirium—bewildered, restless, confused

II. Disturbances of affect
   Euphoria—feeling of well-being inappropriate to apparent events
   Depression—feeling of sadness inappropriate to apparent events
   Grief—sadness appropriate to a real loss
   Anxiety—feeling of apprehension with no known precipitant

III. Disturbance of motor behavior
   Compulsion—uncontrollable impulse to perform an act
   Hyperactivity—restless, aggressive
   Hypoactivity—slowed psychologic and physical function

IV. Disturbances of thinking
   Autistic thinking—thinking that gratifies unfulfilled desires without regard to reality
   Incoherence—run together thoughts without logical coherence
   Flight of ideas—rapid shifting of ideas
   Aphasia—loss of ability to comprehend language (sensory aphasia) or verbalize (motor aphasia)
   Delusion—false belief that cannot be corrected by reasoning
   Hypochondria—exaggerated concern over health not based on real disease
   Obsession—recurrent thought, feeling, or impulse
   Phobia—exaggerated fear of a specific situation

V. Disturbance of perception
   Agnosia—inability to recognize and interpret sensory impressions
   Hallucinations—false sensory perceptions not associated with real stimuli
   Illusions—false sensory perceptions of real stimuli

VI. Disturbances of memory
   Amnesia—partial to total inability to recall past experiences
   Paramnesia—falsification of memory by distortion of recall

VII. Disturbances of intelligence
   Mental retardation—developmental lack of intelligence that interferes with social and vocational performance (IQ less than 70)
   Dementia—loss of intellectual capacity occurring after developmental period and secondarily to organic disease

(Modified from Freedman AM, Kaplan HI, Sadock BJ: Comprehensive Textbook of Psychiatry, III. Baltimore, Williams and Wilkins, 1975, pp. 822–824.)

indicate an abnormal mental state. Symptoms such as anxiety and depression are experienced by almost everyone at certain times. Not infrequently, an individual's symptoms are more troublesome to others than they are to the person with the symptoms.

SPECIFIC DISEASES

The classification of mental illness is based on several major parameters, including intelligence, presence or absence of brain lesions, ability to correctly interpret reality, and duration of the condition.

Subnormal intelligence occurring before age 18 is referred to as *mental retardation;* loss of intellectual capacity after age 18 is termed *dementia.* Mental changes associated with demonstrable brain lesions are termed *organic brain disorders;* all others are classified as "functional disorders." The ability to correctly interpret reality is the fundamental consideration in separating psychoses from nonpsychotic illnesses. Reversible mental disorders of rapid onset are referred to as acute, while prolonged conditions are termed chronic.

The American Psychiatric Association has established criteria for the classification of mental illnesses in its *Diagnostic and Statistical Manual of Mental Disorders,* which is currently in its fourth edition. A modified classification used in this chapter is presented in Table 24–2. Some general comments are in order before proceeding with a discussion of specific categories.

Emotional disorders may be associated with organic disease in several ways. First and most common is the emotional response of an individual to a physical illness or injury. While one would consider this a normal, if not expected, reaction, the severity of symptoms, usually depression and/or anxiety, may vary considerably, depending on such factors as the person's basic personality, prior level of activity, and seriousness of the illness. Often the symptoms are of sufficient intensity and persistence to justify a diagnosis of transient situational disturbance. Second, mental illness may occur secondarily to organic disease of the central nervous system (organic brain syndrome). Third, certain physical illnesses are called *psychophysiologic disorders* because of their frequent association with emotional disturbances. In most instances, the nature of the association between emotional disturbance and physical illness is poorly understood, i.e., the causal relationship is not clear. The

---

TABLE 24–2. CLASSIFICATION OF MENTAL ILLNESS

Organic mental disorders	Personality disorders
Drug-induced disorders	Disorders arising in childhood and adolescence
Dementias	Mental retardation
Mental manifestations of other diseases	Generalized emotional disorders
Schizophrenia	Specific disorders
Affective disorders	Disorders with symptoms suggestive of organic
Bipolar (manic–depressive disorder)	disease
Unipolar (psychotic depression)	Factitious disorder
Intermittent affective disorders	Malingering
Anxiety disorders (neuroses)	Somatoform (psychosomatic) disorder
Phobias	Psychophysiologic disorders
Panic reaction	
Obsessive compulsive neurosis	

combination of emotional disturbance and organic disease can often be best represented by separate diagnoses for the mental and organic components of the illness.

In addition to considering the possible relationship of the patient's mental illness to organic disease, it is also important to judge whether the patient is psychotic or not. Most functional psychoses fall into the categories of schizophrenia and affective disorders; less commonly they fall into the category of disorders of childhood and adolescence. Pure paranoid psychoses are rare, paranoid symptoms more often being a feature of schizophrenia. In addition to the functional psychoses, many organic mental disorders, such as acute drug-induced reactions and chronic dementias, are psychotic reactions.

Another important point to be made about the classification of mental illnesses is that most symptoms are not specific for any particular category. This is particularly true of depression and anxiety, which commonly occur with many different types of mental illness. For example, depression, one of the most common and important symptoms, may be associated with organic brain disease, schizophrenia, affective disorders, alcoholism, and transient situational disturbances. If depression occurs after the onset of one of those disorders, it may be termed a secondary depression. The criteria for and characteristics of the major categories will be discussed below.

Organic Mental Disorders

The essential feature of this category is mental disturbance due to disease of the brain or widespread alteration of brain function due to drugs, toxins, or metabolic products. Acute forms of organic mental disorders often present with psychotic features, while chronic forms more often become psychotic in the late stages of the disease.

Acute organic brain syndromes are usually manifested by delirium and intoxication. Drugs such as alcohol, barbituates, and narcotics are common causes. With some drugs, particularly alcohol, symptoms may be precipitated by withdrawal of the agent. For further discussion of alcoholism see Chapter 30.

Chronic brain syndromes are usually manifested by dementia, loss of recent memory, personality changes, and sometimes by depression, delusions, and hallucinations. Alzheimer's disease, a diffuse atrophy of the brain with loss of neurons of unknown cause, accounts for most cases of chronic irreversible dementia. The course is slowly progressive, with an average survival of about 5 years. See Chapter 23 for further discussion of Alzheimer's disease.

Repeated cerebral infarcts may produce dementia in some individuals but are a much less common cause of dementia than Alzheimer's disease. Multiple cerebral infarcts with dementia are more likely to be associated with a stepwise progression and findings that indicate focal neurologic damage.

Many other brain diseases and metabolic intoxications, such as acidosis, uremia, and hepatic failure, can present with mental disturbances. Organic disease with mental disturbance is usually separable from "functional disease" on the basis of abnormalities found by neurologic examination, laboratory tests, special procedures, and mental status examination. Occasionally, brain lesions, such as a brain tumor, will present as a psychosis and may be misdiagnosed as a "functional disease." Some symptoms and signs of organic psychosis may overlap other psychotic states, but the more common signs and symptoms of organic psychoses are delirium (mental confusion, disorientation, abnormal emotions, and altered consciousness) and dementia (loss of the intellectual processes, such as memory, reasoning, judgment, and problem solving, and the loss of the higher aspects of personality). It is extremely important to recognize organic psychosis because effective treatment is available for many of its causes.

Schizophrenia

Schizophrenia is a common psychosis. It affects approximately 1 percent of the population. Schizophrenia is primarily a thought disorder in which misinterpretation of reality can lead to delusions and hallucinations. Alterations in mood and behavior are also prominent in schizophrenia, and schizophrenic patients are often ambivalent, display inappropriate emotional responses, and become either aggressive or withdrawn. There are several different subtypes of schizophrenia, which are classified on the basis of the predominant symptoms (Table 24–3).

The cause of schizophrenia is unknown. The familial tendency suggests a poorly defined, possible multiple-gene, genetic influence in schizophrenia. General personality characteristics and environmental influences are possible contributing factors in the development of schizophrenia. Numerous biochemical and neurophysiologic abnormalities have been demonstrated in patients with schizophrenia, none of which have been demonstrated to be causative.

Schizophrenia usually has its onset in late adolescence and early adulthood, with fewer new cases occurring in middle age. The severity of the disease varies considerably. Some patients are able to maintain employment while being treated on an outpatient basis, while other require hospitalization. The predominant modes

TABLE 24–3. SUBTYPES OF SCHIZOPHRENIA

Subtype	Characteristics
Disorganized (hebephrenic)	Marked incoherence Flat, incongruous, or silly affect
Catatonic	Stupor or mutism Rigid positioning Inappropriate excited motor activity Inappropriate or bizarre posturing
Paranoid	Persecutory delusions Grandiose delusions Delusions of jealousy Persecutory or grandiose hallucinations
Undifferentiated	Psychotic symptoms prominent (delusions, hallucinations, formal thought disorder, bizarre behavior) Does not fit other subtypes

of medical therapy are medication with phenothiazine derivatives or other antipsychotic (neuroleptic) drugs and psychotherapy. The aim of psychotherapy in the schizophrenic patient is more to aid coping behaviors and improve social adjustment rather than to uncover deep-seated feelings and motivations. Various forms of shock therapy are only rarely utilized today.

Affective Disorders

This group of illnesses is called affective disorders because they are characterized by alteration of mood. *Mania* is a euphoric or irritable, hyperactive state, and *depression* is a sad or melancholy state with either decreased activity or agitation.

Clinically, the term depression is used to describe a cluster of symptoms that may include any of the following: poor appetite or weight loss, sleep disturbance, loss of energy, agitation or retarded motor activity, loss of interest in usual activities or decreased sexual drive, feelings of self-reproach or guilt, complaints of or actual diminished ability to think or concentrate, and thoughts of death or suicide.

Clinically, the manic syndrome includes elevated mood, hyperactivity, involvement in activities without recognizing potential for painful consequences, rapid speech with flight of ideas, inflated self-esteem, decreased need for sleep, and distractibility.

Affective disorders are usually episodic, with one or more episodes clearly distinguished from previous levels of function, or they may be intermittent, with periods of depressed mood having less clear onset and shorter duration. The episodic forms frequently are of psychotic proportions and fall into two major groups, termed *bipolar* and *unipolar*. The intermittent affective disorders are usually nonpsychotic and are similar in nature to personality disorders (see below).

The bipolar form of *manic–depressive psychosis* is inherited, has onset of symptoms in early adulthood, and is characterized by periods of mania and periods of depression alternating with periods of normal function. The inheritance pattern may be partly controlled by an X-linked dominant gene, although father–son transmission has been recorded also. In contrast to schizophrenia, manic–depressive illness occurs in persons of higher socioeconomic classes, and some individuals are highly productive, especially during periods of mania. Manics, however, tend to carry their ambitions beyond reality and end up in difficult situations, such as financial ruin.

The unipolar form of episodic affective disorder is five to ten times more common than the bipolar form and is characterized by depression without episodes of mania. Inheritance also plays a large role in unipolar depression. The cause is not known but aberrations in the levels of the neurotransmitters, norepinephrine or serotonin, have been identified in several patient studies. The onset is usually in middle age or beyond. Approximately 50 percent of patients have more than one episode of psychotic depression.

Anxiety Disorders

The anxiety disorders are characterized by neurotic symptoms such as exaggerated or inappropriate feelings of anxiety, phobia, obsession, or compulsiveness in which one symptom usually dominates and in which there is no evidence of psychosis or of reaction to a transient situation.

Sigmund Freud originally used the term neurosis to indicate a specific disease process but also to indicate unpleasant symptoms in a person with nonpsychotic mental illness. Currently, usage of the terms neurotic disorders or neurotic symptoms indicates a variety of psychiatric symptoms covering several specific disease categories including anxiety disorders.

Neurotic symptoms also occur to some degree in everyone and may be found in patients with other mental illnesses such as psychoses or personality disorders. Thus, neurotic symptoms are not specific. Nevertheless, at times one or more neurotic symptoms become sufficiently dominant as to interfere with an individual's ability to lead a gratifying or productive life. Under such circumstances, a diagnosis of anxiety disorder is often made. Neurotic symptoms may at times be very severe and immobilizing, requiring hospitalization. In anxiety disorders, the patient is aware that the symptoms are irrational but is unable to control them. Thus, this group of mental illnesses does not reflect a thought disorder as do the psychotic illnesses. In addition, persons with anxiety disorders maintain the ability to distinguish external reality from internal processes; thus, reality impairment is isolated to the symptom itself and does not, as in psychotic illnesses, pervade the individual's total awareness.

People with phobias persistently avoid specific objects, activities, or situations because of irrational fears. Patients with agoraphobia avoid being alone, whereas, patients with social phobia avoid specific social situations due to overconcern about humiliation and embarrassment. Common simple phobias involve animals (particularly reptiles, insects, and rodents), tight places (claustrophobia), and high places (acrophobia).

Panic reactions involve sudden, short-lived, severe anxiety reactions without the patient being certain when the attack will occur. Symptoms relate to sudden autonomic nervous system discharge, producing dyspnea, palpitations, chest pain, choking, dizziness, paresthesias, hot and cold flashes, sweating, trembling, and fear.

Obsessive compulsive neuroses involve senseless and repetitive thoughts (obsessions), such as violence, contamination, and doubt, or compulsions to perform an act, such as handwashing, counting, checking, or touching, accompanied by a desire to resist such activity. However, attempts to resist the compulsion are accompanied by tension and anxiety.

It is estimated that 2 to 4 percent of all persons experience an anxiety disorder at some time during their life. Treatment usually consists of supportive therapy, environmental manipulation, and the use of minor tranquilizing drugs such as the benzodiazepines, antidepressant agents, or monoamine oxidase inhibitors.

Personality Disorders

Personality disorders are lifelong patterns of behavior that are usually inconsistent with social norms and are often unacceptable to others. They account for a high proportion of nonpsychotic psychiatric problems. They often include or overlap with such problems as sexual deviation, alcoholism, and drug dependence.

The various types of personality disorders are defined in approximate order of frequency in Table 24–4. Personality disorders may be caused by organic brain disease or may be acquired functional disorders. Occasionally, some types of personality disorders develop into psychotic disorders. The personality disorders commonly associated with psychoses include paranoid personality

TABLE 24–4. TYPES OF PERSONALITY DISORDERS

Type	Characteristics
Histrionic	Excitability, emotional instability, overactivity, self-dramatization
Passive aggressive	Uses passive behavior to express hostility, obstructionism, pouting, procrastination, stubbornness, intentional inefficiency
Dependent	Need overtly expressed, compliant, eager to perform for others, clinging, immature
Avoidant	Withdrawn, overly sensitive, shy, low self-esteem
Antisocial	Lack of loyalty to individuals, groups or society; selfish, irresponsible, impulsive, lack of guilt, violation of rules
Compulsive	Overly conscientious, overly meticulous, perfectionistic
Inadequate	Ineffectual responses to social, psychologic, and physical demands; inept; social instability
Paranoid	Hypersensitive, rigid, unwarranted suspicion, jealous, excessive self-importance
Narcissistic	Grandiose sense of self-importance, exhibitionism, preoccupation with fantasies of unlimited success, power, brilliance, or beauty

with paranoid schizophrenia, cyclothymic personality with manic–depressive psychosis, schizoid personality with schizophrenia, and obsessive compulsive personality with depression.

Disorders Arising in Childhood and Adolescence

Mental illness in children and adolescents differs sufficiently from that of adults to merit separate classification. The many conditions of these age groups will be briefly described below as mental retardation, generalized emotional disorders characteristic of childhood, and specific disorders, such as stereotyped movements and speech disorders.

Mental retardation is subnormal intellectual function (IQ below 70) with onset before age 18 but usually present from birth. Approximately 3 percent of the population of the United States is mentally retarded, and the majority of retarded people live in urban and rural slums. The cause is unknown 80 percent of the time. Known causes include prenatal infectious diseases, such as rubella, toxoplasmosis, cytomegalic inclusion disease, and syphilis; neonatal meningitis or encephalitis; Down's syndrome (mongolism); metabolic disease such as hypothyroidism and phenylketonuria; brain damage from perinatal diseases such as erythroblastosis fetalis, birth injury, and anoxia at birth; and external agents such as trauma, carbon monoxide poisoning, and lead poisoning. Approximately three-fourths of the mentally retarded are mildly so (IQ 55 to 70) and are educable to about the sixth grade level by the time they reach adult age.

Generalized emotional disorders of childhood are likely to affect social and intellectual development. Psychotic disorders are uncommon in childhood, although schizophrenia and manic–depressive psychoses may begin in late childhood and adolescence. Attention deficit disorders are common and become particularly evident in school situations, where they are manifested by an inability to concentrate, hyperactivity, and impulsiveness. Conduct disorders resemble the antisocial personality disorder but are separated, because conduct disorders do not necessarily continue as an antisocial personality disorder. Conduct disorders are common in late childhood and adolescence and may be manifested by persistent lack of concern for others, delinquency, and illegal acts. Anxiety disorders in childhood commonly take the form of separation anxiety, relating to fear of being away from home or of shyness. Specific developmental disorders that are common include defective articulation, poor motor coordination, enuresis (bed wetting), motor tic (rapid spasmodic, involuntary movement), and stuttering.

Disorders with Symptoms Suggestive of Organic Disease

Factitious disorders involve the voluntary production of symptoms that are not real, genuine, or natural for a well-defined goal. These patients create symptoms and inflict physical changes upon themselves in order to receive medical attention and hospitalization. The term *malingering* is used if the symptoms are produced to avoid an obvious external circumstance, such as conscription into the military.

Somatoform disorders are distinguished from factitious disorders and malingering in that the symptoms suggestive of organic illness are not under voluntary control. The complaints, frequently referred to as *psychosomatic,* take the form of headache, fatigue, palpitations, fainting, nausea, loss of sensation, paralysis, blindness, vomiting, abdominal pain, bowel troubles, allergies, and menstrual and sexual difficulties. In its fully developed form, a patient with somatization disorder will have a lifelong pattern beginning in the teenage period of seeking medical evaluation, being hospitalized, and even having unnecessary surgery. The disorder is more common in women and is often associated with use of many potentially addicting prescription drugs obtained from multiple physicians and pharmacies. Pain, often the dominant symptom, accounts for the most common types of drugs obtained by these patients.

It is important to recognize that psychosomatic complaints may be dominant symptoms in other mental illness, especially depression and schizophrenia; however, the other features of these diseases will also be present. Another problem is that the expression of complaints associated with organic illness varies greatly among individuals; thus, the diagnosis of organic lesions may be delayed in persons with frequent psychosomatic complaints or in those that do not readily express pain.

A conversion disorder is a form of a somatization disorder in which there is involuntary loss of function suggestive of physical illness such as paralysis, loss of voice, blindness, anesthesia, or incoordination.

Psychophysiologic Disorders

These are organic illnesses in which emotional factors are felt to play a significant role. These illnesses involve the autonomic nervous system and are usually isolated to a single organ system. Demonstrable organic lesions are present.

Since many emotions are expressed by way of the autonomic nervous system (blushing, sweating, increased heart rate, diarrhea, and nausea), it is not surprising that long-lasting, severe emotions may be expressed as chronic derangements of the autonomic nervous system that eventuate in structural (organic) lesions. The best known examples of psychophysiologic diseases are peptic ulcers, bronchial asthma, migraine, and hypertension. These diseases are all very complicated, and the emotional component in one individual may be much greater than in the next. Other examples of diseases that have variable psychophysiologic overlay are neurodermatitis, psoriasis, obesity, anorexia nervosa, visual disturbances, and sexual disorders such as dyspareunia, frigidity, premature ejaculation, and impotence.

REVIEW QUESTIONS

1. Are all mental or psychiatric disorders functional disorders? Why or why not?
2. How do the roles of psychiatrists, psychologists, and social workers differ?
3. Which six mental illnesses account for the most first admissions to mental hospitals?
4. Which mental illnesses account for 80 percent of long-term hospitalizations in mental illness?
5. At what ages do the major mental illnesses most frequently begin?
6. What factors predispose to suicide?
7. What are the major psychiatric emergencies?
8. How do the psychiatric history and mental status examination differ from the usual medical history and physical examination?

9. What are seven major categories of manifestations of psychiatric illness? What terms are used to define the specific manifestations in each category?
10. What criteria are used to classify the major categories of mental illness?
11. What are the causes, lesions (if any), and major manifestations of the specific diseases discussed in this chapter?
12. How do schizophrenia and manic–depressive illness differ?
13. How do anxiety disorders and personality disorders differ?
14. How do patients with symptoms suggestive of but not due to organic lesions create medical problems?
15. What is the meaning of the following terms?
    Organic brain syndrome
    Chronic brain syndrome
    Mental retardation
    Dementia
    Psychosis
    Depression
    Mania
    Factitious disorder
    Malingering
    Somatoform disorder
    Psychosomatic
    Psychophysiologic

Endocrine System

REVIEW OF STRUCTURE AND FUNCTION

The endocrine system is so named because it includes all of those
organs, or tissues within organs, that secrete their cellular products
(hormones) directly into the bloodstream rather than into viscera,

cavities, or outside the body, as is the case with the exocrine glands. Implicit in this definition is the fact that hormones exert their influence on tissues of the body remote from their site of origin. Major endocrine organs or tissues include hypothalamus, posterior and anterior pituitary gland, thyroid gland, parathyroid glands, adrenal cortex and medulla, islets of Langerhans, ovaries, testes, and placenta (Fig. 25–1). The hormones produced at each of these sites and their effects are listed in Table 25–1.

The anterior pituitary produces trophic hormones, which serve as intermediates by stimulating hormone production in other organs—adenocorticotropic hormone (ACTH), thyroid-stimulating hormone (TSH), follicle-stimulating hormone (FSH), luteinizing hormone (LH), and others that act directly on target tissues (melanocyte-stimulating hormone, growth hormone, prolactin). The pituitary tropic hormones control hormone production by other endocrine glands, but, in turn, the anterior pituitary's hormone production is controlled by hormones from the hypothalamus. The hypothalamus is a portion of the brain that can monitor hormone levels in the blood and secrete releasing or inhibitory hormones to control the production of pituitary hormones. This control mechanism is illustrated in Figure 25–2.

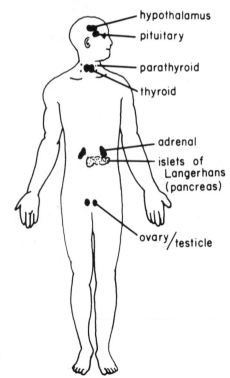

Figure 25–1. Endocrine organs and tissues.

TABLE 25–1. SITE OF ORIGIN AND EFFECTS OF HORMONES

Site	Hormone (Synonyms, Abbreviations)	Effect
Anterior pituitary	Adrenocorticotropic hormone (corticotropin, ACTH)	Stimulates production of glucocorticoids by adrenal cortex
	Melanocyte-stimulating hormone (MSH)	Stimulates pigment production in skin
	Growth hormone (somatotropin, GH)	Promotes growth of body tissues
	Thyroid-stimulating hormone (thyrotropin, TSH)	Stimulates production and release of thyroid hormones
	Follicle-stimulating hormone (FSH)	Initiates maturation of ovarian follicles Stimulates spermatogenesis
	Luteinizing hormone (LH)	Causes ovulation and stimulates ovary to produce estrogen and progesterone Stimulates androgen production by interstitial cells of testis
	Prolactin	Stimulates secretion of breast milk
Hypothalamus	Releasing hormones	Act on anterior pituitary to cause release of specific hormones
	Inhibitory hormones	Act on anterior pituitary to cause inhibition of release of specific hormones
Posterior pituitary	Antidiuretic hormone (vasopressin, ADH)	Causes conservation of body water by promoting water resorption by renal tubules
	Oxytocin	Stimulates smooth muscle contraction in breast to aid in milk ejection
Thyroid	Thyroxine (tetraiodothyronine, T4)	Increases rate of cellular metabolism Nutritional effects on brain and other organs
	Triiodothyronine (T3)	Same as thyroxine
	Calcitonin	Promotes retention of calcium and phosphorus in bone
Parathyroid	Parathyroid hormone (parathormone, PTH)	Regulates metabolism of calcium and phosphorus Promotes resorption of calcium and phosphorus from bone
Adrenal cortex	Glucocorticoids, mostly cortisol-(hydrocortisone) with some corticosterone	Antagonizes effects of insulin Inhibits inflammatory response and fibroblastic activity Many other effects
	Mineralocorticoid, mainly aldosterone	Promotes retention of sodium by renal tubules
	Androgens	Masculinization
Adrenal medulla	Catecholamines (epinephrine and norepinephrine)	Regulation of blood pressure by effects on vascular smooth muscle and heart

(*continued*)

TABLE 25–1. *Continued*

Site	Hormone (Synonyms, Abbreviations)	Effect
Islets of Langerhans and gastrointestinal tract	Insulin	Promotes utilization of glucose and lipid synthesis
	Glucagon	Promotes utilization of glycogen and lipid
Ovaries	Estrogens	Cause development of female secondary sex characteristics Necessary to maintain menstrual cycle and pregnancy
	Progesterone	Preparation of endometrium for implantation and maintenance of pregnancy
Placenta	Human chorionic gonadotropin (HCG)	Maintains corpus luteum and progesterone production in pregnancy
	Human placental lactogen	Stimulates growth of breasts and has growth hormone-like effects
Testes	Testosterone	Causes development of male secondary sex characteristics

Posterior pituitary hormones, like hormones from the remainder of the endocrine organs, act directly on target tissues.

In general, hormones regulate the metabolism of the body and its most important constituents and control sexual maturation and pregnancy. Sex organs and their hormones will not be discussed in this chapter (see Chapters 16, 17, and 18). There are other important hormones, not mentioned in Table 25–1, such as the numerous hormones of the gastrointestinal tract. In addition, there are still other hormones that appear to be of little importance or are

Figure 25–2. Negative feedback control mechanisms for hormones governed by the hypothalamus and anterior pituitary. A stimulating hormone of the hypothalamus causes increased production of the corresponding anterior pituitary hormone, which, in turn, causes the target organ to produce its hormone. The target organ's hormone inhibits the production of stimulating hormone of the hypothalamus, thus shutting off the stimulation until blood levels of the target organ's hormone drop.

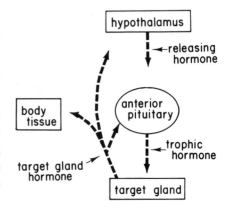

poorly understood at the present time, such as melatonin of the pineal gland.

MOST FREQUENT AND SERIOUS PROBLEMS

Diabetes mellitus is the most common endocrine disease. It is estimated to affect 1 to 2 percent of adults and increases in frequency with age. Hypothyroidism is also quite common, particularly in women. Goiter (enlargement of the thyroid) may be associated with normal, hypo- or hyperfunction. Hyperparathyroidism is not uncommon and is found with parathyroid adenomas and hyperplasia or is secondary to chronic renal disease. Cushing's syndrome, caused by excess corticosteroids (glucocorticoids) is most commonly caused by the administration of corticosteroids for the treatment of various illnesses such as systemic lupus erthematosus, rheumatoid arthritis, and for the prevention of transplant rejections. The syndrome can also be caused by small tumors of the pituitary gland or the adrenal cortex. Hormone-producing neoplasms are relatively uncommon but cause specific syndromes and are often treatable.

SYMPTOMS, SIGNS, AND TESTS

Most endocrine disorders are manifest by hypersecretion or hyposecretion of a hormone. The diagnosis depends upon correctly matching the patient's symptoms and signs with hormone dysfunction and with laboratory confirmation of overproduction or underproduction of a particular hormone. For example, overproduction of insulin will decrease blood sugar levels and the patient will present acutely with measurable hypoglycemia manifested by hunger, pallor, shakiness, and decreased ability to perform mental tasks. All these manifestations result from the body's cells being unable to perform metabolic tasks due to the lack of glucose. On the other hand, too little insulin (diabetes mellitus) leads to elevated blood sugar levels, and the patient presents with excessive urination (polyuria) and excessive drinking (polydipsia). These symptoms occur because the excess glucose along with water is excreted by the kidney. The resultant loss of water leads to dehydration and thirst. Laboratory demonstration of excess glucose in the blood and/or glucose in the urine affords a diagnosis of diabetes mellitus.

The names of the conditions caused by deficiency or excess of the various hormones are given in Table 25–2. The causes and

TABLE 25–2. DISEASES ASSOCIATED WITH DEFICIENCY AND EXCESS OF VARIOUS HORMONES

Hormone	Hormone-Deficiency Diseases	Diseases Associated with Excess Hormone
Adrenocorticotropic hormone	Addison's disease	Cushing's syndrome
Growth hormone	Pituitary dwarfism	Gigantism; acromegaly
Antidiuretic hormone	Diabetes insipidus	Inappropriate ADH-secretion syndrome
Thyroid hormones	Cretinism; hypothyroidism	Hyperthyroidism
Parathyroid hormone	Hypoparathyroidism	Hyperparathyroidism
Glucocorticoids	Addison's disease	Cushing's syndrome
Mineralocorticoids	May occur as part of Addison's disease	Conn's syndrome (hyperaldosteronism)
Insulin	Diabetes mellitus	Insulin shock

manifestations of the most important ones will be discussed under Specific Diseases.

The only endocrine gland accessible to physical examination is the thyroid gland. Generalized enlargement (goiter) can be nodular or diffuse; localized nodules, cysts, and masses can also be felt. Neoplasms of other endocrine organs are detected by the effects of the mass, by the effects of hormones produced by the neoplasm (functional neoplasm), or by hypofunction if the neoplasm replaces the normal glandular tissue.

Many of the major hormones or their breakdown products can be measured directly in the blood or urine by laboratory analysis. These include the thyroid hormones; various steroid hormones of the adrenal cortices, ovaries, and testes; the catecholamines of the adrenal medulla; and growth hormone and prolactin of the pituitary. In other cases, indirect assessment of endocrine function may be accomplished by measuring blood or urine chemicals that are affected by a particular hormone. For instance, the presence of too little insulin may be inferred by finding too much glucose in the blood and urine. The status of the parathyroid glands may be evaluated by measurements of blood and urine calcium and phosphorus levels, because the metabolism of these substances is regulated by parathormone.

SPECIFIC DISEASES

Diseases of the Pituitary Gland

The anterior lobe of the pituitary is composed of epithelial cells that secrete various hormones. The posterior lobe is an extension of the brain composed of nerve fibers with neurosecretory granules. Disease of the anterior lobe consists of destructive conditions and adenomas. Disease of the posterior lobe is caused by destructive lesions.

Panhypopituitarism

Destruction of the anterior pituitary gland leads to panhypopituitarism. The most common causes of this rare condition are neoplasms of the pituitary that are large enough to destroy the gland, postpartum pituitary necrosis, and surgical removal of the pituitary for treatment of tumors. The pituitary is hyperplastic during pregnancy and thus more susceptible to infarction caused by an episode of hypotension as a result of excess bleeding from childbirth. Persons with panhypopituitarism have atrophy of the thyroid, adrenal cortex, and gonads, with variable secondary effects of hypothyroidism, adrenal insufficiency, and decreased libido or secondary sex characteristics, depending on the age of onset. There may be growth retardation from growth hormone deficiency.

CHART 25–1. PANHYPOPITUITARISM

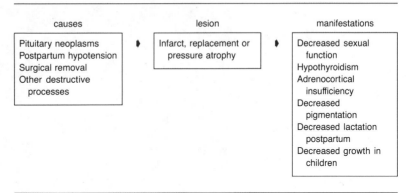

causes	lesion	manifestations
Pituitary neoplasms Postpartum hypotension Surgical removal Other destructive processes	Infarct, replacement or pressure atrophy	Decreased sexual function Hypothyroidism Adrenocortical insufficiency Decreased pigmentation Decreased lactation postpartum Decreased growth in children

Gigantism and Acromegaly

Excess growth hormone leads to enlargement of all tissues; however, bone enlargement is the most prominent clinical finding. In children, bones can grow longer as well as wider, leading to gigantism. In adults, excess growth hormone causes thickening of bones—a condition called acromegaly. Acromegalics have large hands and prominent coarse facial features (Fig. 25–3). Gigantism and acromegaly are caused by pituitary adenomas that secrete growth hormone and are treated by removal of the neoplasm or by radiation to decrease the size of the neoplasm.

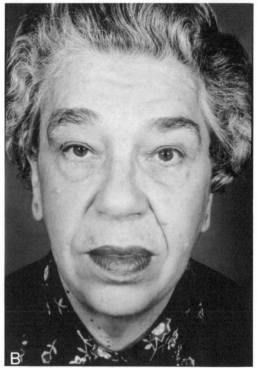

Figure 25–3. Face before (**A**) and after (**B**) development of acromegaly.

CHART 25–2. GIGANTISM AND ACROMEGALY

cause		lesion		manifestations
Growth-hormone-secreting pituitary adenoma	▸	Enlargement of body tissues, especially bones	▸	Rapid increase in height of children (gigantism) Enlarged hands and coarse facial features in adults (acromegaly) Enlarged pituitary fossa by x-ray Increased serum growth hormone

Diabetes Insipidus

Destruction of the posterior pituitary and/or hypothalamus leads to decreased antidiuretic hormone and, thus, to diabetes insipidus. Diabetes insipidus is excessive urination due to failure of water reabsorption in the kidney, resulting from deficiency of antidiuretic hormone. The excess urination (polyuria) leads to excess thirst (polydipsia). This rare condition is caused by infiltrative processes such as neoplasms, and meningitis; by head injury; and, sometimes, by surgical operations in the area. Prognosis depends on the nature of the cause. Hormone replacement therapy is available.

CHART 25–3. DIABETES INSIPIDUS

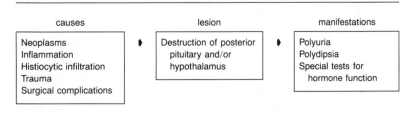

causes		lesion		manifestations
Neoplasms Inflammation Histiocytic infiltration Trauma Surgical complications	▸	Destruction of posterior pituitary and/or hypothalamus	▸	Polyuria Polydipsia Special tests for hormone function

Pituitary Tumors

Adenomas of the anterior pituitary and craniopharyngiomas are the most common types. Most pituitary adenomas do not secrete enough hormone to produce endocrine effects; they more often grow slowly and eventually compress the optic nerve to produce visual field defects. Functioning pituitary adenomas may produce gigantism, acromegaly, Cushing's syndrome, lactation, or amenorrhea, depending on the cell type of the neoplasm. Craniopharyngioma is a benign but locally aggressive tumor containing squamous cell elements that are derived from the embryonic remnant of Rathke's pouch. Craniopharyngiomas are most common in children, while pituitary adenomas are usually found in adults. Pituitary tumors can be cured if they can be removed, but this is often difficult because of their location deep in the skull.

Diseases of the Thyroid Gland

The thyroid gland, which consists of two lobes connected by an isthmus, lies in the neck anterior and lateral to the trachea. It consists of acini that secrete a colloid substance containing thyroid hormones. Calcitonin-secreting C cells are present in the stroma but are not conspicuous. Idiopathic hypothyroidism, Hashimoto's disease, Graves' disease, nodular goiter, adenoma, and papillary carcinoma are the most common conditions of the thyroid gland.

Hypothyroidism

Hypothyroidism is much more commonly due to destruction or atrophy of the thyroid gland than it is to thyroid-stimulating hormone deficiency (hypopituitarism). Destruction of thyroid glands most commonly results from treatment of hyperthyroidism with radioactive iodine. Destruction of the thyroid gland may be necessary to control hyperthyroidism; the resultant hypothyroidism can be treated by administering thyroid hormone to the patient.

The most common natural form of hypothyroidism is associated with atrophy and fibrosis. Atrophy and fibrosis may be idiopathic or may be part of the syndrome called *Hashimoto's disease.* Hashimoto's disease is generally considered to be an autoimmune disease, with destruction of thyroid due to lymphocytes reacting to the body's own tissue. In the early stages of Hashimoto's disease, the thyroid gland is enlarged but usually maintains normal function. It is thought that Hashimoto's disease progresses with time to atrophy and fibrosis, but more patients present with end-stage disease with hypothyroidism than present in the early stage with enlarged thyroid and euthyroidism (normal function). Thus, the most common cause of spontaneous hypothyroidism is variously referred to as idiopathic or Hashimoto's disease. The atrophy and fibrosis are thought to be caused by destruction of the thyroid epithelial cells by lymphocytes sensitized to thyroid epithelial cells. Hypothyroidism is occasionally caused by severe iodine deficiency and other rare conditions of the thyroid.

The results of hypothyroidism are similar regardless of cause. Since the normal action of thyroid hormone is to stimulate cellular metabolism, it is not surprising that patients with hypothyroidism present with slowness in muscular action and slowness in intellectual processes. In addition, they usually feel cold much of the time and have a sallow complexion, dry hair, and dry skin. Patients with prolonged hypothyroidism may develop deposits of mucopolysaccharides in skin, muscle, and viscera, producing a doughy edema called *myxedema*. Diagnosis of hypothyroidism is suspected from the clinical findings and confirmed by one or more laboratory tests that indicate a low level of circulating thyroid hormone, an elevated TSH level, or an abnormally low uptake of radioactive iodine-131.

As with other thyroid diseases, women are much more frequently affected than men. The severity of hypothyroidism varies greatly among patients, and diagnosis depends on careful consideration of laboratory data and clinical observations. In most cases, little can be done about the cause, but replacement therapy with thyroid hormone is effective.

CHART 25–4. HYPOTHYROIDISM

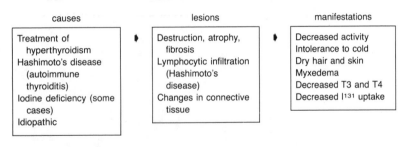

causes	lesions	manifestations
Treatment of hyperthyroidism Hashimoto's disease (autoimmune thyroiditis) Iodine deficiency (some cases) Idiopathic	Destruction, atrophy, fibrosis Lymphocytic infiltration (Hashimoto's disease) Changes in connective tissue	Decreased activity Intolerance to cold Dry hair and skin Myxedema Decreased T3 and T4 Decreased I131 uptake

Hyperthyroidism

Hyperthyroidism may occasionally occur with nodular goiter or a secreting thyroid neoplasm, but the majority of cases are due to Graves' disease.

Graves' disease is a condition of unknown cause but is thought to be due to stimulation of the thyroid by an antibody to some thyroid antigen. Graves' disease is characterized by one or more of the following features: hyperthyroidism with diffuse goiter, ophthalmopathy, or dermopathy (Fig. 25–4). Hyperthyroidism is the most common. In Graves' disease, the thyroid gland undergoes hyperplasia, and there is an associated lymphocytic infiltration. The hyperplastic gland secretes thyroid hormone despite the fact that production of thyroid-stimulating hormone in the pituitary is suppressed by the high serum levels of thyroid hormone.

Common manifestations of hyperthyroidism, regardless of cause, include tremor of extended fingers and tongue, increased nervousness, emotional instability, excessive sweating, heat intol-

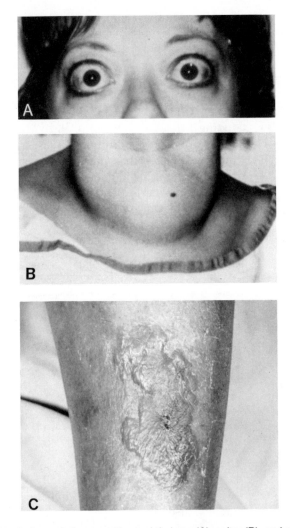

Figure 25–4. Graves' disease with exophthalmos (**A**), goiter (**B**), and pretibial myxedema (**C**).

erance, rapid heartbeat, increased motor activity, and weight loss. The increased metabolic activity associated with hyperthyroidism may lead with time to heart failure and muscle degeneration (thyrotoxic myopathy).

The ophthalmopathy and dermopathy of Graves' disease are not associated with other causes of hyperthyroidism. The eye changes are due to swelling of the soft tissues of the orbit, leading to protrusion of the eye (exophthalmos) and weakness of the eye muscles. The skin changes consist of localized raised thickening of the skin of the dorsum of the legs called *pretibial myxedema*. Graves' disease most commonly begins in young adult women and is progressive, although variable in severity.

Hyperthyroidism must be differentiated from anxiety states, which may mimic many of the nervous symptoms. The measurement of the thyroid hormone blood level is the best test for the demonstration of hyperthyroidism.

Treatment of hyperthyroidism may involve drugs to suppress thyroid function or reduction in thyroid tissue by surgical removal or administration of radioactive iodine. Iodine is selectively taken up by thyroid tissue, so radioactive damage can be safely limited to the thyroid.

CHART 25–5. HYPERTHYROIDISM

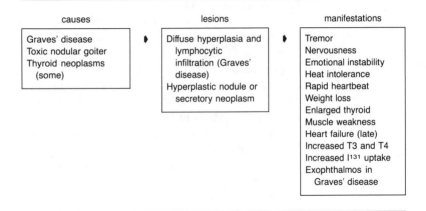

causes	lesions	manifestations
Graves' disease Toxic nodular goiter Thyroid neoplasms (some)	Diffuse hyperplasia and lymphocytic infiltration (Graves' disease) Hyperplastic nodule or secretory neoplasm	Tremor Nervousness Emotional instability Heat intolerance Rapid heartbeat Weight loss Enlarged thyroid Muscle weakness Heart failure (late) Increased T3 and T4 Increased I^{131} uptake Exophthalmos in Graves' disease

Nontoxic Goiter

Generalized enlargement of the thyroid with normal or low thyroid function is termed *nontoxic goiter*. Nontoxic goiter is very common in inland and mountainous areas of the world where there is a deficiency of iodine in the soil and water. Iodine deficiency produces a prolonged, low-grade hyperplasia of the gland that may lead to massive enlargement. The use of iodized salts greatly reduces the prevalence of goiter, but does not eliminate it. Nontoxic goiter in the United States is most often idiopathic or due to Hashimoto's disease.

Idiopathic goiter may be diffuse with increased colloid secretions in the acini (colloid goiter) or nodular due to irregular proliferation with fibrosis (nodular goiter). The nodules of nodular goiter may become large enough to require biopsy to differentiate them from neoplasms. Occasionally, nodules may become functional, resulting in toxic nodular goiter.

CHART 25–6. NONTOXIC GOITER

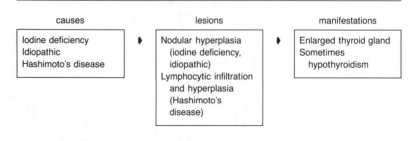

causes		lesions		manifestations
Iodine deficiency Idiopathic Hashimoto's disease	▸	Nodular hyperplasia (iodine deficiency, idiopathic) Lymphocytic infiltration and hyperplasia (Hashimoto's disease)	▸	Enlarged thyroid gland Sometimes hypothyroidism

Thyroid Neoplasms

Adenoma of the thyroid has a variable frequency in different geographic locations. It is a solitary nodule of thyroid acini surrounded by a fibrous tissue capsule. Similar-appearing nodules are found in some nodular goiters. Thyroid adenomas are removed to differentiate them from thyroid carcinoma and occasionally because they are functional. Thyroid carcinoma differs from most other carcinomas in that it occurs predominantly in young adult women and is most often well differentiated, with a papillary growth pattern. Surgical removal often results in a cure. A small percentage of thyroid carcinomas can be causally linked to radiation of the neck region in childhood. A rare type of thyroid carcinoma, called *medullary carcinoma* because the cells occur in solid masses rather than producing glands, arises from the calcitonin-secreting cells of the thyroid.

Diseases of the Parathyroid Glands

Two to six (usually four) tiny glands composed of epithelial cells that secrete parathormone are located adjacent to the inferior and superior aspects of each thyroid lobe. Hyperfunction from adenoma or hyperplasia is the main disease of the parathyroid.

Hypoparathyroidism

Most cases of hypoparathyroidism result from surgical removal of all four parathyroid glands, an uncommon complication of the treatment of hyperparathyroidism or of surgical removal of the thyroid gland. Hypoparathyroidism leads to decreased serum calcium, which, in turn, is manifested by increased irritability of muscle (tetany) and convulsions. Treatment consists of calcium supplements and a synthetic vitamin D derivative.

CHART 25–7. HYPOPARATHYROIDISM

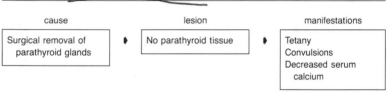

cause	lesion	manifestations
Surgical removal of parathyroid glands	No parathyroid tissue	Tetany Convulsions Decreased serum calcium

Hyperparathyroidism

Hyperparathyroidism is a relatively common and important disease that may be mild and go undetected for a long time. Primary hyperparathyroidism occurs when the cause of increased parathyroid hormone production lies in the parathyroid glands and production is not controlled by normal feedback mechanisms. The causes are adenoma (80%), hyperplasia (18%), and carcinoma (2%). Secondary hyperparathyroidism occurs with conditions associated with low serum calcium, most often chronic renal failure or vitamin D deficiency. The metabolic changes in secondary hyperparathyroidism are quite complex and will not be further discussed.

In primary hyperparathyroidism, the excess parathyroid hormone causes increased breakdown of bone, increased absorption of calcium by the intestine, increased reabsorption of calcium by the kidney, and increased loss of phosphate in the urine. The net effect is to increase serum calcium and decrease serum phosphate. Urine calcium is also increased, because the increase in glomerular filtration of calcium is greater than the increase in reabsorption by the renal tubules.

As osteoblasts attempt to repair the bone destruction, increased amounts of the enzyme alkaline phosphatase are released into the serum. Calcium, phosphorus, and alkaline phosphatase are commonly used as routine biochemical screening tests; thus, there has been a rise in the early detection of hyperparathyroidism. If not discovered by screening, the disease most commonly presents with renal calculi as a result of the increased calcium and phosphorus excretion. Bone destruction with pathologic fracture is a less common mode of presentation and is associated with more advanced disease.

The bone lesions are characterized by the presence of giant multinucleated osteoclasts and fibrosis, sometimes with cyst formation (osteitis fibrosa cystica). The bone resorption is commonly detected by x-rays of the hands and teeth, where subperiosteal resorption of bone and loss of the lamina dura are seen. Primary hyperparathyroidism is treated by removal of the adenoma, hyperplastic glands, or carcinoma.

CHART 25–8. HYPERPARATHYROIDISM

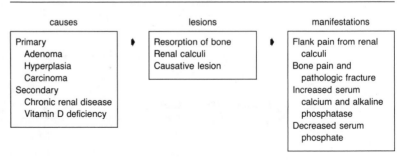

causes	lesions	manifestations

Primary
    Adenoma
    Hyperplasia
    Carcinoma
Secondary
    Chronic renal disease
    Vitamin D deficiency
	Resorption of bone
	Renal calculi
	Causative lesion
		Flank pain from renal
		    calculi
		Bone pain and
		    pathologic fracture
		Increased serum
		    calcium and alkaline
		    phosphatase
		Decreased serum
		    phosphate

Diseases of the Adrenal Glands

The two adrenal glands are located deep within the abdomen embedded in adipose tissue above each kidney. Cortical epithelial cells secrete glucocorticoids, mineralocorticoids, and androgens, while the neural-derived chromaffin cells of the medulla secrete hormones involved in control of the vascular system. Lesions of the adrenal are relatively uncommon and consist of destructive diseases of the cortex, hyperplasias and neoplasms that produce hypersecretion, and two neoplasms of the medulla, neuroblastoma in children and pheochromocytoma in adults.

Addison's Disease

Addison's disease is an uncommon condition characterized by insufficient production of adrenocortical hormones. The most common cause is idiopathic atrophy, thought by some to be autoimmune because of the lymphocytic infiltration of the adrenal cortex, presence of antibodies to adrenal tissue, and association with other presumed autoimmune diseases. Tuberculosis of the adrenal glands was formerly the most common cause. The deficient production of corticosteroids leads to increased release of the stimulatory hormone ACTH. ACTH also causes an MSH-like effect, which accounts for the increased skin pigmentation characteristic of Addison's disease. Other symptoms are often gradual in onset and not specific for Addison's disease, so the diagnosis is easily overlooked. These symptoms include tiredness, anorexia, nausea, weight loss, fainting due to hypotension or hypoglycemia, loss of body hair, and depression. Hormone levels can be measured to confirm the diagnosis. Symptoms may become acute and life threatening, with diarrhea, weakness, and hypotension. Acute insufficiency is likely to be precipitated by stress or infection. Immediate treatment with glucocorticoids and intravenous fluids containing salt and sugar are needed. The chronic condition can be controlled by corticosteroid therapy.

CHART 25–9. ADDISON'S DISEASE

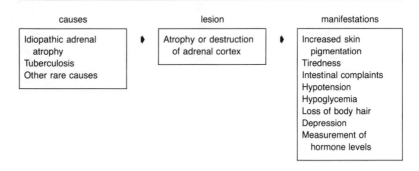

causes	lesion	manifestations
Idiopathic adrenal atrophy Tuberculosis Other rare causes	Atrophy or destruction of adrenal cortex	Increased skin pigmentation Tiredness Intestinal complaints Hypotension Hypoglycemia Loss of body hair Depression Measurement of hormone levels

Cushing's Syndrome

Cushing's syndrome is the result of excess corticosteroids, predominantly glucocorticoids, regardless of whether due to excessive stimulation of the adrenals by ACTH, to primary overproduction by the adrenals, or to exogenous administration of corticosteroids. The most common cause is iatrogenic, the use of corticosteroids in the treatment of other diseases. Noniatrogenic Cushing's syndrome may be caused by adrenal hyperplasia resulting from production of ACTH by pituitary hyperplasia, by corticosteroid-secreting adenomas or carcinomas of the adrenal cortex, or by ACTH-secreting carcinomas such as bronchogenic carcinoma.

The most obvious effect of Cushing's syndrome is a peculiar obesity limited to the face and trunk (Fig. 25–5). Other common findings include purple striae on the skin, easy bruising, hypertension, osteoporosis with spontaneous fractures, hirsuitism (increased body hair), muscle weakness, acne, and hyperglycemia. Some patients become psychotic. In iatrogenic cases, the cause is evident. In other cases, function tests can be used to determine whether the cause is excess cortisol production by the adrenal or excess ACTH production by the pituitary. It would appear that such serious complications would sharply curtail the use of exogenous steroid therapy. Yet, for many diseases, steroids are by far the most effective treatment and the potential positive effects on the patient's health outweigh the negative effects.

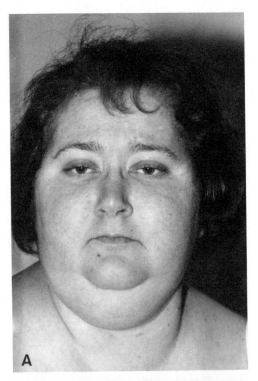

A

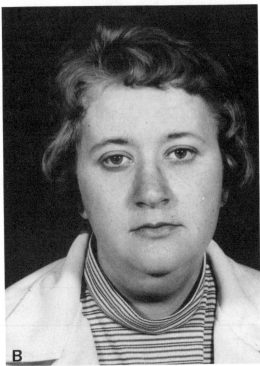

B

Figure 25–5. A. Facial obesity and swelling of Cushing's syndrome. **B.** Same patient after removal of functional adrenal adenoma.

CHART 25-10. CUSHING'S SYNDROME

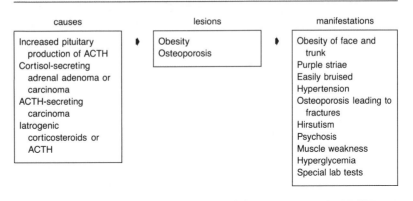

causes	lesions	manifestations
Increased pituitary production of ACTH Cortisol-secreting adrenal adenoma or carcinoma ACTH-secreting carcinoma Iatrogenic corticosteroids or ACTH	Obesity Osteoporosis	Obesity of face and trunk Purple striae Easily bruised Hypertension Osteoporosis leading to fractures Hirsutism Psychosis Muscle weakness Hyperglycemia Special lab tests

Conn's Syndrome

Primary overproduction of mineralocorticoids (aldosterone) leads to Conn's syndrome. The cause is usually an aldosterone-secreting adenoma of the adrenal cortex. There is excess excretion of potassium and retention of sodium. The decreased serum potassium levels are associated with increased urine output and muscular weakness. Salt retention is associated with hypertension. The rare cases of Conn's syndrome must be distinguished from the more common causes of secondary aldosterone overproduction, which include heart failure, cirrhosis of the liver, nephrotic syndrome, renal ischemia, and essential hypertension. Removal of the adenoma is curative in Conn's syndrome.

CHART 25-11. CONN'S SYNDROME

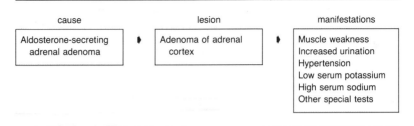

cause	lesion	manifestations
Aldosterone-secreting adrenal adenoma	Adenoma of adrenal cortex	Muscle weakness Increased urination Hypertension Low serum potassium High serum sodium Other special tests

Pheochromocytoma

This is a neoplasm of the chromaffin cells of the adrenal medulla that produces catecholamines with resultant hypertension. It is usually a benign neoplasm but is occasionally malignant. Although it is a rare cause of hypertension, it is important to diagnose, because removal will cure the hypertension. Episodes of sweating and nervousness often accompany the hypertension. Measurement of a catecholamine breakdown product—vanilmandelic acid (VMA)—in the urine is helpful in diagnosis.

Neuroblastoma

Neoplasms of neural cells are particularly common in the adrenal in children under 5 years of age. Poorly differentiated neoplasms are called neuroblastoma, and they usually metastasize widely and kill the patient.

Diseases of the Pancreatic Islets of Langerhans

Myriads of small clusters of neuroendocrine cells that develop from pancreatic ductules are scattered throughout the pancreas, particularly in the tail. The various cell types in the islets secrete different hormones—glucagon (alpha cell), insulin (beta cell), somatostatin (delta cell), and pancreatic polypeptide (PP cell). Diabetes mellitus is the major disease related to insulin secretion of the islets. Islet cell tumors, although uncommon, are of interest because they become manifest due to excessive secretion of any one of several hormones.

Diabetes Mellitus

Diabetes mellitus is a disease condition in which there is more or less persistent hyperglycemia with loss of glucose homeostasis. Since insulin helps determine blood glucose levels, partly by controlling cellular uptake of glucose, diabetes mellitus is sometimes defined as a relative or absolute deficiency of insulin.

Insulin is normally produced by the beta cells of the pancreatic islets of Langerhans. Lack of insulin may result from diminished production of insulin, faulty release of insulin, antibodies to insulin, and abnormalities of the body's cells precluding normal action of insulin. All cases of diabetes cannot be explained by any one of these mechanisms.

Occasional cases of diabetes are due to destructive diseases of the pancreas such as chronic pancreatitis or hemochromatosis, but the vast majority fall into one of two types called insulin-dependent diabetes mellitus (IDDM) or type I (formerly called juvenile-onset diabetes) and noninsulin-dependent diabetes mellitus (NIDDM) or type II (formerly called adult-onset diabetes).

The major differences between type I and type II diabetes mellitus are shown in Table 25–3. Both types have a multifactorial etiology involving both genetic and nongenetic factors. Neither type exhibits a Mendelian-type inheritance pattern. IDDM appears to be less dependent on genetic factors—if one identical twin has

TABLE 25–3. INSULIN-DEPENDENT AND NONINSULIN-DEPENDENT DIABETES MELLITUS

	IDDM (Type I)	NIDDM (Type II)
Age	Usually in children or young adults	Usually after age 25 Frequency increases with age
Onset	Generally abrupt Often triggered by infection	More often insidious Patients often obese
Treatment	Insulin	Often controlled by diet
Complications	Occur early Often severe, especially retinal, renal, and heart disease	Full range of complications Tend to develop more slowly and be less severe

the disease, the other one only has a 50 percent chance of developing it, compared to 90 percent with NIDDM. However, IDDM is associated with certain histocompatibility antigens (HLA types) and antibodies to pancreatic islet cells; whereas, NIDDM is not. A person with NIDDM is more likely to have relatives with the disease than one with IDDM.

Nongenetic etiologic factors are also different for the two diseases. With IDDM, insulin production becomes deficient abruptly over a few days to weeks. At onset, most patients are under age 30, but may be older. An immunologic or viral injury is postulated as the cause in susceptible individuals. The initial lesions involve reduction in the number of insulin-producing beta cells in the islets of Langerhans in the pancreas. Later, some patients have hyalinized islets, others appear relatively normal by light microscopy, but special studies can demonstrate the decrease in insulin-producing cells.

In contrast to the absolute lack of insulin in IDDM, with NIDDM there is only a relative lack of insulin and this is associated with inappropriate insulin release and peripheral cellular resistance to the effects of insulin. Major nongenetic factors producing NIDDM are those that stress glucose metabolism such as obesity, pregnancy, infections, and corticosteroid therapy. NIDDM has a gradual onset, usually in persons over 40 years of age, and responds to treatment that decreases stress on glucose metabolism.

The clinical manifestations of diabetes mellitus are initially due to abnormalities of glucose metabolism. Years later complicating lesions develop. Normally, insulin regulates the uptake and utilization of glucose by cells. The diabetic, therefore, has high levels of blood glucose and deficient cellular glucose. Because of this, the diabetic relies on lipids for energy, with excessive production of acetylcoenzyme A and ketone bodies resulting in acidosis. Coenzyme A enhances the production of cholesterol, which, in turn, may be partially responsible for the increased frequency of severe atherosclerosis in diabetics.

Classically, patients with diabetes present with polydipsia (excessive drinking) and polyuria (increased urination) as the initial manifestations of the disease. This is due to the hyperosmolality of the blood, which results from the increased glucose levels. Generalized weakness, increased tendency to infection, and poor wound healing are also commonly present in the initial stages. These symptoms of diabetes are related to excess blood glucose and also to increased breakdown of lipids, resulting in acidosis. Severe acidosis is characteristic of IDDM and if not treated with insulin the patient will die. After diagnosis, patients may maintain a relatively normal life if glucose metabolism is controlled by therapy (insulin for IDDM, dietary manipulation for NIDDM).

Complications of diabetes mellitus are the same for both types and are predominantly vascular. Complications tend to be more severe with IDDM, probably because the disease is more severe and begins at a much earlier age. The severity and time of development of these complications are not very predictable. They develop over many years.

Most diabetics die from the effects of atherosclerosis, but so do most nondiabetics. Persons with IDDM may die from myocardial infarcts or other atherosclerotic lesions at an early age, which is clearly beyond the norm. Persons with NIDDM die from atherosclerosis at ages that overlap with those of nondiabetics, but statistically they die younger and have more severe atherosclerosis. Myocardial infarct is the most common complication of atherosclerosis in diabetics, but it should be noted that diabetics have a disproportionate amount of vascular disease in the lower extremities leading to gangrene and amputation. The latter is often referred to as "diabetic gangrene," although it is no different than that occurring in nondiabetics.

In addition to accelerated atherosclerosis, diabetics have a more specific microangiopathy characterized by basement membrane thickening and other changes in small vessels. Angiopathy tends to develop in the retina and renal glomeruli at about the same time, leading to blindness and renal failure, respectively (see Chapters 16 and 20). Diabetic neuropathy, which is thought to be due in part to angiopathy involving nerves, has a variable distribution and may result in loss of sensation, neuromuscular weakness, or involvement of the autonomic nervous system with symptoms of delayed emptying of the stomach or bladder and impotence.

Many other conditions have been associated with diabetes. Increased susceptibility to infection is probably the most important. The infections are not different than those that occur in non-diabetics except for frequency and severity.

The diagnosis and treatment of IDDM, which accounts for only 10 percent of patients with diabetes, is relatively straightforward. The abrupt onset calls attention to the disease, urinary and blood glucose levels are distinctly abnormal, and the response to insulin therapy is predictable. Management of the disease requires education of the patient and careful monitoring of therapy. The long-term goal is to prevent the development of complications. It is not entirely clear whether perfect control of glucose metabolism will prevent the late vascular complications or whether these are independent features of the disease. However, implantable devices are being developed to automatically measure blood glucose levels and release the appropriate amount of insulin. It will take a number of years to determine whether such therapy will reduce complications. The possibility of islet cell transplantation is also under investigation.

Diagnosis of NIDDM is more arbitrary and treatment less specific. When a patient develops symptoms of diabetes, spills glucose in the urine, and has a fasting blood glucose level over 140 mg/dl, the diagnosis of NIDDM is not difficult. The diagnostic problem relates to defining normal glucose levels, which vary with age, and deciding whether patients with chemical abnormalities should be labeled as diabetics.

The diagnostic test for borderline cases is the glucose tolerance test. Serum glucose is measured in the fasting state and 1, 2, and 3 hours after ingesting a defined amount of glucose. Table 25–4 shows a point system for making a diagnosis of diabetes developed by the U.S. Public Health Service. Only about half of asymptomatic diabetics diagnosed by these chemical criteria develop overt decompensation of carbohydrate metabolism within 10 years. NIDDM can often be controlled by a reduction in food intake, since most of these patients are overweight. Some diabetics of intermedi-

TABLE 25–4. DIAGNOSIS OF DIABETES BY GLUCOSE TOLERANCE TEST

Time	Plasma Glucose, mg/dl	Points[a]
Fasting	>130	1
1 hour	>195[b]	1/2
2 hours	>140[b]	1/2
3 hours	>130	1

[a] Total of 2 or more = diabetes. Total of 1 or more but less than 2 = possible diabetes.
[b] Add 10 for each decade over 50.

ate severity are treated with sulfonylurea compounds, which enhance the release of insulin from pancreatic beta cells and also lessens the insulin resistance. In any diabetic, treatment consists of careful monitoring of the blood sugar, food intake, and insulin requirements as well as anticipation and treatment of complications of the disease.

Islet Cell Tumors

Insulinomas are uncommon neoplasms of beta cells that secrete insulin. They are manifest by attacks of confusion, stupor, and loss of consciousness associated with low blood sugar and relieved by eating. In most instances, surgical removal of a solitary adenoma effects a cure, but occasional multiple adenomas or malignant islet cell tumors are encountered. Gastrinomas are islet cell tumors that secrete gastrin, a hormone normally secreted by the stomach to increase its acid production. The ectopic hormone production causes Zollinger–Ellison's syndrome, which is characterized by severe, recurrent peptic ulcers. Gastrinomas are more often malignant than benign. Rarer types of islet cell tumors are glucagonoma and somatostatinoma, which produce a diabetic-like illness, and vipoma, which produces diarrhea, hypokalemia (potassium loss), and achlorhydria.

Endocrine Syndromes

Multiple Endocrine Tumors

There are at least two familial syndromes in which tumors of several endocrine organs occur in family members. In type I, there are commonly adenomas of pituitary, parathyroid, and adrenal cortex and carcinoma of pancreatic islets. The pancreatic islet cell neoplasms frequently secrete gastrin, a hormone that causes increased gastric acid production, leading to peptic ulcers. In type II, there commonly are medullary carcinomas of the thyroid, pheochromocytomas, parathyroid hyperplasia, and occasionally neuromas. The inheritance pattern is incomplete dominant, and the clinical presentation may relate to one or more of the neoplasms present.

Ectopic Hormone-Producing Cancers

As has been mentioned earlier, some cancers produce hormones even though they arise from nonendocrine tissue. A poorly differentiated form of bronchogenic carcinoma, called oat cell or small cell carcinoma because of its small, uniform cells, is sometimes associated with ectopic hormone production. Other types of cancer are more rarely involved. Hormone production may be responsible for the initial manifestations of the cancer or may be a problem to be dealt with in the treatment of a known cancer. ACTH, the most common ectopic hormone produced by cancers, leads to Cushing's syndrome.

Certain neoplasms, especially bronchogenic carcinomas of the lung, produce antidiuretic hormone (inappropriate antidiuretic hormone secretion), which leads to water intoxication by causing excess reabsorption of water by the kidneys. Water intoxication is associated with depression, lethargy, mental confusion, irritability, anorexia, nausea, and weakness. Electrolytes (sodium, potassium, and chloride) are lost in the urine while water is reabsorbed, resulting in dilute serum and relatively concentrated urine. Serum sodium levels are very low.

Many other hormones can be produced by nonendocrine cancers, and more than one hormone can be produced by a particular cancer.

ORGAN FAILURE

Failure of each of the major endocrine organs has been mentioned under hypofunction in the preceding sections. Acute failure that is life threatening occurs with lack of parathyroid hormone, adrenocorticosteroids, and insulin. Failure of the pituitary and thyroid lead to more gradual changes that may cause premature death.

REVIEW QUESTIONS

1. What are the most common endocrine disorders?
2. What are the names applied to syndromes of excess or deficiency of ACTH, growth hormone, antidiuretic hormone, thyroid hormones, parathyroid hormone, glucocorticoids, mineralcorticoids, and insulin?
3. Why are hormones sometimes measured directly and other times measured indirectly?
4. What are the causes, lesions, and manifestations of hyposecretion and hypersecretion of the pituitary, thyroid, parathyroid, and adrenal glands and the islets of Langerhans?
5. What are the neoplasms of the endocrine system that are commonly functional? What are the usual manifestations of nonfunctional types?
6. How do insulin-dependent and noninsulin-dependent diabetes differ in cause, presenting manifestations, and complications?

7. What do the following terms mean?
   Craniopharyngioma
   Hashimoto's disease
   Myxedema
   Graves' disease
   Exophthalmos
   Pretibial myxedema
   Goiter
   Ganglioneuroma
   Inappropriate antidiuretic hormone secretion

Section IV

Multiple Organ System Diseases

The purposes of this section are (1) to discuss groups of diseases that frequently involve more than one organ system and (2) to provide a broader overview of some of the diseases that have been encountered in Section III. Chapters 26 through 29 cover many diseases and are organized by cause or causative mechanism (infections, immune reactions, physical injury, chemical injury). Chapter 30 reviews diseases caused by under- and over-nutrition.

Infectious Diseases

INFECTION AND THE BODY'S DEFENSE MECHANISMS
Saprophytes · Normal Flora · Pathogens · Infection · Vectors · Structural Barriers · Inflammatory Process · Immune System

MOST FREQUENT AND SERIOUS PROBLEMS
Overall Impact of Infectious Disease
Respiratory Illnesses · Pneumonia · Secondary (Opportunistic Infections)
Bacterial Infections
Abscesses, Pneumonia, and Wound Infections · Venereal Diseases
Viral Infections
Upper Respiratory Illnesses · Gastroenteritis · Influenza · Cold Sores · Infections of Childhood · Infectious Mononucleosis · Hepatitis · Squamous Papillomas · AIDS
Rickettsial Infections
Fungal Infections
Superficial · Primary Disseminated Infections · Opportunistic Infections
Protozoal Infections
Malaria · Trichomonas Vaginitis
Helminth Infections
Schistosomiasis · Tapeworm Disease · Pinworm Infestation · Ascaris Infestation

SYMPTOMS, SIGNS, AND TESTS
Acute Presentation · Rashes · Cardinal Signs of Inflammation · Cultures · Smears · Immunologic Tests · Nonspecific Tests

SPECIFIC DISEASES
Bacterial Infections in General
Pyogenic Bacterial Infections
  Staphylococcal Infections
  Group A Streptococcal Infections
  Streptococcus viridans Infection
  Streptococcus Pneumoniae (Pneumococcus) Infection
  Meningococcal Meningitis
  Hemophilus influenzae Infection
  Legionellosis
  Enteric Bacterial Infections
  Anaerobic Bacterial Infections
Venereal Bacterial Infections
  Venereal Disease in General
  Gonorrhea
  Nongonococcal Urethritis
  Syphilis
Bacterial Infections Mediated by Exotoxins
  Tetanus
  Botulism
  Gas Gangrene
  Diphtheria

Miscellaneous Bacterial Infections
Whooping Cough
Mycoplasma Disease
Plague
Tularemia
Typhoid Fever
Brucellosis
Leprosy
Tuberculosis
Chlamydial Infections
*Urethritis · Lymphogranuloma Venereum · Trachoma · Inclusion Body
Conjunctivitis · Psittacosis*
Viral Infections in General
Viral Diseases of Childhood
Measles (Rubeola)
Rubella (German Measles)
Varicella (Chicken Pox) and Zoster (Shingles)
Mumps
Smallpox (Variola)
Postviral Encephalitis
Other Common Viral Infections
Influenza
Infectious Mononucleosis
Herpes Simplex Infections
Human Papilloma Virus Infections
Cytomegalic Inclusion Disease
Viral Hepatitis
Human Immunodeficiency Virus Infection
Rickettsial Infections
Typhus · Rocky Mountain Spotted Fever
Fungal Infections
Primary Systemic Fungal Infections
Opportunistic Systemic Fungal Infections
Cutaneous Fungal Infections
Protozoal Diseases
Toxoplasmosis
Giardiasis
Helminth Infections
Pinworm Infestation
Trichinosis
Tapeworm Disease
Ascariasis
Trichuriasis
Filariasis

REVIEW QUESTIONS

INFECTION AND THE BODY'S DEFENSE MECHANISMS

The world is inhabited by myriads of plant and animal species, most of which have no direct contact with man. Some microorganisms, especially certain bacteria, specialize in biodegrading dead animals or other organic material. Such microorganisms are called *saprophytes.* Other microorganisms, again mostly bacteria, live on the skin and in the alimentary tract of man without producing ill effects and sometimes are helpful. These are called *normal flora.* Finally, some microorganisms produce disease when they gain entrance to a host tissue; hence, they are called *pathogens.* The degree to which an organism is a pathogen is termed its *virulence.* Any disease directly caused by pathogens is classified as an infection. Microorganisms that produce infection in man fall into several classes of small plant-like organisms (bacteria, fungi, rickettsiae, and viruses) and animals (protozoa and helminths). It should be noted that the word *infection* is also used to describe the process of organisms gaining entrance to the body regardless of whether disease is produced. In this chapter, *infection* will be used synonymously with *infectious disease.*

Microorganisms may play various roles depending on circumstances. For example, *Clostridium perfringens* is an anaerobic (grows without oxygen) bacterium that is a normal member of the colonic flora. After death of the host animal, it comes a saprophyte by participating in the postmortem degradation of the tissues. *Clostridium perfringens* may also become a pathogen either by releasing necrotizing enzymes and toxins or by invading necrotic tissue.

Many of the common, mildly pathogenic bacteria, such as staphylococcus and *Escherichia coli,* are part of the normal flora of the skin, mouth, or intestines. Strongly pathogenic organisms, including certain viruses, all rickettsiae, and many bacteria, require some method of spreading from one person to another such as direct contact, environmental contamination, or a *vector* (another organism such as an insect that transports the pathogen to the host). When a strong pathogen is present in or on the body without causing disease, the host is said to be a *carrier,* meaning that the host can "carry" the disease to another host.

The body is constructed to provide continual defense against attack by foreign organisms. The most obvious defense mechanism is the structural barrier provided by the epithelium of the skin and internal passageways exposed to external contamination.

The second major line of defense is the inflammatory process. It protects against microorganisms that have penetrated the epithelial barrier. Microorganisms are walled off initially by fibrin deposition and later by fibrous tissue and are engulfed and destroyed by phagocytes (neutrophils and macrophages).

The third major line of defense, the immune system, requires previous or prolonged exposure to the offending agent and operates by enhancing the effectiveness of the inflammatory process. The immune defense mechanism consists of two distinctive systems—the humoral system, in which there are free antibodies in the blood, and the cellular immune system, in which the reaction is mediated by lymphocytes. Both of these systems aid in killing microorganisms and will be discussed in greater detail in the next chapter.

Antibodies are also effective in neutralizing certain bacterial toxins such as tetanus toxin and diptheria toxin, in enhancing phagocytosis of bacteria such as *Streptococcus pneumoniae* and *Bordetella pertussis* (the bacterium that produces whooping cough), and in preventing dissemination of certain viruses such as rabies virus and poliovirus. Routine immunization of infants is used to protect against tetanus, diphtheria, whooping cough, poliomyelitis, measles, rubella, and mumps. Immunization for smallpox (vaccination) has been discontinued because of the remarkable worldwide elimination of this disease. Immune sera are available for selected other diseases when conditions warrant their use. For example, antirabies sera is used after a person is bitten by a rabid animal.

In summary, the major defense mechanisms against infection are mechanical barriers, the inflammatory reaction, and the immune reactions. There are obviously other defense mechanisms, but these are either things we can do little about or are less well understood. For example, viral infection of cells causes release of a substance known as *interferon*, which aids by limiting viral replication, but there is no known deficiency of this substance or way to enhance its action. For unknown reasons, pathogens often show a specific predilection for a part of the body where they can best live. For example, gonococci live on mucous membranes and grow less readily at other sites. In most instances, it is not known why some organisms are more virulent than others.

MOST FREQUENT AND SERIOUS PROBLEMS

Overall Impact of Infectious Disease

As presented in Chapter 2, respiratory illnesses, which are mostly viral infections, account for about 50 percent of all acute illnesses, and other infections account for an additional 10 percent of acute illnesses. Very few chronic illnesses are primarily due to infection, notable exceptions being tuberculosis, leprosy, and syphilis. Pneumonia, which is almost always due to infection, is a common cause of death. Pneumonia may occur as a primary illness, but more often it occurs in persons seriously ill with other diseases, in which case it is termed *secondary* or *opportunistic* because the other disease caused or predisposed to the pneumonia. There are numerous other kinds of secondary infections. Significant predisposing factors to secondary infections are obstructions in body cavities, which enhance secondary bacterial infection, and deficient immunity, such as occurs in patients with chronic debilitating illness.

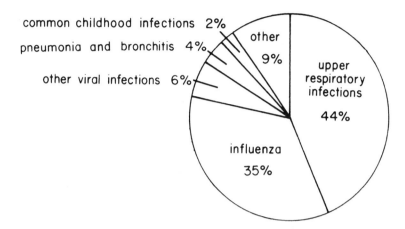

Figure 26–1. Relative frequency of infectious diseases in the United States. *(Data from the National Health Committee: The Killers and Cripplers—Facts on Major Diseases in the US Today. New York, D. McKay, 1976.)*

Patients with terminal cancer are susceptible to various bacterial, fungal, viral, and protozoal infections.

Respiratory infections, including upper respiratory infections, influenza-like infections, pneumonia, and bronchitis, account for over 80 percent of infections (Fig. 26–1). Common childhood infections (mostly viral) and other viral diseases account for about half of the remaining infections. These statistics probably do not include many forms of secondary infection.

Of the primary infections that are on the required list of reportable infections, gonorrhea is by far the most common. Table 26–1 shows the relative frequency of reportable infections in the United States. Unreported venereal infections, however, may be more common (e.g., venereal infections due to chlamydia and human papilloma virus). Human immunodeficiency virus infection is more serious.

Bacterial Infections

Secondary infections and venereal diseases are the most common types of bacterial infection. Common secondary infections include abscesses, pneumonia, and wound infections. One of the most potent predisposing factors to secondary bacterial infection is the presence of dead tissue. Dead tissue is a good growth medium for bacteria and lacks a blood supply to bring in phagocytes and antibodies. The most common predisposing factor for secondary bacterial infection is obstruction of a body passageway. For example, urinary tract obstruction predisposes to pyelonephritis, nasal obstruction leads to sinusitis, pharyngeal obstruction of eustachian tubes leads to otitis media. Obstruction allows normal flora to proliferate to concentrations at which they become pathogenic. Table

TABLE 26–1. NUMBER OF CASES OF REPORTABLE INFECTIOUS DISEASES IN THE UNITED STATES—1982

Disease	No. of Cases	Disease	No. of Cases
Gonorrhea	960,633	Legionellosis	654
Chicken pox	167,423	Thyphoid fever	425
Syphilis	75,579	Tularemia	275
Viral hepatitis	56,773	Leprosy	250
Salmonellosis	40,936	Lymphogranuloma venereum	235
Tuberculosis	25,520		
Shigellosis	18,129	Brucellosis	173
Aseptic meningitis	9,680	Psittacosis	152
Amebiasis	7,304	Rheumatic fever	137
Mumps	5,270	Trichinosis	115
Meningococcal infection	3,056	Leptospirosis	100
Rubella	2,325	Botulism	97
Whooping cough	1,895	Tetanus	88
Measles	1,714	Typhus	58
Encephalitis	1,500	Plague	19
Chancroid	1,392	Granuloma inguinale	17
Malaria	1,056	Poliomyelitis	8
Rocky Mountain spotted fever	976	Diphtheria	2
		Rabies	0

(Data from Morbidity and Mortality Weekly, Report by the Centers for Disease Control. Public Health Service, Vol. 31, No. 54, December 1983.)

26–2 lists the common sites of obstruction and the resulting types of infection.

The classic venereal infections are gonorrhea and syphilis, with gonorrhea being much more frequent and syphilis much more serious; however, papilloma and herpes viruses and chlamydia infections appear to be more common, and human immunodeficiency virus infection leading to acquired immune deficiency syndrome (AIDS) is more serious. Most of the classic bacterial infections that will be presented in this chapter are uncommon or rare and are important because they must be kept in that status.

TABLE 26–2. SITES OF OBSTRUCTIONS AND ASSOCIATED INFECTIONS

Site of Obstruction	Type of Infection
Eustachian tube	Otitis media
Nasal sinus openings	Sinusitis
Bronchus	Pneumonia
Urethra	Cystitis
Ureter	Pyelonephritis
Sebaceous gland	Furuncle and acne
Appendix (fecalith)	Acute appendicitis

Bacterial meningitis is an uncommon but serious problem that must be diagnosed early if treatment is to be effective.

Viral Infections

Upper respiratory illnesses, caused by several types of viruses, far outnumber all other types. Viral gastroenteritis and influenza are less common and occur in epidemics. Cold sores due to *Herpes simplex* are a common type of recurrent viral infection that affect many individuals. Disseminated viral infections of childhood have decreased in frequency as vaccines have been developed for measles, rubella, and mumps. There is no immunization available for chicken pox, and it is still quite common. Infectious mononucleosis and viral hepatitis are moderately common, with peak frequency in adolescents and young adults. Viral encephalitis is a rare but serious viral disease, usually spread by mosquitoes in summer epidemics. Human immunodeficiency virus (HIV) infection, the cause of acquired immune deficiency syndrome (AIDS), has become the most important viral infection because of its chronicity and lethal outcome.

Rickettsial Infections

These are rare but serious disseminated infections caused by organisms intermediate between bacteria and viruses. Rocky Mountain spotted fever occurs sporadically in the United States. Typhus is even rarer.

Fungal Infections

Superficial fungal infections of the skin, such as athlete's foot and ringworm, are quite common but not serious. Superficial infection of skin and mucous membranes by candida, called candidiasis, occurs particularly in diabetics and persons on prolonged antibiotic therapy. Primary disseminated fungal infections, such as histoplasmosis and coccidioidomycosis, are geographically common as asymptomatic diseases but uncommon as significant clinical illnesses. Species of *Candida*, *Aspergillus*, and *Mucor* may produce opportunistic infections in persons debilitated for other reasons.

Protozoal Infections

One of the most prevalent worldwide infections, malaria, is caused by a small protozoan that lives in and destroys red blood cells. In the United States, malaria is uncommon except in those who have returned from tropical and subtropical areas of the world. Possibly the most prevalent protozoan is *Trichomonas vaginalis*, which may cause vaginal and lower urinary tract discomfort but only rarely is associated with significant morbidity.

Giardiasis is an intestinal protozoan infection acquired from contaminated water that causes diarrhea and weight loss if severe and prolonged. Pneumocystis, an organism of uncertain type that has been classified as a protozoan, causes pneumonia in immunosuppressed persons.

Helminth Infections

Helminths comprise the primitive roundworms and flatworms. Worldwide, helminth infestations and infections are extremely common. Schistosomiasis is one of the most important infections in the world, because it produces significant liver and urinary bladder disease in areas of the world such as Asia, Africa, and South America. Many types of helminths infest (live in) the intestinal tract and occasionally produce disease. All types of tapeworm (beef, pork, dog, rat, and fish) can cause disease in man. Pork and dog tapeworm disease may be serious. Pinworm and ascaris are the most common helminths encountered in the United States. In general, helminth infections are uncommon in the United States.

SYMPTOMS, SIGNS, AND TESTS

The long list of diseases produced by microorganisms includes many with distinctive clinical presentations and others that present as nonspecific syndromes. The common syndromes, such as upper respiratory infection, gastroenteritis, cystitis, and meningitis, have been discussed in previous chapters. The clinical presentations of selected specific diseases will be presented in the next section.

Symptoms common to many but not all infectious diseases include rapid onset, short duration, fever, rapid pulse, and a feeling of tiredness and ill health. Physical examination of the lesions may, in many instances, be sufficient for diagnosis. For example, the rash of measles or chicken pox is distinctive, and the parotid swelling of mumps or pus draining from an abscess is usually all that is required for diagnosis. Since most lesions of infectious diseases are inflammatory in nature, they likely will exhibit the cardinal signs of inflammation—heat, redness, swelling, and pain.

The laboratory is critical in the diagnosis of many infectious diseases. Direct culture of organisms from the lesion is the most definitive test, although on occasion contaminants or normal flora may be cultivated while pathogens fail to grow. Culture is most commonly used for bacteria, although it is also the method of choice for diagnosis of many fungal diseases. Many viruses can be cultured, but the lack of effective therapy, time lapse, and high cost limit the use of culture for diagnosis of viral disease. Helminths and most protozoa are not cultured. Rickettsiae are dangerous to culture. The bulk of bacterial cultures are from throat, urine, sputum, and purulent lesions. When indicated, blood and spinal fluid cultures provide critical information for treatment of *septicemia*

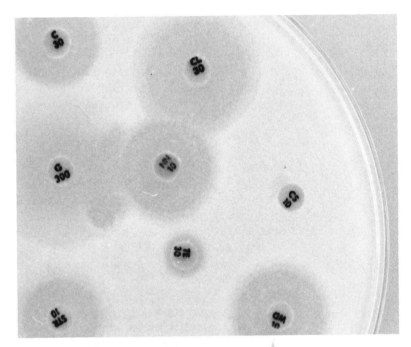

Figure 26–2. Agar plate with antibiotic discs for sensitivity test. Note inhibition of growth on the surface of the medium around some of the antibiotic discs.

(generalized infection involving the blood) and meningitis. Significant pathogens are often tested for their sensitivity to selected antibiotics so that the most appropriate therapy can be administered (Fig. 26–2).

Direct smears with or without staining are occasionally of benefit. Gram-stained bacterial preparations can be used to identify gram-positive (blue-staining) and gram-negative (red-staining) bacteria and to subdivide them into cocci (round) and bacilli (elongated) types. This information obtained from spinal fluid in a patient with meningitis is usually sufficient to start antibiotic treatment while results of culture and sensitivity testing are pending. Gonococci can be identified in pus dripping from the male urethra by gram-stained smear. Smears are generally useful in identifying the larger, more varied types of organisms, including fungi, protozoa, and eggs of the helminths.

In some instances, specific antibodies can be used to detect organisms in samples obtained by swabbing or scraping a lesion. For example, antibodies to group A streptococci attached to latex particles can be mixed with material from a throat swab; agglutination of the particles is interpreted as streptococcal pharyngitis. This is a very rapid and useful test to decide whether a patient with an upper respiratory illness should be treated with penicillin. Another technique involves the use of fluorescent-tagged antibodies mixed with tissue sample and examined under a fluorescent microscope. This technique can be used for rapid diagnosis of herpes simplex type I and II infections and urethritis due to chlamydia.

Immunologic reactivity of the host to an infection can be evaluated by measurement of serum antibodies and skin testing to assess cellular immunity. In either case, the positive result only indicates previous exposure to the organism and may not mean active disease. Furthermore, antibodies or hypersensitivity may not be present in early stages of a disease.

The tests for antibodies in the patient's serum are called serologic tests. Serology is commonly employed for the diagnosis of syphilis, systemic fungal disease, several generalized febrile bacterial diseases (brucellosis, leptospirosis, typhoid fever, tularemia), many viral diseases, rickettsial disease, and some parasitic diseases. Skin tests are most commonly employed as a screening test for tuberculosis.

The most commonly used tests for infectious disease are not really specific for infectious diseases. These tests include measurement of body temperature, white blood cell total and differential counts, and examination of a smear of urine sediment. An elevated body temperature (fever) occurs with many bacterial, viral, and rickettsial infections. An elevated white blood count, particularly when associated with increased numbers of neutrophils, suggests a bacterial infection. Viral infections may have elevated, normal, or low white blood counts, and often the proportion of lymphocytes is increased. Eosinophils are frequently increased with helminth infections. White blood cells in the urine sediment suggest urinary tract infection, and if white blood cell casts are present, it suggests kidney involvement, since casts are formed in renal tubules.

SPECIFIC DISEASES

Bacterial Infections in General

Bacteria are small microbiologic agents distributed ubiquitously throughout nature. Humans normally harbor certain bacteria on the skin, in the oral and nasal cavities, the anterior urethra, the female genital tract, and throughout the intestinal tract. The body provides a home for these bacteria, and in some instances they even provide benefits, an example being the production of vitamin K by intestinal bacteria. At certain times, because of imbalances in either host or bacteria, the bacteria invade body tissues and cause disease by proliferating in an unnatural location. Bacteria invade the body via the lungs, intestine, nasal sinuses, urinary tract, or through breaks or tears in the skin. They can localize in particular areas or organs, or they may gain access to the circulation and be disseminated throughout the body. Once inside the body, most bacteria proliferate in the interstitial tissue, although some prefer to proliferate in the host's cells. The response of the host to bacterial invasion is acute inflammation as outlined in Chapter 4. Different bacteria destroy the host's tissues by different mechanisms; for example, some of the most virulent bacteria, such as *Clostridium botulinum*, secrete exotoxins that are proteinaceous poisons. Some exotoxins actually kill cells directly, others block nerve impulses or inhibit certain intracellular processes. Other bacteria, such as staphylococci, produce enzymes that will enhance their invasiveness by dissolving connective tissue or fibrin. Still other bacteria are surrounded by thick capsules that retard phagocytosis by the host's phagocytes, an example being pneumococcus. Most pathogenic bacteria are *pyogenic,* meaning that they elicit a neutrophilic inflammatory response with purulent exudate. In many cases, the tissue damage in the patient is caused more by release of enzymes from the leukocytes than by the direct action of the bacteria.

Pyogenic Bacterial Infections

Staphylococcal Infections

Of the several varieties of staphylococci, the most virulent is *Staphylococcus aureus. Staphylococcus aureus*, a member of the normal flora, can enter the body by any route and infect any organ; it has a particular propensity to cause abscesses. Most skin abscesses are staphylococcal, and abscesses in other sites are also commonly caused by staphylococci. Abscesses under the skin are referred to as *furuncles* (boils). A collection of furuncles is called a *carbuncle.* Another reason *S. aureus* is such an important pathogen is that it readily develops strains that are resistant to therapy with penicillin and other antibiotics. This can lead to particularly dangerous situations in hospitals, where penicillin-resistant strains of *S. aureus* can infect debilitated patients, causing pneumonia and septicemia. *Staphylococcus* is readily cultured, and sensitivity tests are valuable because of the considerable resistance of many strains to various antibiotics.

CHART 26–1. STAPHYLOCOCCAL INFECTIONS

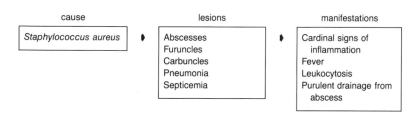

cause		lesions		manifestations
Staphylococcus aureus	▶	Abscesses Furuncles Carbuncles Pneumonia Septicemia	▶	Cardinal signs of inflammation Fever Leukocytosis Purulent drainage from abscess

Group A Streptococcal Infections

Numerous varieties of streptococci exist and are classified into groups according to antigens they possess. Group A streptococci are the most virulent because many possess enzymes that allow them to spread rapidly through tissues via lymphatic vessels—a condition known as *cellulitis*. Scarlet fever is an uncommon systemic streptococcal infection so named because the effect of one of the bacteria's toxins on blood vessels produces a scarlet rash. Acute pharyngitis (strep throat) is a common disease produced by group A streptococci. Impetigo is a superficial skin infection caused by group A streptococci (see Chapter 19). The immunologic sequelae of group A streptococcal infections in terms of production of rheumatic fever and poststreptococcal glomerulonephritis are discussed in Chapters 8 and 16.

CHART 26–2. GROUP A STREPTOCOCCAL INFECTIONS

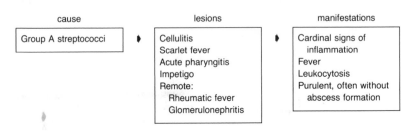

cause	lesions	manifestations

Group A streptococci	Cellulitis
	Scarlet fever
	Acute pharyngitis
	Impetigo
	Remote:
	   Rheumatic fever
	   Glomerulonephritis
		Cardinal signs of
		   inflammation
		Fever
		Leukocytosis
		Purulent, often without
		   abscess formation

Streptococcus Viridans Infection

Streptococcus viridans is part of the normal flora of the mouth and is the most common cause of bacterial endocarditis. These organisms of low virulence occasionally get into the bloodstream, where they can attach to damaged heart valves (usually following rheumatic fever) and slowly destroy the valves, resulting in congestive heart failure (Fig. 26–3). Most strains of streptococcus can be readily cultured and identified in the laboratory, and fortunately, most streptococci are susceptible to antibiotics.

CHART 26–3. *STREPTOCOCCUS VIRIDANS* INFECTION

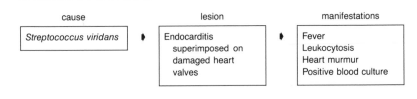

cause	lesion	manifestations
Streptococcus viridans	Endocarditis superimposed on damaged heart valves	Fever Leukocytosis Heart murmur Positive blood culture

Figure 26–3. Heart valves infected with *Streptococcus viridans.*

Streptococcus Pneumoniae (Pneumococcus) Infection

Pneumococcus, a normal member of the throat flora, is classically associated with the production of lobar pneumonia and meningitis and rarely other local or generalized infections. Pneumococcus is quite susceptible to penicillin and related antibiotics; consequently, it is not as dangerous as it was in the past, provided diagnosis is made early enough to institute proper therapy.

CHART 26–4. *STREPTOCOCCUS PNEUMONIAE* INFECTION

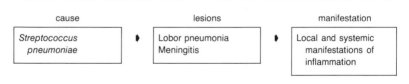

cause		lesions		manifestation
Streptococcus pneumoniae	▶	Lobor pneumonia Meningitis	▶	Local and systemic manifestations of inflammation

Meningococcal Meningitis

Neisseria meningitidis, a small gram-negative intracellular organism, causes epidemic meningitis, which has a very high mortality rate. It is transmitted via the oral route, most commonly in late winter to early spring. The disease usually starts with pharyngitis, fever, and a stiff neck. The myocarditis and disseminated intravascular coagulation caused by these organisms are often more serious than the meningitis.

Hemophilus Influenzae Infection

Hemophilus influenzae causes meningitis in small children and pneumonia in healthy individuals and is also an opportunistic invader in debilitated patients.

Legionellosis

Legionella pneumophilia is a hardy gram-negative bacteria that lives in water reservoirs and cooling units for air conditioners, but requires special media for culture. In retrospect, it was the cause of severe pneumonia in the past, but was only discovered after a severe epidemic at an American Legion convention in Philadelphia in 1976. The lesions are confined to the lung and only occur when large numbers of organisms are inhaled. The disease is more common in summer.

Enteric Bacterial Infections

Enteric bacteria are those bacteria that normally inhabit the intestinal tract of man and includes species of *Escherichia (E. coli), Klebsiella, Proteus, Pseudomonas, Salmonella,* and *Shigella.* Different strains of *E. coli* are found in virtually everyone's intestinal tract, while only a small percentage of people will harbor one or more types of the other organisms. Specific diseases produced by salmonella and shigella are discussed in Chapter 12. The rest of these bacteria can produce disease in man if allowed to invade body tissues, and all are especially pathogenic in debilitated patients. Enteric bacteria generally cause abscesses. *E. coli* is a common cause of meningitis in newborns, and certain strains can cause diarrhea (see Chapter 12). Pseudomonas commonly contaminates

CHART 26–5. ENTERIC BACTERIAL INFECTIONS

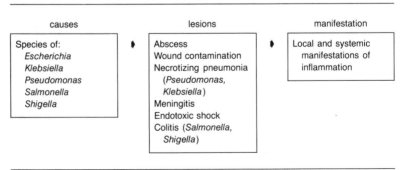

causes	lesions	manifestation

Species of:
Escherichia
Klebsiella
Pseudomonas
Salmonella
Shigella
	Abscess
	Wound contamination
	Necrotizing pneumonia
	  (*Pseudomonas,*
	  Klebsiella)
	Meningitis
	Endotoxic shock
	Colitis (*Salmonella,*
	  Shigella)
		Local and systemic
		manifestations of
		inflammation

wounds and is characteristically associated with green pus. Pseudomonas and klebsiella also cause a severe necrotizing pneumonia. Patients with a bacteremia from gram-negative enteric organisms may lapse into severe shock, which is thought to be caused by a cell wall material called *endotoxin* from gram-negative bacteria. This type of shock is referred to as *endotoxic shock* or *gram-negative shock.*

Anaerobic Bacterial Infections

It is now recognized that abscesses and wound infections may be caused by any of a wide variety of anaerobic bacteria derived from the intestinal tract, genital tract, or mouth. Anaerobes in the lower alimentary tract and in the mouth are very plentiful and outnumber aerobes by more than 1000 to 1. These bacteria cannot grow in the usual culture conditions in which oxygen is present, so they are overlooked unless anaerobic culture techniques are used. The most distinctive feature of infections due to these anaerobes is the foul smell due to the gases produced. Otherwise, the infection will appear similar to that produced by *E. coli* or staphylococcus. Some anaerobic bacteria produce powerful exotoxins; these bacteria will be discussed as a group later in this chapter.

Venereal Bacterial Infections

Venereal Disease in General

Venereal diseases are those diseases spread by genital contact, usually intercourse. However, they may also be spread by oral–genital or genital–rectal activity; oral–oral spread is exceptionally rare. The term *sexually transmitted diseases* should replace the antiquated *venereal disease* designation, but some experts insist that *venereal disease* refers to exclusive transmission via sexual activity, while *sexually transmitted disease* means that the disease is predominantly but not exclusively transmitted by sexual activity. Table 26–3 lists the sexually associated diseases according to this artificial separation.

Neisseria gonorrheae, Chlamydia trachomatis, and *Ureaplasma urealyticum* all produce venereal disease similar to that classically described for gonorrhea. Chlamydia infections are more common than gonorrhea. Ureaplasma infection may also be more common than gonorrhea, but is difficult to diagnose and study. Human papilloma virus and HIV infections are discussed under viral infection.

In large metropolitan areas, venereal disease clinics are set up to efficiently handle the diagnosis of venereal disease and to help identify sexual contacts that should also be treated to limit the

TABLE 26–3. SEXUALLY ASSOCIATED DISEASES

Venereal Disease	Sexually Transmitted Disease
Gonorrhea (bacterial)	AIDS (viral)
Chlamydial infections (chlamydial)	Trichomoniasis (protozoal)
	Candidiasis (fungal)
Ureaplasma infection (bacterial)	Herpes (viral)
Syphilis (bacterial)	Condyloma acuminatum (viral)
Chancroid (bacterial)	Mulluscum contagiosum (viral)
Lymphogranuloma venereum (chlamydial)	Nonspecific urethritis (multiple microbes)
Granuloma inguinale (bacterial)	Scabies (insect)
	Pediculosis (insect)

spread of the disease. However, venereal disease is the least well controlled epidemic disease in the United States today.

Gonorrhea

Gonorrhea, commonly called "clap," "the drip," "the whites," or "the burns," is caused by *Neisseria gonorrheae* and is the number one reported communicable disease in the United States. Gonorrhea outnumbers syphilis by 40 to 1. Gonorrhea affects men and women differently.

Males usually note burning on urination (dysuria) and purulent exudate (drip) within 3 to 5 days after exposure and seek treatment for these symptoms. Therefore, the most severe sequelae are usually avoided. In the asymptomatic male (5 to 7%) or those improperly treated, later sequelae can involve the upper urinary tract, leading to urinary strictures, or involvement of the prostate gland and the epididymus, leading to sterility.

The female seldom has symptomatic urethritis; 90 percent are asymptomatic in the acute stage of the disease. The predominant problem in the female is sterility due to fallopian tube involvement. The gonococci cross the cervical barrier during menstruation and work their way up to the fallopian tubes, where they produce infection (salpingitis) leading to scar formation. Narrowing of the fallopian tubes by scar mechanically interferes with migration of ova, although sometimes sperm can traverse the stricture, resulting in ectopic pregnancy. Another major problem in females is the continuation of purulent infection, with abscess formation at the tuboovarian juncture. Purulent gonococcal infection is commonly referred to as *pelvic inflammatory disease.*

Males and females are equally susceptible to pharyngeal gonorrhea and adult gonococcal conjunctivitis; however, females have disseminated gonococcal infection (septicemia, arthritis, and endocarditis) twice as often as males.

Diagnosis of gonorrhea is by clinical history combined with judicious application of laboratory procedures. The gram-stained smear is the test of choice in the symptomatic male. Smears are of little value in the female, the asymptomatic male, and pharyngeal gonorrhea. In these cases, culture must be attempted. Care must be taken in obtaining specimens for cultural diagnosis, as gonococcus is very susceptible to cold and drying and needs to be cultured in a carbon dioxide atmosphere.

CHART 26–6. GONORRHEA

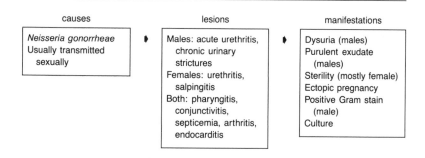

causes	lesions	manifestations
Neisseria gonorrheae Usually transmitted sexually	Males: acute urethritis, chronic urinary strictures Females: urethritis, salpingitis Both: pharyngitis, conjunctivitis, septicemia, arthritis, endocarditis	Dysuria (males) Purulent exudate (males) Sterility (mostly female) Ectopic pregnancy Positive Gram stain (male) Culture

Nongonococcal Urethritis

The minority of cases of acute urethritis in males is due to gonococcus. Most are probably due to chlamydial or ureaplasma infections. The clinical illness is the same as gonorrhea but the treatment is different. Gonorrhea responds to penicillin therapy; tetracyclines are typically used to treat chlamydial and ureaplasma infections. The diagnosis of chlamydial infection has been improved by the availability of tests for antigenic material in exudates and cell culture techniques. The laboratory diagnosis of ureaplasm infection is more difficult and not widely used.

Infections in the female are also similar to gonorrhea. During childbirth, chlamydia may infect the infant causing conjunctivitis or pneumonia.

Syphilis

Syphilis is caused by the spirochete *Treponema pallidum,* a long slender, spiral organism that is a motile bacterium. Syphilis, or the great pox, can be a devastating venereal disease. It produces chronic sequelae that in the past killed many people. Syphilis has a much lower incidence and rate of infection than gonorrhea, because there is a simple diagnostic serologic test and because no antibiotic-resistant strains have emerged. The disease is only transmitted during the first two stages of the disease (about 1 year), while gonorrhea can be transmitted indefinitely. The 3-week incubation period of syphilis allows for therapy to be initiated before the patient can spread the disease. Also, syphilis can be cured by antibiotic therapy given for other diseases, such as streptococcal sore throat, without anyone being aware of its existence.

Unlike gonorrhea, syphilis affects the male and female with equally tragic results. In addition, a woman can transmit the disease to her unborn child when she is in the chronic stage and no longer capable of transmitting the disease sexually. The frequency of congenital syphilis has decreased markedly due to serologic testing of pregnant women. The disease is not transmitted from mother to fetus until after midpregnancy, so that maternal penicillin therapy before this time prevents congenital syphilis.

Syphilis has three stages. In the first stage (primary syphilis), an ulcerated lesion called a *chancre* appears where the spirochete entered the body (Fig. 26–4A). The chancre usually develops on or about the genitalia but is not limited to that area. It is a painless sore resembling a fever blister and can be mistaken for herpes if on the lip. Sometime after the chancre appears, the serologic tests for syphilis become reactive, especially the highly specific fluorescent treponemal antibody test (FTA).

Without treatment, the chancre heals (resolves) in 2 to 3 weeks, and a period of quiescence ensues that can last from 6 weeks to as long as a year. During that time, the spirochete rests for a time, then goes into a rapid growth and multiplication phase, culminating in the production of the secondary stage of syphilis. Again, this stage is basically painless, but a syphilitic rash appears (Fig. 26–4B). The rash may mimic everything from chicken pox to ringworm. The serologic tests for syphilis are at their highest titer during this stage, and serodiagnosis is no problem. The lesions are infective, but not as infective as the chancre. If the patient is not treated, the organisms withdraw into the body for the last time, selecting single or multiple sites within the host where they remain in a dormant state for varying lengths of time (latent stage). The latent stage of the disease is the most potentially dangerous, as the possibility of tertiary syphilitic sequelae exists.

Primary and secondary stages may be mild or hidden and go undetected, only to be inferred by finding a positive serology at a later time. Of those diagnosed in the primary or secondary stage, it is estimated that about one-third would resolve spontaneously, one-third would remain serologically positive without tertiary disease, and one-third would progress to the tertiary stage. Tertiary

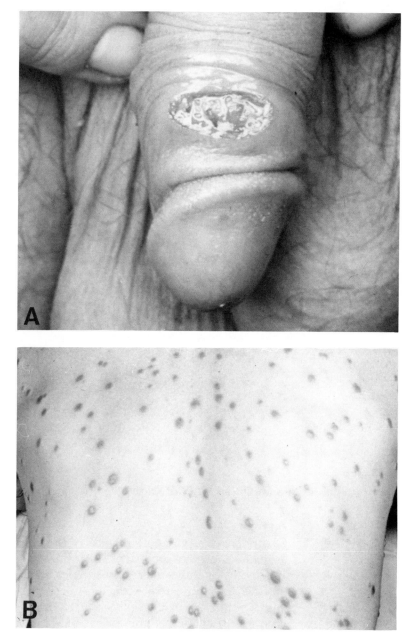

Figure 26-4. A. Syphilitic chancre of the penis. **B.** Rash of secondary syphilis.

syphilis may manifest itself in a variety of ways, involving almost every organ system of the body. Cardiovascular disease, with thoracic aortic aneurysm and secondary heart disease, is most common. Central nervous system disease, with paralysis, insanity, or blindness, is second most common. The basic lesion of syphilis is a vasculitis that commonly involves the aorta, meninges, brain, and spinal cord. Tertiary syphilis, with its severe effects, may occur as early as 3 years or as late as 25 years after the primary stage. Syphilis is diagnosed by the clinical signs or symptoms complemented by laboratory diagnostic procedures, which include darkfield microscopy to visualize the organisms scraped from a lesion, simple screening serologic tests, and specific fluorescent treponemal antibody tests.

CHART 26–7. SYPHILIS

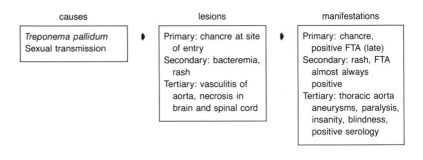

causes	lesions	manifestations
Treponema pallidum Sexual transmission	Primary: chancre at site of entry Secondary: bacteremia, rash Tertiary: vasculitis of aorta, necrosis in brain and spinal cord	Primary: chancre, positive FTA (late) Secondary: rash, FTA almost always positive Tertiary: thoracic aorta aneurysms, paralysis, insanity, blindness, positive serology

Bacterial Infections Mediated by Exotoxins

Tetanus

Clostridia are very important and feared organisms because of the exotoxins they may produce. *Clostridium tetani* is found in the soil, and because it is anaerobic, it will grow well in deep wounds where there is necrotic tissue. Toxin produced in the infected wound diffuses throughout the body, acting on the nerve–muscle junction, with resultant painful contractions of muscles throughout the body. The organisms are confined to the necrotic wound, so debridement of necrotic tissue helps in prevention of the disease. Tetanus, commonly referred to as "lockjaw," has a very high mortality rate. The best treatment of tetanus is prevention by an inoculation of tetanus toxoid to stimulate antibody production against the toxins. Every person should be immunized.

Botulism

Although *Clostridium botulinum* does not produce infection, it does secrete a powerful exotoxin that paralyzes the body's muscles by blocking the nerve–muscle junction, leading to rapid death from respiratory failure. *Clostridium botulinum* proliferates and produces toxin in improperly canned vegetables and meats. Adequate sterilization of canned food prevents the disease.

Gas Gangrene

Several types of clostridial organisms cause gas gangrene, the most common being *C. perfringens.* Clostridia are harbored in the soil and intestinal tract. When introduced into wounds containing necrotic tissue, they proliferate well in the anaerobic conditions of the dead tissue. The organisms ferment carbohydrates in the dead tissue to produce gas. The gas distends the tissue to produce compression of blood vessels, which in turn produces more tissue anoxia, tissue necrosis, and a foul smell. Clostridia also produce enzymes that diffuse through the adjacent tissues to produce more tissue necrosis, thus allowing organisms to proliferate further. Primary treatment is aimed at breaking this perpetual cycle by removal of all dead tissue (debridement). In addition, antibiotics directed against the bacteria, antitoxins directed against the necrotizing enzymes produced by the bacteria, and oxygen therapy to inhibit growth of the bacteria may be helpful. Gas gangrene most commonly occurs after severe wounds but may also occur in the uterus following traumatic nonsterile abortions. Gas gangrene is distinguished from simple gangrene by the production of the necrotizing enzymes (exotoxins). *Simple gangrene* refers to the invasion of dead tissue by a variety of saprophytic anaerobic bacteria that decompose dead tissue. Simple gangrene occurs most commonly in the legs and intestines following vascular occlusion.

CHART 26–8. GAS GANGRENE

cause	lesion	manifestations
Clostridia, especially *Clostridium perfringens*	Gaseous distention of contaminated, necrotic wounds	Bloated, dead tissue Foul smell Local and systemic signs of inflammation

Diphtheria

Corynebacterium diphtheriae produces pharyngitis with secondary toxemia. The bacterium is infected with a virus, which causes it to produce a powerful exotoxin. The toxin causes a local superficial necrotizing reaction in the pharynx called a *pseudomembrane* and is also absorbed into the blood to produce serious systemic disease. Fatal myocardial necrosis and peripheral nerve degeneration occurs in many cases, resulting in congestive failure and paralysis. Although the disease diphtheria is not as dreaded as it once was, there are still occasional outbreaks in this country resulting in loss of life. The disease can be prevented with prior immunization against the toxin.

CHART 26–9. DIPHTHERIA

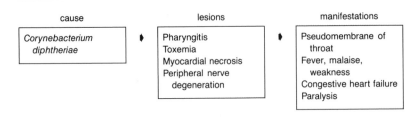

cause	lesions	manifestations
Corynebacterium diphtheriae	Pharyngitis Toxemia Myocardial necrosis Peripheral nerve degeneration	Pseudomembrane of throat Fever, malaise, weakness Congestive heart failure Paralysis

Miscellaneous Bacterial Infections

Whooping Cough

Bordetella pertussis preferentially attacks the ciliated epithelial cells lining the respiratory passages. Whooping cough is no longer as prevalent as it once was, because most children are routinely immunized. The disease derives its name from a characteristic inspiratory whoop as the infected children attempt to breathe. *Bordetella pertussis* elicits a blood lymphocytosis, which is a very unusual response for a bacterial disease.

Mycoplasma Disease

Mycoplasma organisms are bacteria-like but do not possess rigid cell walls. They are the primary cause of an infection referred to as *primary atypical pneumonia,* which may smolder for many months with low-grade symptoms such as coughing, shortness of breath, and fatigue. Primary atypical pneumonia is often referred to as walking pneumonia. Ureaplasma is a similar organism that is implicated as a cause of nongonococcal urethritis.

Plague

This very serious disease caused by *Yersinia pestis* is endemic in wild animals in certain parts of the world but rarely causes disease in the United States. The organism is capable of rapidly spreading in the body and causes extensive necrosis and hemorrhage. At certain times in history, plague annihilated much of the population of Europe.

Tularemia

Tularemia is a disease caused by *Francisella tularensis* and is contracted from a wild animal reservoir, especially rabbits. Tularemia can cause pneumonia as well as a severe systemic infection.

Typhoid Fever

Typhoid is caused by *Salmonella typhosa* but differs from other forms of salmonellosis in that the organism infects only humans and the disease is a more subacute disease than salmonella gastroenteritis. After entry into the gastrointestinal tract from contaminated water or food, the organism spreads throughout the mononuclear phagocytic system, where it grows, causes a histiocytic reaction, and produces bacteremia. The most serious complications are hemorrhage and perforation of lymphoid patches in the small intestine. Carriers harbor the organism in their gallbladder and are the source of outbreaks of the disease.

Brucellosis

Three different species of *Brucella* can cause brucellosis—*B. abortus*, *B. suis*, and *B. melitensis*. Brucellosis is a chronic systemic disease, often referred to as *undulant fever* because of the waxing and waning of febrile episodes in the patient. The bacteria are harbored in cows, sheep, hogs, and goats and can cause human disease by contact with infected meat or milk. The organisms live intracellularly in macrophages, which spill the organisms into the bloodstream every few days to cause the febrile episodes. Brucellosis is still endemic in the midwestern United States because of the large cattle and swine populations. Although the *Brucella* organisms are difficult to culture, patients with the disease will develop antibodies to the organism, which can render a presumptive diagnosis.

Leprosy

Mycobacterium leprae, a tuberculosis-like organism, preferentially invades skin and peripheral nerves, causing palpable lumps of granulomatous inflammation over the nerves. Over the course of many years, much disfigurement results from the bacterial infiltration in the skin and breakdown of tissue that has lost its nerve supply. Leprosy is uncommon but does occur in the United States, particularly in the South.

Tuberculosis

Tuberculosis is discussed in Chapter 12.

Chlamydial Infections

Chlamydia are obligate intracellular parasites, now considered to be bacteria, that lack the ability to survive in the absence of a host-derived energy system. They are susceptible to antibiotics. Various strains of *Chlamydia trachomatis*, a pathogen limited to humans, cause nongonococcal urethritis, lymphogranuloma venereum, trachoma, and inclusion body conjunctivitis. Nongonococcal urethritis has been discussed. Lymphogranuloma venereum is a less common form of venereal disease characterized by lymph node inflammation in males and, much less commonly, by anal strictures as a late lesion in females. Trachoma is spread by direct contact in areas of the world with poor living conditions. It produces a chronic progressive conjunctivitis and is a leading cause of blindness in areas where it occurs. Inclusion body conjunctivitis is a milder disease that is more widespread, but rarely leads to blindness.

Chlamydia psittisi is transmitted from birds, especially imported pet birds, to man and causes pneumonia and occasionally generalized disease.

Viral Infections in General

Viruses are particles of genetic material (DNA or RNA) coated by protein and can only grow inside host cells (*obligate intracellular parasites*). Most are invisible with the light microscope, although some viruses aggregate into inclusion bodies that then become visible in the nucleus or cytoplasm of cells. Viruses cannot reproduce (replicate) by themselves but are capable of controlling the host's cells to produce the ribo- or deoxyribonucleic acid and protein necessary for replication. Some viruses will infect many cells in the body and, after proliferating, exit from those cells into the blood, producing viremia with subsequent infection of other cells. Other viruses lay dormant in cells for years and only appear following appropriate stimuli such as stress. Still others slowly destroy cells over a period of years (Fig. 26–5).

Viruses are known to cause certain neoplasms in animals, and are strongly suspected as the cause of some human cancers. Some viruses probably live a long time in host cells without doing harm. These viruses are called *commensals* and are analogous to normal bacterial flora. Routes of viral invasion are oral, nasal, respiratory, gastrointestinal, and through bites. Most often, the exact route of viral entry cannot be determined but is assumed.

The body's response to viral infection may be inflammatory in

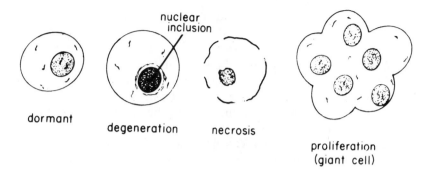

nuclear
inclusion

dormant

degeneration

necrosis

proliferation
(giant cell)

Figure 26–5. Possible reactions of cells to viral infection.

terms of vascular congestion and leakage of exudate, but pus formation seldom attends a virus infection. Immune response to viruses occurs readily, and evidence of this response is the collection of lymphocytes in infected tissues. In addition, the body's cells manufacture a product called *interferon*, which acts to prevent viral replication. Viral diseases do not respond to antibiotic therapy; consequently the treatment must be supportive, as in a head cold, or preventive by immunization. Adequate immunizations in the past have helped to almost eradicate such feared viral diseases as poliomyelitis, yellow fever, and smallpox. Immunization programs in the past few years have helped to control epidemics of influenza and curtail mumps, measles, and rubella.

Viral Diseases of Childhood

Viral diseases of childhood, including measles, rubella, varicella, and mumps, have certain attributes in common. They affect children who have no immunity to them, although they may also readily affect adults who have not had the disease. Once a person has been infected with one of these viruses, antibodies usually confer lifetime immunity. Diagnosis generally is made on the basis of the characteristic clinical picture. In cases where the exact diagnosis is in doubt, serologic tests for antibodies may be performed. Some of these viruses can also be grown in tissue cultures. However, the results of either serologic tests or culture often come too late to aid the individual patient.

Measles (Rubeola)

Measles virus is a small RNA virus transmitted by droplet infection through the mouth and nose. It is highly infectious. After an incubation period of about 10 days, the disease begins with eye pain and photophobia (excessive sensitivity to light). Within a few days, a rash ensues, which is characterized by small red bumps over the face or behind the ears, gradually spreading over the trunk and extremities (Fig. 26–6A). Measles is usually accompanied by a cough and fever, and a severe pneumonia may also result. The symptoms of measles indicate that the measles virus infects many cells in the body, with destruction of some. As the patient begins to produce antibodies, the disease slowly abates. Measles is becoming less common since the employment of measles immunization.

Rubella (German Measles)

Rubella virus is also a small RNA virus transmitted via the respiratory route and causes a very mild disease characterized by a light rash, low-grade fever, and lymphadenopathy. The importance of rubella lies in the fact that it is a major cause of congenital heart defects and other defects in the infant when acquired by the mother within the first trimester of pregnancy. In recent years, massive immunization has lessened tremendously the impact of rubella.

Varicella (Chicken Pox) and Zoster (Shingles)

Varicella–zoster is a DNA virus and is transmitted through the respiratory tract. After an incubation period of 2 weeks, chicken pox starts with malaise and fever followed by a rash that is much different than that of measles, being composed of large red bumps that develop blisters (vesicles) and later form a crust. A patient will show the rash in different stages of development on different areas of the body at the same time (Fig. 26–6B). Chicken pox runs its course in a week or two, and treatment consists only of rest and prevention of secondary bacterial infections in the rash. In some cases, after an attack of chicken pox, or even without a history of chicken pox, the same virus remains dormant in nerve ganglia for many years. If the virus becomes activated in adults, it produces zoster (shingles). In zoster, the virus spreads along the course of a cutaneous nerve causing a painful eruption in the area of skin supplied by the nerve. The difference between chicken pox and shingles, both caused by the same virus, depends on the level of immunity in the patient.

CHART 26–10. VARICELLA AND ZOSTER

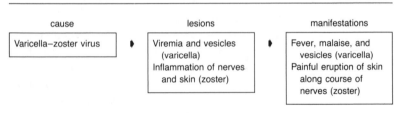

cause	lesions	manifestations
Varicella–zoster virus	Viremia and vesicles (varicella) Inflammation of nerves and skin (zoster)	Fever, malaise, and vesicles (varicella) Painful eruption of skin along course of nerves (zoster)

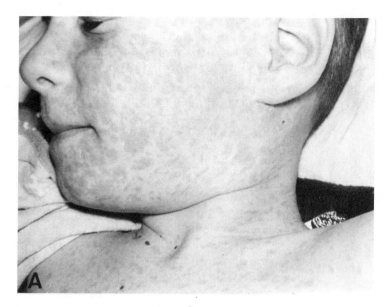

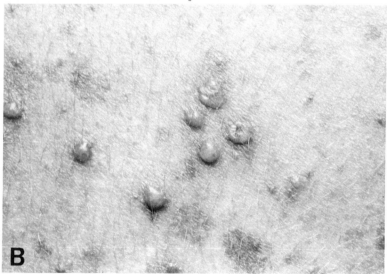

Figure 26–6. A. Macular rash of measles. **B.** Papular rash of chicken pox.

Mumps

Mumps is an RNA virus acquired by the respiratory route that preferentially infects the salivary glands and sometimes the ovaries, testes, and pancreas. After an incubation period of 2 to 3 weeks, the disease starts with malaise and fever and is usually followed by painful swelling of the parotid glands. A few male patients will have swelling and tenderness of the testes (orchitis).

Smallpox (Variola)

This disease is no longer encountered in the United States and probably has been almost eliminated from the world because vaccination has been so successful. Smallpox occupies an important place in history, as epidemics in centuries past have devastated populations. The virus is highly contagious and causes a rash much like that of chicken pox, although when it heals it leaves deep, pitted scars. Focal necrotizing lesions occur in many organs, especially lungs and intestinal tract, leading to death in many cases.

Postviral Encephalitis

All of the viral diseases thus far mentioned may cause encephalitis, which is usually mild and follows the acute phase of the disease. It is attributed to production of antibodies against the virus that cross-react with brain antigens (autoimmunity). Children who manifest this illness have symptoms of altered sensorium and are often said to be "out of their heads."

Other Common Viral Infections

Influenza

There are several strains of influenza viruses that cause a respiratory disease occurring in small or large epidemics. So-called *Asian flu* strains sweep the world every 2 or 3 years, causing disease that has a rather high mortality rate among the elderly and the debilitated. The symptoms of influenza range from those of a head cold with malaise to severe muscular aches and pains and pneumonia. It must be remembered that all the disease states that we call *flu* are not necessarily caused by the influenza virus. There are many viruses that cause various upper respiratory infections, malaise, muscle aches, and colds; some of them are well known, such as adenoviruses and rhinoviruses, but many others are nameless. These viruses do not cause large-scale epidemics of disease as do the influenza viruses. Most of these viruses will probably remain nameless for some time, because they are difficult to culture and the diseases they cause are usually of minor consequence to the individual patient. True influenza, although encompassing a spectrum of severity, can cause severe pneumonia with death of the patient. The mechanism of death is usually that of superimposed (secondary) bacterial pneumonia. Other viruses in this general category of agents that cause flu-like syndromes also cause more or less specific illnesses that can be significant; for example, one type of Coxsackie virus causes a myocarditis, and certain strains of echoviruses cause gastroenteritis.

CHART 26–11. INFLUENZA

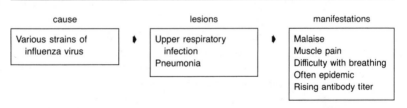

cause	lesions	manifestations
Various strains of influenza virus	Upper respiratory infection Pneumonia	Malaise Muscle pain Difficulty with breathing Often epidemic Rising antibody titer

Infectious Mononucleosis

The agent of infectious mononucleosis has now been identified as the Epstein–Barr virus. This interesting disease characteristically affects teenagers and young adults and generally starts with a sore throat and extreme tiredness. Most victims have splenomegaly and lymphadenopathy, and some may get significant liver disease. Some patients may get over the disease within a few weeks, but others are ill for many months with fatigue. It appears that the older the patient, the more devastating the disease, and physical exertion often exacerbates the symptoms. The peripheral blood in patients with infectious mononucleosis will show increased numbers of lymphocytes, many of which have atypical morphology. Serum from these patients will react on a glass slide with horse red blood cells, causing them to agglutinate. This is the basis of the mono spot test, which is used to diagnose the disease in patients with the appropriate symptoms.

CHART 26–12. INFECTIOUS MONONUCLEOSIS

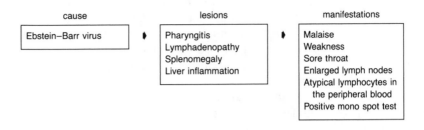

cause	lesions	manifestations
Ebstein–Barr virus	Pharyngitis Lymphadenopathy Splenomegaly Liver inflammation	Malaise Weakness Sore throat Enlarged lymph nodes Atypical lymphocytes in the peripheral blood Positive mono spot test

Herpes Simplex Infections

Herpes simplex is a DNA virus. Two serotypes (I and II) produce a variety of infections. Type I causes most of the oral and skin lesions and is transmitted by direct contact. Type II causes most of the genital lesions and is transmitted as a venereal disease. The vesicular lesions that occur on or near the lips, the most common site of type I lesions, are called cold sores or fever blisters, because they are precipitated by febrile illness or other stress such as overexposure to sunlight. After the initial infection, the virus lies dormant in nerves, particularly the trigeminal ganglion in the head, until another episode of stress causes the virus to spread down the nerve and cause recurrence of the skin lesions. Herpes virus type I occasionally causes a severe and often fatal encephalitis. Herpes simplex virus II may be transmitted during childbirth and produce a fatal disseminated neonatal infection.

Human Papilloma Virus Infections

The 50 or more genetically distinct human papilloma viruses cause or are causally related to proliferative lesions of squamous epithelium including common skin warts (verruca vulgaris), plantar warts (verruca planus), condyloma acuminata (venereal warts), laryngeal polyps, premalignant dysplasias and carcinomas of the cervix and vulva, and squamous carcinoma of skin in immunosuppressed individuals. Different strains are associated with different lesions, e.g., types 2, 4, and 7 with skin warts; 6 and 11 with venereal warts; and 16, 18, and 33 with carcinoma of the cervix. See Chapters 17 and 19 for further details on these lesions.

Cytomegalic Inclusion Disease

This disease is caused by cytomegalic inclusion virus (CMV), which is so named because infection with the virus results in giant cells having large intranuclear inclusions. The disease affects infants in utero or neonatally, causing severe brain destruction with mental retardation similar to toxoplasmosis. Cytomegalic inclusion disease is a rare systemic infection in normal adults, but a common type of infection in immunosuppressed individuals. The diagnosis is presumptively made upon finding the typical large cells with inclusions in brain, kidney, liver, lung, intestinal tract, or other organ and presence of serum antibodies.

Viral Hepatitis

This disease is discussed in Chapter 14.

Human Immunodeficiency Virus Infection

Human immunodeficiency virus (HIV) infection leads to acquired immune deficiency syndrome (AIDS), which is a disease defined as recurrent opportunistic infections, or Kaposi's sarcoma, or a high-grade lymphoma of the brain in a patient with a positive antibody test for HIV. HIV infection is defined as a confirmed positive test, but the virus is present for 3 to 17 weeks before the disease can be detected by a screening antibody test. The disease is characterized by a very long latent period of 2 to 8 years or more before AIDS actually develops. Generalized lymphadenopathy, fever, weight loss, diarrhea, and decreased CD4+ T lymphocytes may precede the development of AIDS.

AIDS was first recognized in the early 1980s and is caused by a retrovirus. It was soon recognized that HIV is transmitted both sexually and by direct inoculation through the use of contaminated needles (mainly by drug addicts) and by transfusion of blood products.

Within 2 to 12 weeks after exposure, an acute clinical illness ensues in some persons and is characterized by fever, night sweats, lymphadenopathy, rash, myalgias, arthralgias, headache, and persistent lethargy. Mood changes, irritability, diarrhea, and anorexia may also be seen. Many HIV-infected persons never develop this initial acute clinical illness. The initial illness is not considered AIDS because the patient usually recovers and has a more or less symptom-free period for years (average 8 years) before developing full-blown AIDS. AIDS is heralded by persistent generalized lymphadenopathy followed by fevers, weight loss, diarrhea, fatigue, night sweats, and encephalopathy with dementia, the latter occurring in 60 to 90 percent of all patients with clinical AIDS. Once full-blown AIDS is developed, the prognosis for survival is considered to be zero. More than half of patients with AIDS develop pneumonia due to *Pneumocystis carinii*, an organism that does not cause disease in normal adults. Other opportunistic infections include toxoplasmosis, salmonellosis, tuberculosis, candidiasis, cytomegalovirus infection, herpes simplex, and histoplasmosis and coccidioidomycosis in endemic areas.

The symptoms of AIDS and its complications are related to a deficiency in the immune system, particularly the cellular immune system. HIV infects cells by attaching to a receptor called CD4. Although found on many types of cells, the CD4 receptor is most prevalent on T helper lymphocytes. Thus, the infection of T helper lymphocytes over a period of years results in the gradual destruction of these cells, which are of paramount importance in orchestrating the immune response of other immunocompetent cells, particularly B lymphocytes. It is this insidious and persistent destruction of the cellular immune system that allows the opportunistic organisms to proliferate.

Within a few weeks to months after initial infection, HIV antibodies appear in the patient's serum. Because the standard serologic test can be falsely positive, a more expensive and more definitive test for the presence of HIV (called *Western blot*) is used to confirm a positive test. Another test that can be used to predict whether the virus is causing significant damage to the immune system is the CD4 helper T cell count—values below 400 cells per microliter are considered abnormal.

The lesions in AIDS are mostly nonspecific or related to opportunistic infections. The lymph nodes may demonstrate a simple increase in lymphocytes. The wasting resulting from the disease is manifest as a loss of adipose tissue and organ atrophy. The opportunistic infections are generally similar to opportunistic infections in patients with other serious illnesses such as cancer. In the brain, however, there is a specific lesion that is virtually diagnostic for AIDS, namely multinucleated giant cells scattered around blood vessels in gray and white matter associated with nonspecific proliferation of glial cells. Some patients with AIDS develop unusual neoplasms, especially lymphomas of unusual type with brain involvement and Kaposi's sarcoma, a neoplastic proliferation of blood vessels that is seen in a relatively high proportion of people with AIDS but rare in other people.

Millions of people in the United States and around the world harbor HIV. Homosexual men and intravenous drug users make up the largest proportion in the United States. Blood testing has greatly reduced the risk for patients with hemophilia who are routinely treated with concentrated blood products or transfusion. Heterosexual transmission is the rule in Africa and rapidly increasing in other areas of the world, especially where there are concentrations of prostitutes.

As yet there is no vaccine to protect an individual against the virus once it has been transmitted. Currently available drugs may slow the replication of the virus and prolong the latent period and severity of the illness initially. Prevention of transmission is quite possible but difficult from an educational and biologic standpoint. The primary means to prevent transmission include testing of blood products before use, elimination of the use of potentially contaminated needles by drug users, and prevention of contact between semen and superficial injuries. The virus is present in macrophages in the semen of male carriers and can gain access to male or female sexual partners through minor tears during intercourse. The transmission rate is higher with anal intercourse. Male-to-female transmission is much more likely than the reverse. Condoms provide protection most of the time, but are not always fail-safe. Intrauterine transmission to the fetus is also possible. Ordinary contact with carriers is not considered to be dangerous, and the risk of health care workers being infected by their patients is small. Transmission of the virus from health care worker to patient has been reported but is very rare.

The AIDS epidemic rages on with no signs of abatement.

Health costs for treatment, prevention, and research continue to escalate.

CHART 26–13. HUMAN IMMUNODEFICIENCY VIRUS INFECTION

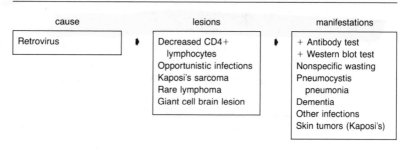

cause	lesions	manifestations
Retrovirus	Decreased CD4+ lymphocytes Opportunistic infections Kaposi's sarcoma Rare lymphoma Giant cell brain lesion	+ Antibody test + Western blot test Nonspecific wasting Pneumocystis pneumonia Dementia Other infections Skin tumors (Kaposi's)

Rickettsial Infections

Rickettsia are very small organisms intermediate between bacteria and viruses. Rickettsia must live inside cells (obligate intracellular organisms), and they preferentially attack the endothelial cells of blood vessels.

Rickettsia cause several types of serious and often fatal infectious diseases in man. Among them are typhus, in which the rickettsia are transmitted to man by the body louse, and Rocky Mountain spotted fever, in which the organism is transmitted to man by the tick. Almost all rickettsial infections cause a skin rash, which is due to small blood vessel involvement in the skin with hemorrhages from these vessels.

CHART 26–14. RICKETTSIAL DISEASE

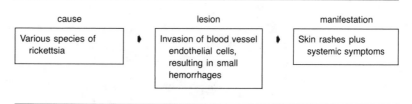

cause	lesion	manifestation
Various species of rickettsia	Invasion of blood vessel endothelial cells, resulting in small hemorrhages	Skin rashes plus systemic symptoms

Fungal Infections

Throughout the world, there are many fungi that cause disease in man. In the United States, there are several important pathogenic fungi that repeatedly cause disease and others that only rarely cause disease in healthy people but may infect debilitated patients (opportunistic fungi). An infection by a fungus is called a *mycosis* or a *mycotic infection*. Some fungi prefer to invade tissues in the yeast (bud) phase, while others proliferate as hyphae (stems) in the body. A very general way to classify fungal disease is into systemic and cutaneous forms.

Primary Systemic Fungal Infections

Certain fungi invade the interior of the body, usually via a respiratory route although other routes are possible. These fungi can be either primary pathogens or they can be opportunistic, which means that the body has some form of diminished resistance. Some of the important primary systemic mycoses are histoplasmosis, coccidioidomycosis, blastomycosis, and cryptococcosis, caused by *Histoplasma capsulatum, Coccidioides immitis, Blastomyces dermatitidis,* and *Cryptococcus neoformans,* respectively. These diseases share the common features of proliferation of fungi in the yeast phase (Fig. 26–7) which usually elicit a granulomatous response from the body. All these diseases can affect the whole body or can be concentrated in one organ. Histoplasmosis is prevalent in parts of the Mississippi and Ohio River valleys and usually causes a subclinical (inapparent) infection. Often, persons get this disease after cleaning chicken coops since the fungal agent is carried in bird

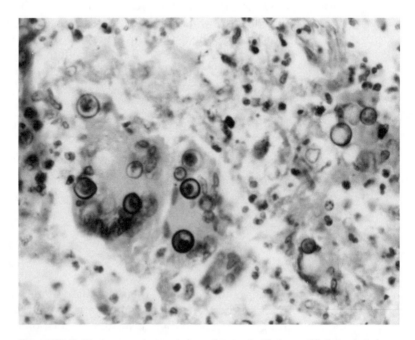

Figure 26–7. Blastomyces in yeast phase (contrast with Aspergillis in hyphal phase, Fig. 26–8).

droppings. Coccidioidomycosis is prevalent in the San Joaquin Valley in California and parts of New Mexico and Arizona. It may also cause either a subclinical or severe systemic infection. The fungal agent is harbored in the soil, and individuals have encountered the disease simply by driving through these endemic areas. Blastomycosis and cryptococcosis can be acquired in most locations. Blastomycosis causes a very destructive lung disease or may present with chronic skin ulcers. Cryptococcosis usually causes a chronic meningitis. These diseases follow a slow, protracted course, and diagnosis depends on demonstrating antibodies to the particular organism in the patient's serum by serologic tests. Treatment is with amphotericin, an antibiotic that can have severe side effects.

CHART 26–15. PRIMARY SYSTEMIC FUNGAL INFECTIONS

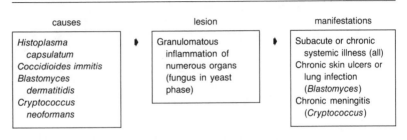

causes	lesion	manifestations
Histoplasma capsulatum *Coccidioides immitis* *Blastomyces dermatitidis* *Cryptococcus neoformans*	Granulomatous inflammation of numerous organs (fungus in yeast phase)	Subacute or chronic systemic illness (all) Chronic skin ulcers or lung infection (*Blastomyces*) Chronic meningitis (*Cryptococcus*)

Opportunistic Systemic Fungal Infections

These fungal infections usually occur with the fungus in the hyphal phase (Fig. 26–8) and usually elicit as an acute inflammatory response rather than a granulomatous one. Some of the more important opportunistic mycoses are candidiasis, aspergillosis, and mucormycosis, caused by species of *Candida*, *Aspergillis*, and *Mucor* respectively. They occur in persons with decreased cellular immune response due to cancer, other chronic infections, or drugs such as steroids. These infections may be responsible for death or may be found incidentally at autopsy. Mucormycosis and candidiasis tend to occur in diabetic patients who are out of control and in patients with leukemia. The infection with any of these fungi may be in one or many organs. Pneumonia is common. Opportunistic infections of all types (fungal, viral, parasitic) have increased in recent years, simply because the powerful drugs used to treat cancer commonly alter or suppress bone marrow and lymphatic tissues, thereby decreasing the patient's ability to produce white blood cells and antibodies.

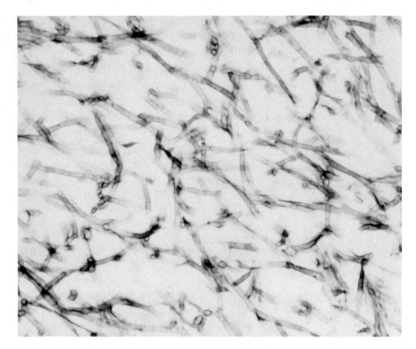

Figure 26–8. Hyphae of Aspergillus in the lung of a patient who died of leukemia.

CHART 26–16. OPPORTUNISTIC SYSTEMIC FUNGAL INFECTION

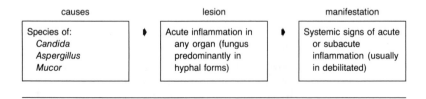

causes	lesion	manifestation

Species of:
  Candida
  Aspergillus
  Mucor
	Acute inflammation in
	any organ (fungus
	predominantly in
	hyphal forms)
		Systemic signs of acute
		or subacute
		inflammation (usually
		in debilitated)

Cutaneous Fungal Infections

These mycoses produce bothersome skin lesions that are not a serious threat to the patient. Examples of cutaneous fungi are the tineas, which cause athlete's foot, and numerous varieties of organisms that cause ringworm.

Protozoal Diseases

Protozoa are unicellular organisms. Pathogenic species are of various types, some invade and live in the blood or tissue cells, while others live on body surfaces. The major blood diseases, malaria and trypanosomiasis, are not endemic in the United States, but are major world health problems. Chagas' disease, found in

South America, is a chronic destructive disease that causes severe damage to the heart and esophagus. Leishmaniasis includes forms with involvement of the mononuclear phagocytic systems and skin, but is not found in the United States. Toxoplasmosis is widespread and will be discussed below. Of the surface diseases, trichomoniasis is a vaginal infection (see Chapter 17); amebiasis is due to an intestinal organism that invades and produces colonic ulcers and occasionally liver abscesses (see Chapter 14); and giardiasis is a diarrheal illness due to intestinal colonization.

Toxoplasmosis

Toxoplasma gondii organisms are found in the soil and are harbored in small animals, particularly the house cat. Toxoplasmosis rarely occurs in healthy adults but sometimes causes chronic inflammation as an opportunistic agent in debilitated patients or causes congenital brain infection in newborns. In the case of infants, the organism is transmitted from the mother to the fetus across the placenta, causing a severe infection with destruction of the brain. It is estimated that many retarded individuals in institutions were victims of toxoplasmosis, but the actual incidence is not known. Subclinical toxoplasmosis is prevalent; 50 percent of people in some geographic areas of the United States have antibodies to the organism.

CHART 26–17. TOXOPLASMOSIS

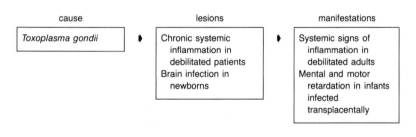

cause	lesions	manifestations
Toxoplasma gondii	Chronic systemic inflammation in debilitated patients Brain infection in newborns	Systemic signs of inflammation in debilitated adults Mental and motor retardation in infants infected transplacentally

Giardiasis

Giardia lamblia is a flagellate that lives in the intestinal lumen of man and animals such as cats and beavers. Man acquires the organism from contaminated water. Most infections are probably asymptomatic, but diarrhea and malabsorption may occur.

Helminth Infections

Helminths are primitive round- and flatworms. Most proliferate in the gastrointestinal tract. Those helminths that cause the more severe diseases are generally those that also invade tissues outside the intestinal tract.

Pinworm Infestation

Enterobius vermicularis (pinworm) can cause annoying anal itching but otherwise does not cause serious disease. These small white worms are rapidly transferred among members of a family by way of bedsheets. They most often cause symptoms in children and can sometimes be seen crawling about the anal orifice where they lay their eggs at night. Cellophane tape is used to dislodge the eggs from the anus for easy diagnosis by viewing the tape under a microscope.

Trichinosis

Trichinella spiralis causes trichinosis. The organism is acquired by ingestion of inadequately cooked meat, especially pork, because the larval stage of this worm encysts in the animal's muscles (Fig. 26–9). After humans eat the meat, the larvae mature and produce more larvae, which disseminate and encyst in the muscles of the human. Other larvae are passed by the human and end up in the garbage, where they may be eaten by hogs, thereby repeating the cycle. Just about any muscle can be affected in humans, including the heart. The brain is sometimes involved. Patients have se-

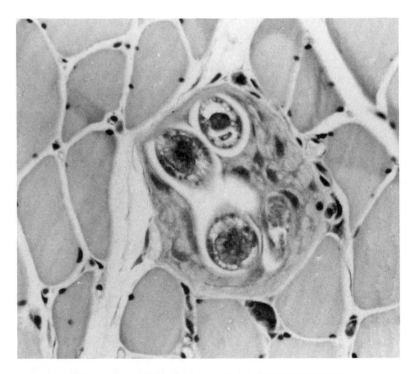

Figure 26–9. Trichinella larvae encysted in skeletal muscle.

vere muscle pain, fever, and marked eosinophilia in the blood. Occasional deaths result from involvement of heart or respiratory muscles.

CHART 26–18. TRICHINOSIS

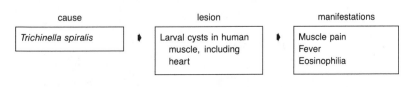

cause		lesion		manifestations
Trichinella spiralis	♦	Larval cysts in human muscle, including heart	♦	Muscle pain Fever Eosinophilia

Tapeworm Disease

Cestodes (tapeworms) include beef, pork, rat, fish, and dog tapeworms. They may cause intestinal disease in man. All are acquired by ingestion of poorly cooked meat or fish. They can cause severe abdominal pain, diarrhea, and even intestinal obstruction. The most important of these cestodes is the pork tapeworm (*Taenia solium*), which causes cysticercosis secondary to encystment of the taenia larvae (cysticerca). Unlike most of the other tapeworms, the cysticerca larvae can invade the intestinal wall and disseminate to other organs, especially the brain. The dog tapeworm (*Echinococcus granulosus*) is also capable of causing a similar larval disease (echinococcosis) in which large cysts, called *hydatid cysts,* develop in the involved organs, especially in the liver. The diagnosis of cestode infection is made by finding parts of the worms in the stool. Diagnosis of disseminated cysts in the case of dog or pork tapeworm disease follows a high index of suspicion. Patients may also have blood eosinophilia.

CHART 26–19. TAPEWORM DISEASE

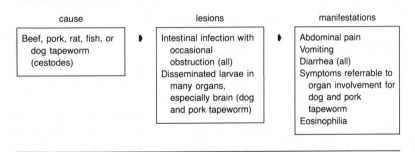

cause		lesions		manifestations
Beef, pork, rat, fish, or dog tapeworm (cestodes)	♦	Intestinal infection with occasional obstruction (all) Disseminated larvae in many organs, especially brain (dog and pork tapeworm)	♦	Abdominal pain Vomiting Diarrhea (all) Symptoms referrable to organ involvement for dog and pork tapeworm Eosinophilia

Ascariasis

Ascaris lumbricoides is a large roundworm measuring up to 30 centimeters long. The worm is acquired by ingestion of ova present in soil contaminated by feces. Larvae enter the blood vessels and migrate through the lungs, where they can cause significant hemorrhage and inflammation. They are then coughed up and swallowed. After maturing in the intestines, adult worms can cause intestinal obstruction.

Trichuriasis

Trichuris trichuria (whipworm) is fairly common in some parts of the United States. The eggs are ingested from fecally contaminated sources and mature into adults in the lower intestinal tract. Heavy infestations of these worms may cause abdominal pain due to distention and anemia due to bleeding.

Filariasis

This is a tropical disease caused by several species of filarial worms that are acquired by mosquito bites and invade and obstruct lymphatics, causing swelling of the affected part of the body. This swelling may be massive and is called *elephantiasis*.

REVIEW QUESTIONS

1. What are the body's major defenses against infection by microorganisms?
2. What are the classes of microorganisms that produce infection?
3. What are the major types of bacterial and viral infections?
4. What are the local and systemic signs of inflammation secondary to infection?
5. How are each of the following tests or procedures helpful in the diagnosis of infectious disease?
   Culture
   Gram-stained smear
   Skin tests
   Serologic tests
   Body temperature
   White blood cell count
   Blood eosinophil count
6. How does the laboratory diagnosis of a bacterial infection differ from that of a viral infection?
7. What are the causes, lesions, and major manifestations of each of the specific infectious diseases discussed in this chapter?
8. What are the common diseases caused by staphylococcus? By streptococcus?

9. How do gonorrhea and syphilis differ in terms of chronicity, complications, and means of diagnosis?
10. Which harmful bacterium does not produce an inflammatory response?
11. How are viral diseases controlled?
12. What are the possible complications of each of the viral diseases of childhood?
13. How does influenza differ from various other illnesses labeled as *flu?* What are some of the causes of these other flu-like illnesses?
14. How do primary and secondary systemic fungal infections differ in terms of common organisms causing each, condition of the respective hosts, and type of inflammatory response elicited by each?
15. Why are pork and dog tapeworm infections more serious than other tapeworm infections?
16. What do the following terms mean?
    Saprophyte
    Normal flora
    Pathogen
    Virulence
    Infection
    Anaerobic
    Vector
    Carrier
    Primary infection
    Secondary (opportunistic) infection
    Septicemia
    Exotoxin
    Pyogenic
    Endotoxin
    Endotoxic shock
    Chancre
    Fluorescent treponemal antibody test (FTA)
    Lockjaw
    Primary atypical pneumonia
    Undulant fever
    Obligate intracellular parasite
    Commensal
    Shingles
    Flu
    Yeast
    Hyphae
    Cysticercosis
    Elephantiasis

Infections Manifest Primarily as Delayed Hypersensitivity Reactions
Graft Rejection
Autoimmune Diseases
Systemic Lupus Erythematosus (SLE)
REVIEW QUESTIONS

REVIEW OF STRUCTURE AND FUNCTION

Immunity is a term that means the resistance to or protection from an individual's environment. Certain forms of protection, such as the skin and the inflammatory response, have been discussed in other chapters. The immune system is an internal chemical system whose purpose is to enhance reactivity to material that is foreign to the body. Material recognized as foreign by the immunologic system is called an antigen. Most antigens are introduced from outside of the body, but some are altered endogenous materials that are treated as if they were foreign. The immunologic system recognizes antigens by producing *antibodies* or specialized lymphocytes specific for each antigen (Fig. 27–1). Large foreign particles, such as bacteria, contain several antigens and may elicit the production of several different antibodies. Also, several different foreign substances may contain the same antigen; for example, some persons have the same antigen in heart muscle that exists in group A streptococci.

Antigens are classified as either complete or incomplete. Complete antigens both induce immune responses and react with the antibodies produced by the immune response. Incomplete antigens can react with antibodies but cannot induce an immune response unless chemically coupled to another antigen. Incomplete antigens are also called *haptens*.

Antibodies are produced within the cytoplasm of plasma cells, which in turn are derived from lymphocytes. Antibodies produced by plasma cells are released into the blood, where they circulate freely as part of the gamma-globulin fraction of serum proteins and

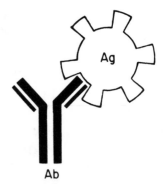

Figure 27–1. Schematic representation of antigen (Ag) and antibody (Ab).

are called *immunoglobulins*. The lymphocytes that are capable of developing into plasma cells to produce immunoglobulins are called *B lymphocytes* or *B cells* after the bursa of Fabricius of chickens. In chickens, the bursa is the primary site of development of B cells. In man, B cells are thought to originate from lymphoid tissue of the alimentary tract and/or bone marrow.

A second type of lymphocyte is the *T lymphocyte* or *T cell,* so named because its production is programmed by the thymus in both humans and lower animals. T lymphocytes and B lymphocytes are similar morphologically in ordinary tissue sections; specialized tests are required to distinguish them. They are both capable of recognizing specific antigens.

There are several known functions of T cells and subtypes of T cells are named to reflect their function. *T-helper cells* physically deliver information about the antigens to B cells in order to aid B cells in the production of antibodies. *T-suppressor cells* suppress B cells in order to prevent the production of excess antibody; they also prevent the production of antibodies to the body's own tissues. *Cytotoxic T cells* can directly kill other cells that process foreign or altered antigens, such as neoplastic cells or cells infected with viruses. Another class of lymphocytes, *natural killer cells* (NK cells) destroy other cells in the absence of any known antigenic stimulation, possibly by reacting with glycoproteins on the target cells' surface. It is hypothesized that the body continually develops neoplastic cell lines but NK cells destroy these neoplastic cells before they can accumulate to form neoplasms. This type of police action by NK cells is termed *immune surveillance.*

Antigens are usually broken-down (processed) by macrophages and then presented to T cells. Once a T cell with a specific receptor for a particular antigen receives the antigen, it will either help program a group of B cells to produce antibody, suppress the production of antibody, or directly kill a foreign cell, depending upon whether it is a helper, suppressor, or cytotoxic T cell. The concept of specific antigenic receptors on T cells implies that there must be countless different T cells in the normal body in order to accommodate the almost infinite number of potential antigens. This indeed appears to be the case and is a very functional aspect of immunity, because, if the possibility of T cell cross-reactivity with different antigens were great, then the possibility of self-reactivity would also be great, resulting in autoimmune phenomena.

In addition to the regulatory mechanisms already outlined, T cells work in cooperation with B cells in other ways to increase the specificity of the immunoglobulin to the antigen. T cells also elaborate nonimmunoglobulin proteins, collectively called *lymphokines,* that may directly lyse an antigen-bearing cell or recruit other lymphocytes or macrophages to help with this reaction.

About 10 days after the first encounter with an antigen, antibodies become detectable. Subsequent encounters with the same antigen are associated with a more rapid production of antibodies called the *secondary response.* Certain lymphocytes serve as *memory cells,* which means that they are long-lived, and are programmed to proliferate rapidly after a second encounter with that antigen.

Immunoglobulins carry out their function in several ways, depending on the structure of the antigen and the class of antibody involved. Immunoglobulins are divided into five classes (IgA, IgG, IgM, IgD, and IgE) based on their biologic properties and major differences in their protein structure. IgG and IgM, the most common, are schematized in Figure 27–2.

IgG, the most abundant immunoglobulin, can combine with antigens such as bacterial exotoxins to neutralize their activity, or it can adhere to antigen on the surface of larger foreign materials such as bacteria to promote their phagocytosis by leukocytes. IgG is particularly important to infants, who because of their age have an immature immune system. IgG antibodies are transmitted across the placenta to the fetus from the mother; thus, the infant is protected during its first 6 months of life from diseases previously

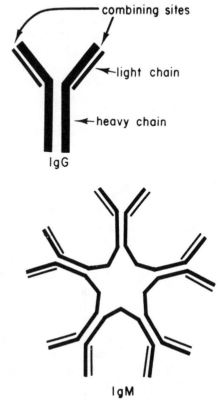

Figure 27–2. Basic structures of immunoglobulins G (IgG) and M (IgM). IgM is the equivalent of five IgG molecules.

encountered by the mother. The combination of IgG with antigen sometimes employs another reaction, called *complement activation.* Complement is a group of special serum proteins that often take part in antigen–antibody reactions. Activation of complement is a complex series of enzymatic reactions that has at least three important effects: proteins are produced that may lyse (rupture) cells, mediators are released that may in turn cause release of histamine from mast cells, and chemotactic agents are produced that may attract leukocytes. The first effect can destroy cells that are recognized as foreign; the other two effects initiate the vascular and cellular components of acute inflammation, respectively.

IgM antibodies are noteworthy for their large size, which is five times that of IgG antibodies (Fig. 27–2). IgM antibodies do not readily pass into the tissues or across the placenta. They develop more quickly than IgG antibodies following antigenic stimulation, and they are important in controlling bacteria that enter the bloodstream and in agglutination (clumping) of large foreign substances such as incompatible red blood cells. IgM antibody reactions may also activate complement.

IgA is noteworthy because it is secreted into body fluids such as tears, saliva, bronchi, and the intestinal tract, where it may interact with antigens before they can enter the body's tissues.

IgE acts in a very specific manner. IgE becomes attached to basophils in the bloodstream and mast cells in tissues. When antigen reacts with IgE on the surface of basophils or mast cells, the cells release vasoactive substances such as histamine. For unknown reasons, IgE reactions are often associated with increased numbers of eosinophils in tissue and blood.

IgD serves as an antigen receptor on the surface of mature B cells (along with IgM).

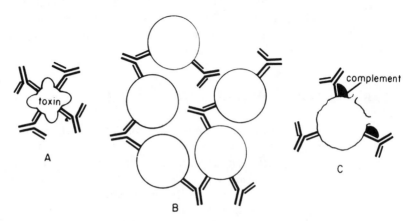

Figure 27–3. Neutralizing (**A**), agglutination (**B**), and lysing (**C**) actions of immunoglobulins.

Thus, immunoglobulins can protect the body from antigenic foreign materials in a variety of ways, including precipitating or neutralizing the chemical action of small foreign materials, agglutinating larger foreign materials so they can be phagocytosed, and lysing foreign cells (Fig. 27–3). Immunoglobulins can also initiate an acute inflammatory reaction to control the spread of and destroy certain foreign agents, and they can combine with foreign substances on body surfaces to prevent their entry into the body.

The major histocompatibility system genetically controls the expression of both self- and foreign antigens on cell surfaces of most cells in the body. This system in man is located on genes in the sixth chromosome and is very similar to the ABO system in which different individuals express different erythrocyte surface antigens. Just as it is important to insure that transfused blood is given to recipients who are of the same ABO type as the donor, likewise, organ transplants such as kidney, skin, or bone marrow must share similar major histocompatibility antigens between donor and recipient. If the antigens are not similar, the recipient's immune system will recognize a transplanted kidney, for example, as foreign and produce antibodies and cytotoxic T cells to destroy the kidney. Conversely, donor lymphocytes, transplanted with an organ into an incompatible host, can, under certain circumstances, mount an attack against the recipient's tissues. This phenomenon is referred to as *graft vs. host disease.*

T cells cannot process antigen unless it is presented to the T cells in conjunction with products of the histocompatibility system. For example, a macrophage presenting antigen to a T cell must carry on its surface with the foreign antigen, certain major histocompatibility antigens which the T cells can recognize as similar to their own. Otherwise, the T cell cannot recognize the antigen as foreign. This is referred to as major histocompatibility complex restriction. The reasons for this restriction are not entirely known, but it must serve as a link in programming B cells to produce antibodies.

CLASSIFICATION OF IMMUNOLOGIC DISEASES

Immune reactions are generally protective and helpful. However, in certain circumstances, they can be more harmful than helpful, in which case they produce disease. Immune diseases fall into two major categories: *immune deficiency diseases*, in which there is too little response to foreign agents, and allergy or hypersensitivity reactions, in which there is too much response to antigens. The term *hypersensitivity* is slightly broader than the term *allergy* and may include some exaggerated responses that are not antigen–

antibody mediated or are of unknown cause. Hypersensitivity reactions to the body's own components mediated by the immune system are called *autoimmune diseases.*

Immune deficiency diseases are subdivided into inherited and acquired forms and into deficiencies of the T-cell and B-cell immune systems or both.

The four major types of allergy (hypersensitivity) are divided on the basis of mechanisms. The first three types involve immunoglobulins, and the fourth type involves T lymphocytes.

Type I, called anaphylactic–atopic allergy, includes those reactions mediated by IgE antibody and involving release of vasoactive chemicals from tissue mast cells or blood basophilic leukocytes (Fig. 27–4). The reaction occurs within minutes; hence, it is sometimes designated as immediate. It is virtually the same reaction as the initial stage of the acute inflammatory reaction.

Type II, called *cytotoxic-type hypersensitivity,* involves destruction of host cells either by agglutination and phagocytosis or by lysis of the cell membrane as a consequence of complement fixation (Fig. 27–5). Red blood cells are commonly targets of this type of hypersensitivity, and the antigens are either the red blood cell membrane itself or a foreign chemical, such as a drug, that adheres to the red cell membrane as a hapten. The reaction may be either immediate or prolonged.

Type III, called *immune complex* or *Arthus-type hypersensitivity* is defined as a complement-mediated reaction to precipitates of antigen and antibody (antigen–antibody complexes). The antigen–antibody complexes lodge in vessel walls, and the activation of certain inflammation-inducing components of complement (C3a,

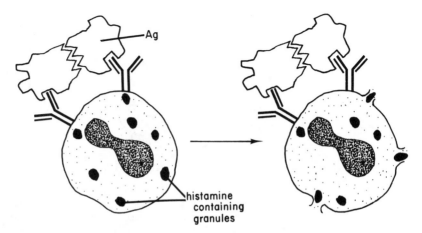

Figure 27–4. Pathogenesis of type I allergic reaction. Antigen reacts with antibody attached to mast cells by cross-linking two IgE molecules (or receptors) to cause release of histamine-containing granules.

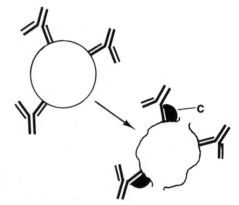

Figure 27–5. Type II allergic reaction with lysis of red blood cells. Antibody reacts with cell surface antigens, resulting in complement fixation (c) and cell lysis.

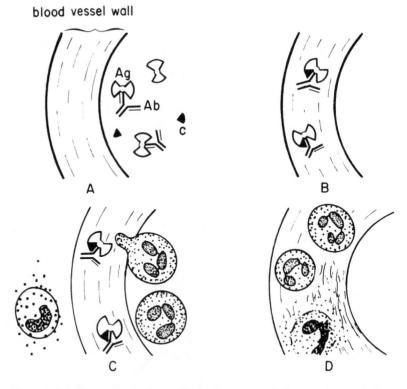

Figure 27–6. Pathogenesis of type III allergic reaction. Antigen and antibody form complexes (**A**) in the presence of excess antigen. The complexes are deposited in the vessel wall and fix complement (**B**). Activated components of complement cause release of vasoactive materials from mast cells and chemotaxis of polys (**C**). The end result is acute inflammation and destruction of the vessel wall (**D**).

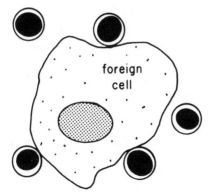

Figure 27–7. Pathogenesis of type IV allergic reaction. Sensitized T lymphocytes attack foreign cells.

C5a) results in release of vasoactive substances from mast cells and attraction of polymorphonuclear leukocytes to the site (Fig. 27–6). The acute inflammatory reaction takes several hours to develop and may be prolonged if the amount of antibody is small. Antigen excess over antibody is required to produce the reaction, because complete binding of all antigens by antibodies inhibits complement activation.

Type IV, called *delayed hypersensitivity* or *cell-mediated hypersensitivity,* is the harmful destruction of tissue by T lymphocytes and macrophages (Fig. 27–7). Lymphocyte-mediated reactions are slow to develop, usually requiring 1 to 2 days to reach a peak, hence the name *delayed hypersensitivity.*

There are numerous other allergic diseases that cannot be subclassified on the basis of mechanism because the mechanism is not known. Many autoimmune diseases and drug reactions fall into the unclassifiable category.

The principle criteria that separate allergic from nonallergic reactions to foreign substances are as follows:

1. The initial reaction in an allergic disease requires approximately 10 days, the time required for antibody production; subsequent responses will recur immediately for immunoglobulin-mediated allergies and in 1 to 2 days for cell-mediated allergies. The reaction time for nonallergic responses to foreign materials will be constant and dependent on the nature of the offending agent rather than on an allergic mechanism. For example, chemical injuries may be immediate, inflammatory reactions may take several hours, and hyperplastic responses may take days to weeks.

2. Allergic diseases tend to occur only in selected susceptible individuals, while other foreign substances tend to affect people without bias. Multiple genetic and environmental factors appear to be involved in the selected susceptibility

to allergens (antigens). Most people have the protective responses, but few have the harmful responses to antigens.

3. In allergic diseases there is usually a poor relationship between dose of allergen and severity of the reactions, whereas with other agents, the dose–injury relationship is usually quite predictable.

MOST FREQUENT AND SERIOUS PROBLEMS

Of type I hypersensitivity reactions, atopic allergies are most frequent, affecting about 10 percent of the population. The most common example is hay fever (allergic rhinitis). Atopic diseases are chronic reactions, because they are caused by exposure to antigens in the environment, e.g., ragweed pollen, house dust, or animal danders. Allergic asthma, another atopic reaction caused by the same antigens, affects approximately 3 percent of the population. Anaphylactic reactions are acute reactions following a distinct reexposure to an antigen. Urticaria (hives) is a common form and may be the result of food allergy or drug reaction. A more severe and immediately life-threatening form of type I hypersensitivity is anaphylactic shock, a condition in which vasoactive substances are released systemically. Drugs and insect stings are the most common causes of anaphylactic shock.

The most common types of cytotoxic (type II) hypersensitivity reactions are transfusion reactions, in which antibodies to transfused blood cells are present in the recipient, and erythroblastosis fetalis, in which maternal antibodies cross the placenta to cause destruction of the infant's red blood cells.

Immune complex hypersensitivities (type III) are common and very complicated and vary considerably in terms of exact mechanisms, tissue involved, and severity. Frequent types are serum sickness, which is usually a drug reaction, and some types of glomerulonephritis. Polyarteritis nodosa is a rare but potentially fatal immune complex hypersensitivity. Rheumatoid arthritis and lupus erythematosus are probably immune complex hypersensitivities.

Allergic contact dermatitis is the most common type of delayed hypersensitivity (type IV) reaction, and poison ivy is the most common specific cause. Graft rejection, following a skin graft or renal transplant, are examples of delayed hypersensitivity in which the donor and recipient tissue antigens are incompatible.

Immune deficiencies are much less common than allergies. Inherited immune deficiencies, whether of the B cell or T cell immune systems, are rare but particularly significant because they lead to repeated, severe infections in the first few years of life and are often fatal. Acquired immune deficiencies are more common than inherited forms and vary in severity from mild to severe. Causes include disseminated hematopoietic diseases such as leukemia or lymphoma, cancer chemotherapy, and immunosuppressive drug therapy, which is used to protect transplanted organs from being rejected as foreign tissue. Acquired immune deficiency syndrome (AIDS) is increasingly frequent, and due to a virus.

SYMPTOMS, SIGNS, AND TESTS

The clinical manifestations of immune disease are quite varied and often unique to the specific types of diseases as well as to the organ system involved. Consequently, clinical manifestations will be discussed in relation to specific diseases.

Many laboratory tests have been devised to aid in the diagnosis of immune diseases. The amount of gamma globulin in the blood can be measured by *electrophoresis of serum* (Fig. 27–8). The five specific types of immunoglobulin can be measured by *immunodiffusion* tests in which prepared antibodies are reacted against IgG, IgA, IgM, IgE, and IgD. The reaction product precipitates in a gel medium and can then be stained and measured. These tests are most important in the evaluation of relative and absolute immune deficiencies.

The components of complement can also be measured by immunodiffusion. A decreased level of complement is inferential evidence that a type II or III reaction has taken place.

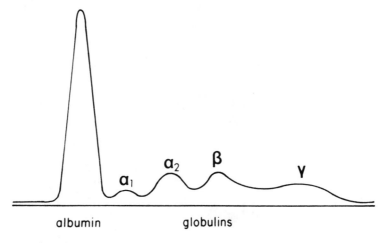

Figure 27–8. Normal serum proteins as measured by serum electrophoresis test.

The most important tests for diagnosing type I anaphylactic–atopic allergies are *skin tests,* in which a suspected antigen, such as a drug, house dust, cat hair, seafood, dander, or ragweed pollen, is injected into the skin. Allergy to a specific antigen is manifested immediately as a small inflammatory reaction at the site of injection. Hundreds of different antigens have been isolated and can be used for such skin tests.

Cytotoxic (type II reactions) antibodies on erythrocytes are demonstrated by the *direct Coombs' test* in which antihuman gamma globulin is added to a test tube of washed erythrocytes. If the erythrocytes have been coated with cytotoxic or other antibodies, they will agglutinate upon addition of the antihuman gamma globulin (Fig. 27–9). The *indirect Coombs' test* will detect serum anti-

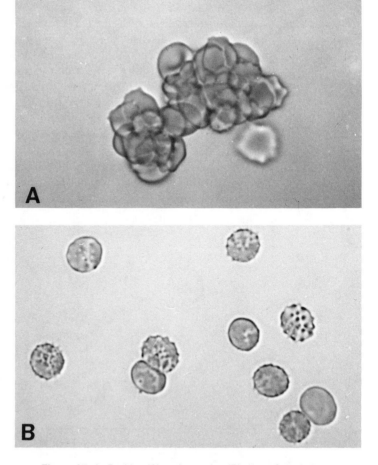

Figure 27–9. Positive (**A**) and negative (**B**) direct Coombs' test.

bodies having the potential to react with erythrocytes. In this test, the patient's serum is first incubated with another person's erythrocytes in a test tube to allow the antierythrocyte antibodies to coat the erythrocytes. The test then proceeds in the same fashion as the direct Coombs' test. Similar tests are less commonly performed to demonstrate antibodies to white cells or platelets.

Immune complex (type III) diseases are often diagnosed on the basis of history plus the clinical picture of the disease. However, the deposition of antigen–antibodies and complement in tissues may be tested for by *immunofluorescence*. In this test, a section of tissue (biopsy) from a diseased organ is overlain with fluorescent-tagged antibodies to specific immunoglobulin or complement. The fluorescent-tagged antibody will form a complex with the deposited immunoglobulin or complement (usually in vessels) and fluoresce a bright yellow-green color under special blue light (Fig. 27–10). This test is used routinely on kidney biopsies in order to separate the various types of glomerulonephritis.

Delayed (cell-mediated) hypersensitivity to microorganisms such as occurs in tuberculosis and fungal diseases can be evaluated

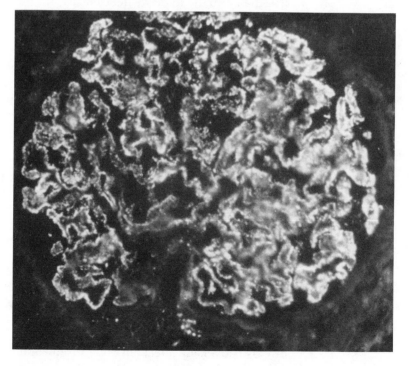

Figure 27–10. Antibodies in renal glomerulus demonstrated by immunofluorescence.

by skin tests in which the antigen (killed microorganisms) is injected into the skin and a delayed (48 to 72 hours) inflammatory reaction ensues. Contact dermatitis is most often diagnosed by good detective work, but suspected allergens may be tested by placing them on the skin (patch test).

Autoimmune diseases are so varied that most are difficult to diagnose as such. An exception is systemic lupus erythematosus (SLE) in which a fairly specific test, the *antinuclear antibody test* (ANA), has been devised. Persons with SLE have antibodies directed against the nuclei of their own cells, and this is demonstrated by the ANA test. In rheumatoid arthritis, the laboratory demonstration of the *rheumatoid factor* (RF) may aid in diagnosis.

SPECIFIC DISEASES

Immune diseases are the result of excessive or deficient antibody or cellular immune response. They can take many forms including developmental, inflammatory, and neoplastic conditions involving various organ systems. Specific diseases in this chapter will be classified into the immune deficiencies and the four basic types of allergy.

Immune Deficiency Diseases

Immune deficiencies are manifest by an increased susceptibility to infections, i.e., more frequent and severe infections. If the deficiency is in the immunoglobulins (B cell system), infections produced by pyogenic bacteria, such as pneumonia, will occur. If the deficiency is in the cell-mediated immune system, infections produced by a variety of weak pathogens, including bacteria, fungi, viruses, and protozoa, will occur. Inherited deficiency of the immunoglobulin system can be substantiated by finding very low levels of serum gamma globulin (*agammaglobulinemia*). In acquired forms, the levels are not usually as low (*hypogammaglobulinemia*). Deficiencies of the T cell immune system are demonstrated by absence of skin reactivity to substances that commonly cause a delayed hypersensitivity reaction. A delayed hypersensitivity skin test is performed by placing the antigen in or on the skin and checking in 1 to 2 days for the typical raised, firm delayed hypersensitivity reaction. A person with a positive skin test for tuberculosis may later revert to a negative skin test following an acquired deficiency of T cell-mediated immunity. T cell deficiencies can also adversely affect antibody production if the deficiency involves T-helper cells.

Acquired Immune Deficiency Syndrome (AIDS)

Acquired immune deficiency syndrome (AIDS) is now the most important and severe form of acquired immune deficiency. It is discussed in Chapter 26.

Anaphylactic–Atopic Allergies (Type I)

Anaphylactic–atopic allergies are caused by the release of chemicals called vasoactive amines. These substances include histamine from mast cells or blood basophils, which produces the most immediate reaction, and other substances which cause slightly more delayed reactions. As a group, these substances produce the following effects: (1) contraction of most nonvascular smooth muscle, producing effects such as bronchial constriction leading to asthmatic breathing; (2) vasodilation, which locally leads to increased blood flow and systemically may lead to shock; (3) increased vascular permeability, which leads to edema such as is seen in the raised wheals of urticaria of the skin or the swollen nasal mucosa of hay fever; and (4) stimulation of secretory activity of some glands, such as the increased mucus secretion in the bronchus in asthma and increased nasal secretions in hay fever. A typical laboratory finding in this group of diseases is an increase in eosinophils in the blood or tissue. Thus, eosinophilia suggests atopic allergy, but it also may be found in some immune complex diseases. Potential atopic allergens may be evaluated by skin testing. A raised edematous lesion with red border (wheal and flare) reaches a maximum about 15 minutes after intradermal injection of the allergen being tested.

Allergic Rhinitis (Hay Fever) and Asthma

Because many allergens are airborne, the respiratory tract is a common site for hypersensitivity reactions to them. The portion of the tract affected presumably reflects individual differences of unknown nature. Both rhinitis and asthma can be triggered by non-allergic mechanisms in susceptible individuals. Asthma involves the lungs and produces three effects: (1) bronchoconstriction, resulting from the contraction of the smooth muscle layers of bronchial and bronchiolar segments of the tract; (2) edema, resulting from vasodilation and increased permeability of bronchial vessels; and (3) increased secretion of thick, tenacious mucoid material. If secretions are not removed by expectoration, their accumulation can impede air flow. The chief mechanical difficulty experienced in bronchial asthma is increased resistance to air flow manifested by wheezing. Generally, attacks are episodic, but the occasional asthma patient may experience a persistence of symptoms for 24 hours or more and fail to respond to medication. This condition of prolonged and unresponsive asthmatic distress is termed *status asthmaticus.* The patient exhibits very labored breathing with great respiratory effort, dyspnea, harassing cough, and sometimes cyanosis. Anxiety is great, and sleeplessness, extreme fatigue, exhaustion, dehydration, and disorientation develop.

When allergic rhinitis is of short duration (days to a few weeks) and seasonal, the term *hay fever* is used. The allergens of hay fever are seasonal plant pollens including ragweed pollen (late summer and early autumn) and tree and grass pollens (spring and early summer). Ubiquitous allergens may produce a chronic condition lasting the year round that may be aggravated by factors such as high humidity, irritating vapors, and upper respiratory tract infections. Allergic rhinitis is marked by edema and hypersecretion by the mucosal lining of the nasopharyngeal cavities that produces partial blockage of the airways and intense nasal and postnasal discharge. The involved nasopharyngeal mucosa appears pale and swollen. Edema may affect the mucosa of the paranasal sinuses, reducing their drainage, and it may close the eustachian tube. Secondary infection and inflammation of the sinuses (sinusitis) and the middle ear (otitis media) may result. After many years, allergic nasal polyps may develop. These are masses of redundant edematous mucosa, which may obstruct breathing, and occur more frequently with nonallergic (intrinsic) rhinitis and asthma.

CHART 27–1. ASTHMA

cause	lesions	manifestations
Type I allergy	Spasm of bronchi Thickened bronchi with increased mucus and eosinophils	Wheezing Dyspnea Cough Positive skin test to allergen (some cases)

CHART 27–2. ALLERGIC RHINITIS

cause	lesions	manifestations

Type I allergy	Edematous nasal mucosa with eosinophils
	Nasal polyps (late)
		Nasal discharge
		Partial obstruction
		Complicated by:
		  Sinusitis
		  Otitis media
		  Nasal polyps
		Positive skin test to allergen

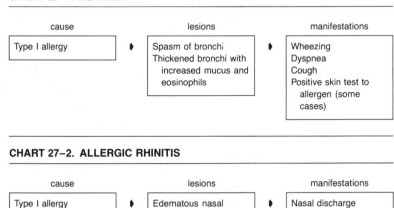

Urticaria (Hives) and Angioedema

Urticaria is a type I allergy that is recognized on the skin by slightly raised, flat, well-demarcated, edematous patches with a congested border (wheal and flare) (Fig. 27–11). Urticaria develops rapidly after exposure to an allergen and is associated with pruritus. Urticaria may be caused by an anaphylactic reaction in the skin to allergens that may have been introduced into the skin (injected drugs or insect stings) or, more often, by allergens that have been ingested and distributed throughout the body after alimentary absorption. A great variety of foods are known to cause urticaria, seafoods, strawberries, and tomatoes being common examples. Some contain histamine or histamine-releasing substances, thus causing urticaria by a nonspecific rather than an IgE-mediated mechanism. Mosquito bites cause a wheal due to nonspecific irritants in the saliva of the mosquito, whereas, stinging insects (mainly *trymenoptera*) inject allergens into the skin with rear stingers rather than mouth parts. Most chronic urticaria is idiopathic. *Angioedema* is a more extreme skin manifestation of immediate hypersensitivity than urticaria. It also is an edematous eruption, but it is more widespread and involves the deep dermis. Often it affects the lips, tongue, face, or even the pharynx, perhaps blocking the airway. Its causes are similar to those of urticaria.

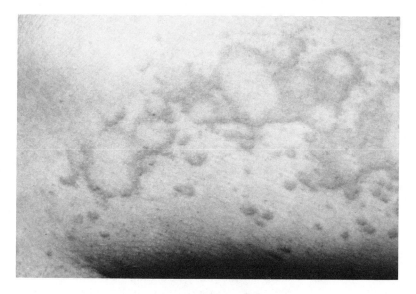

Figure 27–11. Urticaria.

CHART 27–3. URTICARIA AND ANGIOEDEMA

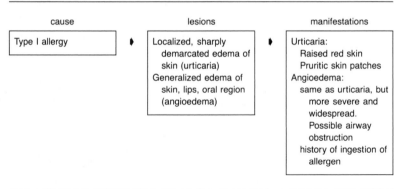

cause	lesions	manifestations
Type I allergy	Localized, sharply demarcated edema of skin (urticaria) Generalized edema of skin, lips, oral region (angioedema)	Urticaria: Raised red skin Pruritic skin patches Angioedema: same as urticaria, but more severe and widespread. Possible airway obstruction history of ingestion of allergen

Systemic Anaphylaxis

This is one of the true medical emergencies. Within seconds to minutes after exposure to the allergen, the patient feels an itching of the scalp, tongue, and throat followed by generalized flushing and headache. Difficulty in breathing begins and is joined shortly thereafter by precipitous drop in blood pressure and body temperature. Shock and loss of consciousness occur within a short time. If early reversal is not instituted, the train of events may lead to death from shock within 15 minutes of allergic exposure. Treatment is immediate subcutaneous administration of epinephrine, which causes vasoconstriction, thereby reversing systemic shock. The more common allergens that cause anaphylaxis are poliens, foods, chemicals, venoms from stinging insects, foreign sera such as diphtheria or tetanus antitoxins, and drugs such as penicillin.

CHART 27–4. SYSTEMIC ANAPHYLAXIS

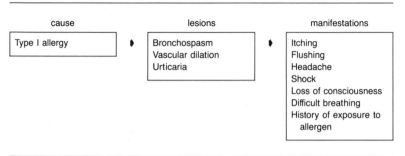

cause	lesions	manifestations
Type I allergy	Bronchospasm Vascular dilation Urticaria	Itching Flushing Headache Shock Loss of consciousness Difficult breathing History of exposure to allergen

Gastrointestinal Food Allergies

Primary allergic reactions in the gastrointestinal tract are less common than skin reactions. A gastrointestinal reaction begins shortly after eating specific foods to which the person is allergic. Symptoms include diarrhea, vomiting, and cramps. Shellfish contain common allergens that can produce this reaction.

Cytotoxic Hypersensitivities (Type II)

Cytotoxic hypersensitivity reactions are usually manifested by low levels of blood cells, because this type of reaction is often directed against blood cells. Antibodies may form against red blood cells, platelets, or white blood cells and produce anemia, thrombocytopenia, or leukopenia, respectively. The direct Coombs' test is used to detect antibodies on the surface of red blood cells.

Erythroblastosis Fetalis

This hemolytic disorder of the newborn is caused by immunologic incompatibility between mother and child and usually involves the Rh antigen of red blood cells. This antigen is expressed as an autosomal dominant trait present on the erythrocytes of 85 percent of the population. When the fetus is Rh positive and the mother Rh negative, the mother will develop anti-Rh antibodies, because some of the fetal erythrocytes will enter the mother's circulation at the time of birth. Consequently, the first pregnancy of a woman with an Rh incompatible fetus is uncomplicated. If the sensitized mother has a subsequent pregnancy with an Rh-positive child, the transplacental passage of IgG immunoglobulins brings anti-Rh to the child's blood. By the time of birth, the child has suffered from continuous hemolysis and may be jaundiced from excess bilirubin, anemic, and edematous. The hemolysis is often accentuated just after birth, at which time the infant no longer has the help of the placenta in removing bilirubin. Consequently, blood of the infant is often exchanged for Rh-negative blood (exchange transfusion). Erythroblastosis fetalis can be prevented by injecting mothers with human gamma globulin containing anti-Rh antibodies within 72 hours after delivery of the first and subsequent Rh-positive children. This binds the antigens on the fetal red blood cells so that the mother's immune system does not recognize them as antigenic and, therefore, does not produce antibodies.

CHART 27–5. ERYTHROBLASTOSIS FETALIS

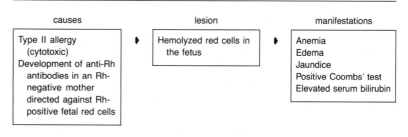

causes	lesion	manifestations
Type II allergy (cytotoxic) Development of anti-Rh antibodies in an Rh-negative mother directed against Rh-positive fetal red cells	Hemolyzed red cells in the fetus	Anemia Edema Jaundice Positive Coombs' test Elevated serum bilirubin

Blood Transfusion Reactions

There are many different antigen systems in red blood cells of which the ABO and Rh systems are most important. An Rh-negative individual does not have anti-Rh antibodies unless previously sensitized. A person with the A antigen on red blood cells has anti-B antibody as a natural phenomenon. Thus, persons with blood group B have anti-A antibodies, those with O blood have both anti-A and anti-B, and those with AB have neither antibody. Because the antibodies are normally present, transfused blood with an ABO incompatibility will produce immediate hemolysis of the transfused red blood cells, resulting in fever, chills, and possible renal failure. These reactions are prevented by typing and cross-matching of blood before transfusion. *Typing* refers to checking for major (ABO and Rh) blood groups to make sure that they are not incompatible, and *cross-matching* refers to mixing samples of donor and recipient blood to see if an in vitro (test tube) reaction occurs due to an unsuspected antibody.

CHART 27–6. BLOOD TRANSFUSION REACTIONS

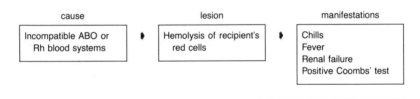

cause	lesion	manifestations
Incompatible ABO or Rh blood systems	Hemolysis of recipient's red cells	Chills Fever Renal failure Positive Coombs' test

Autoimmune Hemolytic Anemia and Thrombocytopenia

Many spontaneously occurring hemolytic anemias and thrombocytopenias are cytotoxic-type hypersensitivity reactions. The reactions may be mild, with agglutinated cells being prematurely removed by the spleen. In mild types, splenectomy may control the disease. Autoimmune hemolytic anemias can be detected by the direct Coombs' test, in which red cells coated with an antibody will be observed to agglutinate in vitro with the addition of antihuman globulin serum. Sometimes drugs attach to the cell surface and become part of the antigen, in which case the drug is a hapten.

Immune Complex, or Arthus-Type, Hypersensitivities (Type III)

Immune complex hypersensitivity reactions produce vasculitis and, as a secondary phenonemon, edema because of the release of vasoactive substances. The frequent involvement of renal glomeruli in immune complex diseases is often associated with loss of protein and red blood cells in the urine and variable degrees of renal failure. Involvement of joint surfaces leads to joint swelling. More severe forms result in a generalized vasculitis with involvement of many organs.

Arthus Reaction

This is the prototype of immune complex hypersensitivity reactions and is an experimental reaction and not a naturally occurring disease. Local injection of soluble antigen in an animal previously sensitized by the same antigen produces an acute inflammation at the site of inoculation. Histologically, the reaction shows evidence of cell necrosis, infiltration with neutrophils, and vasculitis, all sequelae of the acute inflammatory reaction.

Serum Sickness

This is the prototype of a systemic Arthus-type or immune complex reaction. Classically, it occurred after injection of horse serum. The horse serum was used as a source of antibodies to toxins such as tetanus toxin; however, the protective effect was often offset by the harmful effect produced when the patient developed antibodies to the horse serum. The horse serum (antigen) circulates in the patient's blood for a long time. As antibodies begin to develop after about 10 days, antigen–antibody complexes form, lodge in small vessels, and elicit the immune complex reaction at many sites. Although horse serum is rarely used anymore, the same reaction is seen with drugs such as penicillin. The name serum sickness has been retained, although it is no longer appropriate. Symptoms are fever, painful joints, enlarged lymph nodes and spleen, and frequently an allergic urticaria. Usually, after suspending administration of the offending material, the patient recovers with no permanent damage.

CHART 27–7. SERUM SICKNESS

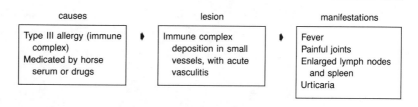

causes	lesion	manifestations
Type III allergy (immune complex) Medicated by horse serum or drugs	Immune complex deposition in small vessels, with acute vasculitis	Fever Painful joints Enlarged lymph nodes and spleen Urticaria

Glomerulonephritis

Some forms of acute and chronic glomerulonephritis are mediated by immune complex reactions due to lodging of antigen–antibody complexes in the basement membrane of glomeruli. One form of glomerulonephritis, poststreptococcal glomerulonephritis, develops in association with the immune response to infection by group A streptococci. The renal disturbance is first seen 1 to 4 weeks after apparent recovery from the acute streptococcal infection. Immune complexes are caught on the glomerular basement membrane, where they fix complement and promote an inflammatory process that compromises the filtering function of the glomerulus. The disease predominantly affects children, but some adults are also affected. Recovery is the rule, probably because the antigenic stimulation of the streptococcal infection subsides. Chronic glomerulonephritis results from a variety of antigens and is often low grade but persistent, eventually leading to renal failure. Glomerulonephritis is also discussed in Chapter 16.

Polyarteritis Nodosa

This is a severe but rare form of immune complex reaction producing widespread, multifocal, necrotizing vasculitis (Fig. 27–12). It usually leads to fatal complications as a result of occlusion or rupture of vessels. The antigen is usually not identified, although hepatitis virus B is implicated in about 35 percent of cases and penicillin is occasionally implicated.

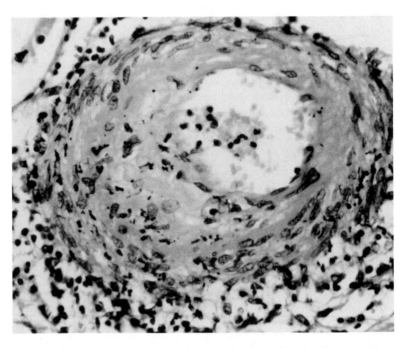

Figure 27–12. Polyarteritis nodosa, with necrosis of vessel wall and intense inflammation.

Delayed, or Cell-Mediated, Hypersensitivities (Type IV)

Delayed hypersensitivity reactions are manifest as subacute or chronic inflammation, with infiltration of the tissue by lymphocytes and macrophages and variable degrees of necrosis. The tuberculin skin test is an example of a typical subacute reaction, with development of a red, firm lump at the site of injection of tuberculin in a sensitized individual. The reaction reaches a peak at 2 days and gradually disappears. Contact dermatitis to a piece of jewelry is an example of a chronic reaction, the appearance being that of a thickened, slightly red irregular lesion of the skin at the site of contact. The lesion persists until the antigen is removed. Internal delayed hypersensitivity reactions are quite variable in appearance but are all characterized by a chronic inflammatory cell reaction, with predominance of lymphocytes and with variable degrees of tissue destruction. These are often seen by the pathologist in tissues invaded by malignant neoplasms. The neoplastic cells carry antigens recognized as foreign that elicit delayed hypersensitivity.

Contact Dermatitis

This is an acute or chronic delayed-type hypersensitive response to allergens placed on the skin surface. A notable example is *poison ivy.* However, the range of agents that cause contact dermatitis is very large and includes many topically applied drugs, cosmetics, paints, dyes, plastics, plants, and jewelry. The lesion varies from simple erythema discretely localized to the area of allergen contact to the more edematous, pruritic, vesicular dermatitis seen with poison ivy. It is sometimes difficult to distinguish the reaction caused by direct irritants from that produced by allergens. Furthermore, it may be difficult to discover the allergen or to remove it from the environment once discovered.

CHART 27–8. CONTACT DERMATITIS

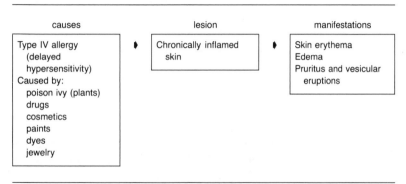

causes	lesion	manifestations
Type IV allergy (delayed hypersensitivity) Caused by: poison ivy (plants) drugs cosmetics paints dyes jewelry	Chronically inflamed skin	Skin erythema Edema Pruritus and vesicular eruptions

Infections Manifest Primarily as Delayed Hypersensitivity Reactions

Some microorganisms tend to stimulate cell-mediated immunity. Examples are the bacteria causing tuberculosis and leprosy, many fungi, and some viruses. A marshalling of sensitized T lymphocytes and macrophages into an infected tissue produces a picture of either chronic or granulomatous inflammation. In tuberculosis and fungal infection, the caseous necrosis is thought to be mediated by sensitized T lymphocytes. Manifestations of viral infections may be due in part to delayed hypersensitivity reactions. Encephalitis occurring after a viral infection may also be a delayed hypersensitivity reaction.

Graft Rejection

The rejection of skin or a kidney grafted from one person to another is in large part due to a delayed hypersensitivity reaction. Different naturally occurring tissue antigens cause the development of sensitized lymphocytes in the person receiving the graft. The lymphocytes then cause necrosis of the graft. Such a reaction does not occur in identical twins because they have the same tissue antigens. Tissue typing, which is analogous to blood typing, is now used to match donor and recipient as closely as possible.

Autoimmune Diseases

Intolerance to self as a disease phenomenon requires an autoimmunization process in which sensitized lymphocytes or antibodies are developed against self-antigens. A number of relatively uncommon diseases appear to evolve as a result of such autoimmunization and its hypersensitive expression. Autoimmunity probably encompasses all four categories of hypersensitivity, but the most common types appear to be immune complex (type III) and hypersensitivity (type IV). Autoimmune hemolytic anemias are good examples of cytotoxic (type II) autoimmunity. A partial list of diseases thought to be autoimmune is given in Table 27–1. One example is discussed.

Systemic Lupus Erythematosus (SLE)

This moderately common systemic disease may affect a number of different organ systems. A characteristic skin lesion, the butterfly rash, is an erythematous dermatitis that covers the bridge of the nose and extends bilaterally onto the cheeks (Fig. 27–13). The skin is generally photosensitive; thus, rashes may appear after excessive exposure to sunlight. The joints may also be involved, producing complaints of arthritis. Muscles may become weak and atrophic. The kidneys develop a glomerulonephritis. Degenerative

TABLE 27–1. AUTOIMMUNE DISEASES

Disease	Organ or Tissue	Antigen
Hashimoto's thyroiditis	Thyroid	Thyroglobulin
Pernicious anemia (vitamin B$_{12}$ deficiency)	Gastric mucosa	Intrinsic factor
Goodpasture's syndrome	Kidney glomeruli and lung	Basement membrane
Autoimmune hemolytic anemia	Red cells	Red cell surface
Idiopathic thrombocytopenic purpura	Platelets	Platelet surface
Myasthenia gravis	Skeletal muscle	Acetylcholine receptors on muscle cells
Rheumatoid arthritis	Synovial membranes	Altered IgG
Systemic lupus erythematosus	Synovial membranes, kidney, skin, blood vessels	Many: DNA, DNA protein, RNA cardiolipin, microsomes

changes in heart valves also may occur. The blood may show hypergammaglobulinemia, anemia, leukopenia, and thrombocytopenia. Lupus erythematosus predominantly affects young to middle-aged women and usually has a protracted course, which ends with renal insufficiency, bacterial endocarditis, cardiac failure, sepsis, or pneumonia. A number of different autoantibodies occur in lupus. The primary tissue damage is probably due to cytotoxic antibodies and immune complexes, as well as to delayed hypersensitivity. Infiltrates of chronic inflammatory cells (lymphocytes and

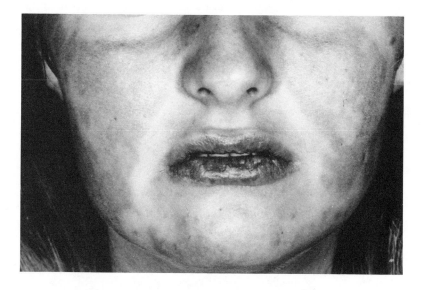

Figure 27–13. Butterfly rash of systemic lupus erythematosus.

plasma cells) are found in many organs. Antinuclear antibody, an antibody directed against the body's own cell nuclei, provides the basis for the diagnostic antinuclear antibody (ANA) test.

CHART 27–9. SYSTEMIC LUPUS ERYTHEMATOSUS

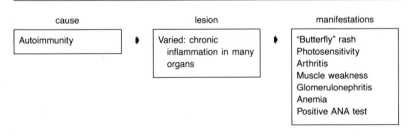

cause	lesion	manifestations
Autoimmunity	Varied: chronic inflammation in many organs	"Butterfly" rash Photosensitivity Arthritis Muscle weakness Glomerulonephritis Anemia Positive ANA test

REVIEW QUESTIONS

1. What are antigens and antibodies?
2. How do B lymphocytes and T lymphocytes differ in terms of origin and function?
3. What are the major functions of each of the five classes of immunoglobulins?
4. What are the four major types of allergy?
5. How are each of the following tests or procedures helpful in evaluation of a patient with allergy or immune deficiency?
   Serum electrophoresis
   Serum immunodiffusion
   Immunofluorescent tests
   ANA test
   Skin tests
   Complement fixation test
   Coombs' test
6. What are the causes, lesions, and major manifestations of each of the specific immune diseases discussed in this chapter?
7. How do deficiencies of the immunoglobulins differ from those of the T cell immune system?
8. What is the relationship between hay fever and asthma in terms of cause, lesions, manifestations, and complications?
9. How are transfusion reactions diagnosed? How are they prevented?
10. What is the basic lesion common to the delayed hypersensitivity diseases?

11. What do the following terms mean?
Hapten
Secondary response
Complement
Agammaglobulinemia
Hypogammaglobulinemia
Rh antigen
ABO red blood cell antigens
Typing and cross-matching
Tissue typing
Butterfly rash

Chapter 28

Physical Injury

MOST FREQUENT AND SERIOUS PROBLEMS

Major external agents causing physical injury fall into three general categories—mechanical, thermal, and radiation. The term *trauma* refers to injury caused by extrinsic forces, particularly when associated with an accident or violence. Overall, trauma is the fourth leading cause of death in the United States and is the most common cause of death in persons under 38 years of age. Trauma accounts for about 7 percent of patient visits to a physician, with

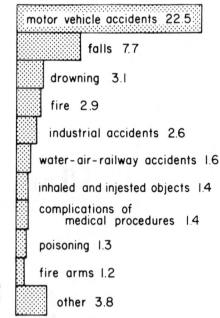

Figure 28–1. Mortality rate (deaths per 100,000 persons per year) from accidents.

half of these visits involving fractures and lacerations. Automobile accidents are the most frequent cause of serious traumatic injury. The mortality rates of various types of accidents is shown in Figure 28–1.

Thermal injuries most commonly occur as burns. Extensive thermal burns produce a serious threat to life and result in a long course of treatment.

Sunlight is the most common source of radiation injury. Radiation from sunlight affects all people, but fair-skinned, outdoor people are more susceptible to sunburn and chronic skin changes. X-ray and gamma-ray injuries are most often encountered as a complication of cancer therapy.

MECHANICAL INJURIES

Types of Simple Wounds

The cause of simple wounds can often be determined by their appearance. Common types include contusion, abrasion, laceration, incision, and puncture wounds.

A *contusion*, or *bruise*, is a crushing, dislocation, or disruption of subsurface tissue caused by a blunt instrument that does not penetrate or break the surface. Blood released from broken vessels has no outlet and pools in the damaged tissue causing gross discoloration. A "black eye" is a typical example.

An *abrasion* is a scraping or scuffing of the skin surface without full-thickness loss of skin.

A *laceration* is a tear in the surface caused by shearing of tissue, leaving irregular, ragged wound margins. Vessels and nerves tend to bridge the gap, because they stretch rather than tear in response to the shearing force.

An *incision* is a clean cut caused by a sharp instrument such as a knife and differs from a laceration in that it has clean, regular edges. The underlying tissue is divided by the edge of the instrument, but peripheral damage is more restricted than with a laceration. Thus, an incision has more bleeding and less tissue injury than a laceration.

A *puncture wound* is a deep, narrow injury caused by a sharp-pointed instrument that penetrates through tissue.

Missile Wounds

A missile wound is a wound caused by a flying object such as a stone, arrow, or bullet. Gunshot wounds are the most common type. Injury by metallic objects thrown from rotary mowers are also common. The nature of the injury produced by missiles depends on the mass of the object and the square of its velocity. Large, slow-moving missiles produce much surface destruction and do not penetrate very deeply. High-velocity, small missiles may have a small entrance wound while producing much internal injury.

Types of Major Body Trauma

Major accidents, such as automobile accidents, often cause characteristic patterns of injury in various parts of the body. Injuries to the abdomen, chest, spine, and head will be discussed. Injuries to the bones and joints are discussed in Chapter 21.

Abdominal Injury

Severe blows, sometimes even without any penetration, can produce significant injury to abdominal viscera. The most important effects of abdominal trauma are hemorrhage, which is often unrecognized and delayed; perforation of the intestines leading to peritonitis; and rupture of the urinary bladder. Hemorrhage is most commonly due to laceration of the spleen but may also be from a lacerated liver or kidney. Because of its muscularity, mobility, and serosal covering, the intestine may slide away from relatively slowly penetrating instruments like needles, swords, and knives. If the intestines are ruptured, spillage of gastrointestinal contents into the peritoneal cavity causes an intense inflammatory reaction (peritonitis) and the motility of the tract is depressed (ileus). Air may leak out of the ruptured viscus and accumulate under the diaphragm, where it may be detected by an abdominal x-ray. Fluid-filled and distended organs, such as the bladder and the pregnant uterus, are prone to rupture with rapid pressure changes such as occur in automobile accidents. Since these serious abdominal injuries are likely to be fatal if not treated promptly, abdominal operations are often performed as a part of the diagnostic work-up when serious abdominal injury is suspected.

Chest Injury

Major effects of chest trauma include fatal hemorrhage and interference with breathing. Fatal hemorrhage occurs from tears in the heart or great vessels produced by stab wounds or gunshot wounds and sometimes by rapid deceleration, as in an automobile accident when the chest hits the steering wheel. Difficulty in breathing is produced by three mechanisms—obstruction of the tracheobronchial tree, fluid in the pleural space compressing the lungs, and multiple rib fractures resulting in an inability to move the chest cage (*flail chest*). Obstruction of the tracheobronchial tree may be due to accumulation of blood, to an inhaled foreign body, or to compression of the trachea. Tracheostomy (making an opening through the neck to the trachea) and suction are used to open the airway. Fluid in the pleural space may be air (pneumothorax), blood (hemothorax), or watery fluid (pleural effusion). Hemothorax is due to rupture of a vessel. Pleural effusion is due to heart failure or pleural inflammation rather than trauma. Pneumothorax is most important because it can be treated but may be fatal if untreated. Open chest wounds must be closed to prevent sucking of air into the pleural space. A tube is placed in the pleural cavity under a water seal so that air is forced out of the pleural space with inhalation and cannot reenter the pleural space during expiration. Multiple rib fractures (flail chest) require fixation of the sternum by wiring it to something overhead in order to prevent collapse of the chest with each inspiration. Artificial breathing may also be necessary. Most deaths following automobile accidents are the result of chest trauma.

Spinal Injury

Fractures or extreme dislocations of vertebrae can shear the spinal cord, interrupting all neural connections below the damaged site. Injury to the lower cord causes *paraplegia*—loss of control over the lower limbs and, usually, loss of bladder function. Injury of the spinal cord in the neck causes *quadriplegia*—loss of control over all limbs. Major long-term complications of spinal cord injury include neurogenic bladder with superimposed cystitis and pyelonephritis. With repeated bouts of pyelonephritis, kidney stones often form, and renal failure may lead to death. Decubitus ulcers (bed sores) form in the severely disabled from pressure of skin against bone. Frequent changes of position through good nursing care prevent these ulcers from occurring over pressure points. Difficulty in breathing and pneumonia are major complications of quadriplegia.

Head Injury

The brain is encased in a nonexpansive bony vault, the skull. If tissue contents of the cranial cavity enlarge, the skull cannot accommodate for the increased mass. The two major mass-producing lesions of trauma are edema and hemorrhage. Edema may be localized or widespread, but, in either case, the increasing mass of the brain causes increased pressure throughout the cranial cavity. The increased pressure may affect neuronal function and compromise blood flow. Herniation of the brain through the foramen magnum into the spinal column causes secondary hemorrhages into the respiratory center of the brain stem, leading to death. See Chapter 23 for further discussion of brain hemorrhage and trauma.

THERMAL INJURY

Burns

Minor small burns are very common and heal by regeneration of the surface epithelium without residual effects. The damage produced by larger burns (Fig. 28–2) depends on two major factors— the depth of the burn and the percentage of body surface involved. The depth of the burn can be estimated by inspection and determined accurately by biopsy. First-degree burns are limited to the epidermis and are red, painful, and dry and may form blisters. Regeneration occurs rapidly from remaining epidermal cells.

Second-degree burns destroy the epidermis and upper dermis and are red and moist. Regeneration of the epidermis can occur because epithelial cells from sweat glands, sebaceous glands, and hair follicles in the dermis can proliferate and, with time, form new epidermis and remodel these skin appendages.

Third-degree burns destroy epidermis and dermis down to the subcutaneous tissue and are charred and dry. Pain sensation is reduced because of destruction of sensory nerve endings in the skin. Since all epithelial cells are destroyed, regeneration of epidermis can only occur from the margins of the burn, and this type of regeneration can only proceed for a few centimeters into the injured area. Thus, third-degree burns cannot regenerate new epithelium, and skin grafting is required if the area is larger than a few

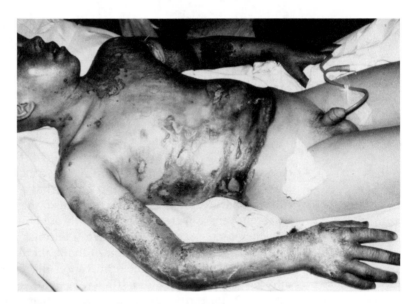

Figure 28–2. Extensive thermal burns of varying severity. Note catheter in urethra to monitor urine output as an index of adequate fluid replacement.

centimeters. After the body has formed granulation tissue under the charred surface and the necrotic tissue is removed, postage stamp-sized pieces of skin are removed from a normal area of the body and placed on the granulation tissue. The graft will grow, and the adjacent grafts will fuse to form a new surface. The donor site will also regenerate, because only a split thickness of skin is removed, leaving skin appendages behind to accomplish the regeneration of the donor site. The grafted sites will not be normal, because the burn will have provoked considerable scarring in the dermis and because the new epidermis will not be capable of regenerating skin appendages. Sometimes excessive scarring occurs (Fig. 28–3).

The amount of body surface involved by second- and third-degree burns correlates roughly with the chances of survival. The general health of the patient and quality of medical care are also very important determinants of survival with extensive burns. As a crude estimate, 30 percent body burns are serious and 60 percent body burns are fatal most of the time. The major complications of extensive burns are fluid loss from the exposed surface leading to shock and infection developing in the burned skin and spreading to the bloodstream. Pseudomonas is a bacterium notorious for its occurrence in burns and its ability to produce fatal septicemia in patients with burns. Severely burned patients should be transferred to major treatment centers, where the skilled personnel and

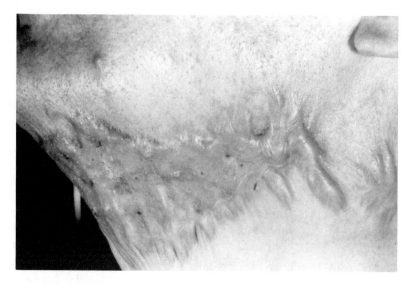

Figure 28–3. Excessive scar formation at the site of a thermal burn.

proper facilities are available to carry out the meticulous care needed to prevent dehydration and infection and to provide skin grafting and follow-up care.

Electrical Burns

Electricity produces two major effects—burns at entry and exit sites and electrical conduction changes, such as cardiac arrest. The amount of local injury depends on the amount of electrical energy and amount of electrical resistance in the tissue. Dry tissue, such as skin, is a poor conductor and, therefore, accumulates the energy and is burned. The effect of the electrical energy as it flows through the body in the shortest pathway from entry to exit site depends on the amount of energy absorbed at the entrance site and the location of entry and exit sites. As the electricity encounters resistance at the exit site, a burn also occurs there. The outcome is likely to be either sudden death or survival with burns at entry and exit sites. Artificial respiration and cardiac resuscitation may be life saving. The burned tissue may be extensive and needs to be debrided to prevent secondary infection.

Excessive Heat Exposure

Prolonged exposure to hot weather can have two possible effects. Excessive sweating can cause dehydration and salt depletion leading to shock. Salt and water replacement are indicated. The other effect is called *malignant hyperpyrexia* and is due to a failure of sweating. The skin is hot and dry and failure to lose heat by vaporization of sweat leads to a progressive rise in body temperature and death.

Frostbite

Frostbite is simply necrosis of tissue due to freezing. It occurs on exposed parts, such as feet, hands, and ears. The dead tissue demarcates from adjacent viable tissue in a few days and is either sloughed or removed surgically so that the remaining viable tissue can heal.

Excessive Cold Exposure

The body can tolerate moderate degrees of hypothermia for a considerable time; however, when hypothermia is excessive, death occurs. Death is much more rapid in water and occurs with temperatures above the freezing level, because water rapidly conducts heat from the body. In air, conduction of heat from the body occurs more slowly.

RADIATION INJURY

Sun Radiations

The acute effects of sun radiations are sunburn, a mild superficial injury of epidermis with vascular congestion. Much of the sunburn effect is due to ultraviolet radiation, which can be reduced by certain types of lotions. The chronic effects of sun exposure include a general deterioration in collagenous tissue of the skin termed *senile elastosis*. This accounts for the wrinkling and drooping of facial and hand skin. Sun exposure also relates to the frequency of premalignant skin lesions (senile or actinic keratoses) and skin cancers (basal cell carcinoma, squamous cell carcinoma, and malignant melanoma).

X-radiation and Gamma Radiation

The basic effect of X-rays and gamma rays on cells is on the nucleus, with resultant death of the cells or interference with its ability to divide. Thus, continuously dividing cell lines are most susceptible to radiation injury. These include bone marrow, small intestinal epithelium, testes, and epidermis. Lymphocytes and ova are also very radiosensitive (easily killed by radiation). Conversely, cells least capable of cell division are least susceptible to radiation injury. Thus, very high doses are required to injure the brain or muscle. Other tissues are intermediate in their radiosensitivity. The relationship of radiosensitivity to proliferative activity also holds for most neoplasms. Rapidly dividing, poorly differentiated malignant neoplasms are radiosensitive; well-differentiated malignant neoplasms are moderately radiosensitive or radioresistant. Benign neoplasms approach their tissue of origin in their degree of radiosensitivity.

In addition to the effects on specific tissues, radiation causes gradual changes in connective tissue that require months to years to develop. These effects include increased density of collagenous tissue, vascular changes leading to decreased blood supply, and ulceration of epithelial surfaces overlying the areas of connective tissue change.

The treatment of cancer by radiation therapy is based on the cancer being more radiosensitive than surrounding normal tissue. The radiologist who performs radiation therapy is an expert at delivering the maximum dose of radiation to the cancer with as little damage to surrounding tissue as possible. Radiation therapy is provided in four forms. High-energy radioactive cobalt sources are available in major treatment centers and are used to deliver penetrating doses to deep-lying cancers. X-ray therapy machines are more widely available and are also used for treatment of solid cancer masses, although the lesser penetration of x-rays results in more tissue damage to the normal tissue lying in the treatment pathway. Radioactive substances that emit less-penetrating gamma rays are selectively used to implant the radioactive source near the cancer. For example, applicators containing radioactive cesium are implanted into the uterus for treatment of cervical and endometrial cancer, and radioactive gold is inserted into the prostate for treatment of prostatic cancer. Finally, radioactive compounds with particular chemical properties may be injected so that they will be localized in the target tissue. For example, radioactive iodine localizes in the thyroid gland and thus can be used to selectively treat hyperplasia of the thyroid or the occasional functional thyroid cancer.

Although most radiation injury is encountered as a complication of radiation therapy for serious disease, the effects of radiation used for diagnosis and of radiation encountered in the environment are of constant concern to society. The damaging effects of diagnostic x-rays appears to be very limited because of careful controls and knowledge gained from previous experiences. Radiologists no longer develop skin cancers on their hands, because protective gloves are worn. Radiation of the chest and neck of children for benign conditions has been stopped because of the high frequency of thyroid cancers that develop some years after such treatment.

Injury from environmental radiation is of much greater long-range concern because of the massive number of people that might be involved and because radiation contamination of the environment might last for years. A background level of atmospheric radiation is present from cosmic rays. The role of cosmic radiations in relation to the spontaneous occurrence of cancer and genetic defects is unknown. In principle, we know that radiation can alter genes and that genetic change is involved in carcinogenesis. There is current concern that alteration of the stratosphere may lead to an increase in the background cosmic radiation. An increase in radioactive fallout from nuclear bombs or accidents would have an effect similar to an increase in cosmic radiation, except that some of the fallout would be in the form of radioactive chemicals that might be selectively concentrated in the biologic chain. From experimental research, it can be predicted that the long-term effects of a generalized increase in environmental radioactivity will be an increase in the frequency of congenital anomalies and of cancer. Many years would be required to gather statistics to prove that an effect had occurred.

Massive high dose exposure to radiation occurs with nuclear bombs and nuclear reactor accidents. Bombs produce injury from mechanical trauma, burns, and superficial injury from beta-particle radiation, as well as from gamma rays. A nuclear accident exposure is more likely to be pure gamma radiation. When the whole body is exposed to gamma rays, the bone marrow is the most sensitive organ involved in fatal injury. About 500 roentgens will destroy the bone marrow beyond its ability to regenerate before fatal infection and anemia occurs in a few weeks from loss of white and red blood cells. Doses of 1000 R or more will produce death in 4 to 7 days from injury to the intestine. Doses of over 50,000 R will produce death in a few hours from brain injury.

REVIEW QUESTIONS

1. What are three major categories of physical injury?
2. What is the most common cause of fatal accidents? What are other important causes?
3. What is the pathogenesis and appearance of contusions, abrasions, lacerations, incisions, and puncture wounds?
4. What factors determine the appearance and amount of damage caused by missile wounds?
5. What are the typical injuries and their complications for abdomen, chest, spine, and head injuries?
6. How do the depth and extent of burns affect the course and treatment of burns?

7. What are the two major effects of electrical injury?
8. What are the effects of frostbite and excessive general exposure to hot and cold environments?
9. What are the acute and chronic effects of sunlight?
10. What are the similarities and differences in the effects of exposure to x-rays and gamma rays?
11. Which tissues and neoplasms are most and least sensitive to injury by radiation?
12. What are four forms of radiation therapy? What are the advantages of each?
13. What are the likely effects of a gradual rise in environmental radioactivity and of a massive nuclear accident?
14. What do the following terms mean?
    Trauma
    Flail chest
    Paraplegia
    Quadriplegia
    First-degree burn
    Second-degree burn
    Third-degree burn
    Radiosensitive
    Radioresistant

Chemical Injury

MODES OF CHEMICAL INJURY

The wide variety of chemical agents that are potentially harmful to man produce diverse types of injury that are difficult to classify. Knowledge of the effects of each chemical agent is needed to provide the best treatment. However, it is impossible for any one person to be familiar with all potential injurious compounds. Application of general principles along with knowledge of common agents is needed to make appropriate initial decisions in cases of chemical injury. Poison control centers provide ready access to information regarding more rarely encountered agents.

The mode of exposure to chemical injury provides the first clue to the likely agents involved. Three modes of chemical injury will be considered: (1) overdose of a drug that a person is purposely taking for its effects; (2) exposure of an individual by accident or because of suicidal or homicidal intent; and (3) potential exposure of many people because of environmental pollution or accident.

Most drugs are potentially harmful if taken in sufficient dose. Often, the harmful dose is close to the therapeutic dose, so that deleterious effects occur as a side effect of therapy. A second mode of drug overdose is miscalculation of the dose or misunderstanding by the patient of how much to take. Drug abuse, which involves self-administration of drugs for their psychological effects, is another mode of drug overdose.

Accidental exposure of an individual to harmful chemicals may occur by ingestion, by inhalation, or by contact with corrosive chemical agents. Agents selected for suicidal or homicidal purposes are likely to be strong agents that can be easily ingested, such as barbiturate drugs.

Environmental accidents and pollution may be either acute or chronic. Carbon monoxide poisoning from car exhaust or fires is an example of acute environmental poisoning. Lead poisoning from eating paint or burning storage batteries is an example of chronic poisoning from environmental exposure.

The likely means of discovery of chemical injury relates to the mode of exposure. Drug overdose as related to therapy will be reported by the patient or observed by the physician. Drug abuse is a social problem that comes to the surface in many ways, sometimes medical, sometimes criminal, and sometimes social in nature. Accidental poisoning brings patients to emergency rooms or poison control centers. Suicide and homicide are investigated by police, medical examiners, and forensic pathologists. Unsuccessful suicide and homicide attempts present as emergency medical problems. Acute environmental exposures are likely to be handled in emergency rooms and may involve enough people to be considered a disaster. Chronic environmental injury is likely to go undetected until pilot cases are identified and the mechanism of exposure is brought to public attention.

MOST FREQUENT AND SERIOUS PROBLEMS

Drug overdose in the form of side effects is very common. Another major and common problem resulting in drug overdose occurs when patients receive prescriptions from multiple physicians and multiple pharmacies for similar or additive drugs, and thus take a drug overdose. This is particularly likely to occur with patients

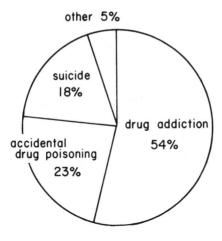

Figure 29–1. Mode of death from drug fatalities. *(Adapted from Statistical Bulletin of the Metropolitan Life Insurance Company, February, 1972.)*

who take tranquilizers, sedatives, and analgesics. Drug abuse with barbiturates, opiates, and other drugs is a major medical and social problem involving millions of individuals and in recent years has caused an increase in fatal poisonings in young adults. Drug abuse is the most common type of drug-related death (Fig. 29–1).

The frequency of poisoning is based on data reported to poison control centers. It gives an approximation of emergency medical care problems relating to chemicals regardless of the mode of injury or potential injury. Most persons seeking medical care for poisoning are asymptomatic and few die (Fig. 29–2). Of the many agents that cause poisoning, drugs are most common (Fig. 29–3). Over 60 percent of reported poisonings occur in children under age 5 (Fig. 29–4). The proportion of cases related to drugs in children

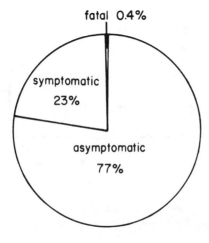

Figure 29–2. Percent of symptomatic and of fatal poisoning among cases of poisoning reported to poison control centers. *(Abstracted from Food and Drug Administration, Poison Control Statistics, 1975.)*

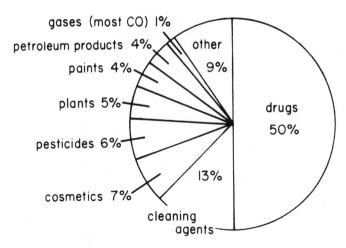

Figure 29–3. Chemical agents involved in poisonings reported to poison control centers, all age groups combined. *(Abstracted from Food and Drug Administration, Poison Control Statistics, 1975.)*

peaks at about 55 percent in 3- and 4-year-olds, whereas, more than 75 percent of cases are drug related in persons age 15 and over.

The importance of environmental injury by chemicals is difficult to assess, because data are difficult to gather and effects may go undetected. Among acute cases of environmental chemicals causing death, carbon monoxide poisoning is the most common. Fatal poisoning from plants, such as mushrooms, is rare in the United States. Pesticides and herbicides are widely used and may produce acute injury when spilled on the skin or ingested. The

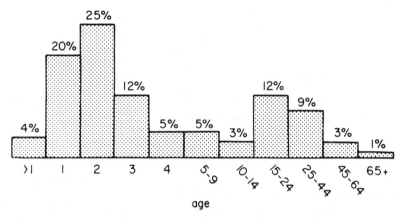

Figure 29–4. Age distribution of reported poisonings. *(Abstracted from Food and Drug Administration, Poison Control Statistics, 1975.)*

long-term effects of many of these compounds in humans are unknown. Air pollutants, particularly hydrocarbons, producing smog in large metropolitan areas, increase the frequency and severity of emphysema and other chronic lung diseases.

Examples of uncommon, but serious exposure to gases include silo filler's disease and ammonia burns. In silo filler's disease, nitrogen dioxide produced in silos by fermentation is inhaled and results in acute and sometimes chronic damage to the lung. Ammonia burns skin, eyes, and respiratory tract when leakage occurs from tanks of liquid ammonia.

Pneumoconioses are chronic pulmonary diseases due to inhalation of particulate matter such as silica, asbestos, and plant fibers (see Chapter 12 for further discussion).

Another large category of chronic environmental chemical injury relates to chemicals that induce cancer. Cigarette smoke is the most well established and has the greatest impact, because it is related to the development of lung cancer, squamous cell cancers of the oral region and esophagus, and transitional cell carcinoma of the bladder. Our knowledge of cancer-causing chemicals is limited, although epidemiologic evidence suggests that they play a role in the development of some cancers.

Probably the most important injurious chemical is alcohol. It does not fit well into the categories discussed above, and, because of its importance, alcohol injury will be discussed separately in Chapter 30.

SYMPTOMS, SIGNS, AND TESTS

The manifestations of chemical injury are obviously dependent upon the route of exposure, the dose, the amount of absorption, the site of metabolism, the degree of excretion, and the specific chemical action of the agent involved. A discussion of these aspects of chemical injury is beyond the scope of this text; specific characteristics of individual agents will be mentioned as needed.

Some broad generalizations can be made about presentation of chemical injuries. From the data presented in the previous section, it is clear that the most common mode of presentation is the child under 5 years of age who has swallowed something and is brought to the emergency room. Examples of signs and symptoms of chemical injury include vomiting, burns, behavioral changes, and unconsciousness. The most important historical information is identification of the chemical agent to which the person was exposed. With knowledge of the agent, the physician can look for the

TABLE 29–1. COMMONLY MEASURED CHEMICAL POISONS

Alcohols (mostly ethyl) and aromatics (such as benzene)

Barbiturates and tranquilizers

Antidepressants

Narcotics

Amphetamine

Salicylates

Iron

Acetaminophen

Carbon monoxide

Lead

specific signs and laboratory abnormalities caused by the agent and select the proper treatment.

Signs of chemical injury are usually not specific but may provide helpful clues. Nausea, vomiting, and diarrhea are observed with intestinal injury. Dyspnea occurs after inhalation of noxious fumes. Careful neurologic examination may detect changes caused by injury to the nervous system. Decreased urine output suggests severe renal injury.

Laboratory tests may be used to suggest whether or not injury has occurred, the likely site of injury, and sometimes the type and amount of the agent involved. General laboratory tests that may suggest whether injury has occurred include elevation of the white blood count, alteration in serum electrolytes, and elevation of serum enzymes, such as aspartate aminotransferase. Elevated blood urea nitrogen and creatinine levels suggest renal injury, high

TABLE 29–2. LABORATORY TESTS AND THERAPEUTIC MONITORING OF DRUGS

Class of Drug	Examples	Mode of Monitoring
Anticonvulsants	Phenobarbital Phenytoin	Chemical assay
Anti-infectives	Gentamycin Chloramphenicol	Chemical assay Microbiologic assay
Bronchodilators	Theophylline	Chemical assay
Antineoplastic agents	Methotrexate Many other agents	Chemical assay White blood count
Anticoagulants	Heparin Coumadin-type drugs	Coagulation time Prothrombin time
Cardiac	Digoxin	Chemical assay
Antipsychotic	Lithium	Chemical assay

aspartate aminotransferase levels are found with liver damage, and altered levels of blood gases are associated with altered oxygen–carbon dioxide exchange in the lungs.

Many chemicals can be measured in samples of blood, urine, feces, or tissues. However, the laboratory needs a clue as to which chemical to look for among the myriads of possibilities. Extensive search for a poison is only indicated when the information will be of value in treatment or in criminal investigation. Table 29–1 lists chemicals that are commonly measured in situations of accidental, suicidal, or homicidal poisoning. Table 29–2 lists drugs for which laboratory tests are commonly performed to detect possible toxicity (therapeutic drug monitoring).

SPECIFIC DISEASES

The emphasis in this section will be on representative specific agents and the type of injury they produce. Agents will be grouped for discussion by mode of injury and/or similarities in patterns of injury. Effects of chemical agents on specific organs that have been previously discussed, such as pneumoconioses, will not be repeated here. Other texts should be consulted for more comprehensive coverage of specific types of chemical injury.

Adverse Drug Reactions

Adverse drug reactions are reactions caused by drugs during the usual course of therapy of a patient. In many instances, the adverse drug reaction is due to hypersensitivity, usually due to an allergic reaction. Other mechanisms include use of drugs with a toxic dose range close to the therapeutic dose range, an enzyme deficiency that accentuates the effective dose of the drug, and renal or liver failure with decreased excretion of the drug.

The most common types of fatal adverse drug reactions are shown in Figure 29–5. Penicillin is a common cause of anaphylaxis although anaphylaxis may be caused by many other drugs. Patients admitted to hospitals and patients given drugs as outpatients should be routinely asked about their sensitivity to drugs, although most cases of anaphylaxis occur in patients not known to be sensitive to a drug. Fatal drug-induced liver injury is most often due to halothane, a commonly used anesthetic. A wide variety of other drugs may cause liver cell injury or bile stasis, but most reactions are of mild degree and reversible upon discontinuance of the drug. Bone marrow suppression leading to leukopenia, thrombocytopenia, and/or anemia may occur as a hypersensitivity reaction to the antibiotic chloramphenicol or as a predictable action of anti-

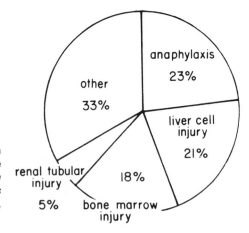

Figure 29–5. Causes of death in patients dying from adverse drug reactions. *(Modified from Irey NS: Adverse drug reactions and death. JAMA 236:575, 1976.)*

cancer drugs. Renal tubular injury is a complication of treatment with several antibiotic agents as well as other drugs. Other specific reactions worthy of special mention include hemorrhage resulting from anticoagulant drug therapy and malignant hyperthermia due to a rare enzyme defect that is manifested when patients are given succinylcholine, a muscle relaxant, in combination with halothane or certain other general anesthetic agents.

Drug Abuse

Agents involved in drug abuse include opiates, barbiturates, tranquilizers, hallucinogens, muscle relaxants, and stimulants such as amphetamines. As mentioned earlier, these agents probably account for about 50 percent of drug-related deaths and for the high incidence of drug overdoses in adolescents and young adults. Death may occur because of overdose as a direct result of the pharmacologic effects of the drug, or it may be due to other complications. The drugs themselves do not produce lesions, and most cause death by their effects on respiration and cardiac function.

Opiates are frequently taken intravenously, leaving telltale lesions on the skin and occasionally leading to severe complications. Local lesions in drug addicts consist of inflammation at the injection sites due to a granulomatous reaction to the talc or starch in the injection mixture or due to infection from failure to use sterile technique. Injected talc or starch is carried to the small vessels of the lung, where a granulomatous reaction occurs in small arterioles and may eventually lead to pulmonary hypertension and heart failure (cor pulmonale) because of the arteriolar obstruction. Allergic reactions may occur to other ingredients in the injection mixture, such as central nervous system damage resulting from a reaction to quinine. Systemic infections, such as bacterial endocarditis and viral hepatitis, are more frequent in addicts, presumably because of direct injection of organisms into the bloodstream.

Accidental Poisonings

Since most accidental poisonings occur in children under 5 years of age, it is not surprising that the agents involved are those that are commonly found around the house. Aspirin is the most common drug involved, although aspirin poisonings have decreased since childproof caps have been used on bottles. The major harmful effects of aspirin are metabolic, with an initial alkalosis due to hyperventilation and later severe acidosis. Accidental or suicidal ingestion of sleeping tablets containing barbiturates produces central nervous system and respiratory depression. Mothers of young children frequently take oral iron medications; accidental ingestion of these iron compounds by a child leads to corrosive damage to the stomach and small intestine, but more importantly, the absorbed iron may lead to fatal shock.

Next to drugs, cleaning agents are the next most common cause of accidental poisoning. Most of these agents cause direct caustic damage to tissue by coagulation of protein. The result is a chemical burn. Benzene and toluene are common ingredients of paint removers, dry cleaning solutions, and glues. Acutely, they cause central nervous system poisoning. Chronic benzene poisoning is noted for its destructive effect on the bone marrow. Carbon tetrachloride is a common solvent that may be inhaled or ingested and leads to severe liver and kidney damage.

Carbon Monoxide Poisoning

Carbon monoxide is produced by incomplete combustion, and when inhaled, it preferentially combines with oxygen-carrying sites on hemoglobin molecules of red blood cells, so that there is a resultant systemic anoxia. The anoxia may be fatal or may lead to permanent brain damage. Automobiles are a common source of carbon monoxide. Most deaths in cases of fire are due to carbon monoxide poisoning rather than burns. Carbon monoxide is particularly dangerous because it is odorless and because initial symptoms of impaired mental function are not apparent to those affected. The effects of chronic low-level carbon monoxide poisoning resulting from smoking, smog, or automobile fumes have not been quantitated. It is possible that mental impairment from mild carbon monoxide poisoning may cause some accidents.

Agricultural Chemical Injury

A number of chemicals encountered in farming cause rather specific types of damage. Nitrogen dioxide, produced by fermentation of silage, may be inhaled to produce severe, acute injury to the lining cells of the pulmonary alveoli. In mild cases of this condition, called *silo filler's disease,* the damage is gradually resolved, but in severe cases, damage to the alveolar walls leads to permanent fibrosis of the lungs and residual decreased breathing capacity.

Fertilizers, herbicides, and insecticides are man-made agricultural poisons. Liquid ammonia, when spilled, may lead to severe burns of the skin, eyes, and respiratory tract. Paraquat is a commonly used weed killer that causes damage to the pulmonary alveoli regardless of whether it is ingested or inhaled. Acute damage may be fatal or may lead to pulmonary fibrosis. Many insecticides contain cholinesterase inhibitors, which may lead to depression of the central and autonomic nervous systems and motor nerve endings.

Injury by Other Agents

Cyanide poisoning may result from inhalation of hydrocyanic acid or ingestion of soluble inorganic cyanide salts or cyanide-releasing substances. Modes of cyanide poisoning include industrial accidents, ingestion of certain plants, suicide, homicide, and execution. Cyanide combines with cytochrome oxidase of cells throughout the body to prevent cellular oxygen utilization. Cyanide is extremely toxic and rapid in its effects.

Methyl alcohol (methanol, wood alcohol) is widely used as a solvent and denaturing agent for ethyl alcohol. The major effect of nonfatal doses is the production of blindness. Methyl alcohol is metabolized to formaldehyde, which in turn causes damage to the retina and optic nerve. Severe acidosis occurs in fatal methyl alcohol poisoning as a consequence of its metabolism to formic acid.

Lead poisoning is usually chronic and often occurs in children who eat paint. It may also follow exposure to burning lead batteries. Lead poisoning affects the brain, peripheral nerves, and bone marrow. Abdominal cramps or colic is a characteristic symptom of peripheral nerve involvement. Encephalopathy of varying degrees, with convulsions and behavioral changes, results from the brain damage. There may be permanent mental incapacity. The mild anemia is recognized as being due to lead poisoning by a peculiar basophilic stippling of the red blood cells. A black line due to deposition of lead sulfide along gingival margins is a characteristic physical finding.

Poisoning with mercuric chloride by ingestion leads to corrosive damage to the oral region and esophagus. The absorbed mercury is excreted by the renal tubules and colonic mucosa, where it produces renal tubular necrosis, leading to acute renal failure, and ulceration of the colon, leading to bloody diarrhea.

REVIEW QUESTIONS

1. What are three general modes of chemical injury?
2. Under what circumstances may drugs lead to injury?
3. What chemical agents account for the most serious and common medical and social problems?
4. What percent of apparent poisonings lead to symptoms? To death?
5. What are the most frequent classes of agents involved in poisoning?
6. What is the age distribution of poisonings?
7. What chemical agent is most important as a cause of cancer?
8. What steps are involved in the evaluation of a person suspected of having injury due to a chemical agent?
9. What roles do laboratory tests play in diagnosis of poisoning?
10. What type of injury is associated with each of the following agents, and under what circumstances do they occur?
      Penicillin
      Halothane
      Chloramphenicol
      Anticancer drugs
      Anticoagulants
      Succinylcholine
      Opiates
      Barbiturates
      Aspirin
      Iron compounds
      Cleaning agents
      Benzene
      Carbon tetrachloride
      Carbon monoxide
      Nitrogen dioxide
      Ammonia
      Paraquat
      Cholinesterase inhibitors
      Cyanide
      Lead
      Mercury
11. What do the following terms mean?
      Drug abuse
      Therapeutic drug monitoring
      Silo filler's disease
      Adverse drug reaction

Nutritional Disorders and Alcoholism

REVIEW OF FUNCTION

Nutrition plays a significant role in health. Other parameters of good health in the individual that are important include heredity, personality, attitude, and environment. Even when these health constituents are normal, a person cannot be healthy if he or she habitually eats too much or too little.

Humans are omnivores and can adapt to a wide variety of diets. Guidelines for good nutrition are difficult to outline because they are subject to various conditions. Infants and young children need relatively more calories for growth and minerals and vitamins to insure normal bone growth. The growth spurt and sexual characterizations at puberty also result in some special dietary needs. Pregnancy and lactation are very demanding situations in the life cycle in which the diet of the mother must be supplemented with nutrients to provide for growth of the fetus, as well as the production of milk. In the elderly, some specific nutritional requirements wane and problems sometimes arise in persuading elderly persons to maintain an adequate diet because of such factors as loss of taste for food and loss of appetite due to loneliness and inactivity. Some general dietary guidelines, good for most every person, would be to (1) maintain ideal weight; (2) eat a variety of foods in order to meet vitamin and mineral needs; (3) avoid excessive fat, saturated fat, and cholesterol; (4) eat food with adequate starch and fiber content; (5) avoid excessive sugar; (6) avoid excessive sodium; and (7) avoid alcohol, or drink only in moderation.

Normal nutritional requirements are a balance of the three major foods: carbohydrates, proteins, and fats (lipids), plus certain vitamins and minerals. Roughly 45 dietary entities are considered essential and these include 9 amino acids (the building blocks of protein), 2 lipids, the vitamins, and several minerals. Note that no carbohydrates are essential.

Energy is derived from carbohydrates, lipids (fats), and protein. There is very little caloric value in vitamins and none in minerals although both of these latter nutritive constituents are necessary for life.

Energy in foods is expressed in calories. A calorie (cal) is the amount of heat required to raise the temperature of 1000 g of water from 15° to 16°C. The caloric value of any food can be directly determined by oxidizing (burning) the food in a calorimeter chamber and measuring the amount of energy released. The energy released by this method is equivalent to the breakdown of the food in the human body although the body utilizes food in a much more prolonged and complicated process than simple oxidation.

The caloric content of each of the three major food categories differ. Overall, carbohydrates yield 4.1 cal per g, protein 4.3 cal per g, and lipids 9.3 cal per g. All three foods are utilized for building structural components of the body's cells, but carbohydrates are the main source of readily available energy in the form of sugars and complex chains of sugars referred to as starches. All carbohydrates can be synthesized in the body from proteins, lipids, or other carbohydrates. Proteins contribute heavily to the structural components of the cells, as well as to enzymes. Proteins are made up of variable-sized aggregates of individual amino acids that are essential, meaning that they are necessary for life. The amino acids can be synthesized in the body to some extent from carbohydrates and lipids, but there are a few essential amino acids that need to be ingested in at least small amounts. Whereas selective deficiencies of carbohydrates or lipids can be compensated by the body by synthesis of the necessary molecules from proteins, a diet selectively deficient in protein will result in a disease called kwashiorkor. Lipids are important components of all cell membranes as well as serving other roles. One or two naturally occurring lipids need to be ingested, but all the others can be synthesized from proteins. Lipids are an important source of energy because their metabolism in the body yields more than twice as many calories as either protein or carbohydrates.

Unlike proteins, carbohydrates and lipids, vitamins do not provide the body with a source of energy. Vitamins are special molecules utilized by the body in small amounts for many and varied specific functions (Table 30–1). Some vitamins are utilized as coenzymes, whereas others contribute to cell and tissue structure. Most vitamins are needed in small amounts and need to be ingested. A few are synthesized by intestinal bacteria, and a few are synthesized in the body by breakdown of other foods.

Of the minerals that are essential for life, large amounts of calcium and phosphorus are incorporated into bone. Iron is needed in relatively large amounts because it is a component of hemoglobin and myoglobin. Blood loss and pregnancy increase dietary iron needs. Magnesium is important for nerve and muscle physiology but normal diets provide adequate amounts. Iodine is a necessary constituent of thyroid hormone, but the body can get along without renewed iodine supplies for relatively long times. Copper, chromium, boron, zinc, and manganese are important but

TABLE 30–1. FUNCTION OF VITAMINS

Vitamin	Function
A	Maturation of various epithelia, necessary for retinal photosensitivity
D	Regulates serum calcium levels
E	Antioxidant (prevents oxidation of certain lipids)
K	Necessary for synthesis of coagulation factors
B_1 (thiamine)	Coenzyme involved in energy metabolism and neural function
B_2 (riboflavin)	Cofactor in cellular respiration
B_3 (niacin)	Coenzyme component for oxidative-reduction reactions
B_6 (pyridoxine)	Coenzyme involved in several metabolic reactions
B_{12} (cobalamine)	DNA synthesis, myelin synthesis
Folic acid	DNA synthesis
C	Collagen synthesis

are used in only trace amounts. Potassium and sodium are necessary electrolytes in cells and blood, but their functions are beyond the scope of this chapter.

MOST FREQUENT AND SERIOUS PROBLEMS

In the United States, overnutrition is recognized as a serious problem. Although there are varying criteria for what constitutes obesity, there is no question that there are millions of Americans in whom obesity contributes to an unhealthy lifestyle in terms of various physical and emotional problems. Obesity may also shorten a lifespan by its contribution to various disease states, the most important of which may be infections, cardiac problems, and atherosclerosis.

Alcoholism is also a serious physical and mental health problem. Alcoholism is found in at least 2 percent of the United States population and is a heavy contributor to accidental death (particularly automobile accidents), loss of work, break-up of families, suicide, and hospitalization for the various secondary effects of alcohol abuse.

Undernutrition is divided into primary malnutrition, in which there is a lack of a balanced diet and decreased total caloric intake, and secondary malnutrition, in which undernutrition results from a disease condition such as cancer, anorexia nervosa, depression, alcoholism, or any chronic disease. As a world-wide phenomenon, undernutrition is extremely significant in terms of scope and individual suffering.

Vitamin deficiencies may result in specific disease states, but they are usually secondary to general undernutrition and are uncommonly seen as isolated deficiencies of specific vitamins.

SYMPTOMS, SIGNS, AND TESTS

Potassium, phosphorus, sodium, and calcium are referred to as electrolytes and their blood levels can be easily and quickly measured. Copper, iodine, and iron can also be measured in blood. Most vitamins can also be measured in the blood if a specific vitamin deficiency is suspected.

Obesity is usually obvious by physical examination, but the degree may be determined by referring to standard charts of body weight and height. The symptoms and signs of alcohol abuse are many and varied. Symptoms of acute alcoholic excess include flushed face, tachycardia, and vomiting. Alcoholism often leads to absence from work, loss of interest in activities, interpersonal difficulties, and divorce. More severe chronic alcohol abuse results in increased incidence of infections, cardiac arrhythmias, peripheral neuropathy, heavy drinking to relieve anger, insomnia, or depression. In its most severe manifestations, alcoholism may result in tremors, withdrawal seizures, delirium tremens, cirrhosis, pancreatitis, gait disturbances, and cardiomyopathy.

Undernutrition manifests with specific signs and symptoms discussed under the specific diseases.

SPECIFIC DISEASES

Many of the disease states resulting from deprivation of essential foods have already been covered in this book because the major manifestations are focused on particular organ systems such as iron-deficiency anemia or rickets due to vitamin D deficiency. This chapter will focus on deficiencies and excesses that have not been adequately covered.

Deficiency Diseases

Generalized Undernutrition
Undernutrition means that the body weight is under that of an accepted standard for height, age, and body build. Undernutrition may result from insufficient diet or from excessive catabolic activity. Undernutrition implies a caloric deficiency of carbohydrates, lipids, and proteins. It is generally stated that up to 75 percent of humanity is undernourished. The true incidence of undernutrition in the United States is not really known. Throughout the world, endemic undernutrition results from several contributing factors, which include poor natural resources, poverty, large families, and lack of knowledge regarding production and preparation of food. Contributing factors to undernutrition include alcoholism, illiteracy, and unstable family situations.

Severe malnutrition in children, such as occurs in many areas of Africa, is divided into two types. *Marasmus* refers to a total caloric malnutrition with severe wasting but without specific effects of serum protein deficiency. *Kwashiorkor* is a severe protein deficiency, with or without caloric deficiency, characterized by generalized edema (due to low serum proteins) and fatty liver (due to insufficient production of lipoproteins).

In general, undernutrition is manifested in children by stunted growth, delayed bone maturation, slowed mental processes, and retarded puberty. Severely undernourished children often have diarrhea, which further complicates the problem. Mental retardation will result from early-onset undernutrition with consequent failure of proper brain development. Children who are undernourished are very susceptible to infections, and most severely undernourished children die by this means.

In adults, undernutrition may not be easily recognizable unless carried over from childhood, in which case short stature will be obvious. Otherwise, a decrease in adipose tissue and muscle mass may be the only obvious physical manifestations. In milder undernourished conditions, there is little desire to improve living conditions, and more primitive forms of gratification surface, such as alcoholic intake and sexual drive. Since the gonads are not involved in milder undernutritional states, the number of pregnancies may increase, much to the detriment of women and their offspring. More severe adult undernourishment is manifested by skin rashes, diarrhea, disappearance of body hair, hepatomegaly, polyneuropathy, and numerous endocrine manifestations. Sexual activity is usually decreased markedly at this stage and is accompanied by gonadal atrophy. Anemia is very common. Superimposed vitamin and mineral deficiencies and infections are common.

Vitamin Deficiencies

Vitamin deficiencies are outlined in Table 30–2. Deficiency of fat-soluble vitamins can be caused by severe fat malabsorption (such as might occur with celiac disease or chronic pancreatitis), but most patients with malabsorption syndrome do not exhibit these deficiencies. A severely restricted diet, such as occurs in some areas of the world, may cause deficiency of vitamins A, thiamine, riboflavin, niacin, and folic acid. Dietary deficiency of vitamins E, K, pyridoxine, and cobalamine are extremely unlikely, although a diet consisting entirely of processed foods potentially may cause a deficiency of pyridoxine. Dietary supplementation with vitamin D has largely eliminated rickets, a disease that occurred in children with reduced exposure to sunlight during the cold season. Vitamin C deficiency is unlikely except with diets deficient in citrus fruits and vegetables for a prolonged period such as used to occur with sailors.

In the United States, dietary vitamin deficiency is uncommon and is most likely to occur in alcoholics, the elderly, poor persons, or persons with severe disease that are living on a very restricted diet. Vitamin B_{12} deficiency is most often due to an autoimmune gastritis (pernicious anemia), a disease that destroys the cells that produce the intrinsic factor needed for vitamin B_{12} absorption. Lifelong injections of vitamin B_{12} prevent further difficulties. Folic acid

TABLE 30–2. VITAMIN DEFICIENCIES

Vitamin	Deficiency	Causes of Deficiency
A	Night blindness, skin changes	Dietary deficiency, malabsorption
D	Rickets, osteomalacia	Lack of sun exposure, dietary deficiency, malabsorption
E	None (lipofuscin and ceroid pigment accumulation)	Malabsorption syndrome
K	Hemorrhagic diathesis	Newborn, prolonged antibiotic therapy, malabsorption syndrome
B_1 (thiamine)	Beriberi (neuropathy, cardiomyopathy)	Dietary deficiency
B_2 (riboflavin)	Skin and mucosal lesions	Dietary deficiency
B_3 (niacin)	Pellagra (dermatitis, diarrhea, dementia)	Dietary deficiency
B_6 (pyridoxine)	Dermatitis, neuropathy, anemia	Diet of processed foods
B_{12} (cobalamine)	Megaloblastic anemia, subacute combined degeneration	Intrinsic factor deficiency, malabsorption
Folic acid	Megaloblastic anemia	Dietary deficiency, malabsorption, cancer chemotherapy, increased need (pregnancy)
C	Scurvy (hemorrhagic diathesis, poor wound healing, retarded bone and tooth growth)	Dietary deficiency

deficiency, most often found in alcoholics with poor dietary intake, mimics pernicious anemia in its hematologic effects but does not cause the neuropathy of pernicious anemia. Folic acid is well absorbed in the face of malabsorption syndrome, so oral therapy followed by adequate diet is sufficient for treatment. In summary, one can say that healthy persons on a balanced diet have adequate vitamin intake, except, perhaps, for the need for vitamin D supplements for children in winter.

Mineral Deficiencies

Dietary iodine deficiency in inland and mountainous areas is easily corrected by use of iodized salt. Iron deficiency usually occurs in situations of increased need (infancy, pregnancy, blood loss) in association with low dietary intake; malabsorption can also lead to iron deficiency. A relative deficiency of calcium may occur, especially during the period of rapid bone growth (childhood). It has been suggested that increased calcium intake may retard the development of osteoporosis. Deficiencies of trace elements, such as magnesium, copper, selenium, and zinc, have been described.

Dietary Excess

Obesity

Obesity means having excess body adipose tissue. Although usually associated with being overweight, a person such as an athlete may be overweight and still not be obese because of increased muscle rather than adipose tissue. Obesity results from an imbalance in calories ingested versus calories expended. Although different persons utilize calories at different rates for a given amount of physical activity, it is a simple fact that obese persons eat too much food relative to their personal metabolic demands and their level of physical activity. Mild obesity is usually tolerated fairly well, but extreme obesity (greater than 20 percent above the standard average body weight at a given age and height) is associated with numerous serious health hazards that collectively reduce life expectancy. Obesity is associated with diabetes mellitus, gout, gallbladder disease, pulmonary emphysema, hypertension, atherosclerosis, and chronic bronchitis.

Treatment of obesity is accomplished by correcting the causative factors. In the cases of habitual overeating and psychogenic problems, the individual must be strongly motivated to decrease caloric intake and exercise properly. Operations that reduce the size of the stomach have been used to treat extreme obesity, but are not without hazard.

Specific Dietary Excesses

Diets high in saturated fatty acids and cholesterol are associated with increased risk for atherosclerotic disease such as myocardial infarct. Excessive protein intake may be deleterious to renal function, especially in patients with chronic renal failure. Excess sugar intake stresses mechanisms that control blood sugar and may lead to overt diabetes mellitus. Complex carbohydrates, such as those found in bread, pasta, rice, and potatoes, help stabilize blood glucose levels.

Vitamin Toxicity

Most vitamins are not toxic, but excessive dietary supplementation with vitamins A and D can lead to problems. Vitamin A toxicity produces a variety of problems with skin, bone, brain, and liver;

mental disturbances mimicking psychoses may occur. Excessive doses of vitamin D may lead to hypercalcemia and to renal calculi.

Mineral Excesses

Excessive exogenous iron from iron tablets, blood transfusions, or production of alcoholic beverages in iron containers can lead to hemochromatosis. Excessive intake of sodium contributes to the development of hypertension. In general, the body regulates the intake of minerals to meet its needs.

Alcoholism

Ethanol is a natural food but may also be considered as a toxic compound when taken in excess over prolonged periods. It must be recognized that psychological and social factors powerfully predispose the potential victim. The term *alcoholism* refers to the pattern of continued alcohol abuse and the biologic and behavioral antecedents and consequences of this pattern. An important ingredient in alcoholism is the poorly understood physiologic dependence (addiction) on ethanol that drives the victim to continued intake. There are probably several different types of alcoholism, each with its own psychophysiologic focus. However, all types of alcoholism have the common denominator of interfering with the person's health, interpersonal relations, and economic and social position. By conservative estimates, over $2\frac{1}{2}$ percent of the population of the United States is alcoholic.

Ethanol absorption starts in the stomach and continues in the duodenum and jejunum. It distributes with body water and rapidly appears in urine, spinal fluid, and pulmonary alveolar air. Alcohol is metabolized by oxidation to acetaldehyde and acetic acid and then to carbon dioxide and water. The main immediate target of ethanol is the brain, where relatively small amounts will depress mental functions, including judgment. Nearly every major organ system in the body is affected, major lesions occurring in the gastrointestinal system (including liver and pancreas), the central nervous system, and the cardiovascular system.

The liver is the most important organ affected by chronic alcoholic intake. Alcoholic changes in the liver caused by ethanol vary from fatty change in hepatocytes, a reversible process, to alcoholic cirrhosis, which is irreversible. Cirrhosis is the most common organic cause of death in alcoholics. The consequences of cirrhosis are discussed in Chapter 14. The acute clinical effects of alcohol intoxication also cause many deaths among alcoholics as well as among innocent bystanders involved in automobile accidents caused by alcoholics. Chronic pancreatitis develops in some chronic alcoholics and may lead to severe disability from pain, malabsorption with weight loss, and diabetes mellitus secondary to destruction of islets of Langerhans. The major, although infrequent, effect on the cardiovascular system by alcohol is cardiomyopathy. Direct toxic injury to heart muscle fibers results in slow accumulation of fibrous scarring and loss of function, first on the right side and then the left. Thiamine deficiency also occurs in alcoholics and causes a similar cardiomyopathy called *beriberi heart disease*.

Excessive alcohol intake may affect the nervous system in several ways. *Wernicke's encephalopathy* involves degeneration of several regions in the center of the brain, producing incoordination and mental confusion. Thiamine deficiency probably causes Wernicke's encephalopathy, as high thiamine doses will reverse the process.

Chronic degenerative changes in the thalamus of alcoholics cause *Korsakoff's dementia,* a condition accompanied by severe memory loss and organic psychosis. Degenerative changes may also occur in the cerebellum, producing ataxia and tremor of the extremities. *Delirium tremens* (DTs) occur in the withdrawal period from acute, excessive alcoholic intake. DTs are manifested by a state of delirium in which the patient experiences frightening visual hallucinations. There is no demonstrable organic lesion in the brain of patients who die after experiencing delirium tremens.

Numerous treatment modalities for alcoholism have been attempted; these include tranquilizers, drugs that make one sick if alcohol is taken following the drug, and criminal incarceration. The only long-term treatment of alcoholism that has proven effective is Alcoholics Anonymous-type organizations, where persons with drinking problems may receive psychological support from other persons with similar problems. In modern society, the issue of alcoholism has shifted from that of a moral problem to one of a disease process and an affliction of society.

REVIEW QUESTIONS

1. How do the manifestations of undernutrition in children differ from those in adults?
2. How do the functions and deficiencies of fat-soluble vitamins differ from those of water-soluble vitamins?
3. What is the usual cause and effect of deficiency for each of the vitamins?

4. What are the major causes and complications of obesity?
5. What is the treatment of obesity?
6. What are the major effects of alcoholism on the liver, pancreas, heart, and brain?
7. How is alcoholism best treated?

Index